A Handbook on Low-Energy Buildings and District-Energy Systems

To Connie and Amanda

A Handbook on Low-Energy Buildings and District-Energy Systems

Fundamentals, Techniques and Examples

L. D. Danny Harvey

Routledge
Taylor & Francis Group

LONDON AND NEW YORK

First published by Earthscan in the UK and USA in 2006

Published 2015 by Routledge
2 Park Square, Milton Park, Abingdon, Oxfordshire OX14 4RN
711 Third Avenue, New York, NY 10017, USA

First issued in paperback 2015

Routledge is an imprint of the Taylor and Francis Group, an informa business

Typesetting by Safehouse Creative
Cover design by Yvonne Booth

A catalogue record for this book is available from the British Library

Library of Congress Cataloging-in-Publication Data

A handbook on low-energy buildings and district-energy systems :
fundamentals, techniques and examples / edited by L.D. Danny Harvey.
     p. cm.
  ISBN-13: 978-1-84407-243-9 (hardback)
  ISBN-10: 1-84407-243-6 (hardback)
 1. Buildings–Energy conservation–Handbooks, manuals, etc. 2. Heating from central stations–Handbooks, manuals, etc. 3. Air conditioning from central stations–Handbooks, manuals, etc. I. Harvey, Leslie Daryl Danny, 1956–
  TJ163.5.B84H364 2006
  696–dc22

                       2005034875

ISBN 13: 978-1-138-96550-8 (pbk)
ISBN 13: 978-1-84407-243-9 (hbk)

Contents

Preface

This is a book about the extraordinary opportunities to reduce our use of non-renewable energy in buildings, both in new buildings and as a result of renovations of existing buildings. It is motivated by the realization, stemming from my own involvement in climate and carbon cycle modelling research over the past 25 years, that unrestrained use of fossil fuels and associated emissions of greenhouse gases (GHGs) pose a serious threat of eventual catastrophic climatic change. Indeed, it is likely that GHG concentrations have already increased to the point that significant negative impacts will eventually occur. It is therefore of the utmost urgency that GHG emissions be reduced as quickly as possible, so as to stabilize concentrations at the lowest possible levels and thereby minimize future damage. This requires simultaneous dramatic improvements in the efficiency with which energy is used in all sectors, so as to reduce total energy demand, combined with a rapid increase in the deployment of renewable sources of energy, leading to an eventual complete phase-out in the use of fossil fuels. Closing the gap between energy demand and that which can be supplied from renewable energy sufficiently rapidly to limit the carbon dioxide concentration to 450 parts per million by volume (ppmv) (the significance of which is discussed in the first chapter of this book) is likely to require overall reductions in energy intensity (the energy use per unit of economic activity) in industrialized countries by a factor of four by 2050. If this reduction is applied proportionally to the industrial, transportation and buildings sectors – each of which accounts for roughly one third of GHG emissions from industrialized countries – then a factor of four reduction is also required from buildings. It is not evident, a priori, that this is possible. In industrialized countries, buildings account for about half of total electricity use. With anticipated improvements in the efficiency of generating electricity, the required reduction in the average energy intensity of buildings (on-site energy use per unit of floor area) is a factor of 2–3 by 2050. This is an average of new buildings and of present buildings still in existence by 2050.

This book began as a comprehensive assessment of the potential to dramatically reduce the energy use of new buildings compared to current conventional practice, and to dramatically reduce the energy use of existing buildings through advanced renovations and retrofits. In order to provide a rigorous assessment, this book systematically examines all uses of energy in buildings and shows, based on documented and accessible case studies, how much better we can do than current practice. The conclusion is that overall energy intensity reductions in the building sector in the order of a factor of 2–3 are indeed possible. To lend credibility to this conclusion, a major focus of the book has been in explaining *how* energy saving technologies and an integrated approach to building design can achieve such large reductions in energy use, without compromising on building comfort or services. This explanatory approach is augmented with the judicious use of equations that present the key relationships between the technical parameters of buildings or equipment, and energy use. At the same time, the major practical issues that can prevent achievement of these savings in reality are addressed. In these ways, this book provides much more of a 'nuts and bolts' level of detail compared to what is found in, for example, the energy scenario literature or in many of the books on 'sustainable' architecture, but provides less 'nuts and bolts' detail than needed for practical implementation by people who actually build or design buildings. An annotated appendix provides sources for this next, higher level of detail.

That having been said, this book will be of great value to architects and engineers because (1) it brings together, in one place, a discussion of all the various ways in which energy is used in buildings, and of the full spectrum of opportunities to reduce the use of fossil fuel and electrical energy in buildings; (2) it provides the reader with a solid grounding in the physical principles underlying energy use and energy savings opportunitip455es in buildings; and (3) it treats buildings as systems in their own right, and as part of larger energy supply and use systems. Much of the energy used in buildings today can be displaced, in new buildings, with various forms of solar energy, generated on-site as part of the building fabric and structure. The remaining energy use can be dramatically reduced through more efficient technology or, in many cases, by an intelligent re-arrangement of the existing components that constitute the various energy-using systems within buildings. The net result is the opportunity, in new buildings, to reduce on-site energy use by a factor of 4–5 compared to the current stock average in industrialized countries or compared to recent, energy intensive, western-style buildings in developing countries, while renovation of the existing building stock can readily achieve reductions of 50 per cent and often much more. Together, this can achieve the required overall reduction in on-site energy intensity by a factor of 2–3.

Almost everything needed to do to this is already known, but it is buried among widely scattered academic journals, conference reports or research institute reports in the various engineering libraries of universities around the world; or in the experience of the limited number of practitioners who have designed and overseen the construction of extremely low-energy buildings. A number of books on 'green buildings' have been published in recent years, and these have included sections on energy use. There are also a number of technical design manuals and guidelines that describe in considerable detail what is needed to achieve deep reductions in energy use in specific areas of building energy use, such as lighting, ventilation and cooling, or heating. What is often lacking is a straightforward explanation of the underlying physical principles, which are generally simple. However, buildings are systems, and systems can behave in ways that might not have been anticipated from the properties of the individual components. Buildings can also be linked to one another through district heating and cooling systems, and can be linked to various sources of waste heat and to systems for large-scale, seasonal storage of thermal energy. Doing so has implications for the design and operation of individual buildings.

This book deals with technologies and practices that are already well established, with new approaches and technologies that have only recently become commercially available, and with technologies that are still under development. In this way the book is deliberately forward looking, documenting what is already established but also providing a 'heads up' for products that may soon enter the marketplace. Only general cost information is given for technologies that have already been commercialized, as costs can change rapidly and can vary significantly with location or, in many cases, with the specific circumstances. Explicit consideration is given to the different challenges facing buildings in cold, hot-dry, and hot-humid climates.

In writing this book, I have drawn upon my own background in computer climate modelling, a background that is rather helpful in writing a book on energy-efficient buildings. In climate modelling, especially the schematic models that I have worked with, one attempts to capture the essence of the phenomenon being modelled in a simple, physically justified manner. A similar philosophy is useful in understanding how buildings and, in particular, passive heating, cooling and ventilation features, operate. This philosophy has guided the explanations given in this book and the various side boxes. In climate modelling, we often find that simple models give answers to within 10–20 per cent of the most complicated, computationally intensive models available, and the same seems to be true with regard to most features of building simulations. This is useful, because it allows one to construct simple models that can be used to guide the design process. Both climate modelling and understanding the heating, cooling, ventilation or lighting of buildings requires a consideration of the basic principles of radiative energy transfer, latent and sensible heat, heat exchange by moving air or water and the buoyancy forces created by temperature differences. So, in the end, the differences between computer modelling of climate and computer modelling of buildings are not great.

L. D. Danny Harvey, November 2005

Technical Preface

Metric units are used throughout this book, as the relationships between power, energy and a host of other related physical quantities are transparent and simple in the metric system, thereby greatly easing calculations or comparisons of different phenomena. More importantly, the metric system is the accepted international standard, is used exclusively in scientific papers and is finding increasing use in English language engineering journals. However, engineers (at least in the US and Canada) still use a large array of bewildering units, such as 'tons' of chilling and 'grains' of moisture, and sometimes mix metric and non-metric units, such as 'Btus per Watt-hour' (Joules per Joule, a dimensionless ratio, is so much simpler!). Rather than cluttering the text with non-metric terms in brackets every time a metric unit is given, a table of conversions is provided in Appendix A at the end of this book.

There is one area where both engineering and non-engineering scientists are sometimes guilty of inconsistent use of units, and this involves the distinction between temperature and temperature change. A temperature of five on the Celsius scale, for example, is five degrees Celsius, and is written as 5°C (the term 'centigrade' was abolished in 1948 at the 9th General Conference on Weights and Measures). A temperature interval or difference of five degrees, on the other hand, is five Celsius degrees, and it is not correct to write this as 5°C. Rather, it should be written as 5C°, except that this would entail introducing a whole new convention. However, on the kelvin scale, absolute temperatures and temperature differences are written in exactly the same way: in the previous example, as five kelvin (5K), that is, without the word degrees (this, along with the convention to write 'kelvin' without a capital 'k' and to use capital 'K' for the symbol, was established in 1967 at the 13th General Conference on Weights and Measures). Since an increment of 1K is the same distance as 1 Celsius degree,

K will be used whenever temperature differences are intended, and °C will be used whenever specific temperatures are intended. Since, in calculations involving temperature, the constants in the equation must be multiplied by either absolute temperature (given by the temperature in kelvin) or by temperature differences, the temperature units of any physical constant or parameter will always be in kelvin. Whether one should use absolute temperature or a temperature difference is guided by an understanding of the process being represented by the equation. For example, the emission of radiation by an object depends on the temperature of that object and not its departure from the temperature of the surroundings, so one uses the absolute temperature. On the other hand, the flow of heat through a window quite intuitively depends on the difference between temperature on the two sides of the window, so one will simply use this temperature difference in the equation used to compute the flow of heat.

Acknowledgements

I would like to thank the people listed below for kindly reviewing parts (or, in some cases, all) of the indicated chapters, in many cases also providing additional information that I would not otherwise have obtained. In addition, numerous people responded to questions about their work and in so doing contributed greatly to the final product. I accept responsibility, however, for any errors, misconceptions, or important omissions that may remain. Comments from readers of this book are certainly welcome.

Chapter 3: Guohui Gan (Nottingham, UK), Ted Kesik (University of Toronto), Eleanor Lee (LBNL, USA), Steve Pope (NRCan, Canada), S. Chandra Sekhar (National University of Singapore), Theodosiou Theodore (Aristotle University of Thessaloniki)

Chapter 4: Andreas Athienitis (Concordia, Canada), Steve Pope (NRCan, Canada)

Chapter 5: Doug Cane (Caneta Research, Canada), John Shonder (ORNL, USA)

Chapter 6: John Andrepont (Cool Solutions, USA), Ben Costelloe (Dublin Institute of Technology), Shashi Dharmadhikari (Ingérop-Litwin, France), Guohui Gan (Nottingham, UK), Srinivas Garimella (Georgia Tech), Soteris Kalogirou (Higher Technical Institute, Nicosia), Maria Kolokotroni (Brunel, UK), Andrew Lowenstein (USA), Steve Pope (NRCan, Canada), Manuel Sanchez (Spain), Theodosiou Theodore (Aristotle University of Thessaloniki)

Chapter 7: Greg Allen (Sustainable Edge, Canada), Fred Belusa (Belnor, Canada), Richard de Dear (Australia), J. Niu (Hong Kong Polytechnic University), Steve Pope (NRCan, Canada), S. Chandra Sekhar (National University of Singapore), Jelena Srebic (Penn State, USA)

Chapter 8: Gary Klein (California Energy Commission)

Chapter 9: Eleanor Lee (LBNL, USA), Russell Leslie (Rensselaer Polytechnic Institute, USA), Danny Li (City University of Hong Kong), Jean-Louis Scartezzini (EPFL, Switzerland)

Chapter 11: Riccardo Battisti (Uniroma, Italy), T.T. Chow (City University of Hong Kong), Soteris Kalogirou (Higher Technical Institute, Nicosia), Nazir Kherani (University of Toronto), B. Sandnes (University of Oslo)

Chapter 12: Doug Hooton (University of Toronto), Graham Treloar (Deakin University, Australia)

Chapter 13: Guliano Todesco (Jacques Whitford, Ottawa)

Chapter 14: Heinrich Feistner (Energy Efficiency Office, City of Toronto)

Chapter 15: John Andrepont (Cool Solutions, USA), Shashi Dharmadhikari (Ingérop-Litwin, France), Recep Yumrutaş (University of Gaziantep, Turkey)

Finally, I would like to thank Katie Myrans for expert assistance in preparing most of the figures found in this book.

List of Acronyms and Abbreviations

ABDS	Automated Building Diagnostic Software
AC	alternating current (pertains to electricity)
ACH	air changes per hour
ACH_{50}	air changes per hour at a pressure difference of 50 Pascals
AFUE	annual fuel utilization efficiency
AIC	advanced insulation component (pertains to refrigerators and freezers)
AIP	advanced insulation panel (pertains to refrigerators and freezers)
ASHP	air-source heat pump
ASHRAE	American Society of Heating, Refrigeration and Air Conditioning Engineers
ASPO	Association for the Study of Peak Oil
ATES	aquifer thermal energy storage
BAS	Building Automation Systems
BAU	business-as-usual (scenario)
BiPV	building integrated photovoltaic
BOS	balance of system
BREEAM	Building Research Establishment Environmental Assessment Method
BTES	borehole thermal energy storage
CAV	constant air volume
CBIP	Commercial Building Incentive Program
CC	chilled ceiling
CDD	cooling degree days
CEPHEUS	Cost-Effective Passive Houses as European Standards
CFC	chlorofluorocarbon
CFD	computational fluid dynamics
CFL	compact fluorescent lamp
CIES	Community-Integrated Energy System
CLTC	climatic limit for thermal comfort
CMH	Ceramic metal halide (pertains to lamps)

COP	coefficient of performance
CPC	compound parabolic collector
CRF	cost recovery factor
CTES	cavern thermal energy storage
DC	direct current (pertains to electricity)
DCV	demand-controlled ventilation
DHW	domestic hot water (i.e. water for showers and washing)
DOAS	dedicated outdoor air supply
DOE	Department of Energy (US)
DT	daylight transmission
DV	displacement ventilation
EAHP	exhaust-air heat pump
EC	electrochromic (pertains to windows)
ECPM	electrically commutated permanent magnetic induction
EIFS	External Insulation and Finishing System
EL	electro-luminescent (pertains to lighting)
EMCS	energy monitoring and control system
EPA	Environmental Protection Agency
EPS	expanded polystyrene (a form of insulation)
ER	energy rating
ESTIF	European Solar Thermal Industry Federation
EU	European Union, consisting of the following 15 countries prior to June 2004: Austria, Belgium, Denmark, Finland, France, Germany, Greece, Ireland, Italy, Luxembourg, The Netherlands, Portugal, Spain, Sweden, the UK. On 1 June 2004, another ten countries joined the EU: Cyprus, Czech Republic, Estonia, Hungary, Latvia, Lithuania, Poland, Romania, Slovak Republic,

	Slovenia
FMS	Facility Management Systems
FSU	former Soviet Union
GHGs	greenhouse gases
GSHP	ground-source heat pump
GWP	global warming potential
HCFC	hydrochlorofluorocarbon
HDD	heating degree day
HERS	Home Energy Rating System
HFC	hydrofluorocarbon
HID	high intensity discharge (pertains to lighting)
HIF	high-intensity fluorescent
HIR	halogen infrared reflecting
HRSG	heat recovery steam generator
HRV	heat recovery ventilator
HVAC	heating, ventilation and air conditioning
ICF	insulating concrete forms
IDP	integrated design process
IEA	International Energy Agency (an agency of the OECD)
IGCC	integrated gasification-combined cycle
LCP	laser-cut panel
LED	light emitting diode
LEED	Leadership in Energy and Environmental Design
LLTD	lower limit of thermal discomfort
LNG	liquefied natural gas
MPP	maximum power point
MSW	municipal solid waste
MV	mixing ventilation
NGCC	natural gas combined cycle
NIR	near infrared (pertains to a part of the solar spectrum)
OECD	Organisation for Economic Co-operation and Development, consisting of the following 30 countries: Australia, Austria, Belgium, Canada, Czech Republic, Denmark, Finland, France, Germany, Greece, Hungary, Iceland, Ireland, Italy, Japan, Korea, Luxembourg, Mexico, The Netherlands, New Zealand, Norway, Poland, Portugal, Slovak Republic, Spain, Sweden, Switzerland, Turkey, the UK, the US

O&M	operation and maintenance
OSB	oriented strand board
PCM	phase-change material
PDEC	passive downdraught evaporative cooling
PEM	personal environmental module
PMMA	polymethyl methacrylate
POU	point of use (pertains to hot-water heaters)
ppbv	parts per billion by volume
ppmv	parts per million by volume
PSC	permanent split capacitor
PV	photovoltaic
PV/T	integrated photovoltaic thermal energy solar collector
RH	relative humidity
RL	recirculation loop (pertains to domestic hot water systems)
RSI	resistance (Système International)
SC	shading coefficient
SESG	Scottish Energy Systems Group
SHGC	solar heat gain coefficient
SIP	structural insulated panel
SPF	seasonal performance factor (pertains to heat pumps)
SSE	Surface Meteorology and Solar Energy
TAP	Tariff Analysis Project
TIM	transparent insulation material
TMY	Typical Meteorological Year
UFAD	underfloor air distribution
UNFCCC	United Nations Framework Convention on Climate Change
USGBC	US Green Building Council
UTES	underground thermal energy storage
UV	ultraviolet (pertains to a part of the solar spectrum)
VAV	variable air volume
VIP	vacuum insulation panel
VOC	volatile organic carbon
VSD	variable speed drive
XPS	extruded polystyrene (a form of insulation)

one

Environmental and Energy-Security Imperatives

This chapter briefly introduces the reasons why reducing energy use in buildings is of the utmost importance: as part of a broader strategy to avert potentially catastrophic climatic change (global warming); to help resolve serious issues of local and regional air pollution (ground level ozone, acid rain, trace toxic chemicals); as a hedge against early increases in the price of fossil fuels as extraction rates peak and then decline; and, in some regions, to address current or pending shortages of natural gas, which is used extensively for heating and for the production of hot water. It concludes with a working definition of 'sustainable' buildings, a term that is widely used without being defined.

The emissions of greenhouse gases (GHGs) (responsible for recent and projected climatic warming, as discussed below) and of almost all atmospheric pollutants are tied to the use of fossil fuels (natural gas, oil and coal) for energy. Buildings rely on fossil fuel energy, either directly through on-site use of natural gas, oil and

(in some parts of the world) coal for space heating and for hot water, or indirectly through the use of electricity that is generated from fossil fuels. Reductions in energy use in buildings will lead to a reduction in the use of fossil fuels, thereby reducing emissions of GHGs and reducing local-to-regional air pollution problems.

1.1 The climatic imperative

A number of naturally occurring gases in the atmosphere – water vapour (H_2O), carbon dioxide (CO_2), ozone (O_3) and methane (CH_4) in particular – make the Earth's climate 33K warmer than it would be otherwise by trapping infrared radiation that would otherwise be emitted to space.[1] This warming effect is popularly referred to as the 'greenhouse effect', and the gases contributing to it are called greenhouse gases (GHGs). Water vapour is the most important contributor to the natural or background greenhouse effect. However, during the past 200 years, humans have directly caused large increases in the concentrations of a number of other GHGs: the CO_2 concentration has increased by over 30 per cent (from 280 parts per million by volume, or ppmv, to 380ppmv by the time this book is published), the methane concentration has increased by more than a factor of 2.5 (from 0.7ppmv to 1.75ppmv), the nitrous oxide (N_2O) concentration has increased by about 15 per cent (from 275 parts per billion by volume (ppbv) to 315ppbv), and the lower atmosphere (or tropospheric) ozone concentration has increased by a factor of 2–5 (depending on location) in the northern hemisphere. In addition, several classes of entirely artificial GHGs have been created and added to the atmosphere, including the chlorofluorocarbons (CFCs) and their replacements, the hydrochlorofluorocarbons (HCFCs) and hydrofluorocarbons (HFCs). The buildup of these gases has strengthened the natural greenhouse effect, trapping an additional 2–3W/m^2 of infrared radiation averaged over the Earth's surface and tending to warm the climate. The increase in CO_2 concentration accounts for about half of this strengthening, while the increase of CH_4 contributes another 20 per cent. The concentration of water vapour in the atmosphere has also increased, not as a direct result of human activity, but due to the warming of climate induced by the increases in the above-mentioned greenhouse gases. This constitutes a strong positive feedback that increases the expected warming by 50–100 per cent.

Figure 1.1 shows the CO_2 and CH_4 increase during the past 200 years in the context of natural variations during the past 400,000 years. Continuous direct observations of the atmospheric CO_2 concentration began in 1958 at Mauna Loa, Hawaii, while systematic observations of the CH_4 concentration began only in 1978. Concentrations prior to these times have been determined by measuring the gas concentrations in air bubbles in Antarctic ice cores, the deepest and oldest of which extend back over 750,000 years. As snow accumulates on an ice sheet, it is gradually transformed into ice, and the air present between the snowflakes eventually becomes sealed from the atmosphere as bubbles, thereby providing samples of past atmospheric composition. Both CO_2 and CH_4 have undergone repeated natural fluctuations, but the increases in

Figure 1.1 Variation in atmospheric CO_2 and CH_4 (methane) concentration during the last 400,000 years as deduced from the Vostok ice core (Petit et al, 1999) (thin lines), and during the past 200 years (heavy line)

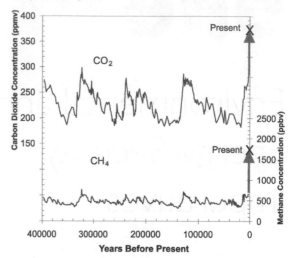

Source: Data in electronic form were obtained from the US National Oceanographic and Atmospheric Administration (NOAA) website (www.ngdc.noaa.gov/paleo)

the concentrations of these two greenhouse gases have been rapid (compared with natural fluctuations) and far outside the bounds of natural variability during the past 400,000 years.

CO_2 is emitted to the atmosphere during the combustion of fossil fuels and from deforestation (commonly referred to as land 'use' changes). Fossil fuel emissions during the 1990s averaged around 6 Gigatonnes (Gt) of carbon per year, while land use changes caused a further emission of 1.6 ± 1.0Gt C per year (giga = billion; see Appendix A for the definition of this and other prefixes). The rate of fossil fuel emission could grow several-fold during the coming century, while long-term land use emissions are constrained by the availability of remaining forests and are unlikely to rise much above the current rate before declining. Thus, in the long term, fossil fuel emissions are of much greater concern than land use emissions as far as the buildup of atmospheric CO_2 is concerned. Methane emissions occur in association with coal mining, extraction of oil and natural gas, distribution of natural gas (which is largely methane) and from rice farming, ruminant livestock (such as cattle), sanitary landfills, sewage treatment plants and biomass burning. Emissions from many of these sources are relatively easy to control, and methane molecules remain on average only 10 years in the atmosphere before being removed (largely through reaction with hydroxyl (OH)), so the methane buildup is largely reversible. In contrast, the natural processes that remove CO_2 from the atmosphere are quite slow (requiring thousands of years to remove most of the emitted CO_2), so the present and future increase in concentration is irreversible for all practical purposes.[2] Furthermore, projected future increases in CO_2 concentration under a typical business-as-usual scenario will dominate further increases in the greenhouse effect. For these reasons, CO_2 emission from the combustion of fossil fuels will be the single largest factor in future human-induced warming of the climate and is of greatest concern.

1.1.1 Past changes of climate

Figure 1.2 shows the variation in global average surface temperature over the period 1856–2005. This curve is a composite of sea surface temperature variations in the oceanic portion of the Earth, and surface air temperature variations over the land portion of the Earth (a detailed discussion of the methods used in constructing this curve can be found in Harvey (2000, Chapter 5)). These data indicate that the Earth's climate has warmed by 0.6–0.8K, on average, during the past 140 years or so. There is an overwhelming body of independent evidence that, collectively, confirms that the Earth's climate is indeed warming: balloon-based observations of temperature in the middle atmosphere, reconstructions of temperature trends in the atmosphere from satellite data, an observed retreat and thinning of Arctic sea ice, reduced extent of snow cover on land in the northern hemisphere, reduced ice cover on lakes, northward and upward movements of climate-sensitive plants and animals, a lengthening of the growing season in the northern hemisphere, warming of the land subsurface as measured in boreholes and a widespread warming of the upper 3000m of the ocean. However, one cannot judge whether the warming seen in Figure 1.2 is unusual or not without a longer record of global-scale temperature variation.

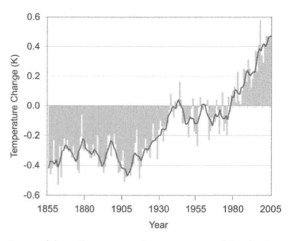

Figure 1.2 Variation in global average surface temperature during the period 1856–2005

Source of data: Climate Research Unit, University of East Anglia (www.cru.uea.ac.uk)

Fortunately, such a record is available, through temperature variations that can be reconstructed from the chemical composition of annual layers in polar ice caps, through the annual variations in tree-ring width (largely at middle latitude sites) and from the chemical composition of annual layers in coral reefs (from low latitudes). Figure 1.3 shows variation in northern hemisphere average temperature from AD 1000 to AD 1980 as reconstructed by Mann et al (1999). Also shown in Figure 1.3 are the directly measured variations in global average temperature from Figure 1.2. Prior to 1900, decadal-scale variations in hemispheric average temperature were generally no more than ±0.1K. Furthermore, there was a long-term downward trend of about 0.2K from AD 1000 to AD 1900. This trend was abruptly reversed, starting around 1900, and the warming during the past 100 years stands out as highly unusual. Methods of reconstructing past global-scale temperature changes and of quantifying the various factors that can potentially explain these changes are reviewed in Jones and Mann (2004). Alternative reconstructions of northern hemisphere temperature variations have been made, some showing greater century time-scale variability than shown in Figure 1.3. However, all scientifically credible reconstructions agree that the late 20th century is the warmest period during at least the past 1000 years (Kerr, 2005). There can

Figure 1.3 Variation in northern hemisphere average surface temperature based largely on ice core, tree ring and coral reef data, as reconstructed by Mann et al (1999) (thin grey line), 20-year running mean of the annual paleoclimatic data (thick black line), and directly observed temperature variation of Figure 1.2 (thin black line)

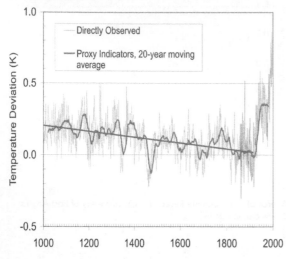

Source: The proxy data of Mann et al (1999) in electronic form were obtained from the US National Oceanographic and Atmospheric Administration (NOAA) website (www.ngdc.noaa. gov/paleo)

be little doubt that this warming is largely due to the concurrent buildup of GHG concentrations during this time period, given that the heat trapping calculated from the measured increase of GHGs is more than adequate to explain the full magnitude of observed warming, while plausible alternative mechanisms (solar variability, volcanic eruptions, heat exchange with the deep ocean) have been comprehensively investigated and found to be capable of explaining only a small fraction of the observed warming (in the case of solar variability) or none at all (in the case of volcanic activity, which has had a net cooling effect, or in the case of heat exchange with the deep ocean, which has acted to passively dampen radiatively driven temperature changes).

1.1.2 Prospective future changes of climate

Given a scenario of CO_2 emissions, the buildup of atmospheric CO_2 concentration can be computed with reasonable accuracy using a computer *carbon cycle model*. Carbon cycle models compute the absorption of a portion of the emitted CO_2 by the terrestrial biosphere (primarily by forests)

and by the oceans. At present, between one half and two-thirds of the annually emitted CO_2 is absorbed by these two sinks, while the balance accumulates in the atmosphere. Given a scenario for the buildup of CO_2 and other GHGs, future climatic changes can be computed using a *climate model*.

The single most important parameter in determining the climatic response to a given GHG buildup is the *climate sensitivity*. This can be defined as the long-term, globally averaged warming for a doubling of atmospheric CO_2 concentration. To a good approximation, the climate response to increases in CO_2 and other GHGs depends only on the net heat trapping of all the gases, and not on the specific gases contributing to the heating. Thus, it is possible to speak of the *equivalent* CO_2 concentration – the concentration of CO_2 alone that would have the same heat-trapping effect as that from the mixture of gases present in reality. A wide variety of evidence (discussed in Chapter 9 of Harvey, 2000) indicates that the climate sensitivity is likely to fall between 2–3K, although a much broader range (from 1K to 5K or more) cannot be ruled out (Andronova and Schlesinger, 2001; Forest et al, 2002; Harvey and Kaufmann, 2002; Knutti et al, 2002). Under most business-as-usual scenarios, GHG concentrations rise far in excess of a CO_2 doubling equivalent (see Prentice et al, 2001, figure 3.12) by 2100. It is evident from these considerations alone that dramatic changes in the world's climate will occur during the next century unless strong preventative action is taken.

Humans have also emitted precursors to a variety of suspended particles, known as aerosols, that have a net cooling effect on climate and have therefore offset a portion of the heating effect due to the buildup of GHGs. Were it not for this aerosol cooling effect, the warming during the last 100 years would have been larger. Aerosol particles remain in the atmosphere for only a few days before being washed out with rain, so they need to be continuously replenished. The primary aerosol – sulphate – is associated with acid rain, but sulphur emissions are being reduced in much of the world in order to reduce the damage caused by acid rain. As emissions are reduced, the aerosol cooling effect will weaken, leading to a temporary acceleration in global warming. The magnitude of the acceleration depends on

Figure 1.4 Comparison of the observed temperature record of Figure 1.2 (up to 2004) and global average temperature changes (up to 2100) as simulated using a simple climate model with prescribed climate sensitivities for a CO_2 doubling (ΔT_{2x}) of 2K and 4K

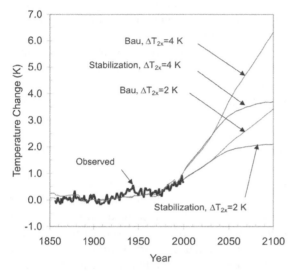

Note: The assumed aerosol cooling effect in 1990 is 30 per cent of the greenhouse gas heating for a 2K sensitivity and 50 per cent of the greenhouse gas heating for a 4K sensitivity. In this way we obtain a reasonable fit to the observed warming for both assumed climate sensitivities. The simulations were extended to 2100 for a scenario in which fossil fuel CO_2 emissions increase by a factor of four while total aerosol cooling decreases by a factor of two (labelled 'BAU' – business-as-usual scenario), and for simulations in which the CO_2 concentration is stabilized at 450ppmv.

Figure 1.5 Proxy temperature variations of Figure 1.3, directly observed temperature variations of Figure 1.2 and business-as-usual model simulations of Figure 1.4

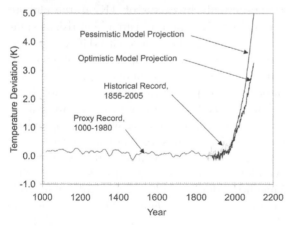

how large the aerosol cooling effect is at present, something that is highly uncertain.

Figure 1.4 shows past and projected future global average warming for a typical business-as-usual scenario of increasing GHG emissions but falling aerosol emissions, for climate sensitivities of 2K and 4K (fossil fuel CO_2 emissions grow from 6.4Gt C/year at present to 20.4Gt C/year by 2100). The globally averaged warming by the end of this century ranges from 3K to 7K (accounting for uncertainties in emissions and in the carbon cycle). Even the optimistic case produces a warming by the end of this century comparable to the difference between the last ice age and the ensuing interglacial climate – a globally averaged difference of 4–6K. However, the warming that occurred at the end of the last ice age occurred over a period of 10,000 years, whereas comparable GHG-induced warming could occur over the next 100 years – 100 times faster. Also shown in Figure 1.4 are results where fossil fuel emissions are reduced to about 1Gt C/year by 2100, which might be sufficient to stabilize the atmospheric CO_2 concentration at 450ppmv. Since this corresponds closely to the equivalent of a CO_2 doubling, the global mean warming by 2100 approaches 2K and 4K for the 2K and 4K sensitivities. The unprecedented and abrupt nature of prospective climatic change during the next century is dramatically illustrated in Figure 1.5, which combines the data shown in Figures 1.3 and 1.4.

1.1.3 The United Nations Framework Convention on Climate Change

The broad principles that should be adhered to in responding to the threat of global warming are laid out in the United Nations Framework Convention on Climate Change (UNFCCC), a document signed and ratified by 186 of the world's countries. The UNFCCC declares its 'ultimate objective' to be:

> stabilization of greenhouse gas concentrations in the atmosphere at a level that would prevent dangerous anthropogenic interference with the climate system. Such a level should be achieved within a time frame sufficient to allow ecosystems to adapt

naturally to climate change, to ensure that food production is not threatened and to enable economic development to proceed in a sustainable manner (UN, 1992, Article 2).

Although the UNFCCC does not specify what constitutes 'dangerous anthropogenic interference', three criteria in determining it – protection of natural ecosystems, food production and sustainable development – are specified.

Inasmuch as the UNFCCC has been accepted and ratified by almost all nations in the world, a number of value judgements have already been implicitly accepted, and do not need further debate. For example, the value judgement has already been made that ecosystems are worthy of protection (by ensuring that rates and magnitudes of climatic change are such as to allow their natural adaptation), irrespective of their economic value to humans. The requirement that GHG concentration caps be such as to enable economic development might appear to place a lower limit on the allowable caps, and possibly conflict with requirements concerning protection of ecosystems and of food production, except for the proviso that economic development be *sustainable*. This rules out the common practice (among economists) of discounting future costs, since the decisions made in that case are anything but sustainable (under discounting, if the interest rate is 3 per cent, a dollar ten years from now is worth only $0.97^{10} = 0.74$ dollars today, while a dollar 100 years from now is worth only 4.8 cents today). Once discounting is rejected, conventional economic analysis indicates that very large emission reductions are justified (Azar and Sterner, 1996), so there is no conflict with the *imperative* of protecting ecosystems and food production.

1.1.4 Threats to ecosystems, food production and human security

Having established the UNFCCC and its goal of avoiding GHG concentrations that endanger ecosystems and food production as the overarching principle, it is worthwhile to examine the risks associated with various degrees of global average warming. As noted above, a CO_2 concentration of 450ppmv roughly corresponds to a doubling of the pre-industrial atmospheric CO_2 concentration when the heating effect of other GHGs is taken into account, assuming *stringent* limitations in emissions of other GHGs (Harvey, 2004). As discussed above, the climate sensitivity – and thus the eventual warming for a mere 450ppmv CO_2 concentration – could be as low as 1K or as high as 5K.

A 1K global average warming is not particularly alarming although, even for this sensitivity, an eventual warming far in excess of 1K would occur for business-as-usual GHG emission scenarios. Furthermore, an increase in atmospheric CO_2 concentration to 450ppmv would have profound impacts on marine life, irrespective of the climate sensitivity, through changes in ocean chemistry. The surface water pH of the ocean would decrease by about 0.2 units, and the supersaturation of surface water with respect to calcite would decrease by about 25 per cent according to simulations with a carbon cycle model by Harvey (2003). Calcite is the structural material of coral reefs, and is the skeletal material of calcareous micro-organisms, at the base of the marine food web. The growth rates and health of these organisms will be adversely affected by these changes, but the full impact is unclear (Kleypas et al, 1999; Riebesell et al, 2000). Thus, the case can be made to limit the atmospheric CO_2 concentration to no more than 450ppmv even if the climate sensitivity is as small as 1K.

However, there is a substantial risk that the climate sensitivity for a CO_2 doubling itself is substantially in excess of 1K. If the climate sensitivity is 5K, then we run a high risk of catastrophic climatic change even if GHG concentrations are limited to a CO_2 doubling equivalent. There is a continuously increasing risk of serious impacts as the projected warming increases from 1K to 5K. A 1K warming (i.e. close to that already experienced) to a 2K warming (optimistic case) risks killing most or all of the coral reefs in the world due to 'bleaching' – the expulsion of symbiotic algae from the coral (Kleypas, 1998; Hoegh-Guldberg, 1999; Wilkinson, 1999; Wellington et al, 2001; Dennis, 2002). Estimates of the number of people at risk from water shortages by Parry et al (2001), as a function of global average warming, show an abrupt increase between 1K and 2K global mean warming. Assessments of the impact of a doubled-CO_2 climate frequently show decreases in agricultural yields in specific regions of 10–20 per cent and more, even after allowing for the beneficial physiological effects of higher CO_2 and for adaptation (see Gitay et al, 2001, table 5.4). Most studies of agricultural impacts assume full realization of the beneficial physiological effects of higher CO_2 (stimulation of photosynthesis and reduced water loss) as measured in experimental settings, although there are substantial reasons for doubting whether the full beneficial effects will occur in practice (Wolfe and Erickson, 1993; Darwin and Kennedy, 2000). A comprehensive assessment by 19 ecologists based in Europe, North and South America, Africa and Australia shows that, for a scenario with a global mean warming of 2K by 2050, between 15 and 37 per cent of terrestrial animal species will be committed to extinction by 2050 (Thomas et al, 2004). A slower warming, even to the same final temperature, would reduce the losses.

A global mean warming of 3–4K over the course of the next century could have highly disruptive effects on many forest ecosystems (Arnell et al, 2002). Leemans and Eickhout (2004) performed a global scale assessment of the impact of 1K, 2K, and 3K warming on terrestrial ecosystems, using a series of computer simulation models. Their simulations indicate that a 1K global mean warming will cause changes in the type of ecosystem over about 10 per cent of the Earth's land surface, increasing to about 22 per cent for a 3K global mean warming. However, even at 1K warming over a period of a century, only 36 per cent of impacted forests can shift in step with the climatic change (that is, 'adapt'), due to their long generation times and relatively slow rate of dispersal. At 3K warming by 2100, only 17 per cent of impacted forests can keep up. Some studies suggest that the issue is not merely one of adaptation, but of the very survival of some forests. Of particular concern are simulations by the HadCM3 atmosphere–ocean model at the Hadley Centre in the UK, in which almost the entire Amazon rainforest is replaced by desert by the 2080s due to a drastic reduction in rainfall in the Amazon basin in association with a global mean warming of about 3.3K (White et al, 1999). In the HadCM2 model (an earlier version of the Hadley Centre model), the Amazon rainforest is still largely intact by 2100 (when the simulation

ends) but with greatly reduced productivity, and a positive climate–carbon cycle feedback results in an atmospheric CO_2 concentration in 2100 of 1000ppmv for an emission scenario that otherwise would have led to 750ppmv by 2100 (Cox et al, 2000).

There is a danger that modest global mean warming could provoke a very large sea level rise, in the order of 10–12m over a period of 1–2000 years, but with much of it occurring on a time scale of a few centuries. Computer simulations indicate that local warming as small as 3K could provoke the irreversible melting of the Greenland ice cap (corresponding to a 7m sea level rise), and that a 10K local warming could provoke the collapse of the West Antarctic ice sheet (producing a 3–4m sea level rise). However, computer models of the West Antarctic ice sheet may be too resistant to warming, as they do not simulate ice streams well (Oppenheimer and Alley, 2005). A critical question is: what amount of global mean warming would be associated with local temperature changes sufficient to provoke the melting of Greenland or the collapse of the West Antarctic ice sheet? It could be as little as 1–2K, implying that current greenhouse gas concentrations are dangerous. Hansen (2004, 2005) notes that the buildup of ice sheets is a slow, dry process, while the collapse of ice sheets is a comparatively fast, wet process, lubricated by glacial meltwater.

It follows from the above that, *if* the climate sensitivity is only 1K, then GHG concentration increases that are collectively equivalent to a CO_2 doubling might be permissible, although there are still important concerns about impacts on coral reefs and the marine food chain through changes in marine chemistry, as well as a small risk of destabilization of the Greenland ice cap. If, on the other hand, the climate sensitivity is close to 5K, then GHG concentrations must be kept substantially below a CO_2-doubling equivalent. Since, at this point in time, we do not know what the true climate sensitivity is, and the repercussions of changes in marine chemistry are uncertain, a CO_2-doubling equivalent (450ppmv) must be regarded as dangerous. It is not necessary to show that 450ppmv CO_2 leads to *certain* impacts that violate the UNFCCC principles of protecting ecosystems and food security; it is simply necessary to show that such a concentration represents dangerous interference in the climate system – that is, that there are non-negligible risks to important ecosystems and to food security. This is clearly the case.

Since the goal of the UNFCCC is to *avoid* dangerous anthropogenic interference in the climate system, and since 450ppmv CO_2 is dangerous (given current uncertainties, and heating effects from other gases), it follows that current policy should be directed at stabilization in the range of 350–400ppmv – a range also advocated by Azar and Rodhe (1997). The mean annual CO_2 concentration will reach 380ppmv by 2006.

1.1.5 Stabilization of atmospheric CO_2 concentration at 450ppmv or less

General pathways that lead to stabilization of atmospheric CO_2 in the 350–450ppmv range are presented in Harvey (2004). They entail completely phasing out the use of fossil fuels by 2075–2100, and some include capturing and injecting into the deep ocean or in deep underground reservoirs some of the CO_2 produced from fossil fuels prior to a complete phase-out in the use of fossil fuels. In every case, we will reach at least 450ppmv CO_2. However, there is the possibility of drawing down atmospheric CO_2 by growing biomass energy crops (which remove CO_2 from the atmosphere as they grow) and then capturing and burying the CO_2 when the energy crops are used, or by building up carbon in soils or in standing biomass (this requires first halting the destruction of existing forests). In order to have any possibility of supplying future energy needs entirely through carbon-free energy, the energy efficiency of the global economy must improve dramatically – eventually leading to improvements in *system* efficiencies by a factor of six to eight. Buildings have an essential role to play in this process, both as loci for improvements in energy efficiency and as collectors of solar energy in the forms in which it can be directly used – as light, as air movement induced by temperature differences, as low-grade (i.e. low-temperature) thermal energy for heating and for hot water – and its conversion to electricity.

1.2 Local and regional air pollution and other environmental impacts of fossil fuels

The use of fossil fuels imposes a number of costs on society and the environment, in addition to rapid and dangerous climatic change, that are not reflected in the prices paid for fossil fuels. These costs include the environmental damage and adverse health impacts caused by pollution from fossil fuel combustion; disability costs and deaths of coalminers; and military expenditures to maintain secure oil supplies. The major forms of atmospheric pollution associated with fossil fuels are as follows:

- ozone ('smog'), produced by chemical reactions involving oxides of nitrogen (NO_x), volatile organic carbon (VOC), methane (CH_4) and carbon monoxide (CO);
- carbon monoxide itself, due to its direct effect on human health;
- particulates (mainly sooty emissions from diesel vehicles);
- acid rain, produced by emissions of NO_x and oxides of sulfur (SO_x);
- trace toxic chemicals (heavy metals such as arsenic, cadmium and mercury; or carbon compounds such as benzene, formaldehyde or toluene) having a variety of carcinogenic, neurological or other effects.

CO_2, the primary greenhouse gas produced by the use of fossil fuels, is not a pollutant in the traditional sense of the word. Indeed, a higher CO_2 concentration has a direct beneficial effect on plants, while having no direct effect on humans over the range of foreseeable future concentrations. Rather, the primary effect of CO_2 is its potential to induce catastrophic climatic change if its emissions are not severely constrained. Reduction of air pollution, on the other hand, will have mixed and relatively modest impacts on climate: ozone, CO and particulate emissions have a modest warming effect on climate, so their reduction will contribute to slowing global warming, while acid rain precursors have a potentially significant cooling effect, so their reduction will lead to a short-term acceleration in global warming. Trace toxic chemicals have no effect on climate.

Societal and environmental costs that are not included in the market price of energy are referred to as *externalities*. Estimation of the magnitude of fossil fuel externalities is fraught with enormous uncertainty and is, to some extent, subjective. This is because the major externalities are health impacts on humans, including premature death. Various techniques have been used to try to quantify these impacts in monetary terms. Mortality-related impacts are often quantified based on the declared willingness of individuals to pay extra money in order to reduce the risk of death by a given amount. Wealthy individuals tend to be more willing than poor people to pay in order to reduce the risk of death, so this approach has the perverse effect of valuing human life more highly in developed countries than in developing countries. Health care costs and the cost of reduced agricultural and forest productivity can be more easily quantified. The external cost in generating a given amount of electricity or in providing heat to a building depends on the technology used to generate the electricity or heat (as this determines the amount of pollution emitted) and on the population density in the region containing the power plant (as this determines the number of people that will be adversely affected). Sundqvist (2004) provides a recent discussion of the differences among various estimates of the external cost of electricity generation.

Figure 1.6 gives the external non-climatic cost of electricity generation from fossil fuels in the north-eastern US, as estimated by Roth and Ambs (2004) for various conventional and advanced fossil fuel technologies (which are explained in Chapter 15, Section 15.1). The top panel gives the breakdown of external costs due to atmospheric emissions of NO_x, SO_x, VOCs, CO and particulate matter; not included are costs due to trace emissions of toxic substances. The costs that are included are based on the cost of controlling emissions, which is evidently a very poor proxy for the damage cost of the emissions themselves. The middle panel includes the external costs due to emissions and a variety of other costs: impacts on land use, water and wildlife; increased dependence on imported energy in the case of oil and natural gas; a premium due to the limitation of future supply options arising from the depletion of coal, oil and natural gas

Figure 1.6 Non-climatic external costs associated with the generation of electricity from fossil fuels, municipal solid waste (MSW) and landfill gas in the north eastern US, (a) related to emissions of pollutants, (b) total emission externality and other externalities, and (c) range of estimates of total externalities.

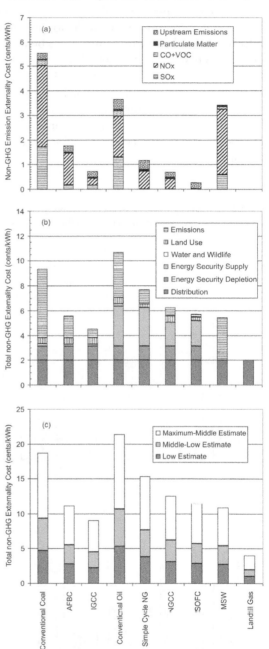

Source: Source of estimates, Roth and Ambs (2004)
Note: (a) Middle estimate of non-GHG emission externality, based on the cost of controlling emissions; (b) middle estimate of total non-GHG externality; and (c) low, middle and high estimates of total non-GHG externality, but with extreme values omitted.

(discussed in Section 1.3); and impacts from electric power transmission. The lower panel gives low, middle and high estimates of the total non-climatic external costs, but with the largest and lowest estimates omitted. For conventional coal-generated power plants with no pollution controls, external costs are estimated to be about 5–18 US cents/kilowatt hour (kWh), compared with 2.5–11 cents/kWh for advanced technologies that are not yet commercially available. External costs for natural gas-generated electricity are 4–15 cents/kWh for conventional technology and 2.5–11 cents/kWh for advanced technologies.

Figure 1.7 Low and high estimates of the non-climatic external cost of electricity generation from (a) coal, and (b) natural gas in various European countries

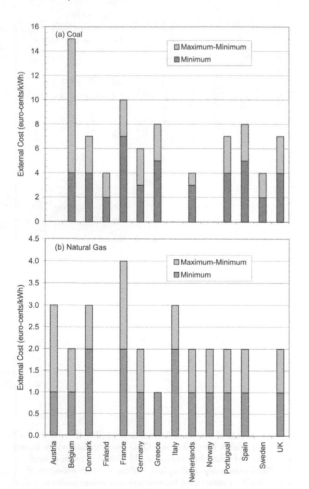

Source: European Commission (2001)

Figure 1.7 gives the non-climatic external costs for present coal- and natural gas-generated electricity in various European Countries as estimated by the European Commission's ExternE project (see www.externe.info). For coal-generated electricity, the minimum estimated non-climatic external costs range from 2 to 7 eurocents/kWh, depending on the country, while the maximum estimated costs range from 4 to 15 cents/kWh. For electricity from natural gas, costs range from 1 to 2 eurocents/kWh (minimum estimate) or 2–4 eurocents/kWh (maximum estimate). These costs are large compared with the market cost of electricity, which ranges from 5 to 16 US cents/kWh (see Chapter 2, Section 2.3). These are real costs to individuals in a society, paid either as consumers or as taxpayers. Nevertheless, these estimates may be too low; as noted by Kammen and Pacca (2004, p326), the cost for two coal power plants in the US is about 50 cents/kWh, based on the estimated number of deaths attributable to air pollution from the plants, and using $5 million as the 'value of a statistical life'. Unless external costs are included in the price of fossil fuels, fossil fuels will often be chosen over renewable energy sources when the latter are less costly, consumers will underinvest in energy efficiency measures and manufacturers will have less incentive to develop more efficient equipment.

Relatively little work has been done on air pollution–health linkages in developing countries. Health effects are dominated by indoor air pollution from biomass fuel, although SO_2 (related in part to coal combustion, both for electricity generation and, in some countries, for space heating) is also an important pollutant. For documents on health impacts in both developed and developing countries, see www.airimpacts.org. Chameides et al (1994) note that there is significant geographical overlap worldwide between major food-producing regions and regional air pollution. They estimate that 9–35 per cent of the world's cropland is exposed to summer ozone concentrations of 50–70ppbv, which reduce crop yields by 5–10 per cent. Overall crop losses due to ozone pollution are likely to be of the order of a few percent.

Estimates of the non-climatic external costs associated with heating of buildings or providing hot water with natural gas are developed in

Table 1.1 Computation of external cost of natural gas heating fuel based on NOx and CO emissions

	Damage ($/tonne)						
	Min	Max					
NOx damage	1050	10,000					
CO damage	500	2500					

	Emissions (gm/GJ)			Emission damage (cents/kWh of fuel)		Total damage (cents/kWh of fuel)	
	Min NOx	Max NOx	CO	Min	Max	Min	Max
Large boilers (>30MWth)	40	60	35	0.02	0.25	1.80	2.03
Small boilers (<30MWth)	13	20	35	0.01	0.10	1.79	1.88
Furnaces (<90kWth)	40	40	17	0.02	0.16	1.80	1.94

Note: Total damage is given by the emission damage plus 1.52 cents/kWh for an energy security premium and 0.26 cents/kWh for land use impacts.
Source: Damage costs, Roth and Ambs (2004); emissions, US EPA (1995)

Table 1.1. Given are the damage costs per tonne for emissions of NO_x and CO, as used by Roth and Ambs (2004) for the estimates of electricity external costs, and emission factors for boilers and furnaces as given by US EPA (1995). The resulting external cost estimate due to emissions – 0.1–0.25 cents/kWh of fuel energy – is small compared to the external cost of electricity due to emissions (0.7–5.5 cents/kWh with existing technology). This is a result of a smaller emission factor for NO_x per unit of fuel energy (no more than 60gm/GJ for heating, compared to 380gm/GJ for electricity generation from uncontrolled coal and 180gm/GJ from natural gas), combined with the fact that three units of fuel are required per unit of electricity for the typical electricity-generation efficiency of 33 per cent. However, there are energy-security related, depletion and land use externalities associated with natural gas. For electricity generated using natural gas combined cycle power plants in the US, Roth and Ambs (2004) adopt a value of 0.53 cents/kWh to account for land use impacts. These costs amount to about 0.27 cents per kWh of fuel energy, and bring the externality for natural gas as a heating fuel to approximately 0.4–0.5 cents/kWh, excluding the energy-security and depletion externalities. For the UK, Gaterell and McEvoy (2005) estimate an externality for natural gas used for heating of 0.8–9.0 eurocents/kWh of fuel – substantially larger than derived here from Roth and Ambs (2004). By comparison, the market price of natural gas for heating ranges from 2 to 8 cents/kWh for a variety of OECD countries (Chapter 2, Section 2.3).

This discussion has deliberately avoided mention of estimates of the cost of climatic change per unit of energy. This has been done for two reasons. First, the potential impacts of global warming are so large, involving the collapse of vital and irreplaceable ecosystems, and death or suffering for hundreds of millions of people, that any monetary cost assigned to global warming will be entirely subjective and largely meaningless. Second, the position is adopted here that the present generation has a *fiduciary responsibility* to preserve irreplaceable and invaluable ecosystem assets, as discussed by Brown (1992). The cost of doing so is not a consideration, except that it must be affordable (which is the case). As discussed above, the UN-FCCC implicitly adopts this principle. The reason for considering non-climatic externalities of fossil fuels here is to show that, having made the decision to drastically reduce the use of fossil fuel for climatic reasons, many if not most of the measures that appear to entail a net cost in fact lead to a net savings for society when full accounting of the costs of fossil fuels is done, as the true cost of fossil fuels is much greater than is reflected in market prices.

1.3 Energy-security imperatives

The rate of extraction of a finite resource grows rapidly at the beginning, since it is relatively plentiful and therefore relatively easy to find. As the resource is depleted, it gets harder, and takes longer, to find further deposits that can be exploited. Thus, at some point the rate of extraction peaks, then gradually declines. This behaviour was predicted by M. King Hubbart in the 1950s, and has been borne out by the production of oil in many regions. For example, oil extraction in the US (excluding Alaska) peaked in the early 1970s and has been on a steady decline ever since.[3] Intensified exploration activity cannot reverse this inexorable trend.

There is general consensus that the peak in the rate of extraction from a given region occurs at about the time that half of the resource that can be extracted has been extracted in the case of oil, and at the time when about 60–75 per cent has been extracted in the case of natural gas. A corollary of this is that the rate of extraction should decline more rapidly, once the peak is reached, for natural gas than for oil. In the case of oil, extraction rates decline about 3 per cent per year once the peak is reached. Colin Campbell and Jean Laherrère have been in the forefront of recent analyses of when the peak in global oil extraction should occur, and place the peak sometime between 2005 and 2010. They have established the Association for the Study of Peak Oil (ASPO), whose websites (www.peakoil.net and www.apsonews.com) contain their key analyses. Bentley (2002) reviews their work and that of others, and concludes that there are reasons for doubting official estimates of recoverable oil or of reserve estimates by oil companies.

With regard to natural gas – the prime fuel used for heating buildings in much of the industrialized world – the cumulative extraction of natural gas to 1996, the remaining reserve (i.e. gas known to exist with near certainty and available at current prices with existing technology) as of 1996, additions to the amount of natural gas that can be extracted from known reserves due to better technology, and undiscovered natural gas as estimated by USGS (2000) for major world regions are given in Table 1.2. The effective date of the assessment is 1 January 1996. Three estimates each are given for reserve additions and undiscovered natural gas – the amounts thought to be equalled or exceeded with 95 per cent, 50 per cent, and 5 per cent probabilities. For North America, the estimate of undiscovered natural gas in the US appears to be unrealistically high compared to the estimate of undiscovered

natural gas in the rest of the continent (Canada + Mexico), given that the US is the most explored part of the continent. Reserve additions were estimated by USGS (2000) for the entire world only, but have been allocated here in proportion to remaining reserves in each region. The last four columns give the cumulative extraction divided by the original endowment. The column labelled 'Certain' uses cumulative extraction plus remaining reserves as the original endowment, while the remaining three columns include reserve additions and undiscovered natural gas with 95 per cent, 50 per cent, and 5 per cent exceedence probabilities. Based on the certain original endowment, the US and Europe have already extracted 83 per cent and 42 per cent, respectively, of the original endowment. Using the 50th percentile for reserve additions and undiscovered natural gas, cumulative extraction in the US and Europe had already reached about 60 per cent and 35 per cent, respectively, of the original endowment by 1996. The global endowment at 95 per cent probability (9914 EJ), as estimated by USGS (2000), is very close to the ultimate global extraction as estimated by Jean Laherrère (9953 EJ) based on fitting a logistical function to the historical variation of cumulative extraction against the cumulative number of gas wells drilled. Not included in either estimate are unconventional sources of natural gas, such as deep offshore areas and coalbed methane. Due to the difficulty in developing these resources, they are very unlikely to affect the timing of peak extraction (at least for the US and Europe), although they can slow the rate of decline in the rate of extraction after the peak. Approximately 75 per cent of the remaining reserves and 60 per cent of estimated undiscovered natural gas are in the Former Soviet Union (FSU), the Middle East and North Africa.

Natural gas can be liquefied at a temperature of −160°C to form liquefied natural gas (LNG) and transported between continents in giant vacuum-insulated pressure vessels in supertankers. LNG currently provides 100 per cent of South Korea's natural gas supply, 97 per cent of Japan's, 86 per cent of Taiwan's, and 53 per cent of Spain's (Jensen, 2003). The process of liquefying natural gas requires an amount of energy equal to 8–13 per cent of the energy content of

Table 1.2 Cumulative extraction of natural gas, remaining reserves, estimated growth of conventional reserves, estimated undiscovered conventional resources and original endowment of natural gas, all as estimated by USGS (2000), and ratio of cumulative extraction to original endowment of natural gas

Region	Cumulative extraction	Remaining reserve	Reserve growth			Undiscovered NG			Original endowment			Ratio (%)			
			F95	F50	F5	F95	F50	F5	F95	F50	F5	Certain	F95	F50	F5
FSU	419	1811	397	1249	2095	462	1586	3496	3088	5066	7819	18.8	13.6	8.3	5.4
ME & N Africa	79	1975	432	1362	2285	458	1375	2806	2944	4792	7145	3.8	2.7	1.7	1.1
Asia Pacific	61	371	81	256	429	118	377	803	631	1064	1663	14.1	9.6	5.7	3.7
Europe	220	301	66	208	348	48	291	789	635	1019	1658	42.2	34.6	21.6	13.3
N America	1010	276	61	191	319	432	591	803	1778	2068	2408				
US only	919	185	41	128	214	423	567	751	1567	1799	2069	83.2	58.6	51.1	44.4
Excluding US	91	91	20	63	105	9	24	52	211	269	339	50.0	43.1	33.9	26.9
C & S America	59	240	52	165	277	104	453	1170	455	917	1747	19.9	13.1	6.5	3.4
Sub-Saharan Africa	8	115	25	79	133	90	237	473	238	439	728	6.5	3.4	1.8	1.1
South Asia	30	68	15	47	78	33	115	268	145	259	443	30.5	20.5	11.4	6.7
Total	1886	5156	1129	3556	5965	1744	5026	10,605	9914	15,624	23,611	26.8	19.0	12.1	8.0

Note: F95, F50 and F5 are the amounts estimated to exist or to be exceeded with a 95 per cent, 50 per cent or 5 per cent probability, respectively. Ratios are given assuming the original endowment to be equal to the cumulative extraction plus remaining reserve only (column labelled 'certain'), or including reserve additions and undiscovered natural gas. All amounts are in EJ (exajoules, 10^{18} joules). FSU = Former Soviet Union; ME = Middle East, South Asia is Pakistan, India, Bangladesh, Nepal, Bhutan and Burma.

the natural gas (depending on the efficiency of the chillers used for liquefaction) (Darley, 2004). During transit, some heat is absorbed, requiring some of the LNG to boil off. The gas that boils off provides about half of the fuel needed to power the ship engines. Over a 14-day, 12,500km journey from Qatar to the eastern coast of the US, for example, about 2.0–3.5 per cent of the LNG energy is lost in this way. Finally, about 1.0–1.5 per cent of the LNG energy is required for regasification at the receiving terminal. Altogether, transportation of natural gas from the Middle East to North America consumes 11–18 per cent of the original energy content of the natural gas. By comparison, the energy required to transport natural gas by high-capacity pipeline is about 2.5 per cent of throughput per 1000km. In this case, natural gas is used to power compressor stations that are located every 60–150km along the gas pipeline.

The 2005 edition of the annual *British Petroleum Statistical Review*, which can be obtained as Excel spreadsheets from www.bp.com, contains data on annual natural gas extraction from 1970 to 2004 inclusive. For some regions, the cumulative extraction from 1970 to 1995 exceeds that given by USGS (2000) for all time up to 1995

inclusive (by up to 15 per cent), which casts doubt on reports of cumulative extraction – presumably the least uncertain of all the data listed in Table 1.2. Table 1.3 compares the cumulative extraction to the end of 1995 and the remaining reserves at the end of 1995, as given by USGS (2000), with the cumulative extraction from 1996 to 2004 inclusive and the remaining reserves at the end of 2004 as given by BP. For some regions, the estimated remaining reserve at the end of 2004 is greater than at the end of 1995. Part of this could reflect real discoveries between 1996 and 2004 and changes in prices. Also given in Table 1.3 are the regional extraction and consumption in 2004 (the difference in global totals reflects changes in inventories or accounting errors).

In light of the estimates presented in Tables 1.2 and 1.3, it is clear that Europe and North America face a growing shortfall between regional natural gas supply and natural gas demand. This growing shortfall can be matched in part through imports of LNG, supplemented by overland imports from the FSU in the case of Europe and development of unconventional natural gas resources in the case of North America, and in part through reduced demand in response to higher prices. On a global scale, three-quarters

Table 1.3 Cumulative extraction of natural gas (to the end of 1995) and remaining reserves (as of the end of 1995) as estimated by USGS (2000), and cumulative extraction between 1996–2004 inclusive, remaining reserves at the end of 2004, extraction in 2004, and consumption in 2004 as given in the 2005 edition of the *BP Petroleum Statistical Review*

Region	Cumulative extraction to end of 1995	Remaining reserves as of end of 1996	Cumulative extraction 1996–2004	Remaining reserve as of end of 2004	Extraction in 2004	Consumption in 2004
FSU	419	1811	232.0	2181	28.2	22.2
ME & N Africa	79	1975	108.3	3066	15.0	11.0
Asia Pacific	61	371	72.8	438	9.5	12.3
Europe	220	301	97.5	252	11.8	18.9
N America US only Excluding US	1010 919 91	276 185 91	260.0 187.3 72.7	278 201 77	29.0 20.6 8.4	29.8 24.6 5.2
C & S America	59	240	33.9	270	4.9	4.5
Sub-Saharan Africa	8	115	6.0	235	1.1	0.8
South Asia	30	68	20.3	102	2.8	2.7
Total	1886	5156	831.4	6822	102.3	102.2

Note: All amounts are in EJ (exajoules, 10^{18} joules). FSU = Former Soviet Union; ME = Middle East. South Asia is Pakistan, India, Bangladesh, Nepal, Bhutan and Burma.

of the endowment could be consumed in 75 years at current rates of consumption, but by 2035 at 2 per cent per year growth in consumption; at which point (if not sooner) real prices will rise essentially everywhere. Consumption of natural gas today therefore imposes a cost on future consumers by driving up the future price of natural gas, and this price increment is an externality imposed on future consumers by present consumers. The magnitude of the externality is larger the longer the time horizon, reaching infinity for an infinitely long time horizon. For oil and a 50-year time horizon, Sabour (2005) estimates this externality to be 40 per cent of the ongoing market price. For natural gas at \$6/GJ (2.16 cents/kWh), a 40 per cent depletion externality is equal to about 0.86 cent/kWh. This brings the total externality to about 1.3 cents/kWh for the particular conditions assumed above.

1.4 Sustainable buildings

There are many books on architecture and on buildings with the word 'sustainable' in their title, but in most cases the term 'sustainable' is not defined. There are also many claims that certain

buildings are 'sustainable', in some cases when the buildings are only slightly less energy-demanding than buildings built according to conventional practice. Others define sustainable buildings in terms of 'minimizing' energy use or the use of natural resources, although there is no logical requirement that some arbitrary level of resource use that is considered to be 'minimal' will be indefinitely sustainable. As discussed by Marshall and Toffel (2005), there are well over 100 different definitions of sustainability more generally, many of which include worthwhile goals such as a higher quality of life and social justice that are separate from the issue of sustainability.

1.4.1 Defining sustainability

The *HOK Guidebook to Sustainable Design* (Mendler and Odell, 2000) lists four principles of sustainability to which 'sustainable' buildings should contribute. These principles are:

- that substances from the Earth's crust must not systematically increase in the ecosphere;
- that substances produced by

society (man-made materials) must not systematically increase in the ecosphere;

- that the productivity and diversity of nature must not be systematically diminished;
- that there must be fair and efficient use of resources to meet human needs (basic needs for all take precedence over providing luxuries for the few).

These four principles, as important as they are, are not sufficient to guarantee sustainability. Furthermore, 'fairness', as desirable as it may be, is not a strict requirement for some system of resource use to be indefinitely sustainable. In any case, at least two more principles are required:

- human energy needs must be met entirely from renewable energy sources, without degrading the long-term capacity of nature to supply energy renewably;
- the local rate of consumption of freshwater must not exceed the local rate at which freshwater is supplied through the hydrological cycle in excess of local ecological needs.

The last point, sustainability of water use, is an issue for buildings because one particular low-energy technique for cooling – evaporative cooling – consumes water. Conventional cooling techniques relying on vapour compression chillers and cooling towers also consume water, both on-site and at the electric power plant, while absorption chilling driven by solar energy would consume even more water while saving energy. Thus, water as well as energy is an important dimension of sustainability for buildings. Efficient use of wood that is harvested sustainably is another dimension of sustainability for buildings, implied by the principle of preserving the productivity and diversity of nature. This dimension is linked to energy use in that wood can displace materials (such as steel or concrete) with a high embodied energy, and because more wood is needed in order to achieve high levels of insulation without using solid foam

insulation (which entails emission of halocarbon gases). The principle of preserving the productivity and diversity of nature also implies efficient use of land, thereby favouring compact, relatively closely spaced multi-storey buildings rather than scattered and sprawling one-storey buildings, but this also has energy implications with regard to heating and cooling loads and the opportunities for wind- and thermally driven passive ventilation, not to mention transportation energy use.

The focus of this book is energy. A society based on fossil fuels is not sustainable for two reasons. The first and most obvious is that fossil fuels are not in infinite supply. Although, technically speaking, we will never run out of fossil fuels, at some point the remaining fossil fuels will become so expensive, and so difficult to extract, that in practical terms they will have been exhausted. The second way in which a society built on fossil fuels is not sustainable is that, as we continue to increase the concentration of greenhouse gases in the atmosphere (largely through the use of fossil fuels), the climate will continue to warm, and with it, the pressure to change course. At some point, long before the fossil fuel resource has been exhausted, this pressure is likely to become overwhelming (although, unfortunately, not before substantial damage has already occurred).

1.4.2 Kaya decomposition of future CO_2 emissions

As argued above, compliance with the UNFCCC requires limiting the atmospheric CO_2 concentration to 450ppmv or less. Stabilization of CO_2 at any concentration ultimately requires reducing anthropogenic emissions to zero, but the rate at which emissions need to be reduced depends in part on the extent to which global warming leads to accelerated natural emissions into the atmosphere (from, for example, forest dieback or decomposition of carbon stored in presently frozen permafrost regions) that are in addition to human emissions. The expected acceleration of natural emissions is a positive *climate–carbon cycle feedback*. According to the climate–carbon cycle model simulations in Harvey (2004), global fossil fuel emission must peak around 2020, drop to 20–40 per cent below present emissions by 2050, and be completely phased out by 2075–2100 in order to stabilize atmospheric CO_2 at 450ppmv, depending on how strong the positive climate–carbon cycle feedback is.

Future CO_2 emissions can be written as the product of population (P), gross domestic product per capita (GDP/P), primary energy used per unit of GDP (PE/GDP), and carbon emission per unit of primary energy (C/PE). That is,

Total emission = P × GDP/P × PE/
GDP × C/PE   (1.1)

where PE/GDP is referred to as the *energy intensity* of the economy and C/PE is referred to as the *carbon intensity* of the energy system. Equation (1.1) is referred to as the *Kaya Identity*, in honour of the Japanese scientist who first presented this equation (Kaya, 1989). This equation is a straightforward mathematical identity, but it hides a series of complex interactions, each term of which will be briefly elaborated on here.

Projections of peak human population have been falling over time. According to the work of Lutz et al (2001), there is a 60 per cent probability that the human population will peak somewhere between 7.8 and 10.8 billion by 2050–2085, then begin to decline, with a 20 per cent probability that the peak will be less than 7.8 billion and a 20 per cent probability that it will be greater than 10.8 billion. However, the future human population is to some extent something that can be determined in advance, through support for family planning and for measures (such as broadly based socioeconomic development and empowerment of women) that reduce the desire for large families.

World average GDP/P, based on estimates of gross world product in terms of constant (2003) dollars as given Mastny et al (2005), grew at the following rates: 2.8 per cent per year from 1950 to 1970, 1.9 per cent per year from 1970 to 1980, and 1.6 per cent per year from 1980 to 2004. Thus, over the long term there has been a steady decline in the rate of growth of GDP/P, although during the four-year period 2000–2004 the growth of GDP/P accelerated to 2.3 per cent per year due to a burst of growth of GDP in China and India (in the order of 6–9 per cent per year) that is not sustainable. The political culture in most parts of the world focuses single-mindedly on the growth of GDP as an end in itself.

However, presumably it is human happiness that is the ultimate goal, with GDP per person at best an imperfect proxy for human happiness. There is a vast policy literature on global warming that implicitly equates happiness with material consumption (in the so-called 'utility' functions), but there is also a whole philosophical and sociological literature that recognizes that, once basic human needs are satisfied, happiness is related to the diversity and richness of human relationships (family, friends, community), health (which will be improved as a byproduct of many of the measures to reduce greenhouse gas emissions) and the preservation of cultural identity, not the accumulation of material wealth (Layard, 2003, 2005). Indeed, there is substantial evidence that, in the US and other developed countries, economic growth has been *negatively* correlated with various measures of human well-being and happiness for the past 2–3 decades (see the Appendix in Daly and Cobb, 1994). This does not in any way imply that significant economic growth is not needed in the developing countries in order to improve the material standard of living. However, at some point, directing increasing labour productivity into more leisure time would likely do more to increase human happiness than further growth of GDP per person. Many would argue that this point has long since been reached in most developed countries.

Energy intensity is much more than energy efficiency, as it also depends on structural changes in the economy (the relative importance of goods versus services, and of energy intensive versus less energy intensive manufactured products) and demographic change (which alters the mix of goods and services demanded, and so is partly redundant with structural change). Carbon intensity reflects the energy sources used – the proportions of fossil fuel and renewable energy, and the proportion of coal, oil and gas within the fossil fuel portion.

To phase out fossil fuels by 2075–2100 requires limiting the growth in total energy demand to levels that can be satisfied with renewable forms of energy. The first three factors in the Kaya identity – population, GDP per person and energy intensity – together determine the growth in energy demand, and all three represent important levers for reducing emissions. To illustrate the relative importance of population, GDP per person, and energy intensity, consider the following scenarios to the year 2100:

- *Scenario 1*: High population (exceeded by only 20 per cent of the scenarios in Lutz et al, 2001), a constant rate of growth in average GDP per person (1.6 per cent per year, the global average of the period 1980–2004) and low rate of reduction in energy intensity (1 per cent per year, the rate of decrease for the world economy over the past few decades).
- *Scenario 2*: High population, constant growth in GDP per person, and 2 per cent per year reduction in energy intensity.
- *Scenario 3*: Low population (exceeded by 80 per cent of the scenarios in Lutz et al, 2001), declining growth in GDP per person (average growth rate decreasing linearly from 1.6 per cent per year in 2000 to 0.8 per cent per year in 2100) and 2 per cent per year reduction in energy intensity.

Figure 1.8a shows the resulting variation in global primary power demand (i.e. the rate of use of energy as it is found in nature, prior to conversion to other forms). The difference between Scenarios 1 and 2 represents the impact of doubling the rate of reduction of energy intensity, while the difference between Scenarios 2 and 3 represents the impact of lower population and GDP per person assumptions. From Figure 1.8a, it appears that doubling the rate of improvement of energy intensity is far more important than the alternative population and GDP per person assumptions. However, this is an artefact of the order in which the changes to Scenario 1 were made. Figure 1.8b shows primary power demand for Scenarios 1 and 3 along with an alternative to Scenario 2, which combines low population and low GDP per person with 1 per cent per year reduction in energy intensity. In Figure 1.8b, low population and low GDP per person appear as the most

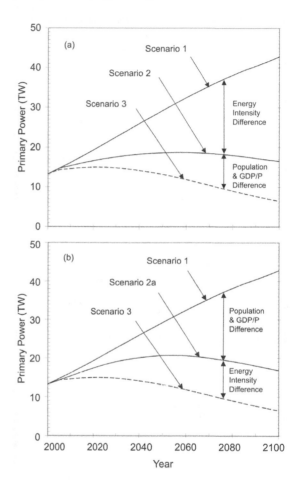

Figure 1.8 World primary power demand for (a) Scenario 1 (high population, high GDP/person, 1 per cent per year reduction in energy intensity), Scenario 2 (high population, high GDP/person, 2 per cent per year reduction in energy intensity), and Scenario 3 (low population, low GDP/person, 2 per cent per year reduction in energy intensity); and (b) Scenarios 1 and 3 (as above) and Scenario 2a (low population, low GDP/person, 1 per cent per year reduction in energy intensity)

important factors, especially during the latter half of this century. This comparison underlines the importance of *both* energy intensity and population/economic growth per person in determining future global energy demand.

The difference between total primary power demand, illustrated in Figure 1.8, and the primary power that can be supplied from fossil fuels without exceeding allowable CO_2 emissions, must be supplied by carbon-free power sources. Figure 1.9a shows the total primary power demand in 2050 as a function of the rate of

Figure 1.9 (a) Primary power demand in 2050, as a function of the rate of decrease of global energy intensity for high population and GDP/person or low population and GDP/person scenarios, and approximate permitted fossil fuel power in 2050 if atmospheric CO_2 is to be stabilized 450ppmv. (b) Amount of C-free power required in 2050 for the same conditions as in (a), and global primary power supply in 2000

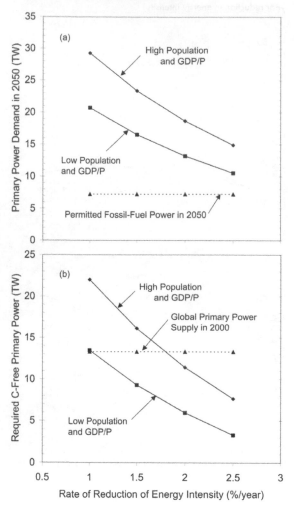

carbon-free sources (of which 0.9TW were nuclear and not likely to be replaced as existing facilities are retired). If energy intensity continues to decrease at 1 per cent per year, global primary demand in 2050 would be 20.7TW and 29.3TW for the low and high scenarios. The permitted fossil fuel use supplies 7TW of primary power, so the required C-free primary power in 2050 is 14–22TW and likely to be impossible on this time frame. If global average energy intensity decreases at 2 per cent per year, global primary power demand in 2050 is 13–19TW and the required C-free power is 6–12TW – still an awesome challenge. At 2.5 per cent per year improvement in energy intensity, the required C-free power in 2050 is 3.3–7.7TW, or 2–4 times the present non-nuclear world C-free power supply of 1.8TW. The required carbon-free power under the low population/GDP scenario is about 40 per cent that of the high population/GDP scenario, even though the primary power demand is 70 per cent that of the high scenario, because a fixed fossil fuel power supply is subtracted in both cases. For the low population/low GDP/P scenario, the required C-free power supply in 2050 ranges from 6.0 to 3.3TW as the rate of decrease of energy intensity ranges from 2.0 to 2.5 per cent per year.

1.4.3 Implications for building energy intensity

Assuming that 3–6TW (3000–6000GW or 3–6 million MW) of C-free primary power by 2050 are feasible, that human population growth is close to the low scenario, and that growth of average GDP per person drops in half over the next century, then energy intensity reductions of 2–2.5 per cent per year until 2050 are consistent with a stabilization of atmospheric CO_2 at 450ppmv; otherwise, faster reductions in energy intensity are required. Global average energy intensities in 2050 would be 36 per cent and 28 per cent, respectively, of the global average energy intensity in 2000 – reductions by factors of 2.8 and 3.6. If the floor space of the global building stock were to grow in proportion to world GDP, and if the same reductions in energy intensity were to occur across all sectors (buildings, industry, transportation, agriculture), then reductions in the primary

reduction in energy intensity, in comparison with the fossil fuel power permitted in 2050 if atmospheric CO_2 is to be stabilized at 450ppmv, assuming either high population combined with high GDP/P (as given above), or low population combined with low GDP/P. Figure 1.9b shows the required C-free power in 2050 in comparison with the global primary power supply in 2000. Global primary power supply in 2000 was 13.3TW (TW = terawatt, 1 TW = 10^{12} watts), of which 10.6TW were from fossil fuels and 2.7TW from

energy intensity of the building stock (annual energy use per unit of floor area) *in the order of a factor of 3–4* are required for stabilization of atmospheric CO_2 at 450ppmv. This condition for the stabilization of CO_2 at 450ppmv is derived from the requirement of an energy use small enough that it can be eventually met with renewable sources of energy, so it is equivalent to long-term sustainability although, strictly speaking, an energy system does not need to evolve toward a carbon-free state fast enough to avoid overshooting 450ppmv CO_2 in order to be moving toward sustainability.

If the building floor area grows more slowly than GDP, the energy intensity of buildings (expressed per unit of area) will not need to fall as fast as the overall energy intensity of the economy (expressed per unit of GDP). As discussed in Chapter 2 (Section 2.2), about 35 per cent of the on-site energy used in buildings in OECD countries is electricity (generated with an average efficiency of 38 per cent) and about 3 per cent is district heat (generated and distributed with an efficiency of only 47 per cent). Allowing for eventual electricity-generation and district heat efficiencies of 60–70 per cent and 80 per cent, respectively, would reduce the primary energy intensity of OECD buildings by 15–17 per cent. To reduce the primary energy intensity of buildings by factors of 2.8–3.6 would require reductions in on-site energy intensity by factors of 2.4–3.0. In non-OECD countries, the level of services is smaller and will therefore grow, but energy is used much less efficiently than in OECD countries, so there should be a comparable potential to reduce energy intensity. To allow for the possibility that floor area could grow more slowly than GDP, the lower bound on the required improvement factor for on-site energy intensity can be reduced to, say, 2.0. *Thus, a reduction in the average building energy intensity (on-site energy use per unit of floor area) by a factor of two to three will be tentatively adopted as a criterion for sustainability in buildings.* This pertains to gross energy demand, that is, excluding on-site generation of electricity by building-integrated photovoltaic modules, as any such power production would contribute to the 3000–6000MW of C-free primary power that was assumed in deducing the need for a factor of 2–3 reduction in building energy intensity. This sustainability criterion

is dependent on specific assumptions concerning future human population and building floor area per capita.

Table 1.4 illustrates combinations of reductions in the energy intensity of new buildings and of the fraction of existing buildings that are replaced, leading to average future on-site gross energy intensities two to three times smaller than the current average on-site energy intensity. Scenarios with small and large growth in total floor area are considered separately for OECD and non-OECD countries, with a factor of two reduction in average energy intensity assumed for the case of small growth in floor area (25 per cent for OECD countries, 100 per cent for non-OECD countries) and a factor of three reduction in energy intensity assumed for the case of large growth in floor area (50 per cent for OECD countries, 200 per cent for non-OECD countries). The larger the increase in total floor area, the greater the amount of the existing building stock that is assumed to be replaced rather than renovated. The greater the reduction in energy intensity of new and replacement buildings relative to the existing stock average in each region, the smaller the reduction in energy intensity required through renovation of existing buildings in order to achieve the factor of two to three reduction in average energy intensity. As can be seen from Table 1.4, a factor of two reduction in average energy intensity can be achieved in OECD countries if the energy intensity of new buildings is reduced by a factor of four and if renovated buildings use 38 per cent less energy compared to the average of existing buildings. A factor of three reduction can be achieved if new buildings use a factor of four to five less energy and renovated buildings 57–51 per cent less energy than the average of existing buildings. In non-OECD countries, a factor of two reduction in average energy intensity can be achieved if the energy intensity of new buildings is reduced by a factor of three and renovated buildings use 20 per cent less energy compared to the average of existing buildings, given a doubling of building floor area. A factor of three reduction can be achieved if new buildings use a factor of three to four less energy and renovated buildings use 67–13 per cent less energy than the average of existing buildings, given a tripling of building floor area. These are of course very rough

Table 1.4 Illustrative combinations of improvements in the energy intensity of new, replacement and renovated buildings as the total building floor area increases by 25 per cent or 50 per cent (OECD cases) or by 100 per cent and 200 per cent (non-OECD cases) while giving a reduction in average building on-site energy intensity by a factor of two (assuming the smaller increase in building floor area) or by a factor of three (assuming the larger increase in building floor area)

Scenario	Relative areas					Energy use					Energy intensities		
	Initial	Renova-tions	Replace-ments	New	Total	Initial	Renova-tions	Replace-ments	New	Total	New	Renova-tions	Aver-age
OECD-1a	100	80	20	25	125	100	51.3	5.0	6.3	62.5	0.250	0.641	0.500
OECD-1b	100	70	30	50	150	100	30.0	7.5	12.5	50.0	0.250	0.429	0.333
OECD-1c	100	70	30	50	150	100	34.0	6.0	10.0	50.0	0.200	0.486	0.333
Non-OECD -1a	100	70	30	100	200	100	56.7	10.0	33.3	100.0	0.333	0.810	0.500
Non-OECD -1b	100	70	30	200	300	100	23.3	10.0	66.7	100.0	0.333	0.333	0.333
Non-OECD -1c	100	40	60	200	300	100	35.0	15.0	50.0	100.0	0.250	0.875	0.333

Note: All areas, energy uses and energy intensities are relative to those for the existing building stock in the corresponding region (OECD or non-OECD).

numbers, dependent on the specific growth and replacement assumptions adopted here, but serve to illustrate the magnitude of energy intensity reductions compared to the existing stock average required for new buildings to qualify as 'sustainable' with respect to energy use: *a factor of 4–5 in OECD countries, and a factor of 3–4 in non-OECD countries.* Compared with conventional practice for new buildings, the required reductions will be less to the extent that new buildings are less energy intensive than the average of existing buildings. Renovations must achieve energy intensity reductions of one-third to two-thirds. With very careful design, high quality workmanship during construction, and intelligent and conscientious operation of the building after it is built or renovated, it is possible to achieve energy intensity reductions of these magnitudes. The purpose of this book is to show how.

Notes

1   As explained in the technical preface, the proper procedure is to refer to temperature differences on the Celsius scale as kelvin or as Celsius degrees (C°) and not as degrees Celsius (°C), which should be used only in reference to actual temperatures. Thus, 28°C–26°C = 2K and not 2°C.

2   As discussed in Harvey (2004), this is not strictly true if we create negative emissions by burying CO_2 produced from the gasification of sustainably grown biomass, used as an energy source in place of fossil fuels. This provides a limited possibility of drawing down atmospheric CO_2 if we overshoot safe levels, as we almost certainly will.

3   Extraction of non-renewable fossil fuels is almost always referred to as 'production', whereas it is nature that produced these fuels over periods of millions of years.

two

Energy Basics, Climate and Meteorological Data, Organizational Strategy

This chapter outlines some basic concepts concerning energy, presents information on the amount of energy used in buildings compared with other uses of energy, and presents an overview of how energy is used today in buildings around the world.

2.1 Energy concepts

Energy as it occurs in nature is referred to as *primary energy*. Examples of primary energy are coal, oil, natural gas and uranium as they occur in the ground. To be useful to humans, these forms of energy need to be extracted and transformed into *secondary energy*. Examples of secondary energy are electricity and refined petroleum products, and processed natural gas ready for use by the customer. However, even these energy forms are not really of interest to us. What we really want are end-use energy services – things like warmth, motion, mechanical power or light. These are referred to as *tertiary energy* or *end-use energy*. The

relationships between primary, secondary and tertiary energy are summarized in Figure 2.1.

Figure 2.1 The transformation from primary to secondary to tertiary energy

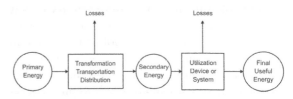

The *first law of thermodynamics* states that energy is neither created nor destroyed. Thus, when transforming energy from primary to secondary, and from secondary to tertiary energy, no energy is lost in the processes. However, the *second law of thermodynamics* states that there will be an unavoidable transformation of some of the energy into unusable forms. In particular, some concentrated or high-quality energy is unavoidably dissipated as low-grade heat, which is random molecular motion. Thus, although no energy disappears when we convert from primary to secondary energy or from secondary to tertiary, there is an unavoidable loss of useful energy. These losses are indicated in Figure 2.1. The efficiency of an energy conversion process is the ratio of the useful output energy to the input energy. For example, the efficiency of a power plant is the ratio of electrical energy produced to the energy of the input fuel. The efficiency of an electric motor is the ratio of mechanical energy created by the motor to the electrical energy input. The fuel energy not converted to electricity, or the electrical energy not converted to mechanical energy, is lost as waste heat. The greater the efficiency, the smaller the losses.

Environmental impacts of energy use are tied to the amount of primary energy that is used. As can be seen from Figure 2.1, there are three general ways to reduce the use of primary energy by buildings:

1   increasing the efficiency of conversion from primary to secondary energy;
2   increasing the efficiency of conversion from secondary to tertiary energy;
3   reducing tertiary energy demand.

Building houses with better insulation and better windows, so that less heat is lost and therefore less needs to be replenished by the furnace, is an example of a measure to reduce tertiary energy demand. Using light coloured roofs along with better insulation to reduce cooling requirements in summer is another example. So is turning off unused lights. Putting in a better furnace, replacing the furnace with an electric heat pump or increasing the efficiency of the heat pump, are all examples of improving the secondary to end-use energy conversion efficiency. More efficient generation of electricity, or smaller losses in extracting natural gas and transmitting it to a building, are examples of increases in the primary to secondary energy conversion efficiency.

The efficiency of a typical power plant is 33–38 per cent, although state-of-the-art natural gas power plants have an efficiency of 55–60 per cent. About 5–10 per cent of the generated electricity is typically lost in transmission from the power plant to the consumer, depending on the distance transported and the nature of the transmission grid. Energy is also required in mining and transporting coal to coal-fired power plants – perhaps 1–3 per cent of the energy content of the fuel. The overall efficiency transforming primary energy into secondary energy at the point of use can be as low as $0.97 \times 0.33 \times 0.9 = 0.29$ (29 per cent). In the case of natural gas, energy is required to power the compressors used to move natural gas (under pressure) from the natural gas fields to the customer. For major markets in North America, this is in the order of 10 per cent of the energy content of the fuel; the efficiency in conversion from primary energy to secondary energy at the point of use is thus about 90 per cent. Because of the much larger losses involved in the generation and transmission of electricity than in the provision of natural gas, measures that reduce overall energy use in a building but with a shift from natural gas to electricity could increase the use of primary energy.

When energy efficiency measures that involve a shift from natural gas (or oil) to electricity are contemplated, the impact on primary energy use should be assessed. This requires determining the appropriate electricity generation efficiency. This, however, is not fixed even for a given jurisdiction. Rather, it depends on which power plants are used to provide an increment of additional (or avoided) electricity use and on the time of day (which determines whether a given generating unit is likely to be operating at full load or part load, something that has a significant effect on efficiency). As well, in the long run, the efficiency of electric power plants will improve (to 55–60 per cent with currently available technology) as older plants are replaced with newer plants or as more of the electricity is produced through cogeneration (see Chapter 15, Sections 15.1.2 and 15.2.5). The computation of primary energy use from building on-site energy use involves some arbitrary assumptions so, as a bare minimum, these assumptions should be explicitly stated and, preferably, results given for a range of assumptions as an aid in decision making.

2.2 Energy use in OECD countries and in buildings

The building sector is often subdivided into residential, commercial and institutional buildings. Commercial buildings include low-rise retail buildings and high-rise office towers, but could also include (depending on one's definition) multi-unit residential buildings – condominiums and apartments. The institutional sector includes government buildings, hospitals, schools and universities. Many buildings in the institutional sector share similarities with commercial buildings. Thus, the distinctions between residential, commercial and institutional buildings are somewhat blurred and arbitrary. For this reason we shall lump institutional buildings with the commercial sector, and shall henceforth refer to just the residential and commercial sectors.

Table 2.1 Primary and secondary energy use in OECD countries (EJ/year), 2002

	Primary energy								Secondary energy		
	Coal	Oil	Natural gas	Nuclear	Hydro	Biomass & waste	Other renew	Total	Elec-tricity	Heat	Total
Primary energy supply	47.3	93.7	54.2	20.4	11.0	7.5	1.4	235.5			
Primary energy for electricity	34.4	4.9	14.6	20.4	11.0	1.7	1.1	88.1			
Primary energy for heat grids	1.6	0.3	1.5	0.0	0.0	0.8	0.0	4.2			
Remaining primary energy	11.3	88.5	38.1	0.0	0.0	5.0	0.3	143.1			
Conversion to electricity or grid heat									35.4	2.3	180.9
	Subtractions from primary energy								Subtractions from secondary energy		
Electric, heat and CHP plants	0.0	0.0	0.0	0.0	0.0	0.0	0.0	0.0	1.8	0.0	1.9
Distribution	0.0	0.0	0.1	0.0	0.0	0.0	0.0	0.1	2.3	0.2	2.6
Petroleum extraction and refining	0.0	4.5	4.0	0.0	0.0	0.1	0.0	8.6	0.6	0.1	9.3
Other energy industry	0.1	0.0	0.2	0.0	0.0	0.0	0.0	0.3	0.3	0.1	0.6
Delivered secondary energy	11.2	83.9	33.8	0.0	0.0	4.9	0.3	134.1	30.4	2.0	166.5
	Uses of delivered secondary energy										
Other industry	10.6	15.2	14.3	0.0	0.0	2.6	0.0	42.6	11.4	0.7	54.7
Transportation	0.0	52.2	1.0	0.0	0.0	0.1	0.0	53.4	0.4	0.0	53.8
Agriculture	0.1	2.3	0.2	0.0	0.0	0.0	0.0	2.6	0.3	0.0	2.9
Commercial and public buildings	0.1	3.6	5.7	0.0	0.0	0.1	0.0	9.6	8.9	0.3	18.9
Residential buildings	0.4	5.4	12.3	0.0	0.0	2.0	0.2	20.3	9.4	0.9	30.5
Total buildings	0.6	9.0	18.0	0.0	0.0	2.1	0.2	29.9	18.3	1.2	49.4
Non-specified	0.0	0.3	0.3	0.0	0.0	0.0	0.0	0.6	0.0	0.1	0.7
Non-energy use	0.0	5.0	0.0	0.0	0.0	0.0	0.0	5.0	0.0	0.0	5.0

Note: 'Other renew' is predominately geothermal, wind and solar.
Source: Computed from IEA (2004a, 2004c)

2.2.1 Breakdown of OECD energy use by sector

Tables 2.1 and 2.2 provide an overview of energy use in OECD and non-OECD countries, respectively, by energy source and by end-use sector.[1] For non-OECD countries, 'non-specified' includes commercial and residential buildings and agricultural uses of energy, but this category is probably overwhelmingly energy use in buildings. The largest component of this category is biomass energy, which is used primarily for heating and cooking and accounts for about half of total secondary energy use in the non-specified category. Figure 2.2 shows the breakdown by end-use sector of electricity use in OECD countries, secondary energy use in OECD countries

and primary energy use in OECD countries. In order to compute the latter, it is first necessary to compute the average efficiency in converting from primary to secondary energy for each of the energy forms. For a given sector, such as residential buildings, the amounts of each kind of secondary energy (e.g. electricity, natural gas, heating oil) are divided by the corresponding primary-to-secondary efficiencies and then added to give the total primary energy use for that sector.[2] Buildings account for half (52 per cent) of total electricity generation in OECD countries, half (47 per cent) of natural gas primary energy (including natural gas used for district heating grids), 29 per cent of secondary energy use and 36 per cent of primary energy use.

Table 2.2 Primary and secondary energy use in non-OECD countries (EJ/year), 2002

	Primary energy								Secondary energy		
	Coal	Oil	Natural gas	Nuclear	Hydro	Biomass & waste	Other renew	Total	Electricity	Heat	Total
Primary energy supply	58.2	57.9	46.2	4.2	15.6	37.5	0.7	220.4			
Primary energy for electricity	30.3	5.7	12.7	4.2	15.6	0.3	0.7	69.6			
Primary energy for heat grids	4.4	1.1	8.0	0.0	0.0	0.2	0.0	13.8			
Remaining primary energy:	23.4	51.1	25.5	0.0	0.0	37.0	0.0	137.1			
Conversion to electricity or grid heat									22.7	9.5	169.3
	Subtractions from primary energy								Subtractions from secondary energy		
Electric, heat and CHP plants	0.5	0.1	0.0	0.0	0.0	0.0	0.0	0.6	1.4	0.1	2.2
Distribution	0.0	0.0	0.9	0.0	0.0	0.0	0.0	0.9	2.9	0.7	4.5
Petroleum extraction and refining	0.0	3.0	4.1	0.0	0.0	0.0	0.0	7.1	0.7	0.7	8.4
Other energy industry	0.8	0.2	1.0	0.0	0.0	0.0	0.0	2.0	0.4	0.1	2.5
Delivered secondary energy	22.1	47.8	19.5	0.0	0.0	37.0	0.0	126.5	17.3	8.0	151.7
	Uses of delivered secondary energy										
Other industry	18.0	11.8	9.6	0.0	0.0	4.1	0.0	43.5	8.4	3.4	55.3
Transportation	0.2	23.4	1.6	0.0	0.0	0.2	0.0	25.5	0.5	0.0	25.9
Non-specified	3.6	10.0	8.3	0.0	0.0	32.7	0.0	54.6	8.4	4.5	67.5
Non-energy use	0.3	2.7	0.0	0.0	0.0	0.0	0.0	3.0	0.0	0.0	3.0

Note: 'Other renew' is predominately geothermal, wind, and solar.
Source: Computed from IEA (2004b, 2004c)

Figure 2.2 Energy use by sector in OECD countries in 2002

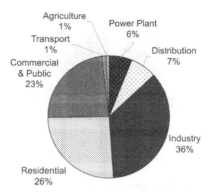

Electricity Use in OECD Countries

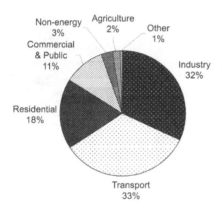

Secondary Energy Use in OECD Countries

Primary Energy Use in OECD Countries

2.2.2 Breakdown of building energy use by end use

Energy is used in buildings for heating, cooling and ventilation; for producing hot water; for lighting; and as electricity to power appliances, and/or consumer goods and/or office equipment. Figure 2.3 shows the breakdown of total residential energy use in the US and in the 15-nation European Union.[3] In both the US and Europe, space and hot water heating account for half to two-thirds of total residential energy use. Figure 2.4 shows the breakdown of energy use in the commercial sector of the US, the EU-15 and Canada, and for representative buildings in Hong Kong, Berlin and Athens. Space heating is a large fraction of total energy use in commercial buildings in Europe but not in the US. Lighting accounts for 15–35 per cent of total on-site energy use, and an even larger fraction of electricity use.

2.2.3 Energy use per unit floor area by country, building type and energy source

The total energy used in a building per year per unit of floor area is a measure of the energy intensity of a building. Although data on electricity use by buildings are readily available, it is not always easy to obtain data on other forms of energy use (heating oil, natural gas, district energy). Furthermore, the definition of floor area differs among different countries (it could be gross floor area, based on the exterior dimensions of the building, or it could exclude common areas such as lobbies, elevators shafts and mechanical rooms). Nevertheless, the International Energy Agency (IEA) and other groups have compiled comparative data on country average building energy intensity and the change over time. Figure 2.5 gives data for commercial buildings in ten

Figure 2.3 Breakdown of residential energy use in (a) the US (b) the EU and (c) in Canada

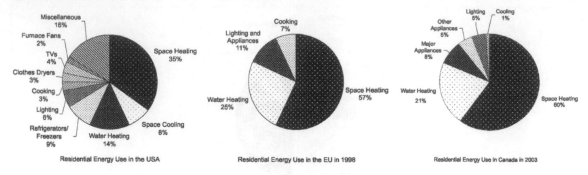

Source: Data from (a) Koomey et al (2000); (b) European Commission (2001); NRCan (2005g)

Figure 2.4 Breakdown of commercial sector energy use in (a) the US in 1999; (b) the European Union in 1998; (c) Canada in 2000; (d) a generic office building in Hong Kong; (e) a 13-storey building in Berlin; and (f) a four-storey building in Athens

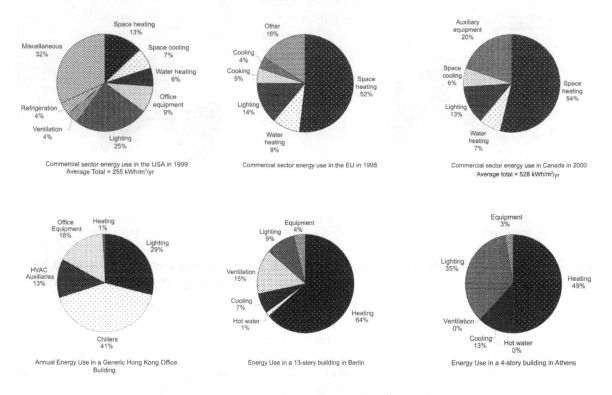

Source: (a) Energy Information Administration, www.eia.doe.gov; (b) European Commission (2001); (c) Anne Auger, Natural Resources Canada (personal communication, 2005); (d) data provided by Joseph Lam, personal communication, 2003, used in Lam and Li (1999) and Lam (2000); (e) and (f) Hestnes and Kofoed (2002)

Figure 2.5 Commercial sector energy intensity (energy use per square metre of floor area per year) in ten countries in 1975, 1990 and 1997 or 1998

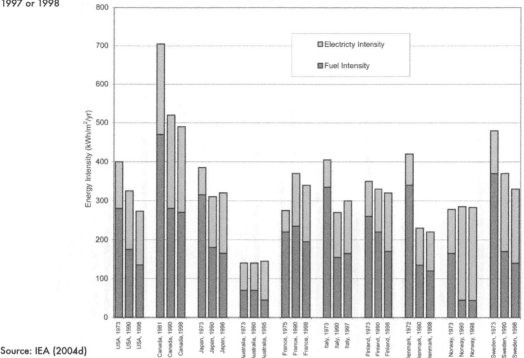

Source: IEA (2004d)

Figure 2.6 Contribution of different end uses and/or energy sources to the energy intensity of commercial buildings in 1990 in six countries

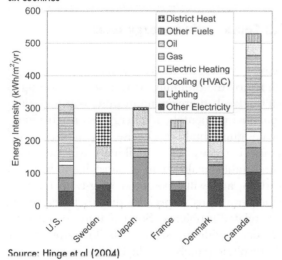

Source: Hinge et al (2004)

Figure 2.7 Comparison of energy intensities of different kinds of commercial buildings in 1990

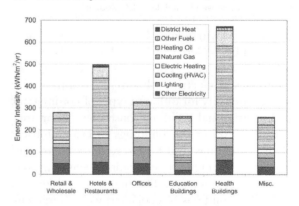

Note: Data are area weighted averages of energy intensities in the US, Canada, Japan, France, Denmark and Sweden. Electric heating includes space and hot water heating.
Source: Adam Hinge, personal communication, 2004

countries. The energy intensities are given in common units of kWh/m²/year, which allows direct comparison and summation of electric and non-electric energy uses, and also allows later comparison with the solar thermal and electric energy that can be produced with solar thermal or photovoltaic collectors. Most commercial energy intensities fell between 300 and 400kWh/m²/year in 1998, although Canada stands out

as having a particularly high energy intensity (500kWh/m²/year) and Australia as having a low energy intensity (about 150kWh/m²/year). Most countries have shown a decline in energy intensity from 1973 to 1998.

Figure 2.6 shows the breakdown of commercial sector energy use by end use and fuel in 1990 for five of the countries shown in Figure 2.5. The aforementioned high energy use in Canada

Figure 2.8 Residential energy intensity (energy use per square metre of floor area per year) in 12 countries in 1973 or 1981 and 1998

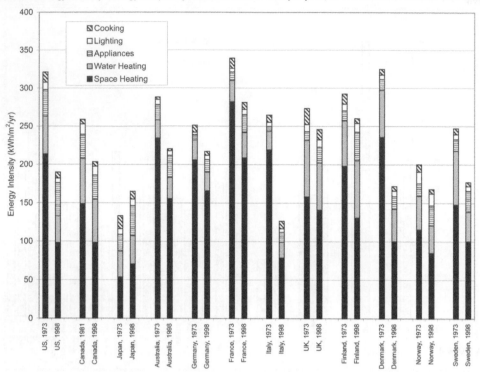

Note: Heating energy use has been normalized to a climate of 2700 heating degree days (HDD).
Source: Figure derived from data in IEA (2004d)

is associated with a large heating energy use compared with Sweden (which also has a cold climate) but also very large energy use for lighting and other uses of electricity. Figure 2.7 compares the energy intensity for different types of commercial and institutional buildings, based on a floor area weighted average of buildings in five countries. Health buildings and hotels/restaurants are the most energy intensive, educational buildings the least energy intensive. The high energy intensity of health buildings is associated with a large energy use for heating, due to the need for 100 per cent outside air for ventilation.

Turning now to residential buildings, Figure 2.8 gives the average residential energy intensity in 1973 (or 1981) and 1998 in 12 countries. Space heating (normalized to a common climate with 2700 heating degree days (HDD)) accounts for the largest single residential use of energy, with energy intensities in 1998 ranging from 70kWh/m²/year in Japan to more than 200kWh/m²/year in France and the US. Total residential energy intensity ranges from 130kWh/m²/year (Italy) to 250kWh/m²/year (UK and Finland).

2.3 Comparative energy costs

One of the justifications for designing buildings to use less energy is the saving in energy costs over the life of the building. As will be seen throughout this book, there are many other reasons (apart from environmental and energy security benefits) for designing buildings to use less energy: if designed correctly, these buildings will be more comfortable, healthier, more satisfying to live and work in, and will have fewer long-term maintenance problems (especially due to fewer moisture-related problems). Some features of energy-efficient buildings, such as building-integrated photovoltaic panels and daylighting systems, can be part of the aesthetic and architectural expression of the building, providing benefits that cannot be expressed in economic terms. Nevertheless, energy costs – in comparison with the cost of efficiency and renewable-energy features – are a factor to be considered in the decision-making process.

Figure 2.9 compares the cost of natural gas to households in 2003 in various countries.

Figure 2.9 Cost of natural gas to households (including taxes) and range in cost of heating equipment amortized over 20 years at a real interest rate of 4 per cent

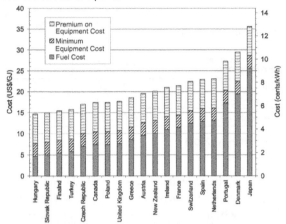

Note: Heating equipment costs are the range (US$3–10/GJ) given in Table 4.5.
Source: Source of natural gas cost data: IEA, (2004d)

Figure 2.10 Retail electricity and heating energy costs in 1998 (except for Canada) in six countries

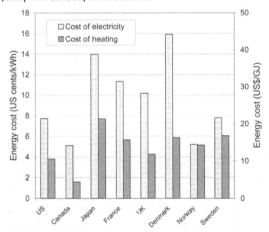

Note: Canadian costs are the costs assumed in developing the Canadian Model National Energy Code in 1997 (NRCan, 1997a).
Source: For all countries except Canada, data from IEA, (2004d)

This cost (including taxes), ranged from a low of US$5/GJ in Hungary to $25/GJ in Japan. The cost of supplying heat includes the cost of the heating system, amortized over its lifetime. Based on Table 4.5, this amounts to $3–10/GJ, bringing the total cost of providing heat with natural gas in residential buildings to $15–35/GJ (about 5–13 cents/kWh) at 2003 natural gas prices. Figure 2.10 compares the average 1997 retail cost of electricity and of energy used for heating in

eight countries. Electricity ranged in price from 5 cents/kWh in Canada and Norway, to 16 cents/kWh in Denmark. The heating energy costs are weighted by the relative proportions of the different energy forms used for heating, which in Norway is largely electricity. For all countries except Norway, average energy costs for heating in 1997 were less than electricity costs.

Average energy costs, however, are not an accurate measure of the benefits of energy saving measures. First, average energy costs will be disproportionately affected by a relatively small number of large consumers, who tend to have lower rates. Costs for most consumers will be greater than the average cost. Second, the *marginal cost* of energy should be considered rather than the average cost. The marginal cost is the total increase in utility bill divided by the increase in energy use. With the exception of small residential customers, most electric utilities have charges based on the amount of electricity used (energy) and based on the peak demand (peak power). Different rates are applicable for different kinds of customers and for customers of different size, and there are often different rates for successive blocks of energy consumption within a billing period. Thus, the price paid for energy savings at the margin depends on:

- the customer size as measured by annual peak demand;
- the season and, in some cases, other features such as local geography or consumer type;
- the total customer consumption and peak demand for the billing period, which determine which rate block applies *at the margin*;
- how the efficiency measure affects both the demand and energy components of the bill.

To assist in calculating the energy cost savings of energy efficiency measures that reduce electricity consumption, the Environmental Technologies Division of Lawrence Berkeley National Laboratory (Berkeley, California) undertook the *Tariff Analysis Project* (TAP). Under TAP, a database of over 400 electricity tariffs from over 100 utilities in 14 countries was created, as well as a web-based interface (available at http://tariffs.lbl.gov). The web interface provides the user with a data entry form (a universal tariff template) that can be configured to mimic the particular structure of a given tariff. Once the form is configured, data values can be entered and edited. Form structures can be stored, allowing relatively quick data entry for multiple tariffs of similar structure. The interface thus allows users to enter and view complex rate structures with a minimum of effort.

As an example of the difference between marginal cost and average cost, the marginal cost of electricity for commercial air conditioners ranges from 6.3 cents/kWh in parts of the south east US, to 12.8 cents/kWh in parts of New England and 16.7 cents/kWh in California according to TAP, compared with an average US price of electricity of 7.7 cents/kWh according to the data in Figure 2.10.

2.4 Emissions of greenhouse gases

This book is motivated by the need to reduce emissions of greenhouse gases (GHGs) in order to limit, and eventually stabilize, human induced warming of the climate. We focus on energy use in buildings because the emission of GHGs is directly tied to the use of fossil fuels, which are the dominant source of energy for buildings (and other sectors). However, there is no direct 1:1 relationship between fossil fuel energy use and GHG emissions. Furthermore, there are GHG emissions associated with air conditioning in buildings beyond those associated with the provision of electricity (namely, through leakage of halocarbon refrigerants) and through the use of halocarbon expanding agents that are used in the manufacture of some types of insulation.

2.4.1 Emissions of CO_2 and CH_4 from the use of fossil fuels

Table 2.3 lists the emission of carbon dioxide per unit of primary energy for coal, oil and natural gas. The emissions factors for coal and oil are about 90 per cent and 30 per cent greater, respectively, than that of natural gas. As noted above, measures that reduce on-site energy use but with a shift from fuels to electricity could entail a significant increase in the use of primary energy, due to the large losses in generating electricity. To

Table 2.3 Carbon dioxide emission factors for combustion of various fossil fuels (given as kg C per GJ of fuel energy) and in the manufacture of cement (given as kg of C per kg of cement)

Source of CO_2 or CH_4	Emission factor	
	CO_2	CH_4
Coal used in electric power plants	22.2–25.9	0.10–0.40
Oil	17–20	<0.03
Natural gas	13.5–14.0	0.18–0.19
Cement	0.08–0.26	–

Source: Harvey (2000)

the extent that electricity is generated from coal, the impact on CO_2 emissions will be amplified further. Conversely, measures that shift from the use of coal-fired electricity (at 30–35 per cent primary energy to consumer efficiency) to natural gas (at 90 per cent primary energy to consumer efficiency) will reduce the CO_2 per unit of on-site energy use by about a factor of five.

Methane is released to the atmosphere through the mining of coal (it leaks from the coal seams), through the extraction of oil (which often occurs in association with natural gas), and through the extraction and distribution of natural gas (which is largely methane). The emission from coal seams is much greater for deep seams than for shallow seams. The emission through the natural gas system depends on how well maintained the distribution system is. Table 2.3 gives emission factors per unit of primary energy. Given an efficiency in generating electricity from coal of about 33 per cent, the emission factor per unit of electrical energy will be three times greater. Although there is considerable uncertainty in the appropriate emission factor in any given region (due to differences in the nature of the coal seams or in the rate of leakage from the natural gas system), it is almost always the case that switching from electricity generated from coal to on-site use of natural gas will entail a reduction in methane emissions.

2.4.2 Emission of CO_2 during the manufacture of cement

The chemical process that produces cement releases CO_2. This is in addition to the CO_2 released through the combustion of fossil fuels that supply the energy needed to manufacture cement. The chemical emission factor is given in Table 2.3. As discussed in Chapter 12 (Section 12.2), the chemical emission of CO_2 can be two to four times that associated with the energy used to make cement.

2.4.3 Emissions of halocarbon gases and global warming potential (GWP)

Halocarbons are compounds containing carbon and one or more halon gases – chlorine, fluorine and bromine. Those containing chlorine and fluorine are *chlorofluorocarbons* (CFCs), those

containing hydrogen, chlorine and fluorine are *hydrochlorofluorocarbons* (HCFCs), and those containing hydrogen and fluorine are *hydrofluorocarbons* (HFCs). All three groups are GHGs, while those containing chlorine (the CFCs and HCFCs) have an ozone-depleting effect as well. CFCs were originally the refrigerant of choice in air conditioners and other cooling equipment, but their use has been phased out in order to protect the stratospheric ozone layer. They have been temporarily replaced by

Table 2.4 Climate-related properties of halocarbon gases and their replacements that are considered in this book

Gas	Uses	Heat trapping (W/m²/ppb)	Lifespan (years)	GWP	Source of data
CFC-11	Centrifugal chillers[a]	0.25	45	4680	Tables TS-1, TS-3
CFC-12	Centrifugal chillers, HP, AC, refrigerators[a]	0.32	100	11726	Tables TS-1, TS-3
HCFC-22[b]	Chillers, HP, AC, insulation (XPS)	0.20	12	1780	Tables TS-1, TS-3
HFC-23		0.19	270	14310	Tables TS-1, TS-3
HFC-32	HP, AC, chillers	0.11	4.9	670	Table 2.6
HCFC-123	Chillers, ejector	0.14	1.3	76	Table 2.6
HCFC-124		0.22	5.8	599	Table 2.6
HFC-125	HP, AC, chiller	0.23	29	3450	Table TS-19
HFC-134a	HP, rotary and centrifugal chillers, ejectors, refrigerators, insulation (XPS)	0.16	14.0	1410	Tables TS-1, TS-3
HCFC-141b	Ejector, insulation (PU, PI)	0.14	9.3	713	Tables TS-1, TS-3
HCFC-142b	Insulation (XPS)	0.20	17.9	2270	Tables TS-1, TS-3
HFC-143a	HP	0.13	52	4400	Table 2.6
HFC-152a	Insulation (XPS)	0.09	1.4	122	Tables TS-1, TS-3
HFC-227ea	Insulation (PU, PI)	0.25	34.2	3140	Table TS-19
HFC-245fa	Insulation (PU, PI)	0.28	7.6	1020	Ashford et al. (2005)
HFC-365mfc	Insulation (PU, PI)	0.21	8.6	782	Table TS-22
R-404a	HP			3862	Table TS-9
R-407c	HP, chillers except centrifugal	HFC mixtures. See Table 5.7 for composition		1600	WK (2001)
R-410a				2060	Table TS-9
R-507a	HP			3925	Table TS-9
HC-290 (propane)	HP	0.0031	0.041	20	WK (2001)
HC-600a (isobutene)	Refrigerators (in Europe)	0.0047	0.019	20	Table 2.6
HC-601 (pentane)	Insulation (PU, PI)	0.0046	0.010	7[c]	Table 2.6
R-717 (ammonia)	HP, chillers		days	<1	WK (2001)
R-718 (water)	Under development			0	
R-744 (CO_2)	Under development			1	

[a] Former uses. Production now phased out.
[b] Phased out in new equipment by 31 December 2003 in Europe, to be phased out by 1 January 2010 in the US.
[c] Inferred
WK (2001) = Westphalen and Koszalinski (2001). AC = air conditioners, HP = heat pumps, XPS = extruded polystyrene, PU = polyurethane, PI = polyisocyanurate
Source: Tables from the technical summary of IPCC/TEAP (2005), except where indicated otherwise

the HCFCs, which in turn will be replaced by the HFCs or non-halocarbon refrigerants. The halocarbons have also been used as blowing agents in various kinds of foam insulation for building exteriors, refrigerators and hot-water tanks.

The impact on climate of the emission of a given mass of gas depends on the effectiveness of the gas in trapping heat, on a molecule-by-molecule basis, and on the average lifespan of molecules of that gas in the atmosphere. As the amount of gas in the atmosphere decreases after a pulse emission, the heat-trapping effect decreases. The integral (or summation) of this heat trapping over some arbitrary time horizon can be computed and compared with that for CO_2; the ratio of the two forms an index called the *global warming potential* (GWP). This is a rough but adequate measure of the relative contribution of equal emissions (in terms of mass) of different gases to global warming (see Harvey, 1993, for a critique of the GWP index).

Table 2.4 provides a consolidated list of the halocarbons and non-halocarbon alternatives considered in various chapters of this book, their heat-trapping ability per molecule, the average lifetime of a molecule in the atmosphere before being removed, and the resulting GWP over a 100-year time horizon. The GWPs given in Table 2.4 are on a mass basis: the mass of non-CO_2 gas emitted times its GWP gives the mass of CO_2 having the same heating effect integrated over a period of 100 years.

Traditional refrigerants and blowing agents (CFC-11 and CFC-12) and their replacements (the HCFCs and HFCs) have GWPs ranging from over 100 to several thousand. For modern and properly serviced cooling equipment, the climatic effect of halocarbon emissions is insignificant compared to the climate effect of the energy used to power the equipment, and compared to potential differences in the efficiency of the equipment using non-halocarbon instead of halocarbon refrigerants or due to other factors. For foam insulation using halocarbon blowing agents, in contrast, the climatic effect of halocarbon emissions can exceed that of the heating energy saved by using the insulation.

2.5 Climatic indices of heating and cooling requirements

A convenient indicator of the heating requirement in a given climate is the number of *heating degree days* (HDD), which is defined as the daily average air temperature departure below some reference temperature for each day when the average temperature is below the reference temperature, and summed over all such days. Normally, the reference temperature is 18°C, because for conventional houses, no heating is required above this temperature. In principle, a lower reference temperature should be used for better-insulated buildings, or for buildings with larger internal heat sources, but the calculated HDD is still a useful rough indicator of climatically-related differences in the potential heating requirements.

Similarly, the potential cooling requirement in a given climate is given by the number of *cooling degree days* (CDD), which is defined as the daily average air temperature departure above some reference temperature for each day when the average temperature is above the reference temperature, and summed over all such days. The reference temperature for CDD should be the outdoor temperature at which internal heat gains maintain the indoor temperature at the desired temperature for comfort. This depends on the magnitude of the internal heat gains, the desired comfort temperature, and whether or not ventilation with outdoor air is permitted. The CCD data given here pertain to a reference temperature of 18°C.

HDD and CDD data for 77 cities of the world are given in Table 2.5 and Table 2.6, respectively. Except for the US and Canada, HDD and CCD data are not widely and easily available. For the US and Canada, the HDD data are taken from published data tables. All other HDD data and the CDD data were obtained from NASA's *Surface Meteorology and Solar Energy* (SSE) dataset (http://eosweb.larc.nasa.gov/sse) by entering the latitude and longitude of the location in question. The data are the average for the period 1983–1993, derived from a combination of meteorological observations and atmospheric model simulations, interpolated to a 1° latitude by 1° longitude grid. The data do not include urban heat island effects and have a stated root mean square error of ±15 per cent. Comparison of the SSE HDD values with published values (which do include urban heat island

Table 2.5 Heating degree days (K-day) relative to a base of 18°C for selected cities in the world, rounded to the nearest 50 degree days

North and South America		Europe and Africa		Asia and Australia	
Canada		Europe		Asia	
Toronto	4000	Athens	1550	Istanbul	2450
Winnipeg	5750	Rome	1900	Ankara	3100
Edmonton	5750	Madrid	1100	Beirut	850
Vancouver	3000	Glasgow	3200	Riyadh	450
Inuvik	11,000	London	3300	Tehran	2550
US		Paris	3200	Kabul	4050
Miami	50	Zurich	4000	New Dehli	900
Atlanta	1550	Berlin	3800	Bombay	0
New Orleans	800	Prague	3900	Madras	0
Albuquerque	2400	Vienna	3900	Katmandu	2950
Phoenix	650	Budapest	3750	Lhasa	7600
San Francisco	1600	Warsaw	4250	Ulaan Baatar	7100
Las Vegas	2750	Minsk	4650	Beijing	4300
Denver	3400	Kiev	4450	Seoul	3500
Chicago	3400	Moscow	5500	Tokyo	2250
New York	2650	Helsinki	5700	Sapporo	4100
Boston	3100	Copenhagen	4250	Shang-hai	1950
Minneapolis	4350	Oslo	3750	Hong Kong	500
Fairbanks	7750	Murmansk	6350	Ho Chi Minh City	0
Latin America		Africa		Singapore	0
Mexico City	450	Casablanca	750	Manila	0
Caracas	0	Cairo	750	Australia	
Bogotá	0	Lagos	0	Alice Springs	750
Brasília	0	Nairobi	50	Cairns	50
São Paulo	0	Salisbury	300	Brisbane	500
Buenos Aires	50	Johannesburg	1000	Sydney	1100
Punta Arenas	7600	Cape Town	900	Hobart	2400

Source: Canada, MSC (2002); US, NOAA (2002); elsewhere, HDD values given by *Surface Meteorology and Solar Energy* (SSE) dataset (http://eosweb.larc.nasa.gov/sse) for the latitude and longitude of the city in question. For US cities, NOAA HDD values are on average 324 K-day lower than SSE values, presumably a reflection of the local urban heat island effect that is absent from the course-resolution SSE data. The appropriate adjustment of the SSE data depends on the length of the heating season and other factors

effects) indicates that about 300 HDD should be subtracted from the SSE HDD values for major cities to account for the urban heat island effect. Other climatic data available from the SSE dataset include monthly mean air temperature, the monthly mean diurnal temperature range, the monthly mean temperature at which condensation would occur (the dewpoint temperature), and the surface air water vapour mixing ratio. Data on the availability of solar energy are also available, and are discussed in Chapter 11 (Section 11.1.2).

2.6 Meteorological data

Meteorological data (i.e. hourly data) can be obtained for any location in the world from the *Meteonorm* dataset (www.meteotest.ch). This commercial dataset is based on a worldwide network of 7400 weather stations. The primary data are monthly means of irradiance (direct and diffuse solar, longwave), temperature, relative humidity, precipitation, luminance, wind speed and direction, and sunshine duration. Hourly data are generated based on a stochastic model (that is, a model of how random variability is distributed around the mean). Data can be interpolated from the weather stations to other points. The output can be exported in a number of different formats, suitable for different building simulation software packages.

For the US, Typical Meteorological Year (TMY) data are available from the National Renewable Energy Laboratory (NREL, www.nrel.gov) for 239 locations. These are hourly solar

Table 2.6 Cooling degree-days (K-day) relative to a base of 18°C for selected cities in the world, rounded to the nearest 50 degree-days

North and South America		Europe and Africa		Asia and Australia	
Canada		Europe		Asia	
Toronto	500	Athens	750	Istanbul	400
Winnipeg	450	Rome	550	Ankara	250
Edmonton	300	Madrid	500	Beirut	1150
Vancouver	0	Glasgow	0	Riyadh	2950
Inuvik	0	London	50	Tehran	600
US		Paris	100	Kabul	200
Miami	2400	Zurich	50	New Dehli	1450
Atlanta	1000	Berlin	100	Bombay	2800
New Orleans	1650	Prague	100	Madras	3000
Albuquerque	700	Vienna	100	Katmandu	300
Phoenix	2300	Budapest	200	Lhasa	0
San Francisco	100	Warsaw	100	Ulaan Baatar	50
Las Vegas	1800	Minsk	100	Beijing	400
Denver	400	Kiev	150	Seoul	400
Chicago	550	Moscow	100	Tokyo	650
New York	650	Helsinki	50	Sapporo	100
Boston	450	Copenhagen	50	Shang-hai	850
Minneapolis	400	Oslo	100	Hong Kong	1900
Fairbanks	50	Murmansk	0	Ho Chi Minh City	1500
Latin America		Africa		Singapore	3050
Mexico City	400	Casablanca	500	Manila	3100
Caracas	2800	Cairo	1500	Australia	
Bogotá	950	Lagos	3050	Alice Springs	1750
Brasília	2750	Nairobi	750	Cairns	2050
São Paulo	2350	Salisbury	900	Brisbane	900
Buenos Aires	1450	Johannesburg	300	Sydney	350
Punta Arenas	0	Cape Town	250	Hobart	50

Source: US, NOAA (2002); elsewhere, CDD values given by *Surface Meteorology and Solar Energy (SSE)* dataset (http://eosweb.larc. nasa.gov/sse) for the latitude and longitude of the city in question. For US cities, NOAA CDD values are on average 232 K-day higher than SSE values, presumably a reflection of the local urban heat island effect that is absent from the course-resolution SSE data. The appropriate adjustment of the SSE data depends on the length of the cooling season and other factors

radiation and meteorological data representing typical conditions (so they cannot be used for designing against extremes). ASHRAE (1997) provides building-oriented datasets of typical hourly weather data for 77 US and Canadian locations, while ASHRAE (2001a) provides typical weather data for 227 locations outside the US and Canada. In both datasets, weather data are supplemented by hourly solar irradiance estimated from hourly earth–sun geometry and weather elements (particularly cloud information).

2.7 Strategy for the remaining chapters

The remainder of this book proceeds as follows: it begins with a discussion of the thermal envelope, heating systems, cooling systems and the

production of hot water for non-heating purposes. A discussion of heat pumps, which can simultaneously provide heating, cooling and hot water requirements, is sandwiched between the discussion of heating and cooling systems. The discussion of cooling and dehumidification is rather long because it is one of the harder energy uses to meet efficiently. It is also complicated: a wide variety of approaches are applicable but in different climates and for buildings with different characteristics. It will become increasingly important as the climate warms during the next few decades. Next, ventilation systems, lighting systems, appliances and office equipment, and the active (as opposed to passive) uses of solar energy are discussed. Full realization of the potential energy savings requires an integrated, team approach to the design of buildings, and a separate chapter deals with advanced designs in new residential and commercial buildings. In both cases, reductions in total energy use in new buildings of 75 per cent compared to current standard practice in OECD countries have been achieved. Commercial buildings are generally more complex than residential buildings, and a very important result, discussed here, is that the design *process* is more important than the use of advanced technology in achieving dramatic energy savings. The chapter on advanced buildings is followed by a chapter that discusses the savings that can be achieved through retrofits of existing buildings. The last chapter considers buildings as part of a 'Community-Integrated Energy System' (CIES).

The approach adopted here in discussing heating, cooling and ventilation requirements is to first examine the extent to which the load can be reduced, then to examine the extent to which each load can be met passively, and only then to discuss options for meeting the residual load as efficiently as possible through mechanical means. For lighting, the sequence is to maximize natural daylighting, then to meet the residual lighting needs as efficiently as possible. Active uses of solar energy (building-integrated PV, solar hot water heaters, solar heat driven chillers) are discussed in a separate chapter, although the distinction between active and passive features is becoming blurred. This is because in the most efficient and cost-effective designs, active and passive solar features will be integrated with each other and with the building envelope, rather than as independent add-on features. Active solar heat collectors can be combined with underground seasonal thermal energy storage, but this is best done at a community scale as part of a district heating system, so yet another layer of analysis and integration arises but is deferred until the CIES chapter.

Notes

1   OECD = Organization for Economic Cooperation and Development. The OECD consists of the following 30 countries: Australia, Austria, Belgium, Canada, Czech Republic, Denmark, Finland, France, Germany, Greece, Hungary, Iceland, Ireland, Italy, Japan, Korea, Luxembourg, Mexico, The Netherlands, New Zealand, Norway, Poland, Portugal, Slovak Republic, Spain, Sweden, Switzerland, Turkey, the UK and the US.

2   In computing the primary to secondary conversion efficiencies, all of the inputs in the row 'Petroleum extraction and refining' were assigned to the Oil column rather than their original column, as these are all assumed to be energy inputs to oil refining. 'Other energy industry' is assumed to be entirely natural gas processing, so all the entries in that row were transferred to the Natural gas column before computing the primary to secondary energy conversion efficiency for natural gas.

3   Prior to June 2004, the European Union consisted of the following 15 countries: Austria, Belgium, Denmark, Finland, France, Germany, Greece, Ireland, Italy, Luxembourg, Netherlands, Portugal, Spain, Sweden and the UK. This group of nations is commonly referred to as the EU-15. On 1 June 2004, another 10 countries joined the EU: Cyprus, Czech Republic, Estonia, Hungary, Latvia, Lithuania, Poland, Romania, Slovak Republic and Slovenia.

three

Thermal Envelope and Building Shape, Form and Orientation

The term *thermal envelope* refers to the shell of the building as a barrier to the loss of interior heat or to the penetration of unwanted outside heat into the building. It refers to the walls, windows, roof and basement floor of the building. The building envelope is more than a thermal barrier, however, as it contributes to the structural integrity of the building and serves as a barrier to moisture, the infiltration or exfiltration of air and noise. These functions are tightly coupled, such that measures that lead to a higher quality thermal barrier will lead to improvements in the envelope as a barrier to wind and rain, thereby reducing maintenance costs and increasing the lifespan of the building envelope. For example, measures to eliminate thermal bridges (places where there are gaps in the insulation or highly conductive materials bridging from indoors to outdoors) reduce the likelihood of condensation and all the problems associated with condensation. Thermally tight buildings will have smaller spatial variations in temperature and lower noise from outside,

contributing to a higher-quality and more comfortable indoor environment.

The effectiveness of the thermal envelope depends on:

- insulation levels in the walls, ceiling, and basement;
- resistance to moisture migration;
- the thermal and optical properties of windows and doors;
- the rate of exchange of inside air with outside air through infiltration and exfiltration;
- the presence of shared walls with other buildings.

A better thermal envelope reduces the amount of heat that needs to be supplied by the heating system in winter, or the amount of cooling that is needed in summer. The heat supplied to, or removed from, a building is an example of tertiary or end use energy demand (see Chapter 2, Section 2.1).

The size and positioning of windows, as well as possible skylights, will influence not only the heat loss from the building in winter, but also the possibilities for passive solar heating and the use of natural daylight in place of artificial light, as well as the amount of unwanted heat gain in summer. Building orientation and overall shape are also important factors in this regard, the former by influencing the receipt of solar energy in summer and winter, the latter by influencing the ratio of surface area to volume. The larger the building, or the more compact the building, the less important the thermal envelope for heat gain and loss, but also the smaller the opportunity for cooling through natural ventilation. Optimization of the trade-off between heat loss, passive heat gains, daylighting, cooling requirements and natural ventilation requires the use of computer simulation models to find the design and choice of materials that minimizes overall energy use for a given climate, site, building function and building occupancy pattern.

This chapter begins with a discussion of heat transfer processes, followed by a thorough discussion of the building envelope as a barrier to the transfer of heat, and closes with a discussion of the role of building shape and form in heat transfer, and the application of life-cycle cost analysis to the choice of alternative envelope systems.

3.1 Processes of heat transfer

Heat can be transferred across the building through molecular conduction, convective mixing, passage of air through leaks in the envelope and through radiative exchange.

3.1.1 Conduction and convection

Conduction involves transferring molecular kinetic energy (heat) to nearby molecules through collisions, while convection involves the movement of air parcels from one place to the next either as turbulent eddies, airflow or a combination of the two. Convection can occur in the air on either side of a wall or window, in cavities inside hollow-core walls or concrete blocks, and in the air gap inside double-glazed windows. In the latter case, an overturning circulation cell will

occur, with air sinking adjacent to the cold glazing and rising adjacent to the warm glazing. Immediately next to a wall, or on either side of a pane of glass, will be a thin layer where heat transfer occurs by molecular conduction only. The heat flux across a wall due to conductive and convective heat transfer, Q_c, is given by:

$$Q_c = \text{temperature difference} \times \text{heat transfer coefficient} = \Delta T \times U \quad (3.1a)$$

or

$$Q_c = \text{temperature difference} / \text{thermal resistance} = \Delta T / R \quad (3.1b)$$

and has units of watts per square metre (W/m²). This times surface area (m²) times time (in seconds) gives one component of the heat loss (joules) that must be supplied by the heating system during that time period. The heat transfer coefficient (U, W/m²/K) for a given layer is given by the *thermal conductivity* (W/m/K) for that material divided by the thickness of the layer. Thus, for a given layer, the heat loss through it will be smaller the thicker it is and the lower its thermal conductivity.

Representation of the heat flow by Equation (3.1) assumes the heat flow to be purely one-dimensional, that is, perpendicular to the wall or window surface. Next to corners or to lateral contrasts in thermal conductivity, there will be a lateral component to the heat flow, and detailed three-dimensional (3D) calculations are needed to compute the heat loss accurately. Otherwise, localized errors of up to 10 per cent in the rate of heat loss can occur (Kośny and Kossecka, 2002).

3.1.2 Exchange of air between inside and outside

The heat loss due to an exchange of warm indoor air with cold outdoor air depends on the indoor to outdoor temperature difference (the greater this difference, the more heat that needs to be added to the incoming outdoor air in order to maintain the same indoor temperature), as well as on the rate of air exchange (F, m³/s), density (ρ, kg/m³), and *specific heat* of air (c_{pa}, J/kg/K).[1] That is:

$$Q_e = \rho c_{pa} F (T_{\text{indoor}} - T_{\text{outdoor}}) = \rho c_{pa} F \Delta T \quad (3.2)$$

where T_{indoor} is the temperature of indoor air (which leaves the building) and $T_{outdoor}$ is the temperature of outdoor air (which enters the building). Q_e has units of watts. Multiplication by time (in seconds) gives the heat loss (joules) due to air exchange over that time period. Exchange of air occurs through unintentional infiltration of outside air and exfiltration of inside air, as well as through deliberate air exchange through the ventilation system in order to provide fresh air.

Given an indoor temperature of 20°C and an outdoor temperature of 0°C, a reduction of 1K in the indoor temperature reduces the inside to outside temperature difference – and the associated heat losses due to conduction/convection and air exchange – by 5 per cent. The heat loss must be made up by supplying heat from the heating system, so the heating energy use also decreases by 5 per cent.

3.1.3 Radiative transfer

All objects above 0K (−273.15°C) emit electromagnetic radiation. The warmer the object, the shorter the wavelengths of the radiation that it emits. The sun (with a surface temperature of 5773K) emits radiation largely between wavelengths of 0.1μm (micron) and 4.0μm (1μm = 10^{-6}m). Solar radiation is divided into ultraviolet (wavelengths of 0.1–0.4μm), visible (0.4–0.7μm), and near infrared (0.7–4.0μm) radiation. Visible solar radiation is what we see as light and different wavelengths correspond to different colours. The peak solar emission occurs at a wavelength of 0.55μm, corresponding to green light. Objects typical of Earth and atmospheric temperatures (200–300K) emit radiation in the longwave or *infrared* part of the spectrum (4.0–50μm), with a peak emission at 10–12μm. The maximum amount of radiation that can be emitted depends on temperature (warmer objects emit more radiation, as well as at shorter wavelengths). This maximum amount of radiation is called *blackbody emission, B*, and is given by:

$$B(W/m^2) = \sigma T^4 \tag{3.3}$$

where σ (= 5.670400 × 10^{-8} W/m²/K) is the Stefan-Boltzman constant and T is in K.

An object that emits the maximum possible amount of radiation is called a *blackbody*. The *emissivity* ε is the ratio of actual emission to maximum possible emission; thus, actual emission E is given by

$$E = \varepsilon \sigma T^4 \tag{3.4}$$

Most solid objects behave almost as blackbodies, that is, ε ≈ 1.0. However, it is possible to achieve quite low emissivities with many metal surfaces if they are highly polished or galvanized, as indicated by the emissivity values given in Table 3.1. The absorption of infrared radiation by a body is given by the incident infrared radiation (emitted by the surroundings) times the *absorptivity* (the fraction absorbed). The absorptivity is always equal to the emissivity. Infrared radiation that is not absorbed by solids (such as window panes or walls) is reflected.

Table 3.1 Emissivity of metals and of common building materials at a wavelength of 9.3μm

Metals			
Aluminium		Iron	
polished	0.04	polished	0.06
oxidized	0.11	galvanized, new	0.23
surface roofing	0.22	galvanized, dirty	0.28
24-ST weathered	0.40	cast, oxidized	0.63
anodized (at 538°C)	0.94	steel plate, rough	0.94
Brass		oxide	0.96
polished	0.10	Silver, polished	0.01
oxidized	0.61	Stainless steel	
Chromium		polished	0.15
polished	0.08	weathered	0.85
Copper		Zinc	
polished	0.04	polished	0.02
oxidized	0.87	galvanized sheet	0.25
Building materials			
Bricks	0.90–0.93	Spruce, sanded	0.82
Concrete	0.94	Walnut, sanded	0.83
Marble	0.95	Oak, planed	0.90
White paper	0.95	Beech	0.94
Paint	0.94–0.96		

Source: Hassani and Hauser (1997) except for concrete and wood, which are from Kreider et al (2002) and pertain to a temperature of 40°C but an unspecified wavelength

Thus,

- a blackbody emits the maximum amount of radiation for its temperature, and absorbs all of the incident infrared radiation from the surroundings (which is why it is called a 'black' body – real black objects absorb all the incident visible light);

- for objects with $\varepsilon < 1$, the emission of radiation is only ε times the blackbody emission (so it cools less), while $(1 - \varepsilon)$ times the incident infrared radiation from the surroundings is reflected and only ε times the incident radiation is absorbed.

When an object absorbs radiation (whether solar or infrared), it gains energy and warms up. Conversely, when an object emits infrared radiation it loses energy, so it cools.

Figure 3.1 shows the distribution with wavelength of solar radiation and of infrared radiation emitted by blackbodies for typical earth surface temperatures. Also given is the variation in the sensitivity of the human eye to radiation. Not surprisingly, the human eye is most sensitive to those wavelengths of solar radiation that are available in greatest abundance, that is, near the peak of solar emission at a wavelength of 0.55μm. Figure 3.1 shows a clear separation between radiation emitted by the sun and by objects at typical atmospheric temperatures, as well as the increase in emission at all wavelengths and the shift of the wavelength of peak emission to shorter wavelengths as temperature increases.

3.1.4 Measuring the resistance to heat flow

The heat-transfer properties of insulation, doors, and windows are rated using three different parameters:

1 *The U-Value* $(W/m^2/K)$: This is the thermal conductivity $(W/m/K)$ divided by the thickness of the material, and is equal to the rate of heat flow (joules per second, or watts) per unit area and per degree of inside-to-outside temperature difference. It is referred to as a heat transfer coefficient. Window manufacturers in the US and Canada tend to use U-values in British units, which are smaller by a factor of 5.678. To avoid possible confusion among North American readers, the units of the U-value will always be specified here. A smaller U-value implies less heat flow for a given temperature difference.

2 *The RSI-Value* $(W/m^2/K)^{-1}$: This is the resistance to heat flow, numerically equal to 1/U when U is expressed in metric units. A larger RSI-value implies less heat flow for a given temperature difference. As the RSI is in metric units by definition, RSI-values will normally be stated here without units.

3 *The R-Value* $(Btu/ft^2/hr/°F)^{-1}$: This is the resistance to heat flow, numerically equal to 1/U when U is expressed in British units. As the R-value is in British units by definition, R-values will normally be stated

Figure 3.1 Variation in the amount of solar radiation with wavelength, in the blackbody emission of radiation with wavelength, and in the sensitivity of the human eye to radiation of different wavelengths. The left scale is for solar radiation and the right scale is for blackbody radiation at –50°C to 50°C.

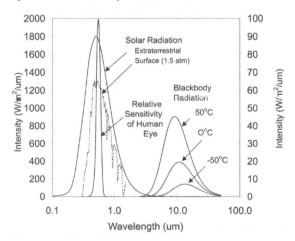

here without units. To convert from RSI-values to R-values, multiply by 5.678.

Each of the above measures of the heat loss tendency has been applied to walls and roofs, while window heat loss tendency is always given as a U-value. An advantage of specifying wall and roof heat properties as a U-value is that it can be directly compared with window heat loss, and in both cases, the rate of heat loss per unit area is directly given by the U-value times the indoor-to-outdoor temperature difference. An advantage of using resistances is that the total resistance of a wall or roof consisting of several layers is given by the sum of the resistances of each layer.

Figure 3.2 shows how the relative heat loss for a given temperature difference varies with the RSI-value for the range of RSI-values that characterize windows (upper panel) or walls and roofs (lower panel). Because the heat flow varies with 1/RSI, there are diminishing returns to adding ever more insulation. The thermal properties of common building structural materials and for different kinds of insulation are given in Table 3.2. The thermal resistance of windows takes into account the exchange of infrared radiation between the window panes (in multiply glazed windows) and between the window and the interior of the room, as discussed later (Section 3.3.4). Radiative heat exchange is not normally a factor in insulation materials, although it can be important for the transfer of heat from one side to the other of hollow-core walls or concrete blocks.

Figure 3.2 Heat flow versus thermal resistance for the range of resistances encountered in windows (upper) and encountered in insulated walls and ceilings (lower). Window heat flow is relative to the heat flow for a single-glazed window, while wall heat flow is relative to the heat flow at R12 insulation (RSI 2.1), which fits into 2″ × 4″ (38 × 89mm) stud walls

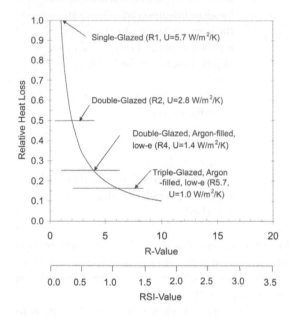

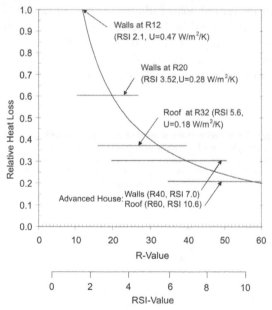

Table 3.2 Thermal properties of building structural materials and of insulation

Material	Conductivity W/m/K	Typical thickness m	inches	U value W/m²/K	RSI value (W/m²/K)⁻¹	R value (Btu/ft²/hr/°F)⁻¹
Structural materials						
Cladding						
Douglas Fir plywood	0.11	0.004	0.2	27.5	0.0364	0.2
Gypsum board (drywall)	0.48	0.0126	0.5	38.1	0.0263	0.1
Particle board	0.14	0.0126	0.5	11.1	0.0900	0.5
Masonry						
Red brick	0.47	0.1015	4.0	4.6	0.2160	1.2
White brick	1.1	0.1015	4.0	10.8	0.0923	0.5
Concrete	2.1	0.2032	8.0	10.3	0.0968	0.5
Hardwoods						
Oak	0.17	0.0507	2.0	3.4	0.2982	1.7
Birch	0.17	0.0507	2.0	3.4	0.2982	1.7
Maple	0.16	0.0507	2.0	3.2	0.3169	1.8
Ash	0.15	0.0507	2.0	3.0	0.3380	1.9
Softwoods						
Douglas Fir	0.14	0.0507	2.0	2.8	0.3621	2.1
Redwood	0.11	0.0507	2.0	2.2	0.4609	2.6
Southern Pine	0.15	0.0507	2.0	3.0	0.3380	1.9
Cedar	0.11	0.0507	2.0	2.2	0.4609	2.6
Metals						
Steel (mild)	45.3	0.0507	2.0	893.5	0.0011	0.006
Aluminium	221	0.0507	2.0	4359.0	0.0002	0.001
Insulation						
Cellulose (blown-in)	0.0389	0.2057	8.1	0.19	5.29	30.0
Fibreglass[a]	0.0421	0.0889	3.5	0.47	2.11	12.0
	0.0433	0.1524	6.0	0.28	3.52	20.0
	0.0407	0.2223	8.8	0.18	5.46	31.0
Rock wool (Roxul)[a]	0.0374	0.0889	3.5	0.42	2.38	13.5
	0.0369	0.1397	5.5	0.26	3.79	21.5
Polystyrene (solid foam)[a]	0.0288	0.0254	1.0	1.14	0.88	5.0
	0.0288	0.0508	2.0	0.57	1.76	10.0
Polyisocyanurate (solid foam)[a]	0.0190	0.0686	2.7	0.28	3.61	20.5
Urea formaldehyde (formed-in-place)	0.0321	0.2032	8.0	0.16	6.33	35.9
Urethane (formed-in-place)	0.0272	0.2032	8.0	0.13	7.47	42.4

[a] Conductivity deduced from RSI-values and thicknesses of locally available products. For cellulose, fibreglass, rock wool and urethane insulation, the thermal conductivity depends on the insulation density, as illustrated in ASHRAE (2001b, Chapter 23), and so will vary with the manufacturer.
Source of conductivity values: Sherman and Jump (1997), except where indicated

3.1.5 Heat loss from basements

As heat is conducted from a warm basement into cool ground, the ground temperature adjacent to and below the basement will increase, thereby reducing the temperature gradient adjacent to the basement wall and floor and reducing the subsequent rate of heat loss. The ground itself provides some resistance to heat flow, in addition to the basement walls or floor. The relationship between the basement interior temperature, the unperturbed ground temperature at a given depth, the basement wall or floor thermal resistance, and the heat loss from the basement is not straightforward, and cannot be represented by Equation (3.1).

Instead, a 3D simulation of the temperature distribution around the basement and the resulting heat flow, taking into account the thermal conductivity of the ground, is required. Figure 3.3 is an example of the temperature field that exists in the ground next to a basement. Solid lines are temperature contours, while dashed lines represent the direction of heat flow (and always intersect the temperature contours at right angles if the thermal conductivity at any given point is the same in all directions). The region between each pair of dashed lines is a heat- flow 'tube', directing the flow of heat. Some of the tubes take the heat deep into the ground and then back up to the ground surface in a broad arc to the side of the house. The resistance to heat flow along each tube is governed by the length of the tube and the ground thermal conductivity. For a given thermal conductivity, the average resistance will be governed by the geometry of the basement – the depth, width and depth to width ratio. Thus, it is possible to compute empirical coefficients for heat loss from basements, based on the kind of simulations that produced Table 3.3. This has been done by ASHRAE (2001b) and the resulting coefficients are given in Table 3.3. The heat loss from the wall over a given depth interval is given by the appropriate coefficient times the length of wall times the internal–external air temperature difference. The heat loss from the basement floor is given by the appropriate coefficient times the floor area times the internal–external air temperature difference.

Figure 3.3 Illustrative temperature contours (solid lines) and heat flow lines (dashed) next to a basement

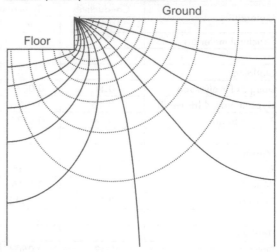

Table 3.3 Heat loss coefficients for below ground basement walls and for basement floors, given a ground thermal conductivity of 1.38W/m/K

	Basement wall		
		Insulated wall coefficient (W/m²/K)	
Depth (m)	Uninsulated wall, U_{soil} (W/m²/K)	RSI = 2.11 m²K/W (R12)	RSI = 3.52 m²K/W (R20)
0.0–0.3	2.33	0.39	0.25
0.3–0.6	1.26	0.34	0.23
0.6–0.9	0.88	0.31	0.21
0.9–1.2	0.67	0.28	0.20
1.2–1.5	0.54	0.25	0.19
1.5–1.8	0.45	0.23	0.17
1.8–2.1	0.39	0.21	0.16
	Basement floor		
Depth of floor below grade (m)	U_b (W/m² /K)		
	Width of building (m)		
	6.0	7.3	8.5
1.5	0.18	0.16	0.15
1.8	0.17	0.15	0.14
2.1	0.16	0.15	0.13

Note: The heat loss coefficient for insulated walls is given by $U = (1/U_{soil} + R_{insul})^{-1}$, where U_{soil} is the coefficient for uninsulated walls and R_{insul} is the insulation thermal resistance. The heat loss coefficient for insulated basement floors can be computed in a similar way.
Source: The un-insulated-basement-wall and basement-floor coefficients are from ASHRAE (2001b, Chapter 28)

The ground thermal conductivity depends strongly on the ground moisture content. The thermal conductivities of dry and wet sand differ by up to a factor of ten, while those of wet and dry loams differ by up to a factor of five. Heat capacities (governing how rapidly temperature changes for a given heat gain or loss) differ by a factor of two to three between dry and saturated states. In addition, heat transfer in moist media can occur in part through downward movement of water (advective heat transfer). Janssen et al (2004) carried out detailed calculations of heat loss from a basement with the climate of Essen (Germany), with and without consideration of ground moisture. They find, for this case, that inclusion of the effect of moisture increases the seasonal heat loss by 10–13 per cent (depending on the soil type, ranging from clay to sand) for basement walls and floors insulated to RSI 1.43 (R8). The relative impact of soil moisture is less important the more highly insulated the basement walls and floor, and it is smaller (about 8 per cent) for slab-on-grade construction.

3.2 Insulation

This section begins with a discussion of conventional insulation materials and their effectiveness, then discusses unconventional and innovative approaches to insulation, and concludes with details on the insulation systems used in energy-efficient houses.

3.2.1 Conventional insulation

As noted above, heat can be transferred by conduction, convection, exchange of air and exchange of radiation. Conduction, involving molecular-scale motion and collisions, requires a substance through which it can take place. Solid materials are relatively good conductors compared to air, so insulation is made with minimal material so as to minimize conduction. This can be done by making the insulation from loose fibres or from porous foam. Convective heat transfer, involving small-scale air motion, can be minimized or eliminated if the air pockets are very small. Insulation can take one of the following forms:

- insulation batts (beds of fibre insulation attached to a paper or some other backing);
- blown-in, loose cellulose;
- rigid, plastic foam panels;
- spray-on foam insulation;
- spray-on adhesive fibre.

Table 3.4 Comparison of the thermal resistance for a given thickness of insulation, for different types of insulation

Type of insulation	Thermal resistance	
	RSI-value per 25mm	R-value per inch
Blown-in cellulose	0.63–0.65	3.6–3.7
Fibreglass batts	0.60–0.62	3.4–3.5
Rockwool batts	0.67–0.69	3.8–3.9
EPS (pentane expanding agent), at 23.8°C (75°F)	0.73	4.1[a]
at –31.7°C (–25°F)	0.91	5.2
XPS (HCFC expanding agent), at 23.8°C	0.87	5.0
at 4°C	0.95	5.4
at –4°C	0.99	5.6
Spray-on polyurethane, HCFC-blown	1.06–1.14	6.0–6.5
HFC-blown	slightly less than above	
water-blown	0.63–0.67	3.6–3.8
Spray-on cellulose	0.61	3.5
Polyisocyanurate	1.00–1.25	5.7–7.1

[a] Ranges from a minimum of 3.7 for Type I (low density) to a minimum to 4.2 for Type III (high density)
Note: Values for foam insulation are stabilized values.
Source: Based on manufacturers' brochures or test reports

The RSI- and R-values per unit thickness for different kinds of insulation, when properly installed, are compared in Table 3.4. The thermal resistance of insulation varies with temperature and, for some kinds of foam insulation, decreases over time (as explained below). Although insulation usually does not serve as an impermeable air barrier, significant exchange of air across the insulation can be prevented if the insulation is installed properly. In many cases, an air gap connected to the outside is placed on the cold side of the insulation so as to vent any moisture passing though the insulation from the inside of the building. If this insulation is not installed snugly against the inner wall surface, and if joints between insulation panels or batts are not carefully sealed, then outside air can flow between the insulation and the inner wall, thereby rendering the insulation much less effective. If there is an air gap on either side of the insulation, then exchange of radiation between the insulation and the adjacent surface can occur. This will be most significant if the gap is on the warm side of the insulation, but can be minimized by using insulation with a low-emissivity backing.

Fibre insulation

Two materials can be used for fibre insulation: glass and mineral matter. The insulation fibres are oriented perpendicularly to the direction of heat flow and with minimal contact with each other, so as to minimize conduction across the insulation. Mineral fibre is made from dirty raw materials, such as boiler slag or blast furnace slag. Both can be purchased as insulation 'batts' with a heavy paper backing, or as insulation boards. Fibreglass and mineral-fibre insulation have comparable thermal conductivities and hence comparable resistance values for a given thickness (Table 3.4). Cellulose insulation consists of shredded newspapers that have been treated chemically to make it fire and insect resistant. It can be blown into cavities or onto attic floors, and has a modestly higher thermal resistance than fibreglass batts for a given thickness.

Sloppy workmanship can seriously undermine the benefit of high levels of insulation. Table 3.5 provides a checklist of good practice for batt and blown-in cellulose insulation, and the rate of compliance as observed in 30 new supposedly energy-efficient houses in California. A

Table 3.5 Measures included in the California Energy Commission Envelope Protocol for housing, and qualitative assessment of the rate of compliance in 30 new 'energy-efficient' houses in California

Protocol item	% of houses where properly implemented
Wall insulation	
Batts correctly sized to fit snugly at sides and ends	70–90%
Insulation cut 2.5cm wider than cavity for non-standard width cavities	70–90%
Rim joists insulated to wall R-value	70–90%
Kneewall batts in contact with drywall	70–90%
Use R-19 batts instead of R-13 batts in 2 × 6 cavities	70–90%
No stuffing of insulation in non-standard width cavities	40–60%
Minimal insulation compression	40–60%
Tub/shower walls insulated	40–60%
Rim joist insulation cut to fit	10–30%
Exterior wall channels insulated	10–30%
Kneewalls/skylights insulated to a minimum of R-19	10–30%
Insulation cut to fit around wiring, plumbing and electrical boxes	0%
Skylight shaft batts in contact with drywall	0%
Attic side of kneewall/skylight batts covered by a facing rated to stop attic air intrusion	0%
Ceiling insulation	
2.5cm free air space between roof sheathing and insulation at eave/soffit vents	100%
Baffles at eave/soffit vents installed to prevent insulation from blocking vents	70–90%
Non-IC fixtures boxed and insulated	70–90%
Insulation covering all IC-rated light fixtures	70–90%
Blown insulation installed uniformly and to required thickness	40–60%
Draft stops in place over all deep drops and interior wall cavities and sufficiently air-tight to stop air movement	10–30%
Attic access insulated with rigid or batt insulation	10–30%
Ceiling batts cut or split to fit around wiring and plumbing	0%
Caulking and sealing procedures	
Top-plate penetrations sealed	100%
Weather stripping at exterior doors	100%
Continuous sealing at bottom plate	40–60%
Weather stripping at attic access	10–30%
Sealing around tub and shower drains to floor	10–30%

Note: IC = insulated ceiling.
Source: Hoeschele et al (2002)

universal shortcoming in these houses was a failure to cut the insulation (rather than compressing it) to fit around wiring, plumbing and electrical boxes (0 per cent compliance). It appears that, in most cases, achieving the potential performance of insulation systems will require additional training of the various subcontractors and close scrutiny during construction.

Sprayed-on fibre insulation consists of cellulose, mineral or glass fibres with its own glue and is sprayed in a moist state into the vertical cavities between open studs. Cellulose fibres are treated with borate to provide fire and insect resistance. After setting, the stud edges are scraped clean with a special milling tool made for that purpose, and the excess can be sent back to a mixing machine for reuse if it is kept clean. Sprayed-on cellulose is a mature technology, but the cavity cannot be closed until the moisture content has dropped to 10 per cent. Sprayed-on fibreglass is less mature but dries quickly (in less than a day). Sprayed-on fibre insulation must be covered after it has dried. Its main advantage is that it can completely fill stud cavities, eliminating air circulation within the wall and providing a continuous air and vapour barrier. It also serves as a substantial sound barrier, particularly sprayed-on mineral wool. Thermal resistance per unit thickness is comparable to that of insulation batts (Table 3.4), but completely fills all irregularly shaped cavities, thereby avoiding gaps that can occur in practice with insulation batts.

Solid foam insulation

There are several solid plastic foam insulation materials: *expanded polystyrene* (EPS) (also referred to as *polystyrene beadboard*), *extruded polystyrene* (XPS), extruded *polyurethane* and extruded *polyisocyanurate* (a polyurethane derivative). EPS is produced from ethylene (a component of natural gas) and benzene (a derivative of petroleum). It begins as small liquid beads combined with pentane as an expanding agent and a fire retardant. The beads are heated and expand, then allowed to set for 24 hours, during which time the expanding agent diffuses through the wall structure and, at some manufacturing facilities, is captured for reuse. The beads are then reheated with steam in a mould, causing the beads to fuse together. The product is the same as the white foam used in many

disposable coffee cups. XPS begins as crystalline polystyrene, which is melted under pressure. An expanding agent and fire retardant are added, then the mixture is squeezed into a mould and expands as the pressure is released. Polyurethane and polyisocyanurate are made from polymeric methylene diisocyanate and polyohydroxyl, both of which are derived from petroleum. Both EPS and XPS have sufficient compressive strength that they can be used beneath basement concrete slabs. A major disadvantage of all foam insulation is that it produces highly toxic smoke if on fire, even if rendered fire-resistant through the use of fire retardants, so it should be generally restricted to exterior applications. It may not be advisable to apply it directly on wooden roof decks, as the combination could burn fiercely.

The expanding agent used in foam insulation can be a halocarbon, pentane, water or, in the near future, possibly CO_2. XPS and polyurethane foam presently use halocarbon expanding agents, while EPS uses pentane. Polyisocyanurate is made with an HCFC in North America but in Europe it is made with pentane – the likely eventual HCFC replacement in North America. Halocarbon blowing agents are greenhouse gases, some quite powerful (see Chapter 2, Section 2.4.3). They leak to the atmosphere from the insulation over time to some extent, but their lower molecular conductivity contributes to a greater thermal resistance and thus lower energy use and CO_2 emissions associated with heating. The trade-off between a warming tendency due to halocarbon emissions versus reduced warming tendency due to reduced heating energy use is assessed in Chapter 12 (Section 12.5). Here, we focus on the extent of leakage and its impact on the thermal resistance of the insulation.

The halocarbon expanding agent is initially trapped as perfectly encapsulated bubbles inside XPS or polyurethane foam. This prevents gas movement by convection or external pressure differences and, along with the low molecular conductivity of the gas compared to air, contributes to the larger RSI-value of XPS and polyurethane insulation compared to fibre or mineral insulation for a given thickness. As the halocarbon gas gradually diffuses out of the insulation over time and is replaced by air, the RSI-value gradually decreases. A number of steps can be taken to reduce the loss of insulation value over time (ASHRAE, 2001b, Chapter 24), so different manufacturers may produce insulation with different stabilized insulation values. EPS begins with air bubbles, as the pentane blowing agent leaks during the manufacturing process, so its RSI-value is smaller than that of other foams but is constant over time (all the resistance values given in Table 3.4 are stabilized values).

Bomberg and Kumaran (1999) have studied the process by which polyurethane loses insulation value over time. Mixing the polyisocyanate and polyhydroxyl components to produce polyurethane foam releases heat, bringing the expanding agent to the boiling point. As the foam cools, the gas pressure falls below atmospheric. The pressure falls further as some of the expanding agent is absorbed into the polymeric matrix. This causes air to gradually diffuse to the interior of the foam, restoring the pressure along the way. Some of the expanding agent also diffuses out of the foam. Figure 3.4 shows the variation in the thermal resistance of a 25mm-thick block of polyurethane foam as simulated by a computer model. Curve 1 illustrates the case in which surfaces are sealed with a coating of epoxy, so that no air can enter or expanding agent escape. In this case, there is no loss of resistance over time. Curve 2 shows the case in which air is allowed to enter, but no expanding agent is absorbed by the matrix or allowed to diffuse out of the insulation. Curves 3 and 4 correspond to the cases in which absorption and absorption plus diffusion, respectively, are allowed to occur. The loss of blowing agent by outward diffusion is not a major factor in the loss of thermal resistance; rather, the entry of air into the insulation is the most important factor. Thus, the presence and quality of an epoxy seal (which can be applied to solid foam panels) is the critical factor in the long-term thermal resistance. Although the loss of expanding agent does not have a large effect on the thermal resistance, the global warming effect of halocarbon agents, if fully released to the atmosphere, can readily exceed the global warming benefit of the insulation (see Chapter 12, Section 12.5).

Sprayed-on foam insulation

Polyurethane foam can also be sprayed on as either internal or external insulation, but the

Figure 3.4 Variation with time of the thermal resistance of a 25mm thick block of polyurethane foam as simulated by a computer model. Curve 1: surfaces are sealed with a coating of epoxy; Curve 2, air is allowed to enter but no expanding agent is absorbed by the matrix or allowed to diffuse out of the insulation; Curve 3, absorption of expanding agent occurs; Curve 4, absorption and diffusion of the expanding agent occurs. Results are for CFC-11 as the blowing agent; using HCFC-141b, the initial and final (for cases 2–4) thermal resistances would be about 15m/K/W and 5m/K/W lower, respectively

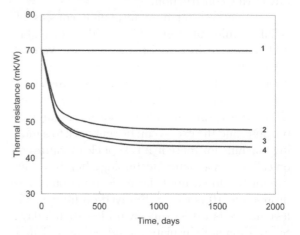

Source: Bomberg and Kumaran (1999)

insulation value depends strongly on the expanding agent used (Table 3.4). Use of an HCFC or HFC expanding agent yields greater thermal resistance (RSI 1.06–1.14 per 25mm, R6–6.5/inch) than XPS, while use of water as an expanding agent yields a thermal resistance (RSI 0.63–6.7 per 25mm, R3.6–3.8/inch) comparable to that of cellulose, fibreglass batts or rock-wool batts. HCFC blown spray-on polyurethane insulation is rigid with a density of about 30kg/m³, and can qualify as an air and vapour barrier, while water-blown insulation is semi-rigid with a density of about 8kg/m³, and does not qualify as an air or vapour barrier.

Inasmuch as an impermeable epoxy coating can reduce both the loss of thermal resistance over time and the escape of halocarbons, solid-foam insulation with a durable epoxy coating is preferable to sprayed-on foam insulation. However, sprayed-on insulation has the advantage of being able to fill irregular shapes and serves as an air barrier (thus reducing infiltration as well as conductive heat losses). The ideal solution might be to rely primarily on solid foam insulation, but to use sprayed-on foam insulation where it is particularly beneficial.

3.2.2 Computation of composite insulation resistance

To compute the total thermal resistance for a uniform section of wall with multiple layers, the resistances of the individual layers are simply added together (similar to combining electrical resistances in series). However, different portions of a wall will have a different total resistance, and the average resistance of the wall has to be computed in the same way that electrical resistances in parallel are combined to give the total resistance of an electrical circuit, as explained in Box 3.1. Heat will preferentially flow through

the elements (such as studs in wood frame construction) with the lowest thermal resistance, so these elements will account for a larger fraction of the total heat loss than expected based on their area. For this reason, a second layer of insulation should be added that covers any thermal bridges within the first layer of insulation. A quantitative example is given in Box 3.1.

External insulation is generally preferable to internal insulation, as thermal bridges can be easily avoided. These arise not only at structural members (such as studs in wood frame walls), but also where internal load-bearing walls or floors

---

BOX 3.1 Determination of the total thermal resistance of a wall

When a wall consists of two or more layers with different resistances, the total resistance through that portion of the wall is obtained by adding the resistance of each layer. Since the resistance is the reciprocal of the U-value, this is equivalent to taking the reciprocals of all the U-values, adding them up, and then taking the reciprocal of the sum to get the total resistance. This is the way that electrical resistances in series are added. If the thermal conductivity and thickness of a given layer are given, the U-value for that layer should be computed first by dividing the thermal conductivity by the thickness.

The average heat flow across a wall or window where a fraction f_1 of the wall area has U-value U_1 and a fraction f_2 ($= 1 - f_1$) has U-value U_2 is given by

$$Q = (f_1U_1 + f_2U_2)\Delta T = \overline{U}\Delta T = \Delta T / \overline{R}$$

(B3.1.1)

where ΔT is the inside to outside temperature difference, $\overline{U}$ is the average U-value, and $\overline{R}$ is the average thermal resistance, given by:

$$\overline{R} = \frac{1}{f_1U_1 + f_2U_2} = \frac{1}{\overline{U}}$$

(B3.1.2)

Given that $R_1 = 1/U_1$ and $R_2 = 1/U_2$, we will define $R_1' = R_1/f_1 = 1/(f_1U_1)$ and $R_2' = R_2/f_2 = 1/(f_2U_2)$. Then a little algebraic manipulation will show that Equation (B3.1.2) is equivalent to:

$$\overline{R} = \frac{R_1'R_2'}{R_1' + R_2'}$$

(B3.1.3)

This is the way that electrical resistances in parallel are added. This formulation assumes that the heat flows through the areas f_1 and f_2 are strictly independent of one another, that is, that the heat flow is one-dimensional. For wood frame construction, the error in the average resistance caused by slight lateral heat flow at the interface between the wood studs and the insulation is less than 2 per cent (Kośny and Kossecka, 2002).

As an example of the application of the above, consider a stud wall with insulation between the studs, as illustrated below.

Figure B3.1 Resistance diagram for heat flow through a stud wall

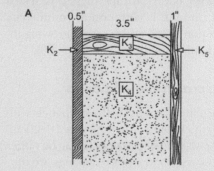

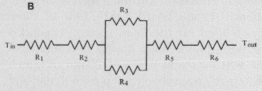

Source: Sherman and Jump (1997)

Resistances R_1 and R_6 arise from the laminar boundary layers attached to the inner and outer surfaces of the wall, with U-values $U_1 = 7.5W/m^2/K$ and $U_6 = 15W/m^2/K$, respectively (both depend on air movement and the wall surface to air temperature difference). The exterior facing is 0.0127m (1/2″) thick Douglas fir and the interior wallboard is 0.0254m (1″) thick gypsum, with thermal conductivities $k_2 = 0.48W/m/K$ and $k_5 = 0.11W/m/K$, respectively. The stud is 0.0889m thick and has a thermal conductivity $k_3 = 0.17W/m/K$, while the insulation has an RSI-value of $2.38(W/m^2/K)^{-1}$ (R13.5). If the studs cover a fraction $f_3 = 0.0972$ of the wall area (corresponding to 2 × 4 studs on 18″ centres), the thermal resistance of the wall is:

$$R_{tot} = R_1 + R_2 + (R_3 R_4)/(R_3 + R_4) + R_5 + R_6$$

(B3.1.4)

where $R_1 = 1/U_1$, $R_2 = 0.0127/k_2$, $R_3 = 0.1016/(f_3 k_3)$, $R_4 = 2.38/(1 - f_3)$, $R_5 = 0.0254/k_5$, and $R_6 = 1/U_6$. The total resistance is $2.24(W/m^2/K)^{-1}$, compared to $2.84(W/m^2/K)^{-1}$ in the region between the studs. The studs, covering 9.7 per cent of the wall area, have reduced the average thermal resistance by 21 per cent. For a wall built with 2 × 6 studs on 24″ centres, with insulation with an RSI-value of $3.79(W/m^2/K)^{-1}$ (R21.5), the wall resistances with and without the stud thermal bridge are 3.55 and $4.25(W/m^2/K)^{-1}$, so the thermal bridge reduces the average resistance by only 16 per cent. That is, thicker insulation requires thicker studs, which allows a greater stud spacing and reduces the total area of thermal bridges, adding to the benefit of thicker insulation. Steel studs, because of their high thermal conductivity, create a particularly strong thermal bridge. This can be reduced through dimples on the edge of the stud (which reduce the contact area between the studs and the interior and exterior sheathing) or through use of insulating foam that is shaped to the stud and covers the stud (Kośny, 2001).

An additional layer of external insulation, spanning the thermal bridges, can be added. The average wall thermal resistance would be simply the average resistance computed above plus that of the external insulation. Using 2″ of solid foam insulation would add a resistance of 1.76 $(W/m^2/K)^{-1}$, increasing the total resistance of the 2×4 and 2×6 stud walls by 80 per cent and 50 per cent, respectively.

join the external wall. Any buildup of condensed moisture can be avoided, and the thermal mass of the wall is exposed to the inside, where it can be used in combination with night-time ventilation to minimize peak summer daytime temperatures in climates with cool summer nights. External insulation, if present, is common in commercial buildings (Section 3.2.5) and is seeing growing interest in residential buildings (Section 3.2.6).

3.2.3 Structural insulated panels

Structural insulated panels (SIPs) are panels with an arbitrarily thick foam core (typically 10–20cm) with a structural (weight bearing) facing on either side. The most common types of facings are drywall and/or structural sheathing such as plywood or oriented strand board (OSB). One of three plastics is used for the foam core: (i) expanded or extruded polystyrene, having insulation values of RSI 0.71 and 0.87 per 25mm (R4/inch and R5/inch), respectively; (ii) polyurethane; and (iii) polyisocyanurate. For the latter two cases, the foam is injected between two wood skins under considerable pressure, and forms a strong bond with the skins when it hardens. In all cases, an HCFC is used as a blowing agent, and gives an initial RSI-value of 1.6 per 25mm (R9/inch). Over time the RSI-value decreases to a stable value of 1.05–1.25 per 25mm (R6–R7/inch), as some of the HCFC escapes. Because of concerns over depletion of stratospheric ozone, HCFCs are only temporary substitutes for CFCs. They are also greenhouse gases.

SIPs provide more uniform insulation compared with most other methods; provide a more air-tight building if properly installed; can be installed quickly; and are almost three times stronger than a 2×4 stud wall. An obvious disadvantage is that electrical wiring, cables and hot-water radiator pipes cannot be placed inside the SIP wall. Assuming R6.5/inch, SIPs offer the prospect of R40 (RSI-7.0) walls at just over 6″ (15cm) insulation thickness (7″ minimum total) and R60 (RSI-10.6) roofs at just over 9″ (23cm) insulation thickness. Manufactured houses using 6″ wall and 8″ roof SIPs in the US are calculated to use 50 per cent of the heating energy of a house built to the HUD (Housing and Urban Development) building code (Baechler et al 2002). Labour savings during construction offset the greater material cost of SIPs. However, the use of HCFC or HFC blowing agents may significantly offset the global warming benefit of greater insulation levels, depending on the rate of leakage of the blowing agent.

3.2.4 Dynamic insulation

Another innovative concept involves *dynamic insulation*, which began with ceilings in Norway in the 1960s and has since been widely used throughout Scandinavia and Austria, with isolated examples in a number of other countries. The concept has also been investigated through experimental test facilities (Baker, 2003). Normally, heat diffuses outward through the building insulation. This can be counteracted by drawing the required outside ventilation air into a building through the wall or roof insulation. The effective resistance value for dynamic insulation is given by (Taylor and Imbabi, 1997),

$$R_{eff} = R_i \exp(AL) + \frac{\exp(AL)-1}{kL} + R_o \qquad (3.5)$$

where

$$A = \frac{v\rho_a c_{pa}}{k} \qquad (3.6)$$

and R_i and R_o are the internal and external laminar boundary layer resistances (see Box 3.1), k is the thermal conductivity of the insulation, L is the thickness of insulation, ρ_a and c_{pa} are the density and specific heat of air, and v is the velocity of air flowing through the insulation. The insulation resistance, R_{ins}, is given by L/k. Figure 3.5 shows how R_{ins} and R_{eff} vary with the thickness of insulation for insulation with $k = 0.042$ W/m/K, with results for R_{eff} given for $v = 0.5$, 1.0 and 1.5m/h (this is the range of velocities that would be required if only the amount of outside air required for ventilation purposes is drawn entirely through the walls and roof for a typical situation). For insulation plus boundary layers sufficient to give R21 (RSI 3.7, about 15cm of insulation), R_{eff} ranges from R30 to R64 (RSI 5.3–11.3). However, as the thickness of insulation increases,

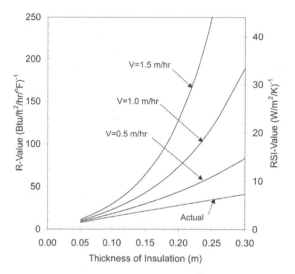

Figure 3.5 Comparison of static wall resistance and the effective resistance for dynamic insulation with airflow velocities of 0.5, 1.0 and 1.5m/s

Computed from Equations (3.5–3.6) with parameter values as given in the main text.

R_{eff} increases at an ever-faster rate, while R_{ins} increases only in proportion to the increase in thickness. On the other hand, a passive system – perhaps with a larger R_{ins} – will always be more reliable than a dynamic system.

Figure 3.6 gives details concerning the construction of a roof with dynamic insulation in a demonstration house in Switzerland (Humm, 1994). Air is drawn into a cavity between the insulation and the wallboard, and distributed through the house via ducts and vents. The overall system is illustrated in Figure 3.7, while temperatures at various points for a time when the outdoor temperature is −10°C are given in Figure 3.8. The incoming air warms to 16°C during its passage through the insulation, and to 19°C as it passes through the ductwork inside the warm floors. The room air temperature is 22°C, which is the temperature at which exhaust air is collected. Heat is extracted from the exhaust air with a heat pump (see Section 5.4), which cools the exhaust air to −2°C and creates water at 45°C. This water is stored in a hot-water tank used for radiant floor heating (see Chapter 7, Section 7.3.1), maintaining a floor temperature of 25°C, which is sufficient to maintain the 22°C room air

Figure 3.6 Details in the construction of a roof using ventilation to create dynamic insulation with an effective U-value of 0.03W/m²/K (about R200)

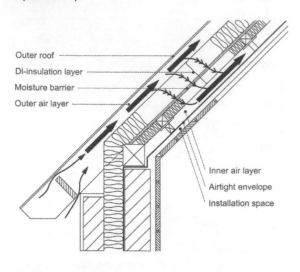

Outer roof
DI-insulation layer
Moisture barrier
Outer air layer

Inner air layer
Airtight envelope
Installation space

Source: Humm (1994)

Figure 3.8 Temperature at different points in the ventilation and heating system shown in Figure 3.6 on a day when the outside air temperature is −10°C

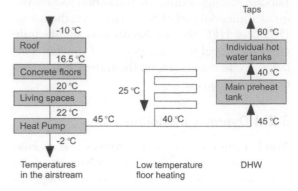

Taps

−10 °C   60 °C

Roof   Individual hot
16.5 °C   water tanks
Concrete floors   40 °C
20 °C   25 °C
Living spaces   Main preheat
22 °C   tank
Heat Pump   45 °C   40 °C   45 °C
−2 °C

Temperatures   Low temperature   DHW
in the airstream   floor heating

Figure 3.7 The heating and ventilation system in a house, built in Switzerland, using dynamic roof insulation (as illustrated in Figure 3.6)

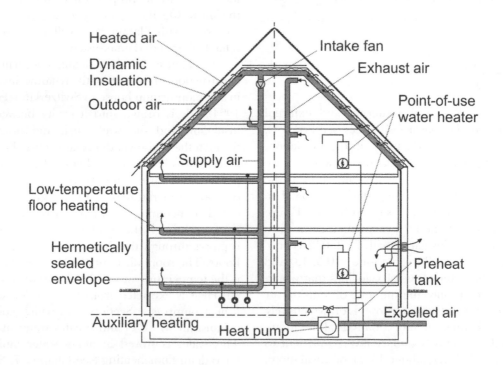

Heated air
Dynamic Insulation
Outdoor air
Intake fan
Exhaust air
Point-of-use water heater
Supply air
Low-temperature floor heating
Hermetically sealed envelope
Preheat tank
Auxiliary heating
Heat pump
Expelled air

temperature. A supplemental heat source is used when heat extracted from the exhaust air is not sufficient, which will be the case when the final exhaust-air temperature is warmer than the outside air temperature.[2] Point-of-use water heaters (which are the norm in much of Europe) are used to boost the temperature to the 60°C required for domestic hot water use (showers, washing). In this particular case, the effective U-value of the roof was reduced from 0.26W/m²/K (RSI 3.85, R24) for the insulation alone, to 0.03W/m²/K (RSI 33.3, R200) due to inflow.

Gan (2000) discusses some practical issues concerning dynamic insulation. Dynamically-insulated walls have surface temperatures lower than the room air temperature during the heating season, which could cause draughts due to cooling of the adjacent air. This can be compensated with slightly warmer floors in a floor radiant heating system. Moreover, solar energy can be used to boost the temperature of the incoming air using wall or roof solar collectors (described in Section 4.1.3). It is essential that a structure relying on dynamic insulation be almost completely air-tight, except for the intentional porosity through the insulation; Taylor and Imbabi (2000) find, through numerical simulation, that the air-tightness achieved by the Canadian R2000 standard is sufficient. Because of variability in the permeability of cellulose insulation, it is common to use a layer of fibreboard – whose properties are more consistent – to control the overall permeability of the wall, so that a dynamically insulated building will perform as designed.

Taylor and Imbabi (2000) discuss the performance of dynamic insulation under combined wind and stack pressures (the latter is explained in Box 3.4). Heat loss increases with increasing wind speed, but the wind speed must reach roughly 10m/s before the heat loss equals that from an air-tight building with comparable amounts of insulation. It is possible that buildings could be constructed so that the stack effect (Section 3.7) can drive adequate airflow through the insulation under calm conditions with large indoor–outdoor temperature differences, but a fan would be needed at other times.

Dynamic insulation can also be used during the cooling season to reduce or eliminate heat gains through the ceiling. In this case,

interior air is forced out through the insulation. Zuluaga and Griffiths (2004) constructed a facility to test this concept. A fan above the insulation draws air through a grill into a gap between the ceiling drywall and insulation (which is supported by a wire mesh), then through the insulation into a gap above the insulation and to the outside. This technique can be combined with evaporative cooling, as it provides a way to exhaust some of the air drawn into the house through the evaporative cooler while reducing the cooling load that must be satisfied by the evaporative cooler (see Chapter 6, Section 6.3.7).

3.2.5 Insulation of concrete and masonry walls

Rigid foam insulation is typically applied to the exterior of concrete slab or concrete block walls in commercial buildings, with an exterior finish separated from the insulation by an air gap. Semi-rigid mineral fibre insulation is equally popular and provides better sound and fire proofing. A 5cm thick layer of polystyrene provides an RSI-value of 1.76 (R10, Table 3.2). Uninsulated concrete walls themselves provide another RSI 0.35–0.53 of resistance to heat flow. Concrete masonry blocks provide RSI 0.18–0.44, the lower value pertaining to the coldest conditions, when the larger temperature difference across the blocks provokes convective heat exchange within the hollow core (Lorente et al, 1998).

There are many ways in which the insulation value of concrete slab and concrete masonry units can be increased. Alternative shapes, illustrated in Figure 3.9, can be used with or without insulated inserts (Kośny, 2001). If built of normal density concrete (thermal conductivity of 1W/m/K), RSI-values of 0.53–1.94 (R3 to R11) can be achieved (Figure 3.9). For lightweight concrete in combination with insulation inserts, RSI values as high as 3.52 (R20) can be achieved.[3] However, due to thermal bridges between the insulation inserts, the effectiveness of the insulation is less than if the same amount of insulation were applied as an external sheathing. If one is restricted to concrete blocks without inserts, an RSI-value as high as 1.41 can be achieved. Pre-cast concrete panels consisting of a layer of insulation sandwiched between two concrete layers, which are joined by plastic ties, have an RSI-value of 2.95 (R16.75) (Tuluca, 1997). Insulating concrete panels and blocks provide one method for achieving relatively high insulation levels, and would be quite adequate in moderate climates where current practice might not entail adding any insulation at all. In cold climates, they could be used to augment currently prescribed insulation levels in new construction.

Figure 3.9 Alternative shapes of concrete blocks (upper) and insulation value as a function of concrete thermal conductivity and presence or absence of insulated inserts (lower)

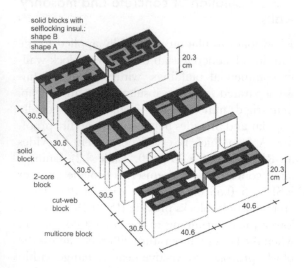

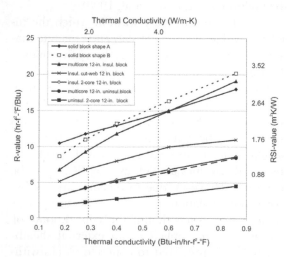

Source: Kośny (2001)

Lightweight insulating structural concrete, with an RSI-value of 1.41 per 30cm, is currently under development (Tuluca, 1997). This has ten times the insulation value of conventional concrete. It could be used where thermal bridges would otherwise occur, such as for columns, beams and balcony floor slabs. An alternative method to eliminate the thermal bridge caused by balcony floors is to use free-standing columns to support the balcony floors, as discussed in Section 3.9.

Another option using concrete is *insulating concrete forms* (ICF). With ICFs, foam insulation serves as the forms into which concrete is poured; the forms stay in place after the concrete has set. Polystyrene, polyurethane and polyisocyanurate foams have been used, either as foam panels held together with pins or as interlocking foam blocks, as illustrated in Figure 3.10. Typical systems have been built with RSI 3.0–3.5 (R17 to R20). ICFs are one of the few new products that have been adopted by the mainstream construction community in North America. Initially, waffle slab or post-and-beam systems were available that had higher insulation values and used significantly less concrete. However, subsequent commercialization and promotion have removed the waffle type systems from the market, leaving concrete use the same. ICFs provide adequate insulation for below-grade walls, and a significant improvement compared to uninsulated walls, but fall short of the insulation levels needed in high-performance houses in cold climates. ICFs made with non-halocarbon blowing agents are available, as are products (such as Durisol, www.durisol.com, and Faswall, www.faswall.com) that are made largely from recycled wood waste, mixed with cement, and treated to be fire and termite resistant.

In the case of hollow-core masonry foundation walls, if the walls are insulated on one side (typically the outside) below grade and on the other side above grade, the insulation will be ineffective no matter how much overlap there is between the inside and outside insulation, unless there is a barrier to vertical air movement inside the wall in the region of overlap. In the absence of an air barrier, vertical air motions provide a thermal bridge between the uninsulated sides of the wall. A single layer of solid concrete blocks in the region of overlap serves as an air barrier.

Figure 3.10 Insulated concrete forms

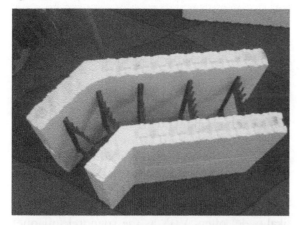

Source: Author

3.2.6 External Insulation and Finishing Systems (EIFSs)

A conventional *External Insulation and Finishing System (EIFS)* consists of four layers, from exterior to interior: a sprayed or towelled acrylic finish that can be chosen to resemble stone or plaster, a fibreglass reinforcing mesh, an insulation layer and an adhesive that attaches the system to the building. EIFSs serve as a barrier against air leakage (see Section 3.7), as well as providing insulation over thermal bridges. The EIFS commonly consists of expanded polystyrene insulation foam board, but because of the toxic smoke produced when polystyrene burns, fire-code regulations in some jurisdictions restrict polystyrene EIFSs to low-rise buildings. Mineral wool based EIFSs can be used in high-rise residential and commercial buildings. As noted in Chapter 12, polystyrene insulation has the largest embodied energy per unit of insulation value of any insulation (depending on where and how it is manufactured), while mineral wool can have a substantially lower embodied energy. Because there is no air gap between the exterior finish and insulation, the EIFS must be well sealed to prevent moisture damage. By the late 1990s, EIFSs had captured about 20 per cent of the wall-cladding market for commercial buildings in the US.

A critical issue with regard to EIFS concerns the details of the interface between the EIFS and other building elements (windows, foundation, other forms of cladding and the roof), the priority being to keep moisture out of the building envelope. Any connection between an upper EIFS and a lower roof element has to be such that the roof can be replaced without tearing apart the EIFS. CMHC (2004) provides diagrams detailing best practice for the connection between the EIFS and other envelope elements. In North America, the EIFS is commonly attached to exterior gypsum board sheathing that is attached to steel studs with screws. This construction can result in the entire cladding system relying on the integrity of the gypsum core, and any moisture-related softening of the gypsum board and/or corrosion of the screw attachments of the sheathing to the studs can cause serious problems. In Europe, the vast majority of EIFSs are installed over masonry or poured concrete.

Figure 3.11 Buildings with External Insulation and Framing Systems (EIFSs)

Source: Sto AG (www.sto.de)

There are many proprietary systems and innovative EIFS products available today. Information on commercially available products can be obtained from the *Advanced Buildings, Technologies and Practices* website given in Appendix E. Examples of buildings using an EIFS are given in Figure 3.11. In the future, the development of nanofibre insulation may lead to a dramatic drop in the thickness required for a given RSI-value and without the use of halocarbons.

3.2.7 Evacuated panels

Vacuum insulation panels (VIPs) have received considerable attention since the 1990s, as they provide the possibility of high thermal resistance at low thickness. The core of the panel consists of a microporous material (such as fumed silica) under a soft vacuum (10^{-4} atmospheres), which yields a thermal conductivity of about 0.004W/m/K (compared with 0.02–0.03W/m/K for solid foam insulation and 0.04W/m/K for cellulose). The evacuated core is wrapped in an airtight envelope that forms a thermal bridge, the extent of which depends on the thickness (d) and thermal conductivity (k) of the envelope. The envelope may consist of aluminium ($d = 0.09\mu m$, $k = 200W/m/K$) or plastic ($d = 3.0\mu m$, $k = 0.24W/m/K$). For $1 \times 1m$, 2cm thick panels, Wakili et al (2004) measured overall thermal conductivities of 0.0047–0.0077W/m/K, and an overall RSI of 6.2 (U-value = 0.16W/m²/K; R35) for a 30mm thick panel. Annex 39 (High Performance Thermal Insulation) of the International Energy Agency's (IEA) *Energy Conservation in Buildings and Community Systems* programme is focusing largely on VIPs, consisting of a microporous core material in a soft vacuum.

The German company Okalux (www.okalux.de) offers VIPs as the spandrel (opaque) portion of curtain wall systems. For 30mm and 60mm thick panels, the U-values are 0.16W/m²/K and 0.08W/m²/K, respectively. Of possible concern is the possibility that such panels could be damaged during the construction process. For this reason, prefabricated assemblies protecting the VIP are preferred. About 10,000m² of VIPs, covered by a protective layer, water barrier and concrete plate, had been installed by 2004 on terrace roofs in Switzerland (Simmler and Brunner, 2005). VIPs are preferred in such applications because the floor height inside and outside can be made equal while providing stringent (RSI > 5) resistance to heat flow on the outside part of the roof. Prefabricated concrete wall slabs are available in the Swiss and German markets with a total wall thickness (including interior finish) of 27cm and an average U-value of 0.15W/m²/K (Binz and Steinke, 2005). An example is shown in Figure 3.12. VIPs have also been used in retrofit applications, as discussed in Chapter 14 (Section 14.1.2).

VIPs are currently quite expensive – in the order of 320€/m² for panels with a U-value of 0.15W/m²/K, compared to 32€/m² for fibre or solid-foam insulation (Erb, 2005). However, the VIP is 8cm thick, compared with 30cm for the alternatives. For a 3.2m floor to floor height, the incremental cost per metre of external wall is 921.6€ and 0.22m² of floor space is saved, at a cost of 4190€/m² of saved floor space. By comparison, the average cost of new residential buildings in Switzerland is about 1800€/m². However, part of the façade will be windows, but the savings in floor space will be the same, so the cost

Figure 3.12 Left: Cross-section of a prefabricated wall unit with a VIP, having a total thickness of 27cm and a U-value of 0.15W/m²/K. Right: photographic example

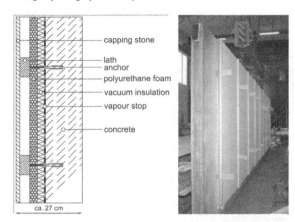

Source: Binz and Steinke (2005)

per m² of saved floor space should be multiplied by the wall fraction. This will bring the cost closer to the market value of saved floor space, but VIPs are still an expensive option at present.

3.2.8 Transparent and translucent insulation

Conventional insulation works by creating pockets of stagnant air (air has a relatively low thermal conductivity). Heat transfer occurs by conduction within the solid insulation matrix, by convection within the air spaces, and by transmission of radiation across the air spaces. Transparent insulation also relies on enclosed air gaps for its insulating power, but is made of a material and has a structure such that it is largely transparent to visible solar radiation. The transparent insulation material (TIM) is placed between enclosing panes of glass that are hermetically sealed and in front of an absorbing surface. A blind can be installed in front of the insulation in order to minimize unwanted heat gain. The most common TIMs are: (i) a silica honeycomb structure with the cell walls perpendicular to the enclosing glass surfaces, and (ii) a homogenous silica aerogel or xerogel (Kaushika and Sumathy, 2003). Companies producing transparent insulation include the Canadian firm Advanced Glazings (www.advancedglazings.com), the German firms Wacotech (www.wacotech.de) and Okalux (www.okalux.de), and the US firm Kalwall (www.kalwall.com).

A honeycomb structure permits most of the solar radiation to traverse the insulation. Cells in the range of 10–15mm in diameter are able to suppress convective air motions through viscous forces, while an air gap between the insulation and absorber surface and use of very thin (50μm thick) cell walls limits conductive heat transfer. Use of a low-emissivity coating on the absorber will reduce radiative heat loss. A protective layer of glass separated by another air gap occurs in front of the TIM, and a roller blind can be placed in this air gap and lowered to reduce overheating when there is excess solar energy and to reduce heat loss during winter nights. The optimal cell size depends on the typical temperature difference across the insulation. A smaller cell size will compensate for the increasing tendency for convection as the temperature difference increases, but increases cost and reduces the transmission of solar energy. Athienitis and Santamouris (2002) present equations that can be used to estimate heat and light transfer across honeycomb TIMs. A silica honeycomb TIM is essentially highly structured fibreglass insulation, and it has a comparable thermal conductivity and density – about 0.04W/m/K and 20kg/m³, respectively. Honeycomb TIMs have also been made from non-silica materials such as polymethyl methacrylate (PMMA), polycarbonate, polyester, polystyrene and polypropylene.

Aerogels and xerogels are similar substances. Silica aerogel can be produced in either monolithic or granular form. Monolithic aerogels have particle sizes generally less than the wavelength of light and so are highly transparent, while granular aerogels have larger particles and are translucent. Silica aerogel is produced from one of two silica containing precursors, tetramethoxysilane (TMOS, $Si(OCH_3)_4$) or tetraethoxylane (TESO, $Si(OC_2H_5)_4$). The precursor is mixed with water, alcohol and various catalysts to produce a mixture called alcosol. After a while the silica precursor forms a porous material called alcogel that contains the alcohol solvent within the pores. The solvent must be removed without causing the collapse of the silica structure, and the initial method used to do this was to heat the material under pressure to above the critical temperature of the solvent (typically to 280°C at a pressure of 90atm).[4] The solvent is extracted as the pressure is slowly reduced. This is an energy-intensive and slow process and runs a risk of explosion (indeed, a pilot plant at Airglass AB in Sweden exploded in 1984). Two alternative approaches are now generally used: (i) to substitute CO_2 for alcohol at a relatively low temperature, then to extract the CO_2 by heating the mixture above the critical point for CO_2 (which is only 31°C), or (ii) to dry the alcogel at moderate temperature (below 100°C) and at atmospheric pressure. Destruction of the gel structure in the latter case is prevented by stiffening the gel by ageing it in various chemical solutions prior to drying. The product of this process is xerogel. It is more dense than aerogel (500kg/m³, compared to about 100kg/m³ for granular aerosol and 180kg/m³ for monolithic aerogel) but much less dense than glass (2200kg/m³). It has twice the thermal conductivity of aerogel at normal atmospheric pressure but a similar conductivity under a vacuum. At normal atmospheric pressure, aerogel has a thermal conductivity of 0.015–0.020W/m/K, but at a moderate vacuum (P < 5000Pa or 0.05 atmospheres), the conductivity drops to less than 0.01W/m/K (Duer and Svendsen, 1998). This is 4–5 times less than the thermal conductivity of fibreglass or mineral wool and 2–3 times less than that of most solid foams (see Table 3.2). Polyurethane aerogel can be produced with a thermal conductivity of 0.012W/m/K at 0.1atm pressure and 0.010W/m/K at 0.01atm pressure (Rigacci et al, 2004).

Table 3.6 presents some data on the properties of a number of transparent or translucent insulation materials. Translucent aerogel skylights and wall panels have been successfully marketed, some under the name 'nanogels'. For 10cm thick panels, $U = 0.284W/m^2/K$ (RSI 3.52, R20), visible transmittance = 0.20, and SHGC (defined in Section 3.3.7) = 0.12. Several examples of buildings using panels are given in Figure 3.13. Silica aerogel can also be used in windows, as discussed in Section 3.3.17, while TIMs more generally have been used with thermal solar collectors, as discussed in Chapter 11 (Section 11.4.2).

The transmission of solar radiation through transparent insulation can more than offset the heat loss through the insulation, leading to effective seasonal mean U-values that are almost always negative. The effective U-value depends on

Table 3.6 Properties of some transparent insulation materials (TIMs)

TIM construction	Solar transmission	U-value (W/m²/K)	Thermal conductivity (W/m/K)
Eight layers of corrugated cellulose acetate sheets between two layers of glass	0.44	1.2	
200mm of capillary material between two layers of 6mm glazing	0.71	0.52	0.104
40mm of granular aerogel between two layers of 6mm low-iron glazing	0.34	0.51	0.020
20mm of evacuated monolithic silica aerogel between 4mm glazings	0.73	0.51	0.010

Source: Hestnes et al (2003)

the solar heat gain relative to the conductive heat loss and on the extent to which surplus heat can be stored for later use. It therefore depends on the physical U-value, climate (as represented by the number of heating degree days), solar irradiance incident on the TIM and the solar transmittance. Hastings (1994) presents data on effective U-values for a wide range of conditions. For equatorward-facing walls with transparent insulation, effective U-values are negative for locations as far north as the Shetland Islands (latitude 60.5°N) if surplus heat can be used. If circumstances do not produce negative effective U-values, then it is better to rely on high levels of conventional insulation (producing U-values as low as 0.1W/m²/K) than to rely on transparent insulation with a much higher U-value (0.7–1.0W/m²/K).

Figure 3.13 Examples of nanogel skylights: upper left: Palisades Center, West Nyack, New York; upper right: University of Central Florida; lower left: Office building in Isle of Jersey, UK; lower right: Orlando International Airport, Florida

Source: Kalwall® Corporation, www.kalwall.com

3.2.9 Radiant barriers

When an attic roof heats up during hot summer days, the hot attic ceiling will emit infrared radiation to the cooler attic floor (where insulation, if present, is normally placed). If the ceiling is lined with a low-emissivity surface, this method of heat transfer will be reduced (see Equation 3.4). An effective method is to drape aluminium foil (emissivity ≈ 0.1) over the roof rafters before the sheathing is installed, or to attach aluminium foil to the underside of the roof rafters (the latter is possible as a retrofit). In tests in the south east US, this simple measure alone can reduce the ceiling heat flux by 25–50 per cent and the total cooling load by 7–10 per cent in single-storey houses (Parker et al, 2001). Alternatively, a low-e layer can be placed on the insulation lying on the attic ceiling. In this case, the low-e surface reflects downward infrared radiation back to the attic roof. In either case, the attic roof will be hotter than otherwise (tending to drive its emission back up), so the effectiveness of the low-e layer will be increased if the roof is ventilated.

3.2.10 Framing and insulation systems in existing advanced houses

A number of advanced houses have been built in various countries around the world during the past decade, as discussed in Chapter 13 (Section 13.2). These houses use as little as 10 per cent of the heating energy of houses built according to the local national building codes. The most highly insulated houses in the world have wall RSI values of 5–7 and roof RSI values of 7–10. Table 3.7 lists the framing and insulation systems that have been used in advanced houses in Canada, along with the resulting U-values, RSI-values and R-values. Wall trusses (two 38 × 38mm framing layers with a 125mm gap) with poured-in-place foam and exterior polystyrene sheathing yield an RSI-value of 8.3–11.1. Standard wood frame construction consists of either 38 × 89mm studs (so called 2″ × 4″ studs) on 406mm (16″) centres, or 38 × 140mm studs (so-called 2″ × 6″ studs) on 610mm (24″) centres. Figure 3.14 provides details concerning the construction of walls with 38 × 140mm studs, and the range of overall RSI values that can be obtained with various kinds of cavity insulation combined with external

insulation (the overall RSI-value takes into account thermal bridges but also the resistance of laminar boundary layers on either side of the wall). With sprayed polyurethane insulation and 26–51mm of external sheathing (depending on the type), an RSI of 6.53 can be obtained. Figure 3.15 provides details and RSI-values for double-wall framing, where an RSI of 7.5 can be achieved using sprayed polyurethane insulation. Wall, roof and basement-slab details for two advanced houses, one in Canada and the other in Germany, are given in Figure 3.16. In the German case, insulation is placed on either side of the foundation as well as below the basement slab. Details for other super-insulated houses around the world can be found in Hestnes et al (2003). The analysis of life-cycle cost, which combines both the initial cost of alternative insulation systems and the energy cost savings over the lifespan of the insulation, is discussed later in this chapter (Section 3.9).

Minimizing the use of wood

High levels of insulation can be combined with advanced framing systems so as to minimize the amount of wood required, and the amount of wood waste generated. An outline of advanced framing systems is found in Baczek et al (2002) and at the Building Science Corporation website, www.buildingscience.com/housesthatwork/advancedframing/default.htm. Optimized wood framing systems eliminate unnecessary studs without compromising strength and, in single-stud walls, can reduce the cost of framing by 40 per cent and reduce the generation of wood waste by 15 per cent.

Engineered I-beams are particularly useful in combination with high levels of insulation. They can be manufactured either from standard 38 × 89mm studs or parallel strand lumber, and joined together with a wood product such as aspenite, as illustrated in Figure 3.17. Parallel strand lumber consists of wood strands glued together and extruded from a rotary press as a continuous piece that can be cut off at any length up to about 20m. Thus, waste wood and 'weedy' tree species can be used, and the resulting product is stronger than conventional studs with the same overall dimensions (see The Engineered Wood Association, www.apawood.org, for more information). Although 38 × 140mm I-beams are

Table 3.7 Framing and insulation systems used in advanced houses in Canada

Framing	Insulation system	U-Value (W/m²/K)	RSI-value	R-value
38 × 89mm (2″ × 4″ framing)	High-density fibreglass batts with exterior polystyrene insulation finishing system	0.17	5.88	R33
38 × 140mm (2″ × 6″ framing)	Blown-in-blanket cellulose with exterior polystyrene sheathing	0.15–0.16	6.25–6.67	R35–R38
38 × 140mm (2″ × 6″ framing)	Blown-in-blanket cellulose with interior strapping and fibreglass batts	0.16	6.25	R35
Wall trusses	Blown-in-blanket cellulose	0.16–0.17	5.88–6.35	R33–R36
Wall trusses	Poured-in-place foam with exterior polystyrene sheathing	0.09–0.12	8.33–11.1	R47–R63
Double wall framing	High-density fibreglass batts	0.11	9.09	R52
Insulated structural panels	Polystyrene foam	0.18	5.56	R32

Source: Mayo (1993)

Figure 3.14 Construction details and RSI values for wood frame stud walls with exterior solid insulation. Below the cavity RSI-values is the type of insulation that will give the indicated RSI-value in the allotted thickness of 140mm. Effective RSI-values account for thermal bridges and the indicated sheathing. The total RSI value is given by the effective RSI-value plus the RSI-values for the exterior and interior finishes

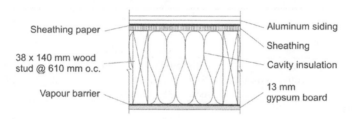

Sheathing	Cavity Insulation			
	RSI 3.25	RSI 3.52	RSI 3.87	RSI 5.90
	glass or mineral fibre batts			sprayed-on polyurethane
	Effective RSI-value (Overall U-value)			
None	3.11 (0.321)	3.29 (0.304)	3.51 (0.285)	4.62 (0.217)
13 mm gypsum	3.20 (0.312)	3.38 (0.296)	3.61 (0.277)	4.74 (0.211)
11 mm plywood or particle board	3.22(0.310)	3.40 (0.294)	3.63 (0.276)	4.76 (0.210)
11 mm waferboard or OSB	3.25(0.308)	3.43 (0.292)	3.66 (0.274)	4.80 (0.208)
11 mm fibreboard	3.32 (0.301)	3.50 (0.286)	3.73 (0.268)	4.89 (0.204)
RSI 0.88 sheathing	4.08 (0.245)	4.28 (0.234)	4.52 (0.221)	5.84 (0.171)
RSI 1.05 sheathing	4.26 (0.235)	4.46 (0.224)	4.71 (0.212)	6.05 (0.165)
RSI 1.14 sheathing	4.35 (0.230)	4.55 (0.220)	4.81 (0.208)	6.16 (0.162)
RSI 1.32 sheathing	4.54 (0.220)	4.75 (0.211)	5.00 (0.200)	6.38 (0.157)
	Exterior Finish			
	13 mm stucco	wood siding	100 mm brick and 25 mm air	19 mm furring behind finish
RSI Adjustment	-0.10	0.04	0.14	0.18
	Interior Finish			
	6 mm wood paneling	16 mm gypsum board	strapped air space	
RSI Adjustment	-0.02	0.02	0.19	

Source: NRCan (1997a)

Figure 3.15 Construction details and RSI values for wood frame double-stud walls. Below the cavity RSI-values is the range of thickness of insulation needed, depending on the type of insulation and its thermal conductivity, as given in Table 3.7 for double-wall insulation. For all except the lowest RSI-value, the range of thickness could be accommodated through various combinations of double studs (in thickness of 64, 89, 140, or 202mm), in some cases with a gap between the stud layers (as shown in the diagram). Effective RSI-values account for thermal bridges. The total RSI-value is given by the effective RSI-value plus the RSI-values for the exterior finish, internal sheathing (if present), and interior finish

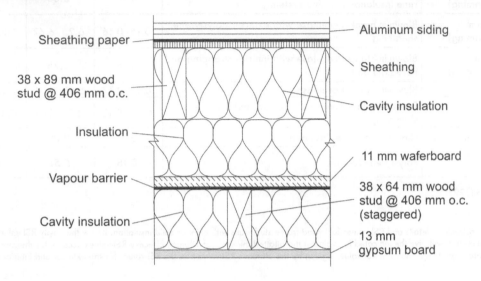

	Cavity Insulation			
	RSI 4.58	RSI 5.63	RSI 7.04	RSI 7.39
	109-178 mm	134-261 mm	168-293 mm	176-308 mm
Structure	Effective RSI-value (Overall U-value)			
38 x 89 mm and 38 x 64 mm, staggered	4.52 (0.221)	5.59 (0.179)	7.01 (0.143)	7.36 (0.136)
38 x 89 mm and 38 x 64 mm, nonstaggered	4.32 (0.231)	5.44 (0.184)	6.91 (0.145)	7.27 (0.138)
38 x 89 mm and 38 x 89 mm, staggered	4.42 (0.226)	5.49 (0.182)	6.92 (0.145)	7.27 (0.137)
38 x 89 mm and 38 x 89 mm, nonstaggered	4.11 (0.243)	5.27 (0.190)	6.77 (0.148)	7.13 (0.140)
	Exterior Finish			
	13 mm stucco	wood siding	100 mm brick and 25 mm air	19 mm furring behind finish
RSI Adjustment	-0.10	0.04	0.14	0.18
	Internal Sheathing			
	13 mm gypsum	11 mm plywood or particle board	11 mm waferboard or OSB	11 mm fibreboard
RSI Adjustment	0.08	0.10	0.12	0.18
	Interior Finish			
	6 mm wood paneling	16 mm gypsum board	strapped air space	
RSI Adjustment	-0.02	0.02	0.19	

Source: NRCan (1997a)

Figure 3.16 Construction details for c super-insulated house in (left) Rottweil, Germany, and (right) Waterloo, Canada

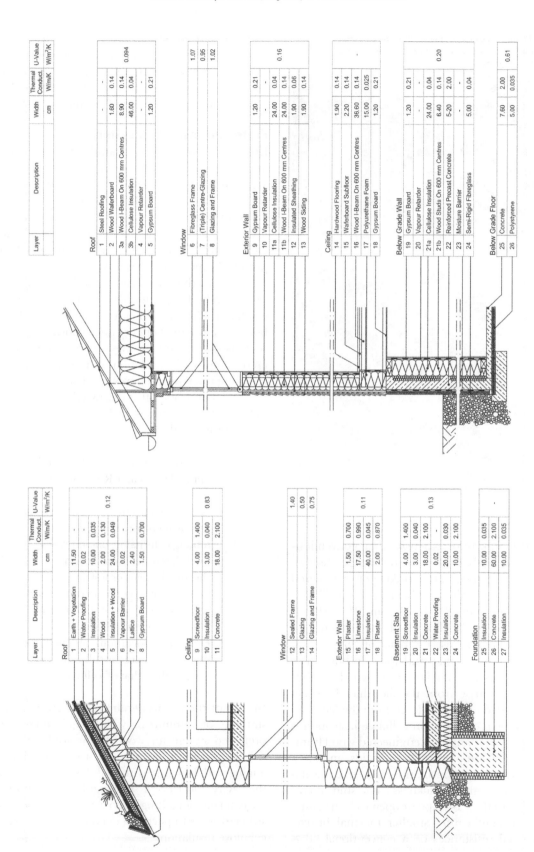

Left (Rottweil, Germany)

Layer		Description	Width cm	Thermal Conduct. W/mK	U-Value W/m²/K
Roof					
	1	Earth + Vegetation	11.50	-	
	2	Water Proofing	0.02	-	
	3	Insulation	10.00	0.035	
	4	Wood	2.00	0.130	0.12
	5	Insulation + Wood	24.00	0.049	
	6	Vapour Barrier	0.02	-	
	7	Lattice	2.40	-	
	8	Gypsum Board	1.50	0.700	
Ceiling					
	9	Screedfloor	4.00	1.400	
	10	Insulation	3.00	0.040	0.83
	11	Concrete	18.00	2.100	
Window					
	12	Sealed Frame			1.40
	13	Glazing			0.50
	14	Glazing and Frame			0.75
Exterior Wall					
	15	Plaster	1.50	0.700	
	16	Limestone	17.50	0.990	0.11
	17	Insulation	40.00	0.045	
	18	Plaster	2.00	0.870	
Basement Slab					
	19	Screedfloor	4.00	1.400	
	20	Insulation	3.00	0.040	0.13
	21	Concrete	18.00	2.100	
	22	Water Proofing	0.02	-	
	23	Insulation	20.00	0.030	
	24	Concrete	10.00	2.100	
Foundation					
	25	Insulation	10.00	0.035	
	26	Concrete	60.00	2.100	-
	27	Insulation	10.00	0.035	

Right (Waterloo, Canada)

Layer		Description	Width cm	Thermal Conduct. W/mK	U-Value W/m²/K
Roof					
	1	Steel Roofing	-	-	
	2	Wood Waferboard	1.60	0.14	
	3a	Wood I-Beam On 600 mm Centres	8.90	0.14	0.094
	3b	Cellulose Insulation	46.00	0.04	
	4	Vapour Retarder	-	-	
	5	Gypsum Board	1.20	0.21	
Window					
	6	Fibreglass Frame			1.07
	7	(Triple) Centre-Glazing			0.95
	8	Glazing and Frame			1.02
Exterior Wall					
	9	Gypsum Board	1.20	0.21	
	10	Vapour Retarder	-	-	
	11a	Cellulose Insulation	24.00	0.04	0.16
	11b	Wood I-Beam On 600 mm Centres	24.00	0.14	
	12	Insulated Sheathing	1.90	0.06	
	13	Wood Siding	1.90	0.14	
Ceiling					
	14	Hardwood Flooring	1.90	0.14	
	15	Waferboard Subfloor	2.20	0.14	
	16	Wood I-Beam On 600 mm Centres	36.60	0.14	-
	17	Polyurethane Foam	15.00	0.025	
	18	Gypsum Board	1.20	0.21	
Below Grade Wall					
	19	Gypsum Board	1.20	0.21	
	20	Vapour Retarder	-	-	
	21a	Cellulose Insulation	24.00	0.04	0.20
	21b	Wood Studs On 600 mm Centres	6.40	0.14	
	22	Reinforced Precast Concrete	5-20	2.00	
	23	Moisture Barrier	-	-	
	24	Semi-Rigid Fibreglass	5.00	0.04	
Below Grade Floor					
	25	Concrete	7.60	2.00	0.61
	26	Polystyrene	5.00	0.035	

Source: Hestnes et al (2003)

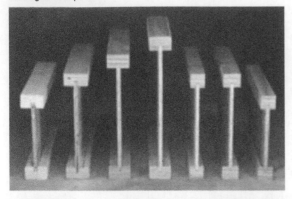

Source: The Engineered Wood Association (www.apawood.com)

140mm stud, a 140mm deep I-stud with insulation between the flanges, and a open-web stud with galvanized metal webs and insulation between the flanges are RSI 1.19, RSI 2.12 and RSI 2.32, respectively (Scanada Consultants, 1995).

Preferred insulation systems

Energy is required to manufacture insulation and transport it to the construction site. This energy is referred to as *embodied energy*, and normally entails emissions of CO_2 and other greenhouse gases. In addition, some solid and spray-on foam insulation involves the use and emission of halocarbon (HCFC and HFC) compounds, which are powerful greenhouse gases (Chapter 12, Section 12.5). Given that there are diminishing additional savings in building energy use as the thickness of insulation increases (Figure 3.2), the trade-off between embodied energy and halocarbon emissions versus energy savings needs to be considered (along with insulation and energy costs) in choosing the type and amount of insulation.

This trade-off is discussed in some detail in Chapter 12 (Sections 12.4 and 12.5). Here, it is sufficient to say that:

- cellulose insulation can be usefully applied to greater RSI values than any other kind of insulation, as it has negligible embodied energy and entails no halocarbons; and
- among external insulation types (useful as a sheathing to span thermal bridges or for direct attachment to concrete or masonry walls), EPS or semi-rigid mineral fibre are preferred because they entail no halocarbons, although their embodied energy is substantially greater than that of cellulose.

In light of these considerations, and given the desirability of minimizing wood requirements, the preferred system for wood frame wall construction at present appears to be a combination of engineered I-beams with cellulose insulation in the cavity and EPS or semi-rigid mineral fibre as an external sheathing. Roofs can often be constructed so as to permit a deep layer of blown-in cellulose insulation.

30–60 per cent more expensive than conventional 38 × 140 studs, this extra cost can be offset by the lack of warping and shrinking of engineered studs, thereby eliminating backing and callbacks to fix popped nails. 38 × 240mm (2″ × 10″) I-beams can be used as floor joists on 610mm (24″) centres, and can be used in place of 38 × 89mm + 38 × 140mm double-stud walls, in both cases using only one-third the volume of wood. Further wood savings arise from the fact that the exact lengths needed can be ordered from the factory.

Blown or spray-on cellulose insulation is the natural companion to engineered wood framing because it completely fills the voids (and has a 5–10 per cent higher insulation value than insulation batts for a given thickness). The studs themselves, due to their I-shape or open web structure, provide a significantly smaller thermal bridge: the thermal resistances of a conventional 38 ×

3.3 Windows

The flow of heat through a window depends on a number of distinct processes:

- *transmission* of solar (short wave) radiation,
- *emission* of infrared (IR) radiation;
- *conduction* of heat through the glass, through the air between the panes, and through the frame and spacers between the panes;
- *convection* between the panes of glass;
- *infiltration* of outside air (especially in poorly built windows).

The thermal resistance of glass is negligible. In a single-glazed window, the resistance to heat flow arises from the thin motionless layer of air on either side of the glass, as motionless air is a relatively good insulator. On a windy day, the thermal resistance on the outside of the glass will be very small, so most of the resistance to heat loss arises from the layer of motionless air attached to the inside surface of the glass. Readily available steps to increase the thermal resistance of windows are discussed below, followed by a discussion of window properties and emerging technologies.

3.3.1 Extra layers of glass (glazing)

In a double-glazed window, a layer of motionless air is created between the two glazings, thereby increasing the thermal resistance. As the thickness (L) of the gap between two panes of glass increases, the thermal resistance to molecular heat conduction (equal to L/k) increases, but if L is large enough for convective motions to begin, the thermal resistance will abruptly drop. Convective motion is more easily triggered the greater the temperature difference between the inside and outside panes of glass. Averaged over a heating season, the heat loss through a double-glazed window increases if the spacing between window panes increases beyond some optimum value that depends on climate. The optimal gap size decreases from 26mm at a temperature difference ΔT of 5K to 14mm at ΔT=33K (Askar et al, 2001). Use of a suboptimal gap size can increase the thermal conductivity of a double-glazed system by 10–30 per cent compared to the optimal size. For any given gap size, the thermal conductivity increases by 80–100 per cent as ΔT increases from 5K to 33K. If one wishes to minimize peak rates of heat loss rather than the total seasonal heat loss, one should choose a gap size that minimizes the heat loss for the coldest expected conditions.

Windows are commercially available with a third and even a fourth pane between the inner and outer panes (a triple- or quadruple-glazed window), which inhibit the development of convective cells in the air space, thereby further increasing the thermal resistance. However, the increased thickness and weight could be a problem in some cases. An alternative to triple and quadruple glazing is to insert one or two thin polyester or mylar films suspended between the inner and outer glazings. These films can be coated with a metallic low-e layer (see Section 3.3.3). Such products are commercially available (for example, Heat Mirror™).

3.3.2 Alternative gases between the window panes

Heavy molecular weight gases, such as argon, krypton, and xenon, are even better than air at inhibiting heat transfer. Table 3.8 compares the molecular thermal conductivity (k) of air and various gases. Argon-filled windows are commercially

Table 3.8 Thermal conductivity (W/m/K) and other properties of different gases that can be used in windows; the optimal thickness of the gas layer is for an average inside to outside temperature difference of 20K

Gas	Conductivity	Molecular mass	Percentage in air	Optimal thickness of gas Layer	Gas-separation energy for a 1.1m² glazed area
Air	0.0250	28.96		20mm	
Argon	0.0161	39.95	0.9	16mm	12kJ
Krypton	0.0096	83.8	0.000114	12mm	508MJ
Xenon	0.0055	131.3	0.0000087	8mm	4.5GJ

Source: McCrae (1997) for thermal conductivity, Muneer et al (2000) for other properties

available in many countries, while krypton-filled windows have been introduced commercially in the US and Switzerland and xenon-filled windows in Germany. Argon has a thermal conductivity two-thirds that of air, while krypton and xenon have thermal conductivities of just over one-third and one-fifth that of air, respectively. However, as explained below, a thinner gap is used in windows with heavier gases, so the contribution of gas-fill conductivity to the overall U-value of the xenon window, for example, is only a factor of two less than that of an air-filled window.

Convective motions are triggered by differences in the density of gas in contact with the warm and cold window panes. The absolute difference in density for a gas at two different temperatures increases with the molecular mass of the gas, so convective motions can be more readily triggered in the gases with a higher molecular mass. As a result, the optimum size of the gap between the two panes of glass is smaller the heavier the enclosed gas, as shown in Table 3.8. This allows triple-glazed windows with argon or heavier gases to fit into sash designs and framing systems used for double-glazed windows.

The last column of Table 3.8 gives the energy needed to separate the amount of argon, krypton or xenon used in an argon-, krypton- or xenon-filled 1.2m × 1.2m window. Due to the fact that argon constitutes almost 1 per cent of the atmosphere, the separation energy is quite small compared to krypton and xenon. These separation energies are compared with the reduction in heat loss over the lifetime of a window in Section 3.3.19.

The amount of gas fill that is lost over time depends on the sealant used. According to Muneer et al (2000), there are a number of sealants for which the loss of argon is less than 10 per cent after 20 years.

3.3.3 Low-emissivity coatings

The emission of infrared radiation (wavelengths of 4–50μm) depends on the temperature and emissivity of the emitting surface (Equation 3.4), and is one of the mechanisms by which heat is transferred from a window to the outside. Standard glass has an emissivity of 0.84. The heat loss can be reduced if the emissivity of the glazing

surface is reduced, and this can be done by coating the glazing with a special film that reduces the window emissivity. Such windows are said to be 'low-e' or 'heat mirror' windows, and are widely available. The term 'heat mirror' is appropriate because a surface that has a low emissivity for infrared radiation also has a low absorptivity for impinging infrared radiation, reflecting the radiation instead. When placed on the inner side of the inner glazing, this serves to keep infrared radiation (heat) inside the room.

Low-e coatings can be produced using the *pyrolytic* technique or the *sputtering* technique (Hollands et al, 2001). In the pyrolytic technique, an aerosol is sprayed from a nozzle above hot (600°C) glass, vapourizing and reacting before reaching the glass. This produces a hard, durable layer 100–400nm thick (nm = nanometer; 1nm = 10^{-9}m). An example reaction, involving tin, is $SnCl_4 + 2H_2O \rightarrow SnO_2 + 4HCl$. The tin oxide ($SnO_2$) forms a coating on the glass. With the sputtering technique, a coating is applied inside a vacuum chamber containing an inert gas at a pressure of 1Pa (10^{-5}atm). The glass passes under a cathode with an attached plate made of raw materials for the low-e coating. A plasma bombards the plate, dislodging atoms that travel at high speed to the glass and stick to it. Typically, 6–9 layers, each 6–12nm thick and of different composition, are created by passing the glass under different cathode plates. These thin layers are soft and fragile. Both techniques can be applied to glass several metres wide, and are fully automated. Pyrolytic coatings are referred to as 'hard' coatings, while sputtered coatings are referred to as 'soft' coatings.

Compositionally, low-e coatings can also be grouped into two categories: *doped-oxide semiconductors* and *metal/dielectric combinations*. Doped-oxide semiconductor films tend to be applied by the pyrolytic technique, and consist of In_2O_3:Sn, SnO_2:F, SnO_2:Sb or ZnO:Al, or combinations of the above. Radiative properties of an In_2O_3:Sn semiconductor with four different free-electron densities are given in the left panel of Figure 3.18. An emissivity of 0.2 can be obtained with a maximum solar transmittance of 78 per cent. Metal/dielectric films are applied using the sputtering technique. Metal layers a few nanometres thick have a low emissivity and but also

Figure 3.18 Transmittance and reflectance for an In₂O₃:Sn semiconductor low-e film (left) and for a silver-based low-e film (right). Thicknesses for the TiO₂ and Ag films are not given in the right panel for the case with 4mm glass, as the information is proprietary

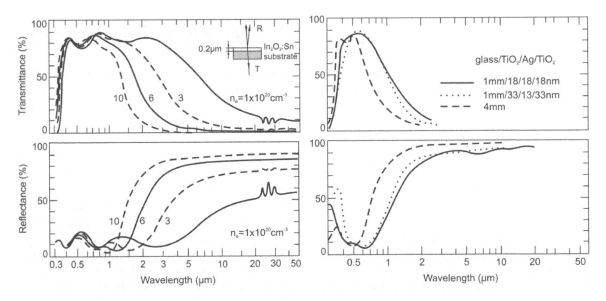

Source: Granqvist (2003)

a low (<60 per cent) solar transmission. Most of the loss of transmission occurs due to reflection at the outer surface, which can be minimized by adding an anti-reflective coating consisting of a dielectric (electrically insulating) layer. The wavelength properties of the low-e film derive from the properties of the individual layers and from their interaction. Silver (Ag) is the most widely used metal, although gold and copper can also be used. Many choices are possible for the dielectric layer(s), including Bi_2O_3, SnO_2, TiO_2, ZnO and ZnS. The radiative properties of a $TiO_2/Ag/TiO_2$ coating with three different thicknesses are given in the right panel of Figure 3.18. Emissivities are approximately given by 1.0 minus the fractional reflectance in the longwave region (wavelengths > 4μm). The reflectivity seen in the NIR region (0.7–4.0μm wavelengths) continues into the longwave region, so low solar transmission is naturally accompanied by a lower emissivity. Thus, a film with ε = 0.15 has solar and visible transmittances of 72 per cent and 87 per cent, respectively, while a film with ε = 0.04 has solar and visible transmittances of 50 per cent and 80 per cent.[5] For both metal and semiconductor low-e films, reduced solar transmittance occurs largely through reduced NIR transmittance, with little impact on visible transmittance. The impact

of low-e films on the transfer of solar heat into a building depends not only on their solar transmittance, but also on their placement within the glazing system, as explained later (Section 3.3.6).

In summary, low-e films can be classified as sputtered (soft) and pyrolitic (hard). Soft coats typically consist of 6–9 thin (6–12nm) layers, have low emissivities (0.04–0.15) but also low solar transmittance (50–72 per cent), consist of metal or metal/dielectric combinations, and are extremely fragile. Hard low-e coatings are thick (100–400nm), have an emissivity of about 0.2, can have relatively high solar transmittance (up to 78 per cent), consist of semiconductors and are durable. Both kinds can be engineered to have a low NIR transmittance and hence low overall solar transmittance while having a negligible effect on visible transmittance.

3.3.4 Vacuum windows

Creating a vacuum between the window glazings completely eliminates conductive (and any convective) heat transfer, except at small, almost-invisible pillars that are needed to prevent the glass layers from collapsing together. If combined with low-e coatings, the two processes of heat transfer through the non-frame part of the window are largely eliminated. The following problems still need to be solved (Muneer et al, 2000):

- the seal must maintain an extremely low pressure (10^{-8}atm) for at least 20 years;
- special attention must be given to the frame, due to the extremely low heat transfer through the rest of the window;
- the temperature difference between the outer and inner panes will be large, producing thermal expansion that could overstress the rigid edge seals.

The manufacture of vacuum glazings normally entails heating the glazings to 450°C so as to fuse the edges together, evacuating the window while baking it to a temperature of 100–250°C so as to remove absorbed and adsorbed gases from the inner surfaces, then melting the end of the pump-out tube to complete the sealing process (Collins and Simko, 1998) This is an energy-intensive process, and produces centre-of-glazing U-values of about 0.6–0.8W/m²/K. Three factors limit the reduction in U-value that can be achieved with vacuum glazing: (i) heat conduction occurs through the pillars separating the glazings (these could be 0.2mm in diameter on a 20–40mm grid, and contribute about 0.4W/m²/K to the U-value); (ii) low-conductivity edge seals cannot be used, as in gas-filled windows, due to the fact that the seal is formed by fusing the glazings together by heating them to 450°C; and (iii) the high temperature normally used during manufacture restricts the low-e coating to hard coatings with an emissivity of 0.25. Griffiths et al (1998) have experimented with an indium-based seal that permits manufacture at <200°C, which in turn permits an $\varepsilon = 0.05$ coating. The predicted centre-of-glazing U-value is 0.36W/m²/K, with a visible transmittance of 0.76.

The gap between glazings in vacuum windows is only about 0.15mm, so vacuum windows are very thin (6–8mm). This makes it possible, in some cases, to retrofit them onto existing windows without having to remove the pre-existing glazing (CADDET, 2001). Nippon Sheet Glass (www1.nsg.co.jp/en) produces a 6mm thick vacuum glazing under the product name SPACIA with a U-value one half that of a conventional double-glazed window.

3.3.5 Readily available low-conductivity frames and spacers

Aluminium frames entail the greatest heat loss of any framing material, followed by aluminium frames with a thermal break, aluminium-clad wood frames, and wood and vinyl frames, with the lowest heat loss from insulated fibreglass frames. The heat loss through the glazed area in high-performance windows is so small that heat loss from the window frame and spacers can be a large fraction of the total heat loss, unless insulated frames and spacers are used. Such frames must not only have a very small thermal conductivity, but must also contract and expand at the same rate as glass if they are to prevent air leakage. They must also be able to support the weight of triple-glazed windows without additional reinforcement, and must be resistant to moisture, chemicals, salty air or acid rain. Pultruded fibreglass satisfies most of these requirements but is also the most expensive option (CADDET, 1999). Saxhof and Carpenter (2003) report the use of a pultruded fibreglass frame with interior cavities that are filled with polystyrene insulation for windows in an advanced house in Waterloo (Canada), and having a frame U-value of 1.2W/m²/K and an overall U-value just over 1.0W/m²/K. Fibreglass is used in many high-performance windows not because of better insulating qualities than wood or vinyl, but because it can be made thinner in profile (due to its strength), thereby providing more glazing area and hence a larger solar heat gain. An example is given in Figure 3.19. For cooling-dominated climates, a low-conductance framing material is useful because it contributes to a slightly smaller solar heat gain than a high-conductance frame with the same area. Fibreglass has a similar thermal expansion coefficient to that of the window glass, so a fibreglass frame maintains the seal against air leakage induced by temperature variations across the window assembly.

Another way to reduce heat loss through the frame is to incorporate the framing system into the wall structure. Figure 3.20a shows the U-values for an evacuated window with a standard frame. The average U-value for the window system is more than twice the centre of glass value

Figure 3.19 Cross section of a triple-glazed window with a pultruded fibreglass frame (upper) and insulating spacers (lower)

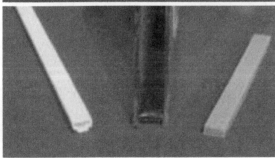

Source: Enermodal Engineering, Waterloo (Canada)

Figure 3.20 U-values of an evacuated window in (a) a standard frame, and (b) a frame overlapping the wall

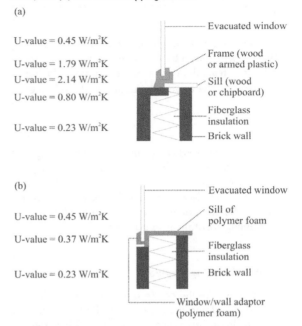

(a)

U-value = 0.45 W/m²K

U-value = 1.79 W/m²K

U-value = 2.14 W/m²K

U-value = 0.80 W/m²K

U-value = 0.23 W/m²K

Evacuated window

Frame (wood or armed plastic)

Sill (wood or chipboard)

Fiberglass insulation

Brick wall

(b)

U-value = 0.45 W/m²K

U-value = 0.37 W/m²K

U-value = 0.23 W/m²K

Evacuated window

Sill of polymer foam

Fiberglass insulation

Brick wall

Window/wall adaptor (polymer foam)

Source: Bredsdorff et al (1998)

(1.05W/m²/K vs. 0.45W/m²/K). Figure 3.20b shows a design in which the window overlaps the wall insulation. The average window U-value is now 0.43W/m²/K, which is slightly *less* than the centre-of-glass value. This strategy has been incorporated in the windows in many high-performance houses built in Europe under the CEPHEUS programme (see Chapter 13, Section 13.2.1). Yet another option, still under development, involves silica aerogel frames and spacers. These are described in Section 3.3.17.

The energy impact of the wall framing necessary to support a window is not normally associated with the window, but can be significant due to the low thermal resistance of wood studs (as discussed in Section 3.2.2). This indirect impact of windows on heat loss can be mitigated through external insulation that spans all wood stud thermal bridges, whether associated with windows or not.

3.3.6 Computation of window U-values and illustrative results

The U-value of a window includes the effect of conductive, convective and radiative heat transfer. It is different for the centre of the glazed portion of the window, along the edges of the glazing and for the frame. The U-value for the separate glazed and unglazed portions of the window must be determined by combining the resistances of successive layers in series (i.e. by adding the resistances), while the U-value for the entire window must be determined by combining the total glazed and unglazed resistances in parallel (i.e. by taking an area-weighting of total U-values for each portion of the window). In the air gap between panes of glass, where separate conductive, convective and radiative heat transfer occurs, these are also combined in parallel. An outline of the steps involved is given in Box 3.2, and complete details can be found in Muneer et al (2000).

BOX 3.2 Computing the U-value of a Window

For a double-glazed window, the overall U-value $(W/m^2/K)$ is given by the area weighting of contributions from the centre of the exposed glazing, the edge of the exposed glazing, and the frame:

$$\overline{U} = f_{cg}U_{cg} + f_{eg}U_{eg} + f_f U_f$$

(B3.2.1)

where U_{cg}, U_{eg} and U_f are the centre-of-glazing, edge-of-glazing and frame U-values, and f_{cg}, f_{eg} and f_f are the corresponding area fractions. Computing the whole-window U-value as an area-average of the component U-values is equivalent to combining resistances in parallel.

Centre-of-glazing U-value

The U-value for the centre of the window is obtained by adding the resistances of each layer (that is, layer resistances are combined in series) and taking the reciprocal of the total resistance. Thus, for a double-glazed window,

$$U_{cg} = \frac{1}{R_{cg}}$$

(B3.2.2)

$$R_{cg} = \frac{1}{h_i} + R_i + \frac{1}{h_s} + R_o + \frac{1}{h_o}$$

(B3.2.3)

where

h_i and h_o are the heat transfer coefficients $(W/m^2/K)$ for the inner and outer laminar boundary layer,

$h_s = h_c + h_r$ is the heat transfer coefficient for the cavity between the panes of glass, h_c and h_r being the coefficients for conductive-convective and radiative transfer, respectively,

$R_i = L_i/k$ is the thermal resistance of the inner glass pane, of thickness L_i,

$R_o = L_o/k$ is the thermal resistance of the outer glass pane, of thickness L_o, and

$k = 1W/m/K$ is the thermal conductivity of glass.

The reciprocals of h_i, h_s and h_o are resistances, so Equation (B3.2.3) is simply the sum of the resistances of each layer that the heat must pass through. Adding h_c and h_r to get h_s is equivalent to

combining the resistances in parallel. The convention is to number the glazing surfaces counting inward from the outermost surface. The circuit diagram for the window, following this convention, is given in Figure B3.2.

Figure B3.2 Resistance diagram for heat flow through a double-glazed window

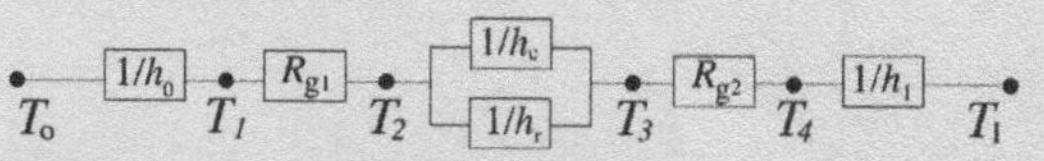

The heat transfer coefficient on the inner surface of the window is given by

$$h_i = h_{ri} + h_{ci}$$

(B3.2.4)

where:

$h_{ci} = 3.0W/m^2/K$ is the contribution from natural convection at a vertical surface,

$h_{ri} = 5.3 \times \varepsilon_{ii}/0.845 W/m^2/K$ is the contribution from the emission of radiation by the glazing, ε_{ii} being the emissivity of the inward-facing side of the inner glazing.

The heat transfer coefficient on the outer surface of the window depends on the wind speed V and angle of attack of the wind. One of many possible relationships that has been used is:

$$h_o = 5.8 + 4.1V$$

(B3.2.5)

Finally, h_r and h_c are given by:

$$h_r = 4\sigma \left[\frac{1}{\varepsilon_2} + \frac{1}{\varepsilon_3} - 1\right]^{-1} T_f^3$$

(B3.2.6)

and

$$h_c = \frac{N_u k}{L}$$

(B3.2.7)

respectively, where:

$T_f = 1/2 (T_2 + T_3)$,

ε_2 and ε_3 are the emissivities of surfaces 2 and 3, respectively,

k is the thermal conductivity of the infill gas,

L is the distance between the two glazing surfaces, and

N_u is the Nusselt number, a function of $T_3 - T_2$, *L*, the height of the window, and the properties of the infill gas (viscosity, diffusivity, expansion coefficient).

Many different formulations for the dependence of N_u on $T_3 - T_2$, *L* and other variables have been proposed, as given by Zhao et al (1999) and Muneer et al (2000). In these formulations, N_u varies with $(T_3 - T_2)^n$, where $n = 0.25$ to 0.33. Representative h_c values are given in Table B3.2.1.

Table B3.2.1 h_c values (W/m²/K) for $T_3 - T_2 = 20$K, taken from Muneer et al (2000, Figure 3.3.2). These values vary approximately with $(T_3 - T_2)^{0.25}$

	Window height		
	0.5m	1.0m	2.0m
Air, 20mm gap	2.09	1.72	1.42
Argon, 16mm gap	1.46	1.21	1.07
Krypton, 12mm gap	1.07	0.88	0.75
Xenon, 8mm gap	0.83	0.69	0.25

For given outside and inside temperatures, the U-value of the window depends on T_3 and T_2 because these temperatures influence h_r and h_c. However, T_3 and T_2 depend directly on h_r and h_c. Therefore, the window U-value needs to be computed iteratively, by choosing an initial h_r and h_c, then using the computed h_r and h_c (along with other parameters) to compute T_3 and T_2, then using these T_3 and T_2 to compute new values for h_r and h_c, and continuing until no further change in any of the computed quantities occurs. In steady state, the heat fluxes between any two pairs of nodes in Table B3.2.1 are equal, and will be proportional to the temperature difference between a pair of nodes divided by the resistance between the two nodes. These facts can be used to derive simple algebraic expressions for the temperature difference between any two nodes *i* and *j*, given the inside to outside difference in air temperature, the total resistance between T_i and T_o, and the total resistance between points *i* and *j*. Illustrative results are given in the main text.

Edge-of-glazing U-Value

The spacer that separates the two panes of glass in a double-glazed window has a significant thermal conductivity, which will lower the temperature along the edge of the inner glazing.

Lateral conduction of heat along the glazing and through the spacer decreases the effective U-value in a band along the edges of the window. One approach is to estimate U_{eg} based on U_{cg} and the type of spacing used:

$$U_{eg} = A + BU_{cg} + CU_{cg}^2$$

(B3.2.8)

where *A*, *B* and *C* are coefficients that depend on the type of spacer used, and U_{eg} is applicable to a band 65mm wide along the outer perimeter of the window. Values for *A*, *B* and *C*, as well as alternative estimation equations, are given in Muneer et al (2000). Table B3.2.2 shows U_{eg} values for representative U_{cg} values and different conventional spacers.

Table B3.2.2 Edge-of-glass U-values associated with different centre of glass values, for metal and insulated spacers

	U_{cg} (W/m²/K)	Type of spacer			
		Metal	Glass	Metal +Ins	Insulated
DG, Air (Case 2)	2.509	3.208	2.902	2.748	2.663
DG, Air, 1S, 1H (Case 9)	1.180	2.222	1.824	1.648	1.546
DG, Krypton, 1S, 1H (Case 15)	0.815	1.934	1.534	1.366	1.265

Note: Case numbers refer to cases in Table 3.9.

From the above table it can be seen that the relative importance of the spacer, and the benefit of spacers with smaller conductivity, increases the smaller the centre-of-glass U-value.

Frame U-value

The U-value of the frame depends on the thermal conductivity of the spacer between the panes of glass (within the frame) and of the framing material. U-values for some conventional frame-spacer combinations are given in Table B3.2.3.

As discussed in the main text, incorporation of the frame into an insulated wall can dramatically reduce these U-values.

Complex windows

Wright (2001) discusses the computation of heat loss through greenhouse windows and other projecting windows and complicated geometries.

Table B3.2.3 U-values of different frames used in operable windows

Frame material	Type of spacer	Glazing system		
		Single[a]	Double[b]	Triple[c]
Aluminium without thermal break	All	13.5	12.9	12.5
Aluminium with thermal break[d]	Metal	6.8	5.2	4.7
	Insulated		5.0	4.4
Wood/vinyl	Metal	3.1	2.9	2.7
	Insulated		2.8	2.3
Insulated fibreglass/vinyl	Metal	2.1	1.9	1.8
	Insulated		1.8	1.5

[a] 3mm glazing unit thickness
[b] 19mm glazing unit thickness
[c] 34.4mm glazing unit thickness
[d] Depends strongly on thickness of thermal break. Value given is for 9.5mm.
Source: ASHRAE (2001b, Chapter 30)

Table 3.9 presents centre-of-glass U-values for 32 different glazing, emissivity and gas fill combinations, computed using the procedure explained in Box 3.2 with coefficients appropriate for a 1m high window, an indoor temperature of 20°C and an outdoor temperature of −10°C. Figure 3.21 shows the temperature profiles for selected cases. The industry convention, adopted here, is to number the glazing surfaces beginning with the outside surface of the outside glazing and counting inward. The percentage change in U-value resulting from the addition of low-e layers or the use of gas fills in the gap between the glazings is always equal to the percentage change in the product of the heat transfer coefficient across the gap (h_s of Box 3.2) and the temperature difference across the gap.

As seen from Table 3.9, a single-glazed window has a centre-of-glass U-value of around $5W/m^2/K$. This value drops in half for clear double-glazed windows, to about $1.0W/m^2/K$ for argon-filled doubling glazing with two low-e coatings, and to about $0.6W/m^2/K$ for triple-glazing with argon fill and two low-e coatings. The thermal resistance of glass itself is almost zero, so the resistance of a single-glazed window arises largely from that of the non-turbulent atmospheric boundary layer next to the glass on each side (heat transfer occurs only by molecular conduction within this layer). These boundary layers also contribute to the thermal resistance of

double- and triple-glazed windows, but they are of less relative importance. The outer boundary layer decreases in thickness with increasing wind speed and angle of attack, causing the window U-value to increase by about 10 per cent in going from calm to severe winds for the best windows shown in Table 3.9, but doubling for single-glazed windows. This amplifies the differences between the windows under those conditions where the U-value matters most, namely, cold, windy conditions.

Careful examination of Table 3.9 and Figure 3.21 reveals the following:

- Adding a low-e coating on the innermost surface makes it colder, although the U-value decreases.
- Otherwise, a lower U-value is associated with a warmer innermost surface (and a colder outermost surface).
- Adding a low-e coating on surface 4 of double-glazed windows has about three-quarters of the impact of adding a coating on surface 3.
- Adding a second low-e coating on surface 2 (Case 5) has less than one-quarter the impact of a first coating on surface 3 (Case 3), while adding a second low-e coating on surface 4 (Case 6) has less than one half of the

Table 3.9 Window centre-of-glass U-values for different gas fills and combinations of hard and soft low-e coatings, computed using the equations and coefficient values given in Box 3.3. Surfaces are numbered beginning with the outermost surface and counting inward, as illustrated in Figure 3.21. For all cases, $\varepsilon_1 = 0.845$, $V = 2.0$ m/s, $T_i = 10°C$, and $T_o = -20°C$. For TG windows, cases are given where Kr and Xe fills one or both gaps. 2H means 2 hard low-e coatings, 2S means two soft low-e coatings

Case	Description	Emissivities, for the indicated surface					U-value (W/m²/K)	
		2	3	4	5	6	Actual	Change
Single-glazed window								
1		0.845					5.036	
Double-glazed windows (DG)								
2	Air	0.845	0.845	0.845			2.509	−2.528[a]
3	Air, 1H	0.845	0.200	0.845			1.719	−0.789[b]
4	Air, 1H	0.845	0.845	0.200			1.935	−0.574[b]
5	Air, 2H	0.200	0.200	0.845			1.554	−0.166[c]
6	Air, 2H	0.845	0.200	0.200			1.415	−0.304[c]
7	Air, 1S	0.845	0.040	0.845			1.395	−1.113[a]
8	Air, 2S	0.040	0.040	0.845			1.350	−0.045[d]
9	Air, 1S, 1H	0.845	0.040	0.200			1.180	−0.215[d]
10	Argon	0.845	0.845	0.845			2.392	−0.117[e]
11	Argon, 1S	0.845	0.040	0.845			1.120	−0.275[e]
12	Argon, 1S, 1H	0.845	0.040	0.200			0.976	−0.204[e]
13	Kr	0.845	0.845	0.845			2.309	−0.199[e]
14	Kr, 1S	0.845	0.040	0.845			0.912	−0.483[e]
15	Kr, 1S, 1H	0.845	0.040	0.200			0.815	−0.366[e]
16	Xenon	0.845	0.845	0.845			2.257	−0.252[e]
17	Xenon, 1S	0.845	0.040	0.845			0.775	−0.620[e]
18	Xenon, 1S, 1H	0.845	0.040	0.200			0.703	−0.478[e]
19	Vacuum	0.845	0.845	0.845			2.180	−0.329[e]
20	Vacuum, 1H	0.845	0.200	0.845			1.044	−0.675[e]
21	Vacuum, 2H	0.845	0.200	0.200			0.932	−0.483[e]
Triple-glazed windows (TG)								
22	Air	0.845	0.845	0.845	0.845	0.845	1.640	−0.869[f]
23	Air, 1H	0.845	0.845	0.845	0.200	0.845	1.236	−1.273[f]
24	Air, 2S	0.845	0.845	0.040	0.040	0.845	1.009	−1.500[f]
25	Air, 2S	0.040	0.845	0.845	0.040	0.845	0.734	−1.775[f]
26	Air, 2S, 1H	0.040	0.845	0.845	0.040	0.200	0.668	−0.066[g]
27	Argon, 2S	0.040	0.845	0.845	0.040	0.845	0.567	−0.167[h]
28	Argon, 2S, 1H	0.040	0.845	0.845	0.040	0.200	0.527	−0.141[h]
29	Kr/Kr, 2S	0.040	0.845	0.845	0.040	0.845	0.451	−0.283[h]
30	Air/Kr, 2S, 1H	0.040	0.845	0.845	0.040	0.200	0.523	−0.145[h]
31	Xe/Xe, 2S, 1H	0.040	0.845	0.845	0.040	0.200	0.359	−0.375[h]
32	Air/Xe, 2S, 1H	0.040	0.845	0.845	0.040	0.200	0.473	−0.195[h]

[a] Relative to Case 1
[b] Relative to Case 2
[c] Relative to Case 3
[d] Relative to Case 7
[e] Relative to corresponding Double, Air case
[f] Relative to Case 2
[g] Relative to Case 25
[h] Relative to corresponding Triple, Air case

impact of a first coating on surface 3 and half the impact than if it is the first coating (Case 4).

- Thus, in double-glazed windows, the single low-e coating with the largest impact is on either of the surfaces facing the inner gap, followed by a coating on the inner surface of the inner glazing.

- Adding an inert gas fill to a window without low-e coatings has about half the effect as when low-e coatings are present, because of the large parallel radiative heat transfer path.

Figure 3.21 Window temperature profiles for selected cases from Table 3.9: (a) Double-glazed windows, Cases 2, 4, 9, 12 and 15; (b) Triple-glazed windows, Cases 22, 25, 26 and 28

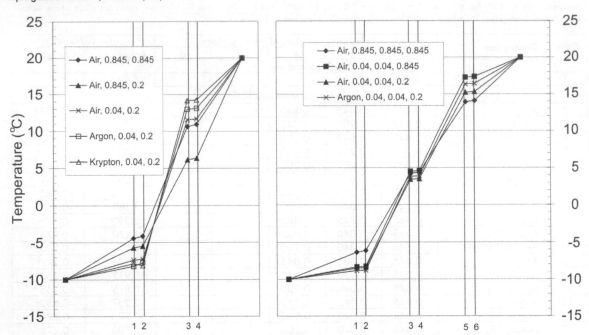

Note: Numbers in the legend are the emissivities on glazing surfaces 3 and 4 (DG) or 2, 5 and 6 (TG).

It follows from Equation (B3.2.6) of Box 3.2 that the effect on the U-value of adding a low-e coating to either surface facing the gap in a double- (or triple-) glazed window is the same. It may seem surprising that applying a low-e coating to the inner surface of the inner glazing reduces the heat loss through the window, as less radiation will be emitted from the window to the interior of the room, so the inner glazing will be warmer and therefore more heat will be available to be conducted to the outer side of the glazing. However, a lower emissivity also means that less radiation from the interior of the room is absorbed by the window and is instead reflected back to the interior. Since the interior walls will (in winter) be warmer than the glazing surface, the net result of a low-e coating on the inside surface is to reduce heat transfer from the room to the window, and to increase the radiant energy impinging on occupants in the room. When a low-e coating is added to a glazing surface, the temperature difference across the gap, and hence h_c of Box 3.2 may increase or decrease, but this changes the U-value by no more than 5 per cent and usually by only

1–2 per cent for the cases considered here.

Placement of a low-e coating on the outside of a window glazing has relatively little impact on the window U-value, because the temperature of the outer glazing surface is closest to that of the surroundings (so the reduction in emission from the window is largely offset by reduced absorption of radiation from the surroundings) and because the impact of a smaller radiative heat transfer coefficient is diluted by the large convective heat transfer coefficient that applies to the outside surface (see Box 3.2). According to Hollands et al (2001), a low-e coating should not be applied to either surface of the inner glazing in triple-glazed windows because the heat from absorbed solar radiation cannot readily escape, so over-heating may occur. In any case, it will not be particularly effective if there is already a low-e coating on the outer glazing, as seen from Table 3.9.

Table 3.10 gives whole window U-values for selected cases from Table 3.9, and for various choices concerning the spacer (which holds the glazings apart) and window frame. With insulated

Table 3.10 Overall window U-values for selected cases from Table 3.9, with the types of spacer and frame indicated below, computed as described in Box 3.2 for a 1.1m × 1.1m window (1.1m² glazed area + frame)

Case	Description	Spacer type	Frame type	U-value (W/m²/K)			
				Centre	Edge	Frame	Overall
2	DG, Air	Metal	Al	2.51	3.21	10.8	3.41
3	DG, Air, 1H	Glass	Wood	1.72	2.26	2.3	1.89
9	DG, Air, 1S, 1H	Insulated		1.18	1.55	1.2	1.26
12	DG, Argon, 1S, 1H	Insulated		0.98	1.39	1.2	1.08
15	DG, Kr, 1S, 1H	Insulated		0.81	1.27	1.2	0.94
18	DG, Xenon, 1S, 1H	Insulated		0.70	1.18	1.2	0.85
20	DG, Vacuum, 1H	Insulated		1.04	1.44	1.2	1.14
26	TG, Air, 2S, 1H	Insulated		0.67	1.16	1.2	0.82

spacers and frames in a moderate-performance glazing system (such as DG air with two low-e coatings), there is little difference between centre of glass and whole-window U-values. When such spacers and frames are combined with the highest-performance glazing systems, whole-window U-values are about 0.10–0.15W/m²/K higher than centre of glass values for 1.1m × 1.1m windows with 5.1cm wide frames. The difference between centre-of-glazing, average-glazing, and whole-window U-values will be smaller the larger the window (assuming a fixed frame width).

The U-value of an inclined window will be greater than that of a vertical window (to which published U-values normally pertain), due to greater convective heat transfer between the glazing layers when inclined and greater boundary layer heat transfer coefficients (h_o and h_i in Box 3.2). For example, Okalux (www.okalux.de) reports a U-value of 1.2W/m²/K for its low-e, Ar-filled, double-glazed windows when vertical but 1.7–1.8W/m²/K when inclined. Sunlite (www.sunlite-ig.com) indicates U-values at a 20° slope that are 10–37 per cent greater than for a vertical window. However, the heat transfer coefficient of transparent insulation (Section 3.2.8) is largely independent of inclination, making it a good choice for skylights. A horizontal skylight of any material still has a larger view factor for exchange of infrared radiation with the cold sky, increasing h_o. However, if the skylight projects from the roof, it will have a reduced view factor of the warm interior space, reducing h_i.

The impact on heat loss of various combinations of low-e coatings must be weighed against the impact of reduced transmission of solar radiation for each additional low-e coating. This is discussed next.

3.3.7 Transmission of solar radiation through windows

The amount of solar radiation transmitted through a window as a fraction of the transmission by a single pane of 3mm thick clear glass is called the *shading coefficient (SC)*. A single pane of 3mm clear glass transmits 84 per cent of the incident solar radiation at a 0° angle of incidence (perpendicular to the window) and 80 per cent at an angle of 60°. A better indicator of the transmission of solar energy is the *solar heat gain coefficient* (SHGC), also known (in Europe) as the *g value*. This is the fraction of the solar radiation incident on a window that passes through the window, taking into account absorption of some solar radiation by the window and the transfer of some of this absorbed energy to the interior. It depends on the angle at which the sun strikes the window and weakly on wind speed (Li and Lam, 2000), the later by influencing the partitioning of absorbed radiation between transfer to the inside and outside. The dependence of solar transmission on the angle of incidence can be used to minimize solar heat gain by tilting the glass from the vertical, as illustrated by Askar et al (2001) for low-latitude sites.

Box 3.3 gives equations that can be used to compute the SHGC for a single pane of glass. To find the approximate SHGC for windows other than single-glazed for a given solar angle, multiply the SHGC for a single pane of glass for that angle times the SC for that window. However, the shape of the variation of SHGC with angle of incidence is different between single-, double- and triple-glazed windows, and for different types of window coatings, and the error in using the angular dependence for a single pane of glass is greatest at high angles of incidence (such as found in summer and at low latitudes). For a more accurate calculation, the polynomial equations of Karlsson and Roos (2000) can be used; these represent a fit to the results of detailed calculations, and require only two input parameters: the number of panes of glass, and the type of coating. For a detailed discussion of window optics, see Rubin et al (1998).

Only about half of the solar radiation reaching the ground consists of visible light, the remainder consisting largely of near-infrared (NIR) radiation (see Figure 3.1). The term *daylight transmission* (DT) is customarily used to refer to the fraction of visible light that is transmitted. Characterizing a window in terms of DT and SC is somewhat awkward, however, because DT refers to the property of a given glazing, while SC refers to the glazing in comparison to standard glass. These terms have been introduced here only to guide the reader through the published data on windows. Here, we shall use the terms *visible transmittance* and *near-infrared transmittance* to refer to the fraction of visible and NIR radiation directly transmitted through the glazing.

As little as 2–5 per cent of the available outside visible light is needed on a sunny day for daylighting purposes (depending on the ratio of window to room area). To minimize cooling loads while allowing for daylighting, one would prefer a window with adequate visible transmittance but minimal NIR transmittance. Conversely, in cold climates we want to maximize the transmittance at all wavelengths. Neither uncoated clear glass nor regular reflective glass are suitable for minimizing the cooling load while maximizing daylighting, because the visible and NIR transmittances are similar in both cases (see Ismail and Henríquez, 2002, for transmittance versus wavelength for a variety of single- and double-glazed systems). However, glazing systems can be designed to have high visible transmittance and low NIR transmittance in two different ways: (i) by engineering a low-e coating to switch from high reflectivity/low transmission and absorption to low reflectivity/high transmission near the visible/NIR boundary rather than near the NIR/infrared boundary, and/or (ii) by varying the iron content of the glass. The former approach was illustrated in Figure 3.18, where NIR

BOX 3.3 Solar heat gain coefficient for a single pane of glass

For a single pane of standard glass, and neglecting dependence on wind speed, the SHGC is given by (Lam and Li, 1998):

$$SHGC = f_{direct}(\tau + 0.267\alpha) + 0.8135 f_{diffuse}$$

(B3.3.1)

where f_{direct} and $f_{diffuse}$ are the fractions of total radiation on the glazing surface as direct and diffuse radiation, respectively (these can be computed using formulae given in Chapter 11), τ is the fractional transmission of direct beam solar radiation, and α is the fractional absorption of direct beam radiation. The latter are given by:

$$\tau = -0.00885 + 2.71235\cos\theta - 0.62062\cos^2\theta$$
$$- 7.07329\cos^3\theta + 9.75995\cos^4\theta - 3.89922\cos^5\theta$$

(B3.3.2)

and

$$\alpha = 0.001154 + 0.77674\cos\theta - 3.94657\cos^2\theta$$
$$+ 8.57881\cos^3\theta - 8.38135\cos^4\theta + 3.01188\cos^5\theta$$

(B3.3.3)

respectively, where θ is the angle between the sun and a line perpendicular to the window surface. The SHGC for diffuse radiation is constant at 0.8135. Polynomial equations for the SHGC for multiply glazed windows and for windows with various films are given in Karlsson and Roos (2000).

The SHGC varies from 0.88 for direct beam radiation striking a single pane of standard glass perpendicularly, to 0.81 for radiation at 60° from the perpendicular, after which it drops off to zero for radiation parallel to the glazing surface. The SHGC for direct beam radiation is thus lower in summer than in winter, which works to slightly reduce summer overheating and increase winter heat gain. For multiply glazed windows, the SHGC drops more rapidly with an increase in the angle of incidence than for a single-glazed window. This acts to further reduce the summer heating.

transmittances ranging from 20 per cent to 80 per cent with relatively little impact on the visible transmittance are seen. The latter approach is illustrated in Figure 3.22, which compares transmittance versus wavelength for standard clear glass and for glass with a low and a high iron-oxide content. Iron is the main absorbing agent in glass, and increasing the iron content reduces the NIR transmittance by increasing NIR absorptance, while having little effect on the visible transmittance. Conversely, decreasing the iron content below that of standard glass gives a uniformly high solar transmittance, increasing the SHGC by about 0.1. At zero iron content, the absorption would be essentially zero, giving 92 per cent solar transmittance and 8 per cent reflectance for a 0° angle of incidence. Addition of an anti-reflective coating can cut the reflection in half, giving a transmission of about 96 per cent for low-iron glass, as also illustrated in Figure 3.22. For a double-glazed window with one hard low-e coating, anti-reflective coatings on both panes increase the visible transmittance from 0.74 to 0.89 (Rosencrantz et al, 2005).

The sun will strike equatorward-facing windows at a lower angle of incidence (closer to perpendicular) in winter and at a larger angle of incidence in summer. Various combinations of thin films can be added to glass to create a strong dependence of solar transmissivity on the angle of incidence. These are referred to as *angular-selective* thin films. Figure 3.23 shows the solar transmissivity for a system of two 12nm films of silver embedded between three films of SiO_2. For 120nm thick SiO_2 films, solar transmittance increases with decreasing angle of incidence (as found during winter). For east- or west-facing windows, however, one might wish to have a smaller transmissivity at low angles of incidence (early morning and later afternoon), when the intensity of solar radiation on the window is largest, and a larger transmissivity at larger angles of incidence. This can be achieved by simply choosing a 170nm thickness for the SiO_2 films, as illustrated in Figure 3.23. The data shown in Figure 3.23 are symmetrical around a perpendicular angle of incidence. Amazingly, thin-films can be produced that have different properties at the same angle on either side of the perpendicular (Granqvist, 2003).

The low NIR transmittance achieved with hard coatings involves 5–10 per cent absorption,

Figure 3.22 Fraction of solar radiation that is transmitted by standard glass, and by glass with a lower or higher iron-oxide content. The latter would have a slight greenish tint

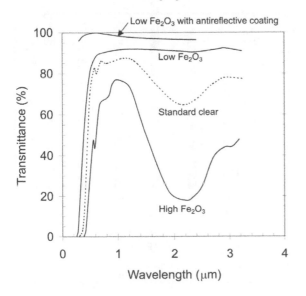

Figure 3.23 Solar transmittance for an $SiO_2/Ag/SiO_2/Ag/SiO_2$ thin film system as a function of the thickness of each SiO_2 layer and of the angle of incidence

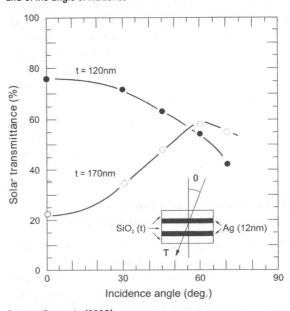

Source: Gombert et al (1998) for low iron with anti-reflective coating, Granqvist (1989) for all others

Source: Granqvist (2003)

as well as substantial reflection (see Figure 3.18 left). Thus, the low e-coating should be placed on the outer glazing of double- or triple-glazed windows if one wishes to minimize heat gain. The inner glazing(s) and gas fill serve to reduce the cooling load in summer, not by reducing conduction of heat from the outside air, but by reducing the transfer of absorbed NIR solar energy to the interior. If the glazing system is placed in a flip window that rotates open for cleaning and can be closed with either side facing outward, then flipping it over in winter can make it an effective trap for solar radiation that is now absorbed by the inner-most glazing, thereby increasing the passive solar heating of the interior. For a double-glazed window with the outer surface of the outer glazing tinted and inner surface of the outer glazing having an $\varepsilon = 0.03$ coating, SHGC = 0.67, while flipping the window around the other way yields SHGC = 0.29 (Feuermann and Novoplansky, 1998). Flipping the same window with a tinted glazing but no low-e coating reduces the SHGC from 0.69 to 0.51. From this, one can see that: (i) the same low-e coating can reduce the SHGC by 0.02 or 0.22, depending on where it is placed, and (ii) flipping a tinted-only window has a relatively small effect on the SHGC.

Flippable windows with low-e coatings will be advantageous from an energy point of view only at locations (such as Albuquerque, New York or North Dakota) with both a significant winter heating and a significant summer cooling load (otherwise, the window can be fixed according to the dominant load). In such locations, the savings in heating energy for the window considered above is around 100–200kWh/m^2 of window area in buildings with heavy construction but only 50–100kWh/m^2 in buildings with light construction (more of the solar heat gain is usable in buildings with larger thermal mass, as there will be less tendency for overheating).

3.3.8 Variation of inner surface temperature with window properties

Windows with smaller U-values will have a higher temperature on the inside surface of the window when it is colder outside than windows with a lower U-value, the only exception being if the lower U-value is due solely to a low-e coating on

the innermost glazing surface. This is illustrated in Figure 3.24, which shows the inside window surface temperature for an outside air temperature of −18°C and an inside air temperature of 21°C. The much higher surface temperature of high-performance windows prevents condensation on the window, prevents drafts and increases the perceived inside air temperature due to the greater emission of infrared radiation from the window to the occupants of the room. The reduction in drafts and greater perceived temperature in turn allow a lower air temperature, thus adding to the energy savings.

The impact of window properties on the window inner-surface temperature during summer depends on the extent to which solar radiation is absorbed or reflected by the window, where within the glazing system it is absorbed, and the ease with which heat can be radiated. As a result, there is not a consistent relationship between window U-value and window inner-surface temperature in summer, as also illustrated in Figure 3.24. Some windows with a low SHGC can have

Figure 3.24 Temperature on the inside of a window surface for an outdoor air temperature of −18°C and an indoor air temperature of 21°C (winter condition, light bars) and for summer conditions (dark bars)

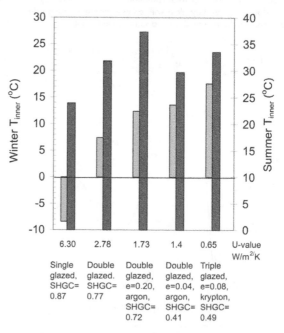

rather warm inner-surface temperatures, implying greater emission of infrared radiation to the inside (which has a heating effect). However, this is accounted for in the definition of SHGC, which accounts for both penetration of solar radiation and re-radiation of solar radiation to the interior as infrared radiation. Thus, a lower SHGC coefficient is beneficial in summer, even if it is associated with warmer inner-surface temperatures.

3.3.9 Air leakage through windows

Air leakage occurs around the frame of a window and where the two parts of operable windows come together. Leakage rates are reported for a standard test condition of 25 or 75Pa pressure difference between the inside and outside of the window and a single temperature (typically 20°C) on both sides of the window. This information is used in computer simulation programs to compute leakage for prescribed hourly winds and for pressure differences that would occur under typical conditions. The pressure differential in turn depends on the difference between indoor and outdoor temperatures, as explained later (Section 3.7). However, the leakage rate also depends directly on the temperature difference between inside and outside, through the fact that different parts of the window will contract by different amounts when the outside temperature drops, but this dependence is not normally accounted for. For a given pressure differential, Henry and Patenaude (1998) report that the leakage rate for an outside temperature of −30°C can be up to five times the leakage rate under isothermal conditions, although in some windows the leakage rate increases by less than 10 per cent. In any case, heat loss due to leakage is only a few percent of total heat loss.

Infiltration of air around a well-built window frame is sufficiently slow that the incoming air is warmed by half to three-quarters of the difference between inside and outside air temperatures by heat conduction from the frame, while outgoing air gives up a substantial amount of heat to the frame before reaching the outside (Hallé and Bernier, 1998). Assuming that infiltration through one window is balanced by exfiltration through another window (rather than balanced by flow through other building-shell elements),

the net result is that the difference in temperature between incoming and outgoing air (and the associated heat loss) is reduced to about one-tenth of the indoor to outdoor temperature difference. That is, air exchange through well-built windows behaves much like dynamic insulation with recovery of heat from the exhaust air (see Section 3.2.4).

3.3.10 Thermal properties of readily available and advanced windows

To summarize the above discussion, there are four thermal properties of windows that need to be considered: the U-value (or heat transfer coefficient, $W/m^2/K$), the shading coefficient (SC) or (preferably) the solar heat gain coefficient (SHGC), the daylight transmission (DT) fraction, and the air leakage rate at a 25Pa pressure differential.

Table 3.11 gives the properties of some high-performance, commercially available windows. Many window products are available with centre of glass U-values less than $1.0W/m^2/K$, with SHGCs ranging from 0.23 to 0.60, and with visible transmittances ranging from 0.38 to 0.72. Windows with low-e coatings have captured more than 40 per cent of the North American market, while triple-glazed windows designed for low or high solar heat gain have captured about 1–2 per cent of the market. In Switzerland, triple-glazed windows with low SHGC had captured about 7 per cent of the market by 2001 (Jakob and Madlener, 2003). Vacuum windows appeared in the European market in small quantities in the mid-1990s.

Table 3.11 Properties of selected high-performance, commercially available windows

Window product	U-value (W/m²/K)			SHGC	Transmittance		Sound insulation
	Air	Argon	Krypton		VIS	UV	
Sunlite (www.sunlite-ig.com), U-values pertain to glazed area only							
DG, Clear	2.689	2.547		0.70	0.79	0.50	
DG, Hard low-e	1.878	1.650		0.67	0.73	0.35	
DG, Soft low-e	1.666	1.407		0.38	0.70	0.14	
TG, low-e	0.903	0.710	0.613	0.32	0.54	0.04	
HM TC-88	1.016	0.835	0.738	0.48	0.63	<0.005	
HM TC-88, low-e	0.948	0.761	0.659	0.34	0.55	<0.005	
HM 44	1.209	1.028	0.937	0.28	0.38	<0.005	
QG, Clear	0.687	0.545	0.454	0.29	0.50	<0.005	
QG, low-e	0.636	0.494	0.397	0.23	0.41	<0.005	
Interpane (www.interpane.net)							
DG, IPLUS ²S		1.1		0.56	0.80		
DG, IPLUS ᶜS			1.1	0.64	0.81		
TG, IPLUS ³S		0.6		0.52	0.72		
TG, IPLUS ³ᶜS			0.5	0.52	0.72		
Steindl Glas (www.steindlglas.com/isog_wd_ws.html#)							
DG	1.7 (1.8)	1.4 (1.5)	1.0 (1.1)	0.58	0.77		30–32dB
TG		0.6 (0.8)		0.44	0.66		32dB
TG			0.7 (0.8)	0.60	0.75		34dB
Nippon Sheet Glass (www1.nsg.co.jp/en)							
SPACIA Normal	1.5 (0.2mm vacuum)			0.76	0.76		30dB
SPACIA Anti-firing	1.5 (0.2mm vacuum)			0.63	0.72		35dB
SPACIA Low-e	1.2 (0.2mm vacuum)			0.50	0.68		30dB

Note: DG, TG and QG = double-, triple- and quadruple-glazed windows, HM = heat mirror. There are likely differences in the test conditions used for calculating U-values, so the results from different companies may not be strictly comparable.
Source: Manufacturers' websites

3.3.11 Optimizing the trade-off between window U-value, SHGC and daylight transmission

A very small window U-value leads to a large reduction in heat loss in winter but only a small reduction in summer air-conditioning load due to the smaller indoor-to-outdoor temperature difference in summer. However, a window with a small U-value achieved through the use of low-e coatings will have a smaller SHGC than uncoated windows. It is the small SHGC in high-performance windows, rather than the small U-value, that contributes to a substantial reduction in glazing cooling load. The reduction in summer cooling load will be larger the lower the emissivity of the low-e coating, since a lower emissivity is associated with a smaller SHGC. For example, a glazing with a single low-e coating has an SHGC of 0.7 if the emissivity of the coating is 0.08, but

an SHGC of 0.3 if the emissivity is 0.03 (Reim et al, 2002). However, for east- and west-facing windows or horizontal skylights, the loss of solar heating in winter due to the smaller SHGC is relatively small, as there is less solar radiation available in winter for these orientations (see Chapter 4, Section 4.1.1). Thus, high-performance windows are often like the best of all possible worlds, simultaneously reducing winter heating loads and summer cooling loads. This is illustrated in a comparison of alternative glazings for atria in Ottawa by Laouadi et al (2002).

Nevertheless, there may be situations where one wants to minimize the U-value while having as large an SHGC as possible (for example, if shading devices are used during the summer, or where summer cooling loads are small and winter heating loads are large). A number of options are available. First, the impact on SHGC of adding a low-e coating to a double-glazed window is minimized if the low-e coating is placed on surface 3 rather than on surface 2 or, in the case of triple-glazed windows, if two low-e coatings are placed on surfaces 3 and 5 rather than on surfaces 2 and 4. This is illustrated in Table 3.12, which compares the SHGC (and corresponding centre-of-glass U-value) for various arrangements of the low-e coatings in a double-glazed window. For a 40° angle of incidence, solar heat gain is reduced by 30 per cent with an $\varepsilon = 0.1$ coating on the outer glazing, but by only 22 per cent if on the inner glazing. In either case, the heat loss (as quantified by the U-value) is reduced by 33 per cent. As noted above and illustrated in Figure 3.22, the use of low-iron glass increases the transmissivity of glass to solar radiation. The increase is large enough to offset the reduction in the SHGC that occurs through the addition of a single low-e coating (Henry and Dubrous, 1998).

Vacuum windows provide an alternative means of achieving a U-value less than 1.0 $W/m^2/K$, but without the large reduction in the SHGC found in triple-glazed windows with multiple low-e coatings. Vacuum windows would be preferred in the coldest climates, while triple-glazed windows might be preferred in less severe climates (especially when avoiding overheating in summer is also important).

The addition of a low-e coating affects the SHGC largely by altering the NIR solar transmittance, with a smaller effect on the visible transmittance (see Figure 3.18). This, in turn, will have an impact on the energy that can be saved through lighting systems that can be dimmed when some natural light is available (see Chapter 9, Section 9.3.2). Altogether then, the choice of the kind and dimensions of a window affects:

- heating energy use;
- cooling energy use;
- lighting energy use.

Figure 3.25 compares the total energy use in a 5m wide perimeter zone of an office building in Chicago (cold-climate case) and Houston (hot-climate case), as calculated by Carmody et al (2004). Results are given for a non-coated double-glazed window and for a high-performance (triple-glazed, low-e) window on walls facing north, south, east and west, and as the window:wall area ratio varies from 0.0 to 0.6. Total energy use increases substantially with increasing window area using the base-case window, but increases only slightly using the advanced window for all orientations and both climates when there are no lighting energy savings. However, when the lighting system can dim in response to daylight, total energy use *decreases* with increasing window area for high-performance windows of any orientation in either climate. Without shading, energy use begins to increase as the window fraction increases beyond 0.4, but is roughly constant beyond a window fraction of 0.4 if fixed shading is provided for east-, west- or south-facing windows. At a window fraction of 0.6, the total energy use for the advanced window with shading and daylight responsive lighting is about half of that for the base case window in the cold climate and about two-thirds that for the base case window in the hot climate. Figure 3.26 directly compares the savings in total energy use with increasing window area for the four orientations, using advanced windows with and without shading.

Table 3.12 Comparison of window centre of glass U-value and SHGC for alternative arrangements of low-e coatings in a double-glazed window

Glazing system	U-value (W/m²/K)	SHGC			
		Angle of incidence			
		0°	40°	60°	80°
No coating	2.73	0.76	0.74	0.64	0.26
$\varepsilon = 0.2$, surface 2	1.99	0.65	0.64	0.56	0.23
$\varepsilon = 0.2$, surface 3	1.99	0.70	0.68	0.59	0.24
$\varepsilon = 0.1$, surface 2	1.82	0.54	0.52	0.44	0.18
$\varepsilon = 0.1$, surface 3	1.82	0.60	0.58	0.51	0.22

Source: ASHRAE (2001b, Chapter 30)

Figure 3.25 Impact of window type and area on total energy use in a 5m wide zone next to the perimeter of an office building in Chicago (cold-climate case) and Houston (hot-climate case) for north- and south-facing, and east- or west-facing walls

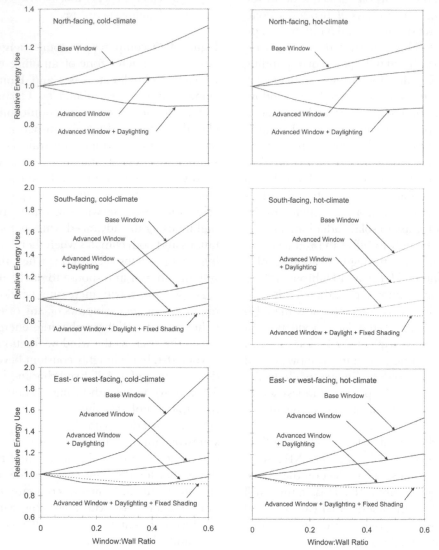

Note: For the cold-climate case, the base case window is double-glazed with $U = 3.41W/m^2/K$, SHGC = 0.6, and visible transmittance = 0.63, while for the hot-climate case the base-case window is double-glazed with a bronze tint, $U = 3.41W/m^2/K$, SHGC = 0.42, and visible transmittance = 0.38. In both case, the advanced window is triple glazed with $U = 1.14W/m^2/K$, SHGC = 0.22, and visible transmittance = 0.37
Source: Data from Carmody et al (2004)

Figures 3.25 and 3.26 understate the full potential for saving energy with advanced windows for a number of reasons:

- the relative energy savings are based on total energy use, rather than heating + cooling + lighting energy use only, which dilutes the relative savings;

- the windows were not optimized based on orientation (for example, a window with a very small SHGC, 0.2, is used for the north-facing window in the cold climate – the same is assumed for south-facing windows in a hot climate);
- fixed shadings (external overhangs or fins) rather than adjustable shading are assumed;

Figure 3.26 Impact on total energy use of increasing the window: wall ratio on north-, south-, east- and west-facing walls using the advanced window of Figure 3.18, and taking full advantage of daylighting to reduce the need for artificial lighting

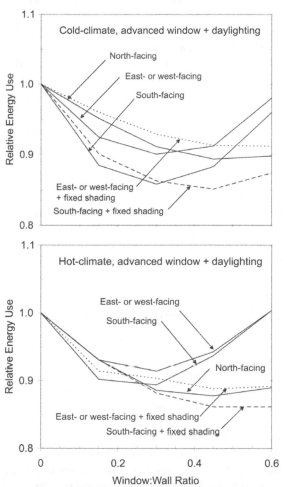

Source: Data from Carmody et al (2004)

• electrochromic windows (discussed later) were not considered.

3.3.12 Windows as net heat sources in cold winter climates

The Canadian Standards Association has developed a single index called the *energy rating* (ER) that combines the effect of usable solar heat gain through the window and the non-solar heat loss (by conduction, emission of infrared radiation, and infiltration of air) (Dubrous and Wilson, 1992; Carpenter et al, 1998). The ER is equal

to the average net heat gain (W/m^2) through a window over the heating season. Not all of the solar heat gain is usable, because at times heat may be absorbed while the room temperature has already reached the thermostat setting. The usable fraction will be smaller if the building heating load is smaller (due to better insulation, for example) and will be larger if there is thermal mass for storing heat with minimal temperature rise. The infiltration heat loss through the window depends on wind conditions but also on house geometry and air tightness in the upper part of the house, as these determine stack-effect infiltration flows (see Section 3.7). In computing the ER for a given window, average climate and sunshine conditions and typical house characteristics in southern Canada are used, and the results are averaged over windows facing the four cardinal directions.

Some representative ratings are given in Table 3.13. A negative rating means that the window loses heat. The best fixed windows serve as a season-averaged heat source (ER > 0) for average conditions and orientation, although not when facing north. Thus, for residential buildings in cold climates, northern-oriented glazing should be kept to a minimum. However, in commercial buildings, the energy value of the admitted daylight will exceed the heat energy lost through high-performance windows, so large glazing areas on north-facing façades can be justified as long as high-performance windows are used and the lighting system is designed to take advantage of daylighting.

Equatorward-, east-, and west-facing high-performance windows are not only a net heat source averaged over the course of the heating season, they can also be a net heat source on the coldest days of the year, given that the coldest days tend to occur under calm sunny conditions, created by stagnant Arctic air masses. As shown in Chapter 4 (Figure 4.1), the minimum (late December) daily average clear-sky solar irradiance on east- or west-facing walls is about 50W/m² at 50°N. With an SHGC of 0.5, a U-value of 0.6–0.8 W/m²/K, an outside temperature of −20°C and an inside temperature 20°C, solar heat gain will balance or almost balance conductive heat loss over a 24-hour period. If an insulating blind (RSI 0.35) is lowered at night, solar heat gain will at least balance conductive heat loss from

Table 3.13 Typical window ER numbers. ER values depend on the average inside to outside temperature difference, the availability of sunlight and the window orientation. Values given here are for average southern Canadian and northern US conditions and are an average over windows facing all four cardinal directions. A positive ER value means that the window is a net heat source to the building

Window characteristics			ER value (W/m²)	
Glazing	Frame	Spacer	Operable	Fixed
Double	TB Alum.	Alum.	−38	−21
Double	Wood/vinyl	Alum.	−25	−15
+ low-e	Wood/vinyl	Alum.	−16	−4
+ low-e/argon	Wood/vinyl	Alum.	−12	+1
+ low-e/argon	Wood/vinyl	Insulated	−8	+5
Triple	Wood/vinyl	Insulated	−7	+2
+ 2 low-e/argon	Wood/vinyl	Insulated	−2	+11
Quadruple+ 2 low-e/argon	Wood/vinyl	Insulated	−5	+8

Note: TB = thermally broken.
Source: Carpenter et al (1998)

east- and west-facing windows on sunny −20°C days in December at 50°N.

3.3.13 Acoustical performance in relation to thermal properties

There are a number of issues involving the acoustical performance of windows in relation to their thermal properties. First, the sound attenuation of a triple-glazed window with two 8mm air gaps is less than that of a double-glazed window with a 20mm air gap, the difference being dependent on the details of the window's manufacture. Second, the sound insulation is relatively constant as the thickness of the air gap increases up to 50mm, increases sharply between a thickness of 50 and 100mm, and more slowly thereafter. In contrast, the optimal air gap thickness from a thermal point of view is 20mm or less. Third, there is negligible difference in the sound insulation of argon and air, but krypton and especially xenon give a marked reduction in sound transmission. These gases are also superior from a thermal point of view, but their use increases the embodied energy of the window (see Section 3.3.19). Further information on the acoustical performance of windows is found in Muneer et al (2000).

3.3.14 Impact of insect screens

Insect screens reduce the window U-value by about $0.2W/m^2/K$ if placed on the outside of the window, and by about $0.4W/m^2/K$ if placed on the inside of the window, given a base value of $2.8W/m^2/K$ (Brunger et al, 1999). This reduction arises through the effect of a screen on the laminar-boundary heat-transfer coefficient (see Box 3.2). Insect screens also reduce the SHGC, by almost half if placed on the outside but by only one-third if placed on the inside (as heat absorbed by the screen is more readily radiated to the inside in this case). During the heating season, the net effect is to reduce the ER rating (i.e. increase the heating requirement) for outdoor screens and to increase the ER rating (i.e. reduce the heating requirement) for indoor screens. In order to maximize passive solar heating, outdoor screens should be removed when the heating season begins.

3.3.15 Electrochromic windows

In an *electrochromic* (EC) *window*, a small voltage (1–5 volts) causes the window to change from a clear to a transparent coloured state, or vice versa. EC windows are also referred to as *smart windows*. Figure 3.27 compares the solar transmission by smart windows in their two extreme states, as well as for two intermediate states. Also given is the voltage required to switch the window from the bleached state to each of the other states. In some EC windows, the transmission drifts less than 1–2 per cent over several days without a new applied voltage. In others, the window must be pulsed a few times per day in order to maintain a given state. In the latter case, electricity use amounts to about 2kWh/year for a 50 × 100cm window (Lee et al, 2002a). Electrochromic glazings have been the subject of research for over 30 years, with numerous demonstrations and ongoing field test programmes in Japan, Europe and the USA. Electrochromic glazing is available in the US through Sage Electrochromics (www.sage-ec.com) and from the German firm Flabeg (www.flabeg.de). The Stadt Sparkasse bank in Dresden, built in 1999, is the world's first operational electrochomic building.

An electrochromic device contains five layers, as illustrated in Figure 3.28. The middle

Figure 3.27 Comparison of the transmission of solar radiation through a smart window in its fully bleached (upper curve) and fully coloured (lower curve) states, as well as for two intermediate states

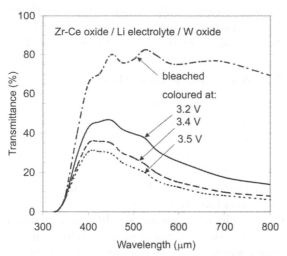

Source: Granqvist et al (1998)

layer is an electrolyte (or ion conductor). On one side is an ion-storage film and on the other side is the electrochromic film. Next to each of these is a transparent conductor. When a voltage is applied between the outer layers, ions travel from the electrochromic film to the ion conductor or vice versa, depending on the polarity of the applied voltage. The most commonly used electrochromic material is tungsten oxide (WO_3); the ion-storage medium can be made of niobium oxide (NbO_3), titanium oxide (TiO_3), vanadium pentoxide (V_2O_5), zirconium oxide (ZrO_2), or cesium oxide (CeO_2); the ion conductor can be made of lithium borate ($LiBO_2$) or lithium aluminium fluoride ($LiAlF_4$); and the transparent conductor can be made of indium tin oxide (ITO). The entire assembly is about 4μm thick (1/50th the thickness of a human hair).

The window can be operated to change from one state to another at a pre-programmed solar intensity, to change gradually from one state to another as the solar intensity varies from one pre-programmed value to another, or can be fully

Figure 3.28 Cross section of an electrochromic coating for smart windows

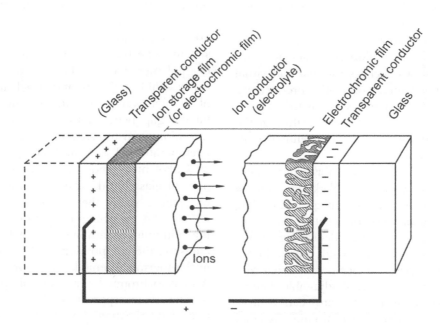

Source: Granqvist et al (1998)

integrated with HVAC and lighting control systems (Lee et al, 2002a; Gugliermetti and Bisegna, 2003).[6] In this way, the window can be tailored to specific climatic regimes (where the relationship between temperature and solar radiation will be different), and even to specific directional orientations. Its properties will frequently vary diurnally. In climates where heating rather than air conditioning is required part of the year, the window can be switched to and left in its transparent state during the heating season. This avoids the trade-off, required at present, between choosing windows with a high solar heat gain to minimize winter heating requirements, or windows with a low solar heat gain in order to minimize summer air conditioning requirements. The time required to switch an EC window from one extreme state to another depends on the area of the window and on temperature. For a 50 × 100 cm window at 5°C, 5 minutes are required, while for a 60 × 175cm window at 21°C, 25 minutes are required (Lee et al, 2002a).

It is expected that significant time will be required to commission and debug EC windows. Cost was about $1000/m² in 2000 but was expected to drop to $100/m² with large-scale manufacturing facilities, not including control systems (Lee et al, 2000). The economic viability of EC windows depends on energy and peak demand costs; heating and cooling system sizing; design, maintenance and operation costs; and less tangible considerations such as thermal and visual comfort and their impact on worker productivity. EC windows will probably serve as the exterior glazing layer, to accommodate wiring through the framing, so replacement of dysfunctional units will occur from the outside.

Detailed simulations for office buildings with the climate of New York state indicate that a savings in combined lighting and cooling electricity use of up to 60 per cent is possible using electrochromic windows, depending on the building characteristics and window area (Lee et al, 2002b). High-performance conventional windows combined with external adjustable shading (including double-façade systems) can achieve comparable performance, perhaps more reliably. Direct measurements of electrochromic skylights indicate a reduction in the SHGC from 0.3–0.5 to 0.1–0.2 (depending on zenith angle) when the skylight is switched from the transparent to opaque states (Klems, 2001). In its opaque state, EC windows may still require interior or exterior shading to prevent glare when the sun is shining directly on the window, depending on the task and the position of the task or eye (Lee et al, 2000).

3.3.16 Thermochromic glazing

Another technology currently under development is thermochromic glazing, which changes (reversibly) from white at higher temperatures to clear at lower temperatures. This eliminates the need for sensors and controls. It could be employed in skylights to automatically permit penetration of solar radiation when heating is desired (i.e. when the outside temperature is cold) but not when cooling is desired. Solar transmissivity switches from approximately 15 per cent to about 70 per cent (Inoue, 2003).

Thermochromic materials involve two components that, below a given temperature, are completely mixed and transparent. Above this temperature, one of the components separates from the other and forms microparticles with a diameter comparable to the wavelength of light and with an index of refraction different from the other component. As a result, scattering of direct beam solar radiation occurs (Goetzberger et al, 2000). There are two kinds of thermochromic systems, one based on hydrogels and the other on polymer blends. The former involves a water-soluble polymer and a gel and, because of its high water content, can only be used to fill the gap between two panes of glass. The transition from transparent to opaque begins at a temperature of 27°C and is largely complete at 30°C. Polymer blends can be coated onto a single pane of glass, but the transition from translucent to opaque spans the temperature range 27–70°C. The switching temperature as well as the minimum transmittance can be altered by adjusting the composition of the materials used in the thermochromic layer (Georg et al, 1998). A thermochromic layer has been applied to the outside of transparent insulation and used to reduce overheating during transitional seasons when some solar heat gain is still desired (Raicu et al, 2002).

3.3.17 Silica aerogel spacers and layers

Solid silica aerogel insulation has been developed for use in skylights and wall panels, as discussed in Section 3.2.8. It is also being developed for window spacers and as a transparent insulating material between window panes. At normal atmospheric pressure, it has a thermal conductivity of 0.015–0.020W/m/K, which is comparable to that of argon, but at a moderate vacuum (P < 5000Pa or 0.05 atmospheres), the conductivity drops to less than 0.01W/m/K, which is comparable to that of krypton (see Table 3.8). Windows with aerogel between the window panes can be designed to have a very small U-value and either a small or large SHGC. As noted in Section 3.2.8, aerogel can be manufactured in either granular or monolithic form.

Reim et al (2002) present data on prototype windows consisting of a 16mm layer of granular aerogel between two thin glass skins, flanked by a 12mm Ar/air or Kr/air gas layer on either side, and finally inner and outer glass panes with a low-e coating on the inside surface of each pane. The solar radiation passing through the window is diffuse and slightly yellowish, so there is no view of the outside. For the Ar/air case, the window achieved (U-value = 0.56W/m²/K, DT = 0.26, SHGC = 0.31) using low-e coatings with ε = 0.08, and (U-value = 0.47 W/m²/K, DT = 0.27, SHGC = 0.20) using low-e coatings with ε = 0.03.

Jensen et al (2004) and Schultz et al (2005) report the development of prototype double-glazed windows using 15mm thick monolithic aerogel in a vacuum between the glazings. Without low-e coatings, the centre-of-glass U-value is 0.41W/m²/K and the overall U-value (centre + edge areas, but excluding the frame) using a specially designed rim seal is 0.6W/m²/K for a 20 × 20cm window and less for larger windows. This is as good as the best triple-glazed windows, but due to the absence of low-e coatings and the use of only two glazings, SHGC = 0.63 (compared to 0.45 for a triple-glazed window with a comparable U-value). With the use of low-iron glass and an anti-reflective coating, SHGC = 0.76. There is almost no distortion of the view through the window, as illustrated in Figure 3.29.

Silica aerogel is even more advantageous than the alternatives if used as a framing

Figure 3.29 View through a double-glazed window with monolithic silicon aerogel between the glazings. The centre of glass U-value is 0.41W/m²/K

Source: Jensen et al (2004)

material, inasmuch as the thermal conductivity of non-evacuated aerogel is four to five times less than that of the best framing material currently used in curtain walls, described in the next paragraphs.

3.3.18 Curtain walls in commercial buildings

A curtain wall is a wall on the exterior of a building that carries no roof or floor loads. It commonly consists entirely or largely of glass and other materials supported by a metal framework, although precast concrete panels have also been used. The frame is referred to as a *mullion*, while the opaque panels between the glazed portions are referred to as the *spandrel*. Mullions can be provided with or without thermal breaks. In the most effective designs, most of the metal frame is on the interior with only a metal cap exposed to the exterior. Table 3.14 gives frame U-values for curtain wall construction. These range from 16.8W/ m²/K for a double-glazed curtain wall with an aluminium frame without a thermal break, to 4.3W/m²/K for a triple-glazed curtain wall using recently available insulated framing systems. For a typical module unit with a 3650mm floor to floor height, a 1525mm width, window head and sill heights above the finished floor of 2750mm and 915mm, respectively, and a 65mm wide mullion, the frame will account for 8–12 per cent of the wall area, depending on the mullion pattern.

Table 3.14 U-values (W/m²/K) of different frames used in curtain walls

Frame material	Type of spacer	Glazing system		
		Single[a]	Double[b]	Triple[c]
Aluminium without thermal break	All	17.1	16.8	16.1
Aluminium with thermal break[d]	Metal	10.2	9.9	9.4
	Insulated	n/a	9.3	8.6
Structural glazing	Metal	10.2	7.2	5.9
	Insulated	n/a	5.8	4.3

[a] 6.4mm glazing unit thickness
[b] 25.4mm glazing unit thickness
[c] 44.4mm glazing unit thickness
[d] Depends strongly on width of thermal break. Value given is for 9.4mm.
Source: ASHRAE (2001b, Chapter 30)

Hence, the large U-values for the frame area are a significant factor in the overall U-value.

Double-glazed and triple-glazed curtain walls are commercially available, with overall U-values (including the frame) for the 7500 Kawneer series as low as 1.8W/m²/K and 0.8W/m²/K, respectively, for large units (see www.kawneer.com). The triple-glazed units represent a 70 per cent reduction in heat loss compared to the glazing value of 2.6W/m²/K that is permitted under the ASHRAE 90.1-2004 commercial building code in moderately cold climates (see Table 13.3). These entail thermal breaks and insulation covering the shoulder of the mullion, thereby reducing the otherwise substantial heat loss through the mullion, so the overall U-value is dependent on the details of the construction. The spandrel portion can achieve U-values of 0.40W/m²/K. In high-performance curtain walls, the spandrel will contain additional insulation on the inside. As noted in Section 3.2.7, vacuum insulation panels are available for the spandrel with U-values of 0.08–0.16W/m²/K. These U-values are a factor of 3–8 smaller than permitted under the ASHRAE 90.1-2004 building code.

3.3.19 Comparison of energy savings and energy cost

With the exception of the frame, the production of high-performance windows requires greater energy than for conventional windows, due to the extraction and processing of the additional raw materials used, and in some cases also during the manufacturing process. This leads to a trade-off between reduced heating energy use and increased embodied energy. High-performance window frames, on the other hand, generally entail both reduced operating energy and reduced embodied energy.

Non-frame window components

Table 3.15 compares the extra primary energy used in adding various energy-saving features to a 1.1 × 1.1m window with the primary energy saved by that feature over a 20-year period, which can be taken as the minimum likely lifespan of the window. The primary energy used to produce a given product is referred to as the *embodied energy*, and is subject to considerable uncertainty, as discussed in Chapter 12. The energy savings are determined from the differences in U-values as computed using the equations and assumptions given in Box 3.2, and take into account edge-effects, which dilute the beneficial impact of lower centre-of-glass U-values. The primary-energy savings are computed from the difference in heating load, assuming a climate with 3000 heating degree days (HDD; see Table 2.5 for typical values), divided by an assumed efficiency of the heating system of 0.9. Not included is increased heating energy use due to decreased SHGC when extra glazings or low-e coatings are added, or reductions in cooling energy use. The energy savings of a given feature depend on which features are already present, so results are given in Table 3.15 in going from one to two, and from two to three panes of glass, as well as for adding a low-e coating to a double-glazed window with no low-e coatings and to a triple-glazed window with two coatings already.

The addition of an extra pane of glass to the 1.1 × 1.1m window increases the energy required to make the window by 242MJ, but saves about 15,800MJ over a 20-year period if it is the second pane of glass and by 5300MJ if it is the third pane of glass. Addition of a single low-e coating requires 8.4MJ to extract and process the materials used in the coating, but negligible energy for the window manufacturing process. The 20-year energy savings is 4844MJ for the first low-e coating added to a double-glazed window, and 394MJ for the third coating added to a triple-glazed

Table 3.15 Incremental primary energy required to produce 1.1 × 1.1m windows (1.1m² glazed area + frame) with the designated energy-saving feature, and annual and 20-year primary energy savings

Energy-saving measure	Cases compared	Energy required (MJ)	Decrease in U-value (W/m²/K)	Energy savings (MJ)	
				Annual	20-year
Change from SG to DG	1 vs 2	242[a]	2.492	789.4	15,788
Change from DG to TG	2 vs 22	242[a]	0.841	266.4	5327
Add 1st hard low-e coating to DG	2 vs 3	8.4	0.764	242.2	4844
Add 2nd hard low-e coating to DG	3 vs 5	8.4	0.159	50.3	1006
Add 1st soft low-e coating to DG	2 vs 7	8.4	1.074	340.4	6808
Add 2nd soft low-e coating to DG	7 vs 8	8.4	0.043	13.6	272
Add hard low-e after soft low-e in DG	8 vs 9	8.4	0.162	51.3	1025
Add 3rd low-e (hard) to TG	25 vs 26	8.4	0.062	19.7	394
Argon instead of air in DG, clear	2 vs 10	0.012	0.114	36.0	720
1S, 1H	9 vs 12	0.012	0.194	61.4	1227
Krypton instead of air in DG, clear	2 vs 13	508	0.194	61.6	1231
1S, 1H	9 vs 15	508	0.346	109.6	2192
Krypton instead of air in TG, 2S, 1H	30 vs 26	508	0.136	43.0	860
Xenon instead of air in DG, clear	2 vs 16	4500	0.245	77.8	1555
1S, 1H	9 vs 18	4500	0.451	142.9	2859
Xenon instead of air in TG, 2S, 1H	32 vs 26	4500	0.183	58.0	1160
DG aerogel instead of DG, clear		685[b] 525[c]	2.045	647.7	12,954

[a] Based on half the glazing embodied energy and half the sash and frame embodied energy of a DG unit
[b] Primary energy from Schramm et al (1994), assuming that electricity is used in the production process
[c] Primary energy from Schramm et al (1994), assuming that fuels are directly used in the production process
DG = double-glazed, TG = triple-glazed. The DG window without low-e coatings or gas fill and with an Al-clad wood frame has a total embodied energy of about 1460MJ.
Source: Incremental production energies are from Muneer et al (2000) except where indicated otherwise, while energy savings are based on differences in glazing averaged U-values (centre and edge regions), assuming 3000HDD and a heating efficiency of 0.9. Case numbers are cases from Table 3.9. See text for further assumptions

window. Not included in this is the reduction in solar heat gain with each extra low-e coating, which adds to the winter heating requirement but subtracts from the summer cooling requirement. For atrium windows in the relatively cool climate of Ottawa (Canada), this trade-off gives an additional net energy savings (Laouadi et al, 2002).

The energy savings when inert gas fills are used are substantially greater if the fill is added to a window that already has low-e coatings. For double-glazed windows with one soft and one hard low-e coating, use of argon in place of air requires an extra 12kJ but saves 1227MJ over a period of 20 years, a payback factor of 84,000. Use of krypton instead of air requires an extra 508MJ but saves only 2200MJ over 20 years, while use of xenon instead of air requires 4.5GJ but saves only 2.86GJ over 20 years. Thus, except in very extreme climates, it would appear that the

use of xenon does not lead to a net energy saving, and that for krypton, about 5 years of energy savings are required before the initial energy input is paid back in a climate with 3000HDD.

However, this analysis does not account for the likely ability to eliminate perimeter heating radiators using high-performance windows (as discussed in Section 3.3.20). As radiators are made of steel or aluminium, both of which have a large embodied energy (see Chapter 12, Table 12.1), elimination of perimeter radiators gives a large reduction in embodied energy that can be credited against the embodied energy of the window. Use of krypton in a 2m high window would entail an extra 706MJ of embodied energy per metre of window width. A finned perimeter radiator has about 135 10 × 10cm aluminium fins per metre of radiator, a 3.8cm diameter hole to accommodate the copper hot-water pipe, and a flange between each fin that is wrapped around the pipe. Assuming a fin thickness of 0.5mm and that the pipe itself has 3mm thick walls, and given densities of aluminium and copper of $2.7\text{gm}/\text{cm}^3$ and $8.92\text{gm}/\text{cm}^3$ (both taken from www.wikipedia.org), the total masses of aluminium and copper per metre of radiator are 1.75kg and 3.20kg, respectively. Assuming primary embodied energies of 230MJ/kg and 120MJ/kg for aluminium and copper, respectively (from the very wide range given in Table 12.1), the embodied energy per metre of radiator is about 950MJ – an estimate that is probably uncertain to at least ±50 per cent. Nevertheless, we come to the important conclusion that, if the use of a krypton fill improves the window U-value to the point that perimeter radiators can be eliminated, then the savings in the embodied energy of the radiators is comparable to the embodied energy in the krypton fill, and is a significant fraction of the 20-year savings in heating energy for moderately cold climates. This justifies the use of krypton fill. Conversely, if argon-filled windows are sufficient to eliminate perimeter radiators, then the saved radiator embodied energy cannot be credited against the krypton embodied energy if the windows are upgraded to krypton fill. A more detailed, region-specific analysis should be carried out in order to assess the trade-off in specific locations, taking into account differences in the embodied energy of krypton and of the radiators or other materials that can be avoided through high-performance windows.

As noted in Section 3.2.8, silica aerogel is produced by heating precursor materials to 280°C at a pressure of 90atm, an energy-intensive process. Schramm et al (1994) estimated that about 70GJ/tonne of primary energy are used in the production of granular silica aerogels, 275GJ/tonne for monolithic aerosol if electricity is used, and 209GJ/tonne if fuels are used directly. For a 15mm thick monolithic aerogol layer (density: $151\text{kg}/\text{m}^3$) in double-glazed windows, the incremental energy required for a window with a 1.1m^2 glazing area is 525–685MJ (depending on whether fuels or electricity are used). However, the reduction in U-value is comparable to that in going from single- to double-glazed and much larger than any other single measure, so the energy payback is short – less than two years.

Window frames

As shown in Table 3.14, aluminium frames have the highest conductivity and hence the greatest heat loss of any framing option. Aluminium is also one of the most energy-intensive industrial products produced (see Chapter 12, Table 12.1), so replacement of an aluminium frame with a fibreglass frame reduces both the embodied energy in the window and the heat lost through the window during its use. An intermediate option is an aluminium-clad wood frame. According to Asif et al (2001), a conventional 1.2 × 1.2m argon-filled, aluminium-clad, wood-frame window requires about 4.6kg of aluminium (3.91kg for the cladding, 0.72kg for other aluminium parts). The total energy required to produce this window in the UK is about 1460MJ, of which about half (835MJ) is for the aluminium. In a conventional wood-frame window without cladding, 1.45kg of aluminium are used in the form of aluminium strips that are not needed in the aluminium-clad window, and 0.15kg of powder coating is not needed, so the amount of aluminium used is reduced by 2.61kg (63 per cent) and the embodied energy is reduced by only 471MJ. Furthermore, some alternative wood finishes are energy intensive, so there may be little difference in the embodied energy of windows with wood frames and aluminium-clad wood frames.

More recently, Asif et al (2005) have estimated the embodied energy for windows with aluminium, polyvinyl chloride, aluminium-clad timber and timber frames. Table 3.16 compares the frame and total estimated embodied energy for a standard double-glazed window with these as well as insulated aluminium and fibreglass frames, excluding possible wood finishes in the case of wood-frame windows. The embodied energy of an aluminium frame – about 5500MJ – is roughly one-third of the savings in operating energy over 20 years in going from a standard double-glazed window to the best window shown in Table 3.15.

3.3.20 Capital cost premium and net capital cost premium of high-performance windows

Quotes of the additional cost of high-performance windows are highly erratic, reflecting an

Table 3.16 Comparison of frame U-values

Frame type	Mass of Al (kg)	U-value (W/m²/K)	Embodied energy (MJ)					Lifespan (years)
			Al	Wood	PVC/FG	Other	Total	
Aluminium	27.8	12.9	5041	0	0	429	5470	50
Insulated Al	19.4	5.2	3529	0	0	429	3958	50
PVC	2.0	2.9	363	0	1357	429	2149	25
Al-clad wood	4.6	2.8	835	195	0	429	1459	50
Wood	2.0	2.8	363	195	0	429	987	50
Fibreglass (FG)	2.0	1.9	363	0	180	429	980	50

Note: U-values are taken from Table B3.2.3, with embodied energies for a clear 1.2 × 1.2m window (0.23m² frame area) with air fill, estimated from or given by Asif et al (2001, 2005), except for fibreglass frames, which are estimated using an energy intensity of 30MJ/kg for fibreglass (taken from Table 12.6), an assumed bulk density of 500kg/m³, and a frame thickness of 5cm. The insulated Al frame is arbitrarily assumed to have 70 per cent the Al of the uninsulated Al frame. PVC = polyvinyl chloride.

immature market in many places and the perception, among window manufacturers, that those seeking high-performance windows are prepared to pay a high price for environmental reasons. Over time, more consistent – and lower – prices should prevail. Table 3.17 provides information on the cost of window upgrades in three locations. Costs given for Toronto are the cost of window glazings to window manufacturers. These costs are substantially less than the cost of the final product (which includes the cost of the frame, assembly, marketing and profit). Thus, the cost of final window products depends in part on the negotiating skill and influence of the potential buyer.

Table 3.18 compares the costs of various window upgrades to window manufacturers, as given in Table 3.17, with the annual savings in heating energy use, as given in Table 3.15 for a climate with 3000HDD. Also given is the annual savings in heating energy cost, assuming a fuel cost of $12/GJ. As seen from Table 3.18, an upgrade from double- to triple-glazing, use of argon fill and the addition of one low-e coating are readily justified by the energy cost savings, although costs to consumers will be greater than indicated here. The use of more than one low-e coating, or of krypton fill, cannot be justified based on heating energy savings. However, additional low-e coatings provide savings in seasonal and peak cooling loads that might justify these upgrades purely in terms of energy-cost savings.

In addition, there is the opportunity with high-performance windows to offset part of the extra cost through avoided perimeter heating units and downsized heating and cooling equipment, ductwork, transformers and electrical

Table 3.17 Costs of windows or window components

Window or window component	Local currency	US$
Costs in Toronto, provided to the author by a supplier to window companies		
Costs to window manufacturers, Toronto		
Integrated DG unit	$40–50/m²	$34–43/m²
Hard low-e coating	$10/m²	$8.5/m²
Soft low-e coating	$20/m²	$17/m²
Argon fill	$1.5–2.0/m²	$1.3–1.7/m²
Krypton fill	$10–15/m²	$8.5–13/m²
Heat mirror film	$70/m²	$60/m²
Selling cost of window (includes frame, assembly, retail markup)		
Basic DG	$250–300/m²	$212–255/m²
Retail costs in the UK (Menzies and Wherrett, 2005)		
Extra glazing	9.5 £/m²	$16/m²
Low-e coating	11 £/m²	$19/m²
Argon fill	4.3 £/m²	$7.3/m²
Krypton fill	29 £/m²	$49/m²
Minimum costs in Switzerland (Jakob, 2006), according to whole-window U-value		
1.4 W/m²/K, 3 m² window	500 CHF/m²	$468/m²
0.85 W/m²/K, 3 m² window	600 CHF/m²	$561/m²
0.75 W/m²/K, 3 m² window	800 CHF/m²	$748/m²
0.65 W/m²/K, 5 m² window	700 CHF/m²	$655/m²

Table 3.18 Incremental costs, annual energy savings, and heating energy cost savings only, for various upgrades to a residential window with a 1.1m² glazed area

Window measure	Cases compared	Annual energy savings (MJ)	Cost (US$)	Annual energy cost savings
TG instead of DG	2 vs 22	266.4	$50	$3.20
Add 1st soft low-e coating to DG	2 vs 7	340.4	$20	$4.08
Add 1st hard low-e coating to DG	2 vs 3	242.2	$10	$2.93
Add 2nd hard low-e coating to DG	3 vs 5	50.3	$10	$0.60
Add 3rd low-e (hard) to TG	25 vs 26	19.7	$10	$0.24
Argon in DG	9 vs 12	61.4	$2	$0.74
Krypton in DG	9 vs 15	109.6	$10–15	$1.32
Krypton in TG	30 vs 26	43.0	$10–15	$0.52

Note: Costs are the cost of the upgrade to window manufacturers in Toronto, energy savings are from Table 3.15 and pertain to a climate with 3000HDD and a heating system efficiency of 0.9, and energy cost savings are for an energy cost of $12/GJ. Case numbers are cases from Table 3.9. The results given can be easily modified by the reader to reflect different climatic and economic conditions.

Figure 3.30 Window U-values below which perimeter heating is not needed for a 2m high window, as a function of the winter design temperature

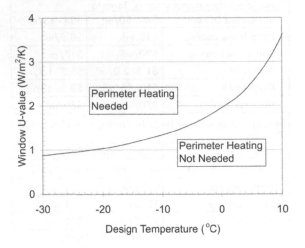

Source: Geoff McDonell, Omicron Consulting, Vancouver

Table 3.19 Comparison of component costs for a building with a conventional VAV mechanical system and conventional (double-glazed, low-e) windows with those for a building with radiant slab heating and cooling and high-performance (triple-glazed, low-e, argon-filled) windows, assuming a 50 per cent glazing area/wall area ratio

Building component	Conventional building	High-performance building
Glazing	$140/m²	$190/m²
Mechanical system	$220/m²	$140/m²
Electrical system	$160/m²	$150/m²
Tenant finishings	$100/m²	$70/m²
Floor to floor height	4.0 m	3.5 m
Total	$620/m²	$550/m²
Energy use	180kWh/m²/year	100kWh/m²/year

Note: Costs are in 2001 Canadian dollars for the Vancouver market in 2001, are given per m² of floor area, and are based on fully costed and built examples over a 3-year period.
Source: Geoff McDonell (Omicron Consulting, Vancouver), personal communication, December 2004, and McDonell (2003)

connections due to the reduced peak heating and cooling loads that occur with high-performance windows. Figure 3.30 shows the window U-value below which perimeter heating is not needed, for 2m high windows, as a function of the coldest winter temperature for which the heating system is designed. Even for a design temperature of −30°C, windows are available that eliminate the need for perimeter heating. When perimeter heating is eliminated, ductwork or hot-water piping can be made shorter, as all the radiators can be located closer to the central core of the

building, with associated cost savings but also savings in fan and pump size and energy use. If the default design involves floor mounted fan-coil units, their elimination will increase the amount of usable floor space. The new courthouse in Denver (US) is an example of a building where triple-glazed windows were used to eliminate perimeter heating in public corridors (Mendler and Odell, 2000). High-performance windows, by reducing peak heating and cooling loads, allow the use of radiant heating and cooling (described in Chapter 7, Section 7.4.4). Table 3.19 gives a breakdown of capital costs for a commercial building with conventional windows (double-glazed, air-filled, low-e with $U = 2.7\text{W/m}^2/\text{K}$ and SHGC = 0.48) and a conventional heating/cooling system, and for a building with moderately high-performance windows (triple-glazed, low-e, argon-filled with $U = 1.4\text{W/m}^2/\text{K}$ and SHGC = 0.24) and radiant-slab heating and cooling. The high-performance building is 9 per cent less expensive than a comparable conventional building, and has just over half the energy use.

3.4 Double-skin façades

A *double-skin façade* is a façade with an inner and outer wall separated by an air space that is not actively heated or cooled. The outer façade consists of a single- or double-glazed glass wall with fixed or adjustable openings, and the inner façade may also consist of a single-, doubled- or triple-glazed glass wall with operable windows, or may be partially opaque. Here, we provide information on the construction, general benefits and costs of double-skin façades. More specific information concerning their heating and cooling loads is given in Chapters 4 (Section 4.1.8) and 6 (Section 6.3.2), respectively.

3.4.1 Types of construction and examples

Oesterle et al (2001) identify the following types of double-skin façades:

- box windows;
- shaft-box façades;
- corridor façades;
- multi-storey façades.

Box windows

These consist of a frame with inward-opening casements. A single-glazed external skin has openings at the bottom and top for the ingress and egress of outside air. Airflow from one box window to another (i.e. from floor to floor or laterally from office to office) does not occur, thereby inhibiting the transmission of sound and smells. Construction details are given in Figure 3.31. These are similar to airflow windows (described in Chapter 4, Section 4.1.7). They can be produced in a factory and lifted into place by crane.

Shaft-box façades

Columns of box windows alternate with vertical shafts and are linked through lateral openings at the top of each box window. The stack effect (described in Section 3.7) draws external air into the bottom of each box window, through the

Figure 3.31 Construction of a box window type double-skin façade

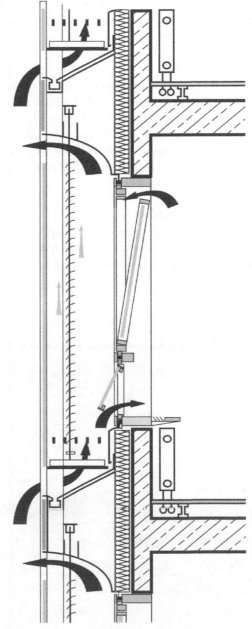

Source: Oesterle et al (2001)

adjacent room if the windows are open, and into and up the shaft. This design requires fewer openings in the external skin, reducing the penetration of outside noise.

Corridor façades

The space inside the façade of a given floor is separated from that of the floors above and below by a partition, but extends without interruption from one side of the façade to the other (vertical partitions will usually be necessary at corners due to large differences in air pressure, and where openings in the inner façade would create uncomfortable drafts from cross currents). Air intakes and outlets occur near the floor and ceiling of each floor, respectively, but to prevent exhaust air from one floor entering the intake of the floor above, these can be staggered from one floor to the next, as in the Centre for Cellular and Biomolecular Research at the University of Toronto, illustrated in Figure 3.32. This will also reduce the spread of smoke in the event of fire.

Multi-storey façades

These are particularly useful when external noise levels are high, as few openings are required, and they can be used as the supply duct for rooms behind the façade that need to be mechanically ventilated. As with corridor façades, the transmission of sound from one office to the next may require attention. The temperature in the façade gap will increase with increasing height, thereby increasing the summer cooling load in upper floors. This requires subdividing the façade gap every few floors in high-rise buildings.

Construction

Motorized louvres or other shading devices are usually placed just inside the outer façade. These will reflect but also absorb sunlight during the summer that would otherwise cause overheating inside the building. The absorption of the sunlight causes the air between the two façades to rise and exit out the top of the façade, while drawing in cooler air from the outside at the base of the façade. The motorized louvres can be programmed to optimize the balance between daylighting (when the adjacent room is occupied, as determined by occupancy sensors) and reduction of the cooling requirements in summer. During winter, the heated air between the two façades can be directed to the inside of the building. The double-skin façade thus serves to preheat the ventilation air. Because the space between the two skins is not actively heated, a larger portion of the solar heat gain can be used than would

Figure 3.32 The corridor double-skin façade at the Centre for Cellular and Biomolecular Research at the University of Toronto

Photographer: Sandy Kiang, Toronto

otherwise be the case. In corridor and multi-storey double-skin façades, the space between the two façades may be as much as 60–100cm wide, with a grille walkway that provides partial shade on the inner windows of the floor below, and access doors from the interior of the building. In some cases, the space has been wide enough to contain stairways.

Condensation on the inner surface of the outer skin can be a problem if (i) the office air is humid and the inner windows are open, and (ii) the outer skin is closed and outside temperatures are low. The solution to this problem is to make sure that inner windows are closed when outside conditions dictate that the outer skin is closed. Exit and entrance slots must be designed to prevent entry of rainwater while also minimizing turbulence, which reduces the effective opening area and hence the rate of passive ventilation. Two examples of such construction are given in Figures 3.33 and 3.34. Figure 3.33 shows a 'fish-mouth' slot, which serves as either an intake vent (in which case the upper curved cover between the two façades is perforated) or an outlet vent (in which case the lower curved cover is perforated).

Most modern examples of double-skin buildings are found in Europe, particularly in Germany, and are amply illustrated by Herzog (1996), Oesterle et al (2001), Wigginton and Harris (2002) and Pasquay (2004). Two early notable examples are the Victoria Ensemble building in Cologne and the Daimler Chrysler Building on Potsdamer Platz in Berlin (the Debis Building), both completed in 1996 and having multi-storey façades. The Victoria Ensemble is illustrated in Figure 3.35 and discussed in Oesterlie et al (2001). It has continuous air inlets at the base of the building and outlets at the top of the façade. The inlets and outlets have automatic flaps that open or close based on outdoor temperature. The Debis Building is illustrated in Figure 3.36 and is discussed by Renzo Piano Building Workshop (1996) and Grut (2003). In the event of fire, all sunblinds in the cavity will be pulled up to reduce the availability of combustible material, and the entire outer façade will open to allow heat to dissipate. Initial analyses indicate that the building consumes 35 per cent of the electric lighting energy and 30 per cent of the heating and cooling energy of a typical naturally ventilated office block in Germany, which in turn is less than that of an air conditioned building. The resulting total energy use is about 75kWh/m²/year – an extraordinarily low energy use for a modern office building, as can be seen by comparison with the data shown in Figures 2.5 and 2.6.

Figure 3.33 Photograph of a mock up of a fish-mouth slot used in a double-skin façade (left) and schematic illustration of its integration into the building structure (right)

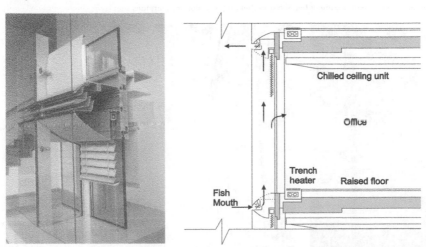

Source: Baird (2001). Reproduced by permission of Taylor & Francis Books UK

Figure 3.34 Construction of a corridor-type double-skin façade, showing the inlet and outlet areas designed to prevent the entry of animals and to minimize turbulence

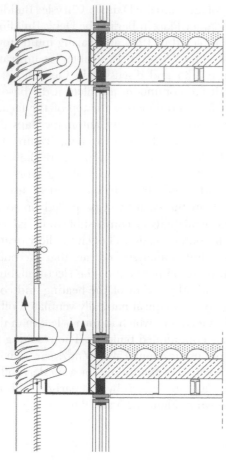

Source: Oesterle et al (2001)

Figure 3.36 The multi-storey double-skin façade on the Daimler-Chrysler (Debis) Building, Potsdamer Platz, Berlin

Source: Author

Figure 3.35 The corridor double-skin façade of the Victoria Building in Cologne, Germany

Architect: Thomas van den Valentyn, Cologne
Photographer: Rainer Mader, Cologne

3.4.2 Energy use and other benefits

The primary energy benefit of double-skin façades is their ability to reduce cooling and ventilation energy use through (i) external sun-shading; (ii) the opportunity to use natural ventilation when outside conditions permit (saving both cooling and ventilation energy use); and (iii) the opportunity to use night ventilation by addressing security concerns and preventing the entry of birds and insects.

All-glass façades have become very popular among architects and their clients. These provide plenty of daylight, but the issues are then to minimize winter heat loss and summer heat gain, and to avoid glare. With regard to winter heat gain, it will never be possible to match the insulative properties of a well-insulated wall with a glass façade. However, glass permits solar energy to enter the building, and for high-performance glazing systems, the glazing is a net heat source during the heating season for most climates and façade orientations (see Chapter 4, Section 4.1.2). This can be achieved without building a second façade over the first, so the double-skin façade does not provide a particular advantage with regard to winter heat loss. However, the main issue with any all-glass façade is to avoid overheating, and this is where the double-skin façade is advantageous (excess heat gain can even be a problem at times on cold but sunny winter days). Avoidance of overheating without excessively decreasing daylighting or rejecting solar heat when it is desired requires adjustable shading devices, but internal shading devices reduce the heat gain by only 50 per cent, compared with 90 per cent for external devices. External devices, especially on tall buildings, are subject to wind damage, and cleaning them is difficult. A major benefit of the double-skin façade is that it permits the installation of external, adjustable shading devices.

In the absence of daylight reflectors inside the double-skin façade, the presence of the façade will reduce the availability of daylight (by 16–20 per cent for regular glass at angles of incidence of 0–60°, and by 4 per cent at a 0° angle of incidence for low-iron glass with an anti-reflective coating) and thereby increase lighting-energy use if a daylight responsive lighting system is in place. This is likely to be a factor only under overcast conditions. So-called 'daylight louvre blinds' have been designed such that the louvres in the upper third of the blind are fixed at a flatter angle, so as to bounce some light off the ceiling when the blind is lowered at times of direct sunlight.

Double-skin façades are most appropriate on buildings subject to large external noise and wind loads, where external shading and natural ventilation would not otherwise be possible. External noise on a busy main road will be 70–75dB, which would preclude natural ventilation using operable windows in a single-skin façade. The recommended maximum sound level for mental work is 50–55dB, with some experts recommending even lower levels. The sound insulation of the external skin will be 3–6dB if the opening area is 10 per cent, and up to 10dB if the opening area is 5 per cent, although the latter may not permit adequate ventilation. A double-skin façade with 7dB of sound insulation would bring the total sound insulation with windows in the inner skin in a tipped position to 21dB, which makes natural ventilation possible. Too much sound insulation will amplify the disturbing effect of internal noises, as they will be less masked by background external noise. If the background noise level drops from 35dB to 30dB, a good-quality wall with sound insulation of 42dB will seem only as good as a wall with 37dB of sound insulation. Sounds from internal sources (especially high-frequency sounds with distracting information content) will be partially reflected back into the building for certain façade designs. This can be mitigated by using sound-absorbing materials within the façade.

Double-skin façades will reduce the effect of wind gusts but not that of steady winds. There should therefore be adequate flow paths between the windward and leeward sides of the building. In the extreme case of no flow pathways and only one door per floor, the entire pressure difference around the building will act on the door, making it difficult to open or causing it to slam in the face of users. At a 10m height, the proportion of time with a windspeed less than 0.5m/s (a 'windless' condition) is usually no more than 1–5 per cent annually.

A possible disadvantage of all-glass (or nearly all-glass) façades, even when they serve as net heat sources in winter and minimize summer heat gain, is the lack of thermal mass to dampen diurnal temperature variations. This can lead to the requirement for early morning heating followed by afternoon air conditioning, although this will be mitigated if the façade has a very low U-value with adjustable shading. Nevertheless, attention should be given to the opportunities to increase the exposed thermal mass of floors and internal partitions and of the possible need to make some portion of the façade non-glazed with greater thermal mass and thermal resistance.

Finally, the embodied energy associated with the additional glazing layers and the metal support structures should be compared with the expected savings in building energy use (see Chapter 12, Section 12.4.1).

3.4.3 Capital cost, net capital cost and payback time

The direct economic cost of double-skin façades depends on a number of factors, including:

- the type and thickness of glass (laminated safety glass ordered in large quantity costs 40–60€/m², compared with 25–40€/m² for standard 12mm glass of identical dimensions, not including possible additional costs for the supporting construction, with these and other costs pertaining to western Europe in 2000 and taken from Oesterle et al, 2001);
- the size (height, width) of individual glass panes;
- the size and type of openings (fixed slits, adjustable slits with flaps, or automatically rotatable glass louvres, the latter costing up to 1000€/m²);
- the type of glass fittings;
- the type of internal shading device (standard mass-produced louvred blinds could be ordered for around 75–100€/m² in 2000, while those with a built-in light-deflecting function for daylighting cost 200–375€/m²);
- the extent to which sound-absorbing material, if any, must be added in corridor and multi-storey type double-skin façades to reduce transmission of sound from one office to the next;
- the extent to which prefabricated, factory-built modules can be used in place of on-site assembly;
- the number of different façade types used in the project;
- the scale of the project.

In Europe, an increasing number of firms are offering prefabricated units (especially for box window and shaft-type double-skin façades) with integrated sunshading systems. These units can be sized and adjusted to fit the needs of a specific project, but this entails one-time retooling costs. Unit costs will be less the greater the number of units. For small projects, the retooling costs will not be justified, so on-site assembly will be more economical but relatively expensive per unit of façade area compared with large projects. Oesterle et al (2001) indicate an overall cost of double-skin façades, including sun-shading devices, ranging from 650 to 1500€/m² of façade area based on experience in Europe up to 2000, with incremental costs of 175–750€/m². They caution that there is a great risk in using lump-sum estimates without knowing the breakdown of the cost factors (as given above) or in using costs from previous projects, as there have been differences of up to 50 per cent between the estimated cost and the final cost. Lang and Herzog (2000) state that double-skin façades cost about twice as much as single-skin façades in central Europe, where there is a good level of experience with this technique, but up to five times as much in the US, where there is little experience. According to Grut (2003), the double-skin façade of the Daimler Chrysler Building in Berlin cost only 20 per cent more than a normal façade (and the façade constituted only 9 per cent of the total building cost). Whether or not such a small cost premium becomes the norm, the double-skin façade option must be carefully compared with other options for reducing building energy use, taking into account the whole range of non-energy benefits of double-skin façade designs.

The indirect costs and savings associated with double-skin façades include:

- the loss of usable internal space if the outside dimensions of the building cannot be increased to accommodate the double-skin façade;
- the increase of usable internal space if the gap between the two skins serves as the supply and exhaust ducts to and from the building interior;
- the possible need to increase the sound insulation of internal partitions in order to compensate for reduced background noise from outside that would otherwise mask internally generated sounds (two layers of drywall on each face of a stud wall instead of one will increase the sound insulation by 5dB at a cost of about €7.5/m^2);
- possible reductions in the sound-insulating capability of the inner façade (if the sound insulation of the inner façade can be reduced from 47dB to 37dB, it will be possible to do without cast-resin panes, saving about 50€/m^2 compared with normal insulating double glazing, while if the required insulation is reduced from 37dB to 32dB, a saving of only €20/m^2 would arise from thinner glass);
- additional fire-protection measures that may be required as a result of the double-skin façade (these include passive measures such as a construction with greater fire resistance; active measures such as enhanced fire detection systems, an enhanced sprinkler plant and systems to extract smoke from the façade gap; and organizational measures such as training of staff in fire-fighting and protection techniques);
- savings in the cost of sunshading devices, assuming that external devices would be used in the absence of a double-skin façade (external shading devices need to be more durable than those placed inside double-skin façades, easily adding 25–50€/m^2 for blinds of standard design);
- possible savings in both cost and space through downsizing of heating, cooling and ventilation systems (a significant downsizing in the cooling equipment is possible if the single-skin alternative does not include external shading).

A double-skin façade will entail differences in annual cleaning, maintenance and energy costs. Oesterle et al (2001) indicate an additional cleaning cost of 0.3–0.6€/m^2/year if the façade is cleaned once per year, which would translate into 0.1–0.4€/year per m^2 of floor area for a building where offices extend 6–9m from the façade on all sides and the floor to floor height is 3m. The double-skin façade will entail greater annual maintenance costs than a single-skin façade, but probably the same percentage of the initial investment cost. The cost of maintaining the chillers and air handlers will decrease due to the reduced annual operating hours. Oesterle et al (2001) suggest an annual cost equal to 2.5 per cent of the investment cost for an air conditioning plant operating 3100 hours per year (as for a single-skin façade in Germany), compared with 1.0 per cent of the investment cost for a plant operating 1400 hours per year (as for a double-skin façade).

As with high-performance windows and curtain walls (Section 3.3.20), part to all of the extra capital cost of a double-skin façade can be offset through savings in HVAC (heating, ventilation and air conditioning) equipment due to downsizing of the equipment. Stec and van Paassen (2005) have computed the cost of the façade and HVAC system for a variety of façade systems in The Netherlands, including double-skin façades. The systems they considered, as well as capital costs and annual energy costs, are presented in Table 3.20. A double-skin façade with operable windows (Case 7) costs roughly twice as much as a double-glazed façade with fixed windows and internal shading according to Table 3.20, while the HVAC cost is reduced by only 10 per cent. Predictive control allows a further slight downsizing of the HVAC system. However, if the double-skin façade serves as the supply duct, then the equipment downsizing plus elimination of supply ductwork completely offsets the incremental cost of the double-skin façade. Also given in Table 3.20 are the annual energy cost savings (based on energy consumption data given in Chapter 6, Table 6.5) and the ratio of incremental cost to annual energy cost savings; this ratio is referred to as the *simple payback time*, as it does not account for interest on the initial investment (nor does it account for any differences in annual maintenance costs). The simple payback time ranges from zero to over 85 years.

Table 3.20 Comparative capital costs, annual energy costs and payback periods for alternative façade systems facing south in The Netherlands. Energy costs are based on the energy use given in Table 6.5, assuming an electricity cost of 0.14 €/kWh and a natural gas cost of 7.9€/kWh (about $9/GJ). DG = double-glazed, TG = triple-glazed, DSF = double-skin façade. Additional information is given in Table 6.5

	Façade system			Capital cost (€/m² floor area)				Energy cost (€/m²/year)	Payback period (years)
Case	Description	U-value (W/ m²/K)	SHGC	Façade	HVAC	Total	Incre-mental		
1	DG, interior shading, fixed windows	1.2	0.26	234	221	455		4.01	
2	DG, external shading, fixed windows	1.2	0.14	242	211	453	–2	3.20	0
3	DG, external shading, fixed windows, mechanical night ventilation	1.2	0.14	242	201	443	–12	2.12	0
4	DG, external shading, operable windows, passive daytime and night ventilation	1.2	0.14	252	191	443	–12	0.98	0
5	TG, internal shading, fixed windows, mechanical night ventilation	1.2	0.20	336	211	547	92	2.93	85
6	DSF, external shading, fixed windows, mechanical night ventilation	1.2	0.095	386	201	587	133	2.12	70
7	DSF, external shading, operable windows, passive daytime and night ventilation	1.2	0.095	465	195	660	205	1.06	70
8	DSF, external shading, operable windows, passive daytime and night ventilation, predictive control	1.2	0.095	405	185	590	135	0.90	43
9	DSF, external shading, operable windows, passive daytime and night ventilation, predictive control, façade gap serves as supply duct	1.6	0.12	354	93	447	–9	1.16	0

Note: Energy costs are based on the energy use given in Table 6.5, assuming an electricity cost of 0.14€/kWh and a natural gas cost of 7.9€/kWh (about $9/GJ). Additional information is given in Table 6.5.
DG = double-glazed, TG = triple-glazed, DSF = double-skin façade.
Source: Data from Stec and van Paassen (2005) and W. Stec (personal communication, 2005)

3.5 Rooftop gardens

Rooftop gardens are discussed in Chapter 6 (Section 6.2.2) as a means of keeping roofs cooler, but it is appropriate here to mention their insulation value. This arises almost entirely from the soil layer, whose thermal conductivity depends strongly on the soil texture and moisture content. Table 3.21 gives the RSI- and R-values for soil layers 100mm, 300mm and 700mm thick (appropriate for small ground plants, shrubs and trees, respectively), and for thermal conductivities of 0.5W/m/K and 1.5W/m/K (appropriate for moderately dry and moderately wet sandy-clay soils, respectively). For wet conditions (which would be more likely in winter in cold regions), the soil will add RSI0.07–0.46 (R0.4–2.6), while for dry conditions, the soil will add RSI0.2–1.4 (R1.1–7.9). Wong et al (2003a) assume that turf, shrubs and trees on roofs in Singapore add another RSI0.35, 1.6 and 0.56 (R2, R9.1 and R3.2), respectively, but Eumorfopoulou and Aravantinos (1998) allow for no additional thermal resistance from the vegetation itself for rooftop gardens in Greece. It seems safe to assume that, in temperate climates, the thermal resistance from rooftop deciduous vegetation is close to zero in winter. In climates with mild winters, where less insulation would be called for than in cold climates, rooftop gardens can make a modest contribution to the roof thermal resistance. However, as far as heating load is concerned, insulating the roof to a high standard is far more effective than adding a rooftop garden.

Table 3.21 RSI values associated with rooftop garden soil layers of various thicknesses and thermal conductivities

Soil thickness	Thermal conductivity (W/m/K)	
	0.5	1.5
100mm	0.2 (1.1)	0.07 (0.4)
300mm	0.6 (3.4)	0.20 (1.1)
700mm	1.4 (7.9)	0.47 (2.6)

Note: R-values are given in parentheses.

3.6 Doors

U-values for common residential doors are given in Table 3.22. A 6cm thick wood slab has a U-value of 2.6W/m²/K, which is greater than that of a double-glazed, low-e window. Thus, the overall door U-value of a thick wood door will decrease with the window fraction if the window is double-glazed with a low-e film. An insulated steel slab in a wood frame has a U-value of about 1.0W/m²/K. Doors containing vacuum insulated panels (VIPs), described in Section 3.2.7, are available in Switzerland. Complete door systems (including fittings, handle, and lock) with two undamaged VIPs have a measured U-value of approximately 0.9W/m²/K, and a measured value of about 1.03W/m²/K when both VIPs are damaged by drilling a hole into them, compared with 1.3W/m²/K for a comparable door with conventional insulation (Nussbaumer et al, 2005). Thus, undamaged VIP door systems have 30 per cent less heat loss compared with doors with conventional insulation, and 65 per cent less loss compared with a 6cm thick wood door, and these savings are only slightly smaller if the VIPs are punctured.

The heat loss through doors ranges from 4 to 10 times that of a modestly insulated (RSI3.52, R20, $U = 0.28$ W/m²/K) wall. As wall insulation increases, the relative importance of doors to total heat loss will increase. For a given glazing fraction, there is almost up to a factor of three variation in the door U-value. Insulated doors and thermally broken frames should be used. Wood doors can be much more aesthetically pleasing than insulated, steel-shell doors, but have little insulating value. However, aesthetically pleasing doors with almost the same insulation value can be achieved by using a wood veneer on a solid-foam insulated core, which can be custom-built by a good woodworking shop.

Table 3.22 U-values (W/m²/K) for common swinging slab doors (970 × 2080mm rough opening)

Door type	No glazing	Single glazing	Double glazing, 12.7mm air space	Double glazing, ε = 0.10, 12.7mm Argon
Wood slab in wood frame[a]	2.61			
6% glazing		2.73	2.61	2.50
25% glazing		3.29	2.61	2.38
45% glazing		3.92	2.61	2.21
Insulated steel slab with a metal edge in metal frame[b]	2.10			
6% glazing		2.50	2.33	2.21
25% glazing		3.12	2.73	2.50
45% glazing		4.03	3.18	2.73
Insulated steel slab with wood edge in wood frame[a]	0.91			
6% glazing		1.19	1.08	1.02
25% glazing		2.21	1.48	1.31
45% glazing		3.29	1.99	1.48

[a] Assumes that the sill is thermally broken. Otherwise, add 0.17W/m²/K.
[b] Sill is not thermally broken.
Notes: Better U-values can be achieved through the use of vacuum insulation panels, as described in the text.
Source: ASHRAE (2001b, Chapter 30)

3.7 Air leakage

Understanding the distribution of pressure inside a building is central to understanding heat loss through infiltration of cold outside air, as well as to understanding how passively absorbed solar energy can be used for heating, ventilation and cooling of buildings. The physical principles behind the pressure distribution are explained in Box 3.4. When air inside a building is heated above that of the outside air, a pressure variation is established such that the interior air pressure in the upper part of the building is slightly greater than the outside air pressure, while the interior pressure in the lower part of the building is less than the outside pressure. The interior air pressure equals the outside pressure at some height z_o. The difference between the inside and outside air pressure at some height z, $\Delta P(z)$, is given by

$$\Delta P(z) = (\rho_e - \rho_i)g(z - z_o) =$$
$$\rho_i g(z - z_o)\frac{(T_i - T_e)}{T_e} \qquad (3.7)$$

where T_i and T_e are the interior and exterior temperatures, respectively. ΔP is positive (greater indoor pressure) above z_o and negative below z_o. As a result, cold inside air is sucked into the lower part of the building through various cracks and openings in the walls and windows, and warm air exits through the upper part of the building. This temperature-induced air exchange is referred to as the *stack effect*. Up to 40 per cent of the heating requirement for houses in cold climates is to heat the outside air that continually replaces the inside air. As seen from Equation (3.7), the pressure differential and hence the rate of air exchange will be greatest when it is coldest outside (i.e. when $T_i - T_e$ is largest), which is the very time that a given rate of air exchange will cause the greatest heat loss (see Equation 3.2).

BOX 3.4 Pressure distribution inside a heated building

The key to understanding the pressure variation within a building and between the inside and outside of a building is the equation of *hydrostatic balance*,

$$\Delta P = -\rho g \Delta z \qquad (B3.4.1)$$

where ΔP is the change in pressure in going up over a distance Δz. The term $\rho g \Delta z$ is the weight of air (per m² of horizontal area) in the layer Δz, so the equation of hydrostatic balance simply states that the difference in pressure between any two heights is equal to the weight of air between those two heights. This depends on the density of the air (ρ), which is related to the pressure and temperature through the *ideal gas law*,

$$P = \rho R T \qquad (B3.4.2)$$

where $R = 287.0 \text{J/kg/K}$ is the gas constant for dry air.

Consider a building with all the windows closed and where the indoor temperature equals the outdoor temperature. The indoor and outdoor pressures decrease at the same rate with height and so are the same at each height. As the building is heated, the overall indoor pressure increases at all heights as required by the ideal gas law. In response to the increase of internal pressure, some air will flow out of the building through leakage points. Suppose that enough air flows out to equalize the pressures at the ground floor level. The changes in pressure in going to a height z above ground level outside and inside the building are given by:

$$\Delta P_e = -\rho_e g z \qquad (B3.4.3)$$
and

$$\Delta P_i = -\rho_i g z = -(\rho_e g z - \Delta \rho g z) \qquad (B3.4.4)$$

respectively, where $\Delta \rho = \rho_e - \rho_i$. Thus, the difference in pressure between the inside and outside is:

$$\Delta P_{i-e} = \Delta P_i - \Delta P_e = (\rho_e - \rho_i)gz = \rho_i \left(\frac{\rho_e - \rho_i}{\rho_i} \right) gz \qquad (B3.4.5)$$

Manipulation of Equation (B3.4.5) using the ideal gas law, and letting T_i and T_e be the internal and external temperatures, respectively,

$$\Delta P_{i-e} = \rho_i g z T_i \left(\frac{1}{T_e} - \frac{1}{T_i} \right) = \rho_i g z \frac{(T_i - T_e)}{T_e} \qquad (B3.4.6)$$

which is essentially the same as Equation (3.7). The pressure difference grows steadily with increasing height, as shown in Figure B3.4. This is the so-called 'stack effect'. If the windows are opened, air will flow out from the upper floor due to the greater pressure inside. As soon as this happens, the mass of air in the building decreases, so the inside pressure at all heights decreases. As a result, the inside pressure in the lower part of the building becomes less than that outside, while the pressure surplus in the upper part becomes smaller. The resulting distribution of inside–outside pressure difference is shown in Figure B3.4, along with the height z_o where the two pressures are the same.

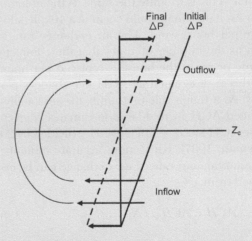

Figure B3.4 Variation in the difference between inside and outside air pressure before adjustment via a net outflow (dashed line) and after adjustment (solid) line

Note: The arrows indicate the steady airflow that is produced in the absence of wind.

It can be seen from Figure B3.4 that opening a window in the upper part of a building will not allow fresh air to enter (in the absence of wind). Rather, to cool the upper part of a building, windows in both the upper and lower levels need to be opened. Cooling of the upper part occurs through fresh air that enters the lower part, rises, and displaces the warm air in the upper part of the building, pushing it to the outside.

3.7.1 Air leakage in houses

The rate of airflow Q through a crack or orifice depends on the difference in air pressure across the crack raised to some power n. That is:

$$Q = c\Delta P^n \tag{3.8}$$

where $n = 0.5$ for fully turbulent flow and $n = 1$ for laminar flow (Walker et al, 1998). The leakiness of houses is computed through a standardized *blower door* test in which a large fan is inserted in the front door opening and the airflow required to produce various pressure drops ranging from 10–50Pa (Pascals) is determined. These are used to determine the values of the coefficients c and n (of Equation 3.8) for that house, from which the rate of air exchange under a variety of natural conditions (normally with ΔP much less than 10Pa) can be computed. The natural leakage rate increases with increasing inside–outside temperature difference (which increases ΔP) and the wind speed. As a metric for comparing different houses, the rate of air exchange is normally given for ΔP = 50Pa. This is roughly the same as the amount of air exchange that would occur in a 40kph (kilometre per hour) wind. The air exchange rate is given as the number of times that the complete volume of air in the house is replaced per hour, and this is designated as ACH (air changes per hour). As a rough rule of thumb, the seasonally-averaged ACH in mid-latitude climates is given by the ACH at ΔP = 50Pa (ACH_{50}) divided by 20 (Sherman, 1987). For a more accurate estimate, the instantaneous rate of air exchange can be estimated from:

$$ACH = ACH_{50} / N \tag{3.9}$$

where

$$N = 14/s \tag{3.10}$$

$$s = (f_w^2 v^2 + f_s^2 \, |\Delta T \, |)^{1/2} \tag{3.11}$$

and f_w and f_s are wind- and stack-infiltration parameters that depend on the building, v is wind speed, and ΔT is the inside–outside temperature difference. Typical values for single-family houses are f_w = 0.13 and f_s = 0.12m/s/K$^{1/2}$ (Sherman, 1987).

The leakage characteristics of houses can also be expressed in terms of the effective leakage area. This is the area of a single orifice that would give the same airflow at ΔP = 4Pa as all the cracks and holes in the house. The relationship used is:

$$ELA = Q_r \sqrt{\frac{\rho}{2\Delta P_r}} \tag{3.12}$$

where Q_r is the airflow at a reference pressure difference ΔP_r of 4Pa.

Figure 3.37 shows the distribution of ACH_{50} values among houses measured in the US, Europe and Canada. US houses tend to be the most leaky (average ACH_{50} = 14.2) and Canadian houses the least leaky (average ACH_{50} = 5.3), with European houses falling midway between the two. Sherman and Dickerhoff (1998) present data on ACH_{50} values for houses in the US, state by state. Average ACH_{50} values and sample sizes (N) for some individual cold-climate states are as follows: Alaska, 40 (N = 2830); Rhode Island, 37.6 (N = 6284); Vermont, 31.2 (N = 1186), Oklahoma, 22.4 (N = 204); New York, 14.6 (N = 282), Illinois, 13.2 (N = 179); and Washington, 8.8 (N = 199). There is an extraordinary geographical variation in the leakiness of US houses, with some of the coldest states having the leakiest houses. In old houses in the UK (Bell and Lowe, 2000), an ACH_{50} of 10–20 is quite common. In a sample of 30 houses in Quebec, ACH_{50} ranges from 1.25 to 11.75, with a broad peak at 4.25 (Parent et al, 1998). In 30 relatively energy-efficient houses in California, ACH_{50} ranges from 2.6 to 8.7, with an average of 5.5 (Hoeschele et al, 2002). However, in only two of these houses was particular attention given to a tight envelope, and in these, $ACH_{50} \approx 2.6$. In one survey of 'Super Good Cents' houses built in the US, ACH_{50} averaged 4.8 while in another the average was 5.5 (Baechler et al, 2002). In a survey of 900 homes built under the Wisconsin Energy Star programme, median ACH_{50} = 2.4, with one-third of the homes falling below 2.0 ACH (Pigg et al, 2002). For new (post-1990) conventional houses in Canada, average provincial ACH_{50} values range from 2.0 (in the cold prairie provinces) to 4.3 (in comparatively mild British Columbia) (Hamlin and Gusdorf, 1997), although P. Parker et al (2000) note that there can be large differences in leakage rates even between adjacent row houses due to inconsistent workmanship. The corresponding average ELAs range from 0.9 to 1.9cm² per m² of envelope area. For houses built to the Canadian R2000 standard (discussed further below), average ACH_{50} ranges from 1.14 in Ontario to 1.44 in British Columbia, which corresponds to about 0.075ACH under average conditions. This is so low that deliberate exchange with outside air must be provided, and this is done with a *heat-recovery ventilator*, in which fresh air is sucked into and circulated through ductwork with fans, then exhausted to the outside. Some of the heat in the outgoing exhaust air is used to heat the incoming air. Heat-recovery ventilators are discussed later (Chapter 7, Section 7.3.3). Finally, the European *Passive House* standard (see Chapter 13, Section 13.2.1) requires an ACH_{50} of 0.6 or less. This represents a factor of 3–30 less leakage compared to average practice, depending on location. The Pirmasens passive house in Germany achieved an ACH_{50} of 0.27 (Badescu and Sicre, 2003), while demonstration houses in Hannover-Kronsberg and Kassel achieved an ACH_{50} of 0.30 and 0.35, respectively (see Fact Sheets for Task 28 of the IEA Solar Heating and Cooling Program, www.iea-shc.org).

Achieving low rates of air leakage requires a continuous impermeable barrier just inside the interior wall and ceiling finish of the entire house, with all breaks (such as for electrical boxes, wiring and plumbing) carefully sealed, and all joints between walls and window or door joists, walls and ceilings, basement and above-grade walls (if made of different material) and between walls and floors sealed. Small gaps can be sealed with caulking, while larger gaps can be sealed with spray-on foam (stuffing insulation into gaps is inadequate). Caulking materials have improved considerably over the past 20 years, with urethane and silicon materials able to stretch by 50–100 per cent without losing their seal. The main factor in achieving low leakage at low cost is the skill of the builder, rather than the cost of materials; experienced builders in Canada have achieved an ACH_{50} of 1.5 at no additional cost compared to those achieving an ACH_{50} of 3–4.

An air barrier is one of four barriers that the building envelope must contain. These are:

- the thermal barrier (insulation and high-performance windows);
- a vapour barrier (to prevent water vapour passing through the insulation from the inside and condensing on the outside of the insulation, or vice versa in hot-humid climates);
- an air barrier;
- a moisture barrier (to prevent water from outside entering the wall).

A clear plastic sheet (polyethylene) can function as both an air barrier and a vapour barrier, as long as all the joints are overlapping and taped or sealed together (preferably caulked). In cold climates, it should be installed on the inside of the stud-cavity insulation, just before the drywall is applied, as illustrated in Figure 3.38. Airtight electrical boxes should be used in place of standard electrical boxes. Particular attention is needed to delineating the responsibilities of the various trades involved in house construction (for example, deciding who is responsible for sealing breaks in an air-tight envelope made by plumbers or electricians). As noted earlier (Section 3.2.1), some forms of spray-on polyurethane insulation

Figure 3.37 Distribution of residential ACH_{50} values in the US, Europe and Canada

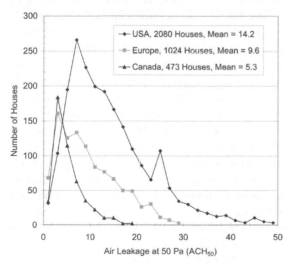

Note: Also given are the mean values of the sampled houses.
Source: Data from ASHRAE (2001b, Chapter 26).

Figure 3.38 Polyethylene air barrier stapled to insulation prior to installation of a drywall

Source: Wulfinghoff (1999), reprinted by permission of the Energy Efficiency Manual by Donald R. Wulfinghoff

can qualify as an air and vapour barrier. Alternatively, external extruded polystyrene insulation with a gasketting system to seal joints can serve as an air barrier. This avoids interfaces with floors and walls and penetrations with electrical outlets. A commonly used moisture barrier is Tyvec, which is wrapped around the outside of the insulation or wall, before the exterior cladding is applied. It can function as an air barrier if joints are taped together. Correct specification of the details concerning air, moisture and vapour barriers requires a sound understanding of how a building envelope behaves; improper specification and installation of any one of these barriers can be a cause of building envelope deterioration.

Recessed pot-lights have become a very popular lighting fixture in North America. The cylindrical 'cans' housing the lamp are quite leaky (having a leakage area of about 4cm² each). They invariably use incandescent lamps, which generate large amounts of heat, heating the air within the can and creating a strong local stack effect that exacerbates the flow of air from the conditioned to the unconditioned space (leakage through the fixtures can be three to five times that which would otherwise occur, and this enhancement is not detected by blower door tests). The US Department of Energy is currently sponsoring the development of close to air-tight (0.21cm² leakage area) recessed pot-light fixtures that are hard-wired to accept only compact fluorescent lamps (McCullough and Gordon, 2002; www.pnl.gov/cfldownlights). These will not only

reduce heating and cooling energy consumption by reducing air leakage, but will also reduce lighting energy use and the associated production of waste heat, thereby indirectly reducing air conditioning energy use (compact fluorescent lamps are four to five times more efficient than incandescent lamps). Note that airtight pot-light fixtures should be used wherever pot-lights are used, and not only where they occur below the roof or attic, as the cavity between the ceiling and an overlying floor is potentially connected to cavities in the outer walls.

A leaky building envelope is a particular problem in high-rise residential buildings, due to the strong stack effect. Apart from energy loss, air leakage can reverse the intended flow of ventilation air, resulting in under-ventilation of some suites and over-ventilation of other suites. Elevator shafts, stairways, garbage chutes and vertical plumbing are ready conduits for airflow. Elevator and stairway doors should therefore be tight-fitting and elevator penthouses well sealed. Alternative elevator systems in low-rise buildings (up to seven floors) eliminate the need for elevator machine rooms (CMHC, 2000). Garbage chute doors and doors to chute rooms should be well gasketed, and all plumbing penetrations inspected and well sealed.

3.7.2 Air leakage in commercial buildings

According to ASHRAE (2001b, Chapter 24), airflow barriers in commercial buildings are often poorly designed and executed. Air leakage in commercial buildings with hollow concrete masonry in particular can be significant, due to the fact that upward airflow in cavities is not always adequately blocked, parallel random leakage paths are found between gypsum wallboard or other finishes and the block face, and masonry cannot form a tight seal with structural steel columns and beams. Wulfinghoff (1999) provides a humorous but sad collection of photographs illustrating common leakage paths in commercial buildings due to shoddy construction practice, incompetence or indifference. Selected examples are shown in Figure 3.39.

As explained above, one of the forces driving natural air exchange is the stack effect arising from the temperature difference between

Figure 3.39 Examples of shoddy workmanship leading to air leakage in commercial buildings. Top: Insulation used to seal the gap between a concrete wall and the rim joist – the wrong material for the job. Middle: Gap around structural steel inside an elevator shaft (the worst place for leaks due to the unimpeded stack effect), caused by masons only partially filling the gap between the steel beams and concrete blocks with clay bricks, which happened to be on hand. Bottom: Block removed by a plumber to gain access to a pipe fitting; as the plumber is not a mason, he did not replace it

Source: Wulfinghoff (1999), reprinted by permission of the Energy Efficiency Manual by Donald R. Wulfinghoff

the inside and the outside. The stack effect increases with increasing height and so will be particularly strong in tall buildings. However, if the upper portion of the building is well sealed, infiltration into (during winter) or exfiltration out of (during summer) the lower portion will be inhibited. Stairwells and elevator shafts require particular attention.

There is very little information available on natural air exchange rates in commercial buildings, but what little there is suggests a range of 0.1–0.6ACH. Two methods have been used to measure leakage in commercial buildings (Bahnfleth et al, 1999). One method is a floor-by-floor blower door test. A blower door is installed in a door leading to a stairwell shaft or in an operable window, and used to pressurize a given floor. To prevent air from leaking to the adjacent floors, two more blower doors are used to maintain the same air pressure in the floor above and below the floor being tested. The measured rate of air leakage is the leakage through the external envelope of that floor only. The second method uses the ventilation fans to pressurize the building in increments. The flow rates needed to achieve various pressurizations are measured, but only the total envelope leakage can be determined in this way. Fennell and Haehnel (2005) indicate the following ranges for typical ACH_{50} values in commercial buildings, based on documented studies: 1.1–1.6 in Canada, 4.9–7.5 in the US, and 6.0–9.1 in the UK. These data imply that there is factor of 6–8 difference between modestly good practice and the typical practice of some countries.

The ventilation systems of most commercial buildings having mechanical ventilation are designed to produce a slightly higher pressure inside the building than outside the building, in order to inhibit infiltration of outside air and associated water vapour (this is done by having the return air fan handle a slightly smaller amount of air than the supply air fan). A primary motivation in creating a tight envelope in commercial buildings is to inhibit infiltration of moisture during the summer rather than transport of heat across the envelope. In leaky buildings, moisture can penetrate at night if the ventilation system is shut down or simply set to a lower flow rate. Even if there are no moisture-related problems, shutting down the ventilation system at night in leaky buildings might not save any energy due to the need to cool and dehumidify all of the air that entered the building during the night.

CMHC (1993) proposes a process for *commissioning* the air barrier of buildings, that is, verifying that the barrier performs as intended while it is still possible to correct any errors, in the same way that mechanical and electrical system are (or should be) commissioned. The proposed steps in the commissioning process for air barriers are:

1   A project brief is written that lays out the expected performance of the air barrier and the winter and summer extreme conditions under which it is to operate.

2   The air barrier is designed; the proposed materials are validated in terms of air permeability, continuity at joints, tolerance of pressure loadings and durability; and the overall design is audited.

3   The performance of the air barrier is tested after it is substantially complete, or individual areas are tested as they are constructed, and/or mock-ups of key details are constructed at the earliest possible stage and tested. Corrective action is taken as needed.

4   A final test of the overall performance of the air barrier is made as the building nears completion.

5   A programme of monitoring, testing and repair after the building is completed is established.

However, this procedure is rarely followed except for major public buildings (hospitals, museums, galleries).

Materials used as air barriers in commercial buildings should have an air leakage rate of less than 0.02 litres/s/m² at 75Pa pressure difference, while building assemblies should have a leakage rate of less than 0.1 litres/s/m². The leakage rate will be larger still for entire buildings. According to Anis (2005), tight commercial buildings in Canada have a leakage rate at 75Pa of about 0.5litres/s per m² of envelope area, where the envelope is the total area enclosing the inside air volume, including below-grade walls and floors. Typical rates in the US range from 6 to 8 litres/s/m² in the north to 10 to 12 litres/s/m² in the south (Anis, 2005).

Minimizing air leakage is important not only because it reduces heating, cooling and de-humidification loads, but also because a relatively air-tight building (except for intentional openings) is necessary in order for many low-energy ventilation techniques to work properly; in particular:

*   natural ventilation (Chapter 6, Section 6.3.1);
*   heat-recovery ventilators (Chapter 7, Section 7.3.3);
*   chilled-ceiling cooling (Chapter 7, Section 7.4.4);
*   displacement ventilation (Chapter 7, Section 7.4.5).

As well, minimization of leakage leads to a more equitable apportioning of energy heating costs between different apartment units in a given building. Finally, construction to a verifiably low rate of leakage will lead to less uncertainty in heating and cooling loads, resulting in less oversizing of HVAC equipment as a precaution, with associated cost savings and efficiency gains (see Chapter 4, Section 4.4, and Chapter 6, Section 6.8).

3.8 Life-cycle cost analysis and the choice of building envelope

High-performance building envelopes entail greater up-front construction costs, but lower energy related costs during the lifetime of the building. The *total* up-front cost of the building may or may not be larger, depending on the extent to which heating and cooling systems can be downsized, simplified or eliminated altogether as a result of the high-performance envelope. Any additional up-front cost will be compensated to some extent by reduced energy costs over the lifetime of the building. Here, we discuss how to combine these costs and benefits into a single life-cycle cost and show how this varies with the insulation level, given reported costs of different insulation levels and a range of future energy costs. We then discuss some principles governing the extent to which better thermal envelopes permit downsizing of heating and cooling equipment, and discuss various ancillary benefits of high-performance envelopes.

3.8.1. Life-cycle cost versus insulation level

The net up-front cost (after any credits for equipment downsizing) and the operating cost can be combined to give the total or *life-cycle* cost of a design choice, if future energy costs are converted to the *present value* of the future costs. The present value is the amount of money that would have to be invested at the prevailing interest rate such that the investment plus accumulated interest would be exactly used up paying for the energy costs by the end of the period used for life-cycle analysis. If C is the annual heating + cooling cost during the first year, n the number of years over which the life-cycle cost is computed, i the interest rate (including inflation), and e the rate of increase of energy costs (including inflation), then the present value PV of the heating costs is:

$$PV = C\left(\frac{1-(1+a)^{-n}}{a}\right) \qquad (3.13)$$

where a is the effective interest rate, given by:

$$a = \frac{i-e}{1+e} \qquad (3.14)$$

The annual heating cost is given by the heating load divided by the efficiency of the heating system, times the cost of heating fuel. The heating load H (Joules per square metre of wall) can be approximated by:

$$H = U \times HDD \times 24 \times 3600 \qquad (3.15)$$

where HDD is the number of heating degree days (see Chapter 2, Section 2.5). HDD is based on the difference between daily average air temperature and a reference temperature of 18°C, while the heat flow across the wall depends on the difference between the internal temperature (typically 20–22°C) and the outdoor air temperature.

However, the extra heat loss associated with the larger temperature difference is offset by internal and passive heat gains and need not be supplied by the heating system. The magnitude of internal and passive heat gains depends on the building and location, but using Equation (3.15) with HDD relative to a fixed reference temperature is still a useful approximation. The annual cooling load can be approximated with an equation similar to Equation (3.15), but using CDD instead of HDD (as long as CDD is computed relative to the lowest outside temperature at which mechanical cooling is needed). The annual cooling energy use would be given by the annual cooling load divided by the average COP (coefficient of performance) of the cooling system.

We will illustrate the application of Equations (3.13) to (3.15) using estimates provided by NRCan (1997a) of the additional cost associated with increasing levels of wall insulation relative to a wall with an RSI of 2.1 (R12), up to RSI6.7 (R38), which is close to the highest amount of insulation used in high-performance building envelopes. The additional construction costs (an extra $33/m² to reduce the U-value by 0.327W/m²/K, to 0.15W/m²/K) are appropriate for conditions in Canada in the mid-1990s and are based on wood-frame construction. Costs using external insulation as reported by Jakob and Madlener (2003) for Switzerland are substantially higher (about $46/m² to reduce the U-value by 0.2W/m²/K to a final value of 0.15W/m²/K). Differences in the life-cycle cost can be compared if additional heating costs are computed for each level of insulation relative to the highest level of insulation considered. This is done using differences in U-values, rather than absolute U-values, in Equation (3.15). The results are given in Figure 3.40 assuming HDD = 5000, $i = 0.06$, $e = 0.03$, and $n = 30$ years, and for heating fuel costs of $10/GJ, $15/GJ and $20/GJ. These heating fuel costs span the range of heating costs found in most countries today (see Figure 2.10). Although the specific incremental construction costs that should be used in any given location will differ from those used in Figure 3.40, a number of conclusions can be drawn from Figure 3.40 that will be generally applicable:

Figure 3.40 Comparison of incremental life-cycle costs of walls with increasing amounts of insulation (successively smaller U-values). The lowest part of each bar is the incremental construction cost relative to the least-insulated wall, the second part of each bar is the incremental heating cost relative to the best-insulated wall for a heating fuel cost of $10/GJ, the third part of each bar is the additional incremental heating cost if the heating fuel costs are $15/GJ instead of $10/GJ, and the top part of each bar is the additional incremental heating cost if the heating fuel costs are $20/GJ instead of $15/GJ

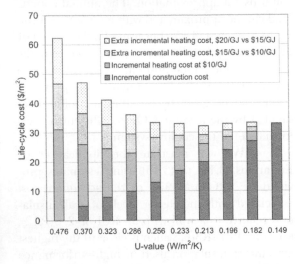

Note: Not included in this cost comparison are the reduction in cooling energy use and the downsizing of heating and cooling equipment that occurs with higher-performance envelopes. Source: Incremental construction cost data are from NRCan (1997a), while incremental heating costs were computed as explained in the text assuming HDD = 5000, i = 0.06, e = 0.03, and n = 30 years

- the minima in life-cycle cost is quite broad, meaning that there is very little difference in the life-cycle cost if insulation levels moderately worse or moderately better than the least-cost insulation level are chosen;
- although the life-cycle cost associated with the highest insulation level (RSI6.7, R38) is not the smallest life-cycle cost, it is not substantially greater than the minimum life-cycle cost when the fuel cost is $15/GJ or $20/GJ, and is less than the life-cycle cost at low levels of insulation (RSI 2-4, which, as discussed in Chapter 13, Section 13.1, are allowed in many jurisdictions with 5000HDD or more).

Figure 3.41 Comparison of incremental life-cycle costs of walls with increasing amounts of insulation (successively smaller U-values) for a fuel cost of $10/GJ, HDD = 5000, i = 0.06, and for alternative time periods and energy inflation rates

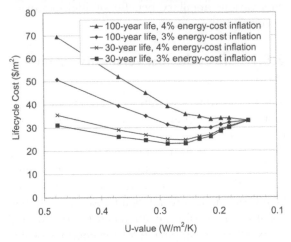

Curves of life-cycle cost versus insulation thickness for the UK (ODPM, 2004, Section 1, Appendix A) have the same shape as those in Figure 3.40, thereby supporting the above conclusions.

Differences in life-cycle costs are influenced by the length of time over which life-cycle costs are computed and by the rate of inflation in energy costs. In the above, a 30-year time frame was chosen because mortgages in North America are typically of this duration. However, much longer mortgages are common in Europe, and in any case, the lifespan of the building should be closer to 100 years. Figure 3.41 compares the incremental life-cycle costs for different levels of insulation for 30- and 100-year lifespans and for e = 0.03 and 0.04. For the 100-year lifespan, the highest insulation level provides the lowest (if e = 0.04) or close to the lowest (if e = 0.03) life-cycle cost.

Whatever the specific life-cycle cost curves in a given jurisdiction, the main conclusion from Figures 3.40 and 3.41 is that it is justified to require insulation levels *substantially* in excess of the level that is calculated to minimize life-cycle cost, particularly if conservative assumptions (30-year time horizon, energy inflation equal to general inflation) are made in doing the calculation and even if credits for downsizing of heating and cooling equipment have been included in the calculation. This conclusion is justified because of uncertainties in future energy costs, because energy costs do not include the environmental damage associated with energy use (most of which cannot be quantified in economic terms), and because of the shallow, concave upward shape of the life-cycle cost curves shown in Figures 3.40 and 3.41. The probability distribution of future energy costs is not symmetrical around the best-guess future cost, but is skewed to the right – there is a greater probability of higher than expected costs than of lower than expected costs (see Chapter 1, Section 1.3). Under such circumstances, the risk minimizing strategy is shifted toward greater rather than lesser insulation levels, as shown formally by Tol (1995) in a different context.

3.8.2 Downsizing of heating and cooling equipment

The conclusion reached in the preceding section – that substantially greater levels of insulation than required at present are justified – would be strengthened if part of the cost of high levels of insulation can be offset by smaller and less expensive heating systems. To assess the extent to which such savings could occur, we need to briefly examine the relationship between insulation level and heating requirements.

The heat loss from a building over the course of a heating season varies in direct proportion to the average envelope U-value. The heating requirement, on the other hand, is the difference between the heat loss and the useable internal and passive solar heat gains. If the useable heat gains are fixed, reducing the envelope U-value by a given fraction will have a disproportionately larger effect on the heating requirement. For example, if seasonal heat losses and gains are 80 and 20 units, respectively, then reducing the heat loss in half reduces the heating requirement by two thirds (this savings will be reduced to some extent by the fact that, with better insulation, the useable solar heat gain decreases). Peak heat loads, on the other hand, occur during times of maximum heat loss and zero solar heat gain, and so will be reduced in proportion to the decrease in the overall building heat loss coefficient. Thus, the permitted downsizing in heating equipment will be smaller, in relative terms, than the reduction in annual heating energy use. As well, the

cost savings due to smaller-capacity residential heating equipment and commercial boilers is diminished by the fact that costs per unit of heating capacity increase with decreasing size (see Chapter 4, Sections 4.2.8 and 4.3.4), and an improved envelope affects only the conductive and leakage but not the ventilation components of heat loss.

Thus, the reduction in the cost of furnaces or boilers due to substantially better thermal envelopes is normally only a small fraction of the additional cost of the better thermal envelope. However, potentially larger cost savings can occur through downsizing or elimination of other components of the heating system – such as ducts to deliver warm air, or radiators. As noted in Section 3.3.20, high-performance windows eliminate the need for perimeter heating. A very high-performance envelope (including non-window elements) can reduce the heating load to that which can be met by ventilation airflow alone, leading to a simpler air-handler in forced-air heating systems, or elimination of radiant floors while still providing the uniform indoor temperatures that are a selling point for radiant floor heating. Finally, high-performance envelopes lead to a reduction in peak cooling requirements, and hence in cooling equipment sizing and costs, and permit use of a variety of passive and low-energy cooling techniques. In a fully integrated design that takes advantage of all the opportunities facilitated by a high-performance envelope, it is indeed possible for savings in the cost of mechanical systems to offset all or much of the additional cost of the high-performance envelope (see Chapter 7, Section 7.4.4, and Chapter 13, Section 13.3.4).

3.8.3 Ancillary benefits of high-performance envelopes

Higher insulation levels provide many benefits in addition to reducing heating loads and associated costs. The small rate of heat loss associated with high levels of insulation creates a more comfortable dwelling because temperatures are more uniform, can indirectly lead to higher efficiency in the equipment supplying the heat (especially when heat pumps are involved, but also for condensing boilers), and permits alternative heating systems – such as radiant floor heating in houses – that would not otherwise be viable but which are superior to conventional heating systems in many respects. Better insulated houses eliminate moisture problems associated with, for example, thermal bridges (because these will be eliminated in well-insulated buildings) and damp basements (insulating the basement floor – not usually done – isolates the basement floor from the cold soil, reducing the tendency for condensation during periods of high humidity in the summer). Increased roof insulation also increases the attenuation of outside sounds such as from aircraft. Mills (2003) explains a number of ways in which high-performance building envelopes reduce the risk of insurance related losses, including:

- eliminating the formation of ice dams behind accumulated snow on sloping roofs by reducing heat loss from roofs that otherwise can periodically melt the accumulated snow;
- the greater length of time required for high-performance windows to break under heat stress during a fire (broken windows accelerate the spread of fire and of toxic fumes);
- the greater resistance of high-performance windows to wind forces;
- the greater resistance of windows with low-e films to shattering.

3.9 Building shape, form and orientation

Building shape, form and orientation are architectural decisions that have impacts on heating and cooling loads, daylighting and the opportunities for passive ventilation, passive solar heating and cooling, and for active solar energy systems. Building shape refers to the relative length of the overall dimensions (height, width, depth), building form refers to small-scale variations in the shape of a building, and building orientation refers to the direction that the longest horizontal dimension faces. Building shape, form and orientation will be more important the smaller the building and for buildings with a lower-performance envelope (less insulation, windows with a higher U-value and SHGC, absence of shading devices). In large buildings,

internal heat sources (people, lighting, equipment) will dominate envelope heat gains (except in the perimeter areas, which will comprise a relatively small fraction of the building), while in highly insulated buildings with high-performance windows, shape and orientation will have only a minor impact on net winter heating requirements, particularly compared with the impact of high levels of insulation or of high-performance windows themselves. Nevertheless, the choice of building shape, form and orientation is one more strategy in the quest to minimize building energy requirements.

It is commonly thought that minimizing the surface area to volume ratio of a building will reduce the heating load for a given insulation system. However, different elements of the building envelope – windows, walls, roof and the floor in contact with the ground – will have a different thermal resistance. Even if the thermal resistance of each element is fixed, a change in the shape of the building will inevitably be accompanied by a change in the relative proportions of the different façade elements, and this can easily outweigh the impact on heat loss of the change in surface: volume ratio when the building shape is altered while keeping the volume fixed. This is illustrated in Table 3.23, which gives the surface:volume ratio, area average envelope U-value, and heat loss coefficient for a building with a fixed floor area of 2500m[2].[7] Case 1 is one storey with horizontal dimensions of 50 × 50m, Case 2 is two stories at 50 × 25m, and Case 3 is two stories at 35.4 × 35.4m. For each case, the floor to floor height is 3.5m and the wall:window area ratio is 0.4. Cutting the floor plan in half and placing one half on top of the other (Case 2 vs Case 1) decreases total envelope area by 31 per cent (excluding parts in contact with the ground) or 39 per cent (including the ground contact area), but increases the wall and window areas by 50 per cent. The net result is a substantial increase in the average envelope U-value but only a 1 per cent increase in the heat loss coefficient if the ground is excluded, or a 6 per cent decrease if the ground is included. Adopting a square rather than rectangular floor plan reduces the heat loss coefficient by about 4 per cent. All of these changes are very small compared to the impact of more insulation, better windows, a more air-tight envelope and the use of

Table 3.23 Illustration of the effect of building shape on the surface area:volume ratio, average envelope U-value and heat loss coefficient for conductive heat loss

Case	Dimensions (m)	Number of floors	Area (m²)		Surface:floor area ratio		Weighted U-value (W/m²/K)	Heat loss coeff (W/K)	
			WW	Roof	Actual	Rel		Actual	Rel
Excluding ground floor as part of the envelope									
1	50 × 50	1	600	2500	1.24	1.00	0.49	1530	1.00
2	50 × 25	2	900	1250	0.86	0.69	0.72	1545	1.01
3	35.4 × 35.4	2	849	1250	0.84	0.60	0.70	1487	0.97
Including ground floor as part of the envelope									
1	50 × 50	1	600	2500	2.24	1.00	0.32	1780	1.00
2	50 × 25	2	900	1250	1.36	0.61	0.49	1670	0.94
3	35.4 × 35.4	2	849	1250	1.34	0.60	0.48	1603	0.90

Note: In every case, the building floor area is 2500m², the floor to floor height is 3m, and the window, wall, roof and effective ground U-values are 2.5, 0.5, 0.3 and 0.1W/m²/K, respectively.
WW = window + wall area. Rel = relative to Case 1.

heat exchangers to recover heat from exhaust air in a mechanical ventilation system. The change from one to two stories would also affect the driving force for air exchange, which varies with the indoor to outdoor pressure difference to the power of 0.5 if the airflow is turbulent (Equation 3.8). The pressure difference varies linearly with distance above the neutral plane for a uniform indoor temperature, so the driving force averaged over the wall will be $\sqrt{2}$ times larger for the two-storey building. If air exchange occurs primarily through the walls rather than the roof, then this factor, combined with the greater wall area in the two-storey building, could lead to a substantial increase in uncontrolled air exchange.

The above example pertains to alternative configurations for a building of fixed volume. The picture is more favourable to multi-storey buildings if one is combining an increasing number of units together into one building, and doing so by increasing the number of floors. Consider living units that are 15 × 6m in plan and 3.5m high, with 6.2m² of window at each end and 10.5m² of windows along one side. Consider three cases: (1) six separate units, (2) two sets of three units side by side (with a loss of 10.5m² of window area, from the middle unit), and (3) two floors of three side by side units each. Wall area drops by 48 per cent between cases (1) and (2), while roof area drops by 50 per cent between cases (2) and (3) with no offsetting increase in

wall area. For the same component U-values as in the previous example, the heat loss coefficient decreases 29 per cent from Case 1 to Case 2, and by 38 per cent from Case 1 to Case 3 (or by 13 per cent from Case 2 to Case 3). Between Cases 2 and 3, the average idealized driving force for air exchange increases by a factor of $\sqrt{2}$, wall area is constant and roof area decreases in half, so there could be a reduction in uncontrolled air exchange (assuming the same leakage:envelope area ratio). Minimizing the surface:volume ratio (and especially the roof area and the area of west-facing walls) will also minimize the summer heat gain from outside, but may adversely affect the opportunity for cooling at night. Readers interested in pursuing the role of building form (and orientation) further may consult Shaviv and Capeluto (1992), Capeluto et al (2003) and Pessenlehner and Mahdavi (2003).

Minimizing the surface:volume ratio can reduce the building cost per unit of floor area, by reducing the relative importance of the external envelope to the total cost. Thus, minimizing the surface:volume ratio can have both energy and investment cost advantages. Conversely, choosing shallow-plan floor designs (narrow buildings) for daylighting and ventilation can increase costs per unit of floor area.

In apartment buildings and condominiums, balconies are a common feature. Unfortunately, the balcony floors serve as effective

Figure 3.42 Free-standing balconies on a residential building in Waterloo, Canada

Source: Duncan Hill, Canadian Mortgage and Housing Corporation

conduits of heat from the interior of the building, as heat is conducted along the concrete floor slab and then radiated to the outside. This is referred to as a 'fin' effect. One solution is to build entirely enclosed balconies (that can be opened during the summer). Other solutions are to use concrete with a greater thermal resistance (Section 3.2.5) or to build free-standing balconies with separate columns to support balconies (Hill and Caruth, 2000). An example of free-standing balconies is shown in Figure 3.42. An added benefit is that the balconies can be cast separately using more durable air-entrained concrete.

Building shape and orientation significantly influence the opportunities for passive solar heating, as discussed in Chapter 4 (Section 4.1.1), and for passive ventilation and cooling. Shading is an important additional consideration with regard to cooling loads. Passive ventilation will occur more readily in multi-storey buildings than in sprawling single-storey buildings, due to a more effective stack effect. Multi-storey buildings (if not too closely spaced) will also be exposed to stronger winds on the upper floors, which can be used for cross-ventilation if appropriately designed. However, compact buildings with small surface:volume ratios, while minimizing envelope heat gains, will reduce the possibility of wind driven passive cooling. These and other considerations are discussed in Chapter 6. Finally,

building shape and form are important with regard to daylighting (discussed in Chapter 9) and with regard to the opportunities to incorporate active solar heating, solar air conditioning and photovoltaic electricity generation as part of the building envelope, items that are discussed in Chapter 11. Building shape and orientation depend in part on the layout of streets, which is an urban planning design choice in new developments. A good discussion of issues pertaining to passive heating, cooling, and daylighting at the community scale is found in Chrisomallidou (2001a, 2001b).

3.10 Summary and synthesis

The critical elements of a building envelope from an energy efficiency point of view are:

- insulation levels in the walls, roof, and floor;
- the thermal and optical performance of windows;
- the air-tightness of the building;
- recovery of heat from mechanical ventilations systems that, for houses and in many commercial buildings, need be operated only when heating or air conditioning is required.

Insulation systems have been built in many high-performance houses that achieve insulation values two to three times that of conventional practice in cold climates. Insulation is available as blown-in cellulose, insulation batts or solid or spray-on foams. Halocarbon blowing agents are generally used with foam insulation. These leak from the foam to a varying extent, thereby causing a decrease in the insulation value over time and contributing to global warming by virtue of the fact that the halocarbon agents used in foam insulation are powerful greenhouse gases. Foams with non-halocarbon blowing agents are available and others are under development; these have a stable but lower insulation value for a given thickness, and the blowing agents have zero or negligible greenhouse effect. A number of alternative insulation systems are available for residential and commercial buildings, including dynamic insulation (in which ventilation requirements are met by drawing outside air through porous insulation), structural insulation panels (SIPs), insulating concrete forms (ICFs), external insulating and finishing systems (EIFSs), vacuum insulation panels and transparent insulation materials (TIMs). The latter can serve as a net heat source during the winter on façades of any orientation at 50°N and on south-facing façades at 60°N if at least some excess heat absorbed during the day is transferred to some storage medium for release at night. That is, solar heat gains can exceed conductive heat loss, corresponding to a negative U-value. To achieve the full potential of advanced insulation systems, as well as to build air-tight structures, requires attention to detail and high-quality workmanship.

High-performance windows (triple-glazed, double low-e coatings, argon or krypton gas fill, fibreglass frames) have one-tenth the conductive heat loss of conventional single-glazed windows or one-fifth the conductive heat loss of conventional double-glazed windows. Window properties can be selected to minimize or maximize solar heat gain while having comparatively little impact on heat loss or on the opportunity for using daylight. Tinted glass is a poor choice for control of heat gain, as it indiscriminately reduces solar transmittance in both the visible and NIR parts of the spectrum. High-performance windows serve as a net heat source on façades facing south, east or west in moderately cold climates (e.g. anywhere in southern Canada) because the solar heat gain exceeds the conductive heat loss during the heating season, thereby eliminating the need to restrict window areas so as to limit heat loss. The window contribution to cooling loads can be dramatically reduced with high-performance windows, not because of their small thermal conductivity but because of the ability to produce windows with a small solar heat gain coefficient (0.30–0.35, compared with 0.60–0.65 for conventional double-glazed windows). The incremental cost of high-performance windows can be partly offset by the elimination of perimeter heating units, the simplification of ductwork or plumbing, and the downsizing of heating and cooling equipment that are possible with high-performance windows.

All-glass façades have become very popular among architects and their clients. These provide plenty of daylighting, but the issues are then to minimize winter heat loss and summer heat gain and to avoid glare. As noted above, high-performance glazing systems are a net heat source during the heating season for most climates and façade orientations, which addresses concerns about the heating load. An effective way to avoid overheating is to construct a second (glass) façade outside the inner façade, with adjustable shading devices between the two façades and openings to permit the flow of outside air into the air space between the two and either back outside (during summer) or into the ventilation system (during winter).

Building shape and form also influence the rate of heat loss in winter and solar heat gain during summer. For a given volume, heat loss is minimized if the surface area:volume ratio is minimized. This in turn requires compact multi-storey buildings rather than sprawling one-storey buildings. Multi-storey buildings also have a smaller roof area in relation to the total floor area, thereby reducing the contribution of solar heating to the cooling requirements. Complex shapes with many protrusions and indentations should be avoided in climates with a significant heating requirement, as should balconies that are not thermally separated from the interior floor slabs. However, excessive compactness and bulk (i.e. very deep floor plans), while further reducing the surface area:volume ratio and associated heat loss in winter, will also reduce the scope for daylighting and for passive ventilation and cooling during the summer. Building shape is less critical for heat loss if the building has a high-performance envelope.

To properly evaluate the difference in life-cycle cost of alternative building envelopes, the downsizing, simplification or elimination of heating and cooling equipment that is possible with a high-performance envelope must be taken into account. Environmental externalities, benefits to power utilities and the value of reduced future uncertainty should also be taken into account. Even in the absence of downsizing credit or environmental externalities, life-cycle cost at very high levels of insulation tends to be only modestly higher than the minimum life-cycle cost (found at intermediate levels of insulation), and is often no higher than the life-cycle cost at the low levels of insulation that are permitted in many jurisdictions.

Notes

1   The specific heat of any substance, c_p, is the amount of heat required to warm 1kg of the substance by 1K. The subscript p denotes the fact that the specific heat used here pertains to addition of heat at constant pressure, and is included for consistency with more specialized textbooks.

2   When the outgoing air temperature is colder than the incoming temperature, the ventilation system is a net source of heat to the building. If this net heat source plus the heat dissipated by the heat pump plus internal heat gains equals the conductive heat loss through the building envelope, then no supplemental heating will be needed.

3   Lightweight concrete attains its lower density through air pockets, implying less embodied energy in its manufacture, which is a benefit in addition to reduced building operating energy.

4   The critical temperature of a substance is the temperature at which it cannot exist as a liquid at any pressure.

5   Solar transmittance means the total transmittance for solar radiation, while 'visible' transmittance means the transmittance averaged over visible wavelengths only. 'Transmittance' is the fraction of incident radiation that passes through the window. The solar transmittance affects solar heat gain, while visible transmittance affects the amount of daylight passing through the window.

6   HVAC = *H*eating *V*entilation *A*ir *C*onditioning.

7   The heat loss coefficient (W/K) is given by the U-value of each façade element times its area, and summed over all elements.

four

Heating Systems

Having reduced the heating load through the use of a high-performance thermal envelope, and possibly through consideration of building shape and orientation, the next step is to provide as much of the reduced heating load through passive solar means as possible. The balance should be satisfied as efficiently as possible, either with external energy sources or through active solar energy with or without seasonal storage. Passive solar heating and mechanical systems other than heat pumps are discussed in this chapter. Heat pumps, which can be used for both heating and cooling, are discussed separately in Chapter 5. Systems for distributing heat (and coldness) are discussed in Chapter 7 (Sections 7.3 and 7.4). Active solar heating systems and district heating systems are discussed in Chapters 11 (Section 11.4) and 15 (Sections 15.3–15.5), respectively.

4.1 Passive solar heating

Passive solar heating occurs when a building is heated by direct absorption of sunlight. CEC (1991) provides a beautifully illustrated, full-colour collection of examples of European buildings from the 1980s that obtained up to 68 per cent of their annual heating requirements in this way, employing only modest levels of insulation and modestly performing windows (by today's standards), and through the use of various combinations of sunspaces, good orientation, thermal mass, passive redistribution of solar heat and shading devices (to avoid overheating). CMHC (1998) and Steven Winter Associates (1998) provide practical design and construction guides for these conventional elements of passive solar design in houses, while Balcomb (1992) and Athienitis and Santamouris (2002) provide a more theoretical discussion. More advanced techniques include the use of *solar collectors*, *airflow windows*, *exhaust-air windows*, *solar walls*, and *integrated PV-solar collector* systems. Technical details concerning conventional and more advanced techniques, real-world examples and data on energy savings are provided in books by Hastings (1994), Hastings and Mørck (2000) and Hestnes et al (2003). All three books are outputs of the International Energy Agency's *Solar Heating and Cooling* Implementation Agreement, which is outlined in Appendix C. Passive solar heating is often regarded as unnecessary or not cost-effective in commercial buildings, due to the presence of significant internal heat gains. However, standard design rules significantly overestimate the magnitude of internal heat gains in commercial buildings, as discussed in Chapter 6 (Section 6.8), so the potential contribution and cost-effectiveness of passive solar heating in commercial buildings are often underestimated.

4.1.1 Solar irradiance on walls or windows of different orientation

Early architectural decisions concerning building shape and orientation strongly influence the ability to implement most passive solar heating features, as the amount of solar radiation striking a given surface and its seasonal and diurnal variation depend strongly on the orientation of the surface. This is illustrated in Figures 4.1 and 4.2, which show the seasonal variation in the daily average intensity of solar radiation (the solar *irradiance*) on walls or windows facing east or west, north and south, as well as on a horizontal surface, for clear skies at the equator, at 30°N, and at 50°N. The diurnal variation of solar irradiance on walls facing east, west, north and south in December and in June at 50°N is shown in Figure 4.3. The curves shown in these figures were calculated using the set of equations given in Chapter 11 (Box 11.1). In a complex urban environment, with partial shading, but also with reflection from nearby buildings, more complex, 3D calculations are required, and some of the software tools available for doing this are reviewed by Compagnon (2004). The calculations presented here show that:

- at the equator, there is little seasonal variation in the diurnally averaged solar irradiance on east- or west-facing windows, but large swings in the irradiance on north- and south-facing windows occur;
- at 30°N, the maximum diurnally averaged irradiance on windows

of any orientation is on south-facing windows in winter, while the diurnally averaged irradiance on east- or west-facing windows is over three times stronger than on south-facing windows in summer but only 40 per cent that on south-facing windows in winter;

- at 50°N, the maximum irradiance on windows of any orientation is on south-facing windows in spring and fall, while the diurnally averaged irradiance on east- or west-facing windows is 30 per cent stronger than on south-facing windows in summer but only 25 per cent that on south-facing windows in winter;
- the diurnally-averaged irradiance on east-, west- or south-facing windows in summer increases steadily as one goes from the equator to at least 50°N;
- during the summer, the solar irradiance is strongest on east-facing windows during mid-morning, on west-facing windows during mid-afternoon, and on equatorward-facing windows at noon, but the peak hourly irradiance on south-facing windows is only half that on east- or west-facing windows at 50°N;
- during winter, the peak hourly solar irradiance is twice as large on south-facing windows as on east- or west-facing windows.

Thus, equatorward-facing windows maximize the heat gain in winter while minimizing it in summer for mid-latitude sites. During transition seasons, an east-facing window can provide heat gain when it is most needed – to warm up a building after it has cooled during the night. However, a moveable shade will be required to prevent overheating. West-facing windows are more likely to cause overheating. Any heat that is collected will be less useful in commercial buildings (arriving as workers begin to leave) but will be more useful in residential buildings. The minimal useful winter heat gain and maximum summer heat

Figure 4.1 Seasonal variation in the daily average solar irradiance for cloudless skies on walls facing east or west, north and south, and on a horizontal surface at the equator, at 30°N and at 50°N. At 50°N, the surface reflectivity is assumed to be 0.25 during the summer half of the year and 0.45 during the winter half

Figure 4.2 Seasonal variation in the daily average solar irradiance for cloudless skies on south-facing (top) and east- or west-facing (bottom) walls at the equator, at 30°N and at 50°N

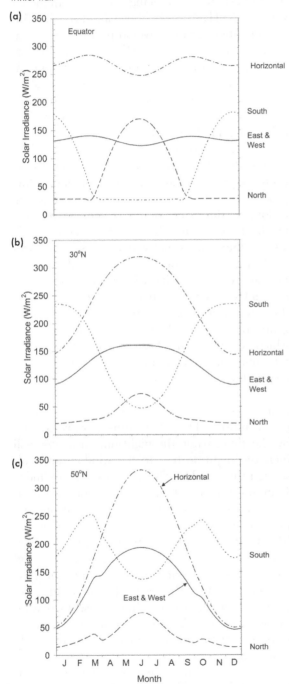

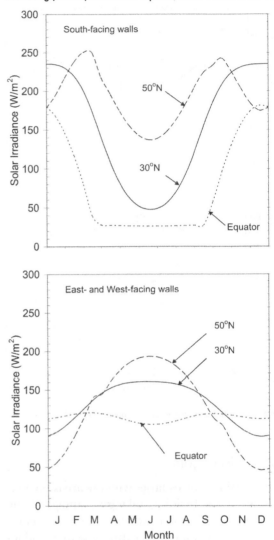

Note: Computed using the equations given in Box 11.1.

Note: Computed using the equations given in Box 11.1.

gain through west-facing windows might imply that the window area on western façades should be kept very small, except that substantial energy savings are possible through daylighting (Chapter 9, Section 9.3). The keys are to choose windows that minimize heat loss and unwanted heat gain, as discussed in Chapter 3 (Section 3.3.11), and to use adjustable external shading, something that can be facilitated through the construction of double-skin façades on tall buildings (Chapter 3, Section 3.4).

Figure 4.3 Diurnal variation in solar irradiance on vertical walls of various orientations for cloudless skies at 50°N in mid-June and mid-December

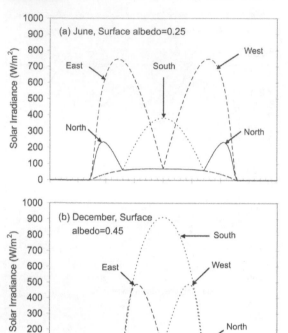

Note: Computed using the equations given in Box 11.1.

4.1.2 Direct gain

Direct gain involves large window areas on sun-lit sides of the building, taking advantage of the distribution of solar irradiance discussed above. It is the simplest passive heating system. Conductive heat loss through the windows needs to be minimized for there to be a net energy gain. Non-residential buildings, such as schools and offices that are used primarily during the daytime, are the best candidates for direct gain, if glare and overheating are avoided. It is possible to design to avoid glare by reflecting sunlight onto the ceiling or onto an interior storage mass (the latter will also reduce overheating).

As noted in Chapter 3 (Section 3.3.10), the best, currently available, advanced windows have centre-of-glass U-values of 0.6W/m²/ K and a SHGC (solar heat gain coefficient) of about 0.5. The heat loss will be 24W/m² when the outdoor temperature is −20°C and the indoor

temperature is 20°C. The coldest winter days are invariably cloud-free, so the clear-sky irradiance given in Figure 4.1c can be used to assess the solar heat gain on −20°C days. The centre-of-glass portion of east- and west-facing windows with the U-value and SHGC given above and no use of insulating blinds at night would provide a net heat gain on any cloud-free −20°C day of the year at 50°N. For south-facing windows, the minimum daily average clear-sky solar irradiance at 50°N is about 175W/m², so the centre portion of these windows will always be a net heat source on days cold enough to be cloud-free.

As discussed in Chapter 3 (Section 3.3.6), the U-value of the frame and glazing next to the frame is greater than the centre-of-glass value, producing a larger overall U-value. However, as long as the centre-of-glass region serves as a net heat source, it will be advantageous to use large rather than small windows, as the centre-of-glass area as a fraction of the total area will increase with increasing window area. Thus, the overall U-value will decrease with increasing window area (as long as spacers are not added to the centre portion). For high-performance windows, which serve as a net heat source beyond some minimum size, increasing the window area will reduce the heating requirements. Offsetting this is the fact that, as the window area increases, the usable fraction of the daytime solar gain will decrease (more will have to be thrown away to avoid overheating), while the night-time heat loss will increase in proportion to the window area (although this can be offset with insulating blinds). From this it follows that the impact of increasing window area on heating requirements cannot be assessed by examining the window U-value and SHGC alone. As discussed in Chapter 13 (Section 13.1), most building codes permit overall window U-values and, therefore, centre of glass U-values (2–3W/m²/K) such that the window often creates a net heat loss. In this case, increasing the window area increases the heating load, irrespective of the impact on the fraction of usable solar gain.

To maximize the utilization of solar heat captured through direct gain, the directly heated space should be ventilated with air from elsewhere in the building and exposed thermal mass should be available. This will prevent or minimize overheating and minimize the associated heat loss through emission of infrared radiation, thereby increasing the net heat gain and distributing it to rooms not directly exposed to solar gain. A high-performance envelope (more insulation, better windows) will tend to reduce the fraction of available solar energy that can be used (since overheating will occur more readily due to the smaller rate of heat loss), thereby increasing the need for effective thermal mass and ventilation. Solar radiation striking a floor can be stored if the floor has significant thermal mass, such as concrete floors. A 20cm thick floor stores only 30 per cent more energy than a 12cm thick floor, while requiring 67 per cent more material (Gratia and de Herde, 1997) and more structural support. Floor colour significantly influences heat storage in floors with thermal mass, dark colours being better. Carpets reduce the ability of a floor in direct sunlight to store heat.

Finally, in assessing the optimal window area for solar heat gain, the impact on summer cooling load should be determined. Increasing the window area will increase the cooling load even if the window is shaded, although this penalty can be minimized if *insulated* external operable shading devices (such as shutters) are used.

4.1.3 Solar air collectors

Solar air collectors can be mounted on a roof or on walls, and integrated into the structure of the roof or wall as part of the building envelope. Hastings and Mørck (2000) provide design guidelines and performance data for a wide variety of solar air collector systems. They recognize six basic system types, which are illustrated in Figure 4.4 and explained below. Solar hot water systems, used for both space heating and domestic hot water, are discussed in Chapter 11 (Section 11.4).

1   *Solar heating of ventilation air.* Outside air is drawn through a rooftop or wall-mounted solar collector, heated and then circulated directly into the space to be ventilated and heated. Because the temperature of the air entering the collector is rather low, very high collection efficiencies are possible (as explained in Chapter 11, Section 11.4.2).

Figure 4.4 Schematic illustration of the six categories of solar air heating systems discussed in the main text

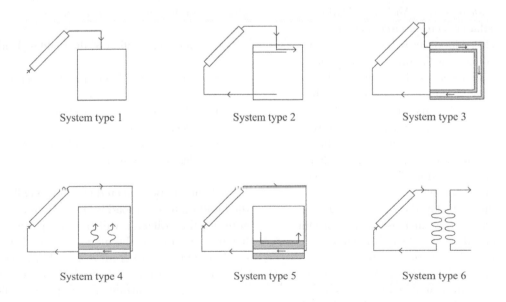

Source: Hastings and Mørck (2000)

2   *Open collection-loop with radiant-discharge storage.* Room air circulates through the solar collector, is heated, then returns to the room, either by natural convection (a solar chimney) or driven by a fan. Heat is stored in the ceiling and re-radiated at night. In the summer, the heated air is vented directly to the outside while ventilation air is supplied through an earth coil or from open north-facing windows.

3   *Double-envelope system.* Air heated by the solar collector is circulated in a closed loop through a cavity in the building envelope. This requires a double-envelope wall, but dramatically reduces the heat loss through the wall. Dampers shut down the airflow at night. In the summer, warm air from the solar collector is used for heating water instead. This system is especially appropriate for retrofitting poorly insulated apartment buildings if a second envelope is built directly over the initial envelope. Hastings and Mørck (2000) provide construction details.

4   *Closed collection loop with radiant discharge storage.* Air heated by the solar collector is circulated in a closed loop through channels inside concrete walls and/or floors. The walls or floors, due to their high mass, are able to absorb heat with minimal temperature increase (and hence, minimal losses) and re-radiate the heat during the night. Fan-forced circulation gives higher solar collection efficiency, since the air passes through the collector more quickly, thereby reducing the temperature increase of the collector and the loss of heat through re-radiation or convection.

5   *Closed collection loop with open discharge loop.* Air heated by the solar collector is circulated in a closed loop through channels inside the storage mass, as

in system 4, but interior room air is circulated through separate channels inside the floor. The storage can be located elsewhere from the room, insulated, and charged to a high temperature, with controlled active discharge. A rock bed rather than the wall or floor mass can be used for storage in this case. On cloudy days when the air is not heated enough to recharge the storage, the heated air would pass through the occupied space (forming an open loop system).

6   *Closed collection loop and hydronic heating loop.* Hot air from the solar collector passes over an air to water heat exchanger. The hot water is then circulated through conventional radiators or radiant floors or walls. This system is useful when the heat has to be transported a considerable distance (due to the smaller heat losses and pumping energy required with hydronic systems) and/or when the time comes to retrofit a building with an existing hydronic system.

In all systems, as the collector area/floor area ratio (A_c/A_f) increases or the building load decreases (due to a better envelope or a warmer climate),

- a larger fraction of the building load is satisfied, but
- heat collected per unit of collector area (Q/A_c) decreases,
- the efficiency of collecting solar heat decreases, and
- the cost of a unit of solar heat is likely to increase.

More heat will be collected by a given collector if the building heating load is larger because the air entering the collector will be cooler (more heat having been lost as the air is circulated through the building) and thus more readily available to absorb heat as it passes through the collector.

A faster airflow through the collector also increases the amount of heat that is collected. This is because, when the airflow is faster, its temperature increase as it passes through the collector is smaller, so less of the absorbed heat is re-radiated to the surroundings.

Table 4.1 summarizes the range of annual heat collection for the six systems described above in different climates, as given by Hastings and Mørck (2000). Solar-collection efficiencies range from 30–70 per cent, which is sufficient to meet 80 per cent or more of the annual heating load for well-insulated buildings with a collector area equal to 16 per cent of the floor area. Solar air collectors cost in the region of 200–400€/m². Further cost and performance data on solar air systems, and on hydronic solar collector systems, are presented in Chapter 11 (Section 11.4.5).

In systems with mechanically forced air movement, the temperature threshold at which the fan turns on can be set such that the delivered heat energy is at least three times the fan energy use. This ensures that there is always a savings in primary energy use if the electricity generation efficiency is only 33 per cent, and will yield a seasonally averaged fan COP (ratio of heat delivered to fan energy use) of 15–20 (Hastings and Mørck, 2000).

Table 4.1 Summary of heating-energy savings associated with different solar-air systems in different climates

| System type | Savings ($kWh/m^2_{collector}$/year) | | | |
| | Temperate | | Mild sunny | Cold sunny |
	Sunny	Cloudy		
1[a]	45–490	30–260	20–250	50–490
2[b]	55–130	40–75	55–115	80–200
2[c]	90–220	50–125	90–200	135–340
3[d]	150–320	100–225	100–225	150–400
4[e]	90–290	60–200	70–270	110–425
4[f]	50–190	45–120	–	50–260
5[e]	30–150	15–95	30–150	–
5[f]	15–140	5–90	8–110	–
6	70–280	60–175	40–290	140–360

[a] Absolute savings per unit collector area for all systems increases with the building heating load, increases with increasing airflow rate and decreases with increasing A_c/A_f.
[b] For natural convection.
[c] For fan-driven airflow of 50m³/h/m²$_{collector}$.
[d] Low savings is for A_c/A_f = 0.04 and low insulation (U = 0.58W/m²/K); high savings is for A_c/A_f = 0.16 and high insulation (U = 0.31W/m²/K for sunny mild climates and U = 0.17W/m²/K for cold mild climates).
[e] Residential buildings.
[f] Office buildings.
Source: Hastings and Mørck (2000)

Details concerning a roof-mounted collector system and the associated circulation of air and the distribution of temperatures in a school in northern Japan are given in Figure 4.5. When snow accumulates on the roof, the fans are reversed so that warm exhaust air warms the roof sufficiently for the snow to slide off, thereby making the roof ready to collect solar heat when the sun shines. In the summer, the airflow bypasses the roof, thereby avoiding unneeded heating of the building. Figure 4.6 shows three modes of operation of any entirely passively driven solar heating system in a house built on a south-facing slope in Santa Fe, New Mexico. This system is similar to System type 5 of Figure 4.4. When the sun is shining and heat is needed, heated air flows from the solar collector into the living space. When heat is not needed, the heated air is directed through a rock thermal storage. When the sun is not shining and heat is needed, warm air is allowed to rise from the warm rock bed and flow through the house.

Hollmuller and Lachal (2001) present data based on extensive monitoring and simulation work for an eight-storey residential building near Geneva with the following features: (i) rooftop solar air heaters; (ii) a ground coil for pre-heating or precooling incoming ventilation air; and (iii) heat exchange between incoming and outgoing ventilation air. The ground coil is placed underneath an underground parking garage, to minimize deviations from the mean annual temperature. Airflow from (i) or (ii) is directed through (iii), depending on the conditions. For a heat exchanger that warms the incoming air by 80 per cent of the temperature difference between incoming and outgoing air (i.e. an effectiveness of 80 per cent, compared with a standard 66 per cent effectiveness), there is little to be gained from ground pre-heating, so it is better to put resources into a better heat exchanger. However, the ground pipes are worthwhile for summer cooling, as they can eliminate the need for an air conditioner in this climate. Winter pre-heating then becomes an added free service. Flow rates are increased in the summer beyond that needed for fresh air, but not in winter, since in winter, additional conditioning (namely, heating) of the incoming outside air is needed. The extra fan energy consumption arising from the ground loop

Figure 4.5 Airflow and temperature distribution in a school in northern Japan with roof-based collection of solar heat

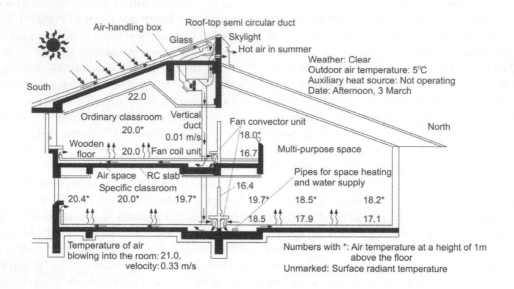

Source: Yoshikawa (1997)

Figure 4.6 Airflow during three modes of operation in a passively heated house in Santa Fe, New Mexico

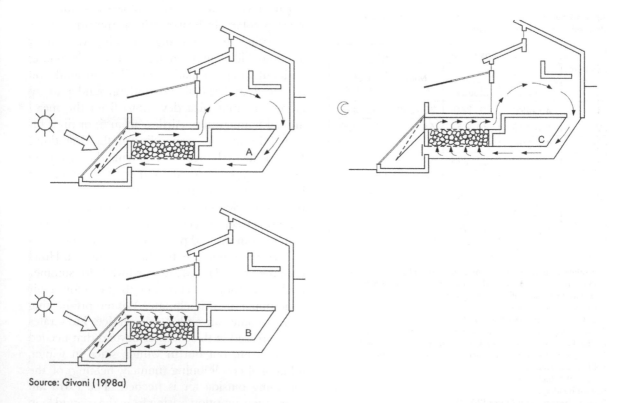

Source: Givoni (1998a)

is estimated to be 0.3–1.1 per cent of the ground loop plus heat exchanger cooling gain in summer, and 0.2–1.5 per cent of the heating gain in winter, for ground pipes of 21–12.5cm diameter (Hollmuller, personal communication, 2004). To avoid possible sanitary problems associated with any infiltration of water into leaky pipes, a closed underground water circuit could be coupled to the fresh air system with an above ground-water-to-air heat exchanger.

Simulation results for a housing development in Scotland indicate that with good insulation (RSI2.3 floors, RSI5.3 walls, RSI7.4 ceilings), triple-glazed windows, 0.3ACH (air changes per hour) without heat recovery and moderate thermal mass, a passive solar heating and ventilation system can meet almost all of the space heating requirements (Imbabi and Musset, 1996).

Zhai et al (2005) used a simulation model to compare the performance of single-pass and double-pass roof solar collectors in heating and ventilation mode for a traditional Chinese-style house. The structure of the double-pass collector and the flow in the two modes are illustrated in Figure 4.7. For a solar irradiance of 500W/m² and an outside air temperature of 0°C, the efficiencies of the single-pass and double-pass collectors in heating mode are 27 per cent and 40 per cent, respectively (in ventilation mode with 400W/m² irradiance, the single-pass and double-pass collectors can induce 12 and 20 air exchanges per hour, respectively).

Figure 4.7 Airflow in (a) heating mode and (b) ventilation mode for a double-pass roof solar collector

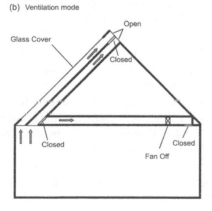

Source: Based on Zhai et al (2005)

4.1.4 Benefits and limitations of thermal mass

Overheating can be avoided either by venting excess heat or through the provision of thermal mass. Use of thermal mass would appear to be better, as heat can be stored and released at night. However, in buildings that are not occupied at night, high thermal mass can lead to increased energy use for heating. This is because the effectiveness of reducing the thermostat setting at night and allowing the building to cool is reduced, while the building will warm more slowly in response to solar heating in the morning, possibly necessitating use of the auxiliary heating system. The net effect of varying thermal mass will depend on a number of factors (building occupancy pattern, envelope characteristics, orientation) that will need to be assessed through computer simulation in order to determine the optimal thermal mass.

4.1.5 Trombe walls

An early passive solar heating system with thermal storage is the *Trombe wall*. The traditional Trombe wall consists of a massive wall behind a glass pane, typically with a 1m air gap. It can be thought of as a variant of the System type 4 system described above, with the entire wall serving as the solar collector. The idea is to collect solar heat that can be supplied after sunset while reducing the risk of overheating. This will complement the direct heat gain through the windows prior to sunset. The traditional Trombe wall is uninsulated so that heat absorbed by the wall can be conducted to the inside surface, but this also allows heat to be lost through conduction at other times. Zalewski et al (2002) have analysed the heat budget of the traditional Trombe wall and three variants using climate data for France. Two of the variants involve insulation behind the wall, with heat removed through airflow rather than by conduction and emission of radiation. The traditional Trombe wall leads to the largest energy savings for the particular case considered. Trombe walls with insulation save 30–40 per cent less energy, but the heat release is controllable and peaks early in the evening, when it is most useful. However, there will certainly be climate-solar irradiance regimes where the insulated Trombe wall will be superior. Insulated walls also reduce unwanted heat gains in the summer, by a factor of approximately three. More importantly, use of low-e double glazing (Chapter 3, Section 3.3.3) rather than standard double glazing dramatically improves the heat collection, such that an insulated Trombe wall with low-e glazing utilizes more solar energy than a traditional Trombe wall with standard glazing.

Greater effectiveness is likely to be achieved through Trombe walls with translucent or transparent insulation (Chapter 3, Section 3.2.8). This allows sunlight to heat the wall, while minimizing heat loss to the outside from the wall or from the building interior. A shading device can be used to minimize heating of the mass wall when heating of the building is not required; this can consist of motorized or manually adjustable blinds, or electrochromic or thermochromic systems, as discussed in Chapter 3 (Sections 3.3.15 and 3.3.16). Conversely, the Trombe wall can be used as a passive solar cooling device, by inducing natural ventilation of outside air through the buildings, as explained in Chapter 6, Section 6.3.1. Finally, instead of using rock or concrete as the thermal mass, a Trombe wall can be constructed with a suitable phase-change material, as discussed below. In some cases, it might be desirable and feasible to construct a building that makes use of direct gain on the east-facing wall, for early morning heating, with Trombe walls on the west face for late afternoon and early evening heating.

In new construction, high-performance glazing with adjustable shading can readily provide greater heating benefits than a Trombe wall unless, of course, the Trombe wall also incorporates relatively high-performance glazing. Trombe walls are potentially of greater value in retrofit situations, as a means of improving the performance of existing walls, particularly if transparent external insulation is applied to the existing wall.

4.1.6 Phase-change materials for heat storage

As already noted, a key to enhancing the effectiveness of passive solar heating is to have a means of storing heat – both to avoid overheating during the day, and to provide heat at night. In most passive solar buildings to date, this has been achieved either through massive walls or through concrete-floor storage. The specific heat and volumetric heat capacity of different conventional materials are compared in Table 4.2. Basement rock beds or water tanks are the best, but still often entail rather large volumes for adequate storage. In existing solar air heating systems, two salt hydrates are widely used to store heat (Hastings and Mørck (2000):

- $Na_2SO_4 \bullet 10H_2O$ (Glauber's salt or decahydrate of sodium sulphate[1]), with a melting point of 32°C and a latent heat of 360MJ/m³;
- $CaCl_2 \bullet 6H_2O$, with a melting point of 29°C and a latent heat of 281MJ/m³.

These salts are commercially available in containers with a variety of shapes. They are two examples of *phase-change materials* (PCMs) – materials that change their state or phase from solid to liquid and back. Figure 4.8 compares the heat stored as a function of temperature for rock (or concrete), water and the two salt hydrates. It is important to have adequate thermal conductivity of the PCM so that heat can be readily absorbed or released, and this can be achieved by the inclusion of various conducting metal fins within the container. Another option is to pack the salts in cylinders that alternate with hollow tubes, through which room air is circulated.

PCMs can also be made of various

Table 4.2 Thermal properties of some common materials at 20°C

Material	Density (kg/m³)	Specific heat (J/kg/K)	Volumetric thermal capacity (MJ/m³/K)
Clay	1458	879	1.28
Brick	1800	837	1.51
Sandstone	2200	712	1.57
Wood	700	2390	1.67
Concrete	2000	880	1.76
Glass	2710	837	2.27
Steel	7840	465	3.68
Gravelly earth	2050	1840	3.77
Magnetite	5177	752	3.89
Water	988	4182	4.17

Figure 4.8 Comparison of the amount of heat stored in rock, water and two salt hydrates as a function of temperature

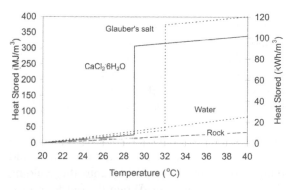

Source: Hastings and Mørck (2000)

organic substances such as polyethylene glycol, fatty acids and paraffin waxes. The latter have the general formula $CH_3\text{-}(CH_3)_n\text{-}CH_3$, where a larger n leads to a warmer melting point and a larger latent heat of melting. An alternative is to store solar heat through PCMs that are incorporated into the walls, floors or ceiling. All three organic PCM groups are thoroughly reviewed by Khudhair and Farid (2004). PCMs can be added to ordinary drywall (wallboard) either by adding micro-encapsulated PCMs during its manufacture, by incorporating an additive into the wet stage of manufacture, or by immersing the wallboard in liquid wax and organic acids after

Figure 4.9 Scanning electron microscope image of micro-encapsulated PCM in gypsum plaster

Source: Schossig et al (2005)

its manufacture. A scanning electron microscope image of drywall with micro-encapsulated PCM is given in Figure 4.9. PCM drywall has about 10 times the heat storage capacity of regular drywall. Bulk encapsulated phase-change materials can also be used and inserted into cavities along with ducts for night ventilation (Yanbing et al, 2003). The specific mixture of acids and waxes used determines the temperature range over which melting occurs, but can be tailored to give a phase change over a finite temperature interval somewhere between 16°C and 21°C with a latent heat of about 300–400kJ/m². In passive solar buildings, this is sufficient to reduce peak daytime temperature by 4K and to greatly reduce the heating load at night (Athienitis et al, 1997). A number of techniques have been developed to render PCM drywall fire resistant.

PCMs can also be used as an alternative to rock or concrete as the thermal mass in a Trombe wall. Stritih (2003) has analysed the thermal properties of a Trombe wall made of a commercially available paraffin wax known as Rubitherm RT30, which melts between 25 and 30°C. The wall consists of a 5cm thick layer of Rubitherm RT30 embedded with metal fins so as to improve the rate of heat transfer into the wall, overlain by transparent insulation and a glass cover on the outside, and separated from an insulated inner wall by an air gap through which ventilation air flows and is heated. The spacing of the metal fins was found to be one of the most important parameters governing the behaviour of the PCM Trombe wall. Another possibility is to incorporate a PCM directly into a roof-integrated solar air collector, as modelled by Saman et al (2005).

PCMs can be used for a wide variety of applications other than storing solar heat for space heating in winter: to limit summer daytime temperature extremes and to increase the effectiveness of night ventilation for cooling (Chapter 6, Sections 6.2.6 and 6.3.3); for storing coldness produced by running air conditioners at night (Chapter 6, Section 6.9.3); and to store solar heat for domestic hot water (Chapter 11, Section 11.4.3). Zalba et al (2003) list 150 organic and inorganic phase-change materials that are used in research, and 45 commercially available products with melting temperatures ranging from $-33°C$ to $112°C$.

4.1.7 Airflow windows and double-skin façades

An airflow window usually consists of an outer double-glazed window, an interior single- or double-glazed window, gaps allowing airflow from the inside or outside of the building into the space between the glazings at the bottom of the window, another gap at the top to allow air to exit the space between the glazings, and an optional adjustable absorbing/reflecting blind within the air gap. Enough light passes through the window for natural lighting. During the winter the blind may be raised, or lowered with the absorbing side turned outward, while during the summer, the blind may be lowered with the reflecting side turned outward. Sir Norman Foster's Commerzbank in Frankfurt, completed in 1997 and profiled in Fischer et al (1997) and Melet (1999), was the first office tower to use airflow windows.

Two variants of the airflow window are found. In the first variant, known as the *supply air* window, air from outside is drawn between the window glazings, then either enters the interior of the building to satisfy some of the ventilation requirements (winter) or is directed back outside (summer). During the winter, the incoming air will pick up some of the heat that would otherwise be lost through the window, much like the airflow through dynamic insulation, as well as picking up heat absorbed between the inner and outer

glazings. During the summer, the blind would be lowered during times of direct sunshine, and the airflow serves to remove heat from the window assembly that would otherwise enter the building.

In the second variant, known as the *exhaust-air* window, indoor air passes through an inner air gap at the base and is either:

- vented directly to the outside at the top;
- recirculated directly through the building;
- passed through hollow-core concrete floor slabs or other thermal storage devices (such as rocks) before being vented;
- directed to a heat exchanger in order to transfer the collected heat to the incoming ventilation air before being vented.

During the winter, one of the last three modes would be preferred rather than direct venting to the outside. In any case, heat conducted to the outside through the window comes largely from the exhaust air rather than from the interior air, thereby reducing heat loss from the building. During the summer, the exhaust air would probably be vented directly to the outside. During both winter and summer, the exhaust-air window would tend to draw fresh outside air into the building through other openings. When the outdoor air is too warm for direct ventilation, the window could be operated as a supply air window with outside air drawn directly from the outside through the lower opening of the window and out the top. These options are summarized in Figure 4.10.

Airflow windows were first used in Finland in the 1950s. Hastings (1994) provides more recent examples from Finland, France, The Netherlands and Switzerland. Figure 4.11 provides details and energy flows for an exhaust-air window on an office building in Finland. Exhaust-air windows have been used in external walls of all orientations in Finland. A supply air window produced by the Finnish company Domlux (www.domus.fi) is illustrated in Figure 4.12.

The Finnish company Tiivi (www.tiivi.fi) produces airflow windows with a whole window

Figure 4.10 Alternative flow configurations for airflow windows and/or Trombe walls and the applicable season

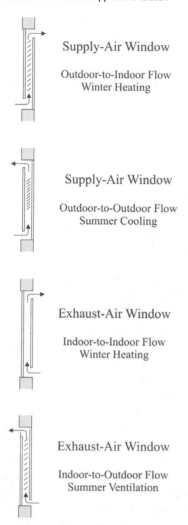

Supply-Air Window

Outdoor-to-Indoor Flow
Winter Heating

Supply-Air Window

Outdoor-to-Outdoor Flow
Summer Cooling

Exhaust-Air Window

Indoor-to-Indoor Flow
Winter Heating

Exhaust-Air Window

Indoor-to-Outdoor Flow
Summer Ventilation

Figure 4.11 Energy flows through an exhaust-air window on an office building in Finland

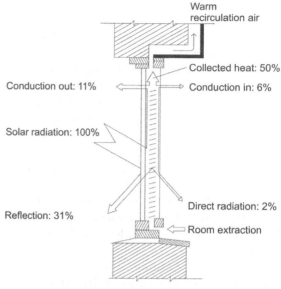

Source: Hastings (1994)

Figure 4.12 A supply air window showing (1) external air inlet; (2) sound attenuating filter; (3) gap between glazings, where the air warms; and (4) adjustable discharge vent

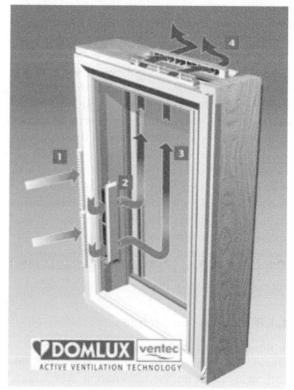

Source: Domlux (www.domus.fi)

U-value of 1.1W/m²/K (0.9W/m²/K centre of glass value), while Domlux produces airflow windows with a centre-of-glass U-value as low as 0.6W/m²/K. Effective U-values, taking into account heat reclaimed by the ventilation air, would be much less in both cases. The effective window U-value depends on the width of the ventilation gap (which should be small enough to prevent turbulent airflow), the placement of a low-e coating, the incident solar irradiance and the ventilation rate. McEvoy et al (2003) report measured effective U-values as low as 0.33W/m²/K, compared with 1.4W/m²/K for a comparable triple-glazed window. Heimonen (2004) reports measured effective U-values of 0.6–0.7W/m²/K in windows with real U-values of $1.6-1.8W/m^2/K$ – a reduction by about 60 per cent.

Haddad and Elmahdy (1998, 1999) computed the difference in monthly heat gain between conventional, supply air and exhaust-air windows facing different directions for the climate of Ottawa. For south-facing windows, the supply air window provides an additional heat gain that can easily be half the heat gain of a conventional window. There is little difference in the winter heat gain between exhaust and supply air windows. During summer, the supply air window slightly increases the cooling load (while providing ventilation), but this increase in cooling load is small because Ottawa's summers are relatively cool. The exhaust-air window causes a very slight reduction in cooling load while providing ventilation.

Airflow windows may use fans, rather than relying entirely on passive driving forces. At low airflow rates, the reduction in heating energy use greatly exceeds the primary energy use associated with the electricity used by the fans. However, as the airflow rate is progressively increased, a point is reached where the additional reduction in heating energy use equals the increase in fan primary energy use. This will occur when the extra fan electricity use reaches about one-third of the savings in heating energy use if the electricity generation efficiency is 33 per cent. At this point, total primary energy use is minimized, so there should be no further increase in flow rate. This optimum flow rate, which can occur at rather small flow rates, should be assessed through computer simulation during the design stage. An example is found in Hastings (1994). If the goal is to minimize CO_2 emissions rather than to minimize energy use, and if electricity is produced by burning coal while heat is supplied from natural gas, then the airflow rate should not be increased further once the increase in fan electricity use reaches approximately one-fifth of the savings in heating energy use (see Chapter 2, Section 2.4.1).

4.1.8 Double-skin façades

A double-skin façade (described in Chapter 3, Section 3.4) functions like an airflow window, but both skins need not be transparent everywhere, and the gap between the two skins can be wide enough to serve as an accessible walkway. The double-skin façade will have a lower overall U-value than the inner skin alone, but heat losses due to ventilation will be larger, particularly since the practice is to use natural ventilation through open façades for external temperatures down to $0-5°C$. The rate of ventilation is likely to be greater than in a mechanical system where the airflow is for ventilation only (with heating and cooling largely through hydronic systems), especially with demand-controlled ventilation. The mechanical system could be equipped with heat recovery, whereas this is difficult with natural ventilation. On the other hand, the heating savings can be much larger than expected based on the change in U-value, due to heating of the air inside the façade by a deployed solar shading device (even a strongly reflecting white or metallized surface will absorb 30–35 per cent of the solar radiation). The shading device should be a minimum of 15cm from the outer skin to avoid excessive heating of this layer, but no further, as this will place it closer to the inner layer and increase the unwanted heat transfer to the interior during the summer. Outside air that is heated as it passes through the double-skin façade could be directed into the mechanical ventilation system when this is useful. The net energy savings will clearly depend on the number and nature of the different operational modes, the temperature thresholds chosen for different operational modes, and the internal heat gain of the building. To fully understand the many interactions between U-value, solar heat gain, sunshading and ventilation, full-scale mockups of the double-skin façades used in many large projects were built prior to finalizing the details for large-scale production.

Faggembauu et al (2003a, 2003b) developed a computer model of a double-skin façade, consisting of an outer glazing layer separated by an air gap from an opaque zone (lower third of each floor) and a double-glazed window (upper two-thirds of each floor), as illustrated in Figure 4.13. They considered a variety of cases for a Mediterranean climate, including cases with a closed air channel, a channel with passive airflow and a channel with forced airflow at a fixed rate (the latter two cases are similar to a supply air window). Figure 4.14 shows the seasonal variation in two quantities of interest for the

Figure 4.13 The double envelope façade simulated by Faggembauu et al (2003a, b)

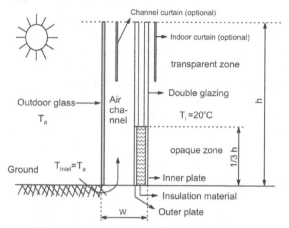

Source: Redrafted from Faggembauu et al (2003a, 2003b)

Figure 4.14 Seasonal variation in the heat flow across the double envelope façade shown in Figure 4.13 (η_i) and into the air ventilating the façade (η_a), as a fraction of the flux of incident solar radiation, for cases with forced ventilation airflow at 2m/s and for natural convection

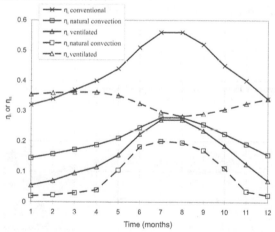

Note: Also given are the heat fluxes for conventional double glazing.
Source: Faggembauu et al (2003b)

cases with an open channel: the heat flow across the façade into the interior of the building as a fraction of the incident solar irradiance (η_i), and the heat flow into the air flowing between the inner and outer envelopes as a fraction of the incident solar irradiance (η_a). The heat flow into the room includes solar radiation as well as the net effect of infrared radiative exchange, conduction and convection. Also shown, for comparison, is η_i for a conventional façade with double-glazed windows.

For a conventional façade, η_i is about 0.32 in winter and 0.54 in summer. The heat flow into the building in summer is reduced by roughly a factor of two using either a passively or mechanically ventilated façade. With natural ventilation, very little heat is picked up by the air in winter and about 20 per cent is picked up in summer. When the façade is ventilated with forced airflow at 2m/s, the air ventilating the façade picks up 35 per cent of the incident solar heat in winter and about 28 per cent in summer. This heated air can be used to supply the building ventilation requirements in winter, but vented to the outside in summer (see also Chapter 6, Section 6.3.2). The total winter heat gain for the ventilated case ($\eta_i + \eta_a$) is 20–25 per cent larger than the total (η_i alone) for the standard case during the winter months. This is a modest impact compared with the factor of two reduction in η_i during the summer months. Hence, in this case, the double-skin façade has a larger impact in reducing the cooling load than in reducing the heating load. The façade, floors, and internal partitions should be designed with sufficient thermal mass to eliminate the need for early morning heating followed by afternoon air conditioning during transition seasons.

4.1.9 Perforated unglazed solar walls

Another option, most applicable to buildings with large window-less walls (such as manufacturing facilities) is the *solar wall* (IEA 1996, 1999a), also known as the *transpired solar collector*. This consists of an exterior metal cladding that performs like a solar panel, but is perforated with millions of small holes. Outside air enters through the holes and is preheated before being fed into the ventilation system. A temperature rise of 20–35K can be achieved on a sunny winter's day. A wide range of colours is possible while maintaining an absorptivity of 0.86 or more. Not only is solar energy captured, but heat loss through the wall is reduced (the two effects can be of comparable importance) and the system can be used to destratify the indoor air. In an automotive manufacturing plant in Canada, the solar collection efficiency is about 70 per cent, with an additional comparable energy benefit due to reduced heat loss and destratification of indoor air (IEA, 1996). Among the limited number of solar air systems used in North America, the solar wall is the most common.

Table 4.3 Technical specifications, cost and energy savings using a transpired solar collector in Toronto

Building description	
Type	Industrial
Floor area	6700m²
Wall insulation	RSl2.1
Roof insulation	RSl3.5
Location	Toronto
Solar collector specifications	
Collector orientation	Vertical, 20° azimuth
Collector area	144m²
Collector colour	Black
Collector absorptivity	0.90
Average efficiency	0.47
Average temperature rise	13.0K
Average airflow	25.1m³/h/m²
Incremental fan power	0.0W
Solar collector costs and credits	
Solar materials	$90/m²
Solar installation	$50/m²
Conventional materials	$45/m² (credit)
Conventional installation	$28/m² (credit)
Solar fan and ductwork, materials	$1.4/m²/h
Solar fan and ductwork, installation	$1.0/m²/h
Conventional fan and ductwork, materials	$1.0/m²/h (credit)
Conventional fan and ductwork, installation	$1.0/m²/h (credit)
Net cost	$15228
Overhead (10%)	$1523
Contingencies (10%)	$1675
Total	$18426
Energy and energy cost savings	
Collected solar heat	56.9MWh/year
Reduced building heat loss	6.3MWh/year
Destratification savings	49.5MWh/year
Total savings	112.7MWh/year
Cost savings for NG at $12/GJ, 0.92 heating efficiency	$5292/year
Simple payback time	3.5 years

Source: Data obtained from the solar air package available from Natural Resources Canada (www.retscreen.net)

The Renewable Energy Technology spreadsheet package (RETScreen) developed by Natural Resources Canada contains a module specifically for the analysis of the performance and cost of transpired plate solar collectors; it can be downloaded for free from www.retscreen.net, is available in 21 languages, and can be applied to any location in the world using the SSE climate dataset from NASA. Technical specifications, costs and energy savings for an industrial building in Toronto are presented in Table 4.3. This analysis indicates a simple payback of 3.5 years if heating is provided by natural gas at $12/GJ with an efficiency of 92 per cent, but almost half of the savings in this case are due to destratification of indoor air, which would not be applicable in multi-unit residential buildings. Conversely, these buildings may face higher retail energy costs than large industrial facilities. Figure 4.15 illustrates the installation and final appearance of a transpired solar wall on an apartment building in Windsor, Canada.

Perforated solar walls have a capital cost comparable to or less than that of the

Figure 4.15 Illustration of a transpired solar collector ('Solarwall') on an apartment building in Windsor, Canada

Source: Solarwall (www.solarwall.com)

conventional exhaust-air heat exchanger that is typically used to preheat ventilation air, but entails a lower pressure drop, so fan energy use will be smaller. However, a conventional heat exchanger will save energy when the sun is not shining. Maximization of energy savings would entail having a heat exchanger that is bypassed when the sun provides adequate preheating. Present transpired solar collectors are made of aluminium, which is costly and has a high embodied energy. Galvanized steel and some plastics could be used with only a few percent reduction in efficiency but significantly reduced cost (Gawlik and Kutscher, 2002).

4.1.10 Integrated PV solar collector systems

Photovoltaic (PV) panels can be integrated with a solar air system. PV panels themselves produce far more heat than electricity, so putting the heat to good use is desirable. An analysis by Kiss et al (1995) indicates that up to three times the heating requirement could be provided in this way for a one and a half storey convention centre in Chicago (when the peak daytime solar irradiance reaches $1000W/m^2$, about $350W/m^2$ would be transferred to the inside of the building, $100W/m^2$ would be converted to electricity, $160W/m^2$ would be reflected, and $390W/m^2$ would be radiated to the sky). Not only is solar energy in the form of heat made use of, but the efficiency of the PV panels in generating electricity increases. This is because the electricity generation efficiency decreases with increasing temperature, and solar panels can be up to 50K above the ambient temperature if heat is not removed.

Charron and Athienitis (2003) analyse the integration of PV panels into ventilated double-skin façades. They consider two configurations, the first with the PV panel forming the outer façade, and the second with the PV panel in the gap between the inner and outer façades (the outer façade being glazed). Outside air enters at the bottom of the façade and can either be drawn into the building (so the ventilated façade serves to preheat the ventilation air) or exhausted to the outside (thereby removing heat that accumulates between the inner and outer façades). In the second case, the PV panel is cooled from both

sides, so more heat is captured and the PV is kept cooler, which tends to increase the electricity generation efficiency, but the outer glazing tends to reduce the production of electricity. With an airflow velocity of 1m/s, a 0.1m gap within the double-skin façade, and for a 1.5m high PV panel outside the lower portion of the façade, the annual heat and electricity production are 331kWh and 155kWh per metre width, respectively. When the PV panel is placed inside the façade, heat production more than doubles, to 692kWh/m, while electricity productions falls about 20 per cent (to 123kWh/m). However, almost the same amount of heat is captured with the first configuration as with the second configuration if 1.5cm conducting fins are attached to the back of the PV panel.

Further discussion of integrated PV/heating systems is found in Chapter 11 (Section 11.6), along with a discussion of the complementary role of PV/double-façade systems in reducing cooling loads.

4.1.11 Seasonal and diurnal variation in thermally driven flow rates

In the absence of fans, the rate of airflow through a Trombe wall, solar wall or airflow window depends on the difference in the density between air in the gap within the window or wall and the outside air (if the air is vented to the outside). This in turn depends on the temperature difference between the two. Although solar heating of the air gap is smallest in winter, the outside air is also coldest then. Thus, it is possible for the induced airflow to be larger in winter than in spring, as found by Alfonso and Oliveira (2000) for a Trombe wall test in Portugal (average flow was approximately 30 per cent greater in January than in March). With regard to diurnal variations in January, Alfonso and Oliveira (2000) obtained maximum flows around 10AM, minimum flow around 3PM and a secondary maximum around 9PM.

4.1.12 Atria and sunspaces

A sunspace attached to a house can provide a net saving in heating energy if used properly, that is, if the door and window between the sunspace and house are opened only when the sunspace temperature exceeds 21°C. For a sunspace with double-glazed windows, the simulated savings for a house in Zurich is 20 per cent (Hastings, 1995). When the windows or doors are never opened, the savings is only 8 per cent, while if the window is always left open, the heating demand is 8 per cent larger than without the sunspace. Use of a small heater with a thermostat setting of 5°C (to prevent the sunspace from freezing) has a negligible effect on annual heating energy use. Mechanical ventilation to exchange air between the sunspace and house is not justified – opening the door or window when the sunspace is warm is a much simpler and almost as effective procedure! Building a sunspace can be an expensive way of providing some of the heating requirements, but it has the added benefit of providing additional living space during part of the year.

In commercial buildings, a savings in heating energy can be achieved through the use of an atrium if the temperature of the atrium is allowed to float to a lower temperature than the rest of the building during winter. This is permissible because atria are normally not continuously occupied by any one person. The heat that is supplied to an atrium can be more than compensated by the reduced heating required in the adjacent spaces. Any heat that is provided should be done radiatively rather than through heated air, as the heated air will rise, thereby reducing its impact on thermal comfort and increasing heat loss. If the atrium temperature is left entirely free to vary, there should be insulation between the atrium and the heated space.

As noted in Chapter 3 (Section 3.3.6), the heat loss from horizontal or slightly inclined glazing is about 50 per cent greater than through the same glazing but oriented vertically. Furthermore, unlike vertical glazing, inclined glazing facing equatorward receives more solar radiation during the summer than during the winter, thereby amplifying the tendency to overheat during the summer. In addition, the greater surface area of a sloped atrium roof increases heat loss through the emission of infrared radiation.

4.1.13 Earth pipe preheating of ventilation air

Given that the subsurface ground is warmer than the outside air during winter, incoming ventilation air can be preheated to some extent by drawing it through pipes that are buried in the ground. In mid-latitude locations in winter, ventilation air can be warmed by 2–6K in this way, the two most important factors being the length of the pipe (30–70m being a typical range) and the depth to which it is buried (1–3m being a reasonable range) (Mihalakakou et al, 1996). Earth pipes can also be used for the cooling of ventilation air in summer, and are discussed further in this context in Chapter 6 (Section 6.3.8). The main value of earth pipes is likely to be in the summer at mid-latitude locations, as the ground is generally cooler than needed for air conditioning in summer but it is not warmer than needed for heating in winter. The cost-effectiveness of such systems compared to alternatives such as heat recovery ventilators, both for heating and cooling, is likely to depend on the severity of the winters and summers.

4.2 On-site non-solar heating systems for single family houses

Having reduced the heating requirement through a good thermal envelope and maximum use of passive solar heating, the key determinant of the amount of secondary energy (fuel or electricity) required to heat a house is the efficiency of the heating system (ratio of heat delivered to secondary energy used). In this section, the efficiency and characteristics of various non-solar options for heating single family houses are discussed, other than on-site cogeneration, which is discussed in Section 4.6.

4.2.1 Furnaces

A furnace, whether fuelled by oil or natural gas, consists of (i) an ignition system; (ii) a burner, where the fuel is combusted; (iii) a heat exchanger, where heat from the hot combustion gases is transferred to flowing air; (iv) a fan (often called a 'blower') for circulating air past the heat exchanger and through the ductwork; and, in some cases, (v) a small fan to assist the flow of the exhaust gases through the chimney (if present) or vent.

Furnaces are available in sizes giving peak heat outputs of $7kW_{th}$ to $94kW_{th}$.[2]

Furnace efficiencies range from about 70 per cent (i.e. 30 per cent of the heat from burning the fuel is lost out the chimney) to 86 per cent using oil and 97 per cent using natural gas (only a 3 per cent loss). The primary factors affecting the efficiency of a furnace are as follows:

- *Use of a pilot light or electronic ignition.* A pilot light is a small flame that burns continuously, summer and winter, so that the furnace is always ready to start. Pilot lights can consume up to 5 per cent of the total energy used by a furnace or 6GJ/year, whichever is less. An electronic ignition avoids this waste.

- *Use of natural draft or fan-assisted draft.* Natural draft refers to combustion gases exiting through the chimney or exhaust vent under the influence of their own buoyancy. A draft hood is required in order to minimize heat losses when the furnace shuts down. Fan-assisted furnaces do not require a draft hood and have smaller off-cycle heat loss.

- *Use of indirect or direct air supply for combustion.* The air needed for combustion can be drawn from the air inside the house, or can be drawn directly from the outside through a pipe. In the former case, the house will be depressurized when the furnace operates, inducing infiltration of cold outside air, thereby increasing the heating requirement. This induced infiltration is avoided if combustion air is drawn from outside directly to the burner, but the efficiency of the furnace will be reduced because the combustion air will be colder. However, this efficiency loss can be avoided if the combustion air is preheated by the exhaust air. A *co-axial* system is commonly used, in which combustion air is drawn through an outer pipe that surrounds

the exhaust pipe, thereby permitting heat to be transferred from the outgoing to the incoming airflow.

- *Effectiveness of the heat exchanger.* A larger or more effective heat exchanger will transfer more heat from the combustion gases to the air that is to be heated. A corollary of better heat transfer is that the temperature of the combustion gases can be reduced enough to avoid the need for a chimney and to permit condensation of exhaust water vapour (see below).

- *Condensation of water vapour in the combustion gas.* The main products of burning a hydrocarbon fuel such as natural gas or heating oil are gaseous CO_2 and H_2O. If the water vapour is condensed, latent heat will be released, and if this heat is used (by preheating combustion air, for example), then the overall efficiency is improved. A larger heat exchanger is needed in order to cool the combustion gases sufficiently for condensation to occur, and the low temperature portion must be made of stainless steel, both of which increase the cost. Condensing units must be connected to a sewer drain in order to get rid of the condensate.

All of the above factors determine the *steady state* efficiency of the furnace, that is, the efficiency under continuous, steady operation. In practice, a furnace will cycle on and off, with the proportion of time that it is off increasing as the heating load decreases. On/off cycling leads to additional energy losses and, therefore, a lower average efficiency. The more that a furnace is oversized (that is, the larger it is compared to what is needed to meet the largest anticipated load), the more frequently it will cycle on and off and hence the lower the average efficiency (not to mention being rather annoying).

An efficiency measure that takes into account on/off cycling losses and energy losses from the pilot light (if present) is the *annual fuel utilization efficiency* (AFUE). Typical AFUE values for natural gas and oil furnaces are given in Table 4.4. Condensing furnaces are 8–9 per cent more efficient than the best non-condensing furnaces, with AFUEs of 90 per cent or better. The best available AFUE values in 2004 for natural gas and oil furnaces were 97 per cent and 86 per cent, respectively (GAMA, 2005).

Table 4.4 Typical AFUE values for natural gas and oil furnaces

Type of natural gas furnace	AFUE (%)
1. Natural draft with standing pilot	64.5
2. Natural draft with intermittent ignition	69.0
3. Natural draft with intermittent ignition and auto vent damper	78.0
4. Fan-assisted combustion with standing pilot or intermittent ignition	80.0
5. Direct vent with preheat, fan-assisted combustion, intermittent ignition	80.0
6. Same as 5 + condensing	90.0
Type of oil furnace	
1. US standard – pre-1992	71.0
2. US standard – post-1992	80.0
3. Same as 2, with improved heat transfer	81.0
4. Same as 3, with automatic vent damper	82.0
5. Condensing	91.0

Source: ASHRAE (2001b, Chapter 28)

Some recent furnaces (and boilers) can alter their heat output by varying the rate of fuel consumption. These are referred to as *modulating* furnaces. A modulating furnace saves energy in two ways: (i) by eliminating on/off cycling losses; and (ii) by permitting a lower thermostat temperature setting (conventional furnaces must have a higher thermostat setting so as to maintain a minimum temperature at the end of the off portion of the on/off cycle). Modulating units are significantly more expensive than non-modulating units due to the cost of the additional valve and controls, and the need for expensive stainless steel in order to protect the heat exchanger from possible condensation at lower heating levels. If the unit is a condensing unit, then protection against condensation is required whether or not fuel flow modulation is employed. An alternative to full modulation is a two-stage furnace, whereby the furnace operates most of the time at lower capacity, with steadier operation, and resorts to the full capacity only when needed.

In addition to using fuel to provide heat, furnaces also use electricity to power the fan that blows the air through a house. This is a significant energy use, with a significant savings potential, as discussed in Chapter 7 (Section 7.3.2).

4.2.2 Fireplaces, wood-burning stoves and wood-burning boilers

In a conventional fireplace, the combustion air is drawn directly from the interior of the house. This air exhausts through the chimney, depressurizing the house and inducing an inflow of cold outside air through various leakage paths. If the house is close to air-tight (as required to minimize heating loads), the depressurization can induce the flow of exhaust air from the furnace or natural gas hot-water heater (if present and running) back into the house, a potentially fatal occurrence. Thus, for both safety and energy-efficiency reasons, fireplaces should be constructed so as to draw combustion air directly from the outside. For energy efficiency reasons, the chimney and air intake should be designed so that the combustion air is preheated by the hot exhaust air. The same principles apply to wood-burning stoves. Hot water can be used to distribute heat from a fireplace or wood-burning stove through a radiant floor heating system (see Chapter 7, Section 7.3.1).

Domestic wood-pellet boilers were introduced in Austria in 1994 and have rapidly grown in popularity. Wood-pellet boilers with efficiencies in excess of 90 per cent produce hot water that feeds a radiant hot-water heating system, and have automatic ignition, electronic throttling down to 30 per cent of peak output, and automatic pellet feeding and metering, heat exchanger cleaning and ash removal. For multi-unit residences, an underground concrete wood-chip storage vessel has been used. Pellets provide a standardized fuel that can be delivered pneumatically by truck. They are made from wood wastes such as sawmill dust. 6–8m^3 of wood waste will be compressed into 1m^3 of pellets, sometimes with a natural bonding agent such as maize starch, yielding a density of at least 650kg/m^3. The production of pellets requires less than 2 per cent of the energy content of the pellets, assuming the wood waste to be already available. The energy density of pellets is 3.3–4.2kWh/litre (4.9–5.4kWh/kg), compared with 9.4kWh/litre for heating oil. The cost of wood pellets is less than half that of heating oil in many countries in Europe, but wood-pellet boilers are considerably more expensive than oil boilers. The economics are more favourable for large boilers (100kW$_{th}$ or more) serving multi-unit buildings. The Austrian company Rika (www.rika.at) exemplifies the types of products available. The *Bioheat Project* of the European Commission is promoting the use of biomass in large buildings; numerous reports and case studies can be obtained from www.bioheat.info.

Pellet heating systems provide a natural complement to active solar heating systems, as the primary reliance on solar energy (combined with a high-performance thermal envelope) reduces the need to transport and store the relatively bulky pellets, while the pellets provide a renewable energy source to meet extreme conditions that are not easily met with solar energy. Fiedler et al (2006) have modelled such hybrid systems, and note that there is a large potential for system optimization.

4.2.3 Dispersed kerosene heaters

In most parts of Japan, kerosene heaters are very popular and used, as needed, to heat individual rooms. The traditional heater, still popular, is the *Kotatsu*, consisting of a low table with a heater underneath and covered with a quilt. Indoor temperatures in bedrooms without heaters are very low. As comfort standards increase, there will be a large increase in heating energy use in Japan unless accompanied by large improvements in the thermal envelope.

4.2.4 Electric resistance heating

The efficiency of electric resistance heating, at the point of use, is 100 per cent – all of the electricity used is converted into heat inside the building. However, electricity in most parts of the world is generated from fossil fuels (largely coal) at rather low efficiency – typically around 35 per cent (see Chapter 15, Section 15.1.1). When losses during electricity distribution are taken into account (typically a few percent of production), the amount of primary energy used is about three

times the amount of electrical energy delivered to the house. For continental-scale distribution of natural gas from natural gas fields to end use customers, in contrast, only about 1.05–1.1 units of primary energy (unprocessed natural gas at the point of extraction) are required per unit of natural gas delivered to the customer (some of the natural gas being transmitted is used to power compressors needed to push the gas through the pipeline network). Even in jurisdictions where coal provides only part of the total electricity supply, it is invariably used as the 'swing' energy source to meet peaks in demand. Since electric resistance heating contributes to peak demand (all the buildings need heating at the same time), it should be counted as coming 100 per cent from coal-fired electricity in such circumstances. Thus, in jurisdictions where electricity is generated from coal at low efficiency, use of electric resistance heating entails more than a factor of two greater primary energy than direct use of oil or natural gas in a high-efficiency (85–92 per cent) furnace. The impact on CO_2 emissions will be greater still, due to the almost double CO_2 emission factor of coal compared with natural gas (see Chapter 2, Section 2.4.1). However, in super-insulated houses with good thermal mass, peak and overall heating loads will be very small, so electric resistance heating could be a reasonable choice where heat pumps are not practical. This is especially the case if and when peak electricity loads can be supplied by renewable energy sources.

4.2.5 Air source, ground source and exhaust-air heat pumps

Heat pumps are discussed extensively in Chapter 5, but are mentioned here for the sake of completeness. Electric air source heat pumps for heating can just barely achieve a COP of 3.0, which cancels the typical factor of three loss in the conversion from primary energy to electricity. A COP of around 4.0 is possible with ground source or exhaust-air heat pumps. Compared with a high-efficiency (90 per cent) natural gas furnace using natural gas supplied with 10 per cent energy loss, primary energy use is reduced by 40 per cent if the electricity is generated and supplied at 33 per cent efficiency, or by 60 per cent if the electricity is generated and supplied at 50 per cent efficiency. However, if the electricity comes from coal, there would be only a 25 per cent reduction in CO_2 emissions in the latter case and a 10 per cent increase in the former case (see Chapter 2, Section 2.4.1).

    Heating-only absorption heat pumps are another possibility (Dieckmann et al 2005). Reversible heat pumps (that can be used for cooling during the summer) can achieve a heating COP of 1.3–1.5 (compared with only 0.7–0.9 for cooling). A heating-only absorption heat pump would cost 25–35 per cent less than a reversible heat pump but \$2000 more than a $29kW_{th}$ (100,000Btu/h) residential furnace. They would use ammonia, which is toxic and therefore requires placement outdoors, and would need to be connected to hot water for heat distribution (LiBr could not be used due to the risk of freezing). By directly using heat from natural gas fuel, they would avoid the factor of three energy loss in generating electricity, but would be 40–60 per cent more efficient than a high efficiency furnace.

    Another option is a gas-engine heat pump. Like the electric heat pump, this is a vapour-compression heat pump, except that mechanical power from a gas engine rather than from an electric motor is directly used to drive a compressor. Waste heat from the gas engine is used along with heat output from the heat pump, giving a COP of 1.3 for heating (see www.aisin. co.jp/life/ghp/english).

4.2.6 Hot-water boilers

A device that directly heats water is called a 'boiler' even if the water is not heated to the boiling point. Small natural-gas fired boilers have been available in Europe for 30 years or more. One unit, manufactured in Italy by Baxi and recently available in North America, is 75cm high, 45cm wide 35cm deep, weighs about 40kg, and can be mounted on a wall (see www.wallhungboilers. com). Heat output can be continuously varied from 10kW to 31kW by varying (i.e. modulating) the rate of fuel flow (i.e. without on/off cycling), with efficiency of the non-condensing model ranging from 79 per cent at minimum output to 84 per cent at full output. A condensing model, with a full-load efficiency of 92 per cent, was introduced in 2005. Heat is extracted from the exhaust gas (and condensation induced) by first pre-heating the return water, then by pre-heating

the combustion air (which is drawn directly from the outside). As adequate hot water for heating can be produced on demand, there is no need for a hot-water storage tank. The hot water produced by the boiler can be supplied to baseboard hot-water radiators, wall-mounted radiators, floor radiant heating loops, or to a heat exchanger in the air handler of a forced air heating system. In the latter case, the hot-water heater can be placed directly above the air handler, and can fit into a small closet. The most efficient residential boiler available has an AFUE of 99 per cent (GAMA, 2005).

4.2.7 Combined space and domestic hot water heaters

A large number of different kinds of integrated space and domestic hot water (DHW) heating systems are available, as discussed in Chapter 8 (Section 8.3.2). These may or may not be more efficient than separate space and DHW heating heaters. Baxi produces a hot-water heater, similar to that described above, that can be used to supply both space heating and DHW, with full-load efficiencies of either 84 per cent (non-condensing unit) or 92 per cent (condensing unit) and no need for storage of hot water. At any given time, the heat output is used either for space heating or for DHW, with DHW having priority over space heating. The temperature of water used for space heating and for DHW can be independently controlled. At the moment, the smallest unit has a rated heat output of $33kW_{th}$ and a minimum output of $9.6kW_{th}$.

4.2.8 Furnace cost versus capacity

Annually updated costs of equipment of different capacities used in buildings (as well as costs of materials and labour used in building construction) are available for North America from the R.S. Means Company. Average costs of furnaces and wall-hung heating units in the US for 2005 are presented in Figure 4.16. These data and other data from R.S. Means are useful in assessing the potential savings in mechanical equipment costs that are possible when heating loads are reduced through an improved thermal envelope, or if on-site mechanical equipment is eliminated altogether through connection to a district heating network. The R.S. Means data do

Table 4.5 Comparison of the costs of condensing and non-condensing furnaces of different size and quality in the Toronto region

Description	Capacity, kW$_{th}$ (kBtu/h)	Cost (Cdn$)			
		Equipment cost		Installed cost	
		Non-condensing	Condensing	Non-condensing	Condensing
Low quality, with PSC motor	14.7 (50)	550	930	1100	1480
	36.6 (125)	675	1070	1225	1620
Middle quality, with PSC motor	14.7 (50)	775	1032	1325	1582
	36.6 (125)	965	1305	1515	1855
Premium, 2-stage with ECPM motor	14.7 (50)	1020	1530	1570	2080
	36.6 (125)	1265	1925	1815	2475
		Installed unit cost (US$/kW$_{th}$)		Cost per unit of heat supplied (US$/GJ)	
Low quality, with PSC motor	14.7 (50)	61.6	82.8	5.70	7.67
	36.6 (125)	27.4	36.3	2.11	2.80
Middle quality, with PSC motor	14.7 (50)	74.2	88.5	6.86	8.19
	36.6 (125)	33.9	41.5	2.62	3.20
Premium, 2-stage with ECPM motor	14.7 (50)	87.6	116.4	8.13	10.77
	36.6 (125)	40.6	55.4	3.13	4.27

Note: Costs per unit of heat supplied were computed assuming a 4% real interest rate, a 20-year lifespan and heat delivery of 20GJ/yr for the small furnaces and 60GJ/year for the large furnace.
Source: Dara Bowser (Bowser Technical), personal communication, 2005

Figure 4.16 Variation with capacity in the average installed cost and installed unit cost of furnaces in the US, as given by Means (2005)

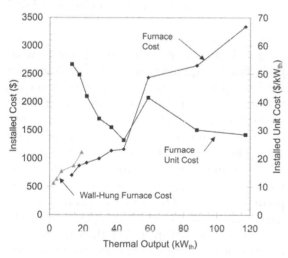

not provide information on equipment efficiency, so the data in Figure 4.16 are supplemented here with the data in Table 4.5 comparing the costs of residential non-condensing and condensing furnaces of the smallest and largest capacities available in the Toronto area. In the 20–60kW$_{th}$ range, there is no large saving when the furnace is downsized, as the cost per kW$_{th}$ increases sharply with decreasing size. The smallest capacities are available for wall-hung units, which can cost more than some larger-capacity furnaces. Unit costs of non-condensing furnaces with standard air handler fans are $30–75/kW$_{th}$.

The cost premium for condensing furnaces, including installation cost, depends on whether the furnace is going into a new building, or is a replacement of a pre-existing non-condensing furnace. In new buildings, there will be little if any difference in the installation cost. Indeed, the total cost will be less with a condensing furnace when the savings from the chimney that is not required is credited against the cost of the condensing furnace. When a condensing furnace replaces a non-condensing furnace, the installation cost will be roughly doubled (from about \$550 to \$1100) because the pre-existing chimney must be blocked, a hole must be drilled through the outside wall above grade and an exhaust vent installed.

4.3 Boilers in commercial and multi-unit residential buildings

A boiler is a pressure vessel in which heat is transferred from combustion gases to a liquid. It can be used either to produce steam or hot water; even though water is not boiling in the latter case, the term 'boiler' is still used. Either kind of boiler is available in sizes having peak heat outputs ranging from 15kW to 15MW. Boilers are classified as either:

- *low-pressure* boilers, producing steam below 1atm pressure or hot water below 10atm pressure and at a temperature no greater than 120°C, or
- *high-pressure* boilers, producing steam below 1atm pressure or hot water below 10atm pressure and at a temperature greater than 120°C.

4.3.1 Full-load efficiency

The *combustion efficiency* of a boiler is equal to the fuel energy input minus heat loss in the exhaust, divided by the energy input. The *overall efficiency* (henceforth, simply the efficiency) is given by the steam or hot-water energy output divided by the energy input. It is smaller than the combustion efficiency, because it includes additional loss due to heat radiated from the boiler and off-cycle energy losses. For a natural-gas-fired boiler with an overall efficiency of 82 per cent, roughly 4 per cent of the energy input would be radiated from the boiler, 4 per cent lost as sensible heat in the exhaust gas, and 10 per cent lost as latent heat (uncondensed water vapour) in the exhaust gas. To obtain high efficiency, water vapour in the exhaust gas must be condensed and the latent heat that is released put to use. This can be done either by preheating the combustion air or the return water with the combustion gases, or both. The former method is more expensive and can increase NO_x emissions. Using the return water requires that the return-water temperature be below the dewpoint of the exhaust gas (about 55°C for natural gas boilers with 10 per cent excess air). Hot water for space heating has traditionally been supplied at about 80°C and returns at about 70°C, which precludes the use of condensing boilers. However, with better thermal envelopes (to reduce peak heating requirements) and radiant heating panels, much lower supply and return temperatures are possible. The colder the water returning to the boiler inlet, the greater the amount of water vapour that can be condensed, and the higher the boiler efficiency. Non-condensing boilers generally have full-load combustion efficiencies of 75–85 per cent (but as low as 68 per cent), while condensing boilers have full-load combustion efficiencies of 88–95 per cent. The latent heat in the exhaust gas of oil-fired boilers is less than with natural gas (6.5 per cent vs 9.6 per cent, due to the smaller H:C ratio in heating oil), so it is less attractive to build condensing oil-fired boilers. The most efficient condensing boilers will have significant amounts of insulation so as to minimize loss of heat from the boiler walls, and will extract heat from any water used to flush water impurities out of the boiler (flushing impurities can account for 1–3 per cent of the energy input).

Condensing boilers require expensive corrosion resistant materials and are generally limited to using clean burning fuels, such as natural gas. Until recently, condensing boilers were prohibitively expensive in the North American market (up to three times the cost of non-condensing boilers, and increasing the total cost of the heating system by 20–25 per cent). As a result, they have captured only 2 per cent of the boiler market in the US (CEE, 2001). The latest models sell for a premium of as little as 25 per cent in the US, and more typically at a 50–100 per cent premium. Costs can be further reduced by installing a small number of condensing units with only one serving as backup, rather than the normal practice of installing two equally sized non-condensing boilers in a building, with one serving as backup. Were the market for condensing boilers to expand in North America, the price premium would decrease through learning-by-doing. Condensing boilers are very common in Europe, and make up over half of the market in Holland.

Figure 4.17 Efficiency of the AERCO 2.0 boiler as a function of return water temperature and output as a fraction of peak output

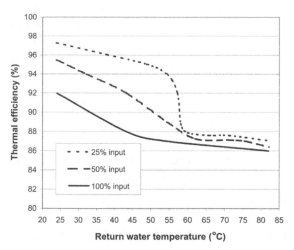

Source: Durkin and Kinney (2002)

4.3.2 Part-load efficiency

Part-load operation of boilers can occur through on/off cycling (with a greater fraction of time off the lower the load, but full output when on), by modulating the fuel flow without modulating the airflow, or by modulating both the fuel and the airflow. Non-condensing boilers generally use on/off cycling, because reduction in fuel flow can cause condensation – something for which they are not designed. On/off cycling is inefficient, as heat is lost during the off-cycle and the boiler has to be reheated to some extent at the start of each on-cycle. Overall efficiency therefore falls with decreasing load in non-condensing boilers, at least below about 80 per cent of full load (where peak efficiency occurs in some cases). Like furnaces, boilers can use natural draft or fan-assisted draft; the former will have a larger standby heat loss. With fuel and airflow modulation, as in modern condensing boilers, efficiency increases at part load. This is because the ratio of heat exchanger surface area to heat flow increases at lower loads. The minimum load that can be achieved in this way ranges from 5 per cent to 33 per cent of peak load. At lower loads, the boiler reverts to on/off cycling, with an attendant sharp drop in efficiency. Because a boiler operates at part load most of the time, the decrease in non-condensing boiler efficiency and the increase in condensing boiler efficiency at part load amplifies the difference in

seasonal average efficiency between condensing and non-condensing boilers. The dependence of boiler efficiency on load and return temperature for a modern boiler is illustrated in Figure 4.17, which shows boiler combustion efficiency versus return water temperature for the AERCO 2.0 boiler at 100 per cent, 50 per cent and 25 per cent of full load. Because condensing boilers have greater efficiency at a relatively low load but an abrupt drop in efficiency at very low loads, energy use will be minimized if there are several small boilers rather than one or two large boilers and if the heat load is spread over some or all of the available boilers (as noted above, this can also reduce capital costs by reducing the amount of backup capacity).

Goldner (2000) has quantified the reduction in efficiency under part-load operation in older, non-condensing boilers based on 14 months of fuel-consumption data from 30 multi-family buildings in New York City. Daily heat load is estimated based on the average temperature departure for that day from a reference temperature of 13°C. Fuel consumption per degree-day of heat load on days with one-quarter the peak heating requirement is about *twice* the fuel consumption per degree-day on the coldest days. That is, the boiler efficiency drops to half of its peak-load efficiency when the load is 25 per cent of the peak load. Fuel consumption per degree-day averages around 50 per cent higher during the fall and spring than during the winter.

If a boiler provides winter heating and year-round domestic hot water, then the use of unequally sized boilers can reduce maintenance and energy costs. Conventional practice for a peak winter heating load of 100 (typical relative units), a night-time summer load of 2.5 and a daytime summer load always less than 10 would be to install two 125-unit boilers, with one serving as backup. During the summer, the boiler always operates at less than 8 per cent of its peak capacity. The addition of a 10-unit boiler would allow the larger boilers to be completely shut down during the entire warm season and would usually improve efficiency.

4.3.3 Constraints on the use of condensing boilers

As noted above, a condensing boiler requires a return water temperature low enough to induce condensation. This in turn requires a high quality thermal envelope in order to permit a relatively low supply water temperature in hydronic heating systems, and thus a lower temperature of the return water. Conversely, the need for low return temperatures largely precludes replacing existing non-condensing boilers in existing buildings with condensing boilers, unless the thermal envelope is upgraded first so that adequate heat can be provided with a lower hot-water supply temperature. New radiators may also be needed.

According to CEE (2001), the potential market for condensing boilers in new apartment and condominium buildings is small because of the inability to bill individual tenants for a central boiler space heating system. However, small wall-hung condensing boilers, suitable for meeting the load of individual apartments, are now available. The market is said to be small in small commercial buildings (<10,000m²) because the predominant load is for space cooling (in the US), which is often provided by rooftop units with integral gas or electric forced air heating. However, it should be possible to integrate the technology of small, wall-hung condensing boilers into rooftop units for commercial buildings.

Figure 4.18 Variation with capacity in the average installed cost and the average installed unit cost in the US for natural gas boilers producing hot water, as given by Means (2005)

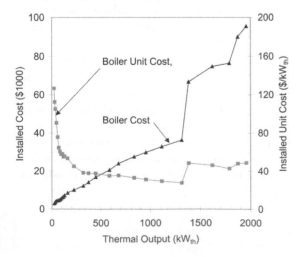

4.3.4 Boiler cost versus capacity

Data from R.S. Means on the cost of conventional boilers in North America in 2005, as a function of heating capacity, are given in Figure 4.18. Unit costs range from over $120/kW_{th} for the smallest unit to about $30/kW_{th} for middle capacity units and about $45/kW_{th} for large capacity units. As noted above, condensing boilers sell for a premium of 25–100 per cent in the US.

4.4 Correct estimation of heating loads to avoid oversizing of equipment

Furnaces and boilers are usually greatly oversized, requiring greater on/off cycling or part load operation. Oversizing is done deliberately, as a precaution to ensure that peak loads are met. More accurate estimate of peak load, with reduced uncertainty, through good quality-control during construction and computer simulation with reliable input data, could reduce the extent of oversizing while ensuring that peak loads are met. Two-stage furnaces address the oversizing problem to some extent in houses, as they operate as smaller furnaces most of the time, but extra capacity can be called upon when needed. They will operate more continuously on low fire. If they are equipped with ECPM motors (Chapter 7, Section 7.3.2), then substantial savings in air handler electricity use will occur (Pigg, 2003). Correct sizing is worthwhile even of a two-stage furnace.

4.5 District heating

Connection of a building to a district heating system provides large energy savings if the heat supplied to the district heating system is waste heat that would otherwise be discarded to the environment. In some cases, some or part of the heat supplied by a district heating system is provided by dedicated centralized boilers, but these can generally be operated more efficiently than individual boilers in each building. An extensive discussion of district heating systems, including energy losses in distributing heat to individual buildings and the opportunities for using heat that would otherwise be wasted, is found in Chapter 15.

4.6 On-site cogeneration

Cogeneration is the simultaneous production of electricity and useful heat. All fossil fuel based electricity generation produces waste heat as a byproduct. In central power plants, this heat is discarded – through the cooling water and in the exhaust gases. In cogeneration, some of this heat is captured at a temperature high enough that it can be put to some use – such as heating nearby buildings or providing hot water. Cogeneration has a long and extensive history of use in industrial facilities and to a lesser extent in district heating systems, as discussed in Chapter 15 (Section 15.2), but has seen very limited use at the scale of individual buildings for a number of reasons:

- electricity generating technology has traditionally been much more expensive and less efficient at small scales;
- conventional generating technology requires a more skilled maintenance staff than is available in many building applications;
- noise and emissions restrict the potential sites for conventional technology.

Cogeneration at the scale of individual buildings or small clusters of buildings can be carried out using reciprocating engines, microturbines or fuel cells. Reciprocating engines are internal combustion engines and, like their automotive counterparts, come in two types: spark ignition and compression ignition engines. Spark ignition engines for cogeneration preferably use natural gas, although they can also use propane, gasoline or landfill gas. Compression ignition engines use diesel fuel or heavy oil. Microturbines are defined as gas engines with an electrical power output of 30–500kW. Fuel cells are electrochemical (rather than combustion) conversion devices, and at present run on natural gas (in the future they may be powered by hydrogen produced, ideally, by splitting water using renewably generated electricity).

4.6.1 Cost and performance of cogeneration equipment

The cost, full-load efficiency, pollutant emissions, and other characteristics of reciprocating engines, microturbines and fuel cells at a variety of different sizes are given in Table 4.6. The critical performance parameters are the overall efficiency, the proportion of the output as electricity and the *marginal efficiency* of electricity generation. The latter is the electricity produced divided by the extra fuel energy used compared to the generation of heat alone. It is given by:

$$\eta_{m\,arginal} = \frac{n_{el}}{1 - \eta_{th}/\eta_b}$$

(4.1)

where η_{el} and η_{th} are the cogeneration electric and thermal efficiencies, and η_b is the boiler efficiency. The more efficient the system for stand-alone heat production, the greater the incremental fuel use for cogeneration, and the lower the marginal efficiency of electricity production. Marginal efficiencies are given in Table 4.6 assuming a boiler at 92 per cent efficiency as the alternative.

From Table 4.6, it can be seen that:

- electrical generation efficiencies are lowest for microturbines (25–29 per cent), intermediate for reciprocating engines (33–41 per cent), and best for fuel cells (36–46 per cent);
- overall efficiencies ($\eta_{el} + \eta_{th}$) are not large (no more than 78 per cent, and only 62–67 per cent for microturbines);
- as a result of the above, marginal efficiencies for electricity production are not large (around 45 per cent for microturbines, 52–62 per cent for fuel cells, and 58–65 per cent for reciprocating engines);
- electrical efficiency and the power: heat ratio tend to increase with increasing size, but overall efficiency shows no consistent trend;
- costs fall dramatically with increasing size;

Table 4.6 Characteristics of cogeneration technologies available for use at the scale of individual large buildings

Capacity	Cost ($/kW)		Efficiency (%)			Power: heat ratio	O&M (cents/ kWh)	Emissions (gm/kWh)		
	Elect Only	CHP	Electrical	Overall	Marginal			NO_x	CO	Hydro-carbons
Microturbines										
30kW	2263	2636	25	67	46	0.52	2.0	0.23	0.63	<0.08
70kW	1708	1926	28	61	44	0.70	1.5	0.20	0.12	<0.08
80kW	1713	1932	27	63	44	0.63	1.3	0.57	0.69	<0.08
100kW	1576	1769	29	62	45	0.73	1.5	0.33	0.20	<0.08
Fuel cells										
200kW PAFC		5200	36	72	59	1.00	2.9	0.02	0.02	<0.01
5–10kW PEMFC		5500	30	69	52	0.79	3.3	0.05	0.03	<0.01
150–250kW PEMFC		3800	35	72	59	0.95	2.3	0.05	0.03	<0.01
250kW MCFC		5000	45	65	58	1.95	4.3	0.03	0.02	<0.01
2000kW MCFC		3250	46	70	62	1.92	3.3	0.02	0.02	<0.01
100–250kW SOFC		3620	45	70	62	1.79	2.4	0.02	0.02	<0.01
Reciprocating engines										
100kW	1030	1350	33	78	65	0.61	1.8	20.9	16.8	1.0
300kW	790	1160	34	77	64	0.67	1.3	2.8	2.8	1.4
1MW	720	945	38	71	59	0.92	0.9	1.4	2.8	1.4
3MW	710	935	39	69	58	1.04	0.9	1.0	3.5	1.8
5MW	695	890	41	73	63	1.02	0.8	0.7	3.4	0.7

Note: For parallel information at the scale of district heating systems, see Table 15.2. PAFC = phosphoric acid fuel cell, PEMFC = proton exchange membrane fuel cell, MCFC = molten carbonate fuel cell, SOFC = solid oxide fuel cell.
Source: Goldstein et al (2003)

Figure 4.19 Variation in the relative electrical generation efficiency (η_{el} divided by η_{el} at full load) with load for different building scale cogeneration devices

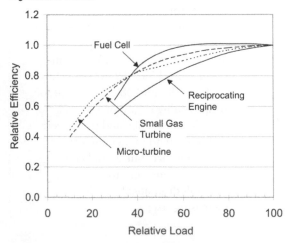

Computed from data in Goldstein et al (2003)

- reciprocating engines are the least expensive but also entail very high pollutant emissions.

The marginal efficiency using microturbines is substantially less than that of state of the art natural gas combined cycle power plants (55 per cent), and the marginal efficiency of all the options listed in Table 4.6 is substantially less than what can be achieved in large-scale combined cycle cogeneration (88 per cent) (see Chapter 15, Section 15.2.5). Thus, although any form of small-scale cogeneration represents a significant improvement in electrical efficiency compared with present typical power plants, none of them represents the most efficient use of natural gas for electricity generation available today. As well, for all systems, the electrical efficiency η_{el} falls with decreasing load, although fuel cells maintain a constant η_{el} down to about 60 per cent of full load before falling. This is shown in Figure 4.19. One of the advantages of cogeneration is that it provides power where it is needed, thereby reducing transmission bottlenecks and energy losses. However, cogeneration at the scale of small district heating systems (25MW and up) provides essentially the same benefits but with much greater savings in primary energy.

Finally, note that reciprocating engines have very high pollutant emissions, the effect of which would be compounded by their close proximity to people when used in buildings for cogeneration. Even microturbines have high emissions compared with community-scale combined cycle cogeneration (see Table 15.2).

4.6.2 Matching heating loads and electricity production

Cogeneration improves overall energy efficiency only if the waste heat that is captured can be utilized. In microturbines, the electricity generation efficiency is only 25–29 per cent at present, meaning that 72–75 per cent of the fuel energy is converted into heat (not all of which is usable). In fuel cells, the electricity generation efficiency is 40–50 per cent (using natural gas as fuel), so a use for less waste heat needs to be found. Since new buildings can be built to require almost no winter heating even in cold climates, the scope for the effective use of cogeneration in new, high-performance buildings is limited. Furthermore, heating loads are seasonal in character, meaning that a cogeneration system sized for the winter heating load will have excess heat during the summer. Hot-water heating loads, on the other hand, are more seasonally uniform and therefore provide a better match for cogeneration (the temperature of the incoming cold water can, however, vary by 10–15K at mid-latitudes, such that energy use for heating is up to one-third less during summer even when monthly hot water use is uniform). According to Braun et al (2004), the annual average hot-water to electricity use ratio is typically about 1.0 in US households, which matches the output ratio of fuel cells.

One option, then, is to size a cogeneration system to match the hot-water load and

design a building to largely or entirely (depending on the climate) eliminate the heating load. The residual heating load could be met with a ground source heat pump, an exhaust-air heat pump, or a high efficiency boiler with low temperature radiant heating. A second option is to size the cogeneration system to meet the combined heating and hot-water load in winter, with both electricity (and heat) production scaled back in summer in order to retain high overall efficiency. If the winter heat load is comparable to or smaller than the hot-water load, the extent to which output will have to be scaled back during summer – and associated financial penalties – will be minimized.

In either case, it will often be possible to entirely meet the hot-water requirements through solar thermal energy during the summer, so there would be no use for the waste heat from cogeneration. Cogeneration is then only attractive where space heating loads are very small and solar heating to produce hot water is not viable. As explained in Box 4.1, is possible that, by implementing solar water heating and rendering cogeneration unviable, primary energy use *increases*. This is because, with solar hot water heating, cogeneration would no longer be available to displace electricity that is inefficiently generated at a central power plant. However, as the efficiency of central power plants increases (from an average of 33–35 per cent at present, to the current state of the art of 55–60 per cent), the savings in power plant energy use through cogeneration will decrease, and it becomes preferable to use solar hot water heating and to forgo cogeneration. This is confirmed by detailed calculations for cogeneration in super-insulated houses (wall and roof U-values of $0.15W/m^2/K$, window U-values of $0.7W/m^2/K$) with fuel cells under Swiss conditions (Dorer et al, 2005): for grid electricity generated at 33 per cent efficiency, cogeneration and solar collectors reduce annual primary energy use by 20–30 per cent and 15 per cent, respectively, while for grid electricity at 47 per cent, cogeneration saves about 12 per cent and solar collectors save 16–20 per cent.

The scope for cogeneration can be increased if thermal storage (beyond that typically used for DHW) is added to the system, as electricity could then be generated at one time and the waste heat used at another time. This

BOX 4.1 Tradeoffs involving solar hot water heating and on-site cogeneration

Cogeneration is attractive from a primary energy point of view because the marginal efficiency of electricity generation (the electricity produced divided by the extra fuel energy used compared to heating only) can be as large as 60–65 per cent for building-scale cogeneration systems. If this electricity displaces electricity that is centrally generated at a coal-fired power plant with a typical efficiency of 35 per cent and 5 per cent loss in transmitting the electricity to the point of use, a substantial saving in primary energy use will occur. However, to achieve a high overall efficiency of cogeneration, a use must be found for the waste heat. Inasmuch as new buildings can be designed to require almost no heating in winter even in cold climates, the only remaining major heat load is for domestic hot water (DHW). Use of solar energy to produce DHW may therefore render cogeneration unattractive and, because inefficiently generated electricity is no longer displaced, can result in an increase in total fossil fuel energy use.

To illustrate, consider a fuel cell that produces 30 units of useful heat for DHW and 45 units of electricity from 100 units of fuel energy (giving an overall efficiency of 75 per cent). If the heat could have been produced with a condensing boiler with an efficiency of 92 per cent, the marginal efficiency of electricity production is 67 per cent. If solar energy is used to provide the heat, and 45 units of electricity are instead generated centrally from coal with a generation times transmission efficiency of 33 per cent, then 135 units of fuel are needed. Total energy use has increased by 35 per cent (and CO_2 emission has increased even more, due to a shift from natural gas to coal for electricity generation). Thus, it is better, in this case, not to use solar energy for water heating if the alternative is cogeneration with fuel cells. On the other hand, if the central power plant is a state of the art natural gas combined cycle plant with 54 per cent electricity generation efficiency, then only 90 units of fuel are used for the same (94 per cent) transmission efficiency, so there is a saving of 10 per cent by using solar energy to make DHW while forgoing cogeneration.

If the cogeneration unit is a microturbine that produces 25 units of electricity and 40 units of useful heat from 100 units of fuel, then producing the heat with solar energy and supplying the electricity centrally at 33 per cent generation times transmission efficiency requires only 75 units of fuel – a saving of 25 per cent using solar energy and forgoing cogeneration. If the alternative is central generation at 50 per cent generation times transmission efficiency, solar heating saves 50 per cent of the heating plus electricity fuel use. In this case, the amount of electricity produced relative to the amount of heat produced is so small that the primary energy that can be saved at the central power plant with cogeneration is less than the fuel that can be saved on-site if solar energy is used for heating, even when the central power plant efficiency is small. Thus, solar heating is preferable to cogeneration for systems (such as microturbines) that have a small electricity to heat ratio.

The above results are altered if solar heat can provide only a portion of the DHW needs. Table B4.1 compares the primary energy used for cogeneration, for the above cases (where solar energy provides 100 per cent of the DHW), and for the above cases where solar energy provides 50 per cent of DHW, with the balance provided by a boiler at 85 per cent efficiency. With microturbines as the cogeneration option, solar DHW is preferable for every case except a 50 per cent solar fraction and 33 per cent central power plant efficiency, where the energy use is almost the same. With fuel cells as the cogeneration option, cogeneration without solar DHW is preferable except when solar energy provides 100 per cent of the DHW and the central power plant times transmission efficiency is 50 per cent.

Table B4.1 Comparison of primary energy use for various cogeneration options without solar heat for DHW, and for solar DHW with central production of electricity at an efficiency of 33 per cent or 50 per cent (including transmission losses) and a boiler at 85 per cent efficiency for any DHW not provided by solar energy

	Microturbine cogeneration $\eta_{el} = 0.25$, $\eta_{th} = 0.40$	Fuel cell cogeneration $\eta_{el} = 0.45$, $\eta_{th} = 0.30$
Cogeneration	100	100
Solar providing 50% of DHW, $\eta_a = 0.33$	99	153
Solar providing 50% of DHW, $\eta_a = 0.50$	74	108
Solar providing 100% of DHW, $\eta_a = 0.33$	75	135
Solar providing 100% of DHW, $\eta_a = 0.50$	50	90

Note: Energy use is relative to the cogeneration option in the same column.

is particularly attractive if electricity rates vary with the time of day, as is increasingly the case. Brown and Somasundaram (1997) analyse various possible configurations with thermal energy storage. The scope is also improved if applied to a cluster of buildings (preferably with different uses) or to multiple housing units, rather than in individual houses, because the peak demand will rise much more slowly than the base demand (in relative terms) due to the fact that peak demands in individual units will not all occur at the same time. Excess heat from summer cogeneration can be stored underground for use during winter, as discussed in Chapter 15 (Section 15.4).

4.6.3 Primary energy implication of cogeneration with absorption chilling

Gas microturbines produce waste heat at a range of temperatures up to in excess of 500°C, which can be used to drive triple-effect absorption chillers (see Chapter 6, Section 6.6.1) for air conditioning. Unlike steam turbine or combined cycle systems (see Chapter 15, Section 15.2.3), putting this waste heat to use does not require reduced electricity output, so it would appear that this is an attractive option for using the excess heat that is available during the summer. However, the fuel that is used to generate electricity through on-site cogeneration in summer could be used in a central power plant with much higher electricity generation efficiency (55–60 per cent for state of the art combined cycle power plants), and the extra electricity so produced could be used in an electric chiller. Assume that the two choices are (i)

a gas microturbine with an electricity generation efficiency of 27.5 per cent and 35.0 per cent useful heat (62.5 per cent overall efficiency) that can be supplied to a double effect absorption chiller with a COP of 1.2; and (ii) a central power plant with an electricity generation efficiency of 55 per cent and 5 per cent transmission loss, so that 52.25 units are delivered with the extra 24.75 units of electricity supplied to an electric chiller. Option (ii) will require less primary energy than option (i) if the electric chiller has a COP of 1.7 or greater.[3] As an alternative to option (i), consider a fuel cell that produces 35 units of electricity and 35 units of heat. In this case, option (ii) requires less primary energy if the electric chiller has a COP of 2.43 or greater. For triple effect absorption chillers (with a COP of 1.6) that are not yet commercially available, the break-even electric chiller COPs are 2.26 and 3.25 for the microturbine and fuel cell cases, respectively.

Since commercial chillers operate with a full load COP of 3.8–7.9 (see Chapter 6, Section 6.5.2), the cogeneration/absorption-chiller option using gas microturbines substantially *increases* the use of primary energy. This unfavourable outcome is a result of the sacrificed electricity production (at the central power plant) being 79 per cent (= $100(55 - 27.5)/35.0$) of the useful heat in the case of microturbines and 57 per cent in the case of fuel cells, and is in contrast to gas combined cycle cogeneration, where smaller amounts of useful heat are captured and the sacrificed electricity production is only 11–28 per cent of the captured heat (see Chapter 15, Section 15.2.3). In this case, use of the excess heat by absorption chillers in summer is an attractive option whenever the electric chiller COP is less than about 5.5–5.8 (see Chapter 15, Section 15.2.6). Gas combined cycle cogeneration, which involves both gas and steam turbines operating together, is not viable at the scale of an individual building, but it is viable as part of a district heating and cooling system – a subject discussed in Chapter 15.

4.6.4 Summary of conditions for reducing the use of primary energy

In summary, cogeneration of heat and electricity at the scale of individual buildings, and scaled to meet the minimum heating or hot-water requirements, provides a net savings in primary energy compared to centralized electricity production and separate space and hot-water heating at present typical efficiencies, but not compared with state of the art systems. Cogeneration is best sized to meet domestic hot-water rather than space-heating loads, as the former is relatively constant during the year. In many cases, it will be possible to entirely meet the hot water requirements with solar thermal energy during the summer, in which case cogeneration will not be viable or useful. If the electricity that is no longer displaced by cogeneration is generated from fossil fuels at a typical efficiency of 35 per cent, the net result of using solar energy will be to increase fossil fuel use. Use of waste heat from gas turbines for chilling with absorption chillers increases primary energy demand compared with efficient centralized electricity production and efficient electric chillers. However, if excess heat would otherwise be discarded, then using it to power absorption chillers is preferable. This could be desirable if summer absorption chilling is part of a package involving winter heating through cogeneration, and if the primary energy savings of the latter exceeds the penalty (compared to efficient centralized electricity generation) of the former.

4.7 Summary and synthesis

In buildings with a high-performance thermal envelope and recovery of heat from expelled ventilation air, passive solar heat can be used to provide a significant fraction of annual heating requirements in cold climates. This fraction will be smaller in larger buildings because the ventilation airflow (air taken directly from outside) and associated heating requirements will increase with the volume of the building, while the potential solar gain will increase with the surface area. Offsetting this will be an increase in the relative importance of internal heat gains, so that passive solar heat combined with internal heat gains can still provide a large fraction of the heating requirements.

Solar heat can be captured through direct gain, solar-air collectors, airflow windows, Trombe walls, perforated unglazed solar walls or through atria and sunspaces. Effective use of passive solar heat gains requires a mechanism (such as circulating air through the passively heated zones) to carry the heat to other parts of the building. The effectiveness of passive solar heating is increased if the building has a large thermal mass, which reduces the extent of overheating during the day and releases heat at night. Thermal mass can be provided through conventional materials such as stone, concrete and bricks, or through phase-change materials stored in tanks or embedded in walls, floors and ceilings. If a building is to be unoccupied at night, a smaller thermal mass can reduce heating energy use by allowing a greater decrease in temperature at night for a given loss of heat, and by reducing the amount of energy (solar or otherwise) needed to reheat the building in the morning.

Residential furnaces can achieve full-load efficiencies of 97 per cent, while boilers used in larger buildings can achieve full-load efficiencies of 88–95 per cent. High furnace efficiency requires that the air returning to the furnace or the incoming outside combustion air induce condensation of water vapour in the exhaust gas, which releases latent heat. High boiler efficiency is achieved when the return water can be preheated by the exhaust gas, with higher efficiency if the return water is cool enough (below about 55°C) to induce condensation in the exhaust gas. The lower the return temperature, the more water vapour that can be condensed, the greater the release of latent heat, and the higher the efficiency. High efficiency also requires a relatively large heat exchanger between the exhaust gases and the air or water flows. Compact, high efficiency (92 per cent at full load) residential boilers are available that can be used for combined space heating and production of domestic hot water.

A high-performance envelope has a smaller rate of heat loss, so radiators do not need to be as hot, allowing both supply and return water temperatures to be lower, which, as noted above, increases the boiler efficiency. If heat pumps are used, the performance of the heat pump will be increased because the temperature lift (from heat source to the heat distribution fluid) does not need to be as large; this effect is particularly strong. Thus, there is a double benefit of a high-performance envelope from an energy use point of view: the heating load is smaller, and the required heat can be supplied more efficiently. Attention must be taken to avoid oversizing heating equipment, as over-sized furnaces will cycle on and off more often (which is inefficient). In the most efficient condensing boilers with modulating burners (preferred here), efficiency can be higher at part load than at full load, although there are still economic reasons to avoid oversizing. Efficiency drops sharply below 5–33 per cent of full load, as the boiler reverts to on/off cycling at such small loads.

Based on the electricity generation divided by the incremental fuel energy use compared to heating only, on-site cogeneration can provide a modest improvement in the efficiency of electricity generation compared with present typical power plant efficiencies (35–38 per cent), but not compared to state of the art power plants (55 per cent efficiency). Any efficiency benefit of cogeneration occurs only if the majority of the waste heat from electricity generation can be used. In high-performance buildings, the space-loading load is so small that cogeneration for space heating cannot be justified unless, possibly, if the building is large. In a typical US household, annual electricity and hot water energy use are roughly the same, as is the electricity:heat output ratio of fuel cells. Thus, the waste heat from on-site residential production of electricity using fuel cells should be well matched to the hot-water load. However, during the summer much of the hot-water load can be met by solar thermal energy, so the waste heat from cogeneration would not be needed, thereby reducing its overall efficiency and economic attractiveness. If solar hot-water heating is implemented instead of cogeneration, inefficiently generated electricity will no longer be displaced, and overall primary energy use

can increase. However, as the efficiency of central power plants increases (from 35 per cent at present to 55 per cent), the savings in power plant energy use through cogeneration will decrease, so it will be preferable to use solar thermal energy for hot water heating and to forgo cogeneration.

Use of waste heat from electricity generation with microturbines or fuel cells to power double-effect absorption chillers increases the use of primary energy compared to efficient (55 per cent) centralized electricity generation. This unfavourable outcome is a result of the sacrificed electricity production at the central power plant being a larger fraction (60–80 per cent) of the useful heat provided by cogeneration, combined with the low efficiency of even double-effect absorption chillers (COP of 1.2) compared with electric chillers (COPs of up to 7.9).

Notes

1   Glauber was a German chemist born in 1604, so this is an extreme case of an early discovery waiting for its calling!
2   The subscript 'th', for 'thermal', is used to distinguish heating power from kW of electrical power.
3   The COP (coefficient of performance) of a chiller is the ratio of cooling provided to energy input.

five

Heat Pumps

Heat pumps can be used for heating, air conditioning and the production of hot water. Residential air conditioners and commercial chillers operate on the same principles as heat pumps, so they can be thought of as heat pumps that operate in only one direction, to cool buildings. Indeed, in some parts of the world (such as Japan), the majority of air conditioners are reversible, so they are really heat pumps. In any case, much of the information presented in this chapter is applicable to air conditioners and commercial cooling equipment, and so forms a complement to Chapter 6, where this equipment is discussed.

5.1 Operating principles

The natural tendency of heat is to flow from warm to cold. A heat pump transfers heat against its natural tendency, from cold to warm, in the same way that a bicycle pump moves air against its natural tendency, from low pressure (outside the tyre) to high pressure (inside the tyre). In both cases, work must be done (requiring energy).

There are two broad types of heat pumps, based on either a vapour compression cycle or an absorption cycle. A vapour compression heat pump can be powered by electricity or by mechanical shaft power. The latter is used in industry and sometimes in the central cooling plants of district cooling systems. An absorption heat pump uses heat rather than electricity or mechanical power as the energy input to drive the heat pump. Absorption chillers (used for cooling purposes only) are much more common than absorption heat pumps (used for cooling and heating). Electric vapour compression heat pumps are widely available and have been mass marketed for decades. Thus, this chapter focuses on electric vapour compression heat pumps, while absorption (and other) chillers are discussed in Chapter 6.

An outline of how an electric vapour compression heat pump works is found in Box 5.1. The transfer of heat from cold to warm is accomplished through a compression–expansion cycle involving a refrigerant or *working fluid*. If heat needs to be transferred from the outside to the inside of a building, the refrigerant must be cooled (through expansion) to a temperature colder than the outside, so that it can absorb heat from the outside. This absorption occurs through a heat exchanger – a coil through which the refrigerant flows as outside air is blown past it. As the refrigerant absorbs heat, it evaporates rather than increases in temperature, so the heat exchanger is called an *evaporator*. Once inside the building, the refrigerant must be warmed (by compressing it) to a temperature warmer than the medium to which the heat is transferred (either air or water), so that it will release heat to the inside. Once again, the

heat transfer occurs through a heat exchanger, which maximizes the contact area between the warmed refrigerant and the air or water to be heated. As heat is released from the working fluid, the working fluid condenses rather than decreases in temperature, so this heat exchanger is called a *condenser*. The difference between the evaporator and condenser temperature is referred to as the *temperature lift*. By reversing the direction of flow of the working fluid, the former evaporator serves as a condenser, and the former condenser serves as an evaporator, and heat is transferred in the opposite direction. Thus, a heat pump can act as a heater in winter and an air conditioner in summer.

BOX 5.1 Principles behind the operation of an electric vapour compression heat pump

Vapour compression heat pumps, refrigerators and air conditioners operate on the basis of two key principles: (1) a gas cools as it expands (or a liquid cools as it evaporates) but warms as it is compressed (or releases heat and thus becomes warmer if condensation occurs), and (2) heat flows from warm to cold. Although a heat pump (or refrigerator or air conditioner) appears to be violating the second principle (by transferring heat from the cold region to the warmer region), this is done by using principle (1) to create smaller scale reversals of the larger scale temperature gradient, such that there is an overall heat transfer from the cold exterior to warm interior (or from a cool interior to a warm exterior) although, at each point in the process, the heat transfer is from warm to cold.

How this is done is shown in the upper part of Figure B5.1. A compressor increases the pressure of a working fluid on the discharge side, and creates low pressure on the suction side. As the working fluid (freon in older systems, CFC replacements in new systems) is compressed, it is heated to a temperature in excess of the indoor air temperature. This allows heat to be transferred to the indoor air stream in an indoor heat exchanger, thereby cooling (and condensing) the refrigerant. The more the gas is compressed (i.e. the greater the pressure), the more it warms up. The liquid refrigerant travels through an expansion valve to a heat exchanger that is connected to the suction side of the compressor. The low pressure there induces evaporation and hence cooling of the refrigerant. The lower the pressure, the greater the cooling that occurs.

The refrigerant must by cooled to below the temperature of the outdoor air in order to absorb heat from the outside air. The cool, low-pressure refrigerant, now in the gaseous state, returns to the compressor, where the cycle is repeated.

By simply reversing the direction of fluid flow, a heat pump can act either as a heating unit (transferring heat from the outside to inside) or as an air conditioner (transferring heat from the inside to outside). This is illustrated in the lower part of Figure B5.1.

Clearly, the colder the outdoor temperature, the more difficult it will be to transfer heat from the outside to the inside because the refrigerant must be cooled more in order to maintain the same temperature difference from the outside air. Thus, more work will be required for a given rate of heat transfer. Indeed, a greater temperature difference would be needed to drive a larger heat flow to match the larger heating demand when it is colder. Similarly, a warmer air distribution temperature or a greater airflow rate, and a greater temperature difference between the condenser and airflow would be needed to meet the larger heat demand when it is colder outside. This requires a warmer condenser temperature and hence greater compression of the refrigerant. Thus, a heat pump will work less efficiently but harder the colder the outdoor temperature while, at the same time, the heating requirement of the building will increase. To handle the heating load on the coldest days, an auxiliary electric heating coil will be used.

Most heat pumps operate on the vapour compression cycle, as described above, with electrically-driven compressors. In industrial applications there are many mechanically-driven compressors (the compressor is driven by a turning shaft from a gas or steam turbine, rather than from a shaft driven by an electric motor). A growing minority of heat pumps use the absorption cycle instead of a compression cycle.

Figure B5.1 Refrigerant flow in a heat pump operating in heating mode (upper panel) and in cooling mode (lower panel).

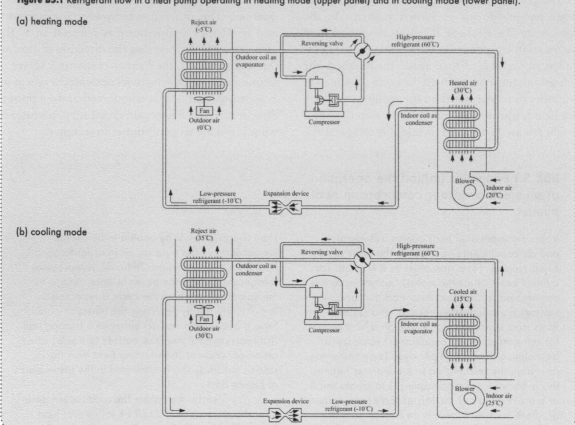

(a) heating mode

(b) cooling mode

5.2 Heat pump performance

The critical parameter measuring the performance of a heat pump is the *coefficient of performance* (COP). When the heat pump is used for heating, the COP is the ratio of heat supplied to energy used.[1] In cooling mode, the COP is the ratio of heat removed from the building to energy used.[2] The performance of a heat pump depends on the performance of the compression–expansion cycle and on the performance of the heat exchangers.

5.2.1 Performance of the compression–expansion cycle

The maximum possible COP of a heat pump is the COP of an ideal heat pump, also known as the Carnot cycle COP. For cooling, it is given by:

$$COP_{cooling,ideal} = \frac{T_L}{T_H - T_L} \quad (5.1)$$

where T_L is the evaporator (or lower) temperature and T_H is the condenser (or higher) temperature. In heating mode, 1.0 is added to the above expression to account for the energy input to the heat pump, which is ultimately dissipated as heat and, along with the heat from outside, is part of the heat supplied to the building. Thus,

$$COP_{heating,ideal} = \frac{T_L}{T_H - T_L} + 1.0 \quad (5.2)$$

In an ideal heat pump, heat is delivered to the evaporator and removed from the condenser isothermally, that is, with no change in the temperature of the working fluid as it evaporates on its way through the evaporator and condenses on its way through the condenser. No heat pump can achieve a COP greater than that of an ideal heat pump, given above. A real heat pump differs from an ideal heat pump, and therefore has a lower COP, for the following reasons (Reay and MacMichael, 1988):

- *Occurrence of superheating.* It is essential that no liquid remains in the vapour when it enters the compressor (to avoid damaging the compressor), so as a safety margin, the refrigerant exiting from the evaporator is heated slightly before it enters the compressor.
- *Non-isentropic compression.* The compressor is where the work needed to drive the compression–expansion cycle is done. In an ideal heat pump, the compressor adds exactly the energy needed to compress the working fluid. Due to heat flow from the compressor to the working fluid and other factors, more energy than is needed will be added, which results in a greater discharge temperature. The ratio of required energy added to actual energy added is called the *isentropic efficiency*, and is typically around 0.7.
- *Mechanical losses.* The mechanical efficiency of a compressor is the ratio of power input to the compressor to work delivered to the working fluid. This efficiency is typically around 0.95.
- *Pressure drops.* As a result of pressure drops through the condenser and evaporator, there is a slight decrease in temperature as the working fluid travels through both the condenser and evaporator.
- *Occurrence of subcooling.* The working fluid exiting from the condenser needs to be cooled slightly since there will be a slight pressure drop between the condenser and the expansion valve, and cooling insures

that no vapour forms prior to the expansion valve (the occurrence of vapour entering the expansion valve would interfere with its performance). However, with a heat exchanger, cold refrigerant from the evaporator can be used to slightly cool the warm refrigerant leaving the condenser while at the same time providing the superheating of the cold refrigerant before it enters the condenser. In this way, the impact on the COP of both superheating and subcooling is minimized.

The ratio of actual COP to that of an ideal heat pump is called the *Carnot efficiency* (η_c). Thus:

$$COP_{\text{cooling,real}} = \eta_c \left(\frac{T_L}{T_H - T_L} \right)$$

(5.3)

The net effect of the above departures from an ideal heat pump is to produce a Carnot efficiency ranging from 0.3 in conventional electric heat pumps, to 0.5 in more advanced residential units and 0.65 in large, advanced electric heat pumps.

Figure 5.1 shows the variation of COP with condenser temperature for an ideal heat pump in heating mode, assuming an evaporator temperature of 0°C. For a given condenser temperature, the COP is smaller the colder the

Figure 5.1 Variation in the COP of an ideal heat pump in heating mode with an evaporator temperature of 0°C for various condenser temperatures

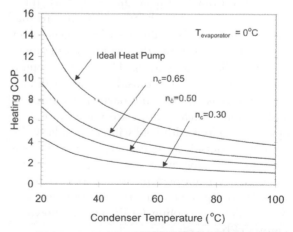

Note: Also shown is the COP for Carnot efficiencies of 0.3, 0.5 and 0.65.

Figure 5.2 Variation of (a) the COP of a heat pump in heating mode with evaporator temperature for various condenser temperatures, and (b) the COP of a heat pump in cooling mode with condenser temperatures for various evaporator temperatures

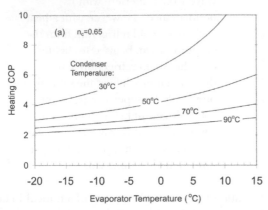

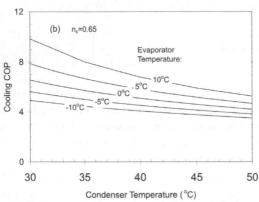

Note: A Carnot efficiency of 0.65 is assumed in both cases.

evaporator temperature (since the denominator in Equation (5.2) is larger). This is illustrated in Figure 5.2a. Since the evaporator must be made to drop to a temperature colder than the outside in order to draw heat from the outside (when the heat pump is used for heating), this means that the heat pump COP decreases when it gets colder (and at the same time, the heating requirement for the building increases). Thus, a heat pump drawing heat from the outside air when the air temperature is −20°C is not particularly efficient, and air-source heat pumps are not an attractive option in regions with cold winters. Similarly, the greater the outside air temperature when the heat pump is used for air conditioning, the greater the condenser temperature required in order to be able to reject heat to the environment, and the smaller the cooling COP. This is shown in Figure 5.2b.

5.2.2 Performance of the heat exchangers

The heat exchangers are critical components of heat pumps. A refrigerant-to-air heat exchanger typically consists of a coil through which the working fluid flows, with metal fins extending into the air. Refrigerant-to-water heat exchangers are concentric without fins, with one fluid flowing in a tube inside the other. To absorb heat from the surroundings, the evaporator must be colder than the surroundings, but how cold it needs to be depends on the characteristics of the evaporator as a heat exchanger. For a given heat exchanger, the greater the temperature difference from the surroundings, the greater the rate of heat flow. Similarly, the condenser must be warmer than the surroundings in order to release heat. Efforts are underway to reduce the size of heat exchangers and to increase the range of temperatures and pressures over which they can operate. Advanced concepts could reduce the volume of a heat exchanger by up to 95 per cent compared with conventional heat exchangers (Reay, 2002).

Figure 5.3 shows the variation of refrigerant temperature as it flows through a condenser. As noted in Section 5.2.1, the refrigerant will have been superheated prior to entering the compressor, so it exits the compressor and enters the condenser above the condensation temperature, and the first step is to cool it to the condensation temperature. During condensation, heat is released with no change of temperature. The refrigerant is subcooled just before exiting the condenser, for reasons explained in Section 5.2.1. Effective subcooling requires that the refrigerant be in contact

Figure 5.3 Variation of refrigerant temperature and air or water temperature as both flow through the condenser of a heat pump

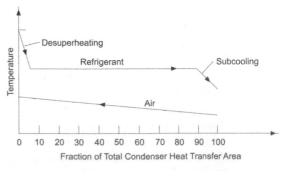

Source: Kreider et al (2002), reproduced with the permission of the McGraw-Hill Companies

with the air (or water) at its coolest, which in turn requires that the air or water flow through the heat exchanger in a direction opposite to that of the refrigerant, as indicated in Figure 5.3.

The heat flow Q across a heat exchanger can be computed as

$$Q = UA \, (LMTD) \tag{5.4}$$

where U is the heat exchanger heat-transfer coefficient (a number of the order of a few 100W/m^2/K), A is the heat exchanger area, and $LMTD$ is the logarithmic mean temperature difference between the refrigerant and the heat source or sink fluid as they flow through the heat exchanger. For a linear variation in the temperature of both fluids through the heat exchanger (i.e. ignoring phase changes), it is given by:

$$LMTD = \frac{\Delta T_1 - \Delta T_2}{\ln(\Delta T_1 / \Delta T_2)} \tag{5.5}$$

where ΔT_1 and ΔT_2 are the temperature differences between the two streams at the inlet and outlet of the heat exchanger, respectively.

Since the evaporator must be colder than the heat source from which it draws heat, and the condenser must be warmer than the air or water of the heat distribution system in order to transfer heat to it, the required temperature lift between the evaporator and condenser is larger than the difference between heat source and sink temperatures, thereby reducing the heat pump COP. Larger heat exchangers, or exchangers that are more effective in transferring heat (having a larger U-value), will minimize these temperature differentials and thereby increase the COP. To illustrate, suppose that the source temperature is 0°C and the distribution temperature in a heating system is 30°C. The apparent Carnot cycle COP in this case would be $273/30 + 1 = 10.1$. If, however, the evaporator is 5K below the source temperature and the condenser 5K above the distribution temperature, in order to achieve adequate rates of heat transfer, then the real Carnot COP is $268/40 + 1 = 7.7$ – one-third lower. The performance of the heat exchangers is thus a critical factor in the performance of the heat pump.

The real Carnot COP, after taking into account the temperature differentials across the heat exchangers, is then multiplied by the product

of the thermodynamic (isentropic) efficiency and the mechanical efficiency to get the final COP. In the above example, and for thermodynamic and mechanical efficiencies of 0.7 and 0.95, respectively, the final COP would be 5.2. This is about half of the apparent Carnot COP.

5.2.3 Impact of the reversing valve

As noted in Box 5.1, a heat pump can be switched from heating to cooling or vice versa simply by reversing the direction of refrigerant flow. This requires a reversing valve. The pressure drop and undesired heat exchange with currently available valves reduce the heat pump capacity by 10–12 per cent and the COP by 6–7 per cent (Fang and Nutter, 1999). If a heat pump is to be used only for heating or only for cooling, then the reversing valve (and associated losses) can be omitted.

5.2.4 Seasonal Performance Factor

Since the COP of a heat pump depends on the difference between the outside and inside air temperature, the average COP over the heating season depends on the climate and will therefore depend on location. The seasonally averaged heat pump COP (which should be weighted by the heating load) is referred to as the *Seasonal Performance Factor* (SPF). A similar average can be computed over the cooling season.

5.3 Air-source and ground-source heat pumps

The evaporator coil can be placed in the air outside the building, with a fan to ensure a steady flow of air past it. The outside air serves as the heat source for the heat pump, so the heat pump is called an *Air-Source Heat Pump* (ASHP). Alternatively, the evaporator coil can be used in a heat exchanger to cool a secondary fluid (usually brine) that flows through a closed pipe loop that is buried underground, so that heat is absorbed from the ground rather than from the air. The ground thus serves as the heat source for the building, so this kind of heat pump is referred to as a closed loop *Ground-Source Heat Pump* (GSHP). Alternatively, the refrigerant can flow directly through the ground loop, eliminating the need for a heat exchanger and secondary fluid; this is referred to as a *direct-expansion* GSHP. The ground loop itself serves as a giant evaporator and heat exchanger. In the closed-loop case, the ground loop serves as a heat exchanger but not evaporator, there being a second heat exchanger between the ground loop and the refrigerant loop in the usual evaporator. The direct expansion system does not require energy to pump a fluid through the ground coil, and does not require as large a compression ratio because there is one less step in the heat transfer chain (so the refrigerant does not need to be as cold at the start). Its disadvantages are that more refrigerant is required, leaks can be catastrophic, trenching requirements are greater and highly skilled installation is required. A third option is an *open-loop* system, in which water is pumped from the ground at one point, used as a heat source for the heat pump, then re-injected at another point. The water flows underground back to the extraction point, being heated along the way. This system can be used only where there is adequate groundwater. A fourth option is a standing column well, which is semi-open. Unless otherwise stated, the term ground-source heat pump will be understood to mean a closed-loop GSHP. GSHPs are sometimes referred to as *Geothermal Heat Pumps.*

A GSHP can also be used for air conditioning, with the ground serving as a heat sink rather than a heat source. If the GSHP replaces air-cooled air conditioners, it will reduce compressor energy use if and when the ground temperature is lower than the air temperature whereas, if the GSHP replaces water-cooled chillers, it will reduce compressor energy use if the ground temperature is lower than that of the cooling water. With water-cooled chillers, the cooling water is cooled through partial evaporation in a cooling tower (Chapter 6, Section 6.10). As the water can often be cooled in this way to be substantially colder than the air temperature, the efficiency gain in using the ground as a heat sink will be smaller than for air-cooled chillers.

The ground loop can be placed horizontally (if adequate space is available) or vertically, as illustrated in Figure 5.4. In vertical systems, a series of boreholes 10–15cm in diameter is drilled to a depth of up to 90m. U-shaped, high-density polyethylene pipes with a typical inside diameter of 30mm are inserted into each borehole and connected at a depth of 1–2m below the ground surface. The boreholes are backfilled with a grout designed to enhance the thermal contact between the pipes and the ground (Carlson, 2000; Zhang and Murphy, 2000), to prevent runoff into groundwater or cross-aquifer contamination, and, in cold climates, to prevent squeezing of the pipes when the ground freezes during heat extraction (Lenarduzzi et al, 2000). A spiral coil can be used instead of vertical or horizontal pumps, and Bi et al (2002) developed a mathematical model of the heat exchange in this case. The required pipe length for air conditioning is about 10–35m per kW of cooling capacity (10–35m/kW_c) for horizontal loops, 20–70m/kW_c for vertical loops, and 45–90m/kW_c (but with less trenching) for spiral loops (US DOE, 2001). In high-rise buildings, vertical loops can be constructed as part of the building foundation piles. Vertical loops tend to be more expensive than horizontal loops (as drilling tends to be more expensive than trenching and twice the length of pipe is needed) and can be subject to heat buildup over time (as explained in Section 5.3.3), but they require less pumping energy (due to the absence of antifreeze, which increases the fluid viscosity) and less ground area, and are subject to smaller seasonal temperature variations.

The four leading countries in the installation of GSHPs are the US (500,000), Sweden (200,000), Germany (40,000) and Canada (36,000) (Lund et al, 2003). In spite of the fact that almost a million GSHP systems have been installed worldwide, with about 45,000 new systems per year, GSHPs are still a relatively unfamiliar technology in many parts of the world. As a result, there have been a number of operational problems that have limited their more widespread adoption. Of 23 buildings using GSHPs in the US studied by Singh et al (2000), nine had problems caused by design or construction errors that were not related to the GSHP technology, 11 had problems due to outdated design, and three had problems due to errors in the construction of the system. All of the problems studied would have been discovered and corrected through proper commissioning (the systematic process of ensuring that building systems operate as designed and intended).

Figure 5.4 Illustration of a ground-source heat pump with (a) a horizontal piping network, and (b) a vertical piping network

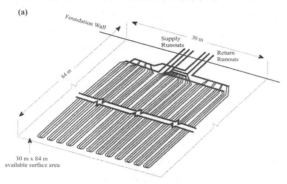

(a)

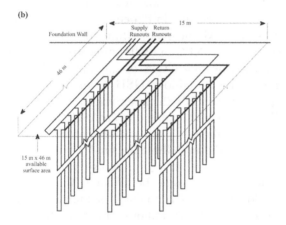

(b)

Source: Caneta Research Inc. (1995)

5.3.1 Subsurface temperature variation

The penetration of heat or coldness into the ground depends on the thermal conductivity k (W/m/K) of the ground material, and the temperature change for a given heat flow into a slab of given thickness depends on the specific heat c_p (J/kg/K) of the ground material and its density ρ. The penetration of the surface temperature variation into the subsoil thus depends on the soil

thermal diffusivity κ ($= k/\rho c_p$, m²/s). Values of soil and rock thermal properties are given in Table 5.1. The amplitude of the seasonal temperature variation at depth z, $\Delta T(z)$, is given by (Oke, 1978):

$$\Delta T(z) = \Delta T(0)e^{-z\left(\frac{\pi}{\kappa P}\right)^{1/2}}$$
(5.6)

where P is the period of variation (1 year) in seconds and $\Delta T(0)$ is the temperature amplitude at the surface. The maxima and minima of the subsurface temperature variation lag behind the surface temperature maxima and minima by an amount that increases with depth according to:

$$\Delta t = \frac{z}{2}\left(\frac{P}{\pi\kappa}\right)^{1/2}$$
(5.7)

As seen from Equations (5.6) and (5.7), the amplitude of seasonal temperature variation decreases exponentially with increasing depth, while the lag between peak surface temperature and peak subsurface temperature increases in direct proportion to depth. A lag of 180 days means that the coldest ground temperature occurs at the warmest time of the year and the warmest ground temperature occurs at the coldest time of the year, which is most useful for both air conditioning and heating. The depth at which the lag equals 180 days ranges from 4.45m to 8.9m as κ increases from 2×10^{-7} m²/s (dry clay soil) to 8×10^{-7} m²/s (saturated sandy soil). For a 15K surface temperature amplitude, the temperature amplitude at the 180-day depth is fixed at 0.65K irrespective of the value of κ.

Figure 5.5 shows the seasonal cycle of surface temperature with a mean annual

Table 5.1 Properties of soil materials related to the downward conduction of surface temperature changes

Material	Remark	Density (10³ kg/m³)	Specific heat (10³ J/kg/K)	Heat capacity (10⁶ J/m³/K)	Thermal conductivity (W/m/K)	Thermal diffusivity (10⁻⁶ m²/s)
Sandy soil (40% porosity)	Dry	1.60	0.80	1.28	0.30	0.24
	Saturated	2.00	1.48	2.96	2.20	0.74
Clay soil (40% porosity)	Dry	1.60	0.89	1.42	0.25	0.18
	Saturated	2.00	1.55	3.10	0.58	0.51

Source: Oke (1978)

Figure 5.5 Illustration of the seasonal cycle of surface air temperature with a mean annual temperature of 10°C and an amplitude of 15K, and of the subsurface temperature at depths of 1m, 2m and 5m for a soil thermal diffusivity of 5×10^{-7} m²/s

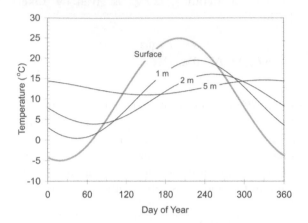

temperature of 10°C and an amplitude of 15K, and of the subsurface temperature at depths of 1m, 2m and 5m for a soil thermal diffusivity of 5×10^{-7} m²/s. The possibility exists to maintain a mean annual surface temperature below the mean annual air temperature by shading the ground surface or by creating a shallow pond over the area to be used for a ground-source heat pump (perhaps as a landscaping element). In this way, the effectiveness of the ground as a heat sink during the summer can be increased.

5.3.2 Energy savings using GSHPs

The energy saving obtained by using GSHPs instead of ASHPs depends on the extent to which the ground is warmer than the air during the heating season and colder than the air or cooling water during the cooling season. We can use the temperature variations in Figure 5.5 and Equation (5.2) for the Carnot cycle COP to compute the relative COPs for different cases. One must take into account that the evaporator temperature must be less than the source temperature in order to draw heat from the source; 3–5K is a reasonable temperature differential. The differences will be greater the larger the heating load and hence the colder the air temperature, since heat will have to be extracted from the source at a greater rate, but the evaporator temperature

will automatically adjust to match the changing heat load. Compared to an air-source heat pump, the annual average heating COP (weighted by the heating load) improves by 10 per cent, 20 per cent and 30 per cent for GSHP pipes placed at depths of 1m, 2m and 5m, respectively.[3] These improvements are somewhat exaggerated, as they do not take into account the fact that, as heat is withdrawn from the ground, the ground temperature and heating COP will fall. Placement of horizontal pipes at a depth of 5m is not practical in most cases, but is included for comparison with more realistic depths. More detailed calculations by Garimella (2003) indicate a savings of up to 20 per cent for cooling in Phoenix, Arizona, and for heating in Minneapolis, Minnesota. For southern Canada, NRCan (2004) indicates seasonally averaged heating COPs of 2.0–2.9 for ASHPs, 2.5–3.3 for closed-loop GSHPs, and 2.9–3.8 for open-loop GSHPs, the latter implying savings of 25–30 per cent compared to an ASHP.

Although the improvement in seasonal mean COP using GSHPs is small for horizontal pipes at depths of 1–2m, short-term extremes in temperature would be avoided. This in turn would allow a GSHP to meet the building's heating needs under conditions where an air-source heat pump would be inadequate, and with greater efficiency. The use of an auxiliary electric heater, with its lower efficiency (COP = 1.0) could be minimized, providing a greater saving in annual energy use than is implied by the difference in COP between air-source and ground-source heat pumps.

W. S. Johnson (2002) measured the energy savings when a GSHP replaced on air-source heat pump for two houses in Tennessee. In one case, horizontal pipes were placed at a depth of 1.4m and tested over four heating and cooling seasons. In the second case, five 4m-deep vertical bore holes were used and the system was tested over three heating and cooling seasons. Measured steady-state and seasonally average COPs are given in Table 5.2. Heating energy use per degree-day was reduced by 50 per cent and 64 per cent using horizontal and vertical pipes, respectively, while cooling energy use was reduced by 37 per cent and 34 per cent, respectively.

Shonder et al (2000) examined the energy

Table 5.2 Average measured COPs for air-source and ground-source heat pumps in Tennessee

| System (GSHP) | ASHP, seasonal | Ground-source heat pump, steady state | | GSHP, seasonal |
		COP	Ground temperature	
Heating season				
Horizontal	1.81	3.6–4.1	9–14°C	3.62
Vertical	1.06	3.2–3.4	0–8°C	2.94
Cooling season				
Horizontal	1.74	3.7–4.4	32–17°C	2.76
Vertical	1.82	2.7–4.0	33–20°C	2.76

Source: W.S. Johnson (2002)

Table 5.3 Comparison of annual electricity and natural gas use (GJ) for heating and cooling of an elementary school in Lincoln, Nebraska (US) using a GSHP with natural gas to preheat ventilation air, and three conventional systems

	GSHP	ACC/ VAV	WCC/ VAV	WCC/ CV
Electricity	1958	2026	1929	3129
Natural gas	795	2389	2403	5172
Total	2743	4415	4332	8301

Note: ACC = air-cooled chiller, WCC = water-cooled chiller, CV = constant volume, VAV = variable air volume.
Source: Shonder et al (2000)

use in four elementary schools equipped with GSHPs in the continental climate of Lincoln, Nebraska. Data from these and other schools were used to develop and calibrate a model for the analysis of alternative heating and cooling systems. Results are given in Table 5.3. Heating and cooling is provided by 54 heat pumps connected with a variable-speed pump to a field of 120 boreholes. Natural gas is used for pre-heating of ventilation air and in some terminal units. Compared to a conventional system with an air-cooled chiller for air conditioning and full reliance on natural gas for heating, there is a negligible change (3 per cent decrease) in total electricity use, but a 67 per cent decrease in natural gas use.

5.3.3 Sustainability of vertical GSHP systems

In vertical GSHP systems, the collector pipes could be 30–50m deep and so will experience negligible seasonal temperature variation except for the top few metres. However, they will be constrained by the limited ability of summer heat to diffuse downward and restore deep ground temperatures before the next heating season (given that the ground at this depth would now be cooled during winter by the operation of the heat pump). Conversely, in climates where the heat pump would be largely or entirely used for cooling, the build up of ground temperature over time is a limiting factor. In some cases, the system is designed to be effective only for a period of 20 years or to operate with lower efficiency and capacity as ground temperatures become less favourable. For example, computer simulations by Chiasson et al (2000a) indicate that when 2.5kW of heat is continuously rejected into a single 76m borehole with an initial ground temperature of 17°C, the ground temperature rises to (and stabilizes near) 45°C after two years for coarse sand, and rises to 29°C in shale (which has a higher thermal conductivity). With a groundwater flow of 60m/year in the coarse sand, the temperature stabilizes at about 36°C.

There are a number of possible solutions to this problem. In some cases, it may be sufficient (and feasible) to increase the spacing of the boreholes (from, say, a 3m grid to a 6m grid). A second solution is to use the ground for heating in winter and cooling in summer, so that winter heat extraction and summer heat rejection will offset to some extent. If this is not sufficient then, for the cold climate case, the ground could be warmed during summer by connecting the ground pipe system to a source of water that is passively heated through contact with the atmosphere or through solar collectors (Chiasson and Yavuzturk, 2003), producing a *solar-assisted ground-source heat pump*. In the warm climate case, there

Table 5.4 Comparison of air conditioning energy use (kWh/year) for systems in which heat is rejected entirely to the ground (a GSHP), and for a hybrid system in which heat is rejected to either the ground or cooling tower using the control strategy that minimizes total energy use

	Houston, Texas		Tulsa, Oklahoma	
	GSHP	Hybrid	GSHP	Hybrid
Chiller or HP				
Year 1	20,399	18,162	17,931	16,648
Year 20	25,904	17,722	20,660	16,432
Ground-pipe pump	16,177	5392	7190	4044
Cooling tower				
Hours, Year 1		5456		5542
Hours, Year 20		4909		5002
Fan, Year 1		2018		2050
Fan, Year 20		1816		1850
Pump, Year 1		330		335
Pump, Year 20		297		302
Total,				
Year 1	36,577	25,903	25,122	23,078
Year 20	42,082	25,229	27,850	22,621

Source: Yavuzturk and Spitler (2000)

may be months when mechanical cooling is not needed (especially if the building is designed to maximize the use of passive cooling techniques). In this case, the ground could be cooled through exchange of heat between the ground-loop fluid and the atmosphere using the cooling tower (Phetteplace and Sullivan, 1998). Alternatively, cooling towers could be used (as in a conventional chilling system) for supplemental heat rejection (see Chapter 6, Section 6.10). The cooling towers would be sized and used such that the annual heat rejection to the ground would approximately balance the annual heat extraction (by the heat pump when used for heating or production of hot water) plus heat dissipated through conduction. Many such *cooling-tower-assisted ground-source heat pump* systems have been constructed. Yavuzturk and Spitler (2000) have calculated the energy use for such a hybrid system for a specified commercial building in Houston (Texas) and Tulsa (Oklahoma). Results are given in Table 5.4 and compared with systems rejecting heat entirely to the ground. During the first year, the hybrid system reduces energy use by 29 per cent in Houston and by 8 per cent in Tulsa. Inasmuch as a more conventional system, relying entirely on a cooling tower, would use greater energy than a GSHP, the savings for the hybrid system would be larger still.

A third possibility, in cooling dominated climates, is to use a shallow pond as a supplemental heat rejecter – a *cooling-pond-assisted ground-source heat pump* system. A portion of the ground loop is placed in the pond, which cools evaporatively. Chiasson et al (2000b) developed a mathematical model to simulate such a system.

In temperate and colder climates, there will be times when heating is needed but the air temperature is warmer than the ground temperature, particularly if substantial cooling of the ground by the heat pump itself has occurred during the course of the heating season. Providing a second, air-based evaporator and using this whenever the air is warmer than the ground would lead to improved efficiency, albeit at greater cost. This could also be part of a strategy, in heating-dominated climates, to balance the removal and replenishment of heat in the ground by avoiding the withdrawal of heat when doing so is less advantageous.

5.3.4 Pumping energy used with GSHPs

Due to the extensive underground pipe network associated with ground-source heat pumps, the energy used to power the pumps can be a significant fraction of the energy used to power the heat pump. Variable speed pumps (discussed in

Chapter 7, Section 7.1.2) are generally used, but the saving that is achieved compared to fixed speed pumps depends on the details of how they are implemented. In four facilities using GSHPs, Henderson et al (2000a) report measured pump energy use ranging from 13 per cent to 53 per cent of the measured heat pump (compressor + supply fan) energy use. However, with best practice, the pump energy use would range from 6 per cent to 14 per cent of the heat pump energy use. For a four-storey office building in Birmingham, Alabama, with a hypothetical ground-source heat pump and 7000m of underground pipe for cooling, Kavanaugh and Lambert (2004) calculate that pump energy use would equal 45 per cent of the compressor and fan energy use if a fixed speed pump is used, but only 7.5 per cent with a variable speed pump.

Kavanaugh and McInerny (2001) evaluated the energy use for four different pump and piping systems for a GSHP in a school in Birmingham, Alabama. Centralized and modular systems are considered; in the centralized system, the total number of ground bores can be reduced by 15 per cent compared to the modular system, because not all parts of the school will need to be cooled at the same time or to the same extent (as solar position changes). In these, the building interior loop is connected to the exterior loop with the same pump. A centralized primary–secondary system was also considered, with separate pumps for the ground and building loops. Results are given in Table 5.5. The baseline system uses a pump that runs at a fixed speed, irrespective of the load. Use of a variable speed drive (VSD) in a centralized system reduces annual pumping energy by an astounding 83 per cent, while increasing the peak load by 6 per cent. The latter is due to the energy losses associated with the electronic drive, which occur even at full motor speed. Pumping energy in a decentralized system with VSD is reduced by a further 30 per cent, in spite of the fact that a large number of small pumps, each with a wire-to-water efficiency of less than 30 per cent, had to be used in the decentralized case. The large energy savings is due to the small pressure heads in the decentralized system and the fact that most of the pumps are off for large portions of the unoccupied periods. The decentralized VSD system also provides a

significant (28 per cent) reduction in peak power demand compared to the centralized fixed speed system.

Table 5.5 Annual heat pump and water pump energy use for a school in Birmingham, Alabama, for different pipe-pump systems

System	Pump energy (kWh/year)	Pump peak power (kW)	Energy Savings (%)
Centralized FSD	108,624	12.4	–
Centralized VSD	18,849	13.2	83
Decentralized VSD	13,132	8.94	88
Dual FSD	65,548	12.9	40

Note: FSD = fixed speed drive; VSD = variable speed drive.
Source: Kavanaugh and McInerny (2001)

5.3.5 Non-energy benefits of GSHPs

GSHPs provide a number of benefits in addition to reduced energy use:

- reduced consumption of water for air conditioning compared to water-cooled chillers, due to the absence of a cooling tower;
- reduced (by about 25 per cent) use of refrigerants (which are halocarbons at present) compared to split-system air-source heat pumps or air conditioning systems;
- lower space requirements, due to the absence of the cooling tower and furnaces or boilers.

5.3.6 Economics of GSHPs

Table 5.6 compares the performance, investment cost and operating costs of a GSHP used for heating and cooling, and the alternative of separate air conditioners and electric resistance heaters or non-condensing boilers. Heat pumps and chillers of the size considered here both cost about $330/kW$_c$, while heating equipment is an order of magnitude less expensive, at $30/kW$_{th}$ (Figure 4.16). The breakdown of the GSHP

Table 5.6 Operating and cost characteristics of a ground-source heat pump and the air conditioner and natural gas (NG) boiler or electric resistance heating system that it can replace for cooling and heating

	Electric base case		NG base case		GSHP	
	Heating	Cooling	Heating	Cooling	Heating	Cooling
Peak load (kW)	30.3	57.5	30.3	57.5	30.3	57.5
Annual load (MWh)	69.7	70.7	69.7	70.7	69.7	70.7
Equipment efficiency	1.0	3.0	0.9	3.0	3.2	5.4
Peak power (kW)	30.3	19.2	0.0	19.2	9.6	10.6
Annual energy (MWh)	69.7	23.6	77.4	23.6	21.8	13.1
Supply efficiency	0.33	0.33	0.9	0.33	0.33	0.33
Annual primary energy (MWh)	282.6		157.5		105.7	
CO_2 emission (tonnes C/year)	6783		2876		2536	
Annual energy cost ($)	11,192		7326		4185	
Investment cost	20,036		20,036		41,253	
Annual financing cost	1475		1475		3035	
Total annual cost	12,667		8801		7220	

Note: Electricity is assumed to be generated from coal with a generation times transmission efficiency of 0.33, and CO_2 emissions are computed assuming emission factors of 13.5 and 24kg C/GJ for natural gas and coal, respectively. Energy costs are computed assuming electricity at 12 cents/kWh and NG at $16/GJ, while base case heating and cooling equipment are assumed to cost $35/kW$_{th}$ and $330/kW$_c$, respectively, based on Figures 4.16 and 6.53.
Source: Source for other data: RETScreen spread sheet for ground-source heat pumps (www.retscreen.net)

system, as given by the RETScreen spreadsheet of Natural Resources Canada (www.retscreen.net) is as follows:

Heat Pump:	55.6kW @ $330/kW	=	$18,341
Drilling and grouting	837m @ $12/m	=	10,043
Ground-loop pipes 1	674m @ $2.5/m	=	4185
Fluid	0.15m³ @ $2600/m³	=	385
Pumps	0.9kW @ $850/kW	=	803
Fittings and valves	55.6kW @ $12/kW	=	667
Internal piping and insulation	55.6kW @ $60/kW	=	3335
Training and contingencies			3494
Total			$41,253

The GSHP costs substantially more than air-source heat pumps in the above example due largely to the cost of the ground loop. However, for typical retail energy costs (electricity at 12 cents/kWh, natural gas at $16/GJ), the savings in annual energy costs more than offsets the additional financing costs (assuming 4 per cent real interest and a 20-year system lifespan), particularly if the GSHP replaces a system with electric resistance heating. However, if the GSHP replaces natural gas heating, there is very little

reduction in overall CO_2 emissions if electricity is derived from coal at the current typical generation times transmission efficiency of about 33 per cent. In some cases, GHSP systems can cost less than conventional systems, as found by Shonder et al (2000) for a GSHP system in four elementary schools in Lincoln, Nebraska. In this case, the scale of the system (630kW$_c$, 388kW$_{th}$) is such that the savings in the cost of mechanical equipment offsets the cost of the ground loop. Kavanaugh (2001) presents further examples where the cost of a GSHP system is no more than that of conventional HVAC systems.

There are two circumstances where the cost of the ground loop can be reduced:

1 Where the building has a number of deep piles required for the foundation, the ground loop can be integrated into the piling system.
2 Where excavation and replacement of the excavated material next to the building is required for structural reasons, the ground loops can be installed as part of this process.

Costs can also be reduced through a variety of methods that allow downsizing of the heat pumps and ground loop, as discussed later (Section 5.9), and through eventual use of microchannel heat exchangers, as discussed next.

It is hard to compare maintenance costs for GSHP systems and conventional systems, since the average age of the former is much less than that of the latter. Data collected by Cane and Garnet (2000) indicate significantly lower (by a factor of two to six) maintenance costs for GSHP systems than for conventional cooling systems, while Martin et al (2000) indicate little difference.

5.4 Exhaust-air heat pumps

The ground is preferred over the ambient air as a heat source or sink because it experiences less temperature fluctuation, being warmer than the air during the winter and cooler than the air during the summer, and subject to less short-term extremes. This allows GSHPs to operate with a higher COP than ASHPs. The stale air exhausted from a building is potentially better still, as it will be at almost the same temperature year round: perhaps 20°C during the winter and 24–28°C during the summer (depending on indoor temperature settings and the difference between the temperature of air supplied to and extracted from the occupied space). The exhaust air will serve as a warmer heat source than either the ground or outside air during winter, although it may not be as cold a heat sink as the ground during summer (depending on the local climate). In addition, the problem of gradual cooling of the heat source during the winter, and gradual warming of the heat sink during the summer, and of the need to ensure a long-term balance between heat additions and withdrawals, is avoided if the exhaust air rather than the ground is used as the heat source and sink. For these reasons, exhaust-air heat pumps (EAHPs) are an attractive option. In heating mode, the EAHP transfers heat from the exhaust air to the supply air. Heat exchangers (discussed in Chapter 7, Section 7.3.3 and 7.4.9) also do this, but can never warm the supply air to a greater temperature than the exhaust air (the best heat exchangers will warm the incoming air by up to 95 per cent of the difference between exhaust and supply air temperatures). However, for an EAHP, if the evaporator in the exhaust

airstream is colder than the outside air and the condenser in the supply airstream is warmer than the exhaust air, then it is possible to warm the outside air by more than the exhaust–outside air temperature difference. In this case, the air leaving the building would be cooled to below the outside temperature, and the ventilation of the building becomes a net source of heat to the building. Today, almost all new single family houses in Sweden are equipped with exhaust-air heat pumps (about 4000/year by 1997), and another 5–10,000/year are installed in Germany (Fehrm et al, 2002). Typically, the exhaust-air is cooled by 20–25K. During the summer, the building exhaust can be used as a heat sink if the heat pump is operated in reverse mode (as an air conditioner).

EAHPs can be used in combination with a heat exchanger and/or an intake pipe that is buried underground so as to moderate the temperature of the air supplied to the heat pump (this system, called an 'earth pipe loop', is discussed in Chapter 6, Section 6.3.8). These may give little improvement in overall energy use (especially when additional required fan energy is taken into account), but serve to moderate extremes in temperature, thus allowing for a smaller heat pump and requiring less backup electric resistance heating. A heat exchanger/EAHP combination was used in demonstration houses in Thening, Austria; Freiburg, Germany; and Ulm, Germany (see Fact Sheets for Task 28 of the IEA Solar Heating and Cooling Program, www.iea-shc.org).

Bühring et al (2003) have developed a model for simulating a system consisting of an earth pipe loop, an exhaust-air to intake-air heat exchanger, an exhaust-air heat pump, a hot-water storage tank, a solar thermal collector, and a backup electric resistance heater. The storage tank permits heat to be upgraded to a temperature of up to 55°C and used for tap hot water, and/or permits heat extracted from exhaust air at one time to be used for space heating at a later time. The heat pump has two condensers, one for recharging the storage tank and the other for directly heating the intake air. The model is programmed in FORTRAN as an additional module for the TRNSYS simulation program (see Appendix D). Its purpose is to aid heat pump manufacturers in the design of heat pumps most suited to this application for various climatic and other conditions. In designing a new heat pump, a manufacturer does not develop new components, but will select commercially available components. The design issue is how best to optimize the selection and integration of the various components for a given application, a task that is aided through the use of a model that incorporates the known characteristics of the separate components. The model has been validated against data from a test facility at the University of Freiburg and from a single family and a multi-unit house in Germany. For a single family high-performance house, the mean annual COP for hot-water storage and direct space heating are 3.0 and 3.3, respectively, and the total electricity consumption for space heating is about 13kWh/m²/year (a typical heating load would be about 200kWh/m²/year).

It appears that, under some circumstances, very high COPs can be obtained with EAHPs. For example, using an EAHP in combination with a ground pre-heating loop and a radiant floor heating system (so that the heat supply temperature need be no warmer than 35°C), Halozan and Rieberer (1997, 1999) calculate seasonal mean COPs of 6–7 for the climate of Graz, Austria. In an experimental setup in Quebec, Minea (2003) measured a seasonal mean COP of 7.5 for heating and 16.3 for cooling.

5.5 Effect of heating and cooling distribution temperatures

During the heating season, the COP of a heat pump will be smaller the warmer the temperature at which heat is distributed by the heating system within a building. This is because the condenser will have to achieve a temperature somewhat greater than the heat distribution temperature in order to reject heat to the distribution system, so the heat pump will have to work harder. In a forced air heating system, which is common in houses in North America, heat is typically supplied at 50–60°C. In Europe it is common to distribute heat with hot water (a *hydronic* system), traditionally with a supply temperature of 90°C and a return temperature of 70°C. As seen from Figure 5.2, this higher distribution temperature reduces the COP by 1.0–3.0 (depending on the temperature of the heat source). If the insulation is improved and air

infiltration reduced in old north-European buildings, the required supply temperature can be reduced to as low as 45–55°C. The lowest distribution temperatures can be achieved through floor radiant heating systems; in new, thermally tight buildings, a distribution temperature of 30–35°C can be used (residential heat-distribution systems are discussed further in Chapter 7, Sections 7.3.1 and 7.3.2). If the condenser temperature can be reduced from 60°C to 40°C, the energy savings by the heat pump is 25–30 per cent for an evaporator at −15°C–0°C (greater relative savings for a warmer evaporator).

There is thus a double benefit from better thermal envelopes: the amount of heat that needs to be supplied decreases, and the efficiency in supplying the required heat with heat pumps increases, because the heat can be distributed at a lower temperature and thus with a higher heat pump COP.

In cooling mode, maximization of the heat pump COP (or that of a chiller) requires that the evaporator temperature be as high as possible, which in turn requires as warm a distribution temperature for cooling as possible. As explained in Chapter 7 (Sections 7.4.3–7.4.4), cooling can be provided through an all-air system, or through a cold water (hydronic system) with an airflow sufficient for ventilation purposes only. Hydronic systems can involve a fan blowing air past a cold-water coil (a *fan coil system*) or chilling the entire ceiling or floor (a *radiant chilling* system). In a fan coil system, a typical supply temperature is around 5–7°C, with a return temperature of around 12–14°C. This requires an evaporator temperature of around 0–2°C. In ceiling radiant-cooling systems, distribution temperatures of 18–20°C have been used, so the evaporator temperature can be much higher.

If humidity is high, dehumidification of the ventilation air will be required whether or not radiant cooling is used. In conventional systems, this is accomplished by cooling the air to a low enough temperature to condense sufficient moisture. For example, to provide air at 60 per cent relative humidity and 16°C (6.81gm water vapour per kg of air) requires cooling the air to 8°C in order to condense any excess water, then reheating it. This is comparable to the chilling required with conventional hydronic cooling, with a comparable COP penalty. However, desiccant dehumidification (discussed in Chapter 7, Sections 7.4.11 and 7.4.12) provides an alternative method of dehumidification, without having to cool the air to below the final distribution temperature. As seen from Figure 5.2, increasing the evaporator temperature from 0°C to 10°C increases the COP by 1.0–3.0. This represents a 25–30 per cent saving in energy use by the heat pump for condensers at 30–40°C.

In summary, lowering the temperature at which heat is distributed or increasing the temperature at which chilled water is distributed leads to significant improvements in the COP of heat pumps. Avoiding the need to overcool ventilation air in order to remove moisture also leads to a significant improvement in COP (as well as saving on energy used to reheat the air, as explained more fully in Chapter 7, Sections 7.4.11 and 7.4.12).

5.6 Part-load performance

As the heating and cooling loads in a building decrease, one needs to reduce the heat or coldness delivered to the building. In conventional systems, this is achieved through periodic on/off cycling; the lower the heating or cooling load, the greater the fraction of time that the heat pump is shut off. On/off cycling reduces the efficiency of a heat pump. It would be better to be able to continuously vary the temperature at which heat or coldness is delivered as the load varies. In heating mode, the rate of heat flow into a room depends on the difference between the radiator temperature and the room temperature, and this in turn must equal the rate of heat loss in order to maintain a constant room temperature. As the outside temperature increases and the heating load decreases, the distribution temperature should decrease. This in turn requires that the heat pump condenser temperature decrease. At the same time, the evaporator temperature must increase (becoming closer to the source temperature), so that there will be a matching decrease in the rate of heat flow from the heat source to the evaporator. Thus, the difference between evaporator and condenser temperature can decrease at part load, thereby increasing the Carnot COP and

tending to improve the actual COP. The increase in evaporator temperature and decrease in condenser temperature at part load can be achieved if the compression ratio is reduced (giving a higher evaporator pressure and a lower condenser pressure), so the compressor needs do less work.[4]

However, heat pumps (like electric chillers) invariably use on/off cycling. Some heat pumps have 2-stage compressor units (with high and low speed), but still revert to on/off cycling. One unit with a reciprocating compressor can run with one or two cylinders, but also with on/off cycling (compressors are discussed in Chapter 6, box 6.3). Others throttle the fluid flow (changing the orientation of vanes that partially obstruct the flow, or partially closing the suction valves). In any case, part-load operation is less efficient than full-load operation, in spite of the larger Carnot COP. In addition, the efficiency of the motors that drive the water pumps and fans is also lower at part load (see Chapter 7, Section 7.1.2). Motor efficiency may drop from 94–98 per cent at full load to 80–90 per cent at 25 per cent load for large motors, and much more for small motors.

However, it is possible to vary the rotation rate of a compressor using variable speed motors. These commonly work by varying the frequency of the input current using a *variable speed drive* (VSD), which is an electronic panel that serves as an interface between the electrical input and the motor and is mounted on a wall up to 150m (with the latest technology) from the motor. The efficiency loss of throttling is thus avoided, although there is an energy loss in the VSD itself (see Chapter 7, Section 7.1.2). Heat pumps with variable speed motors tend to have peak efficiency at about one-third to two-thirds of full load, rather than at full load. This is because the compression ratio decreases at part load in response to a lower compressor speed, resulting in an increase in mechanical efficiency, and because the apparent heat transfer surface of the evaporator and condenser increases in response to the decrease in the volumetric flow rate of the refrigerant (Reay and MacMichael, 1988).

Figure 5.6 shows the variation in COP with load for a screw compressor using three different methods of capacity control: by throttling the fluid flow, by varying the suction volume using a compressor slide valve, and by varying the

Figure 5.6 Variation in the COP of a screw compressor with load when the load is varied by throttling, by adjusting the suction volume, and by varying the shaft rotational speed

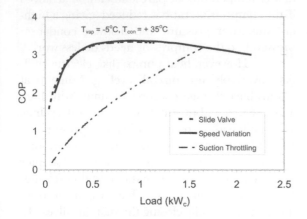

Source: Redrafted from Stosic et al (1998)

compressor rotation speed. The variation of COP with capacity is nearly identical for the latter two methods, being largest at 30 per cent of peak capacity, while the COP with throttling drops to near zero at zero load. Even though the Carnot COP increases at part load (due to the smaller temperature lift), this effect is overwhelmed by the decrease in COP for a given lift when throttling is used. Thus, the incorporation of variable speed motors in the compressors, fans, and pumps of heat pumps (and commercial chillers) leads to a significant increase in COP at part load – the most common operating condition. At present, some heat pumps cannot operate below 50 per cent of full load, and have to resort to on/off cycling in order to achieve lower loads. This, however, is also quite inefficient.

5.7 Hydronically coupled heat pumps

Most residential heat pumps in North America are air-source heat pumps that are directly coupled to the outside air through the outdoor heat exchanger (which serves as a condenser or evaporator, depending on whether the heat pump is in cooling or heating mode), and directly coupled to a forced air heating system through an indoor heat exchanger. These are either round tube or flat plate heat exchangers that are rather large. Automobile air conditioner heat exchangers use microchannel tubes that, if applied to residential heat exchangers, would reduce their size by a

factor of two to three while increasing the heating COP by about 20 per cent (from, say, 2.4 to 2.8), increasing the cooling COP by about 60 per cent (from, say, 2.4 to 3.8), and reducing the refrigerant charge (Garimella, 2003). Reducing the refrigerant charge is desirable because the refrigerant can leak, most refrigerants are powerful greenhouse gases, and some (HCFCs) contribute to depletion of stratospheric ozone (the global warming impact of leakage of the refrigerant is discussed in Chapter 6, Section 6.7).

The main resistance to heat transfer is on the air side of the microchannel heat exchanger. In a hydronically coupled heat pump proposed by Garimella (2003), heat is transferred from the condenser to an intermediate fluid such as an ethylene-glycol solution, and from there to the outdoor or indoor air (depending on whether the heat pump is in cooling or heating mode) using a second, fluid-to-air heat exchanger. A second intermediate fluid loop would couple the evaporator to the other fluid-to-air heat exchanger, as illustrated in Figure 5.7. The main advantages of this system are that (i) the refrigerant-to-hydronic fluid heat exchangers can be reduced in size by a factor of ten compared to conventional refrigerant-to-air heat exchangers, (ii) these heat exchangers can be located within a sealed compact unit, which in turn can be placed anywhere, rather than separated by long refrigerant lines with losses and pressure drops; (iii) the condenser and evaporator do not switch roles when the unit changes from heating to cooling, so each can be optimized for a fixed role; (iv) there is no reversing valve and associated losses (see Section 5.2.3); and (v) integration of space conditioning and water heating is easier. As a result of (i) and (ii), the refrigerant charge can be dramatically reduced. The disadvantage, for an air-source heat pump and a forced air heating system, is that two additional heat exchangers, between the intermediate fluid and the inside or outside air, are needed. As a result, the COP of the hydronically coupled heat pump would be slightly worse than that of the microchannel air coupled heat pump, although still better than a conventional heat pump.

However, with a combination of a ground-source heat pump and a hydronic heating or cooling system, fluid to air heat exchange is

Figure 5.7 Fluid flow in a hydronically coupled heat pump

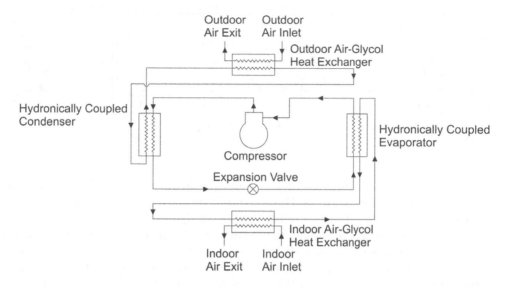

Source: Garimella (2003)

5.8 Working fluids

avoided. It could be replaced with fluid to fluid heat exchangers, or the additional heat exchangers could be avoided altogether if the secondary fluid on the exterior side circulates through the ground coil and the secondary fluid on the interior side circulates through heating or cooling pipes in the building. The factor of ten downsizing with microchannel heat exchangers should lead to a significant reduction in the cost of heat pumps.

5.8 Working fluids

There are two issues of concern with regard to the working fluid (refrigerant) used in heat pumps and other vapour compression cooling equipment: the effect on stratospheric ozone and climate of leakage of the refrigerant to the atmosphere, and the effect on the performance and operating conditions of the equipment. Two classes of halocarbons are used as refrigerants in new equipment – the HCFCs, which deplete stratospheric ozone and warm the climate, and the HFCs, which have no impact on ozone but warm the climate (see Chapter 2, Section 2.4.3).

Table 5.7 lists the working fluids commonly used in vapour compression heat pumps, and their properties. In most cases, mixtures of more than one refrigerant are used. When

mixtures are used, the condensation and evaporation temperatures depend on the heat flux to or from the evaporator or condenser (Reay and Mac-Michael, 1988; Bullard and Radermacher, 1994). When the air is warm relative to the evaporator, the heat flux will tend to be large, which tends to make the evaporator warm (and limit the heat flux). As the air cools in response to heat transfer to the evaporator, the heat flux will decrease, thereby decreasing the evaporator temperature and tending to maintain the original heat flux. The variation in the evaporator temperature is called the *temperature glide*. With a pure refrigerant, as the air flows past the evaporator and cools, the temperature differential will decrease because the evaporator temperature is constant, which limits the overall heat transfer. A similar effect occurs at the condenser. Thus, when the operation of the heat pump leads to a substantial change in the temperature of the air or fluid flowing past the evaporator or condenser, the heat transfer efficiency can be increased using a mixture of working fluids. This in turn allows smaller temperature differentials and hence a larger COP.

Refrigerant emissions occur during the operation of the equipment (due to leakage) and during removal of the remaining refrigerant at the end of the life of the equipment. Lifespan

Table 5.7 Properties of refrigerants used in heat pumps

Refrigerant	R-404a	R-407c	R-410a	R-507a	R-134a	R-290	R-717
Components	44% R125 52% R143a 4% R134a	23% R125 25% R125 52% R134a	50% R32 50% R125	50% R125 50% R143a	R134a	Propane (C₃H₈)	NH₃
Temperature at 25 bar (°C)	53	55–59	41	52	78	68.3	58
Temperature at 1 bar (°C)	–46 to –47	–37 to –44	–52	–47	–26	–42.1	–33.3
Critical temperature (°C)	74.4	86.4	71.8	71	100.6	96.7	130

Note: R32 = HFC-32, R125 = HFC-125, R134a = HFC-134a and R143a = HFC-143a.
Source: Eggen (2000)

emissions of refrigerants have fallen by a factor of ten during the past 30 years and, for HFC refrigerants (which will be the only permissible halocarbon refrigerant by 2010 at the latest), the climatic impact of halocarbon emissions from heat pumps is estimated to be 2–10 per cent of the total impact (including energy use) for a $4kW_c$ unit, and 25–30 per cent for a $56kW_c$ unit, in both cases assuming that only 50–70 per cent of the remaining refrigerant is recovered at the end of a 15-year operating life (Peixoto et al, 2005). As discussed below, non-halocarbon refrigerants can entail COP penalties of 40–50 per cent if the heat pump is not optimized for the alternative refrigerant, but as low as 15 per cent in some circumstances or with a slight COP benefit in other circumstances if the heat pump is fully optimized. Thus, both the performance of the heat pump and the impact of halocarbon emissions need to be considered in evaluating the climatic impact of alternative choices for refrigerants. There is no consistent winner.

5.8.1 Carbon dioxide

Carbon dioxide (R-744) can be used as a working fluid. The key characteristic of CO_2 as a working fluid is that its critical temperature (above which it cannot exist as a liquid at any pressure) is rather low: 31.1°C (critical pressure: 73.8 bar). Since this is cooler than typical desired condenser temperatures, heat rejection must occur above the critical pressure and temperature, while heat absorption occurs below the critical temperature and pressure. This is referred to as a *transcritical cycle*. Condensation is replaced by cooling in a so-called gas cooler. The heat rejection temperature is independent of the heat rejection

pressure in the supercritical region, and there is an optimum pressure that gives the maximum COP (Liao et al, 2000). This is in contrast to traditional subcritical heat rejection, where COP decreases continuously as the rejection pressure (and temperature) increases. CO_2 heat pumps operate at pressures of 80–135 bar and temperatures of up to 90°C on the high-temperature side, and at an evaporator temperature of as low as –6.4°C with pressures of 30–35 bar (White et al, 2002). This makes it more than adequate for making domestic hot water (DHW) and for district heating, where temperatures in excess of 55°C are often needed. Indeed, 17 companies make CO_2 heat pumps for DHW in Japan, with 120,000 units sold in 2004 (Kusakari, 2005). Conventional heat pump water heaters are often limited to a maximum operating temperature of 55°C, as seen from Table 5.7.

The use of CO_2 as a working fluid is not new: the first chillers, in the early twentieth century, used CO_2. This was replaced by ammonia as a working fluid, and later by the CFCs during the 1940s. However, the high required pressures taxed the technologies that existed at the time. The renaissance of CO_2, driven by concerns over the HCFC and HFC replacements for CFCs, is occurring at a time when the high operating pressures do not pose technical problems. This is because *microchannel heat exchangers* (having tubes with a diameter of less than 1mm and a wall thickness of only 0.33mm) are now available that can tolerate the higher operating pressures using CO_2 as a working fluid. Indeed, the high pressures can now be turned to advantage, as they result in smaller volumetric flow rates, so that the dimensions of the compressor, valves, and piping can be smaller (Nekså and Stene, 1997).

Although CO_2 is a greenhouse gas, the

amount contained in heat pumps (which can eventually leak to the atmosphere) is negligible compared to the emissions associated with supplying electricity from fossil fuels to power conventional air conditioners and chillers. At the same time, the GWP of the HCFCs and HFCs that it would replace is up to several thousand times greater than that of CO_2 on a molecule-per-molecule basis (see Table 2.4). Since there can be little impact on COP using alternative candidate working fluids (compared to the other factors influencing COP that have been discussed here), heat pumps using CO_2 as a working fluid are of particular interest.

Performance

The ideal COP using CO_2 is only 50–60 per cent of that using CFC-11 (R-11) due to the high 'condenser' temperature, but the real COP of a heat pump using CO_2 is a larger fraction of the ideal COP, such that the real COP was only about 6 per cent smaller for tap water heating in experiments by Hwang and Radermacher (1999) and slightly smaller to slightly better for space heating in experiments by Richter et al (2003). For a prototype reversible CO_2 heat pump, Aarlien and Frivik (1998) obtained COP values that are 1–15 per cent lower than for an R-22 (CFC-12) heat pump in cooling mode (depending on the initial and final temperature and humidity conditions), and 3–14 per cent higher in heating mode. These and other studies comparing CO_2 and conventional working fluids for space conditioning are not entirely fair, as the chillers using CO_2 are assumed to use high-performance microchannel heat exchangers while the chillers using conventional working fluids are assumed to use conventional heat exchangers.[5] When working fluids are compared assuming the same heat exchangers, the COP using CO_2 is about 43–57 per cent lower according to simulations performed by Brown et al (2002). The key point, however, is that it is possible to achieve a performance comparable to present-day chillers using CO_2 as a working fluid, without the ozone and global warming impacts of HCFC leakage or the global warming impact of HFC leakage. In addition, CO_2 lends itself more easily to the use of microchannel heat exchangers, so the greater COP of conventional fluids with comparable heat exchangers might not be achievable in practice for various reasons. For example,

a major disadvantage of microchannel heat exchanges is their tremendous flow resistance (due to the large surface to volume ratio, the very parameter that makes them so effective as heat exchangers). This is not a problem for CO_2, because it has a very low liquid viscosity (about one-third that of HFC-134a) and a very high vapour density (about seven times that of HFC-134a) (Zhao and Ohadi, 2004). Garimella (2003) discusses a number of other advantages of CO_2 as a working fluid, but cautions that the practical feasibility and economic competitiveness of CO_2 systems remain to be demonstrated.

Proper assessment of the relative efficiencies using CO_2 and conventional working fluids requires consideration of differences in the opportunities for integrating heating and cooling requirements. Here, CO_2 has an advantage because of the high heat-rejection temperatures (70°C for a CO_2 chiller, versus 40°C for a conventional chiller). Nekså et al (1998) designed, built, and simulated a CO_2-based system for refrigeration and heating in a supermarket in Venice. Operating purely as a chiller, the CO_2 unit used 50 per cent more electricity than a conventional chiller. However, when credit was taken for byproduct tap water and space heating, the CO_2 system requires 10–20 per cent less primary energy if the hot water and heating are provided by a high efficiency gas boiler in the conventional system (and saves even more if electricity had been used for hot water and space heating). Adriansyah (2004) has also experimented with a CO_2-based heat pump for simultaneous cooling (0°C evaporator) and production of hot (60°C) water. A unit with a cooling capacity of about 25kW yields a COP of about 3.5 when used only for cooling, but a COP of up to 8.0 when used for simultaneous heating and cooling, with credit taken for the hot water produced.

In summary, the results cited above indicate that heat pumps using CO_2 as a working fluid, once fully optimized for CO_2, are likely to have slightly lower COPs when used for air conditioning but slightly higher COPs when used for heating, compared to heat pumps with halocarbon working fluids. The biggest advantage of CO_2 in heat pumps may be when heat pumps are used for simultaneous cooling and production of hot water.

Restrictions

There are at least two important restrictions on the applicability of heat pumps using CO_2 as a working fluid. First, Chaichana et al (2003) argue that CO_2 will not be suitable for solar-assisted heat pumps, because of the wide fluctuation in operating conditions with a solar input, and instead conclude that ammonia will be best. Second, the large downward temperature glide in the gas cooler (up to 90K) implies that there will be a large temperature gain in the water being heated; in a counterflow heat exchanger, the water will be heated by 50K (Nekså, 2002). The heat distribution system must therefore be designed to have a large temperature drop between the supply and return lines. This can be accomplished with a lower flow rate (saving on pumping energy) and large radiators, but necessitates rather large supply temperatures. In a theoretical study of a retrofit of a typical hydronic heating system in Europe with CO_2 heat pumps, Enkemann et al (1997) propose reducing the flow rate so as to change the supply/return temperatures from 70/50°C to 93/40°C. The seasonal COP increases from 2.8 to 3.2, but there is an increased danger of burns. CO_2 heat pumps could be restricted to dedicated production of domestic hot water, where larger temperature changes are appropriate. To make use of the large temperature glide, a counterflow heat exchanger is required. This precludes using the heat pump for both heating and cooling, as the refrigerant flow will be reversed between the two modes of operation (see Box 5.1) and the flow will be parallel in one of the two modes. This can be overcome with more complex piping and valving, but at a prohibitive cost (Hughes, 2001).

Retrofit opportunity with hydronic heating

A typical hydronic heating system has a supply temperature of 70–90°C. Although a lower temperature is desired in order to enhance the heat pump COP (Section 5.5), this requires first improving the thermal envelope and utilizing larger radiators, which might not always be feasible. With conventional working fluids, these requirements limit the applicability of heat pumps in retrofit situations. This limitation is removed with the use of CO_2 as a working fluid.

5.8.2 Hydrocarbons

Propane (R-290) can also be used as a refrigerant in heat pumps and cooling equipment. It has similar thermodynamic properties as conventional refrigerants (such as HCFC-22) and uses the same lubricant as most HCFC refrigerants. Roth et al (2002a) cite a number of studies that found a 5–10 per cent improvement in COP compared to HCFC-22 for water heating. However, fire safety is a concern with propane. If a secondary loop is used to increase safety, the COP is about 20 per cent lower than when using HCFC-22 or HFC-134a.

5.8.3 Water

Water can be used as a refrigerant, and if used as such, it could also serve as the chilling water and/or the cooling water (in a cooling tower) all in one loop, thereby eliminating the need for an evaporator and/or condenser heat exchanger. Used in such a way, the overall system performance would be better than that of a system using a conventional refrigerant (Roth et al, 2002a). However, equipment using water as a refrigerant might be twice as expensive as conventional equipment, and is at a very early stage of development.

5.9 Heat pump sizing and diurnal thermal storage

Heat pumps are more costly than conventional alternatives but generally have lower energy costs. It is therefore important to minimize the first cost as much as possible, and this can be done if the required peak heating or cooling capacity can be reduced. This can be done in four ways, described below.

5.9.1 Auxiliary electric-resistance heating coil

Rather than sizing a heat pump to meet the heating demand on the coldest days of the year, one can rely on supplemental electric resistance heating. This will decrease the SPF by a few percent but allows downsizing of the heat pump by up to 40 per cent, with a comparable cost savings.

5.9.2 Downsizing heat pumps through diurnal thermal energy storage

If heat or coldness can be stored for a few hours or days, this allows the heat pump system to maintain comfortable indoor temperatures over a greater range of outdoor temperatures without using a supplemental system, or with a smaller heat pump. This can be done by incorporating thermal mass in walls and floors, or through rock beds, water tanks or phase-change materials (Chapter 4, Section 4.1.6). This will minimize the temperature decrease at night and hence the need for artificial heating during the coldest hours. At $4.186MJ/m^3/K$, water has almost three times the volumetric heat capacity of rock or building materials, and provides easy heat transfer (by circulating the water itself). Given that the heat capacity of air is only $1.25kJ/m^3/K$, one cubic metre of water cooled by one degree will warm about 3300 cubic metres of air by the same amount. During the cooling season, coldness can be stored in an ice slurry tank. In the latter case, a heat exchanger between the refrigerant loop and the water freezing loop can be avoided if the evaporator is placed directly in the ice/water tank. This is a direct contact, direct expansion system, also known as a 'flash-freeze' system (Breembroek, 1997).

An added benefit of incorporating thermal storage is that some heating and cooling can be performed at night, when electricity rates are often much lower. The impact of cold water and ice thermal storage on overall cooling system efficiency is discussed in Chapter 6 (Section 6.9). Most heat pump systems in commercial buildings in Japan are coupled to water thermal storage (Breembroek, 1997). A control strategy should be selected such that the storage is almost emptied regularly, as this reduces standby losses.

5.9.3 Downsizing the ground loop through zonation

In mid-size and larger commercial buildings, a separate heat pump should be used for cooling each of several zones. For a GSHP, the most expensive component is the ground loop. If the ground loop is sized based on the total cooling capacity of all of the heat pumps, it will be oversized by up to a factor of 2.5 (Henderson, 1999).

This is because the peak cooling load in different zones will occur at different times, so that the peak cooling load for the entire building (to which the ground loop should be sized) will be less than the total cooling capacity of all the heat pumps.

5.9.4 Downsizing the ground loop by decoupling sensible and latent cooling loads

In cooling dominated applications, especially in climates with humid summers, the ground loop can be downsized if the heat pump system is used for cooling (removing sensible heat) but not for dehumidification (removing latent heat). This can be accomplished if solid or liquid desiccants, regenerated with solar or waste heat, are used for dehumidification (see Chapter 7, Section 7.4.11 and 7.4.12). For a hypothetical GSHP-liquid desiccant system serving an 11-storey tower in northern Italy, Gasparella et al (2005) find that the cost and size of the ground coil are reduced in half compared to a GSHP alone.

5.10 Seasonal underground storage of thermal energy

As noted in Box 5.1, a heat pump can be used for cooling purposes by simply reversing the direction of the refrigerant flow. A GSHP can thus be used for heating in winter and cooling in summer. The ground ends up serving as a seasonal heat storage device: in summer, heat taken from the building is added to the ground by the heat pump, providing additional heat for extraction by the heat pump in the winter. Extraction of this heat during the course of the winter makes the ground colder and thus more suitable for cooling purposes in the summer. The underground pipes may be deliberately placed closed enough together to freeze a large block of ground during the heating season. In effect, summer heat is stored in the ground for use in winter (albeit with some losses), and winter coldness is stored (again, with some losses) for use in summer. As noted in Section 5.3.2, special provisions will be needed in some climates to make sure that winter heat withdrawals and summer heat additions balance over the long term so that the performance of the heat pump does not degrade over time.

A disadvantage of the approach described above is that the ground may have become colder than the air by the end of the heating season (when air temperature is increasing), so that there would be an energy penalty using the ground as the heat source (due to the greater temperature lift). Similarly, the ground might become warmer than the air by the end of the cooling season. Instead, it is preferable from an efficiency point of view to use different areas or depth intervals in the ground as a heat source in winter and as a heat sink in summer, and to preheat the heat source volume in summer with solar energy and to precool the heat sink volume in winter with ambient coldness. The heat pump then serves simply to assist in the extraction of stored heat or stored coldness from the ground. This kind of system is most viable economically at the scale of small district heating and cooling systems, and is discussed further in Chapter 15 (Section 15.4.1).

5.11 Heat pumps for domestic hot water

Domestic hot water (DHW) requires heating water to a temperature of 55–60°C, necessitating an even hotter condenser and a large temperature lift. This in turn will reduce the COP compared to the use of heat pumps for space heating with a low distribution temperature. Nevertheless, energy savings of 40–60 per cent are possible compared to electric resistance water heaters, when supplemental electric resistance heating is accounted for (this is an effective COP of 1.4–1.6) (Zogg et al, 2005). However, if heat is extracted from warm waste water, then the required temperature lift is small, and very attractive COPs are possible. Baek et al (2005) have analysed a hypothetical system whereby wastewater from showers and saunas in a hotel is stored during the day, then used at night (when electricity rates are often low) to recharge the hot-water tank with a heat pump. During the charging, the COP begins at about 9 and drops to 3.5 by the end of the charging, with a mean annual COP of around 5. At present, the cost premium for heat pump water heaters is large ($1400 versus $4–500 for electric resistance heaters in the US), but could drop substantially with large-scale production (10,000 units or more per year).

As noted in Section 5.8.1, heat pumps with CO_2 as a working fluid lend themselves particularly well to DHW. For a heat source at 20°C, the COP in heating water from 10°C to 60°C is 4.3 for a prototype CO_2 heat pump built by Kim et al (2005).

As noted in Section 5.2.1, the refrigerant exiting the evaporator is heated slightly before it enters the compressor so as to ensure that no liquid remains in the vapour when it enters the compressor. The refrigerant is thus too warm for the condenser when it exits the compressor. This excess heat can be removed by the condenser (but reducing its effectiveness in condensing the refrigerant), used to preheat the refrigerant entering the compressor, or used to preheat water for domestic hot water use when the heat pump is used for cooling. The heat exchanger that does this is called a *desuperheater*. It saves on DHW energy use while improving the heat pump COP in cooling mode. Other examples of integrating heating and cooling loads are discussed below.

5.12 Integrated heating, hot water production and air conditioning

Heat pumps provide the opportunity to integrate the heating, cooling, ventilation and hot water requirements of a building in a way that can lead to dramatic savings in energy use beyond that expected based on the COP or SPF of the heat pump. Some examples are given next.

5.12.1 Matching heating and cooling requirements

Large commercial buildings often require cooling in the interior of the building in winter, due to the occurrence of internal heat gains and lack of direct contact with the perimeter. Perimeter regions, however, may require heating at the same time. Heat pumps can be used to simultaneously cool the air or water used for cooling purposes and heat the air or water used for heating purposes. Given a high-performance thermal envelope, net heating may not be needed until the air temperature drops to −5°C, depending on how large the internal heat gains are (this being dependent on the amount, type and efficiency of office equipment and lighting, and the number of occupants). In climates where a building would rarely need all of the heat that can be extracted from a chiller, two condensers are used, or the condenser is split into two sections. One section recovers heat, while the other rejects heat through the cooling tower or to the ground. Heat recovered from the chiller condenser can be used for heating hot water rather than for, or in addition to, space heating. However, in this case a relatively high condenser temperature would be needed, which would reduce the chiller COP.

Suppose that a chiller has a COP of 5, which is not unusual for large units with centrifugal compressors (Chapter 6, Section 6.5.2). One unit of energy input provides 5 units of cooling, and the input energy is ultimately converted to waste heat, so 6 units of heat are delivered to the condenser. The refrigerant-to-water heat recovery from a condenser is about 95 per cent (Tuluca, 1997), so 5.7 units of heat can be recovered. Thus, 10.7 units of cooling or heating are provided for one unit of energy input – equivalent to a COP of 10.7. Such a high COP would apply, however, only to that portion of the heating and cooling loads that balance each other. Furthermore, the heat must be collected at a warm enough temperature to be useful. This, in turn, may require increasing the condenser temperature of the chiller, which would reduce the chilling COP. Of course, the extra energy required by the chiller with a lower COP adds to the waste heat that is available, but this is thermal energy rather than electrical energy. A careful analysis is required to determine the optimum condenser temperature and amount and temperature of waste heat that is recovered.

Hospitals, recreation centres, hotels and holiday resorts lend themselves particularly well to the use of heat pumps for integrating heating and cooling needs. Waste heat from exhaust air, wastewater and refrigeration equipment can be collected, stored and upgraded to be used as needed for space heating and hot water production. Indeed, substantial energy savings can be achieved through judicious use of heat exchangers even without heat pumps. For example, Herrera et al (2003) have shown how the use of just four heat exchangers in a complex in Aguascalientes, Mexico, involving a hospital, laundry centre,

sports centre with a swimming pool and a family health centre could save 38 per cent of the thermal energy. Here, a technique called *pinch analysis* is used to identify the maximum amount of energy that can be saved through heat exchangers, and to identify where within the various flows to place the heat exchangers.

Supermarkets provide another good opportunity for integration of heating and cooling needs. The 30,000 supermarkets in the US account for 4 per cent of the nation's electricity use. Refrigeration systems in supermarkets are large enough to justify collection of the heat from centralized condensers for space and hot-water heating purposes. At present, a small fraction of the heat from supermarket refrigeration systems is captured for use by the HVAC[6] system, by circulating air past the refrigeration condenser coils. A much larger fraction of the refrigeration heat can be collected using a heat pump to extract heat from glycol that is circulated through the condenser coil. Simulations for a supermarket in Washington, DC indicate that the entire heating requirements (presently met by natural gas) can be satisfied in this way, with only a 3 per cent increase in total refrigeration plus HVAC electricity use if the refrigeration system is also reorganized to improve its efficiency (Walker, 2000). At the same time, supermarket refrigeration systems can be redesigned to reduce the amount of refrigerant needed and its leakage to the atmosphere by a factor of three, which provides further global warming benefits (Walker, 2000). A supermarket built in Greenwich (UK) in the late 1990s achieved a reduction in total energy use of 50 per cent compared to a conventional supermarket through such measures as a GSHP for chilling, on-site cogeneration of heat and power, thermal storage through grassed earth walls, daylighting in the sales area and pre cooling of ventilation air by circulating incoming air through the ground first (Rivers, 2000).

Heat pumps can be used for integrating heating and cooling requirements in various industries. Other applications of heat pumps, such as collecting heat from cogeneration, from sewage treatment plants and from airport runways, are discussed in Chapter 15 (Section 15.3.3).

5.12.2 Integrated diurnal heating and cooling thermal-energy storage

When the building thermal inertia and envelope thermal resistance are not sufficient to prevent the need for early-morning heating and afternoon cooling on the same day, a heat pump can be used with thermal energy storage to reduce the combined heating and cooling energy use. This approach has been used in a California university (Goss et al, 1998). During the night (when electricity rates are lowest and fossil power plant efficiencies are largest, as discussed in Chapter 6, Section 6.9.2), a heat pump is used to simultaneously cool a cold-water tank and heat a separate hot-water tank (rather than rejecting heat to the atmosphere with a cooling tower). The stored cold and hot water are used during the day as needed.

5.13 Solar-assisted heat pumps

Solar-assisted heat pumps have been used for space heating and for making hot water for consumptive use (domestic hot water). Passive solar space heating is discussed in Chapter 4 (Section 4.1), while active use of solar energy for space heating and domestic hot water without the assistance of heat pumps is discussed in Chapter 11 (Section 11.4). In many instances, the additional savings in primary energy obtained by combining an ASHP with solar thermal collectors probably does not justify the increased cost and complexity. However, there are cases, discussed below, where the savings may be large enough to justify combining the two approaches. Solar-assisted GSHPs (Section 5.3.3) probably represent a better way to combine solar energy and heat pumps in temperate and colder climates, where feasible.

5.13.1 Space heating

There are two broad approaches to using solar energy with heat pumps for space heating. In the first approach, solar panels serve as the evaporator for electric vapour compression heat pumps. This is referred to as a *direct expansion* solar-assisted heat pump. In the second approach, illustrated in Figure 5.8, water circulates through the collector, is heated and passes through a hot-water storage tank (with or without an appropriate

phase-change material), where some heat is deposited, then continues through the evaporator of the heat pump. When the collector temperature is below some threshold, the collector is bypassed and water circulates between the hot-water tank and the evaporator. An advantage of the direct expansion approach is that a single device serves as collector and evaporator, and no heat exchanger is needed between the collector fluid and the heat pump working fluid. An advantage of the second approach is that solar energy can be stored for later use when in excess, whereas the heat pump using the direct expansion approach reverts to a purely air-source heat pump when the sun is not shining. In either case, coupling the heat pump and solar collectors increases the efficiency of both the collector and the heat pump. The collector efficiency increases because the water flowing through the collector is cooler than it would be without heat having been removed by the heat pump, so radiative and convective heat losses are reduced (see Chapter 11, Section 11.4.2), while the heat pump efficiency increases because the required temperature lift is reduced. Kaygusuz (2000) and Badescu (2002) present a mathematical model of a solar heater/heat pump/thermal energy storage system.

For a storage solar-assisted heat pump used in Turkey, the measured heating COP ranges from 4.2 in February (average outdoor temperature of 7.4°C) to 4.5 in April (average temperature of 11.6°C), with a collector efficiency of 0.62–0.7 (Kaygusuz, 2000). The COP without the solar collector would have been about 1.0 lower, while the collector efficiency without the heat pump would have been about 0.1 lower. The storage efficiency of the hot water tank (diameter: 1.3m, length: 3.2m, U-value: $0.25W/m^2/K$) is 0.63.

Solar-assisted heat pump systems have been installed in some commercial buildings in Japan (Breembroek, 1997). A simpler means of solar-assisted heating is to draw outside air through a space underneath the outer layer of the roof before it passes through the evaporator (which would be placed at the apex of the roof). An average rise of 2.5K was achieved in an early test of this method, leading to an increase in heat pump COP of 8–10 per cent (Reay and MacMichael, 1988).

Figure 5.8 Schematic diagram of a solar-assisted heat pump with thermal energy storage

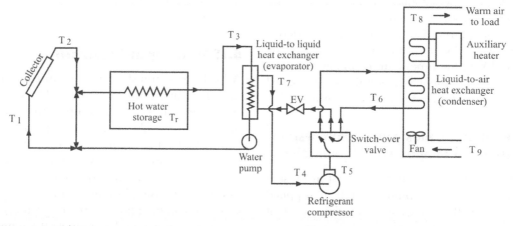

Note: EV = expansion valve.
Source: Kaygusuz (2000)

5.13.2 Domestic hot water

Hawlader et al (2001) developed a prototype so-lar-assisted heat pump for making hot water, and operated it under meteorological conditions for Singapore. They also developed and validated a computer model of their system, which they then used for more extensive testing of the impact of changes in system parameters (such as collector area, storage volume in relation to collector area and compressor speed). The solar collector serves as the evaporator coil, while the condenser coil is placed directly in the hot-water tank. With a larger compressor speed, the evaporation tem-perature in the collector evaporator decreases, thereby reducing the heat loss to the surroundings and increasing the collector efficiency (from, say, 0.65 to 0.70). However, the COP also decreases (from, say, 7.0 to 4.0, the exact values depend-ing on solar irradiance). For any given compres-sor speed, the collector efficiency decreases with time after sunrise (due to increasing water-storage temperature, which increases the temperature of the water returning to the collector) and the COP decreases (because the condenser temperature must increase in order to add heat to the stored hot water as its temperature rises). A larger collec-tor leads to a larger COP because the fluid tem-perature in the collector increases with increasing area, an effect that is stronger the larger the solar irradiance. If the storage volume increases, both the collector efficiency and COP increase because the storage temperature decreases.

The 'free energy' fraction in a solar-as-sisted heat pump can be defined as the energy gained by the collector (Q_u, some of which is so-lar energy and some of which is heat drawn from the ambient air) and not subsequently lost during storage, divided by the delivered end-use energy. The later is equal to Q_u plus the energy input to the compressor, W_c, minus the storage losses, L.

That is:

$$FF = \frac{Q_u - L}{Q_u + W - L} = \frac{Q_c - W - L}{Q_c - L}$$

(5.8)

where Q_c is the heat delivered by the condenser. The COP is given by

$$COP = \frac{Q_c}{W}$$

(5.9)

FF and *COP* increase with collector area and storage volume. For a 4m² collector area (which gives close to the lowest life-cycle cost) and pro-duction of hot water at 50°C, Hawlader et al (2001) found that the COP varies from 6.5 at a solar irradiance of 300W/m², to 9.5 at a solar irradiance of 900W/m². For a daily hot-water demand of 800 litres, the free-fraction ranges from 0.37 with a storage volume of 200 litres to 0.67 with a storage volume of 800 litres. However, comparable or better solar fractions can often be obtained from solar collectors alone (see Chapter 11, Section 11.4.3). Compared to an electric-resistance water heater with an electricity cost of 13cents/kWh, the time required to pay back the additional investment cost for the solar-assisted heat pump is estimated to be about 2 years.

5.14 Heat pump plus power plant system efficiency and comparative CO_2 emissions

For an SPF of 3 (close to the highest at present for an air-to-air heat pump), the heat supplied is three times the energy used (this is equivalent to a furnace with an efficiency of 300 per cent). However, if that energy is supplied as electricity from a coal-fired plant with an efficiency of 33 per cent, there is a factor of three energy loss in going from primary energy (coal) to secondary energy (electricity). Thus, the overall efficiency of primary energy use will be comparable to that of a high-efficiency (88–97 per cent) natural gas furnace or boiler. As the efficiency in generating electricity from fossil fuels and the SPFs of electric heat pumps improve, heat pumps become more favourable from a primary energy point of view. GSHPs permit a higher COP but also require electricity for the ground-loop pumps; for a COP of 4 and pump energy use equal to 15 per cent of the compressor energy use (achievable with current best practice), the effective COP would be $4.0/1.15 = 3.5$. For electricity generated at 45 per cent (the current state of the art using coal) or 55 per cent (the current state of the art using natural gas) efficiency, primary energy use is reduced by a factor of about 1.8 or 2.1, respectively, compared to direct use of natural gas in a high-efficiency furnace. However, the CO_2 emission factor for coal is about 1.8 times that of natural gas (25.0kg C/MJ compared to 13.5kg C/MJ), so in the case of coal-fired electricity generated at 45 per cent efficiency, there would be no reduction in CO_2 emissions. Thus, even with state of the art assumptions, heat pumps used solely for heating are not attractive in systems where the additional electricity that the heat pump would use is supplied with coal. However, if the heat pump is used for coupled heating and cooling, the COP based on the sum of heating plus cooling loads served will be substantially higher, making the heat pump attractive from an emissions point of view.

5.15 Summary and synthesis

A heat pump transfers heat against the temperature gradient, from cold to warm. This can be done either through a vapour compression heat pump, or through an absorption heat pump. The key components of a vapour compression heat pump are the evaporator, condenser and compressor. A working fluid is compressed by the compressor, and as it is compressed, its temperature increases. The working fluid then flows to the condenser, which is a heat exchanger, and as long as the temperature of the working fluid and condenser is warmer than the surroundings, heat will be given off. Condensation of the working fluid occurs during this process. The condensed working fluid then flows to another heat exchanger – the evaporator – where the low pressure created on the suction side of the compressor induces evaporation, which has a cooling effect. As long as the working fluid and evaporator are cooled to below the temperature of their surroundings, they will absorb heat from the surroundings. The key, then, is for the compressor to compress the fluid to a high enough pressure to create a condenser that is warmer than its surroundings, and to create a sufficiently low pressure and hence a sufficiently low temperature through evaporation that the evaporator becomes colder than its surroundings. If the condenser is placed inside a building and the evaporator outside, then the heat pump can be used for heating, while if the condenser is placed outside and the evaporator inside, the heat pump serves as an air conditioner. The roles of the condenser and evaporator can be switched, and the heat pump can change from a heater to an air conditioner, by reversing the flow direction of the working fluid through a simple valve.

The greater the difference between the condenser and evaporator pressure and hence temperature, the more work that must be done by the compressor to transfer a given amount of heat. The ratio of heat transferred to energy used is called the coefficient of performance (COP). The keys to improving the COP of a heat pump are:

- to improve the efficiency of the compressor;
- to make use of variable speed compressors for part-load operation;
- to minimize the difference in temperature required between the evaporator and condenser; that is, to minimize the required temperature lift.

In heating mode, the temperature lift can be minimized if as warm a heat source as possible is used (permitting a warmer evaporator) and as cool a supply-air temperature or hot-water temperature as possible is used for distributing heat through the building (permitting a cooler condenser). In cooling mode, the temperature lift can be minimized if as cool a heat sink (that which receives the rejected heat) as possible is used (permitting a cooler condenser) and as warm a supply-air temperature or chilled-water temperature as possible is used for distributing coldness through the building (permitting a warmer evaporator). In temperate and colder climates, the ground will tend to be sufficiently warmer than the air during winter that it can serve as a relatively warm heat source for winter heating, but it will also tend to be cooler than the air during the summer, thereby serving as a relatively cool heat sink for summer air conditioning. Thus, ground-source heat pumps (GSHPs) tend to be more efficient than air-source heat pumps (up to 50 per cent higher heating COP when electric resistance backup is accounted for). Minimizing the heat distribution temperature and maximizing the coldness-distribution temperature in turn requires that the heating and cooling loads be as small as possible and that radiant ceilings or floors be used for heating and cooling. Minimization of heating and cooling loads in turn requires a high-performance thermal envelope. Thus, the efficiency of heat pumps for heating and cooling depends strongly on the overall design of the building.

A GSHP consists of a long underground coil, either buried horizontally or in a series of vertical boreholes. In either case, a substantial amount of energy can be required to circulate the heat transfer fluid through the coil unless the entire system is designed to minimize the pumping energy requirement. Where a GSHP is used

predominantly or only for heating, the ground temperature could decrease over a period of years as heat is withdrawn unless measures are taken to restore the ground temperature during the summer. One measure is to circulate solar-heated water through the ground loop during the summer, producing a solar-assisted GSHP. Similarly, if a GSHP used predominately or only for cooling, the ground will tend to warm over a period of years. In this case, the GSHP can be combined with a cooling tower and the GSHP used only to the extent that there is not an unacceptable long-term warming of the ground. The result is a cooling tower-assisted GSHP.

There is substantial interest in CO_2 as a working fluid in heat pumps. Heat rejection in this case occurs through a gas cooler rather than a condenser. A key characteristic of CO_2 heat pumps is that they can readily operate at heat rejection temperatures in excess of 55°C, making them ideal for domestic hot water and for district heating systems. Conventional heat pumps are often limited to a maximum operating temperature of 55°C.

Heat pumps can be used in transferring heat in winter from warm exhaust air to cold incoming fresh air in balanced ventilation systems. The heat source (the exhaust air) is comparatively warm, and if the exhaust air is cooled to below the temperature of the fresh air, the ventilation system will serve as a net heat source to the building. Heat pumps can also be used to great advantage whenever there are matching heating and cooling requirements within the same building. The heat pump removes heat from air or water flows that need to be cooled, and adds this heat to the air or water flows that need to be heated. Opportunities for integrating heating and cooling in this way abound in hospitals, recreation centres, hotels and holiday resorts. Finally, it is sometimes advantageous to use solar thermal collectors as a heat source for heat pumps.

Almost all heat pumps use electricity, which is generated at an efficiency of about 33 per cent in older conventional coal power plants, 45 per cent in state of the art coal power plants and 55–60 per cent in state of the art natural gas power plants. An alternative for heating is a high-efficiency (≥90 per cent) natural gas furnace or boiler. Both electricity and natural gas entail losses during transmission and distribution, but these losses are usually comparable (about 5–15 per cent). The CO_2 emission factor for coal is about 25kg C/GJ, whereas that of natural gas is only 13.5kg C/GJ. Consequently, air-source heat pumps (typical COP of 2.5–3.0) powered by electricity from conventional coal-fired power plants do not save primary energy compared to a high-efficiency natural gas furnace or boiler, and will increase CO_2 emissions by almost a factor of two. Ground-source heat pumps (typical COP of 3.0–4.0), with minimal pumping energy use for the ground loop (15 per cent of compressor energy use, giving an effective COP of 2.6–3.5) and a state of the art coal power plant save primary energy but do not reduce CO_2 emissions. The heat pump for heating is attractive from a primary energy point of view only if the product of the effective COP of the heat pump (accounting for ground-loop pumping energy use) times the power plant efficiency exceeds the efficiency of the furnace or boiler (expressed as a fraction). It is attractive from a CO_2 emissions point of view only if this condition is satisfied and the electricity is supplied by natural gas or a CO_2-emission-free energy source, or if coal is used to generate electricity but the saving in primary energy is large enough to offset the greater (almost factor of two) CO_2 emission factor for coal compared to natural gas.

Notes:

1   An alternative measure of heat pump performance in heating mode, used in North America, is the *Heating Season Performance Factor* (HSPF). This is the ratio of total heat output of a heat pump in Btu divided by electricity used in W-hr. This mixture of metric and non-metric units complicates the computation of system-scale energy use. Since 1Btu = 1055 Joules and 1W-hr = 1J s^{-1} × 3600s = 3600 Joules, multiply by 1055/3600 = 0.2931 to get J/J (COP).

2   As with heat pumps in heating mode, mixed Btu-metric measures of performance are often used in North America: the *Seasonal Energy Efficiency Ratio* (SEER), which is the ratio of total heat removed (Btu) from a building to the energy input (W-hr) over the entire cooling season; and the *Energy Efficiency Ratio* (EER), which is the instantaneous, steady-state ratio of the rate of heat removal to the rate of energy use, in Btu/hr per Watt. As with HSPF, multiply by 0.2931 to convert to COP values.

3   These efficiency gains are computed using the difference between air temperature and 17°C as an index of the heating load, and multiplying this times 0.1 to deduce a plausible evaporator-source and condenser-sink temperature differential. The same evaporator-source temperature differential is assumed for both heat pumps.

4   Adjusting the air or water flow past the evaporator (or condenser) will also change the evaporator (or condenser) temperature and hence the evaporator (or condenser) pressure, thereby automatically changing the amount of work done by the compressor, but with partially compensating changes in fan or pump energy use.

5   Microchannel heat exchangers are smaller and more effective than regular heat exchangers (Zhao et al, 2003b). As well, the airflow pressure drop is up to 40 per cent lower across microchannel heat exchangers, which will save on fan power use (Richter et al, 2003) – something not included in the above comparisons.

6   HVAC = *H*eating *V*entilation *A*ir *C*onditioning.

six

Cooling Loads and Cooling Devices

The first step in meeting a cooling load is to reduce the load compared to what it would be under standard practice, then to consider passive techniques for meeting some or all of the load, and lastly to consider mechanical means to meet any remaining load. Reducing the cooling load depends on the building shape and orientation; on the choice of building materials; on window size, orientation and performance characteristics; and on a whole host of other decisions that are made in the early design stage by the architect. The usual procedure today is to ignore climate in the design of buildings, as the engineers are supposed to take care of the problem of making the building habitable through mechanical (and energy-intensive) means. To create buildings that are adapted to the prevailing climate requires more work, because the same universal design template cannot be used everywhere. Highlighting the much greater effort and creative thinking required to design low-energy buildings than conventional buildings, Koch-Nielsen (2002) distinguishes between *active design*, which uses

passive measures to achieve the desired indoor conditions, and *passive design*, which uses active (mechanical) measures to achieve the desired indoor conditions. Because conventional modern designs largely ignore climate, architecture loses its connection to place: the same building forms and designs are now seen in New York, Houston, Hong Kong or Singapore.

6.1 Physical principles

We begin this chapter by identifying the cooling loads related to the temperature and the moisture content of the air and the relationships between them that are relevant to conventional and/or low-energy cooling techniques.

6.1.1 Sensible and latent heat

Heat in air occurs in two forms: as sensible heat and as latent heat. Sensible heat refers to the heat that we can sense, as a warmer temperature. The sensible heat content of dry air per unit mass

(J/kg) is given by:

$$H = c_{pa} T \qquad (6.1)$$

where T is in kelvin and c_{pa} is the specific heat of air ($1004.5 J/kg/K$) – the amount of heat that must be added to warm 1kg of air by 1K.

Latent heat refers to the heat that is released when water vapour in the air condenses. The amount of moisture in air can be represented by its *mixing ratio*, r (kg/kg), which is the ratio of mass of water vapour to mass of dry air. The mixing ratio is related to the *vapour pressure*, e_a, by the relation

$$r = 0.622 \frac{e_a}{P_a - e_a} \qquad (6.2)$$

where 0.622 is the ratio of molecular weights for water vapour and air, and P_a is the total atmospheric pressure. The *saturation vapour pressure*, e_s, is the water vapour pressure when the air is holding the maximum amount of water vapour possible.[1] It increases rapidly with increasing temperature, as shown in Figure 6.1 (this is known as the *Clausius–Clapeyron* relationship). The *relative humidity* of an air parcel is the ratio of the actual vapour pressure (or mixing ratio) to the saturation value, times 100 to give it as a percentage. That is:

$$RH = \frac{e_a}{e_s} 100\% = \frac{r}{r_s} 100\% \qquad (6.3)$$

The latent heat content of an air parcel per unit mass (J/kg) of dry air is given by:

$$L = L_c r \qquad (6.4)$$

where L_c is the *latent heat of condensation* – the amount of heat released when 1kg of water vapour condenses. It depends weakly on temperature, with a value of $2.501 \times 10^6 J/kg$ at a temperature of 0°C.

The combination of sensible heat and latent heat per unit mass of dry air gives the *specific enthalpy* H of an air parcel, where

$$H = c_{pa} T + r(L_c + c_{pwv} T) \qquad (6.5)$$

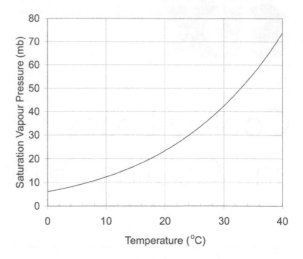

Figure 6.1 Variation of saturation vapour pressure with temperature, computed using the expression given in Lowe (1977)

where c_{pwv} = specific heat of water vapour = $1860 J/kg/K$.

6.1.2 Dewpoint, dry-bulb and wet-bulb temperatures

Suppose that we begin cooling air that is initially unsaturated. Its sensible heat content decreases, but r is constant. As the air cools, the saturation mixing ratio r_s decreases, and at some point the air becomes saturated. Further cooling causes just enough water to condense to maintain r at the saturation value, which continues to fall. This releases latent heat, which must be removed along with the sensible heat required to continue decreasing the air parcel's temperature. The temperature at which condensation begins (i.e. at which the relative humidity is 100 per cent) is called the *dewpoint temperature*, T_{dp}. Since this depends only on how much moisture a parcel of air has, stating the parcel's dewpoint temperature is a convenient way of stating how much moisture it has.[2] It also serves as a warning as to when condensation might occur: if any portion of a cooling system is colder than the dewpoint temperature, condensation can be expected.

If we allow liquid water in an air parcel to evaporate without exchanging any heat with the surroundings (i.e. without adding or removing heat), then the temperature of the parcel will decrease (as heat energy is used to evaporate water) and the humidity of the air will increase. This will continue until the parcel becomes saturated, at which point no further evaporation or cooling will occur. The temperature at which this occurs is called the *wet-bulb temperature*, T_{wb}, because it is given by the temperature of a thermometer bulb with a wet cloth over it, in equilibrium with a steady airflow. The air temperature itself is referred to as the *dry-bulb temperature*, T_{db}, because it is the temperature that a dry thermometer measures. The dry-bulb and wet-bulb temperatures can be easily and simultaneously measured using a *sling psychrometer*. It consists of two side-by-side thermometers, one dry and one whose bulb is surrounded by a cloth that has been moistened with distilled water (to prevent buildup of residue), as illustrated in Figure 6.2. By swinging the sling psychrometer around the handle, a steady airflow past the two thermometer bulbs is created, allowing rapid adjustment to equilibrium temperatures. The sling psychrometer is used in many HVAC laboratories and field services, although alternative setups are also used (see Kreider et al, 2002). Given T_{db} and T_{wb}, the relative humidity of the air can be computed, as explained in Box 6.1.

The wet-bulb temperature is the lowest temperature to which an air parcel can be cooled by evaporating water into the air parcel. Strictly

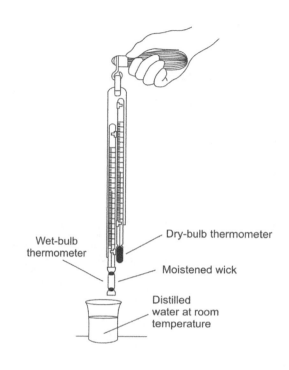

Figure 6.2 A sling psychrometer, used for conveniently and simultaneously measuring the drybulb and wetbulb temperatures

Wet-bulb thermometer

Dry-bulb thermometer

Moistened wick

Distilled water at room temperature

Source: Kreider et al (2002), reproduced with permission of The McGraw-Hill Companies

---

BOX 6.1 Determination of relative humidity from dry-bulb and wet-bulb temperatures

Kreider et al (2002) recommend computing the relative humidity from

$$RH = \frac{e_s(T_{wb}) - e_m}{e_s(T_{db})} \tag{B6.1.1}$$

where $e_s(T_{wb})$ is the saturation vapour pressure at the wet-bulb temperature, $e_s(T_{db})$ is the saturation vapour pressure at the drybulb temperature (the air temperature), and e_m is the water vapour pressure due to the depression of wet-bulb temperature, given by:

$$e_m = P\left(\frac{T_{db} - T_{wb}}{1514}\right)\left(1 + \frac{T_{wb} - 273.15}{873}\right) \tag{B6.1.2}$$

where P is the atmospheric pressure. The saturation vapour pressure can be computed using the equation given in the main text, or from:

$$e_s(T) = e_c(10^{K(1-T_c/T)}) \tag{B6.1.3}$$

where

$$K = 4.39553 - 6.2442\left(\frac{T}{1000}\right) +$$
$$9.953\left(\frac{T}{1000}\right)^2 - 5.151\left(\frac{T}{1000}\right)^3 \tag{B6.1.4}$$

Given RH and $e_s(T_{db})$, e_a can be easily computed from Equation (6.3).

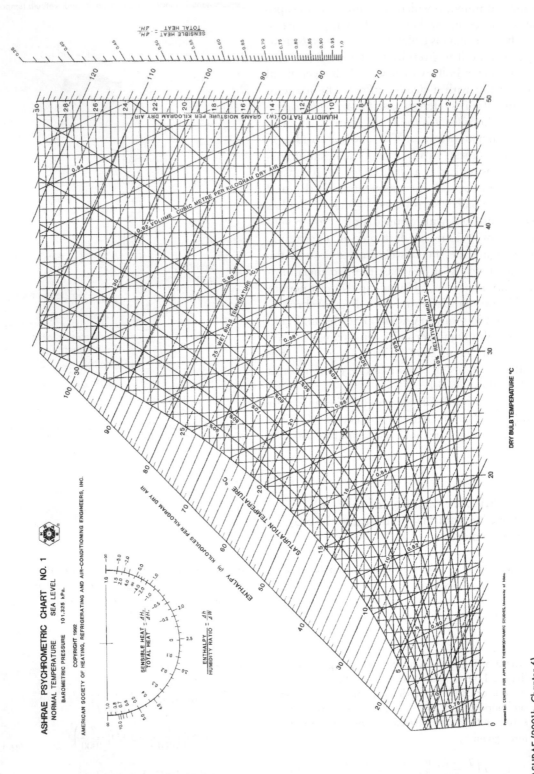

Figure 6.3 The psychrometric chart, showing specific enthalpy (solid lines sloping upward strongly to the left), wetbulb temperature (dashed lines sloping upward strongly to the left), specific volume (solid lines sloping upward weakly to the left) and relative humidity (concave upward lines) as a function of dry-bulb temperature (vertical lines) and humidity ratio (horizontal lines)

Source: ASHRAE (2001b, Chapter 6)

speaking, evaporation does not directly cool an air parcel. Rather, it cools the remaining liquid water. The air then loses sensible heat to the colder liquid water. This cools the air and warms the liquid water. The wet-bulb temperature is the air temperature when the air and liquid have fully converged to a single value at which there is no further evaporation (because the air is saturated). For $T_{db} = 20°C$ and for RHs of 40 per cent, 60 per cent and 80 per cent, $T_{wb} = 12.3°C$, $15.1°C$ and $17.7°C$, respectively (the depression of T_{wb} below T_d increases from 2.3K to 7.7K as RH decreases from 80 per cent to 40 per cent), while for $T_{db} = 30°C$ and the same RHs, $T_{wb} = 20.1°C$, $24.8°C$ and $27.1°C$, respectively (the depression of T_{wb} below T_{db} increases from 2.9K to 9.9K as RH decreases from 80 per cent to 40 per cent). Thus, for a given relative humidity, the difference between T_{wb} and T_{db} increases weakly as T_{db} increases (that is, T_{wb} increases more slowly than T_{db}), while for a given T_{db}, the difference between T_{wb} and T_{db} increases strongly as RH decreases.

However, T_{wb} can be lower in humid temperate regions than in hot arid regions because of the lower initial T_{db} in temperate regions, even though the difference between T_{db} and T_{wb} is greater at the low humidities found in hot arid regions. For example, for $T_{db} = 38°C$ and for a water vapour:air mixing ratio (r) of 10gm/kg (a typical summer afternoon in hot arid regions), RH = 23 per cent and $T_{wb} = 22°C$ (a potential cooling effect of 16K), whereas for $T_{db} = 24°C$ and $r = 12$gm/kg (a typical mid-latitude summer afternoon), RH = 63 per cent and $T_{wb} = 19°C$ (a potential cooling effect of 5K). Thus, lower absolute temperatures can often be achieved through evaporative cooling in humid temperate climates than in hot arid climates, even though the cooling effect is smaller.

Finally, even with a fixed water vapour mixing ratio, T_{wb} will decrease at night as T_{db} decreases, although by a smaller amount. For example, if $T_{db} = 30°C$ by day and falls to 20°C at night with r fixed at 10gm/kg, T_{wb} falls from 19.6°C to 16.2°C – a decrease one-third as large.

6.1.3 Construction and use of the psychrometric chart

The relationships between temperature, mixing ratio, specific enthalpy, dewpoint and wet-bulb temperatures, and relative humidity are given in a specially constructed diagram called a *psychrometric chart*, which is shown in Figure 6.3. Temperature and mixing ratio are given as the vertical and horizontal lines, respectively. The concave upward line to the left gives the variation of saturation mixing ratio with temperature as computed from the Clausius–Clapeyron relationship. The temperature at which a given mixing ratio line intersects this curve gives the dewpoint temperature. Lines of constant specific enthalpy of air slope upward to the left, because a lower temperature can be offset by a larger mixing ratio to give the same total enthalpy. The wet-bulb temperature lines also slope upward to the left but with a slight shift toward lines of greater enthalpy, as these lines represent the process of evaporation, which adds mass as well as latent heat to the air parcel, so that the total enthalpy increases as the air cools. Lines indicating various relative humidity values and specific volumes (the volume for 1kg of dry air) are also found on the psychrometric chart. Given the temperature and mixing ratio of an air parcel, the dewpoint temperature, wet-bulb temperature, and relative humidity can be easily read from the psychrometric chart.

6.1.4 Conventional dehumidification

The commonly used method to remove water vapour from air is to cool the air until the desired amount of water vapour has condensed and fallen out, then to reheat the air to the desired final temperature. If the desired final temperature of air supplied to a room is 16°C with an RH of 50 per cent or 70 per cent (a common practice today), then, as can be seen from the psychrometric chart, the air must be cooled to 5°C or 10°C (unless the initial moisture content is so low that no moisture needs to be removed), then reheated. The psychrometric chart can be used to determine the amount of heat that would need to be removed from an air parcel in cooling it from any initial temperature and mixing ratio to any final temperature and mixing ratio. Suppose that we cool an air parcel at 40°C and $r = 30$gm/kg to 6°C. As the parcel cools, it follows the $r = 30$gm/kg line until it reaches the saturation curve, at which point condensation begins and it moves along the saturation curve to the final temperature of 6°C. The total heat content decreases by 97.2kJ/kg (from 117.4 to 20.2kJ/kg). By disaggregating Equation (6.5), the sensible heat content is seen to decrease by 33.1kJ/kg (from 42.4 to 7.1kJ/kg), which implies that the latent heat content decreases by 64.1kJ/kg. Thus, two-thirds of the heat that is removed from the parcel in this example is as latent heat. However, the parcel is now at 100 per cent relative humidity, with $r = 5.7$gm/kg. Reheating the parcel to 13.5°C has no effect on its water content but raises the saturation mixing ratio sufficient to decrease the relative humidity to a comfortable 66 per cent. This requires adding 5.6kJ/kg back to the parcel, as sensible heat.

6.2 Reducing the cooling load

The first step in reducing the energy required for cooling should be to reduce the cooling load, as this allows a reduction in the size of cooling systems, with savings in upfront capital costs, as well as reduced energy use. Measures, such as increased thermal mass, that decrease peak loads more than average loads will allow cooling equipment to operate at a larger fraction of their peak capacity, thereby improving equipment efficiency. In some instances, loads can be reduced to the point where mechanical cooling equipment can be eliminated altogether. In many developing countries, where much of the population cannot afford air conditioners, reducing the cooling load can do much to improve thermal comfort and

human well-being without requiring increased energy use.

The cooling load of a building can be reduced by decreasing the absorption of solar radiation by the walls and roof; by insulating walls and especially roofs; by reducing the transmission of solar radiation through windows; by reducing the need to cool and dehumidify incoming fresh air by reducing the amount of incoming fresh air; and by reducing the internal generation of heat by equipment, lighting and the ventilation system itself. Analysis of the opportunities for reducing cooling requirements requires identifying the relative importance of these heat sources, which will vary with climate, the building size and shape, the activities carried out in the building, and the properties of the thermal envelope.

Figure 6.4 shows the breakdown of heat gains for a large commercial building in Los Angeles (hot and dry climate) and in Hong Kong (hot and humid climate). Table 6.1 gives the rate of heat generation by people engaged in various activities. For typical indoor activities, this ranges from 40W/person sleeping, to 55–70W/person engaged in various office activities, to 150W/person at a brisk walk. For the Los Angeles building, the three largest heat sources are lighting (28 per cent), windows (21 per cent), and fans (13 per cent), while people account for 12 per cent, conditioning of incoming outside air accounts for 10 per cent of the total cooling load, and heat conduction through walls accounts for only 3

Figure 6.4 Relative contribution of different heat sources to the cooling needs of larger commercial buildings in Los Angeles and Hong Kong

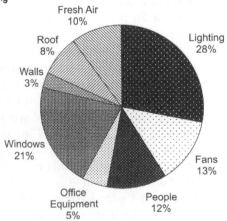

Cooling Loads in a Los Angeles Office Building

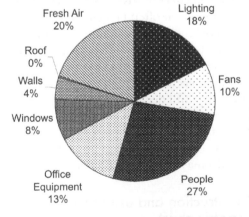

Cooling Loads for a Generic Office Building in Hong Kong

Sources: Data from Los Angeles, Feustel and Stetiu (1995); Hong Kong, Joseph Lam (personal communication, 2003), used in Lam and Li (1999) and Lam (2000)

per cent. For the Hong Kong building, the three largest loads are people (27 per cent), lighting (18 per cent) and conditioning of outside air (20 per cent), with windows and walls accounting for 12 per cent. Most of the cooling load in large commercial buildings comes from internal heat gains (people, lighting, office equipment, fans). In smaller buildings, heat gains through the envelope are relatively more important.

The following subsections discuss techniques for reducing the envelope cooling load, beginning with building shape and orientation. Reduction of cooling loads due to conditioning of outside air, and due to waste heat from fans, lighting and office equipment, is discussed in later chapters.

Table 6.1 Rates of heat generation for various activities

Activity	Watts
Sleeping	40
Seated, reading	55
Seated, typing	65
Standing, filing	80
Walking, 0.9m/s (3.2km/h)	115
Walking, 1.2m/s (4.3km/h)	150
Lifting, packing	120

Source: ASHRAE (2001b)

6.2.1 Building shape and orientation

A rectangular-shaped building will have a greater surface to volume ratio than a square building with the same volume. This will increase the heat loss through the walls in winter, all else being equal, and could increase the heat gain through windows in winter or summer, depending on the building orientation. The critical directions for solar heat gain during the summer are the east- and west-facing sides of a building, as the sun is closer to the horizon when shining on these walls. Thus, the sun's rays are closer to perpendicular to the face of the building, leading to greater solar flux densities than on the equatorward-facing wall in summer (see Chapter 4, Figure 4.1). West-facing surfaces are most critical as the peak in solar irradiance coincides closely with maximum daytime air temperatures. Furthermore, overhangs will not be fully effective in blocking the summer sun on east- and west-facing walls. In winter, there is greater solar energy on the equatorward-facing

wall. Rectangular buildings should therefore be oriented with their shortest sides facing east and west, as this orientation simultaneously minimizes solar heat gain in summer and maximizes it in winter. These considerations imply that residential streets should be oriented east–west rather than north–south, assuming that the houses are spaced sufficiently closely along the street that adjacent houses can shade each other. In this case, the houses will be well exposed to the winter sun but less exposed to the summer sun.

In short, a compact but suitably oriented rectangular building will minimize summer heat gain. However, a compact building also reduces the potential for removing the heat that does accumulate through wind-driven cross-ventilation. The optimal design therefore depends on whether the building will rely on mechanical or passive methods of cooling, whether cross-ventilation will be an important part of the passive cooling strategy, and on the relative importance of internal versus external heat gains. As illustrated in Section 6.3.1, it is possible to have a partially open building form with a breezeway or indented balconies or porches that facilitate cross-ventilation when this is useful, but that can be closed off with insulated shutters when a compact form is need to minimize heat gain. Hirano et al (2006a, 2006b) have investigated the impact of building-scale voids – large 3-D openings that penetrate into the building – in reducing the cooling load in hot and humid climates. When adaptive comfort is taken into account (Chapter 7, Section 7.2.4), building-scale voids in combination with double-skin roofs and walls can reduce the cooling load by up to 50 per cent in Okinawa (latitude 26°N) compared to a building with no voids.

If it is not possible to orient a rectangular building with the long-axis running east–west, due to site constraints, then triangular bay windows on the east- and west-facing walls can be used to maximize winter solar gain but minimize summer solar gain, as illustrated in Figure 6.5. This solution takes advantage of the fact that (in the northern hemisphere) the sun rises and sets south of east or west during winter and north of east or west during summer, with the difference being greater the higher the latitude.

Gómez-Muñoz et al (2003) have analysed the energy budget of hemispherical vaulted roofs. Due to the fact that there will always be

Figure 6.5 Triangular bay windows on the east and west sides of a building that simultaneously maximize winter solar gain and minimize summer solar gain

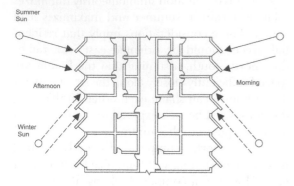

Source: Givoni (1998a)

some self-shading (except when the sun is directly overhead) and the fact that the area of a half sphere is twice that of a circular flat roof of equal radius, the amount of solar radiation intercepted by a domed roof per unit of roof area is always less than that intercepted by a flat roof. At noon, the interception of solar radiation is 30–40 per cent less, while on a daily-average basis, it is 33–37 per cent less at the equator (depending on time of year) and 25–37 per cent less at 20°N. This will reduce the heating of the roof. At the same time, the greater surface area causes the roof to be a more effective radiator of infrared radiation, which also tends to limit the roof temperature. For a rectangular vaulted roof at 23.6°N (Aswan, Egypt), Elseragy and Gadi (2003) compute a reduction in June solar irradiance, compared to a flat roof, by 34 per cent for a building with a north–south long axis and by 25 per cent for a building with an east–west long axis.

6.2.2 Albedo and vegetation

When the sun is shining on a building, the building surface temperature can rise substantially above the air temperature (black roofs can reach 70–80°C). This leads to a conductive heat flow through the walls or roof into the building. The term *albedo* refers to the fraction of incident sunlight on a surface that is reflected. High-albedo surfaces (e.g. white, reflective material) will tend to be cooler than low-albedo surfaces (e.g. black material), as less sunlight is absorbed. High emissivity also contributes to a cooler roof by

increasing the emission of infrared radiation. Cooler roofs might last longer due to reduced thermal stress. Table 6.2 gives albedo and emissivity values for common roof materials. Black asphalt shingles absorb 95–96 per cent of the incident sunlight, and even so-called white shingles absorb 75 per cent of the incident sunlight (reflecting only 25 per cent). Ultra-white

Table 6.2 Albedo of common surfaces

Material	Albedo	Emissivity
General		
Dark gravel	0.08–0.15	0.80–0.90
Light gravel	0.30–0.50	0.80–0.90
Aluminium coating	0.25–0.60	0.20–0.50
Concrete	0.10–0.35	0.85–0.90
Corrugated metal, unpainted dark paint white paint	0.30–0.50 0.05–0.08 0.60–0.70	0.20–0.30 0.80–0.90 0.80–0.90
Coloured paint	0.15–0.35	0.80–0.90
White paint	0.50–0.90	0.80–0.90
Liquid applied coating smooth black rough white smooth white	0.04–0.05 0.50–0.60 0.70–0.85	0.85–0.95 0.85–0.95 0.85–0.95
Shingles		
Black	0.04–0.05	0.80–0.90
Brown	0.05–0.09	0.85–0.90
White (in fact, grey)	0.25–0.27	0.80–0.90
Ultra-white	0.55	0.80–0.90
Cool black	0.18	0.80–0.90
Concrete tiles		
Black	0.04 (0.41)	0.85–0.90
Chocolate	0.12 (0.41)	0.85–0.90
Green	0.17 (0.46)	0.85–0.90
Blue	0.18 (0.44)	0.85–0.90
Grey	0.21 (0.44)	0.85–0.90
Terracotta	0.33 (0.48)	0.85–0.90
Clay tiles		
Red	0.20–0.22	0.85–0.90
White	0.65–0.75	0.85–0.90
Single-ply membranes		
Black	0.04–0.05	0.85–0.95
Grey	0.15–0.20	0.85–0.95
White	0.70–0.78	0.85–0.95

Source: Akbari et al (2004) and Levinson et al (2005). Tile data outside parentheses are for standard tiles, while data in parentheses are for tiles under development with a high NIR reflectivity

shingles are available that reflect 55 per cent of the incident sunlight, by increasing the amount of white pigment (titanium dioxide rutile) on the granules. Conventional concrete tiles reflect 4–33 per cent of the incident solar radiation and clay tiles reflect 20–65 per cent (depending on colour). White coatings can be applied to conventional roofing materials, reflecting as much as 85 per cent of the incident sunlight. Most surfaces have a high emissivity (≥0.80), the exceptions being unpainted corrugated metal and aluminium coatings; for these, the cooling benefit of a modestly higher reflectivity will be largely offset or exceeded by the low emissivity.

One obvious and zero- or low-cost way to reduce cooling requirements is to use light-coloured rather than dark-coloured roofs. A database of reflective roofing materials and their properties is maintained by the Lawrence Berkeley National Laboratory in California (http://EETD/LBL.gov/CoolRoof), and the US Environmental Protection Agency added roofing products to its Energy Star programme in 1999 (www.energystar.gov/products).

Some people may object to lighter-coloured roofs on aesthetic grounds. A partial solution to this subjective obstacle to energy-efficient design is to use roofing materials with inorganic pigments that are highly reflective only in the near-infrared part of the solar spectrum, which accounts for about 40 per cent of the solar energy reaching the ground, while providing a variety of dark shades in the visible part of the spectrum (Miller et al, 2002a). These pigments use various mixtures of 1μm diameter titanium and iron-oxide particles, but none of the spectrally selective roofing materials come close to matching the remarkable spectral properties of green leaves, in which the reflectivity abruptly increases from about 5 per cent at wavelengths less than 0.75μm (where it is useful for photosynthesis) to 90 per cent at wavelengths greater than 0.75μm. Nevertheless, spectrally selective roofing materials can result in roof-surface temperatures up to 30K cooler than standard dark roofing materials. Table 6.2 compares the albedo of conventional and prototype spectrally selective roofing tiles of various colours. The albedo is 0.41–0.48 for a range of colours. In addition to possibly offering higher reflectivity for a given colour, tile roofs are preferable to shingle roofs because of their much

greater durability (50–100 year lifespan instead of 15–20 years), protection against fire, and the smaller rate of heat transfer into the building due to the air gap that occurs under tiles.

Monitoring of six side-by-side houses in Florida that differed only in the reflectivity of the roof, and all having R19 (RSI-3.34) ceiling insulation, showed that white reflective roofs (albedo = 0.66–0.77, depending on the material) reduce cooling energy consumption by about 20–25 per cent and peak power demand by 30–35 per cent compared to a house with dark shingles (albedo = 0.08) (Parker et al, 2002). Hildebrandt et al (1998) applied a white roof coating to the roofs of three test buildings (an office, a museum and a hospice) in Sacramento, California, increasing the albedo from 0.2 to 0.6. The impacts on heating and cooling loads were computed using the DOE-2 building simulation model (see Appendix D), and compared with measured before and after energy use. Assuming the internal heat loads to be the same, the increase in heating energy use (due to less absorption of sunlight during the heating season) was 2–4 times greater than predicted by the DOE-2 model. In two buildings, the measured savings in cooling energy use was more than twice as large as expected, while in the other building (the museum) the measured saving was half as large (perhaps due to difficulty in providing the correct input data to the simulation model). Overall cooling energy use decreased by 20–40 per cent. In two buildings the extra heating energy use was negligible compared to the savings in cooling energy use, while in the third building it offset 40 per cent of the savings. For unventilated test structures in Jodhpur, India, Nahar et al (2003) find that a white glazed-tile roof instead of a conventional reinforced concrete roof reduces the indoor temperature by 11K during times of peak summer heat.

High albedo should also be considered for walls that cannot be shaded. Walls are subject to solar radiation both from above and reflected from the ground; the latter will be particularly important in hot-dry climates, where the surface albedo is large (up to 0.5). Radiation is generally reflected by the ground equally in all directions, and a wall will receive reflected radiation from one half of all possible upward directions. If the atmospheric transmissivity is 0.85, the sun is 30° from the vertical, and the surface albedo is

0.5, then the intensity of reflected radiation on a wall is $0.85 \times \cos(30) \times 0.5 \times 0.5 \times 1370\text{W}/\text{m}^2 = 252\text{W}/\text{m}^2$, compared to $0.85 \times \cos(60) \times 1370\text{W}/\text{m}^2 = 582\text{W}/\text{m}^2$ for radiation directly striking the wall.

Rooftop gardens can also substantially reduce roof temperatures (as well as greatly increase the amenity value of the building while providing a small insulation benefit, the latter as discussed in Chapter 3, Section 3.5). The cooling effect arises in part from the direct shading of the roof by vegetation, and in part from evaporative cooling. Soil adds thermal mass that moderates daily extremes (Theodosiou, 2003). Takakura et al (2000) have calculated that ivy growing in soil on a roof will result in a peak daytime roof-surface temperature of $29°C$ under conditions that would otherwise produce a temperature of $42°C$ with a bare concrete roof. Wong et al (2003a, 2003b) measured a peak soil surface temperature of $26–27°C$ on a roof covered with shrubs in Singapore, compared to a peak temperature of $53°C$ on a bare roof. Bare soil alone limited the peak surface temperature to $43°C$, due most likely to the evaporation of water retained by the soil.

Planting of shade trees around low-rise buildings on the east, west and equatorward sides will directly reduce the cooling requirements of a building. The use of shade trees and high-albedo roofing material, as well as high-albedo pavement, will also indirectly reduce cooling loads, by reducing the ambient air temperature. Rosenfeld et al (1998) calculated the impact on cooling loads in Los Angeles of increasing the roof albedo of all 5 million houses in the Los Angeles basin by 0.35 (a roof area of 1000km^2), increasing the albedo of 250km^2 of commercial roofs by 0.35, increasing the albedo of 1250km^2 of paved surfaces to 0.25 (by using whiter, limestone-based aggregates in pavement whenever roads are resurfaced), and planting 11 million additional trees. Altogether, the albedo of 25 per cent of the surface area of Los Angeles was assumed to be increased by an average of 0.3. The direct effect of higher roof albedo and of shade trees on building energy use was calculated using the DOE-2 building-simulation model. The effect on ambient air temperatures was computed using a high-resolution meteorological model, with the altered air temperatures supplied to the DOE-2 model. In the residential sector,

they computed a 15–17 per cent reduction in air conditioning energy use from higher-albedo roofs alone, a 12–17 per cent reduction from shade trees, and a 23–27 per cent reduction from cooler ambient air temperatures, giving a total savings of 50–61 per cent. The calculated reduction in peak air conditioning loads is 24–33 per cent. These are dramatic savings! For Toronto, Akbari and Konopacki (2004) calculated potential savings in cooling energy use of about 25 per cent for residential buildings and 15 per cent for office and retail buildings, arising from the direct effect of more reflective roofs and more shade trees and the indirect effect of lower ambient temperatures. For a typical house in Shiraz, Iran, Raeissi and Taheri (1999) estimate that appropriately positioned trees can reduce the cooling load by 10–40 per cent.

Vegetation can shade a wall from a distance, but also if it is directly attached to the wall (as in the case of vines). The cooling effect arises in three ways: by displacing the solar absorption plane away from the surface of the building, by reducing the absorption of solar radiation, and through evaporative cooling of the leaves. The absorption of solar radiation is reduced because, as noted above, leaves have a reflectivity of about 90 per cent in the NIR part of the solar spectrum (wavelengths greater than $0.7\mu\text{m}$), which accounts for about 40 per cent of the incident solar radiation. Evaporative cooling requires a supply of soil moisture, which in turn means that rain runoff from roofs should be directed to the ground adjacent to the building rather than into storm sewers. Thus, water and energy management need to be coupled in order to maximize passive cooling opportunities.

Finally, Akbari and Konopacki (2005) compared the impact of low-albedo surfaces and shade trees on both heating and cooling loads at a variety of US locations. Their assessment accounts for both direct effects (reducing the heat gain through the building envelope) and indirect effects (reducing the ambient air temperature). They find that, for climates with about 600 CCD or more, electricity savings from reduced cooling loads more than offset the increase in onsite heating energy use. If the heat is provided by natural gas or oil and the electricity that is saved is produced from coal at 33 per cent generation efficiency, there would be a net reduction in both primary energy use and CO_2 emissions for climates with substantially less than 600 CCD.

6.2.3 Insulation

In low-rise buildings and houses, increasing the amount of roof insulation will reduce air conditioning needs, while the addition of aluminium foil (having low emissivity) above attic insulation can reduce summer heat flow through the ceiling by a further 25–40 per cent for pre-existing attic insulation ranging from R30 to R11 (Medina, 2000). In a simulation of a one-storey factory with no roof insulation or ventilation, the internal air temperature reaches 42°C at noon on a day when the outside air temperature reaches 26°C; the addition of 5cm of roof insulation limits the peak indoor temperature to 34°C, while a combination of insulation and passively driven ventilation limits the peak temperature to 29°C (Rousseau and Matthews, 1996). For a one-storey house in Cyprus, adding 5cm of polystyrene insulation to the roof reduces the cooling load by 45 per cent (and the heating load by 67 per cent), while the addition of 5cm of polystyrene insulation to the walls reduces the remaining cooling load by about 10 per cent (and the remaining heating load by 30 per cent) (Florides et al, 2002a). For a one-storey building in Tehran, Safarzadeh and Bahadori (2005) find that 10cm of insulation on the walls and roof (U-value of $0.38W/m^2/K$) reduces the cooling load by 14 per cent (and the heating load by 55 per cent). For a typical middle-to-high-income house in South Africa, 5cm of roof insulation reduces peak indoor temperatures by about 3K (from 29°C to 26°C for an outdoor peak temperature of 31°C), based on measurements in houses before and after installing insulation (Taylor et al, 2000). Increasing the insulation to 10cm reduces the peak temperature by only a further 0.6K.

Heat gain through the roof in high-rise office towers is a very small portion of the total heat gain, so roof insulation will have a very small effect on the overall cooling load (while benefiting mainly the top floor). In commercial buildings with false ceilings on the top floor (suspended below the roof), it is common to simply lay the insulation above the false ceiling. Such 'lay-in' insulation is ineffective, as the insulation invariably gets displaced and there are gaps in the insulation around ductwork and wiring. It is far more effective to apply insulation directly to the roof, either externally or internally (HMG, 2003). An added benefit is that the plenum space between the false ceiling and roof thus becomes part of the conditioned space, so the ductwork no longer passes through unconditioned space (where it can be subject to heat loss in winter and heat gain in summer). Simply moving the insulation from the ceiling to the roof is estimated to reduce peak cooling requirements by 20–25 per cent in small commercial buildings in California, and to reduce annual cooling energy use by 5–10 per cent (AEC, 2003).

With regard to insulating the walls of a typical high-rise residential building in Hong Kong, Bojić and Yik (2005) find, through computer simulation, that external insulation alone reduces annual and peak air conditioning energy use by about 15 per cent if both external and internal walls are massive, while the application of insulation on both sides of external walls and internal partitions reduces the annual cooling energy use by about 40 per cent while increasing peak energy use by 4 per cent.[3]

Insulation (if applied externally) will be most effective in hot-dry climates, where the best strategy will often be to close up the building during the day and to rely on night ventilation and thermal mass to keep the building comfortable during the daytime. In hot-humid climates, where the diurnal temperature range is smaller, the best strategy will often be to vigorously ventilate the building throughout the day which, as discussed in Chapter 7 (Section 7.2.4), can provide comfortable conditions with temperatures as high as 32°C. In this case, the indoor air temperature will be about the same as the outdoor air temperature and insulation would appear to provide no benefit in reducing the cooling requirements. However, the peak temperature of walls facing the sun will be substantially greater than the outdoor air temperature because of absorption of solar radiation, leading to a transfer of heat to the inside of the wall, which is then radiated to the interior of the building. This can be mitigated with a high-albedo wall, except that growth of fungus on the wall in a hot-humid climate could necessitate repainting the wall every 3–5 years in order to maintain a high albedo. As noted below, direct shading of east- and west-facing façades is difficult. For these reasons, Givoni (1998a) concludes that, in hot-humid climates, insulation of exterior walls to RSI0.6–2.3 is worthwhile, depending on the orientation and albedo of the wall.

External insulation is generally preferred to internal insulation because it isolates the building's thermal mass from daytime heat while coupling it to cool night-time air through night-time ventilation. In hot-humid climates, where the diurnal temperature range is smaller, night-time ventilation can be useful in creating comfortable conditions at night if the building can cool down sufficiently rapidly, but may not be able to extract enough heat from the building mass to sufficiently limit daytime temperatures. In this case, a better strategy is to use minimal thermal mass and to rely on vigorous ventilation to create comfortable conditions. However, many hot-humid regions are also subject to tropical cyclones (hurricanes, typhoons) and so require high-mass buildings for reasons of structural integrity. In this case, it is better to apply the insulation internally since, when ventilated, such buildings will behave as buildings with low thermal mass. As noted by Givoni (1998a), this strategy precludes storing daytime solar heat for night-time use during the winter, so it should be applied only in those hot-humid regions with mild winters.

An alternative to large amounts of roof insulation to reduce the summer heat flux through the roof is the use of ventilated passageways between an inner and outer layer, designed so as to maintain non-turbulent airflow. Ciampi et al (2005) calculate that such microventilated roofs can reduce heat flow from the roof into the room below by up to 30 per cent. In climates with cold winters, however, increased insulation thickness will reduce both summer heat gains and winter heat losses.

6.2.4 Direct shading

Direct shading can be achieved by designing walls or protrusions to shade windows, by designing awnings to maximize the shading of windows in summer (when the sun is high) and to minimize shading in winter (when the sun is low), to include operable shading devices (e.g. Venetian blinds, louvres), and to use PV panels for partial shading, either as awnings, PV windows or atria glazing.[4] External louvres are more effective than internal louvres, as only about 10 per cent of the sunlight

Figure 6.6 Simple shading devices

Source: Koch-Nielsen (2002)

Figure 6.7 Longitudinal cross section of the Mont-Cenis Academy for the state of North Rhine-Westphalia in Germany (upper), showing the roof shade, and view from an open space below the roof shade (lower)

absorbed by the louvre is transferred to the interior of a room in the former case, compared to almost 50 per cent in the latter case (Baker and Steemers, 1999). Louvres should be made of light and reflective materials with low heat storage capacity, and designed to prevent reflection onto any part of the building. It is not possible, however, to design a fixed shading device with zero shading in winter and complete shading in summer, so any fixed device represents a compromise between winter and summer objectives. Emission of longwave radiation from the inner surface of external shading devices can cause glass surface temperatures to be up to 5K warmer than in the absence of longwave emission in hot sunny climates (Kapur, 2004), so a low-e coating on the inner surface of external shading devices is beneficial.

If fixed shading devices are used, then on equatorward-facing walls, a horizontal shading device should be used, while on east- and west-facing walls, a vertical shading device should be used. Horizontal shading devices should have a gap between the device and wall to prevent hot air being trapped under the shading device. A variety of simple shading devices is shown in Figure 6.6. A loose outer roof can be suspended over a building as a sun umbrella, as in the 12,600m² outer roof built over an administrative centre for the state of North Rhine-Wesphalia in Herne, Germany, illustrated in Figure 6.7. This structure is also the world's largest example of building integrated photovoltaic panels, and is discussed further in Chapter 11 (Section 11.3.1). For a one-storey building in Tehran, Safarzadeh and Bahadori (2005) find that the use of Persian blinds on sunny summer days (and winter nights) reduces the cooling load by 27 per cent (and the heating load by 18 per cent). The impact of this and other measures is compared in Figure 6.8. The net effect of insulation, shading, curtains, modestly upgraded windows (double-glazed instead of single-glazed), and a 50 per cent reduction in the size of cracks is to reduce the cooling energy use by about 60 per cent (and the heating energy use by 86 per cent).

Motorized adjustable louvres and Venetian blinds are relatively common in Europe. The AVAX building in Athens, Greece, uses large louvres on a vertical axis that can be completely opened or completely closed, as illustrated in Figure 6.9. The British Research Establishment

Figure 6.8 Cooling and heating energy use for a one-storey administrative building in Tehran as computed by Safarzadeh and Bahadori (2005) for the base case building and with various upgrade measures

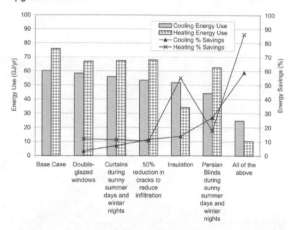

Note: Also shown is the per cent savings for each measure and for the package of measures. The mean daily maximum ambient air temperature is 37°C in July and 8°C in January.

Figure 6.9 Shading with façade-scale louvres on the AVAX building in Athens. Top: Front façade closed, early morning. Bottom: Front façade open, midday

Source: Tombazis and Preuss (2001)

(BRE) building in Garston, in contrast, uses translucent shading panels on a horizontal axis, as illustrated in Figure 6.10. Automated Venetian blinds are discussed in Chapter 9 (Section 9.3.1) as part of an integrated daylighting/cooling-load-reduction strategy. These are most effective if placed outside the glazing, which is possible if they are placed in the gap between inner and outer façades of a ventilated double-skin façade (see Chapter 3, Section 3.4). External light shelves

Figure 6.10 Shading louvres at the British Research Establishment offices in Garston

Photographer: Dennis Gilbert, London

can be designed to shade the lower portion of a window (see Chapter 9, Section 9.3.1). The use of PV panels is becoming an architectural design element in its own right, used to create a more interesting façade and more varied (and hence pleasing) interior light conditions (see Chapter 11, Section 11.3.1).

Traditional houses and streets in many parts of the world used to be laid out in such a way as to provide a significant amount of self-shading. Spacing of buildings close enough to provide significant self-shading will diminish wind strength near the ground, reducing the potential for ventilation, although daytime ventilation is not always useful. Close spacing also reduces solar access in winter, but such access will not be needed in those hot-summer regions with mild winters. Even small-scale, deeply carved patterns on building façades, such as found in Jaisalmer, India, serve to minimize heat gain by providing shading due to texture (Krishan, 1996), while increasing the surface area for cooling at night. Bourbia and Awbi (2004) measured temperatures at a height of 1.5m in traditional (narrow) and contemporary (wide) streets in a city in Algeria. Traditional streets are approximately 5K cooler than contemporary streets, whether oriented north–south or east–west. This is due both to the greater shading in traditional streets (which reduces direct solar irradiance) and the smaller sky viewing angle, which reduces the diffuse solar irradiance.[5]

6.2.5 Windows

In high-rise office towers, total heat gain through the envelope is typically less than the internal heat gain, but is dominated by solar heat gain through the windows. Choosing windows with a low SHGC is therefore important. The SHGC of the commercially available windows listed in Table 3.11 ranges from 0.23 to 0.76 – roughly a factor of three difference. Low SHGCs are achieved through low NIR transmittance, so the visible transmittance – needed for daylighting – is no lower than about 0.4. Electrochromic windows are windows whose solar transmissivity can be actively controlled in response to changing conditions, such that the SHGC can be switched between 0.3–0.5 and 0.1–0.2 (Chapter 3, Section 3.3.15). The product of the window:wall ratio times the window SHGC can be thought of as a *solar aperture*, as the solar heat gain varies directly with its value.[6] Alternatively, windows with laser-cut panels or angular-selective thin films can be used that passively permit more sunlight at those angles of incidence that arise when solar heat gain is more likely to be useful and less at other times, as discussed in Chapter 9 (Section 9.3.1).

6.2.6 Thermal mass

Increasing the thermal mass of a building (through thicker walls, floors, and roof) reduces the cooling load by reducing the peak temperatures reached during the day. This arises because the greater the thermal mass, the smaller the temperature increase for a given absorption of heat. Thermal mass also delays the transmission of heat to the inside surface; for a 30cm thick wall, the delay is 8 hours for heavy-weight concrete and 10 hours for brick (Florides et al, 2002a). Heat absorbed during the day is released at night if the indoor temperature drops below the wall temperature, so early morning temperatures will not be as cold as they would otherwise be. As a result, no or minimal early morning heating is required. This dual benefit is illustrated in Figure 6.11. The slow night-time cooling can be disadvantageous if rapid night-time cooling is desired. It can be counteracted through high rates of mechanical ventilation at night. In residential buildings, high-mass walls can be used in living areas and low-mass walls in sleeping areas. Conversely,

Figure 6.11 Impact of thermal mass on 24-hour temperature variation. A poorly designed building with inadequate thermal mass may require early morning heating and afternoon air conditioning on the same day

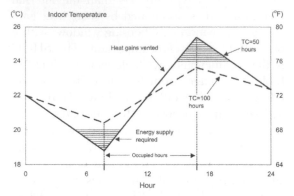

Source: Braham (1998)

if it is desirable to maximize the release of heat at night in order to prevent indoor temperatures from dropping too low, this can be achieved by creating a wall with two or three layers with appropriate thermal properties – a 'thermally tuned wall' (Duffin and Knowles, 1981, 1984).

Thermal mass is effective when there is a large day–night temperature difference, as the heat absorbed during the day can be released at night if there is adequate night-time ventilation. This is the case in hot-dry climates, where the diurnal variation is often 10–20K. Porta-Gándara et al (2002) simulated the cooling load for housing built using adobe bricks and using hollow concrete blocks in Baja California (Mexico), and found the air conditioner load of the former to be one-quarter that of the latter during the hottest months (given a set-point of 26°C). The diurnal temperature variation is much less (frequently only 5–10K) in hot-humid climates, but thermal mass (combined with night-time ventilation) can still be useful in minimizing peak daytime temperatures, depending on the absolute daytime and night-time temperatures. For residential high-rise buildings in Hong Kong (where the diurnal temperature variation averages only 4.5K during the three warmest months), both Bojić and Yik (2005) and Cheung et al (2005) find, through computer simulation, that increasing thermal mass has a negligible effect on yearly cooling energy use, but can substantially reduce (by 30–40 per cent) the peak cooling loads.

For thermal mass to be most effective, it should be in direct contact with the conditioned space. This implies, for example, that false ceilings, carpets, or drywall should be avoided, depending on the building element that is serving as the thermal mass. Insulation, of course, should be on the outside. Alternatively, air channels can be placed inside the thermal mass. The innovative ceilings in the BRE offices in Garston (UK) incorporate both elements through a waveform design (Figure 6.12). Another option is to use hollow pre-cast concrete floor planks, as described by Yorke and Brearley (2003) for the Elizabeth Fry Building at the University of East Anglia. Floor or wall thermal mass will be most effective in areas exposed to direct sunlight. Effective use of thermal mass also requires that the interior temperatures be allowed to vary by several degrees over a 24-hour period. In this way, heat will flow into the thermal mass when the air temperature is warmer than the thermal mass, and will be released when the air temperature is colder. Thus, the benefits of thermal mass and of allowing temperatures to float amplify each another. This is another example of an energy savings synergy in building design.

As noted in Chapter 4 (Section 4.1.6), some fatty-acid phase-change materials (PCMs) melt at a temperature (22–26°C) that would be useful for damping peak temperatures, and can be incorporated into the wall structure or used in phase-change wallboard or spray-on plaster to increase thermal mass. To be fully effective, night-time temperatures should be cold enough (<18°C) for night-time ventilation to fully extract the heat stored in the material during the day. Micro-encapsulated PCM in plaster (illustrated in Figure 4.9) or wallboard has a number of advantages over other approaches (Schossig et al, 2005), namely:

- the capsule shell prevents interaction with the surrounding material;
- no extra work is required at the construction site;
- the capsules are small enough that they do not need protection from destruction;
- they provide a large surface area for heat exchange.

Figure 6.12 Details of the waveform ceilings in the British Research Establishment offices in Garston

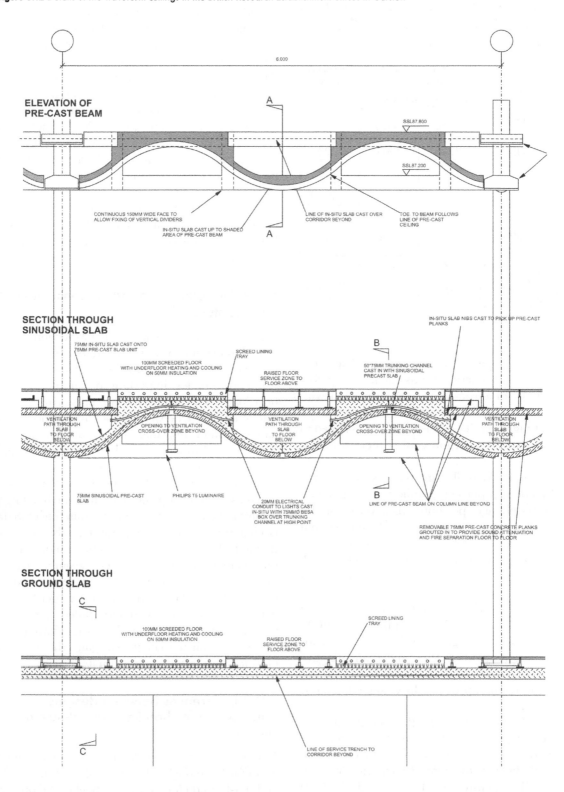

Note: The exposed concrete mass and the hollow channels are shown.
Source: Gething (2003)

Table 6.3 Comparison of annual cooling energy use (kWh) and peak power draw (kW) for a wood-frame house in Ohio (US) with standard or PCM wallboard, with and without night ventilation at four air changes per hour, and for indoor air temperature maintained at an excessively strict 22°C

	Standard wallboard	PCM wallboard	Impact of PCM
Cooling energy use (kWh)			
Without night flushing	4650	4460	–4%
With night flushing	3225	2671	–17%
Impact of night flushing	–30%	–40%	–42% (joint impact)
Peak Power Draw (kW)			
Without night flushing	5.65	5.39	–5%
With night flushing	4.75	4.49	–5%
Impact of night flushing	–16%	–17%	–21% (joint impact)

Source: Kissock (2000)

Table 6.3 compares the peak cooling loads and annual cooling energy use as simulated by Kissock (2000) for a wood-frame house in Ohio, with conventional and PCM wallboard, and with or without night ventilation (the house is assumed to have an unnecessarily low cooling thermostat setting of 22°C). In the absence of night ventilation, the use of PCM wallboard reduces total energy use by about 5 per cent, while night ventilation at four air changes per hour whenever the ambient temperature is lower than 22°C, but without PCM wallboard, reduces energy use by 30 per cent. If night ventilation is already being used, PCM wallboard reduces total energy use by almost 20 per cent. The combination of night ventilation and PCM wallboard reduces energy use by over 40 per cent. Night flushing with PCM wallboard would largely eliminate the need for air conditioning altogether under an adaptive thermal comfort standard, which would permit temperatures of up to 26–30°C (see Chapter 7, Section 7.2).

A factor limiting the effectiveness of bulk PCMs, both in absorbing heat during the day and in releasing it to the ventilation air at night, is the low thermal conductivity of most PCMs. Turnpenny et al (2000, 2001) have overcome this problem by embedding heat pipes into PCMs that are mounted below a ceiling fan.

Heat pipes utilize the latent heat of evaporation of a fluid contained within the pipe to transport large amounts of heat with a minimal temperature gradient, giving them a very high thermal conductivity. This system eliminates the need for summer air conditioning in the UK, and would eliminate the need for air conditioning during the transition seasons in many warmer climates. In climates where air conditioning is needed but the heat load is not excessive, PCMs can be incorporated into chilled-ceiling cooling panels, as discussed in Section 6.9.3. This permits the chillers to run only at night-time, when there can be significant efficiency benefits. Systems with bulk PCM are more complicated than micro-encapsulated PCM in wallboard, so they would have to demonstrate a significant advantage to be justified.

6.2.7 Air tightness to reduce sensible heat and moisture influxes

Air leakage is discussed in Chapter 3 (Section 3.7). In cold climates, it can contribute 40 per cent or more of the total building heating load during winter, but will tend to be less important for summer sensible heat loads. This is because the impact of a given rate of exchange between indoor and outdoor air is larger during winter than during summer, due to a larger typical indoor-to-outdoor temperature, and because the rate of air exchange itself is larger due to the greater pressure difference between the inside and outside at the top and bottom of the building (a result of the greater temperature difference). Safarzadeh and Bahadori (2005) estimate that a reduction in leakage area by 50 per cent for a one-storey, uninsulated and unshaded base-case building in the dry climate of Tehran would reduce the cooling load by about 12 per cent (and by a larger fraction if the building is insulated and shaded) (see Figure 6.8). The major impact of air leakage during summer can be on the latent heat gain (the influx of outside moisture) in humid climates.

6.2.8 Differences between designs appropriate for hot-dry and hot-humid climates

Not all of the strategies discussed above are equally useful for reducing the cooling load in hot-dry and hot-humid climates. In particular, the optimal strategies with respect to urban form, building form and spacing, thermal mass, the nature of the building envelope and openings for daylighting and ventilation are somewhat different between these two climates. These differences are summarized in Table 6.4, based on the extensively illustrated work of Koch-Nielsen (2002). The recommendation to limit thermal mass in hot-humid climates is based on the assumption that the diurnal temperature variation in humid climates is too small to make night-ventilation viable, but, as discussed later in Section 6.3.3, this may not always be the case. The recommendation for widely spaced buildings in hot-humid climates is not fully compatible with the desire for compact and dense urban forms so as to increase the viability of rail-based public transit, so some compromise

will be required. Other strategies, such as the use of low-albedo surfaces, insulation and shading, are applicable to both climates. Philip (2001) notes, with examples, that some architects have created buildings in the tropics that apply a gesture of the devices appropriate for hot-humid climates as part of the aesthetic composition of the building, but with little effectiveness due to rather obvious errors in their application stemming, most likely, from the lack of a genuine desire to create buildings that work with nature rather than against it (the examples include external shading devices that do not shade the main windows, next to adjacent fully enclosed glass-staircases that are completely unshaded).

Table 6.4 Differences in the design of passively cooled buildings in hot-dry and hot-humid climates

Hot-dry climate	Hot-humid climate
Urban form	
• Use compact urban form with narrow (preferably pedestrian!) streets to minimize direct and reflected solar radiation, to provide shade, and to provide shelter from dusty winds • Buildings should be of uniform height to avoid non-uniform winds • Use the courtyard structure with the long axis of the courtyard oriented perpendicular to the prevailing wind direction • Use colonnades, awnings and green gardens or patios	• Buildings should be widely spaced to promote air movement between and through buildings • Trees should be straight to promote air movement, and lined with tall, high-canopy trees to provide shade while not restricting air movement
Building form	
• Buildings should be compact and inward-facing in order to reduce the surface area exposed to solar radiation. • Large surfaces should face north or south, so as to minimize exposure to solar radiation. • High buildings are preferred so as to minimize the roof area for a given volume and to enhance the stack effect.	• Buildings should be open, outward-oriented, and elongated to take advantage of all air movement. • Surface areas should be large compared to the volume, and large surfaces should face north or south. • High buildings are preferred so as to minimize the roof area for a given volume and to enhance the benefit of winds.
Building envelope	
• High thermal mass is needed to reduce daytime temperature peaks, but the envelope must be ventilated with cool night-time air in order to remove heat absorbed during the day. • In composite roofs (having a separate roof and ceiling), the roof should be of lightweight material and the ceiling of dense material • Ventilation openings in roof spaces should be provided to allow hot air to escape. The air outlet opening should be larger than the air inlet opening in order to create pressure differences that will draw more air through the roof space.	• The building envelope should be of low thermal mass in order to minimize storage of heat, so that it can cool down quickly whenever temperatures decrease. • A lightweight outer roof should be provided, which functions as a solar umbrella and which permits unrestricted airflow between it and the main roof. • Ceilings below an outer roof should also be of lightweight construction.

• Roofs sloping toward a courtyard will allow cool air that forms on the roof at night (due to emission of infrared radiation) to flow into the courtyard and cool internal spaces.	• The roof should provide extensive overhangs so as to shade the walls and prevent rainwater from splashing onto the walls.
• Arches, domes and pitched roofs will always have some of their surface in shade, increasing the surface area from which convective and radiative heat loss can occur, and providing a space above the heads of occupants to which warm air can rise.	
• Walls surrounding rooms used only during the day should be massive, while walls surrounding rooms used only at night should be lightweight.	• Solid walls should be minimized. Walls should be designed to maximize internal air movements. Thermal mass should be restricted to construction elements not exposed to solar radiation.
• East-facing walls can cause the most problem, as heat can be released to the interior before the afternoon is over. Insulation and shading can prevent premature release of heat.	• East-facing walls are less critical, due to generally greater humidity and cloudiness during the morning than during the afternoon in humid climates.
• Ground floors should be thermally massive and in direct contact with the ground.	• Ground floors should have no contact with the ground, in order to ensure that the floor does not store heat.
	• Ground floors should be raised above the ground in order to increase exposure of the entire building to wind, to permit ventilation underneath the floor, and to offer protection against moisture and insects.
	• If a solid floor is needed, it should have built-in ducts, as in a hollow concrete slab, in order to reduce the thermal storage capacity and to provide ventilation.
Openings for daylighting	
• Openings should be kept small, and direct sunlight should be avoided.	• Large openings are needed for cross-ventilation, so they should be shaded with overhanging roofs or wide verandas.
• The main view should be directed toward the sky rather than the horizon or ground, as reflected light from the light-coloured ground (characteristic of arid regions) can be a source of glare.	• Openings should not be positioned to face the sky, which is the main source of glare in humid climates.
• Low-level openings are acceptable if they open onto shaded, green areas or non-glare surfaces.	
Openings for ventilation	
• Openings for ventilation and daylighting should be separate and should function independently.	• All openings need to be as large as possible. Large openings are needed in internal as well as external walls.
• Ventilation openings will need to be closed during the day and opened at night.	• Continuous wind-generated ventilation is required, except during midday hours during the hot season.
• Ventilation openings should be located at different heights to make use of the stack effect.	• Openings should be placed on opposing façades to facilitate cross-ventilation.
• Ventilation openings should be located so that air will move over the warmest surfaces (such as the ceiling).	• Openings should be located to allow air movement at body level.

Note: Humidity is being used as a proxy for diurnal temperature variation, the assumption being that this variation will always be too small in humid climates to make effective use of high thermal mass in combination with night ventilation. This should be checked on a case-by-case basis.
Source: Compiled from Koch-Nielsen (2002)

6.3 Passive and passive low-energy cooling

Having reduced the cooling load through the techniques described above, the next strategy in priority is to use passive and/or passive low-energy cooling strategies. A purely passive cooling technique requires no mechanical energy input at all. It includes such techniques as designing a building to maximize natural ventilation. Cooling by maximizing the emission of radiation to the sky at night is another form of passive cooling, but it can be greatly enhanced through mechanical circulation of air or water through a rooftop radiator. Similarly, enhanced ventilation of a building with cool night air or cooling of incoming air by first drawing it through an underground pipe makes use of the natural nocturnal cooling of outside air at night or the coldness of the ground. In this sense they are passive cooling techniques, but they also require a small input of mechanical energy.

Hastings (1994, Chapters 27–32), Givoni (1994, 1998a, 1998b), and Santamouris and Asimakopoulos (1996) discuss many successful examples of passive or passive low-energy cooling techniques in real buildings, and an extensive amount of work has been done in this area under Annex 35 of the International Energy Agency's *Energy Conservation in Buildings and Community Systems* Implementation Agreement (see Appendix C). The European Commission Directorate General for Energy has prepared a design handbook on natural ventilation (Allard, 1998) that also discusses the physics of ventilation, prediction methods and barriers. The following subsections explain how the major techniques work.

6.3.1 Natural ventilation

In the absence of ventilation, the interior air temperature in a building will rise considerably above the outside air temperature, such that the building can be quite uncomfortable even when the outside temperature is pleasant. As the ventilation rate increases, the interior temperature will approach the outside temperature but will remain slightly warmer due to internal heat gains. At the same time that the real interior air temperature decreases, the *perceived* temperature will decrease further due to the greater ability of moving air to remove heat from a warmer body. Finally, with natural ventilation, the *acceptable* air temperature increases due to enhanced psychological adaptation to warmer conditions compared to buildings with mechanical ventilation.

In both mechanically and naturally ventilated buildings, tests with human subjects indicate that the preferred indoor temperature increases with increasing outdoor temperature, due in part to differences in clothing and to differences in expectations (de Dear and Brager, 2002). For an outdoor temperature of 20°C, the preferred temperature is 24°C for both mechanically and naturally ventilated buildings. As outdoor temperature increases to 30°C, the average preferred temperature rises to only 25°C in mechanically ventilated buildings, but to 27°C in naturally ventilated buildings. An air speed of 1m/s increases the neutral air temperature (at which a subject feels neither hot nor cold) by up to 3K, depending on the concurrent radiant temperature (see Chapter 7, Sections 7.2.2 and 7.2.4). Provision of personal control over local air speed through fans further increases the temperatures considered to be acceptable or even preferred. Research in Denmark indicates that a temperature of 28°C with personal control over air speed is overwhelmingly preferred to a temperature of 26°C with a fixed air speed of 0.2m/s (de Dear and Brager, 2002). In Thailand, Busch (1992) found that the maximum temperature accepted by 80 per cent of survey respondents is about 28°C in air conditioned offices and 31°C in naturally ventilated offices. In contrast, most air conditioned buildings are operated to maintain a temperature in the lower part of the 23–26°C range set by the American Society of Heating, Refrigeration and Air Conditioning Engineers (ASHRAE) in its comfort standard (Standard 55), irrespective of outdoor conditions (comfort standards are discussed in Chapter 7, Section 7.2).

Thus, one can envisage the following regime:

- natural ventilation without ceiling fans when this yields an indoor temperature up 26°C;
- activation of ceiling fans when the indoor temperature exceeds 26°C or less and a room is occupied;
- mechanical cooling when the indoor temperature exceeds 28–32°C (depending on the local acceptability of warm temperatures).

To clarify, ventilation refers to the replacement of interior air with fresh outside air. It has a direct cooling effect if the outside air is cooler than the interior air that it replaces. Air movement by ceiling fans is not ventilation, and does not directly cool the interior air. Instead, it reduces the perceived temperature (thereby increasing the acceptable real temperature) by permitting more effective removal of heat from human subjects. It provides no benefit if people are not present, and slightly increases the indoor temperature due to the heat released by the fan motor.

A potential disadvantage of ventilation when outside temperatures are as warm as 32°C is that the entire building mass warms up, making the building hotter and less comfortable at night (and requiring more vigorous night-time ventilation to cool the building back down). This will be less of a problem if the building has low thermal mass, but low thermal mass makes night-time ventilation less effective in limiting daytime temperatures. Clearly, there are several possible strategies in minimizing cooling requirements, and the best strategy depends on the daytime and night-time temperature extremes, the nature of the building, the role of internal heat gains and the occupancy pattern. The evaluation of alternative strategies requires the use of computer simulation models.

In the following paragraphs, various techniques for inducing natural ventilation are described. To be effective, these techniques require that the building envelope be airtight ($1cm^2/m^2$ effective leakage area) except for the ventilation openings. Building form – a key architectural design decision – is critical to most techniques for passive ventilation.

Cross-ventilation

Wherever possible, buildings should be designed with windows that can be opened on both sides, so as to permit cross-flow of air. This can increase ventilation by a factor of five compared to one-sided ventilation (Fordham, 2000). Narrow floor plans, openings on both sides of a building, and unimpeded airflow from one side to the other will facilitate cross-ventilation. Where this is not possible, cross-ventilation can be enhanced through:

- use of alcoves projecting from the main external wall, with operable windows on both sides, as illustrated in Figure 6.13;
- external deflectors (wing walls) projecting from the main wall at right angles next to windows on the downwind side of the windows;
- indented porches or balconies that can be closed off with insulated shutters, as illustrated in Figure 6.14;
- an irregular and less compact floor plan, as illustrated in Figure 6.15;
- a breezeway passage through the centre of detached houses that can be closed off with insulated shutters.

By combining indented porches and breezeways with insulating shutters, the conflicting desires for a compact building form to minimize daytime heat gain and an open form to maximize night-time ventilation can be reconciled. Because of the particular importance of thermal comfort at night, residential units should be laid out in a way that provides the best ventilation in bedrooms while minimizing noise from outside. For the sake of acoustical privacy, each room should be ventilated independently of the others (as in Figure 6.15) or, if that is not possible, the master bedroom can be located next to a study rather than next to a child's room. Where necessary, ventilation louvres can be placed in the walls between rooms, above doors.

The inner courtyard is a feature of traditional architecture throughout the world, as it engenders cross-ventilation and thereby serves as an effective passive cooling device, as documented by Al-Temeemi (1995) for Kuwait, Lee et al (1996) for Korea, Krishan (1996) for India and Oktay (2002) for Cyprus. Many of the blocks of old European cities consist of 5–6 storey buildings lining the streets and an interior courtyard that allows effective cross-ventilation. Tantasavasdi et al (2001) describe many features of traditional Thai houses that enhance natural ventilation, features that can be incorporated in houses worldwide. These include elevating the first floor of a house by half a storey, which allows more wind to flow through the house compared to slab-on-grade construction, and steep roofs with a cathedral attic and venting tower that creates a stack effect (see Chapter 3, Box 3.4). Indeed, two- or three-storey houses will facilitate

Figure 6.13 An example of an alcove projecting from the main wall in order to increase ventilation by wind

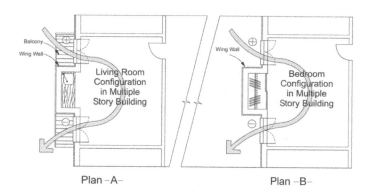

Source: Givoni (1998a)

Figure 6.14 An example of indented balconies that can be closed off with insulated shutters, to create a compact form that minimizes heat gain, then opened to facilitate night-time ventilation to remove the heat that does accumulate

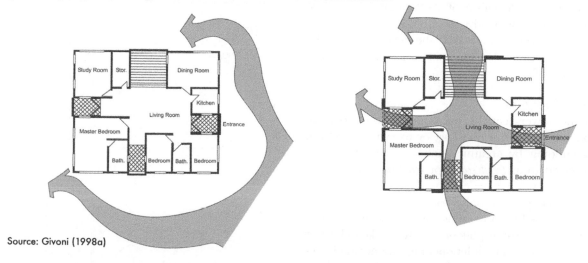

Source: Givoni (1998a)

Figure 6.15 An example contrasting compact and irregular floor plans, the latter serving to greatly increase the cross-ventilation by wind

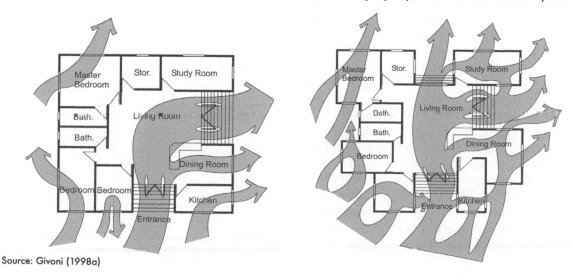

Source: Givoni (1998a)

more ventilation by both wind and stack effects than sprawling bungalows (not to mention contributing to a more compact urban form). Streets and detached houses can be laid out to maximize ventilation by, for example, arranging the gaps within one row of houses such that wind will impinge on houses in the next row, and orienting the long side of houses perpendicular to the prevailing wind direction. Both the unaltered prevailing winds and deflected winds should be taken into account when assessing the prospects for wind-induced ventilation. Optimal orientation from a wind point of view may of course conflict with the optimal orientation from a solar heat gain point of view, in which case some analysis will be needed to determine the best overall orientation. Preference should be given to solar orientation, as a deviation of ±60° from the optimal orientation for wind is possible while still providing good ventilation, while the deviation from the optimal orientation for solar radiation should not exceed ±30°.

Wind towers

Wind towers can be constructed either to force outside air into a building, or to extract air from a building by creating suction. The former are referred to as *wind scoops* or *wind catchers*, and are a traditional passive cooling device in the Middle East. They are usually combined with some device to cool the incoming air, such as water-filled unglazed earthen pots or a water-soaked cloth or wet charcoal. The inlet openings are normally designed to extract the less hot and less dusty air from above the buildings and scoop the air down into the buildings, as illustrated in Figure 6.16a, b and e. The wind tower can be designed to scoop air into the building on one side while acting as an exhaust outlet on the other side (Figure 6.13e), which in turn can permit heat exchange between the supply and exhaust air streams. In order to accommodate variable wind directions, a supply–exhaust wind tower can be divided into four to eight ducts through vertical partitions, a technique commonly used in Iran (Gage and Graham, 2000). Traditional and contemporary examples from Iran and Doha are illustrated in Figure 6.17. Modern wind towers generally utilize both wind and stack (buoyancy) effects, and can be equipped with automatic dampers to control the amount of airflow.

Figure 6.16 Alternative wind tower designs

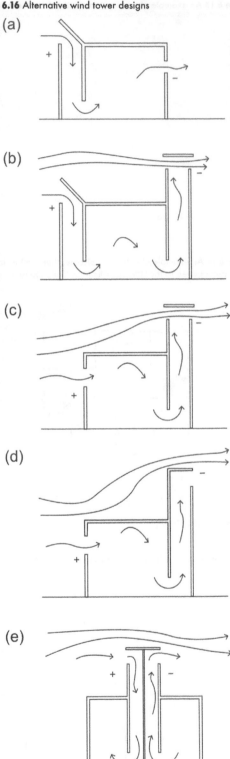

(a)

(b)

(c)

(d)

(e)

Note: Negative signs indicate lower air pressure and a suction effect, positive signs indicate higher pressure.
Source: Koch-Nielsen (2002)

Figure 6.17 A traditional multidirectional wind tower in Yazd, Iran (upper); Multidirectional wind towers at Qatar University, Doha (lower)

Source: Koch-Nielsen (2002)

Suction wind towers are commonly used in north-European residential buildings (Axley, 1999). The stack terminal is designed to maintain a negative pressure (suction) due to wind, and to inhibit rain, insects and animals from entering the stack. Wind normally creates a positive pressure on the windward side of a building and negative pressure on the leeward side, which induces cross-ventilation (see Figures 6.13c and 6.13d). The ventilation stack, if designed and positioned properly, is such that the greatest negative pressure occurs at the stack terminal, forcing air to flow in though all the building inlets and out through the ventilation stack. Airflow over an opening at the top of a domed roof, common in the Arab world, also creates a suction that removes hot air that accumulates inside the top of the dome. Such openings are usually specifically designed to increase the suction effect (Gallo, 1998). Innovative inlet vents developed in The Netherlands and France maintain roughly constant airflow rates over the range of air pressure differences encountered, or reduce the airflow at the higher pressure differences associated with the windiest conditions in order to compensate for the increased infiltration that occurs under these conditions.

Separate wind towers can be constructed for the inflow and outflow of air. Conversely, *balanced wind stack* (BWS) systems provide both inlet and exhaust airflow, either through a co-axial stack (one inside the other), or through vertical partitions running across the tower. BWS systems are being promoted by European manufacturers for use in larger buildings. They allow the possibility of sensible-heat recovery, although the flow

Figure 6.18 Rotatable wind cowl (left) that can be used for air supply or extraction, and a vertical-axis wind catcher (right) that captures the slightest breeze for ventilation

Note: Both are part of the Harare International School in Zimbabwe and are part of a passive ventilation system developed by Ove Arup Consultants.
Source: www.arup.com

resistance will be increased. Heat pipes offer the prospect of modest heat recovery (up to 55 per cent) with minimal flow resistance (a few Pa pressure drop) (Sirén et al, 1997).

Fixed wind towers require a relatively constant wind direction, which is common in coastal areas. For inland regions with variable wind directions, rotating wind towers can be used. The Harare International School in Zimbabwe contains both a rotating wind cowl and a locally developed vertical axis wind turbine that can utilize the slightest wind for ventilation (Figure 6.18).

Wind towers can be combined with solar chimneys (described later) so that both wind and buoyancy forces can be utilized at the same time. Examples of such hybrid systems include the New Parliamentary Building and the School of Architecture at Portsmouth University (UK). Even without solar chimneys, buoyancy forces will be at work, in addition to wind forces. However, in windy regions, wind forcing alone is sufficient or largely sufficient to induce adequate natural ventilation, and there is little if any benefit from additional passive-ventilation techniques. This is the case for San Francisco, where wind-driven natural ventilation was used as the primary means of cooling on levels 6–18 of the new Federal Building (McConahey et al, 2002).

In spite of the many examples of wind towers used in modern buildings, there is still a need for significantly more research in order to define the limits of applicability and to produce the most cost-effective designs. At the moment, there are no design tools available between rough and approximate calculations, and detailed computational fluid dynamics (CFD) simulations. Theoretical work by Li and Delsante (2001) and Li et al (2001) indicates that, when wind catchers are combined with buoyancy forces, more than one flow pattern is possible. The flow pattern occurring for a given set of conditions depends on the previous state of the building; that is, there is a hysteresis behaviour. The more complex the building, the greater the number of possible flows for a given set of conditions. Wind patterns under urban conditions are often quite complex too, requiring physical models or CFD analysis extending two to three blocks from the building on all sides.

Atria

Atria can also be used to induce natural ventilation through the proper placement of air inlets and outlets, along with shading controls to avoid overheating in summer (this can be achieved passively, by using either thermochromic glazing (Chapter 3, Section 3.3.16) or laser-cut glazing (Chapter 9, Section 9.3.1)). Cross-ventilation of air from the outside at night should be allowed, with outside air passing through the space next to the atrium, where it is heated and becomes less dense, thereby flowing up and out through the atrium. In a four-storey office building in England, illustrated schematically in Figure 6.19, this

Figure 6.19 Airflow and temperatures in the Gateway Two building in the UK during (a) summer, and (b) winter

(a) Summer

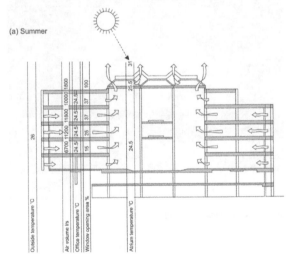

(b) Winter

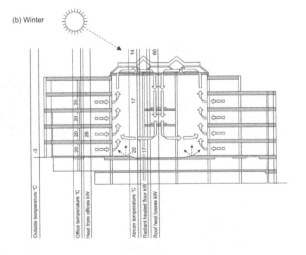

Source: Hawkes (1996)

strategy was sufficient to eliminate the need for mechanical cooling altogether. Natural ventilation induced by the atrium of the Berlin stock exchange is estimated to reduce the air conditioning load by 20 per cent (Herzog, 1996). The 60-storey Commerzbank in Frankfurt, completed in 1997, was the first naturally ventilated high-rise building in the world, with an atrium playing a central role (Shuttleworth, 2003). To avoid too large an airflow, the atrium is sealed every ninth flow and rising air is vented to the outside. A central computer optimizes natural ventilation by controlling the window openings that are part of a double-skin façade, with mechanical air conditioning used only during extreme conditions. Night cooling can also be instigated by fully opening the windows.

To avoid overheating during the summer, adjustable shading devices can be placed over the atrium roof. Conversely, adjustable reflectors can be placed over the roof to increase the daylighting role of the atrium, as in the Genzyme building (see Chapter 9, Figure 9.3). In one office block on Potsdamer Platz in Berlin, shading devices were placed on the outside of the atrium-facing offices to reduce clutter of the atrium roof, to improve daylighting of the lower offices, to improve accessibility and to reduce the maintenance required of the shading devices, and to allow occupants

control of their own shading devices (Richard Rogers Architects, 1996).

Holford and Hunt (2003) have developed computer and physical models of the use of atria to induce natural ventilation through the adjacent space, and have identified design extremes for which an atrium provides no significant enhancement of the airflow. At issue is that fact that, if the ground-level openings in the atrium are too large, air will flow directly through the atrium from the outside rather than through the adjacent space into the atrium.

Solar chimneys

Natural ventilation can also be induced by creating a 'stack' or 'chimney' effect – that is, by creating air that is warmer and therefore less dense than its surroundings, so that it will rise up and out of a building, forcing cooler outside air to enter the lower part of the building, as explained in Box 3.4. As seen in Box 3.4, the outward-directed pressure difference increases with increasing height above the neutral level, so a greater passive ventilation will be induced if a tall tower is constructed. A tower constructed for this purpose is referred to as a *solar chimney*. If solar collectors are integrated into the wall of the tower, even stronger airflows can be induced when the sun is shining. Stairwells can be designed to

Figure 6.20 A traditional house from the hot-humid south-eastern US, with a cupola to facilitate ventilation through the stack effect

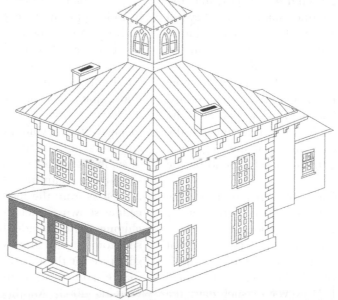

Source: Thomas (1999)

operate as a solar chimney, particularly if they are partially glazed and located next to the edge of the building facing the sun. Solar chimneys can also be designed to provide daylighting for the interior of a building. In detached, semi-detached, or row housing, the stack effect will be larger in taller buildings (so three-storey houses are preferred to one- or two-storey houses) and in buildings having a cathedral ceiling on the top floor with a cupola and windows that can be opened at the apex of the roof, as in the traditional southern US house illustrated in Figure 6.20.

Ong and Chow (2003) present a mathematical model of a solar chimney, which is similar to a Trombe wall (Chapter 4, Section 4.1.5) except that it lacks thermal mass and is intended to induce ventilation rather than storing and later releasing heat. They analyse its behaviour as a function of the geometry, solar irradiance and air temperature. The stack effect depends on the average difference in temperature between the inside and outside of the chimney; heating the air as it leaves will have little effect. Gan and Riffat (1998) developed a simple model to predict the airflow in a solar chimney, given pressure-loss coefficients obtained from measurements or from CFD modelling. They compared flows assuming single, double and triple glazing. The latter prevents condensation on, and reverse flow next to, the glazing surface. Chen et al (2003) experimentally investigated the impact of varying the depth of the chimney gap, the chimney height, and the chimney tilt. If the chimney gap increases beyond some optimal size, the net flow decreases due to backflow at the chimney output. As the chimney is tilted away from vertical, the airflow increases up to an angle of 45°, at which point the flow is about 45 per cent greater than for a vertical chimney. This enhancement occurs because the flow is more uniform in a tilted chimney, so the pressure loss is smaller (the optimum angle and relative flow enhancement might depend on the heating intensity).

Khedari et al (2000a) measured midday summer ventilation rates of 8–15 air changes per hour in an experimental one-room school house in Thailand. On a hot day with a peak ambient temperature of 34°C, the indoor temperature is too high (35–37°C) and the flow velocity too small (4cm/s) for comfort. However, on such days, the use of a solar chimney (with restricted flow) in

Figure 6.21 View of two solar chimneys and intervening glazed section with external shading louvres on the south façade of the British Research Establishment office building in Garston (UK)

Photographer: Dennis Gilbert, London

combination with an air conditioner reduces the air conditioner + ventilation fan energy use by about 30 per cent compared to the case without a solar chimney (Khedari et al, 2003).

Solar chimneys are finding particularly wide application in the UK. An outstanding early example is the Building Research Establishment offices in Garston, built in 1982. The south façade consists of a series of partially glazed solar chimneys that alternate with a glazed façade that contains external motorized horizontal shading louvres. A view of the south façade is shown in Figure 6.21, while design details of the solar chimneys and intervening glazed sections are given in Figure 6.22. The solar chimneys are designed to draw air from ducts in the floor structure as well as from the high points in the waveform ceiling shown earlier in Figure 6.12. The solar chimneys also provide shading against low-angle sun from the east and west and provide a support structure for the external louvres. Further information and construction details can be found in Gething (2003). Another notable UK example is the Inland Revenue Centre in Nottingham, built

Figure 6.22 Cross section of the solar chimney in the British Research Establishment offices (left) and of the glazed portion of the façade between the solar chimneys (right)

Figure 6.23 Inland Revenue Office in Nottingham, UK, showing stairwells as solar chimneys

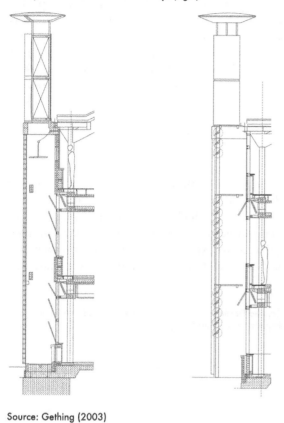

Source: Gething (2003)

Source: Baird (2001) Reproduced by permission of Taylor & Francis Books, UK

Figure 6.24 Solar chimney at the School of Engineering Building at De Montfort University (UK)

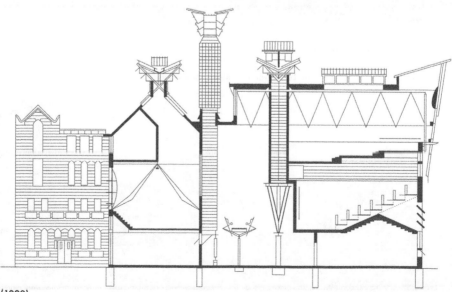

Source: Edwards (1999)

between 1993 and 1995 (Michael Hopkins and Partners, Architect; Ove Arup and Partners, Consulting Engineers). Staircases next to the façade, shown in Figure 6.23, serve as solar chimneys. According to Baird (2001), the building performs as expected in terms of peak indoor temperatures. Other notable UK examples include the School of Engineering Building at De Montfort University (Figure 6.24), designed by Short Ford Associates, London; the library of Coventry University, which combines solar chimneys along the façade with a central solar chimney that also serves as a light well (Smith, 2001); and the Arup Associates headquarters building in Solihull and a student residence in Durham (Hawkes and Forster, 2002).

In Canada, a solar chimney was incorporated into a new building of Sir Sanfred Flemming College (Figure 6.25). In a school in Sweden, passive ventilation was achieved by retrofitting the school with three 6m-high solar chimneys with solar collectors and backup fans (Eriksson and Wahlström, 2002). Large ducts with a minimum of obstructions are helpful, and in the case of the Swedish school, it was deemed acceptable to omit filters from the air intakes. Another Swedish example is Tänga school in Falkenberg (Figure 6.26), designed by Christer Nordstrøm Arkitektkontor AB (www.cna.se) and described in Blomsterberg et al (2002). The stack effect works well in high-rise office buildings, as one might expect, and has been used to induce natural ventilation in the 21-storey Debis Headquarters in Berlin (Herzog, 1996).

A particularly daunting challenge is to incorporate heat recovery in a solar chimney – that is, to extract heat from the rising air that had been heated in the chimney, as it exits, and

Figure 6.25 One of two solar chimneys (assisted with wing forcing) at Sir Sanfred Fleming College, Peterborough, Canada (left) and schematic diagram showing the air intake and exhaust flows under prevailing wind conditions (right)

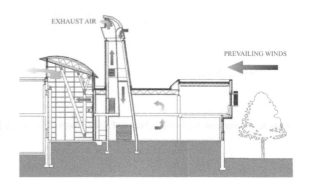

Source: Loghman Azar, Line Architects, Toronto

Figure 6.26 Tänga school in Falkenberg, Sweden. Left: Overview; Right: solar chimney

Source of photographs: Christer Nordstrøm Arkitektkontor AB

to transfer this heat to incoming fresh air. In this way, the solar chimney can be used for ventilation during times when heating rather than cooling is required. Riffat and Gan (1998) and Gan and Riffat (1998) have studied the use of a heat pipe heat exchanger in a solar chimney, both experimentally and theoretically. A heat pipe was chosen among the various heat exchange devices (see Chapter 7, Section 7.4.9) because it entails the smallest flow resistance. The smaller the airflow velocity, the greater the recovery effectiveness and the smaller the pressure drop; at a flow speed of 0.5m/s (at the low end of the feasible range), the pressure drop is about 0.5Pa and the effectiveness 65 per cent for the system studied by Riffat and Gan (1998). Heat recovery reduces the ventilation rate by up to 60 per cent, both by increasing the flow resistance and by cooling the air as it exits the chimney. This can be compensated by building a larger solar chimney, if this is architecturally acceptable.

Cooltowers

The opposite of the solar chimney is the *cooltower*, in which water is pumped into a honeycomb medium at the top of a tower and allowed to evaporate. In so doing, air at the top of the tower is cooled, then falls through the tower and into an adjoining building under its own weight, thereby providing ventilation without fans. This technique is also referred to as *passive downdraught evaporative cooling* (PDEC) and has apparently been used for hundreds of years. It was applied in the new Visitor Centre at Zion National Park, USA (Torcellini et al, 2002), and at the Torrent Pharmaceutical Research Centre in Ahmedabad, India (Ford et al, 1998; Baird, 2001). The Torrent Centre, designed by Short Ford Associates, London, is illustrated schematically in Figure 6.27 and with photographs in Figure 6.28. It contains a central downdraft tower and peripheral solar chimneys. In both cases, relative humidity is low during the warmest period, which permits substantial evaporative cooling, although mechanical air conditioning is still required in two laboratories in the Torrent Centre. Chalfoun (1998) describes other applications of cooltowers in Arizona, South Africa and Saudi Arabia.

In the Torrent Centre, evaporative cooling limits temperatures to 29–30°C on the third (top) floor and to 27°C on the ground floor

Figure 6.27 Schematic illustration of the Torrent Pharmaceutical Research Centre in Ahmedabad, India, showing the central downdraft tower and peripheral solar chimneys

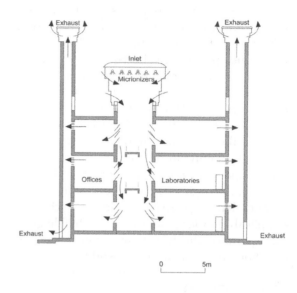

Source: Baird (2001) Reproduced by permission of Taylor & Francis Books, UK

Figure 6.28 The Torrent Pharmaceutical Research Centre in Ahmedabad, India

Source: Baird (2001) Reproduced by permission of Taylor & Francis Books, UK

during extreme outdoor temperatures of 43–44°C, which occur prior to the summer monsoon. The upper-floor temperatures during extreme outside conditions are just within the temperature limits (discussed in Chapter 7 and shown in Figure 7.10) found to be acceptable to most people once psychological adaptation to outside temperatures (i.e. altered expectation) is allowed. Measured average ventilation rates are nine air changes per hour (ACH) on the ground floor and six ACH on the first floor, but much lower on the top floor. Based on measured electricity use, this building uses only 36 per cent of the electricity of a comparable conventional building (Ford et al, 1998). According to the design team, it demonstrates that the passive downdraught technique is applicable to large and complex buildings in many parts of the Indian subcontinent and to other hot and dry areas of the world.

Airflow windows

In Chapter 4 (Section 4.1.7), airflow windows were introduced as a method of passive solar heating. They can also be used, in summer, as a passive cooling device by passing exhaust air through the windows and absorbing some of the solar heat that would otherwise pass through the window (see Figure 4.10). In the TEPCO (Tokyo Electric Power Company) R&D centre, constructed in Tokyo in 1994, this technique reduced the inward heat transfer by one- to two-thirds compared to double-glazed windows with interior or built-in blinds, and eliminated the need for perimeter air conditioning (Yonehara, 1998). The airflow window in this example is combined with monitoring of solar irradiance and the use of computer-controlled blinds, as illustrated in Figure 6.29. The blind-control in turn is coupled to a daylight compensation lighting-control system (automated blinds and daylighting are discussed in Chapter 9, Section 9.3.2). Other examples of computer-controlled blinds inside airflow windows or double-skin façades, programmed to optimize the balance between ventilation, heat gain and daylighting, are found in several of the case studies of 'intelligent' skins presented in Wigginton and Harris (2002).

Erell et al (2004) have proposed an airflow window with clear glass on one side that is sealed and absorbing glass on the side that is open. The window is on a vertically centred hinge, so that it can be flipped from inside to outside. During winter, the absorbing glass is on the inside and it heats interior air, while during the summer the absorbing glass is on the outside, reducing the solar heat gain and heating outside air that is drawn into the airflow window and exits at the top, carrying heat with it. Leal et al (2004) developed a computer simulation module of this window concept for use in whole-building simulation software.

Trombe walls

Trombe walls (Chapter 4, Section 4.1.5) can also serve dual, albeit somewhat different, purposes in winter and summer. Like airflow windows, they can be used to preheat fresh air or reheat recirculated indoor air in winter. In summer, the rising air motion induced by heating of the wall can be used to draw interior air out of the building. Cooler, outside air will enter elsewhere. This can continue into the night, due to the release of heat stored in the mass of the wall during the day, which heats the air next to it while the outside air cools. To be most effective in heating the air between the wall and glazing, double glazing should be used and insulation applied on the inside of the wall (Gan, 1998). The latter will reduce heat loss through the wall during those times in the winter when the air behind the glazing is not heated, and will reduce the inflow of heat into the building during the summer. In climates where the Trombe wall would be used only for cooling, a solar chimney is preferred, because the flow in the latter case exits vertically rather than making a right angle turn in order to exit horizontally. This can increase the flow rate by one-third (Gan, 1998). In addition, there is greater flexibility in the height at which the air exits, allowing the design to make full use of the stack effect (Box 3.4).

Roof solar collectors

Another mechanism to induce natural ventilation is the previously mentioned roof solar collector (Chapter 4, Section 4.1.3), which consists of a solar energy absorbing outer layer, a gap on the order of 14cm, and an insulated lower layer. The lower end of the air gap is connected to the interior air. As the outer layer of the roof and the air in the gap are heated, the air rises, drawing out interior air with it. Experiments on the same one-room school house in Thailand mentioned above

Figure 6.29 Automatic blind control system used in a commercial building in Tokyo

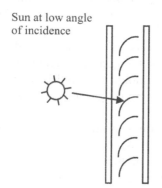

Sun at low angle of incidence

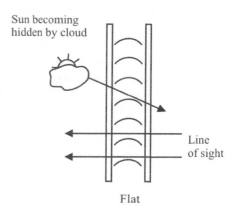

Sun becoming hidden by cloud

Line of sight

Flat

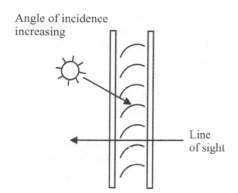

Angle of incidence increasing

Line of sight

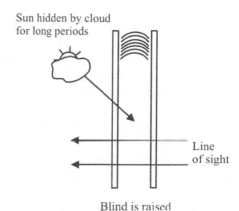

Sun hidden by cloud for long periods

Line of sight

Blind is raised to secure maximum outside views

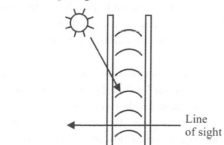

Sun at high angle

Line of sight

To afford outside views as much as possible, slats are controlled at minimum angles required for solar shading

At night

Light

Light reflected by blinds makes lighting more effective

Source: Yonehara (1998)

in the context of solar chimneys, but with a roof solar collector rather than a solar chimney, produced a ventilation rate of 0.08–0.15m³/s per m² of collector area (Khedari et al, 1997), or about 3–6ACH if there is 1m² of collector per 90m³ of house volume. Heat was removed at a rate of 150–300W/m². Interestingly, incorporation of roof solar collectors in the traditional Thai roof architecture – which involves steep, multifaceted gables – gives the largest mass-flow among a variety of roof shapes considered by Hirunlabh et al (2001). Waewsak et al (2003) built and tested a structure in Thailand with a roof solar collector built using traditional clay tiles, except that 15 per cent of the roof area used transparent tiles so that diffuse light could enter the building (the side of the clay tiles facing the air gap was painted white to increase the reflectivity, and aluminium foil was placed on the other side of the air gap). During summer, ACH equalled 13–14 by day and 2–4 by night, while during winter, ACH equalled 5–7 by day and 3–4 by night. Photovoltaic (PV) cladding also functions as a roof solar collector and can be used to induce natural ventilation (Brinkworth et al, 2000). Indeed, electricity from the PV panels can be used to power fans that can double or quadruple the thermally induced rate of ventilation, as in experiments performed by Khedari et al (2002). As noted in Chapter 4 (Section 4.1.3), simulation results by Zhai et al (2005) indicate that a double-pass roof solar collector (illustrated in Figure 4.7) can induce 20ACH under circumstances where a single-pass roof solar collector induces only 12ACH in a traditional Chinese style house.

When a roof solar collector is not employed to induce natural ventilation through the building, the roof itself should still be well ventilated and/or (depending on climate) well insulated in order to minimize the transfer of heat from the roof to the interior of the building.

He et al (2001) report the measured performance of an experimental passively cooled house in the hot and humid climate of southern China. The house features triple-roof and double-wall construction with local building materials. A layer of aluminium foil was installed in the air space within the building envelope to minimize radiant heat transfer (in the same way that low-e film on window panes on the sides facing the air gap minimizes radiative heat transfer across the gap, as discussed in Box 3.2). The house is generally able to maintain peak temperatures 5K colder than the peak outdoor temperature. As the latter sometimes exceed 35°C, mechanical air conditioning is needed to guarantee comfortable conditions, but the amount required is so small that it can be powered through rooftop PV panels.

Allowing natural ventilation on hot and humid days could increase the cooling load, by introducing excess hot and humid air. On the other hand, some outside air must be introduced into a building whether natural or mechanical ventilation is used. Song et al (2002) and Song and Kato (2004) have investigated the net impact on energy use when natural cross-ventilation is permitted in an office building equipped with wall-mounted chilled radiant panels, which provide both cooling and dehumidification. Cross-ventilation is in the upper part of the office, where it can remove heat that rises to the ceiling or is produced by lighting. Wall-mounted radiators that can remove water vapour by inducing condensation are in the lower part of the office, and there is a strong thermal stratification. For an extreme summer day in Japan (30°C, 70 per cent relative humidity), the net effect of natural ventilation on the cooling load is close to zero. The energy benefit is in energy saved by not providing so much mechanical ventilation. During spring and fall, there is a large saving in cooling energy use as well. Clearly, the net impact depends on the climate, characteristics of the building and whether the alternative mechanical cooling system would employ some form of heat recovery on the exhaust air (see Chapter 7, Section 7.4.9).

Natural ventilation could increase heating and cooling loads relative to mechanical ventilation if heat exchangers are eliminated, as would normally be the case with natural ventilation because air inflow and outflow usually occur in different locations. It might be possible in some cases to design a building to be able to fully revert to mechanical ventilation with heat recovery when ambient conditions so merit. As noted above, balanced wind stack systems offer the possibility of heat recovery. Finally, if the incoming air is preheated through an earth pipe loop or by flowing through transpired solar collectors

Figure 6.30 Trickle ventilators

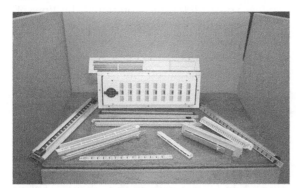

Source: M. White, British Research Establishment, Garston (UK)

Figure 6.31 The glass dome of the Reichstag, Berlin

Source: Author

or double-skin façades, heat recovery would be unnecessary, although solar collectors and double-skin façades would provide preheating only when the sun is shining.

Effective use of natural ventilation requires some control strategy. A common practice is to simply open and close various openings manually on a day-to-day basis or on a fixed seasonal schedule. According to Eftekhari and Marjanovic (2003), the use of automated controls with a negative feedback control loop is difficult with natural ventilation because of the difficulty in defining a representative sensed variable. They propose, instead, a 'fuzzy' rule-based controller. This has rules of the form: IF (Outside temperature is A) AND (Inside temperature is B), THEN Louvre X is Opened (or Closed), where 'A' could be 'very low', 'low', 'average', 'high', or 'very high' (terms which have previously been defined quantitatively), and 'B' could be 'low', 'acceptable', or 'high'.

Some degree of automatic control can be achieved through the use of self-regulating *trickle ventilators*. Trickle ventilators can be constructed with a series of slots having a fixed opening area, or with fixed slots and an exterior flap that moves according to the pressure difference across it. Trickle ventilators are shown in Figure 6.30. In the self-regulating ventilator, the flap moves to create a larger inlet area when the pressure difference falls, thereby maintaining a near-constant airflow. Both kinds of ventilators have an exterior metal mesh to prevent intrusion of dust, rain and insects. Karava et al (2003) present the results of a full-scale experimental investigation of both types of trickle ventilator.

Integrated example: The German Reichstag Building

The refurbished German Reichstag, designed by Norman Foster and Partners and now the new home of the German Bundestag, integrates many of the passive ventilation and cooling concepts described above: a spectacular glass dome, illustrated in Figure 6.31, that functions as a solar chimney and wind suction tower, airflow windows with motor- or manually operated openings, and a heat exchanger at the top to recover heat (Schittich, 2003). Other features of the Reichstag, relevant to other chapters of this book and discussed by Kuehn and Mattner (2003) and Schittich (2003), include displacement ventilation, radiant floor heating and chilled-ceiling cooling, evaporative and desiccant cooling, adjustable external shading, photovoltaics, a central conical daylighting structure inside the glass dome that conveys light into the Chamber of Deputies, seasonal underground storage of excess heat from a biomass-powered cogeneration system in an underground aquifer at a depth of 270–390m, and storage of excess winter coldness for summer cooling in an underground aquifer at a depth of 0–66m.

6.3.2 Double-skin façades for shading and passive ventilation

Details concerning the construction and cost of double-skin façades are given in Chapter 3 (Section 3.4). They normally consist of a second, glass façade constructed over the inner façade of a building. In new high-rise office buildings, the inner façade will normally be a curtain wall, a large portion of which is also made of glass. The Trombe wall is a type of double-skin façade. The benefits of a double-skin façade from a cooling point of view are (i) a shading device can be placed outside the inner façade, where heating of the shading device by the sun does not contribute to heat gain inside the building, and (ii) as air between the two façades is heated, it will rise and can be used to induce passive ventilation of the building, as described above for airflow windows and Trombe walls. In multi-storey buildings, computer-controlled vents can be used to direct the rising air to the outside every one to two stories, which will limit the temperature of the air between the two façades. Deciduous plants have been considered in place of an adjustable shading

device between the two skins (Stec et al, 2005).

Manz (2004) has computed the flow of solar heat through a double-skin façade consisting, from outside to inside, of (i) low-e glass, (ii) an air gap, (iii) a metallized shading screen, (iv) an air gap, and a double-glazed inner façade consisting of (v) low-e glass, (vi) argon fill, and (vii) standard glass. In Case A, the outer low-e glass has a solar transmittance of 0.65 and a solar reflectance of 0.11, while the inner low-e glass has a solar transmittance of 0.35 and a solar reflectance of 0.37. In Case B, the outer and inner low-e glazings are reversed. In Case A with no gaps above or below the shading device but a 10cm high opening to the outside at the top and bottom of the façade for each floor, the solar heat gain (fraction of incident solar energy that passes through the double-skin façade) is 0.065, while in Case B the solar heat gain is only 0.041 – about one-third less. For Case B with air gaps above and below the shading device, the solar heat gain is 0.025 – more than a factor of three lower than Case A without air gaps. In both cases the fraction of solar radiation directly transmitted is the same (0.017), but in Case B less heat is transferred into the building because of the higher reflectance of the outermost glazing, thereby reducing the amount of heat absorbed by the shading device and inner façade, and transferred into the building. This work demonstrates the importance of properly designing glass double-skin façades. Further analysis, involving cases with a combination of fan-driven and buoyancy flows, is presented in Manz et al (2004). Ciampi et al (2003) have analysed various double-skin façades where neither layer is glass.

Gratia and de Herde (2004) compared the heating and cooling loads for a building in Belgium with a south-facing double-skin façade and a building with a normal façade and day and night ventilation for cooling. Heating and cooling loads were computed using a coupled energy balance/computational fluid dynamics model, as described in Appendix D. Compared to the building with a normal façade and ventilation with outdoor air for cooling, the double-skin façade can increase the cooling load. However, day and night ventilation may not be practical in buildings with normal façades, due to exterior noise, dust or the risk of intrusion. The value of a double-skin façade construction, then, is not that it results in the lowest possible energy use, but that it facilitates passive or hybrid ventilation in the first place. Nevertheless, the double-skin façade should be designed to minimize the temperature rise in the façade relative to the ambient air. The temperature rise can reach 12K under central European summer conditions according to Hensen et al (2002), but can be minimized through the choice of an outer glazing with low solar heat gain, the use of reflective (rather than absorbing) shading devices, and by directing relatively cool building exhaust air rather than ambient air into the façade gap during the cooling season (this will also induce ventilation of the building).

Stec and Paassen (2005) simulated the energy use for alternative south-facing façade systems in The Netherlands, including various double-skin façades, integrated with a model of the building HVAC system. The computer model used for the calculations had been validated against observations on an experimental double-skin façade test rig. Results are given in Table 6.5 based on the simulation for one particular year. The cooling loads given in Table 6.5 exclude

dehumidification loads but assume an internal heat gain of 25W/m². Moving the shading device from inside to outside (Cases 1 and 2) alone reduces the cooling load in this climate by about 25 per cent but leaves the shading devices unprotected. This is the result of a 54 per cent reduction in the SHGC when the shading is moved outside (see Table 3.20). Combined with mechanical night cooling, the load is reduced by more than 50 per cent (Case 3), while provision of operable windows with passive daytime and night-time ventilation (Case 4) reduces the load by 85 per cent. The comparable case with a double-skin façade (Case 8) does only slightly better (88 per cent savings), but at much greater expense (Table 3.20). In multi-storey buildings, adjustable external shading and operable windows might not be an option, in which case the double-skin façade provides a significant saving in cooling energy compared to the best feasible alternative – the curtain wall with adjustable shading between glazings (Case 5). Movement of the shading from inside to outside (Cases 1 and 2) increases the heating load. Relative to Case 2, the double-skin façade

Table 6.5 Comparative energy use for alternative south-facing façades in an office building in The Netherlands

Façade system		On-site energy use (kWh/m²/year)			Primary energy use (kWh/m²/year)
Case	Description	Heating	Cooling	Ventilation	
1	DG, interior shading, fixed windows	4.8	74.1	2.94	197
2	DG, external shading, fixed windows	8.4	54.3	2.94	152
3	DG, external shading, fixed windows, mechanical night ventilation	8.1	29.9	3.44	91
4	DG, external shading, operable windows, passive daytime and night ventilation	8.0	10.6	1.76	39
5	TG, internal shading, fixed windows, mechanical night ventilation	5.9	48.6	3.44	136
6	DSF, external shading, fixed windows, mechanical night ventilation	7.3	30.5	3.44	92
7	DSF, external shading, operable windows, passive daytime and night ventilation	7.4	12.6	1.76	43
8	DSF, external shading, operable windows, passive daytime and night ventilation, predictive control	7.1	8.6	2.09	34
9	DSF, external shading, operable windows, passive daytime and night ventilation, predictive control, façade gap serves as supply duct	15.5	10.5	1.46	45

Note: Data computed by Stec and Paassen (2005) based on a computer simulation model that was validated against measurements at a test facility. Primary energy use is computed assuming an electricity generation efficiency of 0.4, and neglecting transmission losses for electricity and natural gas (used for heating).
DG = double-glazed, TG = triple-glazed, DSF = double-skin façade. Additional information is given in Table 3.20.
Source: Stec and Paassen (2005) and W. Stec (personal communication, 2005)

reduces the heating load slightly, except for Case 9, where the gap inside the double-skin façade is used as a supply conduct. This greatly reduces cost (Table 3.20) because some ductwork can be eliminated, but doubles the heating load because of heat losses from the supply air. However, making the outer skin double-glazed (along with or instead of the inner skin), rather than single-glazed (as assumed), could greatly mitigate the increase in heating load. In spite of the increase in heating load, total on-site energy use and primary energy loss for Case 9 decrease by 50 per cent and 77 per cent, respectively, as shown in the last column of Table 6.5. In climates with colder winters, the increase in heating load associated with certain double-skin designs will be a more important consideration. In addition, the relative reduction in cooling load is likely to be smaller in hotter climates, where passive ventilation – facilitated by the double-skin façade – will less often be able to meet the cooling requirements.

Letan et al (2003) discuss the potential use of double-skin façades in multi-storey apartment buildings. A cavity for rising, solar-heated air can be integrated into the south wall, and another cavity for sinking cool outdoor air can be integrated into the north wall. The cool air will enter each storey at floor level on the north side, absorb heat from the interior, and exit at ceiling level on the south side. During winter the system can be used for heating if fresh air enters at the base of the south cavity (rather than at the top of the north cavity) and flows through a roof duct into and down the north cavity, entering each storey at ceiling level on the north side. However, the exhaust should not re-enter the south cavity airflow, as it would be recirculated into other apartments. Instead, the exhaust would have to be collected and vented to the outside, perhaps preheating the air entering the south cavity with a heat exchanger.

Hensen et al (2002) note that passive ventilation in double-skin façade systems depends on many interacting forces, which can often produce highly erratic flows if not properly designed. Simulation models are essential to produce designs that minimize this behaviour. Network models (whereby the airflow is represented by a series of nodes and resistances) can provide much of the required analysis, although there are instances where fluid dynamics models can be

helpful. The work of Faggembauu et al (2003a, 2003b), discussed in Chapter 4 (Section 4.1.8), is also relevant here. Finally, the impact of an adverse wind direction (i.e. wind impinging on the façade from which heated air should be exiting) can be minimized if the double-skin façade extends 5m above the top of the building (Gratia and de Herde, 2004).

Double-skin façades have been widely used in central Europe, which has moderate summers. Cetiner and Özkan (2005) have simulated the impact of double-skin façades in the somewhat hotter climate of Istanbul, Turkey. They find a 33 per cent savings in external cooling loads compared to a double-glazed, single-skin façade with regular glass and no shading control. The double-skin façade is unlikely, however, to be adequate in very hot climates, at least if the inner façade is largely of glass and with minimal thermal mass. That is, we must recognize limits to the appropriateness of buildings with predominantly glass façades. This is borne out by simulations for an office building in the hot-humid climate of Singapore by Hien et al (2005). They find a reduction in total cooling load of about 9 per cent compared to a single-glazed façade, although they also find that the double-skin façade improves the indoor comfort (that is, the energy savings would be larger if the mechanical system in the reference building were sized to meet the same comfort criterion). In hot-arid climates, the traditional concepts of large thermal mass and night-time ventilation are likely to be more appropriate, along with external shading that could be part of a double-skin façade.

6.3.3 Night-time passive and mechanical ventilation

Where the day–night temperature variation is at least 5K but preferably 10K or more, and peak outdoor temperatures are too warm for daytime ventilation beyond the minimal air-quality needs, cool night air can be mechanically forced through hollow-core ceilings, between double-skin walls, or through the occupied space to cool the building prior to the next day. When night-time air is not cold enough but the ground is (i.e. in climates with cold winters), night air could be cooled further by drawing it through an earth pipe (described in Section 6.3.8). Thermally driven passive ventilation and wind induced ventilation, as described in the preceding section, can be used to provide at least some night ventilation but without requiring fan energy. Although passive thermally driven ventilation is driven by temperature differences created by solar heating, these differences will persist to some extent during the night. The diurnal temperature range is largest in hot-dry climates.

Effective use of night ventilation requires:

- a large exposed thermal mass, capable of absorbing heat during the day with minimal temperature gain, and releasing it at night;
- exterior rather than interior insulation, to minimize the heat gained by the thermal mass during the day while exposing it to ventilation airflow during the night;
- minimization of internal heat gains;
- a building designed to maximize stack-induced ventilation;
- a relatively air-tight envelope, so that the air flows as desired;
- automatic control systems in commercial buildings, so that night ventilation does not begin before night-time temperatures have cooled sufficiently to be beneficial, to shut the night-ventilation down if the building will be cooled too much, and to reset ventilation openings at the end of the working day to the proper positions.

The need for external rather than internal insulation stands in contrast to Trombe walls, which are intended to passively induce daytime and night-time ventilation by absorbing and then releasing a large amount of heat. In this case, the best strategy is to apply insulation to the inside of the thermal mass (as discussed in Section 6.3.1).

Thermal mass can play a dual role in night-time ventilation: first, as noted above, by increasing the effectiveness of night ventilation, whether mechanically or passively driven, by storing and releasing daytime heat; and second, by increasing the thermal driving force for passive ventilation through the stack effect by slowing the rate of indoor cooling, thereby creating a larger indoor–outdoor temperature difference than in the absence of thermal mass, and possibly eliminating the need for fans. In the absence of adequate passive driving forces, the energy used by the fans required for night ventilation can exceed the savings in cooling energy use (although peak loads are reduced).

A factor that can limit the applicability of night-time ventilation is the need to avoid producing end-of-night temperatures that are uncomfortably cold, which might be necessary in order to avoid afternoon peaks that are too warm. However, this problem is reduced through the use of high thermal mass, which moderates diurnal temperature extremes in either direction. In residential buildings in hot climates, it can be advantageous to employ greater thermal mass in the walls of the living areas but low thermal mass in the sleeping areas (Tenorio, 2002). In this way, temperature will fall to a lower value during the night in the sleeping area in response to night-time ventilation, while peak daytime temperatures will be more strongly limited in the living areas.

In the following paragraphs, we explore the effectiveness of night-time ventilation in terms of its ability to limit peak daytime temperatures (in the absence of mechanical cooling), and in terms of its ability to reduce energy use and peak loads when mechanical cooling is used as needed to constrain temperature. If daytime peaks are sufficiently limited, and if one is willing to accept occasional periods of warmer-than-desired temperatures, than mechanical cooling equipment can be avoided altogether, with associated savings in capital costs.

Rates of night ventilation and impact on peak temperature without mechanical cooling

Some published examples of the rates of night ventilation used in a variety of different buildings and their impacts are:

- for a high-mass building in California ($0.25m^3$ of concrete per m^2 of floor area), mechanical night-time ventilation at a rate on the order of 30–45ACH is able to limit the maximum indoor temperature to a comfortable 24.5°C on

- isolated days with a peak outdoor temperature of 38°C (which is extremely hot for California) and, more generally, to hold indoor peak temperatures 10K below outdoor peak temperatures (Givoni, 1998b);
- for buildings in Tel Aviv, when a peak outdoor temperature of 29.2°C coincides with a diurnal temperature variation of 7K, passive night ventilation produces 5ACH and reduces the peak indoor temperature 3.2K below the peak outdoor temperature, compared with a 4K reduction with mechanical ventilation at 20ACH (Capeluto et al, 2004);
- based on experiments in and computer simulations of a new building at the Fraunhofer Institute for Solar Energy Systems, Germany, observed passive night ventilation at 3.7–5.3ACH is able to limit peak indoor temperatures to 28.6°C when the outdoor temperature peaks at 29°C in the absence of shading (a reduction of 2–3K compared to the case with no night-time ventilation), and to 25.8°C when shading is used (Pfafferott and Herkel, 2003);
- for a six-storey building in Enschede, The Netherlands that uses night-ventilation and was used to calibrate a computer simulation model, simulations indicate that for Boston weather data, passive ventilation achieves 3–4ACH and is able to limit daily peak indoor temperatures to 26–28°C at the end of a four-day heat wave (peak temperatures of 31°C) (Axley et al, 2002). In this case, mechanical ventilation to 8ACH had little additional effect.

Where air conditioning is still needed by day, external air can be precooled by passing it through the ceiling that has been ventilated at night.

Simulations by Springer et al (2000) indicate that night-time ventilation is sufficient to prevent peak indoor temperatures from exceeding 26°C over 43 per cent of California's geography in houses that include improved wall and ceiling insulation, high-performance windows, extended window overhangs, 'tight' construction, and modestly greater thermal mass compared to standard practice in California. Thus, air conditioners – which account for almost 60 per cent of residential energy use in California – could have been eliminated altogether over almost half of the state while adhering to a rather strict discomfort threshold.

For Beijing, da Graça et al (2002) find that thermally and wind-driven night-time ventilation eliminates the need for air conditioning of a six-unit apartment building during most of the summer (an extreme outdoor peak of 38°C produces a 31°C indoor peak), but there is a high risk of condensation during the day due to moist outdoor air (relative humidity as high as 95 per cent in summer, and usually above 70 per cent) coming into contact with the night-cooled indoor surfaces. A solution would be to close all openings during the day and dehumidify incoming air sufficiently to prevent condensation, which would be easiest with centralized (rather than en-suite) ventilation with desiccant dehumidification (as described in Chapter 7, Section 7.3.8). For Shanghai, night-ventilation is sufficient merely to limit peak indoor temperatures to no warmer than peak outdoor temperatures most of the time (without any ventilation, peak indoor temperatures would exceed peak outdoor temperatures). Theoretical calculations by AboulNaga and Abdrabboh (2000) indicate that a combined roof-wall solar chimney is able to induce night-time ventilation in low-rise buildings in hot and arid climates (where air temperatures drop rapidly at night) sufficient to maintain comfortable daytime conditions in high-mass buildings.

Thermal mass combined with night-time ventilation can be effective in hot-humid climates, depending on the amplitude of the diurnal temperature variation and the absolute peak temperature. Capeluto et al (2004) have simulated the impact of thermal mass and night-time ventilation for various locations in Israel, with daily average relative humidities in August of 71–76 per cent and diurnal temperature variations of 7–10K. Figure 6.32 shows the results for the top floor of a building with minimal insulation (wall

Figure 6.32 Variation of the peak hourly indoor air temperature on an August day in Tel Aviv with a peak outdoor air temperature of 29.2°C as a function of the building thermal mass and the night-time ACH, as simulated by Capeluto et al (2004) for the top floor of a lightly insulated building

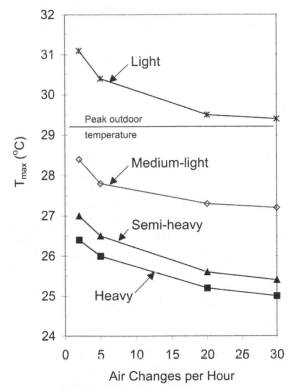

Source: Redrafted from Capeluto et al (2004), American Society of Mechanical Engineers

Figure 6.33 Correlation between the reduction in peak indoor air temperature relative to the peak outdoor air temperature, and the diurnal temperature variation, based on simulations for various cities in Israel by Capeluto et al (2004) for the top floor of a lightly insulated building

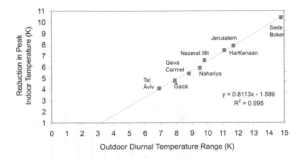

Source: Redrafted from Capeluto et al (2004), American Society of Mechanical Engineers

RSI of 1.32, roof RSI of 1.39) in Tel Aviv. With light construction and no night ventilation, the peak indoor temperature is approximately 2K greater than the peak outdoor temperature, while with heavy construction (concrete floor and ceilings, 20cm concrete block external walls, 10cm concrete block internal walls) and 20ACH at night, the peak daytime temperature is about 4K below the peak outdoor temperature (25.2 versus 29.2°C in this case). The temperature reduction below peak outdoor temperatures is almost perfectly correlated with the diurnal temperature variation, as shown in Figure 6.33.

Energy use

Night-ventilation requires energy to operate fans, while providing some cooling. The ratio of the cooling provided to fan energy use can be thought of as a COP (coefficient of performance), and compared with that of conventional cooling equipment. The fan energy use depends on the extent to which night-time ventilation is at least partly driven by passive forces. In some cases, passive night-time ventilation alone can achieve adequate cooling.

Kolokotroni (2001) systematically investigated the effectiveness of night ventilation in the southern UK by means of a computer simulation model, and found that in conventional buildings (with false ceilings that reduce the exposed thermal mass, 30W/m² internal heat gain, windows with internal blinds covering 60 per cent of the external walls), mechanical night ventilation at rates of 2–10ACH reduces cooling loads by only 3–10 per cent and increases total energy use. Measures to increase the cooling-load savings, and their impact at 5ACH, are as follows:

- heavy-weight construction, 15 per cent savings instead of 5 per cent;
- the above +air tight construction + 30 per cent window area, 20 per cent savings;
- the above + reduction of internal heat gain to 20W/m², 21 per cent savings;
- the above + design to use the stack effect, with 10ACH, 30 per cent energy savings and 40 per cent downsizing of cooling equipment.

A 30 per cent energy saving during the hottest month was also found for a building in Kenya optimized for night ventilation. Where mechanical ventilation is used, there was little additional saving in energy use as the ventilation rate increased beyond 6ACH, and negligible additional savings beyond 10ACH.

The trade-off between increasing fan energy use and decreasing cooling energy use with increasing rates of night ventilation depends on the flow resistance in the given building, the indoor-to-outdoor temperature difference, the thermal mass of the building and its degree of exposure to the night air, and the potential assistance from passive and/or wind-driven circulation. As discussed in Chapter 7 (Section 7.1.2), the energy used by fans varies roughly with airflow rate to the third power, and the fans themselves are a source of heat. Careful analysis is required, as night-ventilation does not always reduce total energy use, even when climatic conditions are favourable. Simulations by Blondeau et al (1997) for one particular building indicate a COP increasing from 4 to 10 (but with large scatter) as the indoor–outdoor temperature difference increases from 4–8K. The lower COP values can be exceeded by vapour-compression chillers (Section 6.5), although there can still be savings in primary energy even if there is no savings in electricity use, because the electricity use is shifted from daytime to night-time, when the efficiency of generating electricity can be larger (see Section 6.9.2). Furthermore, if night ventilation avoids the need for vapour compression cooling equipment altogether, there will be additional benefits through the avoided use and hence leakage of halocarbons (the warming effect of which can be 10 per cent of that associated with supplying electricity to the cooling equipment from coal). Even if there are nights when ventilation rates need to be increased such that the cooling COP is less than for chillers, night-ventilation can be worthwhile if the seasonally averaged COP is superior.

Simulations of the impact of night ventilation combined with design improvements on total heating, cooling and ventilation electricity use in a typical US house in all 16 climate zones of California have been carried out for the California Energy Commission by the Davis Energy Group (DEG, 2004a). The DOE-2 simulation

model was modified and used after calibrating it to measured data from two occupied houses in two climate zones. Results are given in Figure 6.34, which are arranged by climate zone in order of increasing number of cooling degree days (CDD, defined in Chapter 2, Section 2.5). Shown are the CDD values for each zone, the percentage savings compared to the California Title 24 code resulting from design improvements without night ventilation, and the savings resulting from design improvements and night ventilation. The design improvements are (i) installation of a low-e sheathing beneath the roof insulation, (ii) upgrading the windows from vinyl doubled-glazed to vinyl double-glazed with a low-e coating; (iii) use of a single high-efficiency water heater for DHW and space heating (via a heating coil in the air handler) that replaces a standard-efficiency DHW heater and 80 per cent efficient furnace, (iv) upgrading the drywall from ½″ thick to 5/8″ thick, and (v) greater than 50 per cent hard floor surface (to increase thermal mass). The night-ventilation system implemented in the two test houses consists of an air handler with separate heating and cooling coils (for use when needed) with a separate damper designed for residential use that is mounted in the attic above the return-air grill (the damper determines the proportion of outside and return air

Figure 6.34 Comparison of cooling degree day loads in different California climate zones and in the savings in electricity use for cooling, ventilation and air circulation during the cooling season resulting from modest improvements in the design of a typical California house, and resulting from design improvements plus night ventilation

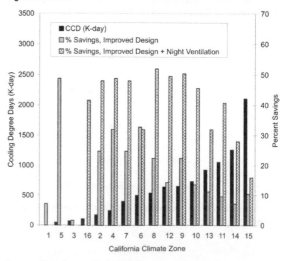

Source: Data from DEG (2004a)

circulated through the house). A user-friendly control system was developed for the project. Despite the very modest design improvements, these alone frequently reduce the afore-mentioned energy use by at least 25 per cent, while the design improvements combined with night ventilation generally reduce energy use by 30–50 per cent.

Where mechanical air conditioning is used in combination with night ventilation, the energy saving from night ventilation depends strongly on the daytime temperature setpoint. For a three-storey office building in La Rochelle, France, Blondeau et al (1997) simulated energy savings due to night ventilation of 12 per cent, 25 per cent, and 54 per cent for setpoints of 22°C, 24°C and 26°C, respectively. More importantly, night ventilation with a 26°C setpoint requires only 9 per cent the cooling energy of the case with a 22°C set point and no night ventilation for this particular building and climate.

A potential problem with night-time ventilation in hot-humid climates is that, although the night-time air may provide a cooling benefit, the moisture content of the building is increased and the added dehumidification requirement during the day may offset the cooling benefit. In this case, Abaza et al (2001) suggest indirect night-time ventilation through an air-to-air heat exchanger. For a house in Virginia (USA), where dehumidification constitutes 55 per cent of the cooling load, indirect night-time ventilation using a heat exchanger with an effectiveness of 70 per cent reduces the cooling energy use by 38 per cent according to computer simulations.

Where passive ventilation is inadequate and fan-energy use for night ventilation would be prohibitive, and where water-cooled chillers with a cooling tower are used, energy use could probably be reduced by circulating cooling-tower water through chilled-ceiling panels at night. The coupling of cooling towers to chilled-ceiling systems is discussed in Section 6.10.2. Since water can be cooled evaporatively to below the air temperature, this would increase the portion of the cooling load that can be satisfied through nocturnal cooling. High thermal-mass elements would be directly chilled, and the pump energy use would be much less than the avoided fan energy use. This is a possibility that remains to be investigated. Another possibility is to increase the effectiveness of night ventilation by

combining a solar-powered adsorption chiller with a solar chimney, as explained in Chapter 11 (Section 11.5.2).

6.3.4 Hybrid natural and mechanical ventilation

Hybrid ventilation systems are systems with primary reliance on wind and buoyancy to provide adequate ventilation, with assistance by fans only as needed. The International Energy Agency's *Energy Conservation in Buildings and Community Systems* Implementation Agreement maintains a website (hybvent.civil.auc.dk) that contains illustrated technical reports on about a dozen hybrid ventilation projects in Europe and Asia, numerous research papers, and conference proceedings. Also available from this website is a comprehensive review of hybrid ventilation systems by Delsante and Vik (2002). The review provides technical information from 22 buildings in 10 countries. Commonly used design elements in these buildings include:

* underground ducts, culverts, or plenums to pre-condition the supply air (in six buildings);
* operable windows and/or ventilation grills, in many cases automatically controlled;
* automatically controlled shading devices;
* solar chimneys or atria, often with backup fans;
* temperature and/or CO_2 sensors.

All the buildings rely on good thermal design and thermal mass with intensive night-time ventilation in order to limit daytime temperatures. The choice of energy-efficient fan systems, whether for night ventilation or for daytime backup, can matter as much or more than the use of passive design features. Many of the buildings have noise attenuators. Some retrofits of purely mechanical systems have been done.

A hybrid mechanical-passive ventilation and cooling system does not necessarily require more energy than a purely passive cooling system, at least for UK conditions (Wright, 1999). This is because, in a hybrid system, greater use of natural light is possible because solar heat gain does not need to be limited as much. The year-round savings in lighting energy use can offset the energy used for mechanical ventilation and cooling summed over those days when it is needed.

Barriers to the use of hybrid ventilation systems have been investigated as part of various European programmes, and the key results are summarized by Delsante and Vik (2002). Among these are the simple fact that design standards and codes in various countries have not been written with the option of hybrid ventilation in mind; this creates uncertainty among designers and developers and thus serves as a barrier to hybrid ventilation. It can also lead to over-design of the system, which reduces the energy savings that would otherwise occur. For example, requirements for filtering have been written for mechanical systems, in which filtering serves to protect fans and other components. The filtering requirement is less in hybrid systems, yet excessive filtering reduces the rate of airflow that can be induced by natural means, thereby increasing the need for supplemental fan power. Another example is heat recovery, which is required in many countries and which also increases the required fan power. In systems in which ventilation air is preheated by flowing through an underground culvert, transpired solar collector, or double-skin façade, heat recovery might not be necessary and so would be redundant. In some countries, minimum airflow rates are clearly excessive (i.e. 16.7 and 16.9 litres/s per person for offices in Germany and Norway, respectively, compared with 7.5 litres/s per person under ASHRAE Standard 62-2001), which requires excessive use of backup fans in hybrid systems.

Fire regulations can also serve as a barrier to hybrid ventilation systems. These regulations generally consist of (i) restrictions on the maximum compartment size, (ii) restrictions on breaks in fire barriers between compartments, and (iii) requirements for the removal of smoke from escape routes. These are barriers to the extent that passive ventilation requires unimpeded airflow between different parts of the building, and may make use of stair wells as solar chimneys or may rely on stack effects in tall buildings, both of which will also facilitate the spread of smoke in the event of fire. These restrictions have been overcome in the case studies examined by

Delsante and Vik (2002) through a combination of careful design and measures, such as the inclusion of water sprinkler systems, that provide alternative ways of satisfying fire regulations but which also increase investment cost. In the case of the 18-storey Federal Building in San Francisco, motorized window actuators supplied from an emergency power circuit are available to close the windows in the event of a fire (McConahey et al, 2002). For manually controlled windows, low-voltage hold-open electromagnets can be used, with gravity-driven closure upon loss of power. If trickle-ventilators are placed at floor level rather than at ceiling level, the amount of smoke leaving the fire floor through the ventilators is unlikely to significantly affect the air quality of adjacent floors. Chow and Chow (2005) discuss further strategies for the reduction of the risk of fire and of the spread of smoke in the event of fire in naturally ventilated atria.

The Panasonic Multimedia Centre in Tokyo, illustrated here in Figure 6.35 and more fully in Ray-Jones (2000), is a particularly interesting (and visually stunning) example of a building with a hybrid ventilation system, particularly since it is located in an area of heavy traffic congestion (so that noise might have been thought of as a problem) and has a large internal cooling load. An atrium creates a solar chimney, through which exhaust air exits the building. Air enters through vents beneath each window, then passes beneath a raised floor, entering the rooms uniformly through floor vents and exiting through slits in the ceiling, where it passes to the atrium (this is a displacement ventilation airflow, a topic discussed extensively in Chapter 7, Section 7.4.5). Fans in the floor assist the airflow when needed, and supplemental mechanical cooling is also available when needed. Ventilation rates are increased at night to cool the concrete slab floors.

Another notable example of a building with a hybrid ventilation system is the mixed office and retail high-rise Eastgate Building in Harare, Zimbabwe, completed in 1996 and shown in Figure 6.36. It is a project of Pearce Partnership Architects and Ove Arup Consulting Engineers, inspired by the natural air conditioning of termite mounds in Zimbabwe. Unlike other office towers in Harare, this building has no mechanical air conditioning. The flow configuration is illustrated

Figure 6.35 Drawing of the Panasonic Multimedia Centre in Tokyo, in which the central atrium serves to induce a displacement ventilation airflow through the offices

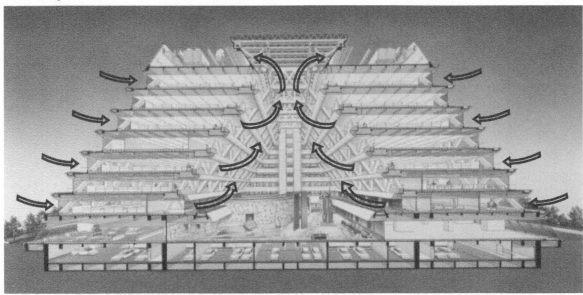

Source of figure: Nikken Sekkei, Japan

Figure 6.36 The Eastgate Building in Harare, Zimbabwe, showing central solar chimneys

Source: Baird (2001) Reproduced by permission of Taylor & Francis Books, UK

in Figures 6.37 and 6.38. Airflow is from a central atrium, underneath the floor of the first office floor, and into a central supply shaft (this is a reversal of the normal flow path with atria, which would be *into* the atrium). Air flows from the central shaft into the hollow-core floor of each office and to the periphery of the office, then passes through the office space, and exits the office at ceiling level next to the supply shaft. The air flows through ducts that traverse the supply shaft, into an exhaust shaft that is within the supply shaft (see Figure 6.38). The airflow is assisted by fans as needed, to give ten air changes per hour at night

and two during the day. For outside temperatures varying diurnally between 12°C and 27°C, office temperatures vary between 19°C and 22°C (Holmes, 2001). It uses 35 per cent less energy than the average consumption of six other conventional buildings in Harare. Other examples include the Tokyo Gas building, Tokyo (Figure 6.39) and Jaer school in Nesodden, Norway, in which passive ventilation is combined with earth pipe cooling (illustrated later).

The design of hybrid ventilation systems is an iterative process, due to the interdependence of the building and HVAC system (Arnold,

Figure 6.37 Passive and fan-assisted ventilation in the Eastgate Building, Harare, Zimbabwe

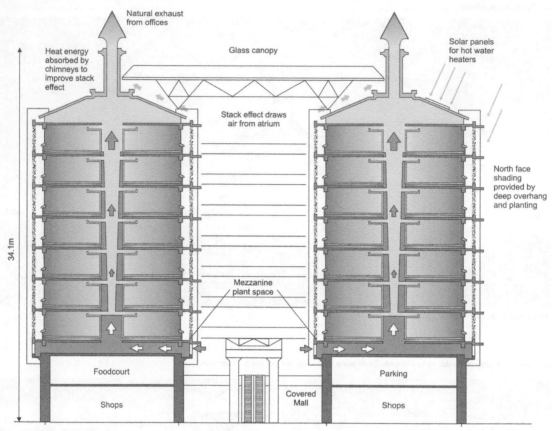

Source: Hawkes and Forster (2002)

Figure 6.38 Detail showing airflow from the central supply air duct, through the hollow-core floor slabs into an office, and then into a central exhaust-air duct in the Eastgate Building

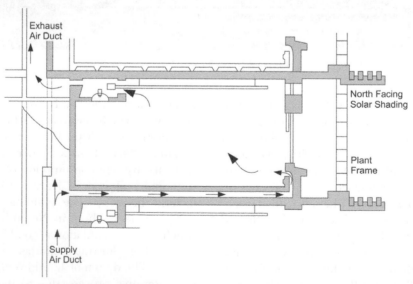

Source: Baird (2001) Reproduced by permission of Taylor & Francis Books, UK

Figure 6.39 The Tokyo Gas (Earth Port) building, Tokyo, designed by Nikken Sekkei Ltd, Architects and Engineers

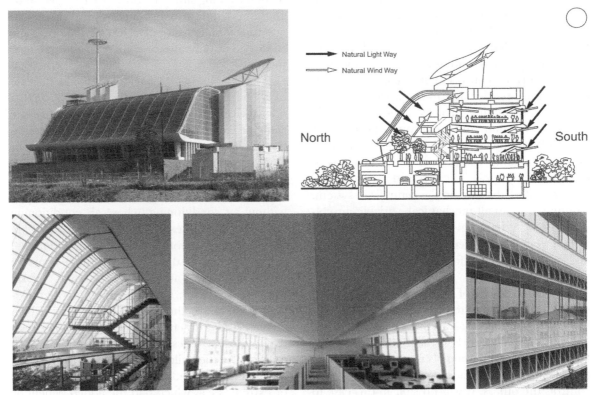

Source: Kato and Chikamoto (2002)

2000). In a conventional building, the building dimensions and materials are specified, the peak heat gain is estimated, and the capacity of the air conditioning equipment is selected to match the peak load. In a hybrid building, an initial design is proposed that includes passive features that reduce peak internal temperatures. Complementary mechanical systems are then selected and sized to reflect the passive features. The original design is then refined in a trial and error approach so as to minimize cost and optimize the passive and active components. This requires the active involvement of all members of the design team. The design of the previously mentioned 18-storey Federal Building in San Francisco, as described by McConahey et al (2002), is a good illustration of the need for an integrated approach to design involving the architects, mechanical and structural engineers, lighting designers, and specialist simulation modellers. This is because some design elements in hybrid systems require alternative arrangements between floor and wall structural elements in order to permit optimal window and

ventilation grill arrangements.

Axley et al (2002) simulated the hybrid ventilation system in a recently completed five-storey office building in Enschede, The Netherlands. Air enters through self-regulating inlet trickle ventilators above the office windows, passes through grills or ductwork into a central atrium, then up and out the top of the atrium. The self-regulating ventilators provide relatively constant airflow over the range of pressure differences commonly encountered, except when wind and buoyancy forces drop to negligible values. The atrium was designed to create the maximum possible suction, and an exhaust fan assists the flow only when needed. Heinonen et al (2002) investigated, through model simulations, the possibility of retrofitting an existing five-storey office building in Helsinki to utilize passive ventilation. A key element is to add an appropriately configured atrium. In one design that was considered, the atrium is connected to controllable openings that are added to the outside walls, as in the Enschede example. In another design, air is drawn in centrally by the atrium suction and distributed through ductwork, with the use of a supply fan as needed. Effective passive ventilation requires that the building be airtight ($ACH_{50} = 0.5$), except for the controllable openings. As noted in other work, complicated and sometimes erratic patterns of airflow can occur in passively ventilated buildings, particularly in the upper stories of multi-storey buildings. For example, air flows in through the atrium and out through the grilles of the 5th-floor offices 50 per cent of the time in the Helsinki building using passive ventilation with decentralized supply. The reversed airflow occurs 12 per cent of the time in the centralized system. Compared to an all-mechanical, variable-air-volume (VAV) system, fan energy use is reduced by 92 per cent for the decentralized hybrid system and by 73 per cent for the centralized hybrid system.

The design of the control system for a hybrid ventilation system is an interesting challenge, as described by da Graça et al (2004) for the aforementioned Federal Building in San Francisco. Here, building occupants control some openings and the building management system controls the others. It was possible in this case to design a control system that is resilient to poor user choices, and where user choices on one side of the building have minimal impact on the conditions on the other side of the building. The key to the control system is to have pressure sensors that determine which side of the building is the windward side and which side is the leeward side, and to orient the control algorithm in terms of windward and leeward sides rather than in terms of fixed sides.

Arnold (2000) gives a real-world example of some of the 'teething' problems associated with a hybrid passive mechanical ventilation system in a new office building in the UK that houses 2200 people with a high level of use of information technologies. These problems generally involved errors in the control software and defects in the control actuators. The problems were exacerbated by an unplanned business need to operate isolated parts of the building for extended hours and in some cases on a 24-hour basis. Because of this and the initial problems, the original building management ignored the original design strategy and operated the building in an air-conditioning mode with heating, cooling, and mechanical ventilation running simultaneously. Over a period of 18 months, the various problems were eliminated and the building restored to its original operating strategy, but only after a change in the management personnel.

Finally, Olsen et al (2004) discuss the case of a four-building complex in Oregon (US) that relies solely on ventilative cooling to moderate summer indoor temperatures, without air conditioning. The occupants tended to be non-profit or environmentally oriented, and thus more disposed to accept a building without air conditioning. The keys to increasing the acceptability of a building without air conditioning (where air conditioning has come to be an expected standard feature) are the occupant control of personal ventilation and the connection to the outside through operable windows that are automatic features of hybrid ventilation systems, and the design of the building to maximize the use of daylighting, which increases the attractiveness of the building relative to those with less daylighting. In the Oregon case, one of the building tenants shifted the work schedule from 9–5:30 to 7–3:30 during a period of extreme heat, which had the added benefit of allowing the employees to be free during more of the summer afternoons.

6.3.5 Acoustical considerations with passive and hybrid ventilation

Increased penetration of outside noise into the building can be a problem with natural ventilation in areas with traffic noise. In addition, exposed heavyweight surfaces (needed for effective thermal mass) will reduce the absorption of sound, and could allow noise levels to build up to unacceptable levels, even in the absence of ventilation paths. Conventional open-plan offices may have highly acoustically absorbent suspended ceilings and carpeted floors, and may rely on the air conditioning system to provide masking noise in order to maintain privacy. General considerations in reducing noise in naturally ventilated buildings are provided in Baker and Steemers (1999), where references to more specialized publications can be found. Information on the sound attenuation provided by various closed and opened windows is given in Table 6.6. Staggered

Table 6.6 Sound attenuation for different windows

	Construction	Mid-frequency sound reduction (dB)
1	3mm glazing, unsealed operable frame	20
2	As in (1) but weather stripped	23
3	As in (2) but double-glazed	25
4	As in (1) but 6mm glazing	27
5	As in (1) with a second pane 150mm away and frame lined with absorbent	49
6	As in (5) but 6mm glazing and 200mm gap	51
7	Typical 225mm masonry wall	54
8	Single casement partially opened window	5–10
9	Double window as in (5) but staggered opening	15–25
10	Acoustic ventilator	47

Source: Baker and Steemers (1999)

Figure 6.40 Influence of window design on the attenuation of outside noise

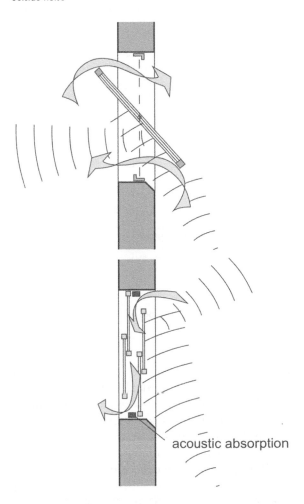

acoustic absorption

Source: Baker and Steemers (1999)

double-window openings provide an attenuation of outside noise (15–25dB) several times that of an unimpeded opening (5–10dB) and comparable to that of a single-glazed unsealed window (Figure 6.40). As discussed in Chapter 3 (Section 3.4.2), double-skin façades with windows tipped open also provide about 20dB of sound insulation. Acoustical absorbers in vents (Figure 6.41) can provide almost five times the sound attenuation of a partially opened window. Barriers placed a short distance in front of the ventilation opening, with absorbent material to absorb sound reflected between that building and the barrier, are another option. A trickle ventilator with acoustically absorbing material can be used in place of an open window. Of course, in some cases external sound will not be a problem, or ventilation openings can be restricted to the quieter side of a building or placed next to interior courtyards. Free-standing absorbent partitions or vertical baffles suspended from ceilings can reduce the internal reflection of sound.

De Salis et al (2002) provide a somewhat more theoretical discussion of sound attenuation. Sound attenuators such as barriers and acoustically absorbing linings inside ducts, or louvres inside tickle ventilators, are most effective at absorbing high-frequency sound, but provide little attenuation of low-frequency sound. However, active noise cancellation can be applied to low-frequency sounds. This involves a microphone near the entrance of a ventilation duct, a microprocessor, and a speaker that produces sound that is out of phase with, and therefore cancels, the incoming sound. Active noise control is widely used to attenuate steady fan noise in building ventilation systems. However, it might not be as easy to apply, or as effective, in attenuating unsteady and intermittent sounds. Nevertheless, Salis et al (2002) hold out the possibility that a 3m inlet duct, lined with absorbent material (to deal

Figure 6.41 A ventilator system providing 47dB of sound attenuation, compared to 10dB for a partially opened window

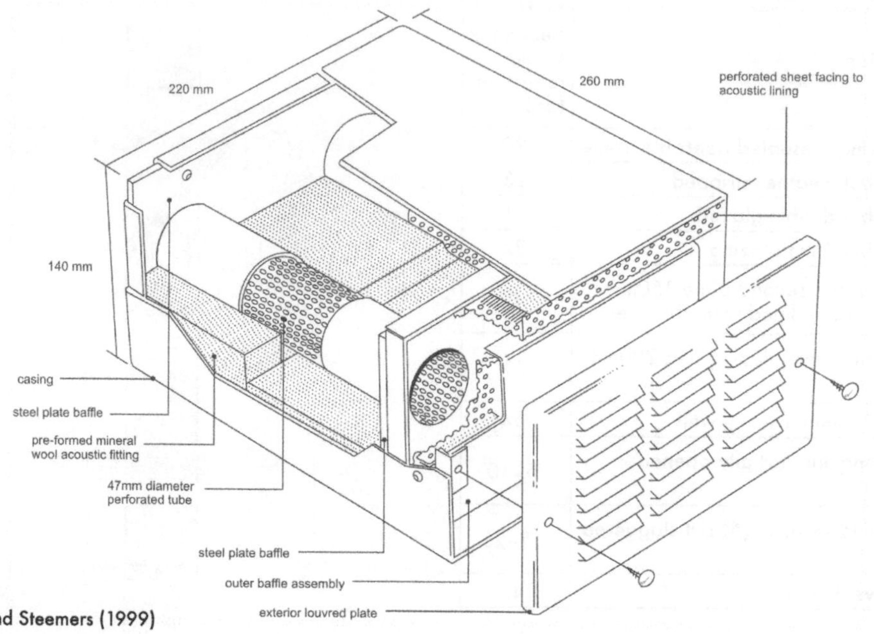

Source: Baker and Steemers (1999)

with high-frequency sounds) and equipped with active sound control (to deal with low-frequency sounds) could be as effective at attenuating sound as a solid masonry wall (an attenuation of 40dB). If only 1m or 2m inlet vents are permitted due to space considerations, attenuations of 23dB and 34dB, respectively, are expected.

6.3.6 Night-time radiative cooling

Givoni (1994) provides a particularly good discussion of radiative cooling techniques, which are most applicable to one- or two-storey buildings. Any painted metallic surface placed over the roof, with an airspace of 5–10 cm beneath it, serves as an effective radiator of infrared radiation (note, from Table 3.1, that unpainted metals generally have a low emissivity and hence will be a poor radiator). An alternative design consists of uncovered copper tubes with aluminium fins (Saitoh, 1999). The temperature drop below the surrounding air temperature that the radiating surface would achieve, in the absence of airflow beneath it, is referred to as the *stagnation temperature drop*. Under clear skies with low relative humidity (which minimizes downward emission of infrared radiation from the atmosphere), it can

reach 5K. In humid regions, a temperature drop of 2–3K can be achieved. It might be possible to achieve greater temperature drops using coatings that have a high emissivity in the atmospheric 'window' but a low absorptivity at other infrared wavelengths.[7] Granqvist (2003) reports a number of materials with these properties. In any case, outside air is mechanically circulated under the radiator at night (being cooled in the process) and then circulated through the building to cool the interior air. Use of a thin (60–100μm) polyethylene film as a screen over the metallic radiator will prevent convective exchange of heat between the radiator and warmer air immediately above it (Mihalakakou et al, 1998). Polyethylene has a very high transmittance for infrared radiation and a very low transmittance for solar radiation, so it does not inhibit loss of infrared radiation and can even sometimes be used for radiative cooling during the day (Al-Nimr et al, 1998). Local building materials, such as clay tiles, can also be used as a radiator if separated from the underlying roof layer by an air gap, but in this case the radiator should be designed to operate as a solar chimney during the day, in order to minimize the storage of heat that needs to be removed during the night (Khedari et al, 2000b).

Radiative cooling is most effective from a flat roof because, in this case, the roof 'sees' only the overlying sky. The net exchange of radiation is given by Equation (B3.2.6) of Box 3.2 times the difference between the roof temperature and the effective radiating temperature of the sky (and using these temperatures to compute T_f). An inclined roof will partially see the warm surroundings and ground. The desire to maximize the radiative cooling conflicts with the possible desire to use the roof as a solar collector at other times, so a trade-off must be made in this case. Radiative cooling is not commonly used, and further research on practical applications is required. The possibility of night-time condensation on the underside of the metal radiator should be evaluated and protected against with a vapour barrier above the roof deck.

In many poor countries in hot climates, corrugated metal roofs are common. As effective radiators, these cool down rapidly at night, but because they are uninsulated and lightweight, they are too hot during the day. Givoni (1998a) suggests centrally hinged interior insulating panels that would be closed during the day and opened during the night during summer (and possibly operated in the opposite sense during winter). To reduce any fire hazard, they could be wrapped in aluminium foil, which would also reduce radiant heat transfer to the interior of the living space.

The HARBEMAN (Harmony Between Man and Nature) house in Sendai, Japan, makes use of night radiative cooling combined with a cold-water storage tank (Saitoh and Fujino, 2001). Weather conditions in Japan during the summer are not suitable for radiative cooling, so sky radiative cooling is used in the HARBEMAN house during the spring to cool water that is stored in a $31m^3$ underground tank, for use during the summer. The sky radiators consist of uncovered copper tubes with aluminium fins. A 600W heat pump is used at nights (when electricity is cheaper) to cool the storage medium down to 4°C while providing space heating and domestic hot water.

6.3.7 Evaporative cooling

As explained in Section 6.1.2, evaporation of water can cool air in contact with the water down to the wet-bulb temperature (T_{wb}). T_{wb} depends on the absolute humidity and the dry-bulb temperature of the air, T_{db}. T_{wb} is always less than T_{db} (unless the air is saturated, in which case the two are equal), but the difference between T_{wb} and T_{db} is greater the lower the absolute humidity. However, the absolute wet-bulb temperature can be lower in humid temperate regions than in hot arid regions because of the lower initial T_{db} in temperate regions. Thus, lower absolute temperatures can often be achieved through evaporative cooling in humid temperate climates than in hot arid climates, even though the cooling effect (the difference between T_{wb} and T_{db}) is smaller.

Although the potential cooling effect of evaporative cooling is greater in arid regions, water is also more limited in such regions. However, as discussed later (Section 6.11), water is also lost through evaporation in conventional cooling-tower based mechanical cooling systems, which are used in arid regions. In addition, where electricity is produced with steam turbines (using fossil fuels or nuclear energy), water is usually used to evaporatively cool the turbine condensers (see Section 6.11.2). Just like the water used to chill the condensers of a chiller, this cooling water is cooled in a cooling tower and is subject to evaporative losses. Thus, on-site evaporative cooling, by reducing the use of electricity for chilling, reduces the water consumption associated with electricity generation whenever the saved electricity is produced with water-cooled steam turbines. To the extent that water is available but in limited supply, the issue becomes the relative efficiency of water use for different cooling systems, something that is discussed in Section 6.11 of this chapter.

The *effectiveness* η of evaporative cooling is defined as the reduction in the temperature of the incoming air as a fraction of the initial difference between the incoming air temperature T_a and T_{wb}. That is:

$$\eta = \frac{T_a - T_f}{T_a - T_{wb}} \qquad (6.6)$$

where T_f is the air temperature after evaporative cooling. In *direct* evaporative cooling systems, water evaporates directly into the air stream to be cooled and the effectiveness is typically 0.8–0.9, but moisture is added to the air (US DOE,

2002a). If this is a problem, the water can evaporate into a secondary air stream that is used to cool the supply air through a heat exchanger, without adding moisture to the supply air. This is referred to as *indirect* evaporative cooling, and η is typically 0.6. The secondary air stream can be either the exhaust air or outside air; if it is the exhaust air, the heat exchanger can be used to preheat the incoming air during the winter. Except in extremely dry climates, most indirect systems require several stages in order to provide sufficient cooling (Tassou and Lau, 1998). However, if indirect cooling is used as a first stage followed by direct evaporative cooling, it is possible to achieve a final temperature *lower* than the initial T_{wb}, that is, to achieve an effectiveness greater than 100 per cent. This is because, for a fixed atmospheric moisture content, T_{wb} is smaller the smaller T_{db}, and the indirect evaporative cooling step serves to decrease T_{db} without altering the moisture content of the air.

Table 6.7 compares extreme ambient conditions for selected cities in the world: the T_{db} that is exceeded only 1 per cent of the time, and the average T_{wb} during these times. Table 6.8 complements Table 6.7 by giving the T_{wb} value that is exceeded only 1 per cent of the time, and the corresponding average T_{db}. Also given in both tables are the supply-air temperature and humidity that can be produced with direct and indirect evaporative cooling, assuming a cooling effectiveness of 85 per cent for direct cooling and 65 per cent for indirect cooling, and for indirect cooling followed by direct cooling. These have been computed using the algorithm given in Box 6.2. With direct cooling, supply-air temperatures are about 18–25°C but relative humidities are rather high (65–90 per cent). Indirect cooling produces higher supply-air temperatures but much lower relative humidities (20–65 per cent), such that the temperature–humidity combination would be acceptable to most people. A combination of indirect and direct cooling produces final temperatures ranging from 0.3K below the ambient T_{wb} (for an ambient RH of 80 per cent) to 2.6K below T_{wb} (for an ambient RH of 9 per cent), but produces a final RH generally greater than 90 per cent, which is unacceptably high. When the direct evaporative cooling step is restricted so that RH does not exceed 80 per cent,

temperatures do not exceed 25°C in most locations for the extreme conditions considered in Table 6.7. Net effectiveness with the RH constraint is as high as 116 per cent. The effectiveness of evaporative cooling can be extended to even the most extreme temperature and humidity conditions if it is combined with liquid-desiccant dehumidification, as explained in Section 6.6.4. A more complete discussion of the climatic conditions under which evaporative cooling can be applied is found in Chapter 7 (Section 7.2.6) in the context of the human perception of thermal comfort.

Having laid out the principles, effectiveness, and climatic applicability of evaporative cooling, the discussion will now turn to specific cooling systems based on evaporative cooling and their associated energy use.

Table 6.7 Drybulb temperature (T_{db},°C) exceeded 1 per cent of the time, average wet-bulb temperature (T_{wb},°C) corresponding to these conditions, and the corresponding relative humidity (RH). Also given are the temperature and RH that would be produced by a direct evaporative cooler with an effectiveness of 0.85, by an indirect evaporative cooler with an effectiveness of 0.65, and by a combined indirect+direct cooler with effectivenesses of 0.65 and 0.85 for the indirect and direct stages, respectively. Temperature and humidity results for the latter case are given with no constraint on RH. The final temperature and overall effectiveness are also given with the RH constrained not to exceed 80% (in which case the overall effectiveness is given). The final column gives the average diurnal temperature variation during the three warmest months of the year

| Location | Ambient conditions | | | Direct cooling | | Indirect cooling | | Indirect+direct cooling | | | | Diurnal ΔT (K) |
| | | | | | | | | Free RH | | RH ≤ 80% | | |
	T_{db}	T_{wb}	RH	T_f	RH	T_f	RH	T_f	RH	T_f	η (%)	
Canada												
Toronto	28.7	20.9	49	22.1	90	23.6	67	20.0	94	21.7	89	11.2
Winnipeg	29.0	19.6	41	21.0	87	22.9	59	18.4	92	19.9	97	11.4
Edmonton	25.6	16.5	39	17.9	87	19.7	55	15.1	91	16.4	101	12.1
Vancouver	23.2	17.6	57	18.4	92	19.6	72	16.8	95	18.6	82	7.8
US												
Miami	32.2	25.1	56	26.2	91	27.6	73	24.5	95	26.6	78	6.3
Atlanta	32.3	23.4	47	24.7	89	26.5	66	22.5	94	24.3	89	9.6
New Orleans	33.1	25.7	55	26.8	91	28.3	73	25.0	95	27.2	79	8.6
Albuquerque	33.9	15.6	10	18.3	75	22.0	21	12.6	82	12.9	114	14.1
Phoenix	42.0	20.9	13	24.1	75	28.3	28	18.3	83	18.8	109	12.8
San Francisco	25.6	16.4	38	17.8	87	19.6	55	15.0	91	16.3	101	9.3
Las Vegas	40.9	18.6	8	21.9	73	26.4	19	15.4	80	15.6	113	13.8
Denver	32.3	15.2	12	17.8	76	21.2	24	12.4	83	12.7	114	14.9
Chicago	31.3	22.8	48	24.1	89	25.8	66	21.9	94	23.7	89	10.9
New York	31.5	22.0	47	24.1	89	25.8	66	21.9	94	23.7	89	8.1
Boston	30.7	21.9	46	23.2	89	25.0	64	20.9	93	22.6	91	8.5
Minneapolis	31.1	21.9	44	23.3	88	25.1	63	20.9	93	22.6	92	10.6
Latin America												
Mexico City	27.9	13.7	16	15.8	79	18.7	29	11.2	85	11.7	113	13.8
Caracas	32.7	28.5	73	29.1	95	30.0	85	28.2	97	30.9	42	7.0
Bogota	20.2	13.5	46	14.5	89	15.8	61	12.4	93	13.6	97	11.5
Lima	28.8	23.2	62	24.0	93	25.2	77	22.6	96	24.8	71	6.4
Manaus	35.1	25.6	47	27.0	89	28.9	66	24.8	94	26.7	88	11.3
Recife	32.7	25.6	56	26.7	91	28.1	74	25.0	95	27.1	78	6.3
Brasilia	30.9	18.2	28	20.1	83	22.6	45	16.4	89	17.5	105	13.0
Sao Paulo	30.9	20.3	37	21.9	86	24.0	56	19.0	91	20.4	98	8.3
Buenos Aires	32.1	22.3	42	23.8	88	25.7	62	21.2	93	22.9	94	12.0
Europe												
Athens	33.0	20.1	29	22.0	84	24.6	48	18.5	90	19.7	103	9.4
Rome	29.8	23.2	57	24.2	92	25.5	73	22.5	95	24.6	79	9.9
Madrid	34.7	20.0	24	22.2	81	25.1	43	18.1	88	19.1	105	16.2
Glasgow	21.6	16.0	56	16.8	91	18.0	70	15.2	94	16.8	86	8.1
London	25.7	17.7	45	18.9	89	20.5	62	16.6	93	18.1	95	9.2
Paris	28.0	19.4	44	20.7	88	22.4	62	18.3	93	19.8	94	10.2
Zurich	26.4	18.1	44	19.3	88	21.0	61	17.0	93	18.4	95	8.9
Berlin	27.9	18.1	38	19.6	87	21.5	55	16.8	91	18.1	100	9.3
Prague	26.8	17.8	41	19.1	87	20.9	58	16.5	92	17.9	98	11.1
Vienna	28.5	19.3	42	20.7	88	22.5	59	18.1	92	19.6	96	9.3

Budapest	30.2	19.9	38	21.4	86	23.5	56	18.6	92	20.0	98	12.2
Warsaw	27.0	19.0	46	20.2	89	21.8	64	18.0	93	19.5	93	11.0
Minsk	25.8	18.0	46	19.2	89	20.7	63	16.9	93	18.4	94	9.6
Kiev	26.7	18.9	47	20.1	89	21.6	64	17.9	93	19.5	92	9.2
Moscow	26.0	18.6	49	19.7	90	21.2	65	17.6	94	19.2	91	8.2
Helsinki	24.1	16.3	44	17.5	89	19.0	60	15.1	93	16.5	97	9.8
Copenhagen	23.2	16.4	49	17.4	90	18.8	65	15.4	93	16.9	92	8.1
Oslo	24.8	16.4	42	17.7	88	19.3	58	15.1	92	16.5	99	8.8
Murmansk	21.2	14.0	44	15.1	89	16.5	59	12.8	92	14.1	99	6.8
Africa												
Casablanca	32.9	21.4	35	23.1	86	25.4	55	20.1	91	21.5	99	11.0
Cairo	36.2	20.5	23	22.9	81	26.0	41	18.5	87	19.5	106	13.3
Niamey	41.2	21.5	16	24.5	77	28.4	33	19.1	85	19.9	108	13.2
Nairobi	28.2	15.8	25	17.7	82	20.1	41	13.8	88	14.7	108	13.5
Harare	29.1	16.3	25	18.2	82	20.8	41	14.3	88	15.2	108	11.7
Johannesburg	27.9	15.6	25	17.4	82	19.9	41	13.6	88	14.5	108	10.4
Cape Town	28.6	19.3	41	20.7	88	22.6	59	18.1	92	19.6	97	8.8
Asia												
Istanbul	29.1	20.8	47	22.0	89	23.7	65	19.8	93	21.5	91	8.5
Ankara	30.2	17.1	25	19.1	82	21.7	42	15.2	88	16.1	107	15.8
Jerusalem	30.2	17.7	28	19.6	83	22.1	45	15.9	89	16.9	106	10.2
Riyadh	43.1	17.8	4	21.6	69	26.7	11	14.1	77	14.1	114	14.0
New Dehli	40.5	22.4	20	25.1	79	28.7	38	20.3	87	21.3	106	12.0
Bombay	34.0	23.3	40	24.9	87	27.0	60	22.2	92	23.8	94	5.2
Madras	37.0	25.2	38	27.0	86	29.3	59	24.1	92	25.8	94	8.1
Lhasa	23.6	10.8	15	12.7	79	15.3	26	8.3	84	8.7	116	11.7
Ulaan Baatar	25.5	14.8	30	16.4	84	18.5	46	13.0	89	14.0	107	9.8
Beijing	32.9	21.8	37	23.5	86	25.7	57	20.5	92	22.0	97	8.7
Seoul	30.1	24.0	60	24.9	92	26.1	76	23.4	96	25.6	74	8.0
Tokyo	31.2	25.1	61	26.0	92	27.2	77	24.5	96	26.8	72	6.2
Sapporo	27.3	21.9	62	22.7	93	23.8	77	21.3	96	23.4	72	6.5
Shang-hai	33.1	27.4	64	28.3	93	29.4	80	27.0	96	29.4	64	6.4
Hong Kong	32.8	26.1	59	27.1	92	28.4	75	25.5	95	27.8	74	4.5
Ho Chi Minh City	34.2	25.2	48	26.6	89	28.4	67	24.4	94	26.3	87	8.2
Singapore	32.3	25.9	60	26.9	92	28.1	76	25.3	96	27.6	73	6.3
Manila	34.1	26.5	55	27.6	91	29.2	73	25.9	95	28.1	79	8.8
Australia												
Alice Springs	38.9	17.7	9	20.9	73	25.1	20	14.6	81	14.7	114	13.7
Cairns	32.1	25.1	57	26.1	91	27.5	74	24.5	95	26.6	78	7.3
Brisbane	30.0	22.4	52	23.5	90	25.1	69	21.6	94	23.5	85	7.6
Sydney	29.4	19.7	40	21.2	87	23.1	58	18.5	92	19.9	97	6.7
Perth	35.1	19.0	20	21.4	79	24.6	36	16.8	86	17.6	108	12.5

Sources: T_{db} and T_{wb} are from ASHRAE (2001b, Chapter 27), while the outputs (including RHs) are computed using the algorithm given in Box 6.2. RHs can also be read from the psychrometric chart (Figure 6.3)

Table 6.8 Wet-bulb temperature (T_{wb}, °C) exceeded 1 per cent of the time, average drybulb temperature (T_{db}, °C) corresponding to these conditions, and the corresponding relative humidity (RH). Also given are the temperature and RH that would be produced by a direct evaporative cooler with an effectiveness of 0.85, by an indirect evaporative cooler with an effectiveness of 0.65, and by a combined indirect+direct cooler with effectivenesses of 0.65 and 0.85 for the indirect and direct stages, respectively. Temperature and humidity results for the latter case are given with no constraint on RH. The final temperature and overall effectiveness are also given with the RH constrained not to exceed 80% (in which case the overall effectiveness is given) in cases where this is possible

Location	Ambient conditions			Direct cooling		Indirect cooling		Indirect+direct cooling			
								Free RH		RH ≤ 80%	
	T_{wb}	T_{db}	RH	T_f	RH	T_f	RH	T_f	RH	T_f	η (%)
Canada											
Toronto	22.2	26.9	66	22.9	94	23.8	80	21.7	96	23.9	64
Winnipeg	20.9	26.8	59	21.8	92	23.0	74	20.2	95	22.2	78
Edmonton	17.7	23.9	54	18.6	91	19.9	69	16.9	94	18.5	86
Vancouver	18.0	22.4	65	18.7	93	19.5	78	17.4	96	19.3	70
US											
Miami	26.1	30.3	71	26.7	95	27.6	84	25.7	97	28.3	48
Atlanta	24.4	30.3	61	25.3	93	26.5	77	23.8	96	26.0	72
New Orleans	26.8	31.3	70	27.5	94	28.4	83	26.4	97	29.0	51
Albuquerque	18.0	28.0	37	19.5	86	21.5	54	16.6	91	17.9	101
Phoenix	23.7	35.4	37	25.5	86	27.8	58	22.5	92	24.1	96
San Francisco	17.2	23.8	51	18.2	90	19.5	67	16.3	94	17.8	90
Las Vegas	21.2	30.4	44	22.6	88	24.4	62	20.1	93	21.8	93
Denver	17.3	26.6	39	18.7	87	20.6	56	16.0	92	17.3	100
Chicago	24.1	29.5	64	24.9	93	26.0	78	23.6	96	25.8	68
New York	24.2	29.3	65	25.0	93	26.0	80	23.7	96	26.0	64
Boston	23.2	28.4	64	24.0	93	25.0	78	22.7	96	24.9	68
Minneapolis	23.4	29.1	62	24.3	93	25.4	77	22.8	96	25.0	72
Latin America											
Mexico City	16.1	23.0	49	17.1	90	18.5	64	15.1	93	16.5	93
Caracas	29.6	31.6	86	29.9	97	30.3	93	29.5	98	32.5	–
Bogota	14.9	18.6	67	15.5	94	16.2	78	14.3	96	16.1	68
Lima	24.0	27.4	75	24.5	95	25.2	86	23.7	97	26.1	36
Manaus	27.6	31.9	72	28.2	95	29.1	84	27.3	97	29.9	46
Recife	26.3	31.2	68	27.0	94	28.0	81	25.9	97	28.3	58
Brasilia	21.3	26.2	65	22.0	93	23.0	78	20.8	96	22.8	68
Sao Paulo	22.3	27.1	66	23.0	94	24.0	79	21.8	96	24.0	65
Buenos Aires	23.8	28.9	65	24.6	93	25.6	79	23.3	96	25.5	65
Europe											
Athens	22.9	29.2	58	23.8	92	25.1	74	22.2	95	24.3	77
Rome	25.4	27.9	81	25.8	97	26.3	90	25.2	98	27.9	–
Madrid	21.0	33.0	33	22.8	85	25.2	53	19.6	91	20.9	100
Glasgow	16.7	20.5	68	17.3	94	18.0	79	16.2	96	18.0	65
London	18.7	23.8	61	19.5	93	20.5	75	18.0	95	19.9	76
Paris	20.3	25.9	60	21.1	92	22.3	75	19.6	95	21.6	77
Zurich	18.9	25.0	56	19.8	91	21.0	71	18.1	95	19.9	83
Berlin	19.2	25.9	53	20.2	91	21.5	69	18.3	94	20.1	86
Prague	18.7	24.7	56	19.6	91	20.8	71	17.9	95	19.7	83
Vienna	20.3	26.8	55	21.3	91	22.6	71	19.5	95	21.4	83
Budapest	20.6	29.2	45	21.9	89	23.6	63	19.6	93	21.2	92

Warsaw	19.9	25.3	61	20.7	92	21.8	75	19.2	95	21.2	76
Minsk	18.7	24.1	60	19.5	92	20.6	74	18.0	95	19.8	79
Kiev	19.9	25.0	62	20.7	93	21.7	76	19.3	96	21.2	73
Moscow	19.5	24.6	62	20.3	93	21.3	76	18.9	96	20.8	74
Helsinki	17.6	22.4	62	18.3	93	19.3	75	16.9	96	18.8	75
Copenhagen	17.4	21.7	65	18.0	93	18.9	78	16.8	96	18.7	70
Oslo	17.4	21.9	64	18.1	93	19.0	77	16.8	96	18.6	73
Murmansk	15.0	19.9	59	15.7	92	16.7	72	14.2	95	15.8	82
Africa											
Casablanca	22.7	30.1	53	23.8	91	25.3	70	21.9	94	23.8	84
Cairo	23.6	30.4	56	24.6	91	26.0	73	22.9	95	25.0	79
Niamey	26.1	34.4	52	27.3	90	29.0	70	25.4	94	27.5	83
Nairobi	18.2	23.2	61	19.0	93	20.0	75	17.5	95	19.4	76
Harare	19.6	24.2	65	20.3	93	21.2	78	19.0	96	21.0	69
Johannesburg	18.1	24.2	55	19.0	91	20.2	70	17.3	94	19.0	85
Cape Town	20.5	26.4	58	21.4	92	22.6	74	19.8	95	21.7	79
Asia											
Istanbul	22.5	26.7	69	23.1	94	24.0	82	22.1	97	24.3	56
Ankara	17.8	28.1	35	19.3	86	21.4	53	16.4	91	17.6	102
Jerusalem	20.5	26.3	59	21.4	92	22.5	74	19.8	95	21.7	78
Riyadh	19.7	36.3	20	22.2	79	25.5	37	17.5	86	18.4	108
New Dehli	27.6	32.6	68	28.4	94	29.4	82	27.2	97	29.8	56
Bombay	27.4	31.3	74	28.0	95	28.8	85	27.1	97	29.8	39
Madras	27.9	32.0	73	28.5	95	29.3	85	27.6	97	30.3	42
Lhasa	12.6	20.0	41	13.7	88	15.2	56	11.3	92	12.4	102
Ulaan Baatar	16.1	23.0	49	17.1	90	18.5	64	15.1	93	16.5	93
Beijing	25.4	29.0	75	25.9	95	26.7	86	25.1	97	27.6	38
Seoul	25.8	28.1	83	26.1	97	26.6	91	25.6	98	28.3	
Tokyo	26.1	30.1	73	26.7	95	27.5	84	25.8	97	28.3	44
Sapporo	22.6	26.5	71	23.2	95	24.0	83	22.2	97	24.5	51
Shang-hai	27.7	31.9	72	28.3	95	29.2	85	27.4	97	30.0	44
Hong Kong	27.3	30.7	77	27.8	96	28.5	87	27.0	98	29.7	28
Ho Chi Minh City	26.9	31.9	68	27.6	94	28.6	81	26.5	97	29.0	58
Singapore	22.1	30.7	47	23.4	89	25.1	65	21.2	93	22.9	90
Manila	27.9	32.3	71	28.6	95	29.4	84	27.6	97	30.2	47
Australia											
Alice Springs	22.2	28.2	59	23.1	92	24.3	75	21.6	95	23.6	76
Cairns	26.6	30.5	73	27.2	95	28.0	85	26.3	97	28.9	41
Brisbane	24.4	27.9	75	24.9	95	25.6	86	24.1	97	26.6	38
Sydney	22.3	26.2	71	22.9	95	23.7	83	21.9	97	24.2	52
Perth	21.2	29.7	47	22.5	89	24.2	65	20.2	93	21.9	91

Sources: T_{db} and T_{wb} are from ASHRAE (2001b, Chapter 27), while the outputs (including RHs) are computed using the algorithm given in Box 6.2. RHs can also be read from the psychrometric chart (Figure 6.3)

BOX 6.2 Determination of the output of direct, indirect, and indirect–direct evaporative coolers

The outputs of direct and indirect evaporative coolers given in Tables 6.7 and 6.8 were computed from the equations given below, using the ambient drybulb temperature (T_{db}) and wet-bulb temperature (T_{wb}) given in these tables as inputs. All equations are taken from ASHRAE (2001b, Chapter 6), except where indicated otherwise.

Given T_{db} and T_{wb}, the saturation vapour pressures $e_s(T_{db})$ and $e_s(T_{wb})$ in Pa are computed using:

$$\ln e_s(T) = C_1/T + C_2 + C_3 T + C_4 T^2 + C_5 T^3 + C_6 \ln T \tag{B6.2.1}$$

where $C_1 = -5.8002206 \times 10^3$, $C_2 = 1.3914993$, $C_3 = -4.8640239 \times 10^{-2}$, $C_4 = 4.1764768 \times 10^{-5}$, $C_5 = -1.4452093 \times 10^{-8}$, $C_6 = 6.5459673$. Next, the saturation humidity mixing ratios $W_s(T_{dp})$ and $W_s(T_{wb})$ (kg water vapour/kg dry air) are computed using:

$$W_s(T) = 0.62198 \frac{e_s(T)}{P_a - e_s(T)} \tag{B6.2.2}$$

where P_a is the atmospheric pressure (taken to be 101350Pa). The mixing ratio W, degree of saturation μ, and relative humidity RH are computed from:

$$W(T) = \frac{(2501 - 2.381 T_{wb}) W_s(T_{wb}) - 1.006(T - T_{wb})}{2501 + 1.805 T - 4.186 T_{wb}} \tag{B6.2.3}$$

$$\mu = \frac{W(T_{db})}{W_s(T_{db})} \tag{B6.2.4}$$

and

$$RH = \frac{\mu}{1 - (1 - \mu)e_s(T_{db})/P_a} \tag{B6.2.5}$$

respectively (temperature inputs to Equation (B6.2.3) are in °C).

During direct evaporation cooling, the air parcel follows a path along a line joining the points $(T_{db}, W(T_{db}))$ and $(T_{wb}, W(T_{wb}))$ on the psychrometric chart. The proportion of the distance between the points that is travelled is equal to the effectiveness of direct evaporative cooling, η_{dir}.

Thus, the final temperature and mixing ratio are given by:

$$T_f = T_{db} - \eta_{dir}(T_{db} - T_{wb}) \tag{B6.2.6}$$

and

$$W_f = W(T_{db}) - \eta_{dir}(W(T_{db}) - W(T_{wb})) \tag{B6.2.7}$$

(evaluation of W_f using Equation (B6.2.3) with T_f as input yields identical results). The resulting RH is computed from Equations (B6.2.4), and (B6.2.5) using $W(T_f)$ and $e_s(T_f)$ as inputs. Computation of the result of indirect evaporation cooling is identical to the above procedure, except that $W_f = W(T_{db})$ and the effectiveness of indirect evaporation cooling, η_{ind}, is used in place of η_{dir}.

In order to calculate the result of direct evaporative cooling following an indirect cooling stage, T_{wb} of the airstream emerging from the indirect evaporative stage (T_{wbf}) must be known and used in place of the original T_{wb} (along with T_f from the indirect stage in place of the original T_{db}) in the algorithm given above for direct evaporative cooling. T_{wbf} can be estimated as the temperature at the intersection of the lines formed by the slope of W_s versus T evaluated at the dewpoint temperature (T_{dp}) and the constant-T_{wf} line passing through T_f (this makes use of the fact that T_{wb} lies at the intersection of the constant-T_{wf} line and the W_s curve, as seen from Figure 6.3). These are the lines with slope magnitudes S_1 and S_2 in Figure B6.2. T_{dp} (°C) is computed from:

Figure B6.2 Finding the wet-bulb temperature (T_{wbf}), given the final drybulb temperature (T_f) and the dewpoint temperature (T_{dp}).

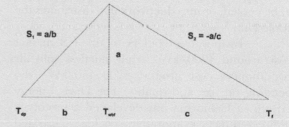

$$T_{dp} = C_7 + C_8\alpha + C_9\alpha^2 + C_{10}\alpha^3 + C_{11}(e_a)^{0.1984} \quad \text{(B6.2.8)}$$

where $C_7 = 6.54$, $C_8 = 14.526$, $C_9 = 0.7389$, $C_{10} = 0.09486$, $C_{11} = 0.4569$, $\alpha = \ln(e_a)$ and e_a is in kPa and is given by:

$$e_a = \frac{P_a W_f}{0.62198 + W_f} \quad \text{(B6.2.9)}$$

with P_a now in kPa. S_1 is computed as:

$$S_1 \approx 0.62198\frac{de_s(T_{dp})/dT}{P_a} \quad \text{(B6.2.10)}$$

where

$$\frac{de_s(T)}{dT} = 0.1*(C_{12} + T(C_{13} + T(C_{14} + T(C_{15} + T(C_{16} + T(C_{17} + TC_{18})))))) \quad \text{(B6.2.11)}$$

de_s/dT is in units of kPa/K, P_a is in kPa, T is in °C, and $C_{12} = 0.4438099984$, $C_{13} = 0.02857002636$, $C_{14} = 7.93805404 \times 10^{-4}$, $C_{15} = 1.215215065 \times 10^{-5}$, $C_{16} = 1.036561403 \times 10^{-7}$, $C_{17} = 3.53242181 \times 10^{-10}$, and $C_{18} = -7.090244804 \times 10^{-13}$.

Equation (B6.2.10) follows from the definition of W_s, while Equation (B6.2.11) is from Lowe (1977). Assuming a constant-T_{wb} line to have exactly the same slope as a constant-enthalpy line, it follows from Equation (6.5) that $S_2 \approx c_{pa}/L_c = 0.000402/$K. From elementary geometry it can be shown that the distance b in Figure B6.2 is given by:

$$b = \frac{T_f - T_{dp}}{1 + S_1/S_2} \quad \text{(B6.2.12)}$$

T_{wbf} is estimated from

$$T_{wbf} = T_{dp} + b \quad \text{(B6.2.13)}$$

To refine this estimate, S_1 is re-evaluated at the midpoint between T_{wbf} and T_{dp}, from which a new b and a new T_{wbf} are computed. This process is repeated until the change in T_{wbf} over one iteration is less than 0.1K. This usually requires 4–5 iterations.

Evaporative coolers

Direct and indirect evaporative coolers are illustrated schematically in Figure 6.42. In a direct evaporative cooler, the air flows through some porous, wetted medium. An indirect evaporative cooler requires some means of exchanging heat between the supply airstream and the cooling airstream, and this can be done by having the supply and cooling air flow through narrow gaps between thin plates and alternating with one another. Two companies manufacture residential evaporative coolers in the USA, and three models are illustrated in Figure 6.43. Energy is required to operate the fans, which draw outside air through the evaporative cooler and directly into the space to be cooled, or into ductwork that distributes the cooled air.

Evaporative coolers in the US market cost around $200/kW$_c$, while ductless split air conditioners cost around $200–450/kW$_c$, depending on size and quality. The Davis Energy Group in California has developed a prototype indirect–direct evaporative cooler, illustrated in Figure 6.44a, that should be able to reduce costs by 30 per cent (DEG, 2004b). The key to cost reduction is the use of plastic plates with

Figure 6.42 Schematic illustration of a direct evaporative cooler (upper) and an indirect evaporative cooler (lower)

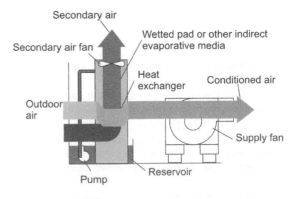

Source: Kinney (2004)

Figure 6.43 Photographs of evaporative coolers from Adobe
Air: Top, Arctic Circle rooftop, ducted; Middle, Alpine window-
mounted; Bottom: Wisper Cool portable

Source: Product brochures at www.adobeair.com

moulded-in spacers, snaps, and air diverters that
provide structure and define the flow paths. The
plates in the indirect-cooling module are shown
in Figure 6.44b. Each plate is folded in half, with
the space between the folded halves serving as the
channels for dry outside air, and the spaces between folded plates serving as the channels for the
secondary airflow (a mixture of exhaust air and a
portion of the indirectly cooled incoming outside
air). About 70 per cent of the indirectly cooled
outside air passes to the direct-cooling module
and from there into the building as ventilation air,
with the remainder directed into the wet passages
of the indirect cooling module. Because the air
entering the wet portion of the indirect module
is cooler than the outside air, its T_{wb} is lower, so
the indirect module can produce greater cooling
of the air in the dry passages. This serves as a
positive reinforcement of the cooling capacity.
Table 6.9 provides measured performance data.
It can be seen that the indirect stage alone produces temperatures that are within the realm of
acceptability if psychological adaptation is accounted for (see Chapter 7, Section 7.2.1), particularly if high-performance windows are used
to maintain a low radiative temperature. The
direct stage produces temperatures much cooler
than needed, with a high relative humidity; being
able to control the unit to produce about half the
cooling and humidification with the direct stage
would be ideal. The COP ranges from approximately 12 when the fan is operating at high speed,
to about 40 at low speed. Simulations for a house
in a variety of Californian climate zones indicate
savings in annual cooling energy use of 92–95 per
cent, while savings are somewhat less (89–91 per
cent) for a modular school classroom. In humid
climates the energy savings would be much less.
However, in humid climates a better approach is
to enhance the evaporative cooling capacity using
liquid desiccants, as explained in Section 6.6.4.

Some evaporative coolers create a net inflow of outside air into the building, which must
be exhausted somewhere. Zuluaga and Griffiths
(2004) suggest directing some of the required exhaust through a dynamic ceiling, which can reduce the ceiling heat gain to zero, thereby amplifying the cooling benefit of the evaporative cooler
(dynamic ceilings are discussed in Chapter 3, Section 3.2.4).

Figure 6.44 (a) Illustration of the residential indirect–direct evaporative cooler developed by Davis Energy Group, and (b) detail showing the folded heat-exchanger plates and the dry and wet air flows in the indirect module

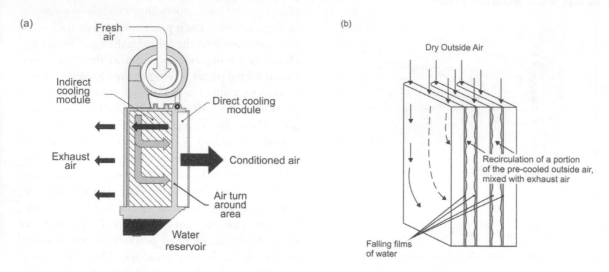

Source: (part a): DEG (2004b).

Table 6.9 Measured operating characteristics of a prototype indirect-direct evaporative cooler for residential and small commercial buildings

	Fan speed		
	High	Medium	Low
Power consumption, W	529	289	81
Airflow to conditioned space, litres/s	732	590	354
Airflow to wetted channels, litres/s	294	226	118
Cooling output, W	6054	5027	3261
Water consumption (litres/hr)[a]	7.4	5.6	4.3
COP	11.6	17.4	40.3
Entering air conditions			
T_d (°C)	40.4	39.8	40.2
T_{wb} (°C)	21.6	21.7	22.8
RH (%)	18.0	19.6	22.2
Between-stage conditions			
T_d (°C)	30.6	29.4	29.2
T_{wb} (°C)	18.3	18.5	19.6
RH	31.0	34.9	40.8
Output conditions			
T_d (°C)	19.9	19.9	20.4
T_{wb} (°C)	18.3	18.5	19.6
RH	86.4	87.8	92.3
Cooling effectiveness			
Indirect cooling unit	0.52	0.57	0.63
Direct cooling unit	0.87	0.87	0.91
Overall cooling unit	1.09	1.10	1.13
Water effectiveness (GJ/m³)	0.78	0.85	0.72

[a] Increase by about 50% to account for periodic flushing so as to prevent mineral buildup.
Source: DEG (2004b)

Passive downdraught evaporative cooling

Passive downdraught evaporative cooling (PDEC) in cooltowers was introduced in Section 6.3.1 as a means of passive ventilation, but of course it is also a means of passive cooling. Cooling is achieved more simply if water is allowed to fall along nylon lines (placed about 10mm apart in two or three staggered rows) in the path of an incoming horizontal airflow, producing a temperature drop of up to 10K for hot (35°C) and dry conditions (Giabaklou and Ballinger, 1996).

Figure 6.45 shows the conceptual design of a PDEC/night-ventilation system that has been developed for a multi-storey office building to be built in Seville (Spain) (Lomas et al, 2004). During the day, the ambient T_{db} and T_{wb} rise to as high as 38°C and 23°C, respectively, but during these hot spells, T_{db} generally drops to 16°C or lower at night. Thus, a combination of night-time ventilation with ambient air and PDEC during the day can satisfy the cooling requirements in this climate. In PDEC mode, hot air enters the upper part of an atrium through inlets that are protected from adverse winds. A fine mist of water is injected into the hot air by micro-ionisers,

at a height well above the top floor to be cooled. The temperature in the capture zone is controlled by varying the rate at which water is supplied to the micro-ionisers, while the rate of airflow is controlled by varying the size of the openings between the capture zone and occupied areas, using a building management system that senses the space temperatures. Air exits the building through outlets that are protected from adverse wind effects by a second façade.

Rooftop evaporation

Night-time evaporative cooling from water on roofs can be used very effectively to reduce cooling loads. In a low-rise building in California, this technique reduced annual cooling energy use by 73 per cent and reduced pump and fan energy use by 66 per cent (Bourne, 2001). An alternative to evaporative enhancement of the cooling is to simply circulate water through a flat-plate radiator as part of a closed circulation loop (Erell and Etzion, 1996). This enhances the cooling power above that obtained by circulating air under a radiator, because the temperature of water falls more slowly than that of air as it loses heat, thereby maintaining a higher average emission of infrared radiation. The same collector can be used for winter heating by circulating the water during the day rather than at night. Meir et al (2002) analysed a system consisting of a 1cm thick polymer panel that is an integral part of the roof, displacing conventional roof cladding. The panel is filled with 2–5mm clay granules so as to increase the rate of heat transfer between the water flowing through the panel and the surface of the panel. A 50m² panel is connected to a 2m³ water reservoir, and used for cooling a house in Oslo with a floor radiant chilling system (which requires water temperatures of 20–26°C). During August, when peak outdoor air temperatures can reach 35°C, this system provides 50–100 per cent of the daily required cooling.

Another alternative is the 'skytherm' system, in which thin plastic bags of water are placed on a flat roof. During the summer, moveable insulation panels are placed over the bags during the day and removed at night. The opposite happens during winter. The tracks and supports for the insulation panels should be designed such that the system forms a tight assembly when closed. Hay

Figure 6.45 Conceptual design of a cooling system for a multi-storey office building in Seville (Spain), using night-time ventilation with exposed thermal mass and daytime passive downdraught evaporative cooling

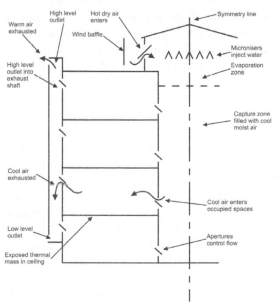

Source: Lomas et al (2004)

and Yellott (1969) first tested this concept on a test room in Arizona, and Niles (1976) reported the performance of the system in a full-scale house in California. In a more recent example from Iran, this system reduced heating demand by 86 per cent and cooling load by 52 per cent (Raeissi and Taheri, 2000). It might be practical to automate such a system.

Free cooling with cooling towers

For the sake of completeness here, it should be mentioned that direct use of the cooling water from a cooling tower in a radiant cooling system is a form of evaporative cooling. It is discussed in Section 6.10.2, as part of a broader discussion of cooling towers.

6.3.8 Underground earth pipe cooling

Outside air can be drawn through a buried coil and cooled by the ground, which, in summer, will be cooler than the air (the same pipe can be used for pre-heating ventilation air in the winter). As discussed in Chapter 5 (Section 5.3.1), the amplitude of seasonal temperature variation decreases exponentially with increasing depth, so that at a depth of 3–5m, the ground temperature will be close to the mean annual surface temperature year round. The lag between minimum ground temperature and minimum surface temperature also increases with increasing depth, with a lag of 180 days (i.e. coldest ground temperature during the warmest part of the year) occurring at a depth of 4–9m (depending on the soil thermal properties).

An example of an air intake and the associated ground coil (during installation) is given in Figure 6.46. The ground coil can range from a few centimetres in diameter (as in Figure 6.46) to

2m in diameter, as in the 250m long corrugated metal pipe that is buried at a depth of 10–12m at the Mercedes Forum in Stuttgart (Thierfelder, 2003, p34), or the concrete pipes at the Jaer school in Norway, illustrated in Figure 6.47. Large-diameter pipes, while more expensive, largely eliminate air resistance and the need for fan power beyond that already required by the ventilation system.

The performance of underground earth pipe cooling can be characterized in terms of peak indoor temperatures compared to peak outdoor temperatures, and by the ratio of the rate of heat removal by the ground to the extra power used by the fans – analogous to the COP of a heat pump. In an experimental house in India, such a system limited the peak indoor temperature to 30–32°C with peak outdoor air temperatures of 42°C, and had a COP of 3.35 (Sawhney et al, 1999). Outdoor relative humidity ranges from about 10 per cent to 30 per cent, resulting in indoor relative humidities of 25–60 per cent. In another experimental study, also from India, the indoor temperature fluctuates between 24 and 31°C as the outdoor temperature fluctuates between 31 and 48°C (Thanu et al, 2001). The measured COP during the summer was 7.9, but much smaller during the monsoon, when the ambient temperature was closer to the ground temperature. By combining earth pipe cooling with solar chimneys or measures to exploit wind suction (Section 6.3.1), it is possible to eliminate the need for fans altogether, at least at some times. By combining earth pipe cooling with liquid desiccant dehumidification (Section 6.6.4), humidity can be reduced with minimal flow resistance.

Underground earth pipe loops have been used in many of the buildings designed by Nikken Sekkei, a multidisciplinary Japanese firm of 1700 planners, architects and engineers whose roots go back over 100 years (Ray-Jones, 2000). Many of these buildings also include rooftop solar collectors and/or atria designed to function as a solar chimney and to induce natural ventilation through the building. An example is the Hokkaido International Centre (Obihiro City), which is served by 230m of 60cm-wide earth pipe. The pipe loop is able to boost outside air from −10°C to −3°C, or to cool the outside air from 34°C (the warmest ever encountered) down to 22°C. Between this and solar air collectors, heating energy use is reduced by about 40 per cent and

Figure 6.46 Earth pipe cooling showing the underground pipes (left) and air intake (right)

Source: www.advancedbuildings.org

Figure 6.47 Jaer school in Nesodden, Norway, showing the solar chimney (top), air intake (bottom left), underground culvert for precooling (summer) or preheating (winter) the air (bottom middle), and fans where the underground culvert enters the basement (bottom right), which are used to assist the flow only when needed (which has rarely been the case)

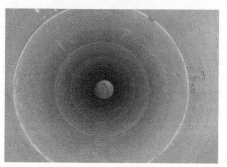

Source: Schild and Blom (2002)

cooling energy use by about 30 per cent.

Al-Ajmi et al (2002) and Hanby et al (2005) have investigated the cooling that can be achieved for an earth pipe in Kuwait, where the mean annual air temperature is 26°C but average afternoon summer air temperatures reach 45°C. For a cooling tube 0.4m in diameter, 60m long, and buried to a depth of 4m, the annual cooling energy use for a residential dwelling is reduced by 30 per cent and the peak cooling load is reduced by about 40 per cent.

In an advanced house in Rottweil, Germany, an underground loop is used in conjunction with a heat recovery ventilator with the following modes of operation (Erhorn et al, 2003b):

<8°C use of ground loop and heat exchanger;

8–15°C bypass of ground loop, use of heat exchanger;

15–25°C fresh air directly supplied to the living space;

>25°C cooling through the ground loop, bypass of heat exchanger.

Two final examples are the Fraunhofer Institute for Solar Energy Systems in Freiburg, Germany, which houses 300 employees and has a 700m long, 25cm diameter cooling loop installed 6m below the ground surface (Reinhart et al, 2000), and the Microelectronics Park in Duisburg, where the atrium was designed to draw outside air through an earth pipe (Compagno, 1999).

In all of the above examples, the ground is used to cool incoming ventilation air.

Alternatively, a fluid could be circulated between the interior space and the ground, where it would be cooled. The fluid would then be able to extract heat from the building when it returns to the building. In an experimental building in Italy, the measured ground loop COP is 5.2 if night-ventilation is not also used, and 4.6 if night ventilation is used (Solaini et al, 1998). The COP is lower in the latter case because there is less heat in the building available to be removed by the ground loop. Peak temperatures with both the ground loop and night ventilation were no more than 30°C while peak outdoor temperatures reached 38°C. The overall COP (heat removed divided by the energy use of the ground loop pump and ventilation fan) was 6.1 in this particular case. These COPs are not better than those that can be achieved with large commercial chillers (where COPs of up 7.9 are possible, as discussed in Section 6.5.1). However, they are substantially better than the COPs of many residential scale air conditioners (which range from 2.3 to 3.7 in Europe and North America, and from 2.9 to 6.4 in Japan) (Table 6.10). Thus, ground-loop precooling will be most advantageous, if at all, at a small scale. Unlike air conditioners or chillers, no HCFCs or HFCs are involved, although this is much less of an issue than in the past due to lower rates of leakage from properly maintained modern equipment compared to old equipment.

Argiriou et al (2004) built and tested an earth pipe that was coupled to a photovoltaic array that directly powers a 370W DC motor. This avoids the 10–20 per cent energy loss in DC to AC conversion normally associated with PV power, and takes advantage of the better efficiency of DC motors. The fan speed increases as the incident solar radiation increases, matching the need for increased cooling. The earth pipe was U-shaped with a total length of 20.18m, the bend buried at a depth of 1.8m, and the inlet and outlet buried at a depth of 2.1m (so that any condensate will drain out). The pipe was packed in sand to insure good thermal contact with the ground. The measured average COP (based on DC power output) for a summer in Athens (Greece) was 12.1. The COP at any given time was statistically related to the difference between the air temperature (T_a) and the ground temperature (T_g) according to the relation:

$$COP = 1.60(\pm 0.01)(T_a - T_g)(r^2 = 0.57) \qquad (6.7)$$

Finally, Flückiger et al (1998) examined the types and concentrations of micro-organisms in the outdoor air, in the air of the pipes, and in the supply air of twelve earth pipe cooling systems in Switzerland. At issue is the possibility that high relative humidity and condensation inside the earth pipe could facilitate the growth of microorganisms, thereby posing a health hazard. They found that, in general, concentrations of fungal spores and bacteria in the air at the end of the earth pipes were smaller than in the outside air, although occasional increases in some microorganisms were found in small, residential systems. To maintain hygienic conditions, the following principles should be followed:

- the air inlet should be placed above the ground and away from bioaerosol sources;
- a screen, a coarse filter, and a fine filter (2–5μm mesh) should be installed;
- pipes should have a flat surface and an inclination to let condensate flow off, and should be buried in well-compressed ground to prevent partial subsidence;
- the air inlet and ducts should be checked periodically and filters washed or replaced.

6.3.9 Impact of projected warming

As noted in Chapter 1, the globally and annually averaged surface air temperature will increase by 3–7K by the end of this century under a business-as-usual scenario of greenhouse gas emissions. Even under stringent emission-reduction scenarios, leading to stabilization of atmospheric CO_2 at 450ppmv, a globally averaged warming of 2–4K is likely. Depending on location, warming during summer could be smaller than or greater than the global annual average warming. Warming of this magnitude would significantly reduce the applicability of the passive and low-energy cooling techniques discussed here. On the other hand, urban centres are already several degrees warmer than the surrounding countryside during summer due to the urban heat island effect. Reduction in the urban heat island through greater provision of trees and greater use of high-reflectivity surfaces could partly offset the increase in regional temperatures due to increasing greenhouse gas concentrations. This predicament – reliance on fossil fuels warming the climate to an extent sufficient to impair the viability of non-fossil fuel alternatives for one of the major uses of fossil fuels (cooling of buildings) – underlines the need to move to a fossil fuel-free energy system as quickly as possible.

An assessment of the impact of future warming on building cooling loads and on passive cooling techniques requires accounting for extreme events. The output from climate models tends to consist of monthly average changes in parameters such as minimum and maximum daily temperature over a very large (200km × 200km or larger) grid square. The process of generating data at higher spatial and temporal resolution from climate model results is a process known as *downscaling*, and is the subject of much research. Belcher et al (2005) present a downscaling approach appropriate to the simulation of building systems.

6.4 Residential air conditioners and dehumidifiers

As the preceding sections have shown, there are many low-energy alternatives or combinations of alternatives to air conditioning that are viable in most climatic regions of the world. This is especially the case in residential and small commercial buildings, where internal heat gains are relatively small. Nevertheless, in North America (even in much of Canada), air conditioners have become a standard feature of new houses, so that they have come to be expected. Having adopted air conditioners as a standard feature, little to no effort is made to minimize cooling requirements in North American housing; air conditioners thereby become another 'need'. Apart from their environmental impacts and contribution to peak power demand (stressing power production and transmission capacity), air conditioners pose significant noise pollution, making it unpleasant for neighbours who would like to rely on passive cooling techniques or enjoy an otherwise peaceful summer evening outside. A similar trend is beginning in Europe and in the affluent parts of Asia. Bucking this trend is difficult, but Hungerford (2004) describes an example of a co-housing development in California where the residents chose, collectively, to forgo air conditioning, in part to reduce capital and energy costs, but also to avoid noise pollution.[8]

An air conditioner can be thought of as a small heat pump that operates only in cooling mode. Like a heat pump, the performance of an air conditioner can be characterized by its *coefficient of performance* (COP), which is the ratio of heat removed from a building to the energy used.[9] The rate of heat removal, like the rate of electricity use, can be expressed as kW. To avoid possible confusion between the two, cooling capacities will be designated here with a subscript 'c'. Thus, a $1kW_c$ air conditioner is one that can remove 1000 joules of heat per second, not one that requires 1kW of electrical power. Air conditioners, heat pumps and commercial chillers (described in Section 6.5) operate on a compression-expansion cycle. A refrigerant or *working fluid* is alternately compressed and allowed to expand. It evaporates at a low pressure in an evaporator, where it becomes colder than the surrounding air, and condenses under pressure in a condenser, where it becomes warmer than the surrounding air.

Air conditioners come in several forms:

- as *single-packaged air conditioners*, in which the compressor, evaporator and condenser are housed in a single unit that is mounted in a window or in an opening in the wall;
- as *single-duct air conditioners*, which consist of a single unit that is placed entirely inside the room to be cooled, with a duct to carry heat from the condenser to the outside;
- as various *split packaged units*, with the evaporator inside the building and the compressor and condenser outside the building (the indoor unit can be mobile or immobile, ducted or non-ducted);
- as *multi-split packaged units*, in which one outdoor condenser serves two or more indoor evaporators.

Air conditioners and commercial vapour-compression chillers use halocarbon refrigerants, all of which are greenhouse gases.[10] The choice of refrigerant affects both the efficiency of the cooling equipment and the contribution to greenhouse trapping of heat when the refrigerant leaks into the atmosphere.

6.4.1 Full-load efficiency of air conditioners

Like a heat pump, the COP of an air conditioner depends in part on the evaporator and condenser temperature, as these set the limit to the theoretical maximum (Carnot) COP (see Equation 5.3). The COP also depends on whether the unit is running at full load or part load. Thus, a proper comparison of air conditioners and other cooling equipment requires that the comparison be made for specified test conditions that correspond to full load. Units with equivalent full-load COP could nevertheless have different part-load behaviour.

For a given cooling capacity, single-duct air conditioners tend to have the smallest COP (due to heat transferred from the duct to the room being cooled), followed by single-packaged (wall or window) air conditioners, split air conditioners, and multi-split air conditioners. COPs of units available in North America, Europe and Japan are compared in Table 6.10. In Europe and North America, residential air conditioners

generally have COPs of 2.5–3.5. Minimum air conditioner COPs in Japan are comparable to or better than the best COPs in Europe and North America, while the maximum COP is 6.4 (on a small, 2.8 kW_c unit). The energy used by an air conditioner includes energy used by the evaporator and condenser fans, as well as by the compressor. For one 10.5kW_c residential air conditioner, the compressor, evaporator fan and condenser fan power consumption are 2584W, 438W and 196W, respectively (Bridges et al, 2001). The operating COP of an air conditioner takes into account fan energy (about 25 per cent of compressor energy alone) as well as compressor energy, whereas the Carnot COP does not account for fan energy.

According to performance data collected from 13,000 residential and commercial air conditioners in California, about two-thirds of air conditioners have the wrong amount of refrigerant and/or too low an airflow speed past the cooling coil (Downey and Proctor, 2002). This causes the energy use of an average air conditioner to be about 15 per cent greater than if these parameters were correct (i.e. more than 20 per cent greater in the units requiring adjustment). Low airflow can be caused by dirty filters, fouled coils, dirty blower wheels, incorrect blower setting or constrictions in the duct system, and is more likely in commercial buildings than in residential buildings.

Table 6.10. Range of air conditioner COPs found in North America, Europe and Japan

	North America[a]	Europe[b]	Japan[c]
Outdoor air test conditions	29.4°C condenser temp?	35°C T_{db}, 24°C T_{wb}	35°C T_{db}, 24°C T_{wb}
Indoor air test conditions	6.7°C evaporator temp?	27°C T_{db}, 19°C T_{wb}	27°C T_{db}, 19°C T_{wb}
Single ducted		2.28–3.09	3.02, all capacities
Single packaged	2.49–3.46	2.62–2.97	2.85, all capacities, window mounted 5.27 for <2.5 kW_c, wall mounted
Single split		2.73–3.56	3.96 for <2.5 kW_c
Multi-split		2.89–3.74	4.12 for <4.0 kW_c

[a] Based on NRCan (2005).
[b] Based on Adnot and Orphelin (1999), which is consistent with current information at www.eurovent-certification.com.
[c] Based on Murakoshi and Nakagami (1999). Given are minimum COPs required for the smallest capacity in each category. Smaller minimum COPs are required at larger capacities. The most efficient air conditioner of any type at 2.8kW_c capacity has a COP of 6.44.

6.4.2 Part-load behaviour of air conditioners

The efficiency of most air conditioners falls dramatically during part-load operation, due to the fact that part-load operation is almost always achieved through repeated on/off cycling. At 20 per cent of full load, efficiency falls to about 80 per cent of the full-load efficiency (Henderson et al, 2000b). Furthermore, some units use power even when they are turned off – 40–120W in central systems according to Henderson et al (2000b)!

The part-load COP of air conditioners can be increased in two ways: (i) by varying the compressor speed and the speed of the evaporator fan rather than relying on throttling or on/off cycling to achieve partial loads; and (ii) by using high-efficiency permanent-magnet motors rather than induction motors. The speed of the compressor in an air conditioner can be adjusted to maintain a given indoor temperature as the cooling load varies, while the speed of the fan that blows indoor air past the evaporator can be adjusted to maintain (or try to maintain) a given humidity. These two speeds can be varied independently. A reduction in the compressor speed of 50 per cent and in fan speed of 60 per cent in response to a given reduction in the cooling load, for example, reduces the energy consumption by about one-third compared to matching the same load through on/off cycling (Andrade and Bullard, 2002). Small air conditioners use fixed-speed induction motors. Electronically commutated permanent magnetic induction (ECPM) motors are used in variable-speed central air conditioners. Two-speed induction motors are also used; they allow the cooling system to operate at two different capacities, thereby reducing on/off cycling and associated losses, but at less cost and energy savings than variable-speed ECPM motors.

The efficiency penalty under part-load operation is exacerbated by the tendency to oversize air conditioning systems, as discussed in Section 6.8.

6.4.3 Impact of thermal pollution from air conditioners in dense developments

In apartment buildings with many closely spaced wall or window air conditioning units in recessed areas, the performance of a given air conditioner will be adversely affected by the heat rejected from the surrounding units. Bojić et al (2001) performed a fluid dynamics simulation of airflow and temperature in the typical recessed space of a 30-storey apartment building in Hong Kong, having four air conditioners per floor, each with a heat rejection of 4kW. When all air conditioners are operating simultaneously, the temperature increase above the ambient temperature grows from zero at the ground floor to 11K above ambient in the inner part of the recessed space on the 30th floor! Although there is a strong rising plume of warm air, the amount of air drawn across the condenser coil decreases by 60 per cent at the 30th floor, which would reduce the cooling capacity and further increase energy use. For a 60-storey apartment building in Hong Kong, Chow et al (2002) compute a 20 per cent decrease in the average COP of the air conditioners, which are placed in recessed spaces, due to heat rejection by the surrounding air conditioners. Xue et al (2004) studied this problem with a 1:20 scale model, and found that the temperature rise over a 16-storey height in a 7m wide recessed space is at least 10K.

6.4.4 Stand-alone dehumidifiers

If the evaporator coil of an air conditioner is below the dewpoint temperature, some condensation of water vapour will occur. That is, the air conditioner will perform some dehumidification. The proportion of total cooling that occurs as removal of latent heat (dehumidification) increases if the flow velocity of air past the cooling coil is decreased (because the temperature of the cooling coil decreases) but is typically about 0.25. This is particularly a problem in hot-humid climates, where 90–95 per cent of the required cooling load can be dehumidification. The problem is exacerbated in insulated and shaded buildings, where the sensible cooling requirements can be greatly reduced with little effect on the latent cooling load. For these reasons, stand-alone dehumidifiers are often used. The effectiveness of these units increases with increasing capacity, ranging from 1.2 litres/kWh at 10 litres/day to up to 2.75 litres/kWh at 50 litres/day. This corresponds to a COP of only 0.83–1.91. This is worse than that of air conditioners. In addition, the heat extracted from the condensed water is added to the room air, as well as the energy input to the compressor, rather than to the outside. This is because the evaporator and condenser are housed together in the same unit (air that has been cooled and dried by the evaporator flows through the condenser, where it is reheated and introduced back into the room). Although wall- or window-mounted air conditioners themselves have no place in new energy-efficient buildings, they are common in existing buildings. If inadequate dehumidification occurs, it is preferable to replace the air conditioner with one having a wrap-around heat pipe, which simultaneously improves the air conditioner COP and increases dehumidification capacity (Chapter 7, Section 7.4.10). The development of solid desiccant systems (Section 6.6.4) for residential systems may be a better solution in the long run.

6.4.5 Alternatives to air conditioners in multi-unit residential and small commercial buildings

Wall- or window-mounted air conditioners generally have a rated COP ranging from 2.5 to 3.5, but often much less under real operating conditions at part loads. Air conditioner COPs are easily two to three times less than that of large commercial chillers, which have COPs as high as 7.9. On the other hand, the best Japanese air conditioners are substantially more efficient, with the most efficient units having COPs that are competitive with those of large commercial chillers. In modern multi-unit residential buildings in south-east Asia, it is standard practice to provide mechanical cooling through air conditioners in each room of each condominium or apartment unit, often with small external pedestals to hold the condenser and compressor of split-type units. This is illustrated in Figure 6.48, which shows a new condominium tower in Beijing. The attraction from a developer's point of view of designing a multi-unit residential building to be cooled in this way is that the purchase, installation, and operation of the air conditioner is the responsibility of the condominium owner. There is no need to provide large, centralized chilling units and the associated chilled-water piping network and

Figure 6.48 One of dozens of similar condominiums under construction in Beijing during the summer of 2004, showing small air conditioners for cooling of individual rooms and the absence of any effort to reduce the cooling load through external shading

Source: Author

cooling towers. However, the approach of having an air conditioner in every room can be two or three times less efficient and produces significant thermal and noise pollution, as noted above. Although the centralized option may entail smaller energy costs and may have lower life-cycle cost, it would require a significant re-orientation of the market and of developers in many parts of the world. As well, it requires installation of separate metering of the chilled water use of each residential unit to provide a financial incentive to avoid waste, and an ability to control the amount of cooling to each room (one advantage of room air conditioners is that electricity use might already be metered, and it is easy to cool only those rooms that are occupied).

There are three alternatives that overcome some of the disadvantages of building-scale chilling systems. One is to build a district cooling system (if one does not already exist) and to connect the building to it. This option is not something that can be undertaken by a building developer alone, as it requires the active involvement of the local municipal government. It overcomes the need for on-site chillers and cooling towers, but still requires an internal chilled-water loop, metering of individual residential units, and controls for individual rooms. District cooling systems and the organizational and institutional barriers to their construction are discussed in Chapter 15 (Section 15.3). The second alternative is to install heat pumps in individual residential units that are coupled to a water circulation loop. This is referred to as a *water-loop heat pump system*. The heat pumps can serve for both heating and cooling. In cooling mode, the circulating water cools the condensers of the heat pumps (improving the COP), with the heat taken from the heat pump condensers removed through evaporative cooling in a cooling tower (described in Section 6.10) rather than warming the surrounding air. It is more efficient than using air conditioners but less efficient than a centralized chilling system, as discussed in Chapter 7 (Section 7.4.2), although it does permit easier control and metering of individual residential units than in a centralized chilling system. The third alternative is a three-way heat pump, tested in prototype form by Ji et al (2003). During air conditioning mode, heat is extracted from the pump and used to make hot water, rather than rejected to the surroundings. The heat pump can be used for air conditioning only (when there is already enough hot water), for making hot water only (when there is no cooling load), and for space heating only. In the latter two cases, the heat pump functions as an air-source heat pump, so it is suitable only in areas with mild winters.

6.5 Commercial air conditioners and vapour-compression chillers

Air conditioning in commercial buildings can be accomplished using central chillers (that produce cold water that is then distributed to radiators throughout the building or to cooling coils in the air handling system), using packaged rooftop air conditioning units where the evaporator and condenser are side-by-side (unitary air conditioner) or separated (with the evaporator inside and near the space to be cooled), or using individual room air conditioners (either wall units, packaged terminal air conditioners, packaged terminal heat pumps, or water-loop heat pumps). Air conditioners and packaged rooftop units use a reciprocating compressor, while central chillers can use reciprocating, rotary or centrifugal compressors. Figure 6.49 shows the breakdown of energy use for cooling in commercial buildings in the USA among these equipment options, where it is seen that rooftop units account for over half of the total cooling energy use.

Figure 6.49 Distribution of primary energy used for cooling of commercial buildings in the US among different types of cooling equipment

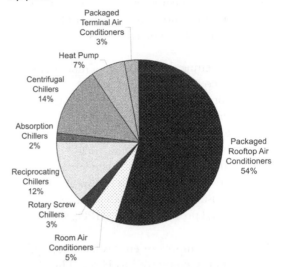

Source: Data from Westphalen and Koszalinski (2001)

The principles governing the operation and efficiency of air conditioning equipment are the same as those governing the efficiency of heat pumps, and are discussed in Chapter 5 (Sections 5.1, 5.2, 5.5 and 5.6). Further information, specific to air conditioning equipment, is presented below.

6.5.1 Packaged rooftop air conditioners

Figure 6.50 is a schematic illustration of a packaged rooftop air conditioner. The condenser is cooled by blowing ambient air past it with a fan. The COP is based on the energy used by the compressor and the fan, namely:

$$COP = \frac{Q_c}{P_{comp} + P_{fan}} \qquad (6.8)$$

where Q_c is the cooling power and P_{comp} and P_{fan} are the compressor and fan power, respectively. P_{comp} depends on (among other things) a factor involving the ratio of condenser to evaporator pressures divided by the motor efficiency (evaporator fan power is not included in the COP because it is part of the ventilation system). A warmer condenser or a colder evaporator (i.e. a greater temperature lift) increases the required compression ratio and hence the amount of work that must be done for a given rate of heat transfer. That is, the COP falls, as implied in the expression for the COP of an ideal Carnot cycle (Equation 5.3). As the cooling load decreases, the rate of heat removal by an air conditioner or air-cooled chiller can be reduced in one of two ways:

- by letting the condenser temperature decrease by reducing the compression ratio, while maintaining the same airflow (condensing temperature control, CTC), or
- by reducing the airflow while maintaining the same condenser pressure and hence a relatively constant condenser temperature (head pressure control, HPC).

CTC saves on compressor energy use but requires more fan energy compared to HPC. The almost universal procedure with air-cooled equipment is HPC – to maintain a near constant condenser temperature (usually around 45–50°C) even as the outdoor temperature and cooling load decrease. This is done in order to maintain an adequate pressure difference between the evaporator and condenser in order to operate a thermostatic expansion valve between the two. This is not necessary with electronic expansion valves. Yu and Chan (2005) evaluated the impact of allowing the condenser temperature to vary with outdoor temperature for the cooling load profile of a typical Hong Kong office building. They find that the compressor energy use drops by 25.3 per cent while fan energy use (initially at 9 per cent of the compressor energy use) increases by 57 per cent. The net result is a decrease in energy use by 18.4 per cent. The CTC strategy is not applicable to window-type air conditioners currently on the market. Water-cooled chillers, discussed below, are similar in that the cooling tower fans are usually operated so that the cooling tower delivers a near constant cooling water temperature as outdoor temperature varies.

As with other vapour compression cooling equipment, humidity control is difficult with rooftop air conditioners. Typically, only 25–30 per cent of the total cooling load can be provided through dehumidification, which is a problem in hot-humid climates or when the sensible cooling load has been reduced through the better design of buildings. Somewhat higher dehumidification can be achieved by decreasing the rate of airflow past the cooling coil. A better solution is to separately handle the latent heat load through desiccant dehumidification; this is already done in some centralized systems (Section 6.6.4). Packaged rooftop air conditioners with built-in solid desiccant wheels are currently under development by Trane and Solar Engineering Company.

According to Davis et al (2002), most rooftop air conditioners have at least one operational deficiency (incorrect refrigerant charge, inadequate evaporator airflow, duct leakage, improper thermostat specification and scheduling). They estimate that routine maintenance can typically provide energy savings of 25 per cent.

Figure 6.50 Schematic illustration of a packaged rooftop air conditioner used in commercial buildings

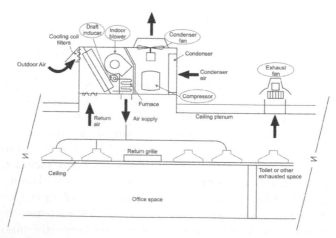

Source: Westphalen and Koszalinski (2001)

6.5.2 Electric chillers

Air conditioners that produce cold water rather than cold air are referred to as 'chillers'. There are three major kinds of commercial chillers, based on the kind of compressor used. Ranging from smallest to largest, these are *reciprocating* chillers; *rotary* (primarily *screw*) chillers; and *centrifugal* chillers. An electric motor normally drives the compressor, but mechanical power from an engine or turbine could also be used to drive the compressor (as discussed in Section 6.5.3). Operating principles and efficiency-related information on compressors is given in Box 6.3.

BOX 6.3 Kinds of compressors and operating principles

Detailed information on compressors, with numerous illustrations and photographs, can be found in Reay and MacMichael (1988) and Wulfinghoff (1999, Reference Note 32). Here, the main points pertaining to the full-load and part-load efficiency of different compressors are summarized.

Reciprocating compressors. These involve pistons with an alternating back and forth motion. Gas enters the piston cylinder at the bottom of the stroke, is compressed as the piston rises, and exits through a valve at the top of the stroke. However, not all of the gas exits before the valve is closed, and the remaining gas re-expands during the suction stroke. Thus, some of the energy used for compression is wasted. In addition, substantial turbulence is generated by the up and down motion, which also wastes energy. As a result, the efficiency of the full-load compressor is less than that of other compressors. However, the reduction in efficiency at part-load is the smallest, such that the part-load efficiency can be larger than for some other compressors. Part-load operation is achieved by on–off cycling in small units and by disabling one or more cylinders in larger (multi-cylinder) units. Reciprocating compressors are the most common type used in small and medium size chillers.

Rotary screw compressors. These work by trapping the refrigerant between the threads of two or more interlocking rotors, at least one of which is screw shaped (and the others screw or star shaped). The best screw compressors are significantly more efficient than reciprocating compressors at all loads, and can be more efficient than centrifugal compressors at low loads. They can have a fixed or variable compression ratio. Part-load operation can be achieved through a slide valve that forms a portion of the housing surrounding the screws, and that can slide away from the rest of the housing to create a gap that deactivates the adjacent portion of the rotors. A variable speed drive can also be used to achieve part-load operation. Frequent on–off cycling is also permitted. Screw compressors can be operated at as little as 10 per cent of full load. They are being increasingly used in place of medium and large reciprocating compressors, and have entered the size domain of centrifugal compressors.

Centrifugal compressors. These have one or more impellers on a single shaft that accelerate a gas. The gas is then slowed down with a diffuser, creating a pressure that compresses the gas along with the aid of centrifugal force. Turbulence and associated energy loss can be minimized through an appropriate shape of the impeller and diffuser, but the shape can be optimized for only one particular flow rate. Guide vanes at the inlet to the impeller are used to reduce the output with the smallest possible efficiency loss, but energy use is almost constant, so the efficiency decreases roughly in proportion to the decrease in load. Alternatively, a variable speed drive can be used to reduce the impeller speed, but there is still a significant efficiency drop below 50 per cent of full load because the impeller shape is no longer optimized at the reduced flow rate. Centrifugal chillers should not be turned on and off at low loads.

In chillers, a discharge valve occurs between the compressor and condenser. If the discharge valve is open to the condenser as the compression takes place, then the discharge pressure never exceeds the pressure needed to condense the refrigerant (except for a small expansion loss between the compressor and condenser). This is the case for centrifugal and reciprocating compressors. The compression ratio adjusts itself to the needs of the system. In some screw compressors, the compression ratio is fixed. Since the compression ratio corresponds to a ratio of condenser and evaporator temperatures, reducing the difference between the condenser and evaporator temperature will not be particularly effective in saving energy if the compressor has a fixed compression ratio.

Reciprocating chillers can have air- or water-cooled condensers, while centrifugal chillers usually have water-cooled condensers. Cooling capacities, full-load COPs and refrigerants used are given in Table 6.11. Within each category, the COP is larger for the largest units. Large centrifugal chillers have a cooling capacity of up to 35MW$_c$ and a full-load COP of up to 7.9. By comparison, residential wall- or window-mounted air conditioners have a maximum cooling capacity of about 6kW and rated COPs of 2.5–3.5 in North American and Europe, and 2.9–6.4 in Japan (Table 6.10). Chillers larger than about 8MW$_c$ are assembled on site. Unless one can acquire the most efficient Japanese air conditioners, large chillers will be a factor of 2–3 more efficient than small, window or split-unit air conditioners (a COP of 5–7.5 instead of 2.5–3.5), so up to a 50–60 per cent energy savings is possible (after accounting for energy used by the cooling tower) if multi-unit residential buildings are designed with a centralized chiller for air conditioning, rather than designed to accommodate a small air conditioner in each room of each apartment unit (as is commonly the case in many parts of southeast Asia). This of course entails constructing a chilled-water piping system and allocating space for the central chilling facility, and it requires metering and billing of individual apartments in order to discourage waste.

The COP of a chiller depends on the fraction of the peak load at which it is running, as well as on the size of the unit. This is illustrated in Figure 6.51, which shows the COP at part-load as a percentage of the full-load COP for various kinds of chillers. The manner in which COP varies with load depends on the type of compressor, the refrigerant used, and whether the compressor motor can operate at only one speed or at variable speeds. Commercial chillers, like air conditioners and heat pumps, are usually equipped with fixed-speed compressors. Part-load operation of such chillers can be achieved through the use of inlet vanes. The inlet vanes regulate the rate of flow of the refrigerant in order to maintain a constant temperature of the chilling water leaving the evaporator, regardless of the cooling load. For fixed-speed centrifugal chillers, the COP falls continuously with decreasing load, dropping to less than 20 per cent of the full-load COP at 20 per cent of full load. For fixed-speed screw chillers, the highest COP occurs around 80 per cent of peak load, but drops sharply below about 50 per cent of full load.

Chiller manufacturers tend to claim that their equipment is more efficient at part load than at full load. In order to capture the differing variation in COP with load for different kinds of chillers and with different refrigerants, the COP weighted over different loads can be calculated

Table 6.11 Types and characteristics of commercial chillers

Chiller type		Capacity range		Full-load ARI COP[a]	Refrigerants used
		MW$_c$	Million Btu/hr		
Vapour-compression chillers					
Reciprocating		up to 1.5	up to 5.1	3.8–4.6	R134a, R717, R407c
Rotary	Screw	0.3–7	1–24	4.1–5.6	R134a, R717, R407c
	Scroll				
	Rolling piston				
	Rotating vane				
Centrifugal		0.3–35	1–120	5.0–7.9	R134a, R717, R123
Absorption chillers					
Single-stage		0.3–6	1–21	0.7	H$_2$O
Double-stage		0.3–8	1–27	1.2	H$_2$O
Direct-fired		0.3–5.2	1–18	1.7	H$_2$O

[a] The COPs for vapour compression chillers are for ARI (Air Conditioning and Refrigeration Institute) standard test conditions, consisting of production of chilled water at 6.7°C (44°F) and cooling water entering the condenser at 29.9°C (85°F). Chiller COPs do not include the energy required to operate cooling tower fans and pumps, which can be 10–30% of the energy that would be used by the compressor of a vapour compression chiller.
Notes: R123 = HCFC-123, H134a = HFC-134a, R717 = ammonia, and R407c is a mixture whose composition is given in Table 5.7.
Sources: IEA (1999b) for vapour-compression chillers, Dharmadhikari (1997) for absorption chillers

Figure 6.51 Variation of COP with load for absorption chillers, electric screw chillers and electric centrifugal chillers

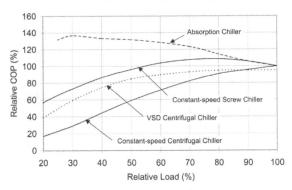

Note: For each chiller except the VSD centrifugal chiller, the COP is relative to its own COP at full load. For the VSD centrifugal chiller, the COP is relative to that of a fixed-speed centrifugal chiller at full load.
Source: For absorption chiller, data from Chow et al (2002); for screw chiller, data from IEA (1999b); for centrifugal chiller, data from Dharmadhikari (1997)

and compared. The *Integrated Part Load Value* (IPLV) is defined as the COP at 100 per cent, 75 per cent, 50 per cent and 25 per cent of full load with a 1:42:45:12 weighting. That is:

$$IPLV = 0.01COP_{100} + 0.42COP_{75} +$$
$$0.45COP_{50} + 0.12COP_{25} \qquad (6.9)$$

IPLV values as high as 9.9 have been claimed. However, as shown by Yu and Chan (2005), the variation of COP with load depends on how outdoor temperature also varies with load, and depends on whether condenser temperature control (CTC) or head pressure control (HPC) is used (see Section 6.5.1). If outdoor temperature is fixed as load decreases, then COP falls with decreasing load whether CPC or HPC is used. If outdoor temperature falls from 35°C to 15°C as load decreases from 100 per cent to 50 per cent, then the COP increases as claimed by the manufacturer for the air-cooled, constant speed chiller that was tested by Yu and Chan (2005), but only under CTC. Under HPC – which is the almost universal procedure – COP is essentially constant. From this it is clear that published IPLVs and even curves of COP versus load are of little value in estimating the relative performance of different chillers or their part-load behaviour

under real operating conditions. Further, even if the part-load COP values used in Equation (6.9) are correct, the IPLV will be an accurate approximation of the average COP of the chiller only to the extent that the distribution of loads matches that assumed in the calculation of the IPLV.

The part-load behaviour of a chiller under CPC or HPC can be improved if the compressor speed decreases as the load decreases. As explained in Chapter 5 (Section 5.6), operation of a motor at variable speeds requires an electronic interface between the input electricity supply and the motor called a variable speed drive (VSD). A VSD can reduce the compressor speed (and hence the rate of refrigerant flow) down to zero and, as noted above, improve the part-load efficiency, but a practical lower limit is 40 per cent of full speed due to the need to circulate lubricating oil. The variation in relative COP for a centrifugal chiller with a VSD is shown in Figure 6.51. The VSD drive itself entails about a 5 per cent energy loss, even at full load, so the COP in this case is relative to the COP of a fixed-speed centrifugal chiller at full load. Even with a VSD, there is still a significant decrease in COP below about 60 per cent of full load. For both fixed-speed and VSD chillers, the variation of COP with load depends on the refrigerant used as well as the kind of compressor, and so can be different from that shown in Figure 6.51 for chillers made by specific vendors. A VSD centrifugal chiller can have a COP as high as 9.9 at typical operating conditions (Smith, 2002). Screw chillers with VSDs have recently become available (Anonymous, 2002).

Lenarduzzi and Yap (1998) retrofitted a variable speed drive onto a 615kW$_c$ centrifugal chiller in an office building in Toronto and monitored its energy use during one cooling season. They found a saving in energy use of 41 per cent at this particular site. The energy saving depends on the amount of time that the chiller operates at various part-load fractions, and thus will vary with climate and the characteristics of the building. As noted by Lenarduzzi and Yap (1998), the use of VSDs in chillers provides other benefits in addition to saving energy. These include:

- providing for a softer start (with less wear);

- eliminating the need for complex system controls, such as inlet guide vanes, dampers, and valves;
- allowing the chiller to follow variations in the load more accurately.

Most chillers operate at a small fraction of full load most of the time, so a major opportunity for reducing energy use arises from using a few large chillers rather than many small chillers to serve a given load, but with the chillers small enough that the cooling load can be split among a small number of unequally sized chillers, which will avoid operating any one chiller at a small fraction of its peak load. If a large centrifugal chiller operates at a small fraction of its peak load, then the advantage of greater full-load efficiency by being larger could be largely or entirely negated. If there are two unequally sized chillers in a cooling system, the smaller chiller can be used at low loads and can operate relatively close to its peak capacity and hence close to its maximum efficiency. When the load is about to exceed its capacity, it can be shut down and the larger chiller started but not at its smallest possible load (where the efficiency penalty would be particularly large). As the load increases further, both chillers can be switched on at some point. Because there is still a significant decrease in efficiency below 60 per cent of full load even with a VSD, correct sizing and staging of multiple VSD chillers to minimize part-load operation is still beneficial. If buildings are linked together in a district cooling system, then it will be possible to serve the combined load with a small number of large chillers.

A commercial chiller is part of a larger system involving air or water flow for distributing coldness through a building, and a mechanism for removing heat from the condenser. In a conventional commercial chiller system, heat from the condenser is transferred to either air, which is blown past the condenser with a fan, or to cooling water that is supplied by a cooling tower on the roof, and from there dissipated to the atmosphere. In the first case, the chiller must concentrate the heat that it removes from a building to a temperature greater than that of the outside air, so that the heat can be transferred to the outside, while in the second case, the heat must be concentrated to a temperature warmer than the outside wet-bulb temperature (see Section 6.10). As outside air temperature increases, the amount of heat that needs to be removed from the building increases and the chiller and/or fans must work harder to reject this heat to the outside air (that is, the COP falls). The heat delivered to the condenser is the heat removed from the building plus the compressor energy. The COP specified for commercial chillers does not include the additional energy needed for delivering the cooled air (or chilled water) throughout the building or for operating cooling tower fans and pumps. As shown later (Table 6.14), the energy used by the cooling tower can be 10 per cent of the energy used by the compressor, while the energy used by the building ventilation fans (in an all-air cooling system) can be up to 60 per cent of the energy used by the compressor.[11]

The 5-Lab working group in the US in its report *Scenarios for a Clean Energy Future* (Koomey et al, 2000) postulates rather modest (10 per cent) improvements in chiller performance by 2020. The COP can be improved if the area of the heat exchangers is increased relative to the compressor size, but this increases the cost by increasing the material requirements. It will increase the energy required for pumping the cooling water and the chilled water unless the flow rates are reduced, so the net effect on energy use of alternative choices needs to be carefully evaluated. The biggest potential improvements lie in the use of ground-source heat pumps (where this is feasible), in the use of large centralized and optimally staged chillers rather than many small and less efficient chillers, by using centralized chilling in multi-unit residences rather than individual air conditioners in every room or unit, and by increasing the chilled-water temperature (from, say, 5°C to 16°C). Increasing the chilled-water temperature improves the COP by reducing the temperature lift, as explained in Chapter 5 (Section 5.5), but requires the use of radiant chilling (Chapter 7, Section 7.4.4) and desiccant dehumidification (Sections 7.4.11 and 7.4.12). That is, the biggest efficiency-improvement potential (often by a factor of two or better) is in the system efficiency rather than the device efficiency.

6.5.3 Mechanically driven chillers

The efficiency in generating electricity is given by the product of the turbine efficiency (ratio of mechanical power produced to thermal power input) times the generator efficiency (ratio of electric power produced to mechanical power input). The efficiency of a chiller is equal to the efficiency of the electric motor (ratio of mechanical power to electric power input) times the efficiency of the compressor. If the shaft of a heat engine or turbine were directly coupled to a compressor, the energy losses represented by the generator and motor efficiencies (typically about 5 per cent each) would be avoided. Whether this leads to a savings in primary energy depends on the efficiency of the heat engine used to drive the compressor compared to the efficiency of the heat engine at the electric power plant. Efficiencies of heat engines tend to increase with increasing size, so the central power plant would have an advantage in this respect. Direct use of mechanical power to drive a compressor is more likely to be advantageous for large heat engines and compressors rather than for smaller units. If heating is also required (say, for hot water), then waste heat from the heat engine that drives the chiller could be used, giving a high overall efficiency.

The variation of COP with load and for different entering condenser-water temperatures (ECWT) is given in Figure 6.52 for engine driven (diesel or gasoline) screw and centrifugal chillers. As with constant speed electric chillers (Figure 6.51), there is a greater efficiency penalty at part load for the centrifugal chiller than for the screw chiller. For an ECWT of 27°C, the peak COP (cooling divided by fuel energy input) is about 2.0 for the screw chiller and 2.5 for the centrifugal chiller. The COP in terms of primary energy is about 1.7–2.1 if there is a 15 per cent loss in conversion from primary energy to secondary energy at the point of use. To provide the same cooling for the same primary energy input using electricity, the electric chiller COP would have to be 5.1–6.3 if the electricity generation × transmission efficiency is 0.33, or 3.4–4.2 if the efficiency is 0.5.[12] As electric chillers in this COP range are readily available, there will normally be little or no efficiency advantage in using mechanically driven chillers. However, mechanically driven chillers can be used to avoid purchasing electricity

at times of peak demand, when it is most expensive (and when electricity generation and transmission efficiencies are the lowest). Mechanically driven chillers have been used as one of several different cooling devices in some district cooling systems (Chapter 15, Section 15.2.7). Electric motors rarely require maintenance, whereas heat engines require regular maintenance, a fact that discourages their use by builder owners but not necessarily by the operators of district cooling systems.

In some situations, the waste heat from the thermal engine that drives the chiller can be used for hot water or to drive absorption chillers for additional cooling. The former is a form of cogeneration, in this case involving simultaneous production of hot and cold water. For

Figure 6.52 Variation of COP with load and condenser cooling water temperature for (a) engine-driven scroll and (b) centrifugal chillers

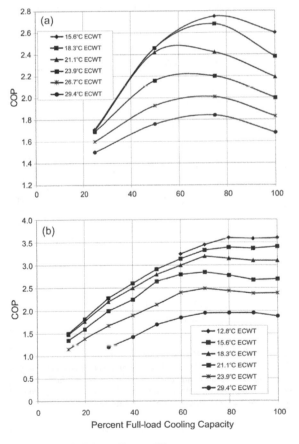

Source: ASHRAE (2000, Chapter 47)

Figure 6.53 Variation with cooling capacity of the cost and unit cost of (a) terminal air conditioners, (b) single-zone rooftop air conditioners, and (c) multi-zone rooftop air conditioners

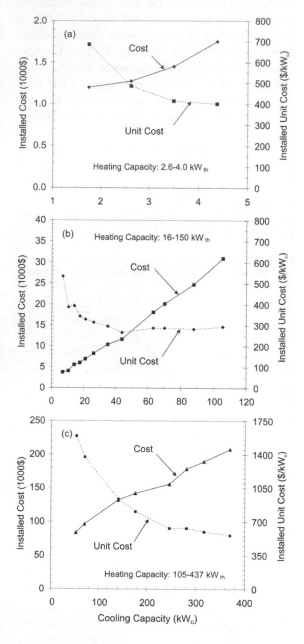

Figure 6.54 Variation in (a) cost and (b) unit cost of reciprocating and centrifugal chillers, as given by Means (1999) for average US conditions

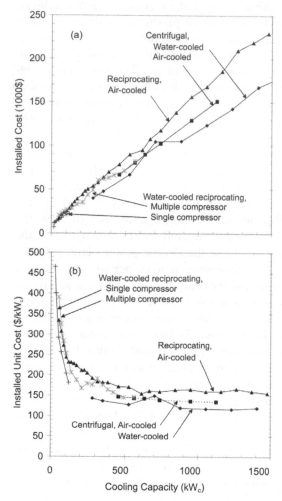

Note: Costs are US 1999 averages as given by Means (1999). The terminal air conditioners also provide electric-resistance heat, while the rooftop air conditioners provide natural gas heat. The heating capacities do not vary in proportion to cooling capacity, so only the heating capacity ranges are given.

electricity generated from fossil fuels with a typical efficiency of 35 per cent and 5 per cent transmission loss, this form of cogeneration will save primary energy. However, if the cogeneration system displaces a natural gas combined cycle power plant (at 55 per cent efficiency), primary energy use could increase or decrease compared to separate chilling with electric chillers and separate production of hot water with a boiler. To illustrate this point, we can make use of experimental data reported by Sun et al (2004). A natural gas engine with a fuel input of 321.1kW produces 142.5kW of heat and 179kW of mechanical power, which drives a compressor with a COP of 2.52, thereby providing 450.6kW of cooling. At 33 per cent generation × transmission efficiency, the electric chiller COP would need to exceed 8.3 in order for separate heating and chilling to use less primary energy (assuming a boiler at 90 per cent efficiency). However, at 50 per cent generation × transmission efficiency, the electric chiller COP need only exceed 5.5 in order for separate heating and cooling to use less primary energy.

6.5.4 Vapour compression chiller cost versus capacity

Representative costs per kW of cooling capacity of various kinds of air conditioners and electric chillers are given in Figures 6.53 and 6.54. Figure 6.53 gives costs for terminal air conditioners and single- and multi-zone rooftop air conditioners. These have high unit costs ($300–1500/kW$_c$), but

also provide heating, and when a credit is applied based on the heating capacity and the boiler unit costs given in Figure 4.18, the cooling unit cost is generally negative. Air-cooled reciprocating chillers range in cost from $330/kW$_c$ at 50kW$_c$ capacity to about $150/kW$_c$ at 1500kW$_c$ capacity. Centrifugal chillers cost around $120–150/kW$_c$ over a very wide range of capacities, dropping to about $80/kW$_c$ for the largest capacity (about 7MW$_c$, not shown).

Figure 6.55 Photographs of an absorption chiller (top left), a 70kW$_c$ adsorption chiller (top right), and a desiccant chiller (lower)

Source: Posters from the AIRCONTEC Trade Fair, Germany, April 2002, available from www.iea-shc-task25.org

Table 6.12 Comparison of various heat-driven cooling options, based on information presented in this chapter, including costs taken from Table 6.13

Technology	Closed-cycle absorption (single-effect)	Closed-cycle absorption	Closed-cycle adsorption	Open-cycle solid Desiccant	Open-cycle liquid Desiccant
Refrigerant	H_2O	NH_3	H_2O	–	–
Sorbent	LiBr	H_2O	silica gel	silica gel	CaCl, LiCl
Chilling carrier	water	water-glycol	water	air	air
Chilling temperature	6–20°C	–60°C to 20°C	6–20°C	16–20°C	16–20°C
Driving temperature	80–110°C	100–140°C	55–100°C	55–100°C	55–100°C
Cooling water temperature	up to 50°C	up to 50°C	up to 35°C	not applicable	as low as possible
Cooling power range	35–7000kW	10–10,000kW	50–430kW	20–350kW	
COP	0.6–0.75	0.6–0.7	0.3–0.7	0.4–0.6[a] 0.5–1.1[b]	0.55–1.8[c] 1.4–2.2[b,d]
Cost ($/kW$_c$)	85–350	410–1800	550–2200	370–700	

[a] Operating as a dehumidifier.
[b] Based on dehumidification and cooling when integrated with an evaporative cooler.
[c] Operating as a dehumidifier, but with achievable capture and use of waste heat in the steam and hot concentrated liquid desiccant streams.
[d] Lower COP pertains to a humid climate, higher COP pertains to a dry climate.

Table 6.13 Costs of absorption, adsorption and desiccant chillers of various cooling capacities

Chiller type	Capacity (kW$_c$)	Conditions	Cost ($/kW$_c$)
Absorption chillers			
Single-effect LiBr	250	110°C input water	240
		80°C input water	350
	1000	110°C input water	125
		80°C input water	210
	2000	110°C input water	85
Single-effect/ Double-lift	250	95–110°C input water	600
	1000		360
	2000		300
	2500		210
Ammonia/water	250	140°C input water	910
		110°C input water	1820
	1000–2000	140°C input water	410
		110°C input water	830
Other chillers			
Silica gel adsorption	70	85°C input water	1380
		71°C input water	2180
	350	85°C input water	550
		71°C input water	955
Desiccant- evaporative cooling	50		700
	200		400
	400		370

Source: IEA (1999c)

6.6 Heat-driven chillers and dehumidifiers

In the vapour compression chiller described above, mechanical power – whether from an electric motor or an engine – is used to drive a compressor, needed to circulate the refrigerant between a condenser and an evaporator. However, heat can be directly used to drive an evaporation–condensation cycle in a variety of ways or to regenerate desiccant based cooling and dehumidification systems. Here, the operating principles and performance of four heat driven cooling and dehumidification devices are presented – absorption chillers, adsorption chillers, ejector chillers and desiccant chillers. Photographs of absorption, adsorption and desiccant chillers are shown in Figure 6.55. The use of solar thermal energy with each of the devices discussed below is discussed in Chapter 11 (Section 11.5). Summary information is presented in Table 6.12, while comparative investment costs are presented in Table 6.13.

6.6.1 Absorption chillers

The principles behind the operation of an absorption chiller are explained in Box 6.4. In an absorption chiller, a refrigerant is alternatively absorbed by an absorbent and driven off with heat. The vast majority of absorption chillers use a lithium bromide (LiBr) water mixture (where water is the refrigerant and LiBr the absorbent). An alternative system uses an ammonia water mixture (where ammonia, NH_3, is the refrigerant and water the absorbent). The heat is normally provided by burning natural gas (a *direct-fired* chiller) or by supplying steam that is produced elsewhere. Commercial absorption chillers have existed since around 1945, and about 5 per cent of the commercial space in the US is cooled today using gas-fired absorption chillers (Herold, 1995). The capacity range of commercially available absorption chillers is given in Table 6.11.

BOX 6.4 Principles behind the operation of an absorption chiller

To explain how an absorption chiller works, it is easiest to first explain how one can use a camp fire and lithium bromide (LiBr) to make beer cold while on a canoe trip on a lake in one of Canada's vast wilderness parks.

Take a mixture of LiBr and water in a bottle. Connect this bottle to another bottle with a tube having a valve that can be closed. Check that the valve is open, and heat the LiBr/water solution near the campfire. This drives off water vapour, which enters the other bottle and condenses, releasing heat. A concentrated LiBr/water solution is left behind. Remove the first bottle from the campfire, close the valve, and let both bottles cool. When the bottles have cooled, open the valve. Water will evaporate from the second bottle as water vapour is sucked into the concentrated LiBr/water solution. As water is reabsorbed, heat is released in the solution, but the bottle where evaporation occurs becomes very cold – sufficiently cold to cool your beer if placed in contact with it (of course, a much easier procedure is to put the beer bottles in the lake and wait one hour!).

The operation of an absorption chiller is illustrated schematically in Figure B6.4.1. The absorption chiller is seen to involve an outer cycle between a condenser and an evaporator, as in the vapour-compression chiller, and an inner cycle between a generator and an absorber in place of a compressor.

Figure B6.4.2 shows the detailed layout of an absorption chiller. In this system, water is the *refrigerant*, and LiBr is the *absorbent*. The LiBr/water solution is heated to 80–95°C in a generator. Water vapour is driven off and flows to a condenser, and is cooled to about 40°C by water coming (indirectly) from a cooling tower. This is sufficient to cause the water vapour to condense. The combination of generator and condenser serves as a water distiller, since it separates pure water from the LiBr water solution, leaving concentrated LiBr behind. The condensed water is sprayed into the upper part of a cylindrical chamber having a pressure of about 1/100atm. This is called the evaporator, as the low pressure causes rapid evaporation, cooling the remaining water to as low as 5°C before

Figure B6.4.1 Schematic representation of an electric vapour compression chiller and an absorption chiller, from which it can be seen that the absorption chiller consists of an outer cycle that is identical to that of the vapour compressions chiller, and an inner subcycle that replaces the compressor of the vapour compression chiller

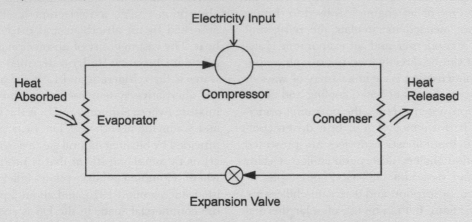

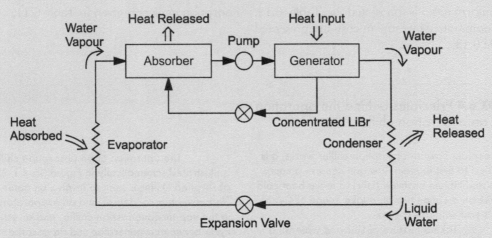

Source: Amplified from Reay and MacMichael (1988)

it lands on the chilled-water coil. Water from the building that needs to be cooled passes through this coil. The hot concentrated LiBr solution that is left behind in the generator is sprayed into the lower part of the cylindrical chamber. Most of this solution forms a thin film on the coil of an *absorber*. The absorber coil is kept cool through the circulation of cooling water from a cooling tower, which reduces the vapour pressure of the LiBr solution. This in turn allows ready absorption of the water vapour produced in the evaporator, and serves to maintain the low vapour pressure in the upper, evaporator portion of the cylinder. As water and concentrated LiBr are mixed, heat is released, but this heat is carried away by the cooling water. The cooled and diluted solution is then pumped back

to the generator, where the cycle begins again. In order to increase the surface area of the diluted solution, so that water can be driven off more readily, the diluted solution is sprayed into the generator. A heat exchanger is used to preheat the absorbed liquid before it enters the generator and to precool the concentrated solution before it enters the absorber.

Most absorption chillers today have at least two generator stages – a high-temperature and a low-temperature stage – which improves the efficiency. Diagrams and photographs of two-stage and other variants can be found in Wulfinghoff (1999, Reference Note 33), while Chua et al (2000) present a general thermodynamic framework.

Figure B6.4.2 Flow diagram of a single-stage single-effect absorption chiller

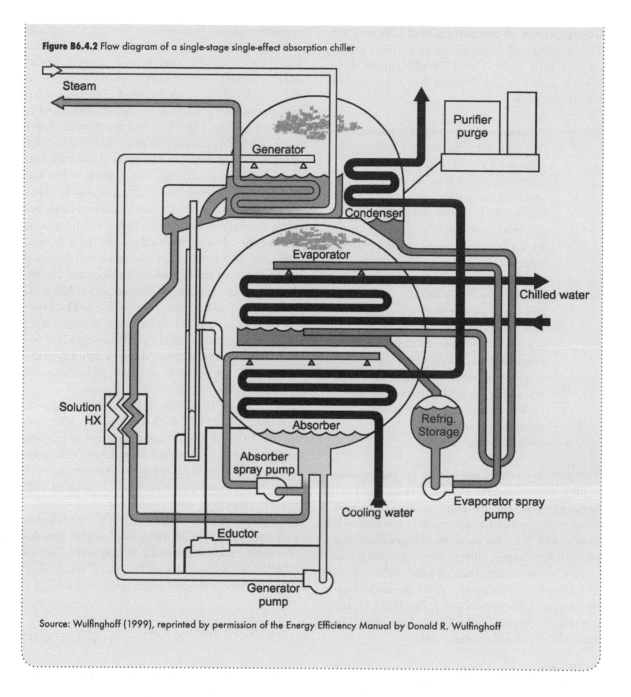

Steam

Generator

Purifier purge

Condenser

Evaporator

Chilled water

Solution HX

Refrig. Storage

Absorber

Absorber spray pump

Cooling water

Evaporator spray pump

Eductor

Generator pump

Source: Wulfinghoff (1999), reprinted by permission of the Energy Efficiency Manual by Donald R. Wulfinghoff

Comparison of ammonia and LiBr chillers

The majority of commercial absorption chillers use the LiBr cycle. The LiBr absorption chiller has a number of advantages over the ammonia absorption chiller:

- it requires a lower input temperature (70–90°C for a single-effect system, compared to 125–170°C for an ammonia chiller if an air-cooled absorber and condenser are used, or 95–120°C with water cooling (Florides et al, 2002b));

- for a given input temperature, it has a higher COP;

- it operates under very low pressure (872Pa evaporator, 7375Pa condenser), while ammonia chillers operate under very high pressure (485kPa evaporator, 1500kPa condenser), which increases the required pumping power;

- it is simpler than the ammonia-water chiller, which requires a rectifier to separate the ammonia;

- ammonia is hazardous.

The primary advantage of the ammonia chiller is that it can produce very low evaporator temperatures (–60°C). By contrast, the LiBr chiller cannot produce an evaporator temperature less than about 5°C, because the refrigerant is water vapour. Ammonia chillers can use a cooling tower or can be air-cooled. Present LiBr chillers require a cooling tower, although efforts are underway to develop a viable air-cooled LiBr chiller by making more compact and efficient heat exchangers (González and Alva Solari, 2002).

COP and energy use

The theoretical maximum COP of an absorption chiller is the Carnot Cycle COP. It is given by (Reay and MacMichael, 1988):

$$COP_{cooling,ideal} = \frac{T_E}{T_C - T_E} \frac{(T_G - T_A)}{T_G} \qquad (6.10)$$

where T_E, T_G, T_C and T_A are the evaporator, generator, condenser and absorber temperatures,

respectively. The first term on the right-hand side can be recognized as equivalent to the Carnot COP for a vapour compression chiller (Equation 5.3), while the second term corresponds to the circulation of concentrated absorbent solution between the generator and absorber and is always less than 1.0. Thus, the absorption chiller Carnot COP is always less than that of a vapour compression chiller with the same evaporator and condenser temperatures. The generator has the highest temperature and the evaporator the lowest temperature, with the condenser and absorber lying at intermediate (and similar) temperatures. With LiBr absorption chillers, the heat output from the condenser and absorber can be used as the input to drive a second absorption chiller. This produces a *double-effect* absorption chiller, although the COP is not quite doubled. The process can be continued to produce a triple-effect chiller, a quadruple-effect chiller and so on, but with diminishing returns. Many variations on the basic cycles are possible, as reviewed by Srikhirin et al (2001).

From Equation (6.10), it would appear that the Carnot COP for a single-effect chiller can exceed 1.0, but the four temperatures appearing in Equation (6.10) cannot be independently varied. In particular, a higher generator temperature tends to increase the COP, but is associated with higher absorber and condenser temperatures, which tend to reduce the COP. The net result is that the Carnot COP of a single-effect absorption chiller cannot exceed 1.0. If α is the Carnot COP of a single-effect chiller, the Carnot COP of an N-effect chiller is given by (Tozer, 2002):

$$COP = \alpha + \alpha^2 + ... + \alpha^N = \frac{\alpha(1 - \alpha^N)}{1 - \alpha} \qquad (6.11)$$

Higher-effect absorption chillers require a higher generator temperature, as given by:

$$T_g = T_c \left(\frac{T_a}{T_e} \right)^N \qquad (6.12)$$

Figure 6.56 shows the variation of COP of real single-, double- and triple-effect absorption chillers as a function of the temperature of the heat source. These three configurations require minimum source temperatures of about 60°C, 95°C and 140°C, respectively (commercial single- and double-effect units operate at somewhat higher temperatures; triple-effect units are not yet commercially available). Within each configuration, the COP initially increases with increasing source temperature, then levels off. For the conditions assumed in Figure 6.56 (29.4°C cooling water temperature and production of chilled water at 7.2°C), the maximum single-, double- and triple-effect chiller COPs are approximately 0.75, 1.25 and 1.75, respectively. Due to the high operating pressure of ammonia absorption chillers, multiple-effect systems are not possible, so the performance is limited to a COP of about 0.7. A factor limiting the commercialization of multiple-effect absorption chillers is the number and size of heat exchangers required. Novel concepts for developing miniature heat exchangers, as reviewed by Garimella (2003), could resolve this impediment to widespread use of advanced absorption chillers.

As shown previously in Figure 6.51, the COP of vapour compression chillers tends to decrease as the load decreases below about 60–80 per cent of full load, whereas the COP of an absorption chiller increases (although this is not always the case; see C. W. Park et al (2004) for a counter example). For the case shown in Figure 6.51, the COP reaches 140 per cent of the full load COP at 30 per cent of full load. The cooling

provided by an absorption chiller can be decreased by decreasing either the temperature or the rate of flow of the hot-water that drives the chiller. This is matched by a decrease in the flow of chilled water at a constant outgoing temperature.

Comparative cost and electrical energy use of absorption and vapour compression chillers

Costs of absorption and centrifugal chillers per unit of cooling capacity are compared in Figure 6.57. The primary results are taken from the RS Means database (Means, 1999), which does not specify the input temperature used to drive the absorption chiller. A higher input temperature gives greater cooling, and thus a smaller investment cost per unit of cooling capacity. Separate costs from Table 6.13 for 80°C and 110°C input temperatures are included in Figure 6.57. Absorption chillers are competitive with regard to cost compared to centrifugal chillers for a capacity of $2.5MW_c$ and larger, and possibly at lower capacities if the input temperature is 110°C.

The removal of one unit of heat from a building requires 1/COP units of energy, either

Figure 6.56 Variation of COP with heat-source temperature for single-, double- and triple-stage absorption chillers, assuming 29.4°C cooling water and production of chilled water at 7.2°C

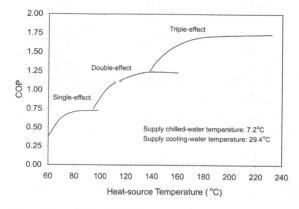

Source: Lee and Sherif (2001)

Figure 6.57 Variation of unit cost with capacity for centrifugal and absorption chillers

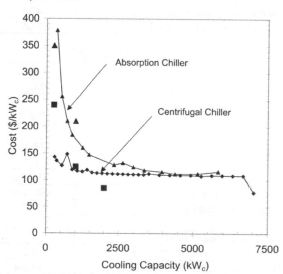

Note: Isolated squares are absorption chiller costs from IEA (1999b) for an input temperature of 110°C, while isolated triangles are for an input temperature of 80°C.
Source: Data in solid lines are from Means (1999)

as thermal energy in the case of an absorption chiller, or electrical energy that is ultimately dissipated as heat in the case of an electric vapour compression chiller. The amount of heat that must be rejected by the cooling tower is therefore $1+1/COP$. For electric and absorption chiller COPs of 6 and 0.7, respectively, 1.17 and 2.4 units of heat need to be rejected. Thus, cooling tower costs and energy use (by pumps and fans) will be substantially greater for absorption chillers. Water requirements will also be larger using an absorption chiller – something that could be a limiting factor in hot and dry climates. The water consumed by the cooling tower decreases the hotter the water is that drives the absorption chiller because the COP of an absorption chiller increases with increasing input temperature. Electric chillers commonly operate without a

cooling tower (albeit at lower efficiency), but such operation is still experimental with absorption chillers.

Energy use and cost data for vapour compression and single-, double- and triple-effect absorption chillers with a cooling capacity of 1750kW, and for the associated cooling tower, are compared in Table 6.14. As noted above, the cooling capacity of an absorption chiller (as well as the COP) decreases with decreasing input driving temperature. The vapour compression and absorption chiller costs are both about 30 per cent less than indicated in Figure 6.57. The single-effect absorption chiller is about 30 per cent more expensive than a vapour compression chiller, and the triple-effect chiller (not yet commercially available) is estimated to be about 120 per cent more expensive, for the driving

Table 6.14 Operating characteristics and costs of absorption and vapour compression chillers with a cooling capacity of 1750kW

	Absorption chillers			V-C chiller
	Single	Double	Triple	
Heat source (°C)	82.2	126.7	176.7	n/a
COP	0.716	1.224	1.671	5
Power input to generator or compressor (kW)	2444	1430	1047	350
Solution pump power (kW)	2.3	1.4	1.0	n/a
Cooling water pump power (kW)	30	22.8	20.0	15.0
Cooling tower fan power (kW)	47.7	36.2	31.8	23.9
Total electrical power (kW)	87.6	69.3	62.8	389
Auxiliary electrical power as a fraction of:				
Cooling power	0.050	0.040	0.036	0.022
Power input to generator or compressor	0.036	0.048	0.060	0.111
Cooling power+power input to generator or compressor	0.021	0.022	0.022	0.019
Chiller cost ($)	177,500	225,000	275,000	125,000
Chiller cost ($/kW,)	101	129	157	71
Cooling tower cost ($)	28,000	22,000	19,000	15,000
Total capital cost ($)	205,500	247,000	294,000	140,000
Total capital cost ($/kW,)	117	141	168	80
Payback period (Years)				
Thermal energy at $0.01/kWh	3.8	2.1	2.3	
Thermal energy at $0.02/kWh	infinite	10.6	4.4	

Note: Payback periods are calculated assuming a 15-year life, 3000 hours of equivalent full-load operation per year, 8% interest, 10 cents/kWh for electricity, and 1–2 cents/kWh for thermal energy. The COP is based on the generator or compressor energy input only. Costs for cooling tower makeup water are not included but are normally small.
Source of energy use and capital-cost data: Lee and Sherif (2001)

temperatures given in Table 6.14. The cooling tower is also more expensive, but the premium is smallest for the triple-effect chiller due to the smaller required heat rejection. Absorption chillers require electricity for pumping cooling water to the cooling tower, for the cooling-tower fans, and for the solution pumps. The first two are also needed by vapour compression chillers, but the energy use by the cooling tower pump and fan is about twice as large for absorption chillers due to the much greater amount of heat that must be rejected to the cooling tower by the absorption chiller. The electrical energy used by a single-effect absorption chiller (to operate pumps and fans) is 5 per cent of the thermal energy input, but about 20–25 per cent of the electrical energy used by the compressor of an electric vapour compression chiller. For the vapour compression chiller system considered here, the energy required to run the cooling tower fans and pumps is about 10 per cent of the compressor energy. Given the total capital cost, electrical and thermal energy use, and prices for electricity and thermal energy, the number of years required to pay back the extra cost of the absorption chiller through reduced energy costs can be calculated. If interest is ignored, this is referred to as the *simple payback*. Simple paybacks are given in Table 6.14 assuming an electricity cost of $0.10/kWh and thermal energy costs of $0.01–0.02/kWh ($2.78–5.56/GJ). Thermal energy this inexpensive is normally available only as waste heat or at times of surplus capacity (see Figure 2.10). At a thermal energy cost of $0.02/kWh, the operating cost of the single-effect absorption chiller exceeds that of the vapour compression chiller, and there is no payback.

Implications for primary energy

Given that fossil fuels can be burned to produce steam that is used to generate electricity that can be used in an electric chiller, or that steam could be used directly in an absorption chiller, the question arises as to which choice requires less primary energy. For dedicated production of either electricity at 55 per cent efficiency (as for state-of-the-art natural gas combined cycle power plants) or steam at 85 per cent efficiency, and with equal distribution losses, the electric chiller option will require less primary energy if the electric

chiller COP is more than 1.55 times the absorption chiller COP. This will always be the case. However, in the production of electricity from fossil fuels or biomass there is some waste heat, which could be used to drive an absorption chiller, except that there can be a penalty in terms of reduced electricity output. Whether an absorption chiller leads to a reduction in primary energy use under these circumstances depends on the ratio of useful heat extracted to electricity production sacrificed, as well as the ratio of the electric and absorption chiller COPs. As discussed in Chapter 4 (Section 4.6.3), if the alternatives are centralized electricity generation at 55 per cent efficiency and electric chillers, or gas turbines or fuel cells for on-site cogeneration of electricity and heat, the latter used in absorption chillers, the centralized/electric chiller option requires less energy. As discussed in Chapter 15 (Section 15.2.7), if the alternatives are dedicated electricity production using a natural gas combined cycle power plant and electric chillers, or natural gas combined cycle cogeneration and absorption chillers, either option could be preferable, depending on the electric chiller and absorption chiller COPs for the particular application under consideration.

6.6.2 Adsorption chillers

The simplest possible adsorption chiller system consists of a single bed of solid material that alternatively releases and then adsorbs a gas. The material that does the adsorbing is called the *adsorbent*, and the material that is adsorbed is called the *adsorbate*. Various adsorbent–adsorbate working pairs have been used, such as zeolite–water, zeolite–organic refrigerants, silica gel–water, salts–ammonia, and activated carbon–methanol. Silica gel–water is the most common pair.

The process of attracting and holding another substance is called *absorption* if a chemical change occurs (as when table salt absorbs water and liquefies) and is called *adsorption* when no chemical change takes place (as when silica gel desiccant adsorbs water – the desiccant does not change except from the addition of the weight of water). In adsorption, a substance is removed from either a gaseous or liquid solution and accumulates in concentrated form on the surface of a solid. In adsorption chillers, the material that is concentrated is the refrigerant. In absorption, one material interpenetrates the other as a solution. In the case of absorption chillers, one of the materials is a refrigerant. In both cases, heat is used to separate the concentrated refrigerant. Thermal energy is thus used to form a compressed liquid refrigerant, rather than mechanical energy through a compressor. The generic term for either absorption or adsorption is *sorption*.

In a single-bed adsorption system, a reactor vessel is heated to drive off the adsorbed gas, which flows to a condenser where it condenses and releases heat. This heat is removed by cooling water, which may flow to a cooling tower. The valve between the reactor vessel and condenser is then closed, and the reactor is cooled with the cooling water. When its temperature drops such that the gas pressure reaches the evaporation pressure of the gas, the valve is opened. The former condenser now serves as an evaporator, and the gas evaporates, flows to the reactor, and is adsorbed (releasing heat in the reactor vessel, which is removed by the cooling water). The evaporator cools sufficiently to chill the chilling water that now circulates through it. When the adsorption is complete, the flow of chilling water through the evaporator stops, and the process begins again with reheating of the reactor vessel.

The main disadvantage of this system is that cooling can be produced only part of the time. More continuous cooling can be achieved with two beds operating out of phase with a separate evaporator and condenser. At any given time, one bed serves as an adsorber and is connected to the evaporator, while the other bed serves as a desorber and is connected to the condenser. Periodically, the roles of the beds and the connections to the evaporator and condenser are reversed. Heat removed from the adsorber provides some of the heat required to drive the desorber, and this partial heat recovery improves the COP. Two-bed adsorption chillers were commercialized in Japan in the early 1990s, and at present, only two Japanese firms produce adsorption chillers. A four-bed chiller has about 70 per cent greater cooling capacity than a two-bed chiller, and a six-bed chiller about 40 per cent greater cooling capacity than a four-bed chiller (Chua et al, 2001). Four- and six-bed chillers may or may not be economically competitive with two-bed chillers. For a given

desorber inlet temperature, the COP falls slightly with an increase in the number of beds because the average regeneration temperature decreases. With an increase in the number of beds, the potential for heat recovery increases, which would increase the COP, but the system becomes more complicated (Wang, 2001).

Three factors limiting the COP of adsorption chillers are: (1) the loss of heat when a bed switches from desorber to adsorber; (2) the existence of an inert mass (the adsorbent); and (3) the low effectiveness of the heat exchangers due to the relatively thick active beds. The COP increases with increasing regeneration temperature and fractional heat recovery, and decreasing inert mass. In an absorption chiller, the inert mass is not important except during startup, because each component operates continuously in the same way. In addition, the solution circuit accomplishes internal heat exchange, and the heat exchange process is more effective. Cerkvenik et al (2001) have developed a model of an absorption chiller that reverts to an adsorption chiller as successive adsorption limitations are enabled. For an absorption chiller with a COP of 0.97, the calculated COP of a simple adsorption chiller (less than 50 per cent heat recovery) is 0.5 if the inertial mass is ignored, and 0.3 if it is accounted for. In all cases, the assumed generator/desorber temperature is 110°C. For a real two-bed chiller with heat recovery, Wang (2001) reports measured COPs of 0.32–0.4, in agreement with calculated values. With modifications that he considers to be realistic, and for an evaporator temperature of 5°C and a condenser temperature of 30°C, the calculated COP ranges from about 0.4 to 0.5 as the desorber temperature increases from 80°C to 120°C. Later simulation studies (Wang et al, 2005) indicate that, for cooling water at 30°C and a chilled-water return temperature of 15°C, the COP will increase from 0.52 to 0.60 and the cooling capacity will increase by two-thirds as the input temperature increases from 65°C to 85°C. A four-bed adsorption chiller is under development in Japan that requires an input temperature of 85°C and a cooling-water temperature of 32°C, and produces chilled water at 9°C with a COP of 0.6 (IEA, 2002a, Appendix 3b; see also Alam et al, 2004).

Although adsorption chillers have a lower COP than absorption chillers for generator temperatures in the 80–95°C range that is best for single-effect absorption chillers, they can operate at driving temperatures too low for an absorption chiller, with little decrease in COP. This is illustrated in Figure 6.58. This would permit operation in cogeneration mode with little or no sacrifice of electricity production. Other advantages of adsorption chillers compared with absorption chillers are that:

- LiBr (which is corrosive, requiring expensive corrosion-resistant steel alloys) and ammonia (which is toxic) are avoided;
- pumps are avoided and the only moving parts are electromagnetic valves;
- much less electricity use for a given cooling load, which reduces both operating cost and the use of primary energy;
- as a result of the absence of moving parts, there is no noise or vibration and a very long operating life;
- for LiBr absorption chillers, there is a risk of crystallization if the generator temperature drops below about 60°C.

However, at unit costs of $550–2200/kW$_c$ in the 50–430 kW$_c$ capacity range (Table 6.12), adsorption chillers are currently more expensive than absorption chillers ($250–400/kW$_c$) and reciprocating chillers ($160–350/kW$_c$) of similar capacity, and even more expensive than small residential air conditioners ($200–450/kW$_c$).

6.6.3 Ejector chillers

An *ejector* is a specially shaped nozzle that is designed to accelerate a primary high-pressure fluid flow (water) while drawing in a secondary low-pressure gaseous flow (the refrigerant). The mixing of the two streams occurs in a mixing chamber within the ejector, then exits through a diffuser (a wider portion of the ejector). A diagram of an ejector and of two alternative ejector cooling systems is given in Figure 6.59. The primary fluid is boiled in a boiler at high pressure and temperature. It flows through Line 1 to the ejector, where it accelerates to high velocity and, in the process, entrains vapour from Line 2. This creates a suction in an evaporator, inducing evaporation of the refrigerant at low pressure and hence at low temperature (a similar effect occurs in the evaporator of an electric chiller, where the evaporator is connected to the suction side of the compressor, and in the evaporator of an absorption chiller, where suction is created as the refrigerant is absorbed into a concentrated solution). This is where the chilling of the chilled water or air stream occurs; evaporator temperatures as low as −25°C can be achieved (Nguyen et al, 2001). The two flows from Lines 1 and 2 mix, exit the ejector through Line 3, and are cooled and condensed in a condenser. Heat is removed from the condenser by a flow of cooling water. The condensed refrigerant is pumped back to the boiler to repeat the cycle. The system can be arranged so that a pump is not needed, relying instead on gravity and the difference in density between the hot and cold refrigerant, as shown in the lower part of Figure 6.59. The first ejector refrigeration system was built in 1910, and it was popular during the 1930s, before being all but forgotten with the advent of mechanical compression refrigeration systems. It is used today in specialized engineering applications.

A number of different refrigerants can be used, with different required heat source temperatures: halocarbons (HCFC-123, HCFC-141b, HCFC-142, HFC-134a, HFC-152), a heat source as low as 60°C; ammonia, a heat

Figure 6.58 Comparison of the variation of COP with generator or desorber temperature for absorption and adsorption chillers

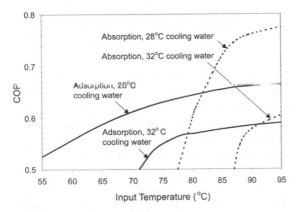

Source: Eicker (2003)

Figure 6.59 Illustration of (a) an ejector, (b) an ejector cooling system with a pump, and (c) an ejector cooling system without a pump

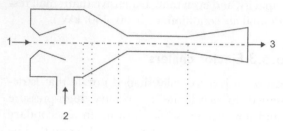

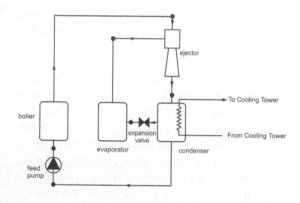

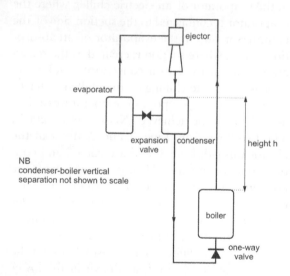

NB
condenser-boiler vertical
separation not shown to scale

Source: Rogdakis and Alexis (2000) for part (a), Nguyen et al (2001) for parts (b) and (c)

source greater than 90°C; water, a heat source of 120–160°C (Chunnanond and Aphornratana, 2004). The low heat-source temperature possible with halocarbons can be supplied with flat-plate solar collectors, but these halocarbons are all strong greenhouse gases (and HCFCs contribute to stratospheric ozone loss), as discussed in Section 6.7. Leakage rates need to be assessed and compared with those of commercial chillers (0.5 per cent per year for best practice). Ammonia is toxic and moderately flammable. Water has no adverse environmental effects, but requires a temperature often too high even for evacuated tube solar collectors (see Chapter 11, Section 11.4.1).

The theoretical maximum COP of an ejector chiller is the Carnot Cycle COP, given by (Yapici and Ersoy, 2005):

$$COP_{\text{cooling,ideal}} = \frac{T_E}{T_C - T_E} \frac{(T_G - T_C)}{T_G} \qquad (6.13)$$

where T_E, T_G and T_C are the evaporator, generator, and condenser temperatures, respectively. This equation is almost identical to that for an absorption chiller (Equation 6.10). Yapici and Ersoy (2005) have used a computer simulation model to compute the COP that can be achieved in practice for a variety of conditions. Using HCFC-123 as the refrigerant, they obtain a COP of 0.22 for $T_E = 5°C$, $T_C = 30°C$, and $T_G = 80°C$. For variations around these operating conditions, they find that:

- the COP increases from 0.12 to 0.30 as T_G increases from 60°C to 100°C;
- the COP increases from 0.11 to 0.68 as T_E increases from 0°C to 15°C;
- the COP decreases from 0.70 to 0.05 as T_C increases from 20°C to 40°C.

These variations assume that the ejector geometry is optimized for each combination of operating temperatures, but given that the ejector will be optimized for only one combination, the COP will not be as high at other conditions as implied by the above ranges. Other workers, such as Alexis and Karayiannis (2005), report smaller ejector COPs. It is particularly important to operate with a relatively high evaporator temperature (which requires displacement ventilation and chilled ceiling cooling systems, with desiccant dehumidification) and a low condenser temperature (which implies use of a cooling tower). The advantages of ejector cooling are (1) the absence of moving parts (except possibly for a pump) and associated long (30-year) lifespan; and (2) the absence of noise and vibration.

6.6.4 Solid desiccant and liquid desiccant dehumidifiers

When air conditioners or chillers are used, dehumidification (removing latent heat) can be achieved only as a byproduct of cooling the air (removing sensible heat). To remove latent heat, the evaporator must be at a temperature below the dewpoint of the air, so as to induce condensation of water vapour. The ratio of latent to sensible heat removal can be increased by lowering the evaporator temperature or by reducing the rate of airflow through the evaporator, but the typical achievable ratio without overcooling the air is only 1:3. However, in humid climates, where dehumidification can be up to 80–90 per cent of the total cooling load, conventional vapour compression chilling systems are usually unable to provide adequate control of humidity. If the sensible heat load is small, an air conditioner will run a small portion of the time, and water that had condensed on the coil will re-evaporate. This not only reduces efficiency, but undoes some of the dehumidification. Improvements in building envelope, shading and reductions in internal heat gain reduce the sensible cooling load while having little effect on the latent cooling load, thereby making it more difficult to achieve adequate dehumidification with conventional cooling equipment. Larger cooling coils permit warmer evaporator temperatures for the same sensible cooling load, thereby increasing chiller efficiency, but also at the expense of reduced dehumidification capability. The alternative is to overcool the air in order to induce adequate condensation, then reheat the air, but at the expense of greater energy use.

Dehumidification without overcooling can be done using solid or liquid desiccants that directly absorb moisture from the air, allowing one to *decouple* the cooling and dehumidification functions. This is particularly advantageous when dehumidification requirements are large. All desiccant dehumidification systems accomplish dehumidification without saturation, using heat as an input to regenerate the desiccant, thereby eliminating the potential for the growth of mould and bacteria and associated health hazards. This is a potentially significant non-energy benefit, as in most HVAC systems, the cooling coil will be wet 90 per cent or more of the time. As well, the growth of mould and bacteria on the cooling coil increases the resistance to airflow (thereby increasing fan energy use) and decreases the effectiveness of heat exchange (thereby increasing the energy use by the compressor), although these energy impacts are not well quantified. Mould and bacterial growth are normally constrained through acid cleaning (which is not completely effective) or through the use of UV radiation (which consumes further energy). Desiccant dehumidification eliminates the need for either while potentially making use of waste or solar heat.

The essence of a solid desiccant dehumidifier is a wheel rotating at 10–60 rpm and packed with porous solid desiccant. The air to be dehumidified flows through a portion of the wheel and moisture is adsorbed. The wheel then rotates into an air stream that has been heated in order to regenerate the desiccant by driving moisture from the desiccant. The essence of dehumidification with a liquid desiccant is to let the desiccant drip down a porous material with high surface area through which the air flows. The desiccant absorbs moisture, collects at the bottom, and is heated in order to drive off the moisture prior to reuse. Solid desiccant cooling is similar to adsorption cooling, and liquid desiccant cooling is similar to absorption cooling, except that desiccant cooling involves open cycles while adsorption and absorption cooling involve closed cycles. Net cooling is achieved with both solid and liquid desiccants by pre cooling a secondary airflow evaporatively and using it to cool the process airflow with a heat exchanger, and/or by direct evaporative cooling of the process airflow if the resulting humidity is acceptable. This requires first overdrying the process air.

Even if the desiccant provides no net cooling, the use of desiccants to handle the latent heat load reduces electricity consumption in two ways: (1) directly, by removing the latent cooling load from the chiller, and (2) indirectly, by allowing a warmer evaporator temperature, since the air needs to be cooled only to the temperature required to handle the sensible cooling load, rather than to the much colder temperature needed to condense water vapour. In chilled-ceiling and displacement ventilation systems (Chapter 7, Sections 7.4.4 and 7.4.5), where relatively warm cooling temperatures are possible, the indirect savings will be particularly large. If electricity is used for reheating air after conventional dehumidification, there is a further saving in electricity use. These factors are sufficient to give a reduction in total (electricity plus thermal) energy use. If waste heat or solar thermal energy is used in the thermal regeneration step, there will be a further savings in total energy use.

Neither solid nor liquid desiccant chillers can be used to cool water, which is disadvantageous because circulating chilled water requires less energy than circulating cool air for a given cooling effect (see Chapter 7, Section 7.1.2). Rather, their value is in the effective dehumidification of outside air.

Cooling and dehumidification with solid desiccants

Silica gel or titanium silicate are used as solid desiccants, the latter having a larger water holding capacity (Vineyard et al, 2002). Molecular sieves with a 3-Ångstrom spacing can be used to allow water molecules (2.8-Ångstrom dimension) but not larger molecules to reach the desiccant. The process by which water is attracted to and retained by the desiccant is *adsorption* (not absorption), as there is no chemical change in the desiccant when it holds water. The adsorptive property of the desiccant arises from its extreme porosity; $1m^3$ of silica gel is estimated to have a surface area of about $3 \times 10^7 m^2$ (Saha et al, 2001). When a solid desiccant adsorbs moisture, heat is released, warming the desiccant and the incoming air stream. The desiccant must be heated further, by a regeneration airstream, in order to drive off the adsorbed moisture.

Gas-fired desiccant dehumidification is widely used in supermarkets, where very low humidity is needed in order to prevent frosting of refrigerated displays. However, the temperature required to drive moisture from the desiccant is sufficiently low (50–150°C) that the required heat can often be provided by flat-plate solar collectors. Solar-powered desiccant systems are discussed in

Chapter 11 (Section 11.5.4).

Zhang and Niu (2002) present a mathematical model of desiccant wheels. Three-quarters of the face area of the desiccant wheel are typically used for dehumidification and one-quarter for regeneration. The desiccant, after regeneration, will be at a temperature of 70–90°C and must rotate into the incoming outdoor air stream through an angle of 5–10 per cent before it cools enough to begin adsorbing moisture. Conversely, a small portion (6–7 per cent) of the dehumidified air can be purged through the wheel immediately after it rotates out of the regeneration air stream, both to minimize heat transfer from the desiccant wheel to the incoming air stream and to maximize moisture removal from the incoming air.

The calculation of the COP of a solid-desiccant cycle under ideal conditions, and as successive idealizations are relaxed, is explained in Box 6.5. The key parameters are the effectiveness of the desiccant matrix (the amount of water vapour adsorbed as a fraction of the maximum possible adsorption for a given set of conditions), the effectiveness of the sensible heat exchanger (the change in the temperature of airflows passing through the heat exchanger as a fraction of the initial temperature difference between the two airflows), and the effectiveness of the evaporative cooler (defined by Equation 6.6). These effectivenesses depend on the physical characteristics of the components themselves but also on variable parameters such as the rotational speed of the desiccant wheel and the velocity of the incoming air stream (Vineyard et al, 2002). Figure 6.60 gives the measured variation in COP and supply air temperature with regeneration temperature for one particular solid desiccant system when the outdoor air is at 32°C and 40 per cent relative humidity. As seen from Figure 6.60 and explained in Box 6.5, the COP is higher the lower the regeneration temperature (decreasing from about 1.1 at a 50°C regeneration temperature, to 0.7 at a 70°C regeneration temperature), although less intense cooling is achieved. This is in contrast to other heat driven methods of cooling, where the COP decreases with decreasing input temperature. The COP is higher than that of a single-effect absorption chiller for regeneration temperatures below about 70–80°C, due to the sharp drop-off in the COP of absorption chillers below this temperature range. Because a smaller cooling load permits less intense cooling and higher COP, the energy benefit of reducing the cooling load is amplified. This is analogous to the positive synergism between heating load and boiler or heat pump efficiency, whereby a smaller heating load leads to greater equipment efficiency in supplying the reduced load (Chapter 4, Section 4.3.1, and Chapter 5, Section 5.5).

BOX 6.5 Principles governing the COP of solid desiccant chilling systems

This box derives the COP of an ideal solid desiccant cycle, and then shows quantitatively how various non-ideal features of real cycles affect the COP that can be achieved in practice. The discussion here is based on Collier (1997). The desiccant cycle to be considered consists of a desiccant wheel, a sensible heat exchanger and an evaporative cooler. For an ideal cycle:

- the latent heat of adsorption (the heat released when water is adsorbed by a desiccant, and which must be added to drive water from the desiccant) is equal to the latent heat of condensation (rather than the heat of condensation plus the heat of wetting);
- the heat capacity of the desiccant and the matrix that holds it is zero;
- the desiccant matrix, heat exchanger and evaporative cooler each have an effectiveness of 100 per cent.

Given the first two conditions, if outside air at point A in Figure B6.5.1 is dried by passing over a solid desiccant, it will move along a line of constant enthalpy to some point B. As there is a limitless supply of outside air, the outside air can be used with a heat exchanger to cool the air being processed back to its original temperature, represented by point C. The decrease in enthalpy of the process air is equal to the decrease in latent heat of the process air. The energy binding the water to the solid desiccant is equal to the heat of adsorption (which equals the latent heat of evaporation, by assumption), so the amount of energy that must be added to the desiccant to regenerate it is equal to the reduction in enthalpy of the process air. If this energy comes from a source external to the cycle, the COP will be 1.0. If, prior to cooling the process air, the outdoor air is cooled evaporatively to the outdoor wet-bulb temperature (point D), then the final temperature that can be achieved is given by point C'. The rate of cooling is proportional to the rate of adsorption of moisture (R_{ads}) by the desiccant plus the rate of evaporation (R_{evap}), while the heat required for regeneration is proportional to the rate of adsorption. The COP is thus given by:

$$COP = 1 + \frac{R_{evap}}{R_{ads}}$$

(B6.5.1)

Figure B6.5.1 Trajectory of an outside air parcel (at point A) as it is dried and heated by a solid desiccant wheel (to point B), then cooled through a heat exchanger with a secondary outside airstream to point C'. The secondary airstream had been evaporatively cooled to point D

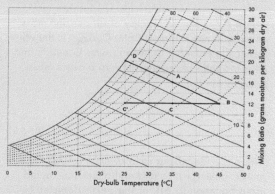

In effect, some free additional cooling is provided by the evaporative cooler.

Now consider the ideal cycle depicted in Figure B6.5.2. This contains a sensible heat exchanger that cools the process air after it passes through the desiccant wheel and warms outgoing indoor air prior to being heated in order to regenerate the desiccant. The indoor air is first cooled evaporatively from point D to point

E in order to provide maximum cooling of the process air. Distance B–C is equal to distance E–F; that is, the heat of adsorption plus the heat removed through evaporative cooling is entirely recuperated by the airflow that will be used for regenerating the desiccant. To regenerate the desiccant, it must be heated to point G, which falls on the line of constant relative humidity passing though point B. The regeneration heat that must be added is equal to the difference in enthalpy between points F and G, times the mass flow rate. The regeneration flow will be cooled as it drives moisture from the desiccant and, ideally, receive moisture until it is saturated (point H). The rate of removal of moisture from the desiccant is equal to the regeneration mass flow rate times the difference in mixing ratio between points G and H′, while the rate of adsorption of moisture by the desiccant is equal to the process mass flow rate times the difference in mixing ratio between points A and B. For a balanced cycle, these two rates must be the same. Inasmuch as the former mixing ratio difference is greater than the latter, only a portion of the airflow at F needs to be heated and passed through the desiccant wheel. The COP is given by:

$$COP = \frac{\Delta E_{A-C}}{FR\Delta E_{G-F}}$$

(B6.5.2)

where ΔE_{A-C} is the enthalpy of air per unit mass as point A minus that at point C, ΔE_{G-F} is the difference in enthalpy between points G and F, and FR is the ratio of regeneration to process mass flows, given by:

$$FR = \frac{r_A - r_B}{r_H - r_G}$$

(B6.5.3)

where r_i is the mixing ratio at point i. The temperatures, mixing ratios and enthalpies (computed using Equation (6.5)) for the points depicted in Figure B6.5.2 are given in Table B6.5.1. From these data it can be seen that the COP is 4.45 if the regeneration air proceeds to point H, and 2.86 if it proceeds to point H′.

We will now consider the impact of relaxing some of the above idealizations. First, assume a desiccant matrix effectiveness of 75 per cent rather than 100 per cent (because adsorption and regeneration proceed as a front from the entrance to the exit of the matrix, the water vapour is not adsorbed or desorbed to the full extent that it could be). This is represented by

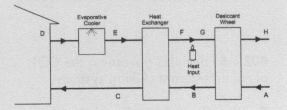

Figure B6.5.2 Trajectory of process and secondary airflows in a system consisting of a solid desiccant wheel, sensible heat wheel, evaporative cooler and heat source to regenerate the desiccant

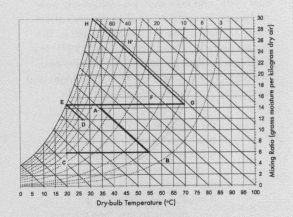

the point B′ in Figure B6.5.3, which, everything else unaltered, leads to final state C′. Regeneration air is warmed to state F′. The resulting COP (calculated from the data in Table B6.5.1) is 2.47 if the flow ratio from Equation B6.5.3 (0.595) is used. Secondly, we allow for a heat exchanger effectiveness of 85 per cent. The process air is cooled and the regeneration air is warmed by 85 per cent of the difference between points B′ and E, yielding states C″ and F″ respectively, and the COP is 1.73. Thirdly, we allow for an evaporative cooler effectiveness of 75 per cent. The regeneration air is initially cooled by 75 per cent of the difference between D and E, yielding state E‴. The process air is cooled to state C‴ and the regeneration air heated to state F‴ by the heat exchanger, producing a COP of 1.70. If FR = 1.0, the COP is 1.01. All of the above assumes that the heat of adsorption is equal to the latent heat of evaporation, whereas it is equal to the latent heat of evaporation plus the heat of wetting. As a result, the trajectory from point A slopes toward lines of increasing enthalpy, while points G and B no longer fall on lines of constant relative humidity. Trajectory A–B is replaced by trajectory A–B‴‴, for example, while the required

Table B6.5.1 States and enthalpies of the points shown in Figures B6.5.2 and B6.5.3

Point	Temperature (°C)	Mixing ratio (gm/kg)	Enthalpy (kJ/kg)
Figure B6.5.2			
A	35	14	71.06
B	55	6	70.85
C	20	6	35.31
D	27	12	57.72
E	20	14.6	57.13
F	55	14.6	93.21
G	70	14.6	108.67
H	31.8	30	108.70
H′	45	24.5	108.47
Figure B6.5.3			
A	35	14	71.06
B	55	6.08	71.06
B′	50	8.06	71.06
C	20	6.08	35.51
C′	20	8.06	40.54
C″	24.5	8.06	45.12
C‴	25.99	8.06	46.64
D	27	12	57.72
E	20	14.83	57.72
E‴	21.75	14.12	57.72
F	55	14.83	93.81
F	50	14.83	88.66
F″	45.5	14.83	83.02
F‴	45.76	14.12	82.26
G	70	14.83	109.28
G‴	69.15	14.12	106.55
H	45	24.82	109.28
H‴	44.15	24.11	106.55

regeneration energy is based on the larger heat of adsorption; the net result is to increase the COP. Radiative heat loss from the desiccant wheel to the surroundings tends to reduce the COP. A more detailed analysis of non-ideal cycles is found in Collier (1997).

An analysis similar to the above can be used to show that the COP of a solid desiccant system increases the lower the regeneration temperature, although the drying power decreases. For the ideal cycle shown in Figure B6.5.3 with FR = 1.0, the COP increases from 2.30 at a 70°C regeneration temperature to 3.91 at a regeneration temperature of 50°C, while the mixing ratio of the supply air increases from 6.1gm/kg to 10.5gm/kg. The COP also increases the warmer and/or the more humid the ambient air. For an ideal cycle with FR = 1.0, the variation of COP with initial conditions is as given in Table B6.5.2.

Table B6.5.2 Impact of initial conditions on the COP of an idealized desiccant cycle

Initial conditions		Final conditions		Regeneration T (°C)	Ideal COP
T (°C)	r (gm/kg)	T (°C)	R (gm/kg)		
25	14	20	8.1	50.2	1.89
35	14	20	8.1	60.0	2.93
35	20	20	8.1	74.4	4.73

Note that the COP increases with increasing initial temperature and humidity even though the regeneration temperature must be increased in order to produce the same final temperature and humidity. An important corollary is that the COP will be lower if the desiccant cycle is applied to outside air that has been mixed with recirculated indoor air. This is another factor in favour of dedicated outdoor air supplies, but these require supplemental hydronic heating and cooling to avoid excessively large ventilation air flows.

Finally, note that additional direct evaporative cooling could be applied to the process air stream after it has passed through the sensible heat exchanger. For example, air at point C in Figure B6.5.2 could be cooled to 16°C and r = 7.3mg/kg (about 65 per cent RH). The result is an indirect/direct evaporative cooler, enhanced by thermally driven desiccant dehumidification. This extends the applicability of evaporative cooling into the hot-humid regions of the world.

Figure B6.5.3 Impact of relaxing various idealizations in the solid desiccant cooling system depicted in Figure B6.5.2

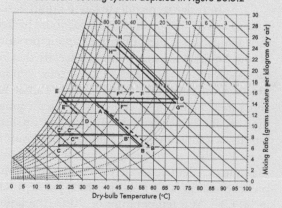

Figure 6.60 Variation of COP and supply air temperature with regeneration temperature for a solid desiccant chiller with incoming ambient air at 32°C and 40 per cent RH

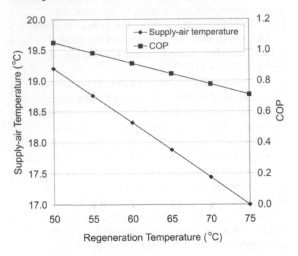

Source: IEA (1999b)

Cooling and dehumidification with liquid desiccants

A liquid desiccant is a liquid solution that has a strong affinity for water vapour. Typical liquid desiccants use triethylene glycol or salts such as calcium chloride or lithium chloride dissolved in water. The more concentrated the desiccant solution, the lower the vapour pressure of the solution, and the more readily it will absorb water vapour from the air. Liquid desiccant systems have been used in hospitals and large installations (Florides et al, 2002b).

Figure 6.61 shows the variation of the water vapour pressure of a LiCl solution of various concentrations as a function of the solution temperature.[13] An air parcel with an initial temperature of 30°C and water vapour mixing ratio of 16gm/kg (60 per cent RH) is plotted on Figure 6.61 as Point A. Condensation of water vapour reduces the mixing ratio but releases the latent heat of condensation, such that the parcel would follow a path parallel to the constant enthalpy lines in Figure 6.61. If water vapour is absorbed by a liquid desiccant, the latent heat of condensation is released as well as the additional latent heat (a further 10 per cent) associated with bonds between water molecules and the desiccant. Thus, as absorption occurs, the desiccant + air parcel follows a path to the right, sloping slightly above the constant enthalpy lines. The temperature at which the air and liquid desiccant come into equilibrium with one another (having equal vapour pressures) is referred to as the *brine-bulb temperature*, and is represented by point B in Figure 6.61 for a 40 per cent LiCl solution.[14, 15] As can be seen from Figure 6.61, absorption of water vapour entails substantial warming of the desiccant, increasing its vapour pressure and reducing its ability to absorb further moisture. It is thus important to provide cooling of the liquid desiccant as it operates. As seen from Point C in Figure 6.61, if enough heat can be removed to maintain a constant liquid desiccant temperature, the potential drying effect is more than doubled. Point D shows the final conditions that could be achieved starting from point A (for which $T_{wb} = 23.6°C$) for a 40 per cent LiCl solution if the air is cooled to within 3K of T_{wb} using cooling water supplied by a cooling tower. The sensible cooling between states A and D in Figure 6.61 is 'free' (as a result of evaporation) but the latent cooling is not; the latter equals the heat required to regenerate the desiccant. Neglecting heat loss to the surroundings, the COP of the liquid desiccant cycle is therefore given by:

$$COP_{ideal} = 1 + \frac{R_{evap}}{R_{ads}}$$

(6.14)

where R_{ads} is the rate of adsorption of moisture by the desiccant and R_{evap} is the rate of evaporation. This is the same expression as Equation (B6.5.1) for the ideal COP of a solid desiccant system with the regeneration heat supplied from outside the cycle.

In order to reconcentrate a liquid desiccant to a given concentration, it has to be heated to a temperature such that the vapour pressure of the liquid desiccant solution exceeds that of the overlying air. The greater the excess vapour pressure, the faster the liquid desiccant can be regenerated. Figure 6.62 shows the water vapour pressure for 30 per cent and 40 per cent LiCl solutions as a function of the solution temperature, as well as the vapour pressure for air with a mixing ratio of 22gm/kg, which is a rather moist ambient condition (it corresponds to 80 per cent RH at 30°C).[16] At a temperature of 70°C, the liquid desiccant vapour pressure would be six times the

Figure 6.61 Variation of liquid desiccant water vapour pressure with temperature for LiCl concentrations of 10 per cent, 20 per cent, 30 per cent and 40 per cent (heavy dashed lines)

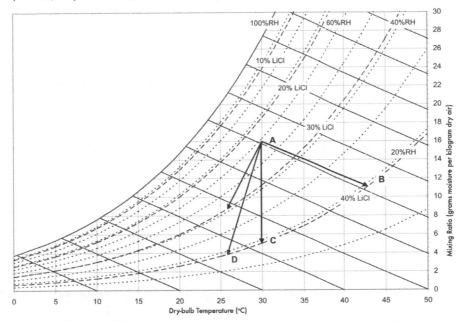

Note: Relative humidity (RH) is given as light dashed lines at 10 per cent intervals. The solid lines sloping upward to the left are lines of constant enthalpy. Also given is an initial ambient state (Point A), the brine-bulb temperature for a 40 per cent LiCl solution (Point B), the final state if enough heat is removed to prevent the desiccant from warming as it absorbs water vapour (Point C), and the final state if the dehumidified air can be cooled to within 1K of the ambient wet-bulb temperature (arrows sloping downward to the left). Water vapour pressure in equilibrium with aqueous LiCl as a function of temperature and LiCl concentration is computed from the analytic expression given in Conde Engineering (2004), with $P_{c,H2O} = 22.064 \times 10^6$ Pa and $T_{c,H2O} = 647.14$ K, as given in Conde Engineering (2002).

air water vapour pressure by the time it has been reconcentrated to 30 per cent, and three times by the time it has been reconcentrated to 40 per cent. This neglects the increase in the mixing ratio of the air as it absorbs water vapour from the liquid desiccant, which will be smaller the greater the rate of airflow past the liquid desiccant. Heat at 70°C can be readily achieved with solar thermal collectors or provided as low-grade waste heat. The greater the regeneration temperature, the greater the liquid desiccant concentration and, as seen from Figure 6.61, the greater the dehumidification potential.

Figure 6.63a is a schematic illustration of a basic liquid desiccant dehumidifier. Concentrated liquid desiccant drips through a porous absorber bed, creating a large surface area of desiccant for the absorption of water vapour from the air. The air to be dehumidified is drawn through the wet bed. Cool water, provided by a cooling tower, circulates through a coil in the absorber. The other major component is the regenerator, where water is driven from the now-dilute solution with heat,

Figure 6.62 Variation of water vapour pressure with temperature for an air parcel initially at 30°C, 101350Pa, and 80 per cent RH; and for liquid desiccant solutions at concentrations of 30 per cent and 40 per cent. The difference between the air and liquid desiccant vapour pressures will drive water from the liquid desiccant

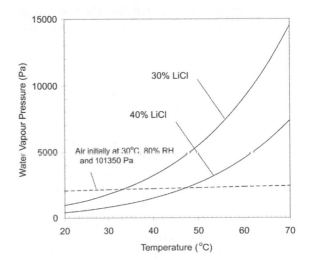

Figure 6.63 Schematic illustration of a liquid desiccant dehumidifier with (a) a single-effect regenerator, and (b) a double-effect regenerator

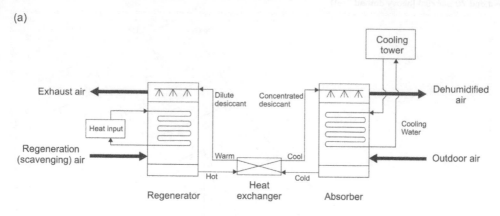

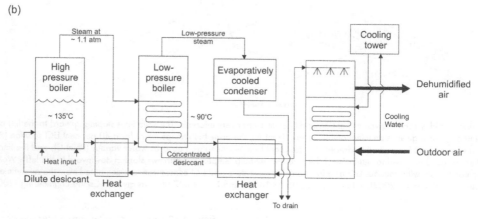

Source: Based on Lowenstein et al (1998)

reconcentrating the desiccant. The absorber and regenerator of a liquid desiccant dehumidifier are analogous to the absorber and generator of an absorption chiller, except that the water that is driven from the desiccant solution leaves the system rather than flowing through a condenser and evaporator, and the absorber is open to the process air. That is, the absorption chiller involves closed cycles (and produces chilled water), while the liquid desiccant chiller is an open system (and directly dehumidifies an air stream).

A key difference between solid and liquid desiccant systems from the standpoint of efficiency is that, in the solid desiccant system, much of the heat that is released during adsorption of water vapour by the desiccant is used for later regeneration of the desiccant, through the sensible heat exchanger depicted in Box 6.5, whereas the heat of absorption is lost with the cooling water and cooling tower in the liquid desiccant system depicted in Figure 6.63a. On the other hand,

some of the heat in the hot, regenerated desiccant is transferred to the dilute desiccant through the heat exchanger shown in Figure 6.63a, where it can be used again. Nevertheless, liquid desiccant systems will tend to have a lower COP than solid desiccant systems. However, the removal of heat by the cooling water allows a much lower humidity mixing ratio for a given regeneration temperature using liquid desiccants. Thus, where there is abundant, low-cost and low-temperature heat available (such as thermal solar collectors sized for winter use), liquid desiccants might be preferred.

Enhancing the performance and acceptability of liquid desiccant dehumidifiers

Three critical issues with regard to the performance and acceptability of liquid desiccant dehumidifiers are:

- to cool the liquid desiccant as effectively as possible while it absorbs water vapour;
- to recapture waste heat flows when the liquid desiccant is regenerated so as to maximize the thermal COP of the regenerator;
- to minimize the rate of flow of liquid desiccant through the absorption bed in order to prevent the generation and entry of liquid desiccant droplets into the supply air.

The use of liquid desiccant dehumidifiers has been limited by cost (several times greater than vapour compression equipment of comparable capacity), the perception of greater maintenance requirements, and the tendency for liquid desiccant droplets to enter the supply air stream unless blocked by a droplet filter or demister. Cost can potentially be reduced through improved performance (and large-scale production). The key to avoiding the generation of liquid desiccant droplets is to achieve effective removal of heat from the absorber bed, so that the liquid desiccant flow rate can be substantially reduced. This will also save electricity use by the pump, which can be 20 per cent that of a comparably sized vapour compression chiller.

There are two options with regard to cooling the absorber bed: (1) to circulate cooling water from a cooling tower through the bed, as in the system shown in Figure 6.63a, or (2) to modify the absorber bed to act as an indirect evaporative cooler, as illustrated in Figure 6.64. In the latter case there are two airflows – the process air and cooling air – which flow through narrow channels that alternate with one another, as in a conventional indirect evaporative cooler (see Figure 6.44b), the only difference being that liquid desiccant flows down the walls of the process air channels. Cooling the absorber bed with water from a cooling tower ultimately relies on evaporative cooling too. The evaporative cooler looks simpler, and the elimination of a cooling tower would make it more amenable to small-scale applications. In addition, the cooling water supplied from a cooling tower will be at least 3K warmer than the wet-bulb temperature (T_{wb}), whereas the water in the evaporative cooler can be maintained at T_{wb} on the cooling side of the plates. However, it is difficult to implement the

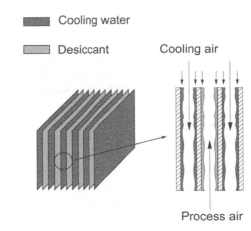

Figure 6.64 A modified indirect evaporative cooler serving as the absorber of a liquid desiccant air conditioner

Source: Lowenstein and Novosel (1995)

Figure 6.65 Diagram of the low-flow liquid desiccant absorber developed by Lowenstein et al (2005)

Note: Cooling water flows through small vertical channels inside the plates, liquid desiccant drips down the sides of the plates, and process air flows horizontally between the plates.
Source: Lowenstein et al (2005)

evaporative cooler in practice, the two biggest problems being keeping the water and desiccant flows separate and preventing carry over of liquid desiccant droplets into the supply air (Lowenstein, personal communication, 2005). Carry over is a problem even with low desiccant flow rates, as it is difficult to maintain a uniform spacing between the plates without using a lot of spacers, and the desiccant tends to bridge where the air gaps are narrow. Nevertheless, groups in the US and Europe are working on the development of evaporatively cooled liquid desiccant absorbers.

Lowenstein et al (2005) have developed an improved method of cooling the absorber bed with cooling-tower water, by circulating the cooling water through very small channels embedded in 2.5mm thick plates. The plates are stacked vertically as shown in Figure 6.65, with process air flowing horizontally in the channels between the plates. This is a parallel plate liquid-to-air heat exchanger. Because it permits very effective removal of heat from the liquid desiccant (in spite of the cooling water being 6K warmer than T_{wb}), desiccant flow rates can be 20–30 times less than in a conventional liquid desiccant dehumidifier, eliminating the generation and carry over of liquid desiccant droplets into the supply air, as well as reducing pump energy use.

The thermal COP (ratio of heat added to the latent heat of the water that is driven off) of most commercially available regenerators is about 0.55–0.75. The COP of a single-stage regenerator is constrained to be less than 1.0 because the heat that must be added to regenerate the liquid desiccant is greater than the latent heat of condensation, due to the additional latent heat of solution, and because of heat losses from the regenerator. In the system shown in Figure 6.63a, some heat is lost in the hot concentrated liquid desiccant (only some of which is transferred to the incoming dilute desiccant through the heat exchanger), some heat is lost in the moist exhaust air, and some heat is radiated to the surroundings. However, if all of the heat lost from the regenerator could be captured and used to preheat the incoming weak desiccant solution, the COP would be about 12. The COP that can be achieved in practice is much lower, but should reach 1.5–1.8 (Lowenstein et al, 1998). One way to do this would be through a *double-effect* regenerator, which is illustrated in Figure 6.63b. The first step involves boiling off some water in a high-pressure boiler to produce steam at a pressure of 1.1atm. This requires a much higher regeneration temperature – roughly 150°C – than used in the basic liquid desiccant dehumidifier. The partly reconcentrated liquid desiccant flows to a second, low-pressure (close to a vacuum) boiler that is heated by the steam from the first step, driving off more water and further concentrating the desiccant. Note that, unlike a solid desiccant system or the basic liquid desiccant system of Figure 6.63a, heat is added directly to the desiccant here. Computer

simulations by Lowenstein and Novosel (1995) indicate that, for an air conditioning system consisting of the evaporatively cooled absorber shown in Figure 6.64 and the double-effect regenerator shown in Figure 6.63b, the seasonally averaged thermal COP would range from a low of 1.44 (in the humid climate of Houston) to a high of 2.24 (in the dry climate of Phoenix) for a cooling thermostat setting of 24.4°C. Performance with the water-cooled absorber would be 15–20 per cent less. A 1.5-effect regenerator (not a true double-effect regenerator) is currently under development that should have a thermal COP of at least 1.0.

Two companies that sell thermally driven liquid desiccant dehumidifiers are Drycor (www.drycor.com) and American Genius (www.geniusac.com). The latter company provides thermally driven and hybrid liquid desiccant/vapour compression units. In the hybrid unit, the liquid desiccant is regenerated solely using waste heat from the chiller condenser. For processing of air at 35°C, the hybrid unit reduces electricity use by 37 per cent (for 40 per cent initial RH) to 55 per cent (for 60 per cent initial RH) compared to a conventional chiller with a COP of 4.5. The conventional chiller produces supply air at 12.8°C and 100 per cent RH, which then needs to be partially reheated, whereas the hybrid unit initially produces supply air at 23.3°C and 25 per cent RH, which is then cooled further evaporatively to 16.7°C and 66 per cent RH. When dehumidification is not needed, the hybrid unit can operate as an indirect evaporative cooler, without using the compressor.

Liquid desiccants have received considerable attention in China. Z. Li et al (2005) have developed a system whereby the evaporator of a chiller is used to cool the liquid desiccant before it absorbs water (rather than as it absorbs water). At the same time, heat from the chiller condenser is used to regenerate the desiccant, which then circulates back to the evaporator while giving off heat to the return flow of dilute desiccant through a heat exchanger. A second liquid desiccant loop is used in heat exchangers to passively precool and pre-dehumidify the incoming fresh air with outgoing exhaust air. During the winter, the system can be used to warm and humidify incoming outside air. Experimental results on a prototype system combined with seasonal simulations indicate a summer COP (based on electricity use by the chiller and solution pumps) of 6.3–7.3 and a winter COP of 4.7–5.0 for Beijing climatic conditions. The system is said to remove pollutants while avoiding carry over of liquid desiccant into the supply air stream.

Primary energy required using solid and liquid desiccants

For a desiccant cooling system with a COP of 1.1 (which is higher than most) and a boiler with an efficiency of 0.9 (also higher than what is commonly chosen), one unit of fuel energy provides one unit of cooling. If electricity is generated and transmitted with 33 per cent efficiency, then one unit of fuel provides one unit of cooling using a chiller with a COP of 3.0, but provides two units of cooling if the chiller COP is 6.0. In the first case, there is no net saving in primary energy using desiccant cooling, while in the second case, the electric chiller uses half the primary energy – or so it would appear. In reality, a net saving can arise because (1) if any residual cooling is required after using a desiccant chiller, the electric chiller can operate at a higher COP because the air does not need to be overcooled, so the evaporator can be warmer; (2) the energy used for overcooling is avoided; and (3) reheating is avoided. Up to half of the heat used for regeneration would be used for reheating in a conventional system, so the effective COP of the desiccant cooler – based on the incremental heating energy use – is up to twice as large. A detailed analysis of the energy use in specific solid and liquid desiccant systems is presented in Chapter 7 (Sections 7.4.11 and 7.4.12), where it can be seen that primary energy savings of 25–30 per cent can be achieved compared to small-scale conventional systems with a chiller COP of 3.0.

If the regeneration heat is waste heat or solar heat, desiccant systems are even better from a primary energy point of view. Because relatively low (50–60°C) regeneration temperatures are possible, heat can be taken from a simple cycle or combined cycle cogeneration power plant with no reduction in electricity generation (see Chapter 15, Section 15.2.3). With absorption chillers, there is a small reduction in electricity production, such that either absorption chillers or electric chillers can minimize primary energy use, depending on their relative COPs (see Chapter 15, Section 15.2.7).

Comparison and relative cost of solid and liquid desiccant systems

Among the advantages of solid desiccant systems compared to liquid desiccant systems are:

- higher COP (at least for present systems), due to recycling of heat used for regeneration;
- more efficient use of water because a cooling tower is not needed (although liquid desiccant systems without cooling towers are under development);
- lower cost.

Among the advantages of liquid desiccant systems are:

- a lower vapour pressure can be achieved for a given regeneration temperature, due to the fact that the heat of sorption is removed during the absorption process (rather than after adsorption, as is the case with solid desiccants);
- the pressure drop in the airflow is smaller (so less fan energy is needed);
- several dehumidifiers can be connected to one generator by pumping liquid desiccant;
- concentrated liquid desiccant can be created at times when excess heat is available and stored for later use (the energy storage density of a 40 per cent LiCl solution is 800–1400MJ/m^3, depending on the water vapour mixing ratio of the process air (Kessling et al, 1998), while the latent heat of ice (which is sometimes used for storing coldness) is only 335MJ/m^3).

As noted in Table 6.13, solid desiccant systems cost on the order of $360–750/kW$_c$ for capacities of 50–400kW$_c$, compared with $160–350/kW$_c$ for rooftop units of comparable cooling capacity. Nevertheless, solid desiccant systems are occasionally used because of their superior dehumidification capability. Liquid desiccant systems are even more expensive, relatively new and little used. However, The US Department of Energy, through the National Renewable Energy Laboratory (NREL) in Golden, Colorado, is funding research into a variety of promising liquid desiccant technologies that have the potential to be 50 per cent smaller and 40 per cent less costly than solid desiccant systems (Slayzak, 2001). Lowenstein (personal communication, 2005) expects that liquid desiccant systems could reach $230–260/kW$_c$ once the technology and market mature.

Comparison of vapour compression, solid desiccant and liquid desiccant processes

The differences in vapour compression, solid desiccant and liquid desiccant processes for cooling and dehumidification can be readily seen by plotting the trajectories for each process on a psychrometric chart. These trajectories are shown in Figure 6.66. The vapour compression approach requires overcooling the air to force condensation of water vapour, then reheating it. Solid desiccants involve heating the air as moisture is removed by the hot desiccant wheel, then recooling it. Only the liquid desiccant approach has the potential to take the air directly to the desired temperature and humidity combination without first heating or overcooling the air.

6.6.5 Hybrid chillers

Various combinations of the above types of chillers are possible, including absorption/vapour compression chillers (Sveine et al, 1998; Swinney et al, 2001), absorption/ejector chillers (Eames and Wu, 2000; Sözen et al, 2004a, b), and ejector/vapour compression chillers (Huang et al, 2001; Hernández et al, 2003). The hybrids improve the overall COP and capacity compared to stand-alone absorption or ejector chillers while allowing heat to serve as the primary energy input. As noted in Section 6.6.4, a hybrid liquid desiccant/vapour compression chiller is already on the market. Solid and liquid desiccant devices have been combined with vapour compression chillers and other equipment in various ways, as discussed in Chapter 7 (Sections 7.4.11 and 7.4.12).

6.7 Trade-offs involving the choice of refrigerant

As noted earlier, vapour compression cooling equipment generally uses halocarbon gases as a refrigerant. The traditional refrigerants were CFC-11, CFC-12 or HCFC-22. The CFCs have been largely phased out in order to reduce damage to the ozone layer and replaced with HCFCs,

Figure 6.66 Comparison of the trajectories of an outside air parcel that is cooled and dehumidified through the conventional approach using a vapour compression chiller, using a solid desiccant wheel and using a liquid desiccant system

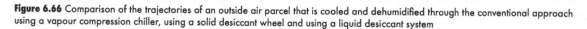

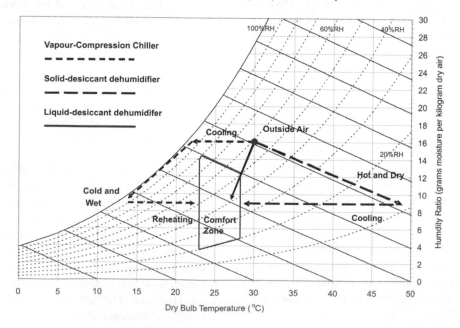

Table 6.15 Traditional and common replacement refrigerants used in different kinds of cooling equipment

Equipment type	Traditional refrigerant	Replacement refrigerants
Room air conditioner	HCFC-22	R-407c, R-410a
Rooftop air conditioner	HCFC-22	R-407c, R-410a
Packaged terminal air conditioners	HCFC-22	R-407c, R-410a
Packaged terminal heat pumps	HCFC-22	R-407c, R-410a
Heat pumps	HCFC-22	R-407c, R-410a
Reciprocating chiller	HCFC-22	R-407c, R-410a
Rotary screw chiller	HCFC-22	R-407c, HFC-134a
Scroll chiller	HCFC-22	R-407c, R-410a
Centrifugal chiller	CFC-11, CFC-12	HCFC-123, HFC-134a
Absorption chiller	R-718 (water)	R-718
Ejector chiller	HCFC-123, HCFC-141b, HFC-134a	

Note: R-407c and R-410a are mixtures whose composition is given in Table 5.7.
Source: Westphalen and Koszalinski (2001), except for ejector chillers, where the list is taken from Section 6.6.3

which will in turn be phased out by 2010 in developed countries and replaced with HFCs or other refrigerants. Traditional and prospective replacement refrigerants used in different kinds of cooling equipment are listed in Table 6.15. The relative warming effect of different gases is represented by an index called the global warming potential (GWP), which is briefly described in Chapter 2 (Section 2.4.3). GWP values of various refrigerants are given in Table 2.4. Note that HFCs are generally several thousand times more powerful as greenhouse gases than CO_2, the exception being HFC-152a, which has a GWP of 122.

However, the climatic impact of air conditioners and chillers is overwhelmingly related to the energy used to power them. For leakage of HFC refrigerants at rates of 1–2 per cent (best practice is about 0.5%/year) and recovery of 85 per cent of the refrigerant at the end of a 15-year life, refrigerant leakage accounts for only 1–2 per cent of the total impact on climate of the cooling equipment (Piexoto et al, 2005). At 50 per cent end-of-life recovery, the refrigerant still accounts for only 10 per cent of the total impact. For a residential split air conditioner with 4%/year leakage of R-410A (a 50:50 mixture of HFC-32 and HFC-125), Barnes and Bullard (2004) find that the global warming effect of refrigerant leakage is about one-ninth that of the CO_2 produced in supplying the air conditioner using average US electricity (having a CO_2 emission of 0.18kgC/kWh). Refrigerant emissions per unit of cooling have fallen by a factor of ten over the past 30 years.

6.8 Correct estimation of cooling loads to avoid oversizing of equipment

In commercial buildings, internal heat gains represent a large fraction of the total cooling load (Figure 6.4). These are often estimated using standardized guidelines that are far in excess of actual cooling loads. For example, typical design guidelines assume the lighting load to be about $20W/m^2$, while measurements in recent systems indicate an average load of about $14W/m^2$ – a number that will drop with further improvements in lighting systems. The average power used by office equipment, and the associated waste heat, is generally only a small fraction of the rated power of the equipment, but it is the latter that is normally assumed in sizing the cooling equipment. Design guidelines overestimate the cooling loads by occupants and by office equipment by 68 per cent and 48 per cent, respectively, based on measured values by Lee et al (2001a) for Hong Kong offices. This results in cooling equipment that is greatly oversized, requiring it to operate at a smaller fraction of its peak capacity than would otherwise be the case. As discussed in Section 6.5.2, vapour compression chillers suffer a significant decrease in efficiency below 50–60 per cent of full load, even when equipped with VSDs. In Hong Kong, many office buildings need only two-thirds of their chillers in order to meet the cooling requirements on the hottest days. Based on a detailed analysis of six buildings in Hong

Kong, Lee et al (2001a) find that more realistic sizing of the cooling equipment alone would have reduced annual cooling energy use by 6–22 per cent and reduced the size of the chilling equipment by 32–46 per cent (thus significantly reducing the capital cost). Hastings (1995) reports that, for typical European office buildings, measured internal heat gains are also much less than the gains commonly assumed during the design process. A corollary of cooling equipment being oversized due to an overestimate of internal heat gains is that the potential contribution and cost-effectiveness of passive solar heating during winter in cold climates will be underestimated.

Oversized air conditioners are also a problem in residential central air conditioning systems. According to field studies reviewed by Neal (1998), central air conditioning systems in the US are oversized by an average of 25–100 per cent. Oversizing is used as a method of minimizing customer complaints, that is, to cover up other problems (Mast et al, 2000). This directly leads to lower average efficiency by requiring frequent on/off cycling. Paradoxically, it creates comfort problems because the run times of the system are so short that the blower does not have time to fully mix the air, causing the occupant to lower the thermostat setting, thereby further increasing energy use. The dehumidification capacity is also reduced because water that condenses on the evaporator coil during the on cycle will partially re-evaporate during the off cycle. At 40 per cent of full load and with on/off cycling, there can be no moisture removal at all (Shirey and Henderson, 2004). The net result of all the above is that an air conditioning system with a rated seasonally averaged COP of 3.5 may have an average operating COP of only 1.9 or less.

To elaborate on the loss of dehumidification capacity due to oversizing, Lstiburek (2002) reports that, for a residential air conditioning system oversized by 20 per cent, only 15 per cent of the total cooling energy under certain part-load conditions is for dehumidification. For a correctly sized system, dehumidification is 30 per cent of the total load under part-load conditions with fixed fan and compressor speed, and 35 per cent using variable speed fans but a fixed compressor speed. Moisture removal would have been even greater with a variable-speed compressor.

The inability of oversized systems to remove moisture under part-load conditions can be a serious problem in humid climates with moderate temperatures, where little cooling is required. Unfortunately, many designers, when seeing that there will be a large moisture load, compound the problem by adding cooling capacity (Harriman and Judge, 2002). As discussed in Section 6.4.4, stand-alone dehumidifiers are even less efficient than residential air conditioners (thermally driven desiccants, discussed in Section 6.6.4, are a better solution).

To avoid oversizing cooling equipment, an accurate estimate of the peak cooling load is required. A number of software tools, usually provided by equipment vendors, exist for this purpose. They do not involve a simulation of the energy flows and temperatures within the building, but instead, are little more than rules of thumb. More accurate estimates can be obtained through building energy simulations. These are needed in order to achieve large energy savings at minimal additional first-cost as part of an integrated design process (see Chapter 13, Section 13.3.2). Thus, capital cost savings achieved through more accurate sizing of cooling (and other HVAC) systems is an additional benefit of the simulation-based, integrated design process.

6.9 Efficiency trade-offs with off-peak air conditioning

Diurnal storage of coldness in large tanks of chilled water is widely used in building cooling systems, with chilled water produced by running chillers during the night when electricity tariffs are lower, and using the stored coldness for at least part of the air conditioning load during the day (electricity tariffs are discussed in Chapter 2, Section 2.3). In the longer run, cold storage tanks could be charged whenever there is excess wind or solar generated electricity available (see Chapter 15, Section 15.5.6). Diurnal storage of coldness in the US has allowed 15GW of electricity demand to be shifted to off-peak periods (Dinçer and Rosen, 2002, p109), out of a total peak demand of about 800GW. Examples of cold water storage tanks are shown in Figure 6.67. Economically successful cold storage systems have been used in hot and humid climates such as Rio de Janeiro (Brazil), Kuala

Lumpur (Malaysia) and Orlando (Florida), in hot and dry climates such as Riyadh (Saudi Arabia), western Australia and Phoenix (Arizona), and in climates where cooling is needed for only a short part of the year (such as the northern US and Edmonton, Canada). The required storage volume is greatly reduced if an ice water slurry is created in the storage tank, this being due to the fact that the latent heat of fusion of water (335kJ/kg) is substantially greater than the heat extracted in cooling water by, say, 10K (41.82kJ/kg). If an ice water slurry (at 0°C) with only 10 per cent ice is created at night, then twice as much coldness can be stored in a given volume compared to chilling water from 17°C down to 5°C. At 50 per cent ice, 4.5 times as much coldness is stored. A

Figure 6.67 Upper: University of Texas (El Paso) 15100m³ cold water storage tank, built in 1999. Cold water is supplied at 4.4°C and returns at 11.7°C, giving a cooling storage capacity of 380GJ. Lower: 15,600m³ cold-water storage tank at the 3M Company in Maplewood, Minnesota, built in 1992. Cold water is supplied at 4.4°C and returns at 12.8°C, giving a cooling storage capacity of 405GJ

Source: John Andrepoint, The Cool Solutions Company

limited amount of storage can also occur simply in the building mass by running the cooling system at night, particularly with concrete core hydronic radiant cooling (Chapter 7, Section 7.4.4).

Ice can be generated either by circulating glycol or brine at 3–6K below the freezing point of water between the evaporator and ice tank, or by placing the evaporator coil directly in the ice tank (a direct expansion system). In an *ice-on-coil* chiller, ice remains on the coil as it forms. This decreases the performance of the chiller as the thickness of ice on the coil increases. In an *ice harvester* chiller, there is less ice-making surface but the surface is specially designed to periodically release the ice. This, however, requires periodically adding heat to the coil, which entails a significant energy penalty. *Ice slurry* systems produce small particles of ice within a solution of glycol and water, resulting in a slushy mixture that can be pumped. They are not subject to efficiency penalties associated with a buildup of ice on the evaporator coil or periodic partial melting of the ice. In *encapsulated ice* systems, water freezes inside spherical plastic containers that are 10–12cm in diameter and immersed in a flowing subfreezing coolant. A cold water system requires a stratified storage tank to prevent mixing between the newly chilled water and the return water, which otherwise reduces the cold storage capacity. A stratified storage tank is not necessary with ice storage. The first cost of a system with chilled water/ice storage is often no greater than an equivalent system without storage, as the storage capital cost is offset by cost savings from significantly smaller chillers (they do not need to be sized to meet short-term peak loads if there is storage), reduced transformer and electricity distribution capacity and smaller pipes, pumps and air handlers, which are permitted due to the colder distribution temperature. Operating costs decrease by up to 20 per cent, due largely to lower electricity rates at night.

6.9.1 Impact of ice thermal storage on on-site energy use

The net effect of storage on the *on-site* energy use depends on the following factors:

- the ambient air temperature is lower at night, which leads in principle to a larger chiller COP;

- the chiller evaporator temperature needs to be lower (as cold as $-10°C$) in order to make ice, which reduces the chiller COP;
- the chiller can be operated at a steady load, which sometimes corresponds to maximum efficiency (the least efficient operation is on/off cycling);
- pump or fan energy use is reduced, since a lower rate of water or airflow will remove heat at the same rate if the water or air temperature is colder;
- when incoming ventilation air is cooled to a very low distribution temperature prior to mixing with the ventilation air in an air-cooling system, the relative humidity of the resulting mixture can be reduced to 30–40 per cent rather than reduced to or kept at 60–70 per cent. A lower relative humidity permits higher room temperatures with the same degree of comfort, thereby providing some air conditioning energy savings if reheating is avoided.

The reduction in ambient temperature at night is particularly beneficial for cooling equipment (such as packaged rooftop air conditioners) with air-cooled condensers as long as they are operated with a variable condenser temperature (see Section 6.5.1). The benefit of night-time operation is smaller for systems with water-cooled condensers and cooling towers, because (as explained in Section 6.1.2), the wet-bulb temperature (which governs the temperature of the cooling water) falls only about one-third as far as the dry-bulb temperature at night.

The distribution temperature is typically around 5–7°C in hydronic cooling systems without storage and the evaporator temperature will be around 0–2°C, while in air-cooling systems, the air will have initially been cooled down to 5–10°C or so for dehumidification (and requiring an evaporator temperature of 0–5°C or less), then partially reheated, as explained in Section 6.1.4. If an evaporator temperature of −10°C is used in an ice-water storage system, the drop in

COP will be 0.7–1.0 in a hydronic chilling system and 1.7–2.7 in an air-cooling system (assuming a condenser temperature of 40–50°C and a Carnot efficiency of 0.65), as can be inferred from Figure 5.2. This acts against use of ice thermal storage. The remaining factors listed above tend to reduce the on-site energy consumption. However, the advantage of operating the chiller at a steady load with ice storage will not be as large if the conventional system (without ice storage) were to use chillers with variable-speed drives or two or more chillers staged so as to maximize overall efficiency. Nevertheless, in a well-designed system, the balance of the above factors can act to slightly reduce on-site energy use with ice thermal energy storage compared to a system without ice thermal energy storage.

For example, Simmonds (1994) found that chiller priority and storage priority ice-based thermal storage systems with air-cooled chillers would save 3 per cent and 14 per cent, respectively, compared to a conventional system in one specific case.[17] Henze et al (2003) have analysed the operating cost savings, energy savings and peak demand savings for ice storage with water-cooled chillers for a variety of buildings in the US state of Wisconsin.[18] All of these savings depend on how the system is operated, and if total costs are to be minimized, this depends on the difference in electricity rates for daytime and night-time electricity (electricity rates for large users generally have a part related to energy use (kWh) and a part related to peak demand (kW)). Henze et al (2003) therefore considered eight different rate structures. They find, for the operating strategy that minimizes total operating costs in a 20-storey office building, that the change in energy use for air conditioning ranges from a decrease of 3.0 per cent to an increase of 1.1 per cent. Peak demand decreases by about 20–30 per cent. Comparable impacts on energy use and peak demand were found for a grocery store.

Chaichana et al (2001) performed a similar analysis for commercial buildings in Thailand. They compared a conventional system with systems using partial and full ice storage. Under partial storage, enough ice is made at night to meet part of the daytime load, so the chillers have to run at part load during the day to supplement cooling from the storage. With full storage, the chillers are run only at night to charge the storage tank. They found that partial storage increased on-site energy use by 18 per cent while full storage decreased it by 5 per cent (while reducing electricity costs by 55 per cent, given the Thai rate structure). The relative energy use for systems with and without storage depends strongly on the part-load efficiencies of the chillers.

These analyses do not consider free cooling (described in Section 6.10.2) and the impact of ice thermal storage on the opportunities for free cooling. To obtain on-site energy savings and minimize the capital cost using ice storage (or to minimize the increase in energy use), the cooling system should be designed to take advantage of the low chilled water temperature. This means that pump sizes and flow rates will be smaller than for a system without ice thermal storage, so that at any cooling load a lower distribution temperature will be needed. This in turn will reduce the number of days when the cooling tower can produce water cold enough to satisfy the cooling load. For many climates, this could be a significant penalty.

Furthermore, if systems using ice thermal storage are compared against advanced alternative systems, ice thermal storage is almost certain to greatly increase the overall electricity use. A chilled ceiling cooling system (discussed later) normally operates with chilled water at 16–20°C (rather than at 5–7°C). Dehumidification can be performed without initial deep cooling using desiccants (Section 6.6.4) and, if combined with a chilled ceiling and displacement ventilation, the ventilation air needs to be cooled down to only 20–25°C (rather than down to 5–10°C). In this case, the COP penalty required to make ice is substantially larger, such that total energy use is likely to increase substantially.

6.9.2 Day–night difference in the energy required to generate and transmit a kWh of electricity

There are significant differences between night and day in the efficiency of generating electricity from fossil fuels and in distributing electricity. Shifting a kW of electricity demand from day to night decreases the peak demand and increases the baseload demand (the level below which demand never falls). Inasmuch as baseload plants

are more efficient than peaking plants, this leads to a reduction in fuel requirements. In addition, less spinning reserve is required (spinning reserve refers to power plants that must be turned on and left spinning even without generating electricity, so that they are able to instantly meet sudden increases in electricity demand when needed). Even when producing electricity but at partial load, the plant efficiency is much less than at full load (fuel consumption per kWh at 30 per cent of full load is 50 per cent larger than at full load). By reducing the difference between baseload and peak demand, smaller total plant capacity is required, and the power plants can operate closer to full load. Finally, electricity distribution losses are smaller at night (by 5 per cent in California) than during daytime peaks in summer, due both to smaller loads and lower air temperatures at night. A detailed analysis by the California Energy Commission (CEC, 1996) indicates that, during summer, the energy required to produce one kWh of electricity at night is 20–43 per cent less than during the daytime peak, depending on the region and method of calculation. Savings due to shifting from daytime to night-time in winter are about two-thirds as large as the savings from shifting during the summer.

Thus, to the extent that the California results are applicable to other jurisdictions, shifting the cooling load to night-time through thermal energy storage can result in significant primary energy savings per kWh of electricity produced. However, as noted above, if systems with ice thermal storage are compared against advanced alternative systems, there is likely to be a large increase in electricity consumed. In order to achieve the savings at fossil fuel power plants by shifting electricity consumption from daytime to night-time, while avoiding or minimizing the on-site energy penalty of ice thermal storage, it is necessary to store coldness at a temperature above 0°C. To do so compactly requires using some alternative to ice that changes from solid to liquid and back at a warmer temperature and which has a reasonably large latent heat of fusion.

6.9.3 Alternative phase change materials

The need for low evaporator temperatures for compact storage of coldness in ice can be avoided if a material with a warmer freezing point is used. *Eutectic salts* are a combination of inorganic salts, water and other elements that form a mixture with a variety of freezing points. The most common is polyglycol, which has a freezing point of 8.3°C. However, its latent heat of fusion is rather low (99.9kJ/kg, compared to 335kJ/kg for water). Other problems with eutectic salts are that (1) they tend to melt into a saturated aqueous phase and a solid phase that consists simply of a lower hydrate of the same salt; (2) they can form a supercooled liquid rather than freezing at the normal freezing point; and (3) they pose a corrosion risk (He and Setterwall, 2002). *Gas hydrates*, which are under development for large commercial systems, have much higher latent heats of fusion, requiring only one-third to half the space and half the weight of systems using eutectic salts (Dinçer and Rosen, 2002). A third option consists of *paraffin waxes*. These have the general formula CH_3-$(CH_3)_n$-CH_3. A larger n leads to a warmer melting point and a larger latent heat of melting. He and Setterwall (2002) discuss a commercially available product, Rubitherm RT5, which is a mixture of C14 to C18 molecules with a freezing point of 7°C, a melting range of 4–6°C, and a latent heat of fusion of 158.3kJ/kg. A melting point anywhere in the range 2–18°C can be obtained by varying the ratio of C14 to C16 molecules (Bo et al, 1999).

As noted above, advanced HVAC systems (chilled ceilings for the bulk of the sensible cooling load, desiccant dehumidification for the latent heat load) eliminate the need for chilled water temperatures below 16–18°C. Phase change materials with a melting point a few degrees colder than this temperature range are thus a good match for advanced HVAC systems, while allowing night-time chilling without the low evaporator temperatures needed with ice thermal storage. However, the only commercial eutectic salt systems that were built had to run in series with the chillers, as they were part of a conventional HVAC system using chilled water at 5–7°C. This provides savings in peak load but does not fully exploit the potential to operate the chillers at a warmer evaporator temperature with eutectic salts.

Paraffin waxes can be incorporated directly into chilled ceiling panels, thereby eliminating the need for storage tanks. Koschenz

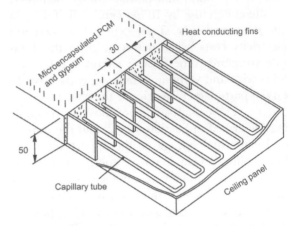

Figure 6.68 Diagram of a chilled ceiling panel with a micro-encapsulated PCM embedded in gypsum

Note: Dimensions are given in mm.
Source: Koschenz and Lehmann (2004)

and Lehmann (2004) have built a prototype chilled ceiling panel that incorporates paraffin as a phase change material and heat conducting fins, as illustrated in Figure 6.68. The paraffin is contained within 10μm capsules that are embedded in gypsum. A 5cm thick panel can store 320Wh/m^2 of heat per day (i.e. 40W/m^2 over an 8-hour period) as the paraffin melts at a temperature of about 22°C (a paraffin with a lower melting point could also be used). This heat is removed and the paraffin refrozen by circulating the chilling water at night. It is possible that such panels could be retrofitted onto existing ceilings in some cases.

6.9.4 Capital costs of systems with and without thermal energy storage

US DOE (2000) provides equations for the cost of chillers, cooling towers and storage systems that can be used to compare the total cost of systems with no thermal storage and with chilled water, ice and eutectic salt storage. The cost equations are given in Table 6.16. The cost equation for eutectic salt storage is somewhat hypothetical, as eutectic salts have not been commercially available for about 10 years.[19] Ice-on-coil and encapsulated ice represent the majority of the ice storage systems, with some ice slurry systems in use but very little if any new ice harvester systems. Air and water-cooled chillers can be used to make ice by freezing water onto the cooling coil. In this

Table 6.16 Costs of chillers, cooling towers and storage systems that can be used with chilled water, ice or eutectic salt storage systems

Component and cost equation	Applicable range
Air cooled reciprocating chiller installed cost = $11900 + $591 (C/3.516)	C = 70–700kW$_c$
Water cooled centrifugal chiller installed cost = $57700 + $307 (C/3.516)	C = 700–5300kW$_c$
Ice slurry generator installed cost = $1000 (C/3.516)	C >= 350kW$_c$ (C is in terms of rate of ice production)
Ice harvesting generator installed cost = $195000+$990 (C/3.516)	C = 700–3500kW$_c$
Cooling tower installed cost = $982 (C/3.516)$^{0.64}$	C = 200–3500kW$_c$
Chilled water storage installed cost = $616 (S/3.516)$^{0.686}$	S = 2100–21000kWh$_c$, ΔT = 8.3 K
Ice-on-coil and encapsulated ice storage installed cost = $70 (S/3.516)	
Ice slurry and ice harvester storage installed cost = $211 (S/3.516)$^{0.686}$	S = 14000–140000kWh$_c$
Eutectic salt storage installed cost = $125 (S/3.516)	

Note: C and S are in units of kW$_c$ and kWh$_c$, respectively. For cooling and storage capacities in tons and ton-hours, delete the factor of 3.516 (1 ton = 3.516kW$_c$).
Source: US DOE (2000)

Figure 6.69 Variation with cooling capacity in the cost of air-cooled reciprocating chillers, water cooled centrifugal chillers, ice-on-coil chillers, ice slurry machines and ice harvester machines. Costs include the cost of the cooling tower, except for air cooled chillers, where no cooling tower is needed. To convert cost per kW$_c$ to cost per ton, multiply by 3.516, and to convert chiller capacity in kW$_c$ to tons, divide by 3.516

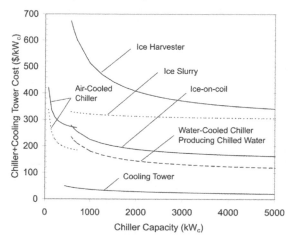

Source: Based on cost equations given in Table 6.16

case, the chilling capacity of a given chiller is reduced (because of the lower required evaporator temperature). A larger chiller is therefore needed, and the impact on cost can be accounted for by multiplying C in the chiller cost equations by 1.5 (but C is unaltered in the cooling tower cost equation, as the peak rate of heat rejection is unaltered, and is unaltered in the cost equations for ice slurry and ice harvester machines, as the latter are designed and rated for making ice). Assuming that the system without thermal storage uses chilled water at 5–7°C, chilled water thermal storage would not use colder water, so no implicit adjustment in the chiller capacity (by altering C) would be necessary with chilled water storage.

Figure 6.69 compares the cost, per kW of peak cooling capacity, of an air-cooled and

water-cooled chiller that produces chilled water (no ice), a water-cooled chiller operating to produce ice, an ice slurry generator and an ice harvester. For all except the air-cooled chiller, the cost of the cooling tower is included in the curves shown in Figure 6.69 (this cost is also given as a separate curve). Not included are differences in the cost of electrical systems (which will be less costly in systems with storage, due to smaller peak loads) or in air handlers or pumps (which can be smaller for systems with ice storage, because the water for chilling will be colder). To compute the cost of a system with thermal energy storage, assume, for illustrative purposes, that the peak chiller capacity with storage is only 0.75 times the peak cooling capacity without storage, and that 6kWh of thermal energy storage is provided for every kW reduction in peak chiller capacity. Figure 6.70a shows the additional cost of the chiller + cooling tower for cases with thermal storage compared to the cases without thermal storage, divided by the reduction in chiller capacity (in order to permit easy comparison with the cost of chiller capacity without storage), with the chiller and cooling tower capacities reduced by 25 per cent for all cases with storage. Both chilled water and eutectic salt systems have lower chiller costs because required peak capacity

Figure 6.70 Variation with system peak cooling capacity in (a) the incremental cost of chillers and cooling towers with various thermal storage options, per kW of chiller capacity that is not needed by using thermal storage, (b) the cost of storage, per kW of avoided chiller capacity, and (c) the system cost, per kW of peak cooling capacity

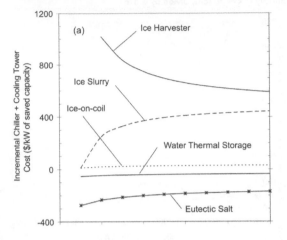

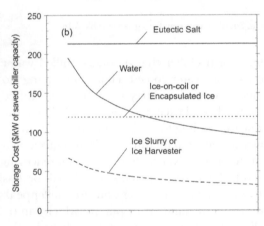

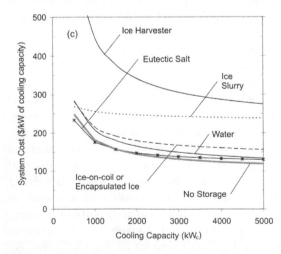

Source: Based on cost equations of Table 6.16 with additional assumptions given in the main text

is 25 per cent lower and there is no derating of the chiller due to colder evaporator temperatures (indeed, the savings would be greater with eutectic salts to the extent that the chiller can operate at a higher evaporator temperature, resulting in a greater capacity for a given machine). Figure 6.70b shows the cost of various storage systems divided by the reduction in chiller capacity. There is a large economy of scale with water storage because the cost is overwhelmingly the cost of the storage vessel, whose surface-to-volume ratio falls with increasing size, whereas there is essentially no economy of scale with eutectic storage because the main cost is the cost of the eutectic salt storage medium. Figure 6.70c shows the overall system cost for the various cases. The net result of reduced chiller cost but expensive storage cost roughly cancels out for eutectic salt, chilled water, ice-on-coil and encapsulated ice storage systems, such that the total capital cost is about the same as for a system without storage. Conversely, ice harvester and ice slurry systems cost substantially more than systems without storage, which explains their lack of popularity.

The time required to pay back the incremental cost with savings in electricity costs depends on the reduction in peak demand, the amount charged per kW of peak demand during each billing period, the impact on total electricity demand, the amount of electricity use that is shifted from daytime to night-time, and the difference between daytime and night-time rates. As noted in Chapter 2 (Section 2.3), these vary from utility to utility, and can even differ for different kinds of customers. The use of thermal storage reduces the required chiller capacity and peak electricity demand, and the reduction in the peak of total building electricity demand could be a large fraction of the original chiller electrical load if cooling is entirely shifted from daytime to night-time and if there is a deep night-time minimum in the non-chiller electricity demand that is filled with the shifted chiller electricity demand. That is, the demand charge savings using thermal storage depends on the diurnal variation in all the other electrical loads in the building. If peak demand is reduced by 0.75 of the original chiller electricity demand, and the demand charge is $10/kW$_c$/month, the COP is 5.0, and the full demand reduction occurs 6 months per year, the average annual cost savings is $9.0/kW$_c$.

Figure 6.71 The time required to pay back the additional cost of various thermal storage options with the savings in demand charges and energy costs, using the assumptions given in the main text

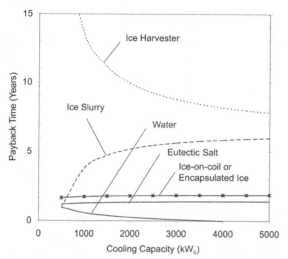

Figure 6.72 Illustration of an indirect contact or closed circuit cooling tower

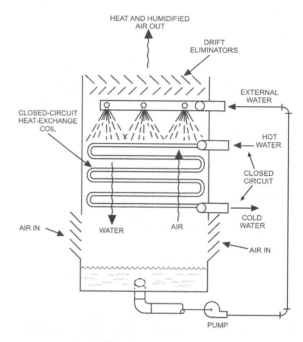

Source: ASHRAE (2001a, chapter 36)

For ice-based storage, if it is assumed that there is no savings in on-site electricity use but that 1095 hours/year (one in eight) of full-load equivalent daytime use (an average of 4 hours per day) can be shifted to night-time with a cost savings of 5 cents/kWh, then the annual energy cost savings is $11/kW$_c$ (however, it is not uncommon in practice to achieve savings of $20–30/kW$_c$/year, so these assumptions are very cautious). The extra cost of each storage option (compared to the case without storage) divided by the annual cost savings gives the time required to pay back the extra cost; this is referred to as the 'simple' payback, as it ignores interest. Simple payback times computed in this way are shown in Figure 6.71, and are less than two years for all options except ice slurry and ice harvester. The payback time for eutectic salts was computed assuming no increase in the cooling capacity of a given machine or reduction in on-site energy use due to a higher evaporator temperature; for modest (10 per cent) changes in these parameters, the payback time for eutectic salts is one year or less. These are very rough calculations, with the correct payback time highly dependent on the electricity tariff structure, the cooling load profile, and the number of months where demand charge savings accrue, but are adequate to indicate that eutectic salt storage can be an attractive option from a private financial perspective.

6.10 Cooling towers

Circulating water can be used to remove heat from the chiller condensers. Based on discussions with a major manufacturer of chillers, Westphalen and Koszalinski (2001) estimate that 40 per cent of the floor space in educational buildings, 50 per cent in office buildings, 55 per cent in public buildings and 70 per cent in lodgings are served by water-cooled chillers in the US, the balance being air cooled. When cooling water is used, the heat removed by the cooling water is rejected to the atmosphere through a *cooling tower*. Chillers of less than 700kW cooling capacity tend to have the compressor and condenser on the roof and use direct air cooling of the condensers, avoiding cooling water and cooling towers but requiring substantial fan energy. The cooling tower cools the cooling water through evaporation; the cooling water can thus be cooled to within a few kelvin of the wet-bulb temperature (T_{wb}) (see Section 6.1.2). Inasmuch as T_{wb} can be 10K or more below the air temperature, the condenser temperature does not need to be as high in order to reject heat. This leads to a larger chiller COP.

For example, Westphalen and Koszalinski (2001) indicate typical COPs of 2.7 and 3.9 for air and water-cooled reciprocating chillers, respectively, and COPs of 3.2 and 4.2 for air and water-cooled screw chillers, respectively (condenser fan energy is included in the COP for the air-cooled case).

In the simplest kind of cooling tower (a *spray filled, direct contact* cooling tower), the cooling water is sprayed into a rising air current. About 5 per cent of the water evaporates, cooling the remaining liquid water as it falls through the air stream. The water can be cooled to within 2–3K of the ambient T_{wb}, or about 20K lower than the condenser of air-cooled systems of reasonable size (ASHRAE, 2001b, chapter 36). The remaining liquid water is collected at the bottom of the tower and returns to the chiller condenser to remove more heat. In an *indirect contact* or *closed circuit* cooling tower, the cooling water circulates

through a closed pipe within the air stream, and a separate water supply is sprayed into the air stream, as illustrated in Figure 6.72. A very wide range of cooling tower designs is possible, as discussed more fully in ASHRAE (2001b, chapter 36). Cooling towers can be placed on the roof, against walls or inside enclosures. In the later two cases, the cooling tower must be placed so as to ensure unobstructed airflow to the fans, to ensure that the discharge air is not deflected back into the cooling tower intake vent, and to ensure that the discharge air does not enter the building.

Altogether then, the cooling system in a large commercial building with hydronic cooling consists of three fluid circulation loops: a refrigerant circulates between the chiller condenser and evaporator, a chilled water loop picks up heat from the building and rejects heat to the evaporator, and a cooling water loop picks up heat from the condenser and rejects it to the atmosphere in the cooling tower. These three loops and a cooling tower are shown schematically in Figure 6.73a.

In addition to consuming water, cooling towers require energy to operate the fan (which induces the airflow) and the pump (which circulates the cooling water). Energy use can be minimized (1) by interlocking the operation of pumps and fans with the operation of the compressors, so that the former shut down when the latter are not operating (in a typical situation, this is not done); (2) through efficient part-load operation; and (3) by facilitating solar induced airflow. The combined chiller cooling tower energy use can be reduced if the system is configured to allow maximum use of 'free-cooling' when outside temperatures are cold enough, as discussed later (Section 6.10.2). Oversizing the cooling tower can reduce energy use in two ways: (1) by reducing the pressure drop experienced by the air as it flows through the tower, due to the fact that a smaller flow velocity is possible with a larger cross section, so that less fan power is needed; and (2) by chilling the cooling water to a temperature closer to T_{wb} (oversizing such that the chilling water comes to within 3–4 per cent of the initial difference from T_{wb}, rather than 10 per cent, is often cost-effective according to US DOE (2002a)). This is a reversal of the situation with regard to chillers themselves, where oversizing increases rather than decreases energy use.

Figure 6.73 (a) Layout of a cooling tower/chiller/chilled water loop system during operation of the chiller (solid lines), with free cooling through bypass of the compressor (dotted lines), or with free cooling through heat exchange between the cooling water and chilling water loops using a heat exchanger (dashed lines). (b) Free cooling through direct interconnection of the cooling water and chilled water loops

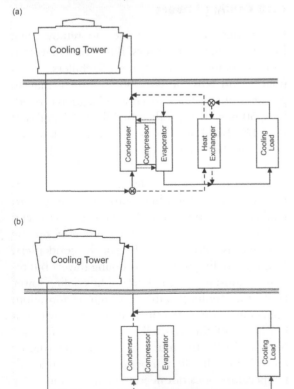

Cooling towers are not needed when ground source heat pumps are used for cooling, as the heat is rejected directly to the ground. Solid desiccant cooling systems also do not require a cooling tower, although water is still consumed during the evaporative cooling step. A district cooling network with a centralized cooling plant eliminates the need for cooling towers in or on individual buildings (see Chapter 15, Section 15.3.7).

6.10.1 Part-load operation

A cooling tower, like other cooling system components, will be sized to meet the peak anticipated load. Thus, most of the time it will be operating at part load. Part-load operation of the fans is achieved through (1) throttling the airflow (partly obstructing the flow path); (2) on/off cycling; (3) use of two-speed fans with on/off cycling for additional modulation of cooling capacity; or (4) use of variable speed fans. Option (1) is the least efficient, while option (2) can cause early motor burnout. Data on fan power versus fan speed (which is proportional to the rate of airflow) presented in ASHRAE (2001b, chapter 36) indicate close to a cubic relationship, implying that very large energy savings are possible through the use of variable speed fans (see Chapter 7, Section 7.1.2). Retrofitting a cooling tower in Toronto, that had relied on throttling, with a variable speed fan yielded a savings in fan energy use of 19 per cent (Lenarduzzi and Yap, 1998).

6.10.2 Free cooling

If the cooling tower water becomes sufficiently cold as the outside air temperature and cooling load decrease, it could be used to directly serve the cooling load, allowing the energy intensive compressor to be turned off. Pumps are still required to circulate the cooling water and chilled water. The operation of the cooling tower in this way is referred to as a *water side economizer*, in contrast to the term *air side economizer*, which refers to the mechanism used to increase the proportion of outside air in an all-air cooling system when the outside temperature is appropriate for meeting at least part of the cooling load (see Chapter 7, Section 7.4.3). Either operation is referred to as *free cooling*.

The compressor can be bypassed in a water side economizer in one of three ways:

1 By circulating the refrigerant directly between the condenser and evaporator, as shown by the dotted lines in Figure. 6.73a. The chiller operates like a heat pipe (Chapter 7, Section 7.4.9): the condenser will be colder than the evaporator when this is done (since the weather will be cold), so refrigerant will migrate to the condenser on its own and condense, releasing latent heat. The liquid refrigerant is pumped back to the evaporator, absorbs heat, evaporates and repeats the cycle.

2 By directing the cooling water through a heat exchanger rather than through the condenser, and by directing the chilling water through the same heat exchanger rather than through the evaporator, as shown by the dashed lines in Figure. 6.73a.

3 By directly interconnecting the cooling water and chilled water circuits, as shown by the solid lines in Figure 6.73b. In this case, contamination of the clean chilling water with impurities from the cooling water can occur if a direct contact cooling tower is used. This problem is avoided with an indirect contact cooling tower.

None of the options, as configured in Figure 6.73, allows simultaneous partial free cooling and operation of the compressor. However, the heat exchanger under option (2) can be configured to pre-cool the return chilled water, with the compressor used to top off the required cooling (Wulfinghoff, 1999, pp373–5). This permits greater use of free cooling.

The coldest temperature that can be achieved through evaporation is T_{wb}, which depends on the ambient air temperature and humidity. As explained in Section 6.1.2, the difference between T_{wb} and the dry-bulb temperature (T_d) is greatest in arid regions, so the temperature *change* achieved with evaporative cooling will be greatest, but the absolute temperature achieved can be lower in humid temperate regions. In practice, the best that one can achieve is to bring the cooling water to within 1.5K of T_{wb}, and if the building chilling water is connected to the cooling tower water through a heat exchanger (Option 2 above), there will be another step of at least 1.5K. However, the warmer that the chilling water is allowed to be while still providing an adequate cooling effect, the more often it will be possible to chill it enough evaporatively using the cooling tower – that is, the greater the opportunity for free cooling.

As discussed in Chapter 7 (Section 7.4.4), radiant chilled ceiling cooling systems operate with a chilled water temperature of 16–20°C, rather than the 5–8°C (with dehumidification) or 11–14°C (without dehumidification) temperatures that are used in conventional systems today (dehumidification without cooling to 5–8°C can be done using solid or liquid desiccants). Assuming a difference between T_{wb} and the chilling water temperature of 3K (which requires the use of enhanced heat exchangers), and assuming the chilling water to be supplied at 18°C, a cooling tower could directly meet the cooling requirements 97 per cent of the time in Dublin and 67 per cent of the time in Milan according to Costelloe and Finn (2003a). If chilling water can be supplied at 20°C, then evaporative cooling in a cooling tower is sufficient 99 per cent of the time in Dublin and 78 per cent of the time in Milan.

Riffat et al (2000) have developed a detailed computer model to simulate the behaviour of a cooling tower coupled to a chilled ceiling. Normally, a cooling tower receives cooling water from a condenser at a temperature of 26–46°C – well above the ambient T_{wb}. However, when the cooling tower is directly coupled to a chilled ceiling, it receives water at a temperature of about 20°C. The effectiveness of the cooling tower – defined as the cooling achieved as a fraction of the difference between the input temperature of the cooling water and T_{wb} (Equation 6.6) – has a maximum value of about 0.7, so if T_{wb} = 16°C, the cooling water will be chilled from 20°C to 17.2°C. This is usually adequate for use in a chilled ceiling. If T_{wb} is warmer than 16°C, as it will frequently be in hot and humid climates, then adequate cooling will not be achieved. The

calculated COP of the cooling tower – defined as the cooling divided by the energy used for pumping the spray water and cooling water – is about 6 for maximum air and spray-water flow rates. This may appear to be no better than the COP of a large chiller, except that the chiller COP does not include energy used by the cooling tower or to circulate chilled water through the building.

If one wishes to produce a constant temperature of cooling water with a cooling tower as the ambient T_{wb} decreases, the cooling water does not need to approach T_{wb} as closely when T_{wb} decreases. This in turn means that the cooling water and airflow rates can be reduced, permitting a large reduction in fan and pump energy use if variable speed fans and pumps are used (see Chapter 7, Section 7.1.2). In an experimental test rig using this strategy, Costelloe and Finn (2003b) measured COPs ranging from 8 to 20 as cooling tower fan and pump power were decreased from 100 per cent to 20 per cent of peak power in response to T_{wb} decreasing from 14°C to 8°C, while always producing cooling water between 15 and 17°C. The best full-load COP of a large vapour compression chiller is about 7.9, but this does not account for the energy used by the cooling tower, which can be 10–30 per cent of the chiller energy use, giving a maximum effective chiller COP of perhaps 5–6.

Strand (2003) used a coupled radiant ceiling/cooling tower model to assess the potential energy savings at four US locations using a hybrid system consisting of (1) a conventional forced air system with an air conditioner, and (2) a chilled ceiling that is cooled solely using water from a cooling tower and only at night-time (9pm–9am). Table 6.17 gives the cooling load that must be supplied by the air conditioner for a conventional forced air system with the indoor temperature fixed at 24°C over the period 1 June to 31 August using average temperatures for each location, and for the hybrid system with the indoor temperature either fixed at 26°C at all times, or allowed to drift at night (the 26°C air temperature combined with the relatively cool radiant panel gives a perceived temperature of 24°C). The hybrid system reduces the air conditioner cooling load by about 50 per cent in the hot-humid climate of Key West, Florida; by about 85 per cent in the hot-dry climate of Phoenix, Arizona and in the mild-humid climate of Illinois;

Table 6.17 Comparison of the air conditioner cooling load (kWh) for a one-storey house over the period 1 June – 31 August with average weather in four US cities, as simulated by Strand (2003)

Location	Climate	Air conditioner cooling load (kWh)		
		Conventional with fixed T_a	Hybrid system Fixed T_a	Variable T_a
Spokane, Washington	Mild-dry	1570	5	4
Peoria, Illinois	Mild-humid	2450	300	108
Phoenix, Arizona	Hot-dry	9480	1400	1350
Key West, Florida	Hot-humid	6020	3140	1540

Note: Results are given for a conventional forced air system in which an air conditioner provides all of the cooling and the indoor temperature is fixed at 26°C, and for a hybrid forced air/radiant ceiling system with the ceiling cooled with water from the cooling tower whenever this is feasible and the indoor temperature either fixed at 26°C at all times, or allowed to drift at night (9pm–9am).

and by about 100 per cent in the mild-dry climate of Washington state, all the while maintaining the same perceived temperature as in the conventional system. When temperatures are allowed to drift at night in the hybrid system, there is almost no further reduction in air conditioner load in the dry climates because the temperature rarely tends to drift above 26°C, whereas there is a substantial further reduction in the humid climates because the night-time temperatures do tend to rise above 26°C, necessitating use of the air conditioner when the air temperature is fixed. Depending on the occupants' tolerance of perceived temperatures greater than 24°C, the hybrid system would reduce the air conditioner load by 50–75 per cent even in the hot-humid climate of south Florida. Although one might question the practicality of cooling towers on individual homes, the system could be viable in multi-unit residences.

A non-energy benefit of using the cooling tower to cool the building chilling water, rather than to cool the condenser of a chiller, is that the possibility of the growth of legionella bacteria is greatly reduced, due to the fact that the water circulating in the cooling tower will normally not exceed 20°C, compared to 26–33°C with condenser water and an optimum temperature of 37°C for the growth of legionella (Costelloe and Finn, 2003a).

An interesting variant of free cooling with a cooling tower is found in an office building in Kitchener, Ontario, Canada (Carpenter and Kokko, 1998). In this case, cooling water from the condenser of a gas-fired absorption chiller is sent

to a cooling pond in front of the building. Heat is rejected to the atmosphere through evaporation, as in a cooling tower. Make-up water is provided by rainwater that is collected from the roof and stored, with groundwater used when needed during extended dry spells. The building heating or cooling needs are meet entirely from free cooling or a heat recovery ventilator for outside temperatures between 0 and 15°C.

6.10.3 Solar assisted airflow

Sánchez et al (2002) and Lucas et al (2003) built and tested a roof mounted solar chimney that creates a natural draft that draws air through a zone beneath the solar chimney. A spray of fine water droplets is added in this zone, which functions as a cooling tower. The system is shown in Figure 6.74. The water droplets are chilled through partial evaporation, and the chilled water is collected at the bottom and can be used to cool a closed cooling water circuit, or can directly serve as the cooling water circuit. A cooling power density of 500–1000W per m² of roof area is achieved for cooling water–wet-bulb temperature differences of 10–15K at the test site (in Catagena, Spain). This is substantially less than the cooling densities of mechanical draft cooling towers (30–90kW/m²), which adds another reason (beyond direct energy savings) for keeping the cooling load as low as possible. With solar assisted, natural draft cooling towers, noise and vibration are eliminated due to the absence of fans.

Figure 6.74 Overview of a solar assisted cooling-tower (top), side view (middle) and schematic diagram (bottom)

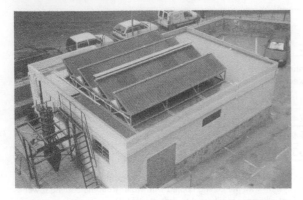

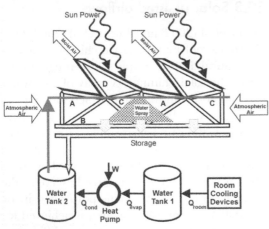

Source: Sánchez et al (2002)

6.11 Water consumption by chillers, evaporative coolers and power plants

As discussed above, the condensers of large commercial chillers are cooled with water that circulates through a cooling tower, where some evaporative losses occur. Steam turbine power plants (whether fossil fuel or nuclear) also have condensers (see Chapter 15, Section 15.1.1) that are usually cooled with water that circulates through a cooling tower, so there is an additional water loss associated with the production of the electricity that powers electric chillers when the electricity is generated at these power plants. Evaporation can be used to directly cool building air or the chilled-water used for cooling purposes (Sections 6.3.7 and 6.10.2). Here, we quantitatively compare the water consumption associated with direct evaporative and mechanical cooling systems. Limited supplies of water could be a constraint in both cases.

6.11.1 Direct water consumption by evaporative cooling and by chillers

In evaporative cooling, evaporating water either directly cools the air stream that enters the building (thereby cooling and humidifying the air), or it cools an air stream that, in turn, is used to cool the air entering the building through a heat exchanger (thus, the building air is not humidified as it is cooled). In a cooling tower, evaporation is used to cool the cooling water from the condenser of a chiller, which includes the heat removed from the building and the additional heat produced by the compressor in the chiller. For a vapour compression chiller with a COP of 4.0, the compressor heat will be 25 per cent of the building heat. For a single-effect absorption chiller with a COP of 0.7, the input heat that drives the generator will be 1.43 times the building heat, so the total required heat rejection will be 2.43 times the building heat load. Thus, one might expect that direct cooling of the building air through evaporation will require less water than vapour compression chilling, which in turn will require less water than absorption chilling.

Given a latent heat of evaporation for water of $2.5 \times 10^6 \mathrm{J/kg}$ and a density of liquid water of $1000 \mathrm{kg/m^3}$, it follows that 2.5GJ of cooling will occur per cubic metre of water that is evaporated. Call this the reference effectiveness of water use, E_{ref}. When evaporation occurs, it is the remaining water that is directly cooled. Evaporative cooling of an air stream will occur through contact between the moving air and the cooled water droplets (a finer mist of droplets will increase the contact area and thereby increase the cooling of the air). The remaining liquid mist falls to a collection tray and is recirculated, so the coldness not transferred to the air during a given circuit is not wasted. A direct evaporative cooler should be able to achieve an effectiveness close to E_{ref}. For indirect evaporative cooling, the effectiveness of water use will be E_{ref} times the effectiveness of the heat exchanger.

In a cooling tower, the water mist is supplied from the condenser water and may be warmer than the air into which it evaporates. In this case, the goal is not to cool the air, but to cool and reuse the cooling water itself. If the cooling water is warmer than the ambient air as it evaporates, it will experience cooling through the transfer of sensible heat to the air, as well as through evaporation. In this case, the cooling achieved per cubic metre of water that is lost through evaporation will exceed the 2.5GJ benchmark given above.

If only the amount of water that is lost through evaporation is replaced in a cooling tower, the concentration of dissolved minerals and particles will continuously increase. To prevent this, additional water must be bled off and replaced. The rate at which water must be bled off (B) is related to the evaporation rate (E), the concentration M of critical contaminants in the make-up water, and the maximum allowable concentration T according to (Wulfinghoff, 1999):

$$B = E \frac{M}{T - M} \qquad (6.15)$$

The need to bleed off water applies to evaporative coolers as well as to cooling towers. The amount of water that needs to be bled and replaced can be reduced if the un-evaporated spray is collected in a container with a sloping bottom, such that solid residues are trapped while water recirculates from the top, and the water in the collector pan is only periodically replaced with fresh water. The water that is flushed out can in turn be used for watering plants surrounding the building or for toilets if the plumbing system is so designed.

Two examples of the rate of water consumption by cooling towers that have been calculated here from published data are as follows:

- For a natural draft cooling tower in Singapore (Hawlader and Liu, 2002): $0.34–0.37 m^3/GJ$, corresponding to a water use effectiveness of $2.7–2.9 GJ/m^3$. This exceeds the $2.5 GJ/m^3$ cooling achievable through evaporation alone, implying about $0.2–0.4 GJ/m^3$ of cooling through the transfer of sensible heat from the water to the ambient air. The cooling water is cooled by $7–12 K$, with evaporation accounting for $84–90$ per cent of the total cooling achieved.
- For a 10m cooling tower in Tucson, Arizona (Chalfoun, 1998): $0.35 m^3/GJ$ for operation during June, 1992, equivalent to $2.8 GJ/m^3$. This water use effectiveness also exceeds that achievable through evaporation alone, presumably due to the transfer of sensible heat.

The water consumption given above is the amount used per unit of heat delivered to and removed by the cooling tower. Table 6.18 gives the water consumption per unit of heat removed from the building, which takes into account the additional water needed to remove the heat produced by the chiller. Desiccant/evaporative cooling has the lowest water consumption, as there is no heat that needs to be rejected beyond that from the ventilation air supplied to the building. Water consumption is $50–60$ per cent larger using a vapour compression chiller with a COP of 4.2, twice as large again using a single-effect absorption chiller, and slightly larger still using an adsorption chiller. The large water use of absorption chillers could be addressed through the development of air-cooled absorption chillers, an effort that is underway (Alva Solari and González, 2002; González and Alva Solari, 2002), although this would probably reduce performance.

Some closed-circuit (indirect) cooling

Table 6.18 Comparative water consumption per unit of heat removed from a building using various cooling methods

Cooling method	Water consumption		Water effectiveness GJ/m³
	m³/kWh$_c$	m³/GJ	
Desiccant evaporative cooling	0.0015–0.0020	0.41–0.56	1.8–2.5
Vapour-compression	0.00274	0.76	1.31
Single-effect absorption chiller	0.00549	1.53	0.65
Single-effect/ double lift absorption chiller	0.00575	1.60	0.63
Adsorption chiller	0.00650	1.81	0.55

Source: Kivistö et al (1996)

towers can operate without a water spray when the ambient air is cold enough, in order to conserve water. In this case, the cooling water loop is often finned in order to enhance the rate of heat loss to the ambient air.

Water use effectiveness calculated here for various evaporation aided cooling systems are as follows:

- For evaporative cooling of the air used to cool the condenser of a residential air conditioner (Hoeschele et al, 1998): 2.14GJ/m³.
- For cooling with a residential direct/indirect evaporative cooler at a variety of sites in south west US (Kinney, 2004): 1.85–1.93GJ/m³.
- At a fast-food restaurant (Young et al, 1999): 1.10GJ/m³ with a direct evaporative cooler, 0.45–0.56GJ/m³ with a direct evaporative pre cooler on the air intake.
- For the DEG prototype indirect/ direct evaporative cooler, 0.72– 0.85GJ/m³ (Table 6.9).

For residential evaporative coolers in the southwest US, operating in houses that have not been designed to minimize cooling requirements, the annual water consumption amounts to 2–5 per cent of the total household water consumption (Kinney, 2004). Water conservation measures (discussed in Chapter 8, Section 8.2) can easily save several times this amount.

Ground source heat pumps do not require cooling towers and hence do not consume water during cooling, as the heat is rejected to the cool ground. The same is true of cooling systems that use lakes, rivers or sewage water as a heat sink, or if heat recovered from the condenser is used to preheat cold incoming water for hot water needs.

6.11.2 Water consumption by thermal power plants

Thermal power plants are power plants in which electricity is generated with a heat engine, such as a steam turbine or a gas turbine (these plants are discussed briefly in Chapter 15, Section 15.1). Cooling water is required, and can be used in a 'once-through' system, in which the cooling water is returned to a surface water body after being used once, or can be used in a recirculating loop system, with the water cooled in a cooling tower before being used again. In the US, cooling of thermo-electric power plants is the single largest user of water (at 48 per cent of total water withdrawals), but 91 per cent of the water that is withdrawn for power plant cooling is used in a once-through system and then returned, while only 9 per cent is used in recirculating loop systems (Hutson et al, 2004). More than half of the water that is withdrawn for recirculating loop cooling systems is consumed through evaporation in cooling towers or through leakage. Recirculating loop systems are used in dry regions of the US because, although significant consumption of cooling water occurs with recirculating loop systems, the amount of water that needs to be supplied is much less. Once-through systems provoke a small evaporative loss due to the fact that water is returned to the river or lake at a warmer temperature than that at which it is withdrawn. Based on data for individual power plants and generating units in the US compiled by the US Department of Energy (US DOE, 2003), the rate of production of electricity for power plants using recirculating loop systems ranges from 0.3GJ/m³ to 1.3GJ/m³. This range corresponds to about 3–12litres per kWh of electricity generated. Torcellini et al (2004b) calculate an average water consumption for all thermal power plants in the US of 1.8litres/kWh (corresponding to 2.0GJ/m³). This average includes once-through power plants.

As for water consumption from thermal power plants elsewhere, a 6MW biomass power plant in India, using primarily rice husks as fuel, requires 50m³ of water per hour for the cooling tower, blow down and drain losses from the boiler, and for ash conditioning (Reddy and Narendra, 2002). This works out to 0.432GJ of electricity produced per cubic metre of water consumed, which is near the low end of the range found for conventional power plants in the US, but the water is used for more than the cooling tower.

A significant amount of water can also be consumed in generating electricity from hydro-electric power plants. In this case, water is lost due to evaporation from the hydro-electric reservoir, beyond that which would occur without the reservoir. For power plants with a very small reservoir area compared to the power output, or for run-of-the-river power plants, the evaporative water loss will be small or zero. Most reservoirs serve other purposes besides power generation, such as flood control, creation of recreational areas or storage of water for irrigation. Thus, not all of the water loss from reservoirs should be attributed to power generation. If, however, one assigns all of the water loss to power generation, the average rate of consumption from hydro-electric power plants in the US is about 40 times that of thermo-electric power plants according to calculations by Torcellini et al (2004b)!

6.11.3 Total water consumption with different methods of cooling

The total water consumed, W, in order to remove H units of heat from a building using an electric chiller, absorption chiller or evaporative cooler can be written as:

$$W = H\left(1+\frac{1}{COP}\right)\frac{1}{E_{ct}}+$$

$$H\left(q_{ct}+\frac{1}{COP}\right)\frac{1}{\eta_t E_{pp}} \quad (6.16)$$

$$W = H\left(1+\frac{1}{COP}\right)\frac{1}{E_{ct}}+H(q_{ct}+q_{aux})\frac{1}{\eta_t E_{pp}} \quad (6.17)$$

and

$$W = \frac{H}{E_{ec}}+Hq_{ec}\frac{1}{\eta_t E_{pp}} \quad (6.18)$$

respectively, where COP is the electric or absorption chiller COP; E_{ct} and E_{ec} are the effectiveness of water use (GJ of rejected heat/m³) by the cooling tower and by the evaporative cooler, respectively; E_{pp} is the effectiveness of water use (GJ of electricity generated/m³) at the electric power plant; q_{ct}, q_{aux} and q_{ec} are the electricity used by the cooling tower fans, absorption chiller pumps and by an evaporative cooler, respectively, as a fraction of the amount of cooling provided; and η_t is the efficiency in transmitting electricity from the power plant to the point of use. The first term on the right hand side of Equations (6.16) and (6.17) is the water consumption by the cooling tower, while the second term in Equations (6.16) to (6.18) is the water consumed in generating any electricity that is used. For air conditioners and air cooled chillers, the first term of Equation (6.16) drops out but, all else being equal, the COP would be smaller than for water-cooled chillers. For power plants that do not use evaporative cooling, the second term of all three equations drops out.

The results of some illustrative calculations using Equations (6.16) to (6.18) are shown in Figures 6.75 and 6.76 assuming an effectiveness of water use at the power plant, E_{pp}, of either 2.0GJ/m³ (the US average) or 0.5GJ/m³ (close to the smallest US value), and for the following additional assumptions: E_{ec} = 2.0GJ/m³, E_{ct} = 2.2GJ/m³, q_{ct} (Equation 6.16) = 0.022 and q_{ct} + q_{aux} (Equation 6.17) = 0.050 (both based on Table 6.13), q_{ec} = 0.0575 (based on the performance of the indirect/direct cooler of Table 6.9 at medium fan speed), and η_t = 0.95. The E_{pp} values are assumed to be applicable to a power plant with an efficiency of electricity generation, η_e, of 0.35.

Figure 6.75 Direct water consumption by a building cooling tower per GJ of cooling provided to a building, and indirect water consumption by the cooling tower at an electrical power plant, as the chiller COP varies from 4.0 to 10.0

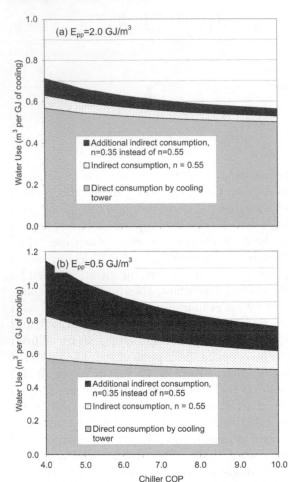

Note: Based on Eq. (6.16) using the assumptions given in the main text. Results are given assuming an effectiveness of water use at the power plant of (a) 2.0GJ/m³ and (b) 0.5GJ/m³.

Figure 6.76 Total (on-site and off-site) water consumption per GJ of cooling provided to a building, for various methods of cooling

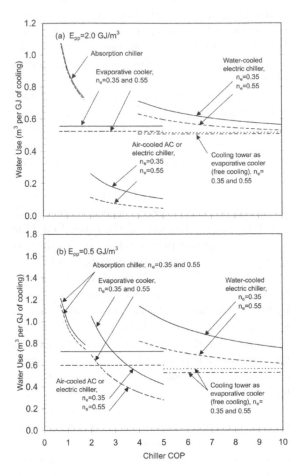

Note: Based on Eqs. (6.16) to (6.18), using the assumptions given in the main text. Results are given assuming an effectiveness of water use at the power plant of (a) 2.0GJ/m³ and (b) 0.5GJ/m³.

Given that the waste heat produced at the power plant per unit of electricity generated is $(1 - \eta_e)/\eta_e$, the water consumption for a power plant with $\eta_e = 0.55$ would be only 0.44 times as large. Using this factor, the total water consumption assuming $\eta_e = 0.55$ is compared with that of the reference power plant ($\eta_e = 0.35$) in Figures 6.75 and 6.76.

Figure 6.75 shows that for a water cooled chiller with $E_{pp} = 2.0\text{GJ/m}^3$, the off-site water consumption is 10–25 per cent of the onsite consumption, while for $E_{pp} = 0.5\text{GJ/m}^3$, the off-site consumption is 65–100 per cent of the on-site consumption.

Figure 6.76 compares the total water consumption per GJ of heat removed from a building for absorption and electric chillers, air conditioners, and for evaporative coolers. The latter are assumed to be applicable only in residential or small commercial applications, where the COP of air conditioner or chiller alternatives would be <5 and the air conditioner or chiller condensers are air cooled. Consequently, the water use of evaporative coolers is shown only up to a COP of 5. In such applications, the indirect water consumption of the air conditioner offsets one-third to two-thirds of the water used by the evaporative cooler (depending on the air conditioner COP) for $E_{pp} = 2.0 GJ/m^3$. However, for $E_{pp} = 0.5 GJ/m^3$, a conventional power plant ($\eta_c = 0.35$) and typical household air conditioners (COP <3 in practice), the evaporative cooler uses less water. For larger applications, the building cooling tower can be used as an evaporative cooler when T_{wb} is sufficiently low (see Section 6.10.2), so it should be compared against the water used by water-cooled chillers. For speculative but plausible adjustments in the parameters to account for the lower temperature of water entering the cooling tower compared to the cooling water from a chiller condenser (namely, $E_{ct} = 2.0$ rather than $2.2 GJ/m^3$ and $q_{ct} = 0.03$ rather than 0.022), the water consumed by a cooling tower is slightly less than that consumed by an evaporative cooler (because q_{ct} remains less than q_{ec}). This is shown by the horizontal lines in Figure 6.76. Water consumption using the cooling tower as an evaporative cooler is less than the water consumption by chillers for an electricity generation efficiency of 0.35 or 0.55, and for either E_{pp} value. A higher chiller COP reduces water use both by reducing the amount of heat that is rejected to the building cooling tower and by reducing the amount of electricity that is used.

It should be stressed that the results shown in Figures 6.75 and 6.76 depend on the values assumed for rather uncertain variables, the correct values of which are likely to vary with climate and from case to case. Rather than providing quantitatively rigorous results, the discussion presented here serves to highlight the significance of the water consumed directly through conventional cooling tower based chillers and indirectly through the provision of electricity, and the potentially significant offset of the on-site water consumed by evaporative coolers through reduced off-site water consumption.

6.12 Summary and synthesis

To minimize the amount of energy used to keep buildings comfortably cool, the first and foremost requirement is to minimize the total cooling load. This can be done through appropriate building shape and orientation, which influence the surface area exposed to sunlight at different times of day; through the use of high-albedo (reflective) surfaces, direct shading and windows with a low solar heat gain; and through minimization of internal heat production by using highly efficient lighting systems and equipment. External horizontal shading elements can also serve as light shelves, while vertical shading elements can also serve as wind deflectors to enhance ventilation. Double-skin façade construction can reduce heat gains through walls by a factor of three by facilitating adjustable external shading devices (which are more effective than shading inside of glazing), as well as by inducing airflow that removes heat from the shading devices and from the inner façade.

Passive and/or low-energy techniques for removing heat consist of the following:

- ventilative cooling;
- radiative cooling;
- evaporative cooling;
- earth pipe cooling.

The opportunity for passive ventilation is maximized through the adoption of narrow and tall rather than low and sprawling building forms; narrow buildings will facilitate greater use of wind driven cross-ventilation, while taller buildings will generate greater thermally driven ventilation, and the upper portions will be exposed to greater wind speeds.[20] Ventilation by wind can be enhanced through the use of wind towers, either as wind scoops or as suction devices. Thermally driven ventilation can be enhanced through the use of atria, solar chimneys, cooltowers, airflow windows, Trombe walls and double-skin walls, all of which can be used to draw inside air out of the building, forcing it to be replaced by an inflow of cooler air from the outside, perhaps preferentially from the cooler side of the building. When outside air is too warm to use for ventilation without mechanical cooling, airflow windows, Trombe walls and double-skin walls can be used to draw outside air through the façade gap and back out, thereby removing heat absorbed by the outer façade and in the air gap that would otherwise enter the building. Alternatively, outside air can be cooled by passively drawing it through a buried earth pipe loop, with fan assistance only when needed. In this way, natural forces can be used for both ventilation and cooling. In order to work properly, any form of passive ventilation requires that the building envelope be close to completely airtight, except for intentional openings. Features to facilitate passive ventilation should not be incorporated at the expense of energy-efficient fan systems for night-time ventilation or daytime backup, as the latter can be just as important for saving energy as the passive design features themselves.

Both wind and thermally driven passive ventilation require operable windows or ample trickle ventilators. Operable windows provide a double benefit in that warmer temperatures are deemed to be acceptable in buildings that are connected to the outside through natural ventilation than in hermitically sealed buildings, as the occupants are given some control over their own environment. In hybrid passive-mechanical ventilation systems, fans are used to enhance the ventilation only when needed. Ventilation is most effective at night if a building has a large exposed thermal mass. Night-time ventilative cooling and thermal mass should be combined with external insulation in order to minimize indoor daytime temperature peaks.

The specific methods employed to satisfy the cooling load with minimal energy use depend on the climate:

- For moderate climates or conditions – utilize passive or hybrid passive-mechanical ventilation to reduce indoor temperatures with minimal energy use while at the same time increasing the acceptable temperature. Where winters are cold enough that ground temperatures are cool during the summer, underground earth pipe loops can be used to cool ventilation air. Evaporative cooling will be effective due to the low summer wet-bulb temperatures characteristic of moderate climates.
- For hot and dry climates (where there tends to be a large diurnal temperature variation), use large thermal mass, external insulation and night-time ventilation, coupled with daytime evaporative cooling

of ventilation air if needed and if water is not scarce. If night-time air temperatures are not quite low enough for adequate cooling at night, but ground temperatures are cold enough, then night-time air can be drawn through a ground loop. Rooftop radiators are most suitable in dry climates in order to reduce the temperature of air used for ventilation at night.
- For hot and humid climates, thermal mass and night-time ventilation will not be very effective during the warmest seasons, but can still be effective during the transition seasons. Thermal mass has little impact on annual energy use for cooling, but can significantly reduce peak cooling loads. However, lightweight buildings can take advantage of brief cool periods. Evaporative cooling will not achieve cool enough temperatures, due to the higher wet-bulb temperature and the need to use indirect rather than direct evaporative cooling. Night-time radiative cooling will also be less effective due to the higher atmospheric emissivity under humid conditions. Mechanical cooling will therefore be needed much of the time. Buildings should be open so as to maximize cross-ventilation. In buildings relying entirely on ventilation for cooling, indoor and outdoor air temperatures will be similar, but a modest amount of insulation is still useful because external wall surface temperatures can rise substantially above the air temperature due to absorption of solar radiation. In buildings that are mechanically cooled, insulation can reduce cooling energy use by up to 40 per cent. If insulation is applied internally in massive buildings (which are required in regions subject to hurricanes), the building will behave as a lightweight building

when ventilated, and will therefore be able to take advantage of brief cooler periods.

In both hot-dry and hot-humid climates, shading devices should be provided and should be:

- placed outside the building (internal blinds are less effective);
- made of light and reflective materials with low heat-storage capacity;
- designed to prevent reflection onto any part of the building.

In much of the world, the wet-bulb temperature (the lowest temperature that can be achieved through evaporation for a given air temperature and humidity) is well below the comfort temperature all the time, so evaporative cooling alone will be adequate. Liquid desiccants can enhance the effectiveness of evaporative coolers, thereby extending the applicability of evaporative cooling to hot-humid regions. Where water is scarce (desert regions), the diurnal temperature range tends to be large, so night-time ventilation and cooling will be adequate in many places where the opportunity for evaporative cooling is limited by water constraints.

In addition to reducing energy use, direct use of water for evaporative cooling will normally consume less water than is consumed with a vapour compression chiller using a cooling tower, particularly if the electricity that powers the chiller is produced from a thermal power plant with recirculated cooling water. Where an evaporative cooler replaces an air conditioner or air-cooled chiller, the savings in water consumed at the power plant due to less need for electricity will offset one- to two-thirds of the water consumed by the evaporative cooler if the power plant uses evaporative cooling. This saving does not occur if the power plant is air cooled or uses once-through water cooling.

Among mechanical cooling equipment, efficiency (as measured by the COP) increases with increasing cooling capacity. Thus, window air conditioners and wall air conditioners (2–6kW cooling capacity) are the least efficient (COP of 2.3–3.5 in North America but up to 6.4 in Japan),

while commercial centrifugal chillers (0.3–35MW cooling capacity) are the most efficient (COP of 5–7.9). Thus, if every room in a multi-unit residential building is to be provided with an air conditioner, it can be much better from an energy-efficiency point of view to supply chilled water from central chillers, with appropriate individual controls and metering so as to discourage wasteful behaviour. An alternative to individual room air conditioners that eliminates the need for on-site chillers and cooling towers is to connect the building to a district cooling system. A third alternative is to install heat pumps in individual residential units that are coupled to a water circulation loop through a heat exchanger. The heat pumps serve both heating and cooling needs, and can be used to redistribute heat gains to the parts of the building needing it, thereby improving overall efficiency if there are frequent simultaneous heating and cooling loads. This approach is more efficient than using separate room air conditioners but less efficient than centralized chilling, and should be coupled with a cooling tower for effective removal of heat when there is a net cooling load.

A variety of heat driven cooling equipment is also available, namely, absorption, adsorption, ejector and desiccant air conditioners. These can utilize waste heat or solar heat. Single- and double-effect absorption chillers are commercially available, require input heat at temperatures of at least 60°C and 95°C, respectively, and have COPs of about 0.7 and 1.2, respectively. The heat can be provided through direct combustion of natural gas, and through production of steam from natural gas. Compared to using the same natural gas to generate electricity that is used in a vapour compression chiller, absorption chillers can increase or decrease primary energy use. Adsorption chillers are commercially available but have seen little use. Their primary advantage is the absence of moving parts. Desiccants are ideal for removing moisture from air (i.e. satisfying the latent heat load), particularly in hot-humid climates, where vapour compression equipment performs poorly. They can save electrical energy both by removing some of the air conditioning load from vapour compression chillers, and by allowing the chiller to operate with a warmer evaporator because overcooling of the air to remove humidity is no longer needed.

Solid desiccants have a larger COP at lower input temperatures, unlike absorption or adsorption chillers, making them better suited to use waste or solar heat. Much of the thermal energy used to regenerate the solid desiccant is recycled via a sensible heat wheel, and net cooling is achieved through evaporative cooling. Liquid desiccants can be readily integrated with an indirect evaporative cooler, or can make use of cooling tower water. Both solid and liquid desiccants extend the range of evaporative cooling to hot-humid climates where it would otherwise be inapplicable.

Where mechanical cooling is needed, correct sizing of the cooling equipment and correct operation of the equipment are important, as grossly oversized or incorrectly operated equipment is significantly less efficient (easily by up to a factor of two). By incorporating thermal mass, and accounting for it and the time lag between peak outdoor temperature and the inward penetration of heat, peak cooling loads will be disproportionately reduced relative to average cooling loads, thereby reducing cooling system capital costs but also allowing the equipment to operate on average at a larger fraction of peak capacity.

There can be a significant saving in primary energy use if chilled water is produced at night and used for cooling purposes during the day. This is because, in jurisdictions where electricity is supplied from fossil fuel power plants, the amount of fuel required to generate a kWh of electricity can be up to 40 per cent less during the night than during the day. To minimize the volume of chilled water that must be stored, the water should be relatively cold or ice should be produced. The later requires an evaporator temperature of −10°C, and this lower temperature reduces the COP of the chiller. This increases the on-site energy use. There can still be a net savings in primary energy if one would be producing cold chilled water (5–6°C) in the absence of night-time chilling and storage. This is often the case in North America. However, in cooling systems using displacement ventilation and chilled ceilings, and where dehumidification is performed with desiccant systems, such low temperatures are not needed. In this case, the COP penalty in conventional night-time chilling is significantly greater, likely eliminating the savings in primary energy. Storage of coldness using phase-change materials with a melting point in the 8–12°C range (such as eutectic salts) could permit effective night-time chilling without a significant COP penalty. Phase-change materials such as paraffin waxes can be embedded in chilled ceiling panels. Illustrative calculations presented here indicate that the capital cost of cooling systems using eutectic salts (8°C freezing point) for thermal storage can be no greater than for systems without thermal storage.

Cooling towers permit more efficient chilling by providing cool water to remove heat from the chiller condenser, rather than relying on outside air to cool the condenser. The condenser does not need to be as warm in order to reject heat, and on hot days – when the temperature difference between air cooled and water-cooled condensers is largest – this can lead to significantly highly chiller COP. Cooling towers can be used to directly cool the chilled water that is used for air conditioning when the outside wet-bulb temperature is sufficiently low. This will be particularly effective if combined with a chilled ceiling cooling system, as relatively warm (16–20°C) chilled water can be used for cooling. The water supplied by a cooling tower is sufficiently cold for chilled ceiling cooling 78 per cent of the time in Milan and 99 per cent of the time in Dublin, under present climate conditions. Alternatively, a pond, supplied with rainwater and designed as a landscaping element, can be used in place of a cooling tower to provide water either for direct chilling or as the cooling water of a chiller condenser.

Notes

1   For temperatures in the range −100–0°C, the vapour pressure in equilibrium with ice is given by $\ln(e_s) = a_1/T + a_2 + a_3 T + a_4 T^2 + a_5 T^3 + a_6 T^4 + a_7 \ln T$, where $a_1 = -5.6745359 \times 10^3$, $a_2 = 6.3925247$, $a_3 = -9.6778430 \times 10^{-3}$, $a_4 = 6.221570 \times 10^{-7}$, $a_5 = 2.0747825 \times 10^{-9}$, $a_6 = -9.4840240 \times 10^{-13}$, $a_7 = 4.1635019$. For temperatures in the range 0–200°C, the vapour pressure in equilibrium with liquid water is given by $\ln(e_s) = a_8/T + a_9 + a_{10} T + a_{11} T^2 + a_{12} T^3 + a_{13} \ln T$,

where $a_8 = -5.8002206 \times 10^3$, $a_9 = 1.3914993$, $a_{10} = -4.8640239 \times 10^{-2}$, $a_{11} = 4.1764768 \times 10^{-5}$, $a_{12} = -1.4452093 \times 10^{-8}$, $a_{13} = 6.5459673$ (ASHRAE, 2001b, Chapter 6). In both cases, T is in K and e_s is in Pa.

2 For dewpoint temperatures in the range 0–93°C, dewpoint can be computed from e_a using $T_{dp} = c_1 + c_2\alpha + a_3\alpha^2 + a_4\alpha^3 + a_5(e_a)^{0.1984}$, where $a_1 = 6.54$, $a_2 = 14.526$, $a_3 = 0.7389$, $a_4 = 0.09486$, $a_5 = 0.4569$, $\alpha = \ln(e_a)$, and e_a is in kilopascals (ASHRAE, 2001b, chapter 6). Conversely, given T_{dp}, e_a can be computed as the saturation value evaluated at T_{dp}.

3 Insulation of internal walls is effective in Hong Kong because of the habit in Hong Kong of air conditioning only the rooms that are occupied at any given time.

4 A louvre is a shutter with horizontal or vertical panels that can be tilted through various angles. Louvres are common in southern Europe.

5 For both traditional and contemporary streets, north–south streets are about 3K cooler than east–west streets, whereas buildings are cooler if oriented east–west so as to minimize exposure to the afternoon sun, which strikes the building façade more directly.

6 The term was first used in reference to the window:wall ratio times the window shading coefficient.

7 A surface that is good at emitting at a particular wavelength is also good at absorbing at that wavelength, if there is radiation available to be absorbed. The term 'atmospheric window' refers to that portion of the electromagnetic spectrum between wavelengths of 8–14μm, where there is relatively little downward radiation from the atmosphere but maximum potential surface emission. Hence, having a high emissivity at these wavelengths allows maximal cooling (through emission) with relatively little counteracting heating through absorption of atmospheric radiation. Outside the atmospheric window, downward radiation from the atmosphere is much larger but potential surface emission is smaller, so it is desirable to have a low emissivity (and hence a low absorptivity) at these wavelengths so as to minimize the absorption of atmospheric radiation.

8 Co-housing is a form of housing development similar to condominiums, except that it is developed and designed by the future occupants themselves.

9 Like heat pumps, the performance of air conditioners is characterized in the US using the SEER and EER values. Elsewhere, the ratio of electrical power consumption to rate of cooling (kW/kW) is sometimes referred to as an 'efficiency', but this ratio (the reciprocal of COP) is an energy intensity.

10 Halocarbons are compounds containing carbon and one or more halon gases – chlorine, fluorine and bromine.

11 Manufacturers of ground source heat pumps, on the other hand, include the energy required to circulate water through the ground coil and cooled (or heated) water or air within the building when specifying the energy performance.

12 As discussed in Chapter 15 (Section 15.1.2), modern natural gas combined-cycle power plants have an electricity generation efficiency of about 55 per cent,

13   The vapour pressure is expressed in terms of the corresponding mass mixing ratio for air using Equation (6.2) with a fixed P_a of 101350Pa in order to permit comparison with atmospheric humidity expressed in terms of mixing ratio.

14   The concept of brine bulb temperature is analogous to that of wet-bulb temperature – the temperature at which an air parcel and liquid water come into equilibrium as evaporation occurs.

15   The equilibrium point would in fact be slightly above point B, as the desiccant solution would be slightly diluted as it absorbs water vapour.

16   From Equation (6.2), $e = rP_a/(0.622 + r)$, where the atmospheric pressure P_a is assumed to be 101350Pa at 30°C and to vary in proportion to the absolute temperature, which implicitly assumes that air density is constant.

17   Under chiller priority, the chillers are used as much as possible during the day. Under storage priority, the melting of ice during the day is maximized in order to minimize electricity use during times of peak electricity cost.

18   As noted above, the potential benefit of night-time operation of a water-cooled chiller is less than for an air-cooled chiller.

19   Note added in proof. Eutectic salt thermal storage systems are again commercially available, through Environmental Process Systems Ltd (www.epsltd.co.uk), Lahren James Inc (www.lahrenjames.com) and perhaps other firms.

20   Narrow buildings also increase the opportunity for daylighting.

seven

Heating, Ventilation and Air Conditioning (HVAC) Systems

The term HVAC refers to the combined *Heating*, *Ventilation* and *Air Conditioning* system in a building. The term is more commonly used in reference to commercial buildings than to residential buildings, as these three functions are closely integrated in the former. In older houses (and many still built), ventilation occurs through uncontrolled air exchange through leaks in the walls and around windows and doors. Energy-efficient houses are almost air-tight (Chapter 3, Section 3.7), so mechanical ventilation has to be provided, often in combination with the heating or cooling system. Thus, it is appropriate to speak of HVAC systems in houses, as well as in commercial buildings.

The preceding chapters have extensively discussed furnaces, boilers, heat pumps and air conditioners – the devices that generate heat or coldness. We now discuss the missing parts – the distribution of heat or coldness throughout a building; the concurrent supply of ventilation air; and the integration of heating and cooling

devices with the distribution of heat and cold, and with ventilation. It is important, from an energy-efficiency point of view, to integrate mechanical HVAC systems with passive heating, cooling and ventilation, so that passive systems can be used whenever conditions permit. Figure 7.1, adapted from Ghiaus (2003), provides a useful overview of the regimes for active and passive ventilation and heating or cooling. In the absence of ventilation or artificial heating or cooling, the temperature of a building will tend to be warmer than the outdoor air temperature. This is referred to as the *free-running* temperature. As well, the upper and lower temperatures that are considered to be comfortable vary with the outdoor temperature. The relationships between outdoor temperature, free-running temperature and the thermal-comfort limits are shown in Figure 7.1. Below about 10°C (depending on the building), heating is required, while above 30°C (or less, depending on the extent of shading, the magnitude of internal heat gains, and the relative humidity (RH)),

Figure 7.1 Domains for mechanical and passive heating, ventilation and cooling

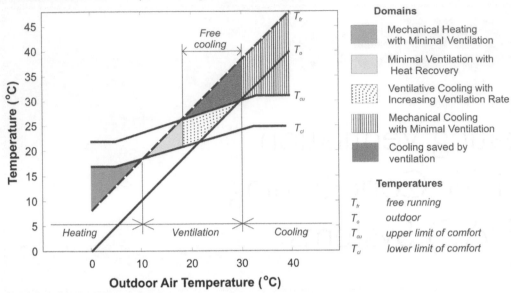

Note: The boundaries between different domains will vary from building to building, depending on the character of its envelope, building geometry and the magnitude of internal heat gains.
Source: Ghiaus (2003)

artificial cooling is required. Between 10°C and 20°C, ventilation with heat recovery only is needed, while between 20°C and 30°C, increasingly vigorous ventilation is needed.

7.1 Physical principles

We begin this chapter with a discussion of physical principles, followed by separate sections dealing with HVAC systems in small residential buildings and in commercial buildings. The relative importance of different heating and cooling loads, and the opportunities for integration of active and passive ventilation systems, are quite different in these two sectors.

7.1.1 Heat transfer by circulating air or water

Heat can be distributed to individual rooms either by supplying warm air through the ventilation system, or by supplying hot water to some sort of radiator. As long as the supply air or radiator is warmer than the room, it will give off some of its heat, thereby returning to the furnace or water heater at a cooler temperature. It is then reheated and completes the circuit again. The

temperature at which the air or hot water leaves the heating system is referred to as the *supply temperature*, while the temperature when it returns to be reheated is referred to as the *return temperature*. Systems based on hot water are called *hydronic* systems. The amount of heat in a given substance at a given temperature is equal to its temperature (in kelvin) times its *heat capacity* (Joules/m³/K). The rate at which heat is given off (in watts) by an air or water flow, Q_H, is given by

$$Q_H = \rho c_p F (T_{supply} - T_{return}) = \rho c_p F \Delta T \qquad (7.1)$$

where F is the volumetric flow rate (m³/s), c_p is the specific heat (J/kg/K), ρ is density, and ΔT here is the difference between supply and return temperatures (compare Equation 7.1 with Equation 3.2). For air, $c_p = c_{pa} = 1004.5$J/kg/K and $\rho = 1.25$kg/m³ at 10°C and 1atm, while for water, $c_p = c_{pw} = 4186$J/kg/K and $\rho = 1000$kg/m³ at 4°C.[1] From this it can be seen that, for a given volumetric flow rate and temperature drop, water delivers 3333 times as much heat as air.

To keep a room at a given temperature, the rate of heat delivered by the heating system must equal the rate of heat loss. For a given flow of a given fluid (air or water), the rate of heating is proportional to the temperature drop of the fluid. A larger temperature drop will occur the warmer the air or water relative to the space being heated (and the slower the flow rate, as this provides more time for the fluid to release heat). If the building has a good thermal envelope, the rate of heat loss is small, so heat does not need to be supplied as rapidly by the heating system, and a smaller supply temperature and/or a smaller flow rate is possible. However, most heating systems are rather inflexible, with a fixed supply temperature and a fixed flow rate, so part load is satisfied with intermittent on/off operation (and often overheating the building, requiring windows to be opened in winter).

The heat output from a radiator is initially close to constant as the rate of water flow decreases, then falls rapidly to zero at zero flow (Figure 7.2). Thus, a 20 per cent reduction in the required heat allows a 60 per cent reduction in the flow rate. This is because the lower flow rate is partly compensated by a larger temperature drop. Pumping energy is reduced by about a factor of eight (as explained in the next section) compared to full flow rather than by 20 per cent (with on/off cycling), and the boiler or furnace efficiency will be larger due to the cooler return temperature.

Figure 7.2 Variation in heat output with water flow rate through a radiator

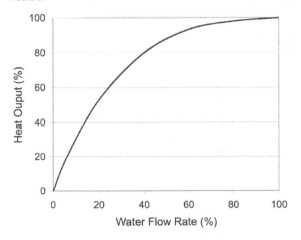

Source: Energy Research Group, University College Dublin, Technology Module 6

7.1.2 Energy used to move air or water

Energy is required for fans or pumps to move air or water. The following discussion applies to both pumps and fans, moving water and air through pipes and ducts, respectively, but to keep the wording less onerous, the term 'pump' will be taken to mean pumps or fans, the term 'pipe' will be taken to mean pipes or ducts, and the term 'fluid' will be taken to mean air or liquid. The rate at which energy must be supplied (watts) to move a fluid can be written as:

$$P_{fluid} = \tfrac{1}{2}\rho AV^3 + \Delta PAV + \Delta\rho AVg\sin\theta \qquad (7.2)$$

where ρ is the fluid density, $\Delta\rho$ is the difference in the density of upward and downward moving flow, A is the pipe cross-sectional area, V is the fluid velocity (so $AV = Q$, the volumetric flow rate), ΔP is the pressure loss during the flow circuit due to friction, g is the acceleration due to gravity, and θ is the slope of the fluid flow. The pressure loss ΔP has to be regenerated as the fluid flows through the pump, and so is referred to as the *pressure head*.

The first term in Equation (7.2) represents the rate at which kinetic energy needs to be supplied, and the second represents the work done creating the pressure needed to drive the flow. The third term is the work needed to overcome the force of gravity in lifting the fluid, and is applicable only to the vertical component of the flow. It will be negative as long as heated air or water is produced in the basement of a building and chilled air or water is produced on the top floor. The kinetic energy term involves V to the third power rather than to the second power because the kinetic energy of a parcel varies with V^2, but the amount of fluid that needs to be moved varies with V. This term would apply only to that portion of the air or water flow that is introduced into the recirculating flow from outside, and therefore accelerated from rest.

Levinson et al (2000) indicate the following as typical airflow velocities: main and branch ducts in commercial buildings, 6–10m/s and 5m/s, respectively; main and branch ducts in residential buildings, 4m/s and 3m/s, respectively. Substituting typical values for the other inputs to Equation (7.2) for airflow ($\rho = 1.25\text{kg/m}^3$, $\Delta P = 100$ Pascals (Pa) for houses and 1000Pa or more for commercial buildings, $\Delta\rho = -0.044$kg/m^3 for a 10K difference between rising air and sinking air) indicates that the second term, representing the power needed to overcome frictional resistance, is the dominant term. The same is true for water flow (for which $\rho = 1000$kg/m^3, ΔP is on the order of 50000Pa (0.5 bar), $\Delta\rho = -0.906$kg/m^3 for a 10K differential centred at 10°C, $v < 1$m/s).

In any case, we shall neglect vertical motions and consider only the power needed to maintain the existing motion. This power is:

$$P_{\text{fluid}} = \Delta P A V = \Delta P Q \qquad (7.3)$$

The electrical power that would be required is given by:

$$P_{\text{electric}} = \frac{P_{\text{fluid}}}{\eta_m \eta_p} = \frac{\Delta P Q}{\eta_m \eta_p} \qquad (7.4)$$

where η_m and η_p are the motor and pump efficiencies, respectively. The motor efficiency is the ratio of shaft power to electrical power, while the pump (or fan) efficiency is the ratio of power supplied to the fluid to shaft power. If the motor speed is varied using a VSD (variable-speed drive), then the required electric power is:

$$P_{\text{electric}} = \frac{P_{\text{fluid}}}{\eta_{VSD} \eta_m \eta_p} = \frac{\Delta P Q}{\eta_{VSD} \eta_m \eta_p} \qquad (7.5)$$

where η_{VSD} is the VSD efficiency (ratio of electrical power out to electrical power in). The rotation rate can be varied in one of three ways using a VSD: by changing the phase lag between different components in a three-phase AC power supply, by changing the voltage of the power supply, or by changing the frequency of AC power.

Dependence of the pressure drop on the piping system and flow rate

The relationship between the pressure drop in the pipe system and the flow rate depends on the diameter, roughness, and length of the pipes, and the number of bends. The pressure drop along a circular pipe for fully turbulent flow is given by the D'arcy–Weisbach equation:

$$\Delta P = \lambda \frac{l}{D} \frac{\rho}{2} V^2 \qquad (7.6)$$

where λ is a friction coefficient that depends on the smoothness of the pipe, l is the length of pipe, D is the diameter of the pipe, and ρ is the density of fluid. From this equation, two critical relationships arise:

- For a given volumetric rate of flow, $V \alpha$ $1/D^2$, so ΔP varies *inversely* with the diameter of the pipe to the *fifth power* and a small increase in the diameter of a pipe will dramatically decrease the energy loss due to friction.
- For a fixed pipe size, $Q \alpha V$, so the power that must be supplied to the fluid decreases in proportion to the decrease in flow volume to the *third power*.

Both relationships are modified slightly by the fact that the friction coefficient λ varies with $D^{0.25}$ and $Q^{0.25}$, so that ΔP varies with $D^{-4.75}$ for fixed Q, while $\Delta P \alpha Q^{1.75}$ and $P_{\text{fluid}} \alpha Q^{2.75}$ for fixed D (Söylemez, 2001). To keep the terminology simple, the relationship $P_{\text{fluid}} \alpha Q^{2.75}$ shall be loosely referred to as the "cubic" relationship. A 50 per cent reduction in the required flow, for example, would reduce the fluid power by a factor of 6.7 (or by a factor of eight if the relationship were strictly cubic). The variation of energy loss with $D^{-4.75}$ applies to that portion of the loss through straight pipe sections. Significant pressure drops occur where there are bends. The keys to reducing the energy required to circulate air or water are thus:

- to *oversize* pipes or ducts, something that will be easier for pipes than for ducts due to space limitations, and which therefore increases the attractiveness that hydronic heating and cooling already have over an all-air heating and cooling system (as discussed later);
- to lay out the heating and cooling elements so as to minimize the length of pipe and minimize the number of turns, which requires

that the HVAC piping contractor be involved throughout the design and construction of the building (as an added benefit, it will be easier to insulate a simpler network, thereby increasing the amount of insulation that is cost-effective);

- to use duct or pipe material with minimal friction and associated pressure drops;
- to use turning vanes (a metal fin) inside any ducts with a 90° turn;
- to minimize the required flow.

Arranging ductwork in a loop, rather than as radial branches to each room (the normal practice) can reduce the required fan power by 20–25 per cent, with comparable savings in the maximum duct diameter (Levermore, 2005). PVC piping has a dramatically smaller roughness than conventional sheet metal ductwork, as illustrated in Table 7.1, leading to smaller frictional losses. Conversely, flexible ducts have large frictional losses, so their use should be kept to a minimum and they should be fully extended. Up to 30 per cent of fan energy is used to push air through filters. Commercially available premium filters, with twice the media surface area per unit volume of conventional filters, can reduce the pressure drop across the filter (for a given flow rate) by 25 per cent with a cost payback in the order of 1 year (Chimack and Sellers, 2000). Further enhancements, produced as a result of a design competition in Norway, reduce the pressure drop by 30–40 per cent compared to conventional filters (Grorud, 2001). Filters should be periodically replaced to avoid excessive pressure drops across the filter.

Variation in energy use with flow rate

As noted above, the power that needs to be supplied to a fluid varies with the flow rate to almost the third power. The electrical power that is required depends on the fluid power divided by the product of the VSD, motor and pump efficiencies. Smaller pumps and motors have smaller full-load efficiencies, so the efficiency at peak flow will be smaller in smaller systems. In addition, all three efficiencies are smaller at part load than at peak load. Thus, the decrease in electric power required as the flow rate decreases is much less than expected based on the cubic relationship.

Figure 7.3 shows the variation of peak-load motor efficiency with the size (power output) of the motor, based on a survey of installed motors in the US and as prescribed under the 1992 US Energy Policy Act (EPAct) and by the National Electrical Manufacturers Association (NEMA) standard for premium motors. Data on peak-load efficiencies for different sized fans and the matching motors are given in Table 7.2, where it can be seen that there is more than a factor of two difference between the smallest and largest systems in the combined efficiency, mostly due to differences in the fan efficiency. As for pumps, a small pump might have a peak efficiency of 55 per cent, while a large pump might have a peak efficiency of 85 per cent (Bernier and Lemire, 1999). Small pumps have a smaller efficiency because internal leakage, drag from seals and packings that prevent escape of the liquid from the pump, and internal roughness are all relatively more important in small pumps (Wulfinghoff, 1999, Reference Note 35).

If an alternative design for a new building leads to a smaller peak airflow rate, not only

Table 7.1 Roughness values for different kinds of pipes and air ducts

Material	Roughness (mm)
PVC piping	0.01
Folded sheet metal ducts	0.15
Smooth sheet metal ducts	0.50
Concrete ducts	1–3
Masonry ducts	3–5
Flexible pipes	0.2–3

Source: Hastings and Mørck (2000)

Table 7.2 Typical fan and motor peak-load efficiencies

Volumetric flow rate (m³/h)	Fan efficiency	Motor efficiency	Combined efficiency
Up to 300	0.4–0.5	0.80	0.32–0.40
300–1000	0.6–0.7	0.80	0.48–0.56
1000–5000	0.7–0.8	0.80	0.56–0.64
5000–100,000	up to 0.85	0.82	up to 0.70

Source: Hastings and Mørck (2000)

Figure 7.3 Variation of peak-load motor efficiency with motor size, as found in a survey of existing installed motors in the USA (US DOE Survey), as prescribed by the 1992 US Energy Policy Act (US EPAct), and as prescribed by the National Electrical Manufacturers Association for premium motors (NEMA Premium)

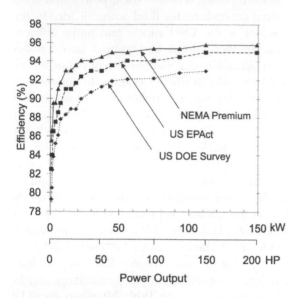

Source: Data from Malinowski (2004)

Figure 7.4 Variation of (a) drive and (b) motor efficiency with speed and torque for an 11.2kW variable speed motor and drive, as determined by Gao et al (2001)

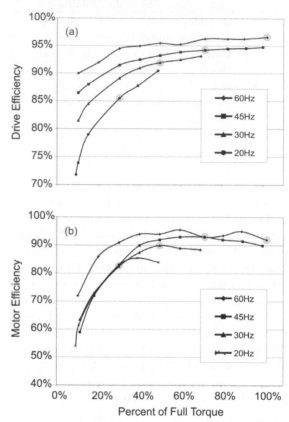

will a smaller and possibly less efficient motor and fan be chosen, but the ducts themselves might by made smaller, in which case the reduction in the required fan output power will be less than implied by the cubic relationship. For example, in comparing all-air and chilled ceiling cooling systems (discussed below), Feustel and Stetiu (1995) assume that the ducts are downsized such that the pressure drop in the duct system is the same for the two cases, with the result that the savings in fan power is proportional to the reduction in airflow (a still-respectable factor of ten). This in turn reduces the required height of each floor by 0.2m, which leads to savings in the embodied energy for concrete, steel and other materials. The absolute savings in fan power in going from 10 per cent to 1 per cent of the base case power (i.e. in not downsizing the ducts) is small compared to the savings in dropping from 100 per cent to 10 per cent, so the energy penalty will in any case be small.

With regard to part-load efficiencies of VSDs, motors and pumps, all three efficiencies depend on the frequency or speed (rotations per minute) of operation and the torque that is produced. The flow rate is proportional to the

frequency, while the torque is proportional to the pressure head. Figure 7.4 shows how the drive and motor efficiencies vary with frequency and torque for one particular VSD motor. The drive and motor efficiencies are both highest at high frequency and torque. Efficiencies are relatively constant between 40 per cent and 100 per cent of full torque, but drop sharply at lower torques. This implies that, below a certain load, it is preferable to revert to on/off cycling rather than a further reduction in the motor speed.

For the experimental setup used to measure drive and motor efficiencies, the speed and torque can be independently varied. In a normal situation, the torque will vary with the speed squared (this follows from the fact that torque is proportional to pressure head, pressure head varies with the flow rate squared, and flow rate varies with speed). Thus, a series of data points corresponding to speed–torque combinations that would occur in reality can be taken from

Figure 7.5 Variation of combined motor-drive efficiency with motor speed based on the data given in Figure 7.4

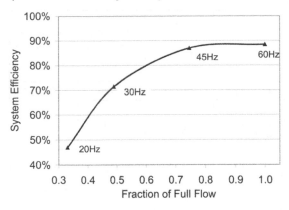

Figure 7.6 Variation of fan or pump power with flow for various ways of reducing the flow

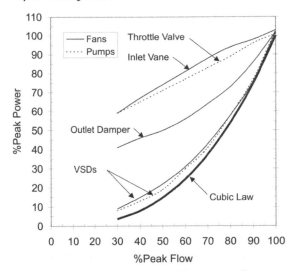

Note: The curve labelled 'cubic law' assumes that power α (flow)$^{2.75}$.
Source: Smith (1997)

Figure 7.4 (indicated by solid circles) and used to give the drive and motor efficiency versus motor speed or flow rate. This is shown in Figure 7.5 for the product of the drive and motor efficiencies, which falls from 88 per cent at full load to 42 per cent at 30 per cent of full load. Bernier and Lemire (1999) and Sfeir et al (2005) provide similar curves along with regression equations of efficiency versus load.

The efficiency of a pump also depends on its speed and torque. Peak efficiency occurs at close to maximum pressure head (torque) and half the maximum flow rate. At lower pressure, efficiency decreases because the impeller produces pressure internally that is not utilized. At lower flow rates, efficiency decreases because the impeller produces turbulence in trying to move the fluid through a system that will not take it, while at higher flow rates the efficiency decreases because flow occurs without the need for all the pressure that is being produced, the excess being dissipated as extra turbulence (Wulfinghoff, 1999, p1341). A pump with a full-load efficiency of 85 per cent might have an efficiency of 75 per cent at 20 per cent of full load.

The change in energy use with flow in a given system depends on how the flow is made to vary. If this is done by throttling (using either inlet vanes or outlet dampers), power consumption falls much more slowly than the reduction in flow. If this is done by varying the rotation rate of a fan or pump using a VSD, the required power decreases several times faster than if throttling is used to control the flow, but not as rapidly as expected based on the cubic relationship, due

to the decrease in VSD, motor and fan or pump efficiencies with decreasing flow. The variation of input power with flow rate using different methods of flow control is illustrated in Figure 7.6, and can be understood with reference to Equation (7.5): when the flow is throttled, Q decreases but ΔP increases, such that power input falls only slightly as the flow is decreased. With a VSD, both ΔP (which varies with Q^2) and Q decrease, as the motor and pump are not trying to generate more flow than needed, but the denominator also decreases. It is important to pay attention to the part-load efficiencies and to avoid oversizing fans or pumps relative to the design flow, because of the substantial decrease in efficiency at part load, but it is normally even more important to use a VSD. Because the VSD itself introduces an energy loss even at full load, a VSD should not be used in those rare circumstances where the pump would need to be operated at or near full load most of the time.

The flow rate by fans can also be varied by changing the pitch of the fan blades. Variable-pitch and variable-speed fans have similar part-load efficiencies. However, noise drops sharply with decreasing fan speed, so variable-speed fans will operate more quietly at part load.

Comparison of hydronic and air systems

The energy required to supply or remove a given amount of heat can be derived from Equations (7.1) and (7.3):

$$E = \frac{\Delta P}{\rho c_p \Delta T}$$

(7.7)

where ΔT is the difference between supply and return temperatures. Using the values of ρ and c_p given with Equation (7.1), and taking ΔP to be 1400Pa in an all-air system and 50,000Pa for an equivalent hydronic system (as computed by Niu et al, 2002, for a given building), it can be seen that heat transfer by water requires about one hundred times less energy for a given ΔT than heat transfer by air, neglecting differences in pump, fan and motor efficiencies. However, if the temperature drop in the air system is four times larger (i.e. 8K rather than 2K), the energy used to transfer a given amount of heat by water would be 25 times smaller than using air. This is still a substantial difference!

Trade-off between parasitic and distribution losses

In addition to the energy required to circulate air or water – so called parasitic losses – there will also be a loss of heat (or coldness) from the piping or duct system as hot or cold fluid flows through it. This distribution loss depends on the temperature difference between the pipe or duct and the surrounding air, the extent of insulation, and the surface area of the pipe. A warmer distribution temperature for heating or a cooler temperature for air conditioning will increase the distribution energy loss but decrease the parasitic loss (the latter because ΔT will be larger, so a smaller flow rate is needed for a given rate of heat transfer). Duct systems are extremely prone to leakage, and often pass through unconditioned spaces, leading to further direct losses of heat or coldness, and indirect losses through the infiltration of outside air into the building that can be induced by a leaky duct. Table 7.3 compares the typical supply and return temperatures, fan or pump energy use, and the parasitic and distribution losses for various residential heat-distribution systems used in Europe and North America.

Table 7.3 Comparison of different heat distribution systems for residential and commercial buildings

System	Supply/ return temperature (°C)	Parasitic energy loss (%)	Distribution energy loss (%)
Forced air	55/45	10–20*	20–55*
Conventional hot water radiators	55–90 45–70	6–9*	3–7*
Floor radiant heating	30–35 25–30	20*	

Note: Parasitic energy loss is the energy used by fans or pumps as a percentage of the heat supplied.
Source: Breembroek and Dieleman (2001) where indicated by asterisk

7.1.3 Matching of fan or pump and system pressure changes

For steady flow, the pressure drop through a piping system will exactly balance the pressure created as the fluid flows through the pump.[2] The rate of flow through a pump varies directly with the speed of the impeller's rotation and its diameter, while the pressure developed varies as the square of the impeller speed and the square of the impeller diameter. These are known as the pump *affinity laws*. Reducing the impeller speed or diameter simultaneously reduces both the fluid flow and the pressure head and thus rapidly reduces the power that must be supplied to the fluid by the pump.

The pressure drop as fluid flows through a piping system varies roughly in proportion to the flow velocity squared (Equation 7.6), with the proportionality constant being a characteristic of the piping system. The pump would, ideally, be chosen so that the pressure head created at the desired flow rate exactly balances the pressure drop in the piping system. However, this does not happen in practice, since there are errors in the calculation of the pressure drop created by the piping network for a given flow rate, and multiple safety factors are routinely built into the calculations, so that the resulting flow is invariably greater than specified. Thus, after the pump has been installed, it has to be operated to determine the actual flow at the specified pump speed. The pump-piping system should then be 'rebalanced' so that the desired flow is created (Egan, 2001). A piping system will typically have many parallel circuits, and the relative flow errors will generally be different for each circuit. A control valve in each circuit will need to be slightly closed (thereby throttling the flow), but this should be done only to bring the ratio of actual flow to predicted flow down to the ratio found in the circuit with the smallest ratio (so no adjustment of the control valve is needed in this circuit). At that point, the pressure head developed by the pump can be reduced, thereby reducing the flow in each circuit and bringing them all simultaneously down to the design-flow. The net result is to achieve the desired flow in all circuits with a minimum of throttling.

The pressure head developed by the pump can be reduced by trimming the impeller inside the pump (which requires disassembling and reassembling the pump), or by reducing the rotational speed of the impeller. However, even if excess flow is recognized, the pump is rarely adjusted to achieve the design flow according to Egan (2001), especially if the system is operating smoothly. Excess flow not only wastes energy, but can cause excess noise and control problems, as documented by Mansfield (2001) in a series of enlightening examples. Wear on equipment and pipe erosion might also be increased.

To understand the implications for energy use by the pump, we need to examine the pressure-head/flow variation for both the pump and the piping system on a single diagram, as shown in Figure 7.7. The convex-upward curves show the variation of pressure head and flow rate for a variety of fixed impeller diameters (these are a characteristic of the particular pump).

For a given impeller diameter, a greater flow rate (due to less resistance in the system to which the pump is connected) is associated with a smaller pressure head. Also shown in Figure 7.7, as dashed lines, are contours of pump efficiency.[3] The concave-upward curves show the variation of piping system pressure drop with flow rate; one curve is the calculated relationship, and the other is the actual relationship for that particular system. Point 'A' is the expected operating point. However, as is typically the case, the pressure

drop for the desired flow is less than the computed pressure drop (due to various safety factors). This is represented by point 'B'. Since the pump generates more pressure than needed, flow increases, the pump pressure head decreases, the system pressure drop increases, and the two come into balance at point 'C', but with excess flow. The fluid power (P_{fluid}), equal to the product of pressure head and flow, is essentially unchanged in this example. However, the pump would have been chosen to have the greatest efficiency near the design flow, and is somewhat less at the actual operating point (the pressure head, flow rate and pump efficiency for all three points are listed in the caption to Figure 7.7). The net result is a 5.6 per cent increase in pump energy use, and most systems would be allowed to operate in this manner. However, at point B (where the pump efficiency is even lower), the required pump power is 24 per cent less than at point C. This is not a negligible saving!

With the widespread availability of VSDs,

Figure 7.7 Intersection of the pressure-flow relationship for a pump with various impeller diameters (convex-upward curves) and for the piping system as originally estimated (upper concave-upward curve) and in reality (lower concave-upward curve)

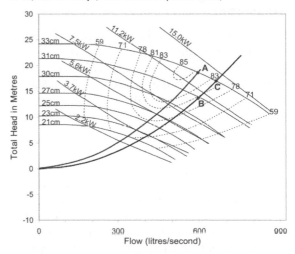

Note: The pump efficiency is given by the dashed curves. The pump power curves are given by the pressure head times the flow rate (in appropriate units), divided by the efficiency. Point A (ΔP = 18.6m, Q = 590l/s, η = 0.846) represents the original operating point, point B (ΔP = 13.5m, Q = 590l/s, η = 0.805) the required point, and point C (ΔP = 17.0m, Q = 662l/s, η = 0.821) the operating point if there is no adjustment of the pump.
Source: A manufacturer's pump diagram, with system curves chosen to roughly match the example discussed by Egan (2001)

there is a tendency not to carefully analyze the pump head required by the system or to dispense with impeller trimming, on the grounds that the VSD will act like an automatic impeller trim (Ahlgren, 2001). However, drive efficiency drops as the rpm decreases (as shown above), whereas a slight reduction in torque from full-load torque has no effect on efficiency, as shown in Figure 7.4. Thus, as noted by Ahlgren (2001), reliance on the VSD rather than an impeller trim to get the correct full-load flow will likely carry an energy penalty, and the energy savings expected using a VSD for rebalancing might not be realized.

As with cooling equipment (Chapter 6, Section 6.8), it is important to avoid oversizing the pump in the first place. In the design process, pumps often need to be specified before the distribution systems and load are finalized, but these characteristics need to be known in order to optimize the pump selection. Thus, an iterative process is called for. Sellers (2005) describes an example where correct sizing of pumps (rather than grossly oversized) reduced pump-first costs by 16 per cent while reducing energy use by 30 per cent.

7.1.4 Cooling-dehumidification trajectories

In a conventional cooling system, moisture is removed when the ventilation air is cooled to below the dewpoint. Condensation of water vapour occurs at the cooling coil, releasing latent heat that is removed by the refrigerant. As warm, moist air flows past the cooling coil, air in immediate contact is cooled and dehumidified while air not in immediate contact is cooled and dehumidified only through mixing with air that came into direct contact with the cooling coil. Thus, some dehumidification occurs without the average air temperature reaching the dewpoint. As a result, the trajectory of an air parcel as it is cooled will not plot parallel to the mixing ratio lines shown in Figure 6.3, but rather, will slope downward from these lines. The ratio of latent heat to sensible heat removed from the air (and hence, the slope of the trajectory) can be altered by adjusting the rate of flow of air past the cooling coil, the rate of flow of refrigerant or chilled water through the coil, and the coil geometry, such that

it is sometimes possible to arrive directly at some desired final temperatures and relative humidities rather than overcooling and reheating the air (Sekhar, 1995).

Air that is cooled and dehumidified at the cooling coil will be distributed throughout the building without dilution if the return air is entirely vented to the outside. Otherwise, it will be mixed with that portion of the return air that is recirculated. In the latter case, the fresh air will need to be cooled and dehumidified more than in the former case, since it will be diluted by the warmer and moister return air. In both cases, the air supplied to rooms will be cooler and drier than the desired room temperature and humidity. The supply air picks up moisture and air from the room and, if it thoroughly mixes with the room air, will exit at the temperature and humidity of the room air. Some or all will be vented to the outside, and the remainder mixes with freshly cooled and dehumidified outside air and repeats the circuit. The temperature–humidity combination maintained in the room is thus different from the temperature–humidity combination created in the supply air. The room temperature–humidity combination that minimizes total energy use depends on the initial temperature and humidity, which implies that systems optimized for one set of climatic conditions may not be optimal for another. As an example, Sekhar (1995) estimates that air conditioning to achieve 26°C and 60 per cent relative humidity in the humid tropics will reduce total cooling+dehumidification energy use by 18 per cent compared to the current practice, in Pacific-rim countries, of air conditioning to 23°C with minimal dehumidification. However, this saving is dependent on being able to avoid overcooling the air to condense the required water vapour and reheating it, but instead, to arrive directly at the temperature–humidity combination that is required for the ventilation air.

A more direct approach is to decouple cooling and dehumidification through the use of desiccants for dehumidification, and to decouple cooling from ventilation so that only the amount of air needed for ventilation purposes need be cooled and dehumidified, with the balance of the cooling load handled through chilled ceilings. Outdoor air passes once through the building without recirculating a portion of it, so the outdoor air need not be overcooled due to subsequent mixing with warm recirculated air. This approach is developed in later parts of this chapter.

7.2 Human comfort and the perception of temperature

The human perception of warmth or coldness depends on four environmental parameters (air temperature, mean radiant temperature, humidity, air movement), two personal parameters (metabolic rate and clothing), and one psychological factor (expectation). The mean radiant temperature (T_{mr}) is the temperature of an isothermal blackbody enclosure that would exchange the same amount of radiation with the occupant as the actual surroundings (see Equation. 3.3). There are two distinct sources of discomfort: thermal sensation (dependent on the perceived temperature), and skin wetness.

7.2.1 Perceived temperature

The perceived temperature is a weighted average of T_{mr} and the air temperature (T_a), with the weighting factor depending on the air speed (v) according to:

$$T_{perceived} = \frac{T_{mr} + 3.17 T_a \sqrt{v}}{1 + 3.17 \sqrt{v}}$$

(7.8)

for sedentary subjects with RH between 40 per cent and 70 per cent (CIBSE, 2006). There is no difference between people from tropical and non-tropical regions in the perception of warmth or coldness. Heat is exchanged between a subject and the surroundings through radiation and through turbulent air motions. When $v = 0$, the perceived temperature is equal to T_{mr}, as there will be no turbulent heat exchange between the subject and the surrounding air. Conversely, as v becomes very large, the perceived temperature approaches the air temperature, as turbulent heat exchange dominates radiative heat exchange. For an air speed of 0.1m/s, the perceived temperature is very close to the average of the air temperature and the radiant temperature. Thus, if the ceiling and floor are kept at a warmer temperature in winter (or at a lower temperature in summer), the air temperature does not need to be

as high in winter (or as low in summer), which will reduce the conductive heat loss (or gain) through the walls and windows and due to exchange of outdoor air with indoor air.

7.2.2 Role of air movement

Whether a given perceived temperature is judged to be comfortable or not depends on the rate of heat loss from the subject and on skin wetness. Air movement shifts the perceived temperature toward the air temperature but also increases the rate of heat loss or gain; heat loss occurs if the air temperature is less than the skin temperature (normally 35–36°C, 1–2K below the normal body core temperature of 37°C). The rate of heat removal varies with the square root of air speed. In winter, air movement should be minimized in order to minimize body heat loss by providing only the fresh air needed to maintain healthy air quality, while in summer air movement should be maximized when cooling is needed through the use of individually controlled ceiling and/or desk fans. Heating and cooling beyond that which can be supplied with the ventilation air should be provided through hot- or cold-water radiators.

Greater airflow also makes it easier for evaporation to occur, so any required evaporative cooling (through perspiration) can occur from a smaller wetted area of skin. If the air temperature is greater than the skin temperature, sensible heat transfer to the skin will increase but so will evaporative cooling, so it is possible that overall cooling will increase. Increasing wind speed when the air is warmer than the skin temperature can increase comfort even though the perceived temperature increases, because skin wetness decreases. At very warm air temperatures, the cooling effect of a given rate of evaporation will decrease as some of the heat required to evaporate the sweat is taken from the ambient air rather than from the skin, and this effect will increase with increasing air speed.

Humphreys (1970) deduced theoretically that the increase in the temperature considered to be comfortable, ΔT_c, varies with air speed according to:

$$\Delta T_c = 7 - \frac{50}{4 + 10 v^{0.5}}, v \geq 0.1 m/s$$

(7.9)

Figure 7.8 Increase with air speed in the temperature considered to be comfortable. For situations where the air temperature and radiant temperature are approximately equal

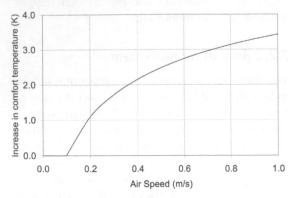

Source: Based on Humphreys (1970)

This relationship is shown in Figure 7.8. A fan will typically produce an air speed of 0.5m/s, which will increase the acceptable temperature by 2K. If $T_a > T_{mr}$ (which will be the case in summer in buildings with chilled ceiling cooling) but is judged to be too warm, increasing air speed will increase the perceived temperature, but the perceived temperature that is judged to be comfortable also increases.

7.2.3 Role of humidity

Relative humidity (RH) has no effect on the rate of evaporation of sweat except under extreme conditions (Givoni, 1998a). Rather, a lower RH can increase comfort by reducing the moist area required to sustain a given rate of evaporation. At low RH, sweat evaporates from within skin pores. As RH increases, sweat spreads over a larger area, tending to maintain the rate of evaporation required to offset the net of all other heat gains and losses at an appropriate skin temperature. We may feel discomfort at higher RH, but this is a beneficial physiological response and there is little if any change in skin temperature.

For seated subjects in light clothing and temperatures up to 30°C, there is no significant difference in preferred temperature between relative humidities of 35 per cent and 70 per cent according to surveys by de Dear et al (1991a, 1991b). Fountain et al (1999) extensively review

previous studies on the effect of humidity on comfort and note that half the studies reviewed found no perceptible relationship between humidity and comfort over a wide range of conditions. Where possible effects occur, they are at and beyond the generally accepted limits of temperature and humidity. For this reason, Fountain et al (1999) performed an extensive set of tests on sedentary subjects with conditions between 60 per cent and 90 per cent RH for temperatures in the range 20–26°C. Although 90 per cent RH was least favourably rated, they found that 80 per cent RH is apparently no less acceptable than 60 per cent or 70 per cent RH. According to Nicol (2004), much of the work reporting an effect of humidity on comfort has not properly separated the effects of temperature and humidity, so it is unclear how large the impact of humidity is in those cases.

7.2.4 Acceptable temperature–humidity–air speed combinations

The range of generally acceptable combinations of temperature, humidity and air speed in a centrally controlled indoor environment as determined by ASHRAE (Standard 55) is given in Figure 7.9. The left panel shows the limits for motionless air, based on surveys of human subjects in test chambers; temperatures below the lower limit were deemed to be too cold by 10 per cent of the subjects, while temperatures above the upper limit were deemed to be too hot by 10 per cent of the subjects. The range considered to be acceptable by 80 per cent of the subjects in typical winter office attire is 20–24.5°C. The upper limit of acceptable temperature in summer varies from 25.5°C at a relative humidity of 60 per cent to 26.5°C at a relative humidity of 30 per cent, while the lower limit is set at around 23°C. As noted above, there is some disagreement concerning how real the effects of humidity are on comfort for sedentary subjects. The acceptable upper limit is likely to be higher than 60 per cent RH at 26°C, in which case substantial amounts of energy are being wasted in many cases to unnecessarily reduce the indoor humidity.

The right panel shows the average air temperature at which sedentary test subjects feel neither warm nor cold for various radiant temperatures and air speeds, all for a relative humidity of 50 per cent and light clothing (this will be called the *neutral* temperature, and is not the same as the preferred air temperature, as subjects may prefer to feel slightly warm or slightly cold). When the radiant temperature is equal to the air temperature (dashed line), the average neutral temperature increases by about 2K (from 25.5°C to 27.5°C) as the air speed increases from 0.1m/s to 0.5m/s and by an additional 1K if the air speed increases to 1.5m/s. This agrees roughly with the relationship shown in Figure 7.8, which can therefore be interpreted as applying only to the case where the radiant and air temperatures are equal. At an air speed of 0.1m/s, a given decrease in the radiant temperature causes a roughly equal increase in the average neutral air temperature, while at a wind speed of 1.5m/s, the increase in neutral air temperature is about one-third as large as the decrease in radiant temperature. Thus, as air speed increases, the impact on the neutral temperature of decreasing the radiant temperature decreases. This is quite reasonable in light of Equation (7.8), which shows that the perceived temperature is more strongly weighted by the air temperature at higher air speed, so that reducing the radiant temperature has less effect on the

perceived temperature. Conversely, as the radiant temperature decreases, the impact of air speed on the neutral air temperature decreases, down to a radiant temperature of 20°C, at which point the average neutral temperature is 30°C irrespective of air speed. At very low radiant temperatures, the neutral air temperature increases with *decreasing* air speed. This is because the cold sensation caused by the low radiant temperature is offset by convective heat transfer from the warm air, but this transfer is less effective the slower the air speed, so a warmer air temperature is required for comfort.

To summarize, for conditions when air conditioning is or might be called for:

- for radiant temperatures down to 20°C, there is a trade-off between reduction in radiant temperature and increase in air speed: the more of one, the less effective is more of the other in increasing the neutral temperature and thereby reducing the required air conditioning;
- at a radiant temperature of 20°C, air speed has no effect on the neutral temperature;

Figure 7.9 Left: Limits of human thermal comfort for people in typical winter and summer clothing engaged in sedentary activity as given by ASHRAE (2001, Chapter 8) for motionless air. Right: Average preferred temperature for sedentary people in summer clothing with 50 per cent relative humidity, as a function of radiant temperature and wind speed, as given by ASHRAE (2001, Chapter 8)

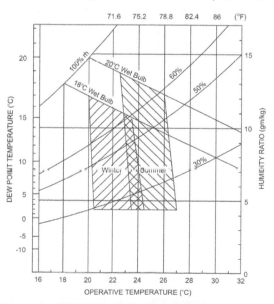
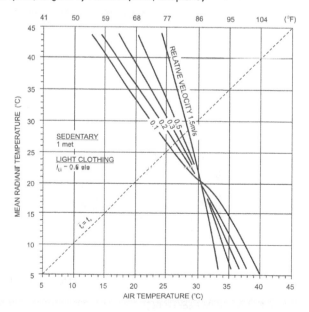

Source: Redrafted from ASHRAE (2001, Chapter 8)

- at a radiant temperature of 20°C or lower, the neutral temperature is 30°C or greater;
- at a radiant temperature below 20°C, the lower the air speed the warmer the neutral air temperature and hence the less that the air needs to be cooled for comfort (if at all).

In addition to these physical effects, there is also a psychological or adaptive component to the temperatures that are considered to be acceptable (de Dear and Brager, 1998, 2002). Warmer interior temperatures are acceptable on hot days and colder interior temperatures are acceptable on cold days if an individual knows what to expect. Many studies indicate that there is no difference in the average preferred temperature for people from different parts of the world, when subject to the same outdoor temperatures. This has lead to the incorporation of a new standard for naturally ventilated buildings in the latest ASHRAE standard (Standard 55-2004), whereby the range of acceptable temperatures inside buildings depends on the outside temperature, as shown in Figure 7.10. According to this standard, the warmest temperature acceptable by 90 per cent of the

building occupants with light clothing increases from 27°C at an outdoor temperature of 20°C, to 31°C at an outdoor temperature of 33°C, compared to the inflexible 23–26°C range adopted under the conventional ASHRAE Standard 55. Acceptable temperatures are of course different from preferred temperatures; at an indoor temperature of 32°C, a majority of occupants may very well prefer a cooler temperature. However, even a temperature setting of 28°C can yield substantial energy savings compared to the common practice of cooling buildings to 23–24°C. For example, computer simulations by Jaboyedoff et al (2004) indicate that increasing the thermostat from 24°C to 28°C will reduce cooling energy use by more than a factor of three for a typical office building in Zurich and by more than a factor of two in Rome, while simulations by Lin and Deng (2004) indicate a factor of 2–3 reduction if the thermostat is increased from 23°C to 27°C for night-time air conditioning of bedrooms in apartments in Hong Kong. As an added benefit, the higher temperature setting for air conditioning that is permitted under an adaptive standard would reduce the thermal shock often encountered in moving between indoor and outdoor spaces on hot days.

Figure 7.10 An alternative standard for acceptable indoor temperatures that takes into account psychological adaptation to different outdoor temperatures

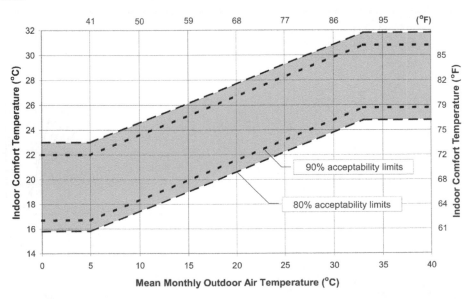

Source: Brager and de Dear (2000)

Part of the psychological adaptation to warmer temperatures is the ability of the occupant to *control* his or her environment by being able, for example, to open or close windows, or to activate or deactivate a fan. As noted in Chapter 6 (Section 6.3.1), the upper limit of acceptable temperatures on hot days is higher in naturally ventilated buildings (up to 31°C, compared to 28°C in air conditioned offices according to one study, assuming suitable summer clothing). This increases the threshold at which mechanical air conditioning needs to be activated, at which point windows will need to be closed. Occasional 'heat holidays' (altered or reduced working hours) should be acceptable in hot regions, so as to avoid the need to design buildings to withstand extreme but rare conditions.

In addition to psychological adaptation, there is a physiological adaptation to warmer temperatures, involving a greater rate of perspiration and a lower core body temperature and heart rate. These adaptations require about two weeks. For this reason, Givoni (1998a) suggests that, with an indoor airspeed of 2m/s, the upper comfort limit be set at 30°C in mid-latitude countries and at 32°C in low latitude countries. From a physiological point of view, comfort at night is more important than comfort during the day, as high heat stress can be tolerated during the day if one has good, restful sleep during the night.

In summary, there is a large body of evidence indicating that temperature and humidity settings in air conditioned buildings substantially warmer or moister than commonly encountered are acceptable. In particular, temperatures from 28°C to as high as 32°C are acceptable on hot days, particularly if individually controlled fans are available to create air speeds of about 0.5m/s and if windows can be opened. Choosing temperatures in the upper part of this range can reduce cooling energy use by a factor of 2–3 in warm climates. The upper temperature limits can be reduced at night, but night-time temperature tends to be lower in any case. There is no convincing evidence that relative humidities below 80 per cent increase human comfort or increase the acceptable temperature for sedentary subjects.

7.2.5 Impact of temperature on perceived air quality

The perceived quality of air is strongly influenced by the temperature and humidity of the air, with lower temperature and humidity resulting in higher perceived quality. This is independent of the impact of temperature and humidity on the thermal sensation. In experiments where clothing is adjusted to give the same thermal effect as temperature is decreased, cooler and/or drier air is perceived to be of higher quality, even when there is no change in chemical composition (Fanger, 2001). A commonly accepted standard for ventilation is 10 litres/s per person, but this is far in excess of actual physiological requirements. Experiments reveal that people perceive the air to be better at 20°C and 40 per cent RH with a ventilation rate of 3.5 litres/s/person than at 23°C and 50 per cent RH with a ventilation rate of 10 litres/s/person. This leads to a trade-off in summer: more energy is required to achieve a lower temperature and humidity, but less outside air needs to be conditioned and less fan energy will be needed. In winter, maintaining a lower temperature and a lower ventilation rate both work to reduce energy use.

Despite the degradation in *perceived* air quality at higher air temperatures in mechanically ventilated buildings, the actual incidence of sick-building symptoms is confined exclusively to air conditioned buildings. There is virtually no known building with sick-building symptoms that is naturally ventilated, despite the higher summer temperatures (de Dear, personal communication, 2004). Indeed, a study of sources of indoor contaminants in 15 air conditioned offices in Copenhagen found that over half (56 per cent) of the contaminants came from the air conditioning system itself (European Commission, Directorate General for Energy, 1998). The major pollution sources from mechanical HVAC systems in European buildings are the filter, oil residues in ducts, poorly maintained humidifiers and poorly installed rotating heat exchangers (Jaboyedoff et al, 2004).

As noted above, a ventilation rate of 10 litres/s/person is not a physiological requirement. The amount inhaled is only 0.1 litre/s/person, but a volume 100 times larger is used to compensate for the ineffectiveness of the typical mixing ventilation system. Displacement ventilation (Section 7.4.5) achieves only a modest improvement compared to the underlying ineffectiveness. The US company Johnson Controls has developed a 'personal environmental module' (PEM) that supplies a small amount of air gently and directly to a person's breathing zone from outlets next to a PC workstation that are connected to an underfloor air distribution system. Use of PEMs reduces overall first costs because HVAC equipment can be downsized, and the supply air can be warmer than is typically the case, thereby saving on cooling energy (Mendler and Odell, 2000).

7.2.6 Building bioclimatic charts and the climatic range of alternative cooling techniques

A *building bioclimatic chart* is the standard psychrometric chart showing the temperature–humidity comfort zone for still air in a closed building, the expanded comfort zone acceptable under daytime ventilation (taking into account both psychological adaptation and the cooling effect of moving air), and the outdoor temperature–humidity conditions under which indoor conditions can be kept within the expanded comfort zone using night-time ventilation with thermal mass, or using evaporative cooling. An example is given in Figure 7.11, which was modified by Stein and Reynolds (2000) from the original building bioclimatic chart of Milne and Givoni (1979). This chart can be used to obtain a quick initial assessment of the viability of various low-energy cooling techniques. For example, if the peak daytime temperature–humidity condition falls within the envelope for night-time ventilation, then night-time ventilation (combined with adequate thermal mass) should be able to maintain comfortable temperatures under peak conditions. By plotting the conditions from the probability distribution for daytime peak weather, the frequency with which night-time ventilation will be inadequate can be estimated. This can be followed by more detailed analysis using, for example, simulation models, if it is thought to be worthwhile. As discussed in Chapter 6 (Section 6.6.4), desiccants can be used to extend the humidity range over which evaporative cooling can provide comfortable conditions. In this way,

non-conventional low-energy cooling techniques can provide comfortable conditions all the time essentially anywhere as long as water is not limited.

Milne and Givoni's (1979) bioclimatic chart is appropriate for residential buildings, where internal heat gains are minimal and where there are large amounts of exposed thermal mass. The boundaries on their chart indicate the conditions for which comfort can be achieved (data points within the boundary), but these will depend on the building thermal mass and internal heat gain. They will also depend on the solar irradiance coincident with a given ambient temperature, as well as on the extent of shading and the reflectivity of the building envelope. Thus, a number of different building bioclimatic charts should be developed to cover the ranges of building types and conditions that occur in practice. With regard to the boundary for evaporative cooling, Lomas et al (2004) have defined two limits: the climatic limit for thermal comfort (CLTC), beyond which evaporative cooling can never achieve comfortable conditions, and the lower limit of thermal discomfort (LLTD), below which evaporative cooling will always achieve comfortable conditions. The CLTC, as its name suggests, depends only on climate and not on the nature of the building, while the LLTD depends on the building, being lower the higher the internal heat gain. Figure 7.12 shows the CLTC, and the LLTD for office buildings in Seville with internal heat gain of $50W/m^2$, as determined by Lomas et al (2004) based on detailed computer simulations. In the region between the LLTD and the CLTC, evaporative cooling sometimes provides comfortable conditions and sometimes does not. This band is wider the larger the internal heat gain.

7.3 HVAC systems in residential buildings

7.3.1 Residential heat and coldness distribution

Most houses heated by natural gas or oil in North America use a *forced-air* system – outside air is heated by burning fuel in a furnace, transferring the heat to the inside air using a heat exchanger, and blowing the heated air through ductwork. Hot

Figure 7.11 A building bioclimatic chart based on Milne and Givoni (1979) showing standard comfort limits, comfort limits with natural ventilation and climatic conditions under which indoor temperatures can be kept within comfort limits through evaporative cooling or by relying on higher thermal mass with or without night-time ventilation

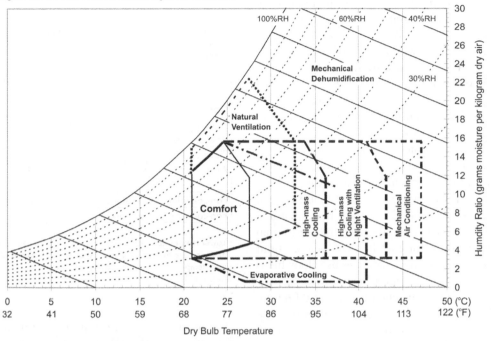

Note: Outside these regions, mechanical dehumidification and/or chilling are required.
Source: Stein and Reynolds (2000)

Figure 7.12 A building bioclimatic chart showing the climate limit for thermal comfort (CLTC), and the lower limit of thermal discomfort (LLTD) for office buildings in Seville with internal heat gains of 10W/m², 30W/m², and 50W/m², as determined by Lomas et al (2004)

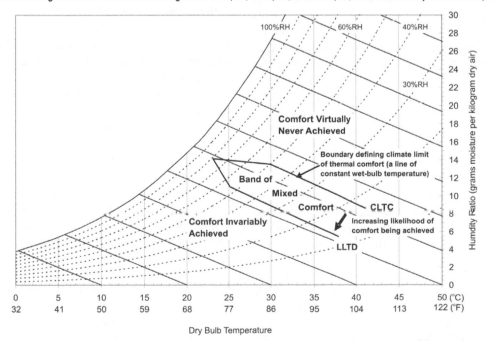

Source: Redrafted from Lomas et al (2004)

water (hydronic) systems, which are the norm in Europe, eliminate the need for ductwork. The energy used for pumping is less than the fan energy used in a forced-air system, and distribution losses tend to be smaller (Table 7.3). Indoor air quality problems associated with the growth of micro-organisms in dust and moisture that accumulate in ductwork are eliminated. However, conventional hydronic systems require a high supply temperature (70–90°C), which leads to inefficiencies and poses a burning hazard.

A better choice is a floor radiant heating system, in which hot-water pipes are embedded in the floor and heat a room by heating the floor. In Germany, Austria and Denmark, 30–50 per cent of new residential buildings have floor heating, while in Korea, where it has been the traditional system of heating for hundreds of years, about 90 per cent of existing residences use floor heating (Olesen, 2002). Radiant heating was used by the Romans 2000 years ago (de Carli and Olesen, 2002). Floor radiant heating is silent and has a number of advantages from an efficiency point of view. First, it requires the lowest distribution temperature of any system, which improves the efficiency of a boiler by a few percent because more water vapour can be condensed from the exhaust by the return water (see Chapter 4, Section 4.3.1), or improves the COP of a heat pump by 1.0 or more because a smaller temperature lift is required (see Chapter 5, Section 5.5). If plastic pipes are embedded in a 7cm floor slab at a density of 10m of pipe per square metre of floor, the required supply temperature is 30–35°C, while if plastic mats with integrated water ducts and a sturdy floor cover are used, a supply temperature of 27–32°C is sufficient (Halozan, 1997). Second, radiant heating allows the air temperature to be 1–2K lower with the same perceived temperature (see Equation 7.8). Third, excess heat on outside walls and windows (arising from perimeter air vents or radiators) is avoided, which will reduce conductive heat loss by 10–30 per cent. Fourth, it avoids warm air rising to the ceiling. Fifth, it provides thermal energy storage in the floor itself, so pumps do not need to run all of the time. A floor radiant system can also be used for cooling; in this case, a warmer distribution temperature (15°C) can be used then needed in a forced-air cooling system (10°C). The typical temperatures at different points in floor radiant heating and radiant cooling systems are given in Table 7.4.

Table 7.4 Temperatures (°C) and temperature changes (K) involved in residential floor radiant heating and cooling, and for an alternative hydronic system using fan coils

	Heating	Cooling
Floor panel inlet temperature	36	15
Floor panel exit temperature	26	20
Temperature change through panel	–10	+5
Average panel temperature	31	18
Floor surface temperature	23	23
Air temperature	18	26
Fan-coil inlet temperature	55	8–15[a]
Fan-coil exit temperature	35	13–18[a]

[a] Lower temperature is with dehumidification by condensing water vapour on the fan coil and its subsequent collection and removal, upper temperature is without dehumidification.
Source: Kilkis (1999).

A floor radiant heating system will provide a heat flux of about 10–11W/m² per degree of temperature difference between the floor surface and air, with just over half of the heating due to radiative heat exchange (de Carli and Olesen, 2002). In well-insulated and air-tight buildings, where the heat load is unlikely to exceed 30–40W/m², a floor temperature of only 23–24°C is needed to maintain an air temperature of 20°C. The required water supply temperature will be a few degrees warmer than the required floor temperature, the difference being greater if a carpet is on the floor or the pipe spacing is wider. Radiant floor heating is not advisable in situations (such as public atria) with relatively large rates of heat loss and concrete floors with the space unoccupied and the heat shut off during the night. Due to the large thermal capacity of the concrete floor, a significant decrease in air temperature (and hence in the cumulative heat loss) will not occur over an 8-hour period, and most of the energy saved at night will have to be put back into the floor in the morning.

If floor radiant heating is used in slab-on-grade floors, then a substantial amount of insulation should be placed below the floor and around the foundation. The heat loss through slab-on-grade floors with radiant floor heating has been simulated by Weitzmann et al (2005). For a square room with a floor area of $36m^2$, heating water supplied at 40°C, and Danish climate conditions, the heat loss through the floor varies from 15–30kWh/m^2/yr as the floor U-value varies from 0.1–0.3W/m^2/K (corresponding to about 30–10cm of solid foam insulation). The heat loss is about 20 per cent greater than the heat loss without floor heating. This will roughly offset the reduction in heat loss due to the fact that the air temperature can be 1–2K colder in houses with radiant floor heating while giving the same perceived temperature.

In well-insulated houses with minimal infiltration heat loss, internal heat loads are a relatively important heat source. This means that there can be a risk of overheating if the internal heat loads abruptly increase. However, this risk is greatly minimized with a floor radiant heating system, because the rate of heat supplied by the system falls much more rapidly with increasing

room temperature than in a conventional system, as shown in Box 7.1. In addition, the rate of water flow to individual rooms can be controlled through individual room thermostats, whereas the heating of individual rooms in a forced-air system cannot be controlled through individual thermostats.

7.3.2 Energy used in distributing heat or coldness

In houses with a furnace and forced-air heating system, the furnace air handler (sometimes called a 'blower') can be the largest or second largest (after refrigerator/freezer units) consumer of electricity. These air handlers are sometimes used to provide ventilation airflow throughout the heating season, but at a lower airflow when the furnace is not operating, and in some cases may operate year-round and in combination with a central air conditioner when cooling is demanded. There

---

BOX 7.1 Negative feedback (damping) of temperature changes

The rate of heat delivery to a space by a mechanical airflow, Q_{af}, can be written as:

$$Q_{af} = \rho c_p F (T_s - T_r) \qquad \text{(B7.1.1)}$$

where F is the airflow rate and T_s and T_r are the supply and return temperatures, respectively. T_r is equal to the room temperature if the air is well mixed. If the room becomes warmer, the rate of heating will decrease, which tends to limit the warming. The strength of this limitation is given by dQ_{af}/dT_r. If F is expressed as the volumetric flow rate per unit of floor area, then Q_{af} has units of W/m^2. For one air exchange per hour and a 2.5m high space (corresponding to 7.0/s/person if there is an area of 10m^2 per person), $F = 0.7 \times 10^{-4}$ m/s and the feedback on temperature changes is given by:

$$\frac{dQ_{af}}{dT_r} = -\rho c_a F = -0.9 \text{W} / \text{m}^2 / \text{K} \qquad \text{(B7.1.2)}$$

This feedback serves to 'dampen' changes in temperature; the stronger the feedback, the smaller the variation in temperature that will occur. The heat flux to a space by a radiant panel, Q_{rad}, per unit of panel area, is given by:

$$Q_{rad} = \sigma T_p^4 - \sigma T_r^4 \qquad \text{(B7.1.3)}$$

where T_p is the panel temperature and T_r is the temperature of non-panel surfaces and objects in the room. The feedback on changes in T_r is given by:

$$\frac{dQ_{rad}}{dT_r} = -4\sigma T_r^3 = -5.7 \text{W} / \text{m}^2 / \text{K} \qquad \text{(B7.1.4)}$$

which is about six times that for heating with airflow.

is a factor of 5–6 variation in the electricity use by furnace air handlers for furnaces of a given size, as shown in Table 7.5. All of the electricity supplied to the air handler is ultimately converted to heat, either immediately through waste heat produced by the motor, or through frictional dissipation of the kinetic energy supplied to the airflow. Thus, the savings in electricity use during the heating season arising from a more efficient air handler is offset by increased heating energy use. However, since electricity is typically generated with an efficiency of 33 per cent, there is still a significant saving in primary energy use when electricity use is reduced.

As discussed in Section 7.1.2, the energy required to circulate air or water through a heating system is given by the volumetric flow rate times the pressure drop (due to friction) through the circuit, divided by the product of the motor and fan or pump efficiencies. Consider a house with a heat load on the coldest day of $12kW_{th}$ (for a typical new house in southern Canada, this heat load would occur at an outdoor temperature of $-20°C$ to $-25°C$). If the heating system is oversized by 25–100 per cent (as is typically the case), the furnace heating capacity will be $15–25kW_{th}$. In a house with a forced-air heating system (common in North America), the airflow would be about $0.7m^3/s$ (700l/s) and the pressure difference about 75Pa. An estimated 90 per cent of such houses use an air handler with a multi-speed *permanent split capacitor* (PSC) fan motor, having an efficiency of 55–67 per cent at high speed (used for air conditioning) and 34–49 per cent at low speed (used for heating) (Sachs et al, 2002). The efficiency of a typical

air handler is given by the efficiency of the motor times the efficiency of the fan times a factor to account for oversizing, degradation over time and inappropriate installation, such that measured efficiencies of real air handlers are typically 10–15 per cent (Walker and Mingee, 2003). This gives a blower power requirement of 350–525W. This estimate is consistent with measured data on motor power draw and airflow rate using PSC motors, as given later in Table 7.6.

In a better-insulated house with a properly sized furnace and air handler, the peak heating load could be reduced to $6kW_{th}$ and the flow rate reduced to $0.2m^3/s$ (which would still provide plenty of ventilation with fresh air, as discussed later in Section 7.3.3). At the same time, there would be no need to locate vents on the outside walls, so they could be located so as to minimize turns and distance travelled. This along with smoother ductwork could readily reduce the pressure difference to 40Pa.[4] More aerodynamic entrance and exit conditions to the blower cabinet, along with high-efficiency filters (which have recently entered the market) would also contribute to a smaller pressure drop. A prototype air handler built by Walker and Mingee (2003) had an efficiency ranging from 17 per cent at full load to 34 per cent at 70 per cent of full load. Adopting an average efficiency of 25–30 per cent, the power use is reduced to 27–32W – a reduction by a factor of 10–20.

A house with a hydronic heating system (common in Europe) sized to a heat load of $15–25kW_{th}$ would have a hot-water flow rate of about $4 \times 10^{-4} m^3/s$ and a pressure drop of 30kPa. The power required to circulate the water is therefore:

Table 7.5 Comparison of electricity use by furnace air handler systems for the most efficient and least efficient air handlers available in North America, and the difference between the most and least efficient air handlers

Furnace heating capacity		Air handler electricity use (kWh/year)		Savings (kWh/year)
kW	Btu/h	Low	High	
7.6–12.3	26,000–42,000	97	459	363
12.6–17.3	43,000–59,000	123	678	555
17.6–22.3	60,000–76,000	167	711	544
22.6–27.3	77,000–93,000	266	806	540
37.6–32.3	94,000–110,000	284	1014	730
32.3–38.1	111,000–130,000	255	1314	1059

Source: Sachs and Smith (2004)

$$(4 \times 10^{-4} m^3/s) \times 30000 Pa = 12W$$

Given a typical motor efficiency (wire to shaft) of 0.40, pump efficiency (shaft to water) of 0.3, and utilization efficiency of 0.7, the electrical power requirement is:

$$12W/(0.4 \times 0.3 \times 0.7) = 142W$$

which is about one-third of that estimated for a comparable house with a conventional forced-air system. Given a 10kW heat load with flow rate of $2 \times 10^{-4} m^3/s$, a pressure drop of 10kPa, and motor and pump efficiencies of 0.5 and 0.4, respectively, the required power for pumping is reduced to 14W. This is about half of a fully optimized forced-air system, so the hydronically heated house still wins. In super-insulated houses, a pump could be eliminated altogether, as gravitationally induced flow (arising from the difference in temperature and hence density of the supply and return water) would be sufficient. Large diameter pipes would be needed in this case. Conversely, a radiant floor heating system would require 2–3 times the pump power of a system with conventional radiators (according to Table 7.3).

Reducing energy use by air handlers in forced-air heating systems

The specific ways in which furnace blower/ductwork system efficiency can be improved are outlined in Kubo et al (2001). They are:

- *Better motors.* A typical furnace blower has a four-speed induction motor. High-end blowers use an *electrically commutated permanent magnetic induction (ECPM)* motor (also called a *brushless permanent magnet* motor). These motors have efficiencies of 74–78 per cent at high speed and greater than 70 per cent at low speed (Sachs et al, 2002), compared to 55–67 per cent and 34–39 per cent for PSC motors.
- *Better fans.* The fan is a typical furnace blower and has an efficiency of less than 70 per cent. More efficient fan blade designs result in larger units (a disadvantage) but

quieter operation (an advantage).
- *Better matching of fans to duct systems.* Blower manufacturers are under pressure from builders to make designs as compact as possible, but this results in suboptimal matching of the blower to the ductwork (this also increases noise). The ductwork itself entails large frictional losses through right-angle turns. According to Proctor and Parker (2000), ductwork is generally undersized relative to the airflow it carries, which further increases frictional losses.

Figure 7.13 gives measured power draw for PSC and ECPM air handlers on two $17.6kW_{th}$ (60000Btu/hr) furnaces. At the minimum PSC airflow, the power draw by the ECPM unit is half that of the PSC unit, and the ECPM unit can extend to lower airflow rates. The difference in the annual electricity use of advanced motor/fan units in residential furnaces and heat pumps and ordinary units is about 800kWh/year for an average US climate, assuming that the unit is also used during the cooling season (Sachs et al, 2002). This difference is greater than the annual electricity use by full-size refrigerator-freezer units meeting the US 2001 standard (479kWh/yr). An ECPM motor of the size considered here costs about $220, compared

Figure 7.13 Comparison power draw versus airflow rate for air handlers with PSC (permanent split capacitor) and ECPM (electronically commutated permanent magnetic) motors, each in a 17.6kW (60000Btu/hour) furnace

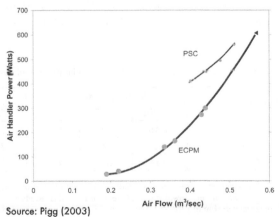

Source: Pigg (2003)

to about \$50 for PSC motors, a premium of \$170. In spite of the four times greater cost of ECPM motors, the extra cost is paid back to the consumer through reduced electricity bills in 2–3 years for typical US electricity costs. ECPM motors have captured only 5–10 per cent of the US furnace market, but with increased market share, the incremental cost of \$170 is expected to eventually drop in half. Advances in induction motors and another technology, the *switched reluctance* motor, may eventually yield efficiencies comparable to existing ECPM motors, but at lower cost.

Pigg and Talerico (2004) find that, in the US, when existing PSC air handlers are replaced with ECPM air handlers, the contractor typically advises the customer to switch from discontinuous to continuous ventilation. This switch will offset some of the savings from the ECPM air handler, and if applied year round, will increase rather than decrease the energy use by the air handler. However, in furnace upgrade or replacement situations, the pre-existing house is probably sufficiently leaky that there is no need for mechanical ventilation during the heating season, when the indoor-to-outdoor temperature difference will be large enough to drive adequate air exchange (see Chapter 3, Section 3.7). During seasons when neither heating nor cooling are required, windows can be opened instead of operating the ventilation system.

Interactions between air handler, heating and air conditioning energy use

As noted above, furnace air handlers can be set to remain operating (but at lower airflow) even when heating is not called for, in order to provide ventilation. They may also be used in conjunction with a central air conditioner to provide mechanical cooling during the summer. An ECPM air handler produces less waste heat, impacting both the heating and air conditioning loads, as well as directly reducing electricity use. The interaction between these different energy uses needs to be determined in order to assess the impact of ECPM air handlers on overall energy use. At the Canadian Centre for Housing Technology (Ottawa), the power consumption by the air handler and heating and air conditioning energy use were compared for two test houses that were identical except that one house had an air handler with a PSC motor and the other had an air handler with an ECPM motor (Gusdorf et al, 2003). The houses had been previously calibrated to have nearly identical energy use when using the same motor. Table 7.6 compares the measured airflow and power draw of the PSC and ECPM air handlers in heating mode and in ventilation mode. The ECPM air handler is able to reduce the airflow to a greater extent in ventilation mode, resulting in a 93 per cent reduction in energy use compared to the PSC air handler. If the ECPM is set to the same flow rate as the PSC air handler, there is still a 46 per cent saving. In heating mode, the saving is 58 per cent.

The impact on heating energy use depends on the heating requirements of the house and the efficiency η of the furnace. A given reduction ΔW in waste heat generation will increase heating energy use by $\Delta W/\eta$. On the other hand, in houses with very low heating loads (due to a high-performance envelope), the waste heat produced by the PSC air handler will exceed the heating requirement of the house during part of the heating season. Thus, not all of the waste heat is utilizable, so not all of the saving in electrical energy use with an ECPM leads to an increase in

Table 7.6 Comparison of ECPM and PSC air handler power and flow rates, as measured in test houses in Ottawa

Mode	Motor power (W)			Airflow (litres/s)			Flow/power (litres/s/W)		
	ECPM	PSC	Ratio	ECPM	PSC	Ratio	ECPM	PSC	Ratio
Heating	246	423	58%	591	658	90%	2.40	1.55	155%
Circulation	22	316	7%	218	486	45%	9.91	1.54	644%
Eq. Circulation	146	316	46%	463	486	95%	3.17	1.54	206%

Note: 'Circulation' mode means ventilation only, with ECPM able to run at a smaller flow rate than PSC. 'Eq. Circulation' means ECPM constrained to run no slower than PSC, even though this results in excess airflow.
Source: Gusdorf et al (2003)

heating energy use. The saving in air conditioning energy use is given by the reduction in waste heat generation divided by the COP of the air conditioner. An additional saving may arise from the fact that, with less heat produced by the air handler, the air conditioner evaporator coil will be more effective in removing moisture from the air, reducing the relative humidity (by up to 5 per cent of saturation in the test house). This might allow a higher thermostat setting for the same level of comfort (see Section 7.2.3).

Using a computer model to extrapolate short-term winter and summer measurements to the entire year, Gusdorf et al (2003) estimate that ECPM air handlers save about 1500–2000kWh/yr (5.4–7.2GJ/year) of electricity in conventional southern Canadian houses without air conditioning, and 2500–3000kWh/year (9.0–10.8GJ/yr) in houses with air conditioning (assuming no adjustment in thermostat setting due to lower humidity in the ECPM house), while increasing heating energy use by 150–220m^3 of natural gas per year (5.6–8.2GJ/year). Thus, the savings in on-site electricity use is roughly offset by the increase in on-site heating energy use for houses without air conditioning. However, electricity is typically generated by coal on the margin at an efficiency of about 33–35 per cent, so there would be a significant saving in the use of primary energy and an even larger savings in CO_2 emissions due to the factor of two smaller emission factor for natural gas compared to coal (see Chapter 2, Section 2.4.1). Furthermore, in houses with high-performance thermal envelopes, or in mild climates, the increase in heating energy use due to the ECPM would be smaller, as explained above.

7.3.3 Ventilation and heat recovery ventilators

In houses that are almost completely air-tight, fresh air needs to be provided by a mechanical ventilation system when the windows are closed. The amount of fresh air that needs to be provided in this way will be substantially less than the uncontrolled ventilation in leaky, poorly built houses (see Chapter 3, Section 3.7.1). Furthermore, a portion of the heat in the outgoing exhaust air

can be used to partially heat the cold incoming air, using an *air-to-air heat exchanger* (to form a *heat recovery ventilator*, HRV). Conversely, HRVs will not be effective in reducing energy use if the building is not reasonably air-tight, as they require an exhaust airflow at least 75 per cent of that of the supply airflow. They reduce peak heating and cooling loads, allowing downsizing of heating and cooling equipment.

The heat exchangers come in two basic types: *flat-plate* and *thermal wheel*. Flat-plate heat exchangers can be of the *counter-flow* or *cross-flow* type, illustrated schematically in Figure 7.14. A cross-flow heat exchanger is shown in Figure 7.15. Counter-flow heat exchangers tend to be more efficient than cross-flow exchangers, but also require larger fans. A thermal wheel consists of a wheel that rotates through the incoming and outgoing air streams. It can transfer both heat and moisture. Thus, in winter, incoming air is

Figure 7.14 Schematic illustration of the two air streams in (a) a counter-flow heat exchanger, and (b) a cross-flow heat exchanger

(a)

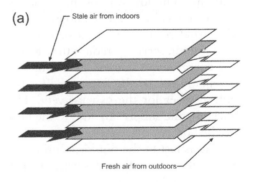

Stale air from indoors

Fresh air from outdoors

(b)

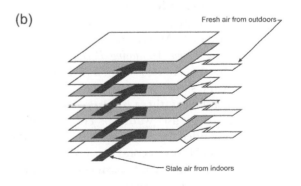

Fresh air from outdoors

Stale air from indoors

Source: Bower (1995)

Figure 7.15 A cross-flow heat exchanger

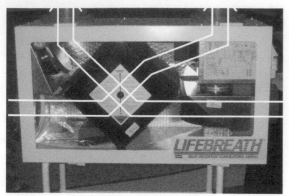

Source: Author

humidified as well as warmed, and this can eliminate the need for a humidifier. However, it is more complicated than a flat-plate heat exchanger, so professional maintenance is necessary. During the summer, the relative coldness and dryness of the outgoing air can be used to partially cool and dry the incoming air. The low rate of air exchange in airtight, mechanically ventilated houses, combined with partial recovery of heat or coldness from the exhaust air, serves to reduce the heating and cooling energy requirements. Offsetting this will be the energy use by the fans – one for the supply vent, another for the exhaust vent. The heat produced by the supply fan adds to the heat available to the house, while the heat produced by the exhaust fan is lost to the outside. During the cooling season, the supply fan adds heat that has to be removed from the house.

Energy use and effectiveness

Data on the energy performance of HRVs available on the North American market are found in the *Certified Home Ventilating Products Directory*, published by the Home Ventilation Institute in the US. For the same volumetric flow rate under given test conditions, there is a large variation in the rate of energy use (wattage). The ventilation rate recommended by NRCan (2002) for a 10-room house is 60litres/second (derived as follows: 10litres/s for an unfinished basement and master bedroom, and 5litres/s each for the living room, dining room, family room, kitchen, two bathrooms, and two other bedrooms) as long as this is not less than 7.5 litres/s per person (ASHRAE

Standard 62-2001). For 60litres/s air exchange, the power consumption of the HRVs on the market varies from 85W to 240W. Note that this rate of air exchange is about one-third that required for heat distribution in a well-insulated house (0.2m³/s or 200litres/s). If the pressure drop across the ventilation system is of the order of 250Pa, then the power that must be supplied to the air is only 15W (0.06m³/s × 250Pa). Actual power consumption of 85–240W is consistent with the motor, fan, and utilization efficiencies given earlier (Section 7.3.2). According to Bower (1995), most HRV manufacturers give flow rates for an external static pressure of 100Pa (i.e. a pressure drop of 100Pa for everything except the HRV itself).

The *effectiveness* of a heat exchanger is defined as the amount by which the incoming air is warmed (or cooled) as a fraction of the indoor-to-outdoor temperature difference after subtracting the heating of the incoming air by the supply fan. The typical effectiveness for a residential HRV is about 0.60–0.85. The Dutch company Brink (www.brinkclimatesystems.nl) produces a HRV with up to 95 per cent heat recovery, an AC/DC transformer, and a DC ECPM motor that uses half the electricity of the motor typically found in HRVs. An HRV saves heating energy through the recovery of heat from the outgoing exhaust air and because the power input to the supply fan (which may be half of the total fan power) serves as a heat source to the building. Offsetting this is the energy consumed by the fans. The amount of heat recovered depends on the indoor-to-outdoor temperature difference (ΔT), the airflow rate, and the heat exchanger effectiveness, while the extra fan energy used because of the heat exchanger depends on the pressure drop across the heat exchanger, the airflow rate and the efficiency of the motor and fan. Below some minimum ΔT, the saving in heating primary energy is less than the extra primary energy used by the fan due to the heat exchanger. An expression for the break-even ΔT in terms of the pressure drop and effectiveness of the heat exchanger and other parameters is derived in Box 7.2. For reasonable parameter values, ΔT need be only about 1–1.5K for it to be worthwhile to use the heat exchanger.

BOX 7.2 Tradeoff between energy saved and energy used by a heat recovery ventilator

If ΔT is the difference between indoor and outdoor temperature, E the effectiveness of a heat exchanger, Q the airflow rate, and FP the power drawn by the fans in a heat recovery ventilator, then the reduction in heating load is:

$$H = \rho c_{pa} QE\Delta T + 1/2 FP \qquad (B7.2.1)$$

where ρ is the air density (about 1.25kg/m³), c_{pa} is the specific heat of air (1004.5J/kg/K), and it is assumed that half of the energy used by the fan supplies heat to the incoming air stream. The savings in primary energy is given by H divided by the efficiency of the heating system (η_f) times the efficiency (η_{PS}) in converting the fuel used for heating from primary energy (in the ground) to secondary energy (at the point of use). The fan power that should be used is not the total fan power, but the additional fan power associated with overcoming the pressure drop (ΔP) through the heat exchanger alone (ventilation would be needed whether or not there is heat recovery through a heat exchanger). That is:

$$FP = \frac{\Delta PQ}{\eta_{mf}} \qquad (B7.2.2)$$

where η_{mf} is the joint motor fan efficiency. The rate of primary energy use associated with FP is given by FP divided by the electricity generation efficiency (η_{pp}) times the electricity transmission efficiency (η_{tr}). From this it follows that the ΔT at which primary energy saved equals primary energy used is given by:

$$\Delta T = \frac{\eta_f \eta_{PS}}{1256\eta_{mf}} \left(\frac{1}{\eta_{pp}\eta_{tr}} - \frac{1}{2\eta_f\eta_{PS}} \right) \frac{\Delta P}{E} \qquad (B7.2.3)$$

Letting $\eta_f = 0.92$, $\eta_{PS} = 0.85$, $\eta_{mf} = 0.40$, $\eta_{pp} = 0.36$, and $\eta_{tr} = 0.95$, the break-even ΔT is given by:

$$\Delta T = 0.00356 \frac{\Delta P}{E} \qquad (B7.2.4)$$

For an HRV where the pressure drop across the heat exchanger is 250–375Pa and the effectiveness is 0.85 (a reasonable combination), there is a net saving in primary energy as long as ΔT > 1.0–1.5K. At such a low ΔT, the HRV can be turned off, the windows can be opened and natural ventilation (if the building is appropriately designed) can be used. Thus, if mechanical ventilation is needed in the first place, it is generally worthwhile to operate it with the heat exchanger. If the flow path in the heat exchanger is made longer to increase the effectiveness, ΔP will increase in proportion to the increase in flow path but E will increase by ever decreasing increments, so $\Delta P/E$ will increase, meaning that the threshold ΔT at which the heat exchanger should be bypassed will increase with increasing effectiveness.

Equations (B7.2.3) and (B7.2.4) are applicable only during the heating season. During the cooling season, a fraction of the fan energy use (assumed to be 0.5) adds to the cooling load. It can be easily shown that the ΔT at which primary energy saved equals primary energy used is given by:

$$\Delta T = \frac{COP}{1256\eta_{mf}} \left(1 + \frac{1}{2COP} \right) \frac{\Delta P}{E} \qquad (B7.2.5)$$

where COP is the COP of the air conditioner or chiller. All of the energy saved or used by the heat exchanger is electricity, so the efficiency of electricity generation and transmission does not affect the breakeven ΔT. For COP = 3 and $\eta_{mf} = 0.40$:

$$\Delta T = 0.00697 \frac{\Delta P}{E} \qquad (B7.2.6)$$

which is a more strict condition for operation of the heat exchanger than during the heating season. That is, a larger ΔT is required in order to justify recovery of coldness than recovery of heat, in part because of the high on-site efficiency of chilling (COP of 4.0) compared to heating and because the extra fan energy adds to the cooling load but subtracts from the heating load.

Factors reducing performance

There are a number of factors that can reduce the performance of an HRV below its rated value. First, some of the exhaust air may re-enter through the intake, reducing the delivery of fresh air. If it is discovered, the flow rate could be increased, but the fan power that must be supplied to the flow varies approximately with the cube of the flow rate. To avoid this, there should be adequate separation between the fresh air intake and exhaust grilles. Second, internal leakage between the incoming and outgoing air streams may occur. Third, operation of the ventilation system might induce exfiltration of interior air through leaks in the building envelope if the supply and exhaust fans are not balanced. In this case, the intake airflow through the heat exchanger will be less than the outgoing airflow. Roulet et al (2001) measured real energy recovery fractions for 13 HRVs. In the best three cases, real heat recovery (taking into account induced exfiltration) was 60–70 per cent for units with a rated heat recovery of 80 per cent, while in the worst cases it was less than 10 per cent. Manz et al (2001) have analysed the performance of single-room HRVs and find that real heat recovery can be over estimated by as much as 63 per cent compared to the efficiency computed from the increase in the temperature of the incoming air. The error in this case is due to the induced exfiltration or infiltration that occurs when the supply and exhaust fans are not balanced. Finally, accumulation of dirt on the heat exchange surfaces will reduce the effectiveness of the heat exchanger, so the heat exchanger should be regularly cleaned in order to maintain its performance.

Integration with the heating system

In houses with a forced-air heating system, the HRV can be connected to its own system of ductwork, or it can be connected to the warm-air heating system. In the latter case, a short duct delivers fresh air from the HRV to the return trunk of the forced-air system. A matching amount of return air is carried outside, and the balance recirculates through the furnace. The furnace fan can serve as the supply fan even when the furnace is not operating, while a separate exhaust fan removes stale air from the house. However, the ductwork will have been sized and laid out under the assumption that it will be delivering a large, warm airflow, and its orientation will be governed by the distribution of the heating loads. This may not be optimal for ventilation only. Second, if the furnace has a PSC air handler (as is normally the case), there will be only modest motor energy savings when operating at the lower flows needed for ventilation. This problem can be obviated through the use of an ECPM air handler. On the other hand, to install separate ductwork and fans for heating and ventilation would be costly and would require excess space. Rather, the preferred solution is to build a ventilation system sized for the ventilation requirements, and to provide heating through a radiant floor heating system. An even better solution is to design the thermal envelope such that the peak heating requirement can be met by the ventilation airflow alone, thereby avoiding the cost of the radiant heating system, with heat provided to the supply air through an exhaust-air heat pump (Chapter 5, Section 5.4) or a small hot-water coil drawn from a combined space-DHW heater (Chapter 4, Section 4.2.7).

Operating strategy

The net energy savings during both heating and cooling are smaller the smaller the temperature difference between the inside and outside air. When heating and cooling are not needed, the HRV should be turned off and windows should be opened to provide free natural ventilation. However, the tendency in houses with mechanical ventilation is for the system to run continuously, year round. Not only should the system be shut down when heat recovery is not needed, but the ventilation rate should be reduced when occupants are sleeping or the house is empty. When occupants are sleeping, the ventilation rate can be safely reduced from the ASHRAE standard of 7.5l/s per person to 3l/s per person (Lin and Deng, 2003). In a test house constructed in The Netherlands, ventilation to individual rooms is controlled by infrared motion detectors. If no motion is detected, the ventilation rate is set to 50 per cent of the normal rate, except that the ventilation rate reverts to the normal level if the relative humidity exceeds 70 per cent (Römer, 2001).

As discussed in Chapter 6 (Section 6.3.3), some techniques for passive cooling require mechanical ventilation of air through interior cavities in the roof, windows and floors at night. If the HRV is operated at night during the summer to provide cooling, the heat exchanger should be bypassed as, in this case, a large difference between outgoing and incoming air temperature is desired.

7.3.4 Interaction between mechanical and uncontrolled ventilation

When a house is mechanically ventilated, the total rate of air exchange can be approximated as (ASHRAE, 2001, Chapter 26):

$$Q_{total} = Q_{bal} + \sqrt{Q_{unbal}^2 + Q_{infiltration}^2} \qquad (7.10)$$

where Q_{bal} is the balanced component of the mechanical ventilation, Q_{unbal} is any unbalanced component, and $Q_{infiltration}$ is the leakage component (see Chapter 3, Section 3.7). If the mechanical ventilator is balanced, then the mechanical and natural rates of ventilation can be added.

If a system with only a central exhaust fan is added to a house, the increase in total ventilation will be much less than the airflow from the exhaust fan. To understand why this is so, suppose that the natural ventilation rate is 20l/s. This means that 20l/s enter the house through one set of holes (the inlet holes) and 20l/s leave the house through another set of holes (the exit holes). If an exhaust fan with a flow rate of 40l/s is turned on, then 40l/s must enter from elsewhere. If this flow is split equally between the inlet and exit holes, and if flow rates through the holes are assumed to be proportional to the pressure difference across the holes, then an extra 20l/s would enter through the inlet holes, giving a total inflow through these holes of 40l/s. The other 20l/s, superimposed on the natural flow of the exit holes, would give a net outflow of 0l/s. The total rate of air exchange is 40l/s, an increase over the natural rate equal to only half (20l/s) the airflow of the exhaust fan. In reality, airflow through an orifice is not linearly proportional to the pressure difference across the orifice (see Equation 3.8). This is accounted for in Equation (7.10), which predicts a total air-exchange rate of 44.7l/s for the above example. An interesting implication of this is that when a bathroom or kitchen exhaust

fan is turned on (an unbalanced system), the increase in total air infiltration (and associated heating load, in winter) is less (potentially much less) than implied by the fan airflow.

7.3.5 Ceiling fans

Ceiling fans are common in hot-climate regions of the US, where up to five fans per house can be found. Moving air creates a cooling effect (by promoting evaporation and transfer of sensible heat away from the skin), thereby allowing a higher temperature to feel comfortable (Figure 7.8). James et al (1996) have examined the trade-off between fan energy use and reduced air conditioning energy use in houses in Florida. A fan set to high speed (0.75m/s airflow) allows the air conditioner thermostat to be set 1.8K higher with the same degree of comfort, as long as the person is within the moving air stream created by the fan. This in turn will reduce the annual air conditioning energy use in Florida by about 20 per cent, after accounting for the energy used by the fan and the waste heat produced by the fan (which needs to be removed by the air conditioner). However, if the thermostat is not adjusted, air conditioning energy use increases due to the waste heat generated by the fan (a ceiling fan uses between 15 and 150W, depending on the model and speed, all of which ultimately appears as heat). In a survey of 400 houses in Florida, James et al (1996) found that there was not a statistically significant difference in the thermostat settings in houses with and without ceiling fans. Furthermore, one-third of the fans in Florida houses run continuously, year-round (Parker et al, 1999). There is an average of 4.3 fans per house in Florida, and they are operated an average of 13.4 hours per day, with an annual electricity consumption of about 800kWh/year (i.e. almost twice that of a refrigerator/freezer unit meeting the US 2001 standard of 479kWh/year) (Sonne and Parker, 1998). The greatest opportunity for energy savings with ceiling fans is through education: to inform the public that fans provide no cooling benefit when the room is not occupied and so should be turned off at such times, and that the air conditioner thermostat setting can be increased if fans will be used when a room is occupied. Some fans on the market are equipped with occupancy sensors that turn the fan off some time after the room is vacated, and at least one model has a photo-optical sensor to prevent the fan from being turned off due to lack of motion in darkness (i.e. while the occupants of a bedroom are sleeping) (Parker et al, 1999).

There are two significant technical measures that can be taken to reduce the energy use by fans. First, most ceiling fans contain lights, and in some units, there are no separate controls for the lights! Ceiling fans typically have a 100W halogen lamp or three 40W incandescent lamps, which means that the lighting electricity consumption (and waste heat generation) is comparable to or greater than that of the fan itself. Ceiling fans with compact fluorescent lamps (CFLs) and separate lighting controls will yield substantial energy savings (CFLs can easily replace incandescent lamps in fans not equipped with CFLs). Second, the electrical-to-shaft (i.e. motor) and fan-blade to airflow (i.e. aerodynamic) efficiencies in ceiling fans tend to be lower than in the fans used in residential HVAC systems (discussed in Section 7.3.2). This is illustrated in Table 7.7, which contains data on the electrical power input to a conventional fan and the resulting power of the airflow (given by $\frac{1}{2}\rho\int V^3 dA$). The aggregate efficiency (which is the product of the motor and aerodynamic efficiencies) ranges from 3.3 per cent (at low speed) to 11.6 per cent (at high speed) for a conventional fan. Sonne and Parker (1998) report similar efficiencies for conventional ceiling fans.

Fan efficiency can be improved through:

- *Better fan motors*. Fan motors are typically the least efficient of the commonly available designs, in order to reduce the upfront cost.
- *Better fan blades*. Most ceiling fans use flat blades with no airfoil characteristics. They are good at creating turbulence but not at moving air. The flat-blade design of existing fans stands in marked contrast to the blades of an experimental high-efficiency fan, illustrated in Figure 7.16.

Figure 7.17 shows power use versus fan speed for a standard fan, a fan with standard blades but a high-efficiency motor, and a fan with efficient blades and motor. The latter reduce average energy use by about 75 per cent (Camilleri, 1995). Table 7.7 gives additional data on the performance of a fan incorporating both a more efficient motor and aerodynamic blades. The efficiency increases by a factor of two (at high speed) to a factor of three (at low speed).

Table 7.7 Comparison of conventional and high-efficiency ceiling fans

Speed		Low	Medium	High
Revolutions per minute		108	204	268
Power output, airflow		0.5W	3.0W	7.3W
Conventional fan	Power input, electrical	15W	38W	63W
	Efficiency	3.3%	7.9%	11.6%
High- efficiency fan	Power input, electrical	4.7W	16W	32W
	Efficiency	10.6%	18.8%	22.8%

Source: Schmidt and Patterson (2001)

Figure 7.16 An aerodynamic ceiling fan referred to as the "Gossamer Wind, developed at the Florida Solar Energy Center and available commercially in North America

Source: Building Design Assistance Center of the Florida Solar Energy Center, www.fsec.ucf.edu/bldg/active/bdac

Figure 7.17 Comparison of power consumption for standard and efficient fans

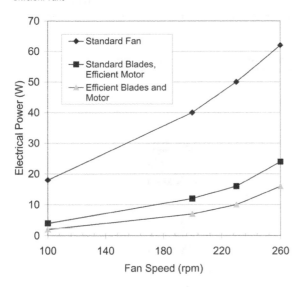

Source: Camilleri (2001)

7.3.6 Energy losses associated with leaky, uninsulated and imbalanced ductwork

In parts of the US (and perhaps in other countries) where houses are built without basements, much of the ductwork passes through unconditioned spaces (attics, crawlspaces). In some cases, the furnace is located in an attached garage. In the northeastern and midwest US, 50 per cent of single-family houses have forced-air heating and cooling systems and 44 per cent of these have ducts in unconditioned spaces, while in the southern and western US, 46 per cent of homes have forced-air systems and 82 per cent of these have ducts in unconditioned spaces (Sherman and Jump, 1997). Heating and cooling energy are directly lost both through leakage of air at duct joints and by conduction of heat through the duct walls. Altogether, duct leakage affects energy use in three ways (Gu et al, 2003):

- through direct loss of energy to the unconditioned space, as noted above;
- by inducing additional infiltration of outside air into the building when the leakage from the supply and return ducts is not equal (as would normally be the case);
- by increasing the average load of the heating or cooling system, which normally acts to slightly increase the system efficiency.

Forced-air heat distribution systems in houses often lose 20–40 per cent of the heat prior to delivery to the intended spaces. Compensating for the reduced air delivery due to leakage by increasing the initial airflow requires a disproportionately large increase in fan power, due to the cubic relationship between airflow and power. Air leakage from ducts induces increased infiltration of outside air due to pressure imbalances. Studies in Tennessee and Florida found 80 per cent and 200 per cent increases, respectively, in the infiltration of outside air when the air handler fan is operating (Sherman and Jump, 1997). In a four-unit apartment building studied by Walker (1999), operation of the leaky forced-air heating system with ductwork in the unheated basement more than doubled the rate of infiltration of outside air. Even when the ducts are inside the thermal envelope, such as between floor joists or inside interior walls, significant leakage to the outside can occur because the spaces housing the ductwork are themselves not air sealed. The extra infiltration induced by the air handler is larger still if internal doors are closed, due to pressure imbalances arising from inadequate return-air pathways. In houses heated with heat pumps, the percentage loss in heat pump + duct system efficiency is up to twice as large as the percentage loss in furnace + duct efficiency for the same fractional air leakage. This is because, in houses with heat pumps, there is greater use of an electric-resistance heating coil for backup purposes when the ducts leak, so not only must more heat be produced, but the average efficiency in producing it decreases. Simulations of houses in Washington state indicate that leaky ducts in houses with heat pumps decrease the seasonally averaged heating COP by about 25–40 per cent (Francisco et al, 2004). In air conditioning systems with ducts passing through an attic at 55°C and 20 per cent relative humidity (a typical condition in the southern US), a mere 10 per cent leakage of attic air into the air returning to the evaporator (also common in the southern US) will decrease the cooling capacity and effective COP of the air conditioner by 30 per cent (O'Neal et al, 2002).

Conductive heat loss or gain through the ductwork walls can also be significant. A study in California found that 23 per cent of the energy delivered to the air at the cooling coil is lost due to conduction prior to reaching the duct outlet. Peak residential duct system heat gains during the cooling season can approach 33 per cent of the cooling capacity for uninsulated ducts in attics in houses in Florida, requiring larger (and therefore more expensive) air conditioners (Parker et al, 1993).

It goes without saying that furnaces and heating and cooling ducts should be placed entirely within the conditioned space. In addition, ductwork for ventilation, heating or air conditioning should be well sealed against leaks and well insulated, particularly if it passes through spaces such as attics. Traditional cloth duct tape does not provide a long-lasting seal, often failing after a few years (Sherman et al, 2000; Walker and Sherman, 2005). Instead, mastic – a wet adhesive – should be used in order to ensure a seal that will last as long as the ducts themselves. Techniques for sealing existing duct systems are described in Chapter 14 (Section 14.1).

7.3.7 HVAC systems in advanced houses

Figure 7.18 shows the heating system in an advanced house built in Montreal that incorporates some of the elements discussed above in a novel way. It is based on a ground source heat pump. Relatively cold water from the ground pipe loop is preheated with heat from the exhaust air with a heat exchanger. The exhaust air includes heat captured from a sun-space within the house. A water-to-air heat pump transfers heat to the ventilation air, and a water-to-water heat pump transfers heat to a floor loop for radiant heating and to the hot water tank for domestic hot water use (showers, washing). As heat is extracted from the water in the groundwater loop, this water is cooled to below the ground temperature, so that it absorbs heat from the ground when it flows through the underground portion of the loop. It is this heat, along with that recovered from the exhaust air, which replenishes the heat extracted by the two heat pumps. Heat is delivered in part through ventilation air and in part through floor

Figure 7.18 The ground source heat pump-based heating system in an advanced house built in Montreal

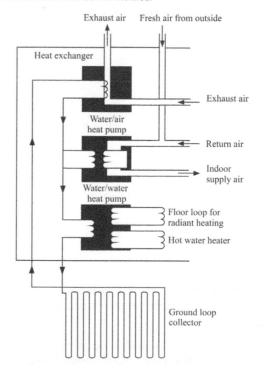

Source: McNeil (1995)

radiant heating. During summer, some of the heat removed from the ventilation air for air conditioning supplies the hot water, with the balance rejected to the ground.

A house built in Brampton, Ontario (Canada), designed by Greg Allen of Sustainable Edge (www.s-edge.com), has probably the most integrated mechanical system ever built (Carpenter 2003a, 2003b). The system is illustrated in Figure 7.19. It provides space heating, space cooling, ventilation, grey-water heat recovery and exhaust-air heat recovery, with a heat pump and separate hot-water and cold-water storage tanks. The heat pump simultaneously cools water in the cold-water tank and heats water in the hot-water tank, which supplies domestic hot water and space heating. The cold-water tank is the direct heat source for the heat pump, and it is recharged by circulating a 33 per cent ethylene glycol-water mixture from the cold-water tank through heat exchangers that pick up heat from grey-water and exhaust air. Being rather cold, this brine is particularly effective in picking up heat from these sources. Incoming fresh air is preheated by passing through thermal mass (a stone wall) inside a two-storey sun-space, which guarantees some preheating of the incoming air even when the sun is not shining (in a more conventional system, heat extracted from the exhaust air is used to preheat the incoming air, but here the exhaust air and grey-water are used to recharge the cold-water storage tank). During summer, the glycol mixture circulates through a cooling coil to provide air conditioning, rather than through the grey-water and exhaust-air heat exchangers. This tends to warm the cold-water tank. In order to maintain adequate coldness in the tank, hot water is discharged from the hot-water tank when needed, which forces the heat pump to run, withdrawing heat from the cold-water tank. In effect, heat extracted from the air for air conditioning provides the domestic hot water requirements, with excess heat dumped to the outside. When hot water is discharged, it is mixed with cold water and used in an underground yard irrigation system.

Dorer and Breer (1998) present information on mechanical ventilation systems in a number of low-energy residential buildings in Switzerland. In these cases, outside air is first pre-warmed (in winter) or precooled (in summer)

Figure 7.19 The integrated mechanical system in an advanced house built in Brampton, Ontario

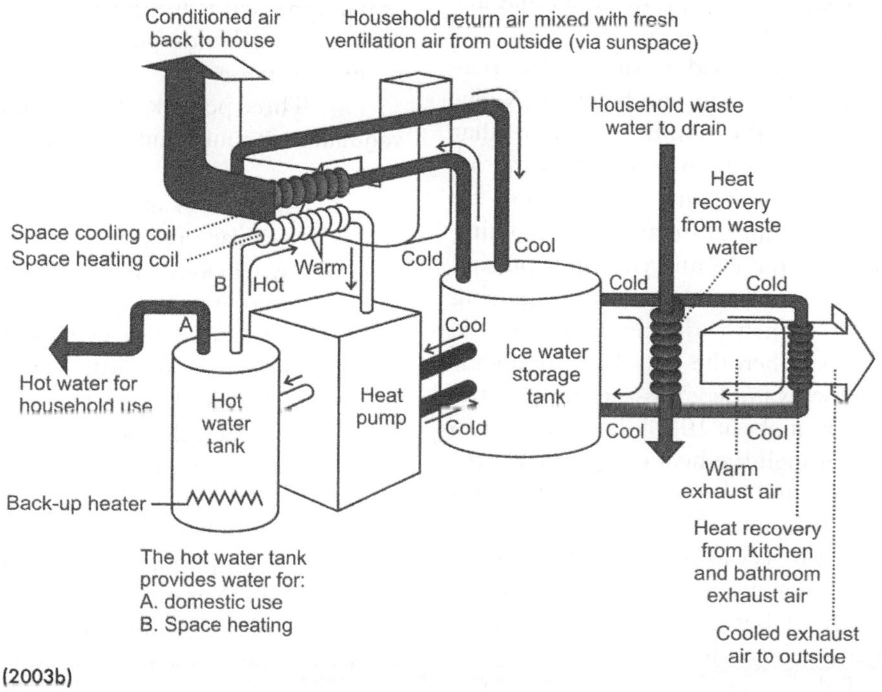

Source: Carpenter (2003b)

Figure 7.20 Temperature duration curves for outdoor air, air after earth pre-conditioning, after passing through the heat exchanger, and after passing through the fans for a 12-unit residential building near Zurich

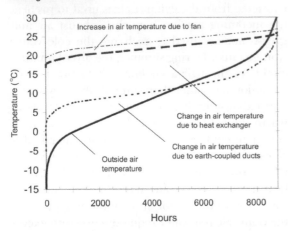

Source: Dorer and Breer (1998)

by passing it through a 30m long, 0.25m diameter coil buried in the ground. It is then passed through a cross-flow plate heat exchanger (with a measured effectiveness of 85 per cent), where it picks up heat from the outgoing air. Apart from the heat released by the fans, no additional heating of the ventilation air is provided. The heat required to offset losses through the building envelope is supplied by hydronic radiators that are connected to a district heating system.

Figure 7.20 provides data on the performance of the system in one building. A given data point gives the number of hours that the temperature is less than or equal to the indicated value. This is shown for the outdoor air temperature, the temperature after the incoming air passes through the ground coil, after passing through the heat exchangers, and after passing through the fan (which is downstream from the heat exchanger). When the outside air temperature is −5°C (for example), passage through the ground (which is at about 10°C) warms the air to 5°C, passage through the heat exchanger warms the air to 19°C, and passage through the fans warms the air to just over 20°C. When the outside air temperature is warmer than 10°C but still in need of warming, passage through the ground cools the air further, but the heat exchanger is still able to warm the incoming air adequately. The main value of the ground loop is that it permits

the heat exchanger to warm the incoming air sufficiently on the coldest days encountered. Its contribution to satisfying the annual heating load is quite small.

7.3.8 Ventilation systems for multi-unit residential buildings

A common conventional approach in ventilating apartment buildings with double-loaded corridors (a central corridor with access to units on both sides) is to over-pressurize the corridor so that air flows under the door into the apartment units, thereby inhibiting the movement of contaminants into the corridor from the apartments. The apartments themselves are ventilated by exfiltration through the external walls and by opening windows. Apart from being energy-inefficient, it does not address apartment air quality, nor is it entirely effective in preventing contaminants from flowing into the corridor, as much of the corridor air bypasses the apartments via shafts, stairwells and other leakage points. A second conventional approach is a centralized, ducted ventilation system. This suffers from a lack of local control and transfer of noise through the ductwork, requires a very high ventilation rate in order to achieve adequate dilution of apartment air, and compromises fire and smoke control. A third conventional approach is to provide en-suite bathroom fans and rangehoods, but these often are not used because of noise.

Three possible alternative approaches to ventilation in multi-unit residential buildings are:

- a balanced and independent ventilation supply and exhaust system for each unit, with heat recovery;
- a balanced floor-by-floor supply and exhaust system with heat recovery;
- a balanced supply and exhaust system with heat recovery serving the entire building.

There are a number of advantages and disadvantages for each option. The first option could be preferred on the grounds that it provides individual control of ventilation rates, with the ability to reduce or shut down the ventilation airflow when it is not needed. It provides excellent ventilation performance if combined with air-tight design of individual suites. However, it is the most expensive option, requires two external envelope penetrations per suite, and available fans at an appropriate size are inefficient (15–20 per cent efficiency). The second and third options are advantageous in that, if radiant cooling is used to provide the sensible cooling requirements, desiccant dehumidification can be easily used to control relative humidity. Maintenance will be easier and more likely to be carried out, especially where residential units are rented rather than owned by the occupants. However, these options require additional space for ductwork and suffer from the cubic-law relationship between airflow and airflow power, only partly offset by the greater efficiency of larger fans (for example, the power that must be supplied to a single airflow of $N \times 30l/s$ is N times greater than what needs to be supplied to N airflows of $30l/s$, whereas the fan efficiency might be only 3–4 times greater – 70–80 per cent instead of 15–20 per cent). Balanced floor-by-floor ventilation with heat recovery during the winter and desiccant dehumidification and heat recovery during the summer might be the best choice.

7.4 HVAC systems in commercial buildings

A variety of methods have been used to ventilate and heat or cool commercial buildings. In an all-air system, air of sufficient coldness and in sufficient volume to remove all of the heat that needs to be removed is circulated through the building. This can be done using either a constant flow rate, or a variable flow rate – referred to as constant air volume (CAV) and variable air volume (VAV) systems, respectively. The air that is circulated through the supply ducts may be taken entirely from the outside and exhausted to the outside by the return ducts, or a portion of the return air may be mixed with fresh outside air and recirculated through the building. The incoming air needs to be cooled and dehumidified in the summer, or heated and sometimes humidified in the winter. In an all-air cooling system, the amount of air that needs to be circulated in order to achieve adequate cooling is so large that fresh-air requirements are easily met; ventilation in this case is a byproduct of the air conditioning.

Alternatively, the ventilation and heating

or cooling functions can be decoupled by circulating only enough air for ventilation requirements, with any additional required heating and cooling provided by circulating hot or cold water through coils. Air in each room is blown past the coil with a fan, producing a *fan-coil* system. This requires three distribution systems – one for air (ductwork), and one each for piping hot and cold water. Since there is a supply and return pipe each for hot and cold water, this is also referred to as a *four-pipe* system. In some buildings, there may be only two pipes, with the system switching from cooling to heating in the fall and from heating to cooling in the spring. Due to the inflexibility of switching between heating and cooling only twice per year, two-pipe systems fell out of favour, although new methods allow daily switching between heating and cooling as needed (Durkin and Kinney, 2002). In some buildings there may be no central ventilation system at all. Heating and cooling is achieved entirely through a fan-coil system, with ventilation provided by opening windows, or through air-to-air heat exchangers in individual rooms.

Existing large commercial buildings typically require simultaneous heating and air conditioning 365 days per year. Air conditioning is required on even the coldest winter days because the core of the building tends to overheat due to all of the waste heat generated by inefficient lighting and office equipment, while the periphery requires heating in winter due to heat losses through windows and walls having minimal thermal resistance. In summer, heat is needed either to reheat air after it has been overcooled for dehumidification purposes (as explained in Chapter 6, Section 6.1.4), or in buildings where ventilation air is precooled to a fixed temperature and then reheated as needed based on the actual cooling load.

In the following subsections, further information on the coupled heating, cooling and ventilation systems used in existing commercial buildings is presented, beginning with old and highly inefficient systems and finishing with a discussion of advanced existing and theoretical systems. Chapter 6 of the *National Best Practices Manual for Building High Performance Schools* (US DOE, 2002a) provides a highly readable introduction to much of the HVAC equipment that is discussed in this chapter, while Tao and Janis (1997) provide more detailed information on the conventional HVAC systems discussed in this chapter.

7.4.1 Heating system component

The heating requirements of a large commercial building can be divided into (i) perimeter heating (to offset heat losses through walls and windows) and (ii) heating and humidification of outside air that is added to the ventilation air. In the majority of large commercial buildings, perimeter heating is provided either by hot water radiators (a hydronic system) or through heating coils that the ventilation air passes through just before entering individual rooms (a *terminal reheat* system). The coils can either be electric resistance or hot water coils (supplied by hot water at up to 60–85°C). Hot water can be provided by centralized gas or oil boilers, by heat pumps, or by connection to a district heating system. Outside air is mixed with recirculating inside air, which has picked up heat from internal heat sources, so heating of incoming outside air is needed in large buildings only when the outside air temperature is less than about 0°C (although perimeter heating is needed at much warmer outside temperatures).

In some buildings, heat from the chiller condensers is used to heat water to about 38°C, which is circulated through the perimeter to meet at least part of the perimeter heating requirements. A separate condenser supplies heat to a cooling tower on the roof when perimeter heating is not needed. In such buildings, the outside air intake unit is sized to meet the fresh air requirements only and is therefore unable to meet the cooling requirements of the interior zones in winter; thus, the chillers are forced to run even in winter, while other parts of the building are being heated. In poorly designed buildings, the lights must be left on at night in order to create a false cooling load, so that the chiller runs and continues to supply heat to the perimeter.

7.4.2 Water loop heat pump systems

Many apartments, condominiums and hotels use through-the-wall (air-source) heat pumps for heating and cooling with auxiliary electric resistance heating. Some new hotels use water-to-air heat pumps with one heat exchanger connected to a common building piping system. In heating mode, the circulating water serves as a heat source, while in cooling mode it serves as a heat sink. When different parts of the building simultaneously need heating and cooling, this permits redistribution of heat gains to the parts of the building needing it, thereby improving overall efficiency. At other times, the circulating water is either chilled centrally with a cooling tower or heated centrally with a boiler, as needed. This system is referred to as a *water loop heat pump system*. This is better than using air conditioners, as the heat pumps are cooled with water that either serves to transfer heat to parts of the building with a heating requirement, or that can be cooled with a cooling tower when there is a net cooling load. However, in warm climates (where there will be few or no occurrences of a need for simultaneous heating and cooling), the water loop heat pump system will be less efficient than a centralized system with large chillers and a cooling tower, as shown by Lian et al (2005). This is due to the larger COP of large centralized chillers (4.0 and larger) compared to that of typical small residential heat pumps (COP of 3.0–3.5).

7.4.3 All-air cooling systems

In many buildings, the ventilation system provides the cooling but the volumetric airflow rate is fixed. Ideally, one would provide less air ventilation (down to the level required to maintain adequate indoor air quality) when the cooling requirements are smaller, and increase the amount of cool ventilation air supplied to each area as the cooling requirements increase. Instead, ventilation air is often precooled and supplied at a fixed flow rate. Just before entering the various offices and working spaces, the air is reheated (sometimes electrically!) by the amount required to maintain the desired workspace temperature. This is the terminal reheat system, mentioned above, which is also used for heating in winter. It used to be quite common (at least in North America), and many HVAC systems using this approach still exist. If combined with over-ventilation or a minimum ventilation rate that provides more cooling than needed, substantial amounts of energy can be wasted. For example, Sellers and Williams

(2000) report the case of a building where the boiler was firing at 50 per cent of its wintertime design maximum on the warmest day of the summer in order to provide the reheating demanded by the system.

Another option for controlling the amount of cooling to various parts of the building is to throttle the airflow (partly close the ducts with vanes in order to reduce the airflow). This is like controlling the speed of a car by varying the amount of braking while the engine is running with the gas pedal to the floor. Most commercial buildings in North American cities still use either terminal reheat or throttling to control the cooling of individual rooms. This is because the motors in the fans of older buildings could run at only one speed (dictated by the frequency of the AC electricity, which is fixed). New buildings use variable speed motors (having a variable speed drive, or VSD), which allow the rate of airflow in the ventilation system to vary with the cooling requirements (see Section 7.1.2).

A third option, which may still be found in old buildings and even in some new buildings in some jurisdictions, is the *dual-duct* system, in which there are separate cold-air and warm-air distribution ducts, with the warm and cold air blended at the terminal unit in the proportions required to maintain the desired temperature. The energy penalty of simultaneously cooling and heating inside air can be reduced if ventilation (fresh air from outside) is supplied through the cool-air supply without mechanical cooling whenever the outdoor temperature is cold enough, and if recirculated indoor air is used as the warm-air supply, with additional heating only when there is a real heating load.

A key to reducing energy use in any all-air system is to make it a VAV system with a VSD. In order to appreciate the energy savings that are possible using VAV rather than CAV systems, it is useful to examine the typical breakdown of energy use by the components of an HVAC system. This is shown in Figure 7.21 for two buildings with CAV systems in Singapore, representing a hot and humid climate, as simulated by Sekhar (1997). Ventilation fans account for 27–30 per cent of the total HVAC system energy use, while operation of the chillers accounts for only about half of the total energy use. For the same

buildings with a VAV system, ventilation fan energy use falls by a factor of two in one case and by a factor of three in the other case. Energy use by all the other components falls modestly. In the case of buildings in Washington, Chicago and Charleston (North Carolina), the switch from a CAV to VAV system lead to a 25 per cent reduction in cooling energy use, a 50–75 per cent reduction in heating energy use (due in part to less terminal reheat), and a 65 per cent reduction in fan energy use according to simulations by Franconi (1998), giving an overall saving of 53–63 per cent.

In some buildings with an all-air cooling system, all of the return air is vented to the

Figure 7.21 HVAC energy use for two buildings in Singapore

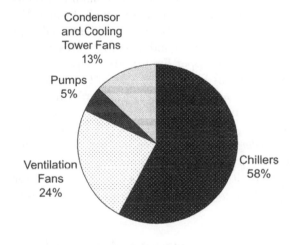

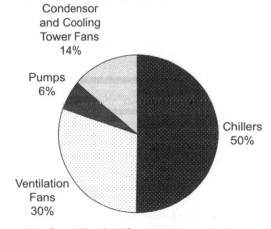

Source: Data from Sekhar (1997)

outside, and the supply air consists of 100 per cent outside air (this is mandatory in hospitals). In most cases, however, some of the return air is mixed with incoming outside air and recirculated through the building. The proportion of outside air will vary with the outside air temperature in the manner shown in Figure 7.22. When no cooling is required, the minimum amount of outside air (needed for ventilation purposes) will be used (Region 1). As a cooling load develops (Region 2, starting at an outside temperature of −12°C to 4°C for buildings with large internal heat sources), an increasing proportion of outside air will be used without use of the chillers, up to the point where 100 per cent outside air is reached (typically at an outside temperature of 12°C). In Region 3, 100 per cent outside air is still used, but the chillers are also in operation. At some point, the outside air is too warm to provide cooling, so the system reverts back to the minimum proportion of outside air. This point is typically at 21°C, but obviously could be warmer if internal heat gains are minimized and if temperatures in summer near the upper limit of the adaptive standard shown in Figure 7.10 (i.e. 28°C rather than 23–26°C) are permitted. The operation of the ventilation system in this way is referred to as an *air-side economizer*. Region 2 is referred to as the *free-cooling* region. However, in small systems with a single chiller, the economizer is usually either entirely closed or entirely open, as the compressor and economizer cannot operate simultaneously

due to problems with on/off cycling in chillers with a fixed compressor speed (AEC, 2003). This reduces the potential savings from free-cooling by 20–30 per cent. Rather than basing the control of the system on outside temperature, it can be based on the outside enthalpy, which combines temperature and humidity (see Equation 6.5). Enthalpy sensors are much more expensive than temperature sensors and most require semi-annual recalibration, but they are needed in humid climates to restrict the use of outside air when humidity would be a problem (Rock and Wu, 1998).

Ventilation air for cooling used to be and often still is supplied at 13–16°C. There is a growing tendency to use lower supply air temperatures (7–11°C), as this permits a smaller airflow for the same cooling effect. This in turn reduces the initial cost of the system, since smaller airflows translate into smaller air handling units, fans, pumps and ductwork. The fan and pump energy use are also reduced, but cooling energy use is normally not increased since the air would need to be cooled to this temperature (or lower) in any case in order to remove moisture. However, as discussed later in Sections 7.4.11 and 7.4.12, alternative methods of dehumidification avoid the need to cool incoming air this low. If these methods are used, chilling down to 7–11°C represents additional work (with a lower chiller COP). In any case, a requirement for chilling down to 7–11°C reduces the opportunity for free cooling.

7.4.4 Chilled ceiling (CC) and chilled beam cooling

An alternative to all-air cooling is to restrict the airflow to that required for ventilation, with no recirculation of return air, and to circulate chilled water throughout the building. Individual rooms are cooled using a *fan-coil* unit, in which a fan blows room air past a series of metal fins attached to the chilled-water pipe. This is referred to as a *hydronic* cooling system. Inasmuch as the energy used by HVAC fans can easily be up to half that used by the chillers, and the energy required to distribute coldness through chilled water is about 25 times less than that required to distribute coldness through cool air (Section 7.1.2), cooling with chilled water can significantly reduce total HVAC

Figure 7.22 Variation with temperature in the proportion of outside air for an air-side economizer using temperature as the controlling variable

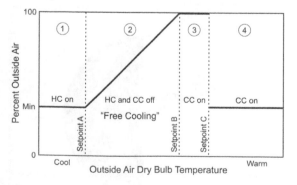

Note: HC = heating coil, CC = cooling coil.
Source: Rock and Wu (1998)

energy use. The ventilation system is referred to as a *dedicated outdoor air supply* (DOAS) system because the air supplied to interior spaces is 100 per cent outside air, and ventilation air is vented directly to the outside. This leads to improved air quality, as well as reduced energy use through the hydronic cooling component.

An alternative to fan coil units placed above a false ceiling or mounted on the floor is to chill the entire ceiling surface. This can be done in one of two ways:

- by circulating chilled water through plastic hydronic piping inside an exposed concrete ceiling;
- by circulating chilled water through metal tubes that are welded to aluminium panels.

Cooling occurs in two ways: through exchange of radiation (the cold panel emits less infrared radiation than it absorbs from the warmer room interior); and through convection (air next to the panel is cooled and sinks). Radiative cooling is the dominant process, so this system is referred to a *radiant chilled ceiling* cooling, or simply as chilled ceiling or CC cooling. Water at a temperature of 16–20°C is circulated through the concrete core or ceiling panels. Concrete-slab cooling costs less than chilled panels (60–75€ versus 150–300€ per m² of active area) but has less cooling power (≤50W/m² versus 60–120W/m²), so a larger active area is needed (Oesterle et al, 2001).

Hydronic concrete-core systems are increasingly used in Europe. As discussed by Koschenz and Dorer (1999), they are cooled during times of low electricity demand at night, when both chillers and power plants operate more efficiently. The concrete slab will absorb heat during the day, resulting in an increase in room temperature during the day. According to these authors, a maximum temperature increase of 3–4K from morning to evening would be acceptable to most HVAC system designers. Night-time cooling is possible with chilled ceiling panels packed with phase-change materials (Chapter 6, Section 6.9.3), although these are not yet commercially available. With concrete-slab cooling, the ceiling must be exposed, so lighting fixtures must be suspended and the light bounced off of the ceiling.

Figure 7.23 Detail showing hydronic pipes for concrete-core cooling fastened to the steel reinforcing of a concrete floor/ceiling slab, prior to pouring of the concrete

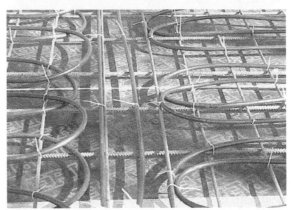

Source: Jim Sawers, Keen Engineering, Calgary

In concrete-slab cooling, the plastic pipes are fastened to the steel reinforcing of the floor/ceiling slab prior to pouring the concrete, as illustrated in Figure 7.23. Floor slabs are 200–300mm thick, so there will be a minimum clearance of 75mm on either side of the pipes, allowing fixtures to be drilled into the slab without risk of puncturing the piping. Additional conduits for electrical and mechanical penetrations will need to be placed at the time the slab is poured, so some coordination between the different contractors is required.

Ceiling radiant cooling has been used in Europe since at least the mid-1970s. In Germany during the 1990s, 10 per cent of retrofitted buildings used panel CC cooling. Of these, 10 per cent relied on natural ventilation, 45 per cent used conventional ventilation in which ventilation air

Figure 7.24 Chilled ceiling cooling panels

Source: www.advancedbuildings.org

is mixed with the room air, and 45 per cent relied on displacement ventilation (which is explained later) (Behne, 1999). An example of installed chilled ceiling panels is given in Figure 7.24. Metallic radiant panels reflect sound waves, so they often need to be combined with some method of sound absorption.

An occupant of a room absorbs heat in two ways: from the infrared radiation emitted by the walls, ceiling and floor within the room; and by convective exchange of sensible and latent heat with the air. The perceived temperature will be roughly the average of the radiative temperature and the air temperature at an air speed of 0.1m/s (Equation 7.8). In an all-air system, the radiative temperature is about 2K warmer than the air temperature, whereas in a CC system the radiative temperature is the same or slightly less than the air temperature (Novoselac and Srebric, 2002). This in turn permits the air temperature to be 2K warmer in CC systems than in all-air systems. The radiant asymmetry associated with a chilled ceiling creates a sensation of "freshness" that is similar to being outdoors under an open sky. This perception increases as the ceiling temperature drops to 14–18°C, then levels off (Hodder et al, 1998).

Because radiation travels at the speed of light, a radiant cooling system responds instantly to the presence of warm objects. If there is nothing in the room that is at a different temperature than the radiant slab, then no heat exchange occurs. However, as soon as a warm object (a human) enters the room, or sunlight enters the

room and heats the floor or furniture, there is an instantaneous exchange of heat between the radiant slab and the warm objects.

Energy use

The impact on energy use of a CC system compared to an all-air system depends on a number of competing factors:

- the energy used by ventilation fans is greatly reduced due to a smaller rate of airflow;
- less outside air needs to be cooled, dehumidified, and reheated compared to an all-air system in which all of the ventilation air is vented to the outside after passing through the building once;[5]
- that portion of the waste heat from ceiling lights or light shelves that enters the ventilation airflow is not recirculated, since the ventilation airflow can be reduced to that needed for ventilation purposes only and vented directly to the outside;
- most of the cooling can be done with water circulating at 16–20°C rather than with air that is initially cooled to 5–7°C, which allows the chillers to operate with a COP approximately 1–3 larger (depending on the condenser temperature; see Figure 5.2b) if separate chillers are used for cooling air and water (which would not be the normal practice) or if the air is also cooled to only 16–18°C (which might require desiccant dehumidification for moisture control);
- because of the relatively high chilled-water temperature used in a CC system, the cooling water produced by a cooling tower will be cold enough for direct use in the CC panels more often than in a conventional cooling system, as discussed in Chapter 6, Section 6.10.2;
- the ventilation air will be 5K below the room temperature in a CC

system, rather than the 8K cooler that is typical of all-air systems, so more energy will be used for reheating after dehumidification in those systems (the overwhelming majority at present) where dehumidification is achieved through overcooling;

- a small amount of energy will be used by the pumps for the hydronic cooling component.

As discussed below, a CC can easily handle 75 per cent of the total cooling load, leaving 25 per cent to be handled by the ventilation system. This would imply a reduction in the required ventilation airflow by a factor of four. If displacement ventilation (described in Section 7.4.5) is used in conjunction with a CC, the difference between supply and return air temperatures will be smaller by a factor of two, but the effectiveness of a given displacement ventilation flow in removing heat is about twice as large as in conventional ventilation, so there is still roughly a factor of four reduction in the required ventilation airflow. Due to the much smaller energy required to remove heat by circulating water compared to circulating air (as discussed in Section 7.1.2), the reduced ventilation airflow acts to reduce the total required HVAC energy.

Stetiu and Feustel (1999) carried out simulations of buildings with all-air and combined air/CC cooling systems for a variety of US climates, assuming the same rate of intake of outside air for the two cases. They found that radiant cooling saves on average 30 per cent of the energy and 27 per cent of the peak power demand used for air conditioning. Results for Seattle and Phoenix are summarized in Table 7.8. Given are the baseline energy use for an all-air system with continuous ventilation and when the ventilation is shut off at night, the energy used with CC cooling, and the percentage energy savings using the latter. In the all-air system, total energy use is slightly larger when the ventilation system is shut off at night, due to the disproportionately large increase in fan energy use during the day that is required to remove the additional sensible heat that accumulated during the night. In the CC system, total energy use is smaller when the ventilation system is shut down at night, especially for Phoenix. Compared to an all-air system, the CC system reduces energy use by as little as 6 per cent in Seattle using continuous ventilation, to as high as 42 per cent in Phoenix when the ventilation is shut down at night. The saving is smaller in hot-humid or cool-humid climates than in hot-dry climates because relatively more of the total air conditioning energy is used for dehumidification, which is not affected by the choice of all-air versus air/CC chilling. Stetiu and Feustel (1999) assumed the same chiller COP in both systems, dictated by the evaporator temperature needed for dehumidification. However, if dehumidification is done without overcooling (for example, using desiccant dehumidification), then a further energy saving with CC would occur.

Jeong et al (2003) simulated the energy use for a chilled-ceiling cooling system that is being installed in a new studio at Pennsylvania State University, and compared this with that computed for a conventional VAV system. The chilled-ceiling system uses 100 per cent outside air at a rate required for ventilation purposes alone, with an enthalpy wheel for recovery of sensible and latent heat (see Section 7.4.9), whereas the VAV has no form of heat recovery. Both systems are assumed to operate between 8am and 7pm only.

Table 7.8 Energy use for an office building in Seattle and in Phoenix using all-air and chilled-ceiling (CC) cooling systems

Mode	City	Energy use (kWh/m²)		Savings (%)
		All-air	CC	
Continuous ventilation	Seattle	15.0	14.3	6
	Phoenix	50.3	32.2	36
Ventilation switched off at night	Seattle	18.3	14.1	23
	Phoenix	51.7	30.0	42

Source: Data taken from simulations by Stetiu and Feustel (1999)

The chilled-ceiling system with an enthalpy wheel is estimated to use about 40 per cent less energy than the VAV system. Although this saving is not entirely due to the chilled-ceiling component, it is another example of the large savings that can be achieved compared to conventional practice using advanced systems.

Risk of condensation

As an air parcel is cooled, its relative humidity rises. If the temperature of the chilled ceiling is less than T_{dp} (the dewpoint temperature) condensation may occur. To avoid this, sufficient moisture must be removed from the ventilation air to bring its dewpoint below the ceiling temperature. If the cooling load can be reduced, a warmer ceiling temperature is possible, so less dehumidification will be needed, thereby amplifying the direct energy savings from the smaller cooling load. However, if relative humidity rises above 80 per cent, human comfort can be adversely affected (Section 7.2.3).

When the ventilation system is running there will be little or no infiltration of air into the building through leaks in the envelope, due to a slight over-pressurization of the building. If the system is turned off, humidity can enter the building, and when the CC component is restarted at the same time as the ventilation component after a weekend, there is an increased risk of condensation. The ventilation component should therefore be started first, to dehumidify the air before the ceiling is chilled. In the hot and humid climate of Hong Kong, for example, dehumidification should begin about 1 hour before the ceiling panels are activated (Zhang and Niu, 2003). In addition, the building should be as airtight as possible in order to minimize infiltration of moist air. Mumma (2001) discusses other condensation issues that have been raised, none of which pose a real problem. The need for dehumidification can be reduced if the need for outside ventilation air is minimized through a displacement ventilation system (described in Section 7.4.5), which is the natural complement to CC cooling.

Cooling capacity and comparison with cooling loads

A rough rule of thumb is that a chilled ceiling surface can provide $11W/m^2$ cooling per degree of temperature difference between it and the room, while a chilled bare floor can provide $6W/m^2$ cooling per degree of temperature difference (for heating, the performance is reversed: a heated ceiling provides $6W/m^2/K$ heating and a heated bare floor provides $11W/m^2/K$ heating). Thus, if the room temperature is $26°C$, panels at $18°C$ would absorb $88W/m^2$ of heat. Given a typical flow rate of 50litres/hr per m^2 of panel area (Roulet et al, 1999), the water would warm by 1.5K as it flows through a 1m² panel. Panels that cover 70 per cent of the ceiling area could remove a heat load of about $60W/m^2$. With an additional $20W/m^2$ handled by displacement ventilation (described below), a total load of $80W/m^2$ could be handled.

In offices with efficient equipment and lighting, the cooling loads will be: lighting, 10–$15W/m^2$; equipment, 8–$15W/m^2$; occupants (at 75W each), 7–$10W/m^2$; total load: 25–$40W/m^2$. In perimeter offices there will be additional heating from sunlight, but this can be minimized by using windows with a low SHGC (see Chapter 3, Section 3.3.7) and through computerized blinds.

If additional cooling is needed in perimeter regions, it can be obtained through *chilled beams* (discussed below). As in floor radiant-heating systems (Section 7.3.1), CC and chilled beam cooling is largely self-regulating, in that the rate of heat removal increases rapidly as the interior temperatures increase, which serves to strongly limit the variation of internal temperature. There is much less self-regulation in an all-air cooling system, as shown in Box 7.1.

Figure 7.25 Use of chilled beams in place of chilled panels in order to create a visually stimulating ceiling

Source: Frenger Cooling, www.buildingdesign.co.uk/mech/frenger2/frenger2.htm

Chilled beams

In chilled beams, the cooling is concentrated in finned coils that are similar to conventional heat exchangers (Alamdari and Butler, 1998). Chilled beams can be either passive (separate ventilation is required, as in chilled panels) or active (ventilation is an integral part of the beam). The beams can be designed to provide several services in addition to cooling: up-lighting, down-lighting, building management system sensors, fire alarm and sprinklers, and cables. They can be deliberately used in place of chilled ceiling panels, rather than as a supplement in areas of high cooling load, in order to avoid monolithic ceilings. An example is shown in Figure 7.25. Conversely, passive chilled beams can be installed above a perforated metal ceiling or above a microperforated ceiling (1.5–2.4mm diameter holes with 22–28 per cent free area), providing up to 130–150W/m² of cooling. There are few situations where this cooling capacity is inadequate, one being the computer rooms in Internet data centres, where cooling loads are on the order of 300–400W/m² (Blazek et al, 2004).

Capital cost

A chilled ceiling eliminates the need for the perimeter fan-coil units that are used in some buildings for summer cooling and winter heating. The elimination of perimeter heating and cooling in turn requires high-performance windows – sufficient to maintain an inner-glazing temperature of at least 16°C during the coldest winter days, and no more than 26–27°C on hot summer days – so as to avoid radiant temperature asymmetry that is too strong. In cold climates, this requires triple-glazed windows with low emissivity coatings and an argon or krypton gas fill. Although high-performance windows cost more than conventional double-glazed windows, the overall capital cost of the glazing plus mechanical and electrical systems can be less than for a conventional building due to savings from avoided perimeter heating units and smaller heating, cooling and air handling equipment.

McDonell (2003) presents data on the cost and energy use of a conventional, all-air HVAC system in an office building in Vancouver (Canada) with a conventional envelope, and that of buildings with radiant slab heating and cooling and a high-performance envelope. His results were summarized in Table 3.19. The buildings with the high-performance envelope and radiant slab heating and cooling cost approximately 10 per cent less and used 45 per cent less energy per unit of floor area. Mumma (2001) compared the cost of a conventional (VAV) HVAC system and a system with chilled ceiling panels and use of 100 per cent outside air for ventilation with sensible and latent heat exchangers, and found the CC system to cost about the same or slightly less than the conventional system.

7.4.5 Displacement ventilation

In a conventional CAV or VAV ventilation system, the ventilation air typically enters a room through an air outlet in one part of the ceiling and returns through a vent elsewhere in the ceiling. Even when the air outlet is placed near the floor, the system still relies on *turbulent mixing* of fresh air with stale air in the lower half of the room in order to provide fresh air to the occupants. An alternative is *displacement ventilation* (DV), in which a gentle flow of fresh air is introduced

Figure 7.26 Diffuser for a displacement ventilation system

Source: Author

from many openings in the floor (where feasible) or from long diffusers at the base of the walls. The air is supplied at a temperature of 16–18°C, spreads laterally, is heated by heat sources within the room, and continuously rises and *displaces* the pre-existing air. This provides superior indoor air quality but with less airflow than in a conventional ventilation system, which shall henceforth be referred to as a *mixing ventilation* (MV) system. Most DV systems provide airflow from the floor, using swirl-shaped diffusers that permit easy individual adjustment of the airflow, something that is not possible with overhead MV systems. An example is illustrated in Figure 7.26. Loftness et al (2002) provide a detailed discussion of underfloor air distribution systems (not all of which are DV), including the design of the floor plenum, alternative diffuser designs, control alternatives and integration with other systems in the building.[6]

DV, like CC cooling, was first applied in northern Europe. By 1989, it had captured 50 per cent of the Scandinavian market for new industrial buildings and 25 per cent for new office buildings (Zhivov and Rymkevich, 1998). The building industry has been much slower to adopt it in North America; by the end of the 1990s, less than 5 per cent of new buildings used underfloor air distribution systems (Lehrer and Bauman, 2003), not all of which would have been DV systems. DV systems are more difficult to design than conventional systems, so both the Federation of European Heating and Air Conditioning Associations and ASHRAE have published design guides (Skistad et al, 2002; Chen and Glicksman, 2003).

With MV, the cooling that can be provided from ventilation air alone is usually not adequate, so a hydronic cooling system is needed (or excessive ventilation must be provided, as in all-air systems). With DV, a hydronic cooling component is more likely to be needed, because a smaller ventilation rate can provide good air quality, and because the ventilation air is supplied at a warmer temperature (16–18°C rather than 7–16°C). Even in winter, the interior of large office buildings needs to be cooled due to the presence of internal heat sources. The ventilation air is supplied at a constant temperature year round, and internal heat sources will be largely the same between summer and winter, so a relatively constant rate of hydronic cooling will be needed. To minimize energy use during winter, the cooling water used in the hydronic system can be chilled by directly passing it through a rooftop heat exchanger or through the cooling tower that is used when the chillers are operating, as explained in Chapter 6 (Section 6.10.2).

Although DV is usually an underfloor air distribution (UFAD) system, not all UFAD systems are DV systems. In a conventional North American UFAD system, air is supplied at relatively high velocity in order to thoroughly mix the supply air and room air, whereas the air in a DV system is supplied at low velocity (0.2m/s) in order to minimize mixing. A DV system normally uses 100 per cent outside air, so the airflow is limited to that needed for ventilation purposes only, whereas a traditional UFAD system mixes outside and return air and provides heating and cooling as well as ventilation. Whether an MV or DV UFAD system is used, careful attention to sealing of the edges of the raised floor is required (Daly, 2002).

The major issues associated with displacement ventilation will now be briefly discussed.

Effectiveness

The effectiveness of a ventilation system in removing air contaminants or heat is given by:

$$\eta = \frac{X_e - X_i}{X_o - X_i} \qquad (7.11)$$

where X is the contaminant concentration or temperature, and subscripts i, o, and e pertain to the incoming air, air at breathing level or anywhere else of interest, and exhaust air. For complete mixing, $\eta = 1.0$ because contaminant concentrations and temperature are the same everywhere in the room. In practice, η is typically about 0.7 for ceiling-level MV (Awbi, 1998). For DV, $\eta = 1.2$–1.8 for CO_2 at breathing level according to simulations by Yuan et al (1999), while for temperature, $\eta = 2.0$ according to calculations by Zhivov and Rymkevich (1998). Based on the standard equation used to compute the ventilation rate required to maintain a particular pollutant below a given concentration (see Awbi, 1998), DV would require only 40–60 per cent the airflow of MV for airflows tied to ventilation requirements.

Cooling capacity

Given a ventilation requirement of 7.5l/s per person and an area of 7.5m² per person, the required ventilation rate is 0.001m³/s per m² of floor area (3m³/hour/m²). For an 8K temperature difference between incoming and outgoing air, this gives a rate of cooling by the DV component of about 10W/m² (using Equation 7.1). However, the airflow with DV can be greater than that required to provide sufficient fresh air, being limited by the need to avoid drafts. In systems where the air is supplied from wall vents at floor level, DV can provide up to 30W/m². With perforated floors, a flow rate sufficient to remove 40W/m² is possible without creating draughts (Loveday et al, 2002). Given the greater energy required for distributing coldness through air rather than through water, one may not wish to increase the DV flow rate beyond that needed for ventilation alone, except that the DV airflow should provide at least 20–25 per cent of the total cooling load in order to maintain good air quality in the breathing zone (see below). This is possible if DV is combined with CC cooling, which, as noted above, can easily handle a cooling load of 60W/m², while the total cooling load will normally not exceed 40W/m² except possibly in perimeter offices, where chilled beams can supplement the cooling capacity if necessary.

Figure 7.27 Vertical variation of temperature in a test facility with displacement ventilation system supplying air at 19°C without a chilled ceiling (CC) or with a CC at various temperatures

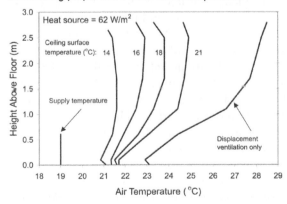

Source: Loveday et al (1998)

Thermal comfort

Thermal comfort requires (i) avoiding displacement airflows that are strong enough or cold enough to produce a draft, and (ii) limiting the temperature difference between heights of 0.1 and 1.1m to 4K (or even less), which in turn limits the difference between incoming and outgoing air temperature to no more than 8K (Behne, 1999). As discussed above, a CC will normally be able to satisfy the entire cooling load without any assistance from DV. On the other hand, in buildings designed to minimize the cooling load, DV alone will be able to satisfy a large fraction of the cooling load without creating drafts. Thus, in high-performance buildings, designers can choose the DV supply temperature and flow rate and the CC temperature that produces a desirable vertical temperature profile while still satisfying the given cooling load. This is illustrated in Figure 7.27, which shows the vertical temperature variation in a test facility with a cooling load of 62W/m² and a displacement ventilation supply air temperature of 19°C, with and without CC cooling. For the case without CC, the 0.1–1.1m temperature difference is 6K. For a CC at 21°C, the difference drops to about 2.5K, and for lower ceiling temperatures, the difference is smaller still.

Noise

The combination of low airflow with DV and the absence of fan-coil units with CC produces a cooling and ventilation system with negligible noise.

Reduced churn costs

Because most DV systems rely on diffusers placed in a raised floor that consists of modular panels, the diffusers can be repositioned if the workspace layout is changed. This in turn will significantly reduce future costs when the tenants in a building or their needs change, and the building space needs to be reconfigured. These are referred to as *churn costs*.

Air quality

There will be three zones in a combined DV/CC system (Novoselac and Srebic, 2002): a lower stratified zone, where fresh air enters; an upper mixed zone, with contaminants (CO_2 from the occupants, emissions from room furnishings);

and a transition zone in between. The convective cooling effect of the CC expands the mixed zone downward, while the upward movement of fresh air tends to push it upward. The lower boundary of the mixed zone should be above breathing level, but where it lies will depend on the relative amounts of cooling provided by the DV and CC components. The DV component should accommodate at least 20–25 per cent of the total cooling load (Behne, 1999). A distinction needs to be made between pollutants (such as CO_2) that are associated with buoyant thermal plumes (emanating from warm objects), and passive pollutants, which are not associated with thermal plumes. In the former case, concentrations next to occupants can be lower than in the absence of a CC. Conversely, if downward convection occurs near the sidewalls, pollutants can be directly transferred from the upper mixed zone to the supply air layer. Concentrated downward convection can be avoided if the CC panels cover a large portion of the ceiling area.

Importance of air tightness

Because the airflow rate is smaller with DV than with MV, particularly compared to an all-air cooling system, there will be less opportunity to counter the effects of a leaky envelope by over-pressurizing the building to prevent infiltration of moisture. It is therefore particularly important to minimize air leakage in buildings with DV in order to be able to maintain the intended interior conditions. Conversely, windows can be opened without compromising the ventilation effectiveness.

Energy use

As noted above, the incorporation of CC cooling in a conventional MV system leads to substantial (30 per cent) energy and peak power savings. The impact on energy use of a DV/CC system compared to an MV/CC system is less clear-cut. It depends on differences in:

- cooling load;
- energy use by ventilation fans;
- energy use by cooling tower pumps;
- energy use by the pumps for the hydronic cooling component;

- reheating of ventilation air after dehumidification;
- the number of hours during which outdoor air will be cold enough to directly supply ventilation air without additional chilling (free cooling);
- the opportunity for evaporative cooling of the ventilation air;
- the opportunity for and effectiveness of increased night ventilation for cooling.

The total cooling load will be less with DV due to the fact that some of the building heat gain occurs above the occupied zone and is directly removed with the ventilation air in a DV system, rather than being mixed with the room air. If DV uses 100 per cent outdoor air (as it should in order to maximize the reduction in the required ventilation rate made possible by DV), then this heat is directly rejected to the outside with the exhaust air, and does not need to be removed by the mechanical cooling system. Calculations by Loudermilk (1999) indicate that, for an office in Chicago, about one-third of the total heat gain (including 50 per cent of the heat gain from electric lighting) can be directly rejected to the outside in this way. Similarly, less of the heat introduced through light shelves (that direct sunlight along the ceiling, as explained in Chapter 9, Section 9.3.1) will reach the occupied space using DV. The amount of energy required by pumps in the CC component of a DV/CC system is very small compared to the potential savings in ventilation-fan and chiller energy use using DV. Savings in cooling load translate into reduced energy use by cooling tower pumps and fans.

Relative to MV/CC, where the airflow is governed by ventilation requirements, DV/CC requires only about half the airflow. Relative to an all-air MV system, where airflow is governed by the cooling load, DV/CC requires only about one-quarter of the airflow. The factor of four difference arises from a reduction in the amount of heat that must be removed by the airflow by a factor of four, and a reduction in the supply–return air temperature difference by a factor of two but offset by DV being twice as effective as MV in removing heat. According to Conroy and Mumma

(2001), the airflow with DV can be as little as one-fifth that of an all-air system, the reduction being constrained by the need for adequate fresh air. Since the power that must be supplied to the air varies roughly with airflow to the third power, this translates into a potential decrease in fan energy use by up to a factor of 125 compared to an all-air system, and by up to a factor of eight compared to an MV/CC system. The saving achieved in practice depends on differences in fan efficiency and in the cross-sectional area of the air ducts, as explained in Section 7.1.2. The smaller fans required with DV might be less efficient, thereby diminishing the savings in fan energy, but under-floor distribution systems (as in DV) permit wider supply ducts, which amplifies the savings in fan energy. As fan energy use decreases, there will be a further decrease in air conditioning requirements, since the fan motors that drive the ventilation system are themselves a significant source of internal heat generation in some cases (see Figure 6.4).

Because of the warmer supply air temperature with DV compared to MV, there will be a greater opportunity for free cooling because the outside air will be cooler than the supply air more often. For the same reason, evaporative cooling will be viable more often using DV than MV, thereby providing a greater relative saving in energy use in a DV system than in an MV system. In moderate climates, such as California's, DV alone can reduce annual HVAC energy use by 30–60 per cent, largely due to increased use of free cooling, while the combination of DV and evaporative cooling can reduce HVAC energy use by 70 per cent. This is illustrated in Figure 7.28 for Oakland and San Diego, which compares the annual HVAC energy use for a baseline VAV MV system with no evaporative cooling, for the VAV MV with evaporative cooling, and for DV with and without evaporative cooling based on rough calculations (using the DOE-2 model) by Bourassa et al (2002). DV alone reduces annual HVAC energy use by 40 per cent in San Diego and by 60 per cent in Oakland, while DV with evaporative cooling reduces annual HVAC energy use by two-thirds in both cities. Howe et al (2003) compute a 50 per cent saving using DV in Colorado, also largely due to the increased opportunity to use outside air for cooling. The warmer supply temperatures with DV will also permit greater use of the ground in temperate climates for cooling of ventilation air.

However, in an all-air cooling system, the warmer supply temperature in a DV system necessitates a greater airflow (and greater fan energy use) in order to provide the same cooling. This can result in negligible overall savings in energy use, as found by Hu et al (1999). However, DV is not suited for an all-air HVAC system and, in any case, an all-air system is a poor choice from an energy efficiency point of view. Novoselac and Srebric (2002) cite studies showing little difference between existing DV/CC and MV/CC energy use, mainly because of greater reheating of ventilation air after dehumidification for DV/CC (due to the warmer supply air temperature required with DV). This might not be the case if some of the heat from the chiller condensers were to be used for reheating, or if the warm return air is used to reheat the over-cooled supply air via a heat exchanger. In any case, it would not be an issue if dehumidification is done without over-cooling and then reheating the ventilation air, using desiccants instead (Chapter 6, Section 6.6.4). Thus, the impact of DV on overall energy use depends very strongly on the nature of the baseline HVAC system, as there are important synergisms between DV and the other HVAC system components.

Figure 7.28 Comparison of annual HVAC energy use for office towers in Oakland and San Diego, as simulated by Bourassa et al (2002) for a VAV mixing ventilation (MV) system with and without evaporative cooling (EC), and for a displacement ventilation (DV) system with and without evaporative cooling

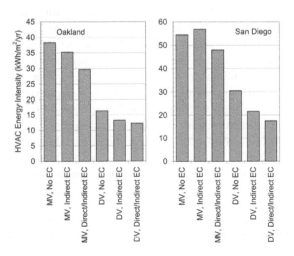

There could be additional savings with a DV/CC system due to the longer retention of coldness if the system is turned off at night due to the fact that the walls, ceiling and floor are 1–2K colder than the air temperature in a DV/CC system, rather than equal to or slightly warmer than the air temperature, as in an all-air system. As a result, the system is less likely to need to be kept running if a single individual wishes to work late.

Capital cost savings

In some new buildings the floors are raised in order to conceal and accommodate cables and services that are laid underneath the floor. The floor can therefore function as a supply air plenum, which can reduce the depth of ceiling plenums. This can lower the height of each floor, or allow more floors for a given total building height (an extra floor for every five according to Conroy and Mumma, 2001), thereby dramatically reducing the cost per floor. In addition, more rentable space is available on each floor, because a smaller area is needed for vertical ducts. Other cost savings arise from the smaller air handling system required with DV, but there are added costs associated with the cold-water distribution and ceiling panels. For buildings in Germany, the net result is that first costs are comparable for DV/CC and all-air systems when the cooling load is $40W/m^2$, but about 25 per cent cheaper for DV/CC for a cooling load of $80W/m^2$ (Sodec, 1999) (the former cooling load is more typical of high-performance buildings). For a six-storey office building in the US, Mumma (2001) finds that a DV/CC system has a lower first cost than a VAV all-air system.

7.4.6 Demand-controlled ventilation

Having decoupled the ventilation and heating or cooling functions of an HVAC system, it is desirable to vary the ventilation rate based on actual and changing ventilation requirements, rather than using a fixed ventilation rate or varying it according to some inflexible schedule. Even in all-air cooling systems, it can be advantageous to vary the ventilation rate or the ratio of outdoor to recirculated air based on the ventilation requirement. This is referred to as *demand-controlled ventilation* (DCV). California's 2001 building code (the Title 24 standards) requires that the ventilation rate respond to changing occupancy in high-density buildings, rather than run at a fixed rate based on maximum occupancy. Depending on the kind of building and occupancy schedule, DCV can save 20–30 per cent of the combined ventilation, heating and cooling energy use (Brandemuehl and Braun, 1999). The total volume of outdoor air circulated through a building during working hours can be reduced by 30–50 per cent (Schell et al, 1998).

DCV generally makes use of CO_2 sensors to determine the required ventilation rate in a given space, with CO_2 used as a proxy for overall air quality and hence of the need for ventilation. CO_2 sensors and various control strategies are discussed in some detail by Schell et al (1998) and by Schell and Int-Hout (2001). As discussed by Roth et al (2003a), there are still only a few control systems that can accept a CO_2 sensor output as in input signal, and CO_2 sensors require a skilled installer in order to work properly. Installed cost is about $400–500 each, with a payback through energy cost savings of 2–3 years (Roth et al, 2002a). Ducarme et al (1998) experimented with the use of IR sensors to control ventilation rates. Where non-human sources of contaminants are of concern, the ventilation rate can be controlled based on a combination of CO_2 and a second, non-occupant-related indoor pollutant (Chao and Hu, 2004).

7.4.7 Task/ambient conditioning

Because DV systems normally rely on floor-based ducts and diffusers, they readily lend themselves to *task/ambient conditioning*, in which air is supplied directly at desktop through flexible ducts that are built into the furniture and connected to the floor diffusers. The desktop units are referred to as *personal environmental modules* (PEM). Task/ambient conditioning is analogous to task/ambient lighting (Chapter 9, Section 9.5.2), and reduces the required ventilation flow while permitting a warmer supply temperature. In the S.C. Johnson Wax commercial products headquarters in Racine, Wisconsin, task/ambient conditioning resulted in a substantial first-cost saving (Mendler and Odell, 2000). Assuming that airflow to the PEM is turned off when not needed, task/ambient conditioning represents a form of demand-controlled ventilation – but without CO_2 or infrared sensors.

7.4.8 Floor radiant cooling

Chilled floors can also be used for cooling purposes, and are particularly effective when there is direct sunlight on the floor. This is often the case in atria, entrance halls and airports. Simmonds et al (2000) analyse the design of a 200,000m² radiant floor cooling system for the concourse of the new Bangkok International Airport. The floor absorbs sunlight penetrating through the atrium and maintains a thermal stratification, with the warmest air in the upper part of the concourse. The peak cooling load is 97W/m², of which 80W/m² are handled by the floor and 17W/m² by the displacement ventilation system, which supplies 4ACH (air changes per hour) of fresh air at 18°C. Of the 80W/m² of heat removed by the floor, 50W/m² are removed through absorption of short- and long-wave radiation and 30W/m² are removed through convective and conductive heat exchange. At full load, the inlet water temperature to the floor is 13°C, the outlet temperature is 19°C, the floor surface temperature is 21°C, and the air temperature at a 1-metre height is 24°C. Compared to the original concept, the energy savings is 31 per cent.

Leigh et al (2005) have investigated the implications for energy use of using the radiant floor heating systems, common in Korea, for summer cooling. Based on laboratory and computer simulation studies, they find that radiant floor cooling combined with dehumidification only when needed to prevent condensation on the floor reduces cooling + ventilation energy use (including the floor pump) by about 67 per cent compared to using an air conditioner (COP = 3.0). In the radiant cooling system, a chiller (COP = 3.0) produces chilled water at 15°C. This is sent directly to a cooling coil in the fresh air intake when dehumidification is needed, but mixed with return water from the radiant floor so as to maintain a floor temperature of 19°C. The large savings between the two systems is due in part to the much greater (but unnecessary) dehumidification done by the air conditioner.

7.4.9 Sensible heat exchangers

Sensible heat exchangers transfer some of the heat or coldness in the outgoing exhaust air to the incoming fresh air. In so doing, they save energy and reduce peak heating and cooling loads, thereby allowing downsizing of heating and cooling equipment, with cost savings that offset part of the cost of the heat exchanger. Not all HVAC designers account for this cost saving when assessing the cost-effectiveness of heat exchangers (Roth et al, 2002a).

Four kinds of sensible heat exchangers are commonly available (Smith, 1997; ASHRAE, 2001b, Chapter 44). Energy performance is characterized in terms of the *effectiveness* (the amount by which incoming air is warmed or cooled as a fraction of the difference in the temperature of the two air streams entering the heat exchanger), while energy cost depends on the pressure drop across the heat exchanger, which depends on the nature of the heat exchanger and on the flow velocity. The latter depends in part on the cross section of the heat exchanger, as a given volumetric flow requires a smaller flow velocity the larger the cross section. Indicative pressure drops are given below.

- *Heat wheel.* A heat-storage material gains heat from the warm air stream, then rotates into the cold air stream, where heat is given off. The material can be impregnated with a desiccant to transfer moisture as well, in which case it is referred to as an *enthalpy wheel* or *total energy wheel*. Effectiveness ranges from 55 to 85 per cent, with a pressure drop of 60–250Pa.

- *Plate type air-to-air heat exchanger.* In most designs, the warm air flows in one direction and the cold air flows in the opposite direction (a counter-flow design). This type of heat exchanger is lighter but more voluminous than the heat wheel, with an effectiveness of 55–85 per cent and a pressure drop of 15–370Pa.

- *Heat pipe.* This uses a refrigerant inside a closed pipe to transfer heat. The refrigerant evaporates in the portion of the pipe traversing the warm air stream, flows to the portion of the pipe traversing the cold air stream, and condenses, releasing heat that is transferred to the cold air stream. The effectiveness is 45 to 75 per cent, with a pressure drop of 100–500Pa.

- *Roundabout system.* Glycol (antifreeze) circulates through a pipe in a grill that picks up heat from the outgoing air, then flows to another grill in the incoming air stream, which it warms. The effectiveness is 55 to 75 per cent, with a pressure drop of 100–500Pa. This system can be used when the supply and exhaust plenums are not physically close.

The presence of a heat exchanger increases the fan energy needed to circulate the air by an amount proportional to the pressure drop times the rate of flow, while the energy saved depends on the amount of heat captured and will be small when the temperature difference between incoming and outgoing air is small. There will be some minimal temperature difference below which more energy is used than is saved. The analysis presented in Box 7.2 indicates that the temperature difference at which the heat exchanger should be bypassed is of the order of 1–2K for a heat exchanger with an effectiveness of 0.85. Bypassing the heat exchanger will significantly reduce the fan energy use only if the fan has a variable speed drive (Figure 7.6). Beyond some effectiveness, designing a heat exchanger to capture more of the available heat will not reduce overall energy consumption when the additional required fan energy is taken into account. Even under circumstances where there is no net reduction in annual cooling + ventilation energy use through the use of heat exchangers, a 40 per cent reduction in peak cooling load can occur (Kavanaugh and Xie, 2000), thereby reducing the required size and cost of the chillers and reducing electric utility demand charges. Sensible heat exchangers were once used primarily in extreme climates or in moderate climates where HVAC systems supply 100 per cent outside air (such as in hospitals and laboratories), but are now common in new large commercial buildings worldwide. In addition, packaged rooftop air conditioners are available with air-to-air heat exchangers; units equipped with heat exchangers can serve a larger floorspace than units of the same size without heat exchangers, such that there is little difference in the purchase cost per unit of floorspace served (Roth et al, 2002a, section 4.5.3). The trade-off between increasing heat exchanger effectiveness and increasing fan energy use should be evaluated before choosing a heat exchanger.

In regions with moderately cold winters, heat exchangers will be more effective during the heating season than during the cooling season, as the temperature difference between incoming and outgoing air will be larger in winter. The effectiveness of heat exchangers in reducing energy use during the cooling seasons is further diminished by the fact that some of the additional fan energy necessitated by the heat exchanger appears as waste heat that needs to be removed by the chiller, whereas during the winter some of it is a useful internal heat gain. The effectiveness of heat pipe exchangers during the cooling season can be enhanced by spraying the pipes with water on the exhaust side, causing evaporative cooling. This will cool the pipes and hence the circulating fluid, thereby allowing greater cooling of the incoming air. This can give a 20 per cent saving in cooling energy use (Tuluca, 1997), but is viable only where the water supply is adequate. It is most effective where atmospheric humidity is low.

A major shortcoming of heat wheels is that they transfer contaminants from the exhaust air to the incoming air, unlike other types of heat exchangers. To minimize this problem, the portion of the wheel that has just passed through the exhaust air stream can be purged with fresh air before it enters the incoming air stream. In effect, the incoming air stream passing through this part of the heat wheel is diverted into the exhaust air stream. Unfortunately, some of the heat carried from the exhaust air stream is lost in this way. The problem of cross-contamination between outgoing and incoming air streams is more severe in enthalpy wheels, as the rough hygroscopic material is more likely to catch contaminants. The solution is to minimize or eliminate indoor sources of contaminants in the first place.

Exhaust-air heat pumps can be used in combination with or in place of sensible heat exchangers, as discussed in Chapter 5 (Section 5.4).

7.4.10 Use of heat pipes to reduce the amount of energy used for dehumidification

As previously noted, humidity (latent heat) is removed from ventilation air in a conventional HVAC system by lowering the temperature enough to condense sufficient moisture, then *reheating* the air. This requires a lower evaporator temperature than if only sensible heat were removed, thereby reducing the chiller COP, and requires additional energy to reheat the air. The energy used for initial cooling of the outdoor air and for subsequent reheating can be reduced if a heat pipe is wrapped around the evaporator coil of an air conditioner, with the lower end (where evaporation of the refrigerant occurs) placed in the warm air stream in front of the evaporator coil, and the high end (where condensation of the refrigerant occurs) placed in the cold air stream behind the evaporator coil. In this way, the air stream is precooled before passing through the evaporator coil and reheated after it has passed through the coil. A vendor of wrap-around heat pipes for air conditioners reports energy savings of 15–30 per cent in Georgia and Florida (Heat Pipe Technology, www.heatpipe.com). In addition, dehumidification capacity can be increased by 25–50 per cent. However, the temperature of the evaporator coil is governed by the required water vapour mixing ratio of the air after passing through the coil, and so cannot be significantly increased, so the COP penalty largely remains. The preferred procedure is to avoid the need for over-cooling altogether, using either solid or liquid desiccants, as discussed next.

7.4.11 Dehumidification and cooling systems using solid desiccants

The use of a solid desiccant wheel in combination with a sensible heat exchange wheel and an evaporative cooler for dehumidification and cooling was introduced in Chapter 6 (Section 6.6.4). The solid desiccant wheel discussed in Chapter 6 is an *active* desiccant wheel, as it makes use of an external heat source (usually applied to the outgoing air stream) to regenerate the wheel.

In a *passive* desiccant wheel, the dryness of the outgoing airflow is the force driving moisture from the desiccant. Heat is also transferred between the incoming and outgoing air streams, so a passive desiccant wheel is also referred to as a *total energy wheel* or *enthalpy wheel*. An enthalpy wheel is thus used for simultaneously drying and cooling or warming and humidifying the incoming air stream (enthalpy is defined in Equation 6.5).

The efficiency of a passive desiccant wheel – that is, the change in enthalpy as the outdoor air passes through the wheel, as a fraction of the difference between indoor and outdoor enthalpies – is usually referred to as the *effectiveness* of the wheel. One can compute a sensible heat and latent heat effectiveness, as well as the effectiveness for total heat or enthalpy. Sensible heat effectiveness is typically around 70–75 per cent, varying only weakly with airflow rate and outdoor temperature. However, the latent heat effectiveness depends strongly on the flow rate and outdoor humidity, varying from 45 per cent to 90 per cent (Simonson et al, 1999). Optimal use of enthalpy wheels in HVAC systems requires effectiveness data appropriate to the range of climatic conditions that will be encountered at a particular site. Calculations by Asiedu et al (2004) indicate that, for the climate of Chicago, passive enthalpy wheels have a payback time of less than one year in new buildings and less than two years in retrofit situations, compared to no heat recovery. These calculations should be regarded only as indicative of the economic benefits of enthalpy wheels, as the economics will be site-specific.

The performance of an active desiccant system is given as its COP – the ratio of the decrease in enthalpy of the incoming air stream to the thermal energy used to regenerate the desiccant. It depends on the effectiveness of the desiccant wheel as well as on the effectiveness of the sensible heat wheel and of the evaporative cooler, as discussed in Chapter 6. An active desiccant wheel, unlike a passive wheel, can drive the humidity of the incoming air to below that of the outgoing air.

Alternative system configurations

We have discussed three kinds of rotating wheels: a sensible heat wheel, which lacks a desiccant and so transfers only sensible heat; a passive desiccant wheel, which transfers both sensible and latent heat and will henceforth be referred

Figure 7.29 Desiccant-based dehumidification systems using alternative arrangements of an enthalpy wheel and/or desiccant wheel and/or sensible heat wheel or heat pipe

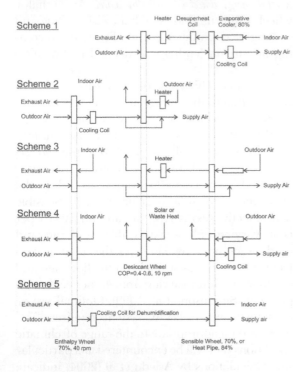

Note: To facilitate easy intercomparison of the different schemes, a given component in different schemes is always drawn at the same horizontal position. The percentage given for enthalpy and sensible heat wheels and for evaporative coolers is the device effectiveness, as defined in the text. Typical rotation rates for enthalpy and desiccant wheels are also given.

to as an enthalpy wheel; and an active desiccant wheel, which removes moisture while adding heat to the incoming air stream, and will henceforth be referred to simply as a desiccant wheel. Sensible heat, desiccant and enthalpy wheels can be combined in various ways, as illustrated in Figure 7.29. Scheme 1 involves a desiccant wheel, a sensible heat wheel, and an evaporative cooler. The ways in which the properties of these components interact to give the overall COP of the system is explained in Chapter 6 (Box 6.5). The sensible heat wheel in Schemes 1 and 3 recycles some of the heat used to regenerate the desiccant wheel back to the regeneration air stream, thereby improving performance. Scheme 1 involves evaporative cooling of indoor air before it passes through a sensible heat exchanger, while Schemes 3 and 4 involve evaporative cooling of

outdoor air and so will be less applicable in hot-humid climates. Schemes 2 and 4 rely on heat completely external to the cycle for regeneration of the desiccant (none of the regeneration heat is recycled); in the case of Scheme 2, this is because the outdoor air is precooled with a chiller prior to passing through the desiccant wheel, rather then after using a sensible heat wheel, while in Scheme 4, this is because solar or waste heat is assumed to be available in excess (the solar collector might be sized for winter heating demand and therefore in excess of summer requirements).

Scheme 1 (without the desuperheat coil) is the standard desiccant dehumidification scheme in use today, and has been analysed by (among others) Vineyard et al (2002), Miller et al (2002b), and Jalalzadeh-Azar et al (2000). To avoid transferring any moisture from the outgoing to the incoming air stream, a heat pipe can be used instead of a sensible heat wheel. The final cooling of the incoming air stream is accomplished with a chiller (through the cooling coil), although direct evaporative cooling could be used if the air passing through the desiccant wheel is sufficiently dry (this might require a higher regeneration temperature). Some heat rejected by the chiller can be supplied to the outgoing air stream through a heat exchange coil (a desuperheat coil) between the compressor and condenser. Like the sensible heat wheel, the desuperheat coil reduces the amount of the heat that must be added prior to regeneration of the desiccant. The desuperheater can provide only 3 per cent of the required regeneration energy, but it also improves the COP of the chiller (Miller et al, 2002b).

Table 7.9 compares the electricity and natural gas use as computed by Mazzei et al (2002) for a commercial building in Rome using a traditional vapour compression chiller with gas reheat and using a desiccant system similar to Scheme 1, with and without evaporative cooling.[7] Electricity energy use is converted to primary energy assuming electricity generation efficiencies of 35 per cent and 55 per cent. Desiccant dehumidification reduces electricity use by shifting some of the cooling load to the desiccant wheel and because the chiller COP increases from 3.0 to 4.0, but it increases natural gas use. On-site energy use increases by 7 per cent without evaporative cooling but decreases by 12 per cent with

Table 7.9 Energy use for alternative dehumidification and cooling systems

	Traditional system	Desiccant system	Desiccant system with evaporative cooling
Chiller COP	3.0	4.0	4.0
Electricity use (kWh)	17,762	12,465	9978
Electricity use (GJ)	63.94	44.84	35.92
Natural gas use (GJ)	32.95	58.51	50.44
Total on-site energy use (GJ)	96.89	103.40	85.36
PE (GJ) assuming electricity at 35% efficiency	182.69	128.21	102.63
PE (GJ) assuming electricity at 55% efficiency	127.89	89.75	71.84
Total PE (GJ) at 35% efficiency	215.65	186.72	153.07
Total PE (GJ) at 55% efficiency	160.84	148.25	122.28
Per cent savings at 35% efficiency		13.4	29.0
Per cent savings at 55% efficiency		7.8	24.0
Per cent savings if solar heat replaces natural gas			
35% efficiency		40.5	52.4
55% efficiency		44.2	55.3

Note: The desiccant system is based on Scheme 1 of Figure 7.29. Results are given assuming efficiencies for electricity generation of 35% and 55%. PE = primary energy.
Source: Based on data in Mazzei et al (2002)

evaporative cooling, while primary energy use decreases by 25–30 per cent with evaporative cooling, depending on the efficiency of generating electricity. If solar heat can be used entirely in place of natural gas for regeneration of the desiccant wheel, primary energy use decreases by 50–55 per cent compared to the traditional system. If water is limited (so that there is no evaporative cooling) but solar energy is available (such as in hot desert climates), the desiccant system reduces primary energy use by 40–45 per cent.

Scheme 2 involves an enthalpy wheel, cooling coil and a desiccant wheel. The incoming air is first cooled and dehumidified by the enthalpy wheel, then it is cooled further if needed with a cooling coil before passing through the desiccant wheel. This is in contrast to other systems, where the cooling coil is after the desiccant

wheel. As a result, the air entering the desiccant wheel is near saturation, so the performance of the desiccant wheel is enhanced while permitting moderate regeneration temperatures. According to an analysis by Fischer et al (2002), energy used for dehumidification is reduced by 45 per cent compared to a conventional overcooling/reheat scheme, or by 75 per cent if the heat required for regeneration (most of which can be supplied as solar heat) is ignored.

Scheme 3 involves an enthalpy, desiccant and sensible heat wheel, as well as an option to bypass the desiccant and sensible heat wheels when the enthalpy wheel alone provides sufficient dehumidification. This scheme has been compared with a conventional scheme by Wong et al (2002a, 2002b), and the computed primary energy use for variants on both schemes for a

Table 7.10 Comparison of relative annual energy use for dehumidification of outdoor air in Atlanta, Georgia (US), using a conventional vapour-compression chiller and using active desiccant dehumidification Scheme 3 of Figure 7.29

	Without enthalpy wheel	With enthalpy wheel
Vapour-compression		
Electric reheat	100	88
Gas reheat	64	52
Free reheat	39	26
Desiccant wheel		
No bypass	94	78
Optional bypass	38	11

Source: Wong et al (2002a)

specific cooling load profile is given in Table 7.10. A standard desiccant system (i.e. without the enthalpy wheel and optional bypass) reduces the total on-site energy use by 6 per cent. Conversely, a desiccant system with an enthalpy wheel as the first step and an optional bypass reduces energy use by almost 90 per cent. These results are for a supply air temperature of 26°C, which is rather warm. A lower supply air temperature can be achieved by placing a cooling coil after the enthalpy wheel, as in Scheme 5 (described below).

Scheme 4 is similar to Scheme 3, except that waste or solar heat rather than heated outside air is used to regenerate the desiccant wheel. Popovic et al (2002) analysed a variant of this scheme in which waste heat from an absorption chiller is used (absorption chillers are discussed in Chapter 6, Section 6.6.1).

Scheme 5 is the only scheme using an enthalpy wheel without a desiccant wheel. When the enthalpy wheel does not provide sufficient dehumidification, a cooling coil is used for additional removal of moisture. This requires reheating the air after sufficient moisture has been removed, which is accomplished by extracting heat from the outgoing air with a sensible heat wheel or heat pipe. Mumma and Shank (2001) have analysed this scheme as an all-air cooling system with no recirculation of return air. Using weather data for Atlanta, and assuming that the air is conditioned to a temperature of 13°C and a dewpoint of 7°C, this scheme reduces annual air conditioning energy use by 44 per cent compared

to a conventional CAV cooling/dehumidification/reheat system. Peak chiller power is reduced by 55 per cent. Compared to a conventional VAV system with partial recirculation of indoor air, the energy saving is only 15 per cent but peak chiller load is still reduced by 50 per cent.

Desiccant dehumidification operates best when the ventilation air passes through the building once and then is vented to the outside, without partial recirculation. With 100 per cent outside air, the airflow should be limited to that needed for ventilation purposes only, which in turn implies that a chilled water system is used to provide the remainder of the required cooling. If chilled ceiling cooling is combined with desiccant dehumidification, one is free to use a relatively warm ventilation air supply temperature without requiring energy wasting reheating. For example, Niu et al (2001) simulated a case with 24°C air supplied at floor level at a velocity of 0.1m/s, a 25°C room air temperature, and a 20.3°C chilled ceiling temperature. The resulting perceived temperature at head level is a completely acceptable 23.5°C.

Niu et al (2002) calculated the energy used for cooling and dehumidifying the intake air for a typical office building in the hot and humid climate of Hong Kong using four different HVAC systems: (i) a constant volume all-air system with an 80:20 mixture of recirculated and fresh air supplied to the chiller for dehumidification by over-cooling, followed by partial reheating; (ii) an all-air system as in (i), except that incoming air is first partly cooled and dehumidified by the cool and dry exhaust using either an energy wheel or a membrane system; (iii) chilled ceiling cooling with treatment of the ventilation air as in (ii); and (iv) a DV/CC/desiccant cooling system, as in Scheme 1 (without the desuperheat coil). In the latter case, a single cooling loop is first used to cool the ceiling, then to cool the incoming air further after it has passed through the sensible heat wheel. In cases (i)–(iii), ventilation air must be cooled to 12°C in order to reduce the mixing ratio to 0.0872kg/kg. It is reheated to 14°C for cases (i) and (ii), and to 17°C in case (iii). In case (iv), all of the required moisture is directly removed by the desiccant wheel, so ventilation air is directly cooled to 18°C. The chiller operates with a higher COP (4.39 instead of 3.31) because the evaporator is at a temperature of 15°C rather than 5°C. On the other hand, heat energy is required to regenerate the desiccant wheel (by driving off moisture) after it passes through the intake air stream, and the required fan pressure increases from 1400Pa in the all-air system to 1600Pa in the desiccant cooling system. The net result is that case (iv) requires only 70 per cent of the energy used for case (i), or only 50 per cent if the desiccant regeneration energy (which can be largely provided with solar heat) is neglected. As well, the desiccant system assumes use of 100 per cent outside air for ventilation, so contaminants from one part of the building will not be spread to another.

Simonson et al (2000a, 2000b) have analysed the part-load operation of enthalpy wheels. The control of enthalpy wheels is complicated because they simultaneously transfer heat and moisture, but the heat and moisture transfer rates cannot be independently controlled. Since the supply air temperature must be less than the room air temperature (which is equal to the exhaust-air temperature), there are temperature conditions when substantial recovery of the sensible heat in the exhaust air would make the incoming air stream too warm, when it would otherwise be cool enough to avoid the need for mechanical chilling. Partial or even no heat recovery is then called for, but this has an impact on the exchange of moisture. The scheduling of full heat recovery, partial heat recovery, or no heat recovery is a somewhat complicated optimization problem. The parameters that can be altered are the rate of supply of outdoor air, the proportion of recirculated air (which would be fixed at zero in a system with 100 per cent outside air), the fraction of outside air that bypasses the enthalpy wheel, and the rate of rotation of the enthalpy wheel. When both sensible and latent loads need to be met, the decision as to whether to give priority to sensible or to latent loads depends on the sensible:latent heat removal ratio of the chiller. Thus, proper analysis of the building and the climatic conditions is required, along with a suitably programmed control algorithm and effective controls in order to maximize the energy savings.

The evaporative coolers shown in Schemes 1, 3 and 4 are indirect evaporative coolers, in that a secondary air stream (exhaust or outside air) is cooled first, then used to cool the supply air through a heat exchanger. If the supply air is dry enough after the desiccant step, it can be cooled further through direct evaporation. In effect, we have an indirect/direct evaporative cooler (as discussed in Chapter 6, Section 6.3.7) supplemented by heat-driven desiccant dehumidification. This approach is considered by Andersson and Lindholm (2001) and Henning et al (2001).

Desiccant dehumidification and evaporative coolers are two relatively new technologies with which many HVAC engineers are not familiar. As a result, they are reluctant to specify this equipment and, when it is specified, operational problems have arisen that can be traced to human errors arising from lack of familiarity with the equipment. An example is provided by Miller et al (2002b), where a desiccant based system was installed in a Florida school with a ventilation system oversized by a factor of 2.5, so it was able to deliver the required cooling and dehumidification in spite of a number of problems that reduced the expected energy savings. These problems would have been undetected if this example had not been part of a carefully monitored test case. Hence, thorough commissioning of the equipment and monitoring of its energy use is needed after installation to ensure that it has been installed properly and that the controls operate as designed.

7.4.12 Dehumidification and cooling systems using liquid desiccants

Liquid desiccant devices were introduced in Chapter 6 (Section 6.6.4). Here, we consider a hybrid vapour compression/liquid desiccant system with indirect evaporative cooling, illustrated in Figure 7.30. The evaporatively cooled outside air is first used to cool the process air after it has left the dehumidifier, then is used to extract heat from the condenser of the chiller that is used for further cooling of the process air. This allows a lower condenser temperature and hence a higher chiller COP. Since over-cooling for dehumidification is not needed, the evaporator need not be as cold, further improving the chiller COP. Since the liquid desiccant should be as cool as possible to absorb moisture from the supply air, and as warm as possible when it is regenerated, heat

Figure 7.30 Flow diagram of a liquid desiccant dehumidification and cooling system integrated with an evaporative cooler and a vapour compression chiller. Arrows in the heat exchangers indicate the direction of heat flow

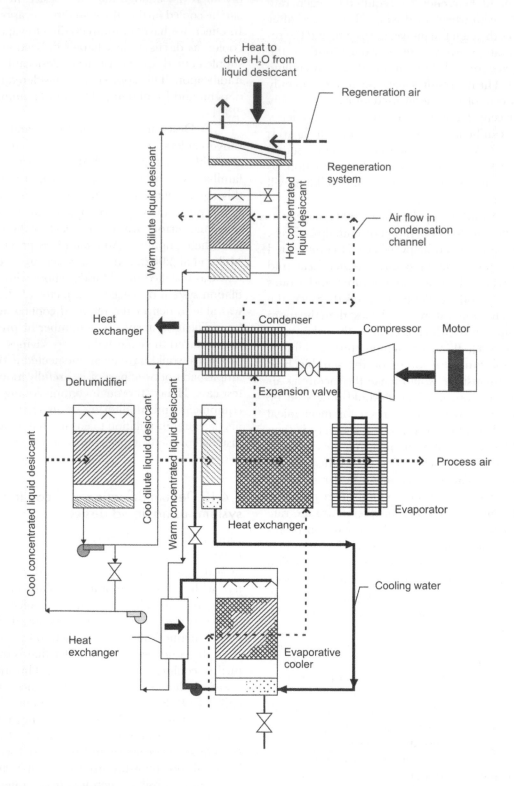

Source: Based on Dai et al (2001)

exchangers are used to cool the desiccant on its way to the dehumidifier and to warm the desiccant on its way to the generator. This reduces the heat input required for regeneration.

Performance

Table 7.11 compares the energy use of the hybrid liquid desiccant/vapour compression system shown in Figure 7.30, for the same system without evaporative cooling, and for a pure vapour compression system, as measured by Dai et al (2001). Results are for outdoor air at 35°C and 40 per cent RH and conditioned air at 18°C and 60 per cent RH. In all cases, the total cooling load (removing both sensible and latent heat) is 5.07kW. For the vapour compression system, this load is removed entirely by the evaporator coil, with a COP of 2.36. For the liquid desiccant systems, the latent load is removed by the desiccant, reducing the evaporator coil load by 34 per cent. For a fixed compressor COP, the electricity use would fall by the same amount but the total load (latent + sensible) is the same, so the system COP based on electricity use would increase by 52 per cent. In reality, there is a slight change in the compressor COP when its load changes, and pumps are required for the liquid desiccant system, with the net result that system COP increases by 49 per cent (from 2.28 to 3.39). When evaporative cooling is added, the COP increases by 67 per cent (to 3.81). This represents a reduction in electricity use by 40 per cent, but the thermal energy

requirement offsets this, giving almost no net change in on-site energy use. However, primary energy use decreases by 14 or 23 per cent (for electricity generation times transmission efficiencies of 50 per cent or 33 per cent, and a boiler efficiency of 85 per cent). The thermal energy input to regenerate the desiccant is at a required temperature of only 56°C, meaning that waste heat or solar thermal energy could be readily used. The load handled by the desiccant component (1.7kW) requires a thermal energy input of about 1.0kW, implying a thermal COP of 1.7 (a thermal COP > 1.0 is possible because of the partial recovery of heat from the hot concentrated desiccant stream and its transfer to the cool dilute desiccant stream; this also causes the sum of evaporator coil heat removal, evaporative cooling and regeneration energy input to be less than the cooling load). Although the baseline chiller COP is small in this case, comparable relative improvements in COP might be possible for larger, more efficient chillers.

Gasparella et al (2005) have evaluated the primary energy needed for heating and cooling two 11-storey towers in northern Italy using a conventional system (natural gas boiler for heating, electric chiller for cooling and dehumidification) with (a) a ground source heat pump (GSHP) for heating, cooling and dehumidification, and (b) a GSHP combined with a liquid desiccant system for summer dehumidification. The liquid desiccant is recharged with heat from a natural gas

Table 7.11 Comparison of vapour compression and liquid desiccant/vapour compression systems for dehumidification and cooling

	Vapour compression System	Liquid desiccant system	
		Without evaporative cooling	With evaporative cooling
Cooling load (kW$_c$)	5.071	5.071	5.071
Heat removal by evaporator coil (kW$_c$)	5.071	3.337	2.953
Heat removal by evaporative cooling (kW$_c$)			0.301
Heat removal by desiccant component (kW$_c$)		1.734	1.817
Electrical input (kW)			
Compressor	2.150	1.392	1.227
Fan	0.070	0.070	0.070
Pump		0.035	0.035
Thermal input		1.008	1.008
Compressor COP	2.36	2.40	2.41
System COP based on electrical input	2.28	3.39	3.81

Note: Outdoor air is at 35°C and 40% RH, while the conditioned air is at 18°C and 60% RH. The liquid desiccant is regenerated at 56°C.
Source: Dai et al (2001)

boiler, which is available to provide some of the winter heating load. Assuming an electricity generation efficiency of 0.33, the saving in primary energy compared to the conventional system is 27 per cent with the GSHP alone, and 43 per cent with the GSHP + liquid desiccant. The size and cost of the ground coil are reduced in half in the latter case.

7.4.13 Membrane sensible and latent heat exchangers

Zhang et al (2000) and Zhang and Niu (2001) present a theoretical analysis of an integrated sensible and latent heat exchanger that uses a membrane that is permeable to water vapour but not to other gases. The fraction of the available latent heat in the incoming air (i.e. the moisture content

in excess of that of the outgoing air) that can be removed ranges from 20 per cent at an RH of 10 per cent (i.e. when both air streams are quite dry to begin with) to 60 per cent at an incoming air RH of 90 per cent. Using meteorological data for Hong Kong, the latent heat removal effectiveness would fluctuate between 54 and 62 per cent during the course of a year, while the sensible heat removal effectiveness would be steady at 67 per cent. The savings in total energy use for cooling and drying incoming fresh air is calculated to be 58 per cent for Hong Kong's hot and humid climate and 24 per cent for the less humid climate of Beijing.

Figure 7.31 shows the flow diagram and temperature mixing ratio variation for the DV/CC/desiccant cooling system in which the incoming fresh air is partially cooled and dehumidified with a membrane heat exchanger prior to passing through the desiccant wheel. The temperature changes in passing through the cooling coil and gas heater are comparable for this system and Scheme 1 in Figure 7.29. Compared to Scheme 1, which lacks a membrane heat exchanger, the energy saving is 10 per cent (for Hong Kong weather data). More important is the fact that a lower regeneration temperature can be used with a membrane heat exchanger. Membrane heat exchangers have been used in Hong Kong schools, but the combination with chilled ceiling cooling has not been reported.

7.4.14 Cogeneration, desiccant dehumidification and radiant-slab cooling

Casas and Schmitz (2005) simulated the heating and cooling energy use for a small office building in Germany, using either a conventional HVAC system or a system consisting of on-site cogeneration with the waste heat used to regenerate a solid desiccant wheel for cooling and dehumidification of outside air. Both the conventional and desiccant system use radiant floor cooling to satisfy the internal sensible cooling load. The desiccant system is similar to Scheme 1 of Figure 7.29, except that there is no evaporative cooler or desuperheat coil. The radiant floor is chilled with ground coldness that is obtained by circulating pure water through a series of eight 100m deep boreholes. The water is chilled to about 18°C. In the conventional

Figure 7.31 (a) Schematic flow diagram and (b) T-q variation for HVAC system 2 of Zhang and Niu (2003a)

(a)

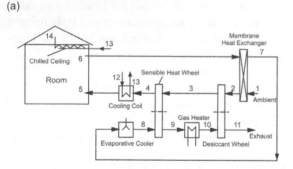

(b)

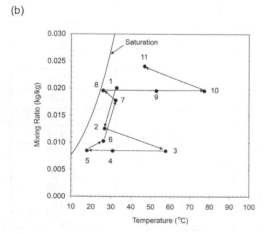

Source: Zhang and Niu (2003)

system, outdoor air is cooled to 10°C for dehumidification, then reheated to 15°C. The chiller is assumed to have a COP of 3.0, which can be obtained assuming an evaporator at 5°C, a condenser at 50°C and a Carnot efficiency of 0.485 (see Chapter 5, Equation 5.3). The annual load for overcooling and dehumidifying outside air is $8962kWh_{th}$, while $2464kWh_{th}$ of heat are used for reheating the overcooled air (so the net change in enthalpy is $8962 - 2464 = 6498kWh_{th}$). In the desiccant based system, the annual heat required for regenerating the desiccant is $5309kWh_{th}$, while the annual load for directly cooling outside air to 15°C is $2723kWh_c$. Thus, the desiccant system provides $6498 - 2723 = 3775kWh_{th}$ of latent cooling, implying a thermal COP of 0.71. The radiant floor satisfies a sensible cooling load of $15200kWh_{th}$. Casas (personal communication, 2005) estimates a pump power draw of 704W while delivering 29.6kW of cooling, for an effective COP of 42. The energy used to pump water through the boreholes depends on the pressure drop (which depends on the distance to the boreholes, and is minimized by the use of pure water) and the flow rate. As a precaution, I shall assume a COP of 20. Casas and Schmitz (2005) make the following additional assumptions: the combined heat and power (CHP) plant has an electricity generation efficiency $\eta_e = 0.24$ and a thermal efficiency (based on the useful heat output) of $\eta_{th} = 0.64$, the central power plant efficiency η_{pp} is 0.4, and the boiler efficiency η_b is 0.9.

The above loads and assumptions will be used here to compute the primary energy required for the conventional system (replicating the results of Casas and Schmitz) and for a series of different desiccant systems based on the system modelled by Casas and Schmitz (2005). The primary energy required by the conventional system is:

$$E = \frac{2464}{\eta_b} + \frac{8962}{COP_{oa}\eta_{pp}} + \frac{15200}{COP_{rf}\eta_{pp}} \quad (7.12)$$

We consider three variants of the conventional system: (i) where the same chiller is used to cool outdoor air and the radiant floor, such that $COP_{oa} = COP_{rf} = 3.0$; (ii) where a separate chiller is used for the radiant floor, with an

evaporator at 13°C rather than 5°C, giving $COP_{rf} = 3.75$ (this higher COP is derived assuming the same Carnot efficiency and condenser temperature as before); and (iii) using the ground loop, giving $COP_{rf} = 20$.

The desiccant/CHP system uses heat from the CHP plant to regenerate the desiccant. The electricity produced displaces centrally generated electricity, and is credited against the primary energy used by the desiccant system. The net primary energy use is thus

$$E = \frac{5309}{\eta_{th}}\left(1 - \frac{\eta_e}{\eta_{pp}}\right) + \frac{2723}{COP_{oa}\eta_{pp}}$$

$$+ \frac{15200}{COP_{rf}\eta_{pp}} \quad (7.13)$$

As in the conventional system, we consider three variants. For the first variant, $COP_{oa} = COP_{rf} = 3.43$ (rather than 3.0, because an evaporator temperature 10°C is assumed). For the second and third variants, $COP_{rf} = 3.75$ and 20, respectively.

The final set of systems, desiccant/boiler, uses heat from a boiler (at 90 per cent efficiency) to regenerate the desiccant. The same three variants for the COPs as in the desiccant/CHP system are considered. The primary energy requirement is:

$$E = \frac{5309}{\eta_b} + \frac{2723}{COP_{oa}\eta_{pp}} + \frac{15200}{COP_{rf}\eta_{pp}} \quad (7.14)$$

In addition to considering systems with a base chiller COP of 3.0, we also consider systems with a base chiller COP of 6.0 (when outdoor air is overcooled to 10°C), which can be obtained assuming a water-cooled chiller with a condenser temperature of 30°C and a Carnot efficiency of 0.539 (instead of 0.485). A cooling tower is assumed, and its electricity use must be accounted for. Based on Table 6.14, it is assumed that the electricity required to operate the cooling tower fans and pumps is 0.022 times the cooling provided.

The results are given in Table 7.12.

Using as a base case, the conventional system with the same chiller to cool outside air and the radiant floor, the desiccant/CHP system with ground loop reduces primary energy use by almost 70 per cent with a low-efficiency chiller in both systems, and by 56 per cent if both systems have a high-efficiency chiller. Even relative to a base case with the ground loop, the saving is 41 per cent. The ground loop alone saves 47 per cent relative to a conventional system with chiller COP of 3.0, and 31 per cent for a chiller COP of 6.0. The desiccant wheel affects energy used to treat outdoor air, and it can be seen that, for a conventional system with a chiller COP of 3.0,

the desiccant/CHP and desiccant boiler systems reduce the primary energy used to condition outdoor air by about 50 per cent and 25 per cent, respectively. Relative to a conventional system with a chiller COP of 6.0, however, there is a negligible saving in treating outside air using the desiccant/boiler system. Schmitz and Casas (2001) estimate that energy cost savings for a small office in Berlin (assuming electricity at $0.11/kWh, demand charges of $150/kW/yr, and natural gas at $0.03/kWh or $8.3/GJ) would pay back the additional cost of Desiccant/CHP-1 compared to a conventional system in approximately four years.

Table 7.12. Comparison of primary energy (kWh) used by alternative systems for conditioning ventilation air and by alternative systems for chilling water to 18°C for use in a chilled floor

Case	Chiller COP		Cooling tower energy	Primary energy used			Savings (%) Relative to		
	Outside air	Radiant floor		Outside air	Radiant floor	Total	Conv-1	Conv-2	Conv-3
Based on low-efficiency, air-cooled chiller									
Conventional-1	3.00	3.00	0.000	10,211	12,676	22,887			
Conventional-2	3.00	3.75	0.000	10,211	10,131	20,342	11		
Conventional-3	3.00	20	0.000	10,211	1900	12,111	47	40	
Desiccant/CHP-1	3.43	3.43	0.000	5301	11,068	16,369	28		
Desiccant/CHP-2	3.43	3.75	0.000	5301	10,131	15,432	33	24	
Desiccant/CHP-3	3.43	20	0.000	5301	1900	7201	69	65	41
Desiccant/Boiler-1	3.43	3.43	0.000	7882	11,068	18,950	17		
Desiccant/Boiler-2	3.43	3.75	0.000	7882	10,131	18,013	21	11	
Desiccant/Boiler-3	3.43	20	0.000	7882	1900	9782	57	52	19
Based on high-efficiency, water cooled chiller									
Conventional-1	6.00	6.00	0.022	6967	7173	14,139			
Conventional-2	6.00	9.07	0.022	6967	5024	11,991	15		
Conventional-3	6.00	20	0.000	6967	2736	9703	31	19	
Desiccant/CHP-1	7.63	7.63	0.022	4360	5816	10,176	28		
Desiccant/CHP-2	7.63	9.07	0.022	4360	5024	9384	34	22	
Desiccant/CHP-3	7.63	20	0.000	4360	1900	6260	56	48	36
Desiccant/Boiler-1	7.63	7.63	0.022	6941	5816	12,757	10		
Desiccant/Boiler-2	7.63	9.07	0.022	6941	5024	11,965	15	0.2	
Desiccant/Boiler-3	7.63	20	0.000	6941	1900	8841	38	26	9

Note: Systems compared were: conventional HVAC, on-site cogeneration (CHP) system with waste heat used to regenerate a desiccant wheel, and using a boiler to regenerate a desiccant wheel. Systems denoted by suffixes, where 1 = using the same chiller as for conditioning outside air, 2 = using a second chiller with a higher COP obtained by using the higher permitted evaporator temperature, and 3 = using a ground loop with a COP of 20. In the conventional systems, outdoor air is overcooled to 10°C for dehumidification, then reheated to 15°C. Cooling tower energy is the energy used by the cooling tower as a fraction of the cooling load.
Source: Based on cooling and heating loads (for reheating or regeneration) as simulated by Casas and Schmitz (2005), with additional assumptions as explained in the main text

7.4.15 Insulating and sealing ducts

Leakage from ducts can reduce the airflow to conditioned spaces by up to 30 per cent in large commercial buildings (Fisk et al, 2000; Xu et al, 2002), while heat conduction through the ducts can cause significant temperature changes in the flow direction. It is therefore important to install ductwork systems that are carefully sealed and well insulated. Ducts can be sealed in situ using duct tape or mastic, but the duct air-tightness is very much dependent on workers' skills. Alternatively, a double sealing gasket can be used, as has long been the practice in Sweden (Carrié et al, 2000). A more recent innovation is aerosol duct sealing. With this technique, one first closes all the vents and HVAC equipment, then pressurizes the ductwork with a fog of small sealant particles. The air can leave only through the leaks but has to make sharp turns, which the sealant particles cannot follow. They stick to the sides of the leaks, sealing them.

In light commercial buildings, ducts are often located between the drop ceiling and the roof deck. In four out of eight buildings studied in northern California, Delp et al (1998) found that the ductwork was functionally outside the building's air and thermal barriers, so duct leakage is an issue here as well. The effective duct leakage area per unit of floor area at a pressure difference of 25Pa was found to be three times that of typical residential buildings in California ($3.4\text{cm}^2/\text{m}^2$ vs $1.3\text{cm}^2/\text{m}^2$). Each of the eight buildings had at least one of the following problems: torn and missing external duct wrap, poor workmanship around duct take-offs and fittings, disconnected ducts and improperly installed duct mastic.

Proper sealing of ducts is a labour intensive process. According to Hamilton et al (2003), the time required for the additional cost of aerosol sealing to be recouped through energy-cost savings in the US is about 10 years. This estimate does not include savings due to reduced peak cooling loads, which (in new construction) allows downsizing of the cooling equipment, and reduced electric utility demand charges (a fee that some utilities charge that depends on the peak power demand, and is in addition to charges for energy consumed).

Internal duct insulation tends to be porous, and airflow into the insulation degrades its thermal performance (Levinson et al, 2000). Encapsulating the surface of internal fibreglass duct insulation with an impervious layer will increase the delivery of heat or coldness by up to 1 per cent.

An obvious advantage of a hydronic heating and cooling system is that there is no leakage of the heat transport medium (water).

7.4.16 Variable flow in hydronic heating and cooling systems

The amount of heat or coldness delivered by a hydronic system depends on the flow rate and the temperature difference between the supply and return water. When ventilation and heating/cooling requirements of buildings are decoupled, as is the tendency now, the flow can be made to vary in response to the changing heating or cooling load.

In variable flow hydronic cooling systems, the system is designed to operate with a fixed temperature difference (ΔT) between the supply and return water. With fixed ΔT, the flow rate must vary in proportion to the cooling load. In most systems, ΔT falls as the load decreases, which necessitates an increase in pumping energy and/or an increase in chiller energy use in order to meet the load. This in turn increases the energy use, and in systems with multiple chillers that should be staged on and off, can result in loss of temperature control. This problem is discussed in some detail by Taylor (2002), and in many cases is linked to incorrectly sized and hence incorrectly matched system components (such as valves and control actuators). Thus, just as it is important to correctly estimate cooling loads and size the chillers in order to minimize energy use (Chapter 6, Section 6.8), so it is important to correctly size all of the components of a cooling system in order to minimize energy use under part-load conditions.

7.4.17 Laboratories and fume hoods

Teaching and research laboratories are energy intensive because of the need for 100 per cent outside air. Fume hoods are used to collect and remove fumes from research laboratories and high technology manufacturing facilities. In the US, each unit is estimated to exhaust $500–1000 per year of heated or cooled indoor air. The energy use associated with fume hoods can be reduced through:

- occupancy related variable airflow;
- redesign such that air is introduced from the fume hood with less turbulence, thereby allowing the exhaust rate to be safely reduced by a factor of four (Rosenfeld, 1999).

The *Laboratories for the 21st Century* programme of the US Environmental Protection Agency and Department of Energy provides case studies of US laboratories that have achieved on-site energy savings of 30-43 per cent, along with design guidelines (all available from www.labs21century.gov/toolkit/design_guide.htm). One lab (at Sandia National Laboratories, New Mexico) makes use of evaporative cooling and fixed sunshades on south-facing walls, in addition to VAV motors, heat recovery and an advanced (digital) control system. Enlightened occupant behaviour (especially with regard to use of fume hoods) can make a substantial difference: in one US laboratory incorporating energy-efficiency features (Pharmacia Building Q in Illinois), simulated electricity use was 38 per cent less than for the reference design, but measured electricity use was only 50 per cent of the expected use. The US case studies highlight the need to be able to lay out and size the ductwork so as to minimize the flow resistance, which is particularly important in laboratories because of the high flow rates. Of particular interest in this respect is the strategy of incorporating *interstitial floors* that house all of the ductwork between the working floors. An example where this strategy was applied is the Fred Hutchinson Cancer Research Center in Seattle, Washington. Working floors 3.2m high alternate with interstitial floors 2.24m high, for a total floor-to-floor height of 5.44m. In spite of the greater height, this design does not necessarily cost more, because of cost savings arising from the fact that construction tasks that must be done consecutively in conventional designs can be done concurrently when there are interstitial floors, because scaffolding and ladders are not needed for mechanical and electrical work, because the duct layout is simpler and with fewer different sizes, because fan motors can be down-sized (by 25–50 per cent)

due to the lower pressure drop facilitated by wide, straight ducts, and because sound attenuators can be eliminated (further reducing the pressure drop). The energy savings arise from the ability to optimize the design of the mechanical system and minimize pressure drops in the ductwork.

Enermodal Engineering (2003) analysed the impact of heat recovery and other efficiency measures on energy use in a generic laboratory in four US locations: Minneapolis (cold winter, humid summer), Denver (cold winter, dry year round), Seattle (mild winter and summer), Atlanta (mild winter, hot and humid summer). The breakdown of energy use for a reference design (with a constant air exchange rate of about 12 ACH and no heat recovery) is given in Table 7.13. Electricity by fans exceeds electricity use by chillers by a substantial margin, even in Atlanta. However, total energy use is dominated by heating requirements, ranging from one half (Atlanta) to three quarters (Minneapolis) of total energy use. At 2000–4000kWh/m²/year, energy use for the reference laboratory is five to ten times that of an air conditioned office building. Implementation of a VAV ventilation system (with airflow dropping to 6ACH during unoccupied periods or in response to reduced fume hood use) reduced fan energy use by about 20–40 per cent and fan + chiller energy use by about 20–30 per cent (chiller energy use either decreases or increases slightly, the latter due to greater part load operation). Heating energy use is reduced by 8–15 per cent. In the advanced case, space heating energy use decreases by 47 per cent (Denver) to 60 per cent (Atlanta). Implementation of heat recovery alone can increase or decrease total electricity use, depending on the trade-off between reduced chiller energy use and increased fan energy use (due to greater flow resistance), while always reducing the heating energy use. Implementation of VAV, redesign of the ductwork to reduce the static pressure drop by 45 per cent, and implementation of an enthalpy wheel with an effectiveness of 0.75 reduces HVAC electricity use by 24–44 per cent and reduces heating energy use by 62–75 per cent. Both electricity and heat energy savings are smaller using a run-around loop, but a run-around loop has the advantage that the supply and exhaust airflows do not need to enter and exit the building adjacent to one another. The last

Table 7.13 Simulated energy use (kWh/m²/year) for a generic research lab in four US cities with a fixed rate of air exchange, and savings in energy use (kWh/m²/year or %) when implementing a variable air volume (VAV) ventilation system alone, or VAV plus redesign of the ductwork to reduce the pressure drop plus heat recovery using either an enthalpy wheel or a run-around loop

Energy use	Location			
	Minneapolis	Denver	Seattle	Atlanta
Equipment	497	497	497	497
Lighting	53	53	53	53
Cooling	86	58	30	175
Pumps	33	18	14	31
Fans	235	235	235	235
Total electricity	904	861	829	990
Natural gas	2716	2498	1360	1142
Total energy	3260	3359	2188	2130
Savings with VAV system				
Fans	76.4	54.9	71.0	98.0
Cooling	5.4	1.1	−5.4	32.3
Natural gas	505	281	309	315
% Fan	33	23	30	42
% Fan+pump+cooling	27	18	23	29
% Natural gas	14	8	14	15
Savings with VAV + reduced pressure drop + enthalpy wheel				
Electricity	118	75	75	194
Natural gas	2040	1600	836	776
% Fan+pump+cooling	33	24	27	44
% Natural gas	75	64	62	68
Savings with VAV + reduced pressure drop + run-around loop				
Electricity	97	54	75	150
Natural gas	1558	1170	767	688
% Fan+pump+cooling	27	17	27	34
% Natural gas	57	47	56	60
Final energy use excl equipment, VAV + reduced pressure drop + enthalpy wheel				
Total electricity	407	364	332	494
Natural gas	675	899	524	366
Total	1082	1263	856	859

Source: Enermodel Engineering (2003)

three rows of Table 7.13 give the energy use for the case with the enthalpy wheel, excluding lab equipment. Heating energy use still constitutes almost half of the total energy use in Atlanta and two-thirds of the total in Minneapolis. Thus, it would be worthwhile to expend further effort to improve the effectiveness of heat recovery without increasing the flow resistance. The reference building has minimal insulation (wall U-value of about 0.5W/m²/K), so improved envelope U-values could be justified in terms of the absolute energy savings even though the relative savings would be small due to the large ventilation heating load.

A recently completed science building at Concordia University, Montreal, with offices, classrooms and 250 fume hoods, achieved a 45 per cent reduction in energy use relative to ASHRAE 90.1-1999 (see Chapter 13, Section 13.1.1) through the following measures (Lemire and Charneux, 2005):

- motion detectors to shut off lights after an adjustable delay when the space is unoccupied, with a signal sent to the building automation system to reduce ventilation rates;
- reduction of lab ventilation rates from 10ACH to 6ACH when unoccupied during the day and to 3ACH when unoccupied during the night;

- reduction of non-lab ventilation rates from 6ACH to 3ACH when unoccupied during the day and to zero when unoccupied during the night;
- combining fume hood exhaust ducts into a manifold;
- heat recovery between exhaust and supply air using a runaround gycol loop;
- use of a low-temperature (30–40°C) water heating loop to recover waste heat from various pieces of equipment;
- use of variable speed drives on all pumps and fans;
- modestly better electric lighting.

These measures increased the total cost of the building by 2.3 per cent ($1,356,000 out of $59,500,000) while yielding an annual energy cost savings of $854,000, for a payback time of 19 months.

7.5 Commissioning, control systems and monitoring

Commissioning is the process of systematically checking that all of the components of an HVAC system are present and functioning properly. It also involves adjusting the system and its controls to achieve the best possible performance. Commissioning costs about 1–3 per cent of the HVAC construction cost, but in the US, only 5 per cent of new buildings are commissioned (Roth et al, 2002a). Consequently, the control systems never operate as intended in many buildings. In a study of 60 buildings carried out by Lawrence Berkeley National Laboratory, half had problems with their control systems. These problems can, for example, cause fans to operate all night that are supposed to shut down at night, cause boilers to start up on summer days when they are not supposed to, or prevent the use of cool outside air for air conditioning when it is supposed to be used (Price and Hart, 2002; Barwig et al, 2002). Air-side economizers, which are supposed to increase the ventilation airflow when doing so can reduce the consumption of energy and reduce the airflow otherwise, frequently fail to operate

correctly. Indications are that only one-quarter of the economizers in commercial buildings work correctly, with perhaps no more than half working correctly when first installed (Financial Times Energy, 2001a). Commissioning typically reduces total energy consumption by 5–20 per cent and sometimes by as much as 30 per cent (Roth et al, 2003b). Commissioning should in fact be an ongoing process, typically eventually yielding savings in annual energy use of 20 per cent (Liu et al, 2003).

Compounding the problem of malfunctioning systems are poorly designed control systems. According to Barwig et al (2002) of the US National Building Controls Program, the reason for pervasive problems with building controls is 'often a lack of understanding of building control systems among the individuals who design, install, and operate them'. They go on to say that, 'This should not be interpreted as an indictment of these individuals' abilities. Instead, it is recognition that building control systems are a complex technology that changes rapidly... Owners often lack an understanding of the attributes and shortcomings of various manufacturers' control systems and control components, so cost becomes the only decision criterion. Designers often do not comprehend the energy impact of various control strategies, so efficient strategies are not specified. Specifiers are often unaware of the importance of strategic sensors. Operators are often uninformed of the intent of control sequences and receive inadequate training on the use of the control system, so when problems arise, they develop a "work-around" and unknowingly create energy waste.'

According to a survey of more than 450 control-related problems in buildings reported in Barwig et al (2002), the two largest causes of problems are software programming errors (32 per cent) and human operator errors (29 per cent). Hardware errors account for a modestly smaller fraction (26 per cent) of the problems. Control systems tend to have large amounts of site-specific code that must be written and implemented (Ardehali et al, 2003). Sensors tend to fail because of the general tendency to select those sensors with lowest first cost. Another common cause of control system problems is a lack of complete documentation about how the system is supposed to work (Esource, 1998a). The design intent should be clearly and fully explained by the designer so that bidders and contractors understand what the system is supposed to achieve, full sequences of operation for every piece of equipment should be supplied in precise and unambiguous terms, and clear requirements for performance verification before turning over the building to its owners need to be prescribed. A typical practice, however, is to create building contract documents that simply state what to install and what the initial settings should be.

Once a building has been commissioned, it is necessary to monitor the HVAC system to make sure that it continues to operate correctly. The key to detecting problems after commissioning is adequate monitoring of energy use. The detection and diagnosis of control system problems can be achieved either through measurement of the energy use by many individual components of the HVAC system, or through the analysis of high temporal resolution (i.e. 15 minute) whole building electricity use data in combination with outside air temperature data. Whole building electricity use data aggregate into a single variable the electricity use of all the different equipment in the building, but a surprising amount of useful information that can identify malfunctions in equipment can be obtained due to the high temporal resolution, as illustrated by Price and Hart (2002). At the other extreme, one may choose to use a state-of-the-art information monitoring and diagnostic system that consists of high-quality temperature, flow and pressure sensors; data acquisition software and hardware; and data visualization software including a web-based remote access system (Piette et al, 2001).

Improved building controls reduce energy costs, reduce the frequency of complaints from tenants, and lead to longer equipment life. In a programme involving over 80 buildings mentioned by Piette et al (2001), improved controls reduced total building energy costs by over 20 per cent and heating + cooling energy costs by over 30 per cent. At issue is not only the proper functioning of the control sensors and actuators, but also the *design* of the control system and the operating strategy. Wulfinghoff (1999) provides many examples to illustrate the principles underlying effective control strategies for

heating plants and cooling plants so as to maximize overall efficiency.

Claridge et al (2001) provide a striking example of how even simple changes in the control strategy can lead to dramatic reductions in energy use, even when the hardware is already working correctly. They reconfigured the control systems for a building on the Texas A&M University campus. The building had been retrofitted, in 1994, from a constant-air-volume to a variable-air-volume system and connected to the campus energy management system so that heating and cooling of non-critical areas (such as classrooms) can be shut down when they are not occupied. This had reduced fan energy use by 44 per cent, chilled water requirements by 23 per cent, and hot water requirements by 84 per cent. However, the control settings were far from optimal, so they were optimized to further reduce the energy use. Table 7.14 shows the major HVAC control settings in this building after the retrofit and after optimization. The pre-optimization settings are typical of North American buildings. Optimization reduced fan, chilling and heating energy use by 27 per cent, 52 per cent and 38 per cent, respectively, of the post-retrofit energy use. These savings arose largely by making duct pressure, and the temperatures at which hot and cold water are supplied, dependent on the outdoor temperature rather than constant. The final fan, chilling and heating energy use were reduced to 41 per cent, 37 per cent and 10 per cent, respectively, of the pre-retrofit energy use. Another example is provided by Plokker et al (2003), who show that optimized controls in an office building completed in The Netherlands in 1998 can save more than 10 per cent of the heating + cooling energy compared to conventional control settings if the internal heat gain is low ($30W/m^2$), and can save 35 per cent if the internal heat gain is higher ($50W/m^2$). Budaiwi (2003) also finds savings in heating + cooling energy use of about 35 per cent with optimized control, in this case for an office building in the hot-dry climate of Saudi Arabia. The optimized control consists of letting the temperature freely float from 4:00pm until one hour before occupancy begins the next morning (i.e. until 6:00am), rather than 24-hour temperature control. Ventilation occurs during occupied

Table 7.14 Major control settings in a university building in Texas before and after optimization

Parameter	Original practice	Optimized practice
Pressure in air ducts	Constant at 625–875Pa	Varies from 250 to 500Pa as outdoor air temperature increases
Cold air temperature	Constant at 10–12.8°C	Varies from 15.6°C to 12.8°C as outdoor air temperature increases
Hot air temperature	Constant at 43–49°C	Varies from 32.2°C to 21.2°C as outdoor air temperature increases
Air flow to rooms	Variable but inefficient	Optimized minimum and maximum flow rates and damper operation
Heating pump	Operated continuously	On only when outdoor air temperature exceeds 12.8°C
Cooling pump	Variable speed with shutoff	Pressure depends on flow

Source: Claridge et al (2001)

hours only (7:00am to 4:00pm) for both the base and optimized cases.

Building Automation Systems (BAS), also known as Facility Management Systems (FMS), were introduced in the mid-1980s to optimize the operation of HVAC equipment through computerized monitoring and control (Sachs et al, 2004). The latest systems have powerful graphics workstations, web-based wireless components and self-adjusting control algorithms. However, their performance has often been disappointing, with a tendency to solve problems by disabling various control loops or equipment schedules. The next generation of BAS/FMS software, Automated Building Diagnostic Software (ABDS), should solve these problems through more advanced self-adjusting control algorithms and automatic data analysis. In effect, ABDS will automatically perform building commissioning on an ongoing basis. Adaptive and Fuzzy-Control algorithms can save energy by enabling control operations that are not feasible with classical controls (Roth et al, 2002a, section 4.1). Levermore (2000) provides a comprehensive discussion of the fundamentals of HVAC control aimed at the practising engineer, with an emphasis on controls appropriate for low-energy HVAC systems and natural ventilation.

7.6 Summary and synthesis

In residential buildings, the first and most obvious rule is to place the heating system and ductwork entirely within the heated space, a rule that is not always followed. Beyond this, the minimal

features of a highly efficient heating and cooling system are:

- to distribute heat through radiant floor heating, in which warm water is used to provide warmth; and
- to use a mechanical ventilation system (fans plus ductwork) to provide only the amount of fresh air needed for ventilation purposes during the heating season or on hot days, when windows are closed, with the system designed to minimize pressure losses (i.e. fat, short, straight ducts to the extent possible), with a high-efficiency variable speed fan, and with a heat exchanger to recover heat or coldness from the exhaust air.

More advanced systems make use of heat pumps to recover heat from exhaust air, and/or make use of earth pipe loops to preheat (in winter) or precool (summer) the ventilation air. Exhaust-air heat pumps and heat exchangers have been used together in some cases, as has the combination of earth pipe loop, heat exchanger and heat pump. To capture the full benefit of earth pipe loops, heat exchangers or exhaust-air heat pumps, the house envelope should have a low leakage rate, and supply and exhaust fans must be adjusted to produce equal airflow. In houses with high-performance thermal envelopes, a very low temperature (30–35°C) can be used in a floor radiant heating system, or the heat supplied to the ventilation air from an exhaust-air heat pump can meet the entire heating load on the coldest day, without the need for radiant floors.

Having minimized heating and cooling loads through the techniques discussed in Chapters 3 and 6, the keys to reducing HVAC system energy use in commercial buildings are:

- to separate the ventilation, cooling and heating functions, by circulating warm or cold water for heating and cooling;
- to provide the minimum amount of ventilation air through demand-controlled displacement ventilation which, being the minimum ventilation airflow, will entail 100 per cent outside air;
- to separate the dehumidification and cooling functions by using solid or liquid desiccants for dehumidification, with chillers for removal of sensible heat only, thereby avoiding the need for overchilling followed by reheating in order to dehumidify air;
- to use chilled ceiling panels to handle that portion of the cooling load that cannot be handled by the displacement ventilation system, with separate dehumidification of ventilation air (where needed);
- to supply ventilation air and chilling water at the warmest possible temperatures, as this will increase the COP of mechanical chillers (as long as overchilling and reheating is not used for dehumidification) and will extend the range of outside temperature conditions where free-cooling can be used;
- to lay out the ductwork and piping in such a way as to minimize the required motor, fan and pump energy requirements;
- to choose the most efficient equipment available, appropriately sized;
- to fully commission all equipment and systems, and to ensure that proper operating procedures are

fully documented and understood by the operational personnel;
- to account for psychological adaptation to seasonal differences in outside temperature when determining the indoor temperatures that the HVAC system will maintain, which presupposes the existence of operable windows and individual control over the ventilation rate in individual workspaces.

The significant energy saving benefits of distributing heat and coldness with water rather than air, with airflow restricted to that needed for ventilation purposes only, follow from the following fundamental relationships:

- the energy that must be imparted to a moving fluid increases with the rate of fluid flow raised to roughly the third power – a cubic relationship; and
- the ratio of thermal energy delivered to energy used by pumps or fans is about 25 times greater using water than using air to distribute heat or coldness.

Under the cubic relationship, if the required flow rate can be cut in half, the energy that must be imparted to the flow drops by a factor of eight. The power that must be supplied to a fan or pump is equal to the energy that must be supplied to the fluid, divided by the product of the motor and pump or fan efficiencies. These efficiencies are less for smaller units or under part-load operation, so the decrease in required pump energy will fall by a factor of four to five instead of a factor of eight.

Many of the measures discussed in this chapter are interlinked. For example, the use of displacement ventilation with airflow no greater than required for ventilation purposes requires a chilled ceiling for additional cooling, which in turn requires a tight envelope (to inhibit infiltration of moist air), but a tight envelope provides its own direct energy saving benefits. Energy recovery heat exchangers, like DV, require a tight envelope, with exhaust airflow equal to a least 75 per cent of the supply airflow (this allows some over-pressurization and some losses through separate bathroom exhaust fans). Effective desiccant dehumidification requires a dedicated outside air supply, which will be provided through DV with minimal airflow.

Some energy-efficient features of HVAC systems provide significant non-energy benefits. DV can be and should be a dedicated outdoor air supply (DOAS) system, which avoids circulation of contaminants from one part of a building to another. Desiccant dehumidification avoids the occurrence of saturated conditions within the HVAC system, thereby inhibiting the growth of mould and bacteria. Advanced control systems provide superior comfort while saving energy.

Notes

1 Air density can be computed from $\rho = P/RT$, where P is the pressure in Pascals, $R = 287.1 \text{J/kg/K}$, and T is the absolute temperature (in K). The density of pure water at 1atm pressure can be computed from $\rho = a_1 + T*(a_2 + T*(a_3 + T*a_4))$, where $a_1 = -7.2169 \times 10^{-2}$, $a_2 = 4.9762 \times 10^{-2}$, $a_3 = -7.5911 \times 10^{-3}$, $a_4 = 3.5187 \times 10^{-5}$, and T is in °C (Friedrich and Levitus, 1972).

2 As in the previous section, the discussion here pertains to fans or pumps but, for the sake of succinctness, the term 'pump' will be taken to mean pumps or fans, and 'piping system' will be taken to mean pipes or ducts.

3 As the flow rate and pressure head depend on the impeller speed and diameter in the same manner, a set of pump curves similar to that shown in Figure 7.7 could be produced for different impeller speeds rather than different impeller diameters. However, the efficiency contours would be different.

4 Needless to say, sharp bends and inadvertent compression of flexible ducts should also be avoided.

5 This is the case in hospitals. The quantity of air circulating through the building, and which needs to be cooled and dehumidified, is much larger than needed for ventilation purposes alone in an all-air cooling system, because of the need to remove sufficient heat without too cold an airflow. In a CC system, the airflow can be reduced to that needed for ventilation purposes alone.

6 The floor plenum is the gap between the floor and raised floor, which serves as the supply air duct in North American buildings. In Europe, air is supplied through ducts that are placed in the plenum, thereby eliminating the possibility of contaminating the air with dirt from the plenum.

7 Technical assumptions (see Box 6.5) include: sensible heat exchanger and evaporative cooler effectiveness of 0.7 and 0.9, respectively; boiler efficiency of 0.8; regeneration airflow rate = process airflow rate; traditional system cools air to 11°C, then reheats it to 15°C; a desiccant system produces air at 15°C.

eight

Water and Domestic Hot Water

Most discussions of building energy efficiency involving water deal exclusively with domestic hot water (DHW) – hot water use for personal hygiene and for washing clothes and dishes. The energy used to heat incoming water to the temperature needed for DHW is significant, but even cold water entails energy consumption – not at the point of use, but in the collection, transport and treatment of the water supplied to consumers, and in the treatment of wastewater. That is, there is *embodied energy* in water. As discussed in Chapter 6 (Section 6.11.2), there is embodied water in most forms of electrical energy. Thus, there is a degree of symmetry between energy and water, as each is to some extent embodied in the other.

We begin this chapter by assessing the amount of energy embodied in water. The energy embodied in a unit of municipal water is normally very small compared to the amount of energy needed to heat that unit for DHW purposes, or compared to the cooling that can be achieved by evaporating the water, but the embodied energy (which occurs mostly as electricity) associated with typical household water consumption can be a significant fraction of the total direct household use of electricity. The bulk of this chapter deals with ways to reduce the amount of non-renewable energy used to make DHW. This can be done by (i) reducing the amount of hot water used; (ii) heating water as much as possible with solar thermal energy; (iii) heating it more efficiently when using non-solar energy; (iv) recovering heat from warm wastewater; and (v) reducing standby and distribution losses. Use of solar energy is discussed in Chapter 11.

8.1 Energy embodied in water

In most parts of the world, water is pumped from the ground or is taken from rivers, lakes, or reservoirs, and pumped some distance to its point of use, all of which requires energy. Energy is also required to treat water before use, for local distribution, and in the treatment of wastewater.

Table 8.1 summarizes estimates of the amount of energy used in the various steps in supplying water in California. Several options are given for the source of water, ranging from recycling of wastewater and pumping of fresh groundwater, to desalination of brackish (slightly salty) groundwater, desalinization of seawater, large-scale diversion from central to southern California by aqueducts, and towing of freshwater at sea in bags. Several options are given for the treatment of wastewater, with and without recovery and use of the biogas energy that is produced at the wastewater treatment plant. The total embodied energy ranges from a minimum of 0.6kWh/m³ (if water can be extracted from an adjacent river with no use of energy) to a maximum of about 5.5kWh/m³, almost all of it as electricity. This estimate does not include the energy used to produce the infrastructure that delivers the water or the plants that treat the water and wastewater, nor does it include the energy used to produce the chemicals used in the water treatment plant. For a household water consumption ranging from 0.5 to 1.0m³/day, the energy used to supply household water amounts to about 100–2000kWh/year. A typical value is probably a few 100kWh per year, plus infrastructure and chemical energy. This is non-negligible compared to the major direct uses of electricity in a household, such as the air handling system (100–1000kWh/year), refrigerator (500–1000kWh/year), and consumer goods (1000–2000kWh/year).

However, the energy embodied in water (equivalent to 0.002–0.02GJ/m³) is very small compared to the energy required to heat water by 50K (0.209GJ/m³) or the cooling that can be achieved through evaporation (2.5GJ/m³).

8.2 Reducing water use

Reducing the use of both cold and hot municipal water saves energy by reducing consumption of the embodied energy associated with that water. Saving hot water saves the additional energy otherwise used to heat the water. The consumption of unheated municipal water can be reduced (i) by recycling 'greywater'; (ii) by collecting and storing rainwater for use where treatment is not required;

Table 8.1 Energy use to deliver and treat water supplied to urban consumers in California

Process	Energy use (kWh/m³)		Energy use (GJ/m³)	
	Minimum	Maximum	Minimum	Maximum
Conveyance to urban centres				
Diversion[a]		2.43		0.00875
Pumping groundwater	0.38	0.60	0.00136	0.00216
Desalinization of seawater	3.57	3.89	0.01284	0.01401
Desalinization of brackish groundwater		1.38	0.00496	0.00496
Recycling of wastewater		?		
Water bags (towed 600km)		0.96		0.00344
Treatment and local distribution				
Water treatment	0.03	0.06	0.00012	0.00020
Local distribution	0.14	0.76	0.00050	0.00274
Wastewater treatment[b]				
Trickling filter	0.18 (0.11)	0.47	0.00066	0.00169
Activated sludge treatment	0.28 (0.18)	0.61	0.00099	0.00219
Advanced treatment	0.32 (0.23)	0.70	0.00117	0.00252
Advanced treatment with nitrification	0.42 (0.32)	0.79	0.00152	0.00286
Representative total				
All steps	0.6	5.5	0.00216	0.0198

[a] Water diverted from central to southern California, with pumping up a vertical distance of about 700m.
[b] Minimum values are for a 400million litre/day facility without energy recovery, or (in brackets) with energy recovery (as biogas). Maximum values are for a 4million litre/day facility without energy recovery.
Source: Cohen et al (2004)

and (iii) through the use of water-efficient fixtures and toilets. Greywater is water such as wastewater from sinks and showers that is not potable but can be used for toilets. With a 50l storage tank, 80 per cent of the toilet flush water can be provided in this way (Dixon et al, 1999). Rainwater can be used for watering outdoor plants or for toilets, although there is little additional gain when rainwater is combined with greywater to supply toilets.

Figure 8.1 shows the breakdown of energy used for DHW by a typical family of four in the US. *Standby loss* is the heat lost from the hot-water tank through conduction and emission of radiation, while *distribution loss* is the heat lost from the pipes as hot water is circulated. Standby and distribution losses account for about one-third of the total DHW energy use in the US, but these losses are poorly quantified (standby losses are likely to be the larger of the two). Nevertheless, they represent a very large fraction of total DHW energy use and therefore merit considerable attention. Options to reduce these losses are discussed later. Showering is another large fraction (about one-third) of total DHW energy use. The consumption of DHW can be reduced through the use of more water-efficient fixtures, appliances and habits, such as:

- low-flow showerheads and faucets;
- more efficient washing machines (which use less water as well as less electricity), or using cold-water washing;
- more efficient dishwashers (which use less water, as well as less electricity), more efficient use of dishwashers (i.e. operation at full load only), or washing efficiently by hand instead.

Table 8.2 compares rates of water use for standard and water-efficient showerheads, faucets and toilets. A reduction in hot water use for showering and washing by at least a factor of two is possible if water-efficient fixtures replace standard fixtures, assuming that showering and washing habits do not otherwise change. However, the savings in DHW energy use will be diluted if hot water is stored in a hot-water tank because a large fraction of the total energy consumption in this case is simply to offset standby losses, which are not altered if total consumption decreases. Indeed, as consumption decreases, the relative importance of standby losses will increase (they are like a fixed overhead).

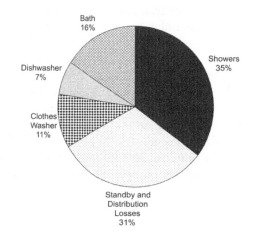

Figure 8.1 Annual energy consumption for domestic hot water by a typical family of four in the US

Total =7500 kWh/year

Source: Data from Holton and Rittelmann (2002).

Table 8.2 Comparison of water use for standard shower heads, faucets and toilets, and for water-efficient varieties

Standard shower head	10–20 litres/minute (2.6–5.3 gallons/minute)
Low-flow shower head	5–10 litres/minute (1.3–2.6 gallons/minute)
Standard sink faucet	10–20 litres/minute (2.6–5.3 gallons/minute)
Aerated bathroom sink faucet	2–4 litres/minute (0.53–1.1 gallons/minute)
Aerated kitchen sink faucet	6–8 litres/minute (1.6–2.1 gallons/minute)
Standard toilet	20 litres/flush (5.3 gallons/flush)
Low-flush toilet	3–6 litres/flush (0.8–1.6 gallons/flush)

Source: City of Toronto water efficiency programme

8.3 Water heating

Water can be heated using electric resistance heaters, natural gas or oil heaters, hybrid solar natural gas systems, heat pumps or by connection to a district heating system. Electric resistance is 100 per cent efficient at the point of use but entails losses in generating electricity (by a factor of three if the electricity comes from a typical coal powered plant), with further losses (typically of 5–10 per cent) in transmitting electricity to the customer. Stand-alone residential natural gas water heaters tend not to be efficient (75–85 per cent overall efficiency) because hot-water loads have not been considered to be large enough to justify the additional expense of high efficiency. However, if a single unit is used to heat hot water for space heating and for domestic hot water, then the high efficiency (85–95 per cent) available for space heating can also be applied to hot-water heating. High efficiency requires being able to condense the water vapour in the exhaust gas, releasing latent heat and using this heat to preheat the supply water or combustion air.

8.3.1 Single-family stand-alone water heaters and storage tanks

Analogously to boilers, we can define the *combustion efficiency* of a natural gas or oil hot-water heater as the fuel energy input minus heat lost in the exhaust, divided by the fuel energy input. The *overall efficiency* (or *recovery efficiency*) is the fraction of fuel energy input converted to hot water under steady operation, and is somewhat smaller than the combustion efficiency due to heat loss from the heater to the surroundings. The recovery efficiency of natural gas heaters in the US ranges from 76 per cent to 85 per cent, and for oil heaters it ranges from 75 per cent to 83 per cent (Lekov et al, 2000). However, if hot water is stored in a tank between periods when it is being used (the normal situation in North America), heat will be conducted through the walls of the water tank and lost; this 'standby' heat loss has to be made up. The fractional standby loss depends on the tank surface area-to-volume ratio, on the temperature difference between the interior of the tank and the surroundings, on the resistance of the insulation filling the cavity between the tank and jacket, and on the hot water load.

Additional standby loss occurs if there is a continuously burning pilot light instead of an electronic ignition. As shown in Figure 8.1, standby energy loss is a large consumer of energy for domestic hot water in the US.

Hot-water tanks typically have 2.5–5cm of polyurethane foam insulation, giving an RSI-value of 0.88–1.76 (R5 to R10). Pre-2003 models used HCFC-141b as the blowing agent, while HFC-245fa or water has been used as a blowing agent since 2003. Table 8.3 gives the insulation resistance for various thicknesses of foam insulation using the above three blowing agents. Although HFC-245fa has no ozone-depleting potential, it is still a strong greenhouse gas (with a global warming potential about 1020 times that of CO_2, on a mass basis). For a given thickness, water-blown insulation provides 73 per cent the insulation value of HFC-245fa. With 10cm of insulation, the insulation value is RSI 4.02 using HFC-245fa and RSI 2.91 using water. Neither is particularly large, given that the temperature difference between the inside and outside of the tank will be 30–40K. Jackets with an insulation value of about RSI 1.0 (R5 to R7) can be used to enhance the insulation already present in a water heater.

The overall efficiency of a gas-fired water heater is measured by the *energy factor*, which takes into account standby losses and combustion efficiency. US 2004 standards require a minimum energy factor for a 150l tank of 0.59 using natural gas and 0.51 using oil (US DOE, 2004). The energy factor depends partially on the total daily hot water use, as this determines how large the standby loss is compared to the amount of hot water supplied, and on the difference in temperature between the tank and ambient air, as this determines in part the absolute value of the standby loss. Thus, the energy factor depends in part on the specific test conditions. Through a variety of improvements to maximize combustion efficiency and minimize standby losses (including up to 7.5cm of insulation), condensing water heaters can achieve a US energy factor as high as 0.86 for small commercial units, while the maximum energy factor for residential units (non-condensing) is about 0.72. The energy factor does not account for the additional space heating load that is created as a result of the extra outdoor air drawn

Table 8.3 Insulative properties of polyurethane foam with different blowing agents and for different thicknesses, used to insulate domestic hot-water tanks

Blowing agent	Thickness (m)	Conductivity (W/m/K)	U-value (W/m²/K)	R-value (Btu/ft²/hr/°F)⁻¹
HCFC-141b	0.025	0.024200	0.97	5.9
	0.050	0.024200	0.48	11.7
	0.075	0.024200	0.32	17.6
	0.100	0.024200	0.24	23.5
HFC-245fa	0.025	0.024922	1.00	5.7
	0.050	0.024922	0.50	11.4
	0.075	0.024922	0.33	17.1
	0.100	0.024922	0.25	22.8
Water	0.025	0.034327	1.37	4.1
	0.050	0.034327	0.69	8.3
	0.075	0.034327	0.46	12.4
	0.100	0.034327	0.34	16.5

Source: Computed from conductivity data given in Lekov et al (2000)

into the house for combustion purposes, which will occur unless the unit is directly supplied with outside air for combustion. Direct use of outside air is rare in water heaters serving household hot-water needs only, but is standard in water heaters serving both space and hot-water needs, the subject of the next subsection.

8.3.2 Integrated space and hot-water heating systems

Up to four kinds of integrated space and hot-water heating systems are commercially available, depending on location. In one, water is heated in a single hot-water tank and stored for both space heating and hot water use. The hot-water tank is connected by pipes to one or more fan coil air handlers, and hot water circulates through the fan coil when space heating is needed. Air is blown past the fan coil and through the ductwork, so this system can easily be retrofitted into a normal furnace/forced air heating system when the furnace needs to be replaced. When hot water is needed, it is drawn from the tank in the usual manner. This is the most common integrated system. A second type of system consists of a boiler and two heat exchanger coils, one being the fan coil of the air handler, and the other being in the hot-water tank. Water or a gycol solution is circulated to the

space-heating or water-heating coil, as needed. A third option is to circulate hot water from the hot-water tank through hot-water radiators located in each room for space heating, with domestic hot water drawn from the same tank as needed. A fourth option is a tankless water heater that feeds a radiant space-heating system as well as supplying DHW. This system is viable only where space heating loads are small.

Not all integrated space and hot-water heating systems are more efficient than stand-alone systems. Energy factors based on water heating alone range from 0.51 to 0.83 (Jones and Bicker, 2004). High efficiency requires the use of a condensing water heater that draws combustion air directly from outside the house so as to avoid inducing infiltration of cold outside air into the house, and a well-insulated storage tank to minimize standby losses. As noted in Chapter 4 (Section 4.2.7), wall-hung tankless boilers are now available for combined space and DHW heating with full-load combustion efficiencies of 84 per cent (non-condensing) or 92 per cent (condensing). In some jurisdictions, the choice of products is limited and it is difficult to find plumbers with experience installing integrated systems.

8.3.3 Tankless and point-of-use water heaters

In Europe and Asia it is common to heat water at the point of use, only as needed. This avoids standby energy losses altogether, as there is no hot-water storage tank. These heaters are referred to as tankless, on-demand or point-of-use (POU) water heaters. We shall refer to them here as tankless water heaters. In Europe, one might find a small tankless water heater at each point of use. However, tankless units are available that can provide up to 30 litres/minute of hot water, which is adequate for a typical household with water-conserving hot-water fixtures (see Table 8.2). A single tankless water heater can therefore serve an entire household, in which case they are no longer POU heaters. The vast majority of tankless water heaters are non-condensing units; condensing tankless units have just recently become available in some jurisdictions. Older tankless water heaters tended to use a pilot light, which can account for half of the total water heating energy use. Tankless and pilotless non-condensing water heaters are available with energy factors in the range 0.64–0.85 (Jones and Bicker, 2004). This is an improvement over the range of 0.54–0.72 for storage type water heaters, and will be better still for condensing units. Condensing, modulating units with a full-load efficiency of 0.92 and independent control of DHW and space-heating water temperatures have recently become available. Periodic maintenance is required for tankless systems in areas with water impurities.

With tankless heaters, some energy is needed to heat up the heat exchanger at the beginning of a hot-water draw, and this energy is lost when the unit cools down between draws. Thus, there is some energy-savings benefit in clustering hot-water draws, above and beyond the reduction in hot water that cools in the pipes between the water heater and faucets (Section 8.5.2). As a result, the published energy factor for instantaneous water heaters depends on the size and clustering of hot-water draws specified in the test procedure.

8.3.4 Commercial boilers

Boilers can be used to heat water for multi-family buildings, athletic facilities or schools. If the boiler is also used for space heating in winter then, due to the low part-load efficiency of most boilers (see Chapter 4, Section 4.3.2), the efficiency for heating water in the summer will be quite low. Averaged over 30 multi-family buildings in New York City, Goldner (2000) measured a summer efficiency of only 42 per cent.

8.3.5 CO$_2$ heat pumps

As noted in Chapter 5 (Section 5.8.1), heat pumps using CO_2 as a working fluid have been introduced to the consumer market in Japan and used to make DHW. The CO_2 heat pump makes hot water largely at 42°C, with a small amount of 60°C water made at night. Using input water at 9°C in the winter and 24°C in the summer, the mean annual COP is 4.1 based on the heat produced by the heat pump, 3.3 after accounting for heat loss from the storage tank (which is applicable to any system with storage), and 3.1 after accounting for energy use by auxiliaries (such as a circulating pump) (Kusakari, 2005).

8.3.6 Solar hot-water heaters

Production of hot water using solar thermal energy is discussed in Chapter 11 (Section 11.4). Typically, 50–90 per cent of annual DHW needs can be met with solar energy.

8.4 Recovery of heat from warm wastewater

The hot-water load in buildings can be classified as either *batch flow*, in which a hot-water flow occurs at one point in time and disposal of the used hot water occurs at a later time, or *simultaneous flow*, in which the two flows occur at the same time. Examples of batch flows would be water used to fill a bathtub, sink basin, clothes washer or dishwasher. A shower is an example of a simultaneous flow. To the extent that the heat in the discarded water can be captured and used to preheat incoming water, the energy needs for water heating can be reduced. This can be done relatively easily for simultaneous flows, but requires a storage tank in order to store the warm wastewater until it can be used to preheat water for the next batch load. A number of studies, summarized by Proskiw (1998), collectively indicate that simultaneous flows account for 30–45 per cent of the total hot-water load in single-family residences in North America.

8.4.1 Types of wastewater heat recovery systems

There are four types of wastewater heat recovery systems:

1   *Storage tank with heat exchanger.* The incoming municipal water passes through a coil that is placed inside a storage tank that holds warm wastewater. Separate plumbing is needed so that solid wastes and perhaps grease (from kitchen sinks) do not enter the storage tank. This increases the cost, but such systems are fairly conventional and well understood. The water that would be supplied to the storage tank is referred to as *greywater.* Storage tanks are generally designed to facilitate thermal stratification (warmest water on top, less warm water below) in order to improve performance. The incoming water can bypass the storage tank coil if it is already warmer than the greywater in the storage tank. The storage tank and heat exchange surface must be periodically cleaned.
2   *Storage tank with heat pump.* This is similar to the above system, except that a heat pump is used to extract additional heat from the stored wastewater.
3   *Heat transfer from the main wastewater line.* The incoming water line is wrapped around the wastewater line, directly capturing a portion of the heat in the outgoing wastewater.
4   *Point-of-use devices.* An example is a shower, in which water heading to the water heater passes through a coil in the shower floor, or a dishwasher in which hot discharged water is temporarily stored.

8.4.2 Specific examples

Direct transfer of heat from the main wastewater line (method 3) is simple, inexpensive and maintenance-free, but waste heat can be captured only during simultaneous flows. The heat recovery efficiency can be enhanced using the *gravity falling film* method. Surface tension and gravity cause falling films of water to spread and cling to the inner wall of a vertical drainpipe. The tension is strong enough to reduce the fall velocity to 0.4–1.2 metres per second and to hold the film thickness to about 0.3–0.7mm at a flow rate of 0.03–0.2 litres/second. This enables a high rate of heat transfer. The cold incoming water passes

Figure 8.2 Gravity falling film heat exchanger for recovery of heat from wastewater. It must be installed almost perfectly vertical so that the water will drain uniformly down the sides of the pipe

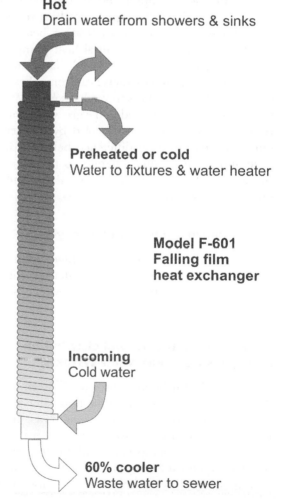

Hot
Drain water from showers & sinks

Preheated or cold
Water to fixtures & water heater

Model F-601
Falling film
heat exchanger

Incoming
Cold water

60% cooler
Waste water to sewer

Source: Vasile (1997)

through a coil that is tightly wrapped around the vertical drainpipe and is warmed as it flows (see Figure 8.2). This system recovers 45–65 per cent of the available heat in the wastewater, depending on water-use patterns, warming the incoming water up to about 20°C (Vasile, 1997). IEA (2002a) reports a simpler system that recovers 30 per cent of the available heat.

De Paepe et al (2003) have developed a simple system that recovers heat from hot water discharged from a dishwater and uses it for pre-heating the inlet water for the next step in the wash/rinse cycle. Hot water is stored in a tank and inlet water passes through a coil inside the storage tank, being heated as it flows through the coil. Since recovered waste heat is useful only if the next step uses warm or hot water, only 25–35 per cent of the total energy required for water heating can be saved in this way.

In the Brampton house discussed in Chapter 7 (Section 7.3.7), greywater heat recovery is integrated into a space-heating heat recovery system that uses a central heat pump, a large phase change storage tank and a circulating brine. The heat pump extracts heat from a brine, and the cooled brine flows through a 1.3cm ($\frac{1}{2}''$) copper tube that is soldered to the greywater line. Because the brine has been cooled by the heat pump, the temperature difference between the brine and the greywater is larger, so more heat can be extracted from the greywater than would otherwise be possible.

8.4.3 Integration with water recycling

Greywater heat recovery with a storage tank is a natural fit with water recycling, as water recycling also requires the separation of greywater and non-greywater flows.

8.5 Standby, parasitic and distribution energy losses in hot-water systems

Standby and distribution losses were previously identified as heat losses from the hot-water tank and from the pipes when hot water flows. If a pump is used to circulate water, the electrical energy used by the pump is an additional energy use called a *parasitic* energy loss. Ways to minimize or eliminate all three energy losses are discussed below.

8.5.1 Minimizing standby and distribution losses in single-family houses

In single-family houses with hot-water storage tanks, distribution losses can be minimized by locating the bathrooms and kitchen close to one another and close to the hot-water tank, something that is easier in multi-storey houses than in sprawling ranch-style houses. Insulating the pipes reduces both standby heat loss (i.e. when there is no water circulation) and heat loss during water flow. The hot-water pipe leaving the hot-water tank serves as a conduit for heat out of the tank, with the heat flow out of the tank proportional to the temperature gradient in the pipe next to the tank. Insulating the first 85cm can reduce total standby heat loss by 50 per cent, but there is very little additional reduction in standby heat loss as the length of insulation exceeds 1m (Stewart et al, 1999). With regard to heat loss due to water flow, for a 2.5cm copper pipe with no insulation, a pipe-ambient temperature difference of 56K, and flow velocity of 3m/s, the heat loss is 44W per metre of pipe. A CPVC (chlorinated polyvinyl chloride) pipe has about 50 per cent less heat loss than copper pipe (Baskin et al, 2004).

Peckler (2003) reports on the total water-heating energy use for the following systems: (1) a conventional, centralized hot-water tank with a branching hot-water pipe network; (2) a centralized water heater that provides hot water on demand, without storage, and a branching pipe network; (3) a centralized hot-water tank with a radial hot-water pipe network (separate hot-water pipes to each point of use); (4) a centralized water heater with a radial pipe network; and (5) tankless, point-of-use water heaters. Options (2) and (4) eliminate standby losses but still entail distribution losses, while option (5) eliminates standby and distribution losses and the waste of energy when hot water in the pipes cools down during long periods of non-use, only to be replaced when hot water is called for again. The relative energy use for the different systems is given in Table 8.4. With low water use, where standby and distribution losses are a large fraction of total energy use with a centralized tank and branching network, a tankless point-of-use water heater saves almost half of the total energy used for hot water. With high water use, the standby and distribution losses are a smaller fraction of total energy use, but the

Table 8.4 Relative energy use for different domestic hot-water heating systems

System	Relative energy use	
	High-use case	Low-use case
Central tank, conventional (branching) arrangement of hot-water pipes	1.00	1.00
Central tankless, branched piping	0.92	0.76
Central tank, radial piping	0.91	0.87
Central tankless, radial piping	0.86	0.66
Point-of-use, tankless	0.72	0.51

Source: Peckler (2003) based on computer simulations

saving is still almost 30 per cent for the point-of-use option.

8.5.2 Minimizing energy and water wasted in houses while waiting for hot water to arrive

The need to drain the hot-water pipes of cold water when hot water is called for after a long period of non-use is a source of waste of both energy and of water. This is done by opening the hot-water tap until hot water arrives at the faucet. The volume of hot water that is wasted is about twice as large as the volume of cold water in the pipes that needs to be displaced, because the pipes themselves are warmed up as the hot-water flow begins, and this roughly doubles the waiting time. The waiting time will be longer the greater the distance from the hot-water heater to the faucet and the slower the flow rate, while the amount of water wasted will be greater the greater the distance between the heater and the faucet and the greater the diameter of the pipe. In pre-1970 houses in the US, the furthest faucet is about 10m from the water heater, there are 5–7 hot-water faucets, and ½″ pipes are used (Klein, 2004a). In more recent houses, the furthest faucet is about 20m from the water heater because most new housing has been in the south, where

basements are rare and the water heater is located in the garage. There are 11–14 hot-water faucets, so a larger-diameter trunk pipe is needed: first 1″ diameter, then stepping down to ¾″ and finally ½″. Figure 8.3 gives an example of a plumbing system with a single trunk line and an example with multiple trunk lines. Altogether, the volume of water that must be replaced before hot water arrives at the furthest fixture is about six times greater in recent houses than in pre-1970 houses. However, as an energy conservation measure, building codes in the US have specified roughly a factor of three reduction in the rate of flow since the 1970s. In both old and

Figure 8.3 An example of hot-water plumbing systems with (a) a single truck and multiple branches, (b) multiple trunks and branches, and (c) a recirculation loop

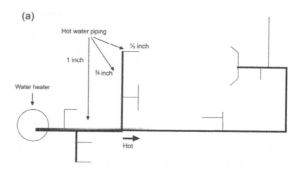

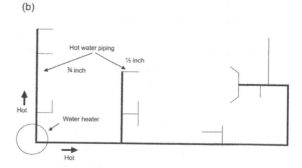

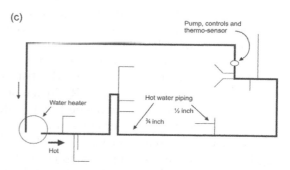

Source: Klein (2004a, 2004c)

new houses, the net result is that it is not uncommon to have to wait 2 minutes before hot water arrives at the faucet. For five waiting times per day of 1–2 minutes each after periods of nonuse long enough for the hot-water pipes to cool down again, with a flow rate of 4 litres/minute, 20–40 litres of water are wasted per day. This is hot water that is wasted, because at the end of a hot-water use, the pipes are filled with hot water which then cools down. The real waste is probably much larger, given that, when the waiting time is long, the user will do other things while the hot-water faucet is open and return some time (perhaps a long time) after hot water has arrived.

Klein (2004b) investigated three possible solutions to this problem: (1) to use more than one water heater, each serving a cluster of hot-water fixtures with short, small-diameter pipes; (2) to install heat trace – a thermostatically controlled resistance heater – between the hot-water pipes and pipe insulation; and (3) to install independent, small-diameter pipes to each hot-water fixture. Option (1) entails greater cost, due to the need for more than one heater, because gas pipes or electrical cables are generally more expensive than hot-water pipes, and because of the need for additional exhaust flues if natural gas heaters are used. Extra space is also required, although this is minimized with tankless water heaters. Option (2) solves the waiting-time problem and saves water, but does not save energy. Option 3 requires a short but thick (1″ diameter) trunk pipe (a manifold), to which the pipes to the individual hot-water fixtures are connected. The volume of water in the manifold system is not necessarily less than in a trunk-and-branch system. If the pipes would have cooled down between uses of hot water, the manifold system can cut the waiting time and waste of water by up to 50 per cent. However, when hot-water uses are clustered together, the manifold system can end up wasting more water over the course of a day.

Klein (2004c) considers a fourth solution: a well-insulated *recirculation loop* (RL) that is periodically primed with hot water. Recirculation loops, in which hot water is continuously pumped through a loop and back to the hot-water tank, are commonly used in commercial buildings. In this way, hot water is instantly available to hot-water fixtures that are attached to the loop through short side branches. Water in a 2cm pipe that is insulated to RSI 0.7–0.9 (R4–5) will remain above 40°C (a typical minimum required DHW temperature) for 45–60 minutes once heated to 50°C. When the temperature drops to 40°C, a temperature sensor can activate a pump that circulates water for 30 seconds, restoring the water in the RL to 50°C (assuming the water in the hot-water tank to be somewhat above 50°C). The RL serves as the immediate source of hot water to the various hot-water fixtures, so small-diameter branch lines can be used. The RL should be kept as short as possible to minimize the amount of hot water needed to prime the loop. An example of a residential RL is shown in Figure 8.3c. The system illustrated in Figure 8.3c can also be retrofitted onto an existing faucet; in this case, the cold-water pipe is used as the return pipe of the recirculation loop (see www.gothotwater.com). The pressure in the hot- and cold-water lines is essentially the same, so the pump sees only the frictional losses in the plumbing at its pressure head.

Table 8.5 compares the costs for water, water heating, and pumping for four systems in a typical US household: a standard trunk-and-branch system, a manifold system, an uninsulated recirculation loop with continuous flow, and an insulated recirculation loop primed only as needed. Both recirculation-loop options reduce the amount of wasted water by about 80 per cent, but a large amount of heat is used for the first option to offset heat lost from the loop, as well as some pumping energy (the primed loop requires negligible pumping energy, as the pump will operate only 10–15 minutes per day). Relative to the standard system, the primed loop reduces total water use 20.6 per cent and heating energy use by 19 per cent. However, it also largely eliminates waiting time – a major selling point – and relative to the continuous recirculation loop (which gives the same reduction in waiting time), total costs are reduced by 60 per cent.

A recirculation loop can be used to save water (and hot water) only if there is a tank to store the water that returns from the loop. This would preclude the use of a tankless water heater. Ideally, one would build a system with a tankless water heater, but add a storage tank large enough to store the warm water from a periodically operating recirculation loop.

Table 8.5 Comparison of representative water and energy costs ($) for different residential DHW systems in the US

	Water cost		Heating cost			Pumping cost	Total cost	Savings
	Used	Wasted	Used	Wasted	Pipes			
Standard	160	73	140	60	–	–	433	
Manifold	160	55	140	45	–	–	400	33
Continuous RL	160	15	140	12	500	25	852	–419
Primed RL	160	15	140	12	10	0	337	96

Note: RL = recirculation loop.
Source: Klein (2004c)

8.5.3 Minimizing losses in multi-unit residential, commercial and educational buildings

Schools, hotels and multi-unit residential buildings in North America (if not elsewhere) commonly use the hot-water recirculation loop (RL) described above. Since the purpose of recirculation is to keep the pipes warm, the required flow can be reduced by insulating the pipes well. Since pumping power varies with the flow rate to the third power (Chapter 7, Section 7.1.2), dramatic reductions in pump energy use are possible along with reduced heat loss. However, wherever possible, tankless POU heaters should be used. Piping losses can constitute 40 per cent to more than 50 per cent of the total hot-water load, even with well-insulated piping (Goldner, 1999; Lutz et al, 2002). Alternatives to continuous circulation, such as shutdown for a few hours overnight, can save 6–10 per cent of total heating energy use (Goldner, 1999). Supplying each unit with pilotless and tankless POU water heaters would dramatically reduce hot water energy use. POU water heaters in multi-unit residential buildings lend themselves more readily to individual metering of energy use, which provides a direct incentive to avoid waste. Such units could be used only to boost the inlet temperature by 10–20K if a low-temperature storage tank is used to store waste heat collected from drainage water or from exhaust air (as illustrated in Figures 3.7 and 3.8) or from an air conditioner condenser (see below).

Hiller et al (2002) monitored the energy use in a new (1997 opening) school in Tennessee using an RL system, and again after it was converted to a POU system. The school had a central gas-fired water heater with an efficiency of 82 per cent, eight separate recirculation loops with a

total length of 257m that served six points of use (the length of the loops was about twice that predicted from a plan view of the school plumbing, due to the large number of ups and downs, and due to some piping that did not show up at all on the plumbing diagrams). All pipes were insulated with 1.27cm of foam insulation. In the original system, water was circulated 24 hours per day, all year long. The original system was monitored for over two months, and it was determined that the system could be completely shut down from 11:50pm to 6:00am. This was done, and the energy use monitored for another month. The RL system was then converted to a POU system with three separate electric water heaters close to the points of water use (electric POU heaters were chosen because of the difficulty of venting gas heaters in a retrofit situation). Energy use for the three systems is compared in Table 8.6. Heat loss from the RL accounted for 74 per cent of the natural gas use in the original system, with standby losses from the hot water tank accounting for another 16 per cent. In the POU system, RL heat losses and pumping energy use are eliminated and tank standby losses reduced by about 75 per cent. The net result is that the hot-water requirements were met with the amount of electricity alone used in the RL system, giving an impressive 91 per cent saving in total on-site energy use. The POU water heaters were operated continuously, but it is estimated that they, like the RL system, could have been shut down at night, on weekends and during school holidays, with a further energy saving of 40 per cent (bringing the total on-site savings to 94.3 per cent). In terms of primary energy, the POU system saves 79–87 per cent, assuming natural gas and electricity to be produced and delivered to the point of use with efficiencies

Table 8.6 Comparison of secondary (on-site) energy use for supplying hot water in a US school using a recirculation loop system operating continuously (RL uncontrolled) or 18 hours per day (RL controlled), and using continuously and intermittently operating point-of-use systems (POU uncontrolled and controlled)

System	Energy use (kWh/year)			Savings (%)		
	Gas	Electricity	Total	Secondary energy	Primary energy	CO_2
RL uncontrolled	26,740	2614	29,354	–	–	–
RL controlled	22,903	2402	25,305	13.8	13.1	12.4
POU uncontrolled	–	2788	2788	90.5	78.5	67.0
POU controlled[a]	–	1672	1672	94.3	87.1	80.2

Note: Also given are the savings in secondary and primary energy and in CO_2 emissions (see text for assumptions).
[a] Estimated.
Source: Hiller et al (2002)

of 85 per cent and 33 per cent, respectively, while CO_2 emissions are reduced by 67–80 per cent if the electricity is produced from coal (see Chapter 2, Sections 2.1 and 2.4.1). If the POU heaters had used natural gas with an efficiency of 90 per cent, thereby completely eliminating electricity use, the savings in primary energy would have been comparable to the savings in secondary energy given in Table 8.6.

8.6 Recovery of heat for preheating water in commercial buildings

Building heat can be recovered and used to make or preheat DHW in two ways:

1   by using heat pumps to extract heat from outgoing ventilation air; or
2   by extracting heat from chillers and refrigeration systems in commercial buildings.

In either case, the energy used for making DHW is reduced, although, as explained below, the energy used for chilling will usually be increased if heat is extracted from chillers. When heat is recovered from chilling systems equipped with VSDs, energy consumption by pumps and cooling tower fans will be smaller compared to the case with no heat recovery, as less cooling water will need to be circulated through the cooling tower. Consumption of water by the cooling tower will also be reduced (if all of the waste heat is recovered, the cooling tower and its water use could be eliminated altogether). For chillers with

air-cooled condensers, there will be less heat pollution of the surrounding environment.

Heat is extracted from the hot refrigerant between the compressor and the condenser using a shell-and-tube refrigerant-to-water heat exchanger called a *desuperheater*. In climates where year-round air conditioning is needed, a backup hot water heater can be avoided if recovered heat is used to fully heat the DHW to the required temperature. In climates where the heat rejection from the cooling system is inadequate during the winter, a backup hot-water heater would normally be needed but can be avoided if part of the cooling tower, that would otherwise be shut down, is operated in reverse (Tan and Deng, 2001). The water normally sprayed onto the cooling coil and evaporated in an indirect contact cooling tower is shut off, while water that has been chilled to below the ambient temperature (as heat is drawn from it to the DHW tank) circulates through the closed loop of the cooling tower. The otherwise unused portion of the cooling tower serves as the outside heat exchanger of an air source heat pump for production of hot water, drawing heat from the ambient air.

8.6.1 Differences between chillers and heat pumps for heating water

It is useful at this point to clarify the distinction between heat pumps and chillers for heat recovery and cooling, as they have the same components and operate according to the same principles. If a heat pump is used for the production of hot water, the evaporator temperature must be less than that of the ambient air or whatever the heat source is, while the condenser temperature must be greater than the desired hot-water temperature. If a chiller is used for air conditioning, the evaporator temperature must be less than that of the air or chilled water that is provided by the cooling system, while the condenser temperature must be greater than the ambient dry-bulb or wet-bulb temperature. To run an existing chiller to produce hot water, the condenser temperature would have to be increased.

If either a heat pump or a chiller is used for simultaneous cooling and production of hot water, the temperature lift (from the evaporator to condenser) will usually be larger than if cooling alone or hot-water production alone were performed. This results in a smaller COP. The temperature lift and the COP penalty will be smaller the warmer the permitted cooling air or chilled water supply temperature. The lower COP means that more energy must be supplied to the compressor. This energy input is part of the recovered heat, so it is not lost, but there is a loss in the quality of the energy – electricity is being converted to heat. Of course, if the recovered heat replaces an electric resistance hot-water heater, there is no net loss in energy quality.

Shiming and Yiqiang (2003) report the operating characteristics of a $36kW_c$ water-to-water heat pump that they considered for simultaneous production of chilled water and of DHW for a hotel. For production of chilled water at 7°C and hot water at 35°C, the heating and cooling COPs are 3.2 and 2.3, respectively, giving a total COP of 5.6. However, for production of hot water at 65°C, the heating and cooling COPs are 1.1 and 0.07. The heat pump in this case is acting largely as an electric resistance heater, with very little cooling capacity. Although this may be an extreme example, it highlights the importance of determining the optimal extent of water heating with a heat pump and by an auxiliary gas heater. Also note that, even for the favourable case of 35°C water, the overall COP is still less than what can be obtained for cooling alone with large chillers (Table 6.11). This underlines the value of connecting small loads to a district cooling system, where chilled water can be produced centrally at much higher efficiencies than are possible with small, on-site chillers.

Durkin and Rishel (2003) measured the performance of a larger, $175kW_c$ chiller. When used solely for chilling with a chilled water supply temperature of 6.6°C and an air-cooled condenser, the COP is about 3.0. When operating as a heat-recovery chiller with hot-water supply and return temperatures of 54.5°C and 43.3°C, respectively, the total COP (based on chilling and heating provided) is 6.7, while at hot-water supply and return temperatures of 32.2°C and 21.1°C, the total COP is 11.5. Applications where the latter temperatures could be useful are to heat ventilation air or to preheat rather than fully heat DHW.

8.6.2 Using chillers to recover heat

There are a number of considerations in using a chiller for heat recovery, as discussed by Wulfing-hoff (1999). Operating a chiller to heat water requires increasing the condenser temperature from perhaps 35–45°C to 60–65°C. This will:

- reduce the COP (as noted above);
- reduce the cooling capacity;
- may adversely affect the compressor and motor reliability;
- may cause surging in centrifugal chillers.

Surging refers to a tendency for the gas in the perimeter of the compressor to flow against the outward flow. Centrifugal chillers specifically designed for heat recovery are available; they create a larger centrifugal force on the gas through either a larger impeller or a greater impeller speed. A heat-recovery chiller is less efficient at normal condensing temperatures than a similar chiller designed to operate at the lower condensing temperature, since it is optimized for the higher temperature.

If the amount of heat that can be recovered for hot-water (or space) heating is less than the heat rejected by the chillers, then more than one chiller should be used. One chiller should be sized to match the heating load and operated with a higher condenser temperature, and the other operated at the normal condenser temperature. Otherwise, the COP penalty applies to the entire cooling load but only a portion of the recovered heat can be put to use.

8.7 Summary and synthesis

Energy, primarily in the form of electricity, is required in order to deliver water to consumers, to treat the water before delivery and for treatment of wastewater. The electrical energy embodied in the water delivered to a typical household in OECD countries is in the order of several hundred kWh per year, which is comparable to the electricity directly used by several other major electricity household end uses. Thus, there are energy benefits from using both cold and hot water more efficiently, through water-efficient fixtures (showerheads, faucets) and toilets and appliances (washing machines, dishwashers), and through less wasteful behaviour. Recapture of greywater and its use in toilets, and capture and storage of rainwater for use in watering outdoor plants, can significantly reduce the use of cold water.

Options to reduce fossil or electrical energy used to produce hot water include: (i) use of low-flow water fixtures, more water-efficient washing machines, cold-water washing and (if used at all) more water-efficient dishwashers (50 per cent typical savings); (ii) use of more efficient and better insulated water heaters or integrated space and hot-water heaters (10–20 per cent savings); (iii) use of tankless (condensing or non-condensing) water heaters, located close to the points of use, to eliminate standby and greatly reduce distribution heat losses (up to 30 per cent savings, depending on the magnitude of standby and distribution losses with centralized tanks); (iv) recovery of heat from warm wastewater; (v) use of air source or exhaust-air heat pumps; and (vi) use of solar thermal water heaters (providing 50–90 per cent of annual hot-water needs, depending on climate). The integrated effect of all of these measures can frequently reach a 90 per cent saving.

The two largest residential DHW loads in systems with centralized hot-water tanks are: (i) standby and distribution losses, and (ii) showering. There are several options available to reduce standby and distribution losses: better insulated hot-water storage tanks and the insulation of the hot-water pipes; tankless water heaters (eliminating standby losses) or small tankless point-of-use water heaters (eliminating both standby and distribution losses); and the use of well-insulated recirculation loops that are primed with hot water only a few times for day (thereby eliminating the waste of hot water while waiting for hot water to arrive at a faucet). Heat from warm wastewater can be used to some extent to preheat cold incoming water. This should be done for showers because showering represents one of the easiest loads for recapturing waster heat and because showering is a large DHW load. It is easy to do in new construction for showers on the ground floor if there is a basement underneath, and for showers on higher floors.

Small, tankless, high-efficiency (>92 per cent) boilers for combined space and DHW heating are now available. These are significantly more efficient than standard oil and natural gas heaters (energy factor of 0.6–0.7 in new US units, less in older units). Heat at a high enough temperature for DHW can be extracted from the condensers of chillers if the chiller is operated at a higher condenser temperature than is normally the case (i.e. 55–60°C rather than 30–40°C). Heat pumps can be used to recover heat from warm exhaust air for use in making DHW when it is not practical or possible to use such heat to preheat incoming ventilation air.

nine

Lighting

Lighting energy constitutes 25–50 per cent of the total electricity used in commercial buildings in OECD countries, and so is an important area for energy conservation. The first strategy with regard to lighting in buildings should be to design the building to make maximum use of natural sunlight, subject to the various other constraints that must be satisfied. This strategy should be complemented by the design of a complementary electric lighting system that provides supplementary light only where it is needed, when it is needed, and in the quantities needed. The third strategy is to choose the most efficient individual components for use in an efficiently designed system. In this way, large (50–90 per cent) reductions in lighting energy use can be achieved in new buildings and frequently in retrofits of existing buildings, particularly when the baseline condition involves lighting in office buildings turned on 24 hours per day, as is often the case in North America. In residential buildings, the role of lighting is much less, but large efficiency gains are easier to achieve.

This chapter begins with a discussion of lighting system components and the definition of efficiency for lighting, followed by a discussion of daylighting, and finishing with a discussion of electric lighting systems and the prospects for further advances in the future.

9.1 Types of electric lighting

The main lighting technologies and the key principle by which electricity is converted to light for each technology are summarized below. Further technical details and common applications are found in Atkinson et al (1997) and in NBI (2003).

9.1.1 Types of lamps

Incandescent lamp
A tungsten filament serves as an electric resistance and, as such, is heated to a temperature of 2100–2800°C, so that about 10 per cent of the electromagnetic radiation that it emits falls within

the visible part of the solar spectrum. This is an inherently inefficient and crude way of producing light. As explained in Chapter 3, the distribution of emitted radiation shifts to shorter temperatures as the temperature of the radiation source increases. The sun, with a surface temperature of about 5800K, emits about 30 per cent of its radiation in the visible part of the spectrum, but due to partial absorption of ultraviolet radiation by ozone and of near infrared radiation by water vapour, about half of the solar radiation reaching the Earth's surface is visible radiation. Incandescent lamps have a typical lifespan of 750–1000 hours.

Halogen lamp

This is similar to an incandescent lamp, except that a small amount of a halogen is mixed with the normal inert-gas fill of an incandescent bulb. In both lamps, tungsten gradually evaporates from the filament, but in a halogen lamp, some is redeposited back on the filament (extending the lamp life). This is because the halogen gas and evaporated tungsten molecules form a compound that is not broken down unless it comes into contact with the hot filament, at which point the tungsten molecule is released directly onto the filament. A quartz rather than a glass envelope is used, permitting a higher temperature. This increases the efficiency, as more of the emitted radiation is at visible wavelengths.

Halogen infrared-reflecting (HIR) lamp

This is a halogen lamp with a coating on the inner surface of the capsule that reflects infrared radiation back to the filament, so that less electricity is needed to heat it.

Fluorescent lamp

This consists of a tube filled with argon and droplets of liquid mercury, with an electrode (cathode or anode) at each end. An electric arc travels between the electrodes, vapourizing the mercury (the discharge source) and causing it to emit ultraviolet radiation. This in turn excites electrons in the *phosphors* coating the inner tube, which then emit light of various wavelengths (depending on the chemical composition of the phosphors) as they drop down in energy level. At present, two kinds of fluorescent lamps are used: two-colour

(di-chromic), in which two phosphors emit blue and yellow light which, when mixed, appear close to white; and three-colour (tri-chromic), in which three phosphors emit red, green and blue light to produce white light that more accurately matches natural light. Fluorescent lamps used in commercial buildings are thin tubes 1–2m in length and 12, 8 or 5 eighths of an inch in diameter (referred to as T12, T8 and T5 lamps, respectively).

Compact fluorescent lamp (CFL)

This is a fluorescent lamp small enough and designed in such a way that it will screw into a regular incandescent light socket. It can be used for residential lighting and for pot lighting. CFLs have a lifespan of about 9000–12,000 hours.

Electrode-less induction lamp

This is a fluorescent lamp without an electrode, introduced commercially in the mid-1990s. It works by converting 50 or 60Hz AC electricity into radio frequency power, which is fed into an electrical coil, where a plasma is excited that releases ultraviolet radiation (Atkinson et al, 1997). Since there are no electrodes to burn out or fail, the lamp can be turned on and off without affecting its life. The lamp's lifespan (up to 60,000 hours) is limited by the degradation of the phosphors.

High intensity discharge (HID) lamp

In an HID lamp, the discharge source directly radiates in the visible part of the spectrum, avoiding the intermediate step involving ultraviolet radiation. There are four kinds, based on the discharge source: mercury vapour, high-pressure sodium, metal halide and sulphur. Until recently, there were only high-output lights, restricted to outdoor lighting (including street lighting and stadiums) and to athletic facilities (such as school gymnasiums). Some low-pressure HID lamps are available for indoor lighting in commercial buildings, with a typical power of 100 watts. Newer HID lamps are dimmable. They require 2–10 minutes to reach full output, and if they lose power, must cool down before they can be restarted. Pulse-start metal halide lamps have a more consistent lumen output than standard metal halide lamps, permitting 25 per cent lower wattage in high bay warehouses and retail and industrial applications where these lamps are used.

Ceramic metal halide (CMH) lamp

The metal halide lamp (one of four kinds of HID lamps) contains a quartz arc tube. Recent advances have allowed replacement of the quartz arc tube with a ceramic arc tube, which in turn permits operation at a higher temperature, with better colour rendering and warmer tones. Compact metal halide lamps are available in reflector lamp configurations that can replace HIR lamps in down lighting, track lighting and retail displays. CMH lamps use less than half the energy for the same light output as HIR lamps and last —three to four times longer (Sachs et al, 2004). CMH lamps are available in the 39–400W range, with a 39W lamp (44W with ballast) replacing a 100W HIR lamp.

Light-emitting diode (LED)

An LED is like a photovoltaic (PV) cell running in reverse. In a PV cell, incident light is converted to electricity using a pair of semiconductors that form a p–n junction (as described in Chapter 11, Section 11.2), while in an LED, a semi-conductor junction is used to convert electricity into light. AC electricity must first be converted to DC electricity. Electrons flow from a cathode into an n-layer or electron-transport layer, while holes (positively charged) are created in a p-layer or hole-transport layer. The electrons and holes travel in opposite directions and combine at the interface between the p and n layers. Energy released by the recombination excites an atom or molecule, which subsequently radiates light through the anode (which is transparent). The colour of the radiated light depends on the chemical composition of the semi-conductors, which involve alloys of gallium, arsenic and indium (as in the more efficient PV cells). The first LEDs produced red light and have been used for traffic lights and exit signs. White light can be produced by mixing light from different LED sources; by using blue LEDs with one or more phosphors (which absorb some of the blue light and re-emit at other wavelengths); or by using UV LEDs with multiple phosphors. Commercial LEDs using the last two approaches are available, but the first approach is likely to be more efficient (Johnson and Simmons, 2002). LEDs using organic materials between the cathode and anode are under development but are at a rather early stage. White LEDs are available with standard screw-in bases that fit into incandescent fixtures. A 1.8W white LED cluster produces as much light as a 25W incandescent lamp, but the cost (including transformer) is about $140 according to information available in 2005 on the National Association of Home Builders technical information website (www.toolbase.org). A 5W desk LED lamp is available for about $90 that can replace a 60W incandescent lamp (see www.tcpi.com) – an energy saving of over 90 per cent. However, white LEDs are mainly used in niche applications where their 100,000-hour lifespan makes them worthwhile.

Electro-luminescent (EL) lamp

EL lights consist of a thin layer of phosphor-impregnated material that is sandwiched between two layers of conducting material, one of which is clear. The phosphor layer produces light when voltage is applied between the two conducting layers. The colour of the light depends on the phosphor that is used. EL lights are ideal for exit signs, the majority of which use incandescent bulbs.

9.1.2 Discussion of the various types of lamps

Fluorescent and HID lamps contain an extra component, called a *ballast*, which converts the 50 or 60Hz (Hertz, 1 Hz = one cycle per second) AC input electricity into AC electricity at a much higher frequency for use by the lamp. The ballast limits the current flow, which would otherwise be too large due to the small resistance once the discharge arc is established. Some of the electrical energy input is lost in the ballast. Fluorescent lights turn on or off with each half of the input current, and the ballast must provide sufficient voltage at the start of each half-cycle to restart the lamp. Older CFLs (and fluorescent tubes) used magnetic ballasts that converted the input electricity to a frequency of about 120Hz. A flicker became noticeable as the lamp aged or if it was dimmed. All new fluorescent lights use electronic ballasts to create a frequency of 20–40kHz. Flicker is eliminated, energy losses in the ballast (as waste heat) are reduced, and lifespan is increased (due to a lower operating temperature, a byproduct of greater efficiency).

Three kinds of electronic ballast are available for fluorescent tubes (McCowan et al, 2002): rapid-start, instant-start and programmed-start. Rapid-start ballasts use slightly more energy than instant-start ballasts because they send a constant warming current to the lamp electrodes, but the lamps have a longer rated lifespan (20,000 hours vs 15,000 hours for instant-start ballasts). Instant-start ballasts should not be used with occupancy sensors due to reduction in lamp life. Programmed-start ballasts use a start procedure that is almost as fast as the instant-start ballast, but is easy on lamp electrodes, extending the lamp life from 18–20,000 starts to 30–40,000 starts. Some ballasts can sense the impending end of the life of the lamp and will shut the lamp off until replaced; such ballasts should be used with T5 lamps because these can disintegrate at the end of their life.

CFLs come in two forms: pin-based and screw-based. Pin-based CFLs have a separate ballast and lamp, with the lamp attached to the ballast with a pin and the ballast screwed into an incandescent lamp fixture. Screw-based CFLs can be directly screwed into an incandescent lamp fixture. For residential lighting, some have argued that only pin-based CFL fixtures should be provided, in order to prevent the homeowner from converting back to inefficient incandescent lamps. Faesy et al (2004) argue in favour of screw-based CFLs because (i) screw-based CFLs are now very inexpensive ($3–4 per lamp); (ii) they are easy to replace if defective or when they wear out, as they are now available everywhere; (iii) one doesn't have to worry about finding the correct ballast/lamp combination after 9–10 years of use (when they will normally burn out); and (iv) there is no evidence that the persistence of the energy savings is any less when screw-based rather than pin-based fixtures and lamps are provided.

Fluorescent lamps require a minimum ambient temperature in order to start. Various CFLs can start at temperatures ranging from −30°C to 0°C. All fluorescent lamps have peak efficiency near a temperature of about 40°C. If the room is kept at normal indoor temperature and the fixture is well ventilated, the lamp will stabilize at a temperature near the optimum temperature. Most CFLs have preheat starters, which take 1–2 seconds to work, after which another few seconds are required to reach full brightness. Some fluorescent lamps combine mercury with another metal in order to maintain high efficiency over a wide range of temperatures by stabilizing the mercury vapour pressure. However, up to a full minute may be required for the lamp to reach full brightness, as the mercury separates from the other metal.

An issue related to CFLs is the need to separate and recycle the mercury that is an inherent part of the technology. Of course, this is also an issue for commercial fluorescent tubes, but in many jurisdictions, systems are in place for the collection and proper disposal of commercial fluorescent tubes. According to one manufacturer (Philips, www.lighting.Philips.com), the loadings are as follows: Series 80 T12 tubes, 2mg/lamp (compared with a stated industry range of 4–8mg/lamp); T5 tubes, 1.4mg/lamp versus an industry benchmark of 2.5–5.0mg/lamp; CFLs, 1.4–2.0mg/lamp, compared with an industry benchmark of 4–5mg/lamp. The mercury loading in CFLs should be compared with the avoided mercury emissions from coal power plants due to reduced electricity use. According to US EPA (1995), mercury emissions from coal power plants are 0.014 tonnes/GW-year and 0.6 tonnes/GW-year from controlled and uncontrolled emissions, respectively. Assuming a lamp lifespan of 7000 hours, replacing a 100W incandescent lamp with a 24W CFL avoids a mercury emission of 0.84mg if the power plant has state-of-the-art controls, and avoids 36mg if the power plant has no controls. These numbers would of course vary with the specific coal being used. Nevertheless, in most cases the mercury loading in the lamp will be less than the avoided emission. This of course does not eliminate the need to recycle the mercury at the end of the lamp life.

9.2 Measuring the efficiency of lamps in producing light

Light is electromagnetic radiation – a form of energy – with wavelengths between 0.4–0.73μm. However, the efficiency of a lamp in producing light cannot be specified as the ratio of watts of light output to watts of electrical energy input. This is because not all watts of light are equal; our eyes are more sensitive to some wavelengths of radiation than to others. The relevant quantity is the *efficacy* of the lamp, which is defined as:

(lumens out)/(watts in)

A lumen (lm) is a measure of light weighted by the sensitivity of our eyes to various wavelengths.

The efficacy of a light source is given by (Lampret et al, 2002):

$$\xi = \frac{683 \int_0^\infty V(\lambda)E(\lambda)d\lambda}{\int_0^\infty E(\lambda)d\lambda} \qquad (9.1)$$

where $V(\lambda)$ is the variation with wavelength in the relative sensitivity of the human eye to radiation (this variation is shown in Figure 3.1, and can be computed using a simple expression given in Lampret et al, 2002), and $E(\lambda)$ is the variation with wavelength in the amount of radiation (W/m^2/μm) emitted by the light source. The factor of 683 times $V(\lambda)$ gives the *spectral efficacy* – the lumens of light per watt for radiation restricted to a very narrow wavelength interval centred at wavelength λ.[1] Since $V(\lambda)$ peaks at a value of 1.0 at a wavelength of 0.556μm in the green part of the spectrum, green light has the highest efficacy.

When calculating the efficacy of natural sunlight, $E(\lambda)$ would be the distribution of solar radiation with wavelength. For extraterrestrial solar radiation, the efficacy is 101lm/W. At ground level, $E(\lambda)$ and hence the efficacy of sunlight depends on cloudiness, the solar altitude and atmospheric conditions, as these influence the relative amounts of solar radiation at different wavelengths. It averages around 120lm/W for total solar radiation (Robledo and Soler, 2000; Vartiainen, 2000) and about 160lm/W for the diffuse component (Robledo and Soler, 2001), in both cases with a large variability.[2] The efficacy of daylight passing through a window can be increased if only the most efficacious wavelengths of solar radiation are admitted. In the extreme case where only light in a wavelength interval centred around 0.556μm (green light) and wide enough to provide 1200 lux is used, the spectral efficacy is 682lm/W. If only visible light (wavelengths of 0.4–0.7μm) is admitted, the efficacy is about 260lm/W, and

if the admitted wavelength interval is restricted slightly further (to 0.45–0.65μm), the efficacy is about 360lm/W.

The efficacy of electric lighting is computed by replacing the denominator of Equation (9.1) with the power used by the lamp, and replacing $E(\lambda)$ in the numerator with the variation with wavelength of radiation emitted by the lamp in question. The efficacy and operating lifespan of the major lighting technologies are given in Table 9.1. The efficacy of a lamp increases as the input wattage increases (so that the luminous flux increases faster than the power input). Table 9.2 compares the wattage, luminous flux and efficacy for different incandescent and CFL lamps. CFL data are shown for electromagnetic and electronic ballasts, the latter being the standard now. Electronic ballasts have a number of advantages over magnetic ballasts, including being rated to start at temperatures as cold as −30°C (making them suitable for outdoor use in all except the coldest climates) and instant starting (although up to one minute is required to reach full brightness).

The key comparisons evident from Tables 9.1 and 9.2 are as follows:

- CFLs are about four times as efficacious as incandescent lamps, two or three times as efficacious as HIR lamps, and three times as efficacious as halogen lamps (all of which they can replace in almost all applications);
- the 80-series T8 and T5 fluorescent tubes are about 60 per cent more efficacious than T12 tubes and 25 per cent more efficacious than standard (70-series) T8 tubes;
- the CMH lamp is about twice as efficacious as the HIR lamp (which it can replace).

A four times greater efficacy implies one-quarter the electricity use for the same light output.

The difference between efficiency and efficacy can be understood by reference to Table 9.3, which compares the disposition of the electrical energy input and the efficacy in four different kinds of lamp. In the 100W incandescent lamp, 10 per cent of the energy input is converted

Table 9.1 Efficacy and operating lifetime of different lighting technologies

Lamp technology	Efficacy (lumens/watt)	Lifetime (hours)	Colour rendering index
Incandescent	10–17	750–2000	98–100
Halogen	12–22	2000–4000	98–100
HIR	20–25[a]	2000–3000	98–100
Compact fluorescent	50–70	10,000	65–88
T12 Fluorescent	62	20,000	62
T8 Fluorescent, 70-series	80	20,000	75
T8 Fluorescent, 80-series	102	20,000	86
T8 Fluorescent, 90-series	62	20,000	95
T5 Fluorescent	105[b]	16,000	85
Electrodeless induction	50	60,000	
HID, Mercury Vapour	25–60	16,000–24,000	50
Metal halide	70–115	5,000–20,000	70
High-pressure sodium	50–140	16,000–24,000	25
Sulphur	80–100	?	
CMH[c]	56–62	9–12,000	
LED[d]	30 (current) 100 (laboratory) 150–200 (projected)	100,000	

[a] Upper value from Turiel et al (1997).
[b] Wisconsin Energy Center (www.ecw.org) fact sheet, *T5 Linear Fluorescent Lamps*.
[c] Sachs et al (2004).
[d] Current efficacy from Johnson and Simmons (2002), laboratory efficacy from Whitaker (2003), projected efficacy from S. Johnson (2000, 2002).
Source: Rows 1–4 and HID data except sulphur, US DOE (2002b); T5–T12 fluorescent lamp data, McCowan et al (2002); other data, Atkinson et al (1997) except where indicated otherwise

Table 9.2 Required power, luminous flux and efficacy of various incandescent and compact fluorescent lamps

Incandescent			CFL, magnetic ballast			CFL, electronic ballast		
Watts	lumens	lm/W	Watts	lumens	lm/W	Watts	lumens	lm/W
25	220	8.8	5	220	44.0			
40	495	12.4	7	400	57.1			
60	855	14.3	13	860	66.2	15	900	60.0
75	1170	15.6	18	1160	64.4	18	1100	61.1
100	1680	16.8	26	1700	65.4	25	1750	70.0

to visible light and the efficacy is 17.5lm/W. In the 40W fluorescent lamp, twice as much of the input energy (20 per cent) is converted into visible light, but the efficacy is over four times larger (79lm/W). This is because that portion of the emission from the fluorescent lamp that is within the visible part of the spectrum is shifted toward wavelengths to which the human eye is more sensitive. Some of the radiation emitted at infrared wavelengths is re-absorbed within the lamp; its re-absorption leads to a higher temperature and greater emission across the electromagnetic spectrum for a given electrical input. Table 9.3 gives the fraction of electrical energy input that is converted into light and infrared radiation (both emitted and re-absorbed) for incandescent, fluorescent, metal halide and high-pressure sodium lights of a particular wattage. Ballast energy losses (as heat) are also given where applicable.

Table 9.3 Distribution of power input for different lighting technologies with the specified power input (the distribution will be slightly different for lamps with different power inputs)

	100W Incandescent		40W Fluorescent	400W Metal hydride	400W High-pressure sodium
	120V[a]	230V[b]			
UV (%)	–	–	–	3	–
Visible (%)	10	5	20	21	30
Total NIR+IR (%)	90	95	63	63	55
Absorbed NIR+IR (%)	18	12	30	31	20
Ballast losses (%)	–	–	17	13	15
Efficacy (lm/W)	17.5	13.8	78.7	100	125

[a] 115–125 volts is the standard socket voltage in North America.
[b] 220–240 volts is the standard socket voltage used outside North America.
Source: McGowan (1989)

The efficacy of the lamp is only one factor in assessing the overall efficiency of a lighting system. Other factors include the efficiency of the luminaire (the reflective or transparent components that partially or entirely surround the lamp) and how much of the light gets to where it is needed. The T5 fluorescent tube has an efficacy comparable to that of the 80-Series T8 fluorescent tube, but because it is a smaller and hence more concentrated light source, it can be combined with a luminaire that transmits more light and directs more of it to where it is wanted, thus giving a greater overall efficiency (up to 34 per cent more light output per watt of input power). Another factor affecting overall efficiency is the tendency, in some lamps, for the lumen output to decrease over time. This is compensated by specifying a greater light output than is needed, so that there is still adequate light by the end of the lamp's life.

The difference between efficacy and efficiency in a broader sense is illustrated by exit lights. Lighting options for exit signs are compared in Table 9.4, where it can be seen that LEDs require only one-tenth the energy of incandescent lamps, even though the efficacy of white-light LEDs is only a factor of two better than that of incandescent lamps. However, an exit sign LED produces only red light, while the incandescent lamp produces light of all colours but the light is filtered so that only red light is seen. EL lamps yield even more dramatic energy savings in applications such as exit signs, as indicated in Table 9.4.

Table 9.4 Lighting options for exit signs

Lighting type	Average power (W)	Savings (%)	Lamp life (years)
Incandescent	32	–	0.20.7
CFL	17	47	1–2
LED	3	91	10–25
EL	0.75	98	8

Note: CFL = compact fluorescent lamp, LED = light-emitting diode, EL = electro-luminescent.
Source: Kubo et al (2001)

9.2.1 Impact of age

The efficacy and output of most lamps decrease as they age. In the case of incandescent lamps, this occurs through gradual evaporation of tungsten from the hot (2100–2800°C) filament and its deposition on the inside of the lamp shell. The filament becomes thinner, increasing resistance and reducing light output more than energy consumption. In addition, the tungsten deposited on the inner shell absorbs some of the light. The efficacy of an incandescent bulb will typically fall to 80–85 per cent of its initial efficacy by the time it burns out. In halogen lamps, much of the evaporated tungsten is redeposited on the filament, as explained above. As a result, the efficacy falls by only about 3 per cent during the life of the bulb (Wulfinghoff, 1999, Reference Note 54). The efficacy of fluorescent lamps decreases by 10–25 per cent over their life due to degradation of the phosphors and deposition on the lamp shell of material evaporated from the electrodes. This is illustrated in Figure 9.1, which compares the drop

Figure 9.1 Decrease in light output over time for various fluorescent lamps

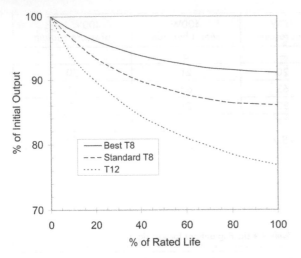

% of Initial Output vs % of Rated Life

Source: Benya et al (2003)

Figure 9.2 (a) Variation of light output and efficacy as the power input is varied for incandescent and fluorescent lamps. (b) Percent energy saved versus percent dimming for incandescent and fluorescent lamps

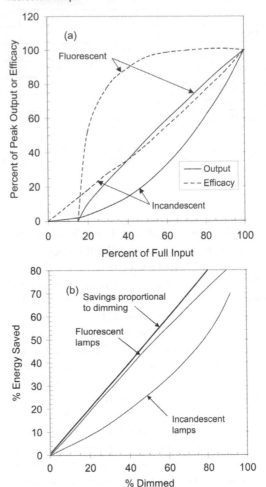

Source: (a) Benya et al (2003), (b), computed from data in (a)

in output over lifespan for T12, standard T8, and advanced T8 fluorescent tubes. Among HID lamps, the lifetime loss of efficacy amounts to about 25–60 per cent for mercury vapour lamps, 25–40 per cent for metal halide lamps, and 15–20 per cent for high-pressure sodium lamps (Wulfinghoff, 1999, Reference Note 56). The output of LEDs falls by 50 per cent during the first 10,000 hours, thereby substantially reducing their useful life compared to their rated life of 100,000 hours (Benya et al, 2003).

9.2.2 Availability and impact of dimming

Incandescent and fluorescent lighting can be dimmed. Early fluorescent systems, using magnetic ballasts, were not particularly good. Recall that fluorescent lights turn on or off with each half of the input current (which is boosted to a frequency of 20–40kHz by the ballast), and that the ballast must provide sufficient voltage at the start of each half-cycle to restart the lamp. Dimmable electronic ballasts simultaneously adjust the voltage and current. High-end ballasts can dim down to 1 per cent of full output, while more common units dim down to 5–20 per cent. Dimmable CFLs are now available – in 15W, 18W and 23W sizes (see www.realgoods.com).

Figure 9.2a shows the variation of output and efficacy as input power is reduced for dimmable fluorescent and incandescent lamps. The efficacy of fluorescent lamps is roughly constant down to about 40 per cent of full output, then drops sharply, whereas the efficacy of incandescent lamps falls continuously as output decreases. As a result, the fractional savings in energy use is roughly proportional to the reduction in output for fluorescent lamps but is about half the fractional reduction in output for incandescent lamps, down to about 40 per cent of peak output, as shown in Figure 9.2b. Lamps of the correct maximum output should be installed in fixtures with dimmers, rather than relying on dimmers to reduce the output, due to the decrease in efficacy with decreasing output.

LEDs can also be dimmed, but with a slight increase in efficacy because they operate at a cooler temperature when dimmed (as discussed in Chapter 11, Section 11.2.2, PV modules – which are like LEDs in reverse – also operate at higher efficiency at cooler temperatures).

With the availability of dimmable fluorescent lighting, there is no longer any need to install separate fluorescent (at full light output) and dimmable incandescent lights in classrooms and conference rooms, as has been the practice. Neither is there any need for incandescent lamps in residential applications where dimming is often used (such as dining and living rooms). For offices, inexpensive desktop controls are available that permit remote dimming.

9.3 Daylighting in buildings

Advances in window technology make it possible to increase window area and hence daylighting opportunities without increasing heat loads in winter for most window orientations (see Chapter 3, Section 3.3.12) and with minimal impact on cooling loads in summer, the latter being greatly aided through double-skin façades with adjustable shading devices (see Chapter 6, Section 6.3.2). Although daylighting is advocated here as an energy-saving measure, it should be pointed out that there are psychological, health and productivity-related benefits of daylighting over electric lighting. There have been many studies claiming to show that daylighting can dramatically increase the rate of learning in schools, increase worker satisfaction and productivity, and increase retail sales. These and other potential benefits of daylighting are thoroughly and objectively reviewed by Boyce et al (2003). While finding fault in the methods used in many of these studies, Boyce et al (2003) nevertheless conclude that there are likely to be positive impacts in each of these areas. These impacts do not have to be large for the economic value of the benefits to exceed the energy cost savings of daylighting. Of particular importance is the *way* in which daylighting is delivered (something that is also important for electric lighting). However large or small the non-energy benefits of daylighting may be, there seems to be no disagreement with the statement that daylighting and associated views of the outside are strongly desired by people and, when provided, have significant psychological benefits.

9.3.1 Techniques

Lam (1986) provides a well-illustrated general discussion of conventional techniques for daylighting using skylights, atria, clerestories, light wells, light shelves and shading devices, along with 25 case studies based on his own work in projects around the world. Particular attention is given to aesthetic and lighting-quality issues. Hastings (1994) provides many examples, photographs and quantitative analyses of buildings that have employed effective daylighting strategies, while IEA (2000a) provides a comprehensive sourcebook of conventional and less conventional techniques and technologies for daylighting. Ternoey (2000a, 2000b) provides a workbook of design approaches, cost and performance data for daylighting in offices and schools. Baker and Steemers (2002) provide much practical information and discuss the strengths, weaknesses and limitations of many techniques. The *National Best Practices Manual for Building High Performance Schools* (US DOE, 2002a) amply illustrates a variety of daylighting techniques used in schools in the US. The Pacific Gas and Electric power company (California) has a series of brochures (available from www.pge.com) on the use of daylighting in retail, industrial, educational, office, restaurant and museum settings. Leslie (2003) provides a listing of other references and design guides. The use of daylight must be coupled with wiring of the electric lighting system in zones parallel to the windows and the use of dimmable lighting systems, with light sensors and occupancy sensors (in private spaces) controlling each zone so that the electric light output can be automatically adjusted up or down as the availability of sunlight changes, and automatically turned off when private rooms or offices are not occupied.

The techniques available for maximizing the use of daylight are outlined below, beginning with the basic design decisions that will be made very early in the design process by the architect, followed by relatively simple, low-cost and well-established techniques for controlling daylight, and proceeding to more experimental but promising techniques. Daylighting systems can be designed to concentrate and redirect diffuse sunlight into the interior of a room, or to reflect direct beam sunlight into a room, with or without shading the window from direct beam sunlight. Conventional shading systems, such as pull-down shades or louvres, often significantly reduce the admission of daylight into a room. Systems that can concentrate and direct diffuse light into a room provide daylighting under overcast conditions.

Building form

Buildings that are narrow enough that light can penetrate to all of the occupied space from both sides will require less electric lighting than thick, massive buildings. This entails long and thin buildings rather than rectangular buildings, which will increase the surface-area-to-volume ratio, possibly increasing the heating and/or cooling requirements and construction cost per unit of floor area. However, if the building is oriented east–west rather than north–south, and if a high-performance thermal envelope is specified, the impact on heating and cooling energy use will be minimal. The north side (in the northern hemisphere) is excellent for daylighting, since there are virtually no problems with glare or excessive heat gain, while the south side is good because the higher-angle sun can be controlled by a variety of techniques.[3]

Window size, shape and position

For a given size, a narrow, high window will admit more light to the back of a room than a wide, lower window. Larger windows will obviously admit more light, but there is little gain in daylighting once the product of window:wall area and visible transmittance (the window 'aperture') exceeds 0.3 (Krarti et al, 2005). Many of the techniques described below rely on a division of the glazed area into a lower viewing component (through which transmission of direct beam radiation is minimized) and an upper daylighting component (which transmits direct beam radiation or concentrated diffuse radiation along the ceiling).

Atria

Provision of an interior atrium can increase daylighting with minimal impact on heating and cooling energy use. Design attributes of an atrium include its shape and orientation to the sun,

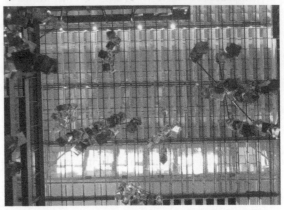

Figure 9.3 View looking up to the roof of the atrium of the Genzyme corporate headquarters building, Cambridge (Massachusetts). Sun-reflecting heliostats are visible in the lower part of the photo, adjustable louvres in the middle part, and operable office windows facing the atrium in the upper part

Note: Also visible is a suspended reflective sculpture that helps to spread light laterally.
Source: Author

the transmittance of the atrium roof, and the reflectivities of the atrium walls and floor. Calcagni and Paroncini (2004) review tools for predicting the distribution of daylight in buildings with atria. The daylighting potential of an inner atrium has been fully exploited in the Genzyme Corporate Headquarters building in Cambridge, Massachusetts, through the use of a roof mounted sun-tracking heliostat and reflecting mirrors, which direct sunlight down into the atrium, and through ornamental and reflective suspended artwork that deflects light laterally (Figure 9.3). Solar chimneys (Chapter 6, Section 6.3.1) are another design element that can sometimes be exploited for daylighting opportunities.

Skylights and roof structure

Roof structures such as sawtooth roofs and *clerestories* (glazed vertical steps in the roof) can be used to distribute natural light throughout a building, as illustrated schematically in Figure 9.4 and with real examples in Figures 9.5 and 9.6. North-facing clerestories can supply adequate daylight while minimizing overheating in summer and avoiding glare. Vertical reflective baffles outside polar-facing clerestories can redirect sunlight into the clerestory. Skylights need cover no more than 5 per cent of the roof area in order to provide

Figure 9.4 Examples of roof configurations that increase the opportunity for daylighting. See Figures 9.5 and 9.6 for photographic examples

Figure 9.5 Daylighting though roof lights and sidelighting; top: Power Gen building, UK; middle: sidelight and roof light; bottom: rooflight in an art gallery

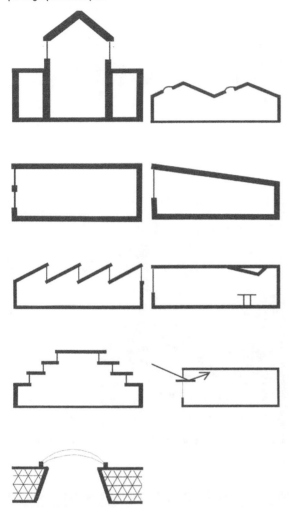

Source: Hastings (1994, chapter 21)

Source: IEA (2000a)

adequate daylight under most daytime conditions. Translucent aerogel skylights will require a greater roof area, but provide a high thermal resistance (up to RSI 3.52 or R20). Domed skylights provide significantly more light at low sun-angles (when it is needed the most) than flat skylights. In a classroom at the University of Wisconsin, skylights are combined with a suspended deflector to create diffuse light that is reflected across the ceiling, and a motorized blackout panel is used when needed with audiovisual equipment (Mendler and Odell, 2000).

Figure 9.6 Exterior and interior views of the Solar Energy Research Facility at the National Renewable Energy Laboratory (NREL), Golden, Colorado

Source: Photography by Warren Gretz, obtained from the Photographic Information Exchange of NREL (www.nrel.gov) and used with permission

Figure 9.7 Examples of a relatively simple (left) and advanced (right) light-shelf system for daylighting. See Figure 9.8 for photographic examples

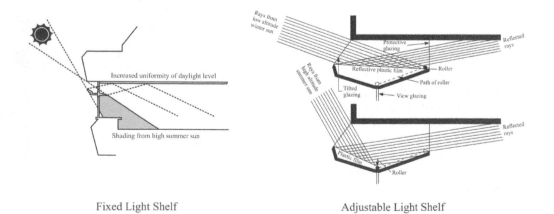

Fixed Light Shelf   Adjustable Light Shelf

Source: Hastings (1994. chapters 21 and 22)

Light shelves

These are horizontal or inclined reflective surfaces that are placed near the upper part of a window, either on the outside or inside, and used to reflect light onto the ceiling and deep into the room. Both fixed and sun-tracking light shelves can be used, as illustrated schematically in Figure 9.7. Fixed interior shelves are illustrated in Figure 9.8. Interior light shelves are perhaps the simplest and least expensive way of making better use of daylight. They are most effective on south-facing walls. Light shelves reduce the light level near the window (where it is usually in excess), thereby reducing lighting contrast and making the light more useful deeper in the room. Light shelves have limited applicability at high latitudes, as additional shading devices will be needed most of the year. Interior light shelves provide negligible reduction in cooling load, while external light shelves reduce cooling loads by providing shading and because the absorption of solar radiation by the shelf occurs outside the building. External light shelves also provide more daylight than internal shelves. Light shelves can sometimes be designed to allow maximal solar heat gain in winter (IEA, 2001). They should be as reflective as possible and combined with highly reflective ceilings.

Light-directing louvres

These are shaped to reflect the maximum amount of light onto the ceiling while permitting very little light to pass horizontally, thus eliminating glare. They are illustrated in Figure 9.9. Existing light-directing louvres substantially reduce the total amount of daylight.

Automatic Venetian blinds

Automatic controls can adjust the tilt-angle of Venetian blinds so as to compensate for changing outdoor light levels and solar position. They should be placed between window glazing layers in order to prevent the accumulation of dust, thereby eliminating the need for cleaning.

Light pipes

Active light pipes consist of sun-tracking mirrors that reflect and possibly concentrate light onto a reflective hollow tube, which carries the light via internal reflections into the interior of large commercial buildings, without glare, as illustrated in Figure 9.10 and Figure 9.11. They are rather expensive, and very few have been installed. Passive

Figure 9.8 Daylighting through interior light shelves. Top left: Semi-transparent double light shelves made out of reflective glass. Top right: Interior light shelf – note light reflected onto the ceiling, while providing some shading below. Middle: Light shelves at Sacramental Municipal Utility District (SMUD) headquarters, Sacramento, California. Bottom: Light shelves in the Michael Capuano Early Childhood Center in Boston

Source: Top and middle, IEA (2000a); bottom, author

light pipes usually consist of a clear dome to collect light on the outside, the light pipe itself, and a diffuser that spreads the light at the other end, as illustrated in Figure 9.12. Over 3 million passive light pipes have been installed worldwide according to Jenkins and Muneer (2004). Recent developments in active and passive light pipes are reviewed in Carter (2004). Rosemann and Kaase (2005) document the design and performance of a sun-tracking system installed at the Technical University of Berlin.

Figure 9.9 Light-directing louvres: (a) the fish system, (b) the Okasolar system

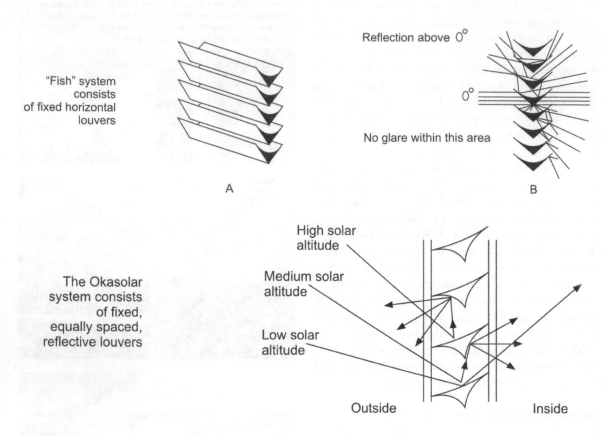

Source: IEA (2000a)

Figure 9.10 Example of a sun-tracking (active) light pipe system used in a building in Toronto

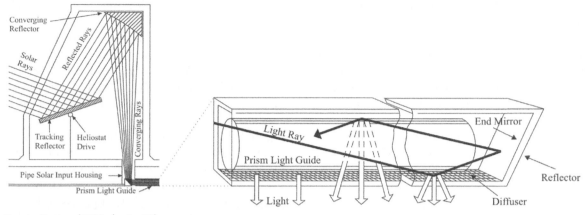

Source: Hastings (1994, chapter 22)

Figure 9.11 A light pipe in a 14-storey shaft

Note: Light is fed into the light pipe from a rooftop sun-tracking heliostat, and refracted laterally by prismatic glass.
Source: Anonymous (2004), from International Association of Lighting Designers

Figure 9.12 Illustration of a passive light pipe

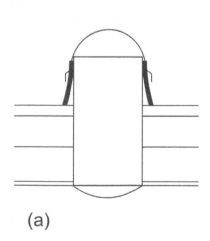
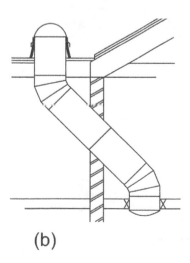

(a)   (b)

Source: Zhang and Muneer (2002). © Edward Arnold (Publishers) Ltd (www.hodderarnoldjournals.com)

Prismatic panels

These consist of an array of acrylic prisms with one surface of the prism aligned to form a plane, as illustrated in the upper left of Figure 9.13. A wide range of optical effects is possible by altering the geometry of the prism faces, the tilt of the panel, and by covering one of the prism surfaces with a reflective coating. They can be used:

- to redirect direct sunlight onto the ceiling while transmitting diffuse light without alteration;
- to direct sunlight horizontally into a room;
- to serve as a fixed sun-shading system, usually inside a double-glazed roof;
- to serve as moveable sun-shading louvres, redirecting direct sunlight to the outside while admitting diffuse light.

Figure 9.13 Daylighting through prismatic panels. Cross section of a linear prismatic panel and visualization of light direction achieved by the panel (upper left), fixed sun-shielding and light-redirecting prismatic panels inside double glazing at Sparkasse, Bamberg, Germany (upper right), and movable sun-shading prismatic panels on the German Parliament Building in Bonn (lower)

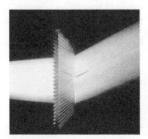

Source: IEA (2000a)

Laser-cut panels

A laser-cut panel (LCP) is a thin panel in which a number of closely-spaced laser cuts have been made, with solid sections around the edge to support the cut sections, as illustrated in the upper left of Figure 9.14. A large portion of the incoming light is deflected through an angle greater than 120° while providing a view to the outside that is similar to that through open Venetian blinds (see Figure 9.14, upper right). Unlike light shelves or prismatic glazing, dust cannot accumulate. LCPs can be mounted directly onto pre-existing glazing, or used as one of the primary glazings in double-glazed windows. Vertical panels will redirect downward light upward, so they should be installed only in the upper portion of windows, possibly in combination with Venetian blinds in the lower part in order to admit just enough light for daylighting while minimizing solar heat gain, as illustrated in Figure 9.15 and in the middle photo of Figure 9.14. Conversely, they can be placed horizontally between the panes of double-glazed windows. In this case, they will redirect to the outside a large portion of the incoming sunlight (thereby reducing solar heat gain) while permitting a view of the outside. They can also be designed as movable louvres, directing sunlight onto the ceiling when vertical and rejecting a large portion of sunlight when horizontal. In normal atria they can be used to deflect most of the solar radiation impinging from a high elevation while admitting most of the low-elevation radiation, as illustrated in Figure 9.16, while in inverted atria they can be used to spread light over the ceiling, as illustrated in the lower photo of Figure 9.14. Laser-cut panels tilted from the upper part of windows can be used to increase daylight levels by a factor of three at the rear of lower-level rooms in highly obstructed residential developments (Edmonds, 2005).

Figure 9.14 Daylighting through laser-cut panels. Top: View through a laser-cut panel (left) and 20mm wide laser-cut panels installed Venetian style between two glass panes (right). Middle: Laser-cut panels in awning windows deflecting sunlight onto the ceiling of a classroom in Brisbane, Australia. Bottom: Laser-cut panels in an inverted skylight, reflecting light onto the ceiling and creating a more uniform distribution of light than would occur otherwise

Figure 9.15 Combination of a laser-cut panel in the upper part of a window and Venetian blinds in the lower part of a window to admit just enough light for daylighting without glare. See Figure 9.14 for photographic examples

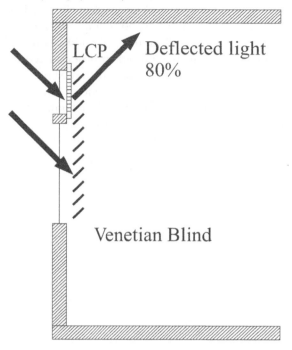

Source: Edmonds and Greenup (2002)

Source: IEA (2000a)

Figure 9.16 Use of laser-cut panels in a skylight to impede penetration of solar radiation from high elevations but not from low elevations. See Figure 9.14 for photographic examples

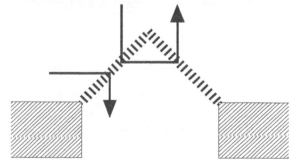

Source: Edmonds and Greenup (2002)

Light-guiding shades

A light-guiding shade simultaneously shades a window while directing the light incident on it inside and onto the ceiling, as illustrated in Figure 9.17. They are designed to improve the daylighting in rooms of subtropical buildings where external shading is required in order to reduce heat gain through windows, so they should be compared against shaded windows. Exterior and interior views are provided in Figure 9.18. Light guiding shades must be made with precisely shaped and highly reflective surfaces. They are based on non-imaging optics, as are anidolic systems (described below).

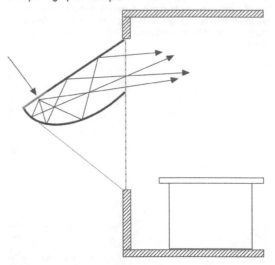

Figure 9.17 Illustration of a light-guiding shade, which shades a window while directing a portion of the incident light inward along the ceiling to provide daylighting without glare. See Figure 9.18 for a photographic example

Source: Edmonds and Greenup (2002)

Figure 9.18 Daylighting through light-guiding shades. Exterior view of a light-guiding shade on a school in Brisbane, Australia (upper), and interior view with a conventional shade (lower left) and a light-guiding shade (lower right). The latter directs sunlight onto the ceiling, and is intended for use at low latitudes where shading of the windows is needed to reduce heat gain

Source: IEA (2000a)

Sun-directing glass

Sun-directing glass consists of a doubled-glazed unit with concave acrylic elements stacked vertically between the glazings. Sunlight from a wide range of angles is directed onto the ceiling, as illustrated in Figure 9.19. Examples are shown in Figure 9.20. Under overcast conditions, sun-directing glass can provide a modest improvement in lighting within the first 3m of the window but has negligible effect at greater distances, while under clear-sky conditions, lighting is improved over a depth of 5m. Tilted reflective elements in the ceiling can be used to concentrate reflected light to specific task areas.

Figure 9.19 Illustration of sun-directing glass. See Figure 9.20 for photographic examples

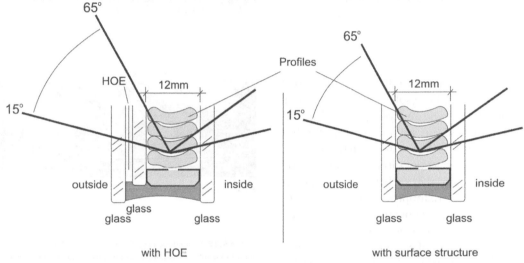

Note: HOE = holographic optical element.
Source: IEA (2000a)

Figure 9.20 Daylighting through sun directing glass. Left: Sun-directing glass above a normal viewing window in the ADO office building in Cologne, Germany. Right: sun-directing glass in the upper windows of the Geyssel Office Building, Cologne, Germany

Source: IEA (2000a)

Zenithal light-directing glass

A holographic film is sandwiched between two glass panes, which are tilted at an appropriate angle and used to redirect diffuse light from a range of angles into the building.[4] This makes them appropriate for daylighting on building façades that do not receive direct sunlight, either because of the façade orientation or because of shading by other buildings.

Holographic transparent shading

Holographic film can be used to redirect or reflect solar radiation from within a small angular region, while transmitting radiation from all other angles. Thus, if the glass incorporating holographic layers rotates to follow the sun, direct beam radiation (from the solar angle and a small region around it) can be always rejected while diffuse beam radiation (i.e. radiation from all other angles) is transmitted. The system can be designed to concentrate the rejected direct beam radiation onto some other surface, such as a PV panel or solar water heater. Examples of buildings with shading and shading + concentrating systems are given in Figure 9.21. Holographic film can also be applied to the tops of vertical windows, causing light striking the top of the window to be deflected along the ceiling while diffuse light passes through unimpaired (Tuluca, 1997; IEA, 2001).

Anidolic systems

The term 'anidolic' means 'non-imagining', from the Greek (*an* = without, *eidolon* = image). Light from an angular range of about 90° is collected and redirected in another direction with a much smaller angular range. They can thus be used to direct diffuse light (i.e. under overcast conditions) deep into a room. Three kinds of system are possible (Courret et al, 1998; Scartezzini and Courret, 2002):

- *anidolic ceilings*, consisting of an anidolic optical concentrator on the external façade above a window, a highly reflective duct (reflectivity of 0.9) above a false ceiling, and a parabolic reflector to deconcentrate the light at the back of the room, illustrated schematically in Figure 9.22;

Figure 9.21 Daylighting through holographic optical elements. Top: REWE Headquarters, Cologne, Germany. Bottom: Sunlight-concentrating system at the IGA rowhouses in Stuttgart, Germany

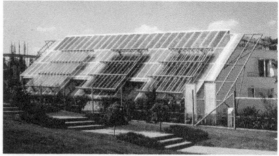

Source: IEA (2000a)

Figure 9.22 An anidolic ceiling daylighting system, which is well-suited to overcast conditions. See Figure 9.25 for photographic examples. All dimensions are in metres

(a)

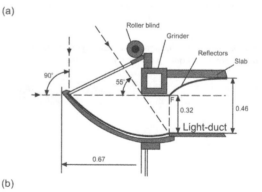

(b)

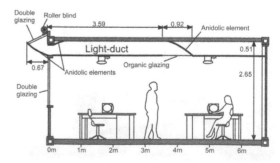

Source: Scartezzini and Courret (2002)

- *anidolic zenithal openings*, illustrated schematically in Figure 9.23;
- *anidolic solar blinds*, which have a honeycomb structure of small anidolic elements that admit very little light from near vertical but permit passage of progressively more of the diffuse light from progressively lower altitudes.

The penetration of sunlight into a room with conventional double glazing and an anidolic ceiling is compared in Figure 9.24, and photographs of real systems are given in Figure 9.25. Anidolic ceilings and zenithal openings work well under overcast conditions, when sunlight arrives rather uniformly from all directions. They can also be used in buildings with limited access to direct sunlight due to obstructions.

Luminescent solar concentrators

Still at an early stage of development, these utilize three different coloured fluorescent dyes to produce a concentrated near-white light source that is transferred onto flexible polymer sheets (Earp et al, 2004a). A remote room can be illuminated with over 1000 lumens of natural light with a luminous efficacy of over 300lm/W (Earp et al, 2004b).

Discussion of lighting techniques

The suitability of the above techniques for use with direct and diffuse radiation, and their commercial status as of 2000, are given in Table 9.5.

In many instances, light shelves, light pipes or other devices are not needed to bring daylight into a room, because adequate light ($\geq$300 lux) can penetrate several metres from a window. The problem, instead, is too much light (and associated solar gain) near the window, as illustrated in Figure 9.26 for standard double-glazed windows. As shown by Lorenz (2001) and illustrated in Figure 9.26, prismatic panels can be designed to greatly attenuate the light intensity within the first 1–2m of the window, while having negligible effect beyond 4m. At middle latitudes, the design passively limits transmission of direct beam solar radiation during the summer and transitional seasons but has minimal effect during the winter.

Figure 9.23 Illustration of an anidolic ceiling opening (lower) and light-admission sector (upper)

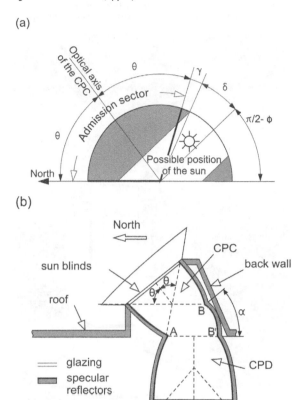

Source: IEA (2000a)

Figure 9.24 Comparison of light transmission for a double-glazed window and an anidolic system for overcast conditions with an obstruction to a height of 30°.

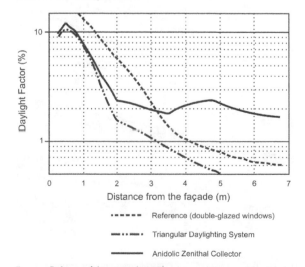

Source: Baker and Steemers (2002)

Figure 9.25 Daylighting through anidolic ceilings. Test facility, Switzerland (top) and renovated south façade of the Solar Energy and Building Physics Laboratory (LESO-PB) of the École Polytechnique Fédérale de Lausanne (EPFL), Switzerland (bottom)

Source: IEA (2000a)

Table 9.5 Summary of daylighting techniques

Technique	Effective with		Status as of 2000 based on IEA (2000a)
	Direct radiation	Diffuse radiation	
Light shelves	x		A
Light pipes	x		A
Light-directing louvres	x		A
Prismatic panels	x		A
Prismatic panels in skylights	x		T
Laser-cut panels	x		T
Light-guiding shades	x		T
Sun-directing glass	x	x	A
Zenithal light-directing glass		x	A
Holographic transparent shading	x		A
Anidolic ceilings		x	T
Anidolic zenithal openings		x	T
Anidolic solar blinds	x		T

Note: Status categories are: A, commercially available; T, still being tested.

Figure 9.26 Variation of solar illumination (lux) with distance from a standard double-glazed window (solid lines) and from a prismatic window (dashed lines)

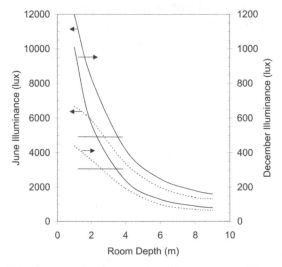

Note: Fluxes are given for overcast conditions at noon on 20 December and for clear sky conditions on 20 June, both at 50°N latitude. The horizontal lines intersect the December curves at 300 and 500 lux, representing typical lower and upper limits for offices.
Source of data: Lorenz (2001)

Figure 9.27 Comparison of solar irradiance through regular glazing (dashed lines) and through laser-cut panels (solid lines) at 27°S for (a) an east-facing window, (b) a north-facing window, and (c) a pyramidal skylight

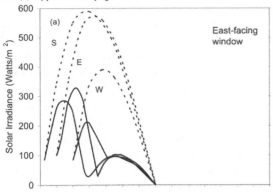

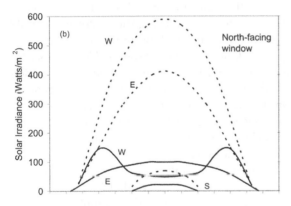

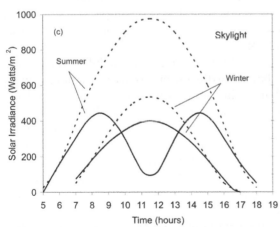

Note: In (a) and (b), S = mid-summer, E = equinox, and W = mid-winter.
Source: Edmonds and Greenup (2002)

Edmonds and Greenup (2002) discuss daylighting systems appropriate for tropical and subtropical latitudes, where the sun is closer to directly overhead than at higher latitudes and where minimizing unwanted heat gain is essential. These include light-guiding shades and LCPs. The solar irradiance admitted by regular glazing and by LCPs on a vertical east-facing window, north-facing window, and through a pyramidal skylight is compared in Figure 9.27 for Brisbane (latitude 27°S). Information on the ways in which the light-deflection and transmission properties of LCPs can be altered by adjusting the geometry of the panels is given in Reppel and Edmonds (1998).

Oakley et al (2000) and Chirarattananon et al (2000) discuss the use of passive light pipes in mid-latitude and tropical locations, respectively, while Jenkins and Muneer (2004) review the available design tools (used to predict light output). As noted above, peak outdoor light flux densities can exceed 100,000 lux. Light pipe systems can be designed to deliver 300–1000 lux under clear sky conditions, implying a reduction in light intensity – and associated transmission of heat to the workspace – by more than a factor of 100 compared to outside light (resulting in a heat gain of 3–10W/m² for 300–1000 lux). Even the darkest windows will transmit 20 per cent of the incident solar energy (given a shading coefficient of 0.23, as indicated in Table 3.12). However, the light pipe itself can get quite warm, due to the absorption of solar radiation by the walls of the pipe. Given a wall reflectivity of 96 per cent (typical of aluminium) and 20 reflections of a beam of radiation from one side of the pipe to another, the fraction transmitted to the end of the pipe will be $(0.96)^{20} = 0.44$; that is, over half of the radiation entering the light pipe is absorbed by the pipe. Heating of the building can be minimized by placing as much of the light pipe on or just under the roof as possible, by insulating the pipe except where exposed to the outside, and by collecting only the amount of sunlight needed (this can involve collecting only visible light, or – perhaps – automatically adjusting the collector based on the solar flux intensity so that adequate light is available for overcast conditions while avoiding excess heating during clear conditions). A recently developed film with a reflectivity of 98–99 per cent could be used to line light pipes.

9.3.2 Control systems

In order for daylighting to save energy, it has to be possible to reduce the output of the electrical lighting system so as to provide just the required total light. This in turn requires photosensors and dimmable lighting. Inasmuch as clear windows will provide more light than needed at times, causing overheating and discomfort and/or an increase in the cooling load, full optimization of the daylighting system also requires some form of automatically adjustable shading system, such as motorized roller blades, Venetian blinds or louvres. Systems involving automatically adjusting Venetian blinds, light sensors and dimmable lights that optimize the balance between daylight admission and avoidance of solar heating were built and tested in real office buildings in the 1990s, with a high degree of user-satisfaction (Vine et al, 1998). Current state-of-the-art, commercially available daylighting and shading control systems involve the following components (Lee et al, 2004a):

- a central computer, connected to the various sensors and controls, and to the building energy management system;
- automatic dimmable lighting controls, which change the electric light output in response to changing daylight levels;
- a manual override via remote control or a wall-mounted switch or touch screen;
- schedules for occupancy or day- or night-time operating conditions;
- an automated shading system, with automatic limits on operation for external shading devices if there is ice, snow or high wind;
- a time delay to prevent the shading system from changing too often in response to changing weather conditions;
- fault detection and automated diagnostics that help to troubleshoot hardware failure, enable software-based commissioning of zones and provide real-time plots showing the control history;

- dimmable ballasts that enable reconfiguration of the dimming zones in the software.

Automatic blind systems can use DC or AC motors. Systems with DC motors appear to be more reliable, have more intelligence built into them, and include self-diagnostics as well as some self-correction features (Lee et al, 2002a). Software refinements allow quiet and precise tilt action. To keep costs down, a single transformer is usually dedicated to a group of DC-motorized shades. Sequential rather than simultaneous control of an entire bank of shades on a single transformer reduces costs further. DC motors are normally quieter than AC motors, although motor noise can be reduced by placing the motor in the ceiling plenum and providing sound insulation.

Guillemin and Molteni (2002) describe a genetic-algorithm automatic-control system for blinds that takes the user's wishes into account, but ignores energetically bad wishes if expressed only once during a given time period. This system could lead to wide acceptance of automatic blind control. Future systems with automatic blinds inside double-skin walls (described in Chapter 3, Section 3.4) may be controlled by an optimization routine with an embedded simulation model which is used to predict the impact of alternative system settings, as described by C. S. Park et al (2004).

Price and limited availability are currently barriers to the use of motorized shading systems that are integrated with occupancy sensors, light sensors and dimmable lighting, although Lee et al (2004b) propose various concepts for reducing the networking and integration costs. As well, dimmable ballasts (currently at $60–120 each) should fall to $20–30 each (Lee et al, 2004a). There are also ongoing concerns about reliability. To address both cost and reliability concerns, Leslie et al (2005) propose a simpler system in which blinds are opened or closed only *once* per day (with the timing dependent on the façade orientation), and in which lights are turned on when the luminance drops below 400 lux and off when it rises above 600 lux. There is no dimming system, and blinds can be raised or lower manually if desired. Computer simulations indicate savings of 34 per cent in private offices and 24 per cent in open plan offices in Albany, New York, compared to savings of 57 per cent and 50 per cent with fully automatic blind control and dimming. The savings with the proposed system arise from the fact that, with manual blinds, users often close the blinds when glare becomes a problem, and then leave them closed even when the sun has moved to a position where glare would not be a problem. Similarly, lights are turned on when needed, but not turned off when no longer needed.

9.3.3 Energy savings

Rubinstein et al (1998) report the results of detailed measurements of lighting energy savings over a 6-month period in a San Francisco office building, in which light sensors and dimmable ballasts were installed in one entire floor. Annual energy savings between 6am and 6pm were calculated to be 41 per cent and 30 per cent for the outer rows of lights on the south and north sides of the building, respectively. The economic benefit of daylighting is enhanced by the fact that it reduces electricity demand most strongly when the sun is strongest, which is when the daily peak in electricity demand tends to occur. Jennings et al (2000) analyse the benefits of combined daylighting controls and occupancy sensors as measured over a 7-month period in the same office building; the combined savings approached 50 per cent during working hours even in areas with high minimum lighting requirements. Detailed simulations by Bodart and de Herde (2002) indicate savings of 50–80 per cent over the hours 8am to 6pm for a building in Belgium with offices along a single corridor, depending on the window orientation and internal wall reflectivity. Li and Lam (2003) measured a mean annual saving between 7am and 7pm of 70 per cent along a daylight corridor in a Hong Kong building, while Atif and Galasiu (2003) measured an annual saving between 7am and midnight of 46 per cent (or 60 per cent between 8am and 6pm) after the retrofit of the lighting system in a large atrium in Canada (the atrium itself was unchanged and therefore not optimized for daylighting).

Reinhart (2002) performed detailed simulations of daylight distribution in commercial offices for 1000 different combinations of internal design parameters (such as workstation size, partition height, partition reflectance, floor reflectance) and, for many office designs, 160 different combinations of façade orientation (4 choices), glazing transmission (2 choices), blind control strategy (4 choices) and solar radiation regime (25 Canadian and 161 US cities were first clustered into five different regimes). For a south orientation and with automatic control of blinds, the lighting electricity saving ranges from 50 to 70 per cent for peripheral offices. Façade orientation alters the electricity saving for peripheral offices by only a few percent. Electricity savings for second row offices facing south varies from 25 to 50 per cent as the partition height decreases from 1.8m to 1.2m. These savings are for buildings that have not been optimized for daylighting (i.e. without light shelves or light pipes, as would be the case in a retrofit situation).

Daylighting can lead to a reduction in cooling loads if solar heat gain is managed. This is because the luminous efficacy of natural light averages around 105–128lm/W (Section 9.2). The best fluorescent lighting systems provide about 63lm/W (90lm/W from the lamp/ballast system, times a luminaire light-transmission efficiency of 0.7) (Muhs, 2000), so replacing artificial light with just the amount of natural light needed reduces internal heating. Offsetting this theoretical benefit is the fact that (i) not all the waste heat from artificial lighting adds to the cooling load (see Section 9.5.3) and (ii) natural light cannot be distributed as uniformly as electric light. If electrical dimmable lighting is wired in rows parallel to the outer wall, with separate controls for each row, then the lighting distribution can be made more uniform and the use of daylight optimized. If non-visible solar radiation is excluded (which can be done in part with non-absorbing selective filters), then the luminous efficacy of natural light would be about 260lm/W, leading to greater savings in cooling loads.

Lee et al (1998) measured savings in lighting and combined lighting + cooling energy use of 22–86 per cent and 23–33 per cent, respectively, in a full-scale testbed of an office building with automated lighting and blind control in Oakland, California, compared to an adjacent otherwise identical testbed without daylighting controls and a static horizontal blind. The peak cooling load in the former case was reduced by 18–32 per cent, which would permit a comparable downsizing of the cooling equipment and associated cost savings (with manual blinds and lighting controls, the mechanical engineer would size the cooling equipment based on the worst-case combination of blind and light settings).

Ullah and Lefebvre (2000) report measured savings in cooling + ventilation energy use of 13–32 per cent using automatic blinds in a building in Singapore, depending on the orientation of the external wall. The control algorithm was rather crude in this case, with the blinds either fully closed or fully opened, depending on a threshold illumination. Tzempelikos and Athienitis (2003) simulated the savings in lighting energy expected for offices with automated dimming and shading control on the south-west and south-east façades of a new building planned for Montreal. Continuous dimming with a minimum of 500 lux illumination saves 83 per cent of the lighting electricity use, while three-level switching saves almost as much (72 per cent) but at a much lower cost. Peak cooling load for these offices (excluding cooling of fresh air used for ventilation) is reduced by about 10 per cent. In Hong Kong offices, D. Li et al (2005) estimate that optimized daylighting will reduce peak cooling loads by 5 per cent and peak electrical loads by about 9 per cent.

Daylighting systems should be designed with full consideration of the users of the building. Otherwise, they will start to tamper with the system if they do not like it, and this may increase energy use. Simply maximizing the window area can counteract the energy-saving possibilities by creating highly contrasting luminance ratios within the space, thereby requiring more artificial lighting to balance the lighting environment. For automatic shading systems, a manual over-ride is essential, although it is not used often (Lee et al, 2002a).

9.3.4 Current design practice

The design of a building with daylighting, and the design of the daylighting system itself, involves a large number of decisions but a number of different players. These decisions and players are summarized in Table 9.6. Successful daylighting systems require a high degree of coordination among the different players. The location and design of the daylight sensors is particularly critical, as discussed by Littlefair and Motin (2001). Vaidya et al (2004) provide eight examples of daylighting systems that did not work as intended because of mistakes made during the design or installation of the system. According to these authors, fully

Table 9.6 Players and decisions involved in the design of buildings with daylighting and in the design of daylighting systems

Player			
Architect	Interior designer	Lighting designer	Commissioning agent
Building orientation	Interior space planning	Lighting illuminance	Building operator education
Building shape	Interior partitions	Lighting fixture type	User awareness
Ceiling height	Interior colours	Lighting lamp type	Problem-reporting protocol
Window area		Lights to be controlled	Performance monitoring
Window location		Control sequence	Performance reporting
Glazing type		Lighting switch/dim control	
Exterior shading		Ballast type	
Interior shading		Photosensor type	
		Photosensor location	
		Photosensor, number of	
		Controller dials available	
		Controller location	
		Calibration	
		Relamping guidelines	

Source: Based on Vaidya et al (2004)

successful daylighting systems are still the exception.

Turnbull and Loisos (2000) discuss the current practice of daylighting design among architectural firms, based on research carried out for Pacific Gas and Electric (PG&E), a Californian utility that has been actively involved in research and promotion of daylighting in buildings. They sent a questionnaire to 76 architectural firms in California, of whom 58 replied. Based on this response, they classified the ability of firms to use daylighting as (i) Level 1, rudimentary, 34 firms; (ii) Level 2, intermediate, 21 firms; and (iii) Level 3, sophisticated, 2 firms (one firm could not be classified). Level 1 firms do not fully understand advanced glazing systems and their use in daylighting, do not understand dimming systems and are not aware of such basic things as alternative window heights as a design option for daylighting. They see 'increased cost' as a primary barrier to the use of daylighting. Level 2 firms are able to incorporate client requests for daylighting into their overall design and are aware of various advanced daylighting techniques. Among Level 3 firms, lighting and daylighting decisions do not simply impact building design, they are fundamental design drivers. They consider multiple design options, and, unlike Level 2 firms, use physical and computer models to predict the impact of alternative design options in terms of lighting quality and energy use.

An impediment to more widespread use of daylighting, according to Turnbull and Loisos (2000), is the lack of 'off-the-shelf functionality' for daylighting dimming systems. Dimming systems produced by different manufacturers do not at present conform to any standard performance and communications protocols, so it is difficult to correctly specify and install this essential component of a daylighting system. Although there are dozens of examples of successful and well-documented buildings with good dimming systems, it takes a significant (and often commercially nonviable) amount of expert time to create such systems. However, the prospects for rapid improvement in this area are good according to Turnbull and Loisos.

Another impediment to the effective use of daylighting in buildings is the sequential or linear nature of the typical design process, a problem that is discussed more fully in Chapter 13 (Section 13.3). Based on a survey of 18 lighting professionals, Turnbull and Loisos (2000) found that architects normally make a number of irreversible decisions at an early stage that adversely impact daylighting, *then* pass on their work to the lighting consultants and electrical engineers to do the lighting design, rather than involving lighting consultants from the very beginning. As a result, the lighting system becomes, *de facto*, a strictly electric design.

Ubbelohde and Loisos (2002) discuss in some detail the process of designing multi-storey office buildings to use daylight. Daylighting in multi-storey buildings is particularly challenging, and requires careful integration between the window design and the structural and mechanical designs. They illustrate this using six state-of-the-art buildings in the US and Europe as case studies. The design process involves detailed computer simulations of daylight distribution in the buildings for different design options and construction of a physical model to test and calibrate the computer model under clear sky and overcast conditions. Physical models can be tested using the real sun or any parallel beam light source combined with a chart to determine the sun's position at any time of the day and year for any location. Artificial light characteristic of overcast skies can be also created in a controlled and replicable manner.

Finally, Lee at al (2004a) discuss the process by which the *New York Times* came to adopt advanced daylighting in a major new building that it will be constructing. This process involved constructing a full-scale outdoor mockup of a portion of one floor, complete with interior furnishings! Potential vendors of daylighting systems were invited to install their products in the mockup, with monitored data collected over a 6-month period spanning the full range of solar conditions. Manufacturers were permitted to tune their systems to obtain optimal performance and improve their designs. The monitored data were supplemented with computer simulations using the RADIANCE programme (see Appendix D) in order to explore various design issues. The monitoring and computer simulations allowed the building owner to develop a detailed performance specification that would then be open for bid by any vendor. Of particular interest is the observation by Lee et al (2004a) that the building owners found the quality of daylight to be palpably different between morning and afternoon, with subtle shifts of colour, intensity, sparkle and mood throughout the day that, along with the rise and fall of the automated shades, tuned the occupant's awareness to the varying outdoor solar conditions. In other words, the daylighting system provided a sense of connection to the outside and visual stimulation, in contrast to the monotony of unchanging electric lighting in spaces that do not make use of daylighting.

9.4 Residential lighting

Most residential lighting uses incandescent bulbs. The vast majority of these can be replaced with compact fluorescent lamps (CFLs), with a factor of 4–5 reduction in light output for the same electricity use. CFLs have now dropped in price to $3–4 per lamp, which means they are comparable in price to the 6–10 incandescent lamps that each one replaces (due to their 6–10 times longer operating life).

Halogen lamps are used in track lighting, due to their ability to direct their light into specific directions. HIR lamps provide a modest reduction in energy use compared to standard halogen lamps, but are several times less efficacious than CFLs. Small CFLs with appropriate reflectors are now commercially available that can replicate the directional characteristics of conventional track lighting. CFLs can also be used in ceiling downlighting (potlighting). Indeed, CFLs are now available that can replace almost any residential use of incandescent or halogen lighting (see, for example, the product listing at Technical Consumer Products Inc., www.tcpi.com).

A factor limiting the use of CFLs in potlighting is the buildup of heat from the lamp. Although the heat buildup is far less than with incandescent lamps (due to the much smaller power draw), it will reduce the life of electronic ballasts by 50 per cent for every 10K temperature increase. Existing CFL downlight products have either used magnetic ballasts (which are inferior) or have used low-wattage lamps (generally 13W). The installation of CFL downlights is more expensive and difficult than the installation of incandescent downlights, involving cans, lamps, sockets and control switches that must work together but are often sold separately. For these reasons, Lawrence Berkeley National Laboratory (California) commissioned the development of a prototype CFL downlighting system with the following features: a single electronic ballast serving four to six lamps, with the ballast in thermal contact with a rectangular pan that in turn is attached to the downlight pot; 'plug-and-play' connecting wires that permit the simple connection of the various CFL lamps to the ballast; and a CFL optimized for temperatures that are higher than in non-downlight applications (Sminovitch and Page, 2003). The system is illustrated in Figure 9.28. The metal pan conducts heat away from the ballast, preventing excessive temperatures, but costs are kept low as it involves only one of several

Figure 9.28 Prototype CFL downlighting system, with a master ballast attached to a metal pan in order to dissipate heat, and 'plug-and-play' connecting wires

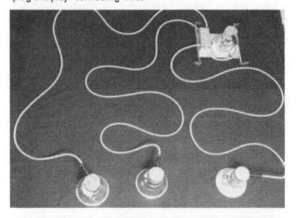

Source: Sminovitch and Page (2003)

Figure 9.29 Energy-efficient residential lighting, with small linear fluorescent tubes and CFLs

Source: Banwell et al (2004)

CFL downlights. It is expected that commercial products based on the prototype will follow. As noted in Chapter 3 (Section 3.7.1), it is important that potlight cans be airtight.

Other options to avoid incandescent lamps involve T8 or T5 fluorescent tubes hidden in recessed areas, and used for indirect lighting – bouncing light off ceilings or walls rather than providing direct illumination. This provides a soft, even illumination without glare and with a minimal number of downlight interruptions in the ceilings, as illustrated in Figure 9.29. The connected lighting power density in a US house is typically 22–28W/m², compared to 9.8W/m² in the house illustrated in Figure 9.29 (Banwell et al, 2004). Energy efficient lighting decisions must be made early in the construction process in order to minimize additional first cost, particularly if the lighting is to be integrated into the architectural features.

9.5 Advanced lighting systems in commercial buildings

Lighting in a building is a system, involving individual devices and controls but also their *arrangement*. However, the dominant paradigm in building lighting in North America is to provide uniform illumination using a rectangular grid, irrespective of the spatial variation in lighting needs. As discussed by Wulfinghoff (1999, pp1019–20), every profession involved in the design of buildings has its own set of design procedures, which are formalized and promoted by the profession's leading organization. According to Wulfinghoff: ' the official lighting design doctrine has grown in a direction that is fundamentally in conflict with energy efficiency. It wastes energy to an extent that ranges from moderate to extreme, depending on application ... the basic tool of contemporary lighting design is a rectangular grid, and lighting layout is usually guided by non-lighting constraints, such as the spacing of ceiling tiles'. Wulfinghoff (1999) provides many photographs of lighting systems in contemporary buildings that can be charitably described as thoughtless, along with examples of carefully thought out and efficient systems. Here, the energy savings that can be achieved through a systems and task-oriented approach to daylighting is documented along with the impact on heating and cooling energy use.

9.5.1 Lighting system savings

Fluorescent lighting in commercial buildings is a lighting system consisting of much more than the fluorescent lamps. In retrofits of fluorescent lighting systems, a 30–50 per cent electricity saving can be routinely achieved. With considerable effort, a 70–75 per cent saving in retrofits is possible). In new construction, 75 per cent and larger savings compared to recent standards can be readily achieved.

The mix of measures that lead to these large savings, and the percentage savings from individual measures, is as follows:

- more efficient lamps and ballasts (15–30 per cent saving);
- use of occupancy sensors and scheduling controls (25–50 per cent saving);
- use of daylight dimming (25–85 per cent saving);
- use of specular reflectors behind the lamps (up to 25 per cent saving);
- use of lighter-colour finishes and furnishings (10–20 per cent saving);
- electronic compensation for gradual lumen depreciation over time, rather than oversizing at the beginning (15 per cent saving).

Older, T12 lamps, have an efficacy of 65–70lm/W. More recent T8 and T5 lamps have efficacies of approximately 100lm/W and provide more accurate rendering of colours, as represented by the Colour Rendering Index (CRI) (Table 9.1). The 90-series T8 lamps provide even better colour rendering, but at the expense of reduced efficacy (no better than T12 lamps). The T5 lamps allow for slimmer fixture designs, which in turn permit better optical control. This is useful when light needs to be projected further. However, T5 lamps were not initially available in low-mercury versions (so disposal is a greater problem), and ballast options were initially not as extensive as for T8 lamps (McCowan et al, 2002).

Advances have also occurred in occupancy-sensor technology. Ultrasound sensors can detect motion behind obstacles that mask the occupant's presence when infrared sensors are used. 'Smart' occupancy sensors can adjust the time delay (between the last detected motion and when the lights are turned off) to changing activity levels during the course of a day, giving another 5 per cent energy savings (Garg and Bansal, 2000). According to McCowan et al (2002), however, many contractors simply install the sensors with the factory default (or random) setting and 'walk away'. Dissatisfaction with poorly adjusted sensors leads to a large percentage of sensors being disabled. Occupancy sensors must be commissioned after installation to ensure that they are at the correct sensitivity for the given application.

It is often claimed that special reflectors placed behind fluorescent tubes reduce a requirement for four tubes to two tubes – a 50 per cent savings. According to Wulfinghoff (1999, p1069), this claim is false because (i) conventional fixtures already reflect 90 per cent of the incident light; (ii) the tubes themselves absorb some light; and (iii) most of the loss of light is due to absorption by the diffusers placed in front of the lamps. What reflectors can do is concentrate the light output more directly downward, which in some cases may lead to more effective use of the light, thereby permitting a reduction in light output and in associated energy use.

9.5.2 Task/ambient lighting and light requirements for different tasks

The above subsections show that up to 75 per cent savings in the energy use by electric lighting systems can be achieved while providing *uniform* illumination. However, not all tasks require the same amount of light. A simple strategy to reduce energy use is to provide a relatively low background lighting level (say, 200–300 lux instead of 500–600 lux), and to provide local levels of greater illumination (up to, say, 750 lux) at individual workstations. This strategy is referred to as *task/ambient lighting*, and is popular in Europe. Not only can this alone cut lighting energy use in half, but individual control over personal lighting levels is possible. The Illuminating Engineering Society of North America (IESNA) has classified hundreds of tasks into nine general categories, and provides minimum, average and maximum required illumination levels. These categories and the corresponding illumination levels are given in Table 9.7. For most categories, there is a

Table 9.7 Illuminance categories and low, medium and high recommended illuminance as given by the Illuminating Engineering Society of North America

Type of activity	Illuminance code	Illuminance (lumens/m² or lux)
Public spaces with dark surroundings	A	20–30–50
Simple orientation for short, temporary visits	B	50–75–100
Working spaces where visual tasks are only occasionally performed	C	100–150–200
Performance of visual tasks of high contrast or large size	D	200–300–500
Performance of visual tasks of medium contrast or small size	E	500–750–1000
Performance of visual tasks of low contrast or very small size	F	1000–1500–2000
Performance of visual tasks of low contrast and very small size over a prolonged period	G	2000–3000–5000
Performance of very prolonged and exacting visual tasks	H	5000–7000–10,000
Performance of very special visual tasks of extremely low contrast and small size	I	10,000–15,000–20,000

Source: Rea (2000)

factor of two difference between the minimum and maximum recommended lighting level. These reflect legitimate differences between individual lighting needs. To provide the maximum rather than the minimum recommended illumination everywhere has major energy implications. With task/ambient lighting, very high local illumination levels can be supplied if desired, with negligible impact on overall energy use, because it is supplied only where it is needed. If ambient light is provided at 200 lux and 500 lux provided over 20 per cent of the floor area, instead of 500 lux everywhere, then lighting energy use is reduced by almost a factor of two.

It appears that most people require less light at night and on cloudy days, because of less window glare but also because requirements are conditioned in part by expectations (Torcellini et al, 2004a). This parallels the finding that the acceptable temperature range depends in part on expectations, which vary with outside conditions (see Chapter 7, Section 7.2.4). An added benefit of task/ambient lighting, then, is that it allows the lighting level to be adjusted to changing preferences, as well as to differences between users.

9.5.3 Interaction with the HVAC system

Lighting is one of the internal sources of heat that needs to be removed by the cooling system in summer in perimeter zones and year round in core zones (Figure 6.4), and which reduces the winter heating requirement in perimeter zones. However, not all of the waste heat from the lighting system contributes to a cooling load or reduces the heating load, since only some of the waste heat is released to the occupied space and some of the remainder is released directly into the return air, which is exhausted at least in part to the outside. The proportion released to the occupied space ranges from 0.3 for recessed vented fixtures to 1.0 for suspended fixtures. The reduction in cooling electricity use, E_c, as a fraction of the reduction in lighting energy use, is given by:

$$E_C = L_C / COP$$
(9.2)

where L_c is the reduction in the cooling load as a fraction of the reduction in lighting energy use, and COP is the COP of the chillers. For $L_c = 0.3–1.0$, year-round cooling, and a COP of 3.0, the savings in cooling energy use is 10–30 per cent of the savings in lighting electricity use. The increase in heating energy use, E_H, as a fraction of the reduction in lighting energy use, is given by

$$E_H = L_H / \eta$$
(9.3)

where L_H is the increase in heating load as a fraction of the reduction in lighting energy use, and η is the efficiency of the heating system. If all of the heat is released into the occupied space, L_H will be equal to the fraction of the year where heating is required. In core zones, this can be zero even in cold climates. The net energy savings (E_N) from a lighting retrofit, as a fraction of the direct energy savings, is given by $1 + E_C - E_H$. Simulations by Zmeureanu and Peragine (1999) for an office building in Montreal indicate that E_N could range from 0.63 to 1.12. For the same building in Phoenix, E_N would range from 1.04 to 1.22. Thus, for Montreal, the net energy saving can be less than or greater than the direct energy savings, while for Phoenix, the net energy saving is always greater than the direct energy saving. In the Montreal case, there is almost a factor of two variation in the net effect energy saving, depending on the nature of the baseline lighting system.

The above discussion pertains to on-site energy use. If $E(1 + E_c)$ is the saving in electrical energy (where E is the saving directly attributable to reduced lighting energy use) and η_c is the efficiency in generating and transmitting electricity, then $E(1 + E_c)/\eta_e$ is the saving in primary energy (η_e is typically 0.33–0.35). If heating is provided by oil or natural gas and η_f is the efficiency in transforming oil or gas primary energy (unprocessed fuel in the ground) into secondary energy (heating fuel at the point of use), then the net savings in primary energy, E_{PE}, expressed as a fraction of the direct savings due to reduced lighting energy use, is:

$$E_{PE} = 1 + E_c - E_H\left(\frac{\eta_e}{\eta_f}\right) \tag{9.4}$$

Inasmuch as $\eta_f \approx 0.75$–0.85, the net saving in primary energy use is larger than the net saving in on-site energy use. Indeed, there can be a saving in primary energy at the same time that there is an increase in on-site energy use, if heating is supplied by fuel rather than electricity.

Extending the above analysis to CO_2, if F_{coal} is the emission factor for coal (kg C per unit of coal energy) and F_{NG} is the emission factor for natural gas, then the net saving in CO_2 emissions

expressed as a fraction of the direct saving due to reduced lighting use is:

$$E_{PE} = 1 + E_c - E_H\left(\frac{\eta_e}{\eta_f}\right)\frac{F_{NG}}{F_{coal}} \tag{9.5}$$

for electricity from coal and heating from natural gas. As F_{NG} is about half that of coal (see Chapter 2, Section 2.4.1), the impact of increased heating energy use due to more efficient lighting is further diminished.

9.6 Athletic and industrial facilities

Lighting in athletic facilities, including in schools, colleges and universities, is generally provided by HID lamps (usually metal halide) and is designed for sanctioned competitive events. These lamps are inflexible, so most other activities are over-illuminated. HID lamps have a long warm-up time, and so are generally turned on in the morning and left on all day, often spanning long unoccupied time periods (particularly in schools). HID lamps are also used for high-ceiling spaces in retail stores and manufacturing and warehousing facilities.

High-intensity fluorescent (HIF) lamps can now provide the same lighting that indoor HID lamps provide but with as little as one-third or less of the energy use, with essentially instant start-up, a dimming capability and better colour rendition. Both HID and fluorescent lamps produce light from an electric arc through vaporized metal. In HID lamps, the light is produced from such a small volume that they can be considered to be 'point sources'. The light is directed using a parabolic reflector that surrounds the light source. Fluorescent lamps traditionally produce diffuse light from a long glass tube – the opposite of a point source. However, more intense T5 fluorescent lamps and specially designed reflectors allow them to act almost as point sources.

The T5 lamp has an efficacy (105lm/W) comparable to or slightly better than that of new metal halide lamps (90–103lm/W). High-pressure sodium lamps have greater efficacy (up to 124lm/W) but poor colour rendition. However, even if left on all the time at full power, HIF lamps save energy in two ways: (i) they suffer as little as 5 per cent loss in illumination or *lumen depreciation* over their lifetime, compared to 20 per cent for the best metal halide lamps and up to 33 per cent for standard metal halide lamps; and (ii) HIF lamps come with reflectors with 80–98 per cent efficiency (only 2–20 per cent of the light emitted by the lamp is absorbed by the reflector), whereas HID reflectors are only 70–85 per cent efficient (Financial Times Energy, 2001b). Because of lumen depreciation, HID lamps must be oversized in order to provide enough light near the end of their lifetime. Between avoided oversizing and better reflectors, HIF lamps can save up to 25 per cent in energy compared to the best metal halide lamps and up to 40 per cent compared to standard metal halide lamps.

Most HIF fixtures have multiple lamps, which can be wired with multiple circuits that can be independently switched on or off (McCowan et al, 2002). Significant additional energy savings are possible through the dimming and instant start-up capabilities of HIF lamps. HID lamps can also be dimmed but with very little saving in energy and with a noticeable change in colour if dimmed below 60 per cent. With their dimming capability, HIF lamps in sports facilities can be at a lower light level when full light output is not needed (as is often the case). In many warehouses, aisles are vacant for long periods of time. HIF lamps could be at their minimum light output except when motion detectors sense activity. Because of their instant start-up ability, HIF lamps in school gymnasiums (for example) can be shut off during midday periods of non-use.

9.7 Integrated daylighting photovoltaic energy systems

Mention must also be made of the emerging use of window photovoltaic panels, which combine generation of electricity with partial shading and daylighting. These use ruthenium-based nanocrystalline dye cells (see Chapter 11, Section 11.2.1).

Another hybrid daylighting/PV system is a variant of the sun-tracking light pipe system illustrated in Figure 9.10. In the hybrid system, a two-axis tracker concentrates solar radiation onto a mirror that splits the radiation into visible light and near infrared (NIR) radiation. The

visible light is distributed through optical fibres and used for daylighting, while the NIR radiation is converted to electricity using GaSb (gallium antimonide) modules that have a bandgap in the NIR part of the spectrum (Schlegel et al, 2004). For an incident flux of 1000W, 850W can be collected and split into 490W of visible solar energy and 360W of non-visible energy (Muhs, 2000). Because the efficacy of visible solar energy is at least 200lm/W (see Section 9.2) and that of artificial lighting about 63lm/W, 900W of electrical energy are displaced when a 15 per cent credit for reduced cooling load is included. The 360W of near infrared radiation generate 70W of electricity, giving a total net electricity savings of 970W. This is equivalent to close to 100 per cent efficiency in the utilization of solar energy.

9.8 Prospects for yet further advances in lighting technology

As demonstrated above, the opportunity exists to reduce lighting energy using by at least 75 per cent compared to recent standard practice in commercial buildings in many jurisdictions through careful attention to the design of the entire lighting system (involving light fixtures, ballasts, reflectors and lenses, controls, daylighting). Replacement of incandescent lights with CFLs in high-use residential applications gives a 75–80 per cent energy savings for that portion of the lighting that is replaced. These remarkable energy savings could be increased yet further through advances in the efficiency and performance of the individual components of the lighting system. Some of the potential advances are outlined below, based on Rubinstein and Johnson (1998).

Improved phosphor efficiency in fluorescent lamps

The efficacy of fluorescent lamps is limited by the fact that (i) one UV photon produces only one photon of visible light; and (ii) the energy of a UV photon is about twice that of a photon of visible light (due to its shorter wavelength). Thus, 50 per cent of the energy input is lost in the UV-to-visible conversion step. The development of two-photon phosphors or the ability to excite the phosphor with longer wavelength (lower energy) UV radiation could lead to a large improvement in the efficacy of fluorescent lamps.

Electrodeless fluorescent and HID lamps

The chemical composition of the material that emits UV or light radiation (in fluorescent and HID lamps, respectively) is linked to the chemical reactivity of the electrodes. Electrode-less lamps, introduced in the early 1990s, remove this limitation, opening up the possibility of a wider range of discharge materials that in turn could be selected to give higher efficacy (at present, the efficacy is lower than for standard fluorescent lamps). They use a high-frequency magnetic field to excite the mercury in the lamp.

Very low-power HID lamps

HID lamps have very high efficacy, exceeding that of compact fluorescent lamps (Table 9.1) but are currently available only at medium and high power. The development of lower-power (1–35W) and small HID lamps could result in a replacement for the incandescent lamp that is up to eight times more efficient (125lm/W versus 14–18lm/W).

More efficient LEDs

The efficacy of white-light LEDs is presently rather low, no better than 30lm/W (Johnson and Simmons, 2002). However, LEDs promise eventual efficacies of 150–200lm/W, that is, approaching that of natural light (S. Johnson, 2000, 2002). At present, LEDs are capable of converting up to 90 per cent of the supplied energy to visible radiation (depending on the wavelength) at the p–n junction (this is referred to as the *internal quantum efficiency*). Overall performance is limited, however, by the tendency of much of the light to be internally reflected (and eventually absorbed) through refraction. Improvement in the lamp efficacy is thus dependent on chip geometries and other enhancements that minimize refraction, something that is achievable.

Improved dimmable electronic ballasts

Dimmable electronic ballasts are used to enable matching of electric light output with daylight, via light sensors and controls, but have lower efficacy at full power compared to non-dimmable ballasts, and lose efficacy with dimming below 40 per cent of full output. Improving dimmable ballasts is a potentially important advance.

9.9 Summary and synthesis

In commercial buildings, lighting energy usually constitutes 25–50 per cent of the total electrical energy used. The single most important step in reducing lighting energy use in commercial buildings is to design the building to make maximum use of daylighting opportunities, with appropriate sensors and controls so that the output of the electric lighting system can be adjusted in response to changing amounts of natural light, and completely turned off when the room or zone is unoccupied. Daytime energy savings from daylighting as high as 80 per cent have been documented. The provision of daylighting has significant beneficial effects on the well-being and productivity of the occupants of a building, so its benefits extend well beyond direct energy cost savings and reduced environmental impact. Building shape, the window:wall ratio on walls of each orientation, the height of the windows, and the presence or absence of skylights, are the earliest design decisions that will impact the extent to which daylighting can provide a building's lighting requirements. High-performance windows that are optimized according to orientation have a beneficial impact on heating energy use (as solar heat gains exceed heat losses during the winter, except on polar-facing windows in cold climates) and have minimal impact on cooling energy use if combined with adjustable shading and especially if part of a double-skin envelope. Window dimensions and placement can be chosen so as to yield significant reductions in the sum of heating, cooling and lighting energy use.

The daylighting benefits of windows can be greatly enhanced through a variety of techniques, ranging from relatively simple and low-cost methods such as internal light shelves and aerogel skylights of various shapes and orientations, to increasingly sophisticated methods such as automatic Venetian blinds, light-directing louvres, light-directing shades, light pipes, prismatic panels, laser-cut panels, various holographic systems and anidolic systems that can concentrate diffuse light. Some of the latter techniques are still being tested. However, there is a widespread lack of awareness and misunderstanding of even the most basic daylighting techniques within the design profession. Many techniques make use of light that is reflected off ceilings. It is often useful to distinguish between windows for viewing and windows for daylighting.

With regard to electric lighting, energy use can be reduced through the use of efficient T8 or T5 lamps and efficient ballasts and luminaries in commercial buildings, compact fluorescent lamps in residential buildings, LEDs for exit signs, and multi-lamp fluorescent fixtures in athletic facilities. Subdividing lit areas into separate zones, controlled by occupancy sensors, and provision of task lighting with reduced general lighting intensity can easily provide a 50 per cent or greater saving in the lighting energy needs that remain after full utilization of daylight. Ongoing research should yield further significant opportunities to reduce lighting energy use.

Notes

1 See Appendix A for the origin of the factor 683.

2 With a peak solar radiation intensity of $1000W/m^2$, this implies that peak outdoor illumination can exceed $100,000lm/m^2$ (100,000 lux). For reading and writing, a light intensity of 300–500 lux is required.

3 Glare on computer screens occurs when the intensity of reflected light is large compared to the intensity of light emitted by the computer screen. The maximum brightness of a computer screen is about 500 lux, so a luminance on the screen of 5000 lux with 5 per cent reflection will cause problems.

4 A holographic film is created by exposing a film to the interference pattern of two laser beams.

ten

Appliances, Consumer Electronics, Office Equipment and Elevators

Although appliances are not, strictly speaking, part of a residential building, they are features that are often chosen by the developer or builder. If one anticipates meeting part of the residential electricity load with building-integrated photovoltaic panels, then the efficiency of any electricity using equipment in the residence is important. Office equipment will be chosen by the occupant of the office, but must be anticipated and taken into account when designing the cooling and heating systems, since office equipment can be a modest source of internal heat gain (Figure 6.4).

10.1 Appliance energy use

Table 10.1 gives the range of energy use, under standard test conditions, for the refrigerator/freezer units, freezers, ovens, ranges, cooktops, dishwashers, washing machines and clothes dryers that were commercially available in North America in 2001. Within a given size category, there is up to almost a factor of two difference in the energy used by refrigerator/freezer units,

dishwashers and clothes dryers; up to a factor of 1.5 difference in the energy used by freezers; and up to a factor of four difference in the energy used by different washing machines. For ovens and cooktops, the range in energy use is much smaller (up to 34 per cent and 24 per cent difference, respectively, within a given size category). Table 10.2 gives peak power and typical annual energy use for a variety of other household appliances.

10.1.1 Refrigerator/freezer units

Figure 10.1 shows the variation in average energy use by new refrigerators sold in the US from 1947 to 2000. Energy use by new US refrigerators peaked at 1800kWh/year in 1975, shortly before the first standards came into effect. The newest standard, which came into effect in 2001, saw the energy consumption drop to 478kWh/year for a refrigerator with a top freezer unit, no through-the-door ice service, auto defrost and an adjusted volume of 20.7ft^3 (582l, the current average size

Table 10.1 Energy use for household appliances in North America

Item	Smallest models		Largest models	
	Capacity (litres)	Energy use (kWh/year)	Capacity (litres)	Energy use (kWh/year)
Refrigerator/freezer units				
With top-mounted freezer	465–520	480–697	690–740	620–732
With side-mounted freezer	525–575	782–934	800–850	765–1132
With bottom-mounted freezer	465–520	650–906	580–630	537–785
Freezers				
Upright, manual defrost	<200	347	666–720	644
Upright, automatic defrost	<200	504–516	546–600	800–896
Chest	<200	215–291	666–720	620–629
Other				
Oven/range unit	24″ wide	690–816	30″ wide	639–858
Single oven	<60 litres	309–324	>80 litres	405–456
Cooktop	Conventional	329–384	Modular	362–408
Dishwasher	Compact	425–496	Standard	381–698
Top-loading washing machine	<50	243–537	80–90	337–1000
Front-loading washing machine	<50	189–322	80–90	282–362
Clothes dryer	<100	394–398	>200	906–950

Source: NRCan (2001)

in the US).[1] However, a 20ft³ fridge has been designed and demonstrated by Oak Ridge National Laboratory (in the US) that uses only 365kWh/year (Brown et al, 1998). There is no reason not to expect that further reductions in energy use, down to perhaps 200kWh/year, cannot eventually be achieved in mass-produced units. Conversely, greater energy use is permitted under the 2001 standards for alternative refrigerator/freezer configurations: 542kWh/year with a top freezer and through-the-door ice service, 574kWh/year for a bottom freezer without through-the-door ice service, 631kWh/year with a side freezer without through-the-door ice service, and 692kWh/year with a side freezer and through-the-door ice service (US DOE, 2004). Figure 10.2 compares the average energy use of the refrigerator stock in various OECD countries at various times. Although significant reductions in the energy use of new refrigerators in the US and Canada have occurred, the average energy use by refrigerators in both countries still exceeds that of most other countries by 50–100 per cent and will exceed that of most other countries even after full turnover of the existing stock. This may be partly due to larger refrigerators in the US and Canada, but

Figure 10.1 Variation in the annual energy use by new refrigerators sold in the US, 1947–2001

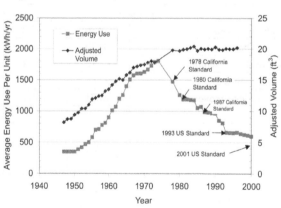

Note: During this period, the average volume of refrigerators sold in the US increased from 8ft³ to 20ft³. The dark line is the sales-weighted average of the refrigerators sold, while for the light line, all of the consumption data have been adjusted to correspond to a 20ft³ fridge.

Source: Rosenfeld (1999). Reprinted, with permission, from the *Annual Review of Energy and the Environment*, Volume 24 © 1999 by Annual Reviews www.annualreviews.org

Table 10.2 Power and annual energy use by common household appliances

Electric appliance	Power (Watts)	Typical use (kWh/year)
Food preparation		
Blender	300	1
Broiler	1140	85
Carving knife	92	8
Coffee maker	1200	140
Deep fryer	1448	83
Dishwasher	1201	363
Egg cooker	516	14
Frying pan	1196	100
Hot plate	1200	90
Mixer	127	2
Oven, microwave	1450	190
Range with oven	12,200	700
self-cleaning oven	12,200	730
Roaster	1333	60
Sandwich grill	1161	33
Toaster	1146	39
Trash compactor	400	50
Waffle iron	1200	20
Waste dispenser	445	7
Laundry		
Clothes dryer	4856	993
Iron (hand)	1100	60
Washing machine	512	103
Water heater	2475	4219
quick recovery	4474	4811
Fans		
Fan (attic)	370	291
Fan (circulating)	88	43
Fan (rollaway)	171	138
Fan (window)	200	170

Source: Smith (1997)

Figure 10.2 Average energy use of the refrigerator stock in different OECD countries in 1973, 1980, 1990 and 1998

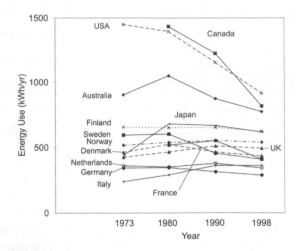

Source: Data from IEA (2004d)

Table 10.3 Average effective thermal resistances of a refrigerator door using different insulation technologies

Refrigerator door configuration	RSI-value (R-value)
CFC blown foam w/ conventional steel outer shell	1.59 (9.03)
Gas-filled AIP/foam composite w/ conventional steel outer shell	1.71 (9.71)
Evacuated AIP/foam composite w/ conventional steel outer shell	1.96 (11.14)
Gas-filled AIP/foam composite w/ polymer outer shell	1.96 (11.15)
Evacuated AIP/foam composite w/ polymer outer shell	2.31 (13.09)
Gas-filled polymer-barrier AIC	2.38 (13.50)
Evacuated-powder polymer-barrier AIC	3.31 (18.80)

Note: AIP = advanced insulation panel; AIC = advanced insulation component.
Source: Griffith and Arasteh (1995)

also partly due to design features that increase energy use.

The energy used by a refrigerator/freezer unit is highly influenced by the thermal resistance of the casing, and by the COP of the compressor. It is instructive to briefly compare these parameters with those typical of building systems. Table 10.3 compares the thermal resistance of the door of a conventional fridge with that of various advanced insulation panels (AIPs) and advanced insulation components (AICs) that are under development. A conventional fridge shell has a thermal resistance of RSI 1.6 (R9), compared to RSI 3.3 (R19) for the best AICs (and RSI 5–10 for the walls of high-performance houses in cold climates). A conventional fridge/freezer uses a rotary fixed-speed compressor with a COP of 1.1–1.6 (compared to 2.3–3.5

for residential air conditioners, and up to 7.9 for large commercial chillers). The freezer is directly cooled to about −1°C, and the fridge indirectly cooled (to 4°C) through exchange of air with the freezer. This means that when cooling of the fridge only is needed, the compressor must work harder than necessary, as it is working against the much greater temperature difference between the freezer compartment and room air rather than between the fridge compartment and room air. Use of separate compressors for the fridge and freezer would directly reduce energy use by 30 per cent, with additional savings due to the need for less defrosting (Gan et al, 2000).

10.1.2 Clothes washers

There is a dramatic difference in the energy use for clothes washing between vertical-axis machines (normal in North America) and horizontal-axis machines (normal in Europe), and between the maximum energy use permitted under the current (1987) US standard and that of the best machines, as shown in Figure 10.3. Vertical-axis machines are top loading. Horizontal-axis machines can be top- or front-loading, but most are front-loading. The best horizontal-axis machines use less than half the energy of the best vertical-axis machines and less than one-fifth the energy permitted under the US standard for top-loading machines (about 200kWh/year vs 1000kWh/year for the assumed annual load). The main reason for the lower energy use by horizontal-axis machines is that much less water is used. In vertical-axis machines, the clothes are fully immersed in water and agitated; in horizontal-axis machines, the drum only partially fills and the clothes tumble in and out of it as the drum turns. Other benefits of horizontal-axis machines are (i) faster spin (1500rpm instead of 750rpm), such that the clothes are drier after washing, thereby reducing clothes dryer energy use; (ii) use of less laundry detergent; and (iii) less wear on clothes. The reduction in energy needed to manufacture detergent adds another one-quarter to one-half to the direct energy savings. Horizontal-axis machines cost substantially more than vertical-axis machines but are generally the lower-cost option over their lifetime, when energy and detergent cost savings are accounted for (FEMP, 2000b).

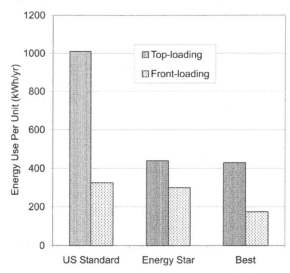

Figure 10.3 Energy use by vertical and horizontal-axis clothes washers

Source: Data from FEMP (2000a)

10.1.3 Clothes dryers

The energy required to evaporate water amounts to 0.73kWh/kg. Measurements on a typical dryer indicate that about 1.5kWh are used per kg of water removed (Smith, 1997). Thus, about half of the energy supplied to a dryer is dissipated as waste heat, in part through the vent and in part to the surroundings. Airflow through the vent induces infiltration of outside air into the building, and during the heating season this adds to the net energy use by a clothes dryer. Ideally, outside air could be directly fed to the clothes dryer with a heat exchanger to recover heat from the exhaust air. Saving up loads and doing several loads sequentially will save energy by avoiding cooling of the drier between loads. Operating at one-third to one-half load will increase the energy use per unit of clothes dried by 10–15 per cent (Smith, 1997). Typical annual US energy use is about 1000kWh/year (US DOE, 2004).

Extracting water with mechanical energy uses 70 times less energy than if done by heating. High-spin clothes washers (discussed above) therefore lead to significant reductions in energy used by dryers. Increasing the spin speed from 550rpm to 850rpm, for example, reduces moisture retention from 65 per cent to 41 per cent, thereby reducing the dryer energy use by 40 per cent (McMahon et al, 1997). Models are available with a sensor to automatically turn off the dryer when the clothes reach a predetermined dryness. And one must not forget the option of drying clothes on a clothes line using free energy from the sun, or even indoors, something that will be much quicker if one has a high-spin washing machine!

10.1.4 Dishwashers

The keys to improving dishwasher energy efficiency are (i) to reduce the use of hot water, and (ii) to allow air drying after the wash cycle. In one study (reported by Smith, 1997), pre-washing and washing accounts for 26 per cent of the total energy used (0.16kWh per load, through the heating of 72l of water), while drying accounts for 74 per cent (0.45kWh).

Minor reductions (20 per cent total) can be achieved through (i) more efficient food filters, which minimize redeposition during washing and rinsing, thereby reducing the rinsing requirement; (ii) improved spray arm geometry; and (iii) improved controls over the amount of water used. Eliminating the pre-wash by rinsing with cold water will save about one-quarter of the hot water otherwise used. European dishwashers typically use cold inlet water that is heated by the dishwasher itself only for part of the wash/rinse cycles. This would allow a lower temperature setting on the hot water tank, common in North America, as well as reducing the energy used for heating. Up to 35 per cent of the heat in wastewater can be recovered and used by the dishwasher itself, and more if stored and used for other purposes, as discussed in Chapter 8 (Section 8.4.2).

Even the most efficient dishwashers are energy intensive compared to water-efficient washing by hand, by virtue of the fact that considerably less water is needed with hand-washing if done efficiently (for example, rinsing the dishes first, then dabbing a sponge into a dish filled with soapy water for washing rather filling the sink or running hot water the entire time). Dishwashers use anywhere from 35–75l of hot water per load. Water mixed with soap does not need to be hot in order to disinfect dishes, as it is the soap that disinfects.

10.1.5 Ovens

Oven efficiency is rated at around 12–15 per cent (Marbeck Resource Consultants, 1992), so it would appear that there is scope for large improvements in the efficiency of ovens. Factors influencing the efficiency of an oven for cooking food are: the amount of insulation (standard ovens have about 5cm of fibreglass insulation, self-cleaning ovens about 10cm); the presence of forced-air convection (which improves heat transfer to the food, giving about 20 per cent energy savings); improved door seals to reduce leakage (10 per cent savings potential); and oven interior-walls that reflect heat into the interior of the oven (40 per cent savings potential). Consumer habits also have a significant influence on oven energy use. A study by the US National Bureau of Standards (Fechter and Porter, 1979) found a difference of 50 per cent in the energy use between non-professional cooks preparing the same recipe using the same oven.

Smith (1997) gives examples of the amount of energy used for cooking with different kinds of ovens. Four baked potatoes require 2.3kWh and 60 minutes in the oven (5.2kW peak) of an electric range, 0.5kWh and 75 minutes in a toaster oven (1.0kW peak), and 0.3kWh and 16 minutes in a microwave oven (1.3kW peak). Cooking a meal consisting of canned ham (2.3kg), frozen peas (0.23kg), four yams, and a pineapple upside-down cake requires 5.2kWh when done separately with a toaster oven and electric stove; 2.5kWh when cooked together in an oven; and 1.2kWh when cooked separately using a microwave oven. Note that microwave ovens use about half the energy or less of a regular oven and at peak power draw one-quarter as large for the examples given here. Energy use by microwave ovens themselves could be reduced by up to 12 per cent through use of high-efficiency transformers (McMahon et al, 1997) (Transformers are used to increase the voltage from 120V to 4000V in order to operate the magnets inside a microwave oven.)

10.1.6 Cooktops (stoves)

Gas cooktops have either a continuously burning pilot flame, or an electronic spark ignition system. Spark ignition uses a negligible amount of electricity, but the pilot flame typically uses as much energy as is used for cooking (McMahon et al, 1997). Elimination of the pilot flame therefore reduces gas cooktop energy use by about 50 per cent.

A commercial induction range is reported to have a cooking efficiency of 92 per cent, compared to 72 per cent for an electric range, 47 per cent for a residential gas range, and 30 per cent for a commercial gas range (Anonymous, 2003). The cooking vessel must be ferromagnetic, and when placed on a ceramic glass panel above the inductor, is heated by a magnetic field. The cooking vessel thus serves as the heating element found in a conventional cooktop. In restaurants, an induction range contributes to keeping the kitchen cooler.

10.2 Consumer electronics

Although the consumer electronics goods that fill a house are chosen entirely by the occupant, the choice of consumer goods, like that of appliances, has an important influence on total energy use by the house. A large energy-savings potential lies in the choice of items with low standby power consumption. Many household items consume power even when turned off, often in excess of 10W. As shown in Table 10.4, for many items, the annual energy consumption in standby mode is much greater than the consumption when the item is being used. However, the standby power is 1.0W or less for the most efficient make for almost all of the items listed in Table 10.2. Standby power is also consumed by low-voltage lighting, such as bedside lamps and office tabletop lamps, which have transformers that consume an average of 2W even when the lamp is turned off. In these applications, new technology can reduce standby power consumption by a factor of 1000 (Röing and Avasoo, 2001).

Total standby power and total annual standby energy use as estimated for households in a large number of different countries are given in Table 10.5. In many countries, total standby power consumption is comparable to or greater than that of recent North American full-size refrigerator-freezer units (i.e. 500kWh/year or more). In individual households in California, standby power consumption represents as much as 26 per cent

of total household electricity consumption (Ross and Meier, 2002). Replacing existing household equipment with equipment using 1W or less of standby power would reduce total standby power use by 59 per cent in Halifax (Canada) according to measurements by Fung et al (2003) and by 68 per cent in California according to measurements by Ross and Meier (2002), in both cases reducing total residential electricity use by about 7 per cent. Machines with low standby power generally entail no greater cost than machines with high standby power use (all else being equal).

The AC–DC power converters that accompany most electronic equipment (either as external or internal adaptors) are another significant source of energy waste. Most power supplies have conversion efficiencies of 30–90 per cent (depending in part on the load), while efficient units have efficiencies of 75–93 per cent (Foster et al, 2004). The efficiency of inefficient and efficient power converters is compared in Figure 10.4. Efficient power converters are much smaller and generate less waste heat than inefficient converters (permitting smaller and less noisy cooling fans in equipment such as computers).

Figure 10.4 AC–DC conversion efficiency as a function of load for efficient and inefficient zip-drive power converters

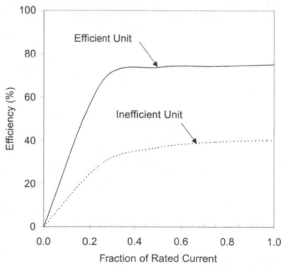

Source: Computed from input and output data given in Foster et al (2004)

Table 10.4 Active power, standby power and annual energy use for household and office equipment

Appliance	Active power (W)			Standby power (W)			Hours used per year	Annual energy use (kWh)	
	Min	Ave	Max	Min	Ave	Max		Total	Standby
Answering machine	2.5	3.7	5.2	0.2	2.2	4.5	60	19.4	19.1
Caller ID				0.1	1.5	3.3	60	13.1	13.1
Cellular telephone recharger	2.1	6.1	10.0	0.0	0.2	0.7	300	3.5	1.7
Cordless telephone	2.4	4.4	5.5	0.1	1.2	3.5	720	12.8	9.6
Colour TV (<26")	29.0	60.8	124.1	0.0	2.7	9.6	1260	96.9	20.3
Colour TV (26–36")	55.4	86.2	145.0	0.0	5.5	11.8	1260	149.9	41.3
TV/VCR combo	33.5	54.4	88.6	0.9	8.4	16.2	1460	140.7	61.3
Black & white TV	21.8	27.9	38.4	0.0	0.3	1.2	920	28.0	2.4
VCR	6.8	15.1	37.1	0.6	4.9	19.4	185	44.8	42.0
Cable receiver	4.0	6.7	10.2	3.2	5.7	9.0	1260	51.2	42.8
Digital cable receiver	17.1	18.1	19.6	16.7	17.8	19.0	1260	156.3	133.5
Satellite receiver	14.1	15.8	17.9	14.1	15.8	17.9	1260	138.4	118.5
Game console	5.9	22.2	85.1	0.0	0.4	1.0	720	19.2	3.2
Computer (desktop/tower)	25.0	58.5	165.6	0.0	1.6	28.0	1000	70.9	12.4
Laptop computer	12.4	20.7	28.9	0.0	1.7	5.6	1000	33.9	13.2
Computer monitor (≤15")	16.4	55.0	77.7	0.0	0.1	1.8	1000	55.8	0.8
Computer monitor (17")	51.0	71.9	115.2	0.0	0.7	3.2	1000	77.3	5.4
Computer monitor (19")	72.8	77.7	86.5	0.0	1.1	3.3	1000	86.2	8.5
Computer peripherals	1.9	4.5	7.1	0.0	4.9	13.8	1000	42.9	38.0
Computer speakers	83.2	83.2	83.2	0.0	1.3	9.1	1000	93.3	10.1
Inkjet printer	2.3	12.4	22.5	0.0	3.2	11.4	60	28.6	27.8
Laser printer	23.0	34.0	45.0	0.0	1.8	6.3	60	17.7	15.7
Fax	7.1	7.1	8.2	2.1	7.2	13.5	60	63.1	62.6
Baby monitor	0.2	1.8	4.4	0.0	0.6	1.6	3000	8.9	3.5
Central vacuum cleaner	730.0	1026.7	1425.0	0.0	0.6	1.3	100	107.9	5.2
Clock	0.2	5	84.5	0.1	0.9	5.2	8760	43.8	0.0
CO detector	–	–	–	2.8	3.6	5.5	0	31.5	31.5
Microwave	1000.0	1359.5	1578.0	0.0	2.1	7.3	50	86.3	18.3
Portable audio	0.4	4.9	12.5	0.0	0.9	3.0	650	10.5	7.3
Radio	1.0	3.1	4.3	0.0	0.3	0.6	500	4.0	2.5
Desktop audio (1 disc)	1.8	11.2	29.9	0.0	2.1	13.0	400	22.0	17.6
Desktop audio (> 1 disc)	11.2	21.7	28.5	0.0	7.7	10.5	400	73.1	64.4
component stereo system	13.1	37.1	57.6	0.0	2.1	16.4	400	32.4	17.6
Stereo CD layer	5.5	8.8	19.2	0.0	2.2	6.2	400	21.9	18.4
Stereo receiver	12.1	28.4	46.5	0.0	3.1	24.3	400	37.3	25.9
Stereo tape player	3.6	8.8	20.0	0.0	0.6	4.5	400	8.5	5.0
Stereo tuner	0.0	18.0	50.0	0.0	0.5	3.4	400	11.4	4.2
Stereo turntable	0.0	2.4	5.1	0.0	0.5	6.2	400	5.1	4.2

Note: Hours used per year are the author's guesses.
Source for power data: Fung et al (2003)

Table 10.5 Estimates of household standby energy use in different countries

Country	Number of homes in survey	Year of survey	Estimation method	Standby power (W)	Standby energy consumption (kWh/year)
Australia	65	2000–2001	W	87	763
Australia		2000	B	86	
Bulgaria	30	2001	B	33	288
Canada	79	2001	W	38	329
Canada		2001	B	41	
China	157	2001	W	34	50–200[a]
Denmark	100	2001	W	60	530
France	179	1999	W	38	237
France		2000	B	38	
Greece	100	2001	W	50	440
Germany		2001	B	52	
Hungary	39	2001	B	30	259
Italy	100	2001	W	57	500
Japan	79	1997–2000	W	52	462
Netherlands		1995	B	37	
New Zealand	30	1999–2001	W	101	887
Portugal	100	2001	W	46	400
Romania	30	2001	B	14	124
Sweden	1	1997	W	80	475
Switzerland		1999	B	19	
UK	32	2000	W	32	277
US	19	2000–2001	W	72	630
US		1996	B	50	

[a] Based on a survey of 28 urban homes by Meier et al. (2004)
Note: W = whole house measurement, B = bottom-up estimate.
Source: Bulgaria, Romania, Hungary: Ürge-Vorsatz et al (2002); all other countries: summary tables in Bertoldi et al (2002) except where indicated otherwise

10.3 Office equipment

Figure 10.5 gives the estimated breakdown of energy use by office equipment in the US. Personal computers, workstations and monitors account for about 40 per cent of the energy used by office equipment. Servers, photocopiers and printers account for another 26 per cent. In cooling-dominated climates, this energy use contributes to the cooling load. Table 10.6 gives the energy use of present office equipment for different modes of operation.

In addition to the direct energy used by electronic equipment, its effect on the electrical *power factor* needs to be considered. If an electrical device acts like a resistor (as does an incandescent light bulb), the voltage and current of AC electricity are perfectly in phase (their peaks and troughs coincide). If an electrical device acts like an inductor, the voltage and current are out of phase to some extent (for a perfect inductor, the peaks in the current occur one-quarter cycle after the peaks in the voltage). This happens with motors and electronic ballasts. The power produced and consumed (given by the average of the product of voltage and current) decreases with increasing lag, so there must be a compensating increase in current, which increases transmission and distribution losses. For most office equipment, the power factor is about 0.6 (but site dependent and changing rapidly), so the transmission and distribution losses will be multiplied by a factor of $1/0.6 = 1.67$.

Table 10.6 Power consumption (W) by office equipment

Item	Rated power	Mode of operation			
		Active	Standby	Suspend	Off
Personal computers					
Desktop		55.0		25.0	1.5
Laptop		15.0		3.0	2.0
CRT[a] monitors					
14–15″		61.0	53.0	19.0	3.0
17″		90.0	26.0	9.2	4.3
19″		104.0	31.0	13.0	4.0
21″		135.0	43.0	14.0	4.7
LCD[b] monitors					
13″	7.5	2.5	0.7	0.2	0.1
14″	20.0	6.7	1.9	0.7	0.3
15″	35.0	11.7	3.4	1.2	0.6
17″	50.0	16.7	4.8	1.7	0.8
18″	75.0	25.0	7.2	2.5	1.2
20″	95.0	31.7	9.2	3.2	1.6
21″	107.0	36.0	10.4	3.6	1.8
Workstations					
Dell 330	330.0	110.0			
Compaq APP 550	375.0	125.0			
Compaq SP 750	475.0	158.0			
SGI 230	428.0	143.0			
Servers[c]					
Compaq low-end	250.0	125.0			
Compaq work-horse	1300.0	650.0			
Compaq midrange	2450.0	1225.0			
IBM high-end	5040.0	2520.0			
Printers					
Inkjet		53.0	13.0	6.0	0
Laser, small desktop		130.0	75.0	10.0	
Laser, desktop		215.0	100.0	35.0	
Laser, small office		320.0	160.0	70.0	
Laser, large office		550.0	275.0	125.0	
Photocopiers					
<12 cpm[d]		778.0.	56	2.2	1.1
12–30 cpm		1044.0	179	42.0	0.5
31–69 cpm		1354.0	396	68.0	0.6
70+ cpm		2963.0	673	300.0	2.3
Average of 378 copiers		660.0	74	50	0
Scanner		150.0	15		0
FAX Machine		30–175	1535		0

[a] Cathode-ray tube, used with desktop personal computers.
[b] Liquid crystal display, used with laptop computers.
[c] Given are maximum and average power consumption.
[d] cpm = copies per minute.
Source: Roth et al (2002b)

Figure 10.5 Relative energy use by office equipment in US commercial buildings (excluding telecommunication networks)

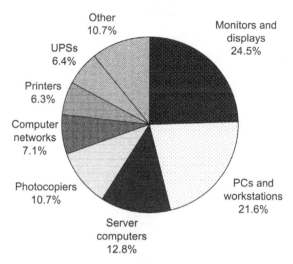

Source: Data from Roth et al (2002b)

10.4 Voltage transformers

Large buildings often purchase electricity directly from the medium-voltage distribution grid (3-phase AC at 4–35kV). They have their own transformers to step the voltage down to that used in buildings. Typically, 3–5 per cent of the incoming electricity is lost in this step. Table 10.7 shows the potential for reducing transformer energy losses through use of advanced transformer design. Up to one-third of current transformer losses (around 1–2 per cent of total energy use) can be avoided using energy-efficient transformers. This is not large but, according to Kubo et al (2001), is economically justified for first-time purchases or at the normal time for transformer replacement (the lifespan of a transformer is typically 35 years). The US Energy Star Program includes energy-efficient transformers (www.epa.gov/appdstar/transform). In comparing different transformers, particular attention should be paid to part-load efficiencies, since transformers operate most of the time at one-third or less of their rated load, and even peak loads are often no more than 50 per cent of the rated load (Thomas et al, 2002). In a survey of 43 buildings in the US, Korn et al (2000) measured an average transformer load of 16 per cent of the rated load.

Table 10.7 Efficiency in transforming from high voltage to low voltage

Initial voltage (kV)	Baseline efficiency	'Energy-efficient' efficiency	Reduction in losses (%)
15	95.7	97.0	29.4
30	96.5	97.5	29.0
45	97.0	97.7	23.4
75	97.3	98.0	24.0
150	97.9	98.3	19.3
300	98.3	98.6	18.1
750	98.6	98.8	13.0

Source: Kubo et al (2001)

Figure 10.6 Transformer efficiency versus load for two conventional transformers (with permitted temperature rises of 80K and 150K) and for an Energy Star transformer

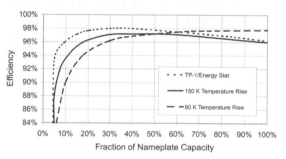

Source: Korn et al (2000)

Figure 10.6 shows the variation of efficiency with load for two conventional transformers and for an Energy Star transformer. At 16 per cent of the rated load, the transformer efficiency can be as low as 94 per cent, compared to 97.5 per cent for the Energy Star transformer.

10.5 Elevators

State-of-the-art elevators reduce energy use in two ways: through a more efficient elevator, and through the elimination of a machine room above the elevator shaft (thereby reducing the surface area for heat loss and air leakage). Greater efficiency is obtained through the use of permanent-magnet motors, a smaller sheave that turns at a faster, more efficient speed, and a gearless design that eliminates the energy loss of geared systems. The energy use of an advanced gearless elevator is about 40 per cent less than a geared system at full load and about 60 per cent less at 50 per cent of full load according to Otis Elevator Company (www.otis.com).

10.6 Summary and synthesis

Energy use by household appliances and consumer electronics is an important fraction of total household electricity use. The most efficient appliances require a factor of two to five less energy than the least efficient appliances available today. Standby use by consumer electronics (i.e. energy used when the machine is turned off) in a typical household in many countries typically exceeds the energy used by a refrigerator/freezer unit complying with the latest US standards. Electricity use by office equipment may not be large compared to electricity use by the HVAC system, but can be a modest source of internal heat gain. The biggest savings potential is to choose equipment that shuts down when inactive (increasingly, this is the only equipment available), and to turn equipment off during non-working hours.

Note

1   This adjusted volume is derived from a fridge volume of $14.5ft^3$ (411litres) plus a freezer volume of $3.8ft^3$ (108litres) multiplied by an adjustment factor of 1.63 to account for the greater work required to cool the freezer.

eleven

Active Collection and Transformation of Solar and Wind Energy

We have discussed passive uses of solar energy on many occasions so far in this book, as part of passive heating (Chapter 4), passive ventilation and cooling (Chapter 6) and as daylighting (Chapter 9). Many of the passive heating and cooling systems rely on solar collectors that heat air and are part of the building cladding. In this chapter we focus on more active solar energy systems – photovoltaic panels (PV) that are part of the building cladding or windows, and hot-water solar panels that can be used for domestic hot-water needs or to power solar air conditioners. Solar air collectors and PV panels can be combined to form a single system, so the distinction between passive and active solar energy systems is somewhat arbitrary. We have also discussed passive use of wind-energy, as a means of cooling through cross ventilation, through wind catchers or through passive suction devices or building designs that create a suction effect. We complement that discussion here with brief mention of wind turbines integrated into a building structure – the active transformation of wind energy into electricity.

11.1 Availability of solar energy

The intensity of solar radiation (that is, the solar irradiance) on a plane perpendicular to the sun's rays, outside the Earth's atmosphere and at the average Earth–sun distance, is referred to as the *solar constant*. It has a value of $1370 W/m^2$ but, despite the name, varies by about $\pm 1.5 W/m^2$ during the course of the 11-year sunspot cycle. When averaged over day and night, and from North Pole to South Pole, the average incoming solar irradiance is only one-quarter of this – about $342 W/m^2$. On average, 30 per cent of this is reflected back to space (largely by clouds and the surface), leaving a global average rate of absorption of solar energy of about $240 W/m^2$.

Figure 11.1 shows the geographical distribution of measured or estimated annual mean solar irradiance at the Earth's surface, taking into account observed cloudiness. The best locations receive an average of 250–275W/m^2 (2200–2400kWh/m^2/yr). This is a low power density compared to the concentrated power demands in cities of the industrialized world. For example, a typical thermal power plant (coal fired or nuclear) would produce 2GW of electric power. For a solar-power density of 200W/m², an area of 10km² would be required to intercept this much power. The area required to generate this much electric power would be several times larger, since the efficiency of conversion from solar energy to electricity will likely not rise much above about 10–20 per cent for economically competitive modules. This implies that it will be difficult to meet a large fraction of the world's energy needs with solar energy through conversion of solar energy into electricity (although certainly not impossible, as the total amount of solar energy intercepted by the Earth is about 12,000 times the total present world primary power demand of 14.5TW). On the other hand, and as demonstrated in this book, a large portion of the energy needs of buildings (lighting, ventilation, heating, cooling) can be met directly with solar energy through passive means, with the building itself serving as the collector and transformer of solar energy. The energy needs that can be met *only* by electricity are vastly smaller than the total energy use of current buildings, such that a significant fraction of the reduced requirement could be met through PV panels integrated into the building structure (as discussed in Section 11.3).

11.1.1 Illustrative calculations

To calculate the amount of radiation incident on a solar collector of any orientation, we need to know the intensity perpendicular to the sun's rays at ground level and the angle between the sun's rays and a line perpendicular to the module (such a line is called the 'normal'). The latter in turn depends on the sun's location, as specified by its *zenith angle* (the angle measured from the vertical) and its *azimuth* (the horizontal direction measured from north, positive to the east and negative to the west). We have to consider three separate components of solar radiation: *direct beam radiation*, which has not been scattered by molecules, airborne particles or clouds and so arrives from

Figure 11.1 Geographical variation in observed mean annual solar irradiance (W/m²) on a horizontal surface at ground level

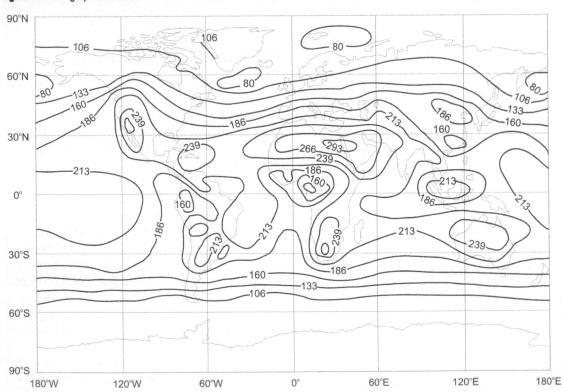

Source: Henderson-Sellers and Robinson (1986)

one direction only (this is the component that produces shadows); *diffuse beam radiation*, which arrives from all directions due to scattering; and *reflected radiation*, which is reflected from the surrounding ground and buildings onto the solar collector. Diffuse radiation can be divided into radiation that has passed through the atmosphere only once before striking the solar collector, and radiation that has been scattered back and forth between the surface and atmosphere before striking the collector. The reflected component depends on the tilt of the collector and on the reflectivity or *albedo* of the surroundings. The albedo depends on the type of surface and increases with increasing solar zenith angle; albedo values for a zenith angle of 60° range from 0.15 for a dense vegetation canopy, to 0.25–0.35 for a sandy surface, to 0.8 for fresh snow. As the albedo of the surrounding landscape increases, the contribution of reflected radiation (and of multiply scattered diffuse radiation) increases.

The set of equations needed to compute the total flux of solar radiation on a solar collector under clear skies is given in Box 11.1. For a solar collector with a fixed orientation, the mean annual solar irradiance is maximized if the collector is inclined at an angle equal to local latitude (i.e. for a location at 45°N, incline the collector to the south at an angle of 45°). This might be possible for some roof-integrated collectors. The equations presented in Box 11.1 can be readily used for calculating the irradiance on an unobstructed wall of any orientation under clear skies (by setting δ = 90° and choosing the appropriate azimuth angle). The seasonal variation of diurnally averaged clear sky solar irradiance on vertical walls of various orientations at various latitudes, as calculated using the equations in Box 11.1, was presented in Figues 4.1 and 4.2 of Chapter 4, and the diurnal variation in irradiance in mid December and mid June at 50°N was presented in Figure 4.3. The largest irradiance in June occurs on east- and west-facing walls rather than on south-facing walls, due to the fact that the sun is closer to the horizon early and late in the day and thus closer to perpendicular to a vertical surface.

BOX 11.1 Intensity of solar radiation on a flat surface

The intensity of incident solar radiation at the top of the atmosphere is given by:

$$Q(\theta, t) = \left(\frac{Q_s}{d(t)^2} \right) \cos\theta \qquad (B11.1.1)$$

where Q_s is the solar constant, $d(t)$ is the Earth–sun distance at time t as a fraction of the mean distance, and θ is the zenith angle (the angle of the sun from the vertical). $Q_s = 1370$ W/m² but is not strictly constant, as it varies by about ±0.07 per cent over the 11-year sunspot cycle. $d(t)$ is given by (Berger, 1978):

$$d = \frac{1 - \varepsilon^2}{1 + \varepsilon \cos\left((\gamma - P)\frac{\pi}{180} \right)} \qquad (B11.1.2)$$

where ε is the orbital eccentricity (0.01672), P is the longitude of perihelion (282.04°), and γ is the orbital angle (in degrees) measured from the vernal equinox (20 March). $\gamma = P$ on 3 January, when the Earth is closest to the sun. The solar zenith angle is given by (Sellers, 1965):

$$\cos\theta = \sin\phi \sin D + \cos\phi \cos D \cos h \qquad (B11.1.3)$$

where ϕ is latitude, D is declination (the angular distance of the sun north (positive) or south (negative) of the equator), and h is the solar angle (the angle through which the Earth must turn to bring the longitude in question directly under the sun). The declination can be computed as:

$$D = -\arcsin(0.397901\cos(2\pi(DN-1)/365.25)) \qquad (B11.1.4)$$

where DN is the day number within the year ($DN = 1$ on 1 January) and D will be in radians (this equation has been slightly modified from Kreider and Kreith, 1981, to give a maximum declination of ±23.447°, equal to the obliquity of the Earth's axis at present). If $\cos\theta < 0$ the solar irradiance is zero.

In the case of a surface inclined at an angle δ from the horizontal with an azimuth of a′, the angle θ* between the surface normal and the sun is given by:

$$\cos\theta^* = \cos\delta\cos\theta + \sin\delta\sin\theta\cos(a-a') \qquad \text{(B11.1.5)}$$

where a is the solar azimuth and is given by:

$$\cos a = \frac{\sin\phi\cos\theta - \sin D}{\cos\phi\sin\theta} \qquad \text{(B11.1.6)}$$

For the sun, the azimuth is the horizontal direction of the sun at a given point in time, while for a module, the azimuth is the direction that it is facing if it is tilted. The azimuth is zero for directions due south, is positive when directed west of south up to a maximum value of π when due north, and negative when directed east of south up to a maximum value of $-\pi$ when due north. Equations (B11.1.5) and (B11.1.6) are derived in Sellers (1965). For a sloping surface directed due south, $a' = 0$, so:

$$\cos\theta^* = \cos\delta\cos\theta + \sin\delta\left[\frac{\sin\phi\cos\theta - \sin D}{\cos\phi}\right]$$

$$\text{(B11.1.7)}$$

The solar irradiance I_T incident on an inclined plane at the Earth's surface is given by:

$$I_T = I_{Dr} + I_{Df} + I_R \qquad \text{(B11.1.8)}$$

where I_{Dr}, I_{Df} and I_R are the directly transmitted, diffuse and reflected components. These are given by:

$$I_{DR} = \left(\frac{Q_s}{d^2}\right)(1-O_3)(1-A_w)D_R\cos\theta^* \qquad \text{(B11.1.9)}$$

$$I_{Df} = D\downarrow\left\{D_R\frac{\cos\theta^*}{\cos\theta} + 0.5(1-D_R)(1+\cos\delta)\right\} \qquad \text{(B11.1.10)}$$

$$I_R = \left(\frac{Q_s}{d^2}\right)(1-O_3)(1-A_w)D_R\cos\theta\left\{\frac{1}{2}\alpha_s(1-\cos\delta)\right\}$$

$$\text{(B11.1.11)}$$

where O_3 and A_w are the fractional absorption of solar radiation by ozone and water vapour, D_R is the zenith angle-dependent fraction of radiation transmitted as direct beam, α_s is the zenith angle-dependent surface albedo for direct beam radiation, and $D\downarrow$ is the diffuse beam irradiance on a horizontal surface. Equation (B11.1.10) is based on Hay (1986), with the first and second terms representing the circumsolar and uniform

background components of diffuse radiation, respectively (as D_R decreases, the diffuse radiation is more strongly weighted toward the uniform component). The factor $0.5(1-\cos\delta)$ in Equation (B11.1.11) is a view factor that arises by assuming the reflecting surface to be semi-infinite and the reflected radiation to be isotropic. $D\downarrow$ is computed as the sum of singly and multiply scattered components D_1 and D_2, which are given by:

$$D_1 = \left(\frac{Q_s}{d^2}\right)(1-O_3)(1-A_w)S_F\cos\theta \qquad \text{(B11.1.12)}$$

$$D_2 = \left(\frac{Q_s}{d^2}\right)(1-O_3)(1-A_w)\left\{(D_R\alpha_s\overline{S}_B + \right.$$

$$\left. + S_F\overline{\alpha_s S_B})\cos\theta/(1-\overline{\alpha_s S_B})\right\} \qquad \text{(B11.1.13)}$$

respectively. S_F is a zenith angle-dependent forward scattering fraction, while $\overline{S}_B$ is the back scattering fraction for upward reflected diffuse radiation, evaluated at $\cos\theta = 0.5$

D_R, S_F and $\overline{S}_B$ can be evaluated using polynomial functions of the form:

$$x = a_0 + OAM(a_1 + OAM(a_2 + OAM$$

$$(a_3 + OAM(a_4 + a_5 OAM)))) \qquad \text{(B11.1.14)}$$

where x is D_R, S_F and $\overline{S}_B$, and OAM is the optical air mass and is given by:

$$OAM = \frac{1}{\cos\theta + 0.15/(57.3(1.6386-\theta))^{1.253}} \qquad \text{(B11.1.15)}$$

where θ is in radians. Equations (B11.1.14) and (B11.1.15) are from Thompson and Baron (1981). Values of the a_i coefficients for evaluation of D_R, S_F and $\overline{S}_B$ for non-polluted conditions are given in Table B11.1 (under hazy pollution, D_R will decrease while S_F and $\overline{S}_B$ increase).

The fractional absorption of solar radiation by water vapour can be computed as:

$$A_W = \frac{2.9W_a}{(1.0+141.5W_a)^{0.635} + 5.925W_a} \qquad \text{(B11.1.16)}$$

where $W_a = e_a OAM$ and e_a is the surface water vapour pressure in mb. Finally, an adequate formulation for the zenith angle dependence of the surface albedo for direct beam radiation is:

Table B11.1 Values of the coefficients used in Equation (B11.1.14) for the computation of D_R, S_F and $\bar{S}_B$

	D_R	S_F	S_B
a_0	8.9788×10^{-1}	2.519×10^{-2}	$1.73548478 \times 10^{-2}$
a_1	-1.1908×10^{-1}	6.582×10^{-2}	4.8979660×10^{-2}
a_2	7.3649×10^{-3}	-4.3399×10^{-3}	$-3.02510749 \times 10^{-3}$
a_3	-2.4966×10^{-4}	1.2697×10^{-4}	$8.87911913 \times 10^{-5}$
a_4	4.4260×10^{-6}	-1.3608×10^{-6}	$-9.65852864 \times 10^{-7}$
a_5	-3.1921×10^{-8}		

$$\alpha_s = \bar{\alpha}_s \frac{1.7 - 0.751\cos\theta}{1.0 + 0.649\cos\theta} \qquad \text{(B11.1.17)}$$

where $\bar{\alpha}_s$ is the surface albedo at $\theta = 60°$, which is used as the albedo for diffuse radiation. More detailed treatments of solar radiation can be found in Clarke (2001) and Eicker (2003), but the equations presented here are entirely adequate for illustrative purposes while being easy to implement.

Further results, applicable to roof-mounted or roof-integrated solar panels, are shown in Figures 11.2 and 11.3. Figure 11.2 shows the seasonal variation in diurnally averaged solar irradiance on panels inclined at a slope of 0°, 30° and 50° at latitudes 0°N, 30°N and 50°N. For panels at a slope of 30° and 50°, results are given for the panels pointing due south and pointing 45° west of due south. The largest mean annual solar irradiance occurs for panels pointing due south and inclined at an angle equal to the latitude. Inclining the panel at this slope results in a more seasonally uniform irradiance (that is, the irradiance is reduced in summer and increased in winter). Orienting an optimally tilted panel 45° to the west decreases the mean annual irradiance by about 10 per cent, while shifting the diurnal peak to 1–2 hours after solar noon, when it might be more useful.

11.1.2 Surface Meteorology and Solar Energy dataset

Further information on the availability of solar energy anywhere in the world can be obtained from the *Meteonorm* dataset (www.meteotest.ch) and from NASA's *Surface Meteorology and Solar Energy* (SSE) dataset (http://eosweb.larc.nasa.gov/sse). The Meteonorm dataset is a commercial dataset based on a worldwide network of 7400 weather stations. The primary data are monthly means of irradiance (direct and diffuse solar, longwave), and sunshine duration (along with temperature, humidity, precipitation and wind data) that can be interpolated from the weather stations to other points. Hourly data are generated based on a stochastic model (that is, a model of how random variability is distributed around the mean). The SSE dataset includes estimates of the monthly and annual average solar irradiance on horizontal surfaces, on surfaces tilted toward the equator at various fixed angles, and on surfaces tilted at the optimal angle for each month (but pointing toward the equator). Also given are separate estimates of the diffuse and clear sky irradiance on horizontal surfaces, the monthly mean cloud cover averaged over daylight hours, and many other parameters useful in the design of solar energy systems. The estimates are derived from a combination of satellite and meteorological observations and atmospheric model simulations for the period 1983–1993, interpolated to a 1° latitude by 1° longitude grid. Estimates for a given location can be obtained by entering the latitude and longitude of the location in question, or by zooming in on a map. The estimates do not include the effects of small-scale variations in average cloud cover. Temperature and humidity data are also available. This dataset is also available through the RETScreen package of spreadsheets that is available free of charge from Natural Resources Canada (at www.retscreen.net) in 21 languages. These spreadsheets provide preliminary estimates of the costs and benefits of solar thermal and photovoltaic energy systems in buildings.

Figure 11.2 Seasonal variation in diurnally averaged solar irradiance (W/m²) for clear skies on a horizontal panel and on panels tilted at 30° and 50° and pointing due south or to the southwest. Results are given for panels at the equator (top), 30°N (middle), and 50°N (bottom)

Figure 11.3 Diurnal variation of solar irradiance (W/m²) for clear skies on 21 June on a horizontal panel and on panels tilted at 30° and 50° and pointing due south or to the southwest. Results are given for panels at the equator (top), 30°N (middle), and 50°N (bottom)

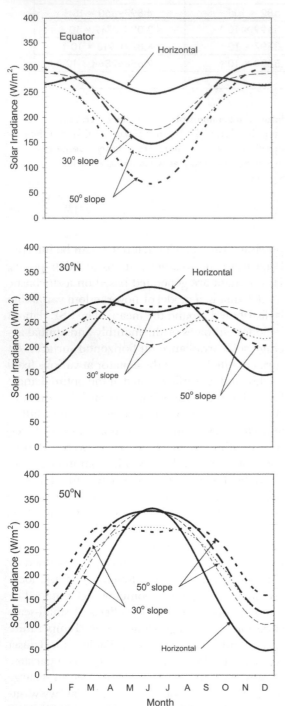

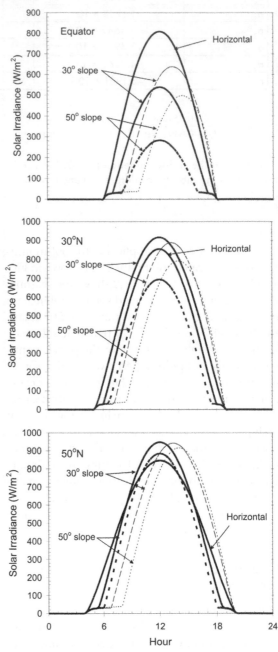

Note: For tilted panels, the heavy lines are for panels facing due south, while the light lines are for panels tilted 45° west of due south.
Source: Computed using the equations given in Box 11.1

Note: For tilted panels, the heavy lines are for panels facing due south, while the light lines are for panels tilted 45° west of due south.
Source: Computed using the equations given in Box 11.1

Table 11.1 Annual solar irradiance on a horizontal surface or on a surface pointing toward the equator and at the tilt from the horizontal that maximizes the annual irradiance, based on a ten-year average of a mix of observed and model-simulated data

City	Irradiance (kWh/m²/year)		Irradiance as a percentage of maximum listed irradiance		Tilt (°) with maximum Irradiance
	Horizontal	Tilted	Horizontal	Tilted	
North and South America					
Canada					
Toronto	1256	1383	57	60	28
Winnipeg	1245	1478	57	64	34
Edmonton	1164	1427	53	62	38
Vancouver	1281	1610	59	70	49
Inuvik	887	1351	41	59	53
US					
Miami	1694	1756	77	76	25
Atlanta	1540	1639	70	71	33
New Orleans	1580	1650	72	72	30
Albuquerque	1668	1821	76	79	35
Phoenix	1843	2015	84	87	33
San Francisco	1584	1730	72	75	37
Las Vegas	1748	1913	80	83	36
Denver	1570	1756	72	76	39
Chicago	1314	1438	60	62	26
New York	1434	1595	66	69	40
Boston	1365	1507	62	65	27
Minneapolis	1307	1493	60	65	30
Fairbanks	938	1424	43	62	49
Latin America					
Mexico City	1821	1894	83	82	19
Caracas	1978	2008	90	87	10
Bogotá	1650	1661	75	72	19
Brasília	1854	1920	85	83	15
São Paulo	1653	1701	76	74	23
Buenos Aires	1628	1726	74	75	34
Punta Arenas	865	1000	40	43	38
Europe and Africa					
Europe					
Athens	1507	1606	69	70	23
Rome	1686	1907	77	83	26
Madrid	1475	1613	67	70	25
Glasgow	942	1146	43	50	40
London	975	1095	45	48	36
Paris	1110	1252	51	54	33
Zurich	1201	1372	55	60	32
Berlin	1000	1142	46	50	37
Prague	1022	1153	47	50	36
Vienna	1146	1296	52	56	33
Budapest	1223	1387	56	60	32
Warsaw	1022	1197	47	52	37
Minsk	989	1190	45	52	38
Kiev	1110	1303	51	57	35
Moscow	989	1219	45	53	40
Helsinki	931	1205	43	52	45
Copenhagen	1026	1237	47	54	40
Oslo	1007	1132	46	49	35

City	Irradiance (kWh/m²/year)		Irradiance as a percent of maximum listed irradiance		Tilt (°) with maximum Irradiance
	Horizontal	Tilted	Horizontal	Tilted	
Murmansk	748	1234	34	54	54
Africa					
Casablanca	1865	2000	85	87	33
Cairo	1924	2004	88	87	15
Lagos	1792	1829	82	79	21
Nairobi	2022	2048	92	89	16
Salisbury	1971	2051	90	89	17
Johannesburg	1902	2008	87	87	26
Cape Town	1898	2018	87	88	18
Asia and Australia					
Asia					
Istanbul	1398	1515	64	66	26
Ankara	1529	1668	70	72	24
Beirut	1876	1989	86	86	33
Riyadh	2113	2164	97	94	24
Teheran	1675	1792	77	78	20
Kabul	1690	1799	77	78	34
New Delhi	1869	2000	85	87	28
Bombay	2190	2303	100	100	19
Madras	1938	1982	89	86	13
Katmandu	1821	1975	83	86	27
Lhasa	1774	1942	81	84	29
Ulaan Baatar	1394	1697	64	74	47
Beijing	1573	1829	72	79	39
Seoul	1445	1628	66	71	37
Tokyo	1361	1515	62	66	35
Sapporo	1263	1398	58	61	28
Shang-hai	1438	1522	66	66	31
Hong Kong	1478	1518	68	66	22
Ho Chi Min City	1869	1902	85	83	25
Singapore	1716	1730	78	75	16
Manila	1712	1712	77	74	14
Australia					
Alice Springs	2168	2245	99	97	23
Cairns	1967	2026	90	88	16
Brisbane	1785	1865	82	81	27
Sydney	1573	1686	72	73	34
Hobart	1325	1475	61	64	27

Note: Also given is the tilt that maximizes the annual irradiance (among the small number of tilts that can be chosen).
Source: Surface Meteorology and Solar Energy (SSE) dataset (http://eosweb.larc.nasa.gov/sse)

Irradiance values on horizontal and tilted modules (pointed toward the equator) for selected world cities, as obtained from the SSE website, are given in Table 11.1. Also given is the irradiance on horizontal and tilted modules at each location as a fraction of the maximum irradiance reported in Table 11.1 (which is for Bombay). For horizontal modules, these fractions are less than 0.4–0.50 for the high-latitude sites listed in Table 11.1, but for inclined modules, the worst sites have (with one exception) at least 50 per cent of the irradiance of the best site (Bombay). This is because tilting the module toward the equator partly compensates for the effect of higher latitude, thereby reducing the spread in annual irradiance.

11.2 Photovoltaic solar energy

Solar energy can be converted to DC electricity using *photovoltaic panels*. Sunlight can be thought of as a stream of energetic particles called *photons*. A *semiconductor* is a material in which electrons occur in two energy bands or levels, the lower referred to as the *valence band* and the upper as the *conduction band*. The energy difference between the top edge of the valence band and the bottom edge of the conduction band is called the *bandgap*. Silica is a common and inexpensive semiconductor. When an electron absorbs a photon of energy equal to or greater than the bandgap, it will jump from the valence band to the conduction band. In order to induce a flow in one particular direction – that is, to generate DC electricity – two separate layers with different impurities added are joined together, one of which tends to attract electrons and the other of which tends to give up electrons. These are referred to as *n-type* and *p-type* layers, respectively, and the juxtaposition of the two is referred to as a *p–n junction*.

The layout of a solar cell is shown in Figure 11.4. On either side of the p–n junction are metallic contacts, and when an electron near the p–n junction absorbs a photon it is induced by the electric field around the p–n junction to flow out of the semiconductor through one of the metallic contacts. The power produced by a solar cell is given by the product of voltage and current, the voltage being provided by the internal electric field at the p–n junction.

Figure 11.4 Layout of a silicon solar cell

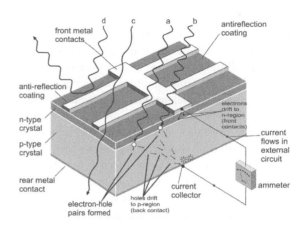

Note: Photon a has an amount of energy just sufficient to dislodge an electron from its orbit next to a silicon atom near the p–n junction, causing it to enter an energy state called the conduction band. It then migrates into the n-type layer under the influence of the electric field straddling and set up by the p–n junction. A hole is left behind, which is filled by an electron from the p-layer, creating a hole in that layer, which progressively 'migrates' to the bottom electrical contact. The electron in the n-layer finds it way to the top electrical contact, travels through the external circuit to the bottom contact, and fills a waiting hole. Photon b has more energy than needed to dislodge an electron, and the excess is dissipated as heat. Photon c has less energy than the bandgap and passes right through the material as if nothing were there. Other photons are reflected by the surface or absorbed by the current collectors on the cell surface.
Source: Boyle (1996)

Photons with energy equal to or greater than the bandgap will be able to bump electrons from the valence to the conduction band. However, for photons with energy greater than the bandgap, the excess energy is converted to heat. To maximize the conversion of sunlight to electricity, the bandgap should coincide with the photon energy where the peak emission occurs. Photons at longer wavelengths will not be able to bump electrons, while photons at shorter wavelengths have excess energy that is not used in bumping electrons. Because most photons either do not have enough energy to bump an electron, or have excess energy that is wasted, the maximum possible efficiency for converting sunlight to electricity using a single bandgap is limited to around 40 per cent (Luque and Marti, 1999). However, if cells with different bandgaps are stacked on top of each other, then the maximum possible efficiency is 85–90 per cent.

11.2.1 Kinds of PV solar cells and their efficiency

Solar cells can be made from a single crystal of silicon, from multiple crystals, from amorphous (non-crystalline) silicon, or from other materials. For details on the manufacture and structure of alternative solar cells, see Green (2003a). Single-crystal cells (x-Si) are produced by first growing a single large crystal of silicon under carefully controlled conditions, then cutting it into thin (0.1–1.0mm thick) wafers. A solar cell has a typical dimension of 10cm × 10cm (an area of 0.01m²), and many cells are placed next to each other and wired together in series to form a module with a typical area of 0.4–1.0m². The best efficiencies (ratio of electrical energy produced to incident solar energy) as of 2004 achieved for laboratory cells with a minimum area of 1cm² and for laboratory modules with a minimum area of 800cm² conditions are summarized in Table 11.2 for different technologies. Also given are the efficiency ranges for commercially available products. Module efficiencies are less than cell efficiencies (about 75–90 per cent as large) because slight differences between adjacent cells cause some of their

electrical power output to cancel out, because there are inactive regions between cells, and due to resistance losses in the wiring between cells and in the diodes used to protect cells from short circuiting (Boyle, 1996).

Crystalline and polycrystalline cells

As of 2004, the best efficiency obtained for x-Si cells in the laboratory was 24.7 per cent, and the best module efficiency was 22.7 per cent. Polycrystalline silicon (p-Si) consists of numerous individual grains made from lower-grade silica and under less stringent conditions than for single-crystal silicon. Due to the presence of crystal variations and transfer boundaries, electron conduction is more frequently impeded than in an x-Si cell, so the best cell and module efficiencies are somewhat lower – 20.3 per cent and 15.3 per cent, respectively.

The wafers used to make crystalline cells are typically one half millimetre (500μm) thick. Recent advances in micro-machining techniques have made it possible to cut silicon wafers to a thickness of 35–70μm, without compromising on efficiency. Although substantially thicker than amorphous silicon cells (described below), these cells are three times as efficient while using as little as one-thirtieth the silicon of conventional crystalline cells.

Thin-film cells

PV cells can be made through direct deposition of a semiconductor onto a substrate (usually glass), rather than by making wafers. Typically, less than 1μm (micron) of semiconductor material is used, reducing the material requirements by up to a factor of 1000 compared to crystalline silicon. This in turn allows almost any semiconductor to be economically competitive with silicon. Thin-film modules can be made using amorphous (non-crystalline) silicon (a-Si) (either with a single junction or by stacking a number of p–n layers on top of each other, each with slightly different properties), using a-Si and x-Si combinations, or using p-Si.

Pure amorphous silicon (a-Si) cells have very low efficiency (1 per cent for the first ones made, in 1973), but the efficiency has been dramatically improved by adding up to 10 per cent hydrogen by volume (giving a-Si:H). However,

Table 11.2 Efficiencies of the best laboratory PV cells and modules achieved as of 2004 for various technologies at a temperature of 25°C, and of commercially available modules

Technology	Laboratory efficiency (%)		Efficiency (%) of commercial modules
	Cell	Module	
x-Si	24.7	22.7	7–16
p-Si	20.3	15.3	7–14
a-Si	10.1	–	4–8[a]
a-Si/CIGS	14.6	–	5–6
a-Si/a-SiGe/a-SiGe	12.4	10.4	5–6
CdTe	16.5	10.7	6–10[a]
CIGS[c]	18.4	13.4	7–9
GaInP/GaAs/Ge	32.0	–	
Dye-sensitized	10.8[b]	4.7	

[a] Upper limit from SOLRIF PV Inroof System (www.solrif.ch).
[b] Grätzel (2001).
[c] CIGS = CuInGaSe$_2$.
Note: The electrical output decreases by 0.5% for each Celsius degree warming for all cell types except a-Si, where the temperature effect is close to zero.
Source: Green et al (2005), except where indicated otherwise

exposure to strong sunlight undoes some of the benefit of adding hydrogen, leading to a decrease in cell efficiency during the first 1000 hours of use. To overcome this problem, a cell design different from crystalline cells has been adopted (Green, 2000). The required cell thickness is less than that needed to absorb all of the usable incident sunlight, so several cells are stacked on top of one another so that light not absorbed by the upper cell can be absorbed by a lower cell.

Stabilized multi-junction a-Si cell and module efficiencies have reached 13.5 per cent and 10.4 per cent, respectively, under laboratory conditions (Table 11.2). The best stabilized efficiency in commercial modules is only 6–7 per cent (Green, 2000), and it had been hoped that eventual efficiencies of 10 per cent would be achievable. However, it now appears that a breakthrough will be needed to achieve a stabilized efficiency of 10 per cent (von Roedern, 2003). Amorphous silicon cells are suited to continuous production, and large areas of cell can be deposited onto a wide variety of substrates, including building materials such as steel and glass.

A variety of other kinds of solar cells have been created, involving rare and/or toxic materials such as cadmium (Cd), indium (In), telluride (Te), selenium (Se), gallium (Ga) and arsenic (As). These cells are listed in Table 11.2, where it is seen that cell efficiencies as high as 32 per cent have been achieved. However, there has been an understandable market resistance to PV products based on toxic materials. Furthermore, there are severe resource constraints with these materials, such that it is highly unlikely that they could ever be used to supply a significant part of global energy needs (Andersson, 2000). However, that does not prevent them from playing a niche role, in applications where their high efficiency justifies what might be a greater module cost. This may include concentrator PV modules, where mirrors are used to concentrate sunlight onto the PV module with 3–30 times its normal intensity (in effect, partially replacing expensive solar cells with inexpensive mirrors, but requiring a solar tracking capability).

Nanocrystalline dye cells

A nanocrystalline dye-sensitized solar cell consists of a porous matrix of titanium oxide crystals that is coated with a one-molecule thick layer of a ruthenium-based organometallic dye. A platinum catalyst is required. The solar cells can be printed on foils, making them light and flexible with low manufacturing costs, and allowing for special design features of architectural interest. The crystals are of the order of 10nm in size (nm = nanometer; 1nm = 10^{-9}m). The dye absorbs sunlight, creating an excited electron that is transferred to the titanium oxide matrix, and through the matrix to the cell contacts and an external circuit. Electrons transferred from the dye to the matrix are replaced with electrons transferred from an electrolyte to the dye. The matrix structure and the dye cell are illustrated in Figure 11.5. The dye cell mimics photosynthesis, which also uses an organometallic compound (chlorophyll) to absorb sunlight, creating excited electrons that are then transferred to a transport medium. The dye absorbs only a band of photon energies, rather than all photons above a given energy. This raises the possibility of creating windows that convert near infrared solar energy into electricity while letting visible light through. Platinum is a rare element and would face severe supply constraints if there were a worldwide move to automotive fuel cells (Andersson, 2000). Ruthenium

Figure 11.5 Structure of a dye-sensitized solar cell

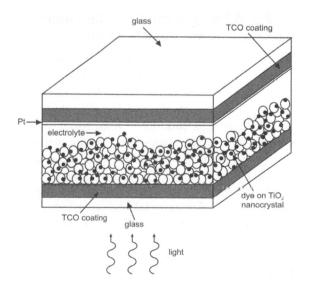

Note: TCO = transparent conducting oxide.
Source: Green (2000)

in significant quantities would be available as a byproduct of mining platinum for automotive fuel cells. Since the pioneering work on dye-sensitized cells in 1991 (O'Regan and Grätzel, 1991), an efficiency of 10.8 per cent has been achieved on small (<1cm²) active areas (Grätzel, 2001) and 6 per cent on large (68cm²) active areas (Späth et al, 2003).

11.2.2 Effect of temperature on module efficiency

For crystalline silicon modules, the efficiency in generating electricity decreases as the module temperature increases, for reasons explained in detail by Green (2003b). Each Celsius degree increase in module temperature decreases the electricity output by 0.04–0.9 per cent, with most systems showing a decrease of 0.4–0.6 per cent/K. Since unventilated PV panels can reach temperatures of up to 80°C, this can result in a significant reduction in electrical output. As noted above, a-Si modules show a decrease in performance during the first 1000 hours of their operation. This performance decrease is reduced when the modules are heated. Thus, a-Si modules tend to show a smaller decrease (Scheuermann et al, 2002) or even a slight increase in electrical output with increasing temperature, by perhaps 0.2 per cent/K (Bazilian et al, 2001).

11.2.3 Voltage/current variation and maximum power point tracking

An example of the variation of the current and voltage produced by a PV module for various solar irradiances is shown in Figure 11.6. Electrical power (W) is given by the product of voltage and current, and this product is maximized at one voltage-current combination called the *maximum power point* (MPP). The MPP for each value of irradiance is indicated in Figure 11.6. The current, and hence the voltage and power, can be adjusted through a variable external resistance using one of several possible control algorithms (Hohm and Ropp, 2003). An electronic device that continuously adjusts the resistance so as to maximize the power output as the solar irradiance changes is called a *maximum power point tracker*. Since the tracking process is not perfect, there will be some loss compared to the maximum possible power

Figure 11.6 Variation of current with voltage in a given PV module for various solar irradiances, and the maximum power point (MPP) for each irradiance

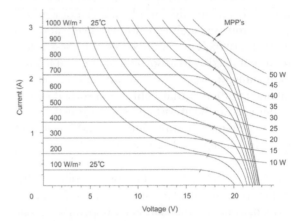

Source: Hastings and Mørck (2000)

Figure 11.7 The combination of PV modules in series (within each row) and in parallel (separate rows)

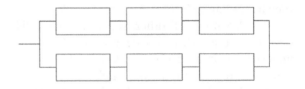

Note: Here, the voltage would be three times that of a single module and the current twice that of a single module.

output, but this loss is normally less than 5 per cent. The DC output from a PV module or an array of modules must be converted to AC electricity, a process called *power conditioning*. An *inverter* is the device that performs both DC–AC power conversion and MPP tracking. Efficiencies today are 90–95 per cent (van Zolingen, 2004). Information on commercially available products can be obtained through the California Energy Commission at www.consumerenergycentre.org/cgi-bin/eligible_inverters.cgi.

To increase the voltage of a PV system, individual PV modules are connected together in series (end-to-end). To increase the current produced by a PV system, individual PV modules are connected in parallel. A desired voltage-current combination can be achieved through a mixture of series and parallel connections, as illustrated in Figure 11.7. If any cell or module is shaded, it will produce a smaller current, and this will limit the output of *all* of the other cells to which it is connected in series. Thus, there are merits in limiting the extent of series connections if partial shading will occur.

11.2.4 System efficiency

As noted above, module efficiencies tend to be about 75–90 per cent of cell efficiencies, due to slight mismatches between the cells that make up a given module. The ratio of AC system output to DC module output is known as the *performance ratio*, and reflects losses due to wiring losses, module mismatches and DC–AC conversion losses. The performance ratio ranged from 0.46 to 0.86 in 19 systems that were monitored in California, with an average of 0.62 (Scheuermann et al, 2002). A similar range was found in a survey of 18 systems in Europe (Nordmann and Clavadetscher, 2003). For 333 residential systems in Japan, Nakagami et al (2003) report a range from less than 0.5 to better than 0.85, with an average of 0.71. For systems installed in Germany between 1991 and 1994, Jahn and Nasse (2004) report performance ratios of 0.30–0.87, with an average of 0.64 but falling over time for any given building, whereas for systems installed after 1996, the performance ratio ranges from 0.52 to 0.91, averages 0.74, and is constant over time.

11.2.5 Illustrative costs of PV electricity

The equations that can be used to calculate the cost of electricity from solar energy are given in Box 11.2, while Table 11.3 gives illustrative electricity costs for the range of present and possible future component costs for $1100kWh/m^2/year$ irradiance (close to the lowest found in Table 11.1), $1650kWh/m^2/year$, and $2200kWh/m^2/year$ (close to the highest found in Table 11.1). The key inputs are the module and balance of system (BOS) costs ($/m^2$), the power conditioning costs ($ per kW of peak output), the module and balance of system efficiencies, insurance, maintenance and contingencies; and the real interest rate (after inflation) and system lifetime. The interest rate and system lifetime determine the *cost*

recovery factor – the fraction of the total investment cost that must be paid per year as principle and interest. BOS costs include any support structures and connections between modules. At a module cost of $400/m^2$ (a representative present cost for x-Si), a BOS of $100/m^2$, $800/kW DC for power conditioning, an additional 25 per cent for installation (net of any credits), a module efficiency of 15 per cent (giving $6.89/W_p$ capital cost), and a BOS efficiency of 85 per cent, the cost of electricity ranges from 24cents/kWh (for a solar irradiance of $2200kWh/m^2/year$ and 4 per cent real interest rate) to 66cents/kWh (for a solar irradiance of $1100kWh/m^2/year$ and 8 per cent real interest rate). If modules and BOS costs can be reduced to $100/m^2$ and $50/m^2$, respectively, while maintaining the module efficiency at 15 per cent and power conditioning at $800/kW DC (giving $3.00/W_p$ capital cost), electricity cost ranges from 11 to 29 cents/kWh. If the net module cost is zero (because it displaces other envelope materials of equal or greater cost), electricity costs 8–22cents/kWh for modules with 10 per cent efficiency and 6.8–18.7cents/kWh for modules with 15 per cent efficiency. By comparison, typical average retail electricity costs in OECD countries are 5–16cents/kWh (Figure 2.10), although costs at times of peak demand (when solar energy is often available) can be substantially higher.

BOX 11.2 Economics of PV electricity and solar thermal energy

The levelized cost of electricity ($/kWh) from a power plant is given by the annual revenue requirements divided by the number of kWh produced per year. The annual revenue requirements are given by (i) the fixed fraction of the initial capital cost that must be recovered each year, such that the initial investment plus interest is exactly paid off at the end of the lifetime of the power plant; (ii) the fuel cost; and (iii) the operation and maintenance costs. For solar energy there are, of course, no fuel costs. The fraction of the original investment that must be paid each year is called the *cost recovery factor* (CRF). It depends on the real interest rate (excluding inflation) expressed as a fraction, *i*, and the power plant lifespan in years, N:

$$CRF = i/(1 - (1 + i)^N)$$

(B11.2.1)

For a fossil fuel power plant, capital costs are expressed in terms of $ per kW of capacity. We can express capital costs per kW of peak DC power output *assuming* the peak solar irradiance to be 1000W/m². Given module and BOS costs initially specified in terms of $ per m², the cost in terms of $ per kW$_p$ is given by:

$$C_m(\$/kW_p) = C_m(\$/m^2)/\eta_m I_p$$

(B11.2.2)

where η_m is the module efficiency (solar energy to DC electricity) and I_p is the assumed peak insolation of 1kW/m². This (and other) capital costs are then divided by the number of hours in a year times the capacity factor, as in the cost equation for electricity derived from fossil fuel. Here, the capacity factor would be given by the average irradiance on the panel, I_a, divided by the assumed peak flux (I_p). To convert from a cost per DC kWh to AC kWh, one must divide by the efficiency (η_{BOS}) in converting from DC to AC power. Given a capital recovery factor CRF, the contribution of the module cost to the levelized cost of electricity from solar polar is

$$CRF\left(\frac{C_m(\$/kW_p)}{8760(I_a/I_p)\eta_{BOS}}\right) = CRF\left(\frac{C_m(\$/m^2)}{8766\eta_m\eta_{BOS}I_a}\right)$$

(B11.2.3)

where the denominator on the right hand side is the number of kWh of AC electricity produced per year per m² of module area and I_a is in kW.

Altogether, the levelized cost of producing PV AC electricity is given by:

$$C(\$/kWh) =$$
$$\frac{\left\{(CRF + INS)(1 + ID)(C_m + C_b + C_p + C_{ins}) + OM\right\}I_p}{8760 I_a \eta_{BOS}}$$

(B11.2.4)

where C_m, C_b, C_p and C_{ins} are the module, balance of system, power conditioning costs and installation costs, respectively (all as $/kW$_p$ DC), OM is the operation and maintenance cost (as $/kW$_p$(DC)-year), INS is the annual insurance as a fraction of the capital costs, and ID is an indirect cost factor to account for site engineering, maintenance of inventories and contingencies. In some cases, C_b and C_{ins} might be lumped together as C_b.

An equivalent expression for the levelized cost of AC electricity is:

$$C(\$/kWh) = \frac{(CRF + INS)C_{total} + OM}{R\eta_m\eta_{BOS}}$$

(B11.2.5)

where C_{total} is the total investment cost ($/m²), OM is now in $/m²/year, and R is the annual irradiance (kWh/m²/year).

A similar expression applies to the cost of thermal energy from solar thermal collectors, except that the system thermal efficiency (useable thermal energy divided by incident solar irradiance) replaces $\eta_m\eta_{BOS}$.

Table 11.3 Illustrative costs of PV electricity for the indicated solar irradiance and real interest rate, a 25-year lifespan, 1%/year insurance, 25% indirect costs, $800/kW(DC) or 800€/kW (DC) power conditioning, 75% BOS efficiency, $1.1/m² or 1.1€/m² per year operation and maintenance costs, and other assumptions as indicated below

Cost ($/m² or €/m²)		Module efficiency	Electricity cost (cents or eurocents/kWh(AC))								
			1100kWh/ m²/year			1650kWh/ m²/year			2200kWh/ m²/year		
			Interest rate			Interest rate			Interest rate		
Module	BOS		0.04	0.06	0.08	0.04	0.06	0.08	0.04	0.06	0.08
400	100	0.10	66.4	78.9	92.4	44.2	52.6	61.6	33.2	39.4	46.2
		0.15	47.2	56.1	65.8	31.5	37.4	43.9	23.6	28.1	32.9
100	100	0.10	32.7	38.8	45.3	21.8	25.8	30.2	16.4	19.4	22.7
		0.15	24.8	29.4	34.4	16.5	19.6	22.9	12.4	14.7	17.2
	50	0.10	27.1	32.1	37.5	18.1	21.4	25.0	13.6	16.0	18.7
		0.15	21.1	25.0	29.2	14.0	16.6	19.4	10.5	12.5	14.6
50	50	0.10	21.5	25.4	29.6	14.3	16.9	19.7	10.8	12.7	14.8
		0.15	17.3	20.5	23.9	11.6	13.7	16.0	8.7	10.2	12.0
0	100	0.10	21.5	25.4	29.6	14.3	16.9	19.7	10.8	12.7	14.8
		0.15	17.3	20.5	23.9	11.6	13.7	16.0	8.7	10.2	12.0
	50	0.10	15.9	18.7	21.8	10.6	12.5	14.5	8.0	9.4	10.9
		0.15	13.6	16.0	18.7	9.1	10.7	12.5	6.8	8.0	9.3

Note: A module cost of zero is the net cost after credit for displaced expensive cladding materials that would otherwise be used for aesthetic purposes. BOS = balance of system.

11.2.6 Global PV power market

Total production of PV modules worldwide reached 1195MW$_p$/year in 2004, having grown by an average of 35 per cent per year since 1996. Of all PV module sales in 2004, 29 per cent were made of monocrystalline silicon wafers, 56 per cent were lower-quality multicrystalline silicon wafers, 8.4 per cent were multicrystalline silicon ribbons and self-supporting silicon sheets, 3.9 per cent were amorphous silicon, 1.1 per cent were CdTe, and 0.3 per cent were based on copper indium diselinide (Maycock, 2005).

Table 11.4 gives the peak power production from PV modules in reporting International Energy Agency (IEA) countries as of the end of 2004, and the new power added in 2004. The total installed capacity in IEA countries was 2596MW$_p$. The leading countries in terms of existing capacity and rate of installation of new capacity are Japan (with almost half the cumulative total), Germany (with almost one third of the cumulative total and almost half the installation in 2004), and the US. In Japan and Germany, the PV power is overwhelmingly grid-connected and distributed.

Table 11.4 Cumulative installed PV peak power capacity (kW) in reporting IEA countries as of the end of 2004

Country	Off-grid		Grid-connected		Total	Per capita (W)	Total Installed in 2004
	Domestic	Non-domestic	Distributed	Centralized			
Australia	15,900	29,640	5410	1350	52,300	2.6	6670
Austria	2687		15,340	1153	19,180	2.37	2347
Canada	5291	8081	476	36	13,884	0.44	2054
Czech Rep	2810	290	18,440	1560	23,100	3.12	2100
Denmark	65	190	2035	0	2290	0.43	400
Germany	26,000		768,000		794,000	9.62	363,000
France	12,500	5800	8000	0	26,300	0.44	5228
Israel	653	210	9	14	886	0.13	353
Italy	5300	6700	12,000	6700	37,000	0.53	4700
Japan	1136	83,109	1,044,846	2900	1,131,991	8.87	272,368
Korea	461	4898	4533	0	9892	0.21	3454
Mexico	14,169	4003	10	0	18,182	0.17	1041
Netherlands	4769		41,830	2480	49,079	3.01	3162
Norway	6438	375	75	0	6888	1.5	273
Portugal	1657	569	417	0	2643	0.25	574
Spain	14,000		23,000		37,000	0.87	10,000
Sweden	3070	602	194	0	3866	0.43	285
UK	193	585	7386	0	8164	0.14	2261
USA	77,900	111,700	153,600	22,000	365,200	1.24	90,000
Total	170,730	281,021	2,064,201	79,593	2,595,545		770,270

Source: IEA Photovoltaic Power Systems Program (www.iea-pvps.org)

11.3 Building-integrated PV panels

Early uses of PV panels to generate electricity involved setting up large arrays of panels in sunny, dry locations, or have involved small-scale generation in remote locations where the alternatives are very expensive. This requires support structures and, if the electricity is fed into the power-transmission grid, conversion to AC electricity with the appropriate frequency and voltage. However, solar modules can now be built as part of structural materials for buildings such as roof shingles, wall siding, curtain walls and windows. This reduces the module cost once a credit is given for the structural materials that it replaces, although energy output will be smaller on average since the orientation of the modules will be constrained by the building. PV modules that are part of the structure of a building are referred to as *building-integrated PV* (BiPV).

The technical characteristics, costs and benefits of BiPV are discussed in Boyle (1996), Sick and Erge (1996), Thomas and Fordham (2001), Schoen (2001), and at the international demonstration centre and website of the IEA's Implementation Agreement concerning PV power systems (see Appendix C). van Mierlo and Oudshoff (1999) discuss the solutions pertaining to a large number of non-technical problems associated with the introduction of BiPV: financing, administration, architectural integration, communication, marketing and environmental considerations. The September 2004 issue of *Progress in Photovoltaics: Research and Applications* is devoted to BiPV. Eiffert and Kiss (2000) and Prasad and Snow (2005) provide design details and examples of particular interest to architects. Summary information is presented below.

11.3.1 Styles of integration

PV modules can be attached to conventional roof and façade materials, or they can replace conventional materials. They can serve as skylights in atria, or as fixed or adjustable external shading devices. Examples, advantages and disadvantages of each style are briefly discussed here.

Mounting onto a sloped roof

This can often be the cheapest method of mounting PV modules, due to its simplicity (there is no need for a water-tight structure). Modules can either be attached to clamps that can be attached to roofing tiles with screws, or attached to special tiles having built-in plates. Another advantage is that an air gap is provided underneath, which limits the module temperature and loss of output (which amounts to 0.4–0.6 per cent per K of temperature rise for x-Si modules). This is the style that has been used in most national programmes

Figure 11.8 PV panels mounted onto a sloping roof

Source: Klöber GmbH & Co. KG, Germany (www.kloeber.de)

to encourage roof-top PV systems. An example is provided in Figure 11.8.

Integration into a sloped roof

Modules integrated into a sloped roof displace conventional roofing material, must form a watertight layer, and tend to be more aesthetically pleasing than modules mounted onto the roof. They can take the form of overlapping tiles or panels, or non-overlapping units, as illustrated in Figure 11.9. Bahaj (2003) reviews the state-of-the-art concerning photovoltaic roofing tiles and shingles. Examples of houses with BiPV roofs are given in Figure 11.10. Membrane and drain trays provide weather barriers as part of the PV system. The modules will get hot unless a ventilation gap is provided underneath, with an inlet at the bottom and an outlet at the top of the roof. The modules thus serve as the outer layer of a double-skin roof, which can be used to preheat ventilation air or to induce ventilation for cooling purposes (see Chapter 4, Section 4.1.10, and Chapter 6, Section 6.3.1). Alternatively, PV panels can be cooled by circulating water through tubes that are attached to the back of the panel, thereby preheating water for domestic hot-water use in the process (see Section 11.6). If a-Si modules are used instead of x-Si modules, the loss of efficiency with increasing temperature is much less, so it is less important to provide ventilation. Recently, ceramic PV tiles have been developed that have a PV-free area at the top of the tile, thereby ensuring overlap, and there are significant ongoing research and development activities.

Tilted modules on a flat roof

This requires a support structure, but provides for cooling of the modules by airflow underneath. The building cooling load will also be reduced due to shading of the roof. Tilting modules toward the sun will increase the mean annual irradiance on the modules, but less total power will be produced because not all of the roof area can be used, due to the need to provide space between the modules so as to reduce shading of modules by adjacent modules. Gravel on roofs can be used as ballast to hold the modules in place, along with tension cables, as illustrated in Figure 11.11.

Figure 11.9 PV panels integrated into a sloping roof, either as shingles on the Eco-Energy House, University of Nottingham (top and middle), or flush panels (lower)

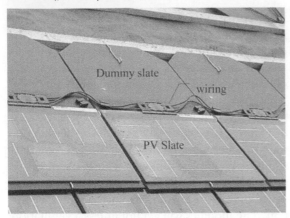

Source: Top and middle, Omer et al (2003); lower, Peter Toggweiler, Enecolo AG, Switzerland (www.solarstrom.ch)

Figure 11.10 BIPV in the residential sector. Top: the single-family house in Pietarsaari, Finland, from Hestnes (1999); bottom, a single-family house in coastal Maine.

Figure 11.11 Tilted PV modules mounted onto a flat roof

Source: Peter Toggweiler, Enecolo AG, Switzerland (www.solarstrom.ch)

Horizontal modules on a flat roof

Modules can be used in place of conventional roofing material, perhaps with attached interlocking insulation, as illustrated in Figure 11.12. Alternatively, modules can be elevated above the roof for ventilative cooling.

Modules on façades

Modules can be mounted on vertical façades, and may be physically integrated as in the Kyocera building, Austria (Figure 11.13, left), or visually but not physically integrated into the façade, as in the façade of the Brundtland Centre in Denmark (Figure 11.13, right). When used in curtain walls, PV modules can alternate with window glazing, as in the Condé Nast Building, New York (Figure 11.14). A Japanese firm produces a window with laser-etched a-Si laminated between two layers of glass, with 4.1 per cent efficiency (www. msk.ne.jp). It has 10 per cent light transmission, which optimizes the balance between heat gain and daylighting in Japan. PV modules can be used as fixed sunscreens (as in Building No. 31 of The Netherlands Energy Research Foundation (ECN), Figure 11.15), or as adjustable louvres (as in the SBIC building in Japan, Figure 11.16), or the outer skin of a double-skin façade, as in the Telus Building (Chapter 14, Figure 14.14). In the latter case, PV modules have been used to power DC fans that maximize the airflow in the façade during summer. Whether used for shading, as part of a double-skin façade for heating or cooling, or for preheating of water for consumptive use, the PV panel serves multiple purposes, so that orienting it for maximum electricity production is less critical.

Modules as skylights in atria

BiPV is now being used as a design element in its own right as a part of skylights (Figure 11.17). When used as skylights, the overall transmittance and appearance of PV modules can be altered by varying the proportion of the skylight with and without opaque crystalline cells. In thin-film modules, a wide variety of patterns can be laser-etched. For semi-transparent PV modules that form the roof of an atrium, air inlets and outlets should be placed to facilitate the flow of outside air under the atrium roof when cooling is needed (Gan and Riffat, 2001). Use of PV modules as

Figure 11.12 Illustration of PV modules separated by an air gap from interlocking insulation

Source: www.powerlight.com

Figure 11.13 PV panels physically integrated into the façade (Kyocera commercial office building, Austria, left) or visually integrated (Brundtland Centre, Denmark, right)

Source: Left, Stromaufwärts, Austria (www.stromaufwarts.at); right, BEAR Architecten, The Netherlands (www.bear.nl)

Figure 11.14 BiPV panels integrated into the curtain wall instead of conventional spandrel panels in the Condé Nast Building, New York. The BiPV panels (dark) are mounted in exactly the same way as the glazing (light)

Source: Eiffert and Kiss (2000)

Figure 11.15 PV panels as fixed sunscreens on Building No. 31 of The Netherlands Energy Research Foundation (ECN)

Source: BEAR Architecten, Case Study (www.bear.nl)
Photographer: Marcel von Kerckhoven

Figure 11.16 PV panels as adjustable louvres in the SBIC East head office building in Tokyo

Source: Shinkenchiku-Sha and www.oja-services.nl/iea-pvps/cases jpn_02.htm

Figure 11.17 PV providing partial shading in an atrium: Upper, Brundtland Center, Denmark, KHR A/S (architect); Lower, Kowa Elementary School, Japan.

Source: Upper, Henrik Sørensen, Esbensen Consulting; lower, www.nsk.ne.jp

Figure 11.18 Illustration of the way in which alternative geometries in a sawtooth roof with translucent PV affect the balance between electricity generation, daylighting and heat gain

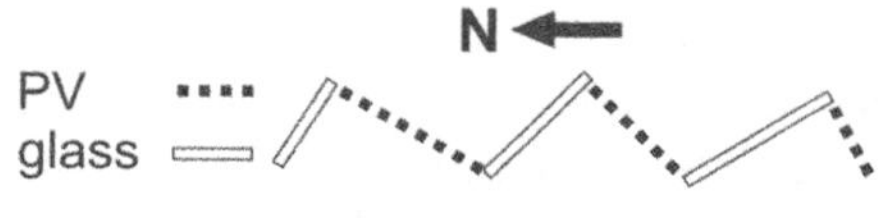

electric power	+	0	-
daylighting	-	0	+
indoor temperature	+	0	-

+ = favourable,   - = unfavourable,   0 = neutral

Source: Prasad and Snow (2005)

skylights requires optimizing the balance between electricity production, daylighting and cooling or heating load. Figure 11.18 shows how alternative geometries in a sawtooth roof affect this trade-off. The quantitative trade-offs as the geometry is changed will depend on latitude and climate.

Two large examples of BiPV are the Nieuwland development in Amersfoort (The Netherlands), which involves about 500 houses and 1MW peak power (Schoen, 2001), and the outer roof over the Herne administrative complex in Germany, illustrated earlier in Figure 6.7. This roof serves as both a shading device and power source, with modules laid to a density of 86 per cent in the roof area over buildings and to a density of 58 per cent in the transitional zones to areas of clear roof glazing between buildings, thereby softening the contrast in brightness. PV panels occupy 9300m^2 out of a total roof area of 12600m^2, and an additional 780m^2 are integrated into the south-west façade. Peak power output is 1MW.

11.3.2 Cost and benefits of BiPV electricity

IEA (2005) presents data on the cost of PV modules and the installed cost of PV systems in IEA countries. Typical module costs in 2004 ranged from a low of US$2.9/W$_p$ in Israel to a high of $7.4/W$_p$ in Australia, with 11 out of 15 reporting countries having typical costs of $3–5/W$_p$. For some countries, the best reported module price is significantly lower than the typical price. Typical installed prices for grid-connected systems smaller than 10kW peak power range from $5/W$_p$ to $15/W$_p$, with most countries having median

prices of $6–9/W$_p$. In most countries, the unit cost of larger systems is slightly less. For up-to-date retail costs of solar modules, invertors and electronic controls in the US and Europe; for lists of suppliers in the US and Europe; and for information on funding programmes in the US and Europe, consult www.solarbuzz.com.

The price of PV modules, like that of other technologies, tends to fall with cumulative production by the entire industry. This is referred to as a *technology learning curve*. Specifically, prices have been observed to be multiplied by a fixed fraction called the *progress ratio* with each doubling in cumulative production. For PV modules, the progress ratio was 0.80 from 1981 to 1990 and 0.77 from 1991 to 2000 (Parent et al, 2002), after prices are adjusted for inflation. The prices used for this calculation are based on deals closed and not on catalogue prices, and reflect a mixture of crystalline and a-Si products. Thus, from 1991 to 2000, the sales-weighted price of PV modules fell by about 23 per cent for each doubling in cumulative production, which has been occurring in a little more than every two years. More recently, prices have remained constant or even increased in some markets, as demand has grown faster than supply and due to government incentive programmes combined with bottlenecks in the supply of silicon to manufacturing plants (Haas, 2004; Jones, 2005; Krampitz, 2005)

Amorphous silicon modules may offer the best prospect for low-cost photovoltaic electricity, in spite of their low efficiency. Payne et al (2001) have estimated the costs of manufacturing two different kinds of a-Si modules – one rigid (on a glass substrate) with 8 per cent sunlight to DC conversion efficiency, and the other flexible with 10 per cent conversion efficiency. They assume a manufacturing facility with an output 10–20 times larger than present plants. Total installed costs for rooftop systems are given in Table 11.5 and amount to $2.67–2.95 per watt of electrical output for a solar irradiance of 1000W/m^2 (this output is referred to as a peak watt, W$_p$, since 1000W/m^2 is about the largest solar irradiance found anywhere). This is one-half to one-third the year 2002 costs of installed systems in most countries. However, present a-Si modules have achieved an efficiency of only about 6 per cent.

For the installed costs given in Table 11.5, and solar irradiance corresponding to southern California or northern Illinois (US), the retail cost of electricity amounts to 9.4–13.2cents/kWh (including an 18 per cent government subsidy). This is comparable to the retail cost of electricity to residential customers in these regions. Thus, with net metering (where the customer's meter runs backwards when there is excess electricity to export to the grid), rooftop PV could become competitive in these locations. These results are roughly consistent with the analysis of Poponi (2003), who finds that BiPV systems need to fall to an installed cost of $3.2–4.5/W$_p$ (a reduction by a factor of 2–3 in most countries) in order to be competitive with 15cent/kWh electricity without a subsidy.

The power output of PV systems tends to coincide with peaks in electricity demand, because air conditioning energy use and the availability of solar energy are highly correlated. Peak demand is satisfied by generation units that operate only a small part of the year, so the cost of supplying electricity at times of peak demand can be two or three times the average cost of electricity. Thus, a kWh of PV electricity is more valuable than implied by the average retail cost of electricity (by 30–50 per cent in California according to calculations by Borenstein (2005) and by 22–63 per cent and 7–27 per cent in Alberta and Ontario, respectively, according to Rowlands (2005)). If the greater cost of electricity at time of peak demand is reflected in the electricity rate structure, the economic attractiveness of PV electricity from a private perspective will be greatly enhanced. Utility rate structures can reflect the greater cost of peak electricity in two ways:

- through a demand change, which is proportional to the peak power demand (kW) by the customer during any given billing period, and is in addition to the charge for energy (kWh) consumption;
- by charging different rates per kWh for electricity consumed at different times.

Table 11.5 Lifecycle costs of 4kW$_p$ PV systems on rooftops of new residences, once the module cost reaches $160/m^2

		Cost for 10% efficient, flexible (stainless steel) triple-junction modules		Cost for 8% efficient, rigid (glass) double-junction modules	
		$	$/W$_p$	$	$/W$_p$
Module price		6640	1.66	6200	1.55
Installation		1640	0.41	3164	0.79
O & M (NPV)		840	0.21	1040	0.26
Government subsidy[a]		(1800)	(0.45)	(1880)	0.47
Total	DC output basis	7320	1.82	8520	8520
	AC output basis[b]	7320	2.37	8520	2.65
Power conditioning		919	0.30	966	0.30
Total		8240	2.67	9490	2.95
Cost of electricity (cents/kWh)		9.4–11.9		10.3–13.2	
Assumed irradiance (kWh/m^2/year)		1600–2000		1600–2000	

[a] In the US, this arises automatically from the fact that interest on mortgage payments are tax deductible, and that the system is assumed to be financed as part of the homeowner's mortgage.
[b] Assumes a power conditioning efficiency of 76.6 per cent for flexible modules (flush on the roof with no ventilation underneath) and 80.5 per cent for rigid modules.
Note: The assumed solar irradiance corresponds to northern Illinois (low value) or southern California (high value). NPV = net present value of lifetime maintenance at a 5.1 per cent real discount rate.
Source: Payne et al (2001)

Information on rate structures can be obtained through the *Tariff Analysis Project*, discussed in Chapter 2 (Section 2.3). BiPV not only reduces loads at times when electricity tariffs are the greatest, it resolves a growing problem in many urban areas of transmission bottlenecks by providing electricity at the point of use.

Perez et al (2003) analyse the reduction in peak loads that can be expected for PV systems on an office building in New York City and on department stores in Long Island and Hawaii. PV power output does not exactly match the daytime variation in building power demand, and short-term cloud cover could largely eliminate the savings in peak demand altogether. However, if the power use by and output of the cooling system is temporarily reduced in response to short-term reductions in PV power output, then reductions in peak power demand can be achieved. Even where there are no savings in demand charges, time varying electricity rates can still enhance the economic attractiveness of PV power.

BOX 11.3 National PV roof solar programmes

At least four countries have national programmes to encourage the installation of PV or BiPV systems in buildings. These are Korea, with a target of 30,000 systems; Japan, with a target of 70,000 systems; Germany, with a target of 100,000 systems; and the US, with a target of 1 million systems (including solar thermal systems).

Korea launched its *Solar Land 2010* programme in 2002, with the goal of delivering 30,000 rooftop PV systems of 3kW capacity by 2010 (IEA, 2003).

Japan began a subsidy programme for PV power for residential buildings in 1994, as part of an overall target of reaching 4800MW of peak PV power (4000MW$_p$) by 2010. The rate of installation of PV power has steadily increased since then, resulting in a cumulative installation of 637MW$_p$ by the end of 2002, of which 421MW$_p$ were residential (Tsurusaki et al, 2004). This represents about half of the total of IEA countries (see Table 11.4). Installed capital costs fell from $33.6/W$_p$ in 1993 to $6.6/W$_p$ in 2002, while electricity costs fell from $2.36/kWh to 0.45/kWh.

On 1 April 2000, the German *Law for the Priority of Renewable Energy* (REL) was enacted (Weiss and Sprau, 2002). This law provides for the buyback of rooftop PV-generated electricity at an initial rate of 50.62 eurocents/kWh (about $0.60/kWh), with the rate declining by 5 per cent per year beginning in 2002. Low-interest (1.91 per cent) loans are also involved. The buyback is referred to as a *feed-in tariff*. The goal was to stimulate the installation of 100,000 rooftop PV systems with an average peak power of 3kW, for a total of 300MW$_p$. By July 2003, this goal was achieved, with 314MW$_p$ of installed capacity and over 1.2 billion euros of credit provided. The German PV industry consists of 30 companies making PV modules, mostly as roof tiles; 300 installation companies; and 100 PV module and component distributors. At the end of 2000, there were 3800 full-time jobs in the German PV industry and an undisclosed number of part-time jobs. Fifteen companies manufactured about 300 different kinds and sizes of inverters for PV systems.

Installed-costs in 2001 ranged from an average of about 7000 euros/kW$_p$ for systems smaller than 2kW$_p$, to 6000 euros/kW$_p$ for systems larger than 20kW$_p$ (Dobelmann, 2003). Costs fell 50 per cent during the past 10 years, and are expected to fall by 50 per cent again during the next 10 years if annual production grows by 20 per cent per year.

After a brief hiatus following the closing of the 100,000-roofs program, new rates were established for subsequent systems beginning on 1 January 2004: a base rate of 45.7 eurocents/kWh, plus an extra 5 eurocents/kWh for façade systems, 9.3 eurocents/kWh for rooftop systems over 30kW, and 11.7 eurocents/kWh for rooftop systems under 30kW (*Renewable Energy World*, Jan–Feb 2004, p12). The rates for new facilities going into operation after 2004 will decrease by 5 per cent per year. It is expected that 200MW$_p$ will be added in 2004. In a similar move, Portugal has established a fixed feed-in tariff of 51 eurocents/kWh, with a goal of installing 100MW$_p$ by 2010.

The benefits of BiPV extend beyond the mere provision of electricity, and so the decision as to whether or not to incorporate PV into a building depends on more than the cost of electricity from BiPV compared to the cost of purchased electricity. Other benefits include the role of BiPV as a façade element, replacing conventional materials and providing protection from UV radiation; providing thermal benefits such as shading or heating; augmenting power quality by serving a dedicated load; serving as backup to an isolated load that would automatically separate from the utility grid in the event of a line outage or disturbance; and (as already noted) reducing power-transmission bottlenecks and the need for peaking power plants. These benefits are discussed in some detail by Eiffert et al (2002). According to Eiffert et al (2002), system cost can be reduced from $5.83/$W_p$ when PV modules are directly mounted onto a curtain wall, to $4.21/$W_p$ when fully integrated into the building design – a saving of 28 per cent. The grid and load saving benefits to the power utility are worth $107–180 per year per kW of peak power produced.

The economics of BiPV is further improved when a credit is applied to the building envelope materials that BiPV replaces. Inasmuch as BiPV is used as part of an aesthetic design element in a building, it can replace rather expensive building materials. The cost of PV modules ranges from $400 to 1300/$m^2$. By comparison, the costs of envelope materials in the US that PV modules can replace range from $250 to 350/$m^2$ for stainless steel, $500–750/$m^2$ for glass wall systems, to at least $750/$m^2$ for rough stone and $2000–2500/$m^2$ for polished stone (AEC, 2002).

Several countries (Germany, Japan, Korea) have aggressive programmes to promote the installation of rooftop photovoltaic systems, as outlined in Box 11.3, and these have been largely responsible for the 35 per cent/year growth in the installation of PV systems since 1996.

11.3.3 Direct coupling to fans

The power required to drive an airflow varies with the volumetric rate of airflow to the third power (Chapter 7, Section 7.1.2). Thus, if BiPV modules are used to drive ventilation fans in a building, the airflow rate will vary with the solar irradiance to the cube root. That is, the relative variation of ventilation rate with solar irradiance will be much less than the relative variation of the irradiance. Furthermore, it will be largest when the solar irradiance is largest, which may or may not be advantageous.

Costs can be reduced and the overall efficiency can be improved if DC output from a PV module is used to drive a DC motor, thereby eliminating the inverter (which contributes up to 25 per cent of total system costs, and is the largest source of failure). DC motors are somewhat more efficient than AC motors, and DC/AC conversion losses are avoided. However, MPP tracking ability is also lost, so there can be no net improvement in module efficiency.

11.3.4 Potential electricity production from BiPV

Gutschner and Task-7 Members (2001) have estimated the potential power production from BiPV in member countries of the IEA. They first estimated building ground floor areas from population data, from which roof and façade areas were estimated. They then took into account architectural suitability (based on limitations due to construction, historical considerations, shading effects and the use of the available surfaces for other purposes) and solar suitability (taking into account the relative amounts of solar radiation on surfaces of different orientation). Based on architectural considerations, they find that the suitable roof area on average is 60 per cent of the ground floor area and the suitable façade area is 20 per cent of the ground floor area. Restricting the usable roof and façade area to those elements where the solar irradiance is at least 80 per cent of that on the best elements in a given location, defined separately for roof and façade elements, they find that the suitable roof area is 40 per cent of the ground floor area and the suitable façade area is 15 per cent of the ground floor area.

Table 11.6 gives the resulting estimates of the roof and façade areas available for BiPV, the potential generation of electricity from roofs and façades, the total electricity consumption in each country in 1998, and the percent of total 1998 electricity demand that could be provided by BiPV (assuming 10 per cent sunlight-to-AC

Table 11.6 Potential area and electricity generation from BiPV in selected IEA countries, and potential BiPV-generated electricity as a per cent of total domestic electricity demand in 1998

Country	Potential BiPV area (km²)		Potential solar electricity (TWh/year)			National electricity consumption (TWh/year)	Potential BiPV portion (%)
	Roofs	Facades	Roofs	Facades	Total		
Australia	422.3	158.3	68.2	15.9	84.1	182.2	46.1
Austria	139.6	52.4	15.2	3.5	18.7	53.9	34.7
Canada	963.5	361.3	118.7	33.1	151.7	495.3	30.6
Denmark	88.0	33.0	8.7	2.2	10.9	34.4	31.6
Finland	127.3	33.0	11.8	3.1	14.6	76.5	19.4
Germany	1295.9	486.0	128.3	31.8	160.0	531.6	30.1
Italy	763.5	286.3	103.1	23.8	126.9	262.0	45.0
Japan	966.4	362.4	117.4	29.5	146.9	1012.9	14.5
Netherlands	259.4	97.3	26.7	6.2	31.9	99.1	32.2
Spain	448.8	168.3	70.7	15.8	86.5	180.2	48.0
Sweden	218.8	82.0	21.2	5.5	26.7	137.1	19.5
Switzerland	138.2	51.8	15.0	3.4	18.4	53.2	34.6
UK	914.7	343.0	83.2	22.2	105.4	343.6	30.7
US	10096.3	3786.1	1662.4	418.3	2080.7	3602.6	57.8

Source: Gutschner and Task-7 Members (2001)

power conversion efficiency). This ranges from about 15 per cent (Japan) to almost 60 per cent (US). Chaudhari et al (2004) independently estimated the available roof area in the US to be 5800km², compared with 10,100km² estimated by Gutschner and Task-7 Members (2001). Although one may question the specific assumptions used, there is in any case a significant potential for electricity production from BiPV. Thus, systematic incorporation of BiPV into new buildings, and into old buildings when they are renovated and whenever this is feasible, can make an important contribution to electricity supply.

11.3.5 Reduction in cooling loads

The presence of PV panels separated from the main building façade by an air gap can substantially reduce the penetration of heat through the façade, thereby reducing the air conditioning load. This occurs both through direct shading of the inner façade, and through passive (or mechanically forced) ventilation of the air gap with outside air, which removes the heat that accumulates in the gap. Based on calculations with an energy balance model for a summer day, Yang et al (2001a) obtained the following reductions in heat transfer through a wall with PV panels separated

by a passively ventilated air gap, compared to a wall without PV panels: for east- and west-facing walls, reductions in heat flow of 43 per cent, 49 per cent and 53 per cent for buildings in Hong Kong, Shanghai and Beijing, respectively; for a south-facing wall, reductions of 33 per cent, 41 per cent, and 50 per cent. The reduction in average heat transfer is on the order of $20W/m^2$, which is comparable to the amount of electricity generated (the savings in cooling electricity use, however, is this amount divided by the COP of the air conditioning system). Simulation of a ventilated PV roof at 22°N by Yang et al (2001b) indicates a reduction in the roof contribution to cooling load by two-thirds compared to a conventional roof.

The *Energy-10* simulation software package can be used to simulate the impact of BiPV on the thermal properties of a building envelope. Walker et al (2003) provide an example of this use of Energy-10. Energy-10, like all user-friendly software packages, is dependent on coefficients that are derived from detailed fluid dynamics calculations for the calculation of heat transfer, and the most appropriate values of these coefficients should depend on the type of building and the prevailing climatic conditions (Mei et al, 2002).

11.3.6 Window-integrated PV

Miyazaki et al (2005) used the *EnergyPlus* building-simulation tool (discussed in Appendix D) to optimize the design of window-integrated PV for an office building in Tokyo. The window-integrated PV consists of two panes of glass with a semi-transparent PV layer between them. The layer is rendered semi-transparent by making microscopic holes. The reduction in power output is almost equal to the transmittance of the PV layer. Optimization involves taking into account the impact of the semi-transparent PV layer on the heating load, cooling load, lighting load (with dimmable, daylight responsive lighting) and on electricity generation. For an HVAC system where heating and cooling are done with heat pumps with heating and cooling COPs of 4.3 and 3.3, respectively, the optimized PV layer reduces electricity demand from the grid for lighting + HVAC loads by 55 per cent compared to a single-glazed window with no lighting control and no electricity production.

11.4 Solar thermal energy for heating and hot water

Solar thermal energy can be collected, stored in hot-water tanks, and used for domestic hot water or to provide space heating. The collectors can be mounted on the roof or, in the case of flat collectors, can form part of the building fabric. Here, we discuss the key efficiency and operating characteristics of the kinds of solar collector systems that can be used in buildings. The uses of solar energy for passive space heating and in solar-assisted heat pumps are discussed in Chapters 4 (Section 4.1) and 5 (Section 5.13), respectively. More detailed technical information can be found in Reddy (1997), Peuser et al (2002), Andrén (2003), Weiss (2003), and Kalogirou (2004). Peuser et al (2002) present a synthesis of 30 years of research and operating experience with solar thermal systems for domestic hot water and space heating in Germany, including a discussion of issues pertaining to the interconnection of solar panels, solar thermal storage tanks, auxiliary

Figure 11.19 Types of solar thermal collectors and associated temperature rises

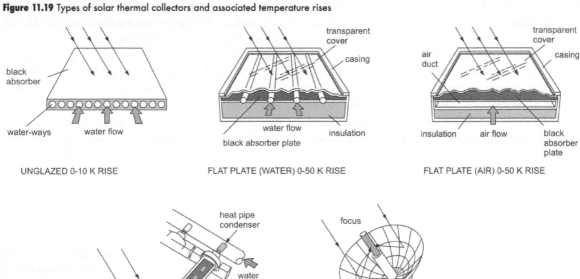

Source of figure: Everett (1996)

water heaters and possible auxiliary storage tanks; the sizing of various components; and control sequences. Andrén (2003) provides a nuts-and-bolts treatment of solar thermal collectors and related components for space and water heating, with an emphasis on examples and performance in Sweden, one of the world's northernmost countries. Weiss (2003) presents a collection of papers on the design, cost and performance of 19 different systems for combined space heating and domestic hot water in various countries in Europe based on research carried out as part of the International Energy Agency, Solar Heating and Cooling Programme Task 26 (Solar Combisystems). Kalogirou (2004) discusses solar thermal collectors more in terms of the underlying physics and mathematical representation.

11.4.1 Types of solar collectors

The main types of solar collectors are illustrated in Figure 11.19 and analysed in great detail in Morrison (2001a). The main types that can be used on or as part of buildings are: *flat-plate* collectors, *evacuated-tube* collectors and *compound parabolic* collectors.

A flat-plate collector contains pipes, through which water flows, embedded in a black absorbing surface and underlain by insulation that assists in retaining heat. Unglazed flat-plate collectors give a temperature rise of up to 10K,

Figure 11.20 Evacuated-tube thermal collector

Source: Posters from the AIRCONTEC Trade Fair, Germany, April 2002, available from www.iea-shc-task25.org

Figure 11.21 All-glass evacuated-tube thermal collector from China, integrated with a storage tank

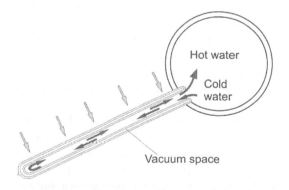

Source: Morrison et al (2004)

Figure 11.22 The Power Roof™ tracking with a curved collector for the production of solar thermal energy, developed by Solargenix Energy (formerly Duke Solar)

Source: Gee et al (2003), American Society of Mechanical Engineers

Figure 11.23 Thermal solar collectors integrated into a building façade (top and middle) and building roof (bottom)

Source: Top, Sonnenkraft, Austria; middle and bottom, AEE INTEC, Austria

while transparent flat-plate collectors with one or two layers of glass give a temperature rise of up to 50K and can be used to heat water or air. The absorber plate should have a very high absorptivity (up to 0.95, compared to 0.9 for normal black paint, which still reflects 10 per cent of the incident radiation), a high temperature conductivity to effectively transfer heat to the water, and a low infrared emissivity (less than 0.15) to minimize radiant energy losses (emissivity is discussed in Chapter 3, Section 3.1.3). The insulation must be able to withstand temperatures of up to 150°C, which may occur during a summer day if the water flow is interrupted.

Among the most efficient solar collectors is the *evacuated-tube collector*, which consists of a series of tubes with a vacuum inside in order to eliminate convective heat losses. There are several kinds of evacuated-tube collector, but the only two with demonstrated long life under fluctuating outdoor conditions are the heat pipe collector and the all-glass collector. The heat pipe collector has a black absorber plate inside each vacuum tube, and within each absorber plate, a pipe that carries a liquid at a pressure such that it will boil where it absorbs solar heat and condense onto a pipe carrying the water that is to be heated (Figure 11.20). In this way, very effective heat transfer occurs, with an effective thermal conductivity much greater than can be achieved with a solid. A temperature rise of up to 150K can be achieved, but the tubes must be inclined at a slope of at least 25°. The all-glass collector is illustrated in Figure 11.21. The solar absorbing surface is placed on the vacuum side of the inner glass tube. Heat is conducted through the glass and heats water, which rises into a cylindrical storage container, to which the evacuated tubes are attached. China is the largest manufacturer and user of all-glass evacuated-tube collectors, with an estimated production of 20 million tubes in 2001 (Morrison et al, 2004).

In a parabolic collector, a parabola-shaped reflecting surface concentrates sunlight onto a linear absorber tube. In a *compound parabolic collector* (CPC), an additional concentrator is integrated into a glass envelope that surrounds the absorber tube. The gap between the glass envelope and the absorber tube may or may not be evacuated. The US firm Solargenix (formerly Duke Solar) has developed a system referred to as Power Roof™ that is mounted on the roofs of buildings. It consists of a fixed curved (non-parabolic) reflector that reflects sunlight onto an evacuated tube that moves during the day as the sun's position changes. The reflector, receiver and curved tracking guides are illustrated in Figure 11.22. The receiver consists of a glass cylinder with an anti-reflective coating to improve transmission and an inner absorber tube with a solar absorptance greater than 0.96 and an emissivity less than 0.10. The space between the glass tube and receiver is a vacuum and contains a silver-coated secondary reflector. The primary reflectors are arranged in rows adjacent to one another in such a way that some of the light intercepted by one reflector is reflected onto the backside of an adjacent reflector and can be used for daylighting.

As with PV modules, flat-plate solar thermal collectors can be integrated into the building façade, rather than forming an independent add-on feature. A number of factory-produced roof units, that include structural support, insulation, the thermal solar collector and waterproofing, are available in Europe, as illustrated by Kovács et al (2003). Some examples of the integration of solar thermal collectors into the building roof and building walls are illustrated in Figures 11.23 and 11.24, respectively. For aesthetic reasons and also to minimize the potential of water leakage at transitions between different roofing materials, it is usually desirable to have the complete roof or natural divisions of the roof covered by a solar collector array. If this results in too large a collector array, a part of the roof can be covered with dummy collectors. Wall-integrated solar collectors are also available. An advantage of equatorward-facing wall-integrated collectors is that the seasonal variation of solar irradiance peaks during the winter or early spring/late fall and is minimal during the summer (see Figure 4.1). This

Figure 11.24 Installation of solar thermal collectors

Source: www.socool-inc.com

provides a better match to the seasonal variation in space heating requirements than horizontal or sloping modules.

11.4.2 Efficiency of solar collectors

As a solar collector absorbs solar energy and heats up, heat losses through emission of infrared radiation and through convection will increase. This increase offsets some of the absorption of solar energy, thereby limiting the net efficiency of the collector in converting solar energy into heat. A lower collector temperature, and hence a lower outlet temperature, will lead to greater efficiency for a given collector, but the heat is less useful due to its lower temperature. The efficiency of a solar collector, η_T, is represented as:

$$\eta_T = F_R \tau \alpha - F_R U_L \frac{T_i - T_a}{I}$$

(11.1)

where τ is the transmissivity of the cover plate, α is the absorptivity of the absorber plate, U_L is a heat exchange coefficient that depends on the collector, T_i is the temperature of the fluid entering the collector, T_a is the ambient air temperature, I is the solar irradiance and F_R is an empirical coefficient that differs from collector to collector. The first term represents the fraction of incident solar radiation that is absorbed, while the second term represents the effect of radiative and convective heat losses. Physically, this term should depend on the average temperature of the absorber plate, but as this is not easy to measure in practice, the inlet temperature is used instead. The effect of this difference is incorporated in the value of U_L. Atmaca and Yigit (2003) suggest the following parameter values: for the product $F_R\tau\alpha$, 0.9, 0.75 and 0.70 for single-glazed flat-plate, double-glazed flat-plate and evacuated-tube collectors, respectively; for the product $F_R U_L$, 10.0W/m²/K, 6.5W/m²/K and 3.3W/m²/K for the same three collector types. Thus, for a 25K temperature difference (giving $(T_i - T_a)/I$=0.036 for $I = 700$W/m²), a typical flat-plate collector efficiency will be 0.50–0.55, dropping to 0.2–0.3 for a 50K temperature difference. The collector efficiency will be highest in early morning, and decrease steadily during the course of the day, as the collector temperature increases. For the Solargenix Power Roof, Gee et al (2003) find that $F_R\tau\alpha = 0.501$ and $F_R U_L = 0.116$W/m²/K for temperatures 0–300K above ambient temperature, with I the solar irradiance in the plane of the curved receiver. This gives an efficiency that decreases from 0.50 to 0.46 as the temperature rise increases from 0 to 300K for $I = 800$W/m². For flat-plate thermal collectors, the U-value will be smaller if the collector is vertically mounted, due to the reduction in convective heat loss. This gives a collector efficiency that is a few per cent more efficient at large $T_i - T_a$ (Kovács et al, 2003).

The thermal efficiency of a solar collector can be increased in two general ways:

- through changes in the design of the collector itself;
- through changes in the system of which the collector is a part, so as to reduce the temperature (T_i in Equation 11.1) of the water returning to and entering the collector.

Design of the collector

Specific examples of changes to the collector include:

- use of an anti-reflective coating on the glass cover, which increases the efficiency by a rather uniform amount (4–6 per cent, depending on the angle of incidence) for all collector temperatures (Furbo and Shah, 2003);
- application of transparent insulation (described in Chapter 3, Section 3.2.8), such that the efficiency of a conventional flat-plate solar collector is doubled at a temperature of 60°C (Saxhof, 2003);
- improving the absorption of heat by the circulating water. This has been done by one manufacturer with an absorber having channels filled with small porous spheres. The surface area of the working water is large due to capillary forces, increasing the absorption of heat (see www.solarnor.com).

Figure 11.25 Variation of solar collector thermal efficiency with $(T - T_a)/I$ for flat-plate collectors with one or two glazings, for an evacuated-tube collector, and for a flat-plate collector with a 5cm encapsulated TIM (transparent insulation material)

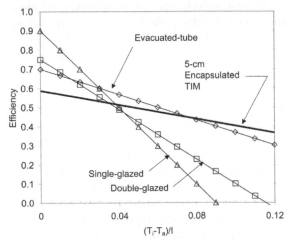

Source: Based on equations for a TIM collector in Kaushika and Sumathy (2003) and in Atmaca and Yigit (2003) for other collectors

Figure 11.25 shows the efficiency of flat-plate solar collectors with one and two cover glazings, an evacuated-tube collector, and a flat-plate collector with 5cm of transparent insulation. Adding an additional glazing reduces the efficiency when $(T_i - T_a)/I$ is small, due to partial reflection of sunlight by the additional glazing, but increases it when $(T_i - T_a)/I$ is large due to the reduced heat loss. Evacuated-tube collectors have a smaller efficiency when $(T_i - T_a)/I$ is close to zero, but also have a very small decrease in efficiency with increasing $(T_i - T_a)/I$. The efficiency of a flat-plate collector with transparent insulation is similar to that of an evacuated-tube collector.

Minimizing the collector inlet temperature

In systems that provide domestic hot water (DHW), the temperature of the water that returns to the collector inlet can be minimized if (i) the hot-water tank is stratified, with cooler water that flows to the collector for heating at the bottom of the tank, and (ii) the peak storage temperature is capped through the use of phase change materials. Use of a heat pipe to transfer heat from the solar collector to the storage tank for DHW (discussed below) also increases the collector efficiency by reducing the collector inlet temperature. In systems that provide space heating, use of solar heat as it is collected can be preferable to accumulating it in a storage tank through a continuously increasing storage temperature. This, however, requires buildings with adequate thermal mass and low rates of heat loss, so that making full use of sunshine to heat the building when the sun is shining avoids overheating during the day while maintaining comfortable temperatures during the night.

11.4.3 Solar water heating

Active solar heat is a sensible option for meeting at least part of domestic hot water requirements in most parts of the world. In southern Canada, rooftop solar collectors can provide half of annual hot water requirements, an amount that rises to 80 per cent of hot water needs in Cyprus (Kalogirou and Papamarcou, 2000) and to 97 per cent in Darwin, Australia (Aye et al, 2002). Solar water heaters can be classified as *open* loop (or direct) and *closed* loop (or indirect). In an open-loop

system, the potable water circulates through the collector from a hot-water tank. In a closed-loop system, a fluid containing anti-freeze (in cold climates) circulates between the collector and a heat exchanger inside the hot-water storage tank, but does not mix with the potable water. In a passive or *thermosyphon* system, the hot-water tank is placed above the solar collector. As water is heated in the collector, it rises naturally into the storage tank as colder and heavier water enters the bottom of the collector. The collector and storage tanks are sold as a single unit, as illustrated in the evacuated-tube system shown in Figure 11.21. A thermosyphon system is viable only where there is high solar irradiance throughout the year. Active systems use pumps to transfer the fluid between the collector and storage tank, so the tank can be located anywhere that is convenient. Morrison (2001b) provides a detailed discussion of solar water heating systems.

Phase change materials (PCMs) in the upper portion of the storage tank permit a greater density of heat storage for the same tank temperature, thereby reducing the conductive heat loss associated with a given amount of heat storage (Mehling et al, 2003). Conversely, a PCM can be distributed throughout the storage tank in small canisters and used to recharge the water temperature after a withdrawal of hot water. Metal fins can be inserted into the PCM to increase the rate of heat transfer. Possibilities include sodium thiosulphate pentahydrate ($Na_2S_2O_3 \cdot 5H_2O$, 48°C melting point, $L_f = 187kJ/kg$, $c_p = 1.5kJ/kg/K$; Kiatsiriroat et al, 2000) and paraffin (58–59°C melting point, $L_f = 214.4kJ/kg$, $c_p = 0.9kJ/kg/K$; Velraj et al, 1999). Hasnain (1998) provides data on many other PCMs suitable for domestic hot water (and other applications). However, PCMs are not yet mainstream features of solar DHW systems.

Figure 11.26 presents data collected by Sandnes et al (2000) comparing the energy stored versus water temperature for a 1000l tank with and without 250l of a commercially available PCM. The PCM consists of a mixture of eutectic salts and hydrate enclosed in plastic modules with an air gap to allow for expansion of the PCM when it melts. This PCM has a latent heat of fusion of 210kJ/kg, solid and liquid densities of 1500kg/m³ and 1340kg/m³,

Figure 11.26 Total energy stored as a function of temperature for a tank with 1000l of water, or 720l of water and 250l (solid) or 280l (liquid) of PCM

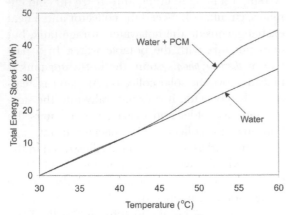

Source: As measured by Sandnes et al (2000)

Figure 11.27 Variation of water temperature during charging and discharging using the PCM/water system of Figure 11.26

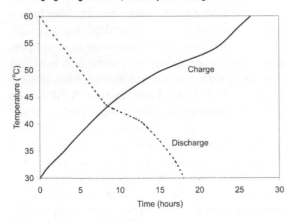

Source: Sandnes et al (2000)

respectively, and solid and liquid specific heats of 1.85kJ/kg/K and 3.1kJ/kg/K, respectively, while water has a density of 1000kg/m³ and a specific heat of 4.19kJ/kg/K. This gives the PCM a heat capacity of 2.28kJ/l/K in the solid state and 4.15kJ/l/K in the liquid state, compared to 4.19kJ/l/K for water. Thus, one should avoid operating the system at temperatures where a phase change will not occur, as the heat storage in that case will be less than for an equivalent volume of water alone. This PCM melts over a range of 47–50°C, which is ideal for hot water needs, and freezes more abruptly, at about 47°C. Over the temperature range from 30°C to 60°C, the

cumulative thermal storage for a tank filled one-quarter with PCM modules is almost 40 per cent greater than for water alone. This allows a smaller storage volume, reducing the cost and heat loss, or permits a smaller peak temperature, thereby improving the solar collector efficiency. Figure 11.27 shows the variation in storage water temperature during charging and discharging. Even while the PCM is melting, the water temperature continues to increase, although more slowly, creating a slight plateau in the curve of temperature versus time. Similarly, the temperature continues to fall, although more slowly, when the PCM is freezing. The plateau during charging occurs at a warmer temperature than during discharging, because, during melting, the water temperature must be warmer than the PCM (in order to transfer heat to the PCM), while during freezing it must be colder than the PCM (in order to extract heat from the PCM).

The better the heat exchanger for heat transfer between the collector and the storage tank, and the larger the storage tank for a given collector area, the more heat that can be collected for a given collector area (Belessiotis and Mathioulakis, 2002). As a minimum, the collector to storage tank heat-transfer coefficient should be larger than the collector to atmosphere heat-transfer coefficient (U_L in Equation 11.1). Effective heat transfer can occur if a heat pipe is an integral part of the solar collector (Mathioulakis and Belessiotis, 2002). The solar collector serves as the evaporator, where a refrigerant boils. The refrigerant vapour flows to a condenser, which is placed inside the lower part of the storage tank. Condensation and heat release occur in the condenser. Much of the heat is transferred to the storage tank in latent form, rather than entirely as sensible heat, as in a conventional system. The result is a higher collector efficiency (up to 60 per cent for a flat-plate collector) and operation even at small temperature differences between the collector and tank.

Existing solar water heating systems are not optimized as integrated systems. The solar collector is heated by solar radiation only on one side (the side facing the sun). Mills and Morrison (2003) analysed a design whereby reflectors at the top and bottom of a solar collector reflect sunlight onto a portion of the underside of the collector. One reflector is oriented such that it operates only during months on the winter side of the spring and autumn (fall) equinoxes, thereby reducing the seasonal asymmetry in collected energy. This and other simple design changes increase the fraction of annual hot water demand that can be met with solar energy from 50 per cent to 80–90 per cent in Sydney, Australia. The solar fraction is limited by occasional occurrences of several days of continuously cloudy weather; increasing the collector area does not increase the solar fraction, but rather, increases the amount of collected energy that has to be dumped.

Rather than using a solar collector directly to produce domestic hot water, the solar collector can serve as the evaporator of a solar-assisted heat pump. In this way, hot water can be produced during times of low solar irradiance if necessary. Solar-assisted heat pumps as applied to the production of domestic hot water are discussed in Chapter 5 (Section 5.13.2), where it is suggested that the added cost and complexity compared to thermal collectors alone might not be justified.

Diab and Achard (1999) assessed the potential to meet hot water demand in the French city of Chambéry using rooftop solar collectors. Based on data on total floor area and the ratio of roof area to floor area for different building types, they estimated the total roof area available for each building type. These numbers are then multiplied by three factors: (i) an orientation factor (1.0 for flat roofs, 0.25 for tilted roofs) to account for some tilted roofs not having the correct orientation; (ii) an availability factor, to account for some of the roof area being taken up by equipment; and (iii) an arrangement factor (0.4 for flat roofs), to account for the spacing needed to avoid self-shading by tilted collector modules on flat roofs. They found that the available roof area could meet 109 per cent of the city's hot water requirements, ignoring temporal mismatches between solar supply and demand. How much could be supplied in practice depends on the extent of thermal storage.

11.4.4 Solar combisystems

'Combisystems' are solar systems that provide space and water heating. They are inherently more complex than systems for producing domestic hot water, since they involve an additional heat distribution system (normally some low-temperature system, such as radiant floor heating). In addition, the heating load undergoes a marked seasonal fluctuation, in contrast to the hot-water load, and is largest when there is the least amount of solar energy available. During the period 1975–1985, many complex and non-standard combisystems were designed by engineers. After 1990, simple and inexpensive systems were designed by solar-collector companies. Current designs are based on field experience but have not yet been carefully optimized, so it is thought that there is a great potential for cost reduction, improved performance and improved reliability. Modular collector panels for combisystems are available that replace conventional roofing or wall materials. Even when the collectors do not deliver usable heat to the storage tank, they reduce heat loss through the wall or roof element. Incorporation of solar collectors into equatorward-facing walls increases the heat collection in winter compared to roof collectors, due to the high solar zenith angle, while decreasing it in summer (see Figure 4.1).

In 2001, the total collector area installed for solar combisystems in eight European countries was about 340,000m^2, two-thirds of which was in Germany and most of the rest in Austria (Weiss, 2003). Assuming an average collector area of 15m^2, this translates into 22,600 solar combisystems installed in 2001 alone. Problems to date have been mainly due to an improper control strategy when combining the solar and auxiliary parts of the system. Through design improvements, the number of pipe connections required in an optimized residential system has fallen from 17 to 8, the required collector area has decreased by half, and the total mass of the system has fallen from 250kg to 160kg. In The Netherlands, a typical solar combisystem consists of 4–6m^2 of solar collector and a 300litre storage tank, so solar energy meets a small (10 per cent) share of the space + water heating demand. In Switzerland, Austria and Sweden, common systems for a single-family house consist of 15–30m^2 of collector and a 1–3m^3 storage tank, so 20–60 per cent of the heating demand is met by solar energy. In Gleisdorf, Austria, a system was installed in 1998 for an office building and six terraced houses. The collectors have been integrated into the roofs of the winter gardens and cover 80 per cent of the annual hot water demand and 60 per cent of the heating demand. Combisystems have been built with the tank for domestic hot water (at about 60°C) placed inside the tank for space heating (which is at a lower temperature), or with two separate tanks at different temperatures, or as one large stratified tank (with hotter water, for domestic hot water, in the upper portion of the tank) (Suter and Letz, 2003).

Active solar heating becomes more viable in district heating systems combined with underground seasonal heat storage and heat pumps (Lindenberger et al, 2000; Yumrutaş and Ünsal, 2000). In Austria, individual modules of up to 15m² area for large systems have reduced overall costs by 30 per cent. The volume-to-surface area ratio for the storage tank is much larger than for a tank sized to store the heat needed for a single house, thereby greatly reducing the thickness of insulation required in order to give a high heat-storage efficiency. More information on seasonal heat storage is found in Chapter 15 (Section 15.4).

It is not economically attractive to build poorly insulated houses and then try to heat them through active solar heating. It is less expensive to build a very well-insulated house and to rely on passive solar heating, rather than active solar heating. To collect enough heat in winter using collector panels, the panels would have to be very large. In the summer, much of the collected heat would be wasted.

11.4.5 Comparative cost and performance of solar thermal systems

In this section, the investment cost, cost of solar heat and two performance characteristics of various kinds of solar thermal systems under German economic and climate conditions are compared. The two performance characteristics are:

- *Degree of utilization*. This is the ratio of captured solar heat that is used to solar energy arriving on the collector field (it is the same as an efficiency). The *collector degree of*

utilization is based on the amount of heat delivered by the collector to the storage tank, while the *system degree of utilization* is based on the amount of heat delivered from the storage tank to the heat loads, after subtracting circulation and standby losses. The difference between the two can be significant if the conventional system consists of tankless water heaters next to the point of use, while the solar component involves storage tanks far from the points of use, so that substantial distribution losses can occur for the solar component but not for the conventional component of the delivered heat.

- *Solar fraction*. This is the ratio of heat supplied by the solar collectors to the total heat supplied (solar plus non-solar) for that purpose. Thus, if solar heat is used only for DHW, the solar fraction should be computed based on the total energy input to the DHW system (this will include standby and distribution losses), whereas if solar heat is used for both space heating and DHW, the solar fraction should be based on the total space and DHW heating load.

In general, the larger the solar fraction, the smaller the degree of utilization and the greater the cost of solar heat. A reduction in energy demand (or smaller energy demand than expected when the solar system was designed) increases the solar fraction but decreases the degree of utilization, decreases the collector efficiency and increases the cost of solar heat. With a smaller energy demand, heat is drawn off the storage tank less often or to a smaller extent, so that the average storage tank and hence collector temperature is greater, which reduces the collector efficiency and the amount of solar heat that is collected and delivered to where heat is used. Thus, a halving of energy demand increases the solar fraction by less than a doubling. A reduction in energy demand reduces the degree of utilization because the proportion of the time when the availability of solar heat exceeds the heat requirements increases, so that less of the available solar heat can be used. These trade-offs are illustrated in Figure 11.28; the upper panel gives the variation in solar heat output from the collector (kWh/m²/year), collector solar utilization, and solar fraction with DHW demand for a 5m² system located in Germany; the middle panel gives the same information for a large-scale system consisting of solar collectors oriented within 30° of due south and tilted at 30–40° at a location with 1200kWh/m²/year solar irradiance on a horizontal plane; and the lower panel gives the same information for a combisystem in Frankfurt serving a 100m² living area with a 15m² collector area and a 1000litre storage tank. System utilization ranges from 20 per cent to 55 per cent.

Henning (2004a) indicates the following costs for solar collectors, support structures and piping (but excluding storage systems, heat exchangers and pumps):

- solar-air collectors, 200–400€/m²;
- flat-plate or stationary compound parabolic collectors, 200–500€/m²;
- evacuated-tube collectors, 450–1200€/m².

Table 11.7 gives illustrative costs of solar thermal energy for total system costs of 400–1200€/m², solar irradiance of 1100–2200kWh/m²/year (which spans the majority of values given in Table 11.1), solar utilization (system efficiency) of 20–60 per cent (which roughly spans the range encountered in practice), and real interest rates of 4 per cent, 6 per cent and 8 per cent. In all cases, a 25-year system lifespan and annual insurance and maintenance costs of 1 per cent and 2 per cent of the investment cost, respectively, are assumed. Solar utilization will tend to be larger in regions with smaller solar irradiance, thereby partly offsetting the impact on cost of lower irradiance. Thus, solar thermal energy costs the same at 1100kWh/m²/year irradiance and 40 per cent system utilization as at 2200kWh/m²/year irradiance and 20 per cent system utilization. For mid-European conditions (1650kWh/m²/year irradiance), 4 per cent financing cost, and total system costs of 800€ per m² of collector, the cost of thermal energy ranges from 7.6 eurocents/kWh at 60 per cent solar utilization (not yet achieved

Figure 11.28 Top: variation in solar heat output from a 5m² collector located in Germany, collector solar utilization and solar fraction for a system serving various DHW loads. Middle: variation in solar heat output from medium to large DHW systems serving various DHW loads, collector solar utilization, and solar fraction for a system with collectors oriented within 30° of due south, tilted at 30–40°, and at a location with 1200kWh/m²/year solar irradiance on a horizontal plane. Bottom: variation in solar heat output from a 15m² collector array serving DHW and space heating loads for a living unit of 100m² floor area in Frankfurt with a 1000 litre storage tank, as well as system solar utilization and solar fraction. DHW and space heating loads in the bottom panel are per m² of living area, while solar yields are per m² of collector area

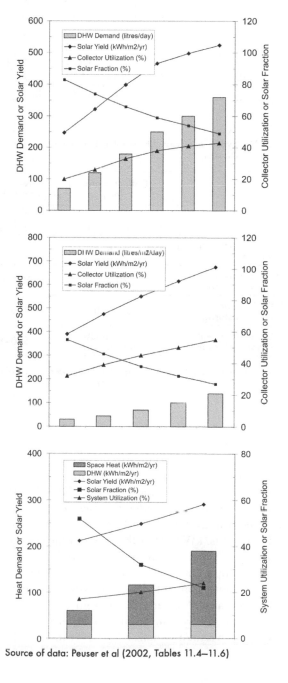

Source of data: Peuser et al (2002, Tables 11.4–11.6)

Table 11.7 Illustrative costs of solar thermal energy for the indicated solar irradiance and real interest rate, a 25-year lifespan, 1%/year insurance, 2%/year maintenance cost and the given system cost and efficiency

System cost ($/m² or €/m²)	System efficiency	Cost of thermal energy (cents or eurocents/kWh)								
		1100kWh/m²/year			1650kWh/m²/year			2200kWh/m²/year		
		interest rate			interest rate			interest rate		
		0.04	0.06	0.08	0.04	0.06	0.08	0.04	0.06	0.08
400	0.2	17.1	19.7	22.5	11.4	13.1	15.0	8.5	9.8	11.2
	0.4	8.5	9.8	11.2	5.7	6.6	7.5	4.3	4.9	5.6
	0.6	5.7	6.6	7.5	3.8	4.4	5.0	2.8	3.3	3.7
800	0.2	34.2	39.4	45.0	22.8	26.2	30.0	17.1	19.7	22.5
	0.4	17.1	19.7	22.5	11.4	13.1	15.0	8.5	9.8	11.2
	0.6	11.4	13.1	15.0	7.6	8.7	10.0	5.7	6.6	7.5
1200	0.2	51.3	59.0	67.5	34.2	39.4	45.0	25.6	29.5	33.7
	0.4	25.6	29.5	33.7	17.1	19.7	22.5	12.8	14.8	16.9
	0.6	17.1	19.7	22.5	11.4	13.1	15.0	8.5	9.8	11.2

Table 11.8 System costs, cost of heat, solar utilization and solar fraction for various kinds and sizes of solar thermal DHW or space heating systems in Germany

System	Collector area (m²)	System cost (€ per m² of collector)	Cost of heat (€/kWh)	Solar utilization	Solar fraction
Small DHW	4–5	800–1300	0.13–0.62	40–20%	50–80%
Large DHW	100–1600	400–900	0.09–0.23	55–25%	20–60%
Combisystem, diurnal storage	15		0.40–0.50	25–18%	20–50%
Combisystem, seasonal storage	20–80	900–1900		23–12%	70–100%
District heat, no seasonal storage	100–1000	400–500	0.10–0.13		7–10%
District heat, with seasonal storage	3000–6000 (540–6000)	620–800	0.18–0.30 (0.16–0.42)	25–28%	50% (30–62%)

Source: Peuser et al (2002), except for data in brackets for cost of district heat with seasonal storage, which are a summary of Table 15.11 that in turn is based on Schmidt et al (2004)

in practice) to 22.8 eurocents/kWh at 20 per cent solar utilization.

Table 11.8 summarizes cost and performance data for a variety of solar thermal systems in Germany. The cost achieved for the large DHW systems constructed under the *Solarthermie 2000 Program (Part 2)* during the 1990s ranged from 9–23 eurocents/kWh. After the programme evaluation, some systems were optimized, reducing the cost. Comparable costs have been achieved for large systems connected to district-heating grids without seasonal storage. To go from a DHW system to a combisystem requires a larger collector area, which is then oversized for DHW in summer.

Solar usability decreases, thereby contributing to a greater cost (40–50 eurocents/kWh). Solar systems connected to district heating networks with underground seasonal energy storage have provided up to two-thirds of the total DHW + space heating load in well-insulated buildings, at a cost of 16–42 eurocents/kWh. Without seasonal thermal energy storage, the system primarily serves summer loads, with an annual solar fraction of 7–10 per cent and costs of 10–13 eurocents/kWh. A factor tending to reduce the efficiency of solar district heating compared to solar domestic water heating is the fact that a district heating network will be designed to have return temperatures no lower than 30–50°C, compared to 15–20°C for domestic water heating (solar energy can be captured only when the storage buffer is warmer than the return temperature). If the district heating return temperature is higher than intended (as has sometimes been the case), the solar utilization decreases and the cost increases. Solar district heating with seasonal energy storage is discussed more fully in Chapter 15 (Sections 15.4 and 15.5).

Energy costs should fall with ongoing decreases in the costs of individual system components, and with better optimization and design. For example, Furbo et al (2005) show that better design of solar domestic hot-water storage tanks when combined with an auxiliary energy source can improve the utilization of solar energy by 5–35 per cent, thereby permitting a smaller collector area for the same solar yield.

11.4.6 Global solar thermal market

Table 11.9 gives the solar thermal collector area in operation by the end of 2003 in countries thought to represent 85–90 per cent of the world market. Total installed collector area was over 132 million m² by 2003, corresponding to a peak heat production capacity of about 93GW. This is greatly in excess of the total installed PV electrical capacity of 1.3GW at the end of 2002 (Table 11.4), and the global wind power capacity of 48GW at the end of 2004 (Cameron, 2005). The collectors are overwhelmingly (98.7 per cent) water collectors rather than air collectors. Of these, 41 per cent are glazed, 35 per cent are evacuated-tube, and 24 per cent are unglazed collectors (the latter are used predominantly in the US and Australia for heating swimming pools). Also given in Table 11.9 are the collector areas installed in 2003. The leading countries are China, the US and Japan in terms of existing collector area, and China, the US and Turkey in terms of installation in 2003. The rate of installation has been growing by almost 30 per cent/year in China, and constitutes 12 per cent of the national water heater market according to Jones (2005). However, in terms of per capita use, Cyprus is likely the leading country in the world, with one operating solar water heater for every 3.7 inhabitants (Kalogirou, 2002). CPCs have seen essentially no use to date.

The European Solar Thermal Industry Federation (ESTIF) estimates the technical potential for solar thermal energy in the 15-nation EU for heating, hot water, air conditioning and industrial processes to be a collector area of 1.41 billion m² (3.74m² per capita) and an energy output of 682TWh/year (ESTIF, 2003). It has developed a strategy for reaching an installed area of 100 million m² by 2010–2015, compared to an installed area of 12.3 million m² by the end of 2002 and 1.2 million m² installed during 2002. In 1999, Barcelona City Council unanimously passed a 'solar ordinance' that requires the installation of solar thermal collectors sufficient to meet 60 per cent of the hot water needs on all new buildings and on larger buildings during renovation. Other municipalities in Spain are considering similar laws.

Price can be a barrier to the widespread adoption of solar thermal water heating systems. However, the National Renewable Energy Laboratory in the US is currently funding the development of two integrated collector storage systems using polymer materials that could break the current price barrier (Lutz et al, 2002). Polymers resist corrosion and mineral buildup, are cheaper to manufacture than glass and copper components, and are lightweight and less fragile, making them cheaper to transport and assemble. However, they must be able to withstand high temperature and pressure for at least 10 years and resist ultraviolet-induced degradation. Raman et al (2000) review various polymer materials that could be used for solar thermal collectors. If a single storage tank is used for a hybrid solar/natural gas water heating system, the incremental cost of the solar component is about half of the cost when bought as a separate add-on system (see www.iea-shc.org/task24, Case Study 1). In this case, the heating coil from the auxiliary heater would be placed in the upper portion of the storage tank, in order to avoid heating all of the water in the tank when the sun is not shining or is inadequate. Care is needed to maintain temperature stratification within the storage tank.

Table 11.9 Total solar thermal collector area (m²) in operation by the end of 2003 in reporting countries thought to represent 85–90% of the world market, and the collector area added in 2003

Country	Water collectors			Air collectors		Total	Total added in 2003
	Unglazed	Glazed	Evacuated	Unglazed	Glazed		
Australia	2,767,000	1,444,000				4,211,000	521,000
Austria	594,823	2,066,145	32,309			2,693,177	176,820
Barbados		71,870				71,870	2731
Belgium	25,232	36,348	2783			64,363	10,920
Brazil		2,233,000				2,233,000	33,000
Canada	565,988	75,220	1000	65,156		707,364	37,896
China	600,000	6,800,000	44,000,000			51,400,000	11,400,000
Cyprus		677,000				677,000	15,000
Czech Rep.		31,930	3100			35,030	7006
Denmark	21,870	294,570	550			316,990	8000
Finland		15,000	150	70,000		85,150	1500
France	111,500	574,113	60			685,673	44,876
France Terr.		86,350				86,350	45,000
Germany	715,000	4,274,000	677,000		40,000	5,706,000	720,000
Greece		3,248,000				3,248,000	126,000
Hungary	29,400	1,740	2560	2500	37,200	73,400	1500
India		800,000				800,000	100,000
Ireland		4199	1011		200	5410	1200
Israel		4,720,000				4,720,000	400,000
Italy	12,000	365,000	23,000			400,000	60,000
Japan		12,363,636	318,869			12,682,505	280,482
Mexico	399,493	157,210				556,703	75,303
Netherlands	175,599	267,504	3190	4794		451,087	41,686
New Zealand	2400	76,183	460			79,043	6300
Norway	1299	9450	200	410,000	1200	422,149	7200
Poland	1150	65,185	4334	3000	2000	75,669	26,520
Portugal	1000	258,210	500			259,710	9210
Slovenia		99,076	575			99,651	1124
South Africa	140,500	152,099	20			292,619	57,409
Spain	6112	569,106	17,343			592,561	70,000
Sweden	41,756	229,316	5214			276,286	23,697
Switzerland	209,450	292,460	23,760	830,000		1,355,670	25,520
Turkey		9,500,000				9,500,000	800,000
UK	99,000	177,000	26,730			302,730	22,000
US	24,989,264	1,545,711	552,580		227,487	27,315,042	1,063,476
Total	31,509,836	53,580,631	45,697,198	1,385,450	308,087	132,481,202	16,222,376
Total capacity (MW$_{th}$)	22,057	37,506	31,988	970	216	92,737	11,356

Note: Also given is the total world heat production capacity corresponding to the reported collector areas.
Source: Weiss et al (2005), with their 2003 capacity additions converted to area by dividing by their conversion factor of 0.7kW/m²

11.5 Solar thermal energy for air conditioning

Solar thermal energy can be directly used for cooling and dehumidification. There was a substantial effort to develop solar cooling systems in the late 1970s and early 1980s in the US, but this effort subsided and the centre of activity shifted to Japan and Europe. There are about 45 solar air conditioning systems in Europe, with a total solar collector area of about 19,000m² and a total capacity of about 4.8MW chilling power (Henning, 2004b). About 50 per cent of these use absorption chillers, 30 per cent use rotating desiccant wheels and 20 per cent used adsorption chillers. The largest solar air conditioning system is at a cosmetics factory in Greece, consisting of 2 350kW$_c$ adsorption chillers and a 2700m² collector field. Task 25 of the International Energy Agency's *Solar Heating and Cooling Programme* (www.

iea-shc-task25) was directed toward improving conditions for the market introduction of solar-assisted cooling systems. This task, which concluded in 2004, reviewed and evaluated international projects, prepared a database of projects, carried out 11 demonstration projects and developed design tools, detailed simulation models and advanced control strategies. A design handbook was prepared as part of Task 25 (Henning, 2004a). A number of projects have been carried out as part of the *Solar Air Conditioning in Europe* (SACE) programme, which has been supported by the European Commission. Complete solar air conditioning systems and services are available commercially in Europe through *Solar Installation and Design* of Austria (www.solid.at) and in the US through its US distributor, *Solar Capital Partners* (www.ussolid.com). Integrated systems providing air conditioning, space heating and hot water have been built.

Table 11.10 Characteristics of solar air conditioned buildings in Germany

Building use	Technology	Cooling capacity (kW)	Solar collector type	Collector area (m²)	Project status
Offices	Absorption; chilled ceiling	70	Vacuum tube	176	In operation since 1985
Offices	Absorption; chilled ceiling and displacement ventilation	35	Vacuum tube	22	In operation since 1998
Offices	Absorption	70	Vacuum tube	244	In operation since 2000
Chilled water network	Absorption	70	Flat plate, high efficiency	209	In operation since 2000
Offices, laboratory	Absorption; cooling baffles	7	Vacuum tube	20	In operation since 1999
Offices, slab cooling	Absorption	143	Vacuum tube	260	In operation
Offices	Absorption (ammonia/ water); ventilation system	15	Vacuum tube	79	In operation since 1999
Offices	Adsorption; silent cooling	105	Flat plate	150	In operation since 1999
Offices	Adsorption; chilled ceiling	71	Flat plate	156	In operation since 1996
Offices	Adsorption; chilled ceiling	70	Flat plate	70	In operation since 1996
Offices, computer and seminar room	Adsorption; cooling baffles and process chilled water	352	Flat plate	2000	In operation since 2000
Hospital	Adsorption	105	TIM-collector	116	In operation since 2000
Laboratory	Adsorption; ventilation system	70	Vacuum tube	170	In operation since 1999

Offices	Adsorption and desiccant; ventilation system	70	Vacuum tube	120	In operation since 2000
Seminar room, foyer	Desiccant; ventilation	30	Flat plate air collector	100	In operation since 1998
Meeting room	Desiccant (LiCl) and heat pump	18	Flat plate with volumetric absorber	20	In operation
Meeting room	Desiccant; ventilation	18	Flat plate	20	In operation since 1997
Exhibition room	Desiccant; ventilation	18	Flat plat air collector	20	In operation
Factory	Desiccant	108	Flat plat air collector	100	In operation since 2000
Meeting room	Desiccant; ventilation	60	Flat plat air collector	100	In operation since 2001
Test plant	Desiccant	24	Flat plat air collector	40	In operation since 2000

Source: Balaras et al (2003)

Balaras et al (2003) provide summary information on 21 solar air conditioned buildings in Germany. Information on these systems is given in Table 11.10. Costs per unit of energy will be minimized if solar collectors are designed to be used for winter heating and summer cooling, rather for only one or the other. Information on different kinds of solar air conditioning is given below. Using solar energy for home air conditioning does not make sense if the roof colour has not first been changed to a light colour (if not already so), since this costs little to nothing at the time the roof is built or needs to be replaced and can reduce cooling loads by up to 40 per cent (Chapter 6, Section 6.2.2). In the discussion to follow, three different COPs arise: the chiller COP (heat energy removed divided by the thermal energy input), the solar COP (chiller COP times solar collector efficiency), and system COP (solar COP with account taken of losses during transfer and storage of solar heat). The system COP is equal to the thermal energy removed from the building divided by the solar energy incident on the collectors.

11.5.1 Solar-powered absorption chillers

The principles behind the operation of an absorption chiller are thoroughly explained in Chapter 6 (Section 6.6.1 and Box 6.4). Absorption chillers require a refrigerant and an absorbent, either an ammonia (NH_3)-water mixture (where NH_3 is the refrigerant and water the absorbent) or a lithium bromide (LiBr)-water mixture (where water is the refrigerant and LiBr the absorbent). Absorption of the refrigerant by the absorbent lowers the vapour pressure of the remaining refrigerant, provoking its evaporation in an evaporator. Heat is released in the absorbent but cooling of the evaporator occurs. Heat must be added to the absorbent–refrigerant solution in a generator in order to drive off the refrigerant and regenerate the absorbent, and this is done through the circulation of dilute and concentrated solutions and refrigerant in such a way as to permit continuous operation. The heat input can be supplied by a natural gas burner, by steam or by hot water. Table 11.10 lists seven solar-power absorption chiller projects in Germany using flat-plate or evacuated-tube solar collectors. Single-effect absorption chillers require an input heat at 70–90°C. In California, a double-effect absorption chiller powered by a compound parabolic collector began operation in 1998 (Duff et al, 2003; Bergquam et al, 2003). The chiller has a COP of 1.0 operating on a hot-water input at 150°C, but the goal is to reduce the input temperature to 120°C with no reduction in COP or cooling capacity. Solar absorption chillers are commercially available from Yazaki Energy (www.yazakienergy.com).

Overcoming weaknesses in single-effect LiBr chillers

Absorption chilling is a widely used technology for cooling purposes, with the majority of commercial absorption chillers using the LiBr-water cycle rather then the ammonia-water cycle. The LiBr chiller is a better match to solar energy because it requires a lower input temperature (70–90°C for single-effect systems), such as can be produced by relatively low-cost flat-plate solar collectors (more expensive evacuated-tube collectors would be needed with an ammonia chiller). However, conventional single-effect LiBr absorption chillers suffer from two weaknesses that are critical to their coupling with solar collectors:

- there is a sharp drop in the production of chiller water and in the COP if the input temperature drops below a threshold ranging from 60°C (for an evaporator at 10°C) to 90°C (for an evaporator at –20°C);
- there is little increase in COP as the input temperature increases, so the decrease in the efficiency of the solar collector with increasing temperature is not compensated by increased cooling efficiency.

A solution to these shortcomings might be the use of a *two-stage* absorption chiller, which is different from a double-*effect* chiller (described in Chapter 6, Section 6.6.1). The double-effect chiller is essentially two single-effect chillers coupled together by supplying heat from the condenser of the first chiller to the generator of the second chiller. The input temperature to the first chiller must be greater than for a single-effect chiller, so that heat from the condenser is sufficiently warm to drive the second chiller. In a two-stage chiller, illustrated in Figure 11.29, there is only one condenser and one evaporator but two generators and absorbers at different pressures. Heat from the low-pressure generator is supplied to the high-pressure absorber. The two-stage absorption chiller has been used in industrial applications to make use of low-grade waste heat.

Efforts are also underway to develop an ammonia chiller that can be driven with solar

Figure 11.29 Flow diagram for a two-stage absorption chilling system

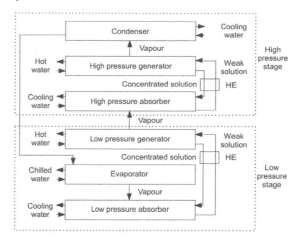

Source: Sumathy et al (2002)

heat at 80°C while producing an evaporator temperature of –20°C, as well as to develop a solar-powered ammonia chiller suitable for residential applications that would operate with an input temperature of 93–102°C and produce chilled water at 10–11°C with a COP of 0.54–0.62 (IEA, 2002a).

COP

The variation of COP and cooling capacity of commercially available two-stage and single-effect chillers are compared in Figure 11.30 for production of chilled water at 9°C. As can be seen from this figure, the two-stage chiller allows for lower input temperatures without a sudden drop-off in performance as the input temperature decreases. At a common input temperature of 80°C, the two-stage unit provides 2.5 times the cooling at almost twice the COP. The lower input temperature allowed with the two-stage system leads to an increase in the solar collector efficiency by 40–50 per cent (Sumathy et al, 2002). A 100kW solar-powered two-stage absorption chiller with a collector area of 500m² began operation in China in January 1998 (Li and Sumathy, 2000). Disadvantages of this system are (i) the complexity of the chiller; (ii) a lower COP (somewhat over 0.4) than a single-effect chiller (0.6) at the nominal generator temperature; and (iii) a need for twice as much cooling water and, hence, a larger cooling tower. However, the lower COP of the

Figure 11.30 Comparison of cooling output and COP of commercially available single-stage and two-stage LiBr absorption chillers

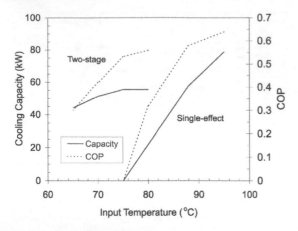

Note: Chilled water is produced at 9°C in both cases, while cooling water is assumed to be supplied at 31°C for the single-stage chiller and 32°C for the two-stage chiller.
Source: Data from Sumathy et al (2002)

two-stage chiller is offset by the greater solar collector efficiency, leading to a solar COP in both cases of 0.20–0.22.

Impact of chilled-water temperature

Figure 11.31 shows the variation in the COP and cooling capacity with chilled-water temperature and generator temperature for a commercially available absorption chiller. At chilled-water temperatures typical of most HVAC systems (7–11°C), the COP and cooling capacity drop sharply as the generator temperature decreases from 100°C to 75°C (this is also seen in Figure 11.30). However, at a chilled-water temperature of 16°C (suitable for chilled ceiling cooling and displacement ventilation if overcooling is not used for dehumidification), the COP is almost completely independent of generator temperature, ranging from 0.65 to 0.68. Cooling capacity still decreases with decreasing generator temperature, but to the extent that a lower generator temperature is due to smaller solar irradiance, less cooling capacity will be needed when the generator temperature is low.

Figure 11.31 Impact of chilled water temperature and generating temperature on the COP and capacity of a commercially available absorption chiller, as measured by Li and Sumathy (2001b) for a cooling water temperature of 29.5°C

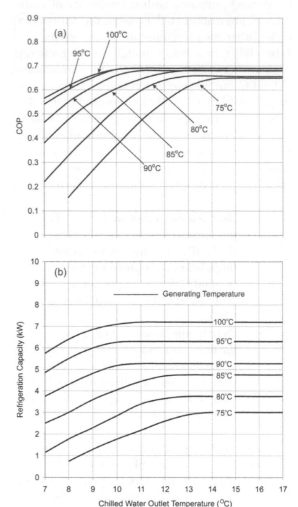

Source: Li and Sumathy (2001b)

Extending the duration of cooling through separate low- and high-temperature thermal storage, cold storage and/or evacuated-tube collectors

The operation of an absorption chiller is more continuous if a hot-water storage tank is used to buffer fluctuations in the rate of production of hot water from the solar collectors. However, this can delay the start-up of the chiller until noon or later if a single, well-mixed storage tank is used. If a large and small tank are used, with the smaller tank collecting early morning heat, chilling can begin two hours sooner (Li and Sumathy, 2001a, 2002b). The low-temperature tank would be charged by early morning energy and store heat at 50–70°C, while the high-temperature tank would store heat at 85–95°C. Separation of the storage into high- and low-temperature tanks can increase the amount of heat collected by a given array by 30–50 per cent and increase the seasonally averaged chiller COP by 15 per cent (Li and Sumathy, 2000). Together, this would reduce the collector area needed to cool a building by 30–40 per cent.

A disadvantage of hot-water storage is heat loss to the surroundings. An alternative is to store chilled water. There will be less heat gain with chilled water than heat loss with hot water because of a smaller temperature difference between the water and surrounding air. If the chilled water is stored in an area to be air conditioned, the heat gain will assist in cooling that area.

Use of evacuated-tube solar collectors rather than flat-plate solar collectors permits collection of solar energy at lower irradiance (beginning at $140W/m^2$ rather that at $250–300W/m^2$), thus providing energy for cooling (and heating) earlier in the morning and later in the evening

Cooling towers

Solar-powered absorption chillers used today utilize wet cooling towers, which provide cooling water for heat rejection at a temperature close to the outdoor wet-bulb temperature. This makes them inapplicable to residential applications, since the average homeowner will be unable to provide the necessary maintenance, and they are generally considered to be unattractive. In addition, water is scarce in many parts of the world where solar energy is plentiful. In a centralized district cooling system (see Chapter 15, Section 15.3.6) serving a residential neighbourhood, a wet cooling tower would be easier to implement. Use of lower temperature solar heat requires a single-stage absorption chiller, with a relatively low COP, which increases the total amount of heat that needs to be rejected by the cooling tower.

Options where water is scarce are: (i) to use an air-cooled single-effect chiller; and (ii) to use a *dual-cycle* chiller (Li and Sumathy, 2000). Use of a dry cooling tower with a single-effect chiller will require a warmer generator temperature

(100°C rather than 78°C if the evaporator is at 5°C). This in turn will reduce the fraction of total cooling provided by solar energy by 10–20 per cent. The dual-effect chiller requires evacuated-tube collectors in order to provide a sufficiently hot heat input, is rather complicated and has a low COP.

Impact of the backup component on overall energy use

If a backup fossil fuel heat source is used to power an absorption chiller when solar heat is inadequate, the benefit of solar heat could be offset by the low COP of the absorption chiller. If the baseline system is an electric chiller with a COP of 4.0 and an electricity generation efficiency of 33 per cent, and the solar system is an absorption chiller having a COP of 0.75 and a boiler for backup heat with an efficiency of 85 per cent, solar energy must provide 40 per cent of the required heat input in order to break even from a primary energy point of view if differences in auxiliary electricity use (for pumps and fans) between the two systems are ignored. If the electric chiller has a COP of 6.0 and electricity is generated at an efficiency of 55 per cent (the current state-of-the-art), solar energy must provide 60 per cent of the heat input in the hybrid system in order to require no more primary energy than the electric system. Thus, in order for a solar-powered single-effect absorption chiller to reduce the use of primary energy, either the solar collectors must be sized to guarantee the required minimum solar fraction, or an electric chiller should be used as a backup, or there should be no backup at all (which implies accepting occasional temperatures warmer than preferred). Electric chillers in the $50–200kW_c$ range cost $150–300/kW_c$ (Figure 6.53b), while boilers in the $75–300kW_{th}$ size range cost $30–60/kW_{th}$ (Figure 4.18), so electric chiller backup would be considerably more expensive than a backup boiler. Alternatively, a double-effect absorption chiller, requiring evacuated-tube solar collectors but having a COP of 1.2, can be used with a natural gas boiler as a backup heat source. The attraction of using a boiler as backup is that it can also serve space heating loads in winter. A $70kW_c$ hybrid natural gas–solar absorption chiller has been in operation in California for several years (Gee et al, 2001).

Another option is to design the chiller to operate as a double-effect chiller when fossil fuel heat (at high temperature) is available, and to operate as a single-effect chiller when solar heat (at lower temperature) is available. Lamp et al (1998) worked out the design of such a chiller, which could operate with simultaneous solar and natural gas heat as well as with either heat source alone. The primary energy used by this chiller with a solar fraction of 0.6 is the same as that of an electric chiller with a COP of 5 and an electricity-generation efficiency of 55 per cent, but for a smaller efficiency or COP, or a larger solar fraction, there is a net saving in primary energy.

Impact of auxiliary electricity use on savings in primary energy

Electricity is required to operate the cooling tower pumps and fans associated with an absorption chiller. If a solar-powered absorption chiller replaces an air-cooled rooftop air conditioner, this is an additional energy load. Even if the solar-absorption chiller replaces a water-cooled electric chiller, the amount of heat that will need to be rejected by the cooling tower for a given building cooling load – and the associated pump and fan energy use – will be greater due to the lower COP of the absorption chiller, so the auxiliary electricity use will increase. This increase should be subtracted from the savings in chiller electricity use to give the net savings in electricity. Henning (2004a) indicates that cooling tower fan energy use is about 6–20W per kW rate of cooling tower heat rejection (kW_{ct}). Inasmuch as kW_{ct} is $(1 + 1/COP)$ times the rate at which heat is removed from the building (kW_c), the fan energy use per kW_c is $(1 + 1/COP)$ times larger. The total auxiliary electricity use (E_{aux}), including pumps, could be 10–40W/kW_{ct} (19–22W/kW_{ct} is given in Table 6.13). For an absorption-chiller COP of 0.7, this amounts to 24–97W/kW_c. For an air-cooled chiller with a COP of 3.5, electricity use is 285W/kW_c. Thus, one-tenth to one-third of the savings in chiller electricity use would be offset by cooling tower electricity use.

For a water-cooled chiller, the electric power per kW_c is $1000/COP + E_{aux}(1 + 1/COP)$. If an absorption chiller with a COP of 0.7 replaces a water-cooled chiller with a COP of 6.0, the savings in total electricity use decreases from

Table 11.11 Comparison of energy use and performance for a conventional electric chiller and for an absorption chiller driven by heat from flat-plate solar collectors, as calculated for a hotel in Athens

	Conventional system	Solar absorption system
System components		
Chiller capacity (kW_c)	29.4	29.4
Chiller COP	3.0	0.7
Cooling Tower Power (kW_c)	0	74.6
Collector area (m²)	0	200
Heat storage volume (m³)	0	18
Energy flows and water use		
Chilling load (kWh_c/year)	67,164	67,164
Electricity use (kWh/year), chillers pumps and fans total	22,388 2423 24,811	0 10,977 10,977
Peak electricity demand (kW)	10.0	2.9
Solar heat on collector (kWh/year)	0	314,980
Solar output (kWh/year)	0	128,803
Useful solar output (kWh/year)	0	103,611
Water use (m³/year)	0	368.3
Effectiveness of water use, GJ of building heat load per m³ GJ of cooling tower heat load per m³	0 0	0.66 1.76

Source: Henning (2004a)

86 per cent to 55 per cent as E_{aux} increases from 10 to 40W/kW_{ct}. It is thus very important to minimize energy use by the cooling tower associated with solar-powered absorption chillers.

System performance

The effect of solar collector efficiency, chiller COP and auxiliary electricity use on the overall use of primary energy is illustrated by a detailed design for a solar absorption cooling system for a hotel in Greece, and its comparison with a conventional cooling system. Results are summarized in Table 11.11 for a single-effect $29.4kW_c$ absorption chiller without backup. Direct electricity use for chilling is eliminated, but electricity use by fans and pumps increases by a factor of four, giving an overall savings in electricity (and primary energy) of 56 per cent. The mean annual solar collector efficiency is 41 per cent, but only 33 per cent of the solar energy incident on the collectors is ultimately used. About 0.66GJ of building heat is removed per m^3 of water consumed, an effectiveness comparable to that given in Chapter 6 (Table 6.18) for absorption chillers.

System cost

The cost of solar-powered absorption chilling systems is dominated by the cost of the solar collectors. Grossman (2002) estimated the cost of a single-effect system using flat-plate solar collectors, double-effect systems using flat-plat or compound parabolic collectors, and a triple-effect system using evacuated-tube collectors. His results are summarized in Table 11.12, based on somewhat more favourable collector costs than given earlier from Henning (2004a). Evacuated-tube collectors cost almost four times as much as flat-plate collectors, but only half of the collector area is needed, due to the higher COP of the chiller to which they are attached. The triple-effect chiller (not yet commercially available) is assumed to be 15 per cent less expensive than the single-effect chiller, due to its higher COP. Nevertheless, the triple-effect chiller with evacuated-tube collectors is almost twice as expensive as either the single- or double-effect systems. If the cost of evacuated-tube collectors can be substantially reduced, then the smaller required area for this system could allow its widespread use.

Bergquam Energy Systems built, installed and tested a complete solar absorption cooling system for an $800m^2$ office building in California, consisting of a $5.7kW_c$ natural gas double-effect chiller modified to use hot water, an integrated compound parabolic solar collector, and a 3800litre storage tank (BES, 2002). The original natural gas-fired absorption chiller cost about $480/kW_c$, while it is expected that a solar-powered chiller could be produced for less than $400/kW_c$, and a complete system installed for about $1400/kW_c$. This is comparable to the overall cost estimated by Grossman (2002) for a solar system with a double-effect absorption chiller using flat-plate or compound-parabolic collectors, although the estimated costs of the various components differ. The chiller cost ($400/kW_c$) is substantially greater than that given in Table 6.13 and Figure 6.54 for large ($250–2000kW_c$) commercial absorption chillers ($85–300/kW_c$), so it might be possible to further reduce costs with larger systems. According to Table 6.14, large systems consisting of a vapour-compression chiller and cooling tower can be built for around $100/kW_c$.

A detailed breakdown of costs for the two cooling systems featured in Table 11.11 is given in Table 11.13. These costs should be regarded as illustrative and approximate only. The comparison is likely biased in favour of the solar/absorption chiller option, in that the same chiller, pump, installation and control costs are assumed for both cooling systems. At the moment, costs are prohibitively expensive (about $3000€/kW_c$), with a 10-year simple payback even with a 50 per cent subsidy. Over half of the cost is due to the solar collectors, at $250/m^2$. Water costs and additional maintenance costs for the solar option offset just over half of the savings in energy costs, thereby more than doubling the payback times. The economics of solar air conditioning are more favourable in China than in Europe

Table 11.12 Comparative costs and other characteristics of different solar collector-absorption chiller systems, as calculated by Grossman (2002)

| | Typical COP | Input T (°C) | Collector type | Collector cost ($/m²) | Heat collected per day (kWh/m²) | Collector area (m² per kW of cooling capacity) | Cost ($ per kW of cooling capacity) | | |
							Solar	Chiller	Total
Single-effect	0.7	85	Flat-plate	110	1.53	7.48	1234	200	1434
Double-effect	1.2	130	FP/CPC	160	1.31	5.07	1216	175	1391
Triple-effect	1.7	220	Evac tube	400	1.05	4.49	2694	165	2859

Note: Chiller cost data ($/kW_c) can be compared with comparable data in Tables 6.12 or 6.13, 6.14, 15.7 and Figure 6.57.

Table 11.13 Comparison of investment and operating costs (euros) for 29.4kW$_c$ vapour compression and solar absorption chiller systems serving a hotel in Greece, and of the simple payback time for the solar system assuming subsidies of various magnitudes

	Conventional system	Solar absorption system
Investment costs		
Chiller	11,760	11,760 (400€/kW$_c$)
Solar collectors	0	50,000 (250€/m²)
Heat storage unit	0	9000
Cooling tower	0	2611
Pumps	750	750
Control system	8000	8000
Installation	10,000	10,000
Total	30,510	92,121 (3133€/kW$_c$)
Annual operating costs		
Electricity (at 20 euro cents/kWh)	4962	2199
Electricity demand charge	1000	290
Maintenance and inspection	837	1594
Water	0	1105
Total	6799	5188
Simple payback time (years)		
With 0% subsidy	–	38.2
With 25% subsidy	–	23.9
With 50% subsidy	–	9.7

Source: Henning (2004a)

due to the availability of lower cost flat-plate and evacuated-tube collectors in China – in the range of 300–350 Yuan/m² (30–35€/m²) according to Yattara et al (2003), compared to 200–500€/m² for flat-plate collectors in Europe.

11.5.2 Solar-powered adsorption chillers

The principles behind the operation of an adsorption chiller were outlined in Chapter 6 (Section 6.6.2). A more detailed discussion of adsorption chilling as applied to solar adsorption refrigerators, which have been used to make ice and to store vaccines and drugs in tropical regions without access to conventional energy sources, is found in Anyanwu (2003, 2004). Adsorption systems have the advantages of operation with no moving parts except magnetic valves, mechanical

simplicity, high reliability and long life, and very low noise and vibration. A single-bed solar-powered adsorption chiller is illustrated schematically in Figure 11.32. The bed is either open to a condenser or open to an evaporator, but is shown in Figure 11.32 as two beds, one with each option. During the day, the adsorbate (water, organic refrigerants, ammonia or methanol) is driven from the adsorbent (zeolite, silica gel, salts or activated carbon) with solar heat and condenses in a condenser, where the heat of condensation is rejected to an external water loop. This external water loop can flow through a cooling tower (as shown in Figure 11.32), or can be used to produce domestic hot water or, during the heating season, for space heating. The condensed adsorbate flows to an evaporator. During the night, the valve to the condenser is closed, the valve to the evaporator opens, and water from the cooling tower rather than the solar collector flows through the adsorbent bed, thereby lowering its vapour pressure. This induces evaporation of the adsorbate, producing a cooling effect in the evaporator. Heat that is released during adsorption is removed by the cooling tower loop. Water in another secondary loop flows through the evaporator and is chilled in the process. Continuous cooling can be achieved by using two or more adsorbent beds operated out of phase.

When the regeneration temperature is less than 100°C, the COP of an adsorption chiller decreases significantly as the evaporator temperature decreases from 10°C to −10°C, or as the condenser temperature increases from 25°C to 40°C. To maximize the efficiency while using the comparatively low regeneration temperatures available with solar energy, solar adsorption chillers should be combined with cooling systems that use as high a chilled-water distribution temperature and as low a cooling-water temperature as possible. For a 10°C evaporator temperature and a 30°C condenser temperature, the COP will range from about 0.25 at a 75°C regeneration temperature, to about 0.45 at a 100°C regeneration temperature. Li et al (2004) calculate system COPs for solar-powered adsorption chillers of 0.08–0.19.

Seven buildings in Germany with solar adsorption chillers are listed in Table 11.10, with cooling capacities ranging from 70kW$_c$ to

Figure 11.32 Schematic diagram of a single-bed solar-powered adsorption chiller. The adsorption bed is shown as two separate beds in order to illustrate the flow and valve positions during daytime operation (right) and night-time operation (left)

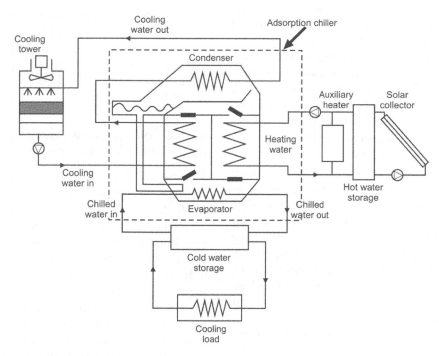

Source of figure: Papadopoulos et al (2003)

353kW$_c$, and using either flat-plate or vacuum tube collectors. A 100kW Mycom adsorption chiller has been installed for municipal services in Remscheid, Germany, with the intention of providing 75 per cent of the driving heat from solar collectors and the balance from the district heating system (Dieng and Wang, 2001). The chiller requires a hot-water input at 75°C and cooling water at 29°C, and reportedly produces chilled water at 9°C with a COP of 0.6. This is much greater than the COP expected based on the performance information given above.

Adsorption chillers normally cannot operate at an input temperature of less than 65°C; however, with the use of a mechanical pump to reduce the pressure inside the adsorption bed, solar heat down to a temperature of 50°C can be used. Such a hybrid system for use with solar energy is under development in Japan (IEA, 2002a). The COP based on the electrical input only ranges from 5 at an input temperature of 50°C to about 13 at an input temperature of 70°C,

while the COP based on the heat input ranges from 0.52–0.56 over the same input temperature range.

Dai et al (2003) propose an innovative system for a house that combines a solar chimney and a single-bed adsorption chiller with a separate condenser and evaporator. As in other single-bed solar-powered adsorption chillers, cooling occurs at night and regeneration during the day. During the day, a solar chimney induces natural ventilation (but limited to that required for fresh air if outside conditions are hot). During the night, heat rejected by the condenser warms air in the solar chimney, thereby sustaining a high ventilation rate during the night. At the same time, the evaporator is used to cool the outside air that is drawn into the house. This increases the effectiveness of night ventilation (Chapter 6, Section 6.3.3) in cooling the house.

11.5.3 Solar-powered ejector cooling

The principles behind the operation of an ejector chiller are summarized in Chapter 6 (Section 6.6.3). Nguyen et al (2001) successfully assembled and operated a system at Loughborough University in the UK using solar energy as the driving force. Their system used an evacuated-tube collector with a boiler designed to operate at 90°C, a condenser at 35°C, and an evaporator at 10°C. The ejector COP was 0.27–0.32 at the test location. Pridasawas and Lundqvist (2004) present a detailed mathematical analysis of a solar-driven ejector refrigeration system based on the measured performance of the solar and ejector components. For incident solar radiation of 700W/m² on double-glazed flat-plate collectors, a solar collector outlet temperature of 90°C, a generator temperature of 80°C, a condensing temperature of 37°C and an evaporator temperature of 10°C, the solar collector efficiency and ejector COP are about 0.41 and 0.27, respectively, giving a solar COP of about 0.11.

Hernández et al (2003) used a mathematical model of an ejector to investigate the performance of a hypothetical hybrid solar-ejector/vapour compression chiller. The compressor is used to boost the pressure of the working fluid between the outlet of the evaporator and the inlet to the ejector. The system COP is based on the thermal and electrical energy inputs. If E_R is the ratio of electrical to thermal energy inputs, then the COPs based on electrical and thermal energy alone are:

$$COP_e = COP_s\left(1 + \frac{1}{E_R}\right)$$

(11.2)

and

$$COP_{th} = COP_s(1 + E_R)$$

(11.3)

respectively, where COP_s is the system COP. For an ejector using HFC-134a as a working fluid, an 85°C generator temperature, and a 30°C condenser temperature, system COP ranges from 0.24 to 0.47 as E_R increases from 0.028 to 0.70 (due to greater compression of the working fluid). This corresponds to a COP_e of 8.8–7.2 and a COP_{th} of 0.26–0.50. The former is comparable to or better than the most efficient (and large)

vapour compression chillers, while the latter is substantially better than what can be obtained with an ejector without the assistance of a compressor. Hybrid solar-powered ejector absorption and ejector adsorption systems have also been investigated, with some operational benefits compared to either system operating alone (Zhang and Wang, 2002; Li et al, 2002; Sözen et al, 2004a, b)

11.5.4 Solar-powered desiccant dehumidification and cooling

The use of solid and liquid desiccants for dehumidification and/or cooling is discussed in Chapter 6 (Section 6.6.4) and Chapter 7 (Sections 7.4.11 and 7.4.12). These provide an alternative to dehumidification by overcooling and then reheating the ventilation air, and would be particularly advantageous in buildings with a chilled ceiling and displacement ventilation cooling system, as the chilling water and ventilation air need to be cooled only to 16–20°C in the absence of dehumidification. As desiccant dehumidification systems are driven by low-grade heat (at 50–100°C), they lend themselves to the use of solar energy. Smith et al (1994), Halliday et al (2002), and Mavroudaki et al (2002) present a theoretical analysis of a solar-powered desiccant dehumidification system. The main factor limiting the use of solar energy is that, as the latent heat load increases, the required regeneration temperature increases, and in humid climates can exceed that which can be provided by flat-plate collectors. In this case, more expensive evacuated-tube collectors or a supplemental boiler are needed.

Henning et al (2001) built an experimental solar-powered solid desiccant air conditioning system at a technology centre in Saxony, Germany. The system was designed using the TRNSYS software package (see Appendix D), with two new subroutines added to model the desiccant components and their control. It is shown in the upper part of Figure 11.33. A more extensive system with a vapour compression chiller and cooling tower for supplemental chilling and direct use of the auxiliary heater for regeneration of the desiccant or for heating in winter is shown in the lower part of Figure 11.33. It is similar to Scheme 1 of Figure 7.29. There are several possible different

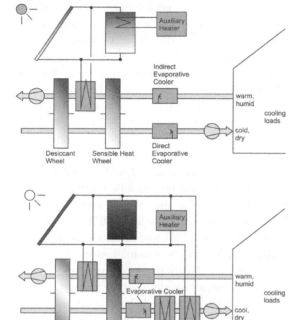

Figure 11.33 (a) Layout of a solar-powered solid-desiccant cooling and dehumidification system, as constructed at a test site in Saxony, Germany. (b) Layout of a solar-powered system with a supplemental electric chiller and cooling tower, and direct connection of the auxiliary heating to the desiccant wheel, as needed for use in hot and humid climates

Source: Modified from Henning et al (2001)

modes of operation for desiccant cooling: direct use of solar energy; storage of solar energy for later use; use of stored energy only; direct use of the auxiliary heater; or simultaneous use of some combination of stored energy, the auxiliary heater, and solar energy. Control systems and sensors are therefore important in optimizing the overall performance of the system. The control of a solar desiccant cooling system is a demanding task, due both to a variable heat source (solar collector) and the need to control the amount of cooling supplied (Balaras et al, 2003).

Kessling et al (1998) discuss solar-powered liquid desiccant cooling and dehumidification systems. A conventional liquid desiccant cooling system was presented in Chapter 7 (Section 7.4.12). The elements of a solar-powered liquid desiccant system are shown in Figure 11.34. A concentrated liquid desiccant solution is sprayed into the incoming air stream, or flows as a thin film on a porous structure through which the incoming air flows. Water vapour is absorbed from the air by the desiccant, but the desiccant is warmed in the process. This would limit the absorption of water vapour. To enhance the dehumidification, and because the incoming air may need to be

Figure 11.34 Outline of a solar-powered liquid desiccant cooling and dehumidification system, as proposed by Kessling et al (1998)

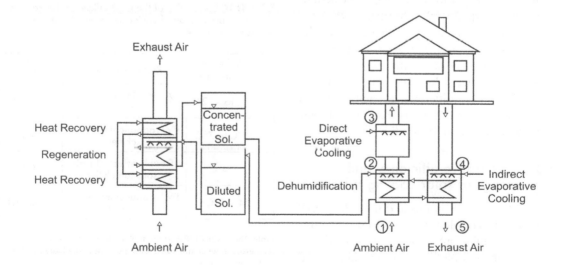

Source: Redrafted from Kessling et al (1998)

cooled in any case for thermal comfort, indirect evaporative cooling can be used (the outgoing air stream is cooled through evaporation, and the coldness is transferred to the incoming air with a heat exchanger but without adding moisture). If the incoming air has been overdried, it can be cooled further through direct evaporative cooling, without exceeding the allowable humidity. The diluted desiccant is stored until solar heat is available to regenerate it, which is accomplished by spraying the desiccant into a hot air stream, provoking evaporation of water. The concentrated desiccant is stored until needed. Heat from the air used to regenerate the desiccant is captured and used to preheat the regeneration air. One advantage of a liquid desiccant system over a solid desiccant system is that dehumidification capacity can be stored as regenerated liquid desiccant.

Gommed and Grossman (2004) simulated the behaviour of a solar powered liquid desiccant dehumidification system with a 20m² collector area designed to dehumidify air for an office building in Haifa, Israel (a heat pump is used to cool the air after it is dehumidified). A combination of a 120l tank for storing concentrated (43 per cent) LiCl and a 1000l hot-water tank allows 4 hours of operation without solar radiation. The thermal COP – defined as the latent heat removed divided by the heat added to the desiccant to regenerate it – ranges from 0.85 at a regeneration temperature of 35°C and a cooling-water temperature of 30°C, to about 0.7 at a regeneration temperature of 80°C. The system COP would be equal to the thermal COP times the solar collector efficiency (which is lower the warmer the regeneration temperature) times the hot-water thermal storage efficiency. Although the COPs are smaller at higher regeneration temperature, the dehumidification capacity is much larger (solid desiccant systems exhibit a similar behaviour, as explained in Box 6.5).

11.5.5 Comparative efficiency and energy savings of solar cooling systems

Syed et al (2002) have computed the solar COP for PV panels coupled to an electric vapour compression chiller and for various kinds of solar collectors coupled to various kinds and configurations (i.e. number of stages) of absorption

cooling systems or to various adsorption systems, for cooling temperatures ranging from −40°C to 20°C (any thermal storage losses in heat driven systems are not included). The vapour compression chiller spans the range −30°C to 20°C, while the various sorption systems are effective within much smaller ranges. The COP of all systems is lower the deeper the desired cooling, but the dependence on cooling temperature is generally weaker for sorption systems. Table 11.14 lists the solar cooling system with the highest solar COP for various cooling temperature ranges, assuming a 15 per cent PV sunlight-to-AC efficiency and using an evacuated-tube collector or compound parabolic collector for heat driven systems. Also given is the associated solar COP. A double-effect LiBr chiller gives the largest solar COP for chilling to about 7–15°C, the single-effect ammonia chiller gives the largest solar COP for chilling to about −12°C to −17°C, and a vapour compression chiller coupled to PV modules gives the largest solar COP for all other temperature ranges. The solar COP of active desiccant dehumidification cooling is in the range of 0.4–0.7 while achieving cooling down to 16–18°C, a region where the PV/electric chiller is best if the PV module efficiency is 15 per cent. However, desiccant cooling easily allows independent control of the temperature and humidity without resorting to reheating.

A potential advantage of solar thermal air conditioning systems over PV vapour

Table 11.14 Solar COP of the most efficient solar cooling option for production of chilled water within different temperature ranges

Chilled-water temperature	Preferred system	Solar COP
−17°C to −12°C	Ammonia absorption chiller	0.4–0.45
−12°C to 7°C	PV/vapour-compression chiller	0.4–0.65
7–14°C	Double-effect LiBr absorption chiller	0.7–0.9
14–20°C	PV/vapour-compression chiller	0.85–1.1

Note: The lower COP pertains to the lower chilled-water temperature, while the upper COP pertains to the higher chilled-water temperature.
Source: Based on calculations by Syed et al (2002)

Table 11.15 Electricity and thermal energy (kWh) used in vapour-compression and various thermally driven cooling systems, including electricity used by the cooling tower and air handler. Thermal energy is energy supplied to the airflow, to the absorption or adsorption chiller, or to the desiccant wheel

	System			
	Conventional	Absorption	Adsorption	Desiccant
Copenhagen, cooling load = 5446kWh				
Electricity	3661	2223	2239	2363
Thermal energy	3164	12,141	13,465	15,654
Freiburg, cooling load = 13,659kWh				
Electricity	5623	2380	2415	2631
Thermal energy	6361	25,623	28,577	26,675
Trapani, cooling load = 85,958kWh				
Electricity	28,351	6349	6571	10,670
Thermal energy	24,496	162,021	180,504	81,779

Source: Henning and Hindenburg (1999)

compression systems is that the collectors that are used to collect solar heat for air conditioning in the summer can be used for active space heating in the winter.

Henning and Hindenburg (1999) present an analysis of the performance of solar-assisted cooling systems using absorption, adsorption, and desiccant chillers. Table 11.15 compares the annual thermal and electrical energy use for these systems with no solar contribution and for a conventional system with overcooling and reheating for dehumidification and a heat recovery wheel (effectiveness of 80 per cent). Results are for a typical office room in three different climates with an all-air cooling system. Cooling towers are required for the conventional, absorption, and adsorption systems. In Copenhagen, where cooling loads are small, the electricity used by the absorption or adsorption chiller pumps and the additional electricity used by the cooling towers is about two-thirds of the electricity used in the conventional system, so the net electricity savings is only about one third. For the desiccant cooling system, two thirds of the savings in chiller electricity use is offset by the increased fan energy due to the additional flow resistance (a pressure drop of 1700Pa instead of 1100Pa). In Freiburg (Germany), the fractional savings in electricity use with the thermally driven systems is 40–50 per cent, while in Trapani (Italy), the savings is about 60–80 per cent. Not all the heat used in the thermally driven systems is an additional energy

requirement, as some heat is used by the conventional system too.

From the data given in Table 11.15, the primary energy use for each system has been computed here assuming an electricity-generation efficiency of 0.35 and a boiler efficiency of 0.9. When solar energy is used to supply a portion of the required heat, the boiler primary energy use is accordingly reduced. Figure 11.35 gives the ratio of primary energy used by various solar-cooling systems with various solar fractions, to the primary energy used by the conventional system. For Trapani, results are also given assuming an electricity-generation efficiency of 0.55. The solar fraction must reach at least 0.3–0.5 (depending on the location and system) just to break even in terms of primary energy use. For a 50 per cent saving in primary energy, the solar fraction must reach 70–85 per cent, depending on the kind of system and location. The desiccant system saves more primary energy than the other systems at low (50–70 per cent) solar fraction. For desiccant cooling in Trapani, the solar fraction required to save 50 per cent of primary energy is about 70 per cent for an electricity-generation efficiency of 0.35 and about 85 per cent for an efficiency of 0.85. However, a large solar fraction requires a large collector area, as illustrated in Figure 11.36 for solar-desiccant cooling with flat-plate collectors and a 10m³ hot-water storage capacity to buffer the variable solar energy input. At 85 per cent solar fraction, the required collector area

Figure 11.35 Ratio of primary energy used by various thermally driven cooling systems, to the primary energy used by a vapour compression system with reheating for humidity control and a sensible heat wheel. Results are given for three cities in Europe for solar fractions ranging from 0.0 to 1.0, assuming an electricity generation efficiency of 0.35 and (in the case of Trapani) 0.55. Ratios are computed from the data given in Table 11.15

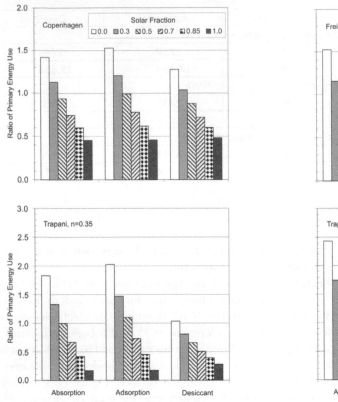
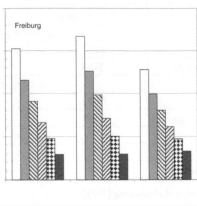
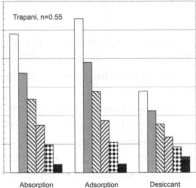

Source: Granqvist et al (1998)

Figure 11.36 Areas of flat-plate solar collector required for a desiccant chilling system in Copenhagen, Freiburg and Trapani when solar energy provides various fractions of the total thermal energy input. The capacity of the cooling system is indicated below the name of the city. The collector areas are based on simulations by Henning and Hindenburg (1999) and assume a hot-water thermal storage reservoir of 10m³

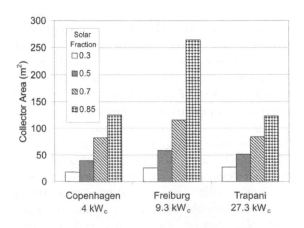

in Copenhagen and Freiburg is about 30m² per kW of cooling capacity, and about 4.5m²/kW$_c$ in Trapani. Given flat-plate collector costs of at least $200–400/m² in Europe and the US, and chiller costs of $300/kW$_c$ (conventional) to $700/kW$_c$ (desiccant) in the capacity range considered here, it can be seen that the total cost will often be dominated by the collector cost. This large cost component would be eliminated if solar collectors are installed for winter space heating and year-round domestic hot water, and the cooling load is such that only excess thermal output is needed during the summer. As noted in Section 11.5.1, collector costs are substantially lower in China.

11.5.6 Hybrid electric thermal cooling using parabolic solar collectors

Gordon and Ng (2000) proposed a hybrid electric thermal system for chilling using solar energy that should be able to achieve a substantially higher solar COP than is possible using a PV electric chiller or a thermal collector-absorption chiller combination alone. The key is to begin with high temperature (1000°C) thermal energy and to cascade the energy flow to the maximum possible extent. The proposal makes use of small (0.2m diameter), roof-mounted, mass-produced (and therefore inexpensive) parabolic dishes, each of which would concentrate sunlight onto a fibre-optic cable. A high solar collection efficiency (0.8) would be achieved as the mini-dishes themselves would remain at a low temperature. The fibre-optic cables would direct energy to a central receiver facility, where a hot gas at a temperature of 1000°C would be created. After heat exchange, hot gas at a temperature of 800°C would be supplied to a micro-turbine. Such turbines have an efficiency (at present) of about 0.23 if ambient air is drawn through the combustion chamber and there is no heat recovery from the exhaust gas. However, by using waste heat to preheat air to 185°C, the thermal-to-electricity efficiency would be increased to 0.35 for present turbines and to 0.42 for future, more efficient turbines. Electricity from the turbine drives a vapour compression chiller with a COP of 3.0–4.0, while waste heat from the turbine at 360°C (minus 10 per cent losses) drives a double-effect absorption chiller with a COP of 1.35. Both the electric and absorption chillers would produced chiller water

at 5°C. The net solar COP would be 1.4–1.7 with present turbines and 1.5–1.8 with future turbines. The energy flow in the latter case is shown in Figure 11.37.

During night-time, an auxiliary natural gas burner could be used to drive the micro-turbine. The chilled water from the absorption chiller would be used to cool the condenser of the electric chiller, which could then be used to make ice for daytime chilling. The electric chiller would operate at evaporator and condenser temperatures of –20°C and 5°C, respectively, which would yield an electric chiller COP of 5.6. Given a 90 per cent natural gas-to-heat efficiency and a turbine heat-to-electricity efficiency of 0.35–0.42, the primary-energy COP is 1.8–2.1.

11.6 Solar cogeneration: Integrated PV modules and thermal collectors

PV panels and solar thermal collectors can be combined to form an integrated PV/thermal (PV/T) collector. This can be done in a number of different ways, including:

- by using a PV module as the outer cladding of an airflow thermal solar collector (as noted in Chapter 4, Section 4.1.10);
- by pasting solar cells directly onto the black absorbing surface of a solar water heater.

In the first case, heat is carried away by air flowing underneath the PV modules, while in the second case, heat is carried by flowing water. As of 2002, eight companies worldwide produced commercial PV/T products, five supplying air-cooled systems and three supplying water-cooled systems (IEA, 2002b). The PV and thermal components of an integrated PV/T system share the same supporting system, and so will achieve some cost savings compared to separate components. There will also be a savings in total collector area compared to separate PV and thermal collectors. Bazilian et al (2001) provide information on buildings with integrated PV/T collectors and on commercially available products. The PV/T collector is likely to experience larger temperature swings than standard PV modules. All materials used,

Figure 11.37 Energy flow in a hybrid electric thermal solar cooling system proposed by Gordon and Ng (2000)

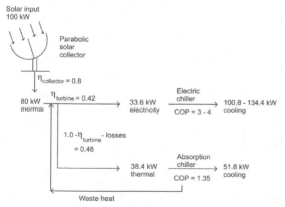

including the encapsulation and adhesive material, must withstand these temperatures without significant degradation of properties such as optical transparency and elasticity. The solar cell material is brittle, so a very elastic adhesive must be used when there are differences in thermal expansion between the cells and absorber plate. In addition to being elastic, the adhesive should have a high thermal conductivity for effective cooling of the solar cells (Sandnes and Rekstad, 2002).

Water is much more effective than air in removing heat, but whatever is used to remove heat, the PV module can be uncovered, or it can be covered by one or two glazing layers (with air gaps) in order to increase its effectiveness in collecting thermal energy. The impact on electricity production of combining PV and thermal energy collection depends on two factors: (i) the additional reflection and absorption of solar radiation by any glazing layers; and (ii) the effect of the glazing layers and of heat removal by air or water flow on the temperature of the PV modules (efficiency decreases as temperature increases, at least for x-Si modules). The first effect dominates if glazing layers are present, so the electrical efficiency of a glazing-covered PV/T collector will be less than that of uncovered stand-alone PV module. However, most commercial PV modules have a glass cover (but without an air gap) for protection of the cells and to provide rigidity, so the PV/T collector will experience additional electricity loss only if it has more than one glazing layer. The thermal efficiency of a PV/T collector is less than that of a stand-alone thermal collector because (i) PV modules have a lower absorptance than the usual collector plate; (ii) there is poorer thermal contact with the heat-carrying fluid; and (iii) some energy is extracted to generate electricity.

The relative efficiencies of stand-alone PV modules, stand-alone thermal collectors, and several different designs for integrated PV/T systems are illustrated in Table 11.16, based on measurements made by Zondag et al (2003). As expected, the integrated PV/T units have a lower thermal efficiency than a stand-alone thermal collector, and a lower electrical efficiency than a stand-alone PV unit except for the uncovered sheet-and-tube design. In the latter case, a metal sheet is placed underneath the PV module, and tubes through which water flows are attached to the metal sheet. Electrical output is almost one-tenth greater than for a stand-alone PV unit (due to the cooling effect of the water flow), but the thermal output is less than half that of a stand-alone thermal collector (due largely to the absence of a cover). Also given in Table 11.16 is the

Table 11.16 Measured annual efficiency of different PV/T systems

System	Annual thermal efficiency	Annual electrical efficiency	Displaced primary energy
Stand-alone PV	–	0.072	0.131
Sheet-and-tube PV/T, uncovered	0.24	0.076	0.404
Sheet-and-tube PV/T, 1 cover	0.35	0.066	0.509
Sheet-and-tube PV/T, 2 covers	0.38	0.058	0.528
Glass/air/glass/water/PV panel	0.38	0.061	0.533
Glass/air/opaque PV/water	0.35	0.067	0.511
Glass/air/transparent PV/water	0.37	0.065	0.529
Glass/air/water/PV panel	0.34	0.063	0.492
Glass/air/glass/water/transparent PV/air/absorber/water	0.39	0.061	0.544
Glass/air/glass/water/transparent PV/air/transparent insulation/ absorber/water	0.37	0.061	0.522
Stand-alone thermal collector	0.51	–	0.567

Note: The displaced primary energy (as a fraction of the incident solar energy) has been computed assuming the efficiency of the furnace or boiler it displaces to be 90% and the efficiency of central electricity generation to be 55%, corresponding to the current state-of-the-art. For smaller efficiencies, the savings of primary energy would be greater.
Source for thermal and electrical efficiencies: Zondag et al (2003)

Figure 11.38 Variation in (a) thermal, and (b) electrical efficiency with water mass flow rate for a covered tube-and-sheet PV/T collector, as simulated by Chow (2003)

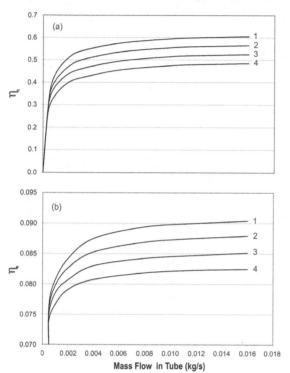

Source: Redrafted from Chow (2003)

Table 11.17 Comparison of annual efficiency of a stand-alone thermal collector, stand-alone PV unit and integrated PV unit

	Thermal Collector	PV Unit	PV/T unit
Heat	0.462	–	0.320
Electricity	–	0.107	0.106
Total	0.462	0.107	0.426

Source: Saitoh et al (2003)

a detailed dynamic model by Chow (2003). As the flow rate increases, the thermal efficiency increases, which implies that the collector temperature decreases (see Equation 11.1), which in turn leads to an increase in the electrical efficiency.

Saitoh et al (2003) compared the energy production of PV, thermal and integrated PV/T collectors in an experimental setup at Hokkaido University, Japan. The PV/T collector has a single glazing layer and a tube-and-sheet collector below the solar cells. Results are summarized in Table 11.17. In this case, the loss in electricity output in a PV/T unit compared to a PV unit is negligible, while the drop in thermal output for a PV/T unit compared to a stand-alone thermal collector is close to the electricity output. Thus, total energy output for the PV/T and thermal units is almost the same, but with about one-quarter occurring as electricity for the PV/T units. Stand-alone PV units could be installed along with PV/T units to match the electricity:heat demand ratio of the building in question.

Chow et al (2003) compared the electricity output and building heat gain for BiPV, PV/T and PV/C units on a 30-storey hotel at 22° latitude in south China. The PV/T collects heat through a warm air stream behind the solar cells, which enters at the bottom and exits at the top, and so functions as a solar chimney. The 'PV/C' unit is a PV/T unit except that the sides are open so as to permit free ventilation with ambient air by wind. The purpose in this case is not to collect heat, but to keep the PV panel as cool as possible. The analysis is based on simulations with the ESP-r computer model (see Appendix D). The annual electricity output is almost identical for the PV/T and PV/C options, and is always slightly larger than for the BiPV unit. This is in contrast to the empirical study of Zondag et al (2003),

savings in primary energy as a fraction of the incident solar radiation, assuming that the collected thermal energy displaces heat that would be generated in a furnace or boiler at 90 per cent efficiency, and that the electricity displaces electricity that would be generated in a fossil fuel plant at 55 per cent efficiency. These efficiencies pertain to the current state-of-the-art. If the PV/T collector displaces heat and electricity that is produced less efficiently, the primary energy savings would be larger. The PV/T units save less primary energy than a stand-alone thermal collector when high efficiency for fossil fuel generation of electricity is assumed, but would save more primary energy in most cases assuming an efficiency typical of current power plants (35–40 per cent).

The variation of electrical and thermal efficiency with water flow rate for one particular design – a covered tube-and-sheet collector – is shown in Figure 11.38, based on calculations with

and could be a result of the equipment input assumptions or of the hotter assumed climate. More significantly, the BiPV unit reduces the heat transfer through the wall into the building by about 30 per cent compared to a conventional building envelope, with almost another 20 per cent savings for either the PV/T or PV/C units. The difference in heat gain between PV/T and PV/C is negligible, so a PV/T unit can be used to collect heat for preheating of water for consumptive use without noticeably increasing the cooling load (compared to PV/C) in hot climates.

Krauter et al (2001) have carried the integration process one step further, having constructed, studied and simulated a system consisting of a PV module, thermal collector and insulation. For a case with 700W/m^2 solar irradiance and 20°C air temperature, the following cell temperatures were obtained: for a PV module mounted directly to insulation, which is attached directly to the exterior of a wall, 77°C; for a PV module with a naturally ventilated air gap, 56°C; for a PV module with a mechanically ventilated air gap, 40°C; for a PV module with insulation and water cooling, 36°C. The electricity production for cases two, three and four is 110 per cent, 119 per cent and 120 per cent, respectively, that of case one. With a mechanically ventilated, 1.2m long and 5cm deep gap, the ventilation air is warmed by 7.4K. For a water-cooled module, the water is warmed by only 1.9K. There are few circumstances where this heat would be useful (given the 20°C ambient conditions).

Because electricity is the more valuable output, one will want to maximize its output by minimizing the temperature of the thermal output (i.e. by using only one glass cover-plate rather than two). This is possible in low-temperature heat distribution systems, but domestic hot water requires warmer temperatures. If heat is accumulated over the course of a day in a thermal storage tank, then the temperature of the water flowing from the tank to the thermal collector will increase over the course of a day. If, on the other hand, the heat is used as it is collected (for example, for radiant floor heating, where the floor slabs can store heat for night-time), then the average collector temperature will be lower. This will increase both the collector and PV efficiencies. Simulations by Sandnes and Rekstad (2002)

indicate that integration of a PV/T unit with floor radiant heating can increase the daily electricity output by up to 10 per cent compared to the case where the heat is accumulated.

If tilted solar collectors are arranged in parallel rows on a flat roof, with adequate spacing to avoid self-shading most of the time, the placement of diffuse reflectors between the rows can increase the electrical and thermal output by about 15 per cent in winter and by about 20 per cent in summer (Tripanagnostopoulos et al, 2002). Alternatively, solar energy can be concentrated with mirrors onto PV/T modules in order to decrease the required module area. For a system with concentration by a factor of 37, Coventry (2005) measured electrical and thermal efficiencies of 11 per cent and 58 per cent, respectively, while producing hot water at 65°C.

Bakker et al (2005) investigated the performance of the hypothetical system for a one-family house in Holland consisting of 25m^2 of PV/T collector, a 200l hot-water tank, a heat pump, two 35m deep ground loops, and radiant floor heating. The PV/T collectors regenerate ground temperatures after winter draw down, and preheat water, which is then warmed to 55°C with the heat pump and once per week to 65°C with an electric-resistance heater to control legionella bacteria. The system provides 100 per cent of the space and tap-water heating loads (19.4GJ/year), with the electricity output from the PV/T covering 95 per cent of the electricity required by the heat pump. At present, the total cost of the system is 27830€, of which 3000€ are for the heat pump, 2750€ for the ground loops, and 19750€ total installed cost for the PV/T collectors (790€/m^2). At 3 per cent interest and a 20-year lifespan, the cost of heat is 0.35€/kWh.

11.7 Energy payback time for PV, PV/T and thermal modules

The process of producing x-Si cells is energy intensive, as a key step involves heating sand to the point where it melts. The energy-intensity of a-Si cells is much less, due to the vastly smaller amounts of material used. In the future, the energy intensity of both kinds of cells will drop due to better integration of the various steps used in the production process and due to other

Figure 11.39 Energy payback times for present and possible future solar PV systems using (a) polycrystalline silicon, and (b) amorphous silicon

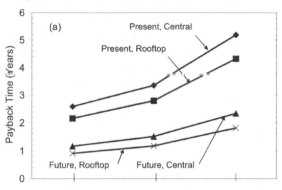

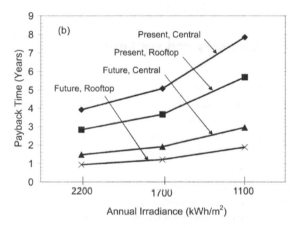

Annual Irradiance (kWh/m²)

advances. In stand-alone systems, there is a significant amount of energy embodied in the steel used for the support structures, something that is avoided in BiPV. The energy payback time for a PV module is the primary energy required to make the module (and any support structures) divided by the primary energy used to produce the electrical energy that is now generated by the module per year. The primary energy savings thus depends on the efficiency with which avoided grid electricity is generated, which varies with the time of day (see Chapter 6, Section 6.9.2). Most of the manufacturing energy is electrical. The payback time will be shorter the less energy that was required to produce the module, the greater its efficiency, and the greater the amount of sunlight striking the module. The latter depends on its location and orientation.

Figure 11.39 gives energy payback times for stand-alone and roof-integrated p-Si and a-Si modules with an assumed efficiency of 10 per

cent, for solar energy incident on the panel of 2200kWh/m²/year (typical of the Sahara desert and south-west US), 1700kWh/m²/year (typical of southern Europe and much of the US), and 1100kWh/m²/year (typical of middle Europe). These payback times do not include credits for any displaced roofing materials (as can be seen from Chapter 12, Section 12.3, such credits would be small compared to the embodied energy of x-Si or a-Si modules, taken here to be 5500MJ/m² and 2900MJ/m², respectively). For present manufacturing energy intensity, the payback periods for roof-integrated panels range from just over two years under south-west US conditions to over four years under middle European conditions, and they are worse for stand-alone power plants due to the need for separate support structures. However, in the future, when embodied energies are lower, the payback times could drop to the 1–2 year range for roof-integrated systems.

Keoleian and Lewis (2003) have computed the ratio of electrical energy produced over a 20-year period to the primary energy input required to manufacture triple-junction a-Si roof shingles and metal roofing, what they call production efficiencies. For various US cities, this ratio ranges from 3.6 to 5.9. The time required to save an amount of primary energy equal to that needed to manufacture the modules depends on the primary energy that would be required to produce the electricity that can now be produced with the PV modules. This in turn depends on the efficiency of the power plants that would supply the electricity that is no longer needed. Since PV electricity will mostly be available during times of peak demand (at least in jurisdictions where the peak load is during the summer), the appropriate efficiency to use is that of the power plants that supply peak (rather than baseload) demand. These efficiencies can be as low as 25 per cent. The energy payback times corresponding to 20-year production efficiencies of 3.6 and 5.9 are 1.4 and 0.85 years, respectively, assuming a 25 per cent efficiency for the displaced electricity. These payback times do not include a credit for the displaced conventional building materials.

Tripanagnostopoulos et al (2003, 2004) have computed the energy payback for PV modules in Patras, Greece, on a horizontal roof with no required support structure but using reflectors (as discussed in Section 11.6), and compared this with that of a water-cooled PV/T collector on a flat roof with the same reflectors. The calculated payback time for PV modules is 3.9 years, compared to 1.7–3.6 years for the PV/T (the thermal and electrical energy output is lower, and the payback times longer, the greater the operating temperature of the collector). For air-cooled collectors, the payback period is about 1.2 years, assuming that the heated air is used for space heating for six months of the year and for pre-heating water for DHW the other six months of the year. For roof-integrated tilted modules in Rome, Battisti and Corrado (2005) compute energy payback times of 3.3 years for monocrystalline PV modules, 2.8 years for PV/T modules with the thermal output used for space heating, 2.3 years for PV/T modules displacing natural gas for DHW, and 1.7 years for PV/T modules displacing electricity for DHW.

With regard to complete solar DHW systems, the energy payback time requires accounting for any difference in the size of the hot-water storage tank compared to the non-solar system and the energy used to manufacture the tank. Conventional DHW systems may or may not have a storage tank. Nevertheless, Crawford et al (2003) find that the energy payback time for a solar gas system in southern Australia is 2–2.5 years, in spite of the embodied energy being 12 times that of a tankless gas system.[1] Ardente et al (2005) compute an energy payback time of 1.3–4.0 years for an integrated thermosyphon flat-plate solar collector and storage device operating in Palermo, Italy.

11.8 Building-integrated wind turbines

Wind energy has emerged as one of the most cost-effective forms of renewable energy, with 8145MW of capacity installed worldwide in 2004, bringing the cumulative capacity to nearly 48,000MW (Cameron, 2005). However, this capacity has been overwhelmingly as large (1–3MW) turbines situated outside urban areas or offshore, often as large (100–200MW) wind farms, where economies of scale reduce the cost. In sites with good wind, electricity can be generated at a cost of 4–6 cents/kWh. The wind turbines invariably involve three blades rotating on a horizontal axis

at 15–20 revolutions per minute, with rotor diameters of the order of 60–100m.

There has been some interest in smaller, vertical axis turbines mounted on top of buildings or integrated into the corners of buildings. Vertical axis turbines would be less affected by changes in wind direction and would produce less vibration than horizontal axis turbines. In some situations, wind speeds could be enhanced by the presence of buildings, but turbulence may also be increased. Full integration of wind turbines into the built environment would require optimizing the shape and layout of buildings. Vertical axis wind turbines suitable for use with buildings are still under development.

11.9 Summary and synthesis

The active use of solar energy by buildings involves roof-mounted photovoltaic (PV) panels or building-integrated photovoltaic (BiPV) panels, thermal solar collectors for space heating and/or the production of domestic hot water (DHW), and thermal solar collectors coupled to one of several heat-driven chillers for solar air conditioning. Combined photovoltaic and thermal (PV/T) panels are also available, simultaneously generating electricity and providing useful heat.

A PV module consists of a number of PV cells wired together in series, usually with a metal frame and a glass cover for protection and rigidity. There are three broad categories of PV cells: crystalline (and polycrystalline), thin film (usually amorphous), and nanocrystalline dye cells. The vast majority (93 per cent) of PV modules produced in 2004 were crystalline, with sunlight to AC electricity conversion efficiencies of 10–15 per cent for presently available commercial models, but eventually reaching 20 per cent. Thin-film modules have much lower efficiency (6–7 per cent) but also lower production costs and material requirements, and thin-film materials can be directly deposited onto a variety of building materials, including roof shingles, wall siding and windows. The use of such BiPV panels has become a design element in its own right, opening new architectural and artistic possibilities. When compared against the cost of alternative cladding materials that are used for purely artistic and design purposes, the cost of BiPV is very attractive even at present. The energy payback time (the time for the primary energy saved by a BiPV module to equal the primary energy required to produce it) ranges from just over two years in sunny locations to just over four years under middle European conditions, once credit is taken for the avoided conventional building materials. Nanocrystalline dye cells mimic the process of photosynthesis and might be used to create windows that are transparent to visible solar radiation while converting near infrared solar radiation into electricity. In OECD countries, there is enough wall and roof area available that BiPV could meet anywhere from 15 per cent to 60 per cent of total current domestic electricity demand.

There are three broad categories of solar thermal collector: flat-plate, evacuated-tube and parabolic. Flat-plate collectors can produce hot water at a temperature of 60–95°C and are less expensive than evacuated-tube collectors. The latter can produce hot water (under pressure) well in excess of 100°C. Solar thermal collectors can be used in converting solar energy into useful heat with an efficiency typically ranging from 30 to 60 per cent. For a given hot-water temperature, the efficiency is greater for evacuated-tube than for flat-plate collectors and for collectors with a protective glass layer. Efficiency is greater the lower the temperature of the water entering the collector, which in turn depends in part on the design of the system of which the collector is a part. Solar thermal energy can be used to meet the majority of annual DHW needs in most parts of the world. Combisystems, which provide DHW and space heating, have been designed to meet 20–60 per cent of the annual heating needs in central and northern European locations. Compound parabolic collectors have seen very little use in buildings at present, but could be used in the future in combination with fibre-optic cables for efficient collection of high temperature (1000°C) thermal energy that could be used in microturbines for on-site generation of electricity.

Heat from solar thermal collectors has been used to drive a variety of different heat-driven chilling systems: absorption chillers, adsorption chillers, ejector chillers and both solid and liquid desiccant dehumidification and cooling systems. The solar coefficient of performance (solar COP) of these systems is given by the COP of the

chiller times the efficiency of the solar collector. The system COP takes into account heat losses transferring and storing hot water. It is equal to the cooling provided (heat energy removed) divided by the solar energy incident on the solar collector. Solar COPs tend to be rather low, in the order of 0.05–0.2 for adsorption chillers, 0.08–0.32 for ejector systems, 0.4–0.7 for desiccant systems, and up to 0.9 for double-effect absorption chiller systems. The solar COP for a system consisting of a PV panel and an electric chiller is given by the sunlight to AC electricity conversion efficiency (0.1–0.2) times the electric chiller COP (3–6, depending on size), and can be larger or smaller than for thermal systems. In solar thermal air conditioning, as in applications using thermal collectors for space heating, the most expensive components are the collectors. An advantage of solar thermal air conditioning over solar electric air conditioning is that the same solar collectors that are used for heating in winter can be used for air conditioning in summer, thereby reducing costs and making better use of the available roof area. Hybrid solar electric thermal systems using roof-mounted parabolic solar collectors and small turbines could eventually yield solar COPs of 1.5–1.8.

If natural gas used to provide heat to drive a single-effect absorption chiller when solar heat is inadequate, there can be an increase rather than a decrease in primary energy use compared to exclusive use of an electric vapour compression chiller, depending on the COP of the electric alternative and the fraction of cooling provided by solar heat. This is due to the low COP of absorption chillers. Two solutions to this problem are (i) to operate the absorption chiller as a double-effect chiller when using natural gas and as a single-effect chiller when using solar energy; (ii) to operate a double-effect chiller at all times; (iii) to use an electric vapour compression chiller as backup; or (iv) to accept occasional periods when solar chiller is inadequate with no backup.

Finally, integrated PV and thermal collectors have been used for solar cogeneration – the simultaneous production of electricity and useful heat. The electrical and thermal efficiencies of such combined collectors tend to be lower than the efficiencies of stand-alone PV or thermal collectors, but the overall efficiency is usually higher, as both electricity and useful heat are produced from the same collector area. PV/T collectors can be designed to significantly reduce the heat flow into a building, thereby reducing the cooling load in hot climates at the same time that useful energy is provided.

Note

1   The data for the gas storage and gas instantaneous systems in Table 6 of Crawford et al (2003) were inadvertently interchanged.

twelve

Comparison of Embodied Energy, Operating Energy and Associated Greenhouse Gas Emissions

The *embodied energy* of a building is the energy used to manufacture and transport the materials used in the building, as well as the energy used during the construction process itself. The *operating energy* refers to the energy for heating, cooling, ventilation, lighting and appliances or office machines, which has been the subject of this book so far. The embodied energy depends on the energy intensity of the industries involved in producing building materials, while transportation energy depends on the energy intensity of transportation and the distances transported. There is tremendous scope for reducing industrial and transportation energy intensities (i.e. energy used per unit of output or per tonne-km of transport), and as these energy intensities improve, the embodied energy in new buildings will decrease. Nevertheless, it is appropriate here to briefly identify areas where alternative designs can reduce the embodied energy of buildings, and to examine instances where decreased building operating energy might be associated with increased embodied energy, in order to determine the sign and magnitude of the net change in energy use.

12.1 Determining the embodied energy of building materials

To determine the embodied energy of a building requires determining the embodied energy per unit mass or per unit volume of all the materials that go into a building, multiplying by the amounts of each material used, accounting for energy used during construction and adding all these terms. The energy needed to produce, for example, a tonne of steel involves the energy used by the steel mill, which is the *direct* or *zeroeth-order* embodied energy in the steel. However, energy was also required in extracting the raw materials used by the steel mill and in bringing them to the mill, and in manufacturing the machinery used by the mill. But the facilities where the equipment used by the mill was manufactured, or where the transportation vehicles were produced, also required inputs from an array of other industries, each of which required energy input. In other words, there are a number of *indirect* energy inputs of first order, second order, third order and so on.[1] Fully accounting for the embodied energy in building materials requires accounting for a very large succession of linkages. This can be done in two ways: by explicitly measuring each of a large number of linkages until it is felt that the excluded higher orders are not important, and through an input–output analysis of the entire economy.

An input–output analysis requires dividing the economy into a number of sectors, which could be in the order of 100. Any one sector could, in principle, have inputs from all the other sectors and its output could be an input to all of the other sectors. All of the possible input–output relationships can be represented by an $N \times N$ matrix, where N is the number of sectors. Through manipulations of this input–output matrix, the full cascade of effects on the embodied energy of products from any one sector can be accounted for. The input–output analysis will identify a large number of linkages between individual sectors that all directly or indirectly contribute to the embodied energy of building materials. For example, the energy path tree accounting for 90 per cent of the total embodied energy in Australian residential buildings involves 592 energy paths, while analysis up to and including 12th order involves over 20 million energy paths (Treloar et al, 2001).

There are two problems with this approach: (i) much of the basic data used to construct the matrix are in terms of the dollar-value (or other currency) of the inputs and outputs, and so have to be converted to energy intensities based on sector average energy use (since energy data at a more disaggregated level are usually not available); and (ii) different products from a given sector, or different uses of the same product, are not distinguished. However, the energy use at each step in each path is, for that step, a direct energy use – something that can be evaluated in physical terms. The direct energy use at each step of a given path deduced from the input–output matrix is independent of the energy uses deduced for all the other links. Thus, the problems associated with input–output analysis can be mitigated to some extent by replacing the most important energy intensities as deduced from the input–output matrix with the correct intensities as calculated physically. The amounts of inputs into an output at a given link in the input–output analysis as determined by the input–output analysis also need to be replaced with physically determined quantities. This is referred to as a hybrid approach by Treloar et al (2001), and in their analysis of residential buildings in Australia, 52 per cent of the energy paths in the input–output model were replaced with case-specific process based energy intensities. For the building analysed by Treloar et al (2001), the unaltered input–output analysis indicated an embodied energy of 834.8GJ. Alteration of a portion of the energy intensities increased this to 1154.4GJ, while alteration of a portion of the energy intensities and product quantities increased the total to 1762.4GJ – a factor of two greater than indicated by the original analysis.

On the other hand, to rely entirely on a process based approach will underestimate the true embodied energy, since interactions beyond some relatively low order will be omitted (that is, there is a truncation error). Lenzen and Dey (2000) find that, in the case of iron and steel in Australia, terminating the analysis after considering only zeroeth and first-order terms leads to an estimated energy intensity of 26.05GJ/t, whereas use of an input–output matrix with a process-based analysis for the zeroeth and first-order terms leads to an embodied energy of 40.05GJ/t. According to Lenzen and Treloar (2002), going as far as fourth-order inputs for wood-frame and concrete-frame construction still captures only 75 per cent of the total embodied energy; convergence to the full embodied energy is not reached until 12th-order inputs are included. Since most analysis of embodied energy are process-based analyses that stop after first-order or second-order interactions, these analyses likely significantly underestimate the true embodied energy of buildings.

Apart from these methodological issues, which cast doubt on the accuracy of the vast majority of published studies of embodied energy, there is a large range in published embodied energy values for most building materials. This is partly due to real differences in the embodied energy of the same material produced in different countries, due to differences in the energy efficiency of the manufacturing plants and the generation of electricity, and in the distances that raw materials or final products are transported.

Table 12.1 Embodied energy (GJ/tonne) for building materials

Material	Embodied energy	Material	Embodied energy
Very high energy		Medium energy	
Aluminium	200–500	Lime	3–5
Plastics	50–100	Clay bricks and tiles	2–7
Copper	100+	Gypsum plaster	1–4
Stainless steel	100+	Concrete	
High energy		In-situ	0.8–1.5
Steel	30–60	Blocks	0.8–3.5
Lead, Zinc	25+	Precast	1.5–8
Glass	12–25	Sand-lime bricks	0.8–1.2
Cement	5–9	Timber	0.1–5
Plasterboard	8–10	Low energy	
		Sand, aggregate	<0.5
		Flyash, volcanic ash	<0.5

Source: Thomas (1999)

However, some of the discrepancy is likely due to uncertainties and hence errors in some of the inputs used in the calculations.

In spite of these concerns, it is useful to have a rough idea of the relative and absolute energy intensities of different building materials. Table 12.1 provides such information, with building materials classified as: (i) very high energy intensity (aluminium, plastics, copper, stainless steel); (ii) high energy intensity (steel, glass, cement, plasterboard); (iii) medium energy intensity (clay bricks and tiles, concrete, timber); and (iv) low energy intensity (sand, aggregate, flyash).

12.2 Special considerations with regard to cement and concrete

Concrete is a mixture of cement (anywhere from 8–23 per cent by mass), aggregates (crushed stone and gravel), and various supplementary cementing materials (up to 50 per cent replacement of cement) such as flyash and blast furnace slag. Although the energy intensity of concrete is low compared to that of steel and plastic, enough of it is used in many buildings that it can readily account for 40 per cent or more of the total embodied energy. Furthermore, the chemical reaction that produces cement releases CO_2, which is in addition to the CO_2 emitted while supplying the energy used to produce cement.

Cement is mainly produced from limestone or chalk, and clay. Limestone and chalk are calcareous ($CaCO_3$-containing) materials. The first step in the production of cement is the grinding of the raw materials into a fine powder. This is followed by the *calcination* reaction at a temperature of 700–1000°C:

$$CaCO_3 \rightarrow CaO + CO_2. \qquad (12.1)$$

This is followed by *clinkerization* (a fusion process) at 1400–1450°C to produce *clinker*, which is then ground after cooling and mixed with various additives, such as gypsum, or in the case of blended cements, with blast furnace slag or fly ash. The most common cement is Portland cement, so-called because the first Portland cement (1824) resembled natural stone from the peninsula Portland, in England. It consists of 95 per cent clinker and 5 per cent gypsum. The direct energy use in making cement is about 4–6GJ/tonne, and 80–90 per cent of this energy is fuel for heating. Almost all cement plants burn pulverized coal, but many also burn waste. Assuming the fuel to be coal, the CO_2 emission associated with the production of cement is about 100–150kg of carbon per tonne of cement. By contrast, the CO_2 released during calcination is 203kg of carbon per tonne of Portland cement. This is equivalent, in terms of CO_2 emission, to an additional 8GJ of energy as coal per tonne of cement produced.

The energy used to make concrete involves the energy in extracting, transporting and crushing the aggregates, in producing the cement and in processing the input materials. Additional energy is used in transporting the concrete to the work site. Estimates of the embodied energy of various concrete products in Canada are given in Table 12.2, based on zero- and first-order inputs only (i.e. excluding the energy used to build the manufacturing plants and transportation vehicles). The embodied energy is given for hollow-core concrete slabs and double-T beams (illustrated in Figure 12.1) as well as for solid concrete slabs with the same dimensions as the hollow-core slab, and for a double-T beam scaled to the same width. Also given is the percent of cement by mass in the various forms of concrete, and the mass of cement in solid slabs with the same overall dimensions as the hollow-core slab, in the scaled double-T slab and in the hollow-core slab itself, given a 40 per cent void fraction. In terms of strength, the hollow-core slab would be equivalent to a 40MPa solid slab, which would have a CO_2-equivalent embodied energy of about 1.19GJ/m. By comparison, the CO_2-equivalent embodied energy of the hollow-core slab is 1.16GJ/m, and that of the double-T slab is 0.81GJ/m. The hollow-core slab also requires less steel, but steel accounts for only about one-sixth of the embodied energy of the concrete. Augmenting the benefit of the hollow-core slab is the potential savings in heating and cooling energy through its ability to store mid day solar heat in winter or to make good use of

night ventilation for cooling in summer. Another consideration, for both hollow-core and double-T slabs, is their smaller weight compared to solid slabs, which has ramifications for the strength of materials required in other structural parts of the building, and thus in the associated mass and embodied energy.

The amount of concrete and hence cement used in walls can be reduced through the use of waffle-shaped foundation walls, as in the advanced house in Waterloo, Canada (Carpenter, 2003c). The below-grade walls are precast concrete panels that are flat on the outside but waffle-shaped on the inside. The waffle cavities are filled with cellulose insulation, and an additional 50mm of rigid insulation is applied to the outside, while wood-stud walls with additional insulation are applied on the inside. This system uses only half the concrete of conventional poured basements in Canada.

It is not uncommon to reduce the use of clinker by using up to 50 per cent granulated blast furnace slag or flyash. Blast furnace slag concrete has good compressive strength, making it suitable for mixing into cement used for columns. Cement with a greater fraction of flyash requires a longer curing time, is a more workable mixture and requires different finishing technique compared to

Table 12.2 Cement content and embodied energy of different kinds or uses of concrete as computed for Canada

Type of Concrete	Density (kg per m³ of solid)		% Cement	Embodied energy			Results for slabs with same dimensions as hollow-core slab		
	Concrete	Cement		Fuel	Total	Units	Embodied energy (GJ/m)	Mass of cement (kg/m)	CO_2-Equiv embodied Energy (GJ/m)
15 MPa Ready Mix	2303	191	8.3	1.19	1.36	GJ/m³	0.34	47.3	0.72
20 MPa Ready Mix	2334	218	9.3	1.35	1.54	GJ/m³	0.38	53.9	0.82
30 MPa Ready Mix	2324	319	13.7	1.79	2.04	GJ/m³	0.51	78.9	1.15
60 MPa Ready Mix	2386	352	14.8	2.02	2.29	GJ/m³	0.57	87.1	1.27
Hollow-core	2201	505	22.9	0.478	0.551	GJ/m	0.55	75.0	1.16
Double-T	2201	505	22.9	0.834	0.962	GJ/m	0.39	52.4	0.81
Block	1943	189		0.020	0.023	GJ/block			
Mortar	1277	307		1.45	1.67	GJ/m³			

Note: The hollow-core slab is 4 feet wide (1.219m) and 8″ (0.203m) thick, for a volume of 0.247m³ per m of length, and has a void fraction of 0.4. The double-T slab is 10 feet (3.048m) wide. The difference between fuel and total embodied energy is electrical embodied energy. The final column is the real embodied energy plus the additional embodied energy that would yield the same total CO_2 emissions as produced during the manufacture of the concrete, given a chemical production of CO_2 of 203kg C per kg of cement and converting this to the amount of coal with the same emission using a coal emission factor of 25.0kg C/GJ.
Source for concrete density, cement content, and embodied energy: Venta, Glaser, and Associates (1999)

Figure 12.1 A hollow-core concrete slab (left) and a double-T slab (right)

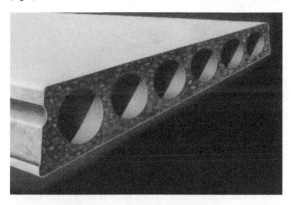

Source: Canadian Precast/Prestressed Concrete Institute, www.cpci.ca

Portland cement. In the case of the Environmental Technology Centre at Sonoma State University (California), the amount of Portland cement used in construction was reduced by 50 per cent by partially substituting flyash and rice hull husk (Beeler, 1998).

12.3 Special considerations for residential roofing materials

Residential roofing materials (asphalt shingles, clay and concrete tiles) deserve special consideration, as all three are energy intensive products, asphalt shingles require substantial amounts of oil products as a feedstock, and concrete tiles entail CO_2 emissions associated with the chemical reaction that produces cement, as well as through the energy used to manufacture them. Asphalt shingles account for over 80 per cent of the residential roofing in the US and 99 per cent in Canada, while clay and concrete tiles are dominant in Europe and elsewhere.

Asphalt shingles begin with a base consisting of either an organic (paper) felt or a fibreglass mat, the former made from a combination of recycled and virgin fibres. Paper felts are saturated with a less viscous asphalt, then coated with a more viscous asphalt, while glass mats require the viscous asphalt only. Asphalt itself is derived from the heavy residue remaining after refining of petroleum into transportation or heating fuels. The asphalt is coated with mineral granules that provide the visually attractive surface, fire resistance and protection of the asphalt from ultraviolet radiation. A fine mineral coating is applied to the backside of the shingles so that they do not stick together while in storage or in transit. Saturated felt shingles consist only of organic felt and saturating asphalt.

The production of clay tiles is similar to the production of clay bricks. Surface clay or shale is extracted, crushed (in the case of shale), mixed with water, deaerated, passed through a dye to create the desired shape, dried over a period of 2–4 days using exhaust heat from firing kilns, then fired to 1100–1200°C over a period of 40–150 hours, The key to successful firing is to induce incipient fusion and partial vitrification but without viscous fusion. Concrete tiles are made with 6–12 per cent cement and 5 per cent water by weight, and rely on consolidation of the input mixture by vibration in order to reduce both the amount of water and of cement that must be used.

Table 12.3 presents estimates for Canada of the embodied energy in various kinds of asphalt shingles, based on the direct embodied energy of the input materials, the energy used at the shingle manufacturing facility, and the average amount of energy used to transport the materials to the manufacturing facility and to transport the product to construction sites (transportation energy is only a small fraction of the total embodied energy). The embodied energies thus represent zero- and first-order terms (energy used at the manufacturing plant and embodied in the inputs), but would neglect most higher-order terms (such as the energy used to produce the manufacturing plant or the vehicles that transport the materials to the plant). Also given in Table 12.3 is the energy equivalent of the oil-based feedstocks that are used in making asphalt shingles. The feedstock energy is several times the energy used in making asphalt shingles. If one is interested in greenhouse gas and other emissions of alternative roofing materials, then the feedstock energy can be ignored, but if one is interested in alternative roofing materials from the standpoint of energy supply, then the feedstock energy should be included.

Also given in Table 12.3 are the embodied energies for clay and concrete tiles. For asphalt shingles and clay tiles, the embodied energies have been broken into electric and fuel energy inputs, as these have very different implications for primary energy use. For concrete tiles, the embodied energy is broken into that used to make cement and that used in other ways. As noted in Section 12.2, the chemical reaction that produces cement releases CO_2 that is in addition to the CO_2 associated with providing energy for the manufacturing process. Assuming the cement embodied energy to be 6GJ/tonne and given a chemical emission of 203kg C/tonne, the amount of natural gas that, if burned, would produce the same CO_2 emission as is produced chemically can be calculated. These amounts are given in Table 12.3, and roughly double CO_2 emissions (most of the

energy used to make concrete tiles is heat, which can be provided by natural gas or other fuels). The final column of Table 12.3 gives the loading (tonnes/100m²) for different kinds of shingles.

Figure 12.2 gives the embodied energy in Canada, per 100m² of roof area, for different kinds of roofing materials. Neglecting (or, in some cases, including) feedstock energy, asphalt shingles have the lowest embodied energy, while clay tiles (especially medium- and heavy-weight tiles) have a particularly large embodied energy. Clay tiles also have a large electricity input, which is often generated at only 33–38 per cent efficiency from coal. Partly offsetting this will be the much longer lifespan of clay tiles (50–100 years) compared to asphalt shingles (13–15 years). Concrete tiles also have long lifespans, but with substantially less emission than clay tiles. Emissions for concrete tiles can be reduced further through the use of fibre-cement composites, which have the advantage of lighter weight. Buchanan and Honey (1994) estimated the primary embodied energy of asphalt shingles in New Zealand to be 28GJ/100m² for strip shingles and 8.5GJ/100m² for rolled shingles (assuming 33 per cent electricity

Table 12.3 Data on the embodied energy of shingles and roofing tiles in Canada, used in computing the results given in Figure 12.3

Type of shingle	Embodied energy (GJ/tonne)				Loading tonnes/100m²
	Fuel	Electricity	Feedstock	Total	
Asphalt shingles					
Felt mat	2.74	0.38	10.32	13.44	1.141
Glass mat	1.99	0.29	7.89	10.17	1.076
#15 felt	8.60	1.42	26.37	36.39	0.073
#30 felt	8.52	1.38	26.92	36.82	0.144
Mineral surface roll	2.82	0.41	10.30	13.53	0.585
Clay tiles					
Light					2.832
Medium	2.66	1.92	0.00	4.58	4.862
Heavy					8.984
Concrete tiles					
	Energy to make cement	Other energy	Natural gas energy equivalent of chemical CO_2 emission	Total equivalent energy input	
Light					2.64
Medium	0.60	0.69	1.54	2.81	4.28
Heavy					5.13

Source: Venta and Nisbet (2000)

Figure 12.2 Embodied energy of different kinds of residential roofing materials in Canada

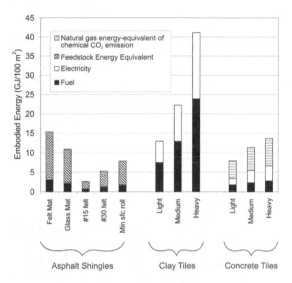

Source: Computed from the data given in Table 12.3

generation efficiency). These energies are substantially in excess of those given above for Canada.

Steel has an embodied energy of 30–60GJ/tonne (Table 12.1), compared to about 1GJ/tonne for cement tiles, 5GJ/tonne for clay tiles and 1.3–10GJ/tonne for asphalt shingles (excluding feedstock energy). In spite of being thinner and hence lighter-weight, and in spite of a credit from recycling at the end of its life equal to about three-quarters of the initial embodied energy, steel roofing is likely to entail greater embodied energy than concrete tiles. It is likely to be competitive with clay tiles and some asphalt shingles in terms of embodied energy.

12.4 Comparison of embodied energy and operating energy

There are many published analyses of the embodied energy of buildings, and many published analyses of the operating energy, but very few in which embodied and operating energy are compared for the same building. Results for three such studies are summarized in Table 12.4. For office buildings in Vancouver and Toronto, the embodied energy is equivalent to only a few years of operating energy, even if the embodied energy

is underestimated by a factor of two. Thus, over a 50-year time span, reducing the operating energy is more important than reducing the embodied energy. For well-insulated detached houses in Sweden, the embodied energy is equivalent to 6–7 years of operating energy. For an advanced house in Gothenburg (Sweden), the embodied energy is equivalent to 43 years of the very low operating energy. However, about one-third of the embodied energy could be recovered through recycling of material when the house is renovated. For a single-family house in Michigan, the embodied energy is given (i) using only the energy required to process wood (taken to be 10.7MJ/kg); and (ii) including the energy value of wood if used as a fuel (17.7MJ/kg), along with the process energy. The conventional house is a 2 × 4 wood-frame structure with R15 (RSI 2.64) fibreglass insulation and asphalt shingles, while the well-insulated house has a double-wall construction with R35 (RSI 6.16) cellulose insulation and recycled plastic/wood composite shingles. The embodied energies of the conventional and well-insulated houses are equivalent to 3–5 years and 10–18 years, respectively, of the operating + maintenance energy.

Trusty and Meil (2000) compared both the embodied energy and the operating energy for a 13-floor office building as constructed according to ASHRAE 90.1 and according to the Canadian C-2000 code for commercial buildings. The C-2000 had slightly greater embodied primary energy, 1.58GJ/m² versus 1.53GJ/m², but 54 per cent smaller operating energy: 82kWh/m²/year versus 180kWh/m²/year. The operating energies are almost all electrical, and if generated and transmitted with an overall efficiency of 33 per cent, correspond to 0.89 and 2.16GJ/m²/yr of primary energy.[2] If a building complying with ASHRAE 90.1 were demolished and rebuilt according to the C-2000 standard, the embodied energy of the new building would be paid back in 1.5 years through the savings in operating energy. If the structure of an existing building were saved and everything else replaced, the saved structural embodied energy would be 1.17GJ/m² and the incurred new embodied energy would be 0.41GJ/m², which would be paid back within a few months.

Table 12.4 Comparison of building embodied and operating energies

Case	Initial embodied energy (GJ/m²)	Renovation embodied energy over 50 years (GJ/m²)	Operating Energy (GJ/m²/year)	Years of operating energy required to equal initial embodied energy	Reference
Conventional offices in Vancouver	4.5–5.1	6.6	1.05	4–5	Cole and Kernon (1996)
Conventional offices in Toronto	4.5–5.1	6.6	1.76	2.5–3	Cole and Kernon (1996)
Conventional offices in Japan	7–13	1.2–2.6[a]	1.2–1.6	6–8	Suzuki and Oka (1998)

Office buildings in Germany
  Typical
  Good practice
  Potsdamer Platz, Berlin
	10
	5
	7[b]
		16.0
		7.0
		15.3
			0.9
			0.45
			0.26
				11
				11
				8[c]
					Richard Rogers Architects (1996)

Well-insulated houses in Sweden	2.9–3.7		0.46–0.51	6–7	Adalberth (1997)
Advanced house in Gothenburg	7.03		0.164	43	Thormark (2002)
Conventional house in Michigan	6.62 3.97		1.27	5.2 3.1	Keoleian et al (2001)
Well-insulated house in Michigan	7.32 4.13		0.41	18 10	Keoleian et al (2001)

[a] Renovation energy computed over a 40-year cycle.
[b] Higher than for good practice due to choice of materials.
[c] Payback time if a typical building is demolished and rebuilt with 'good practice' embodied energy and Potsdamer Platz operating energy.

From this it follows that, for existing buildings with high operating energy, the payback – in energy terms – of demolishing the building and rebuilding it from scratch as an ultra-low-energy building is only a few years. Demolition in such cases is therefore worthwhile from an energy point of view, although there are other considerations that may still rule out demolition, such as the desire to preserve historical buildings or to minimize waste.

12.4.1 Potential trade-offs involving advanced buildings

Our concern here is to identify instances where a change in building design that reduces the lifetime operating energy leads to an increase in the embodied energy. Where such a trade-off occurs, we need to determine in what cases or under what circumstances the increase in embodied energy is greater than the decrease in operating energy, such that the total operating + embodied energy increases rather than decreases. Some potentially unfavourable trade-offs are briefly discussed below.

Concrete versus wood-frame construction

Greater reliance on concrete increases the thermal mass of a building, allowing one to minimize peak daytime temperatures and to take fuller advantage of night-time ventilative cooling in summer, and to maximize the use of passive solar heating with minimal temperature drop at night in winter. The difference in energy use between wood-frame and concrete buildings depends on a large number of assumptions concerning how the wood and concrete are produced and delivered to the site, as discussed by Börjesson and Gustavsson (2000). As a very rough estimate, the difference in embodied energy amounts to about 1GJ per m² of living space in a 20-unit, four-storey residential building in Sweden, with the concrete-frame building having greater embodied energy. To break even over a 50-year period, the operating energy would need to be smaller by $20MJ/m^2/year$ for the concrete-frame building. With operating energies of $120–800MJ/m^2/year$ for Swedish housing, there would appear to be ample opportunity to achieve the required savings through passive-design features that are facilitated by the concrete construction. The longer-term energy and greenhouse gas balance depends on a wide range of additional assumptions concerning what is done with wood waste at the end of the building's life, the extent of recycling of both wood and concrete, and the extent of carbonization of concrete over its lifetime (concrete absorbs back some of the CO_2 released by calcination during the production of concrete, as discussed in Richardson (1988)). Also note that new phase change materials can be used to increase the thermal mass of a building without using concrete, as discussed in Chapter 4 (Section 4.1.6), but the embodied energy in these materials has not been reported.

Alternative façade systems in commercial buildings

Commercial buildings can be constructed with masonry walls, lightweight metal walls, heavyweight concrete walls, curtain walls or double-skin façades. The latter will contain an adjustable shading device between the inner and outer façade, while shading devices of some sort can be optionally applied to the other kinds of walls. Table 12.5 gives the embodied energy (per unit wall area) in the UK for these alternative walls as estimated by Kolokotroni et al (2004), along with the embodied energy for horizontal overhangs, glass louvres and metal louvres, as well as for vertical metal louvres. The wall embodied energy ranges from $770MJ/m^2$ for lightweight cladding to $1610MJ/m^2$ for standard curtain walls (no shading) and $2120MJ/m^2$ for a double-skin façade with horizontal glass louvres. For higher-performance curtain wall or double-skin façades, the embodied energy would be greater. Shading-device embodied energy ranges from $150MJ/m^2$ (vertical metal louvres) to $430MJ/m^2$ (horizontal metal louvres).

High-performance versus conventional windows

This potential trade-off has already been discussed in Chapter 3 (Section 3.3.19). Measures up to and including triple glazing, argon fill and three low-e coatings are well justified by the lifetime savings in operating energy. Where the use of krypton fill permits elimination of perimeter heating units, the savings in heating-unit embodied energy is comparable to the window embodied energy.

Table 12.5 Comparison of the embodied energy of different façades systems for a commercial building in the UK

Façade type	Construction	U-value (W/m²/K)	Embodied Energy (MJ/m²)
Light cladding	Metal, insulation, metal	0.3	769
Masonry	Brick, air cavity, insulation, block, plaster	0.3	1028
Heavy cladding	Concrete, insulation, concrete	0.3	1041
Curtainwall	10mm toughened glass, 12mm air cavity, 10.6mm laminated glass	1.6	1610
Double-skinned	12mm toughened glass, 0.8m cavity, 8mm float glass, 12mm argon, 8.4mm laminated glass	1.1	2120
Additional embodied energy due to shading			
75cm deep horizontal overhang			220
40cm deep horizontal glass louvres			340
10cm deep horizontal metal louvres			430
75cm deep vertical metal louvres			150

Source: Kolokotroni et al (2004)

Ground source heat pump versus water-cooled chillers

Genchi et al (2002) have estimated the energy payback time associated with a ground source heat pump instead of a chiller and cooling tower combination. They account for the embodied energy of the ground pipe loop and the energy required for excavation, as well as the avoided cooling tower embodied energy. For a high-density system (in Tokyo), they compute an energy payback time of less than two years.

Minimal versus maximal insulation thickness

Chen et al (2001) provide estimates of the embodied primary energy in different kinds of insulation in Hong Kong, while Adalberth (1997) and Lenzen and Treloar (2002) provide estimates for Australia. These are summarized in Table 12.6, and indicate considerable differences between different estimates of the embodied energy. Foam insulation is made from petroleum feedstocks that can also be used as an energy source. Table 12.7 compares the energy used in making polyisocyanurate insulation with the feedstock energy; the energy in the feedstock is comparable to the energy used to make the insulation. Table 12.8 gives embodied primary energy on a volumetric basis for different insulation materials (ignoring feedstock energy in the case of petroleum-based products). Along with thermal conductivity values, this can be used to compute the embodied energy for a 1m²

Table 12.6 Embodied primary energy (GJ/t) of different insulation materials

	Chen et al (2001)	Adalberth (1997)	Lenzen and Treloar (2002)
Cellulose	3.3		
Fibreglass	30.3		
Polyester	53.7		
Glass wool	14.0		93.1
Mineral wool		19.2	93.1
Polystyrene	105	106.7	
Urea formaldehyde	78.2		

Table 12.7 Energy and feedstocks used to produce a 1kg of polyisocyanurate insulation in Canada

Input	Used as feedstock		Used as energy (MJ)	Total embodied energy (MJ)
	Mass (gm)	Energy equivalent (MJ)		
Oil	482	20.2	4.0	24.2
Natural gas	203	10.1	19.3	29.4
Coal			9.3	9.3
Other			3.4	3.4
Total	685	30.3	35.9	66.2

Note: Feedstocks have been converted to energy equivalents using 1gm oil = 42kJ and 1gm natural gas = 50kJ.
Source: Franklin Associates (2001)

Table 12.8 Embodied primary energy for a 1m² insulation panel with an RSI value of 1, computed from embodied energy per unit volume and thermal conductivity

Type of insulation	Density (kg/m³)	Embodied energy		Conductivity	Area (m²) of RSI-1 panels made from 1m³	Embodied energy (MJ/m²/RSI)
		(MJ/kg)	(MJ/m³)	(W/m/K)		
Cellulose	40–70	0.9	36–61	0.045	22.2	1.6–2.8
Fibreboard	190–240	11.2–11.8	2124–2826	0.053–0.045	18.9–22.2	113–127
Polystyrene	15–30	127	1900–3780	0.032–0.030	31.3–33.3	61–113
Polyurethane	30–35	137	4104–4788	0.035–0.020	28.6–50.0	144–96
Mineral wool	20–140	18	360–2520	0.045–0.035	22.2–28.6	16–88
Mineral wool						40
Fibreglass						16.5

Note: Embodied energies (as MJ/m³), densities and thermal conductivities are from Petersdorff et al (2002) and are used to compute other entries, except for polystyrene thermal conductivities (which are from BASF (2003)) and embodied energies in the last two rows, which are from Norris (1998).

panel with an RSI-value of 1, a quantity that is also given in Table 12.8. As the density increases, the material requirements and hence embodied energy increases, while the thermal conductivity decreases, but by less than the proportional increase in density. Thus, pushing a given insulation type to a lower thermal conductivity by increasing the density (in order to reduce the space requirements) increases the embodied energy for a given RSI increment (as well as increasing the cost). The polystyrene thermal conductivities given in Table 12.8 are for a pentane-blown commercial product called Neopor® that contains a low-e coating to minimize infrared radiative transfer within the insulation matrix, and is available in small quantities in Europe. This reduces the thermal conductivity by 10–15 per cent (BASF, 2003).

Payback times for insulation can be computed based on the savings in heating energy and the embodied energy in the full thickness of insulation, or can be based on the additional savings in heating energy and the additional embodied energy when an extra increment of insulation is added. In analogy to economic cost/benefit analysis, the latter will be called the *marginal payback time*. Because the absolute reduction in heat loss with successive increments of insulation decreases as more insulation is added (as seen from Figure 3.2), the marginal payback time is longer for a given increment of insulation the greater the pre-existing insulation level. At any level of insulation, the marginal payback time is longer than the overall payback time. The overall payback time is a useful indicator of the value of a given amount of insulation, while the marginal payback time is useful in deciding when (on a lifecycle energy basis) to stop adding more insulation: the amount of insulation should not be increased beyond the point where the marginal payback time equals the expected lifespan of the insulation.

Figure 12.3 gives overall and marginal payback times based on the primary energy needed to manufacture (but not transport or install) various kinds of insulation for a climate with 4000 heating degree-days (HDDs) and a heating-system efficiency (η) of 0.9. As indicated in Table 2.5, a climate with 4000 HDD pertains to regions with moderately cold winters. The saving in heating energy is computed as:

Figure 12.3 Variation in the overall energy payback time as the total RSI is increased (a) from 0.5 or (b) from 3.0 to values as large as 10.0, and (c) variation in the marginal energy payback time when the RSI value is increased by 1.0 to the indicated final value

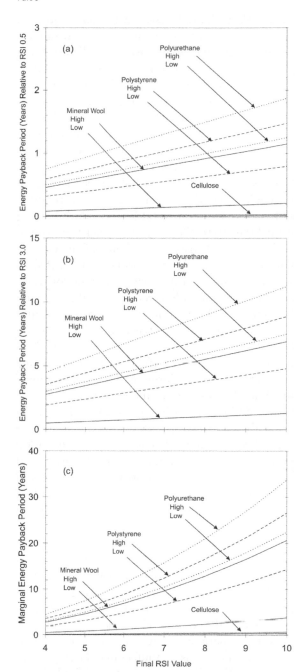

Note: Energy payback times are for a climate with 4000 HDD and a heating system efficiency of 0.9, and are based on the embodied energies given in Table 12.8.

$$Saving(J) = HDD \times 24 \times 3600 \times \Delta U / \eta$$

$$(12.2)$$

where ΔU is the difference in U-values with and without foam insulation (or with and without a given increment of foam insulation). Figure 12.3a gives overall payback times relative to an uninsulated brick or masonry wall with an RSI value of 0.5. Even for the most energy-intensive insulation, the overall payback time is less than 2 years for an RSI of up to 10. An alternative base case is a wood-frame structure with internal insulation giving an effective RSI value (including thermal bridges) of 3.0. Figure 12.3b gives the overall payback times when external insulation is added sufficient to bring the total RSI to values ranging from 4.0 to 10.0. The overall payback time for the added external insulation when sufficient external insulation is added to bring the total RSI to 6.7 (about the most that has been done for walls) is 3–6 years for polystyrene, depending on the efficiency of the manufacturing process. Figure 12.3c gives marginal payback times, based on the last RSI increment of 1.0 added. When the RSI is increased from 9 to 10, the payback time for this increment is as large as 34 years, but inasmuch as this is less than the expected lifetime of the insulation, increasing the RSI to as high as 10 using the most energy-intensive foam insulation is justified on a lifecycle energy basis with a 4000 HDD climate. However, RSI values as large as 10 have been used only in roofs which, if flat could be insulated to this level using a continuous bed of cellulose insulation, which is much less energy intensive.

    To compare the savings in CO_2 emissions resulting from additional insulation with the CO_2 emissions associated with manufacturing the insulation, the kinds of energy used to make the insulation, the kind of energy used for heating and the sources of any electricity used for heating or manufacturing need to be considered. If the insulation is manufactured largely using electricity that is generated at coal-fired power plants, while the heating energy that is saved is natural gas, then the time required to offset the CO_2 emitted during the manufacturing process will be almost twice as large as the energy payback time, because the CO_2 emission factor for coal is almost twice as large as that for natural gas (Chapter 2, Section 2.4.1). Savings in cooling energy use due to greater insulation also need to be considered, along with the downsizing of heating and cooling equipment and radiators that is possible with greater insulation and the savings in the embodied energy of this equipment. Since radiators involve energy-intensive products such as aluminium or steel, consideration of the savings in radiator embodied energy works in favour of greater amounts of insulation. Finally, for solid-foam and sprayed-on foam insulation, the global warming impact of any halocarbon gases used as an expanding agent needs to be considered, an issue that is discussed in Section 12.5.

Building-integrated PV panels (BiPVs)

These were discussed in Chapter 11 (Section 11.3), where it was concluded that BiPVs provide a net energy saving but require two years (southwest US, southern Europe) to four years (middle Europe) for the energy that they produce to equal the energy required to produce them. As BiPVs improve, the energy payback time should drop considerably. Alsema and Nieuwlaar (2000) estimate that the production of roof-integrated PV modules requires about $2900MJ/m^2$ of primary energy (assuming that the electricity used in the manufacturing process is generated at 35 per cent efficiency). The embodied primary energy of displaced shingles is small by comparison.

12.4.2 Modern versus traditional building materials

Modern building materials – concrete, concrete blocks, glazed bricks, steel – tend to contain substantially more embodied energy than traditional materials. Tiwari et al (1996) and Tiwari (2001) have examined the impact on cost and CO_2 emissions of partial substitution of some modern building materials with traditional and locally available materials for housing in India. The latter include compressed mud blocks, used in place of bricks; lime, surkhi, and flyash instead of cement and sand in mortar; lime and surkhi in place of cement and sand for plastering; and filler slabs with Mangalore tiles in place of concrete slab roofs and floors. The net result is a 60 per cent reduction in CO_2 emissions associated with the construction of housing, and a 45 per cent

Table 12.9 Blowing agents used for polyurethane and extruded polystyrene insulation. For HFC blowing agents, the global warming potential (GWP) relative to CO_2 over a 100-year lifespan is given

Type of insulation	Original blowing agent	HCFC blowing agent	HFC blowing agent	HFC GWP	Non-halocarbon blowing agents
Polyurethane, polyisocyanurate	CFC-11 GWP = 4680	HCFC-141b GWP = 713	HFC-245fa	1020	Cyclopentane, CO_2
			HFC-365mfc/ HFC-227ea	950–1100	Various isomers of pentane
Extruded polystyrene (XPS)	CFC-12 GWP = 10720	HCFC-142b	HFC-134a	1410	Normal pentane, CO_2
		HCFC-22 GWP = 1780	HFC-152a	122	cyclopentane/isopentane blends

Source: Ashford et al (2005)

reduction in costs. A number of materials can be impregnated into mud bricks in order to increase their durability and resistance to water (Ren and Kagi, 1995). Soil-cement blocks (with 8 per cent cement) have about one-third the embodied energy of burnt clay bricks (Reddy and Jagadish, 2003).

12.5 Halocarbon greenhouse gas emissions associated with foam insulation

Halocarbon gases are normally used as expanding agents in the production of solid-foam insulation, and in the application of sprayed-on foam insulation (Chapter 3, Section 3.2.1). Some of the gas is released during the manufacture of the insulation, but the majority is embedded in the foam pores and can gradually leak over time. The original expanding agents were CFC-11 and CFC-12, whose use has been phased out as part of the Montreal Protocol and subsequent accords to protect the stratospheric ozone layer. HCFCs have been introduced as temporary replacements for the CFCs, but these too will be phased out. They can be replaced by a variety of HFC and hydrocarbon compounds. The HFCs pose no harm to the ozone layer but, like the CFCs and HCFCs, are powerful greenhouse gases. Table 12.9 lists the various expanding agents that have been used for polyurethane, polyisocyanurate and extruded polystyrene insulation, along with the global warming potential (GWP) of the HFC agents. The GWP is a rough measure of the relative heating effect of a gas, compared to the heating effect of CO_2, integrated over some

arbitrarily determined lifespan (see Chapter 2, Section 2.4.3). The impact on climate of the use of halocarbons in insulation depends on the amount of halocarbon used and the fraction that leaks. For extruded polystyrene and spray-on polyurethane insulation, most to all of the blowing agent ultimately leaks, whereas for solid-foam polyurethane insulation, about two-thirds of the blowing agent is believed to be retained after 50 years (Ashford et al, 2005).

With regard to solid polyurethane foams, some currently use HCFC-141b as the blowing agent, while others use HFC-365mfc or n-pentane. Table 12.10 gives the GWP, amount of blowing agent used, and the foam thermal conductivity for these three blowing agents.[3] From these parameters, and given a foam density of $32.5kg/m^3$, the embodied energy and mass of blowing agent used in $1m^2$ of foam with an RSI-value of 1.0 can be computed, and is given in Table 12.10. Assuming the energy used to manufacture the foam to be natural gas, with an emission factor of $55kg\ CO_2/GJ$, the CO_2 emission associated with the manufacture of the insulation are as given in the second last row of Table 12.10. To compute the equivalent in CO_2 emissions of the blowing agent emission to the atmosphere, the fraction of the blowing agent that is emitted must be determined. Based on Harnisch et al (2003), it is assumed that 8 per cent is emitted to the atmosphere during the blowing and processing of the foam insulation, and another 0.5 per cent per year of the remaining blowing agent is emitted (out of a possible range of 0.2–1.0 per cent per year). At the end of 50 years, it is assumed that the building is demolished, the foam waste transferred to a

Table 12.10 Input and computed intermediate parameters used to compute the time required for savings in heating energy CO_2 emissions to completely offset the GHG emissions associated with the manufacture of polyurethane solid-foam insulation and associated with the leakage of three different blowing agents into the atmosphere

Parameter	Blowing agent (BA)		
	HCFC-141b	HFC-365mfc	n-pentane
GWP	713	782	7
kg of BA used per kg of foam	0.100	0.123	0.050
BA density (kg/m³ foam)	3.200	3.936	1.600
Insulation conductivity (W/m/K)	0.021	0.022	0.024
Area of RSI-1 panel (m²) produced from 1m³ of foam	47.619	45.455	41.667
Foam mass (kg/m²/RSI)	0.672	0.704	0.768
BA mass (kg/m²/RSI)	0.067	0.087	0.038
Foam embodied energy (MJ/m²/RSI)	92.06	96.45	105.22
CO_2 from manufacture (kg/m²/RSI)	5.06	5.30	5.79
CO_{2eq} of BA emission (kg/m²/RSI)	26.58	37.56	0.14

In all cases, the foam density is assumed to be 32.5kg/m² and the embodied energy 137MJ/kg. Foam density and embodied energy are taken from Table 12.8, while foam conductivities and blowing agent GWP and loading are taken from Harnisch et al (2003).

landfill, and 20 per cent of the remaining blowing agent emitted to the atmosphere. These assumptions result in 42 per cent of the original blowing agent being emitted to the atmosphere over a period of 50 years. This number is quite uncertain, but is useful for the illustrative calculations that follow. This emission fraction times the GWP and blowing agents loadings given in Table 12.10 produce the CO_2 equivalents given in the last row of Table 12.10.

Figure 12.4 gives the number of years required for the savings in heating energy CO_2 emission (resulting from the application of the polyurethane foam insulation) to completely offset the equivalent CO_2 emissions associated with the manufacture of the insulation and with leakage of the blowing agent. Heating energy savings are computed using Equation (12.2), assuming 4000 HDD and a heating efficiency of 0.9, then converted to CO_2 emissions assuming the energy used for heating to be natural gas. Results are given for foam insulation sufficient to give a total RSI ranging from 4.0 to 10.0, beginning with an initial RSI-value of either 0.5 (Figure 12.4a) or 3.0 (Figure 12.4b), as well as based on RSI increments of 1.0 up to a final RSI-value of 10.0 (Figure 12.4c). The latter are marginal payback times, analogous to the marginal payback times given in Figure 12.3c for embodied energy.

As seen from Figure 12.4, payback times assuming an initial RSI-value of 0.5 and a final RSI-value of 6.5 are 4 and 5 years using

HCFC-141b and HFC-365mfc, respectively. These payback times increase to 22 and 39 years, respectively, if the initial RSI-value is 3.0 and the final RSI-value is 6.5. This would correspond to the case where external foam insulation is added to a wood-frame structure with cellulose or fibreglass insulation in the wall cavities. For a final RSI-value of 6.5, marginal payback times are in excess of 40 years, while for a final RSI-value of 10.0 (obtained in the roofs of some low-energy houses), the marginal payback times are in excess of 100 years. If the fractional leakage of the blowing agent is greater than assumed here, the payback times would be even longer. However, for polyurethane blown with n-pentane, the marginal payback at an RSI-value of 10.0 is only 25 years, and average paybacks (based on the total thickness of foam insulation) are 8 years and 1.4 years if the starting RSI-values are 3.0 and 0.5, respectively. Based on these results, it is seen that the net climatic benefit of halocarbon-blown foam insulation can be quite small, and that when used to build the total RSI value up to the levels of insulation used in low-energy houses (RSI 6.5 in walls, 10.0 in roofs), halocarbon-blown foam insulation is *counterproductive* from a climatic point of view.

HFC-152a, with a GWP of only 122, would appear to be an attractive blowing agent for extruded polystyrene insulation. However, its leakage rate is several times faster than for HFC-134a, leading to a faster degradation in the

Figure 12.4 Variation in the overall payback time for equivalent CO_2 emissions as the total RSI is increased (a) from 0.5 or (b) from 3.0 to values as large as 10.0, and (c) variation in the marginal emission payback time when the RSI value is increased by 1.0 to the indicated final value

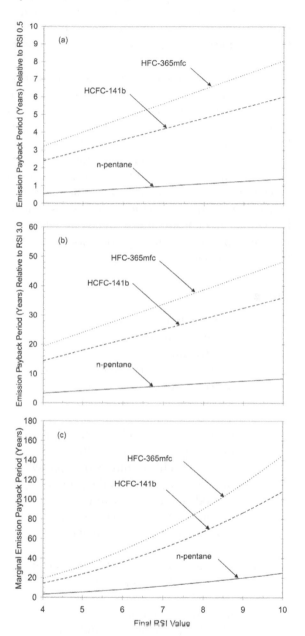

Note: Emission payback times are for polyurethane insulation using either HCFC-141b, HFC-365mfc, or n-pentane as blowing agents, for a climate with 4000 HDD, for a heating system efficiency of 0.9, and additional assumptions as given in Table 12.10.

effectiveness of the insulation over time. Hydrocarbon blowing agents have negligible greenhouse effect, but pose a number of problems of their own, the most significant being that they pose a fire hazard during the manufacturing process, so expensive flame-proof equipment is needed. Carbon dioxide can also be used as an expanding agent for extruded polystyrene and for spray-on polyurethane; the CO_2 in this case is generated by the reaction of H_2O with excess isocyanurate. The thermal conductivity of the resulting insulation is greater, so a greater thickness is needed to give the same RSI value. However, in HCFC and HFC insulation, the higher thermal conductivity is due in part to the low molecular conductivity of the blowing agent gas, which is initially trapped in the foam bubbles. As the blowing agent leaks, the thermal conductivity advantage of the insulation using HCFCs or HFCs declines. Thus, over time, the difference between insulation blown using CO_2 and using other blowing agents will decrease.

12.6 Summary and synthesis

The embodied energy in a building is the energy that was used in constructing the building and in making and transporting to the building site all the material used. Two broad approaches have been used to compute the embodied energy of building materials: (i) a physical approach based on the energy used on-site during construction (zeroeth-order embodied energy), in the factories where all materials used in the building are produced (first-order embodied energy), and in transporting them to the building site (another first-order embodied energy); and (ii) based on an input–output matrix that includes the major outputs of an economy and the major inputs to each of the outputs, where all of the inputs to a given output consist of all (or some) of the other outputs. Both approaches suffer from severe limitations. The physically or process-based approach is limited by the fact that the energy embodied in a building includes not only zeroeth and first order energy inputs, but second order inputs (energy used to make the materials that are inputs to the factories where building materials are made, and energy used to make the machinery in these factories), third order inputs, and so on. Full accounting of the embodied energy requires carrying the analysis up to about 12th order, but most process-based analyses stop after 2nd or 3rd order (thereby accounting for as little as half of the full embodied energy). The input–output approach accounts for the full cascade of energy flows embodied in any given product, but does so on the basis of currency flows and the average relationship, within a sector, between currency flows and energy flows. This can lead to substantial errors for individual products. A hybrid approach begins with an economically estimated input–output matrix, but then replaces the energy intensities for the first few orders with physically based energy intensities.

Despite the uncertainties in the calculation of building embodied energy, due to the shortcomings in the methods used, it is safe to conclude that the new embodied energy required to demolish an old building and construct a new building is equal to only a few years of energy savings if the new building achieves a substantial (factor of 2–4) reduction in annual energy use compared to the old building. Demolition in such cases is therefore worthwhile from an energy point of view, although there are other considerations that may still rule out demolition, such as the desire to preserve historical buildings or to minimize waste.

With regard to potential trade-offs between increased embodied energy and reduced operating energy, (i) use of concrete in place of wood in order to increase the use of passive heating and cooling will generally be favourable, although phase change materials may give similar benefits but with less increase of embodied energy; (ii) increasing the glazing area can reduce embodied energy and can reduce operating energy for most glazing orientations in most climates; (iii) upgrading window performance to triple glazing with argon fill and three low-e coatings is well justified in temperate and colder climates, as is an upgrade to krypton-fill if this allows perimeter radiators to be eliminated; (iv) for a modestly cold climate (4000 degree-days heating load), the additional embodied energy when the amount of insulation is increased from RSI 9 to RSI 10 is paid back with heating energy savings in less than 10 years for fibreglass insulation, and in one year for cellulose insulation; and (v) building-integrated PV modules pay back the energy required to produce them in two and four years for south-west US/southern European and north European conditions, respectively. The time required to pay back the additional energy required to manufacture high-performance windows or extra insulation is reduced if savings in the embodied energy of mechanical equipment (due to less need for mechanical heating and cooling) is accounted for.

For solid-foam or spray-on foam insulation, the global warming impact of halocarbons used in the manufacture of the insulation can be several times that associated with the energy used to manufacture the insulation. This renders the use of more than minimal (RSI 1-2) foam insulation counterproductive from a climatic point of view if halocarbon expanding agents are used in the manufacture or application of foam insulation. High levels of foam insulation can be justified in cold climates if the insulation is produced using hydrocarbons or CO_2 as an expanding agent.

Notes

1   For a building, the direct energy is the energy used on site, during the construction process. The direct energy for the production of steel is first-order energy for the building, and so on.

2   The savings in operating primary energy will be smaller in the future, as electricity generation efficiency increases. This would be only partly offset by a decrease in the primary embodied energy, because much of the energy needed to produce steel and cement is thermal energy produced directly from fossil fuels.

3   The same amount of blowing agent in gaseous form is needed for all three blowing agents, but the greater mass loading required for HFC-365mfc compared with the others arises from differences in molecular weight (about twice as large for HFC-365mfc than for n-pentane) combined with the fact that 20–30 per cent of HFC-365mfc dissolves into the foam matrix, requiring a greater dose.

thirteen

Demonstrated Energy Savings in Advanced New Buildings and the Role of the Design Process

The preceding chapters have discussed in some detail the dramatic reductions in energy use that are possible through improved building envelopes; more efficient heating and cooling devices; through maximizing the use of passive solar heating, cooling and ventilation; through daylighting and advanced electric lighting systems; through clever ways of combining the components of residential and commercial HVAC systems; through better use and supply of hot water; and through more efficient household appliances, consumer goods and office equipment. In this chapter, the net effect of these improvements on total building energy use, when incorporated together in a systems approach to building design, is illustrated through a discussion of recent, exemplary buildings.

In Chapter 1 (Section 1.4), 'sustainable' buildings were defined (in energy terms) as buildings whose on-site energy use is so small that the remaining energy use can eventually be met entirely through renewable energy, either through

active building-integrated solar and wind energy, or from energy grids. From this it was argued that the gross on-site energy intensity (kWh per m^2 of floor area per year, excluding building-integrated active solar and wind energy) of new buildings needs to be reduced by a factor of four to five in OECD countries and by a factor of three to four in non-OECD countries, compared to the average of existing buildings. As the energy intensity of new buildings is somewhat less than the stock average in most jurisdictions, the required reductions compared to current practice for new buildings are somewhat less.

We begin this chapter with a discussion of the energy provisions of current building codes, which govern new buildings. This is followed by a discussion of exemplary buildings in the residential and commercial sectors, with a comparison of the technical specifications of the building components with the local building codes in some cases and of the energy intensity of these exemplary buildings compared to those that minimally comply with the local building code. Through greater insulation levels, high-performance windows, more efficient equipment and innovative HVAC systems, energy savings of 75 per cent or better have sometimes been achieved. Savings of this magnitude are consistent with the energy dimension of sustainability given above. The early fundamental decisions by the architect, and adherence to an integrated rather than a linear design process involving the architect, engineers, contractors and subcontractors, are crucial factors in achieving savings of this magnitude. That is, in order to achieve dramatic energy savings, major changes in the usual *process* by which buildings are designed are required. For that reason, this chapter contains a substantial discussion of the design process behind exemplary buildings.

13.1 Energy-related provisions of building codes

The first comprehensive energy code for commercial buildings was developed by the American Society of Heating, Refrigeration and Air Conditioning Engineers (ASHRAE) in 1975, in the aftermath of the first OPEC oil embargo (Crowder and Foster, 1998). This standard, ASHRAE 90.1-1975, was revised in 1980, 1989, 1999, 2001 and 2004. The various versions of ASHRAE 90.1 form the basis for building energy codes in many parts of the US and in many countries throughout the world. A standard for residential construction, ASHRAE 90.2, was published in 1993. ASHRAE standards are developed by a number of separate committees through a consensus process. For this reason, they are quite cautious with regard to energy efficiency and do not reflect what can be achieved with high-performance building components, much less that which can be achieved through integrated design. Nevertheless, even today, substantial training and technical assistance is required to enable many design and engineering firms to comply with ASHRAE 90.1 (Epstein et al, 2002).

   Building codes are under state jurisdiction in the US, and under provincial jurisdiction in Canada. To assist states and provinces in developing their own building codes, both federal governments have developed their own voluntary national building codes. These were referred to simply as the *Model Energy Code* (MEC) in the US, with different versions published in 1992, 1993 and 1995. After 1995, the MEC has been referred to as the *International Energy Conservation Code* (IECC), although it is a purely US product. Editions were published in 1998, 2000 and 2003. In Canada, there is the *Model National Energy Code for Houses* (MNECH) and the *Model National Energy Code for Buildings* (MNECB), which replaced earlier model energy codes in 1997. In the EU, the *European Directive for Energy Performance of Buildings* was adopted in 2002 and came into force in 2005. This directive provides a consistent template for unifying the diverse national building codes in Europe, but does not itself set any energy-related standards.

13.1.1 US and Canadian building energy codes

Table 13.1 compares the energy-related provision of selected US building codes and the Canadian MNECH. Table 13.2 gives the characteristics of operable windows required to meet the window U-values specified in the various codes shown in Table 13.1. The performance required by the 2003 IECC for the northern US plains is close to the highest that can be achieved in commercially available products. Table 13.3 lists the envelope and lighting power density requirements of the ASHRAE 90.1-1989 and 90.1-2004 codes and the Canadian MNECB, while Table 13.4 lists the performance of air conditioners, chillers, heat pumps, and boilers as required under ASHRAE 90.1-1989 and 90.1-2004. ASHRAE 90.1-1989 provided a 20-30 per cent improvement in energy use compared to the prevailing practice at the time. According to an assessment conducted by the US Department of Energy, based on a weighted average of different building types and in different US climate zones, ASHRAE 90.1-1999 (not shown here) provides only a 3.9 per cent savings in on-site energy use compared to ASHRAE 90.1-1989 (this is composed of a 7.3 per cent reduction in electricity use and an 8.6 per cent increase in natural gas use) (see Table 7 at www.energycodes.gov/implement/determinations_com_exp.stm). The main difference between ASHRAE 90.1-2001 and older versions is a decrease in the permitted lighting power density, from 16-20 W/m² to 14 W/m². The 2004 version aims to achieve a 20 per cent savings relative to the 1999 version, and includes further tightening of the lighting energy standard. The main differences between the 2004 and 1989 versions are (i) a significantly lower roof U-value and modestly better or no better window U-value, depending on climate, (ii) better lighting (11.8W/m² vs 16.2-20.5W/m² connected power density), and (iii) more efficient equipment (for example, COPs of 4.2-6.1 instead of 3.8-5.2 for water-cooled chillers) for the 2004 version. California has developed its own energy code (Title 24, updated in 2001). In California's climate, buildings constructed under Title 24-2001 use about 12 per cent less energy on average than those constructed under ASHRAE 90.1-1999, based on a sample of 940 buildings that were simulated by computer as built, then upgraded or downgraded

to be in minimum compliance with both codes (Eley Associates, 2001). A further strengthening of Title 24 came into effect in late 2005, with an emphasis on reducing peak loads (see www.energy.ca.gov/title24/2005standards/index.html). The City of Seattle has developed its own code that saves about 16 per cent compared to ASHRAE 90.1-1999 (Hogan, 2005). Table 13.5 lists the US states where various building codes are mandatory, and those states where there are no mandatory building codes, while Table 13.6 indicates which countries in the Pacific-rim region have mandatory energy provisions in their building codes, voluntary energy provisions codes, or no energy provisions at all. Note that a number of northern US states have already adopted the 2003 IECC, with its relatively stringent window U-value requirement.

Table 13.1 Comparison of residential building energy codes in the US and Canada

Envelope element	1993 US MEC	2003 IECC	Location	Building America upgrade	Location	1997 Canadian MNEC	Location
Attic-type roof RSI value (U-values, W/m²/K, in brackets)							
	4.3 (0.23)	4.3 (0.23)	Houston	8.45 (0.12)	Phoenix		
	5.5 (0.18)	5.5 (0.18)	Baltimore	8.45 (0.12)	Orlando		
	6.8 (0.15)	6.8 (0.15)	Fargo	8.45 (0.12)	Denver	5.4 (0.19)	BC <3500 HDD
				8.45 (0.12)	Minneapolis	5.6 (0.18)	Ontario <5000 HDD
						7.0 (0.14)	Ontario >5000 HDD
						10.6 (0.094)	Yukon north
Other-roofs RSI value (U-values, W/m²/K, in brackets)							
						4.3 (0.23)	BC <3500 HDD
						4.3 (0.23)	Ontario <5000 HDD
						4.3 (0.23)	Ontario >5000 HDD
						7.1 (0.14)	Yukon north
Above-grade wall RSI value (U-values, W/m²/K, in brackets)							
	2.1 (0.48)	2.1 (0.48)	Houston	5.81 (0.17)	Phoenix		
	3.0 (0.33)	3.0 (0.33)	Baltimore	5.01 (0.17)	Orlando		
	3.8 (0.26)	3.8 (0.26)	Fargo	5.81 (0.17)	Denver	2.0 (0.50)	BC <3500 HDD
				5.81 (0.17)	Minneapolis	2.9 (0.34)	Ontario <5000 HDD
						3.3 (0.30)	Ontario >5000 HDD
						4.7 (0.21)	Yukon north
Below-grade wall RSI value (U-values, W/m²/K, in brackets)							
	1.0 (1.0)	1.0 (1.0)	Houston	None	Phoenix		
	1.8 (0.56)	1.8 (0.56)	Baltimore	None	Orlando		

2.9 (0.34)	2.9 (0.34)	Fargo	2.6 (0.38)	Denver	2.1 (0.48)	BC <3500 HDD
			2.6 (0.38)	Minneapolis	3.1 (0.32)	Ontario <5000 HDD
					3.1 (0.32)	Ontario >5000 HDD
					3.1 (0.32)	Yukon north
Basement floor/slab-on-grade RSI value						
None	None	Houston		Phoenix		
0.7 (1.42)	0.7 (1.42)	Baltimore		Orlando		
1.3 (0.77)	1.3 (0.77)	Fargo		Denver	1.08 (0.93)	BC <3500 HDD
				Minneapolis	2.6 (0.38)	Ontario <5000 HDD
					2.6 (0.38)	Ontario >5000 HDD
					2.4 (0.42)	Yukon north
Window U-value (W/m²/K)						
unspecified	2.67	Houston	1.87	Phoenix		
unspecified	1.70	Baltimore	1.87	Orlando		
unspecified	1.48	Fargo	1.87	Denver	3.2	BC <3500 HDD
			1.87	Minneapolis	2.6	Ontario <5000 HDD
					2.6	Ontario >5000 HDD
					2.4	Yukon north
Window SHGC						
unspecified	0.40	Houston				
unspecified	0.68	Baltimore				
unspecified	0.68	Fargo				

Note: HDD and CDD (K-days, relative to 18°C and rounded to the nearest 50) of cities named below are as follows: Houston, 750 and 1600; Baltimore, 2750 and 650; Fargo, 5150 and 350; Phoenix, 800 and 4350; Orlando, 300 and 4500; Denver, 3400 and 1450; Minneapolis, 4500 and 450.
Sources: 1993 US MEC and 2003 IEEC, Hendron et al (2004); Building America upgrade, Rudd et al (2004); 1997 Canadian MNEC, (NRCan, 1997a); HDD and CDD data, ASHRAE (1989)

Table 13.2 Characteristics of operable windows required to meet the U-values specified in the residential-building energy codes of Table 13.1

Code	HDD (K-days)	U-value (W/m²/K)	Window Characteristics
MNEC, BC	<3500	3.20	Double-glazed, clear, air fill, Al-clad wood frame
2003 IECC, Houston	750	2.67	Double-glazed, $\varepsilon = 0.6$ coating on surface 2 or 3 and argon fill, wood/vinyl frame
MNEC, Ontario	>5000	2.60	
MNEC, Yukon north	up to 11000	2.40	Double-glazed, $\varepsilon = 0.4$ coating on surface 2 or 3 and argon fill, wood/vinyl frame
BA Upgrade, all	all	1.87	Double glazed, $\varepsilon = 0.05$ coating on surface 2 or 3 and argon fill, insulated fibreglass frame
2003 IECC, Baltimore	2750	1.70	
2003 IECC, Fargo	5150	1.48	Triple glazed, $\varepsilon = 0.2$ coatings on 2 surfaces, argon fill, insulated fibreglass frame

Source: Based on U-values given in ASHRAE (2001b, Chapter 30)

Table 13.3 Comparison of the ASHRAE 90.1-1989, ASHRAE 90.1-2004 and 1997 Canadian MNEC for commercial buildings

Building Characteristic	ASHRAE 90.1-1989			ASHRAE 90.1-2004			1997 Canadian MNEC		
	HDD (K-days)			HDD (K-days)			HDD (K-days)		
	2500	4500	7500	2500	4500	7500	<3500	4500	11,000
Building Envelope Assembly U-values (W/m²/K)									
Roof	0.51	0.36	0.25	0.19–0.36	0.15–0.36	0.15–0.28	0.47	0.47	0.25
Above-grade wall	0.61	0.40	0.28	0.51–0.86	0.51–0.64	0.29–0.54	0.81	0.55	0.27
Below-grade wall	0.72	0.54	0.39				1.11	0.77	0.32
Slab-on-grade	0.95	0.95	0.95						0.67
Window	4.1	2.6	2.6	2.6	2.6	2.0	3.2	3.2	1.2
Other characteristics									
Office lighting power density	20.5W/m² (smallest offices) to 16.2W/m² (largest offices)			11.8W/m²			22.6–29.1W/m², depending on the kind of office		

Source: ASHRAE (1989, 2004) and NRCan (1997b)

Table 13.4 Minimum seasonally averaged COP (under ARI (American Refrigeration Institute) test conditions) required by ASHRAE 90.1-1989 and ASHRAE 90.1-2004 for various kinds of cooling equipment, and minimum AFUE or combustion efficiency required for boilers

	ASHRAE 90.1-1989			ASHRAE 90.1-2004	
	Capacity (kW$_c$ or kW$_{th}$)	COP		Capacity (kW$_c$ or kW$_{th}$)	COP
		<1/1/92	>1/1/92		
Air conditioners					
Room air conditioners	All	2.34	2.64	All	2.93, 3.52ᵘ
Air-cooled air conditioners				65–135	2.96
	≤223	2.40	2.49	135–240	2.78
	>223	2.34	2.40	240–760	2.73
				>760	2.64
Water-cooled air conditioners	All	2.76	2.81	≤65	3.55
				65–135	3.31
				135–240	3.17
				>240	2.96
Heat pumps					
Air source heat pumps, cooling				19–40	2.90
	≤223	2.40	2.49	40–70	2.67
	>223	2.34	2.40	>70	2.64
Air source heat pumps, heating	8.3°C T$_o$	2.80	2.90		
	–8.3°C T$_o$	1.85	2.00		
Water source heat pump, cooling	≤19	2.64	2.73	≤5	3.28
	19–40	2.78	3.08	>5	3.52
Ground water heat pump, cooling	21°C ewt	2.93	3.22	25°C ewt	3.93
	10°C ewt	3.08	3.37	15°C ewt	4.75
Ground water heat pump, heating				0°C ewt	3.1
Chillers					
Air-cooled chiller, unspecified type	<527	2.6	2.7		
	≥527	2.4	2.5		
Water-cooled chiller, unspecified type	<527	3.7	3.8		
	527–1055	3.7	4.2		
	≥1055	4.6	5.2		
Water-cooled chiller, reciprocating				All	4.20

Water-cooled chiller, rotary & screw				<527	4.45
				527–1055	4.90
				≥1055	5.50
Water-cooled chiller, centrifugal				<527	5.00
				527–1055	5.55
				≥1055	6.10
Absorption chiller, SE, air-cooled					0.6
Absorption chiller, SE, water-cooled					0.7
Absorption chiller, DE, water-cooled					1.0
Boilers					
Gas (AFUE)	< 87.9	0.68	0.80	No change	
(combustion efficiency)	≥ 87.9	0.75	0.80		
Oil (AFUE)	< 87.9	0.78	0.80		
(combustion efficiency)	≥87.9	0.78	0.83		

ᵃ Before and after 23/01/2006.
Note: ewt = entering water temperature.
Source: ASHRAE (1989, 2004)

Table 13.5 Implementation of building energy codes in the US

Code version	States where implemented
Energy codes for residential buildings	
2003 IECC or IRC	Kansas, Nebraska, New Mexico, Pennsylvania, Rhode Island, Utah
2000 IECC or IRC, adopted or under review	Alabama, California, District of Columbia, Georgia, Florida, Idaho, Kentucky, Maryland, New Hampshire, New York, North Carolina, Ohio, Oregon, South Carolina, Texas, Virginia, Wisconsin, West Virginia
1998 IECC	Oklahoma
1995 MEC	Connecticut, Maine, Massachusetts, Minnesota, New Jersey, Vermont, Hawaii
1995 MEC, partial	Louisiana
1993 MEC	Delaware, North Dakota, Montana
1992 MEC	Arkansas, Iowa, Indiana, Tennessee
<1992 MEC	Maine, Michigan
None	Alaska, Arizona, Colorado, Illinois, Mississippi, Missouri, Nevada, South Dakota, Wyoming
Energy codes for commercial buildings	
2003 IECC/ ASHRAE 90.1-2001	Georgia, Kansas, Maine, Nebraska, New Mexico, Montana, Pennsylvania, Rhode Island, Utah
ASHRAE/IESNA 90.1-1999	Alabama, California, Florida, Idaho, Kentucky, Massachusetts, Michigan, New Jersey, New York, North Carolina, Ohio, Oregon, Texas, Vermont,, Washington, Wisconsin
ASHRAE/IESNA 90.1-1989	Arkansas, Connecticut, Delaware, District of Columbia, Hawaii, Iowa, Louisiana, Maryland, Minnesota, New Hampshire, North Dakota, Oklahoma, South Carolina, Virginia, Wisconsin
<ASHRAE/IESNA 90.1-1989	Indiana, Mississippi
None	Alaska, Arizona, Colorado, Illinois, Missouri, Nevada, South Dakota, Tennessee, Wyoming

Notes: ASHRAE = American Society of Heating, Refrigeration, and Air Conditioning Engineers, IECC = International Energy Conservation Code, IESNA = Illuminating Engineering Society of North America, IRC = International Residential Code, MEC = Model Energy Code.
Source: Building Codes Assistance Project, www.bcap-energy.org

Table 13.6 Asia-Pacific countries with building energy codes

Country	Commercial energy code?	Residential energy code?
Australia	Mandatory	Mandatory
Canada	Mandatory	Mandatory
Chile	Mandatory	
Hong Kong	Mandatory	
Indonesia	Voluntary	
Japan	Mandatory	
Korea	Mandatory	Mandatory
Malaysia	Planned	Voluntary
Mexico	Planned	Mandatory
New Zealand	Mandatory	Mandatory
Papua New Guinea	Mandatory	Mandatory
Peru	Voluntary	Voluntary
Philippines	Mandatory	
Russia	Mandatory	
Taiwan	Mandatory	
US	Voluntary/Mandatory	Voluntary/Mandatory
Viet Nam	Planned	Planned

Source: APERC (2001)

Both the US and Canadian building energy codes allow some flexibility in how the code requirements are met. Generally, there are some basic *mandatory* requirements that cannot be bypassed. Beyond that, there are further *prescriptive* requirements, but small trade-offs are allowed (for example, reducing thermal resistance in one part of the envelope but increasing it in another part so that the total heat loss does not increase). A third route is the *performance-based* route: one can design a building with any desired thermal characteristics, provided that the calculated overall energy use is no greater than that of a building designed in strict conformity with the prescriptive requirements. This requires the use of computer simulation models, which are now adequate for simulating the impact of standard upgrade features but not, as discussed later, for simulating the impact of some of the innovative features discussed in this book. Thus, credit will not be given for some measures that could bring the building into compliance, and for those striving only to meet code requirements, the lack of credit will serve as a disincentive to implement certain measures (the capabilities and limitations of some popular simulation models are given in Appendix D). On the other hand, conscientious design of buildings will result in a performance that far exceeds any

regulatory requirements, so the lack of credit for some measures should not discourage their use.

A voluntary residential building code called the *R2000 Standard* was established in Canada in the 1980s. This is a performance-based standard, with a maximum permitted annual space heating energy consumption of:

$$Q_s = S * (49 * HDD / 6000) * (40 + V / 2.5) \tag{13.1}$$

where S = 4.5MJ for fuel-fired space heating systems and S = 3.6MJ (1kWh) for electric space heating systems, HDD = heating degree days relative to 18°C, and V is the interior heated volume (m³), including the basement (further information can be found at http://oee.nrcan.gc.ca/residential/personal/index). Figure 13.1 shows the permitted heating-energy use in kWh/m²/year for HDD ranging from 1000–11,000 K-days and for floor area (assumed to equal V/2.5) ranging from 100m² to 400m². For an average-sized Canadian house (270m² floor area), the permitted heating energy use ranges from 40–80kWh/m²/year for HDD of 3000–6000 K-days. This is about 30 per cent less than permitted under the mandatory provincial building codes.

Figure 13.1 Annual heating energy use (kWh/m²/year) permitted under the voluntary Canadian R2000 building code, as a function of the number of heating degree days and floor area (including the basement)

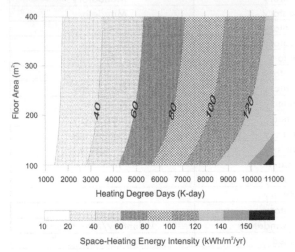

Space-Heating Energy Intensity (kWh/m²/yr)

13.1.2 European building energy codes

The European Commission published the *Energy Performance of Buildings Directive* (EPBD) on 4 January, 2003. This directive requires member states to set minimum energy performance standards for new and existing buildings and for different categories of buildings, and to review the standards at least once every five years in the light of technological progress. The standards are to be set by individual member states but shall be performance based. Standards are to be implemented by 5 January, 2006. Performance standards are to be enacted for major renovations of buildings (>25 per cent of the initial cost of the building) with a usable floor space of 1000m² or more. Member states are to require that, before construction begins, consideration be given to district heating and cooling system (including systems using renewable energy), to combined heat and power, and to the use of heat pumps. Most European countries already have building energy-related standards, but the EPBD provides an impetus to strengthen these standards. Links to the building codes in force in many European countries can be found at the *Energy Performance Norm* website that is maintained by the Fraunhofer Institute for Solar Energy Systems (www.enper.org/index).

Table 13.7 gives the insulation U-values and RSI-values, and the window U-values, that are currently required by or are consistent with residential building codes in the UK, Switzerland, Denmark, Italy (Rome), Finland (Helsinki) and Sweden (Stockholm). Also given for the UK are specifications that would be consistent with a 25 per cent improvement in building energy performance, as is being contemplated for the UK in response to the EPBD (ODPM, 2004). Table 13.8 compares the heating energy use of different kinds of housing in Germany under successively stricter building regulations.

Swedish buildings are among the most highly insulated in Europe. The 2002 revision of the building code prescribes a maximum permitted average U-value for the entire building (see www.boverket.se/English) that, for residential buildings, is given by:

$$U_{max} = 0.16 + 0.81 \frac{A_w}{A_{total}}, W/m^2/K \qquad (13.2)$$

where A_w is the window area and A_{total} is the total envelope area (ground floor, walls, roof) (for non-residential buildings, the 0.16 term is replaced by 0.22). For purposes of determining if the building complies with the regulated maximum U-value, the window U-value can be reduced by 1.2W/m²/K when calculating the average U-value if the window faces south-east to south-west, by 0.7W/m²/K if the window faces south east to north-east or south west to north-west, and by 0.4W/m²/K if the window faces north east to north west. Windows whose U-value is less than the above adjustments act as a net heat source to the building due to passive solar gain. The U-value of building elements in contact with the ground can be multiplied by 0.75 for purposes of assessing compliance. For $A_w/A_{total} = 0.25$, the permitted average U-value (after adjusting the U-values of windows and floors as indicated above) is 0.36W/m²/K. This is consistent with the average value given in Table 13.7 for Stockholm, once the above adjustments are taken into account. The permitted air leakage amounts to about $ACH_{50} = 1.0$ for residential buildings and $ACH_{50} = 2.0$ for non-residential buildings (for a discussion of ACH_{50}, see Chapter 3, Section 3.7).

Table 13.7 Comparison of envelope U-values (W/m²/K) required in Rome, the UK, Switzerland, Denmark, Rome, Helsinki and Stockholm, and possible values that might be required in the UK in response to the EU EBPD

| | Rome (1700 HDD) | UK (3000 HDD) | | Switzerlanad (3700 HDD) | Denmark (4100 HDD) | Helsinki (5400 HDD) | Stockholm (5400 HDD) |
		Current	EBPD				
Roof	0.6	0.25	0.13–0.20	0.2	0.15–0.20	0.22	0.28
Wall	0.7	0.35	0.22–0.27	0.2	0.2–0.3	0.28	0.30
Floor	0.7	0.25	0.20–0.22	0.2	0.2	0.36	0.21
Windows	5.0	2.2	1.3–1.8	1.2	1.8	2.0	1.7
Weighted average	1.4	0.63	0.38–0.51	0.37	0.46–0.51	0.57	0.52
Curtain wall		1.6					
Air permeability		10	37				

Note: Air permeabilities (m³/m²/h) are also given for the UK. For Switzerland, lower values are required in some Cantons.
HDD = heating degree days (K-days). Weighted averages are computed as a 1:3:1:1 weighting of the roof, wall, floor and window values for illustrative purposes only.
Source: UK, Switzerland, Denmark: ODPM (2004); Rome, Helsinki, Stockholm, Eicker (2003)

Table 13.8 Heating performance standards (kWh/m²/year) for different kinds of housing in Germany under successively more stringent regulations

| Standard | Housing type | | |
	Detached houses	Row houses	Apartment blocks
Pre-1982 building stock	260	190	160
1982 upper limit	150	110	90
1995 upper limit	100	75	65
Low-energy house	<70	<60	<55

Source: Gauzin-Müller (2002)

13.1.3 Japanese and Chinese building energy codes

Table 13.9 presents information on the energy provisions of residential building codes in Japan. The legally binding requirements are written in terms of performance. This includes an overall building heat loss coefficient (which includes the effect of heat losses due to conduction through the envelope, and due to air exchange) and maximum permitted energy intensities (per unit of floor area) for heating + cooling.

Table 13.10 presents the U-values permitted under the 1995 regulations for new residential buildings in China. In climate zones with 2000 HDD, a larger U-value is permitted for walls if the window U-value is smaller. Smaller roof and wall U-values are required in buildings where the ratio of envelope area to floor area is greater than 0.3, compared to buildings where this ratio is less than 0.3. Chinese residential energy efficiency standards are discussed further in Lang (2004). Permitted envelope U-values are two or three times greater than in affluent countries in similar climates. For residential buildings in southern China, the maximum permitted window shading coefficient is regulated. For buildings with walls having minimal thermal resistance, the maximum permitted shading coefficient decreases from 0.8 for a window-to-wall ratio of 0.25, to 0.4 for a window-to-wall ratio of 0.45 (Huang et al, 2003).

Table 13.9 Performance standards and prescriptive guidelines for housing in Japan according to the 1999 standard (unless otherwise indicated)

	Climate zone and HDD (K-days)					
	I	II	III	IV	V	VI
	>3500	3000–3500	2500–3000	1500–2500	500–1500	<500
Overall building heat loss coefficient (W per m² of floor area per K)						
1980	3.26	4.19	5.12	5.58	7.91	–
1992	1.74	2.67	3.14	3.95	4.30	6.40
1999	1.6	1.9	2.7	2.7	2.7	3.7
Maximum heating + cooling energy use (kWh/m²/year)						
1999	108	108	128	128	97	81
Prescriptive guidelines						
Window U-values (W/m²/K)						
Centre of glass	2.33	2.33	3.49	4.65	4.65	6.51
Frame	1.91	1.91	2.91	4.0	4.0	–
Maximum window SHGC						
Facing ±30° from N	0.52	0.52	0.55	0.55	0.55	0.60
Other directions	0.52	0.52	0.45	0.45	0.45	0.40
Building-frame U-values (W/m²/K) for buildings with low thermal mass						
Roof	0.17	0.24	0.24	0.24	0.24	0.24
Wall	0.35	0.53	0.53	0.53	0.53	0.53
Floor	0.53	0.53	0.76	0.76	0.76	–
Insulation RSI-values (for buildings with low thermal mass)						
Roof	6.6	4.6	4.6	4.6	4.6	4.6
Wall	3.3	2.2	2.2	2.2	2.2	2.2
Floor	1.2	1.2	0.5	0.5	0.5	0.5
Leakage area per unit floor area (cm²/m²)						
1999	0.08	0.08	0.07	0.07	0.07	0.06

Table 13.10 U-values (W/m²/K) required for the envelopes of new residential buildings under the 1995 regulations in China

	JGJ 26-95		
Envelope Element	HDD (K-days)		
	2000	4000	6000
Roof, SC ≤0.3	0.80	0.60	0.40
SC >0.3	0.60	0.40	0.25
Wall, SC ≤0.3	0.92 (1.20)	0.65	0.52
SC >0.3	0.60 (0.85)	0.50	0.40
Window	4.70 (4.00)	3.00	2.00
Door	1.70	1.35	1.35
Floor over			
Air	0.60	0.40	0.25
Unheated basement	0.65	0.55	0.45
Ground, perimeter	0.53	0.30	0.30
Non-perimeter	0.30	0.30	0.30

Note: SC = shape coefficient (ratio of exterior surface area to floor area). Alternative U-values, involving a trade-off between window and wall values, are given in brackets for the 2000-HDD case.

13.2 Advanced residential buildings

This section summarizes published reports on the characteristics and measured performance of advanced houses around the world, on computer analyses comparing the energy use of advanced designs and conventional designs, and on programmes to promote widespread construction of residential buildings with significantly lower energy use than under current conventional practice. Examples of advanced HVAC systems in houses were presented in earlier chapters, incorporating either dynamic insulation (Chapter 3, Section 3.2.4) or a heat pump (Chapter 7, Section 7.3.7); advanced framing and insulation systems in existing advanced houses were discussed in Chapter 3 (Section 3.2.10); and advanced techniques to reduce envelope and duct leakage were discussed in Chapters 3 (Section 3.7) and 7 (Section 7.3.6), respectively. Energy-efficient HVAC systems rely on low distribution temperatures and therefore require advanced building envelopes consisting of high-performance windows, high degrees of insulation, very low rates of air leakage and heat recovery when mechanical ventilation is used.

13.2.1 Characteristics and performance of built and simulated advanced houses

Hamada et al (2003) summarize the characteristics and energy savings for 66 advanced houses in 17 countries. The majority of the houses surveyed have external-wall U-values of 0.1–0.2W/m^2/K (RSI 5-10, R28 to R57), and one-third of the houses have a window U-value $\leq 1.0W/m^2$/K. For the 28 houses where the saving in heating energy use is reported, the savings compared to the same house built according to conventional standards ranges from 23 per cent to 98 per cent, with eight houses achieving a saving of 75 per cent or better. Total purchased energy use ranges from 33–100kWh/m^2/year in 28 out of 37 houses where total purchased energy use is reported.

Deep reductions in residential energy use have been achieved under two voluntary standards in Europe. The first is the *Passive House Standard* – a house with an annual heating requirement of no more than 15kWh/m^2/year irrespective of the climate, and a total energy consumption of no more than 42kWh/m^2/year. By comparison, the average heating load of the existing residential building stock is about 60kWh/m^2/year in Rome and Stockholm and 90kWh/m^2/year in Helsinki (according to Eicker, 2003), 165kWh/m^2/year and 65kWh/m^2/year in old (pre-1970) and new (post-1990) housing in Zurich (Zimmermann, 2004), 140–170kWh/m^2/year for the average of existing buildings in Finland (according to the Tuusniemi fact sheet of Task 28 of the IEA Solar Heating and Cooling Program), 220kWh/m^2/year in Germany according to the CEPHEUS website given below (although only 65–100kWh/m^2/year under the 1995 regulations), and 250–400kWh/m^2/year in central and eastern Europe. Thus, Passive Houses represent a reduction in heating energy use by a factor of four to five compared to new buildings, and by a factor of 10–25 compared to the average of existing buildings. CEPHEUS (Cost-Effective Passive Houses as European Standards) is a programme whereby about 250 houses in five countries (Sweden, Germany, France, Switzerland and Austria) were constructed to the Passive House standard between 1998 and 2001. 'Cost-effective' houses are defined as those where the energy–cost savings pay back the additional investment cost in 30 years or less. Technical details, measured performance, design issues and occupant response to Passive Houses in various countries can be found in Krapmeier and Drössler (2001), Feist et al (2005), and Schnieders and Hermelink (2006), with full technical reports available at www.cepheus.de. Over 1000 Passive Houses have been constructed in Germany and several hundred each in Austria and Switzerland outside the CEPHEUS programme (Gauzin-Müller, 2002). Averaged over 13 projects in Germany, Sweden, Austria and Switzerland, Schnieders and Hermelink (2006) report that the additional cost is 8 per cent of the cost of a standard house, but that when amortized over 25 years at 4 per cent interest and divided by the saved energy, the cost of saved energy averages 6.2 eurocents/kWh (the range is 1.1–11 eurocents/kWh).

The other European standard is the Swiss *Minergie Standard,* which requires a heating energy consumption of <45kWh/m^2/year in new housing and <40kWh/m^2/year in new office buildings, or <90kWh/m^2/year in retrofitted pre-1970 housing and <70kWh/m^2/year in retrofitted pre-1970 offices. Electricity consumption cannot exceed 15kWh/m^2/year in either type of building, of any vintage. By 2000, about 750 Minergie-rated projects had been completed (Gauzin-Müller, 2002).

Table 13.11 gives indicative insulation and window U-values required to meet the Passive House standard in Rome, Stockholm and Helsinki; typical wall and window U-values in the coldest climates are 0.08W/m^2/K and 0.7W/m^2/K, respectively. Other features of Passive Houses are (i) a rate of air exchange at 50Pa pressurization not exceeding 0.6ACH, an exceptionally stringent standard (see Chapter 3, Section 3.7.1), and (ii) mechanical ventilation at a rate of 8.3litres/second per person and a heat exchanger effectiveness of at least 0.8. These houses cost 0–9 per cent more than conventionally built houses. The heating load is so small that, in some cases, passive solar heat gain, internal heat gains

Table 13.11 Indicative U-values (W/m²/K) and RSI-values required to meet the European Passive House standard of no more than 15kWh/ m²/year of heating energy

	Rome (1700 HDD)		Helsinki (5400 HDD)		Stockholm (5400 HDD)	
	U-Value	RSI-value	U-Value	RSI-value	U-Value	RSI-value
Roof	0.13	7.69	0.08	12.5	0.08	12.5
Wall	0.13	7.69	0.08	12.5	0.08	12.5
Floor	0.23	4.35	0.08	12.5	0.10	10.0
Window	1.4		0.70		0.70	
Overall	0.33		0.16		0.17	

Note: HDD = heating degree days (K-days).
Source: Eicker (2003)

and heat recaptured from ventilation exhaust air are sufficient to meet the entire heating requirements, while in other cases ground preheating of the supply air, an exhaust-air heat pump and/or a small supplemental heat source are required on the coldest days. The required heat can be distributed entirely through the small ventilation airflow. Badescu and Sicre (2003) provide detailed information on one passive house, the Pirmasens Passive House in Germany. Solar thermal energy from a 9m² collector with a 600litre stratified storage tank provides two-thirds of the DHW load, while the small heating load is met by a wood-pellet stove.

Hastings (2004) provides technical information on three multi-family buildings in Switzerland and two in Germany, all of which have a peak heating load of 10W per m² of living area or less when the indoor temperature is 20°C and the outdoor temperature is −10°C. Based on the characteristics of these buildings, Hastings recommends the following standards in cold climates:

- wall U-value ≤0.15W/m²/K (RSI 6.67, R38);
- whole-window U-value ≤0.8W/ m²/K and window SHGC (g-value) ≥0.5;
- ventilation heat recovery effectiveness ≥0.75.

So little heat is required in these houses that it can be difficult to provide the required heat without overheating. A regular wood-burning stove risks overheating the room where it is located. An alternative is an insulated wood-pellet oven that distributes heated water to other rooms. Other options are to use a heat pump to extract heat from the exhaust air to the ventilation air, or to place a small electric-resistance heating coil in the ventilation duct. In one project, bathroom temperatures are separately controlled by using the towel racks as radiators. Reducing the thermostat setting at night saves very little energy because the house is very slow to lose heat, while a very long time can be required to reheat the house if the temperature is allowed to drop substantially (to, say, 10°C) while on vacation because of the low capacity of the heating system. Thick insulation, air-tight construction, and night ventilation are effective strategies in maintaining comfortable summer temperatures without air conditioning in Germany and Switzerland, even during periods of extreme heat such as the summer of 2003, as long as the windows are shaded.

Figure 13.2 gives the annual heat budget for two Passive (or almost Passive) Houses in Germany. In the Berlin house, the heating system involves a large, two-storey tall, central insulated hot-water tank that stores summer heat for winter use. The heat supplied by the heating system offsets only one-quarter to one-half of the total heat loss, the balance being offset by internal heat gains and passive solar gains (in houses with typical heat losses, these gains would be insignificant by comparison). In other words, in superinsulated houses, the heat required from the heating system is a relatively small difference between heat losses and passive and internal heat gains. As a result, a change in the heat loss will disproportionately affect the heating requirements. For example, if the heat loss in the Rottweil house were to double (due to less insulation), the heating requirement would increase by more than a factor of

Figure 13.2 Heat gains and losses in two energy-efficient houses

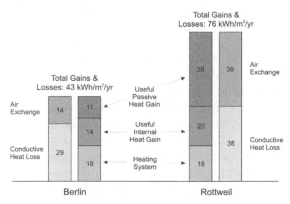

Source: Data from Erhorn et al (2003a, 2003b)

five (from 18kWh/m²/year to 94kWh/m²/year) (in fact, somewhat less, because the usable passive and internal heat gains would increase).

Ten advanced houses that have been built and monitored in Canada have achieved a 75 per cent reduction in energy use compared to typical house construction. A specific example is the advanced house built in Brampton in the early 1990s (discussed in Chapter 7, Section 7.3.7). It has:

- RSI-10.7 (R60) ceilings, RSI-7.0 (R40) above-grade walls, RSI-6.5 (R37) below-grade (basement) walls, and an RSI-1.23 (R7) basement floor;
- triple-glazed, low-e, argon-filled windows;
- a continuous air barrier, that prevents air leakage and protects the building fabric;
- a two-storey passive-solar sun space;
- an energy-efficient fireplace (with combustion air drawn from outside rather than from the interior of the house) and heat storage in masonry;
- a heat pump to recover heat from greywater, ventilation exhaust and excess solar gains, which is stored in a hot-water tank that replaces the furnace (additional energy is used to maintain the hot-water tank temperature at the required level);

- a single piece of equipment to replace the furnace, hot-water tank, air conditioner, and ventilation system;
- energy-efficient lighting throughout;
- energy-efficient appliances.

A comparable house built to the provincial standards at the time (which have not improved much since then) would have used 40,880kWh per year of energy (including energy for appliances and lighting), an R-2000 house would use 27,567kWh, while the advanced house uses only 11,221kWh (31kWh/m²/year) – a saving of 73 per cent. Peak load is 10kW, compared to 20kW for the conventional house.

The added component costs at the time were:

- windows: +30 per cent (3–6 year simple payback);
- integrated mechanical system: no extra cost;
- extra insulation (minor), extra framing: +15 per cent to +25 per cent;
- air-tight construction: +1 per cent of total cost of house.

It should be noted that, in residential buildings with high-performance envelopes, the indoor temperature setting and occupancy have a large relative impact on the heating energy requirements. This is because, with very small heat loss, passive solar and internal heat gains are a large fraction of the heat loss, and the heating requirement is a small residual. Thus, for mid-European conditions with a heat load of 32kWh/m²/year at 20°C indoor temperature with no people, increasing the thermostat setting to 24°C will double the heat load to 64kWh/m²/year (Streicher et al, 2003). Conversely, adding two people with the thermostat at 20°C reduces the heat load to about 12kWh/m²/year (given a floor area of 140m²). Thus, the behavioural conditions need to be carefully defined in order to properly compare different standards for high-performance residential buildings. European standards most likely assume a 20°C indoor temperature, whereas North American standards tend to assume a 22°C indoor temperature.

Parker et al (1998) show how a handful of very simple measures (attic radiant barriers; wider and shorter return air ducts; use of the most efficient air conditioners with variable speed drives; use of solar hot-water heaters; efficient refrigerators, lighting and pool pumps) can reduce total energy use by 40–45 per cent in single-family houses in Florida, compared to conventional practices. These savings are achieved while still retaining black asphalt shingle roofs that produce roof surface temperatures of up to 82°C! Further measures, such as use of a white tile roof instead of asphalt shingles, can increase the savings during the warmest month to over 50 per cent (D. Parker et al, 2000).[1] Replacement of standard, single-glazed windows with moderately advanced windows (double glazing, low-e, SHGC of 0.39 instead of 0.85) alone saves 17 per cent of cooling energy use in standard Florida houses and permits a 25 per cent downsizing of the cooling

equipment (Anello et al, 2000). Farrar et al (1998) report a 40 per cent savings in a prototype house built in the hot and dry climate of Arizona, again using only a handful of simple measures.

Holton and Rittelmann (2002) used the DOE-2 building simulation model to assess the impact of various upgrade measures on the total energy use of a medium size (246m² floor area), two-storey house in four US cities (Chicago, San Antonio, Minneapolis and San Francisco). Table 13.12 gives the simulated heating, cooling, hot water, lighting and appliance energy use for the base case house (built according to the MEC) and for the highest upgrade considered, and gives the technical characteristics of the base and upgraded houses. The latter reduce energy use in every case by at least 50 per cent. The maximum upgrade still leaves room for improving the wall insulation, windows and air tightness. If these were to be improved, and passive solar energy

Table 13.12 Energy use (kWh/m²/year) and characteristics of base case and upgraded two-storey houses in four US cities, as simulated by Holton and Rittelmann (2002)

	Chicago		San Antonio		Minneapolis		San Francisco	
	MEC	Adv	MEC	Adv	MEC	Adv	MEC	Adv
Energy use								
Heating	231.9	105.5	74.4	30.1	307.1	140.8	93.4	42.0
Cooling	7.9	3.6	35.3	11.4	5.9	2.2	0.2	0.1
DHW	30.2	12.2	24.4	9.2	32.1	13.2	30.2	12.2
Lighting	14.3	5.5	14.3	5.5	14.3	5.5	14.3	5.5
Appliances	24.2	19.2	24.2	19.2	24.2	19.2	24.2	19.2
Total	308.6	145.9	172.6	75.4	383.7	180.8	162.3	78.9
Saving		53%		56%		53%		65%
Measures								
Roof RSI-value	6.7	8.6	4.4	5.3	6.7	8.6	5.3	5.3
Wall cavity RSI-value	1.9	2.6	0	2.3	2.3	3.3	1.9	3.3
Sheathing RSI-value	0.4	0.9	0	0	0.9	0.9	0.9	0.0
Basement wall RSI-value	1.4	1.8	n/a	n/a	1.4	2.3	n/a	n/a
Floor RSI-value	3.3	3.3	3.3	3.3	3.3	3.3	3.3	3.3
Window COG U-value	2.78	1.48	2.78	1.65	2.78	1.48	2.78	1.82
Window SHGC	0.76	0.65	0.76	0.29	0.76	0.65	0.76	0.65
Infiltration ACH	0.5	0.1	0.5	0.1	0.5	0.1	0.5	0.1
Ventilation ACH	0.0	0.25	0.0	0.25	0.0	0.25	0.0	0.25
HRV efficiency	n/a	0.8	n/a	n/a	n/a	n/a	n/a	n/a
Duct efficiency	0.8	1.0	0.8	1.0	0.8	1.0	0.8	1.0
Heating AFUE	0.78	0.92	0.78	0.92	0.78	0.92	0.78	0.92
Cooling COP	2.9	4.1	2.9	4.1	2.9	4.1	2.9	4.1
Water heater EF	0.54	0.86	0.54	0.86	0.54	0.86	0.54	0.86
Hot-water use (lpd)	254	154	235	148	257	163	254	159

Note: MEC = US Model Energy Code.

were incorporated, such that heating energy use were reduced to 20kWh/m²/year (i.e. approaching the Passive House Standard), total energy use could be reduced by about 75 per cent in Chicago and Minneapolis. Use of solar thermal energy to meet a portion of hot water needs, and further measures to reduce air conditioning energy use could achieve a 75 per cent reduction target for San Antonio and San Francisco as well.

Gamble et al (2004) assess the economics of building low-energy (50 per cent savings) houses in three locations in the US: Phoenix (Arizona), Springfield (Missouri) and Albany (New York). These involve a combination of envelope and HVAC-system upgrades, energy-efficient lighting and appliances, and PV panels sized to produce as

much electricity as is consumed (with export to or import from the electric grid, as needed). Overall cost is minimized when the incremental cost (per unit of energy saved) of further reductions in energy demand equals the cost of PV energy. This optimum was calculated assuming a 30 per cent subsidy of the cost of PV systems (assumed to have an installed cost of $10,000/kW before the subsidy) and a subsidy of $750 for solar water heaters. Upgrade costs are converted into a monthly cost assuming a 30-year mortgage at 6 per cent interest. Opposing this are reductions in monthly energy costs. The specifications for these houses, the reductions in energy demand achieved, the net energy demand after subtracting PV energy production, and the net monthly cash flow, are given

Table 13.13 Characteristics, energy savings and net cash flow of hypothetical close-to-zero net energy houses in the US

House characteristic	Base case	Upgrade case		
		Phoenix	Springfield	Albany
Window area	18%	12%	12%	12%
Window U-value (W/m²/K)	2.7, 2.3, 1.6	1.7	1.7	1.7
Window SHGC	0.40, 0.68, 0.68	0.30	0.35	0.35
Attic insulation, RSI-value	4.40, 5.81, 7.22	7.75 (R44)	7.75 (R44)	7.75 (R44)
Wall insulation, RSI-value	2.11, 3.52, 4.23	3.70 (R21)	3.70 (R21)	4.23 (R24)
Wall sheathing, RSI-value	None	1.41 (R8)	1.41 (R8)	1.41 (R8)
Slab insulation, RSI-value	None, 0.53, 0.70	None	1.41 (R8)	1.41 (R8)
Roof solar absorption	0.75	0.35	0.35	0.35
Air infiltration	0.39, 0.54, 0.52	0.35	0.15+ERV	0.15+HRV
Air conditioner seasonal COP	2.9	5.6	5.6	4.4
Gas furnace AFUE	0.78	0.78	0.92	0.92
Heat pump seasonal COP	2.0	2.5	2.5	2.5
Duct leakage	15%	1%	1%	1%
Hot water	0.54 EF gas	Solar DHW	Solar DHW	Solar DHW
Exterior shading	None	2-m porch	2-m porch	2-m porch
Orientation of front	West	North	North	South
Lighting	Standard	Energy Star	Energy Star	Energy Star
Appliances	Standard	Energy Star	Energy Star	Energy Star
Gross energy savings		74%	56%	59%
Net energy savings		89%	63%	65%
Monthly cash flow		−$8	$1	−$42
Gross energy use (kWh/m²/year)		36.8	79.9	92.7

Note: ERV = energy (sensible and latent heat) recovery ventilator, 75 per cent assumed effectiveness. HRV = heat recovery ventilator, 75 per cent assumed effectiveness.
Source: Gamble et al (2004)

in Table 13.13. The reductions in gross energy demand range from 56 to 77 per cent and the reductions in net energy demand range from 63 to 89 per cent (averaged over three different house sizes and single and two-storey houses, for a total of six configurations). These energy savings are relative to a house based on the 2003 International Energy Conservation Code, a relatively stringent code (Table 13.1). The resultant gross energy use ranges from 40 to 90kWh/m²/year. For an electricity cost of 11.3 cents/kWh, one house achieves a positive cash flow while the other two have negative cash flows. However, installed costs of PV systems in the US are already sometimes less than assumed by Gamble et al (2004) and can be expected to continue to fall in the long run (see Chapter 11, Section 11.3.2). Note that these savings are achieved through modest incremental improvements in individual components of the house (more insulation, slightly better windows, more efficient equipment). The same is true in a similar study by Rudd et al (2004) of the measures needed to achieve 30–40 per cent savings in gross energy use in single-family houses in a variety of US climates. In both studies, alterations in building form to facilitate passive solar heating, use of thermal mass combined with night ventilation to meet cooling requirements (where applicable), or use of features such as earth pipe cooling, evaporative coolers or exhaust-air heat pumps are not considered.

Demirbilek et al (2000) find, through computer simulation, that a variety of simple and modest measures can reduce heating energy requirements by 60 per cent compared to conventional designs for two-storey single-family houses in Ankara, Turkey. These measures include increasing the thickness of insulation from 4.5 to 7cm for the floor, from 10 to 15cm for the roof, and from 1.2 to 2.4cm for walls; applying curtains at night to decrease the window U-value from 2.9 to 2.0W/m²/K and using a simple roller shade in summer; incorporating a small (7.7m²) Trombe wall on the south façade with summer shading and night insulation during the winter; and designing the stairwell to act as a solar chimney. The resulting heating energy use of about 66kWh/m²/year is still quite high, particularly given the mild winters (2700 HDD) in Ankara. The shading and passive ventilation features maintain comfortable (≤26°C) conditions with minimal energy use (5–10kWh/m²/year) by conventional mechanical cooling equipment.

Advanced buildings do not always perform as expected. Branco et al (2004) evaluated the energy performance of a multi-family building in the *Cité Solaire de Plan-les-Ouates*, near Geneva in Switzerland. This building contains rooftop hot-water solar collectors, an earth pipe loop for precooling ventilation air in summer, a heat exchanger between exhaust and supply air for use in the winter, and a modest thermal envelope (wall and window U-values of 0.25W/m²/K and 1.30W/m²/K, respectively). However, the measured natural gas use was about 50 per cent higher than predicted (68kWh/m²/year vs. 44kWh/m²/year) and the active solar contribution to the heat load smaller than predicted (19 per cent vs. 31 per cent). Part of the discrepancy arises from the fact that, because the energy use was expected to be so low, individual meters were not installed. As a result, winter temperatures in the apartments averaged 22.5°C rather than 20°C. Other important factors include greater heat loss through balcony floors than expected due to the fin effect (see Chapter 3, Section 3.9), 25 per cent higher measured window U-values than specified, and failure to maintain thermal stratification in the hot-water tank due to turbulence as hot water enters the tank. Branco et al (2004) recommend that mechanical systems be kept as simple as possible.

13.2.2 Moving beyond isolated examples

It has been difficult to induce construction of significantly more efficient houses beyond the scale of isolated demonstration homes, even when these homes are cost-effective from the purchaser's point of view and do not incorporate any features that change the appearance of the house or its operation, and in spite of the fact that an energy-efficient house is a more comfortable house due to more uniform temperatures and the absence of drafts. The Energy Star program of the US Environmental Protection Agency (EPA) has been extended to houses, and requires a minimum improvement of 30 per cent compared to the US Model Energy Code (MEC). As of the end of 1999, 10,525 houses had been certified as meeting the Energy Star criterion, with an average energy saving of 40 per cent compared to the MEC (Werling et al, 2000), while 397,822 had been certified by the end of 2004. The US Department of Energy (DOE) created the 'Building America' programme, which brings together integrated teams of architects, engineers, builders and equipment manufacturers to produce homes on a community scale that use 30–50 per cent less energy while also reducing construction time and waste (this is possible because framing systems using 2×6 studs on 24inch centres require less wood and assembly than systems with 2×4 studs on 16inch centres, while accommodating more insulation). So far, five teams involving more than 180 companies have been assembled and 1500 homes built. They have typically achieved a 50 per cent saving in heating energy and a 30 per cent saving in cooling energy by using simple measures with negligible impact on the final construction cost (Edminster et al, 2000). In Canada, the R2000 programme requires an energy savings of about 30 per cent compared to the Model National Energy Code. Although this programme, which includes training sessions for designers and contractors, has been in existence since the 1980s, only 8000 certified R2000 homes had been built as of 2000 (P. Parker et al, 2000). As noted above, over 1000 Passive Houses have been built in Germany and several hundred each in Austria and Switzerland, saving 75 per cent or more in heating energy compared to new houses.

Possible factors discouraging more widespread use of advanced designs include: (i) the need for thicker walls (to accommodate more insulation), which will subtract from the living space if outside dimensions are fixed; and (ii) the greater cost of more insulation and better windows. In principle, it should be possible to offset some of the cost of a more expensive building envelope with savings due to smaller heating and cooling equipment. In practice, smaller capacity heating equipment can be *more* expensive than larger capacity equipment at present due to the fact that only small quantities of small units are produced, due to the limited market for low-capacity equipment. In other cases, equipment (such as furnaces and air handlers) small enough to match the heating load of high-performance buildings are simply not available. In the long run,

manufacturers may provide lower-capacity equipment, and in larger quantities (bringing down costs) as the demand for such equipment increases. In both the short and long term, however, it is preferable to consider alternative systems altogether (such as small, wall-hung, integrated space and DHW boilers with some form of hydronic heating). Vacuum insulation or nano-fibre insulation may eventually allow thinner walls for the same insulation value, while the number of high-performance window products on the market is growing rapidly, and costs should fall. Incorporation of sunspaces and a large window area on south-facing walls, which can be done without increasing energy use if high-performance windows are used, can be used to increase the marketability of high-performance houses.

13.3 Advanced commercial buildings and the pivotal role of the design process

A commercial building is a complex system, with the energy use and performance of any one part of the system affecting the energy use of the building as a whole through a complex cascade of interactions. However, the typical design process for commercial buildings is a linear, sequential process that precludes the analysis and design of the buildings as an integrated system. In order to achieve deep savings in energy use, an integrated and iterative design process, involving all members of the design team, is required. A checklist of measures that need to be considered in the design of high-performance buildings is presented in Box 13.1.

BOX 13.1 Checklist for the design of commercial buildings with low-energy use

As documented in the main text, the best commercial buildings typically use one-quarter of the total energy use (heating, cooling, ventilation, lighting) of comparable existing buildings, and about one-half or less of the energy use that would result from following the building codes in force in North America. In this box, a checklist of the measures that lead to these dramatic energy savings is presented, based on the New York City and Pennsylvania guidelines for high-performance buildings (NYC DDC, 2000; PDEP, 1999). Additional recommendations pertaining to heat pumps, displacement ventilation, chilled ceilings and solar-powered desiccant dehumidification, not found in these guidelines, have been added here. The guidelines also discuss non-energy characteristics of high-performance buildings, which the interested reader may consult.

Siting

- Optimize the building shape and orientation to maximize the benefits of the solar and wind characteristics of the site, including cool breezes in summer and avoidance of cold winds in winter;
- landscape for energy conservation;
- plan for on-site food production and gardens;
- incorporate ground source heat pumps where feasible.

Envelope

- Optimize the thermal envelope before relying on building space conditioning systems;
- specify high-performance windows (U-value, spectral solar properties);
- specify high wall and ceiling insulation and light roof colour;
- use building mass to moderate temperature swings;
- use brick-cavity walls or pressurized curtain walls in preference to solid masonry;
- implement procedures to ensure very low envelope air leakage.

Interior layout of space

- Configure occupied spaces so as to maximize natural ventilation and daylighting;
- connect occupants to the natural environment whenever possible via operable windows and daylight – make sure that this can be accommodated by the HVAC design;
- design public areas and circulation zones to serve as thermal collectors and buffers – these spaces can accept a wider range of temperature swings, based on limited duration of occupancy;
- provide inviting, pleasant staircases to encourage use of the stairs rather than elevators in low-rise buildings;
- group similar room functions together in order to concentrate similar heating and cooling requirements and to simplify the HVAC zoning controls – determine the optimum locations within the building in order to take advantage of microclimatic conditions and building orientation;
- locate HVAC equipment so as to minimize HVAC loads;
- where choices exist, place non-windowed spaces on the northern side of the building (in the northern hemisphere) against an exterior wall so as to create a thermal buffer.

Lighting

- Maximize the use and penetration of natural light, via light shelves, fibre-optic light pipes, and low partitions in offices next to windows;
- use light sensors in conjunction with dimmable lights in order to capture the benefits of partial daylighting;
- use occupancy sensors to turn lights off when a room or area is unoccupied;
- use high-efficiency lighting systems (luminaires, ballasts, reflectors);
- use interior colours and finishes that minimize the need for artificial light;
- provide lumen maintenance controls so that excess light output is not needed when a lighting system is first installed in order to compensate for degradation in light output that otherwise occurs over time;
- provide task lighting.

Mechanical systems

- Use a variable air volume, demand-controlled, displacement ventilation system with 100 per cent outside air;
- use desiccant dehumidification (latent heat exchangers) of incoming ventilation air;
- use sensible heat exchangers between incoming and outgoing ventilation air;
- use separate heating and cooling systems for zones with different thermal requirements;
- give building occupants control over individual room environments.

Heating and cooling systems

- Wherever simultaneous heating and cooling are required, use heat pumps to integrate these needs;
- consider chilled ceiling radiant cooling, particularly in combination with displacement ventilation;
- consider exhaust-air heat pumps, in addition to exhaust/supply air heat exchangers, as a means of providing heating whenever heat from the cooling system condensers is not adequate to meet the heating and hot water needs;
- consider ground source and solar assisted heat pumps for heating when waste heat from the cooling system and recovered from the exhaust air is not adequate, and consider ground source heat pumps for air conditioning in place of conventional chillers with rooftop cooling towers;
- minimize heat distribution temperatures and maximize chilled water distribution temperatures;
- insulate pipes and ducts, and ensure minimal duct leakage.

Electrical systems

- In large buildings, consider distributing power at 480/277 volts rather than at 208/120 volts;
- specify energy-efficient office equipment, including computers, printers and copy machines;
- improve the power factor by specifying appropriate equipment as required;
- use K-rated transformers to serve non-linear equipment;

- utilize direct current (DC) from photovoltaic or fuel cell power sources for use in applications where DC is more appropriate than AC.

Energy load management

- Provide an energy management and control system that collects and displays data in graphical form and can automatically operate HVAC equipment;
- use only those control systems that allow the desired temperatures to be adjusted without complete reprogramming.

Materials

- Choose materials with the lowest possible embodied energy that will accomplish the task;
- specify locally manufactured materials and products to minimize transportation energy use;
- use salvaged materials whenever possible.

The guidelines enumerated above pertain to reducing a building's energy needs through efficient design, including incorporation of passive cooling and ventilation and of natural daylighting. Additional, active use of solar energy through photovoltaic panels that are part of the building fabric or mounted on the roof, and through roof mounted solar panel water heaters, is discussed in detail in Chapter 11.

There are a number of factors that inhibit designing commercial buildings as integrated systems. These are discussed in *Scenarios for a Clean Energy Future* (Koomey et al, 2000) and summarized below.

- First, the prevailing fee structures of building-design engineers are explicitly or implicitly based on a percentage of the capital cost of the project (more costly projects result in greater fees to the designers).[2]
- Second, and compounding this problem, dramatically more efficient designs reduce the energy costs *and* (frequently) the upfront capital costs (reducing designers' commissions), but require more work.

- Third, heavy reliance is placed, in the design process, on 'rule-of-thumb' equipment sizing. The result is that HVAC equipment is typically oversized by a factor of two to three, greatly increasing energy use and capital costs.
- Fourth, legal and liability concerns result in 'defensive design and institutionalized conformity' (Lovins, 1992). Unfortunately, there is no such thing as state-of-the-art technology with a 20-year track record.

In technical terms, however, it is quite easy to achieve energy savings of 25–50 per cent relative to current practice, such as the ASHRAE

Table 13.14 Comparison of specifications and energy use in four buildings as built according to the Canadian building code and with proposed energy saving alterations

Location	Winnipeg, Manitoba		Vancouver, British Columbia		Halifax, Nova Scotia		Ottawa, Ontario	
Type	Office		Office/retail		School		Retail	
Design	Code	Proposed	Code	Proposed	Code	Proposed	Code	Proposed
Wall RSI-value	2.88	4.98	1.24	1.52	2.10	2.56	1.84	6.21
Roof RSI-value	4.29	5.32	2.08	5.29	2.39	3.72	2.08	7.25
Window U-value (SI)	2.09	1.93	3.20	2.17	3.21	1.93	3.25	1.60
LDP (W/m²)	12.3	12.6	24.5	13.2	16.6	13.4	30.0	14.6
SWH efficiency (%)	100	100	100	100	80	80	80	94
Pump type	Fixed	Fixed	Fixed	Variable	Fixed	Fixed	Fixed	Fixed
Heating efficiency (%)	300	327	80	80	80	87	80	82
Cooling COP	2.50	5.10	3.80	3.25	n/a		2.50	2.64
Energy savings (%)		44		33		21		33

Note: LDP = lighting power density, SWH = service water heating.
Source: Mottillo (2001)

90.1-1999 code. Mottillo (2001) performed an analysis of the sensitivity of building energy use to variations in a number of basic building characteristics. The specifications and energy use for four of the ten buildings that she analysed are given in Table 13.14. Many of the proposed specifications are not particularly stringent, and well-tested features such as advanced daylighting, and passive ventilation, heating and cooling, are not part of the proposed designs. Nevertheless, the projected energy saving ranges from 21–44 per cent.

An analysis of almost one full year of energy use data from two new buildings (six and eight stories high) in Oakland, California, further illustrates how little effort is needed to beat existing building codes by a substantial margin (Motegi et al, 2002). In this case, the buildings were subject to a performance contract, whereby the developer would be rewarded if the building beat the California energy code by more than 25 per cent, and penalized if the building did not beat the code by 25 per cent. The buildings managed to beat the code by more than 25 per cent, in spite of:

- using a standard central VAV HVAC system with hot-water reheat;
- using grossly oversized chillers (average load 20 per cent of

capacity, peak load 60 per cent of capacity), with an average operating COP of 4.4 instead of the manufacturer's specification of 8.0;
- operational control problems, such that the boilers started to heat the building every morning during the summer at 6am, then shut down at 11am when the chillers began to operate![3]

ASHRAE, in conjunction with the American Institute of Architects, the Illuminating Engineering Society of North America, the New Buildings Institute, and the US Department of Energy, released an energy design guide for small office buildings in 2004 to help designers achieve a 30 per cent energy saving compared to ASHRAE 90.1-1999. The latest code, ASHRAE 90.1-2004 is expected to achieve a 20 per cent energy saving relative to the 1999 version, largely through better thermal envelope, lighting and mechanical equipment.

A number of national governments are developing programmes to encourage the building industry to achieve even better improvements. The US Department of Energy, through the National Renewable Energy Laboratory (NREL) in Golden, Colorado, has an 'Exemplary Buildings Program' that works with building design and construction teams to produce low-energy buildings, with a goal of eventually producing buildings that use 70 per cent less non-renewable energy than a building built according to ASHRAE 90.1-1989 (which differs little from 90.1-1999), with minimal or no increase in construction cost (Hayter et al, 1998). As of late 2002, 19 high-performance buildings had been designed under this programme and 12 constructed (four residential and eight commercial) (Murphy, 2002). Two of NRELs own buildings save 45 per cent and 63 per cent in annual energy use, with lighting energy use reduced by 75 per cent. A home-improvement store and warehouse recently built in Silverthorne, Colorado with design assistance from NREL uses 54 per cent less energy compared to ASHRAE 90.1-2001, not counting the contribution from building-integrated PV panels (Hayter et al, 2000; Torcellini et al, 2004a). In the Silverthorne case, efficient electric lighting and daylighting reduced lighting energy use by 93 per cent in the warehouse and by 67 per cent in the office and retail areas.

In 1993, Natural Resources Canada (NRCan) launched its own programme for advanced commercial buildings, the C-2000 Programme, to encourage the construction of buildings using 50 per cent or less of the energy required under standard practice. Initially, the benchmark was ASHRAE 90.1-1989, but later this was changed to the 1997 MNECB. To encourage industry participation in the programme, other characteristics of high-performance buildings – such as health and comfort, functional performance, longevity, adaptability of building systems, and operation and maintenance issues – were targeted, in addition to energy use. Initially, the C-2000 programme provided financial assistance to cover both the design and construction of advanced buildings, but it emerged that the most important ingredient in producing high-performance buildings is the design process, so in the second phase (C-2000 Design Facilitation, or C-2000 DF), support was provided for only the design process. Eleven buildings were built under the C-2000 programme and achieved their energy-performance target. Details of the C-2000 programme and information on the 11 completed buildings can be found by searching for C-2000 at oee.nrcan.

gc.ca/commercial. Initial demonstration projects ended up costing 4–14 per cent more than the base-case design (while cutting energy use in half), while later projects came in at the same or *less* cost than the base-case design. In order to broaden industry participation, the C-2000 programme was merged with another NRCan programme, the Commercial Building Incentive Programme (CBIP), which had been established in 1997 with a minimum required savings of 25 per cent. As in C-2000 DF, support is provided only for the design phase in order to compensate for the extra cost and time required to design more energy-efficient buildings. Although the energy savings required by this phase of the C-2000 programme is very modest, participation has been good (about 300 projects by 2001) and a sample of 43 had an average predicted energy saving of 37 per cent (Larsson, 2001).

The following material is excerpted directly from the NRCan C-2000 website, and underlines the importance of the design process in achieving significant energy savings:

Based on the experience in Phase 1, the C-2000 DF process relies on the following elements to increase the likelihood of high performance:

- *an insistence on close teamwork by all members of the design team from the beginning of the design process, so that the performance and cost implications of various design options are considered in a holistic way, and at an early stage,*
- *the involvement of one or more design facilitators in most design meetings, to act as a guide to performance options and a link to various contracted specialists,*
- *augmentation of normal design team expertise with an energy engineer, an environmental specialist and a cost consultant,*
- *the availability of a roster of specialized technical experts who can be called in at short notice to assist the design team in issues such as daylighting, thermal storage or other specialized technical areas,*
- *the use of a clear and comprehensive technical guideline document, such as that*

developed for the C-2000 programme,

• *the development of short written performance strategies by the design team, so that performance targets can be established within each performance area and the costs and benefits of each option examined, and*

• *during the construction phase, commissioning of all major systems including the building envelope to ensure that systems are properly installed and perform to designed levels.*

... there is a consensus amongst participants that the largest single factor appeared to be the strong teamwork amongst architects, engineers, energy analysts and others that begins early in the design process... Thus, a preliminary but significant result of the first phase of the C-2000 program was the apparent importance of process as compared to technical wizardry in achieving high performance ... it appears to be teamwork and the careful integration of a number of relatively conventional technologies into the process that allowed C-2000 designers to achieve relatively high performance levels with minimal costs.

Results to date indicate that, if an integrated teamwork process can be implemented in the context of well-articulated guidelines and specialized support we may expect a 35–50 per cent improvement in current energy performance levels with only modest increases in design budgets. The combination of such an approach with truly advanced technologies could have spectacular results. However, there is an undoubted reluctance within the design profession, particularly amongst architects, to change their established relationships, and this will require years of patient prodding to overcome.

These findings are consistent with two recent projects in The Netherlands, where implementation of an improved design process achieved reductions in projected energy savings of greater than 30 per cent in one case and greater than 35 per cent in another case (Poel, 2004).

In recognition of the importance of the design process, two other programmes – the California *Savings By Design* programme and the German *Solar optimized building – SolarBau* programme – are used to fund the extra cost of designing and monitoring (but not constructing) highly energy-efficient buildings (see www.savingsbydesign.com and Reinhart et al, 2000). The *Savings by Design* programme also provides assistance to design teams in the use of building energy simulation tools. Its goal is a minimum energy savings of 30 per cent compared to the California Title 24 standards, although savings as large as 60 per cent have been achieved. The *SolarBau* programme has as its target a primary energy consumption not exceeding 100kWh/m²/year (excluding office equipment), compared with 300–600kWh/m²/year for typical German and Swiss office buildings. Measured performance information on ten buildings in the programme where at least one year of data were available by 2003 is given in Wagner et al (2004). Five of the ten buildings achieved the 100kWh/m²/year target, but no building uses more than 140 kWh/m²/year of primary energy. Additional costs are reported to be comparable to the difference in cost between alternative standards for interior finishings. Two of the ten buildings have cogeneration and five have some PV power (supplying one-third to two-thirds of total electricity demand). Heating energy alone ranges from 20 to 70kWh/m²/year, a result of mean envelope U-values of 0.21–0.43W/m²/K and ACH_{50} values not exceeding 1.2. Passive cooling techniques – either mechanical night ventilation with exposed thermal mass or hydronic cooling integrated with groundwater – are used in combination with measures such as external shading, reduced glazing area and minimal internal heat gains in order to reduce the cooling loads. This strategy permitted the elimination of mechanical air conditioning for most of the floor space. Earth-to-air heat exchangers save very little energy but are essential to maintaining comfortable indoor temperatures using passive and low-energy cooling techniques. Installed lighting power density is generally in the range of 6 to 14W/m².

Table 13.15 provides summary information on commercial buildings from around the world that have achieved 45–85 per cent savings in energy use compared to typical new buildings in each region.

Based on the experience summarized above and in Table 13.15, the following conclusions can be reached:

Table 13.15. Summary of exemplary (in terms of energy use) new commercial buildings where baseline and reference energy use have been published

Building and location	Energy use	Energy savings	Reference for comparison of energy use	Key features	Reference
US examples					
NREL offices and labs, Golden, Colorado		45% & 63% (two buildings)	ASHRAE 90.1		Murphy (2002)
Federal Courthouse, Denver		50%	ASHRAE 90.1-1989	Triple glazing, modest insulation, sunshading, daylighting, T5 lamps, VAV displacement ventilation, direct and indirect evaporative cooling, VSD on all air handlers and pumps, BiPV	Mendler and Odell (2000)
Home improvement store, Silverthorne, Colorado		54%	ASHRAE 90.1-2001	Daylighting and efficient electric lighting	Torcellini et al (2004a)
SC Johnson Wax Headquarters, Racine (Wisconsin)	<218kWh/m²/year total	54% / 69%	Ave new buildings Existing SCJ buildings	Daylighting with automatic controls, fixed and adjustable shading, demand-controlled desktop personal air supply	Mendler and Odell (2000)
Academic building, U. of Wisconsin, Green Bay		60%	Wisconsin energy code	Wall U-value 0.16W/m²/K Roof U-value 0.11W/m²/K Skylights with suspended reflectors and motorized blackout panels BiPV	Mendler and Odell (2000)
Center for Health and Healing at the Oregon Health and Science University, River Campus		60%	ASHRAE 90.1-1999	Hybrid ventilation, solar preheating of ventilation air, heat recovery, radiant heating/cooling, demand-controlled displacement ventilation, PV modules as exterior shading, commissioning	Interface Engineering (2005)
Federal Reserve Bank, Minneapolis	<134kWh/m²/year total 9.1W/m² connected lighting load 7.0W/m² average lighting load	74%	ASHRAE 90.1	Window U-value 0.74W/m²/K Wall U-value 0.2W/m²/K Conventional VAV HVAC	Mendler and Odell (2000)
Environmental Technology Centre, Sonoma State University, California		80%	California Title 24		Beeler (1998)
WRI Headquarters, Washington	8.2W/m² connected lighting load 4.3W/m² average lighting load	83%	ASHRAE 90.1	Daylighting with automatic controls	Mendler and Odell (2000)

Canadian examples					
Green on the Grand (offices), Kitchener, Ontario	81.2kWh/m²/year (design total) Natural gas 43.1kWh/m²/year Electricity 38kWh/m²/year	50.4%	ASHRAE 90.1-1989	Double-stud manufactured wood-frame wall; fibreglass-frame, triple-glazed, double-low-e, argon- filled, insulating-spacer windows; reduced lighting power densities; radiant heating and cooling panels, DOAS w/ ERV, natural gas fired absorption chiller; outdoor pond replaces conventional cooling tower	C-2000 Internal Programme Report[a]
Crestwood Corporate Centre Building No.8, Richmond, BC	62.6kWh/m²/year (design total) Natural gas 14.2kWh/m²/year Electricity 48.4kWh/m²/year	51.7%	ASHRAE 90.1-1989	Tilt-up concrete walls with upgraded air tightness and insulation; thermally broken Al- framed DG low-e windows; reduced lighting power densities; high efficiency boiler and chiller; 4-pipe fan coil system w/ DOAS	C-2000 Internal Programme Report[a]
MEC Retail Store, Ottawa	202.8kW/m² annual (design) Natural gas 110.3kWh/m²/year Electricity 92.5kWh/m²/year	56%	MNECB	Upgraded wall and roof insulation, DG low-e argon-filled, warm edge spacer windows in clad wood or TB Al frames; TG low-e windows on north faces; roof monitors for daylighting and greatly reduced connected lighting power; high efficiency boiler, mid efficiency rooftop ventilation unit, ERV ventilation heat recovery, upgraded chiller efficiency, CO_2 DCV, variable speed fan drives	CBIP Internal Technical Review Report[a]
MEC Retail Store, Winnipeg	101.5ekW/m² annual (design) Natural gas 41.9 ekWh/m² Electricity 59.6kWh/m²	55.5%	MNECB	Upgraded insulation in walls and roof, low fenestration-to-wall ratio, DG low-e argon-filled warm edge spacer windows in TB aluminium frames; daylighting w/ occupancy sensor controls and reduced connected lighting power; mid-efficiency boiler, DOAS, radiant slab and panel heating with groundwater cooling	C-2000 Internal Programme Report[a]
SC3 Smith Carter Office, Winnipeg	142.8kW/m²/year (design) Electricity 142.8kWh/m²/year	54.9%	MNECB	Upgraded insulation in walls and roof; DG low-e argon-filled warm edge spacer windows in TB Al frames; daylighting w/ wireless digital and occupancy sensor controls; exterior solar shading, reduced connected lighting power, combination boiler and ground source heat pump w/ GSHP sized for cooling, DOAS w/ UFAD	CBIP Internal Technical Review Report[a]
MEC Retail Store, Montreal	147.3kW/m²/year (design) Electricity 147.3kWh/m²/year 159.0kW/m²/year (actual 2004)	68%	MNECB	High-performance envelope, daylighting, GSHP, DOAS, radiant slab heating and cooling, earth coupled OA tempering	Genest and Charneux (2005)

Centre for Interactive Research on Sustainability, Vancouver (proposed design)	56kWh/m²/year without BiPV and solar thermal (47kWh/m²/year with solar)	84%	Typical existing building (353kWh/m²/year)	High-performance envelope, adjustable atrium shading, hybrid ventilation, daylighting, VSDs, DCV, 90% heat recovery effectiveness	Hepting and Ehret (2005)
European examples					
Brundtland Centre, Denmark (Figure 11.17)	50kWh/m²/year	70%	Typical comparable building (170kWh/m²/year)		Prasad and Snow (2005)
Center for Sustainable Building, Kassel, Germany	16.5kWh/m²/year heating 32–42kWh/m²/year total energy use	73% 76–82%	1995 German Building Code Typical office building	Wall U-value 0.11W/m²/K, window U-value 0.8W/m²/K, radiant slab heating and cooling, ground heat exchanger (COP = 23), hybrid ventilation, daylighting	Schmidt (2002) and Schmidt (personal communication, 2006)
Debis Building, Potsdamer Platz, Berlin (Figure 3.36)	75kWm/m²/year total	80%		Double-skin façade and passive ventilation	Grut (2003)
Ionica Building, UK	64kWm/m²/year total	46%	Good-practice air conditioned building	Hybrid ventilation	Hybvent website (http://hybvent.civil.auc.dk)
Solar Bau programme, ten buildings in Germany	25–140 kWh/m²/year primary energy excluding office equipment	50–90%	Typical office buildings, 300–600kWh/m²/year primary energy	Mechanical night ventilation with exposed thermal mass or hydronic cooling integrated with groundwater, external shading, reduced glazing area, minimal internal heat gains, efficient lighting	Wagner et al (2004)
Solar Office, Doxford International Business Park, UK	85kWh/m²/year	80%	Typical new air-conditioned buildings in the UK (400kWh/m²/year)	Passive ventilation and cooling; BiPV functioning as partial shading devices	Prasad and Snow (2005)
Asian examples					
Kier Building, South Korea	68kWm/m²/year electricity, 18kWm/m²/year heat			Double-skin façade, ground coupled heat exchanger, solar thermal and PV	Prasad and Snow (2005)
Liberty Tower, Meiji University, Japan		48%	Japanese building code	Hybrid ventilation	Hybvent website (http://hybvent.civil.auc.dk)
Tokyo Gas (Earth Port) (Figure 6.39)	380kWm/m²/year primary energy	45%	Typical office building in Japan	Hybrid ventilation, daylighting, external shading	Baird (2001)
Torrent Pharmaceutical Research Centre in Ahmedabad, India (Figures 6.27 and 6.28)	Electricity use 36% of a conventional modern building			Evaporative cooling and hybrid ventilation (passive downdraught cooling)	Ford et al (1998)

[a] Available from Stephen Pope, Natural Resources Canada.

1   with nothing more than a better design *process* but using conventional technology in conventional buildings, savings in total energy use by commercial buildings of 35–50 per cent can be achieved, compared to the ASHRAE 90.1-1989 standard (or compared to the ASHRAE 90.1-1999 standard, which is not much different), at no or little additional upfront cost;

2   by combining a holistic, systems-oriented design process with advanced technologies and less conventional approaches to meeting the building energy needs, substantially larger energy savings (up to 75–80 per cent) can be achieved.

Corresponding to these two achievement levels are two levels of design integration: basic and advanced. These integrated design processes, how they differ from the conventional design process, and the tools available to assist in the integrated design process, are explained in the following subsections.

13.3.1 Integrated design process

Lewis (2004) defines the integrated design process as 'a process in which all of the design variables that affect one another are considered together and resolved in an optimal fashion'. The traditional design process proceeds in a linear fashion, as a sequence of steps, with no reconsideration of the initial steps in light of the results during the first pass through the latter steps. To worsen matters, the sequence of linear steps in the traditional design process is often opposite to the natural flow of the design logic. For example, and as explained by Lewis (2004), HVAC capacity is usually calculated (and equipment selected) before the lighting design is done or the choice of glazing is finalized. The cooling equipment ends up being oversized, but usually enough space is not provided for ductwork, which has to be undersized in order to fit, resulting in higher pressure drops and increased noise. This is not to say that there is no integration or teamwork in the traditional design process, but rather, that the integration is not normally directed toward minimizing total energy use through an iterative modification of a number of alternative initial designs and concepts so as to optimize the design as a whole (rather than optimizing individual subsystems).

Two steps in the design process can be said to be 'integrated' if the effect of the first design decision on the second design decision is taken into account when making the first design decision. The greater the number of design decisions that are linked in this way, the more 'integrated' is the design. Integrated design means, as a minimum, to consider the impact of a change in early design decisions such as building shape and orientation, envelope characteristics and the inclusion of heat recovery on subsequent decisions, such as the nature and sizing of mechanical systems. A number of early alternative choices should be considered with the goal of minimizing some objective criterion such as lifecycle cost, net present value of capital and operating costs, or energy cost alone subject to some upper limit concerning the acceptable payback time. If this is complemented by the choice of the most efficient equipment available, optimization of the equipment operation, and commissioning to ensure that everything works as intended, then savings of the order of 35–50 per cent

Figure 13.3 Steps in the basic integrated design process (IDP)

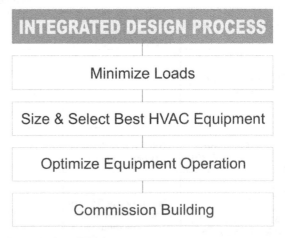

Source: Todesco (2004)

can be achieved relative to ASHRAE 90.1-1989 or − 1999, as shown above. This is the basic IDP, and it is illustrated in Figure 13.3. It is a simple 'back-to-basics' approach that stresses a quality envelope, quality equipment, and quality sizing and operation of mechanical equipment. Buildings produced in this way are still conventional buildings, but computer simulation will be needed if one wants to optimize the design choices.

To push the savings beyond the 35–50 per cent achievable through the 'back-to-basics' approach illustrated in Figure 13.3, an increasing number of unconventional measures will need to be carefully combined – measures such as passive ventilation, heating, and cooling involving perhaps double-skin façades and airflow windows; thermal mass with night ventilation; chilled ceiling, displacement ventilation and desiccant dehumidification; daylighting; and adaptive thermal comfort. Computer simulation involving simulation specialists who serve as a liaison between the architects and engineers is essential. This design process is compared with the conventional design process in Figure 13.4, with further elaboration in Box 13.2. The integrated design process entails two-way interactions between the client and design team, between the architect and engineers, and between the design team and the contractors. Once the design is complete, the design team must be available during construction to explain details that are not clear, because no matter how thorough the plans and specifications, some details that affect the energy use by the building will be overlooked.[4] The integrated design process in a number of buildings in Europe and North America is discussed in some detail, along with lessons learned, in IEA (2000c, 2002c).

In order to foster good working relationships within the design team, Esource (1998b) recommends the use of a coordinator. The coordinator needs to be, 'an effective communicator and a good negotiator' who may be called upon 'to challenge the design team to innovate or to find a better solution to a given problem' and 'to call upon independent analyses …in order to convince sceptical architects and engineers to change their approaches'. The extent to which the latter is required depends, of course, on the experience and competence of the other members of the design team. The coordi-

Figure 13.4 Conceptualization of the design process (a) as commonly practiced when the client will not occupy the building, (b) as commonly practiced when the client will occupy the building, and (c) for the highly integrated approach needed to produce a high-performance building

Level 1:

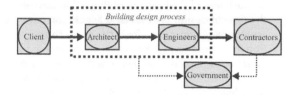

Level 2:

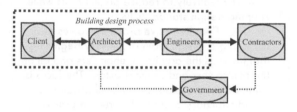

Level 3:

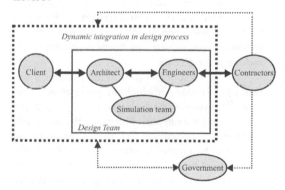

Source: Hien et al (2000)

BOX 13.2 The importance of process in designing high-performance buildings

The *Guidelines for Creating High Performance Green Buildings* published by the Pennsylvania Department of Environmental Protection describe the typical design process as a linear, compartmentalized progression from design to construction to occupancy, with most decisions driven by initial costs, time requirements and the quality of the product desired. Each party completes its task on a 'need-to-know' basis and passes it on to the next party. It is a system in which design professionals and consultants are paid a percentage of the project costs, thereby providing little incentive to work creatively to reduce project costs while keeping design standards high.

The alternative process involves establishing a design team at the very beginning that involves the architects, engineers and all of the contractors and subcontractors. The focus is shifted from a compartmentalized process to a multidisciplinary process. This design process is more expensive than the traditional design, but the greatest opportunities to reduce capital costs, energy use and operating costs will be identified during the early team-building and design-optimization stages. This is described as a 'front-end loaded, human energy-intensive exercise'. It requires a fee structure that rewards solutions that reduce project costs.

An important part of this process is design optimization, in which the performance and cost-effectiveness of all elements of a project are refined and maximized. Normally, the emphasis is placed on optimizing design choices based on budget or time considerations, without evaluating their impact on total building performance. Components such as window glazing systems, thermal envelope and mechanical systems and their distribution need to be considered along with the breakdown of the building into different zones (with different heating, cooling and ventilation requirements) and the shape, orientation and layout of individual rooms and common areas. This can be done only through computer simulation of the entire building. Commonly used programmes (among high-end design teams) are the DOE-2 and ENERGY-10 programmes. In the words of the Pennsylvania *Guidelines*:

> Decision makers must encourage project team members to vigorously undertake the optimisation process. Team members must be capable and willing to evaluate each optimisation suggestion or opportunity as it presents itself throughout the design and construction process. A commitment to continuous refinement and design optimisation is the best insurance of a cost-effective, high-performance green building.

An effective high-performance design team will plan education and orientation sessions into the bidding and negotiation sessions. The design team will establish an atmosphere in which information is shared and questions and recommendations are encouraged. This will help to ensure that construction is carried out in an informed and responsible manner; otherwise, the performance of the building will be undermined, in spite of good design. Commissioning – the process of ensuring that the building systems are properly installed and function as designed – is another critical step in the creation of high-performance buildings. A study of 60 commercial buildings by the US DOE found that:

- more than half had control problems;
- 40 per cent had problems with HVAC equipment;
- one-third had sensors that were not operating properly;
- 15 per cent were missing specified equipment!

nator should be hired before the design competition begins, and used as a resource for each of the competing firms so that they can prepare more innovative submissions than they would otherwise submit.

Torcellini et al (1999) identify and describe nine steps in an integrated design process for low-energy buildings. The steps that they identify are:

1   Simulate a base-case building that meets the energy efficiency requirements of ASHRAE 90.1 or 90.2.
2   Simulate the energy use for a building in which various loads (such as conductive losses, solar gains, internal heat gains) are eliminated in order to determine which factors are most important.
3   Have a brainstorming session with the entire design team to develop possible ways to minimize energy use.
4   Perform simulations with variants of the base-case building that incorporate the solutions identified in step (3).
5   Architectural team prepares a preliminary set of drawings.
6   Determine the HVAC system needed to meet the predicted loads.
7   Finalize plans and specifications, ensuring that the building is properly detailed to avoid questions during construction. Ensure that specifications are correct. The final design should achieve a 50–70 per cent energy savings.[5]
8   When unforeseen circumstances require design changes during construction, rerun the simulation to ensure that proposed changes do not adversely affect energy use.
9   Commission all equipment and controls, and train building operators in the efficient use of the equipment.

de Wilde and van der Voorden (2004) present a strategy for the selection of energy saving components during the building design process (steps 3, 4 and 6 from above). This strategy draws upon principles of systems engineering and decision theory. The elements of the strategy are:

1   To define an option space, consisting of different energy savings combinations that are to be considered.
2   To identify all the performance characteristics (dimensions) of all the design options, in order to determine a set of criteria for the selection of the preferred option.
3   To specify objectives, constraints and performance indicators.
4   To predict the performance of the design options, primarily using computer tools but possibly also using experimental set-ups.
5   To evaluate the predicted performance, based on how well each option performs in the various performance dimensions, with some weighting of the different performance dimensions.

Scofield (2002) documents the many flaws in the design of what was supposed to be an exemplary 'green' building at a US college that will not be named here. Measured energy consumption was found to be three times that expected. Most of the errors documented by Scofield will hopefully appear as rather obvious to anyone who has read this book this far. Scofield (2002) lays much of the blame on a lack of integration during the design process. As he states:

Rather than a design team that interacted regularly, the team was assembled from geographically scattered, highly paid consultants with limited time to devote to this project. Each consultant focused on his/her area of specialty and paid little or no attention to the activity of others. The building owner provided inadequate technical oversight of the project, relying instead on the architect to manage the project.[6]

The design process described by Scofield (2002) stands in sharp contrast to the process used for the design of the Environmental Technology Centre at Sonoma State University (California), described in some detail by Beeler (1998). In this case, a fully integrated team was involved from the very beginning in the design of a building expected to use less than 20 per cent of the energy that would be used if it had been built to the California energy code (one of the more stringent in the US).[7] Details on the design process in five other high-performance buildings, in Canada, Denmark, Germany and The Netherlands, can be found in Poel et al (2002).

A study of barriers to the construction of energy-efficient buildings in the UK also highlights the need for an integrated, teamwork-based approach to building design in place of the linear approach that is the norm (Sorrell, 2003). The construction industry in the UK (and elsewhere) is fragmented, dominated by subcontracting, and commonly sees new design teams, constructors and suppliers assembled for each project. In the words of Sorrell: 'This inhibits co-ordination and learning and prevents the development of skilled and integrated teams who may be able to deliver green buildings'. Notwithstanding the fact that competently designed energy-efficient buildings should cost no or very little more than standard buildings, perceived capital constraints and the pressure to finish on budget often lead to the exclusion of energy-efficient features at the last minute; as a building is nearing completion and approaching its budget limit, anything regarded as an 'extra' tends to be deleted, and this frequently includes energy efficiency measures with extremely attractive paybacks. Sorrell (2003) points out that *time* constraints can be as serious an impediment as the perceived capital constraint. By the time the design phase reaches the building mechanical services, not only have many decisions already been made that restrict the scope for energy-efficient mechanical systems, but also the design might be running behind schedule. This inhibits creative work that, of necessity, requires ample time. Frequently there is also inadequate time when bids are prepared, leading to the widespread practice of using bids from previous work and modifying them only slightly to meet the new project.

Procedural tools

The International Energy Agency, through Task 23 of its *Solar Heating and Cooling Programme*, has prepared a number of reports and guides pertaining to the integrated design process. Among these is the *Multi-Criteria Decision-Making* guide and software, MCDM-23 (Balcomb et al, 2002). MCDM-23 is a formalized step-by-step procedure to aid in the process of designing a building, with an accompanying computer program that automates many of the tasks involved and produces worksheets, bar charts and star diagrams (on which multidimensional performance indicators can be plotted). MCDM-23 does not provide the 'right' answer, but rather, provides a framework within which to carry out the tasks inherent in the decision-making process. The first step is to decide on the criteria that the design team wants to use and how they will be measured and weighted, a step that users of MCDM-23 have described as an extremely valuable activity in and of itself. Next, the performance of alternative designs are predicted (using other tools) and then presented within the MCDM-23 framework for discussion and decision making. The Task-23 working group has also produced an integrated design process guide (Löhnert et al, 2003), an online interactive information source (*IDP Navigator*), a report detailing the design process used in specific buildings and lessons learned (Poel et al, 2002), and a blueprint for the kick-off workshop for the design team, a crucial first step in the process.

Green Globes Design is an online tool developed for Natural Resources Canada that promotes the integrated design process by providing a structured, step-by-step process throughout all the stages of a project (project initiation and goal definition, option/site analysis, schematic design, detailed design, production of construction documents, commissioning). There is a separate module for each stage, each with about 120 questions with 'pop-up' explanations. Most questions require a simple 'yes' or 'no'. A report is generated that gives estimated energy use and CO_2 emissions, and compares the proposed building against various benchmarks. The tool interfaces with other web-based tools and information sources. *Green Globes Design* is available at www.greenglobes.com at a cost of $250 CDN per building assessment. Metric and imperial versions are available through the Canadian and US versions of the site, and a free trial version is available online.

13.3.2 Building simulation software

Computer simulation is needed when designing a high-performance building in order to be able to assess, in advance, the impacts on energy use of alternative designs and in order to be able to determine the optimum design for a given building in a given climate. Appendix D presents an overview of the different approaches to building simulation and of some of the frequently used building simulation tools. Here, we briefly identify some of the issues and problems (at present) pertaining to the use of computer simulation tools in the design of high-performance buildings.

Computer assisted design or CAD software is widely used in the architectural profession, but CAD software is used as a visualization and communication tool, not for the analysis of energy use. Many software tools are available for the separate analysis of energy use, HVAC systems, air quality and ventilation, lighting and acoustics. Hien et al (2000) provide an extensive listing of the available tools in each category, but what is lacking are integrated tools. Rivard et al (1999) discuss the role of industry fragmentation, archaic systems of data exchange and computer design software in reinforcing the fragmentation of the construction industry; and the multiple and conflicting representations of a building by different specialists as barriers to integrated, energy-efficient design. The use of building simulation software appears to be more prevalent in North America than elsewhere, and the majority of the available tools have been developed in the US. However, building software is largely restricted to CAD software and to computation of peak loads for purposes of sizing the HVAC system components (Ellis and Mathews, 2002).

Many of the design tools have been developed by vendors of HVAC equipment and are of a specific and piecemeal nature. The use of more general and at-least partially integrated models, such as the EnergyPlus, ESP-r and TRNSYS tools described in Appendix D, is quite limited. This is because generic models require substantial data inputs, which can take days to prepare,

and require some familiarity with the inner workings of the model. In a competitive bidding environment, few firms feel that they can invest the required time or convince potential clients that it is worthwhile. For this reason, these models tend to be used largely by university-based researchers. Indeed, the EnergyPlus (or its predecessors, DOE-2 and BLAST), ESP-r and TRNSYS models were used in a large fraction of the academic simulation studies cited in this book. Even these, because of their generic nature, are limited in their ability to incorporate truly advanced design features. These require full-scale fluid dynamics models in order to develop design guidelines that can be incorporated in more practical software tools.

In any case, the next generation of building design tools will be more fully integrated than most of the tools that are available and used at present (Ellis and Mathews, 2002). This will allow an analysis of the effect that various sub-components of the building have on the building as a whole. An example is the EnergyPlus software package, released by the US Department of Energy in April, 2001 (Strand and Pederson, 2002; Strand et al, 2002). Within the first four months of its release, 3000 people had downloaded this model. In spite of many improvements compared to previous models intended for use by practising designers, however, there are still many features that cannot be simulated with EnergyPlus, including:

- combining radiant cooling with free cooling through cooling towers or ground loops;
- combining radiant cooling with dehumidification of ventilation air;
- night-time ventilation of hollow slabs.

Olsen and Chen (2003) provide a specific example of the effort required to simulate advanced systems (displacement ventilation, chilled ceilings and beams, night cooling, variable airflow) and natural ventilation in buildings using a software package as recent as EnergyPlus. They evaluated the energy use for alternative systems in a recently completed building in the UK, but in order to use EnergyPlus, they had to make many modifications. Although radiant heating and cooling can be simulated with EnergyPlus, it cannot be combined with other processes (such as dehumidification of ventilation air) in the way that the set of processes would operate together in reality (Strand and Baumgartner, 2005). This is because radiant systems, although conceptually simple, are rather complex from a modelling standpoint.

A coordinated effort is underway to develop a common computer representation of building components that can be used by different types of building simulation software (Hitchcock, 2003). At the moment, building data have to be separately entered in different software tools, such as CAD and energy simulation software. *Software interoperability* is the ability to share data among a variety of related but different software programs. The common data model now under development is called the *Industry Foundation Classes* (IFC). The interoperability process began in earnest in 1998, and is rapidly developing (Bazjanac, 2004). Over 20 IFC-compliant software tools are now available, including CAD, cost estimation, energy simulation and HVAC design. Efforts are underway to link CAD and energy simulation tools together into single software packages, as outlined in Appendix D.

The next step will be the creation of building software packages that automatically find optimal designs – that is, designs which minimize some weighting of building energy use, number of hours of discomfort, and other indicators of poor performance. Optimization involves systematically varying a number of design parameters, within prescribed limits, computing the differences in the outcome, and converging on the optimal solution through a series of iterations. Some initial efforts at automated optimization are found in Holst (2003) and Zhou et al (2003).

13.3.3 Use of building simulation models within the design profession

In conventional practice, simulation is used only for the final confirmation of the performance of the mechanical system, rather than as an integrated element of the design process. For building simulation software to lead to improved designs, it must be easy to use in a comparative mode, that is, to be able to systematically generate alternative designs by varying input parameters and to generate an *n*-dimensional design performance space. Current simulation environments do not support this process (Mahdavi and Gurtekin, 2001).

Morbitzer et al (2001) outline how building simulation can be integrated within the design process, utilizing the three design stages identified by the Royal Institute of British Architects: the *outline design stage, scheme design stage* and *detailed design stage*. The decisions that are made during each stage are outlined in Table 13.16. A single simulation tool should be used for all three stages, but different interfaces with the users are required for different stages due to the differing backgrounds and time constraints of the users at different stages and the difference in issues being addressed. As noted above, CAD software can be used to define the building geometry, which can then be converted to a format suitable for import into the building simulation software.

de Wilde et al (2001) surveyed the architects and consulting firms involved in 70 low-energy buildings in The Netherlands. They identified five design stages: *feasibility study, conceptual design, preliminary design, final design* and *construction-drawing and building specification*. They found that over 70 per cent of the energy saving components that were used were selected during the feasibility and conceptual design stages, with computational tools used largely in later stages, and then only for optimization and verification rather than to aid in the selection of energy saving components. Instead, selection was based on past experience and reference buildings, but, among architects, in only 11 per cent of the cases was there a real design decision between at least two options. de Wilde and van der Voorden (2004) have identified a number of reasons for the limited use of building simulation tools by the design profession:

- unavailability of appropriate tools (tools that can answer specific design questions are either hard to find or non-existent);
- lack of trust in the results;
- high level of expertise needed to

Table 13.16 Division of the design process into three stages, and decisions made at each stage, as proposed by the Royal Institute of British Architects

Outline design stage	Scheme design stage	Detailed design stage
Orientation (appraisal)	Glazing area (detailed analysis)	Different heating systems
U-values (opaque and transparent)	Glazing type	Different heating control strategies
Heat recovery systems	Shading and/or blinds	Different cooling systems:
Light/heavy construction	Blind control	• mechanical
Air change rate (appraisal)	Orientation (small adjustments)	• free
Space usage	Air change rate (detailed analysis)	Different cooling control strategies
Glazing area (appraisal)	Material adjustment in overheating areas	Different ventilation strategies
Floor-plan depth	Lighting strategy	
Fuel type	Cooling required: yes/no?	

Source: Morbitzer et al (2001)

fully utilize the tools;
• cost (time and money) associated with the tools;
• problems of data exchange between design and simulation software.

The Scottish Energy Systems Group (SESG) at the University of Strathclyde (Glasgow) has perhaps the only university-based outreach programme in the world that is directed to the design community. This outreach programme consists of (McElroy et al, 2001):

• loan of fully configured computers with a variety of building simulation tools already installed;
• provision of in-house training and an experienced operator to the design firm for the use of the programmes in real projects.

As a result, firms are able to explore technical issues that they could only guess at before. Thermal, lighting, structural, and cost-analysis methods become an integral part of the design process. Access to modelling engenders the confidence to implement innovative solutions that would otherwise not be considered using conventional design methods. Firms that have participated in the outreach wanted to develop an in-house simulation capability, as they see that as something that will be increasingly demanded by clients. The challenge is to make the transition from old to new practices while still meeting day-to-day requirements and deadlines.

Augenbroe (2001) reviews the history and current status of building simulation among design firms in the US. The widespread and instant accessibility of design experts and their specialized analysis tools has lead to an increasing delegation of many analysis tasks to outside experts through the Internet, with less import of 'designer friendly' tools into the nucleus of the design team. The challenge instead has become one of 'sustaining complete, coherent and expressive communications between remote simulation experts and other design team members'. The development of 'interoperable' design tools (Section 13.3.2) is a prerequisite to this process. Full reliance on outside simulation experts lies at one end of the spectrum of possibilities, and development of full in-house simulation capability, as represented by the SESG programme, lies at the other end. The appropriate choice depends on the size and nature of the design firm.

There are two different areas in which the building simulation profession can be improved (Augenbroe, 2001): *tool-related*, involving the development of more functional tools, and *process-related*, involving better integration of analysis tools into the design process – that is, the use of outputs from simulation analysis in making design decisions, rather than as an incidental and unstructured process. Two tool-development paths are possible: (1) the development of increasingly sophisticated but user-friendly tools that can be used by the non-specialist (such as the architect), or (2) the increasing use of specialized tools by specialists but better integration of these specialists and their tools into the design team. Augenbroe (2001) supports the latter, and I have to agree: the most aggressive energy saving measures cannot be simulated in any currently available generic model. Simulation specialists are needed. As Augenbroe (2001) states, no architectural firm would take the risk of relying on designer friendly analysis tools for complex projects, as it would take a high degree of expertise just to judiciously apply simplified analysis in such cases.

With regard to process, it is important to involve simulation tools and experts at the early stages of the design process. However, as noted above, the current generation of design tools generally plays no significant role in the early stages. Instead, the first few critical decisions about building form and technologies are based on 'expert' judgement, which of necessity is biased toward past practice, without adequate evidence (backed up by simulation) of their performance in a particular case. In the words of Augenbroe (2001):

> 'once the decision for a certain energy saving technology is made on the grounds of overall design considerations or particular owner requirements and cost considerations, the consultants' expertise is invoked later for dimensioning and fine tuning. By that time the consultant is restricted to a narrow "design space" which limits the impact of the performance analysis and follow-up recommendations'.

The role of analysis can be made explicit through the development of a conceptual *design project model*. Again in the words of Augenbroe (2001) such a model:

> 'contains information about how the project is managed and how a particular domain consultant interacts with other members of the design team. It captures what, when, and how specific design analysis requests are handed to a consultant, and it keeps track of what downstream decisions may be impacted by the invoked expertise. Design iterations are modeled explicitly, together with the information that is exchanged in each step of the iteration cycle'.

The development of such a design model is a joint upfront responsibility of the design team. According to Augenbroe (2001), there has been a new wave of research in engineering on process integration and workflow control, but little activity in this area so far when it comes to the building industry.

For further information on building energy simulation tools and their use in the design profession, the interested reader can consult the detailed review by Jacobs and Henderson (2002). For information on 291 building software tools, see the US DOE Energy Efficiency and Renewable Energy website (www.eere.doe.gov/buildings/tools_directory). The energy tools listed at this website include databases, spreadsheets, component and systems analyses, and whole-building energy performance simulation programs. A short description is provided for each tool along with other information including expertise required, users, audience, input, output, computer platforms, programming language, strengths, weaknesses, technical contact and availability. An overview of the most frequently used tools and their capabilities is presented here in Appendix D.

13.3.4 Greater energy savings can cost less than smaller energy savings

It is possible that, through an integrated design, very large (50 per cent or more) energy savings will cost less than more modest energy savings, due to the savings from downsizing of mechanical equipment. This is illustrated in Table 13.17, which compares the technical characteristics, equipment sizing and cost, and simulated energy use for a 20-storey office building (18,700m² floor area) in Chicago with a window-to-wall ratio of 0.36. Results are shown if built according to ASHRAE 90.1-2001, or built with progressively better envelopes and mechanical equipment (cases IDP-1 to IDP-3). In IDP-1, a modestly better envelope and more efficient lighting reduce total energy use by over 27 per cent. The improvements entail an extra $302,000 in costs (1.5 per cent of the total construction cost), but downsizing of mechanical equipment saves $152,000, for a net incremental cost of $150,000, which is paid back in 1.6 years through energy cost

Table 13.17 Comparison of design parameters, equipment sizing and energy use for a conventional design of an office building in Chicago, and for three designs using an Integrated Design Process (IDP)

	Base case	IDP-1	IDP-2	IDP-3
Envelope measures				
Wall U-value (W/m²/K)	0.48	0.35	0.35	0.35
Window U-value (W/m²/K)	2.61	2.04	1.53	1.2
Window SHGC	0.26	0.19	0.19	0.19
Lighting and equipment				
Connected lighting load (W/m²)	14.0	8.07	8.07	8.07
Cooling plant COP	6.4	7.05	7.05	7.05
Heating plant efficiency	0.80	0.88	0.90	0.90
HVAC size				
Cooling plant (kW$_c$)	1494	1125	1020	1020
Cooling tower (kW$_c$)	1740	1287	1160	1160
Heating plant (kW$_{th}$)	1289	1026	967	879
Peak airflow (m³/s)	103.8	82.6	77.9	23.6
Fan power (kW)	167.9	111.9	111.9	37.3
Peak water flow (litres/s)	64.3	48.6	44.2	44.2
Pump power (kW)	11.2	7.5	7.5	7.5
Energy use				
Electricity (MWh/year)				
Lighting	737.9	425.7	314.8	314.8
Chillers	258.1	154.6	141.0	136.7
Cooling tower	51.8	29.5	26.4	25.5
Pumps	42.2	32.3	30.1	30.1
Fans	457.5	267.0	243.8	81.8
Plug load	600.1	600.1	600.1	600.1
Total	2147.6	1508.9	1356	1188.9
Peak electrical load (kW)	826	572	486	351
Electricity (GJ/year)	7731	5432	4882	4280
Natural gas (GJ/year)	7505	5666	4901	4264
Energy intensity (kWh/m²/year)	228.2	165.8	146.4	127.7
Savings (GJ/year)		4138	5453	6692
Savings (%)		27.4	35.8	43.9
Savings (%) excluding plug load, relative to a base case with lighting at 18W/m² instead of 14W/m²		35.4	44.4	53.9
Capital cost savings				
Cooling plant		$42,500	$53,125	$53,125
Cooling tower		$14,000	$21,000	$21,000
Heating plant		$15,000	$15,000	$21,000
Perimeter heating system		–	$27,600	$38,500
Air handler		$75,000	$75,000	$297,500
Savings from elimination of VAV Boxes		–	–	$856,000
Pumps		$4,000	$6,000	$6,000
Drop ceiling savings		–	–	$120,000
Total		$150,500	$197,725	$1,413,125
Additional costs and net incremental cost				
Insulation		$45,000	$45,000	$45,000
High-performance glazing		$175,000	$420,000	$525,000

T8 lighting		$30,000	$60,000	$60,000
High-efficiency chiller		$22,500	$22,500	$22,500
High-efficiency boiler		$30,000	$120,000	$120,000
Radiant cooling system		–	–	$960,000
Total		$302,500	$667,500	$1,732,500
Net		$152,000	$469,775	$319,375
Energy costs and savings, payback on incremental capital cost				
Energy cost	$332,953/year	$235,996/year	$208,524/year	$173,580/year
Energy cost savings		$96,957/year	$124,429/year	$159,373/year
Payback time (years)		1.6	3.8	2.0

Note: Energy cost savings are computed here from the energy and peak demand data assuming costs for electricity and natural gas of $0.10/kWh and $0.025/kWh ($6.94/GJ), respectively, and electricity demand charges of $80/kW/year.
Source: Todesco (2004) for the Base, IDP-1 and IDP-2 cases, with IDP-3 calculations kindly performed and provided for this table by G. Todesco (Jacques Whitford, Ottawa) along with additional IDP-2 data

savings (assuming electricity at $0.10/kWh, natural gas at $7/GJ ($0.025/kWh), and electricity demand charges of $80/kW/year). IDP-2 entails further improvements in the windows and use of daylighting in perimeter zones, with a further downsizing in mechanical equipment, for a net incremental cost of $575,000, a 36 per cent savings in total energy use, and a 4.6-year payback. IDP-3 involves a displacement ventilation/chilled ceiling HVAC system with use of 100 per cent outside air for ventilation, and a further window upgrade (as required with DV/CC systems in cold climates). Energy savings increase to 44 per cent, but the net cost decreases from $575,000 for IDP-2 to $320,000 for IDP 3 and the payback time drops to 2.0 years. The incremental cost and payback time would be cut in half if credit (about $150,000) is given for the CC system displacing a conventional sprinkler system for fires. Other potential cost savings, such as elimination of perimeter heating with high-performance windows (Chapter 3, Section 3.3.20), have also not been considered here.

These energy savings are relative to ASHRAE 90.1-2001, which requires a smaller lighting power density than earlier versions (14W/m² instead of 16–20W/m²), and is diluted by the plug load, which is assumed to be unchanged. Relative to a base case with lighting power density more typical of recent practice (18W/m²),

and excluding plug load, the savings are 35–54 per cent for IDP-1 to IDP-3. In the above calculations, it is assumed that ducts and pipes are downsized when the required flow decreases, as this is normal engineering practice, so that the energy savings expected through the cubic law are not realized (see Chapter 7, Section 7.1.2). Further savings would be possible through passive ventilation and desiccant dehumidification and cooling (Chapter 6, Section 6.6.4, and Chapter 7, Section 7.4.11).

Another indication that larger energy savings can cost less than smaller energy savings is provided by a survey of the incremental cost and energy savings for 32 buildings in the USA by Kats et al (2003). These buildings meet various levels of the LEED standard, which is discussed in Section 13.4. Summary results are given in Table 13.18. The energy savings are broken into reductions in gross energy demand, and reductions in net energy demand including on-site generation (by, for example, PV modules), which tends to be expensive. The cost premium is the total cost premium required to meet the various LEED standards and so includes the cost of non-energy features as well. Nevertheless, average costs are less than 2 per cent of the cost of the reference building, and are smaller on average for buildings with 50 per cent savings in net energy use than for buildings with 30 per cent savings.

Table 13.18 Energy savings relative to ASHRAE 90.1-1999 and cost premium for buildings meeting various levels of the LEED standard in the US

| LEED Level | Sample Size | % Energy savings, based on | | Cost Premium |
		Gross energy use	Net energy use	
Certified	8	18	28	0.66%
Silver	18	30	30	2.11%
Gold	6	37	48	1.82%

Source: Kats et al (2003)

Table 13.19 Economics of the new Oregon Health and Science University building.

Item	
Total project cost	$145.4 million
Energy efficiency features	$975,000
PV system	$500,000
Solar thermal system	$386,000
Commissioning	$150,000
Total	$2,011,000
Savings in mechanical systems	$3,500,000
Value of saved space	$2,000,000
Net cost	-$3,489,000
Estimated annual operating cost savings	$600,000

Source: Interface Engineering (2005)

The final example presented of the beneficial economics of energy-efficient buildings is one of the first buildings to be built on the new Oregon Health and Science University, River Campus, and completed in 2006. This 16-storey building is expected to achieve energy savings of 60 per cent relative to ASHRAE 90.1-1999 through such measures as hybrid ventilation using the stack effect in stairwells, solar preheating of office ventilation air, heat recovery with laboratory ventilation, radiant heating/cooling, demand-controlled displacement ventilation, PV modules as exterior shading, accurate equipment sizing and commissioning. Incremental costs or upfront savings are given in Table 13.19. Cost savings due to downsizing of the mechanical systems permitted by the efficiency measures exceeded the cost of the efficiency measures. A further credit arises from the space saved due to more efficient and downsized mechanical systems. The net result is a cost savings of about $3.5 million out of an original budget of $145.4 million and estimated operating cost savings of $600,000 per year.

13.4 Building rating systems

In an effort to promote the design and construction of more efficient buildings, a number of government agencies and trade associations have developed various building rating systems. Some of these will be briefly described here.

13.4.1 BREEAM

The BREEAM (Building Research Establishment Environmental Assessment Method) was developed in 1992 by the UK Building Research Establishment, formerly a branch of the British Ministry of the Environment. The rating system has been applied to over 1000 commercial buildings in Europe, Asia and North America, but adapted to local conditions (Skopek, 1999). Buildings earn points based on the building envelope and systems, and based on the operation and management of its systems. In this way, an older building that may be inherently inefficient can make up for this deficiency (under the rating system) through excellent management. Buildings gain points in the areas of energy conservation, water conservation, transportation (i.e. location with respect to public transportation) and indoor air quality. The rating system is used as part of a 'before/after' procedure. The building is assessed and rated compared to a baseline best practices building, and a report with a series of recommendations is presented to the building management. Management is given several months to implement the recommendations, then the building is revisited and a final report is prepared.

13.4.2 LEED Standard

In 1993, the US Green Building Council (US-GBC) was created. It released Version 1.0 of the *Leadership in Energy and Environmental Design* (LEED) standard for new buildings in late 1998. A building earns credits by satisfying requirements in five categories: sustainable site (i.e. part of a compact development served by public transit), water

efficiency, energy efficiency, materials and resources, and indoor air quality. The total number
of credits determines the building's rating: certified, silver, gold or platinum. As noted in Section
13.3.4, the average energy savings compared to
ASHRAE 90.1-1999 ranges from 18 per cent in
sample of certified buildings to 37 per cent in a
sample of gold buildings, at an average additional
upfront cost (including non-energy features) of 2
per cent compared with conventional designs. Average lifetime operating cost savings are equal to
20 per cent of the incremental construction cost
(i.e. a benefit:cost ratio of 10:1).

Adherence to an appropriate design
process is an important part of the LEED rating system. Hiring a qualified LEED coordinator
and a qualified commissioning agent are desirable steps. A minimal commissioning process is
required to receive credits. Additional credits can
be earned if the project developers: (i) commission a focused design review by a qualified third
party other than the design team prior to development of construction documents; (ii) conduct a
review of construction documents by a third party
when close to completion; (iii) conduct a selective
review of contractor submittals of commissioned
equipment; (iv) develop a recommissioning management manual; and (v) have in place review
near the end of warranties or after occupancy.

As discussed by Eijadi et al (2002), actions with vastly different impacts on energy use
sometimes earn an equal number of points under
LEEDs, whereas some actions that greatly reduce
energy use are not counted at all. For example,
use of a low-albedo roof is awarded one point
irrespective of climate, even though the savings
in annual energy use is large in hot climates and
close to zero or even negative in cold climates.
On the other hand, measures such as the use of
natural ventilation, good window design and appropriate building orientation and thermal mass
cannot be counted in the energy efficiency calculations for the LEEDs rating. This is because
the energy efficiency calculations are based on
an ASHRAE procedure that does not account
for these options. Thus, a higher score under the
LEEDs system does not necessarily imply a more
energy-efficient building, and a very high score
does not imply that the building has come close
to exhausting the opportunities to reduce its use

of energy. However, the effort to earn points under the LEEDs system does require changes to
and improvements in the conventional building
design and construction process, which is one
of its underlying purposes. Indeed, the LEEDs
stamp of approval is becoming a prerequisite for
an increasing number of building projects. The
danger is that the accumulation of LEEDs points
becomes a goal in its own right, with investments
made in those measures that generate the most
points per dollar spent rather than those measures that maximize energy savings.

However, the LEED standard is not
static, and many shortcomings in Version 1.0
were addressed in Version 2.0 (released in March
2000). Further improvements can be expected in
Version 3.0.

13.4.3 HK-BEAM

The Hong Kong Building Environmental Assessment Method (HK-BEAM) was released in December 1996 by the Hong Kong Real Estate Developers Association. Buildings are rated based
on the characteristics of the thermal envelope,
the lighting load, the HVAC system, metering
and monitoring, operations and maintenance of
systems (Lee et al, 2001b).

13.4.4 HERS and EnerGuide for Houses

The Home Energy Rating System (HERS) was
established in the 1980s in the US by members
of the building industry but was active in only
six states until the mid-1990s, when a national
industry council, the HERS Council, was established by the Department of Energy (DOE). A
rating system ranging from 0 to 100 was established, with 1 representing an open tent, 80 representing housing built according to the national
Model Energy Code, and 100 representing a
house that does not require any utility electricity or fossil fuels for heating, cooling or hot water (Rashkin, 2000). Thus, each point above 80
represents a saving of 5 per cent compared to
the MEC. Shortly thereafter, the Environmental
Protection Agency (EPA) decided to extend the Energy Star certification system to homes, with a home
qualifying as an Energy Star home if its energy
use is 30 per cent less than that for the MEC,

that is, with a HERS rating of 86 or better. The HERS system has been largely used for rating new houses, as part of a procedure to secure 'energy-efficient' mortgages (that is, granting of mortgages with higher carrying costs than would otherwise be permitted, in recognition of the lower energy costs of energy-efficient houses). During the period 1993–8, 63,165 homes were rated and 8533 energy-efficient mortgages granted (Judkoff and Farhar, 2000). More recently, it is being marketed as a means of quality control that reduces the risk of litigation arising from moisture or comfort problems (which are eliminated as a byproduct of properly constructed energy-efficient houses).

A similar rating system, EnerGuide for Houses, was launched in Canada in April 1998, with the difference between EnerGuide and HERS largely reflecting differences in climate. A score of 80 approximately represents an R2000 house, while a score of 100 represents a house with zero use of utility energy (including energy used for lighting and appliances). Unlike HERS, the EnerGuide system has been largely used for rating existing houses rather than new houses. Recommendations for improving the energy performance of the house are presented to the home owner. Those who implement the recommended changes are entitled to a post-implementation re-evaluation. By the end of April 2000, over 14,000 houses had been rated, recommendations that would save 20-25 per cent of total energy use were made, and over 1100 houses were re-evaluated (Parekh et al, 2000). In these houses, customers installed about two-thirds of the recommended measures and saved 10–15 per cent in total energy use (some of the data generated by this programme are presented in Chapter 14). A follow-up survey indicates that about 70 per cent of customers implemented at least one recommended action but without requesting a re-evaluation.

13.5 Summary and synthesis

An achievable but ambitious target for new high-performance housing is to reduce the heating and cooling energy use by a factor of four compared to conventional practice, whatever the country and climate under consideration. There are many examples of high-performance houses

around the world that demonstrate that this is possible using presently-available and proven techniques and equipment. Similarly, an achievable but ambitious target for new high-performance commercial buildings is to reduce the heating, cooling + ventilation, and lighting energy use by a factor of four compared to conventional practice, whatever the country and climate under consideration. There are many examples of high-performance commercial buildings around the world that demonstrate that this is possible using presently available and proven techniques and equipment. Inasmuch as current conventional practice is somewhat better than the average of existing buildings, the demonstrated savings in the best new buildings satisfy the criterion for sustainability (in energy terms) adopted in Chapter 1 of a factor of four to five reduction in the energy intensity of new buildings compared to the stock average.

The design of high-performance buildings requires integration of (or back-and-forth information flow between) the various design steps and among the various members of the design team, with an explicit goal from the very beginning of minimizing energy use. A basic integrated design process (IDP) focuses on a high-performance thermal envelope, minimization of lighting loads, and highly efficient, appropriately sized, properly commissioned, and optimally operated mechanical equipment. This alone can achieve a 35–50 per cent savings in total energy use compared to current practice for new buildings. The basic IDP stands in contrast to the usual design process, which is a linear process in which the design proceeds as a series of linear steps, without back-and-forth iteration and especially without optimization of the overall design. The basic IDP requires computer simulation models in order to optimize the design, although experienced designers can achieve large savings without using computer simulation models (computer tools are used, instead, for sizing of equipment once the building form, orientation, and envelope characteristics have been selected).

More advanced IDP will incorporate as-yet unconventional features, such as passive ventilation, heating, and cooling involving perhaps double-skin façades and airflow windows; thermal mass with night ventilation; chilled ceiling, displacement ventilation and desiccant dehumidification; daylighting; and adaptive thermal comfort. Professional facilitators and computer simulation specialists are also required from the very beginning of the design process, as it is essential to be able to objectively compare a number of alterative design and system options before firm decisions are made. Savings of up to 80 per cent have been achieved in this way.

A distinction must be made between an incremental approach to building energy efficiency, often involving the optimization of an inherently bad design through the use of slightly more efficient equipment, and a holistic approach that examines the building as an entire system. The first approach invariably results in more efficient buildings being more expensive, with the cost premium increasing the more efficient the building (up to the limit achievable in this way), while the holistic approach can often result in buildings that are no more expensive or only slightly more expensive than conventional buildings, but with vastly greater efficiency than can be obtained through the incremental approach.

Notes

1   An added benefit of tile roofs is that they are more durable than asphalt shingle roofs. The use of asphalt shingles for roofs is largely a North American phenomenon.

2   Eley et al (1998) explain how an alternative fee structure, based on the energy performance of the building, can be structured. They also outline procedures for verification.

3   In some buildings, it seems that the absurdity of heating the building in the morning and air conditioning it in the afternoon is intended. For example, Price and Hart (2002) discuss how whole building electricity use data at 15-minute resolution was used to detect the 'problem' of heating beginning at 3:30am rather than close to the time when occupancy begins, on days when the cooling

system begins at 1:30pm.

4   Hayter et al (1998) provide examples of errors during construction that caused the energy savings in one building to be only 63 per cent (compared to ASHRAE 90.1-1989) instead of 70 per cent (had the design intent been followed).

5   Although Torcellini wrote in 1999 and his target would be relative to ASHRAE 90.1-1989 in the case of commercial buildings, and ASHRAE 90.1 has since been strengthened (by 20–25 per cent relative to the 1989 version), technologies have also improved since Torcellini wrote, so a 50–70 per cent savings target relative to the latest version of ASHRAE 90.1 is still a reasonable target.

6   Adding insult to injury, the architect was paid a fee of at least $1 million!

7   The better-than-80 per cent energy savings expected for this building is achieved largely through the use of passive ventilation, cooling and heating; and through daylighting.

fourteen

Energy Savings through Retrofits of Existing Buildings

The preceding discussion demonstrates the possibility of achieving energy savings of up to 80 per cent in new residential and commercial buildings, relative to current practice in most countries. However, there is a large stock of existing, inefficient buildings. Our long-term ability to reduce energy use depends critically on the extent to which energy use in these buildings can be reduced when they are renovated. In Chapter 1, a rough criterion for sustainability of a factor of two to three reduction in the energy intensity of existing buildings was developed. In this chapter we show that reductions in the energy use of existing buildings of this magnitude are indeed possible.

14.1 Conventional retrofits of houses and apartments

The equipment inside a house, such as the furnace, water heater, appliances, air conditioner (where present) and lighting is completely replaced over time periods ranging from every few years to every 20–30 years. The building shell – walls, roof, windows and doors – last much longer. There are two opportunities to reduce heating and cooling energy use by improving the building envelope: (i) at any time prior to a major renovation, based on simple measures that pay for themselves through reduced energy costs; and (ii) when renovations are going to be made, including replacing windows and roofs.

The energy intensity of the housing stock has decreased dramatically during the past few decades in many countries, but still remains much higher than what can be cost-effectively achieved in new housing. Upgrading old housing to the energy standards of new housing or even better would lead to a reduction in housing energy use by a factor of three to four. Figure 14.1 shows the heating energy intensity for houses of various ages in the Waterloo region of Ontario (Canada).[1] Canadian houses built in the 2000s use about one-third the energy for heating as houses built early in the 20th century and about half that of houses built in the 1960s. Figure 14.2 gives the space heating, hot-water and electricity energy use for existing and new residential buildings in Germany.[2] New residential buildings use about one-quarter the heating energy of existing buildings and the most advanced new buildings (Passive buildings) use about one-third that of conventional new buildings. Figure 14.3 compares the overall energy intensity of single-family and multi-unit residential buildings in Austria built from before 1919 to after 1990. A particularly large decrease – by more than a factor of two – is seen for multi-family buildings. Figure 14.4 gives the floor area, energy intensity for space heating and hot water, and year of construction for residential buildings in the Canton of Zurich. The most recent buildings use just over one-third the energy of buildings constructed before 1970 (which account for about three-quarters of the total floor area). The high energy intensity of old buildings, combined with the large stock of such buildings in many countries, implies a large energy savings potential by renovating old buildings. Conversely, in the case of Sweden, old buildings are only 20–35 per cent more energy intensive than recent buildings, as seen from Figure 14.5.

Figure 14.2 Energy intensities of existing and new residential buildings in Germany (about 3500 HDD)

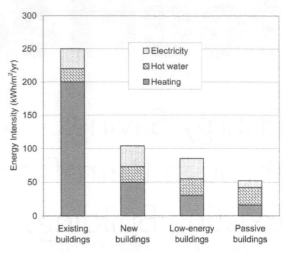

Note: Alternative estimates of heating energy use in existing German residential buildings range from 150 to 220kWh/m²/year.
Source: Eicker (2003)

Figure 14.3 Heating energy intensities of single-family and multi-family dwellings of different ages in Austria

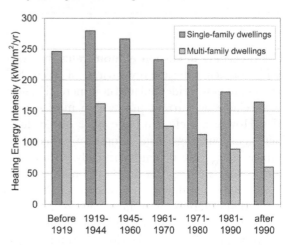

Source: Haas et al (1998)

Figure 14.1 Heating energy intensity of houses of different ages in the Waterloo region of Ontario, Canada (about 4000 HDD)

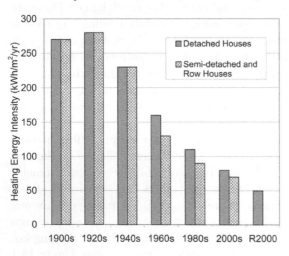

Note: Based on a non-random sample of 5700 houses that might be biased high.
Source: Data from Paul Parker, Faculty of Environmental Studies, University of Waterloo

Figure 14.4 Floor area and energy intensity for space and water heating for residential buildings in the Canton of Zurich built during various time intervals in the past, and projected for the near future

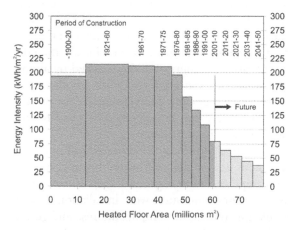

Source: Zimmermann (2004)

Figure 14.5 Heating energy supplied to multi-family residential buildings in Sweden according to time period of construction

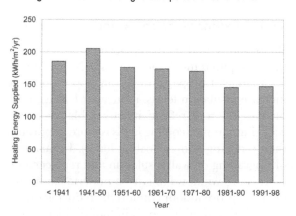

Source: Data from Erlandsson and Levin (2005)

14.1.1 Simple, low-cost measures

Cost-effective measures that can be undertaken without a major renovation include simple improvements to the thermal envelope, sealing and insulating ductwork, and upgrading the heating system and hot-water heater when they are due for replacement.

Simple envelope improvements include sealing points of air leakage around baseboards, electrical outlets and fixtures, plumbing, the clothes dryer vent, door joists and window joists;

weather stripping of windows and doors; and adding insulation in attics or wall cavities. Figure 14.6 shows the average total heat loss for houses in Canada built during the 1890s and during each decade of the 1900s, as well as the energy savings that can be cost-effectively saved through simple measures such as those listed above.[3] Also given is the cost-effective energy savings as a percentage of the total heating energy use. Houses built in the 1990s in Canada lose about half as much energy as houses built in the 1890s, although the average floor areas for the sampled houses are almost the same. The cost-effective energy savings potential ranges from 25–30 per cent for houses built before the 1940s, to about 12 per cent for houses built in the 1990s.

Rosenfeld (1999) refers to an 'AeroSeal' technique that he estimates is already saving $3 billion/year in energy costs in the US. Without proper sealing, homes in the US lose, on average, about one-quarter of the heating and cooling energy through duct leaks in unconditioned spaces – attics, crawl spaces, basements (see also Chapter 7, Section 7.3.6). This not only does no good, it does harm – blowing cold air out of a duct in the attic creates a partial vacuum in the house, which sucks in warm air from outside. With the AeroSeal technique, one first closes all the vents and HVAC equipment, then pressurizes the ductwork with a fog of small sealant particles. The air can leave only through the leaks but has to make sharp turns, which the sealant particles cannot follow. They stick to the sides of the leaks, sealing them. Studies summarized by Francisco et al (1998) indicate that air-sealing retrofits alone can save an average of 15–20 per cent of annual heating and air conditioning energy use in US houses. Additional energy savings would arise by insulating ductwork in unconditioned spaces.

Replacement of older furnaces or boilers (with a typical efficiency of 60–70 per cent) with condensing units (90 per cent efficiency or better) can yield a 20–30 per cent saving in fuel use. Natural gas and oil-fired water heaters tend to be even less efficient, so if both the furnace and water heater are replaced with a condensing unit that provides both heat and hot water (such as the small, wall-hung boilers that are now available), a large saving in energy use for hot water can also be achieved. These units can be easily retrofitted

Figure 14.6 (a) Heat loss from houses in the Waterloo area of Canada according to decade built, (b) near-term energy saving potential, and (c) energy saving potential as a percentage of heating energy use

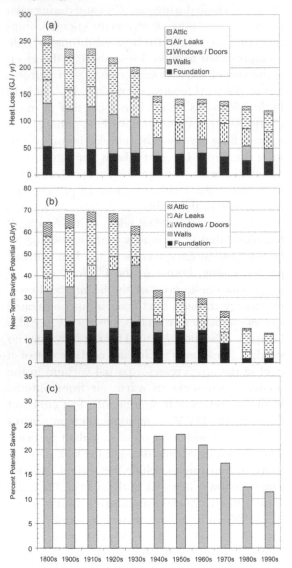

Note: Houses in the sample had an average floor area of 276m² in the 1800s, decreasing to 186m² in the 1940s and then increasing to 270m² in the 1990s.
Source: P. Parker et al (2000)

into a pre-existing forced air heating system, by providing a heating coil inside the main supply duct, next to the blower. In houses with basements that are heated by the forced air system, it will be advantageous to fully insulate the basement walls,

remove the ductwork to the basement, and install baseboard hot-water radiators whose heat output can be controlled independently of the forced air system, which heats the above-grade living space. If the existing water heater is electric and the electricity comes from coal at a typical 33 per cent efficiency (after transmission losses), then the use of primary energy will be reduced by almost a factor of three using a condensing boiler. Condensing units permit side venting, so the pre-existing chimney can be sealed, which will reduce heat loss due to air leakage.

In a retrofit of 4003 homes in Louisiana, the heating, cooling and water heating systems were replaced with a ground source heat pump system. Other measures were installation of attic insulation and use of compact fluorescent lighting and low-flow showerheads. Space and hot-water heating previously provided by natural gas was supplied instead by electricity (through the heat pump), but total electricity use still decreased by one-third (Hughes and Shonder, 1998).

14.1.2 Extensive retrofits and renovation

Renovations and roof replacement (something that is required much more often in North America than in Europe) provide an opportunity to increase insulation levels, reduce air leakage and increase the roof reflectivity (advisable whenever reducing summer cooling loads is more important than reducing winter heating loads). Installation or upgrading of wall insulation and replacement of existing windows might be undertaken in order to improve indoor comfort (by reducing drafts and creating more uniform indoor temperatures) even if the walls and windows are not due for a renovation because of deterioration. The savings in energy use that can be achieved depends on the initial insulation levels and climate, the former depending on the age of the building. Table 14.1 gives estimated wall, roof and floor U-values for buildings constructed between 1975–90 and after 1990 in three different climate zones in Europe. Older buildings often have no insulation at all, so wall and roof U-values will be of the order of 1–2W/m²/K. Upgrading such buildings to post-1990 standards would reduce wall and roof heat loss by a factor of two to four. For the case of residential buildings in Osaka, Japan, Shimoda

Table 14.1 Envelope U-values (W/m²/K) for European buildings built between 1975 and 1990 and after 1990

Envelope element	Mild climate (1800 HDD)		Moderate climate (3500 HDD)		Cold climate (>4500 HDD)	
	1975–90	After 1990	1975–90	After 1990	1975–90	After 1990
Roof	0.8	0.5	0.5	0.4	0.2	0.15
Wall	1.2	0.6	1.0	0.5	0.3	0.20
Floor	0.8	0.5	0.8	0.5	0.2	0.15

Source: Taken from estimates by Petersdorrf et al (2002)

et al (2003) estimate the current heating energy use to be 3470TJ/year, but that if all buildings could be upgraded to the 1999 building standard, the energy use would be 855TJ/year – a reduction of 75 per cent. Many uninsulated old buildings have lath-and-plaster walls, and Katrakis et al (1994) provide technical details concerning techniques for insulating and sealing old multi-unit residential buildings with such walls when they are ready for a major renovation (involving replacement of the lath-and-plaster finish with a new finish). To the extent that winter temperatures in existing buildings are colder than desired, some of this potential saving will instead lead to improved comfort, as residents increase the thermostat setting. In Austria, this rebound is estimated to be around 15–30 per cent of the savings that would otherwise occur (Haas et al, 1998).

The time when the building envelope is upgraded is a good time to replace the heating system, as this provides an opportunity for downsizing the system, or for switching to a more efficient heating system. If the old furnace or boiler is kept, not only is one keeping an inefficient system, but the efficiency will fall further due to the fact that it will be operating at a lower average load (see Chapter 4, Sections 4.2.1 and 4.3.2).

In a carefully documented retrofit of four representative houses in the York region of the UK, installation of new window and door wood frames, sealing of suspended timber ground floors, and repair of defects in plaster reduced the rate of air leakage by a factor of 2.5–3.0 (Bell and Lowe, 2000). This, combined with improved insulation, doors and windows, reduced the heating energy required by an average of 35 per cent. Bell and Lowe (2000) believe that a reduction of 50 per cent could be achieved at modest cost using well-proven (early 1980s) technologies, and a further 30–40 per cent reduction through additional

Table 14.2 Example of energy savings from retrofitting an apartment block in Switzerland. Given is energy use in units of MJ/m² per year

	Prior to retrofitting	After retrofitting
Conductive heat loss	473	216
Ventilation heat loss[a]	119	76[b]
Internal heat gain	114	108
Net heating requirement	478	184
Hot water requirement	108	100
Energy for space and hot water heating	586	284
Heating efficiency	0.85	3.2
Electricity demand		
Heat pump		89
Mechanical air circulation		6
Photovoltaic system (gain)		8
Secondary energy demand	690	87
Primary energy demand, MJ/m²/year	863	193
kWh/m²/year	240	54

[a] Associated with an exchange rate of 0.4/h.
[b] Reduction due to installation of heat recovery from the exhaust air stream.
Source: Humm (1997)

measures. Conversely, improper installation of new windows can largely negate the benefits of high-performance windows.

Table 14.2 compares the energy balance of an apartment block in Switzerland before and after retrofitting. The largest savings in secondary energy use resulted from replacing an oil-fired boiler, at 85 per cent seasonal average efficiency, with an electric heat pump having a seasonal average COP of 3.2. In the future, heat pumps with better COPs should be available. Nevertheless, assuming an eventual electricity generation

Figure 14.7 Details showing the application of solid-foam insulation over a pre-existing sloping roof in a retrofit application

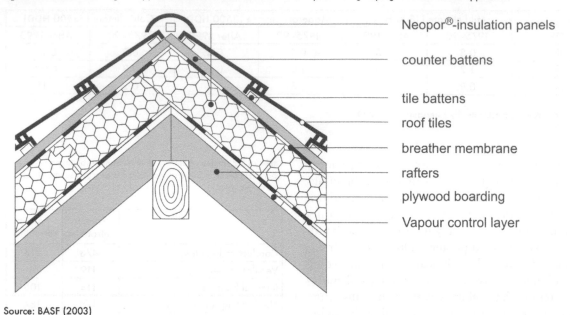

Neopor®-insulation panels

counter battens

tile battens

roof tiles

breather membrane

rafters

plywood boarding

Vapour control layer

Source: BASF (2003)

efficiency of 50 per cent, an efficiency in refining and delivering oil of 80 per cent, and 5 per cent energy loss in delivering natural gas to the electric power plant and a further 5 per cent loss in transmitting the electricity to the apartment block, the total primary energy requirement decreased by a factor of 4.4.

External Insulation and Finishing Systems (EIFSs), described in Chapter 3 (Section 3.2.6) provide an excellent opportunity for upgrading the insulation and improving the airtightness of single- and multi-unit residential buildings, as well as institutional and commercial buildings. This is because of the wide range of external finishes that can be applied, ranging from stone-like to a finish resembling aged plaster. It is important, however, to specify insulation that is produced without halocarbon blowing agents (including HFCs), as explained in Chapter 12 (Section 12.5). The German company BASF (Badische Anilin Und Soda Fabrik), which manufactures some of the components used in EIFSs, undertook a major renovation of some of its own 1930s multi-unit residential buildings in Ludwigshafen (see www.3lh.de). The EIFS in combination with other measures achieved a factor of eight measured reduction in the

Figure 14.8 View of a multi-unit residential building owned by BASF (a) before, and (b) after a retrofit that reduced heating energy used by a factor of eight

Source: Wolfgang Greifenhagen, BASF

Figure 14.9 Installation of external insulation (upper) and of spray-on plaster with micro-encapsulated phase change material (lower) in the building illustrated in Figure 14.8

Source: Wolfgang Greifenhagen, BASF

Figure 14.10 Photographs of a 100-year old residential building on Magnusstrasse in Zurich before and after a renovation (upper panel) that reduced total energy use by a factor of six (from 233kWh/m²/year to 37kWh/m²/year). The renovation entailed removal of the original roof and its replacement with prefabricated roof elements with built-in solar thermal collectors, one of which is shown being lowered by crane in the lower panel

Source: Zimmermann (2004)

heating energy use. The external insulation consisted of 20cm of pentane-blown polystyrene on walls and 30cm on the roof (16cm between rafters and 14cm on top). Figure 14.7 shows details of the roof construction. The other measures included replacement of the existing windows with triple-glazed, krypton-filled windows having a U-value of $0.8W/m^2/K$, reduction in air leakage from $ACH_{50} = 4.5$ to $ACH_{50} = 0.6$, installation of controlled ventilation with heat recovery (85 per cent effectiveness), and use of spray-on plaster with micro-encapsulated phase change material (paraffin wax in 50μm bubbles) in combination with night ventilation in order to limit daytime temperature peaks during the summer. Figure 14.8 presents photographs of one building before and after the retrofit, while Figure 14.9 shows the installation of the external insulation and spray-on plaster. The potential for similar savings is large wherever there is a large stock of poorly insulated buildings.

An even more radical renovation for old buildings involves replacement of the existing upper floor and roof with prefabricated units that are lowered into place by crane. Figure 14.10 shows a roof unit being lowered into place on a 100-year old residential building in Zurich, along with photographs of the building before and after renovation. The roof units contain a total of 15.5m² of solar thermal collectors, which supply heat to a 2600litre hot-water tank for space heating, supplemented by individual wood-burning stoves when the outside temperature drops below −2°C (Viridén et al, 2003). In this case, heating energy use as measured over a 2-year period was reduced by 88 per cent (from 165kWh/m²/year to 19kWh/m²/year) and total energy use by 84 per cent (from 233kWh/m²/year to 37kWh/m²/year). Single-glazed windows (10 per cent of the façade area) were replaced with windows with

Figure 14.11 Use of vacuum insulation panels in prefabricated dormer units used in a building renovation in Zurich, showing (a) fabrication, (b) mounting, and (c) final result

Source: Binz and Steinke (2005)

a U-value of $0.75W/m^2/K$, 70 per cent of the façade was insulated to a U-value of $0.15W/m^2/K$, and 30 per cent of the façade could not be touched.

Finally, Figure 14.11 illustrates the use of prefabricated dormer units containing vacuum insulation panels (VIPs) in the renovation of old buildings in Zurich (VIPs are described in Chapter 3, Section 3.2.7).

14.2 Conventional retrofits of institutional and commercial buildings

Commercial and institutional buildings afford numerous opportunities for deep reductions in energy use, but are generally more complex than residential buildings. The measures that can be taken in retrofits of commercial buildings are first enumerated here, followed by examples of retrofits that have achieved energy savings of 50 per cent or more.

14.2.1 Measures for commercial buildings

The first condition for energy efficiency in commercial buildings is to have heating, ventilation and cooling systems that can be turned off when they are not needed, or at least adjusted to match actual needs. Closely associated with this is the ability to open windows in order to take advantage of even minimal passive ventilation and cooling when conditions are favourable. However, even these most basic conditions are not satisfied in many commercial buildings, particularly in North America.

In addition to the above, the ideal HVAC system in a commercial building would:

- separate ventilation from cooling/heating requirements by using a hydronic system for the latter, with airflow set to ventilation requirements only;
- separate sensible from latent cooling loads by using desiccant dehumidification for the removal of water vapour;

- use demand-controlled displacement ventilation with 100 per cent outside air for fresh air requirements, rather than invariant mixing ventilation and partial recirculation;
- use some sort of heat exchanger to recover thermal energy from the exhaust air;
- use a chilled ceiling for sensible cooling, so as to permit the maximum possible temperature for cooling;
- use variable-speed drives on all pumps, fans and chillers.

Instead,

- one might be stuck with an all-air, constant volume, 100 per cent outside air system (which, in this case, would be highly wasteful);
- if there is a variable-air-volume ventilation system, the airflow is throttled when it needs to be reduced, and the amount of outdoor air varies in proportion to the total airflow (that is, the proportion of outdoor and recirculated air is fixed);
- dehumidification would be achieved by overcooling, followed by electric or gas reheat;
- if there is partial recirculation of indoor air rather than using 100 per cent outside air, dehumidification

might be done to the airflow after mixing rather than applied directly to the outside air before mixing, so that the recirculated air cannot be used for reheating the supply air after overcooling it for dehumidification purposes;
- there might be no option to increase the proportion of outside air when it is cool enough to satisfy the cooling load, or if there is, the system probably does not work properly.

The building envelope is likely to be poorly insulated if it is insulated at all, and consists of single-glazed (or, at best, double-glazed) windows with a relatively high solar heat gain coefficient, no spectral selectivity and no external shading devices. Windows cannot be opened, nor are there trickle ventilators, so it is impossible to take advantage of outdoor coolness when cooling is required or to provide natural ventilation. However, leakage points abound, thereby undermining the HVAC system when it is running (which is all the time) and increasing the latent heat load in hot-humid climates. The lighting is entirely electric and inefficient, provides uniform intensity even though uniform lighting is not needed, has no dimming capability, is not controlled by occupancy sensors and is invariably left on when not needed.

Specific measures that can be used to address the shortcomings outlined above are listed in Table 14.3, grouped as envelope, HVAC, domestic hot water and lighting measures.

Table 14.3 Energy retrofit measures that can be undertaken in commercial buildings

Envelope measures		
	Relatively inexpensive	
		Reduce air leakage (especially effective in the upper stories of tall buildings)
	Expensive but provides significant non-energy benefits	
		Addition of extra insulation (external or internal) Upgrade of windows (smaller U-value and/or smaller SHGC if cooling is dominant) More reflective façade and roof to minimize summer heating Operable external shading, such as sliding or pivoting shutters Construction of a second (perhaps curtain wall) façade over the first façade (can deal with moisture and deterioration problems with existing façade)
HVAC systems		
	Replacement of heating and cooling equipment	

	When replacing boilers, replace non-condensing units with condensing units (may require first reducing building heat loss through envelope measures, and increasing the size of the radiators or reducing flow rate so as to induce a greater temperature drop, so that return water is cold enough to induce condensation of water vapour in flue exhaust) Replace a single boiler that supplies both winter space heating and summer DHW loads with two unequally sized units, the smaller being appropriate for the summer load When replacing chillers, choose the most efficient equipment, with a variable-speed drive Replace rooftop air conditioners with units having desiccant dehumidification (in addition to sensible cooling) Replace rooftop air conditioning/heating units with exhaust-air heat pump (EAHP), or install EAHP as an add-on to existing rooftop units
Modest overhaul, generally feasible	
	Conversion from CAV (constant-air-volume) to VAV (variable air volume), with use of variable-speed drives rather than throttling to control the rate of airflow (to save both fan energy and energy used for heating and cooling outside air) If already VAV with throttling, install variable-speed drives If stuck with an all-air system with partial recirculation of indoor air, install an economizer to vary the proportion of outdoor air so as to permit 'free-cooling' If pre-existing system has hydronic cooling and chillers with a cooling tower, reconfigure to allow direct use of cooling-tower water when this is adequate for cooling purposes. Ensure that replacement equipment is properly sized (rather than grossly oversized, as is typically the case) Eliminate terminal reheat when conversion to VAV occurs Install energy recovery between exhaust air and supply air Install desiccant dehumidification wheel on outside air intake (along with sensible heat recovery) Install indirect evaporative cooling Replace inefficient pumps, fans and motors with premium (highest-efficiency) units Insulate all pipes, and insulate and seal all ductwork (or repair or upgrade existing insulation) Install state-of-the-art controls and sensors, using wireless systems to reduce cost and disruption Use energy saving control system settings
Major overhaul, sometimes not feasible	
	Replace single-zone systems with multi-zone systems Implement demand-controlled ventilation (DCV) in a DOAS system Convert from mixing ventilation to displacement ventilation (would require vents at base of wall rather than in floor in a retrofit situation) Convert from purely mechanical to hybrid passive/mechanical ventilation by replacing inoperable windows with operable windows or adding trickle ventilators, and creating suitable internal paths and exits for airflow if necessary
Domestic hot water	
	Install water-conserving fixtures Install timer on recirculation loop systems to shut down flow during periods of non-use Install point-of-use water heaters in recirculation loop systems where justified Insulate hot-water pipes where accessible Implement recovery of warm wastewater (such as from showers or dishwashers) where loads are significant and wherever feasible Upgrade to high-efficiency heaters or boilers Use desuperheater on chiller or heat pump to provide some DHW heat requirements
Lighting	
	Replace inefficient ballasts, lamps and luminaries with efficient models Implement simple measures to enhance daylighting (light shelves, various types of light-directing glazing) and convert non-dimmable to dimmable electric lighting Replace uniform lighting with task/ambient lighting Install occupancy sensors

14.2.2 Examples of deep energy saving retrofits in commercial buildings

In the early 1990s, the Pacific Gas and Electric Company (PG&E) in California sponsored a $10 million demonstration of advanced retrofits. The project was entitled the 'Advanced Customer Technology Test' (ACT²). In six of seven retrofit projects, an energy saving of 50 per cent was obtained with a modest payback time; in the seventh project, a 45 per cent energy saving was achieved. For Rosenfeld (1999), the most interesting result was not that an alert, motivated team could achieve savings of 50 per cent with conventional technology, but that it was very hard to *find* a team competent enough to achieve these results! (The first team selected had to be fired.) Unfortunately, by the time the ACT² project was finished, deregulation and restructuring of the power industry in California removed the incentive for electric utilities to be involved in reducing their customers' energy needs, and the ACT² team was dismantled.

Liu and Claridge (1999) show how the conversion of a dual-duct CAV system to a VAV system in a Texas office building *alone* saved 41 per cent in total heating + cooling + ventilation energy use. Fisher et al (1999) find that optimizing the design of a restaurant fume hood and reducing the exhaust airflow during times when cooking is not being done can save more than 50 per cent of heating and cooling energy for restaurants in cities throughout the US. The following measures achieved a 74 per cent reduction in annual cooling energy use in a one-storey commercial building in Florida (Withers and Cummings, 1998): sealing ducts achieved an 80 per cent reduction in leakage (31 per cent savings in cooling energy use); replacing an old air conditioner (seasonal COP of 1.8) with a new air conditioner (seasonal COP of 3.5) (43 per cent saving of remaining energy use); and turning off the attic fan (36 per cent saving). A survey of 80 office buildings in Toronto, ranging in age from very recent to more than 36 years old, indicates that lighting system improvements alone could reduce the average total energy use from 402kWh/m²/year to 280kWh/m²/year (Larsson, 2001) – a saving of 30 per cent. The March 1997 issue of the *CADDET Energy Efficiency Newsletter* (published by the International Energy Agency, as explained in Appendix C) gives several examples of retrofits of schools in Europe and Australia where total energy or heating-only savings of 50–70 per cent were achieved. These and many other examples that could be cited indicate that energy savings of 50–70 per cent can be consistently achieved for existing buildings with aggressive retrofits.

As noted in Chapter 7 (Section 7.4.15), leaky ducts can be a major source of energy waste in commercial buildings. Fortunately, this source of energy waste can be significantly reduced with the aeroseal technique, described above for single-unit houses, as it has been successfully applied to the much larger ducts (and cracks) found in large commercial buildings. In one test building, 66 per cent of the leakage area was sealed after 2.5 hours of injection of aerosolized particles into the ductwork, while in another test building, 86 per cent of the leakage area was sealed after 5 hours (Modera et al, 2002). However, if the ducts are poorly constructed, aerosol sealing may provide only a temporary fix until structural repairs are performed.

The European Commission funded an analysis of energy use and retrofitting opportunities in commercial buildings throughout Europe (Hestnes and Kofoed, 1997, 2002; Dascalaki and Santamouris, 2002). The location, height, year built and occupancy level for each of the buildings are given in Table 14.4. The characteristics of the HVAC systems in the buildings are as follows (Jagemar, 1997): eight of the buildings use a constant air volume HVAC system, of which four use 100 per cent outside air (i.e. no recirculation in order to reduce the intake of fresh air to that required for ventilation alone); only one building has heat recovery from the exhaust air, and only five make use of free cooling; three buildings have an all-air cooling system, four have a fan coil hydronic system (of which one also has chilled ceilings), and three have no mechanical cooling; three buildings have single-glazed windows, two have triple-glazed windows, and the remainder double-glazed; only one building has windows with a low-e coating; the lighting energy use in the office spaces is 15–20W/m² in four buildings, 10–15W/m² in three buildings, <10W/m² in one building, and <20W/m² in one building.

Computer models were constructed for ten buildings and validated against detailed

Table 14.4 Location and characteristics of the ten buildings studied as part of the European OFFICE Project

Case	Location		Characteristics		
	City	Country	No of floors	Year Built	No of Occupants
1	Nordboorg	Denmark	6, 11	1961	1000
2	Gothenburg	Sweden	6	1991	1000
3	Berlin	Germany	23	1971	800
4	Trondheim	Norway	13	1961	300
5	London	UK	1	1870	70
6	Bern	Switzerland	13	1971	200
7	Lausanne	Switzerland	1	1982	25
8	Florence	Italy	7	1976	85
9	Athens	Greece	4	1975	350
10	Lyon	France	7	1987	160

Source: Hestnes and Kofoed (1997)

measurements. The models were then used to assess the impact of a variety of standard retrofitting activities (such as improved insulation, simple air-sealing measures, replacement of poor windows, lighting and HVAC system upgrades) as well as passive solar measures (such as passive solar heating, night ventilation, daylighting and use of shading devices). Table 14.5 lists the measures that were considered for each of the buildings. Simulations were performed to determine the impact of individual measures and of a package of non-redundant measures. Figure 14.12 compares the simulated thermal and electrical energy use before and after retrofits for all ten buildings (office equipment is not included in the electrical energy use, since no changes were assumed). The percentage saving in thermal + electrical energy use ranges from 36 per cent to 77 per cent. The investment cost, simple payback period (i.e. investment cost divided by the annual energy cost saving), and net present value of the investment (future energy cost saving, discounted to the present, minus the investment cost) are given at the bottom of Table 14.5. In some cases, the investment does not pay for itself in terms of energy cost savings. However, building retrofits are performed for a wide range of reasons other than reducing energy costs, such as improved comfort and control, or to replace components that are wearing out.

Dascalaki and Santamouris (2002) classified the buildings in the European project in one

Figure 14.12 Comparison of measured thermal and electric energy use in the ten buildings studied under the European OFFICE project, and simulated energy use after implementation of recommended renovation measures

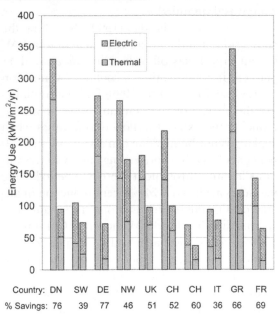

Note: Office equipment is excluded from electric energy use, as no changes were assumed.
Source: Data from Hestnes and Kofoed (1997)

Table 14.5 List of measures evaluated for buildings in the European OFFICE Project, economic costs when a package consisting of most measures is implemented (some measures are redundant and thus excluded from the package), and percentage savings in total non-office-equipment energy use

Retrofit measures in the OFFICE Project	Case study building									
	1	2	3	4	5	6	7	8	9	10
Improve thermal insulation of opaque elements	X		X	X	X	X	X	X	X	X
Replace window frames in bad condition	X		X	X		X	X	X	X	X
Improve insulation standard of window panes	X		X		X	X	X	X	X	X
Reduce infiltration rate	X		X	X		X		X	X	X
Use of mass walls, solar walls, atria or double façades	X							X	X	X
Use of reflective blinds or light shelves				X			X	X		
Use of effective solar shading systems	X	X	X	X	X	X	X	X	X	X
Use of night ventilation	X	X	X	X	X	X	X	X	X	X
Use of exposed thermal mass	X		X							X
Use of ceiling fans					X		X	X	X	X
Use of indirect evaporative cooling			X					X	X	X
Use of natural ventilation	X		X			X		X		
Use of solar protective glazing	X			X						
Daylight responsive control of artificial lighting	X	X	X	X	X		X	X	X	
Control of artificial lighting in response to occupancy	X	X			X					
Use of HF-ballasts	X	X	X		X		X			
Use of modern luminaries or reflectors				X	X			X	X	
Use of task lighting					X				X	
Reduction in installed power for lighting								X	X	
Use of heat recovery on ventilation air	X		X			X			X	X
Reduce ventilation rates	X		X			X				
Reduce running time of mechanical ventilation	X	X								
Improve fan efficiency	X	X	X			X				
Improve mechanical cooling machines						X		X	X	
Use of heat recovery from condenser coil			X			X				
Use of water reservoir for free cooling or for heat sink		X			X					
Use of efficient boiler system						X		X	X	
Use of heat recovery from flue gases								X	X	
Improve HVAC distribution system						X				
Reduce heating set-point	X	X	X							
Increase cooling set-point		X	X							
Package investment cost (euros/m²)	179	2	119	139	183	369	75	210	138	340
Simple payback period (years)	12	1	10	31	52	56	15	9	15	194
Net present value (euros/m²)	77	45	341	−48	−71	−362	182		161	−356
Percent savings	76	39	77	46	51	52	60	36	66	69

Source: Hestnes and Kofoed (1997)

of five categories (ranging from free-standing tall office towers with a high volume-to-surface-area ratio and massive walls and ceilings, to buildings with a smaller volume-to-surface-area ratio and light construction) and one of four climatic zones (Mediterranean, continental, mid-coastal and north-coastal). They performed additional simulations for all 20 possible building type–climate combinations, and found potential energy savings of 46–80 per cent for specific sets of measures, which, if implemented solely as energy saving measures, give simple payback periods of 10–15 years.

Balaras (2001) describes the retrofit of an 1865 two-storey office building in Athens, where low-energy was achieved through some passive technologies that required the cooperation of the occupants (such as closing windows when mechanical cooling was called for). In part of the building, a control system was installed that would, for example, shut down the local fan coil unit if the window is open, and activate a ceiling fan rather than the fan coil units when warranted. Balaras (2001) documents the learning process required on the part of the occupants, initial resistance to some features and non-cooperation, and eventual acceptance and satisfaction with the system. The need to maintain the historical identity of the building served as a constraint on the available options. Nevertheless, total annual energy consumption dropped from 210kWh/m² before the retrofit to 130kWh/m² after the retrofit and adjustment period – a saving of about 40 per cent.

Finally, Sekhar and Phua (2003) have computed the energy savings from the retrofit of an institutional building complex in Singapore. In this complex, the chilled water and cooling water pumps and the cooling tower fan and air handling unit all ran at a fixed speed regardless of the cooling load. Replacement with variable-speed units, upgrading the chillers to high-efficiency units (COP = 7.3), sequencing of the chiller operation to minimize inefficient part-load operation, and increasing the thermostat setting from cooling from 21°C to 22°C (still unnecessarily cold!) would reduce the energy use by 51 per cent.

14.3 Solar retrofits of residential, institutional and commercial buildings

The retrofit examples described above, while achieving dramatic (35–75 per cent) energy savings, rely on making incremental improvements to the existing building components and systems. More radical measures involve reconfiguring the building so that it can make direct use of solar energy for heating, cooling and ventilation. The now-completed Task 20 of the International Energy Agency's *Solar Heating and Cooling (SHC)* implementing agreement was devoted to solar retrofitting techniques.

Solar renovation measures that have been used are:

- installation of roof or façade-integrated solar air collectors;
- installation of roof mounted or integrated solar DHW heating;
- installation of transpired solar air collectors;
- advanced glazing of balconies, integrated with the ventilation system of apartments;
- installation of external transparent insulation, with or without external shading;
- a transparent-insulation wall in place of glazing for daylighting in an industrial facility (also eliminating problems of glare);
- construction of a second-skin façade over the original façade.

Figure 14.13 gives an example of a solar renovation involving glazed balconies, while Figure 14.14 gives an example of the construction of a second, glass façade over the original façade of a commercial building. Table 14.6 provides information on a number of solar renovation projects completed in Europe under IEA SHC Task 20. Further information can be found in the references given in Table 14.6, or in Voss (2000b). The glazed balconies include a solar collector mounted on the balcony wall below the balcony windows, so that air can be preheated before entering the balcony at the base of the window. Mechanical ventilation then draws the air into the apartments. This air is centrally vented to the

Figure 14.13 View of a multi-unit residential building in Zurich (a) before, and (b) after a retrofit that reduced total energy used by a factor of four

Source: Zimmermann (2004)

Figure 14.14 View of the Telus Building in Vancouver, Canada, (a) before and (b) after construction of a second, glass façade over the first façade

Source: Terri Meyer-Boake, School of Architecture, University of Waterloo (Canada)

outside so that heat can be recovered and used to heat water with a heat pump (this is an alternative to transferring heat from outgoing air to incoming ventilation with a heat exchanger, which is not possible nor necessary in this case). Energy savings in the projects are as high as 70 per cent, while the cost of solar heat (if all of the expenditure is treated as an energy-savings measure) is in the range of 20–40 eurocents/kWh. However, this is an artificially high cost, as the renovations would be eventually needed even if there were no energy savings.

Construction of a second façade can be used to save buildings that would otherwise need to be eventually demolished or undergo future expensive repairs due to water damage arising from poor workmanship at the time they were built (of which there are many examples, some rather recent!). It can also be used to improve the appearance of ugly buildings. The gap between the original and second façade can be used for preheating ventilation air, for heating exhaust air with the heat transferred to supply air via a heat exchanger, and/or used to induce passive ventilation. The original wall must have sufficient thermal mass for a second façade to be effective. In addition to reducing energy use, a second façade can solve air and water leakage problems in existing metal/glass façades or reduce noise from traffic. It can permit night cooling without the security risk of intrusion, and simple movable shading devices can be incorporated into the space between the inner and outer façade.

Transparent insulation can also be applied to old walls with substantial thermal mass but lacking interior insulation, with an air gap between the insulation and original wall that would be used in the same ways as with a glass façade. In walls without significant thermal mass, an opaque, perforated solar wall can be installed (see Chapter 4, Section 4.1.9). Translucent aerogel can be used effectively in window replacement if some loss of vision glazing is permitted, as in the examples illustrated in Figure 14.15.

Figure 14.16 compares the energy loss through the south- and north-facing façades of a school in Switzerland as calculated for the base case (no renovation), a standard renovation and for various second façades (differing in the number of glazing units and emissivity).

Figure 14.15 Examples of the use of translucent aerogel windows in retrofit applications

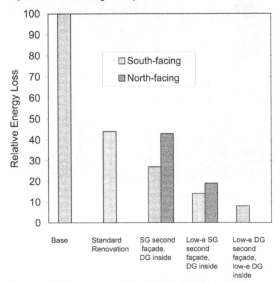

Source: Kalwall® Corporation, www.kalwall.com

Figure 14.16 Comparison of energy loss through a south- and north-facing façade of a school building in Switzerland, relative to the loss prior to any renovation, for a conventional renovation and for renovations involving construction of a second, transparent façade around the original façade

Note: SG = single-glazed, DG = double-glazed.
Source: Data from de Herde and Nihoul (1997)

Table 14.6 Summary information on solar renovation projects performed as part of IEA SHC Task 20

Location	Building description	Dates Built	Dates Renovated	Measures implemented	Space heating energy use (kWh/m²/year) Before	After	Investment (euros/m² floor area)	Cost of saved energy (euros/kWh)	Reference
Jambes, Belgium	8-storey apartment	1976	1990s	Glazed balconies	64	47			Boonstra and Thijssen (1997)
Perwez, Belgium	Single-family rowhouses	1800s	1990s	Added 2-level greenhouse on SE façade	Almost 40% savings				Boonstra and Thijssen (1997)
Aalborg, Denmark	8 apartments in 4-storey building	1900	1996	Preheat ventilation air in ventilated solar walls, roof-integrated DHW solar collectors, demand-controlled ventilation, low-e glazing	230	70	2780ᵃ		Boonstra and Thijssen (1997)
Freiburg, Germany	Multi-family (8 units)	1950s	1989	Standard insulation of all façades, roof, and cellar; low-e blinds, 120 m² TI on SE and SW façades with adjustable shading	225	43			Haller et al (1997)
Freiburg, Germany	Multi-family Villa Tannheim	1912	1995	Insulation, new windows, 53m² TI on west façade for space heating and DHW, no shading.	225	75ᵇ	633	0.22	Haller et al (1997); Voss (2000a)
Salzgitter, Germany	Industrial hall	1940	1995–7	7500m² TI glazed façade	300	225			Haller et al (1997)
Zaandam, Netherlands	384 apartments in 14-stories	1968	1997	Solar DHW, glazed balconies, TI walls	145	80	59		Boonstra and Thijssen (1997)
Gardstensbergen, Gothenburg, Sweden	11-storey residential	1975	1990s	Insulation, new windows, roof-integrated DHW solar preheating, glazed balconies, TI	270	160	60		Boonstra and Thijssen, 1997
Rannebergen, Gothenburg, Sweden	188-unit apartment building	1975	1990s	Roof-mounted solar air collector	40% savings				Boonstra and Thijssen (1997)
Hedingen, Switzerland	11-unit multi-family	1971	1994	Standard insulation of all façades, roof, and cellar. New low-e windows, 63m² TI on south façade with adjustable shading	245	140	86	0.36	Boonstra and Thijssen (1997); Haller et al (1997)
Niederurnen, Switzerland	12 apartments in 4 stories	1971	1996	Insulation, new windows, TI on SW façade with external blinds	175	105	79	0.43	Boonstra and Thijssen (1997)
Wollerau, Switzerland	Multi-family	1965	1996	Insulation, new windows, TI with fixed horizontal lamellae as sunshades in summer	70kWh/year savings per m² of TI		610 euros per m² of TI		Boonstra and Thijssen (1997)

ᵃ High cost due to intensive demonstration of advanced technologies.
ᵇ A 75% saving in space heating, along with a 50% saving in DHW energy use.
Note: TI = transparent insulation.

Photovoltaic panels can also be part of a solar retrofit, although they tend to be one of the more expensive measures at present. They can serve multiple purposes, such as shading or insulation. PV modules mounted on extruded polystyrene are available and are well-suited to the renovations of large flat roofs. They are interlocking and do not require roof penetrations (see Figure 11.10). PV models can be mounted relatively easily onto existing sloping roofs, with an air gap underneath for ventilation.

14.4 Assessment tools

A number of tools have been developed to facilitate easy preliminary assessments of the costs and energy savings of retrofitting existing buildings. Among these are *Green Globes* tools (Chapter 13, Section 13.3.1), developed in Canada, and the European assessment tools TOBUS and EPIQR (Flourentzos et al, 2000; 2002).

14.5 Summary and synthesis

Conventional but aggressive retrofits of residential and commercial buildings that are in need of refurbishment should achieve at least a 50 per cent saving in annual energy use. In many cases, a saving of 75 per cent can be achieved. There are documented cases where savings of 80 per cent or more have been achieved.

Solar retrofits involve measures such as construction of a second, transparent façade over the existing façade of a building, integrated with the ventilation system of the building; installation of building-integrated solar collectors; advanced glazing of balconies and integration with the ventilation system; and use of transparent insulation, either for heating only or for heating and daylighting. Solar retrofits provide additional savings beyond whatever could be achieved through conventional retrofits alone. Construction of a second façade can be used to protect buildings with water problems that might otherwise need to be demolished or undergo expensive future repairs, or to improve the appearance of ugly buildings.

Notes

1 The data are based on over 5700 houses where heating load was measured, but the sample is not random. Rather, it pertains to houses where the homeowner requested an evaluation of energy use, and so might be biased high, as those with heating problems are more likely to request an evaluation.

2 The data presented here are not entirely consistent with Table 13.8, which comes from a different source.

3 This should not be confused with the energy used for heating, given in Figure 14.1. The latter is equal to heat losses minus heat gains, divided by the furnace efficiency.

fifteen

Community-Integrated Energy Systems

Up to this point we have discussed buildings as isolated systems. However, in most contexts, buildings are part of a larger community, and if the heating, cooling and electricity needs of the larger collection of buildings can be linked together in an integrated system, then significant savings in the use of primary energy are possible and significant new opportunities for the use of renewable energy arise. These opportunities have implications for the design and operation of individual buildings, and especially for the planning of developments involving more than one building. The key elements of an integrated system are: (i) a district heating network for the collection of waste or surplus heat and solar thermal energy from dispersed sources and its delivery to where it is needed; (ii) a district cooling network for the delivery of chilled water to individual buildings; (iii) central production of steam and/or hot water in combination with the generation of electricity (cogeneration) to serve the district heating network; (iv) central production of chilled water

to serve the district cooling network, using either waste heat from electricity generation to drive absorption chillers and alongside production of hot water or steam for district heating (trigeneration), or as a separate process using electric chillers; (v) production of electricity through building integrated photovoltaic (BiPV) panels; (vi) possible diurnal storage of heat and coldness produced during off-peak hours; and (vii) possible seasonal underground storage of summer heat and winter coldness.

A district heating system consists of a network of insulated underground pipes carrying hot water and/or steam, with the hot water or steam produced at or extracted from a limited number of sites. District heat can be economically transported several tens of kilometres, depending on the relative steam and natural gas costs, the pipeline energy loss (typically a few per cent), the pipeline cost (which is highly location-dependent), and the density of the heat load (Karvountzi et al, 2002). A district cooling system consists of a network of insulated pipes carrying cold water (typically at 4–6°C) or an alternative thermal-transfer fluid, or carrying an ice-water slurry that is produced centrally.

District energy pipes are often pre-insulated and installed in trenches, as illustrated in Figure 15.1. The easiest heat source to capture for district heating is the waste heat from electricity generation, through cogeneration (discussed below). Many cities in the world already contain district heating systems, but many of these have dedicated central heating plants rather than co-generation plants. Table 15.1 names the cities and gives the annual heat load for the ten largest district heating systems in the world. Figure 15.2 gives the share of the total space heating load that is meet by district heating in western Europe, Canada and the US. A number of cities also have district cooling networks. Most of the new district energy systems that have been built during a recent 15-year period in the US have been district cooling rather than district heating systems (14 district cooling systems, six district heating systems, and one joint system) (Spurr, 1998).

Figure 15.1 District heating pipes about to be installed in a trench

Source: Løgstør Rør (www.logstor.com)

Figure 15.2 Share of total space heating load supplied by district heating in selected countries

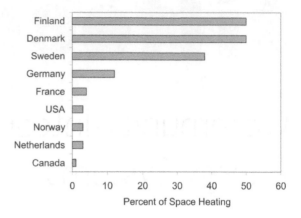

Source: Data from Maker and Penny (1999)

Table 15.1 The world's ten largest district heating systems

Location	Annual heat delivery	
	PJ	GWh
St Petersburg	237	66,000
Moscow	150	42,000
Prague	54	15,000
Warsaw	38	10,600
Bucharest	37	10,200
Seoul	36	10,000
Berlin	33	9247
Copenhagen	30	8000
New York	28	7800
Stockholm	27	7500

Source: www.energy.rochester.edu/dh/largest.htm

The term 'Community-Integrated Energy System' has been coined to refer to heat and cooling sources and sinks linked together by district heating and cooling networks, with local production of electricity through cogeneration or trigeneration. Since energy losses occur during the distribution of heat and coldness through underground pipes, these losses can be reduced – and the overall efficiency of the system increased – if the distances that hot and cold water need to be distributed are minimized. This in turn provides another reason for building compact cities rather than sprawling, low-density suburbs (the first and stronger reason being to make public transit, cycling and walking viable alternatives to the private automobile, as detailed in Cervero (1998) and Newman and Kenworthy (1999), among others). In a fully integrated urban system, the primary mechanized mode of transport would be electric, rail-based public transit using electricity produced in whole or in part by the community co- or tri-generation system, which in turn would be supplied by renewable primary energy.

Cogeneration of heat and electricity at the scale of individual buildings was discussed in Chapter 4 (Section 4.6), where it was noted that there is no efficiency benefit in producing electricity this way if a use cannot be found for the waste heat. To the extent that buildings are superinsulated and some of the hot water needs are met through solar thermal collectors, the scope for cogeneration will be greatly reduced. In jurisdictions where electricity is already produced from non-fossil-fuel energy forms, cogeneration will increase rather than decrease greenhouse gas emissions unless biomass is used as the input fuel. On-site generation of electricity through BiPV panels will sometimes exceed on-site electricity needs, requiring export of the excess electricity to the electrical grid, while at other times there will be a shortfall. Minimization of building energy use at the community scale requires optimization of measures to reduce on-site heating, hot water and electricity demand, alongside on-site production of electricity and thermal energy and its integration with the electricity grid, community-scale cogeneration, community-scale chilling plants, and district heating and cooling networks. Born et al (2001) present a modelling system that incorporates most of these considerations. For UK conditions, if on-site energy needs can be reduced by a factor of three, then over half of the remaining energy needs can be met through on-site active forms of renewable energy, thereby reducing the need for utility energy by more than a factor of six.

In jurisdictions where electricity is produced from fossil fuels, the most significant benefit of a CIES is likely to be the ability to integrate the production of heat and electricity. To understand how these benefits arise, and the limitations on the benefits, we need to digress briefly and discuss how electricity is generated today from coal and natural gas, and how it will be generated in the future from these fuels.

15.1 Generation of electricity from coal and natural gas

Today, 39 per cent of the world's electricity is generated from coal, 17 per cent from natural gas and 8 per cent from oil (UN, 2001) – a total of 64 per cent from fossil fuels. Another 18 per cent is generated with nuclear energy, 17 per cent from hydropower and 1 per cent from biomass.

15.1.1 Pulverized coal power plants

Electricity is generated using coal by first pulverizing it to a consistency of talcum powder and pneumatically injecting it through the burners into a furnace. Combustion takes place almost entirely while the coal is suspended in the furnace volume. The furnace heats water under pressure to produce pressurized steam (much like a pressure cooker). The steam is carried to a turbine, where it expands and cools, causing the turbine to rotate. There is a shaft along the axis of rotation of the turbine that is attached to the rotor in a generator, where electricity is produced. The low-pressure and low-temperature steam exiting from the steam turbine is carried to a condenser, where water is produced. This water is pumped back to the boiler, where the cycle begins again. A separate piping loop runs through the condenser, with cooler water entering and warmer water exiting, thereby carrying heat away from the condenser. The cooling water runs through a heat exchanger in a cooling tower, whereby heat is dissipated to the atmosphere, much in the same way that cooling towers are used to dissipate the heat rejected from commercial chillers in large buildings (see Chapter 6, Section 6.10), albeit on a much larger scale. Alternatively, river or lake water may flow through the condenser once for cooling purposes before being returned to its source.

The heat produced from burning the fuel ends up in one of two forms: as electricity and as heat dissipated to the surrounding environment, both through the cooling water and as hot exhaust gases. The maximum possible efficiency of a generic thermal cycle is the efficiency η_c of a *Carnot Cycle*, given by:

$$\eta_c = \frac{T_{in} - T_{out}}{T_{in}} \tag{15.1}$$

where T_{in} is the turbine inlet temperature and T_{out} is the turbine outlet temperature.[1] The turbine inlet temperature is limited by the strength of the steel used to make the boiler, since the pressure of steam increases exponentially with increasing temperature. In a typical steam turbine, $T_{in} \sim 860K$ and $T_{out} \sim 360K$, so $\eta_c = 0.58$. Real efficiencies are much less than the Carnot efficiency, ranging from less than 30 per cent in smaller turbines (common in developing countries), to 33–35 per cent typical of many developed countries, to 45 per cent in the newest, state-of-the-art units. Efficiencies of 48–50 per cent should eventually be achieved once steel alloys permitting $T_{in} \sim 1000K$ are developed (McMullan et al, 2001). Efficiencies at part load are smaller than the efficiencies given above, which pertain to full load. For example, a power plant with a full-load efficiency of 38 per cent would have a part-load efficiency as low as 28 per cent (Brown et al, 2000).

15.1.2 Electricity from natural gas

Electricity can be generated using natural gas in a gas turbine. Hot combustion gases rather than steam are directly used in a gas turbine, so the limitation on the turbine inlet pressure for steam due to the exponential increase of steam pressure with increasing temperature does not apply to gas turbines. This allows gas turbines to operate at higher temperatures – around 1250°C. This would increase the efficiency, except that about half of the shaft power is needed to drive a compressor, which draws air in and compresses it to 30 times the outside air pressure, then directs the air to a combustor. The electricity generation efficiency of a typical gas turbine ranges from 25–31 per cent, a state-of-the-art turbine has an efficiency of about 36 per cent, and turbines currently under development will have an efficiency of about 41 per cent (Kim, 2004). These efficiencies are less than those of corresponding steam turbines. As with steam turbines, the electricity generation efficiency falls sharply at part load (see Section 15.2.5).

However, the exhaust from the gas turbine is hot enough that it can be used to make steam in a *heat recovery steam generator* (HRSG). The steam can then be used for industrial processes or as input into a steam district heating system (giving *simple cycle cogeneration*), or it can be fed into a steam turbine for production of additional electricity (giving *combined cycle* power generation, so-called because gas and steam turbines are combined). Or both can be done: some of the steam in the steam turbine of a combined cycle system can be drawn off and used for heating, giving *combined cycle cogeneration*.

The best efficiency for electricity generation today using natural gas combined cycle (NGCC) is 50 per cent with air-cooled gas turbines and 55–60 per cent with steam-cooled gas turbines. An efficiency of 60 per cent is achieved through technological advances that permit turbine inlet temperatures of up to 1500°C and through such things as injecting a mist of fine water particles into the compressor to cool the air during compression, thereby greatly reducing the compressor work (Pilavachi, 2000; Williams et al, 2000). The fraction of fuel input that is converted to electricity or useful heat is typically 75–85 per cent in simple cycle or combined cycle cogeneration but can approach 92 per cent if water returning to the HRSG is cool enough to condense most of the water vapour in the exhaust and use it to preheat the returning water before it enters the HRSG. Natural gas power plants can be deployed at scales ranging from a few megawatts to hundreds of megawatts, with the lowest cost (per kW) at the largest scale. However, combined cycle power plants, with their greater efficiency, are economically viable only at about 25MW and larger.

In the long run, hybrid fuel cell turbine power plants might be developed. Waste heat from a high temperature ($\geq 650°C$) fuel cell (either molten carbonate or solid oxide) would be used in a steam or a gas turbine uses, with an overall electricity generation efficiency of 70–75 per cent (Srinivasan et al, 1999). This is about twice the current average in industrialized countries, and more than twice the current average in developing countries.

15.1.3 Integrated gasification combined cycle

An emerging technology for producing electricity from coal is the *integrated gasification combined cycle (IGCC)* power plant. Coal is gasified to a mixture of CO_2, CO and H_2 by heating it to high temperature (around $1000^\circ C$) at high pressure (up to 25 atmospheres) in the presence of pure oxygen. The CO can be converted to CO_2 and H_2 by reaction with water. The H_2 and CO (if present) are then burnt in a gas turbine (which typically operates at $1250^\circ C$). The hot exhaust gases from the gas turbine are used to produce steam for use in a steam turbine, as in natural gas combined cycle power plants. Efficiencies of IGCC plants using pure oxygen rather than air are expected to range from 42 per cent to 48 per cent, depending on the specific technology used (McMullan et al, 2001). These efficiencies are, at best, only slightly better than the best expected pulverized coal efficiency.

However, there are a number of reasons why oxygen-blown IGCC is preferred on environmental grounds. Among these benefits are the fact that emissions of some conventional pollutants (NO_x, SO_x, particulate matter) will be significantly lower than for even the cleanest pulverized coal plants, IGCC lends itself more readily to the capture of CO_2 from the exhaust gases and subsequent injection deep underground or into the deep ocean (something that might be needed as a complement to a rapid phase out of fossil fuels, as discussed in Harvey (2003, 2004)), the cooling water requirement of IGCC is less than 40 per cent that of conventional steam turbine plants (Neathery et al, 1999), and – most importantly in the present context – IGCC lends itself to more efficient generation of electricity in cogeneration mode than do pulverized coal power plants (64 per cent versus 48 per cent; see Section 15.2.2, below). Although coal would be the first fossil fuel to be phased out in a world in which all fossil fuels are phased out by 2075–2100, there will still be many decades during which coal will be needed as we make the transition to a carbon-free energy system, so we do need to consider the cleanest and most efficient ways in which coal can be used and how, where there is no near-term alternative, it can be made part of a CIES. At the very least, IGCC could form the future baseline for central production of electricity using coal, and so is relevant to determining the long run impact of changes in electricity use in buildings where coal is used to generate electricity. However, IGCC is still at a pilot stage.

15.2 Cogeneration and trigeneration

Cogeneration at the community scale can occur in a number of different ways:

- using a simple gas turbine combined with an HRSG;
- as part of a combined cycle power plant, in which exhaust heat from a gas turbine is first used to produce steam for a steam turbine, and some of the steam from the steam turbine is used to provide useful heat;
- using a simple steam turbine when there is no upstream gas turbine, as in coal or nuclear power plants.

Cogeneration in district heating systems occurs at a scale of several 10s to 100s of MW of electrical and thermal output. Cogeneration can also be done at the scale of individual buildings (5–100kW electrical or thermal output) using micro-turbines or fuel cells but, as discussed in Chapter 4 (Section 4.6), may or may not be advantageous in economic and energy efficiency terms compared to separate production of heat and electricity or compared to centralized cogeneration in a district heating system.

Figure 15.3 gives the proportion of total electricity generation that is produced decentrally in the ten countries with the largest decentralized fraction. The decentralized production consists of cogeneration and PV power, but the latter constitutes less than 3 per cent of the decentralized portion. Cogeneration accounts for just over 50 per cent of total electricity production in Denmark, and about 7 per cent of global production (247GW of capacity at the beginning of 2003).

Critical issues involving cogeneration include the impact of extracting useful heat on electricity generation, the ratio of electricity to useful-heat production, and the overall efficiency in the use of primary energy. In order to understand the design choices relevant to these issues, further information on steam turbines and the process of heat recovery is needed. This information is provided next.

Figure 15.3 Fraction of total electricity supply that is generated decentrally in various countries

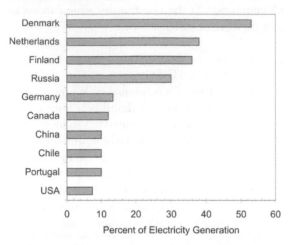

Percent of Electricity Generation

Note: Decentralized electricity production is overwhelmingly as cogeneration.
Source: Data from World Alliance for Decentralized Energy, www.localpower.org

15.2.1 Condensing, back-pressure and extraction steam turbines

In a conventional *condensing steam turbine* – the workhorse of coal-fired power plants – the steam leaves the turbine at low pressure (a high vacuum) to a condenser, where it condenses to water (at 40–50°C), which returns to the boiler. The high vacuum allows maximal extraction of mechanical energy from the steam (to rotate the turbine), but the exhaust heat is of too low a quality (pressure and temperature) to be useful in most cases. As in water-cooled chiller condensers, cooling water is needed in order to extract heat from the condenser and maintain the condenser at a cool enough temperature to provoke condensation of the water vapour leaving the steam turbine. In a *back-pressure steam turbine*, all of the steam is expanded through the turbine, but it exits above atmospheric pressure. Instead of passing to a condenser, the steam is used for heating elsewhere (where it condenses on its own), then returns to the boiler. A condenser and cooling water are not needed. Both types of turbines can be equipped with extraction ports so that some of the steam can be extracted from the steam turbine at pressures between the inlet and exit pressures, giving either a *condensing extraction turbine* or a *back-pressure*

extraction turbine. In the former case, the condenser (and cooling tower, if used) can be reduced in size if a minimum amount of steam will always be extracted. Most or all of the heat that is recovered in cogeneration is from the extracted steam, but the boiler exhaust can also be a source of recovered heat in some cases.

15.2.2 Efficiency of heat recovery in combined cycle and/or cogeneration plants

In most HRSGs used with gas turbines, turbine exhaust flows up through the device and exits at the top, while water that is converted into steam flows in the opposite direction. As illustrated in Figure 15.4, this leads to the juxtaposition of the coolest water next to the coolest exhaust and the hottest steam next to the hottest exhaust, thereby ensuring that the exhaust is always warmer than the adjacent steam or hot water and maximizing the heat recovery. The lower temperature exhaust is used to preheat the water, intermediate temperature exhaust is used to vaporize the water to form saturated steam, and the highest temperature water is used to superheat the steam. The temperature difference between the exhaust gas and the water where the water first starts to vaporize is called the *pinch point* temperature difference. The smaller this difference, the more of the exhaust gas thermal energy that will be utilized (Caton and Turner, 1997). A typical difference is 20–45K, while a high-efficiency unit may have a difference as low as 15K. However, a smaller difference results is a smaller rate of heat transfer, so a larger heat exchanger surface is required, which increases the capital cost. Another factor affecting the overall system efficiency is the pressure drop across the HRSG (usually at least 2500Pa). A smaller pressure drop leads to greater efficiency, but also increases the cost of the unit.

The lower the temperature of the water entering the HRSG, the more heat that can be extracted from the exhaust. If the water is returning from a steam turbine, this means having as low a steam turbine condenser temperature as possible. If the water is returning from a district heating system, it should do so at as low a temperature as possible. If the return water is cool enough, some of the water vapour in the gas turbine exhaust

Figure 15.4 The temperature of the exhaust gas and water/steam passing through an HRSG in opposite directions

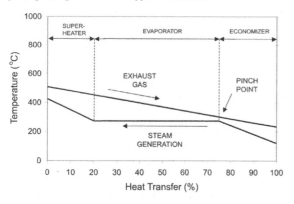

Source: Modified from Caton and Turner (1997)

Figure 15.5 Loss of electricity generation when thermal energy is extracted from a steam turbine, as a fraction of the amount of energy extracted, as a function of the extraction temperature

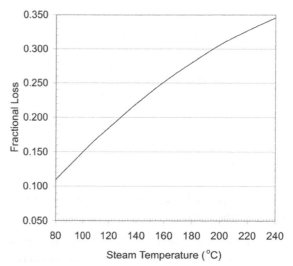

Source of figure: Bolland and Undrum (1999)

will be condensed, adding to the heat recovery. However, condensation leads to the formation of sulphuric acid if sulphur-containing fuels (such as coal) are used, in which case condensation must be avoided. This in turn requires a minimum exhaust gas outlet temperature, which limits the heat recovery and overall efficiency.

15.2.3 Trade-off between increased useful heat and reduced electricity production

For cogeneration using a gas turbine, the withdrawal of useful heat from the turbine exhaust has no impact on the production of electricity. For cogeneration using a steam turbine, an increase in the withdrawal of useful heat leads to a decrease in the production of electricity. However, the ratio of electricity loss to heat gain is smaller the lower the temperature at which useful heat is withdrawn. This is illustrated in Figure 15.5, which shows the loss in electrical energy output per unit of steam energy withdrawn from one particular steam turbine, as a function of the steam temperature. The loss in electrical output ranges from 11 per cent of the thermal energy recovered if it is recovered at a temperature of 80°C, to 35 per cent for recovery at 240°C. The amount of heat that can be transferred to district heat water will be greater the lower the return temperature from the district heat network, and the electricity:useful heat ratio will be greater the lower the supply temperature.

Because the electricity loss:heat gain ratio is less than 1.0, the total production of useful energy will always be greater when useful thermal energy is extracted. However, whether this increases the *system* efficiency depends on how efficiently the electricity that has been sacrificed would have been used. In the present context, the alternative to withdrawing useful heat would be to use the additional electricity in an electric heat pump. However, since the capture of one unit of thermal energy at 80°C sacrifices the production of only 0.11 units of electricity and may entail a loss of 5 per cent during its transport from the power plant to nearby buildings, the COP of the heat pump would have to be about 9.6 in order to break even. Since heating COPs are never this large, it is always advantageous from a primary energy point of view to produce useful heat and sacrifice a small portion of the electricity production that would otherwise be used for heating with a heat pump.

We can now see an interesting synergy. Well-insulated buildings will have a small heating load, which means that the input water temperature in a radiant heating system will not have to be as warm (only 30–35°C for a floor radiant system) as for less well-insulated buildings. This in turn allows the district heating distribution temperature to be lower, which in turn minimizes the loss in electricity production, reduces the distribution heat loss, and permits a lower return temperature and thus greater extraction of heat from the HRSG and a greater overall efficiency. On the other hand, the temperature drop as the water flows through the radiant floors will be smaller (since the supply temperature is close to the room temperature), so the difference between the supply and return temperature of the district heating system will be smaller. This results in a smaller heating capacity for a given flow rate, necessitating more pumping energy.

15.2.4 Ratio of electricity to useful heat in cogeneration

Typical ratios of electricity to useful heat output are shown in Figure 15.6 for different cogeneration systems producing steam at 10 atmospheres pressure (T = 180°C). A simple back-pressure steam turbine produces only 0.25 units of electricity for every unit of heat. With a gas turbine using an HRSG – which can be used with natural gas or with coal and biomass fuels if they are gasified first – the electricity:useful heat ratio rises to 0.8:1.0. With fuel cells (an emerging technology), the ratio is 1:1. With a combined cycle system (gas

turbine and back-pressure steam turbine), the ratio is 1.25:1, and in advanced systems still under development involving fuel cells, gas turbines and back-pressure steam turbines, an electricity:useful heat ratio of 2:1 is foreseen. With adjustable back-pressure turbines, the electricity:heat ratio can be varied with a constant rate of fuel input. Electricity production can be maximized and heat production minimized during the day (by operating the turbine with minimum back pressure), and vice versa at night.

The electricity:heat ratio depends on the temperature at which useful heat is withdrawn from the turbine. Figure 15.7 gives the electricity generation efficiency, total efficiency

Figure 15.7 Variation in electricity generation efficiency, total efficiency and electricity:heat production ratio as the temperature of heat supplied to a district heating system varies, for (a) a 40MW back-pressure steam turbine, and (b) a gas combined cycle system

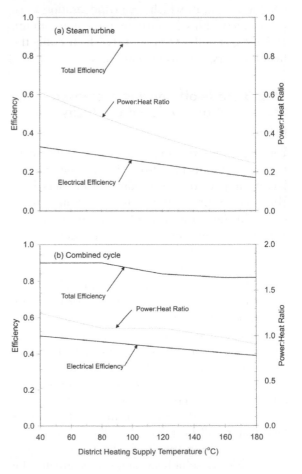

Figure 15.6 The ratio of electricity to useful heat produced in various cogeneration systems

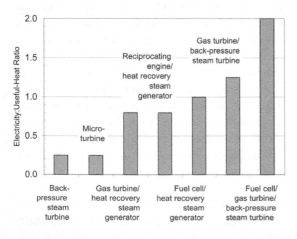

Note: Where a steam turbine is involved, the steam turbine is a back-pressure turbine without a steam condenser. The ratios are for steam produced at a pressure of 10 atmospheres.
Source: Data from Williams et al (2000)

Source: Spurr and Larsson (1996)

and electricity:heat ratio as a function of the temperature at which heat is supplied to a district heating system, for a back-pressure steam turbines and for gas combined cycle systems. As noted above, the electricity production increases as heat is extracted at a lower temperature. The overall efficiency is constant or increases modestly, such that the electricity:heat ratio increases. For the back-pressure turbine, the electricity:heat ratio varies from 0.22 to 0.61 as the district heating supply temperature decreases from 180°C to 40°C, while for the combined cycle system, the ratio increases from 0.9 to 1.25. The electricity:heat ratio decreases at part load operation (as does the overall efficiency).

A consideration of cogeneration for the production of electricity presupposes that fossil fuels are already at least part of the local electricity supply system. If electricity is generated entirely from carbon-free energy sources, then, from a climatic point of view, it is better to build systems that exclusively produce heat and hot water. In a cogeneration system, it is best to have as large an electricity:useful heat ratio as possible, given that electricity is the higher value product, and more centrally and less-efficiently generated electricity can be displaced for a given thermal load the higher this ratio (the efficiencies of centrally generated and cogenerated electricity are compared below). Furthermore, in well-designed or renovated buildings, the heating loads will be very small, and cogeneration provides an efficiency gain only to the extent that waste heat can be used. Designing a building to require the waste heat produced from cogeneration of electricity does not represent an improvement in efficiency! For these reasons, it is desirable to develop a system with as high an electricity:useful heat ratio as possible. Microturbines are being promoted for cogeneration within individual buildings and even considered by some to be an element of supposedly 'green' buildings, but they have a very low electricity:useful heat ratio (about 0.25, as shown in Figure 15.6) and so cannot be recommended.

15.2.5 Matching heat and electricity loads using heat pumps in a district heating system

If waste heat from a cogeneration plant is used to supply a district heating network, the usual approach would be to size the cogeneration plant to meet the minimum heat load, with supplemental boilers to supply additional heating requirements. Lucas (2001) discusses an alternative strategy, which yields higher overall efficiency:

- the cogeneration plant can be sized to produce more heat than needed at times of low heat load, but at these times, excess heat goes into thermal storage (perhaps underground, as discussed in Section 15.4);
- at a range of intermediate loads, use all of the heat from cogeneration plus additional heat provided by heat pumps;
- at high loads, use all the heat from cogeneration plus heat pumps plus heat from storage.

When the heat load increases without a matching of the electric load, the heat pumps can bring the two back into balance by reducing the net heat load and by increasing electricity demand.

15.2.6 Efficiency and cost in generating electricity through cogeneration

A critical performance parameter for cogeneration is the *marginal efficiency* of electricity production, which was defined in Chapter 4 (Section 4.6) as the electrical energy produced divided by the additional fuel energy used compared to production of heat only. It can be computed using Equation (4.1), and is smaller the greater the efficiency of stand-alone heat production in the non-cogeneration alternative. Table 15.2 gives the investment cost, marginal efficiency and other performance parameters for simple cycle and combined cycle gas cogeneration, assuming that the alternative heating system has an efficiency of 90 per cent ($\eta_b = 0.9$). The marginal efficiency is only 44–62 per cent for simple cycle cogeneration (increasing with the size of the plant) but is 88-90 per cent for combined cycle cogeneration. Costs are substantially lower for combined cycle than for simple cycle plants, as well as substantially less than building-scale cogeneration with microturbines or fuel cells (see Table 4.6). Investment and O&M (operation and maintenance) costs decrease with increasing plant size, while the

Table 15.2 Characteristics of natural gas cogeneration technologies available for use at the scale of district heating systems

Capacity	Cost ($/kW_e)		Efficiency (LHV, %)			Power: heat ratio	O&M (cents/ kWh)	Emissions (gm/kWh)		
	Elect only	CHP	Electrical	Overall	Marginal			NO_x	CO	Hydrocarbons
Gas turbines (simple cycle cogeneration)										
1MW	1403	1910	24	65	44	0.51	1.0	1.09	0.32	2–3
5MW	779	1024	30	67	50	0.68	0.6	0.50	0.27	2–3
10MW	716	928	32	69	54	0.73	0.6	0.50	0.23	2–3
25MW	659	800	38	70	58	0.95	0.5	0.41	0.18	2–3
40MW	592	702	41	72	62	1.07	0.4	0.36	0.18	2–3
Gas and steam turbines (combined cycle cogeneration)										
20–50MW		860	47	90	88	1.09	0.5	0.33	0.15	1–2
50–100MW		770	49	90	88	1.20	0.5	0.30	0.15	1–2
> 100MW		600	55	90	90	1.57	0.5	0.13	0.08	1–2

Note: For parallel information at the scale of individual large buildings, see Table 4.6. LHV = lower heating value
Source: Goldstein et al (2003) for gas turbines, Lemar (2001) for combined cycle plants except for emissions, which have been estimated from the simple-cycle emissions based on relative efficiencies and equipment size

power:heat ratio increases. Thus, economic considerations strongly favour larger cogeneration plants.

Table 15.3 compares the fuel use, efficiency and cost for stand-alone electricity generation, stand-alone steam production and for cogeneration of steam and electricity using industrial boilers for stand-alone steam production and using NGCC, IGCC or pulverized coal power plants for electricity generation. Also given is the marginal efficiency, computed from the fuel use data given in Table 15.3, assuming $\eta_b = 0.9$. The marginal efficiency in this case is 84 per cent for NGCC, but only 65 per cent for IGCC and 48 per cent using pulverized coal. The latter is no better than what can be expected in future central coal fired power plants (48–50 per cent). The IGCC marginal efficiency is substantially better than for stand-alone electricity generation using IGCC (about 50 per cent) but is worse than potential efficiencies using natural gas of 70–75 per cent in stand-alone electricity generation and 85–90 per cent in cogeneration.

As can also be seen from Table 15.3, the capital cost of natural gas combined cycle cogeneration, at the 400MW scale, is two or three times less than either alternative (the IGCC cost is the projected future cost). At a scale of 10MW, however, gas turbines for cogeneration are about twice as expensive as shown in Table 15.3 ($964/ kW compared to $485/kW), increasing to $1600/

kW at a scale of 1MW (Fischer, 2004). Electricity from cogeneration is 0.3–0.9 cents/kWh less expensive than from dedicated electricity power plants, depending on the kind of power plant.

At present, cogeneration provides a marked improvement in the efficiency of electricity generation (from 45 per cent in the best coal fired power plants and 55 per cent in the best gas combined cycle power plants, to 85–90 per cent marginal efficiency in the best cogeneration power plants). There is, however, very little if any gain in the efficiency of providing heat, as the efficiency for stand-alone heat production is around 80–90 per cent in properly operated systems. In the future, central power plants should attain electricity generation efficiencies of 70–75 per cent through fuel cell–steam turbine hybrid power plants powered by natural gas, in which case the efficiency benefit of cogeneration will be much smaller than today when natural gas is available. If coal is available and natural gas is not, then cogeneration – using IGCC – is clearly preferable to stand-alone electricity generation.

All of the efficiencies given in Table 15.3 are efficiencies at full load. They would be applicable to a cogeneration system that is sized to meet the minimum annual heat load, and which therefore operates at a constant electricity and heat output. However, if a cogeneration system is operated with varying heat and electricity output, then a proper assessment of the efficiency

Table 15.3 Comparison of cogeneration of electricity and steam with the separate production of electricity and steam, using natural gas combined cycle, integrated coal gasification combined cycle and pulverized coal combustion

	Separate production of electricity and steam			Cogeneration Facility	Marginal efficiency of electricity generation
	Electricity	Steam	Total		
Power generation (MW$_e$)	400	–	400	400	
Steam production at 10–15atm (MW$_{th}$)	–	400	400	400	
Natural gas combined cycle					
Fuel use (MJ/s)	739	492	1231	967	
Efficiency (%)	54.1	81.1	64.9[a]	82.8	84.2
Capital cost ($/kW-electric)	415	120	535	485	
Electricity cost (cents/kWh)	2.72			1.96	
Steam cost (cents/kWh)		1.47		1.47	
Integrated coal gasification combined cycle (IGCC)					
Fuel use (MJ/s)	889	458	1347	1078	
Efficiency (%)	45.1	87.2	59.4[a]	74.3	64.6
Capital cost ($/kW-electric)	1133	493	1625	1343	
Electricity cost (cents/kWh)	3.31			2.44	
Steam cost (cents/kWh)		1.50		1.50	
Pulverized coal combustion					
Fuel use (MJ/s)	942	458	1400	1300	
Efficiency (%)	42.4	87.2	57.1[a]	61.6	47.5
Capital cost ($/kW-electric)	1133	493	1625	1530	
Electricity cost (cents/kWh)	3.36			3.06	
Steam cost (cents/kWh)		1.50		1.50	

[a] This efficiency is the harmonic rather than the arithmetic mean of the individual efficiencies.
Note: The marginal efficiency of electricity generation is computed from the basic data, as described in the text. Fuel costs are assumed to be $2.7/GJ for natural gas and $1/GJ for coal.
Source: Basic data from Williams et al (2000)

benefits of cogeneration compared to stand-alone electricity and heat production requires a consideration of the part-load efficiency in generating electricity for both stand-alone generation and cogeneration.

Consider first the dedicated production of electricity with a gas turbine. The power output of a gas turbine can be varied in two ways: by varying the rate of fuel flow with constant airflow, or by modulating the airflow into the compressor using guide vanes, while maintaining constant fuel flow (Kim, 2004). In either case, the efficiency falls dramatically as the load decreases, as illustrated in the upper portion of Figure 15.8. At 50 per cent of full load, the efficiency is 80–95 per cent of the full-load efficiency, while at 20 per cent of full load, the efficiency is only 55–60 per cent of the full-load efficiency for state-of-the-art

turbines (older turbines, with a smaller full-load efficiency, exhibit an even larger drop in efficiency at part load). The lower part of Figure 15.8 shows the variation in relative efficiency with load for a combined cycle system using both control methods. With fuel flow control, the temperature of the gas turbine exhaust decreases as the load decreases, making it less useful for the steam turbine. With guide-vane control, the gas turbine exhaust temperature is held roughly constant down to about 50 per cent of peak power output. Hence, the efficiency drop in a combined cycle system with guide-vane control is smaller than with fuel flow control, or for a gas turbine alone. In a cogeneration system, the total efficiency drops much less than the electrical efficiency, so the electricity: heat ratio decreases as the heat output decreases.

Figure 15.8 (a) Variation with load in the efficiency of producing electricity with a gas turbine as a fraction of the full-load efficiency, using airflow modulation. Cases A, B and C correspond to a mid-performance gas turbine, a state-of-the-art gas turbine, and a soon-to-be-available gas turbine, respectively, with full-load efficiencies of 30.9 per cent, 36.1 per cent and 40.9 per cent. (b) Same as (a), except for a combined cycle system using turbines B and C with airflow (or guide-vane, GV) and fuel flow (FF) modulation.

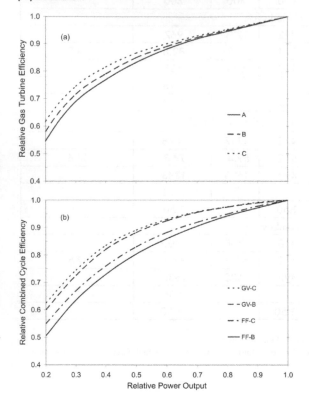

Notes: Full-load efficiencies are 54.6% and 58.1% for turbines B and C.
Modified from Kim (2004)

15.2.7 System efficiency when chilled water is produced through trigeneration

The term trigeneration refers to the production of electricity, heat and chilled water as part of a single, *integrated* process. This can occur in two ways at present: using the mechanical power from a gas or steam turbine to drive a vapour compression chiller, or to use waste heat from a gas or steam turbine to drive an absorption chiller (the operation of which is explained in Chapter 6, Section 6.6.1). In both cases, some electricity production is sacrificed. The alternative is to maximize the production of electricity and use the extra electricity to drive an electric vapour compression chiller. This is a *sequential* process. Waste heat can be used for chilling with absorption chillers in two ways: (i) centrally, using absorption chillers that take heat directly from the electric turbine, and with a separate district cooling grid in order to distribute chilled water; or (ii) at individual building sites, using absorption chillers powered with hot water or steam from the district heating grid. This avoids the need for a separate district cooling grid, but would entail greater distribution losses, and might not be considered to be trigeneration. However, the analysis presented below is still applicable.

If ΔH is the heat extracted from a steam turbine and ΔE is the electricity production that is lost, a COP for heat extraction (COP_{ext}) can be defined as $\Delta H / \Delta E$. As ΔH times the absorption chiller COP is the amount of chilling produced, the electrical equivalent COP of an absorption chiller can be defined as:

$$COP_{e\_abs} = COP_{ext}COP_{abs} = \frac{\Delta H}{\Delta E}COP_{abs}$$

$$= \frac{chilling \quad provided}{electricity \quad lost} \tag{15.2}$$

This is analogous to the COP of an electric chiller, given by (chilling provided)/(electricity used). In both cases, there is additional electricity used by the cooling tower pumps and fans and the chilled water pump – so-called auxiliaries. If q_{aux} is the electricity used by auxiliaries per unit of cooling, then the total electricity used, q_{tot}, by absorption and electric chillers per unit of cooling is:

$$q_{tot} = \frac{1}{COP_{e\_abs}} + q_{aux-abs} = q_{abs} + q_{aux-abs}$$

$$\tag{15.3}$$

and

$$q_{tot} = \frac{1}{COP_e} + q_{aux-e} \tag{15.4}$$

respectively, where $q_{abs} = 1/COP_{e-abs}$. The difference in auxiliary energy use is significant, as $q_{aux-abs} \approx 0.04$–0.05 while $q_{aux-e} \approx 0.02$ (see Table 6.13).

To clarify, the choices are to extract some thermal energy at a high enough temperature to be used in an absorption chiller, or to maximize electricity production with no thermal energy useful for chilling, and to use the extra electricity in an electric chiller. In both cases, chilling water is produced centrally, so there is no difference in distribution losses between the two choices. The data required to compute the electric chiller COP at which primary energy use is the same are given in Table 15.4. For single-, double- and triple-effect absorption chiller COPs of 0.7, 1.2 and 1.7, respectively, and corresponding input temperatures of 80°C, 120°C and 180°C (see Figure 6.56), the electricity loss:useful heat ratios are 0.11, 0.185 and 0.28, respectively (according to Figure 15.5). The COP$_{ext}$ values are given by the reciprocal of these ratios. Neglecting differences in auxiliary energy use, the break-even electric chiller COPs are 6.4, 6.5 and 6.1, respectively. However, $q_{aux-abs}$ is about one-quarter to one-third of the electricity that is sacrificed per unit of chilling with an absorption chiller but is much smaller for electric chillers. Taking into account this difference, the break-even electric chiller COPs are 5.3–5.8.[2] For smaller electric chiller COPs, absorption chillers require less primary energy, while for larger electric chiller COPs, the electric chiller requires less primary energy. Since large centrifugal chillers have COPs in the range 5–7.9 (see Chapter 6, Section 6.5.2), trigeneration with absorption chillers in place of dedicated electricity production will either approximately break even or increase total primary energy use. Actual energy differences will depend on the part-load performance of each alternative, the extent to which each kind of chiller can be operated close to its peak efficiency through careful on/off staging of differently sized chillers and the frequency distribution of different loads.

Schweigler et al (1998) report on the operation of two single-effect/double-lift absorption chillers in district heating networks, one a 300kW$_c$ unit in Düsseldorf and the other a 400kW$_c$ unit in Berlin. A single-effect absorption chiller normally has only one stage, that is, one generator, condenser, absorber and evaporator. In a single-stage/double-lift chiller, the driving hot water flows in series through three generators in two subcycles, but there is still only one condenser and one evaporator.[3] This machine offers the prospect of making better use of heat from a district heating system, particularly under part-load conditions, and was developed specifically for use in district heating networks. Part load can be achieved in two ways: by reducing the temperature of the input hot water, from 95°C at full load to 60°C at 25 per cent of full load, with a constant mass flow; or by decreasing the mass flow with a constant input temperature. In the first case, the COP decreases from 0.62 at full load to 0.45 at 25 per cent load and the temperature of the return water decreases from about 60°C to about 50°C. However, the efficiency of the boiler or steam turbine supplying the heat increases. In the second case, the COP increases from 0.62 at full load to 0.70 at 25 per cent and the return temperature drops even lower, to 40°C. A lower return temperature increases the amount of heat that can be extracted from a boiler or steam turbine, but

Table 15.4 Computation of the electric chiller COP at which total primary energy consumption is the same using electric chillers (with q_{aux-e} = 0.022) or using absorption chillers with the properties given below and driven by waste heat taken from a steam turbine at the expense of reduced production of electricity

# Effects	Input T	COP	COP$_{ext}$	COP$_{e-abs}$	q_{abs}	$q_{aux-abs}$	Break-even electric-chiller COP
Single	80°C	0.7	9.09	6.36	0.157	0.050	5.34
Double	120°C	1.2	5.41	6.49	0.154	0.040	5.75
Triple	180°C	1.7	3.57	6.07	0.165	0.036	5.54

Note: For electric chillers with a larger COP than the break-even COP given here, operation of electric chillers with maximization of electricity production uses less primary energy. The $q_{aux-abs}$ and q_{aux-e} values used here are taken from Table 6.14, and could vary substantially from case to case. See text for a full explanation.

the constant supply temperature (at 95°C) locks in the penalty in electricity production that arises if the heat is supplied from a steam turbine.

Lin et al (2001) have analysed the prospect of using hot water from the district heating system in Beijing for cooling using an absorption chiller. Addition of cogeneration to the district heating system is being contemplated. Lin et al (2001) have identified an important system interaction, as explained now. At present, the Beijing system supplies 100MW of domestic hot water in summer using supply and return temperatures of 75°C and 50°C, respectively. Supplying water at this temperature through cogeneration would entail no penalty in terms of electricity production. However, operation of an absorption chiller requires supplying water at 95°C, which would entail an electricity production penalty that would apply to *all* of the hot water produced, whether for absorption chilling or for domestic hot water. When this interaction between the production of chilled water and hot water is taken into account, the absorption chiller option is decidedly less attractive than using an electric chiller.

For a year-2010 trigeneration district energy system powered by a solid oxide fuel cell/gas turbine combination, with chilled water produced either by a vapour compression chiller or by an absorption chiller supplemented with a vapour compression chiller as needed, Burer et al (2003) conclude that use of an absorption chiller does not make sense, either economically or in terms of CO_2 emissions. This is because of the high electricity generation efficiency (up to 70 per

cent) and low waste heat production anticipated for such systems.

In the case of using the mechanical shaft power from a gas or steam turbine to directly drive a vapour compression chiller, an efficiency advantage is expected. This is because two steps – conversion of shaft power to electricity with a generator, and conversion of electricity back to shaft power with a motor – are omitted. The minimum energy loss for each of these steps is about 5 per cent (i.e. generator and motor efficiencies are around 0.95). However, smaller turbines have lower efficiencies than larger turbines, and if the central chiller plant uses a significantly smaller turbine than the electric power plant (as would most likely be the case), the gain from avoiding generator and motor losses might be negated. However, if the choice is between on-site power generation and chilling with electric vapour compression chillers, or on-site chilling with a directly driven vapour compression chiller, using the same turbine in both cases, then it is preferable to drive the vapour compression chiller directly. At any time, the mechanical work supplied by a heat engine can be distributed between an electric generator and a compressor at variable and complementary rates (d'Accadia, 2001).

Figure 15.9 shows the components found in the Trigen-Peoples District Energy system in Chicago. The system makes use of three kinds of chillers (electric, absorption, direct drive) for cooling, heat recovery steam generators and dedicated boilers for heating, gas turbines for electricity generation, and diurnal storage of chilled water

Figure 15.9 Energy flows and equipment used in the Trigen-Peoples trigeneration system in Chicago

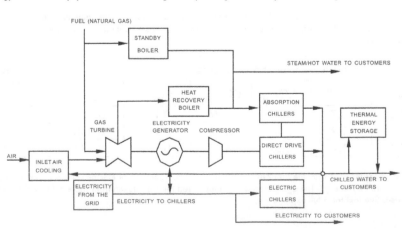

with anti-freeze. The system can provide $60MW_c$ of peak cooling, $11.4MW_{th}$ of peak heating, and 3.3MW of electricity, and can store 1556GJ of coldness with a $-1°C$ supply temperature and a $12°C$ return temperature (equivalent to about 7 hours of chilling at the peak rate) (Andrepont, 2000a). About 1 per cent of the cooling capacity of the system is used to cool the air entering the gas turbines (down to $4-10°C$), as this increases the efficiency and power output of the turbines sufficient to give a 5 per cent saving in fuel use (Andrepont, 2000b).

A trigeneration system built for the Expo '98 project in Lisbon also made use of about 1 per cent of the chilled water output to cool the combustion air, in this case increasing electricity production by 17 per cent and reducing fuel use by 7 per cent (Dharmadhikari et al, 2000). However, chilling does not need to occur as trigeneration – that is, using heat or mechanical power integrated with electricity production – in order to achieve this gain. Instead, chilled water can be produced centrally, using electric chillers next to a cogeneration plant. As shown above, trigeneration can increase rather than decrease primary energy use compared to the use of electric chillers, and this should also be true when part of the cooling output is used to cool the combustion air.

Alternative to absorption chillers

The above discussion of trigeneration has focused exclusively on absorption chillers which, as noted above, require a rather hot input and thus result in less electricity production, as well as requiring a non-negligible amount of electricity for the operation of pumps and of the cooling tower. Trigeneration using adsorption chillers (Chapter 6, Section 6.6.2) is likely to be attractive from a primary energy point of view, inasmuch as adsorption chillers can utilize cooler input heat (at as low at $50-60°C$) with only a modest loss of COP, particularly if cooling water is available at $28°C$ rather than at $32°C$ (see Figure 6.58). Adsorption chillers at present are substantially more expensive than absorption chillers (Table 6.12). Another alternative would be to distribute relatively low-temperature heat through the district heating system, to be used at individual building sites to regenerate solid or liquid desiccants that are used for dehumidification of ventilation air.

15.3 District heating and district cooling

A district heating system can utilize waste heat from cogeneration, or it may distribute heat that is produced centrally and separately from the generation of electricity. In the preceding section we showed that, under most circumstances where fossil fuels are part of the supply mix for electricity, production of useful heat in conjunction with the production of electricity leads to a reduction in primary energy use compared to independent production of heat and electricity. However, there can be efficiency gains through centralized production of heat compared to use of on-site boilers or furnaces even in the absence of cogeneration and in spite of distribution losses. Similarly, chilled water supplied to a district cooling network can be produced through trigeneration, or it can be produced through a centralized chilling plant independently of the generation of electricity. The co-production of chilled water (with absorption chillers) and electricity can either increase or decrease primary energy use compared to a system using electric chillers where electricity production is maximized and waste heat is discarded. As with district heating, there can be efficiency gains with centralized production (of chilled water, using electric chillers) compared to on-site production due to economies of scale, even when not integrated with the production of electricity. Furthermore, centralized systems can make it easier to shift to renewable energy for heating and cooling purposes, as explained later.

From the perspective of the owner or operator of a building, there are a number of reasons why hooking up to a district heating or cooling system might be preferable to on-site production of hot and chilled water:

- there are savings in the upfront capital cost since boilers and chillers do not need to be purchased as part of the building;
- there are savings in space, maintenance costs and insurance by virtue of having no on-site heating or cooling equipment;
- in the case of district cooling, there is no need for on-site cooling towers and condenser piping;

- noise and vibration associated with heating and/or cooling equipment are eliminated;
- the absence of roof-mounted cooling towers eliminates the risk of legionnaires disease, particularly if air intakes are also located on the roof;
- the absence of roof-mounted cooling towers increases flexibility in the design of the building, including an increased opportunity for the incorporation of PV or solar thermal collectors.

The difficulties with district heating and cooling occur at the societal scale: district energy projects are typically complex and involve a large number of institutional, technological, legal and financial issues (MacRae, 1992). The required feasibility studies are expensive and time consuming. District energy systems are capital intensive, with a large upfront investment that is paid back over a long period of time. Strong and determined leadership is required to bring about these systems.

15.3.1 Hot water versus steam district heating systems

Heat can be distributed as either steam or hot water. Steam systems have been classified as high pressure (>8.6 bar), medium pressure (2.0–8.6 bar) and low pressure (<2.0 bar), while hot-water systems have been classified as high-temperature (>150°C), medium-temperature (90–150°C) and low-temperature (<90°C). Water is typically supplied at 8–16bar and returns at 5bar. The advantages and disadvantages of steam and hot-water distribution systems are summarized in Table 15.5. Many older systems use steam, whereas most modern systems use hot water. Steam has a much higher thermal energy density than hot water, so much less mass needs to be circulated in a steam system than in a hot-water system.[4] On the other hand, hot water provides a number of efficiency advantages compared to steam: less electricity production is sacrificed if the heat is supplied through cogeneration, lower-temperature heat sources can be used and distribution losses are smaller due to the smaller temperature difference between the pipe and the surroundings. The supply temperature can be adjusted with the seasons, or there may be more than one piping network, carrying water at different temperatures, thereby allowing both low-grade (low temperature) and high-grade (high temperature) heat to be captured and utilized. As discussed later (Section 15.5.3), solar-assisted district heating systems are based on hot water, with supply temperatures as low as 60–70°C and return temperatures of 30–40°C.

Table 15.5 Advantages and disadvantages of steam and hot-water district heating systems

Steam – Advantages
Pumps are not required – steam flows under its own pressure
High thermal energy density
Steam – Disadvantages
If extracted from a steam turbine, a substantial amount of electricity is sacrificed
Can be transported only 5km, more in newer systems with improved insulation
Poor match to end-use heating requirements in most cases
Steel pipes are required; these are expensive and can corrode
Water must be treated to prevent mineralization
High heat loss during transport (15–45%)
Not as safe as hot water, requiring stringent boiler, piping and personnel codes
Metering of energy use is difficult
Maintenance costs are greater than for hot-water systems
Difficult to operate under conditions of varying load
Very susceptible to the loss or mis-sizing of a single large customer

Hot water – Advantages
If extracted from a steam turbine, relatively little electricity is sacrificed
Can be readily transported 25km and sometimes up to 100km
Good match to end-use heating requirements in most cases
Plastic pipes can be used; these are less expensive and do not corrode
Water need not be treated
Small heat loss during transport (5–15%)
Safe
Metering of energy use is easy
Low maintenance costs
Easy to operate under conditions of varying load
Can be stored
Direct flow of the district heating water through the building hydronic heating system is possible, eliminating the need for a heat exchanger and associated heat losses
Low-grade thermal energy sources can be upgraded to low-temperature hot water with a heat pump
Low-temperature hot-water systems can use solar energy
Hot water – Disadvantages
Must be pumped
System balance is required in order to ensure adequate flow to all customers
Cannot provide high-pressure steam if a customer requires it – can only perform preheating

Source: Based largely on MacRae (1992)

15.3.2 Efficient operation of centralized heating facilities

District heating systems (DHSs) can provide an efficiency advantage compared to on-site heating even in the absence of cogeneration. This is because (i) a heating system is sized to meet the heating requirement on the expected coldest day of the year, and so will be running at part-load most of the time, and (ii) non-condensing boilers have peak efficiency at full load, with reduced efficiency at part load, particularly below 20–50 per cent of full load (see Chapter 4, Section 4.3.2). During the summer, when the only heating load is for domestic hot water, the heat load may be only 10 per cent of the peak load. In a DHS, the heat load will be satisfied by many boilers rather than by one giant boiler. As the load varies, so will the number of boilers that are operating. The number and size of boilers operating at any given time can be selected to provide maximum efficiency. Furthermore, boilers are more likely to be properly maintained in a centralized heating system than in small, individual heating systems. Altogether, the annual energy saving for a centralized heating system can be 10–20 per cent

compared to a decentralized system. This efficiency gain will be offset by the heat losses from the pipes that connect the buildings to the central heat plant. These heat losses are of the order of 5–10 per cent, depending on the distance transmitted, how well insulated the pipes are and whether the system is based on hot water or steam. However, the heat loss can also be much larger, especially if the distribution system has not been maintained, thereby annulling the benefit of more efficient heating plant operation.

The efficiency benefits of a DHS will be reduced or eliminated if the buildings served by the DHS would utilize condensing boilers (where efficiency improves at part load down to 5–33 per cent of peak load) in the absence of the DHS, or if they would have a small boiler designed to carry the summer domestic hot water load.[5] To ensure that there is a net efficiency gain compared to the best alternative, the DHS should be combined with cogeneration whenever this will displace electricity that is generated with fossil fuels and should be designed to tap other sources of heat. As with building scale heating systems, the water returning to the heating plant should be as cool as possible in order to maximize the plant efficiency or to maximize the amount of heat that can be extracted from other heat sources. This in turn requires well-insulated buildings (to minimize the required supply temperature) with large radiators (to maximize the temperature drop).

15.3.3 Tapping into other sources of heat

The district heating network that supplies heat to the buildings in a community could be used to collect heat from scattered sources. Examples include sewage treatment plants (sewage water will be several degrees above freezing even in winter), bakeries, some manufacturing facilities and electrical transformer stations. Since the heat from these sources will usually be at a lower temperature than that at which heat is distributed by the district heat system, heat pumps will be needed to transfer heat from these heat sources to the heat distribution grid. In Tokyo, sewage has a stable temperature of about 16°C in February, and is used as a heat source for production of hot water at 47°C with a COP of 3.9 in one small district heating network (Yoshikawa, 1997). Over 20 per cent of the heat for district heating in Gothenberg, Sweden, is extracted from waste water (Balmér, 1997). Heat that is rejected by chillers in ice arenas or district cooling systems could also be supplied to the district heating system (to the extent that the timing of the loads matches). In effect, the district heating system would serve as a *heat energy broker*, collecting it where there is an excess and supplying it where there is a deficit. About half of the total heat output from heat pumps in Sweden comes from units in district heating systems, usually to upgrade the heat from lake water and sewage water (Berntsson, 2002). In addition, steam from incineration plants is used with 6MW absorption heat pumps to extract heat from the exhaust gas for district heating; the exhaust gas is cooled from 70°C to 30°C in the process.

Low-grade geothermal heat, left over after higher-temperature heat has been used for heating or hot water purposes, is another example of a heat source that can be put to use with a heat pump if a district heating system is present. In Tianjin, China, geothermal 'waste' heat is available at 40°C but should not be discharged to waterways at a temperature above

30°C. A heat pump with an appropriate mixture of working fluids can be used to upgrade the return-water of the district heating system from 65°C to the supply temperature of 80°C while cooling the discarded geothermal water to 30°C (Zhao et al, 2003a).

15.3.4 Accommodating low distribution temperatures

In order to utilize low-temperature heat sources, a relatively low distribution temperature (<70°C) is needed. The heat flow from a radiator depends on the radiator temperature, size and hot-water flow rate. In new developments, the building envelope can be designed to require smaller peak heating rates. However, if a high-temperature district heating system is converted to a low-temperature system in order to utilize low-grade heat sources, special provisions will need to be taken to assure adequate heating of existing buildings. Wherever possible, the thermal envelope in these buildings should be upgraded prior to conversion so that the amount and temperature of the required heat can be reduced. Other options are:

- to increase the flow rate (necessitating larger pumps and, in rebuilt systems, larger-diameter pipes); or
- to install larger radiators; or
- to add a peaking plant to boost the temperature during coldest conditions.

All of these options entail added cost, so the split between the different options should be optimized, as discussed by Kilkis (1998). For a system in which the supply temperature is reduced from 90°C to 60°C, the radiators could be oversized by 60 per cent and the distribution temperature boosted to 76°C during peak conditions. Flow oversizing will not be the prime method of restoring equipment capacity, but can be used for fine tuning.

Another option is to use on-site heat pumps to boost the internal distribution temperature for those buildings that require higher temperatures (Curti et al, 2000). This avoids the need to provide heating water at the temperature required by the most demanding building. In buildings where a low internal distribution temperature is permitted, heat pumps can be used to extract heat from the return branch of the district heating system rather than from the supply branch. This will increase the efficiency of the central heating plant, and also reduce the required water flow in the district heating loop, because the temperature differential between supply and return water will be increased.

15.3.5 Opportunities for cascading of heat flow

To maximize the temperature drop between supply and return water in the district heating system, heat should be cascaded to the greatest extent possible (Kilkis, 2002). If heat is supplied near 70°C, it can be cascaded to progressively lower temperatures by being used first to supply domestic hot water at 55–65°C, then used in a fan coil to heat incoming ventilation air and, finally, used in a radiant heating panel to supply the heat load that cannot be satisfied through ventilation air. If heat is supplied at a temperature closer to 35°C, an on-site heat pump can be used to boost the water temperature for domestic hot water or to meet peak heating loads. The lowest-temperature systems require a larger or more effective heat exchanger between the power plant and district heat water, and between the district heat water and building heating water. This is because the permitted temperature drop across each heat exchanger is smaller due to the lower starting temperature.

15.3.6 Efficient operation of centralized chilling facilities

Constant-speed screw chillers and centrifugal chillers with variable-speed compressors maintain close to their full-load efficiency for loads down to about 60 per cent of full load, while fixed-speed centrifugal chillers suffer from a continuous drop in efficiency as load decreases (see Figure 6.51). A central cooling plant serving several nearby buildings can be built from many chillers that would otherwise go into individual buildings. The system can be designed with the number of chillers and size mix chosen to give the greatest operational flexibility, and the number of

chillers in operation and the distribution of the cooling load among the chillers can be chosen so that most of the chillers operate at peak efficiency at any given time. Variable-speed drives, which entail about a 5 per cent efficiency loss even at full load, could be eliminated. However, clever equipment sizing will not yield an energy savings if the start and stop sequencing of the various chillers is not arranged for maximum efficiency. This in turn requires a knowledgeable and highly trained operating staff, or a sophisticated control system – something more likely and more economical in a centralized cooling system. Dharmadhikari et al (2000) present the example of the Expo '98 project in Lisbon, where a central plant reduced chilling energy use by 45 per cent compared to meeting the same loads with chillers in each building.

A large efficiency gain is possible if a centralized chilling facility is used in place of individual, small air conditioners in multi-unit residential buildings, as air conditioners are less efficient than commercial chillers (COPs generally of 2.5–3.5 but higher in Japan, compared to 5.0–7.5 for large commercial chillers). This can be done at the scale of an entire building, but requires finding space for chillers and installing the cold-water piping, radiators and a cooling tower, and then maintaining the system. Connection to a district cooling system provides a convenient alternative to separate chillers and cooling towers in multi-unit residential buildings, although an internal cold-water distribution system and radiators are still needed. Indeed, in some cases the only practical alternatives will be connection to a district cooling network or use of inefficient wall-mounted air conditioners. As with district heating, it is essential that the cooling provided by a central facility be metered so that customers can be billed for their actual use, thereby providing an incentive to minimize their cooling requirements.

A growing number of district cooling systems are using a cold aqueous fluid, or even an ice-water slurry, rather than cold water to distribute coldness. The aqueous fluid allows generation, storage and distribution at reduced supply temperatures and increased supply-to-return differentials. This reduces the size and cost of distribution pumps and piping, and can reduce the size and cost of heat exchangers and customer air-side fans and pumps. For an ice-water slurry, as the percentage of ice in the mixture increases up to 10 per cent, the friction and heat transfer coefficients decrease by about 10 per cent due to a reduction in turbulence, so pumping energy and heat loss decrease (Knodel et al, 2000). This is in addition to the benefit arising from the much greater cooling capacity for a given flow rate due to the latent heat of ice. Ice-water slurry remains an issue of concern in multi-customer piping networks. The disadvantage of both types of cold thermal storage is that the chiller evaporators must be at a lower temperature, reducing the COP of the chilling equipment. Furthermore, the supply temperature will be colder than needed for displacement ventilation (DV) or chilled ceiling (CC) cooling systems (see Chapter 7, Sections 7.4.4 and 7.4.5). If some buildings in the district cooling system use DV and CC systems and others use conventional cooling systems, then the district cooling system must satisfy the supply water temperature requirement of the conventional buildings, so the potential COP gain of DV and CC systems will not be realized.

The vast majority of district cooling systems work as designed. However, Elleson and Dettmers (1999) document a case where basic errors in the design and operation of a system that serves five buildings were made. They discuss design flaws, shortcuts that violated the original (flawed) design, and the resulting problems and inefficiencies in the operation of the system. Correct sizing of each component of the system, and an adequate number of control sensors in appropriate locations, are essential features of an efficient system. The system evaluated by Elleson and Dettmers (1999) used a cold storage tank which was supposed to be naturally stratified, with 4°C water from the chillers at the bottom and return water from the buildings at the top, but due to mismatches in the system components, stratification did not occur and the storage capacity of the tank was much less than anticipated. Unplanned frequent starts and stops of the system caused excessive wear and tear. However, the more common problems with district cooling systems involve logistical, marketing and financing difficulties.

15.3.7 Utilizing dispersed heat sinks

In a conventional cooling system, the heat extracted from a building is rejected to the atmosphere through a cooling tower on the roof of the building (see Chapter 6, Section 6.10), or through the air-cooled condensers of rooftop chillers. Air-cooled condensers must be warmer than the ambient air in order to reject heat. This is more difficult the warmer the air, so the efficiency of the cooling system decreases (at the same time that the amount of heat that needs to be removed from the building increases). If cooling towers are used, the cooling water supplied to the condenser will be 2–4K warmer than the ambient wet-bulb temperature, which depends on ambient temperature and relative humidity, and the condenser must be slightly warmer than the cooling water in order to reject heat. If the heat in the cooling system can be rejected to a colder medium, the system efficiency can be improved. Potential media for receiving the heat from a cooling system include sewage water, lake water or sea water, or the ground itself (in the latter case, long-term heating of the ground could be a constraint, as discussed in Chapter 5, Section 5.3.3). However, the majority of buildings in a region will likely not be situated so as to be able to use these heat sinks. By linking the buildings in a district cooling network, it may become possible to utilize heat sinks that could not otherwise be used.

In Japan, sewage water is used as a heat sink for cooling purposes in summer (and as a heat source, with heat pumps, for the heating network in winter). During extraordinary conditions in Tokyo in 1995, with a peak outside temperature of 37°C, the sewage temperature rose to 29.4°C, and the monthly average COP was 4.3 for the production of chilled water at 7°C (Yoshikawa, 1997).

Yik et al (2001) analysed the impact on energy use and cost of systems using alternative heat sinks for air conditioning in Hong Kong. At present, most building chillers have air-cooled condensers, due to restrictions on the use of freshwater in cooling towers. Alternatives include (i) use of sea water in a closed loop cooling tower; ii) direct use of sea water to cool the chiller condensers (the condensers are in contact with sea water); (iii) indirect use of sea water to cool the condensers (the condensers are cooled by a freshwater loop, that exchanges heat with the sea water through a heat exchanger); and (iv) a district cooling system with large centralized, water-cooled chillers and cooling towers. For a wet-bulb temperature of 28°C, the cooling tower delivers cooling water to the condenser at 32°C. Indirect use of sea water (at 27°C) also provides cooling water at 32°C, whereas direct use of sea water provides cooling water at 27°C. The annual energy use for cooling a group of buildings (including pumps to supply sea water or chilled water to individual buildings) and the full-load COP of the chillers used in different systems are compared in Table 15.6. The differences in energy use depend on differences in pumping energy and the opportunity to optimize the staging of the chillers for different systems, as well as on differences in COP. The district cooling system requires about 25–30 per cent less energy than the use of air-cooled chillers in individual buildings, and 17–22 per cent less energy than the use of water-cooled chillers in individual buildings, with a larger district cooling system being slightly more efficient due to the use of larger and more efficient chillers. With the most efficient chillers (COP = 7.5), the energy saving is 45 per cent compared to air-cooled chillers. District cooling is now being actively considered by the Hong Kong government (Parsons Brinckerhoff Asia, 2003; Ove Arup and Partners Hong Kong, 2003).

Where lake or sea water is sufficiently cold, it can be used directly for cooling buildings (rather than as a medium for cooling the condensers of chillers). In Stockholm, this is done using water from the Baltic Sea that is coupled to the building chilled water loops through heat exchangers (so that sea water itself does not circulate through the buildings). When the water it

not cold enough, it is cooled with a heat pump, with the ejected heat added to the district heating network. The use of deep cold sea water is being pursued in Hawaii (see Section 15.5.5 for more information). In Toronto, cold water (at 4°C) from below the 80m depth of Lake Ontario will be used in place of electric chillers in an eventual 183MW$_c$ (52,000 ton) district cooling network, while a 63MW$_c$ (20,000 ton) district cooling system is in operation at Cornell University (New York state) using cold water from Cayuga Lake.

15.3.8 Opportunities for centralized storage of coldness (Coolth)

The benefits of diurnal storage of coolth in individual buildings, using chilled water or an ice/water slurry, were discussed in Chapter 6 (Section 6.9). Coolth can also be stored centrally, as part of a district cooling system. For example, the 1994 Chicago Trigen-Peoples District Energy system has what was then the largest welded steel tank cool storage system (capacity: 1556GJ). In 2003, the Orlando University district cooling system in Orlando, Florida, began operation with an even larger steel tank for cold storage (capacity: 1818GJ based on the initial chilled water service, convertible to 2840GJ in the future with low-temperature fluid service). Another example is at Cosmo Square, Osaka (371GJ). At Yale University, reinforced concrete tanks are located beneath parking lots and playing fields, and have a total capacity of 11 million litres (11,000 m^3 or 22.2m × 22.2m × 22.2m if a single cube). If the stored water warms up by 5K each day and is cooled by 5K each night, the storage capacity of the Yale system is 230GJ (11,000m^3 × 0.004186GJ/m^3/K × 5K). Large-scale, centralized storage is

Table 15.6 Energy use (kWh/m²/year) for alternative systems for cooling buildings in Hong Kong

System	COP	System energy use	Savings
Air-cooled air conditioners	2.8	161	
Cooling tower for cooling condensers	4.7	147	8.7%
Indirect use of sea water for cooling condensers	4.7	143	11.1%
Direct use of sea water for cooling condensers	5.2	134	16.8%
District cooling system, 40MW$_c$	5.7	122	24.2%
200MW$_c$	5.9	115	28.6%

Source: Yik et al (2001)

advantageous compared to storage for individual buildings because, as the scale of the storage system increases, the volume to surface area ratio increases. Since energy losses depend on the surface area, this tends to reduce the loss per unit of thermal storage. Unit costs also fall dramatically as the scale of the storage system increases, which favours district cooling applications of thermal storage.

District heating and cooling systems also lend themselves to large-scale, seasonal storage of heat and coldness underground, as discussed later.

15.3.9 Integrated heating and cooling

When there is a need for simultaneous heating and cooling, such as for air conditioning and domestic hot water in summer, a hot pump could be used to produce chilled water while rejecting heat at a high enough temperature to serve DHW requirements. The advantages and disadvantages of this method of producing hot water from an energy efficiency point of view are discussed in Chapter 8 (Section 8.6.1) in the context of individual buildings. It might be more easily implemented in centralized district heating and cooling systems, but a full analysis is still required in order to determine if, to what extent, and under what circumstances there is a saving in primary energy use.

15.3.10 Capital cost and economies of scale

Centralized district heating and cooling systems entail a substantial upfront capital cost to install the pipe network. Offsetting this are a number of factors that tend to reduce the total capital cost of the heating and cooling system:

- total heating and cooling capacities do not need to be as large as the sum of the capacities that would occur in individual buildings, because peak demands would not all occur at the same time;
- unit costs of electrical generators, heat exchanges, boilers and chillers all tend to fall with increasing capacity. This is illustrated in Figure 15.10 for market conditions in Japan.

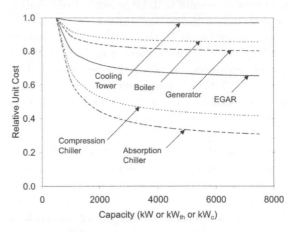

Figure 15.10 Variation in the unit cost of a reciprocating gas-engine electrical generator and of heating and cooling equipment with capacity for market conditions in Japan

Note: EGAR = exhaust heat gas absorption chiller/heater.
Source: Based on cost equations in Yamaguchi et al (2004)

Boiler operation and maintenance costs will be lower due to economies of scale, while the energy operating cost of the cooling plant will be less due to the greater efficiency of large chillers compared with small chillers, the better opportunity to schedule multiple chillers so as so maximize overall efficiency at part-load operation, and the relatively smaller cost of maintenance and supervisory personnel.

The capital costs of providing chilling through individual on-site chillers, through on-site absorption or desiccant chillers supplied by district heat, or through a district cooling network with centralized vapour-compression or absorption chillers, are compared in Table 15.7. The savings in the cost of the chilling equipment with a district cooling system more than offsets the cost of the network for the network costs presented in Table 15.7, resulting in a smaller total capital cost. The smaller cost of the chilling equipment is due in part to smaller costs per unit of cooling capacity for larger chillers, and also because the total capacity need not be as large in a centralized plant due to the fact that the peak loads in individual buildings will not all occur at the same time. With individual chillers in each building, each building must be able to supply its own peak cooling load. The total plantroom floor area occupied by the

Table 15.7 Comparison of capital costs, energy use and water use for individual vapour compression chillers (VCCs), for various systems driven by heat from a district-heating system, and for centralized chillers serving a district-cooling system that serves a 25 per cent larger total load due to individual peak loads not exactly coinciding

Cost component	System						
		District heat supplied to:			District cooling chilled water from:		
	500kW$_c$ VCCs in individual buildings, peak COP = 4.2	500kW$_c$ SE/DL absorption chillers	500kW$_c$ adsorption chillers	500kW$_c$ desiccant cooling units[a]	5MW$_c$ of vapour compression chillers serving 6.25MW$_c$ load	4 1.25MW$_c$ Li/Br absorption chillers serving 6.25MW$_c$ load	2.5MW$_c$ of NH$_4$/H$_2$O absorption chillers serving 6.25MW$_c$ load
Chiller, $	76,470	214,120	431,760	94,830	413,000	416,000	1,529,400
Chiller, $/kW$_c$	153	428	864	190	83	83	
Cooling Tower, $	41,650	26,120	62,590	0	110,000	196,500	
Mechanical Installation, $	60,060	82,350	87,710	included	275,000	508,700	81,060
Electrical Installation, $	35,700	39,000	40,180	included	137,500	126,500	88,410
Building, $	57,060	57,060	57,060	26,470	76,000	258,800	258,820
Other, $	101,120	181,830	247,070	59,880	308,000	531,600	1,014,890[b]
Network, low,[c] $	0	0	0	0	674,100	674,100	427,650[d]
Network, high,[e] $	0	0	0	0	2,195,300	2,195,300	1,133,530[d]
Total, $	372,000	600,350	926,350	181,180	1,993,600–3,514,800	2,754,700–4,275,900	3,400,120–4,106,000
Total, $/kW$_c$ of load served	744	1200	1853	362	319–562	441–685	544–657
EFLH/year	800	800	800	800	1000	1000	2000
Chilling provided (MWh/year)	400	400	400	400	5000	5000	5000
Electricity use (MWh/year)	140	84	66.5	46.8	1220	737.5	1016
Heat use (MWh/year)	0	635	770	334	0	736	8700
Average COP	2.9	0.63	0.52	0.84	4.1[f]		0.57
Water use (m³/year)	1098	2300	2600	800	1020[g]	22,032	30,604
Water use (m³/GJ)	0.73	1.60	1.81	0.56	0.71	1.22	1.70

[a] Provide chilled and dehumidified air only.
[b] Includes $264,710 for an ice-slurry storage tank.
[c] 100m of network per MWh/year of cooling load.
[d] Network costs for an ice-slurry system.
[e] 400 m of network per MWh/year of cooling load.
[f] Assumed here.
[g] Reduced from first case in proportion to reduction in required heat rejection.
Note: EFLH = equivalent full-load hours.
Source: IEA (1999b) for all cases except the district-cooling system with vapour compression chillers, where all costs except network costs are taken from Spurr and Larsson (1996) and multiplied by 1.1 in order to be roughly comparable with the more recent costs in IEA (1999c). Chiller cost data in $/kW$_c$ can be compared with data in Tables 6.12, 6.13 and 6.14, and Figure 6.57

central chiller plantroom is much smaller than the total plantroom floor area occupied by all the chiller plants in individual buildings, in part due to less total capacity and in part due to more efficient use of space in a larger system.

15.3.11 Use of district energy systems today and targeted future use

In the US there are around 6000 district energy systems (approximately 2000 at universities, 2000 at medical facilities and 2000 at other facilities, including district energy utilities, private industry and airports), serving more than 10 per cent of the country's commercial floor space. Only 10 per cent of these are combined with cogeneration of electricity, however, representing 3.5GW of generation capacity (Kaarsberg, et al 1999). Table 15.8 gives the proportion of electricity produced through cogeneration in the European Union and other countries in central and eastern Europe. In both the US and Europe (as well as

Table 15.8 Electricity production through cogeneration as a fraction of total production or of thermal electricity production alone

Country	Proportion of electricity from cogeneration (%)	Comment
Austria	77	This is the fraction of thermal electricity (66% of total electricity is hydro)
Belgium	n/a	
Denmark	50	Mostly district heating cogeneration
Finland	32	Cogeneration provides 75% of heating energy
France	2	Largely industrial cogeneration
Germany	10	Equal district heating and industrial cogeneration
Greece	2.5	
Ireland	2.1	
Italy	16	Dominated by large industrial cogeneration
Luxembourg	58	This is as a fraction of electricity production; 95% of electricity is imported
Netherlands	38	
Portugal	16	Dominated by industrial cogeneration
Spain	n/a	
Sweden	6	Equal district heating and industrial cogeneration
UK	5	Mostly industrial cogeneration
EU-15 average	10	
Bulgaria	16	
Czech Republic	18	
Estonia	n/a	
Hungary	10	
Latvia	n/a	
Lithuania	11	
Poland	n/a	Half of the district heat output is through cogeneration
Romania	40	Almost all is district-heating cogeneration
Slovak Republic	n/a	
Slovenia	n/a	
Switzerland	50	This is the fraction of thermal electricity, which is 5% of total electricity production
Non-European Countries		
Brazil	3	
Canada	11.3	Almost exclusively industrial
China	9.7	
India	2.5	
Japan	3	
Russia	30	
US	8	

Source: ESD (2001) for European countries, WADE (2004) for non-European countries

elsewhere), there is a huge potential for cogeneration, even without building more district heating systems. The European Union and the US each have a target of doubling the use of cogeneration for electricity production by 2010. To get the highest possible efficiency from a cogeneration system, it is necessary to make use of as much of the waste heat as possible.

15.4 Seasonal underground storage of thermal energy

The term 'thermal' energy storage will be used here to refer to the storing of both heat and coldness, the former by raising the temperature of a storage medium, the latter by lowering the temperature of a medium or by freezing water or some other substance (that is, storing coldness in the latent heat of melting). Diurnal thermal energy storage is already widely used in building heating and cooling systems, as discussed in Chapter 6 (Section 6.9). Seasonal thermal storage requires larger storage volumes and/or greater changes in storage temperatures than diurnal storage. A greater storage volume of artificial storage containers (such as water tanks) entails greater cost, while a greater storage temperature (in the case of heat storage) will lead to greater heat loss from the storage container prior to withdrawal of the heat. However, the ground itself provides a natural, large-scale medium for storing heat and coldness between summer and winter. The ground has been used for storing excess summer solar thermal energy and excess heat from summer cogeneration, as well as winter coldness. In the following pages, underground thermal storage systems are discussed, while later (Section 15.5.3), further information on solar-assisted district heating with underground storage is presented.

15.4.1 Kinds of underground thermal energy storage

Underground thermal energy storage (UTES) can occur as aquifer thermal energy storage (ATES), as borehole thermal energy storage (BTES) or as cavern thermal energy storage (CTES). In the case of CTES, artificial openings in the rock are created. However, all three involve the pre-existing subsurface rock or sediment as the heat-storage medium, and so will be referred to here as

'natural' UTES. These are in contrast to 'artificial' UTES options, which are: (i) to build a large steel or concrete tank partially or entirely underground, with or without an inner stainless steel lining, and filled with water; or (ii) to construct an underground, water-tight gravel pit that is saturated with water. A third option is a combination of the two, with a central, high-temperature, steel or concrete storage vessel surrounded by a gravel pit or a BTES system, where heat is stored at a lower temperature and extracted with heat pumps. In this way, heat lost from the central storage vessel to the surrounding ground is recaptured.

Heat can be stored either as low-temperature heat (<50°C) or as high-temperature heat (>50°C). Low-temperature storage requires advanced heating systems with a low distribution temperature so that the stored heat can be used, or requires heat pumps (and additional energy input) in order to upgrade the stored heat to higher temperature. High-temperature storage does not require heat pumps. Heat storage in the range of 10–40°C has been demonstrated successfully on many occasions, but high-temperature natural storage – up to 150°C – caused many problems in experimental and pilot projects in the 1980s (Sanner, 1999). These problems have been solved, at least for less extreme storage temperatures.

The major options are summarized in Figure 15.11 and discussed below, while technical characteristics of BTES, underground tanks and gravel/water pits are summarized in Table 15.9.

ATES

In an ATES system, heated water is injected into the ground at a series of injection wells and later withdrawn. The same well(s) can be used for injection and withdrawal, or separate wells can be used for injection and withdrawal. In the latter case, water flows in the aquifer between the injection and withdrawal well. ATES is a *direct* water heat exchange system, in that the water circulated through the buildings is taken directly from the ground. Well depth has normally not exceeded 400m. Schaetzle et al (1980, Chapter 7) describe a number of piping arrangements for distributing thermal energy in a community energy system with injection and withdrawal wells. Cool water can be extracted from one well to use for cooling purposes in summer (perhaps with a heat pump).

Figure 15.11 The major types of seasonal underground storage of thermal energy

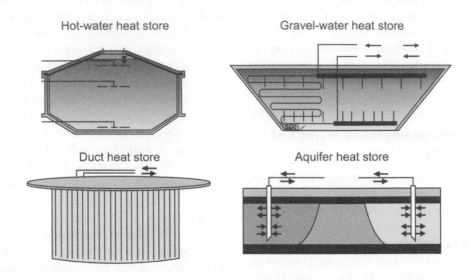

Source: Schmidt et al (2004)

Table 15.9 Characteristics of some alternative options for storing heat underground

Storage system	BTES	Hot-water tank	Gravel/water pit
Storage concept	Hot water circulates through pipes in the ground, to a maximum depth of 150m	Water-tight container buried underground	
Construction	U-shaped coaxial plastic pipes with a 1.5–3m separation	Reinforced concrete, steel- or fibreglass-reinforced plastic, or a pit with a cover and lid, stainless steel cover	Pit with waterproof lining, filled with water and gravel without a load-bearing cover
Maximum/minimum volumes	>100,000m³ due to high lateral heat loss	Max. 100,000m³, the largest store designed so far being 28,000m³	-
Insulation	Only in the covering layer, 5–10m from the surface	15–30cm on top and at the sides, and also underneath if the pressure can be withstood	As with hot-water tank
Storage volume/flat-plate collector area	8–10m³/m²	1.5–2.5m³/m²	2.5–4
Cost (€/m³) at 20,000m³ storage volume	25	70–80	65–85
Other characteristics	Easily constructed	Container store is costly	With a gravel proportion of 60–70% by volume, the total storage volume is about 50% larger than for a water store

Source: Eicker (2003)

The water is warmed in the process and injected into another well. In winter, the warm water is extracted and used for heating. It is cooled in the process, and injected into the cold well. This is a cyclic process that can be repeated indefinitely while reducing cooling costs by 80 per cent and heating costs by 40 per cent (Dinçer and Rosen, 2002). According to Schaetzle et al (1980), 60 per cent of the surface area of the US is underlain by aquifers adequate for use as thermal energy storage systems, and a similar percentage is expected for other continental areas. To the extent that settlement is concentrated along waterways, the fraction of the population with aquifer storage is larger (about 75 per cent in the case of the US). There are some issues related to changes in water chemistry associated with large temperature variations, leading to clogging, scaling, corrosion and leaching, but the expertise now exists to build reliable ATES systems (Dinçer and Rosen, 2002).

BTES

In a BTES system, U-shaped or coiled pipes are placed in a series of vertical boreholes, and a fluid is circulated through the pipe system. Different possible layouts of the pipes and a sample installation are shown in Figure 15.12. The pipe system could serve as the ground coil of a heat pump, if a heat pump is used to inject and extract heat, or could be connected to a source of hot water (such as a cogeneration power plant or solar collectors) for unassisted heat transfer. The water circulating through the underground coils is isolated from the groundwater, so BTES is referred to as an *indirect* water heat exchange (in contrast to the direct water exchange of ATES). It is sometimes also referred to as *duct heat storage*. The depth of borehole heat exchangers in BTES systems normally ranges from 30m to 100m (Sanner, 1999). With deep aquifers and deep boreholes (>1000m), the surrounding ground temperature is greater, so the heat loss associated with a given storage temperature is smaller. This blends into the use of geothermal energy for heating purposes, except that heat withdrawn in winter is replaced in summer, thereby overcoming a major disadvantage of geothermal energy – that it is not renewable.

Steel or concrete tanks

Buried stainless steel tanks with a storage volume of up to at least 2000m³ and concrete tanks (some

Figure 15.12 Different arrangements of pipes in a borehole heat exchanger (left) and a sample installation (right)

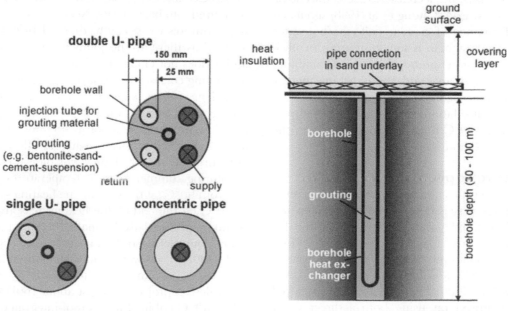

Source: Schmidt et al (2004)

with a steel lining) up to at least 12,000m³ have been used for seasonal storage of solar thermal energy (Fisch et al, 1998), with storage temperatures varying from 40°C to 90°C seasonally. Bermed tanks (i.e. partially underground, with earth piled up against the above-ground portion) can be an economically attractive alternative to complete burial (Rosen, 1998). An advantage of using hot-water tanks is that the water can be stratified, with the hottest water on top. Hot water from solar collectors or cogeneration is supplied at the top of the tank and returns to the collectors from the bottom of the tank, while hot water to the heating network is withdrawn from the top of the tank and returns to the bottom of the tank. This results in a greater collector efficiency and greater supply temperatures compared to a well-mixed storage tank. In order to maintain the temperature stratification, diffusing water inlets are needed. For a prototype system built at the University of Calabria (Italy) using 91.2m² of vacuum tube solar collectors and a 500m³ concrete tank with 0.2m of foam insulation, the average collector and storage efficiencies are each about 55 per cent, giving an overall efficiency in the use of solar energy of about 30 per cent (Oliveti et al, 1998). The temperature difference between the top and bottom produced by thermal stratification reaches a maximum of about 15K (out of a peak temperature of 72°C) at the start of the heating season, and decreases to 5K by the end of the heating season. Chung et al (1998) simulated the performance of a solar collector/hot-water tank storage system in Korea for heating an office building and greenhouse, for various combinations of collector area and storage volume and for flat-plate and evacuated tube collectors. For 0.5m of urethane insulation, the storage temperature is greater than 90°C for 4000 hours per year (one year has 8760 hours).

Gravel/water pits

Gravel/water pits are cheaper than concrete/steel storage vessels beyond some minimum size. An early example was built by the 'Institut für Thermodynamik und Wärmetechnik' (ITW) of the University of Stuttgart on the university campus in 1985, and has been in operation ever since (Hahne, 2000). Heat is added to or removed from the gravel pit using both a direct water

exchange system and indirect heat exchange using plastic tubing, aided by a heat pump. Heat is accumulated during the summer for winter heating and coldness is accumulated during the winter for summer cooling, so the gravel pit is relatively cold (near 0°C) as a heat source for heating at the end of the winter season, and relatively warm (up to 34°C) as a heat sink for cooling at the end of the cooling season. As a result, the combined heating + cooling COP (i.e. annual heating or cooling provided to the buildings divided by the energy used by the heat pumps) is only 2.9–3.2 (depending on the year). However, the system has been extensively monitored over a period of many years, and much practical information has been gained.

A more recent example of a gravel/water pit for seasonal storage is the one built for a solar housing development in Steinfurt, Germany, in the late 1990s (Pfeil and Koch, 2000). Working from the outside inward, the structure of this pit consists of (i) insulation; (ii) aluminium-polyethylene foil; (iii) water-tight lining; (iv) fleece; and (v) the gravel fill. Granulated recycled glass was used as the insulation on the top and sides, while 0.15m thick foam glass plates were used at the bottom The granulated glass has a thermal conductivity of 0.07W/m/K, compared to 0.019–0.029W/m/K for solid-foam insulation used on building exteriors (Table 3.2), but it can be transported by silo trucks and either blown in by an air stream or poured into bags on site. Needless to say, it has high compressive strength, thermal bridges are avoided as there are no joints in the insulation, and it does not entail emissions of halocarbons. The lining consists of two layers of flexible polypropylene whose water tightness was checked during construction using a vacuum test. The aluminium-polyethylene foil is used as a barrier against diffusion of steam, while the fleece provides protection to the lining. The pit is used only for storing summer heat, at temperatures ranging from 90°C (at the start of the heating season) to 30°C (at the end of the heating season). The Steinfurt pit is relatively small (1650m³, 1100m³ water-equivalent), so the unit cost was relatively high (490DM/m³ water-equivalent, about 245€/m³), but a system scaled to 16,000m³ water-equivalent is projected to cost about 160DM/m³ (80€/m³). Combined with a roof-integrated solar collector cost of 350DM/m³ (175€/m³), which has been reached in Germany once the savings in conventional roofing material is accounted for, the cost of solar thermal energy would be about 25DPf/kWh (0.125€/kWh).

15.4.2 Historical development and current examples

The subsurface has been used as a heat source and sink for heat pumps since the late 1940s. A number of large ATES systems began in China in the 1960s, for cold storage. Underground heat storage was considered outside China in the early 1970s and demonstrated in practice in the early 1980s in combination with solar thermal energy, waste heat and heat pumps. A project in Toronto began in 1985. A comprehensive guide on seasonal energy storage was published in 1988 in French (Hadorn, 1988) and translated into several languages, the English translation (*Guide to Seasonal Heat Storage*) having been published by Public Works Canada in 1990. Since 1990, cold storage has been used at a number of sites

in Canada, Sweden, The Netherlands, and elsewhere. Bakema et al (1995) produced a then state-of-the-art report on UTES. At the University of Massachusetts at Amherst in the US, solar heat collected during the summer is stored in subsurface clay and used for heating an athletic centre (Tomlinson and Kannberg, 1990).

The International Energy Agency (IEA) Implementing Agreement *Energy Conservation through Energy Storage* (ECES) consists of a number of subactivities called 'Annexes' (see Appendix C). Annex 12, established in December 1997, addresses high temperature underground thermal storage (HT-UTES). A report on preliminary findings, including a review of previous experience, has been published (Sanner, 1999). Table 15.10 lists the operating HT-UTES systems given in Sanner (1999). The longest operating high-temperature ATES system dates back to 1991, at Utrecht University in The Netherlands, while the longest operating HT-BTES system dates back to 1983 in Sweden. The Reichstag project involves two aquifers, one at a depth of 0–66m, and the other at a depth of 270–390m. The lower aquifer

Table 15.10 High-temperature natural underground thermal energy storage systems in operation as of 1999

Location	Start Date	Energy source	Loading temperature (°C)	Unloading temperature (°C)	Number and depth of wells or boreholes, or cavern volume	Capacity
ATES systems						
Utrecht University, the Netherlands	1991	Cogen	90		2 @ 260m	<7200GJ
Reichstag, Berlin, Germany	1998	Cogen	70	60–20	2 @ 320m	
Houge Bourch Hospital, the Netherlands	1998	Cogen				
BTES systems						
Kullavik, Sweden	1983	Solar	60	50–40	200 @ 8m	14–29GJ
Vaulruz, Switzerland	1983	Solar	54	40–5	horizontal pipes, 1.2–1.6m deep	600GJ
Groningen, the Netherlands	1984	Solar	60	50–30	360 @ 20m	3600GJ
Neckarsulm, Germany	1998	Solar	80		168 @ 30m	
CTES systems						
Avesta, Sweden	1982		115	70	15,000m³	2900GJ
Lyckenbo, Sweden	1983	Solar	80–90	80–55	104,300m³	20,000GJ

Source: Sanner (1999)

is supplied with excess heat from summer cogeneration through one injection well and reaches a maximum temperature of 70°C. This temperature drops to about 25°C by the end of the heating season, and is upgraded by absorption heat pumps, as a 45°C heat distribution temperature is used. During the winter, the upper aquifer is chilled by ambient cold through three injection wells whenever the ambient temperature is cold enough, reaching a minimum temperature of 10°C. The temperature rises to 30°C by the end of the cooling season.

More recent UTES projects, involving district heating systems, are discussed in Section 15.5.3.

15.4.3 Computer simulation for optimization

Pahud (2000) analysed the cost and performance of the system depicted in Figure 15.13, consisting of solar collectors, a short-term storage unit (the buffer tank), seasonal underground heat storage in boreholes, a supplemental boiler and a district-heating network. The short-term surface storage tank is used to buffer short-term variations in direct heat supply and demand in winter. The top of the ground storage volume is insulated. Water circulates between the solar heat exchanger and the buffer tank whenever the collector outlet temperature exceeds the water temperature at the bottom of the buffer tank. The return

water from the district heating system circulates through the buffer tank when the return water temperature is less than the water temperature at the top of the tank. The lowest possible flow rate through the district heating system is used at any given time, so as to give the lowest return water temperature and therefore the maximum possible heat extraction.

Pahud (2000) systematically investigated the impact of five system parameters on the cost and performance of the system shown in Figure 15.13: the collector area per GJ of annual heat demand; the buffer tank volume, ground storage volume and borehole length per unit of collector area; and the shape of the ground storage volume. A total of 720 combinations were considered for nine different heat loads, characterized by the absolute heat load (i.e. the scale of the district heating system), the distribution temperature (moderate, 55–60°C, or low, 25–30°C), and the hot-water:space-heating ratio. Low-temperature distribution is feasible only if there is no hot-water load. Design parameters were determined such that solar energy provides at least 70 per cent of the annual heating plus hot-water load. For a relatively large system (18,000GJ/year heat load, sufficient for 300–600 single-family houses, of which 25 per cent is for hot water), the cost computed for Swiss conditions is around 0.3 Swiss Francs per kWh of heat (about US$0.24/kWh). For a load of 4000GJ/year, this rises to 0.4SF/kWh (about US$0.32/kWh). A control system that optimizes

Figure 15.13 Components and layout of a centralized system for collecting and storing solar thermal energy on a short-term and on a seasonal basis

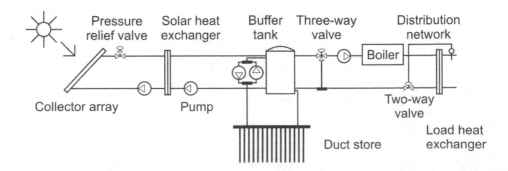

Pressure relief valve Solar heat exchanger Buffer tank Three-way valve Distribution network

Boiler

Collector array   Pump   Duct store   Two-way valve   Load heat exchanger

Source: Pahud (2000)

the transfer of heat between the short-term and long-term storage based on short-term weather forecasts increases the solar energy yield by 10 per cent. Adjusting the flow rates in the ground loop so as to avoid disrupting the thermal stratification in the buffer tank whenever possible increases the solar yield by 4 per cent.

Finally, Gabrielsson et al (2000) report simulations of seasonal storage of solar thermal energy in soft clay for 200 dwelling units using low-temperature floor radiant heating. The calculated cost of heat is US$0.126/kWh and the calculated solar fraction is 0.7. The storage system would be fully charged from mid-April to mid-August to a temperature of 60–70°C, then used to provide heating until the end of February, at which point the storage temperature would have decreased to 30°C. Thereafter, heat would be supplied directly from the solar collectors and from the backup heat source.

Figure 15.14 Fuels used for district heating in Sweden, 1980–2004

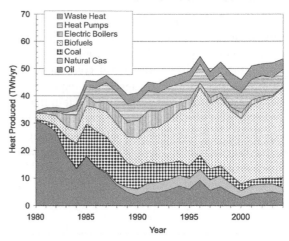

Source: Data from Swedish Energy Agency (2005)

15.5 Utilization of renewable energy with district energy systems

The prior existence of a district heating and cooling system in a city makes it much easier to directly utilize renewable energy forms, as they become available, and will make it easier to eventually use hydrogen produced from renewable energy sources for heating of individual buildings.

15.5.1 Use of biomass energy for district heating

Sweden has been able to switch a large fraction of its building heating requirements to biomass energy (plantation forestry) by switching from fossil fuels to biomass as the fuel for its district-heating systems (Figure 15.14). To directly heat individual buildings with biomass energy would not have been easy (as discussed in Chapter 4, Section 4.2.2, building-scale biomass boilers are available, but there are logistical difficulties in supplying the biomass unless the building is designed to accommodate it). Similarly, Denmark has been able to utilize biogas from farm manure for heating purposes in some of its district-heating systems (Ramage and Scurlock, 1996). By the end of 1997, there were 369 district heating plants in Austria powered by biomass (Faninger, 2000). Maker and Penny (1999) present

a guidebook with case studies on how municipal communities can assess the potential for biomass district energy, and on how to proceed in the event of a positive assessment.

15.5.2 Use of geothermal energy for district heating

Where geothermal energy is available, a district-heating network is required in order to distribute the heat to individual buildings. Over 97 per cent of the inhabitants of Reykjavik and 90 per cent of the population of Iceland receive geothermal district heat. Some municipalities in the Paris region rely largely on geothermal energy to heat buildings through its many local district heating networks (Figure 15.15). By 2000, Turkey had connected about 52,000 residences to geothermal district energy networks, with plans to connect about 500,000 residences (30 per cent of the total in the country) by 2010 (Lund, 2002). Other countries using or developing geothermal district heating include the US (18 systems), Hungary, Romania, Poland, China, Denmark and Sweden (Bloomquist, 2003). In the latter case, heat pumps are used to boost 23°C geothermal water to 80°C. The US systems are concentrated in the states of California, Nevada, Utah and Hawaii (Lund, 2003).

15.5.3 Solar-assisted district heating

As discussed in Section 15.4, solar thermal energy, collected during the summer months, has been successfully stored underground and used for heating in winter in several countries. Seasonal thermal energy storage is less expensive at the scale of a small district heating system than at the scale of individual buildings or houses. This is because, as the scale of seasonal energy storage reservoirs increases, the surface-to-volume ratio will decrease. This will decrease the fractional loss of stored thermal energy. At the same time, for underground storage in concrete or steel tanks or

in gravel pits, the capital cost depends largely on the surface area, so the cost per unit of storage capacity will decrease with increasing size (unless the size increases to the point where scaffolding needs to be erected during construction). Finally, larger storage and district heating systems will require larger heat pumps, which have a lower unit cost and higher COP than smaller heat pumps.

European examples

The pioneering work on solar-assisted district heating was carried out in Sweden during the 1970s and 1980s (Dalenbäch, 1990). Solar-assisted district heating systems with storage can

Figure 15.15 The Creil district heating system, installed north of Paris in 1976

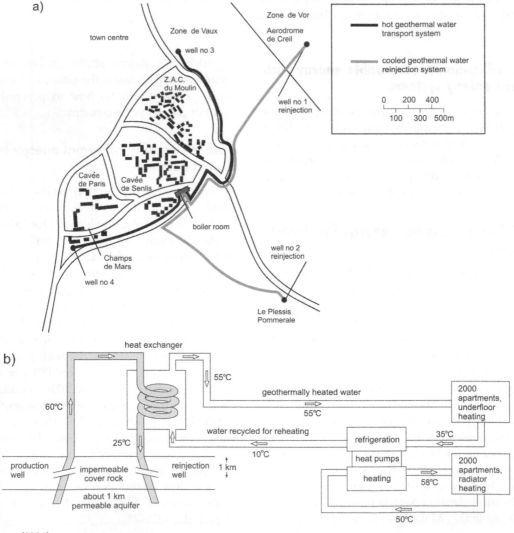

be designed such that 32–95 per cent of total annual heating and hot water requirements are provided under German conditions (Lindenberger et al, 2000). By 2003, eight solar-assisted district heating systems had been constructed under the German *Solarthermie-2000* programme that begin in 1993 to promote solar systems both with and without seasonal storage (Lottner et al, 2000; Schmidt et al, 2004). Technical information and costs are given in Table 15.11. The largest of these systems are in Hamburg (serving 124 single-family homes), Friedrichshafen (serving 570 apartments in eight buildings), and in Neckarsulm (serving six blocks of flats, a school and a commercial centre). The fraction of the heating load supplied by solar energy in these systems ranges from 30 per cent to 62 per cent, but about 4 years of operation are required to reach the long-term solar fractions, as the ground surrounding the storage tanks is warmed. The cost of solar heat ranges from 16 eurocents/kWh to 42 eurocents/kWh. A number of municipalities and housing companies in Germany have apparently committed themselves to using solar

energy by constructing low-energy housing with solar-assisted district heating. Another eight existing and six planned solar-assisted district heating systems in Sweden, Denmark, The Netherlands and Austria are reviewed by Fisch et al (1998). The largest of these, in Denmark, involves 1300 houses, a 70,000m³ gravel-pit for storage, and a 30 per cent solar fraction. Additional information on Danish systems is found in Heller (2000). The *European Large-Scale Solar Heating Network* maintains a website (linked to www.hvac.chalmers.se) on solar heating plants with collector areas ranging from 500m² to 10000m². The construction of solar-assisted district heating plants has been facilitated by the development of large collectors, with an absorber area of 15m², such that a collector area of 1250m² can be installed within 3 days.

General guidelines for the sizing of solar-assisted district heating in central and northern Europe are given in Table 15.12, along with cost data. Seasonal solar heating could reach costs of 13–25 euro-cents/kWh, compared to 4.3 cents/kWh fuel cost for natural gas at $12/GJ and another 1–3 cents/kWh for the purchase

Table 15.11 Technical data concerning solar-assisted district heating systems constructed in Germany under the Solarthermie-2000 programme

Location	First year of operation	Housing type	Heated area (m²)	Annual heat demand (MWh)	Solar collector area (m²)	Storage volume (m³)	Solar fraction	Solar heat cost (€/kWh)
Hamburg	1996	124 single-family units	14,800	1610	3000	4500 (hot water)	0.49	0.256
Friedrichshafen	1996	570 apartments in 8 buildings	39,500	4106	5600	12,000 (hot water)	0.47	0.158
Chemnitz	1996	1 office building, hotel, and warehouse	4680	573	540 vacuum tubes	8000 (gravel/ water)	0.30	0.240
Steinfurt	1998	42 apartments in 22 buildings	3800	325	510	1500 (gravel/ water)	0.34	0.424
Neckarsulm	1999	6 blocks of flats, school, commercial centre	20,000	1663	2700 (5000, phase II)	20,000 (duct) (63,400, phase II)	0.50	0.172
Attenkirchen	1999	30 apartments	6200	487	800	500+9350 (hot water + duct)	0.55	0.170
Rostock	2000	108 apartments	7000	497	1000	20,000 (aquifer)	0.62	0.255
Hannover	2000	106 apartments	7365	694	1350	2750 (hot water)	0.39	0.424

Source: Schmidt et al (2004)

and maintenance of on-site heating equipment (see Figure 2.9). Because solar heat tends to be expensive, houses supplied with solar heat should be very well insulated so that the total heat load is very small, thereby compensating for the high unit cost of solar heat. In a recent project in Anneberg that was fully operational in March 2002, 100 houses are supplied with solar heat stored using 99 65m deep boreholes. Of the total annual heating load of about 4000GJ, about 2500GJ will be supplied by solar energy. The cost of heat is 1000 SEK/MWh (about 11.3 cents/kWh), compared with 1100 SEK/MWh for a small district heating system using biomass fuel (common in Sweden) and 920 SEK/MWh with individual ground source heat pumps (Nordell and Lundin, 2001).

The annual energy flows for the solar district heating system in Attenkirchen (Germany) are given in Figure 15.16. Twenty single-family and five semi-detached houses will be supplied with district heat at 45°C (two houses were finished in March 2002). Heat from 800m² of solar collectors on the roof of a sports facility will be stored at up to 80°C using 110 25m

deep boreholes with a 2m spacing, surrounding a 500m³ uninsulated concrete water tank (Reuss and Mueller, 2000). The solar collector has an average collection efficiency of 43 per cent. Solar heat flows either directly into the water tank or into the borehole field. In the latter case, it is upgraded to a higher temperature and transferred to the water tank with a heat pump. About 60 per cent of the heat lost from the concrete tank is recovered by the borehole heat exchanger. The total annual heat load will be about 1750GJ, of which 20 per cent will be supplied directly from solar energy and/or the hot-water tank, and 80 per cent via heat pumps and ground heat storage. As can be seen from Figure 15.16, the overall COP (heat delivered to the district heating network divided by electricity use) is 4.5. Ground source heat pumps can attain a comparable COP without the solar input, but not on a sustainable basis unless excessive ground volumes are used, due to the long-term withdrawal of heat. The solar component of the Attenkirchen system serves to maintain a high heat pump COP by recharging the ground thermal reservoir every summer.

Table 15.12 Design guidelines for solar-assisted district heating systems in central and northern Europe

System type	Small solar system for domestic hot water (for comparison)	Central solar heating plant with diurnal storage	Central solar heating plant with seasonal storage
Minimum system size	–	More than 40 apartments or more than 100 persons	More than 100 apartments (70m² each)
Collector area	1–1.5m²/person	0.8–1.2m² per person	1.5–2.5m² per MWh annual heat demand
Storage volume (per m² of collector area)	50–80 litres/m²	50–60 litres/m²	1.5–2.5m³/m²
Solar net energy	350–380kWh/m²/year	400–430kWh/m²/year	280–320kWh/m²

Solar fraction with new building code
  domestic hot water
  total heat demand
	
	50%
	10%
		
		50%
		20%
			
			–
			40–50%

Investment cost per m² of heated area	20–25€	30–50€	90–150€
Solar heat cost in Germany	0.2–0.4€/kWh	0.08–0.15€/kWh	0.17–0.25€/kWh

Note: Flat-plate collectors are assumed.
Source: Mangold (2001), except for investment costs, which are from Mahler (2002)

Figure 15.16 Annual energy flow in the solar-assisted district heating system in Attenkirchen, Germany

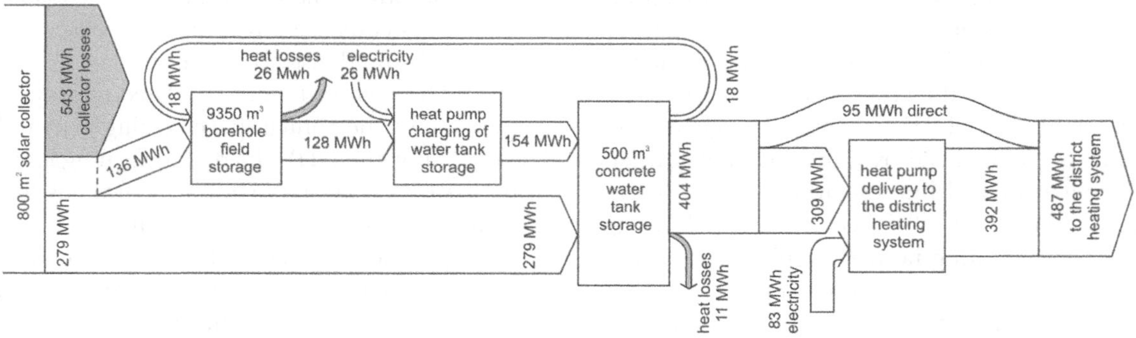

Source: Reuss and Mueller (2000)

North America's first solar-assisted district-heating system

The first solar district heating system in North America began construction in 2005 in the town of Okotoks, near Calgary (Alberta, Canada) (see www.dlsc.ca). The project will serve 52 small (145m² floor area) detached houses, and will consist of a large (250m³) water tank that will serve as a short-term thermal energy storage, a field of 144 37m-deep boreholes covered with insulation to RSI 7.04 (R40) for longer term UTES, a district heat supply loop with a variable-speed pump, 2300m² of solar collectors, and a high efficiency (90 per cent) backup boiler. The space heating and domestic hot water (DHW) systems are kept separate, as the DHW load is relatively uniform seasonally but requires a higher supply temperature. Separation of the two allows the space heating system to operate at higher efficiency than if the two were combined. The water tank stores heat from the collectors for immediate use, without the need to cycle the heat through the underground heat store. It also allows some daytime heat to be stored and dumped into the BTES overnight, thereby reducing the cost of the BTES by reducing the rate at which it must absorb heat. The decision was made to keep the system as simple as possible in order to (i) reduce costs; (ii) to more readily match the ability of installers and maintenance contractors; and (iii) to match the operating abilities of homeowners. In order to reduce the required size and hence cost of the solar energy system, the design of the houses was upgraded to meet the R2000 standard (see Chapter 13, Section 13.1.1).

The BTES can reach 80°C at the end of the summer, but the district heating system distributes heat at 35°C to heating coils inside a regular (for North America) forced air heating system in individual houses. Use of on-site forced air heating systems with such a low-temperature coil required a specially designed heat exchanger and a low flow velocity in order to avoid draughts associated with 'cold heat'. At the time of writing, the possibility of increasing the temperature of the short-term storage tank to 55°C in anticipation of forecasted extreme cold, using the backup boiler, was being considered.

Table 15.13 presents preliminary investment cost and performance estimates for the project, based on computer simulations (SAIC, 2004). These are used here to compute annualized costs using the cost-recovery factor computed as in Box 11.2. The base case house has space heating and DHW loads of 49.6GJ/year and 13.1GJ/year, respectively. Upgrading the envelope to the R2000 standard, and implementating heat recovery on hot waste water (see Chapter 8, Section 8.4) and other measures to conserve hot water, reduce these loads to 34.2GJ/year and 7.7GJ/year, respectively (on an areal basis, the reduced heating load is 66kWh/m²/year, which is still over four times that of the Passive House Standard discussed in Chapter 13 (Section 13.2.1)). These upgrades add $5980 to the cost of each house (onto a base price of about $240,000), but are offset slightly by savings in furnace and DHW heater costs due to downsizing. The solar district heating system is estimated to add $29,600 to the cost of each house, which is in addition to the upgrade cost of $5980 per house (this excludes one-time costs associated with the design of a system that is the first of its kind in North America). The solar costs are divided in Table 15.13 into the costs of solar collectors and individual solar DHW systems, which are assumed to be financed by individual homeowners at 4 per cent real interest over a 25-year mortgage, and the costs of the central thermal storage and district heating network, which are assumed to be financed at 3 per cent real interest over a 50-year period. These assumptions result in the annualized capital costs given in Table 15.13. The costs for the Okotoks projects are given in the column labelled 'Solar-1' in Table 15.13. These costs could fall over time; for illustrative purposes, two other cases are considered: Solar-2, where

the costs paid by individuals fall by 25 per cent, and Solar-3, where these costs fall by 50 per cent. Storage costs could be lower in other projects if ATES can be used instead of BTES.

The solar system is calculated to provide 95 per cent of the annual space heating load and 50 per cent of the DHW load. The heating loads in the base case are met with a natural gas furnace at 80 per cent efficiency, while heating loads in the upgraded houses and the residual heat loads for the case with solar heat are assumed to be met by a natural gas furnace or a central condensing boiler at 90 per cent efficiency. The resulting annual fuel costs are given in Table 15.13 assuming unit costs to the customer of $12/GJ and $24/GJ ($12/GJ/year is a typical retail price in Europe but is much greater than past prices in

Table 15.13 Investment costs per house, annualized costs and heating costs (including amortization of equipment) for the Okotoks project in Alberta, Canada

	Case				
	Base	Upgraded	Solar-1	Solar-2	Solar-3
Capital costs ($/house)					
Furnace+DHW tank	1840[a]	2360[b]	0	0	0
House thermal upgrade	0	5980	5980	5980	5980
Solar cost, individual	0	0	15,792	11,844	7896
Solar cost, collective	0	0	13,805	13,805	13,805
Solar cost, total	0	0	29,598	25,650	21,701
Total capital cost	1840	8340	35,577	31,629	27,681
Heating load and fuel use (GJ/year/house) and heating efficiencies					
Non-solar space-heating load	49.60	34.20	1.70	1.70	1.70
Non-solar DHW load	13.10	7.70	3.90	3.90	3.90
Space-heating efficiency	0.80	0.90	0.90	0.90	0.90
DHW efficiency	0.74	0.74	0.74	0.74	0.74
Fuel use	79.70	48.40	7.10	7.10	7.10
Annualized costs ($/year/house)					
Capital	118	534	1930	1677	1425
Fuel (at $12/GJ)	956	581	85	85	85
Fuel (at $24/GJ)	1913	1162	170	170	170
Total (at $12/GJ)	1074	1115	2015	1763	1510
Total (at $24/GJ)	2031	1696	2101	1848	1595
Cost of heating (cents/kWh)					
Based on the initial heating load and including upgrade costs where applicable					
For fuel at $12/GJ	7.80	8.10	14.60	12.80	11.00
For fuel at $24/GJ	14.70	12.30	15.20	13.40	11.60
Based on heating load after the envelope upgrade but excluding upgrade costs					
For fuel at $12/GJ			17.20	14.50	11.90
For fuel at $24/GJ			18.10	15.40	12.80

[a] 29.3kW$_{th}$ (100,000Btu/hr) non-condensing furnace at $940 and 180 litre DHW heater at $900 assumed here.
[b] 23.4kW$_{th}$ (75,000Btu/hr) condensing furnace at $1460 and 180 litre DHW heater at $900 assumed here.
Source: Investment cost data are from SAIC (2004)

North America; however, the natural gas supply in North America is much more precarious than in Europe (as discussed in Chapter 1, Section 1.3), and as of September, 2005, the wholesale cost of natural gas was $15–16/GJ). At $12/GJ retail fuel cost, the cost to heat the upgraded house is comparable to that of the base case house, while at $24/GJ, the cost of supplying 95 per cent of the space heating load with solar energy (Solar-1) is comparable to that of the base case house. If the Solar-2 cost assumptions can be reached, solar district energy would be competitive with the base case at a fuel cost of about $18/GJ.

The final lines of Table 15.13 show the cost of heat energy, including amortization of heating system capital costs and of thermal upgrades to the base case house. These costs are computed based on the heating load of the base case house, rather than on the reduced heating load after the thermal upgrade (that is, the thermal upgrade and solar energy are treated as a single package to satisfy the heating load of the base case house). For the solar cases, costs are also given based on the cost of the solar components alone, divided by the heating load of the upgraded house (that is, the upgraded house rather than the base case house is used as the reference for computing the cost of solar heat). The latter is a more typical calculation procedure, and yields solar heat costs of 10–15 cents/kWh. However, the former procedure is justified until the upgraded house without solar heat becomes standard practice, and yields a heating cost of 9–12 cents/kWh.

Minimizing the distribution temperature

In order to use solar energy in district heating, the required storage and distribution temperatures must be kept as low as possible in order to minimize the heat loss during storage and distribution, while the return temperature should be no more than 40°C if possible so as to permit a relatively high solar collector efficiency. To provide an adequate temperature difference between supply and return hot-water flows and a minimal supply temperature, radiant floor heating should be used, or a very efficient heat exchanger in a low-velocity forced air system is needed. In the aforementioned systems in Hamburg and Friedrichshafen, the building heating systems are designed for 60/30°C supply/return and 70/40°C supply/

return temperatures, respectively, while storage temperatures reach 85–95°C in the upper portion of gravel/water pits. Heat is transferred from the district heat loop to hot-water loops within each building through a heat exchanger. The minimum permitted temperature on the district heating side of the heat exchanger is 65°C, which guarantees a minimum temperature of 50°C on the building side of the heat exchanger. For the system in Anneberg (Sweden) solar heat will be collected at 60°C using 3000m² of flat-plate collectors and stored at 30–45°C in 60,000m³ of underground rock using an array of 99 65m deep bores with a 3m spacing (Nordell and Hellström, 2000). Floor heating coils will provide heat at 25–32°C. As noted above, district heat at 35°C is supplied to forced air systems in the Okotoks project in Canada.

If district heating is to provide domestic hot water, then a somewhat warmer supply temperature (75°C) is needed. In summer, when there is no space heating load, it is not possible to induce a large enough temperature drop to produce return water no warmer than 40°C. This will diminish the efficiency in collecting solar heat during the summer. In the Anneberg project, auxiliary electric heaters will be used to boost the water temperature to at least 55°C for domestic hot water needs.

Optimization

Lindenberger et al (2000) have analysed alternative configurations for a solar-assisted district heating system to supply 100 well-insulated houses in Germany with heat at 65°C. The total peak heating load is 753kW$_{th}$. The systems considered consist of 600–1500m² of solar collectors, a 720–6300m³ stratified underground water tank, backup condensing boilers, and either (i) no other heating devices; (ii) an electric heat pump; (iii) a gas-fired absorption heat pump; (iv) a gas motor heat pump; or (v) gas-fired cogeneration. Some properties of the different systems are summarized in Table 15.14. The solar fraction ranges from 32 per cent to 95 per cent. The boiler efficiency is lowest (91 per cent) for the largest solar collector area and storage volume because this gives the warmest minimum temperature at the top of the tank (52°C), and it is this water that is used to extract heat from the boiler flue gas. With a smaller solar collector area and storage volume, the minimum tank-top temperature is colder (31°C), so more heat can be extracted from the flue gas, giving an efficiency as high as 98 per cent. Use of a heat pump, by lowering the storage temperature, improves the collection of solar energy and the boiler efficiency. When the storage temperature is coldest, the boiler rather than the heat pump is used, because this is when the boiler efficiency is highest and the heat pump COP the lowest. With a larger solar fraction, less heat is needed from cogeneration, so less electricity is produced. If this electricity displaces inefficient centrally produced electricity, then the net savings in primary energy when the solar component is added to a system with cogeneration is substantially reduced.

Nordell and Hellström (2000) used the MINSUM and TRNSYS computer models to simulate a solar-assisted district heating system, the latter of which is discussed in Appendix D. These are ready-to-use models. Meliß and Späte (2000) used the same models to simulate the Solar-campus Jülich.

15.5.4 Hybrid biomass/solar district heating systems

As noted above, biomass energy is being increasingly used as a fuel for district heating systems in some countries, and solar energy complements biomass energy well. Because the heat load is relatively small during the summer and transitional seasons, the boilers (whether fuelled by biomass or fossil fuel) would operate with frequent on/off operation, which is inherently inefficient. However, this is the time when solar energy is most readily available, so the boilers can be shut down altogether during part of this time. The world's first hybrid solar/biomass district heating plant was installed in the village of Deutsch Tschantschendorf in Austria in 1994, and by August 1998, 12 hybrid solar/biomass plants were in operation in Austria with solar collector areas ranging from 225m² to 1250m² (Faninger, 2000). In these particular cases, a relatively small thermal storage tank was constructed (adequate for 3–5 days of hot-water demand), so the solar fraction is relatively small: about 15 per cent of the total space + water heating load.

15.5.5 Use of deep lake water and deep sea water for district cooling

As discussed in Section 15.3.7, district cooling systems in mid-latitude locations such as Toronto

Table 15.14 Characteristics of possible alternative solar-assisted district heating systems serving 100 well-insulated houses with heat at 65°C in Germany

System	Collector area (m²)	Storage volume (m³)	Solar fraction	Boiler fraction	Heat pump COP	Boiler efficiency
No additional heating devices	600	720	0.32	0.68	–	0.97
		2520	0.43	0.56	–	0.96
	1500	1800	0.55	0.45	–	0.95
		6300	0.85	0.15	–	0.91
Electric heat pump	600	720	0.41	0.52	3.38	0.98
		2520	0.57	0.31	3.51	0.98
	1500	1800	0.71	0.16	3.57	0.98
		6300	0.95	0.00	4.39	–
Gas fired absorption heat pump	900	1620	0.56	0.06	1.61	0.97
		3780	0.69	0.02	1.67	0.97
	1500	1800	0.63	0.03	1.65	0.97
		6300	0.89	0.01	1.71	–
Gas motor heat pump	600	720	0.35	0.58	2.10	0.98
		2520	0.49	0.41	2.10	0.98
	1500	1800	0.62	0.27	2.14	0.97
		6300	0.93	0.00	2.19	–

Note: Each system consists of solar collectors, an underground stratified hot-water tank, a condensing boiler and additional equipment as indicated.
Source: Lindenberger et al (2000)

and Cornell utilize cold deep lake water as the source of chilled water for cooling purposes. This is a form of renewable energy, as the coldness of the deep lake water is restored every winter. Freshwater is densest at a temperature of 4°C, so as surface water cools to this temperature it sinks, to be replaced with slightly warmer water from below which in turn cools and sinks. The result is to establish a nearly uniform temperature of 4°C throughout the depth of the lake. During the spring and summer, only a relatively shallow layer of water is heated, but as it is less dense than the deep water, it does not mix to any substantial extent with the cold underlying water.

In the case of Lake Ontario, the water at a depth of 80m and greater is always at a temperature of 4°C. The cold-water intake for the Toronto system occurs at a depth of 83m and a distance of 5km from shore; installation was completed in August 2003 and is really an extension to deeper and cleaner water of the intake for the city's drinking water supply. Coldness from the drinking water is transferred to the district cooling system through heat exchangers, so there is no physical mixing between water used for drinking purposes and for cooling purposes. At full capacity (183MW$_c$), the deep lake water-cooling system will be able to supply the air conditioning needs for 1.8million m^2 of office space (about 100 office towers or 40 per cent of the air conditioning load of the downtown core), displace 35MW of peak power demand, and reduce the amount of electricity used for air conditioning in the buildings served by 75 per cent. The city's water supply would be warmed to 12.5°C as it cools the return water from the district cooling loop, thereby reducing slightly the energy needed to make domestic hot water. Cooling of all the buildings in all of the cities surrounding the lake would have a negligible effect on the lake's thermal balance, while projected warming of the climate by 2100 might warm the water at the 80m depth from 4°C to 5°C or 6°C, as shown by Boyce et al (1993). This would still be adequate for cooling purposes.

Cornell University began using deep water from nearby Cayuga Lake for cooling beginning in July 2000 (Peer and Joyce, 2002). Water at 4°C is piped a distance of 8km, providing 56.3MW of cooling with a net energy savings of 86 per cent compared to chiller based cooling.

Deep sea water is also cold year round, even in tropical locations, as cold water forms in polar regions, spreads throughout the global ocean, and upwells. However, water at 4°C or colder is generally not reached until a depth of 1km. Along the south shore of the island of Oahu, Hawaii, water at 7°C is reached at a depth of 500m and a distance of 4–6km from shore (Andrepont and Rezachek, 2003). The technical feasibility of using deep sea water in Hawaii has been demonstrated, but initial upfront costs remain a major hurdle. Sea water would not circulate through the district cooling loop, but instead, would transfer its coldness to the district cooling loop through heat exchangers in order to avoid the need for protection against corrosion by salt in the district cooling loop. As noted in Section 15.3.7, various ways of indirectly using sea water in a district cooling network (by cooling the chiller condensers) have been considered for Hong Kong.

15.5.6 Wind energy buffer

Wind energy has already become competitive with fossil fuel electricity in regions with very good winds (average hub-height wind speed $\geq$7.5m/s), where electricity can be generated at a cost of less than 5 cents/kWh. The cost of wind turbines has been falling about 20 per cent for each doubling in cumulative production, which has been occurring every 3–4 years recently (IEA, 2000b). Thus, the competitiveness of wind energy will continue to increase. However, due to the fluctuating and somewhat unpredictable nature of wind energy, it will not be practical to supply more than 20 per cent of total electricity demand with wind energy without some mechanism for storing wind energy. One way to do this is through integration with a district heating and cooling system. If the wind energy component of a local power system is sized such that it sometimes provides more energy than needed, the excess power can be used with heat pumps to supply hot or chilled water in place of fossil fuel or biomass energy, or to charge thermal storage reservoirs (either the above ground cold storage tanks illustrated in Figure 6.67, or the underground hot-water tanks discussed in Section 15.4.1, or smaller above ground hot-water tanks).

This allows the wind system to be sized to meet a larger fraction of total electricity demand with less wastage of excess wind energy. For a scenario in which 50 per cent of Danish electricity demand in 2030 is met by wind, integrated operation of the electricity and heating systems cuts the wasted wind energy potential in half (Redlinger et al, 2002, chapter 2). This in turn improves the economics of wind energy.

15.5.7 Making the transition to the hydrogen economy

In order to make the final transition off fossil fuels – something that is needed before the end of this century if critical ecosystems and food production are to be safeguarded (Harvey, 2004) – it will be necessary to replace fossil fuels with hydrogen fuel in applications where renewable energy cannot be directly used. Hydrogen would be produced from a variety of renewable forms of energy (primarily wind and solar energy), in locations where renewable energy is plentiful, and transported largely by pipeline to where it is needed, stored, and used when needed. Such a system, where hydrogen serves as the bridge between renewable energy resources and energy demand, is referred to as the 'hydrogen economy'. Fischedick et al (2005) provide a recent overview of an all-renewable energy system using hydrogen and electricity as interchangeable energy currencies.

The switch from fossil fuels to hydrogen for space and water heating will be much easier, and less expensive, in those communities that are served by district heating systems. In buildings that are individually heated with natural gas, it is likely that a whole new pipeline infrastructure will have to be built in order to utilize hydrogen. In buildings served by district heating systems, the only infrastructure that will need to be rebuilt are the pipelines supplying the central power plant. Furthermore, if boilers or furnaces need to be modified or replaced in order to use hydrogen fuel, this will be easier and probably less expensive if done at a single, centralized facility rather than in each individual building.

15.6 Summary and synthesis

Cogeneration is the simultaneous production of electricity and useful heat. A district heating system is a system of underground pipes that distributes hot water or steam from a centralized heating plant to individual buildings connected to the system. This heat can be used for domestic hot water or for space heating. A typical existing coal fired power plant generates electricity with an efficiency of 33–35 per cent. Advanced pulverized coal power plants should attain an efficiency of about 48 per cent, while possible future integrated coal gasification combined cycle power plants might reach an efficiency of 50 per cent. Existing state-of-the-art natural gas power plants have an efficiency of 55–60 per cent. In contrast, the marginal efficiency of electricity generation (i.e. the electricity produced divided by the extra fuel used compared to heat production only) through cogeneration using natural gas is 80–90 per cent. Cogeneration requires a use for the waste heat from electricity production, and district heating systems provide an ideal use.

Trigeneration – the simultaneous production of electricity, hot water and chilled water – is possible if some of the waste heat from electricity production is used to produce chilled water using a single- or double-effect absorption chiller. This can be done centrally, at the site of power generation, in which case a separate district cooling grid is required, or it can be done using absorption chillers at each building, using heat distributed by the district heating grid. In either case, operation of absorption chillers requires reducing the electricity output, electricity that in most cases could have been used with an electric vapour compression chiller to produce greater cooling than by using waste heat in an absorption chiller. Thus, it is normally better to maximize electricity production and to throw away some of the waste heat, rather than producing chilled water with absorption chillers using waste heat from electricity production. Adsorption chillers can also be used to make chilled water in a trigeneration system, but can make use of lower-temperature heat, such that little if any electricity production is sacrificed. If no electricity is sacrificed, the energy is free, but the adsorption chillers cost about twice as much as absorption chillers.

Even when not combined with power production through cogeneration or trigeneration, centralized production of hot and chilled water for heating and cooling through district heating and cooling systems can be more efficient than on-site heating and cooling, even after accounting for energy losses during the distribution of heat and coldness. For district heating, these savings arise from the ability to avoid low part-load efficiencies by having many boilers that can be staged so that, at any given time, all except at most one will be operating at their peak efficiency. For district cooling systems, the savings arise from the factor of two or more difference in the COP of large chillers (6–7.9) compared with the COP (2.5–4) of the small chillers that would be found in residential and many commercial buildings, as well as from the advantage of the ability to stage the operation of many centralized chillers to avoid low part-load efficiencies. When energy use by pumps is considered, a district cooling system for Hong Kong is estimated to reduce energy use by about 30 per cent using central chillers with a COP of 5.9.

Seasonal underground storage of heat is an effective way of storing solar thermal energy, plentiful during the summer, and using it during the winter for heating purposes. Solar thermal energy can be stored in underground aquifers, in subsurface rock using pipes in a series of vertical boreholes, in insulated steel or concrete tanks (with waterproof liners), or in gravel-water pits with waterproof liners. Winter coldness can also be stored in underground aquifers or in subsurface rocks. Many projects have been built in Sweden, Germany and elsewhere in Europe, in which solar thermal collectors collect heat that is stored underground and used for space heating and domestic hot water purposes. Such systems are most viable economically at a scale of many buildings, connected by a district-heating network. Delivered solar heat in recent German systems costs 16–42 eurocents/kWh while providing 30–64 per cent of the total heat load. Solar heat from a project in Alberta (Canada) will cost about 10–15 cents/kWh while supplying 95 per cent of the annual space heating load, or 9–12 cents/kWh if packaged with upgrades to the thermal envelope. A number of hybrid biomass/solar district heating systems have been built in Austria. In these cases, relatively small storage tanks – sufficient for 3–5 days hot-water demand – have been used.

In solar-assisted district heating systems, the lowest possible distribution and storage temperatures should be used. Distribution supply temperatures of 35–50°C are adequate if space heating alone is provided. This in turn requires that buildings have a high-performance thermal envelope and large enough radiators that heat at this temperature is adequate for space heating. If DHW is to be provided, then a higher distribution temperature (75°C) or supplemental on-site heating will be needed. Use of solar thermal energy for district heating reduces the opportunity for cogeneration. In jurisdictions where electricity is generated at inefficient (35–40 per cent) central coal fired power plants, this will this result in increased CO_2 emissions and primary energy use, as less coal fired electricity will be displaced by cogenerated electricity. However, it is preferable to take a long-term view, and in the long term, the savings from cogeneration will decrease because the efficiency of alternative electricity sources will improve, and electricity will increasingly be generated from renewable energy sources. The loss of cogeneration opportunity by implementing solar-assisted district heating is therefore less critical.

Apart from facilitating the use of solar and biomass energy, district heating (and cooling) systems combined with heat pumps can be used as a buffer to make use of excess wind energy. This in turn will allow wind energy systems to be sized to meet a larger fraction of total electricity demand with minimal economic penalty due to capping the power output when the wind electricity potential exceeds total demand. Finally, district energy systems will facilitate the eventual transition to a hydrogen economy, as only one or a small number of central heating and power plants will need to be converted to hydrogen fuel, rather than each individual building.

Notes

1   This expression is the reciprocal of the Carnot efficiency for an ideal heat pump (Equation 5.3), which is reasonable given that an ideal heat engine generates electricity while transferring heat from a high-temperature source to a low-temperature sink, while a heat pump consumes electricity while transferring heat from a low-temperature source to a high-temperature sink.

2   This result is independent of the efficiency with which grid electricity is already generated, because the trade-off being considered here is between more or less electricity production in a new cogeneration facility, with electric or absorption chillers feeding a district cooling network. For the trade-off involving using fossil fuel energy for dedicated heat production and an absorption chiller, or dedicated electricity production and an electric chiller, the efficiency of electricity production does matter. This case is considered in Chapter 4 (Section 4.6.2).

3   This is in contrast to a double-effect chiller, where the heat output from the first condenser and absorber drives the generator of a second chiller unit.

4   For example, steam that is cooled from 170°C and 690kPa to 80°C at atmospheric pressure (101kPa) releases 2400kJ/kg, while cooling water from 175 to 120°C releases only 240kJ/kg.

5   There is a sharp drop in the efficiency of condensing boilers below 5–33 per cent of peak load, as the boiler reverts to on/off cycling.

sixteen

Summary and Conclusions

In this chapter, tables categorizing and summarizing the technologies available to reduce energy use in buildings, summarizing all quantitative references in this book to energy savings potential, and summarizing all quantitative references to costs are presented. We then return to the question that motivates this book: Can energy use in new and existing buildings be reduced to levels low enough, in the context of future population growth and in the growth of floor area per capita, such that the remaining energy use can eventually be met by renewable sources of energy? In other words, is sustainability in the buildings sector feasible?

16.1 Summary of technologies, development status and applicability

Table 16.1 summarizes the design features, technologies and systems available to reduce energy use in buildings that have been discussed in this book. These have been categorized into the following broad categories:

- simple and well understood, primarily architectural decisions;
- simple and well understood, pertaining to mechanical systems and lighting;
- moderately complex.

These categories have been further divided into zero-to-low cost, low-to-moderate cost, or moderate-to-high cost, either with partially to entirely offsetting savings in mechanical equipment through downsizing, with attractive payback periods in spite of high initial cost, or without attractive payback periods except in special circumstances. The vast majority of the technologies and techniques listed in Table 16.1 already exist and have been used successfully in at least one or more countries, although some very promising techniques – such as liquid desiccant dehumidification systems for hot-humid climates, advanced daylighting systems, building-integrated photovoltaics, solar-assisted district heating, and solar-driven cooling systems require further development to bring down costs and to improve performance.

Table 16.1 Summary of design features, technologies and systems available to reduce energy use in buildings. Items have been subjectively classified as simple and well understood, or moderately complex and further classified according to capital cost. High-cost items are classified here as 'cost-effective' if they represent the lowest-cost choice on a lifecycle basis

Simple and well understood, primarily architectural decisions			
	Conventional, applicable everywhere		
		Zero-to-small cost	
			• Building form, shape, glazing area and shape on each façade, and choice of operable instead of fixed windows: Impacts heating load and cooling load through impact on solar heat gain and on surface-area: volume ratio. Impacts opportunities for passive ventilation and daylighting. Optimal choices depend on type of building, occupancy pattern and climate. Requires simple analysis tools for optimization of these choices. • Choice of high-reflectivity (low albedo) building envelope to minimize summer cooling requirement • Upgraded construction details to reduce air leakage through the envelope
		Low-to-high cost, partly to entirely offset through savings in mechanical systems and/or other benefits	
			• Insulation (provides heating and/or cooling benefits, depending on climate, and peak load benefits) • High-performance windows (low heat loss coefficient to reduce heating requirements and/or low solar heat transfer to reduce cooling requirements, depending on climate) • Thermal mass (to maximize use of solar heat in winter and the benefit of night-time ventilation in summer where this is useful) (less applicable in hot-humid climates due to smaller diurnal temperature range) • Fixed external shading devices (applicable to equatorward-facing walls) • Adjustable external shading devices (manual or automatic, most useful on east- and west-facing walls) • Simple design features (beyond building form) to facilitate passive and hybrid ventilation (trickle ventilators, wider air ducts, internal airflow pathways)
	Unconventional, widely applicable		
		Low-to-moderate cost, often with quick payback	
			• Evaporative cooling (commercial products available in some markets, but requires further development of commercial products everywhere; not applicable in hot-humid climates unless combined with desiccant dehumidification; not applicable in hot-arid climates with severe water shortages, where night ventilation with thermal mass is often effective)
		Moderate-to-high cost, partly to largely offset through savings in mechanical equipment	
			• Earth pipe cooling (less applicable where ground temperature and peak summer temperatures differ little)
Simple and well-understood, pertaining to mechanical systems and lighting			
	Modest additional capital cost with short payback times		

			• Variable air volume ventilation systems instead of constant air volume systems • Variable-speed drives on all pumps and fans • Variable-speed (modulating) air conditioners and chillers • Basic hydronic heating and cooling (2-pipe or 4-pipe systems with fan coils) to separate heating, cooling and ventilation requirements • Heat recovery between fresh air and exhaust air • Basic automation systems (such as shutting off or reducing heating, cooling and ventilation when not needed) • High performance but fully electric lighting systems
	Simple (because packaged), expensive in most jurisdictions due to immature markets, potentially or already cost-effective on a lifecycle basis		
			• Condensing furnaces and boilers • Condensing combisystems for integrated space heating and domestic hot water • Domestic and small commercial biomass boilers
Moderately complex			
	Well developed, widely used in some jurisdictions		
		Modest or no net capital cost when properly integrated into the building design	
			• Advanced design features to facilitate passive ventilation (solar chimneys, wind cowls) • Displacement ventilation • Chilled-slab heating and cooling • Dehumidification using solid desiccants • Demand-controlled ventilation system and other more advanced control systems • District heating and cooling systems (no net capital cost possible compared to heating and cooling plants in individual buildings)
		Modest-to-large capital cost, but cost-effective in many jurisdictions	
			• Solar thermal collectors for domestic hot water • Transpired solar collectors for preheating ventilation air • Ground source heat pumps
	Requiring further development and refinement through practical demonstration		
		Potentially low-cost	
			• Liquid desiccant dehumidification (overcomes limitation of evaporative cooling in hot-humid climates) • Simple daylighting systems (involving simple components such as light shelves, passive light pipes, or sun-reflecting glass) (cost and complexity arise through controls, sensors and dimmable ballasts in the electric component, and in automatic control of shading devices) • Diurnal storage of coldness using eutectic salts (storage with ice is well established and cost-effective, but unlike eutectic salts, is not suited for high temperature (16–20°C), efficient displacement ventilation and chilled ceiling cooling systems)
		Likely to remain high cost but potentially cost-effective in some jurisdictions	
			• Advanced glazing (electrochromic and thermochromic windows) • Transparent insulation • Double-skin façades • Advanced daylighting systems (such as light pipes with sun-tracking reflectors) • Solar-powered absorption, adsorption or ejector chilling • Solar-powered desiccant (solid or liquid) dehumidification and cooling • Building-integrated PV panels • Solar cogeneration • Seasonal underground storage of thermal energy (solar, or from cogeneration)

16.2 Summary of energy savings potential

Table 16.2 summarizes the quantitative references to energy savings potential that are found in this book, along with the section where each reference is found. All of the technologies listed in Table 16.2 already exist and their performance has been measured in numerous field tests. Envelope measures alone can reduce heating requirements in both residential and commercial buildings by up to 90 per cent, while envelope measures combined with passive ventilation or combined with night ventilation and thermal mass can frequently reduce energy requirements for cooling and ventilation by 75 per cent and can sometimes eliminate these energy uses

altogether. Many individual HVAC measures can reduce HVAC energy use by 25–50 per cent, a savings that would be applied to the reduced loads created through envelope measures. Low-energy cooling techniques in one form or another (night ventilation combined with thermal mass, earth pipe cooling, evaporative cooling, desiccant cooling) are applicable under almost any climatic conditions that are encountered almost anywhere in the world, including both hot-arid and hot-humid climates. Advanced electric lightings systems have frequently achieved 50–75 per cent reductions in lighting energy use compared to typical recent practice in new buildings in many jurisdictions, with further savings through advanced daylighting systems.

Table 16.2 Summary of most references to energy savings potential in this book.

Item	Savings potential	Section
Chapter 3, Thermal Envelope		
Structural insulated panels	50% reduction in heating energy use compared to standard US housing	3.2.3
Dynamic insulation	Up to a factor of eight reduction in effective U-value	3.2.4
Attic low-e foil	25–50% reduction in heat flux, 7–10% reduction in cooling load in US SE	3.2.9
Advanced insulation systems	Provide 2–3 times the resistance to heat loss as conventional insulation systems in most jurisdictions	3.2.10
High-performance windows	As little as 1/5 the heat loss of non-coated double-glazed windows, serving as a net heat source in winter in most climates for most orientations	3.3.6 and 3.3.12
High-performance windows	SHGC as low as 0.23 with visible transmittance of 0.41	3.3.10
High-performance curtain wall, glazed portion	70% reduction in heat loss compared to ASHRAE 90.1-2004 in 4500 HDD climate	3.3.18
Vacuum insulated panels in curtain wall spandrel (opaque) sections	Factor of 4–8 less heat loss compared to ASHRAE 90.1-2004 in 4500 HDD climate	3.3.18
High-performance glazing + radiant slab heating and cooling in Vancouver buildings with 50% glazed wall area	45% savings in total energy use	3.3.20
External shading devices	Reduce heat gain by 90% (compared to 50% by internal devices)	3.4.2
Vacuum insulated panels in doors	30% less heat loss for complete door system compared to door with conventional insulation, and 65% less heat loss compared to a 6cm thick wood door	3.6
Air leakage, residential	Factors of 3–30 less leakage achieved in housing through quality construction, compared to average practice, depending on location	3.7.1
Air leakage, commercial	Factor of 6–8 difference between modestly good practice and typical practice of some countries	3.7.2

Chapter 4, Heating Systems		
Solar air collectors	Can meet 80% of annual heating load in well-insulated buildings with collector area equal to 16% of floor area	4.1.3
Airflow windows	60–75% reduction in effective U-value	4.1.7
Furnace efficiency	70–97% (up to 28% savings)	4.2.1
Boiler efficiency, full load	68–95% (up to 28% savings)	4.3.1
Boiler efficiency, part load	Can be as little as half the full-load efficiency at 25% of full load	4.3.2
On-site cogeneration	62–78% overall efficiency, 45–65% marginal efficiency of electricity generation	4.6.1
Chapter 5, Heat Pumps		
GSHPs	Can save 25–65% in heating energy and about 35% in cooling energy compared to ASHPs	5.3.2
Reduction in heat-distribution temperature (through envelope improvements) so that condenser temperature can be reduced from 60°C to 40°C	25–30% energy savings for an evaporator at −15°C to 0°C	5.5
Increase in chilled-water temperature (through alternative system design) so that evaporator temperature can be increased from 0°C to 10°C	25–30% energy savings for a condenser at 30°C to 40°C	5.5
Heat pumps with CO_2 as a working fluid	Slightly lower COP for cooling, slightly higher COP for heating if fully optimized	5.8.1
Integration of heating and cooling loads in a sports and health complex, Mexico	38% savings in thermal energy	5.12.1
Chapter 6, Cooling Loads and Cooling Devices		
Building-scale voids and double-skin roofs and façades	Can reduce cooling load by up to 50% in Okinawa (latitude 26°N)	6.2.1
White reflective roofs	Reduce cooling loads by 20–25% in Florida and 20–40% in California	6.2.2
Shade trees + higher albedo roofs and roads	50–60% savings in cooling energy use in Los Angeles	6.2.2
5cm of polystyrene roof insulation	Reduce cooling load by 45% in Cyprus	6.2.3
10cm of roof and wall insulation	Reduce cooling load by 14% in Tehran (and heating load by 55%)	6.2.3
Raising insulation from false ceiling to roof	5–10% savings in annual cooling load in California (and 20–25% reduction in peak load)	6.2.3
Insulation on external and internal walls in Hong Kong apartments	40% reduction in annual cooling energy use	6.2.3
Thermal mass combined with passive night ventilation	100% reduction in cooling loads in locations with cool nights; negligible reduction in Hong Kong but 30–40% reduction in peak cooling loads	6.2.6
Double-skin façades (DSF)	Can reduce solar heat gain to as little as 2.5%	6.3.2
DSF with passive daytime and night-time ventilation	About 80% reduction in primary energy used for heating, sensible cooling and ventilation energy for south-facing façades in The Netherlands	6.3.2
Thermal mass combined with mechanical night ventilation	30% net energy savings for optimized buildings in the UK and in Kenya during the hottest month	6.3.3
Evaporative cooling	92–95% savings in California homes	6.3.7
Evaporative cooling from roofs	73% reduction in cooling load, 66% reduction in fan energy in California; 52% reduction in cooling load in Iran	6.3.7
Earth pipe cooling	Cooling COPs of 5–8 over a range of sites	6.3.8
Residential air conditioners	Best Japanese models twice as efficient as typical North American and European models	6.4.1

Rooftop air conditioners	Most have at least one operational deficiency that can save 25% if corrected	6.5.1
Large, centralized chillers	Up to 2–3 times more efficient than window and wall-mounted air conditioners	6.5.2
Variable speed drive on centrifugal chiller	41% annual savings in a Toronto case	6.5.2
Absorption chillers	Can increase or decrease primary energy use	6.6.1
Correct sizing of cooling equipment	Can reduce annual cooling loads by 6–22% in Hong Kong office buildings. Real seasonal COPs of residential air conditioners in the US as little as half the rated COP due to oversizing	6.8
Off-peak cooling	Energy to generate a kWh of electricity 20–43% less at night than during the day in California	6.9.2
Free cooling with cooling towers and chilled ceilings	Cooling tower meets cooling load 67% (Milan) to 97% (Dublin) of the time. COP of 8–20 based on fan and pump energy use. Reduction of annual cooling load by 50% in Key West (Florida) to 85% in Arizona	6.10.2
Chapter 7, HVAC Systems		
Reduce required airflow by 50%	Reduces fan energy use by a factor of six if duct size is not changed	7.1.2
Loop rather than radial duct design	Reduces fan energy use by 20–25%	7.1.2
Rebalancing pump systems after installation	Up to 25% savings in pump energy use	7.1.3
Proper sizing of pumps	Up to 30% savings in pump energy use	7.1.3
Optimal combination of cooling and dehumidification in the humid tropics	Up to about 20% savings in air conditioning energy use	7.1.4
Furnace air handlers	Up to 93% savings in annual energy use using ECPM motors instead of PSC motors	7.3.2
Heat recovery ventilators	Can recover up to 95% of the heat in ventilation exhaust air, but require proper installation and maintenance	7.3.3
Ceiling fans	75% savings through more efficient motors and aerodynamic blade design	7.3.5
Leaky ducts	Reduce effective COP of air conditioners and heat pumps by 25–40%	7.3.6
VAV instead of CAV systems in commercial buildings	50–65% savings in combined cooling, heating and ventilation energy use in various US cities	7.4.3
Chilled ceiling cooling	6–42% savings in cooling energy use in various US cities compared to all-air systems	7.4.4
Displacement ventilation	40–60% savings in cooling energy use in various US cities compared to a standard VAV system	7.4.5
Demand-controlled ventilation	20–30% savings in combined cooling, heating and ventilation energy use in various US cities	7.4.6
Floor radiant cooling	31% savings in cooling energy use at Bangkok International Airport	7.4.8
Sensible heat exchangers	Recovery of up to 85% of the heat in exhaust air	7.4.9
Heat pipes around air conditioners	15–30% savings (and 25–50% increase in dehumidification capacity)	7.4.10
Solid desiccants	25–30% savings in primary energy use for cooling and dehumidification, 40–50% savings if waste heat or solar heat can be used	7.4.11
Liquid desiccants	15–25% savings in primary energy use for cooling and dehumidification, 40% savings if waste heat or solar heat can be used	7.4.12
Membrane heat exchangers (experimental)	About 25% computed savings for cooling and dehumidification in Beijing, 60% in Hong Kong	7.4.13

Earth pipe loop + cogeneration applied to desiccant dehumidification in Germany	70% savings relative to vapour compression chilling with a COP of 3.0, 56% savings if the baseline COP is 6.0	7.4.14
VAV in laboratories instead of CAV; reduced pressure drop, enthalpy wheel in US cities	24–44% savings in fan + pump + chiller electricity use 62–75% savings in heating energy use	7.4.17
Fume hoods with variable flow (based on occupancy) and design to minimize turbulence	45% savings in a university laboratory in Montreal.	7.4.17
Commissioning	20% savings in total HVAC energy use	7.5
Optimal control settings	20–35% savings in total HVAC energy use	7.5
Chapter 8, Domestic Hot Water		
Water savings fixtures	50% savings in water use	8.2
Condensing, tankless water heater	15% savings in overall efficiency + elimination of standby losses (which account for up to 1/3 of DHW energy use)	8.3.2
Heat recovery	25–35% savings	8.4.2
Residential circulation loop	Saves 80% of the hot water wasted waiting for hot water to arrive at the faucet	8.5.2
School recirculation loop	94% savings in pump energy use through point-of-use water heaters	8.5.3
Chapter 9, Lighting		
Daylighting with dimmers	40–80% savings in lighting energy use in perimeter offices, 20–33% savings in combined lighting + cooling energy use	9.3.3
Advanced lighting (without daylighting)	50–75% savings	9.5.1
Task/ambient lighting	Up to 50% savings	9.5.2
High intensity fluorescent lamps	40% savings in athletic facilities relative to standard metal halide lamps	9.6
Chapter 10, Appliances, Consumer Electronics, Office Equipment and Elevators		
Refrigerator/freezer units	Almost factor of 2 difference in energy use, best vs worse, within a given size category	10.1.1
Freezers	Almost factor of 1.5 difference in energy use, best vs worse, within a given size category	10.1.1
Ovens, cooktops	25–35% savings, best vs worse. Within a given size category	10.1.1
Washing machine	80% savings, best front-loading machine vs US standard for top loading machines	10.1.2
Clothes dryer	40% savings due to dryer clothes from front-loading washing machine	10.1.3
Dishwasher	20–35% savings potential	10.1.4
Standby energy use by consumer electronics	60–70% potential reduction at little or no additional cost	10.2
Advanced gearless elevators	40–60% savings	10.5
Chapter 13, Advanced New Buildings		
Passive houses in Europe and Advanced houses in Canada	75% savings in heating energy use relative to conventional new housing	13.2.1
Modest measures in US houses	50% savings in total energy use	13.2.1
Modest measures in Turkey	60% savings in heating energy use	13.2.1
Integrated design process (IDP) for commercial buildings	35–50 % savings in total energy use compared to conventional new buildings	13.3
IDP combined with advanced features	50–80% savings in total energy use compared to conventional new buildings	13.3

Chapter 14, Retrofits of Existing Buildings		
Simple, cost-effective envelope measures	25–30% savings in old residential buildings	14.1
Extensive retrofits	Up to 90% savings in heating energy and 80% savings in total energy in old residential buildings	14.1
Deep retrofits of commercial buildings	50–75% savings demonstrated in a wide variety of types of buildings in many different countries	14.2
Chapter 15, Community-Integrated Energy Systems		
Simple cycle cogeneration	65–72% overall efficiency, 44–62% marginal efficiency of electricity generation	15.2.6
Combined cycle cogeneration	90% overall efficiency, 88–90% marginal efficiency of electricity generation	15.2.6
Centralized heating	10–20% savings compared to decentralized heating, at least partly offset by distribution losses	15.3.2
Centralized chilling	Up to factor of three improvement in chilling COP, depending on the size and part-load efficiency of decentralized equipment that is replaced, partly offset by distribution losses and pump energy use	15.3.6
Seasonal underground storage of solar energy	Provides up to 95% of annual space heating load	15.5.3
Heating and cooling systems as a wind energy buffer	Can allow wind to supply 50% of total electricity demand with minimal spillage of wind energy	15.5.6

16.3 Summary of costs

Table 16.3 summarizes all references to cost found in this book, along with the section where each reference is found. The cost-effectiveness of envelope measures is dependent on, or greatly enhanced by, savings in the cost of mechanical systems (heating and cooling plants, ductwork, radiators) and associated electrical systems through the downsizing that is possible with a better envelope. In order to realize these savings, a fully integrated and iterative design process is essential. Many high-efficiency measures involving HVAC systems (such as chilled ceiling cooling, displacement ventilation and sensible and latent heat exchangers) also permit downsizing of heating and cooling equipment and of ventilation fans, thereby offsetting some to all of the additional cost of the energy efficiency measures. Promising low-energy cooling techniques, involving enhancement of evaporative cooling with solid or liquid desiccants, have the potential to become competitive with conventional cooling systems on a first-cost basis. Advanced lighting systems involving

daylighting controlled dimming remain expensive and require skilled designers and installers, but are still at an early stage of commercialization. Electricity from building-integrated photovoltaic panels (BiPV) is 1.5 to 3 times the average cost of electricity in sunny locations ($2200kWh/m^2$/year) with favourable financing (i.e. 24–33 cents/kWh) and 3–5 times the average cost of electricity in less sunny locations ($1650kWh/m^2$/year) with favourable financing (30–45 cents/kWh). However, if costs can be cut in half, BiPV will be competitive with grid electricity at times of peak demand in many locations, and will serve to relieve growing transmission bottlenecks in urban areas by providing electricity where it is needed. Solar thermal energy with seasonal storage is already competitive with conventional heating systems on a lifecycle cost basis in some locations. Solar driven absorption and desiccant chilling systems, on the other hand, are likely to remain expensive unless they can make use of excess solar thermal heat from collectors that are installed primarily to meet winter heating requirements.

Table 16.3 Summary of all references to capital cost (i.e. upfront cost) in this book

Item	Cost	Section
Chapter 3, Thermal Envelope		
Structural insulation panels	Labour savings can offset greater material costs	3.2.3
Vacuum insulation panels	Cost in Switzerland of 320€/m² for U = 0.15W/m²/K panel, compared with 32€/m² for comparable fibre or solid-foam insulation. Incremental cost when applied to walls is about (4000€ × window area fraction) per m² of saved floor space	3.2.7
Engineered I-beams	Lack of warping and shrinking, eliminating backing and callbacks to fix popped nails, can offset greater material costs	3.2.10
Electrochromic windows	$1000/m² in 2000, expected to drop to $100/m² with large-scale manufacturing, not including control costs	3.3.15
Components of high-performance windows	Heating-cost savings in 3000 HDD climate may justify triple glazing, two low-e coatings and argon fill. Further measures may be justified through savings in mechanical equipment and radiators due to smaller heating and cooling loads	3.3.20
High-performance windows + radiant slab cooling	Overall cost, Vancouver office buildings: $550/m², versus $620/m² for conventional façade and HVAC system (while reducing energy use by 45%)	3.3.20
Double-skin façades	120–200% the cost of a normal façade for builders experienced with double-skin façades; downsizing of heating and cooling equipment can offset much of the incremental cost; elimination of vertical ducts with the gap serving as a supply/exhaust vent can result in no net cost. Costs are highly site-specific	3.4
Air leakage of 1.5 ACH at 50Pa instead of 3–4 in houses	No additional cost for experienced builders in Canada.	3.7.1
Lifecycle costs of higher levels of insulation	The minimum in lifecycle costs is quite broad, with lifecycle cost at maximum implemented levels of insulation less than lifecycle cost at levels that are prescribed by building codes in many jurisdictions, even at present energy costs and without consideration of externalities	3.8
Chapter 4, Heating Systems		
Transpired solar collectors	Economics is highly favourable in large-scale applications (at least), with a simple payback of the order of 3–4 years in Toronto (4000 HDD) for natural gas at $12/GJ. Future galvanized steel and plastic collectors might be significantly less costly	4.1.9
Furnaces	Condensing units more expensive than non-condensing furnaces (about 30% greater installed cost) due to need for corrosion-resistant heat exchanger, and modulating units more expensive still due to additional valves and controls. Larger premium in retrofits if condensing furnace replaces non-condensing furnace due to need to block chimney	4.2.1 & 4.2.8
Wood pellet boilers	Considerably more expensive than oil boilers, but less so at larger sizes	4.2.2
Heating-only absorption heat pump	$2000 more than a 29kW$_{th}$ residential furnace	4.2.4
Commercial condensing boilers	25–100% more expensive than non-condensing boilers in North America due to immature market, partially offset where possible by choosing a greater number of smaller units so that there is less backup capacity and hence less total capacity. Typical non-condensing cost: $30–75/kW$_{th}$	4.3.1

Equipment for building-scale cogeneration	Reciprocating engines, \$900–1400/kW_e installed; Microturbines: \$1800–2600/kW_e installed; Fuel cells, \$3200–5500/kW_e	4.6.1
	Chapter 5, Heat Pumps	
Ground source heat pumps	Will often have greater upfront cost but with a favourable payback in many cases. Upfront costs can be comparable to first cost of conventional systems in some cases. Ground loop is the most expensive part, but its cost can be reduced if the ground loop is integrated into building piles (if present) or into excavation and replacement of excavated material next to the building (if needed for structural reasons), and/or if the loop can be downsized in various ways	5.3.6 & 5.9
Hydronically coupled heat pumps	Factor-of-ten downsizing in size of heat exchangers through microchannel heat exchangers should lead to a significant reduction in the cost of heat pumps	5.7
Heat pump water heater	\$1400 compared to \$400–500 for comparable electric resistance heaters, but cost premium could drop substantially.	5.11
Solar-assisted heat pump for domestic hot water	Improved performance generally not likely to justify increased cost and complexity compared with solar DHW heater alone. 2-year payback in Singapore compared to electric resistance water heater and 13 cents/kWh electricity	5.13.2
	Chapter 6, Cooling Loads and Cooling Devices	
Low-cost measures to reduce cooling load (building shape and orientation; albedo; added insulation; external shading; windows with low SHGC; thermal mass)	Cost largely or entirely offset by downsizing or elimination of mechanical cooling equipment	6.2
Features (atria, solar chimneys, airflow windows and double façades, trickle ventilators, wind cowls) to facilitate passive, hybrid, and night-time ventilation	Cost partly offset by downsizing or elimination of mechanical cooling equipment	6.3
Evaporative coolers	\$200/kW_c in the US market, compared with \$200–450/kW_c for ductless split air conditioners, and likely to fall at least 30% in cost	6.3.7
Vapour compression chillers	Terminal air conditioners, \$400–700/kW_c at 4.4–1.7kW_c capacity; multizone rooftop air conditioners, \$300–1500/kW_c at 400–600kW_c capacity; air-cooled reciprocating chillers, \$150–330/kW_c at 1500–50/kW_c capacity; centrifugal chillers, \$80–130/kW_c at 7000–300kW_c capacity	6.5.4
Absorption chiller	\$85–350/kW_c, competitive in cost with centrifugal chillers of 2.5MW_c capacity and larger. Greater cost per unit of cooling capacity the lower the regeneration temperature	6.6.1
Adsorption chiller	\$550–2200/kW_c for capacities of 50–430kW_c	6.6.2
Solid desiccant/evaporative cooling	\$400–700/kW_c for capacities of 20–350kW_c	6.6.4
Liquid desiccant dehumidification	More expensive than solid desiccant cooling, but may drop to \$230–260/kW_c	6.6.5
Correct sizing of cooling equipment	Potentially significant reduction in capital cost and modest energy savings	6.8
Ice thermal storage	No net increase in capital cost possible	6.9.4
Eutectic salt thermal storage	No net increase in capital cost possible	6.9.4

Chapter 7, HVAC Systems		
High-performance filters	1-year simple payback on incremental cost	7.1.2
Correct pump sizing	Potentially modest reduction in capital cost and significant savings in pump energy use	7.1.3
Personal environmental modules	Reduced overall HVAC cost due to downsizing of equipment and warmer permitted supply temperature	7.2.5
Premium furnace blower (with ECPM motor)	Four times the cost of standard motor, but 2–3 year simple payback at US electricity costs; should drop to 2–3 times cost of standard motor with volume	7.3.2
Chilled ceiling cooling combined with high-performance building envelope	Same or smaller cost than conventional VAV cooling system and envelope	7.4.4
Displacement ventilation	Same or smaller cost than conventional VAV cooling and ventilation system. Reduced churn costs	7.4.5
Demand-controlled ventilation with CO_2 sensors	2–3 year simple payback through energy cost savings	7.4.6
Task/ambient conditioning	Reduced capital cost due to equipment downsizing	7.4.7
Sensible heat exchanger between exhaust and supply air	Reduce peak cooling and heating loads, allowing equipment downsizing that offsets part or all of the cost of the heat exchanger	7.4.9
Cogeneration/desiccant dehumidification/radiant-slab cooling	4-year simple payback compared to conventional all-air HVAC system in one study	7.4.14
Duct sealing	10-year simple payback through energy cost savings, shorter if cooling and air-handling equipment can be downsized	7.4.15
Occupant responsive fume-hood ventilation with heat recovery	Less than 2-year simple payback	7.4.17
Commissioning	1–3% of HVAC cost	7.5
Chapter 9, Lighting		
Daylighting and shading control systems	Expensive, but costs reduced through use of a single transformer for several DC-motorized shades and sequential rather than simultaneous operation Dimmable ballasts currently $60–120 each but could fall to $20–30 each. Simplified systems can reduce cost significantly with minimal impact on energy savings. Part of cost will be offset by downsizing of cooling equipment and electrical system	9.3.2
Air-tight, CFL downlights	Prototype concepts reduce costs significantly	9.4
Chapter 10, Appliances, Consumer Electronics, and Office Equipment		
Horizontal-axis washing machines	Substantially more expensive than vertical-axis machines but generally lower-cost over their lifetime	10.1.2
Consumer electronic products with low standby power draw	Generally no more expensive than standard products	10.2
Chapter 11, Active Collection and Transformation of Solar and Wind Energy		
PV electricity	24–33 cents/kWh (high solar irradiance, 4% interest) to 30–45 cents/kWh (mid solar irradiance, 4% interest) at present. Cost could drop in in half, which would make it competitive with costs of grid electricity at the time that PV power is most available (i.e. at times of peak load)	11.2.5
Solar thermal energy, mid-European conditions	9–23 eurocents/kWh for DHW 40–50 eurocents/kWh for combisystems 10–13 eurocents/kWh for district heat without seasonal storage, 16–42 eurocents/kWh with storage	11.4.5

Flat-plate solar collectors	200–500€/m² in Europe; 30–35€/m² in China	11.5.1
Solar thermal absorption chiller cooling	$1400/kW$_c$ for a complete system with compound parabolic collectors and double-effect absorption chiller in California; 3000€/kW$_c$ for a system in Greece; economics much more favourable in China	11.5.1
Solar thermal desiccant dehumidification and cooling	High cost if sized to meet 70–90% of cooling load in Europe (required in order to reduce primary energy use for cooling by 50%), less costly if collector cost is assigned in whole or in part to winter heating	11.5.5
Energy self-sufficient house with PV/T in The Netherlands	35 eurocents/kWh for heat, including cost of radiant floor distribution system	11.6
Chapter 12, Embodied Energy		
Appropriate traditional building materials	Up to 45% less expensive than modern materials in India (with 60% less embodied CO_2 emission)	12.4
Chapter 13, Advanced New Buildings		
CEPHEUS (Cost-Effective Passive Houses as European Standards) houses	Passive Houses have 10–20 times less heating energy than conventional houses; 'cost-effective' is defined as houses where the energy-cost savings pay back the incremental investment in 30 years or less. Cost of saved energy is 6.1 eurocents/kWh averaged over 13 cases (range: 1.1–11 eurocents/kWh)	13.2.1
Measures to reduce energy use by 56–77% in new US houses	Energy cost savings largely cancel incremental financing costs in simulation study	13.2.1
Building America programme, residential housing	50% savings in heating energy, 30% in cooling energy at negligible upfront cost	13.2.2
C-2000 (commercial building) projects in Canada	35–50% savings over ASHRAE 90.1-1989 at little or no additional cost	13.3
6 LEED buildings in US	Average cost premium <2% while saving an average of 50% in energy use	13.3.4
Solar Bau (commercial building) programme in Germany	33–85% savings at a cost comparable to the difference in cost of alternative interior finishings	13.3.4
Integrated design of Chicago office	1.6-year payback for 27% savings, 4.6-year payback for 36% savings, 2-year payback for 44% savings	13.3.4
16-storey building, Oregon Health & Science University	60% savings over ASHRAE 90.1-1999 with reduced first cost	13.3.4
Chapter 14, Retrofits of Existing Buildings		
ACT² project, California	45–50% savings with modest payback	14.2
Proposed retrofits of ten European offices	46–80% estimated savings with 10–15 year simple payback	14.2
Chapter 15, Community-Integrated Energy Systems		
Natural gas power plants for cogeneration	$700–1900/kW for simple cycle cogeneration; $400–860/kW for combined cycle cogeneration	15.2.6
District heating and cooling systems	Economies of scale for heating or cooling units, and reduced total capacity due to peak loads in individual buildings not coinciding, can more than offset the cost of pipes in high-density developments, thereby reducing total cost	15.3.10
Eight solar-assisted district heating systems completed in Germany (30–62% solar fraction)	16–42 eurocents/kWh	15.5.3
Large-scale seasonally stored solar heat in Sweden	11 eurocents/kWh	15.5.3
Medium-scale seasonally stored solar heat and district heating in Alberta, Canada	15 cents/kWh for the initial project	15.5.3

16.4 Policy directions

A number of conclusions clearly emerge from the summary tables presented here and from the broader discussion throughout this book:

1   Most of the technologies needed to achieve levels of energy use in buildings consistent with sustainability and with stabilization of atmospheric CO_2 at 450ppmv already exist and are well understood, at least in some jurisdictions.

2   Although there is a role for technology transfer between so-called 'developed' and 'developing' countries (i.e. between OECD and non-OECD countries), there is at least as great a need for technology transfer *between* various OECD countries as well as *within* individual OECD countries (i.e. from the few architectural and engineering design firms that are on the cutting edge, to the rest).

3   A corollary of (2) is that there is no need to wait for technological development to progress before beginning ambitious programmes targeting deep reductions in energy use and CO_2 emissions in the buildings sector.

4   Although significant reductions (factors of two to ten) in energy use can be achieved in retrofitting existing buildings, it is generally not possible to achieve as low an absolute energy intensity as can be achieved in new buildings, and such reductions as can be achieved entail greater cost than if buildings are designed from the beginning to require minimal use of energy.

5   A corollary of (4) is that, by delaying programmes to dramatically reduce the energy intensity of new buildings, or the energy intensity of old buildings when they require major renovations, significant windows of opportunity will be lost (irreversibly in the case of new buildings).

The primary barriers to achieving deep (factors of 3–5) reductions in CO_2 emissions in new buildings are not technological (because technologies and, more so, system designs, already exist that can provide deep reductions) nor even economic (because fully integrated designs entail very little and often no additional upfront cost compared to current conventional practice). Rather, the barriers are behavioural in nature: they involve the fragmented nature of the design process and resulting lack of optimization, lack of awareness, time constraints during the design process, and an over-reliance on established ways of doing things. There is a widespread if not universal perception that low-energy buildings must, *of necessity*, entail greater capital costs. This leads to a lack of desire on the part of the client to have a low-energy building, and without a committed client, architects and engineers will usually not undertake the additional design effort required to produce low-energy buildings.

Lack of awareness of energy savings opportunities among practising architects, engineers, lighting specialists and interior designers is a major impediment to the construction of low-energy buildings. This in part is a reflection of inadequate training at universities and technical schools, where the curricula often reflect the fragmentation seen in the building-design profession. There is a significant need, in many countries, to create comprehensive, integrated programmes at universities for training future architects and engineers in the design of low-energy buildings, with parallel programmes at technical schools for training technical specialists. The value of such programmes would be significantly enhanced if they have an outreach component to upgrade the skills and knowledge of practising architects and engineers and to assist in the use of computer simulation tools as part of the integrated design process

The purpose of this book has been to assess how and to what extent energy intensity in new and existing buildings can be reduced. It is beyond the scope of this book to deal extensively with the policies needed to achieve these reductions. Nevertheless, it is appropriate to outline some of the key actions required:

1 Upgrading the teaching of building sciences at universities (in architecture and engineering departments), community colleges and vocational schools with, in particular, the creation of comprehensive, integrated programmes that combine all of the elements needed to create truly sustainable buildings.

2 Developing university-based outreach programmes to improve the design process among practising design firms (architectural and engineering)

3 Undertaking 'market transformation' programmes so that high-performance buildings, designed using the integrated design process, are increasingly what the market expects. This could entail financial support for high profile projects that demonstrate large savings at little to no incremental cost (many examples of which already exist and have been documented in this book), incentives to support the additional cost of design using the integrated design process, and training in the integrated design process.

4 Upgrading building codes, providing training in meeting upgraded building codes, and providing enhanced inspection ability to increase the degree of compliance with the upgraded building codes.

5 Providing financial support for continued improvement and reduction in the cost of a wide array of promising advanced technologies that further increase the potential for reducing the energy intensity of buildings, particularly of the large stock of existing buildings.

16.5 Is sustainability feasible?

A sustainability energy system is a system that can continue indefinitely, which implies that it is based entirely on renewable energy sources. Although the energy flow from the sun that is intercepted by the Earth is enormous, solar energy is diffuse, intermittent and not always available where or when it is needed. This places limits on the rate at which renewable energy sources can be supplied to major population centres. Current levels of per capita energy consumption in OECD countries are clearly not sustainable. Buildings themselves can serve as collectors and transformers of solar energy, but for this intercepted energy to meet a large fraction of the building energy needs (with the balance eventually provided by renewably based grid energy), total energy demand must be kept small.

Sustainable new buildings were defined in Chapter 1 (Section 1.4.3) as buildings that, among other things, achieve a factor of four to five reduction in energy intensity (energy use per unit floor area) compared to recent new buildings in OECD countries. This book has presented several examples of buildings that have achieved energy-intensity reductions of this magnitude (see Table 13.15), although factors of two to three reduction are more common. To achieve a factor of four to five reduction is pushing the energy-efficiency measures considered here to the limit. There is no room for half-measures if true sustainability is to be achieved. The criterion presented here for sustainability in new buildings assumes that the energy intensity of that portion of the existing building stock that is not eventually demolished can be reduced by about 40–60 per cent, which is readily achievable through extensive and comprehensive retrofits (as discussed in Chapter 14). If larger savings in the existing stock can be achieved through retrofits, then there will be less pressure to reduce the energy intensity of new buildings.

Thus, we may tentatively conclude that sustainability in the buildings sector is achievable, but it requires making use of every available opportunity to reduce the energy demands of buildings, and it requires that we begin immediately so as to reduce the buildup of inefficient building stock that serves as a future liability. This book is directed to those who are in a position to do that and, it is hoped, will contribute to this urgent process. The well-being of future generations, and the preservation of at least some portion of our priceless ecological and evolutionary heritage, is dependent on it.

Appendices

Appendix A

Units and Conversions

Prefixes

The meaning of the various prefixes used in dealing with energy, power, mass or length, are as shown in Table A.1.

Table A.1 Prefixes used with physical units in the metric system

Prefix	Factor	Common examples
nano	10^{-9}	nm (nanometre)
micro	10^{-6}	µm (micrometre)
milli	10^{-3}	mm (millimetre)
kilo	10^3	kW (kilowatt), km (kilometre), kg (kilogram)
mega	10^6	MW (megawatt), MJ (megajoule)
giga	10^9	GW (gigawatt), GJ (gigajoule), Gt (gigatonne)
tera	10^{12}	TW (terawatt), TJ (terajoule)
peta	10^{15}	PJ (petajoule)
exa	10^{18}	EJ (exajoule)

Fundamental units

All physical quantities not involving electric charge can be expressed in the fundamental units of distance, time and mass. In the metric system, two sets of units can be used: the mks system (metres, kilograms, seconds), and the cgs system (centimetres, grams, seconds). Any other unit can be expressed as a combination of the fundamental units of distance, mass and time. Some of the key derived units are shown in Table A.2.

Table A.2 Fundamental units in the mks metric system

Quantity	Definition	Units Metric	Units Fundamental
Velocity	change in distance/time	m/s	m/s
Acceleration	change in velocity/time	m/s^2	m/s^2
Force	mass × acceleration	Newtons	kg m/s^2
Energy, work done	force × distance	Joules	kg m^2/s^2
Power	energy/time	Watts (J/s)	kg m^2/s^3
Pressure	force/area	Pascals	kg/m/s^2

Conversion factors between and among metric and British units

Table A.3 Conversion factors between and among metric and British units

British to metric	Metric to British
Energy and power	
1 British Thermal Unit (Btu) = 1055Joules	1000Joules = 0.9486Btu
1 horsepower (Hp) = 746Watts	1000Watts = 1.341Hp
1 ton of cooling = heat removal at a rate of 3.516kW (3.516kW$_c$)^a	1kW$_c$ = 0.2844tons
1000Btu/h (used for heating) = 0.2931kW (0.2931kW$_{th}$)	1kW$_{th}$ = 3412Btu/h
1 kilowatt-hour (kWh) = 1000Watts × 3600 seconds = 3.6 megajoules = 0.0036gigajoules	
Pressure	
1 atmosphere (atm) = 101.325kPa = 1.01325bars = 14.696 pounds per square inch (psi)	
1Pa (Pascal) = 0.102mm water = 4.02 × 10^{-3} inches water = 7.5 × 10^{-3}mm Hg	

0.1IWC (inches water column) = 25Pa	1mb (millibar) = 100Pa = 0.402IWC
Distance, area, volume, mass and weight	
1ft (foot) = 0.30480m (metres)	1m (metre) = 3.28083ft = 39.37in (inches)
1mile = 5280ft = 1.6093km (kilometre)	1km (kilometre) = 0.6214 miles
1ft^2 = 0.092904m^2	1m^2 = 10.7638ft^2
1ft^3 = 28.32l (litres) = 0.02832m^3	1m^3 = 1000ll = 35.3145ft^3
1 Imperial gallon = 4.546l	1l = 0.2198 imperial gallons
1US gallon = 3.785l	1l = 0.2642 US gallons
1cfm (cubic foot per minute) = 0.4720l/s	1l/s = 2.119cfm
1 pound (a unit of weight, equal to 16 ounces) corresponds to 0.4536kg on Earth	1kg (a unit of mass) corresponds to 2.2046 pounds on Earth

ᵃ The term 'ton of cooling' originates from the time when cooling was measured in terms of the amount of cooling from the melting of a given amount of ice. A ton of cooling is the rate of cooling provided by melting 2000 pounds of ice over a period of 24 hours, which absorbs 12,000Btu of heat per hour.

Energy equivalents of heating fuels

1m^3 of natural gas = 36–39MJ (depending on composition)

1 litre of heating oil = 38.7MJ

Lighting

Luminous intensity is the 'force' that generates the light that we see, analogous to pressure or voltage. The metric unit is the *candela*, and British unit is the *candlepower*.

Historically, a source of 1 candela (1 cd) was defined as the light flux per unit of solid angle (steradian, sr) that is emitted from a surface area of 1/60cm^2 by a particular blackbody emitter-platinum at its melting point (2044.9K). The power output from this source is 1/683W per steradian. This is the origin of the constant 683 that appears in the definition of efficacy (the factor has changed as the calculated power of the light source has changed). Today, a candela is defined as the luminous intensity from a monochromatic source radiating at a wavelength of 0.555μm with a power output of 1/683 W/sr.

Luminous intensity times the solid angle gives *luminous flux*, which has units of *lumens* (lm). Thus,

$$\text{lumens} = \text{cd} \times \text{sr}$$
$$\text{candela} = \text{lm/sr}$$

Illuminance is the luminous flux crossing a plane per m^2, that is, lumens/m^2, and has units of *lux*. Given that a sphere represents a solid angle of 4π steradians, a source of 1 candela, radiating in all directions, produces 4π lumens (2π lumens in the half sphere). The luminous flux per m^2 on a sphere of radius 1m (area: 4πm^2) is therefore 1 lumen, and the illuminance on the sphere is 1 lux. The British unit of illuminance is the *footcandle*, equal to lumens/ft^2. Thus,

$$1 \text{ footcandle} = 10.764 \text{ lumens/m}^2 =$$
$$10.764 \text{ lux, because } 1\text{m}^2 = 10.764\text{ft}^2$$

Temperature

Temperature is not a fundamental quantity, but rather, is an index of heat content (heat being a form of energy). Temperature on the Celsius scale can be converted to temperature on the Fahrenheit scale as follows:

$$°F = °C \times 9/5 + 32$$

Conversion factors from frequently encountered Celsius temperatures to Fahrenheit temperatures are given in Table A.4. Note that, starting with 32°F at 0°C, the Fahrenheit temperature increases by 18 degrees for every 10 Celsius degrees.

Table A.4 The Fahrenheit equivalents of some Celsius temperatures that are frequently encountered in this book

0°C	32°F	26°C	78.8°F
10°C	50°F	30°C	86°F
16°C	60.8°F	40°C	104°F
20°C	68°F	55°C	131°F
22°C	72°F	80°C	176°F

Appendix B

Physical Constants

Below are the constants used in the calculation of sensible and latent heat, enthalpy and the emission of electromagnetic radiation.

Table B.1 Physical constants at 0°C and 1atm pressure (Standard temperature and pressure, STP)

Constant	Value at STP
Specific heat of water vapour, c_{pwv}:	1850J/kg/K
Specific heat of liquid water, c_{pw}:	4218J/kg/K
Specific heat of ice, c_{pi}:	2106J/kg/K
Specific heat of air, c_{pa}:	1005J/kg/K
Latent heat fusion of water, L_f:	0.3337×10^6 J/kg
Latent heat of condensation of water, L_c:	2.5008×10^6 J/kg
Stefan Boltzman Constant, σ:	5.670400×10^{-8} W/m²/K⁴

Source: Iribarne and Godson (1973)

Appendix C

Activities of the International Energy Agency Related to Buildings

The International Energy Agency (IEA) is an autonomous agency of the OECD (Organisation for Economic Co-operation and Development), based in Paris, that was established in 1974 in the wake of the first OPEC (Organization of Petroleum Exporting Countries) oil embargo. Its original goal was to deal with oil-supply disruptions, which is still one of its concerns. However, its mandate has been broadened to include programmes involving renewable energy and energy efficiency, motivated in part by concerns over climatic change. It has 26 member countries, which are listed in Table C.1. The IEA issues a biannual *World Energy Outlook*, regular reports on the energy policies of member countries and of some non-member countries, and annual energy statistics. It also develops cooperative relationships with non-member countries, industry and international organizations.

The IEA has coordinated a vast amount of collaborative research and dissemination of information pertaining to non-renewable energy, renewable energy and energy efficiency among its member countries. This work has taken the form of a number of *implementing agreements*, which fall into the following five categories:

1   information centres and modelling;
2   fossil fuels;
3   renewable energy;
4   energy end-use technologies;
5   fusion power.

Several programmes exist under each of these categories. Each programme in turn is divided into a number of *tasks* or *annexes*. Different groups of countries are involved in each of the tasks or annexes, through the involvement of national experts that may come from government research institutes, universities or industry. The major programmes pertaining to energy use in buildings are outlined below, while participating countries involved in some of the recent tasks or annexes under the programmes of interest are indicated in Tables C.1 and C.2.

Table C.1 Participation of IEA-member and non-member countries in various IEA implementing agreements and tasks or annexes. Years of accession to the IEA are given for countries that joined after 1974

Country	Solar Heating and Cooling, Tasks													Energy Conservation in Buildings and Community Systems, Annexes														
	18	19	20	21	22	23	24	25	26	27	28	31	5	26	27	28	30	31	32	33	34	35	36	37	38	39	40	42
Australia, 1979				X	x						X	X	X											x	x	x		
Austria			x	x		x					x	x										x	x		x	x		
Belgium				x			x	x		x	x	x					x		X			x			x		x	
(Brazil)											x																x	
Canada				x	x	x	x			x	x	x	X	x		x		x	x	x	x	x		x	x	x	x	X
Czech Republic, 2001																						X	x	X				
Denmark				X	x	x	x	x	x	x	x	x		x			x	x	x		x	X	x	x	x	x	x	x
Finland				x	x	x		x	x	x	x	x	x	x	x	x	x	x	x		x	x	x	X	x	x	X	x
France, 1992				x	x	x		x	x	x	X	x	x	x	x	x	x				x	x	x	x	x	x	X	x
Germany				X		x		X	x	X	X	x		x		x				x	x		X	x	x	x	x	x
Greece, 1977								x					x									x						
(Hong Kong, PRC)																						x					x	
Hungary, 1997																		x						X			x	
Ireland																												
Italy, 1978				x				x	x	x	x	x	x	x	x					x	x	x		x	x	x		x
Japan						x		x			x	x		x	x			x			x	x		x			x	
Korea, 2002																												
(Mexico)								x																				
Luxembourg																												
Netherlands				X			x	x	X	x	X	x	x	x	x	x		x	x	x	x	x		x	x	x	x	x
New Zealand, 1977				x		X				x	X	x		x				x										
Norway				x		X				x	X	x	x	x			x			x			x	x	x	x	x	
(Poland)																							x					
Portugal, 1981						x		x		x						x												
Spain						x		x																				
Sweden			X	x	x	x	X	x	x	X	X	x		X	X	x	x	x	x		x	x		x	x	x	x	x
Switzerland		X		x	x	x	x	x	X	x	x	x		x	x	x	x	x	x	x	x		x		X	X	x	x
Turkey, 1981																				x								
UK	X			x	X	x				x	x	x		x	x		x	x			X	x	x	x			x	x
US				x	X	x	x	x	x	x		x	x	X	x	x	x	x	x	x	x	x	x	x	X	x	x	x

Note: Countries in brackets are not IEA members but have participated in one or more tasks or annexes. Lead countries for individual tasks or annexes are indicated with X.

Source: Websites for the IEA and for individual implementing agreements

Table C.2 Participation of IEA-member and non-member countries in various IEA implementing agreements and tasks or annexes

Country	CADDET	PV power systems				Fuel cells	DSM			DHC	Heat pump programme								Energy storage					
		1	5	7	10		1	7	10		5	8	15	16	24	25	27	28	1	3	10	12	14	17
Australia, 1979	x	X	x	x	x																			
Austria		x	x	x	x	x							x	x				x						
Belgium								x												x				x	
Canada	x	x	x	x	x		x	x		x		x	x	x	x			x			x	x	x	
Denmark	x	x	x	x	x		x			x									x	x				
Finland		x		x		x		x	x	x	x								x	x	x			
France, 1992	x	x	x	x	x	x	x		x							X		x						
Germany		x	x	x	x	x				x	x	x						x	x		x	X		
Greece, 1977																								
Hungary, 1997																								
Ireland																								
(Israel)		x																						
Italy, 1978	x	x	x	x	x	x	x		x		x				x									
Japan	x	x	X	x	x	x	x		x	x			x	x	x		x	x		x		x		
Korea, 2002	x	x					x	x																
(Mexico)		x																						
Luxembourg																								
Netherlands	x	x	x	X	x	x	X	x	x	X				X	x	x		x						
New Zealand, 1977																								
Norway	x	x			x	x	x	x	x	x				x	x		X	x						
(Poland)																					x			
Portugal, 1981		x																						
Spain	x	x	x				x				x													
Sweden	x	x		x		x	X	x	x	X	x				X		x	x	x	x	X	x	x	X
Switzerland		x	x	x		x						x					x	X	X	X	x		X	
Turkey, 1981								X													x		X	
UK	X	x	x	x	x			x				x	x	x	x	x		x	x	x	x	x		
US	x	x	x			x	x	x		x		x	x	x	x	x		x	x	x	x			x

Note: Years of accession to the IEA are given for countries that joined after 1974. Countries in brackets are not IEA members but have participated in one or more tasks or annexes. Lead countries for individual tasks or annexes are indicated with X.

Source: Websites for the IEA and for individual implementing agreements

1 Information Centres and Modelling

The relevant implementing agreement in this category established the *Centre for the Analysis and Dissemination of Demonstrated Energy Technologies* (CADDET) in 1988 in Harwell, UK. In 1993 the agreement was expanded into two annexes: *CADDET Energy Efficiency* and *CADDET Renewable Energy* (both at www.caddet.org). National teams in each participating country collect information on the latest energy saving or renewable energy technologies used in their country and submit it to the appropriate centre, where it is pooled and analysed. Several different information products are produced at the two centres:

- *InfoStore* – a database on over 2000 full-scale energy demonstration projects (many of which involve buildings) from around the world
- Brochures – which provide more detailed information on selected items within the InfoStore database
- Reports – which focus on individual technologies
- *InfoPoint* – a quarterly newsletter, originally one 8-page newsletters, then two 8-page newsletters (one for each of the two CADDET branches), but recently combined into a single 16-page newsletter.

The vast majority of past newsletters and other information products are available online, many of which have been used in writing this book. In 1996, GREENTIE was added as a third annex. GREENTIE provides information on suppliers of greenhouse gas abatement technologies and services. It has 33 member countries, many of whom are not IEA members, including several large industrializing countries such as Brazil, China, India and Mexico. Its web site URL is www.greentie.org.

2 Renewable Energy

There are two implementing agreements under this category that are relevant to buildings:

Photovoltaic Power Systems (www.iea-pvps.org)

This implementing agreement has produced a large number of books (including design guides and source books of built examples), has sponsored conferences and has published conference proceedings. An 8-page newsletter (*PV Power*) is produced twice per year. Ten tasks have been completed or are underway. Those of greatest interest here are:

Task 1: Trends in photovoltaic applications in IEA countries (published annually)

Task 2: Database on the performance and cost of 431 monitored PV systems, distributed worldwide (updated June 2005)

Task 5: Grid interconnection of building-integrated and other dispersed photovoltaic power systems

Task 7: Photovoltaics in the built environment

Task 10: Large-scale application of grid connected photovoltaics in the built environment.

Task 7 sponsored a design competition, with winners announced in 2001. An international demonstration centre (open to the public) for photovoltaic building elements was established in Lausanne, Switzerland (www.demosite.ch), and a database of buildings with PV systems is maintained from The Netherlands (www.pvdatabase.com). *Designing with Solar Power* (Prasad and Snow, 2005) is an output of Task 7.

Solar Heating and Cooling (www.iea-shc.org)

Thirty-five tasks have been established so far under this implementing agreement. Among those of greatest interest here are:

Task 11: Passive and hybrid solar commercial buildings

Task 13: Advanced solar low-energy buildings

Task 14: Advanced active solar
systems

Task 16: Photovoltaics in buildings

Task 18: Advanced glazing materials
for solar applications

Task 19: Solar air systems

Task 20: Solar energy in building
renovation

Task 21: Daylighting in buildings

Task 22: Building energy analysis
tools

Task 23: Optimization of solar
energy use in large
buildings

Task 24: Active solar procurement

Task 25: Solar assisted air
conditioning of buildings

Task 26: Solar combisystems

Task 27: Performance of solar
façade components

Task 28: Solar sustainable housing

Task 31: Daylighting buildings in the
21st century

Task 32: Advanced storage concepts
for solar thermal systems in
low-energy buildings

Task 35: PV/Thermal systems

A large component of Task 23 was the development of an integrated design *process*, utilizing experts from 12 countries. An interactive CD was produced with all of the reports, design tools and software developed by the task members. Task 24 was directed toward facilitating cooperation among large buyer groups for solar water heating. Task 26 has produced an impressive and comprehensive set of technical documents, including design tools and a discussion of practical considerations, available at www. iea-shc.org/task26/index.html. Task 27 covers the full spectrum of window products, shading devices, transparent insulation, PV windows and solar collectors. Task 28 is directed toward helping participating countries achieve significant market penetration of sustainable solar housing by 2010, defined as housing with a factor of four reduction in heating energy use and a factor of two reduction in hot water energy use compared to national building codes. Interestingly, because current building codes tend to be more stringent in colder climates, application of these reduction

factors results in approximately the same absolute heating + hot-water energy use in mild, temperate and cold climates: about $60kWh/m^2/$ year $(0.22GJ/m^2/year)$.

A number of books have been published under this implementing agreement, including *Passive Solar Commercial and Institutional Buildings: A Sourcebook of Examples and Design Insights* (Task 11, Hastings, 1994), *Solar Energy Houses: Strategies, Technologies, Examples* (Task 13, Hestnes et al, 2003), *Photovoltaics in Buildings: A Design Handbook for Architects and Engineers* (Task 16, Sick and Erge, 1996), *Solar Air Systems: A Design Handbook* (Task 19, Hastings and Mørck, 2000), *Daylighting in Buildings: A Sourcebook on Daylighting Systems and Components* (Task 21, IEA, 2000a), *Solar-assisted Air-conditioning in Buildings: A Handbook for Planners* (Task 25, Henning, 2004a), and *Solar Heating Systems for Houses: A Design Handbook for Solar Combisystems* (Task 26, Weiss, 2003).

3 Energy End-Use Technologies

There are five implementing agreements of interest here:

Advanced Fuel Cells (www.ieafuelcell.com)

The task most relevant to this book is:

Task 19: Fuel Cells Stationary
Applications, which
includes subtasks
devoted to residential
and large buildings.

Energy Conservation in Buildings and Community Systems (www.ecbcs.org)

Under this implementing agreement, 43 tasks (called annexes) have been completed or are underway or ongoing. Among the completed tasks are:

Annex 26: Energy-efficient
ventilation of large
enclosures

Annex 27: Evaluation and
demonstration of
domestic ventilation

systems
Annex 28: Low-energy cooling
systems
Annex 29: Daylighting in buildings
(in partnership with Solar
Heating and Cooling
Task 21)
Annex 30: Bringing simulation to
application
Annex 32: Integrated building
envelope performance
assessment
Annex 33: Local energy planning
Annex 34: Computer aided
evaluation of HVAC
system performance
Annex 35: Control strategies for
hybrid ventilation in new
and retrofitted office
buildings (HYBVENT)
Annex 36: Retrofitting educational
buildings
Annex 37: Low-exergy systems for
heating and cooling
Annex 38: Solar sustainable housing
Annex 39: High-performance
insulation
Annex 40: Commissioning of
building HVAC systems
for improved energy
performance

Among the ongoing or incomplete tasks (as of
December 2005) are:

Annex 5: Air Infiltration and
Ventilation Centre
Annex 41: Whole-building heat, air
and moisture response
(MOIST-EN)
Annex 42: The simulation of
building-integrated
fuel cells and other
cogeneration systems
(COGEN-SIM)
Annex 43: Testing and validation
of building energy
simulation models
Annex 45: Energy-Efficient Future

Electric Lighting for
Buildings
Annex 46: Holistic Assessment Tool-
kit on Energy Efficient
Retrofit Measures for
Government Buildings
(EnERGo)

Annex 5 involves the establishment of a centre
with its own quarterly newsletter (*Air Information
Review*) and an extensive set of technical documents that can be accessed online (at www.aivc.
org). Many books on simplified and detailed design tools for low-energy cooling, and case studies
on operation and performance in practice, were
published by the British Research Establishment
(BRE) as part of Annex 28. As well, a number of
new detailed design tools for low-energy cooling
systems were developed under this annex, and are
outlined in Table C.3.

Annexes 5, 35, 40 and 42 have their own
websites (www.aivc.org, http://hybvent.civil.auc.
dk, www.commissioning-hvac.org, and http://co-
gen-sim.net, respectively). One of the most useful
books produced under this implementing agreement is *Hybrid Ventilation: State-of-the-art Review*
(Annex 35, Delsante and Vik, 2002). Many books
can be ordered through the ECBCS website.

Demand Side Management (http://dsm.iea.org)

This implementing agreement deals largely with
policies and programmes to encourage more efficient use of energy, and only secondarily with
technologies. A newsletter (*DSM Spotlight*) and a
quarterly news summary are produced. Among
the Tasks most relevant to this book are:

Task 1: International database
of energy efficient
demand-side management
technologies and
programmes (INDEEP)
Task 7: International collaboration
in market transformation
Task 10: Performance contracting
Task 12: Cooperation in energy
standards

Table C.3 Detailed design tools developed under Annex 28 of the IEA Implementing Agreement *Energy Conservation in Buildings and Community Systems*

Detailed design tool	Summary	Source code
Direct and indirect evaporative coolers	Models of direct and indirect evaporative coolers based on laboratory studies and published literature	FORTRAN
Evaporative cooling control strategy	Strategy for control of direct and indirect evaporative coolers in conjunction with heating and cooling coils	Turbo Pascal
Desiccant + evaporative cooling	Model of a desiccant dehumidification wheel based on manufacturer's data plus theoretical models of other system elements	DOE-2 Function
Desiccant cooling	Spreadsheet to simulate the performance of desiccant cooling systems based on manufacturer's isotherm performance characteristics	Excel Spreadsheet
Evaporative cooling post-processor for impact assessment	Spreadsheet for processing outputs from simulation to estimate the effect of introducing evaporative cooling on internal conditions, loads and energy and water consumption	Excel Spreadsheet
Water cooled slab	Theoretical models of a water cooled slab validated using monitored data plus a theoretical model of a wet cooling tower	FORTRAN
Air cooled slab	Theoretical resistive capacitance network model of a hollow core slab	Excel Spreadsheet
Night cooling control strategies – commercial	Three rule-based control strategies	–
Night cooling control strategies – residential	Estimation of ventilation rate as a function of free opening area, outdoor noise, security, occupation and solar shading using look-up tables based on monitored data	FORTRAN
Absorption cooling machine	Model of an absorption chiller based on manufacturer's data	Basic
Displacement ventilation	Multi-node air-flow model based on published data and simulation results	HSLIGHTS
Ground cooling – water	System performance information based on measured data	–
Ground cooling – air	Theoretical model of an air/ground heat exchanger validated using monitored data	Quick Basic

Source: Law (1998)

District Heating and Cooling (www.iea-dhc.org)

Work under this implementing agreement has been divided into a series of 3-year annexes, the most recent being Annex 7 (2002–2005). Among the projects most directly related to energy efficiency are:

Annex 4 (1993–1996):
• Integrating district cooling and combined heat and power
• Efficient substations and installation

Annex 5 (1996–1999):
• Optimizing operating temperatures
• Balancing the production and demand in CHP

Annex 6 (1999–2002):
• Optimization of a DH system by maximizing building system temperature differences
• Optimization of cool thermal storage and distribution

Annex 7 (2002–2005):
• Detailed comparison of large-scale and small-scale DHC systems
• Improvement of operational temperature differences in DH systems

Annex 8 (2005–2008):
• Cost benefits and long-term behaviour of a new all plastic piping systems
• Assessing the Actual Annual Energy Efficiency of Building-Scale Cooling Systems

- New materials and constructions for improving the quality and lifetime of district heating pipes including joints – thermal, mechanical and environmental performance.

Heat Pump Programme (www.heatpumpcentre.org)

The most significant activity under this implementing agreement has been the establishment of the *Heat Pump Centre* in Sittard, Netherlands (Annex 16), with its quarterly newsletter (*IEA Heat Pump Centre Newsletter*). This newsletter served as an invaluable source of information in writing the chapter on heat pumps in this book. Other annexes of particular interest here are:

Annex 5: Integration of large heat pumps into district heating and large housing blocks

Annex 8: Advanced in-ground heat exchange technology for heat pump systems

Annex 15: Heat pump systems with direct expansion ground coils

Annex 24: Absorption machines for heating and cooling in future energy systems

Annex 25: Year-round residential space conditioning systems using heat pumps

Annex 27: Selected issues on CO_2 as a working fluid in compression systems

Annex 28: Test procedures and seasonal performance calculation for residential heat pumps with combined space and domestic hot-water heating

Annex 30: Retrofit Heat Pumps for Buildings

Annex 32: Economical Heating and Cooling Systems for Low-Energy Houses

Energy Conservation Through Energy Storage (www.iea-eces.org)

Most of the activities under this implementing agreement pertain to storage of thermal energy, and so are applicable to buildings. Among the completed or ongoing annexes are:

Annex 1: Large-scale thermal storage systems evaluation

Annex 3: Aquifer storage demonstration plant in Lausanne Dorigny

Annex 10: Phase-change materials and chemical reactions in thermal energy storage

Annex 12: High-temperature underground thermal energy storage (HT UTES)

Annex 14: Cooling with thermal energy storage in all climates

Annex 17: Advanced thermal energy storage through phase change materials and chemical reactions – feasibility studies and demonstration projects

Annex 20: Sustainable Cooling with Thermal Energy Storage (ongoing as of Dec 2005)

Annex 14 has its own website (http://cevre.cu.edu.tr). This implementing agreement has produced a number of books and provides a significant amount of information on storage technologies and on successful applications. Works cited in this book are *Underground Thermal Energy Storage, State-of-the-Art 1994* (Annex 8, Bakema et al, 1995) and *High Temperature Underground Thermal Energy Storage, State-of-the-art and Prospects* (Annex 12, Sanner, 1999).

Appendix D

Building Simulation Software

The energy used for heating or cooling a building depends in part on the distribution of temperature within the building, as well as on the properties of the thermal envelope and the rate of air exchange between the inside and outside. The temperature distribution is important because the rate of heat flow across a window or wall, for example, depends on the difference in temperature between the window or wall surface and the air temperature adjacent to the surface. It also depends on the heat transfer coefficient, which in turn depends on the rate of air movement and the degree of turbulence. The detailed pattern of air movement depends of course on the mechanical ventilation system (if activated) but also on the temperature distribution itself. The temperature distribution depends on the air movement and on the temperatures on all the room surfaces (the boundary conditions). Thus, there is a feedback loop: the temperature distribution and boundary conditions depend in part on the air movement, which depend (in part) on the temperature distribution and boundary conditions. To compute this interaction for an entire building requires solving the equations of fluid dynamics and temperature on a fine grid that corresponds to the geometry of the building in question. This approach is referred to as *computational fluid dynamics* (CFD), and requires substantial computing power and memory. CFD is used for assessing whether a building will be comfortable (by computing temperature distributions) and can also be used to assess indoor air quality, by simulating the distribution of indoor pollutants along with temperature. However, the computed velocity fields do not depend strongly on the boundary conditions, so the heat fluxes – which are linked to the boundary condition and adjacent air temperature through the specified heat transfer coefficient – are not accurately modelled. For this reason, and because of its computer intensive nature, CFD has largely not been used for calculating the energy flows into or out of a building. As well, the effective use of CFD requires an expert user and a substantial amount of time to set up the problem (Maliska, 2001). The CFD solutions are not always stable and do not always converge to a consistent answer as the grid is refined.

Instead, a much simpler approach has been used. This involves treating the rooms in a building as a series of well-mixed boxes. 'Well-mixed' means that the temperature inside the box is assumed to be uniform (the same everywhere), so only one temperature needs to be computed for each box. The heat flow across each surface bounding each box is computed using Equation (3.1), based on the temperature difference across the surface and a specified resistance to heat flow. Heat sources such as lighting, fans and solar heat gain can be included, as well as the net heating or cooling due to differences in the ventilation supply and return air temperature and due to other heating or cooling systems. Changes in temperature over a given time period depend on the net of all the fluxes into or out of the box during that time period. This is referred to as an *energy simulation* (ES) approach, and is the approach that is almost always used when computer simulation models are used as part of the design process. However, ES programs cannot accurately predict energy use for systems that produce non-uniform temperatures within the building, such as displacement ventilation systems. Furthermore, most ES programs cannot provide information on the airflow entering a building through natural ventilation, but such airflow is important for predicting the interior air temperatures and the heating or cooling loads in naturally ventilated buildings.

With the growth in the power of desktop computers, efforts have been made to couple CFD and ES models (Maliska, 2001; Zhai et al, 2001; Zhai and Chen, 2004). The key challenges are the accurate computation of the internal heat transfer coefficients and the calculation of highly resolved flows near walls. The heat transfer coefficient is the link between the CFD and ES calculations. The existing tools that integrate CFD and ES tools rely largely on empirical correlations for computing the heat transfer coefficients. However, the available correlations have large disagreements among themselves. The airflow (i.e. the CFD part of the calculation) responds to changes in the boundary conditions on a timescale of seconds, so these calculations must be performed with time steps of a few seconds length. Heat flow through the thermal envelope, on the other hand, is a slow process, so envelope temperatures (one of the CFD boundary conditions) respond with a timescale of hours. The most computationally efficient way to couple CFD and ES programs, therefore, is to run the CFD program to a steady state for a given set of boundary conditions, then to hold the flow field constant while running the ES component long enough for a noticeable change in the boundary conditions to occur, then computing a new flow field, and so on. The CFD simulation is thus a series of quasi-steady states. The CFD model can be used to compute the interior heat flux coefficient, and the resulting value can be different from that assumed in the ES model. In one case where a CFD and ES were coupled together, Zhai et al (2001) found a discrepancy of 9.4 per cent, with the CFD model giving a larger heat transfer coefficient and hence a larger heating load than the original ES model.

For the new National Library Board building in Singapore, two different hybrid passive/mechanical ventilation schemes that relied on an atrium and wind for the passive component had to be compared. CFD programs were used to simulate the velocity and temperature fields in the building, with the velocities at the edge of the building (a required input to the CFD simulation) determined by constructing a physical model of the building and its surroundings and experimenting with it in a wind tunnel (Lam et al, 2001).

For a detailed discussion of numerical methods used in writing building simulation programs (that is, how the mathematical equations that the program uses are translated into computational procedures that can be carried out on a computer), see Clarke (2001).

Examples of energy simulation tools

Crawley et al (2005) have recently published a comprehensive comparison of the simulation capabilities of 20 building energy simulation programs. Table D.1 compares the simulation capabilities of the seven programs with the greatest number of energy efficiency and renewable energy options. There is considerable ambiguity as to what the tools can do in reality, with many nuances of 'capability' that will require further work to resolve. Thus, Table D.1 only provides a rough indication of what the models can do. All of the models listed in Table D.1 are compatible with some other software tools, have fully interactive interior conditions and HVAC systems with floating indoor temperatures, account for adsorption/desorption of water vapour by building materials, and have built-in weather data but can also read weather data in a variety of formats. A brief description of each model is given below.

Table D.1 Comparison of some of the leading building simulation tools in terms of building properties affecting energy use that the user can specify (such as envelope properties and equipment efficiency) or in terms of energy saving features that can be incorporated

Possible model feature	Simulation tool					
	BSim	Energy Plus	ESP-r	IDA ICE	IES <VE>	TRNSYS
Building envelope						
Envelope reflectivity	x	x	x	x	x	x
Envelope insulation levels	X	x	x	x	x	x
Envelope leakage	x	x	x	x	x	x/o
Green roofs	?		e	?	?	o
Dynamic insulation	?		e	?	?	e
Moveable transparent insulation	x	x	x	x	x	x
Radiant barriers	?		e	?	?	x
Window U-value	x	x	x	x	x	x
Window SHGC	x	x	x	x	x	x
Electrochromic glazing		x	x	x	x	x
Thermochromic glazing			x	x		e
Passive and low-energy heating and cooling						
Internal thermal mass	x	x	x	x	x	x

Trombe walls		x	x	x	x	x
Phase change materials			i	o		e
Transpired solar walls			x			x
Airflow windows		x	x	o	x	x
Controllable window blinds	x	x	x	x	x	x
Blinds between glazings		x	x	x	x	x
Natural ventilation	x	x	x	x	x	x
Hybrid ventilation	x			x	x	x
Night-time ventilation		x	x	x	x	x
Air-side economizer		x	x	x	x	X
Earth pipe coupling		x	x			X
Direct evaporative cooling		x				O
Indirect evaporative cooling		x				O
Water side economizer		x		x		x
Pond heat exchanger		x				x
Heating and cooling equipment						
Boiler		x	i	x	x	xo
Ground source heat pump		x	r	o	x	x
Centrifugal chiller with VSD		x				o
Steam absorption chiller		x				o
Hot-water absorption chiller		x				x
Accounts for part-load performance of heating and cooling equipment		x	x	x	x	x
HVAC system						
Chilled ceiling cooling	?	x	x	?	?	x
Radiant heating	?	x	x	?	?	x
Displacement ventilation		e	e	x	x	o
Demand-controlled ventilation	x			x	x	x
Solid desiccant dehumidification		x	e			x
Water/ice diurnal thermal storage		p				xo
Underground thermal energy storage						o
Variable-speed pumps		x	x	x	x	x
Variable-speed fans	x	x	x	x	x	x
Centrifugal chiller with VSD		x				o
Variable-speed cooling tower fans and pumps		x	x			X
Duct leakage		e	x	x	x	x
Duct insulation		e	x	x	x	x
Pipe insulation	?		x	?	?	
Heat recovery ventilation	x	x	x	x	x	x
Daylighting						
Computes interior solar illumination	x	x	x	x	x	
Glare simulation and control		x	x		x	
Stepped or dimmable electric lights	x	x	x	x	x	x
Light shelves	x	x	x		x	x
Passive light pipes		x	x		x	
Active solar energy						
Solar thermal collectors		x	x		x	x

Complex arrangement of storage tanks						x
Integrated collector/storage tank						x
Building-integrated PV	x	x	x			x
Economic analysis						
Simple energy and demand charges	x	x	x	x	x	x
Complex energy tariffs	x	x			x	e
Estimate of upfront capital cost		x			x	

Note: x = standard feature, o = optional feature, e = feature requiring expert knowledge, r = feature intended for research use only, i = feature requiring difficult-to-obtain input data, p = partial.
Source: Crawley et al (2005)

BSim (www.bsim.dk) was developed by the Danish Building Research Institute and has been used extensively over the past 20 years, largely in Denmark. Its strengths lie in the analysis of the distribution of daylight in rooms, in the analysis of natural ventilation and in calculating the electrical yield of PV systems. It computes transient indoor temperature and humidity conditions, taking into account the role of the building and its furnishings as a thermal and moisture buffer. Simulations of the penetration of sunlight into the building can be saved as animations of the movement of sunspots in the building spaces. It is available at a cost of about 2680€ for one to five users, plus an annual fee of about 670€. A demonstration version is available for free.

EnergyPlus (www.eere.energy.gov/buildings/energyplus) merges and improves the DOE-2 and BLAST models, two widely used programs that had been developed by the US Department of Energy and Department of Defence, respectively. In these models, heating or cooling loads were calculated on each time step (typically every 15 minutes) using a heat balance module. The loads were then passed to the mechanical systems module to calculate energy use by the mechanical systems. However, the earlier models assumed that the output from the mechanical system is able to meet the calculated loads, so there is no feedback between the heat balance and mechanical system modules in terms of loads not met. This deficiency is corrected in EnergyPlus: the output from the mechanical systems modules it used to calculate the change in temperature and humidity over the time step, and the simulated temperatures and humidities at the end of the time step affect the load calculated at the beginning of the next time step (Crawley et al, 2001). EnergyPlus contains airflow (COMIS) and fenestration (WINDOW-5) modules. The latter allows the user to enter a layer-by-layer description of a window or to choose from a library of windows. Moveable interior and exterior shades and electrochromatic glazing can be simulated (Winkelmann, 2001). It was used in the design of the passively ventilated Federal Building in San Francisco (Haves et al, 2004) that is discussed in Chapter 6 (Section 6.3.1 and 6.3.4). It is available for free but requires a high degree of expertise to use. Free weather data from about 1150 locations worldwide (as of December 2005) can be downloaded for use in the model.

ESP-r (Environmental Systems Performance – research) (www.esru.strath.ac.uk/Programs.ESP-r.htm) was developed by the Energy Systems Research Unit (ESRU) of the University of Strathclyde, Glasgow over a period of 25 years. It has been extensively used for research purposes, but also as a consulting tool by some architects and engineers. It was used as the starting point in the development of the HOT2000 and later models used as the analysis tool in the Canadian EnerGuide-for-Homes program (Haltrecht et al, 1999). The ESRU website has a tutorial that guides the user through the main features of the program as part of the ESRU's online courseware. It operates on UNIX systems, is written in FORTRAN and is available for free but requires some expertise to use.

IDA Indoor Climate and Energy (IDA ICE) (www. equa.se/ice) is a commercial product developed by a European industrial consortium and released in 1998. Its special strengths include features common in northern European buildings, such as displacement ventilation, active chilled beams, and water- and air-based slab heating and cooling systems. Another strength is the realistic modelling of controls. It is available at four levels: a free online version that leads the user through the steps in building a model from component modules; a standard version; an advanced version where the user can browse and edit the mathematical equations that represent the model; and a programmable version for those who would like to develop further program components.

IES <Virtual Environment> (IES <VE>) (www. iesve.com) permits modelling of natural ventilation and infiltration, as well as detailed shading and solar penetration analysis, and detailed design and simulation of electric lighting systems. It combines many design-oriented capabilities within a single package. It is a commercial product with reduced rates for academic and student users.

TRNSYS (TraNsient SYstem Simulation) (http:// sel.me.wisc.edu/trnsys) was first released commercially in 1975 by the Solar Energy Center at the University of Wisconsin. Version 16 (TRNSYS 16) was released in 2004. A large number of groups have contributed modules (written in FORTRAN) for individual building components, which can then be linked together within the TRNSYS modelling framework. However, FORTRAN programming knowledge is not required if the standard package is used. As of 1999, over 300 institutes and consulting firms worldwide were equipped with the then-latest version of TRNSYS (Pelletret and Keilholz, 1999). TRNSYS is much more difficult to use than EnergyPlus or Energy-10, but also permits the simulation of many features that cannot be simulated with the latter models, including (Hiller et al, 2001):

- heating and cooling radiant slabs;
- double-skin façades and airflow windows.

TRNSYS can call external programs and external DLLs (Dynamic Link Libraries), and can read ASCII input files. It contains a standard library of window and glazing system data, including angular-dependent transmission, reflection and absorption. It sells for $4000 to commercial users and $2000 to educational institutions, although a free demonstration diskette is available.

Also worthy of mention is the Energy-10 model (www.sbicouncil.org) that was developed by the National Renewable Energy Laboratory (NREL) in Golden, Colorado. Although less capable than the models mentioned above in terms of the range of features that it can simulate, it can be used to evaluate BiPV options based on simulated building electrical loads and PV performance for each hour of the year (Walker et al, 2003). The PV component uses a module from TRNSYS. Various routine energy efficiency strategies can be assembled and automatically compared with a baseline design. The software and manual are available for $50, and the package can be used be a novice.

For detailed computations of heat flow through complex building envelopes, THERM can be used. This is a freely available (from http://windows.lbl.gov/software), user-friendly, two-dimensional heat transfer model that can be used for calculating the impact of thermal bridges in building elements such as windows and doors, and for computing the net radiative energy transfer associated with projecting building elements, where self-viewing reduces the heat loss (by about 5 per cent averaged over a window and frame) (Curcija et al, 2001).

A hierarchy of models is available for free from Lawrence Berkeley National Laboratory (US) for detailed computations of the optical properties of window glazing materials (Optics5), for the computation of the optical and thermal properties of window systems (WINDOW), and for computing heating and cooling energy use associated with windows in residential buildings (RESFEN). These can be obtained from http:// windows.lbl.gov/software. A glazing library (http://windows.lbl.gov/materials/optical_ data/default.htrm) contains optical data for 1153 glazing products that can be used with the WINDOW program. FRAMEPlus is a free Windows-based package available from Natural Resources

Canada (www.frameplus.net) to compute the thermal properties of windows, doors and other building envelope elements. It permits a very high degree of user customization. WIS (Advanced Window Information System) is a free software tool developed in Europe that can be used to compute the thermal and optical properties of windows. It contains databases with component properties. Unlike the US and Canadian tools, the effect of various shading systems (such as slatted blinds and louvres, pleated binds, roller blinds and diffusing panes and other scattering devices not belonging to one of the other categories) on daylight and heat gain can be simulated, as well as airflow windows. It can be accessed from www.windat.org.

A number of tools are also available for the calculation of the distribution of daylight within buildings. These range from simple, free and Excel-based packages (such as SkyCalc, available from www.energydesignresources.com), to commercial windows-based products (such as Lumen Micro, www.lighting-technologies.com, and Lightscape, www.lightscape.com), to the UNIX-based RADIANCE software that can be used for research purposes (developed at Lawrence Berkeley National Laboratory and available for free to non-commercial users from http://radsite.lbl.gov/radiance/framew.html).

Validation of building simulation models

'Validation' is the process of verifying that a model produces an accurate simulation. There are three types of validation for building-simulation models (Neymark et al, 2001):

1   *Empirical Validation.* This involves comparing the model results with measured data from a real building. For the test to be valid, the inputs (such as envelope properties) to the model must conform to the properties of the building against which the model is compared.

2   *Analytical validation.* The output of a numerical (computer-based) model are compared against an analytical (that is, mathematically derivable) solution for an isolated heat transfer mechanism under very simple and highly defined conditions.

3   *Comparative testing.* The results of a model are compared against other, perhaps more detailed models, which have been validated in some way. The comparison might be done for situations more complex than those under which the reference model was validated.

ASHRAE has developed Standard 140-2001, which provides a standard method for testing building simulation software. As part of Task 22 of the IEA Solar Heating and Cooling Programme, researchers at NREL developed the Building Energy Simulation Test and Diagnostic Method (BESTEST). A number of test cases were developed for building envelopes, with input and verification data, against which model simulations can be compared (i.e. using the third validation method, listed above). As part of Task 22, energy use as simulated by eight different simulation programs for a variety of buildings and building system components were compared. During the course of the testing, a number of errors were found in the various programs, and corrected. Several hundred copies of the test procedure have been distributed to energy software developers, energy standard making organizations, researchers and universities, where it is being used as a teaching tool for simulation courses. The original test procedure has been extended for testing mechanical system simulation models (Neymark et al, 2001). As with the envelope BESTEST protocol, a number of errors where found in mechanical system models (causing errors in calculated energy use of up to 50 per cent) and subsequently corrected. The mechanical system test includes some analytical solutions.

CAD-Energy simulation model interfaces

Three-dimensional CAD (Computer-Assisted Design) software is used in the architectural profession to define building geometry and as a visualization tool. Energy simulation (ES) software is available to analyse energy use in buildings, as outlined above. Efforts are underway to develop tools that couple CAD and ES software. Prominent among these is SimCAD, which is an object-oriented CAD tool designed for generating building description data that can be imported into TRNSYS (Pelletret and Keilholz, 1999; Hiller et al, 2001). It has been written in C⁺⁺. The National Renewable Energy Laboratory (Golden, Colorado), Lawrence Berkeley National Laboratory (Berkeley, California), and Pacific Gas & Electric (California) are working together to bring the powerful RADIANCE daylight simulation capabilities into practical CAD software.

Appendix E

Web Sites Pertaining to High-Performance Buildings

This appendix lists websites from which documents in English pertaining to high-performance buildings, and potentially of interest to academics and professionals, can be downloaded at no cost or can be purchased. Not included on this list are many local and regional associations that act to promote more efficient buildings by, among other things, providing reports and brochures aimed at the general public.

Government Research Centres and Information Services

Advanced Buildings, Technologies and Practices. This is a website established by Natural Resources Canada (NRCan), Public Works and Government Services Canada, and the Canadian Mortgage and Housing Corporation (CMHC). Intended for architects, engineers and building managers, it provides information on over 90 energy saving technologies, generally under the categories of definition, description, benefits, limitations, applications, experiences, example buildings, costs, example commercial suppliers, contact information for the contributing expert and sources of further information. Website: www.advancedbuildings.org.

Buildings Research Group, Natural Resources Canada. Developed and administers the commercial C-2000 program and building code, the residential R-2000 program, the EnerGuide-for-Homes rating system, and carries out research involving computer simulation of buildings, advanced windows, advanced insulation techniques, and energy efficient HVAC systems for houses and commercial buildings. Numerous research reports can be freely downloaded. Website: www.buildings-group.nrcan.gc.ca.

Canadian Mortgage and Housing Corporation (CMHC). Conducts research covering a wide range of topics pertaining to energy efficiency and durability of single-family residential, multi-unit residential and commercial buildings, with numerous downloadable reports of interest to readers of this book. Many reports focus on construction details needed to ensure that envelope systems perform as designed. Website: www.cmhc-schl.gc.ca.

California Energy Commission. A source of numerous reports covering many facets of energy efficiency in buildings. Includes a link to the *Public Interest Energy Research* (PIER) program. Website: www.energy.ca.gov.

Carbon Trust. The Carbon Trust is an independent, non-profit company set up and funded by the UK government. It holds the largest online collection of free publications on energy efficiency in the UK. The August 2005 publication list contains 784 titles covering energy efficiency in all sectors, including buildings. Website: www.thecarbontrust.co.uk.

Federal Technology Alerts (FTAs). These contain up-to-date information on US-developed emerging technologies in the areas of energy efficiency, water conservation and renewable energy; their applications; and a list key of contacts and information sources pertaining to the technologies. Website: www.pnl.gov/fta.

Fraunhofer Institute for Solar Energy Systems (Freiburg, Germany). This is the largest solar energy research institute in Europe, with a staff of about 400. Building-related research includes advanced window films and solar retrofits of multi-unit residential buildings. It is associated with the University of Freiburg. Web site: www.ise.fhg.de.

Lawrence Berkeley National Laboratory (LBNL) (US). Located near San Francisco, LBNL focuses on advanced windows, computer and office equipment and household appliances, advanced lighting systems and daylighting, mitigation of the urban heat island, minimization of energy loss due to infiltration and leaky duct systems, and building simulation. Loosely affiliated with the University of California, Berkeley, this national lab dominates US research pertaining to energy efficiency in buildings. Numerous research reports can be freely downloaded. Website: http://eetd.lbl.gov and http://buildings.lbl.gov.

National Renewable Energy Laboratory (NREL) (US). Located in Golden, Colorado, NREL is active in the *Building America Program* (which produces houses that use up to 70 per cent less energy than under standard practice), high-performance commercial buildings, desiccant cooling and dehumidification, BiPV, electrochromatic windows and passive solar buildings. Also codeveloped the *Energy-10* computer simulation software and the *BESTEST* method for evaluating building energy simulation software. The website contains many technical reports and a library of photographs that can be downloaded covering all forms of renewable energy and its use in buildings. Website: www.nrel.gov.

Oak Ridge National Laboratory (ORNL) (US). Located in Oak Ridge, Tennessee, ORNL focuses on cogeneration, ground source heat pumps, testing and development of desiccant dehumidification systems, development of energy efficient refrigerator/freezer units, development of 'drop-in' heat pump water heaters to replace existing water heaters, and development of hybrid PV/daylighting systems. Numerous research reports can be freely downloaded. Website: www.ornl.gov/sci/eere/buildings.

US Department of Energy. This site is the gateway to programs and information involving energy efficiency and renewable energy provided by the US Department of Energy. It contains a worldwide directory of almost 300 building software tools, and links to the EnergyPlus and other DOE-sponsored tools. Website: www.eere.energy.gov/buildings.

Industry and Trade Associations, Advocacy Groups

American Council for an Energy Efficient Economy. Publishes numerous detailed studies, reports and fact sheets pertaining to energy efficiency in all sectors of the economy, as well as listings of the most efficient models for furnaces, boilers, air conditioners, various household appliances and office equipment in the North American market. Website: www.aceee.org.

Building Research Establishment (BRE). BRE is a research establishment wholly owned by the BRE Trust, a registered UK charity whose members consist of firms, professional bodies and other groups related to the design, construction and operation of buildings. BRE created the Building Research Establishment Environmental Assessment Method (BREE-AM) rating system for buildings. Publications are available for purchase only. Website: www. bre.co.uk.

Consortium for Energy Efficiency. An advocacy group with a number of publications pertaining to energy efficient equipment in buildings. Website: www.cee1.org.

Energy Design Resources group. This is a consortium of California energy utilities that maintains an outstanding website with many (generally 28-page) design briefs, design guideline source books, and case studies, along with free software (eQuest, eVALUator, and SkyCalc). Website: www.energydesignresources.com.

European Solar Thermal Industry Federation. Covers heating and cooling applications of solar thermal energy, with market data and position papers. Website: www.estif.org.

International Building Performance Simulation Association (IBPSA). Organizes bi-annual international conferences pertaining to building simulation. Information on local conferences by regional chapters is also available. Conference proceedings can be downloaded. Website: www.ibpsa. org.

International Initiative for Sustainable Built Environment (IISBE). With an international board of directors from almost every continent and a small secretariat in Ottawa, this site provides links to *Advanced Building News* (the IISBE Newsletter), the *Sustainable Buildings Information System* (which provides access to 1200 downloadable documents), the *Green Building Challenge* (an international project to develop the theory and practice of environmental assessment of buildings), the *Green Building Policies Network*, *GBTool* software (designed to provide a method of assessing building performance across national boundaries in widely differing site conditions, using regional benchmark comparisons without bias to any particular climatic, economic or cultural context), and information about past and future IISBE conferences. Website: www.iisbe.org.

National Alliance of Home Builders Research Center Toolbase Services. Provides timely information on new energy-efficient technologies. Website: www.toolbase.org.

New Buildings Institute (California). Promotes demonstration projects, and has produced a number of detailed guidelines pertaining to various elements of high-performance buildings that can be freely downloaded. Website: www.newbuildings.org.

SolarBuzz, San Francisco (US). Provides up-to-date (within one month) retail costs of solar modules, invertors and electronic controls in the US and Europe; lists of suppliers in the US and Europe; and provides information on funding programmes in the US and Europe. Website: www.solarbuzz.com.

Sustainable Buildings Industry Council, formerly the *Passive Solar Industries Council*. The developer of the Energy-10 Software tool and a source of occasional valuable reports that can be downloaded or purchased. Website: www.sbi-council.org.

Usable Buildings Trust (UBT). The Usable Buildings Trust is an independent charity, registered in the UK, that promotes better buildings through the more effective use of feedback on how they actually work. UBT is also a home for approaches that are not quite ready for widespread application and an incubator for their development. Has several downloadable reports. Website: www.usablebuildings.co.uk.

Research Institutes Affiliated with Universities

Centre for Energy Policy and Economics (CEPE). Part of the Swiss Federal Institutes of Technology, CEPE has a large number of downloadable research reports, several of which pertain to buildings. Website: www.cepe.ch.

Florida Solar Energy Center. Affiliated with the University of Central Florida, this centre performs research related to the hydrogen economy, buildings (primarily dealing with reducing cooling loads in hot-humid climates), BiPV and building solar thermal energy. Includes the Building Design Assistance Center, directed toward architects and engineers. Numerous reports, including the results of laboratory and field tests, are available. Website: www.fsec.ucf.edu.

Lighting Research Center, Rensselaer Polytechnic Institute. Contains an extensive bibliography of papers published by research staff at the Lighting Research Center, although they are not available for download. Short technical facts sheets and past newsletters are available. Website: www.lrc.rpi.edu.

Swiss Federal Institute of Technology, Lausanne, Switzerland. The *Solar Energy and Building Physics Laboratory* has been on the cutting edge of work on solar architecture and passive ventilation. Website: http://lesowww.epfl.ch.

University of California Energy Institute (UCEI). Has a broad spectrum of downloadable research reports, including several relevant to buildings. Website: www.ucei.org.

Private Research Companies and Associations

Architectural Energy Corporation (Boulder, Colorado). Has produced many 28-page design briefs pertaining to the efficient use of energy in buildings, case studies and research reports that can be freely downloaded. Many of these documents were produced for the *EnergyDesignResources* group. Website: www.archenergy.com.

Athena Institute (Merrickville, Ontario, Canada). Provides software, databases and publications pertaining to lifecycle assessments of the environmental impacts and embodied energy of materials accounting for 90–95 per cent of the structural and envelope systems in residential and commercial buildings. Quantitative data are applicable to various regions in Canada and the US, but background information and methods are generally applicable. Website: www.athenasmi.ca.

Caneta Research Inc (Mississauga, Ontario, Canada). Performs research, develops software and carries out building energy simulations. Has a particular expertise in ground source heat pumps. Software tools for sizing ground source heat pumps have been developed and can be purchased, as well as a number of reports pertaining to heat pumps. Website: www.canetaenergy.com.

Ecofys (The Netherlands). Performs research pertaining to renewable energy and energy efficient buildings, much of which can be downloaded. Website: www.ecofys.com.

Energy Center of Wisconsin (Madison, Wisconsin). Commissions and publishes research that includes a number of topics relevant to buildings, much of which can be downloaded. Website: www.ecw.org.

Enermodal Engineering (Waterloo, Ontario, Canada). Performs research, develops software and carries out building energy simulations. Has a particular expertise in windows, and is involved in design projects pertaining only to low-energy buildings. Reports on individual design projects can be downloaded. Website: www.enermodal.com.

Passivhaus Institut (Darmstadt, Germany). Founded in 1996 as an independent research institution, the Passivhaus Institut provides consulting to architects and engineers on the design of passively heated buildings, provides support to manufacturers for the development and optimization of components suitable for passive construction, and provides energy and CFD simulations. A worksheet in German to compute the heating requirements of passive houses can be purchased for 87€. Website: www.passivehouse.com.

The Twenty-First Century Research (21CR) (Arlington, Virginia). 21CR is a private–public sector research collaboration of the heating, ventilation, air-conditioning and refrigeration (HVAC/R) industry, with the mission to identify, prioritize and undertake precompetitive research that focuses on decreasing energy consumption, increasing indoor environmental quality and safeguarding the environment. It is associated with the Air Conditioning and Refrigeration Technology Institute, a non-profit organization established in 1989 and based in Arlington, Virginia. Contains numerous research reports that can be freely downloaded. Website: www.arti-21cr.org.

Valentin Energy Software (Berlin, Germany) provides software packages for the design of solar thermal systems (T*Sol), PV systems in buildings (PV*Sol) and cogeneration systems. Demonstration versions can be downloaded for free. It also provides a meteorological database package (Meteonorm) with data from 2400 stations worldwide in a format that can used by T*Sol and PV*Sol. Website: www.valentin.de.

Appendix F

University Programs Pertaining to Low-Energy Buildings

Based on the affiliations of the first authors of the scientific papers, reports and books cited in this book, the leading university programmes pertaining to energy use in buildings have been identified. Information from the web pages of universities with programmes on energy use in buildings or where there is some integration of architecture and engineering is summarized below, in alphabetical order by country. In addition to the universities identified below, there are many universities where individual researchers have made important contributions to the scientific literature on buildings. Important work is sometimes done by university-based researchers on a contractual basis for government research centres, work that does not appear in the open, peer-reviewed literature and therefore has not been used in compiling the list given below.

Concordia University, Montreal, Canada: The *Centre for Building Studies* has courses and research on computer simulation, HVAC systems, and the use of solar energy. Website: www.bcee.concordia.ca/cbs/Home.htm.

City University of Hong Kong, Hong Kong, China: The *Department of Building and Construction* does work on computer simulation of buildings, HVAC systems, and solar radiation and daylighting. Website: www.bc.cityu.edu.hk.

Hong Kong Polytechnic University, Hong Kong, China: The *Department of Building Services Engineering* is said to be the largest academic unit in the world devoted exclusively to building services engineering, with research in building energy simulation, solar-powered dehumidification and cooling systems, energy efficient mechanical ventilation, and BiPV. Website: www.bse.polyu.edu.hk.

Technical University of Denmark, Lyngby, Denmark: The former *Department of Buildings and Energy* has been rolled into the Department of Civil Engineering, while the *International Center for Indoor Environment and Energy* has been established. Research has focused on passive and active solar heating, thermal energy storage, windows and solar shading devices, the integrated design of building envelopes and services, and the design process. Website: www.adm.dtu.dk.

Dublin Institute of Technology, Dublin, Ireland: The *Department of Building Services Engineering* in the *Faculty of the Built Environment* offers an integrated programme on buildings and energy use in buildings. Website: www.dit.ie.

University College, Dublin, Ireland: The *Energy Research Group* in the *School of Architecture* does research and consulting pertaining to passive solar heating, passive ventilation, daylighting and integrated design. The group explicitly tries to integrate architects, engineers and urban planners. A mid-career teaching package pertaining to office equipment, passive solar energy and natural ventilation is available over the internet. Website: http://erg.ucd.ie.

Kumamoto University, Japan: Like many universities in Japan and Korea, Architecture and Civil Engineering are often combined into their own department within the Faculty of Engineering. The *Department of Architecture and Civil Engineering* here offers students two streams: one combining architecture and engineering, the other combining architecture and urban/regional planning. 'Harmony of interactions' is stressed as a design principle in both cases in the website information. Website: www.civil.kumamoto-u.ac.jp.

Nagoya University, Japan: Here, the *Department of Architecture* is within the *School of Engineering*. Some work has been done on solar assisted district heating. Website: www.nuac.nagoya-u.ac.jp.

Norwegian University of Science and Technology, Norway: Here, the *Department of Building Technology* is within the *Faculty of Architecture*. Although not evident from the web pages (which are largely in Norwegian), important work on low-energy and solar housing has been done here. Website: www.ntnu.no/arkitekt/eng/.

National University of Singapore, Singapore: The *Department of Building, School of Design and Environment* and the *School of Building and Estate Management* carry out research pertaining to energy-efficient cooling and dehumidification in tropical climates, energy efficient HVAC systems, rooftop gardens, and the integrated design process. Website: www.bdg.nus.edu.sg.

Chalmers University of Technology, Gothenburg, Sweden: The *Department of Building Services Engineering*, with a staff of 20, does work in district solar heating, daylighting, evaporative and desiccant cooling, building retrofits, and computer simulation of buildings. Website: www.hvac.chalmers.se.

Lund Institute of Technology, Lund, Sweden: The *Department of Building Science* has a number of projects pertaining to solar energy in buildings and the design of low-energy buildings. Architecture is a division within this department. Website: www.bkl.lth.se.

University of Applied Science, Horw, Switzerland: The *School of Engineering and Architecture* does work on energy efficient HVAC systems. Website: www.hta.fhz.ch.

King Mongkut's University of Technology, Thonburin, Thailand: The *Building Scientific Research Center* and the *Energy Technology Division, School of Energy and Materials* has done work on passive ventilation and cooling in hot and humid climates. Website: www.arch.kmutt.ac.th..

Loughborough University, UK: The *Centre for Renewable Energy Systems Technology (CREST)* seems to offer one of the most comprehensive programmes (at the MSc level) on renewable energy anywhere. Research related to the direct use of renewable energy in buildings involves integrated solar facades, BiPV, solar adsorption cooling and electrical system-scale issues related to the use of BiPV in buildings. The *Department of Civil and Building Engineering* has foci on building thermal loads and the design and modelling of low-energy buildings. Website: www.lboro.ac.uk.

University of Nottingham, UK: The *Institute of Building Technology* in the *School of the Built Environment* is notable for its work pertaining to passive ventilation and dynamic insulation. Website: www.nottingham.ac.uk/sbe.

University of Strathclyde, Glasgow, UK: The *Energy Systems Research Unit* (ESRU) is one of the leading university-based groups involved in the development of building energy simulation software (through the ESP-r model), has an impressive outreach programme involving the professional architectural and engineering communities, and has offered internet-based teaching modules. Website: www.esru.strath.ac.uk.

Carnegie Mellon University, USA: The *School of Architecture* hosts the *Centre for Building Performance and Diagnostics*, a 700-square-metre laboratory that provides hands-on experience in the design of integrated building systems, involving innovative envelope, HVAC, lighting, and structural systems. The School of Architecture has developed the *Building Investment Decision Support tool* (BIDS), which contains succinct summaries of advanced building designs and their economic benefits. Website: http://cbpd.arc.cmu.edu.

Pennsylvania State University, USA: The *Department of Architectural Engineering* does work and/or offers courses pertaining to daylighting, radiant cooling, solar thermal energy in buildings and advanced HVAC systems. Website: www.engr.psu.edu.

Texas A & M University, USA: The *Energy Systems Laboratory* does research pertaining to fans, evaporative coolers, cooling towers, air conditioners and heat pumps, and the continuous building commissioning process. It also serves as the official testing laboratory of the Home Ventilating Institute. Website: www-esl.tamu.edu.

University of California Berkeley, USA: The *Center for Environmental Design Research* contains a *Building Science Group* and a *Center for the Built Environment* that does work on HVAC systems, alternatives to compression cooling, daylighting, adaptive standards of human thermal comfort and building computer simulations. Website: www.cedr.berkeley.edu.

University of California Los Angeles (UCLA), USA: The *Department of Architecture and Urban Design* has a focus on climate responsive design and the incorporation of passive and active uses of solar energy into buildings. Website: www.aud.ucla.edu.

University of Colorado and Colorado State University, USA: The *Joint Center for Energy Management* does research pertaining to energy efficient HVAC systems and controls, computer simulation of building energy systems and lighting. Courses include topics such as active and passive uses of solar energy, daylighting, HVAC systems and computer simulation of buildings. Website: http://civil.colorado.edu/Graduate_Programs/Jcem/jcemmain.html.

University of Maryland, USA: The *Center for Environmental Energy Engineering* does buildings-related research involving refrigeration (both vapour compression and absorption), heat pumps and alternative working fluids (particularly CO_2). Website: www.enme.umd.edu/ceee/.

University of Nebraska, USA: The *Program in Architectural Engineering* includes courses in advanced HVAC systems, incorporation of solar energy in buildings, the integrated design of buildings, and the use of computer simulation of energy use in buildings. Website: www.ae.unomaha.edu.

University of Wisconsin, USA: The *HVAC&R* (Heating, Ventilating, Air-Conditioning, and Refrigerating) *Center* (www.hvacr.wisc.edu) was established in 1989 and provides a specific focus on HVAC systems, including computer modelling of energy use. The *Solar Energy Laboratory* (http://sel.me.wisc.edu) – the first of its kind in the world – was established in 1954 and is the source of the widely used TRNSYS computer program for simulating energy use in buildings.

Appendix G

Annotated Guide to Books for Further Reading

This book has introduced and explained a wide spectrum of techniques for reducing energy use in buildings but, because of its comprehensive nature, has not been able to go into great depth in any one area. However, entire books have been written about each of the areas touched upon in this book. This appendix provides an annotated guide to the most useful of these books, divided into three categories. Those that are outputs of the International Energy Agency (IEA) Solar Heating and Cooling (SHC) or Photovoltaic Power Systems (PVPS) Programme are indicated.

Pictorial galleries

These books are full of photographs (usually in colour) of buildings that have achieved low-energy use (and low environmental impact in other respects too), with descriptive essays and sometimes with historical information, but usually with very little technical information. These will be of greatest interest to architects.

The Architectural Expression of Environmental Control Systems, George Baird, Spon Press, London, 2001, 264 pages. This book is concerned with the aesthetics of environmental control systems, be they passive or mechanical. Contains numerous diagrams and photographs of 15 case study buildings from around the world and additional information based on the author's visit to each building and interviews with the designers and building operators.

CEPHEUS: Living Comfort Without Heating, H. Krapmeier and E. Drössler (eds), Springer-Verlag, Vienna, 2001. Contains several colour photographs of nine residences built in Austria under the CEPHEUS (Cost-Effective Passive Houses as European Standards) programme. Construction details of the insulation of walls, roofs and around windows and where floors and walls are joined are also given for each project, along with a very brief description of the ventilation system and the system to provide the residual heating requirement.

Daylight Performance of Buildings, Marc Fontoynont (ed), James & James, London, 1999, 304 pages. Presents 60 case studies of daylighting in buildings in Europe, spanning glazed streets, transportation terminals, churches, museums, offices, educational buildings, libraries, houses and others. The book is the outcome of a three-year monitoring campaign funded by the European Commission. There are many colour photographs of the daylighting in each building, along with charts showing the distribution of light. Energy savings were not assessed.

The HOK Guidebook to Sustainable Design, Sandra Mendler and William Odell, John Wiley, 2nd edition, 2005, 412 pages. Features the work of the design firm Hellmuth, Obata + Kassabaum (HOK), founded in St. Louis in 1955 and now employing 1600 worldwide. It begins with a section that provides guidance on the integrated design process (team formation, goal setting, gathering information, design optimization, documents and specifications, construction, and operations and maintenance). This is followed by a Design Guidance section that covers planning and site work, energy, building materials, indoor air quality, water conservation, recycling and waste management. The majority of the book consists of 16 case studies of projects designed by or designed and completed by HOK. Each case study includes a brief description of the energy saving features included in the design, and some quantitative information on either the amount of energy used by the building or the savings in energy used compared to standard practice.

Intelligent Skins, Michael Wigginton and Jude Harris, Butterworth-Heinemann, Oxford, 2002, 176 pages. Presents 22 case studies of buildings with intelligent skins. The intelligent skins in these buildings modulate the amount of daylight, solar heat gain, and passive ventilation in response to changing internal and external conditions, through computer control systems that may or may not include some built-in learning capability. Each case study is illustrated with colour photographs, schematic diagrams showing details of the façade and/or floor plan, an outline of the intelligent features in the building, and a summary table of building data (including energy use).

Solar Architecture: Strategies, Visions, Concepts, C. Schittich (ed), Birkhäuser, Basel, 2003, 176 pages. Contains photographs and construction details of façades, floor plans, and ventilation flow in a number of buildings (mostly in Germany, and including the Reichstag) with double-skin façade construction and/or passive ventilation and/or use of daylighting.

Sustainable Architecture and Urbanism, Dominique Gauzin-Müller, Birkhäuser, Basel, 2002, 255 pages. Deals with several dimensions of sustainability in buildings, including energy use. Contains superb photographs of a number of buildings with double-skin façades.

Sustainable Architecture in Japan: The Green Buildings of Nikken Sekkei, Anna Ray-Jones (ed), Wiley Academic, Chichester (UK), 2000, 190 pages. Nikken Sekkei is a multidisciplinary Japanese firm of 1700 planners, architects and engineers whose roots go back over 100 years. The book contains an introductory essay on 'The Nikken Sekkei approach to green buildings', ten case studies of Nikken Sekkei buildings, and eight technical essays, all superbly illustrated with colour photographs and graphs and/or bar and pie charts. Among the low-energy techniques profiled are daylighting, shading, natural ventilation, earth pipe cooling and the use of thermal mass.

Sustainable Architecture: Towards a Diverse Built Environment, Ed Melet, NAI Publishers, Rotterdam, 1999, 191 pages. Contains a brief but enlightening discussion of glazed thermal buffers, smart buildings, buildings as energy generators and compact cities. The discussion of each of these topics is followed by photographs, diagrams and brief explanations of several buildings that employ the techniques discussed. The examples are largely from The Netherlands and Germany, with a few examples also from France, Austria and the United Kingdom. Many of the buildings illustrated in the book also represent outstanding examples of the incorporation of nature in buildings, through extensive indoor gardens that, in addition to creating a pleasant indoor environment, also humidify, cool and purify the air (but they do not enrich the oxygen supply, contrary to repeated claims otherwise, because the cumulative oxygen production is proportional to the buildup in plant and soil carbon and would be swamped by the effects of ventilation).

Transsolar Climate Engineering, Anja Thierfelder (ed), Birkhäuser, Basel, 2003, 229 pages. Presents an impressive portfolio of projects by the engineering firm Transsolar Energietechnik in Stuttgart. The projects are organized around the four elements of the ancient Greeks: Fire (corresponding to utilization of atria, daylighting, solar collectors, solar protection and shading), Water (evaporative cooling, groundwater cooling, water walls), Earth (boreholes, underground ducts, thermal storage, night ventilation, rooftop gardens), and Air (displacement ventilation, double-skin façades, solar chimneys, wind-enhanced ventilation, thermal stratification).

Design and construction guides

These books provide practical technical information – details concerning the design of specific features of low-energy buildings, and/or details concerning their construction. Many of these books are also extensively illustrated with photographs, but of a more technical nature than the photographs in the pictorial galleries.

Daylight Performance and Design, 2nd Edition, G. D. Ander, John Wiley, 2003, 335 pages. Begins with a discussion of the principles of daylighting, human visual comfort, and glazing properties, followed by a discussion of three kinds of design tools: hand calculations using graphical information, computed simulation models, and physical models. Several worked examples of hand calculations are provided. About half of the book is taken up with a series of detailed case studies.

Daylighting in Buildings: A Sourcebook on Daylighting Systems and Components, 2000 (IEA SHC Task 21), available from Lawrence Berkeley National Laboratory, Environmental Energy Technologies Division (http://eetd.lbl.gov) as LBNL-47493. The most comprehensive single source of information on daylighting techniques, performance, control systems, and design tools. Accompanied by a CD-ROM of case-study buildings.

A Design Guide for Energy-Efficient Research Laboratories, US EPA, Washington, 81 pages. Identifies advanced energy efficiency features in laboratories. It focuses on a systems approach to the design of laboratories, drawing information from direct facilities engineering experience and over 230 research papers, conference proceedings, design texts, case studies, recommended practices and manufacturers' experience. Available for free from www.labs21century.gov/toolkit/design_guide.htm.

Designing with Solar Power: A Source Book for Building Integrated Photovoltaics (BiPV), D. Prasad and M. Snow (eds), Earthscan, London, 2005, 252 pages (IEA PVPS Task 7). Contains, among other things, a chapter on design issues related to BiPV and different methods of integration into the building roof or façade, a chapter with 13 case studies from around the world, a chapter on assessment and design tools, and a chapter on market trends and marketing issues. It is illustrated with about 350 colour photographs, including many showing the process of installing PV modules, plus additional diagrams and charts.

Energy Efficiency Manual, Donald Wulfinghoff, Energy Institute Press, Wheaton, Maryland, USA, 1999, 1531 pages. The core of this book is a set of 400 measures covering all the mechanical, energy-using equipment in commercial buildings, the building envelope and lighting. Contains technical and practical information about each measure, with over 800 photographs and illustrations. Discusses new buildings and retrofit situations.

Hybrid Ventilation: State-of-the-art Review, A. Delsante and T. A. Vik (eds), 2002 (IEA ECBCS Annex 35). Discusses techniques and control systems for hybrid passive-mechanical ventilation systems, as well as barriers to the use of hybrid systems and how these barriers can be overcome. Contains 2-page fact sheets on 22 office and educational buildings located in different climate zones.

Natural Ventilation in Buildings: A Design Handbook, F. Allard (ed), James & James, London, 1998, 356 pages. Discusses methods for predicting and monitoring natural airflows in buildings, barriers to natural ventilations and how to overcome them, design guidelines, and case studies. Contains a CD with a software tool for predicting natural airflow in buildings.

Natural Ventilation in the Urban Environment, C. Ghiaus and F. Allard (eds), James & James, London, 2005, 241 pages. An outcome of the European Project 'URBVENT: Natural Ventilation in the Urban Environment', this book pays particular attention to noise and outdoor and indoor pollutants as they pertain to naturally ventilated buildings. Contains chapters on the physics of natural ventilation, strategies for natural ventilation, specific kinds of openings and optimization of the openings. Contains a CD with software for assessing the potential of a given site and for sizing the openings.

Passive Solar Commercial and Institutional Buildings: A Sourcebook of Examples and Design Insights, S. R. Hastings (ed), John Wiley, Chichester, 1994, 454 pages (IEA SHC Task 11). Well illustrated with photographs and diagrams, this book contains lots of data on the performance of a variety of passive solar features. Separate chapters are devoted to direct gain systems, air collector systems, airflow windows, mass wall systems, transparent insulation systems, absorber walls, daylighting, cooling (avoiding overheating) and atria. Each chapter contains sections dealing with basic principles, system variations, examples, analysis tools and conclusions.

The Passive Solar Design and Construction Handbook, Steven Winter Associates, M. J. Crosbie (ed), John Wiley and Sons, New York, 1998, 291 pages. Provides many diagrams detailing the construction of passively heated houses, covering direct gain, thermal storage walls, sunspaces, convective loops, and materials.

Planning and Installing Solar Thermal Systems: A Guide for Installers, Architects and Engineers, German Solar Energy Society and Ecofys, James and James, London, 2005, 298 pages. This book contains chapters outlining the various choices for the components and layout of solar thermal systems for space heating and domestic hot water, swimming pools and solar air conditioning.

Solar Air Systems: A Design Handbook, S. R. Hastings and O. Mørck. (eds), James & James, London, 2000, 256 pages (IEA SHC Task 19). Focuses on the sizing, design and performance of systems for using solar energy to provide space heating in buildings. There is an extensive set of charts and tables documenting the performance of a wide variety of systems under a variety of climatic conditions, based on built examples or detailed computer simulations. There is also a brief discussion of alternative uses of parts of systems intended primarily for space heating, namely, production of domestic hot water, cooling and as integrated PV/thermal collectors.

Solar-Assisted Air-Conditioning in Buildings: A Handbook for Planners, H. M. Henning (ed), Springer, Vienna, 2004, 150 pages (IEA SHC Task 25). It provides technical information concerning the various components of solar air conditioning systems: solar collectors (flat-plate and evacuated tube, compound parabolic concentrators, solar air collectors), heat-driven cooling equipment (absorption and adsorption chillers, solid and liquid desiccant dehumidifiers), cooling units (such as chilled ceiling panels and fan coil radiators), and storage vessels for hot and cold water. Also given are details concerning several different possible system configurations.

Solar Heating Systems for Houses: A Design Handbook for Solar Combisystems, W. Weiss (ed), James & James, London, 2003, 313 pages (IEA SHC Task 26). Contains chapters on 19 generic solar combisystems, building-related considerations, performance, durability and reliability, dimensioning, built examples, and testing and certification.

Solar Energy Houses: Strategies, Technologies, Examples, 2nd edition, A. G. Hestnes, R. Hastings and B. Saxhof (eds), James & James, London, 2003, 202 pages (IEA SHC Task 13). Contains a series of short chapters on technologies for creating houses with a sufficiently low heat load in cold climates that much of the required heating can be supplied by passive and active solar energy, followed by a series of short chapters with information on the design and performance of specific buildings in North America, Europe and Japan.

Solar Thermal Systems: Successful Planning and Construction, F. A. Peuser, K.-H. Remmers, M. Schnauss, James & James, London, 2002, 364 pages. Summarizes 30 years of experience with solar thermal systems in various research projects in Germany. As such, there are no equations or calculations, but plenty of information concerning practical matters such as reliability, unexpected problems, and how they were resolved. Some quantitative performance and cost data are presented graphically. Separate chapters deal with solar collectors, collector assembly types, collector loops, storage units, heat exchangers, control and characteristic parameters of various types of solar systems.

Technical and theoretical books

These books develop the main ideas and principles in considerable theoretical detail, with, in some cases, hundreds of equations and dozens of problems and worked-out solutions. These will be of greatest interest to engineers.

Energy Efficient Motor Systems: Handbook on Technology, Program, and Policy Opportunities, S. Nadel, R. N. Elliott, M. Shepard, S. Greenberg, G. Katz, and A. T. de Almeida, American Council for an Energy-Efficient Economy, Washington, 2002, 494 pages. Contains chapters on motor technologies, motor-control technologies, system considerations, motor applications, motor markets and policies to promote more efficient motor and use of motors.

Heating and Cooling of Buildings: Design for Efficiency, J. F. Kreider, P. S. Curtiss, and A. Rabl, McGraw Hill, New York, 2002, 877 pages. Covers conventional HVAC systems in commercial buildings, with plenty of examples of the calculation of solutions to various mathematically posed problems, and a large number of problems at the end of each chapter.

Solar Technologies for Buildings, Ursula Eicker, John Wiley, Chichester, 2003, 323 pages. Contains a detailed technical discussion of the distribution of solar radiation, and of solar thermal energy, solar cooling, grid-connected PV systems, passive solar energy and daylighting. Provides worked problems involving quantitative design and the calculation of performance. Contains several hundred equations, dozens of diagrams, a few photographs of equipment but no photographs of buildings.

Thermal Analysis and Design of Passive Solar Buildings, A. K. Athienitis and M. Santamouris, James & James, London, 2002, 288 pages. Provides a mathematical treatment of transient heat transfer and thermal storage; transfer of solar radiation through windows, transparent insulation and light shelves; the passive response of buildings to solar radiation; small auxiliary heating and cooling systems; and the control and utilization of solar thermal energy in buildings.

References

Aarlien, R. and Frivik, P. E. (1998) 'Comparison of practical performance between CO_2 and R-22 reversible heat pumps for residential use', in *Natural Working Fluids '98, IIR – Gustav Lorentzen Conference*, 2–5 June 1998, Oslo, International Institute of Refrigeration, Paris, pp388–398

Abaza, H., Beliveau, Y. and Jones, J. (2001) 'Dehumidification through indirect nighttime ventilation', in *2001 International Solar Energy Conference*, 21–25 April 2001, Washington, DC, pp51–56

AboulNaga, M. M. and Abdrabboh, S. N. (2000) 'Improving night ventilation into low-rise buildings in hot-arid climates exploring a combined wall-roof solar chimney', *Renewable Energy*, vol 19, pp47–54

Adalberth, K. (1997) 'Energy use during the life cycle of single-unit dwellings: examples', *Building and Environment*, vol 32, pp321–329

Adnot, J. and Orphelin, M. (1999) 'Hungary cooling: Room air-conditioners', *Appliance Efficiency*, vol 3, no 3, pp6–10

Adriansyah, W. (2004) 'Combined air conditioning and tap water heating plant using CO_2 as refrigerant', *Energy and Buildings*, vol 36, pp690–695

AEC (Architectural Energy Corporation) (2002) *Energy Design Brief: Building Integrated Photovoltaics*. Available from www.archenergy.com

AEC (Architectural Energy Corporation) (2003) *Small HVAC System Design Guide*. California Energy Commission Report 500-03-082-A12. Available from www.consumerenergycenter.org/bookstore/index.html

Ahlgren, R. C. E. (2001) 'Why did I buy such an oversized pump?' *ASHRAE Transactions*, vol 107 (Part 2), pp566–572

Akbari, H. and Konopacki, S. (2004) 'Energy effects of heat-island reduction strategies in Toronto, Canada', *Energy*, vol 29, pp191–210

Akbari, H. and Konopacki, S. (2005) 'Calculating energy-saving potentials of heat-island reduction strategies', *Energy Policy*, vol 33, pp721–756

Akbari, H., Berdahl, P., Levinson, R., Wiel, S., Desjarlais, A., Miller, W., Jenkins, N., Rosenfeld, A. and Scruton, C. (2004) 'Cool colored materials for roofs', in *Proceedings of the 2004 ACEEE Summer Study on Energy Efficiency in Buildings*, vol 1, American Council for an Energy Efficient Economy, Washington, DC, pp1–12

Al-Ajmi, F., Hanby, V. I. and Loveday, D. L. (2002) 'The potential for ground cooling in a hot, arid climate', *Climate Change and the Built Environment*, Paper 119, Manchester, UK, Tyndall Centre for Climatic Change/CIB Task Group 21

Alam, K. C. A., Akahira, A., Hamamoto, Y., Akisawa, A. and Kashiwagi, T. (2004) 'A four-bed mass recovery adsorption refrigeration cycle driven by low temperature waste/renewable heat source', *Renewable Energy*, vol 29, pp1461–1475

Alamdari, F. and Butler, D. (1998) 'Chilled ceilings and displacement ventilation – do they work?', in Tassou, S. (ed) *Low-Energy Cooling Technologies for Buildings: Challenges and Opportunities for the Environmental Control of Buildings*, IMechE Seminar Publication 1998–7, Professional Engineering Publishing, London, pp69–83

Alexis, G. K. and Karayiannis, E. K. (2005) 'A solar ejector cooling system using refrigerant R134a in the Athens area', *Renewable Energy*, vol 30, pp1457–1469

Alfonso, C. and Oliveira, A. (2000) 'Solar chimneys: simulation and experiment', *Energy and Buildings*, vol 32, pp71–79

Allard, F. (ed) (1998) *Natural Ventilation in Buildings: A Design Handbook*, James & James, London

Al-Nimr, M. A., Kodah, Z. and Nassar, B. (1998) 'A theoretical and experimental investigation of a radiative cooling system', *Solar Energy*, vol 63, pp367–373

Alsema, E. A. and Nieuwlaar, E. (2000) 'Energy viability of photovoltaic systems', *Energy Policy*, vol 28, pp999–1010

Al-Temeemi, A. D. (1995) 'Climatic design techniques for reducing cooling energy consumption in Kuwaiti houses', *Energy and Buildings*, vol 23, pp41–48

Alva Solari, L. H. and González, J. E. (2002) 'Simulation of an air-cooled solar-assisted absorption air conditioning system', *Journal of Solar Energy Engineering*, vol 124, pp276–282

Andersson, B. A. (2000) 'Materials availability for large-scale thin-film photovoltaics', *Progress in Photovoltaics*, vol 8, pp61–76

Andersson, J. V. and Lindholm, T. (2001) 'Desiccant cooling for Swedish office buildings', *ASHRAE Transactions*, vol 107 (Part 1), pp490–500

Andrade, M. A. and Bullard, C. W. (2002) 'Modulating blower and compressor capacities for efficient comfort control', *ASHRAE Transactions*, vol 108 (Part 1), pp631–637

Andrén, L. (2003) *Solar Installations: Practical Applications for the Built Environment*, James & James, London, 130pp

Andrepont, J. S. (2000a) 'Long-term performance of a low temperature fluid in thermal storage and distribution applications', presented at *International District Energy Association (IDEA) 13ᵗʰ Annual College/University Conference*, Vancouver, BC, 24 February 2000

Andrepont, J. S. (2000b) 'Combustion turbine inlet air cooling (CTIAC): benefits, technology options, and applications for district energy', presented at *International District Energy Association (IDEA) 91st Annual Conference*, Montreal, Quebec, 13 June 2000

Andrepont, J. S. and Rezachek, D. (2003) 'Potential and benefits for district cooling using sweater air conditioning (SWAC) integrated with thermal energy storage (TES)', presented at *International District Energy Association (IDEA) 94th Annual Conference*, Philadelphia, Pennsylvania, 23 June 2003

Andronova, N. G. and Schlesinger, M. E. (2001) 'Objective estimation of the probability density function for climate sensitivity', *Journal of Geophysical Research*, vol 106 pp22605–22611

Anello, M. T., Parker, D. S., Sherwin, J. R. and Richards, K. (2000) 'Measured impact of advanced windows on cooling energy use', in *Proceedings of the 2000 ACEEE Summer Study on Energy Efficiency in Buildings*, vol 1, American Council for an Energy Efficient Economy, Washington, DC, pp29–38

Anis, W. (2005) 'Commissioning the air barrier system', *ASHRAE Journal*, vol 47, no3, pp35–44

Anonymous (2002) 'Carrier unveils variable speed screw chiller', *ASHRAE Journal*, vol 44. no 5, p7

Anonymous (2003) 'A cooler way to cook', *ASHRAE Journal*, vol 45, no 10, p10

Anonymous (2004) 'Let there be daylight', *ASHRAE Journal*, vol 46, no 1, pp6–8

Anyanwu, E. E. (2003) 'Review of solid adsorption solar refrigerator 1: an overview of the refrigeration cycle', *Energy Conversion and Management*, vol 44, pp301–312

Anyanwu, E. E. (2004) 'Review of solid adsorption solar refrigerator 2: an overview of the principles and theory', *Energy Conversion and Management*, vol 45, pp1279–1295

APERC (Asia Pacific Energy Research Centre) (2001) *Energy Efficiency Indicators: A Study of Energy Efficiency Indicators in APEC Economies*, APERC, Tokyo. Available from www.ieej.or.jp/aperc

Ardehali, M. M., Smith, T. F., House, J. M. and Klaassen, C. J. (2003) 'Building energy use and control problems: an assessment of case studies', *ASHRAE Transactions*, vol 109 (Part 2), pp111–121

Ardente, F., Beccali, G., Cellura, M. and Brano, V. L. (2005) 'Life cycle assessment of a solar thermal collector: sensitivity analysis, energy and environmental balances', *Renewable Energy*, vol 30, pp109–130

Argiriou, A. A., Lykoudis, S. P., Balaras, C. A. and Asimakopoulos, D. N. (2004) 'Experimental study of a earth-to-air heat exchanger coupled to a photovoltaic system', *Journal of Solar Energy Engineering*, vol 126, pp620–625

Arnell, N. W., Cannell, M. G. R., Hulme, M., Kovats, R. S., Mitchell, J. F. B., Nicholls, R. J., Parry, M. L., Livermore, M. T. J. and White, A. (2002) 'The consequences of CO_2 stabilization for the impacts of climate change', *Climatic Change*, vol 53, pp413–446

Arnold, D. (2000) 'Thermal storage case study: combined building mass and cooling pond', *ASHRAE Transactions*, vol 106 (Part 1), pp819–827

Ashford, P., Wu, J., Jeffs, M., Kocchi, S., Vodianitskaia, P., Lee, S., Nott, D., Johnson, B., Ambrose, A., Mutton, J., Maine, T. and Veenendaal, B. (2005) 'Foams', in *IPCC/TEAP Special Report on Safeguarding the Ozone Layer and the Global Climate System: Issues related to Hydrofluorocarbons and Perfluorocarbons*, chapter 7, Cambridge University Press, Cambridge

ASHRAE (American Society of Heating, Refrigerating and Air-Conditioning Engineers) (1989) *ASHRAE Standard 90.1-1989, Energy Efficiency Design of New Buildings Except Low-Rise Residential Buildings*, American Society of Heating, Refrigerating and Air-Conditioning Engineers, Atlanta, 147pp

ASHRAE (American Society of Heating, Refrigerating and Air-Conditioning Engineers) (1997) *WYEC2 (Weather Year for Energy Calculations 2) Data and Toolkit CD-ROM*, American Society of Heating, Refrigeration and Air-Conditioning Engineers, Atlanta

ASHRAE (American Society of Heating, Refrigerating and Air-Conditioning Engineers) (2000) *2000 ASHRAE Handbook, Heating, Ventilating, and Air Conditioning Systems and Equipment, SI Edition,*

American Society of Heating, Refrigeration and Air-Conditioning Engineers, Atlanta

ASHRAE (American Society of Heating, Refrigerating and Air-Conditioning Engineers) (2001a) *International Weather for Energy Calculations (IWEC Weather Files) Users Manual and CD-ROM*, American Society of Heating, Refrigeration and Air-Conditioning Engineers, Atlanta

ASHRAE (American Society of Heating, Refrigerating and Air-Conditioning Engineers) (2001b) *2001 ASHRAE Handbook, Fundamentals, SI Edition*, American Society of Heating, Refrigeration and Air-Conditioning Engineers, Atlanta

ASHRAE (American Society of Heating, Refrigerating and Air-Conditioning Engineers) (2004) *ASHRAE Standard 90.1-2004 – Energy Standard for Buildings Except Low-Rise Residential Buildings, SI Edition*, American Society of Heating, Refrigerating and Air-Conditioning Engineers, Atlanta, 182pp

Asiedu, Y., Besant, R. W. and Simonson, C. J. (2004) 'Wheel selection for heat and energy recovery in simple HAVC ventilation design problems', *ASHRAE Transactions*, vol 110 (Part 1), pp381–398

Asif, M., Davidson, A. and Muneer, T. (2001) 'Technical note, embodied energy analysis of aluminium-clad windows', *Building Service Engineering Resources Technology*, vol 22, no 3, pp195–199

Asif, M., Muneer, T. and Kubie, J. (2005) 'Sustainability analysis of window frames', *Building Service Engineering Resources Technology*, vol 26, pp71–87

Askar, H., Probert, S. D. and Batty, W. J. (2001) 'Windows for buildings in hot arid countries', *Applied Energy*, vol 70, pp77–101

Athienitis, A. K. and Santamouris, M. (2002) *Thermal Analysis and Design of Passive Solar Buildings*, James & James, London

Athienitis, A. K., Liu, C., Hawes, D., Banu, D. and Feldman, D. (1997) 'Investigation of the thermal performance of a passive solar test-room with wall latent heat storage', *Building and Environmenti*, 32, pp405–410

Atif, M. R. and Galasiu, A. D. (2003) 'Energy performance of daylight-linked automatic lighting control systems in large atrium spaces: report on two field-monitored case studies', *Energy and Buildings*, vol 35, 441–461

Atkinson, B., Denver, A., McMahon, J. E., Shown, L. and Clear, R. (1997) 'Energy efficient technologies: energy efficient lighting technologies and their applications in the commercial and residential sectors', in Kreith, F. and West, R. E. (eds) *CRC Handbook of Energy Efficiency*, CRC Press, Boca Raton, FL, pp399–427

Atmaca, I. and Yigit, A. (2003) 'Simulation of solar-powered absorption cooling system', *Renewable Energy*, vol 28, pp1277–1293

Augenbroe, G. (2001) 'Building simulation trends going into the new millennium', in *Seventh International IBPSA Conference*, Rio de Janeiro, Brazil, 13–15 August 2001, pp15–27. Available from www.ibpsa.org

Awbi, H. B. (1998) 'Chapter 7: ventilation', *Renewable and Sustainable Energy Reviews*, vol 2, pp157–188

Axley, J. (1999) 'Passive ventilation for residential air quality control', *ASHRAE Transactions*, vol 105 (Part 2), pp864–876

Axley, J. W., Emmerich, S. J. and Walton, G. N. (2002) 'Modeling the performance of a naturally ventilated commercial building with a multizone coupled thermal/airflow simulation tool', *ASHRAE Transactions*, vol 108 (Part 2), pp1260–1275

Aye, L., Charters, W. W. S. and Chaichana, C. (2002) 'Solar heat pump systems for domestic hot water', *Solar Energy*, vol 73, pp169–175

Azar, C. and Rodhe, H. (1997) 'Targets for stabilization of atmospheric CO_2', *Science*, vol 276, pp1818–1819

Azar, C. and Sterner, T. (1996) 'Discounting and distributional considerations in the context of global warming', *Ecological Economics*, vol 19, pp169–184

Baczek, S., Yost, P. and Finegan, S. (2002) 'Using wood efficiently: from optimizing design to minimizing the dumpster', Building Science Corporation. Available from www.buildingscience.com/resources/misc/wood_efficiency.pdf

Badescu, V. (2002) 'Model of a solar-assisted heat-pump system for space heating integrating a thermal energy storage unit', *Energy and Buildings*, vol 34, pp715–726

Badescu, V. and Sicre, B. (2003) 'Renewable energy for passive house heating Part 1. Building description', *Energy and Buildings*, vol 35, pp1077–1084

Baechler, M., Hadley, D., Spatkman, S. and Lubbliner, M. (2002) 'Pushing the envelope: a case study of building the first manufactured home using structural insulated panels', in *Proceedings of the 2002 ACEEE Summer Study on Energy Efficiency in Buildings*, vol. 1, American Council for an Energy Efficient Economy, Washington, DC, pp29–41

Baek, N. C., Shin, U. C. and Yoon, J. H. (2005) 'A study on the design and anlysis of a heat pump heating system using wastewater as a heat source', *Solar Energy*, vol 78, pp427–440

Bahaj, A. S. (2003) 'Photovoltaic roofing: issues of design and integration into buildings', *Renewable Energy*, vol 28, p2195–2204

Bahnfleth, W. P., Yuill, G. K. and Lee, B. W. (1999) 'Protocol for field testing of tall buildings to determine envelope air leakage rate', *ASHRAE Transactions*, vol 105, pp27–38

Baird, G. (2001) *The Architectural Expression of Environmental Control Systems*, Spon Press, London. 264pp

Bakema, G., Snijders, A. and Nordell, B. (eds) (1995) *Underground Thermal Energy Storage, State-of-the-Art 1994*, IEA ECES Annex 8, IF Technology, Arnhem, The Netherlands

Baker, N. and Steemers, K. (1999) *Energy and Environment in Architecture: A Technical Design Guide*, E & FN Spon, London, 224pp

Baker, N. and Steemers, K. (2002) *Daylight Design of Buildings: A Handbook for Architects and Engineers*, James & James, London, 250pp

Baker, P. H. (2003) 'The thermal performance of a prototype dynamically insulated wall', *Building Service Engineering Resources Technology*, vol 24, no 1, pp25–34

Bakker, M., Zondag, H. A., Elswijk, M. J., Strootman, K. J. and Jong, M. J. M. (2005) 'Performance and costs of a roof-sized PV/thermal array combined with a ground coupled heat pump', *Solar Energy*, vol 78, pp331–339

Balaras, C. A. (2001) 'Energy retrofit of a neoclassic office building – Social aspects and lessons learned', *ASHRAE Transactions*, vol 107 (Part 1), pp191–197

Balaras, C. A., Argiriou, A. A., Michel, E. and Henning, H. M. (2003) 'Recent activities on solar air conditioning', *ASHRAE Transactions*, vol 109 (Part 1), pp251–260

Balcomb, J. D. (ed) (1992) *Passive Solar Buildings*, MIT Press, Cambridge, MA, 534pp

Balcomb, J. D., Andresen, I., Hestnes, A. G. and Aggerholm, S. (2002) *Multi-Criteria Decision-Making MCDM-23: A Method for Specifying and Prioritising Criteria and Goals in Design*, International Energy Agency, Solar Heating and Cooling Program, Task 23 (Optimization of Solar Energy Use in Large Buildings), Subtask C (Tools for Trade-Off Analysis), Paris. Available from www.iea-shc.org/task23.

Balmér, P. (1997) 'Energy conscious waste water treatment plant', *CADDET Energy Efficiency Newsletter*, June 1997, pp11–12

Banwell, P., Brons, P. J., Freyssinier-Nova, J. P., Rizzo, P. and Figueiro, M. (2004) 'A demonstration of energy-efficient lighting in residential new construction', *Lighting Resources Technology*, vol 36, no 2 pp147–164

Barnes, P. R. and Bullard, C. W. (2004) 'Minimizing TEWI in a compact chiller for unitary applications', *ASHRAE Transactions*, vol 110 (Part 2), pp335–344

Barwig, F. E., House, J. M., Klaassen, C. J., Ardehali, M. M. and Smith, T. F. (2002) 'The National Building Controls information program', in *Proceedings of the 2002 ACEEE Summer Study on Energy Efficiency in Buildings*, vol 3, American Council for an Energy Efficient Economy, Washington, DC, pp1–14

BASF (2003) 'Neopor® Expandable Polystyrene (EPS) Insulation Innovation', brochure.' Available from www.3lh.de

Baskin, E., Lenarduzzi, R., Wendt, R. and Woodbury, K. A. (2004) 'Numerical evaluation of alternative residential hot water distribution systems', *ASHRAE Transactions*, vol 110 (Part 1), pp671–681

Battisti, R. and Corrado, A. (2005) 'Evaluation of technical improvements of photovoltaic systems through life cycle assessment methodology', *Energy*, vol 30, pp952–967

Bazilian, M. D., Leenders, F., van Der Ree, B. G. C. and Prasad, D. (2001) 'Photovoltaic cogeneration in the built environment', *Solar Energy*, vol 71, pp57–69

Bazjanac, V. (2004) 'Building energy performance simulation as part of interoperable software environments', *Building and Environment*, vol 39, pp879–883

Beeler, A. G. (1998) 'Integrated design team management within the context of environmental systems theory', in *Proceedings of the 1998 ACEEE Summer Study on Energy Efficiency in Buildings*, vol 10, American Council for an Energy Efficient Economy, Washington, DC, pp19–30

Behne, M. (1999) 'Indoor air quality in rooms with cooled ceilings. Mixing ventilation or rather displacement ventilation?', *Energy and Buildings*, vol 30, pp155–166

Belcher, S. E., Hacker, J. N. and Powell, D. S. (2005) 'Constructing design weather data for future climates', *Building Service Engineering Resources Technology*, vol 26, pp49–61

Belessiotis, V. and Mathioulakis, E. (2002) 'Analytical approach of thermosyphon solar domestic hot water system performance', *Solar Energy*, vol 72, pp307–315

Bell, M. and Lowe, R. (2000) 'Energy efficient modernisation of housing: a UK case study', *Energy and Buildings*, vol 32, pp267–280

Bentley, R. W. (2002) 'Global oil and gas depletion: an overview', *Energy Policy*, vol 30, pp189–205

Benya, J., Heschong, L., McGowan, T., Miller, N. and Rubinstein, F. (2003) *Advanced Lighting Guidelines*, New Buildings Institute, White Salmon, Washington, DC, 445pp. Available from www.newbuildings.org

Berger, A. (1978) 'Long-term variations in daily insolation and Quaternary climatic change', *Journal of the Atmospheric Sciences*, vol 35, pp2362–2367

Bergquam Energy Systems (BES) (2002) *Design and Optimization of Solar Absorption Chillers*, California Energy Commission Report P500-02-047F. Available from www.consumerenergycenter.org/bookstore/index.html

Bergquam, J., Brezner, J. and Jensen, A. (2003) 'Design and testing of an augmented generator for solar fired absorption chillers', in *2003 International Solar Energy Conference*, Kohala Coast, Hawaii, USA, 15–18 March 2003, pp351–358

Bernier, M. and Lemire, N. (1999) 'Non-dimensional pumping power curves for water loop heat pump systems', *ASHRAE Transactions*, vol 105 (Part 2), pp1226–1232

Berntsson, T. (2002) 'Heat sources – technology, economy and environment', *International Journal of Refrigeration*, vol 25, pp428–438

Bertoldi, P., Aebischer, B., Edlington, C., Herschberg, C., Lebot, B., Lin, J., Marker, T., Meier, A., Nakagami, H., Shibata, Y., Siderius, H. P. and Webber, C. (2002) 'Standby power use: how big is the problem? What policies and technical solutions can address it?', in *Proceedings of the 2002 ACEEE Summer Study on Energy Efficiency in Buildings*, vol 7, American Council for an Energy Efficient Economy, Washington, DC, pp41–60

Bi, Y., Chen, L. and Wu, C. (2002) 'Ground heat exchanger temperature distribution analysis and experimental verification', *Applied Thermal Engineering*, vol 22, pp183–189

Binz, A. and Steinke, G. (2005) 'Applications of vacuum insulation in the building sector' in Zimmermann, M. (ed) *7th International Vacuum Insulation Symposium*, EMPA, Duebendorf, Switzerland, 28–29 September 2005, pp43–48. Available from www.empa.ch/VIP-Symposium

Blazek, M., Chong, H., Woonsien, L. and Koomey, J. G. (2004) 'Data centers revisited: assessment of the energy impact of retrofits and technology trends in a high-density computing facility', *Journal of Infrastructure Systems*, vol 10, pp98–104

Blomsterberg, A., Wahlstrøm, A., Sandberg, M. and Eriksson, J. (2002) *Pilot Study Report: Tånga, Falkenberg, Sweden*, International Energy Agency, Energy Conservation in Buildings and Community Systems, Annex 35. Available from http://hybvent.civil.auc.dk

Blondeau, P., Sperandio, M. and Allard, F. (1997) 'Night ventilation for building cooling in summer', *Solar Energy*, vol 6, pp327–335

Bloomquist, R. G. (2003) 'Geothermal space heating', *Geothermics*, vol32, pp513–526

Bo, H., Gustafsson, E. M. and Setterwall, F. (1999) 'Tetradecane and hexadecane binary mixtures as phase change materials (PCMs) for cool storage in district cooling systems', *Energy*, vol 24, pp1015–1028

Bodart, M. and de Herde, A. (2002) 'Global energy savings in office buildings by the use of daylighting', *Energy and Buildings*, vol 34, pp421–429

Bojić, M. and Yik, F. (2005) 'Cooling energy evaluation for high-rise residential buildings in Hong Kong', *Energy and Buildings*, vol 37, pp345–351

Bojić, M., Lee, M. and Yik, F. (2001) 'Flow and temperatures outside a high-rise residential building due to heat rejection by its air conditioners', *Energy and Buildings*, vol 33, pp737–751

Bolland, O. and Undrum, H. (1999) 'Removal of CO_2 from gas turbine power plants: evaluation of pre- and postcombustion methods', in Riemer, P., Eliasson, B. and Wokaun, A. (eds) *Greenhouse Gas Control Technologies*, Elsevier Science, New York, pp125–130

Bomberg, M. T. and Kumaran, M. K. (1999) *Use of Field-Applied Polyurethane Foams in Buildings, Construction Technology Update No. 32*, Institute for Construction Research, Ottawa. Available from http://irc.nrc-cnrc.gc.ca

Boonstra, C. and Thijssen, I. (1997) *Solar Energy in Building Renovation*, James & James, London

Borenstein, S. (2005) *Valuing the Time-Varying Electricity Production of Solar Photovoltaic Cells*, CSEM WP 142, University of California Energy Institute. Available from www.ucei.org.

Börjesson, P. and Gustavsson, L. (2000) 'Greenhouse gas balances in building construction: wood versus concrete from life-cycle and forest land-use perspectives', *Energy Policy*, vol 28, pp575–588

Born, F. J., Clarke, J. A., Johnstone, C. M., Kelly, N. J., Burt, G., Dysko, A., McDonald, J. and Hunter, I. B. B. (2001) 'On the integration of renewable energy systems within the built environment', *Building Service Engineering Resources Technology*, vol 22, no 1, pp3–13

Bourassa, N., Haves, P. and Huang, J. (2002) 'A computer simulation appraisal of non-residential low energy cooling systems in California', in *Proceedings of the 2002 ACEEE Summer Study on Energy Efficiency in Buildings*, vol 3, American Council for an Energy Efficient Economy, Washington, DC, pp41–53

Bourbia, F. and Awbi, H. B. (2004) 'Building cluster and shading in urban canyon for hot dry climate. Part 1: air and surface temperature measurements', *Renewable Energy*, vol 29, pp249–262

Bourne, R. C. (2001) '"NightSky" natural cooling system saves energy', *CADDET Newsletter*, 16–18 August 2001, International Energy Agency, CADDET Energy Efficiency, Harwell, UK

Bower, J. (1995) *Understanding Ventilation: How to Design, Select, and Install Residential Ventilation Systems*, Healthy House Institute, Bloomington, Indiana

Boyce, F. M, Hamblin, P. F., Harvey, L. D. D., Schertzer, W. M. and McCrimmon, R. C. (1993) 'Response of thermal structure of Lake Ontario to deep cooling water withdrawals and to global warming', *Journal of Great Lakes Research*, vol 19, pp603–616

Boyce, P., Hunter, C. and Howlett, O. (2003) *The Benefits of Daylight Through Windows*, US Department of Energy, Washington, DC. Available from www. daylightdividends.org

Boyle, G. (1996) 'Solar photovoltaics', in Boyle, G. (ed) *Renewable Energy, Power for a Sustainable Future*, Oxford University Press, Oxford, pp89–136

Brager, G. S. and de Dear, R. (2000) 'A standard for natural ventilation', *ASHRAE Journal*, vol 42, no10, pp21–28

Braham, G.D. (1998) 'Slab cooling and the termodeck system', in Tassou, S. (ed) *Low-Energy Cooling Technologies for Buildings: Challenges and Opportunities for the Environmental Control of Buildings*, IMechE Seminar Publication 1998–7, Professional Engineering Publishing, London, pp45–57

Branco, G., Lachal, B., Gallinelli, P. and Weber, W. (2004) 'Predicted versus observed heat consumption of a low energy multifamily complex in Switzerland based on long-term experimental data', *Energy and Buildings*, vol 36, pp543–555

Brandemuehl, M. J. and Braun, J. E. (1999) 'The impact of demand-controlled and economizer ventilation strategies on energy use in buildings', *ASHRAE Transactions*, vol 105 (Part 2), pp39–50

Braun, J., Klein, S. A. and Reindl, D. T. (2004) 'Considerations in the design and application of solid oxide fuel cell energy systems in residential markets', *ASHRAE Transactions*, vol 110 (Part 1), pp14–24

Bredsdorff, M., Bezzel, E. and Moltke, I. (1998) 'The centre U-value is less than half the truth', *CADDET Energy Efficiency Newsletter*, December 1998, pp7–8

Breembroek, G. (1997) 'Systems and controls: an international overview', *IEA Heat Pump Centre Newsletter*, vol 15, no 2, pp10–15

Breembroek, G. and Dieleman, M. (2001) *Domestic Heating and Cooling Distribution and Ventilation Systems and Their Use with Residential Heat Pumps – Considerations in designing and selecting*, IEA Heat Pump Centre, Sittard, The Netherlands

Bridges, D. B., Harshbarger, D. S. and Bullard, C. W. (2001) 'Second law analysis of refrigerators and air conditioners', *ASHRAE Transactions*, vol 107 (Part 1), pp644–651

Brinkworth, B. J., Marshall, R. H. and Ibarahim, Z. (2000) 'A validated model of naturally ventilated PV cladding', *Solar Energy*, vol 69, pp67–81

Brown, D. R. and Somasundaram, S. (1997) 'Recuperators, regenerators, and storage: thermal energy storage applications in gas-fired power plants', in Kreith, F. and West, R. E. (eds) *CRC Handbook of Energy Efficiency*, CRC Press, Boca Raton, FL, pp511–532

Brown, G. (1996) 'Geothermal energy', in Boyle, G. (ed) *Renewable Energy, Power for a Sustainable Future*, Oxford University Press, Oxford, pp353–392

Brown, J. S., Kim, Y. and Domanski, P. A. (2002) 'Evaluation of carbon dioxide as R-22 substitute for residential air conditioning', *ASHRAE Transactions*, vol 108 (Part 2), pp954–964

Brown, M. A., Levine, M. D., Romm, J. P., Rosenfeld, A. H. and Koomey, J. G. (1998) 'Engineering-economic studies of energy technologies to reduce greenhouse emissions: opportunities and challenges', *Annual Review of Energy and the Environment*, vol 23, pp287–385

Brown, M. A., Short, W., Levine, M. D., Braitsch, J., Hadley, S., McIntyre, T. and Nicholls, A. (2000) 'The longer-term and global context', in *Scenarios for a Clean Energy Future, ORNL/CON-476*, Oak Ridge National Laboratory, Oak Ridge

Brown, P. G. (1992) 'Climate change and the planetary trust', *Energy Policy*, vol 20, pp208–222

Brunger, A., Dubrous, F. M. and Harrison, S. (1999) 'Measurement of the solar heat gain coefficient and U value of windows with insect screens', *ASHRAE Transactions*, vol 105 (Part 2), pp1038–1044

Buchanan, A. H. and Honey, B. G. (1994) 'Energy and carbon dioxide implications of building construction', *Energy and Buildings*, vol 20, pp205–217

Budaiwi, I. M. (2003) 'Air conditioning system operation strategies for intermittent occupancy buildings in a hot-humid climate', in *Eighth International IBPSA Conference*, Eindhoven, The Netherlands, 11–14 August 2003, pp115–121. Available from www. ibpsa.org.

Bühring, A., Schmitz, G. and Voss, K. (2003) 'Modeling and development of compression heat pumps in integrated ventilation and heat supply units for passive solar houses', *ASHRAE Transactions*, vol 109 (Part 2), pp207–218

Bullard, C. W. and Radermacher, R. (1994) 'New technologies for air conditioning and refrigeration', *Annual Review of Energy and the Environment*, vol 19, pp113–152

Burer, M., Tanaka, K., Favrat, D. and Yamada, K.

(2003) 'Multi-criteria optimization of a district cogeneration plant integrating a solid oxide fuel cell-gas turbine combined cycle, heat pumps and chillers', *Energy*, vol 28, pp497–518

Busch, J. F. (1992) 'A tale of two populations: thermal comfort in air-conditioned and naturally-ventilated offices in Thailand', *Energy and Buildings*, vol 18, pp235–249

CADDET (Centre for the Analysis and Dissemination of Demonstrated Energy Technologies) (1999) *Saving Energy with Advanced Windows. Maxi Brochure 12*, International Energy Agency, Paris

CADDET (Centre for the Analysis and Dissemination of Demonstrated Energy Technologies) (2001) *Vacuum Glazing with Excellent Heat Insulation Properties*, International Energy Agency, Paris

Calcagni, B. and Paroncini, M. (2004) 'Daylight factor prediction in atria building designs', *Solar Energy*, vol 76, pp669–682

Cameron, A. (2005) 'Steady as she goes: BTM's world market update', *Renewable Energy World*, vol 8, no 4, pp134–145

Camilleri, S. (1995) 'Development of a silent, high efficiency ceiling fan', *CADDET Energy Efficiency Newsletter*, June 1995, pp6–8

Camilleri, S. (2001) 'Development of a silent, high efficiency ceiling fan', *CADDET Energy Efficiency Newsletter* No. 2, 2001. Available from www.cadet.org

Cane, D. and Garnet, J. M. (2000) 'Update on maintenance and service costs of commercial building ground-source heat pump systems', *ASHRAE Transactions*, vol 106 (Part 1), pp399–407

Caneta Research Incorporated (1995) *Commercial/Institutional Ground-Source Heat Pump Engineering Manual*, American Society of Heating, Refrigerating and Air-Conditioning Engineers, Atlanta

Capeluto, I. G., Yezioro, A. and Shaviv, E. (2003) 'Energy, economics and architecture', in *Eighth International IBPSA Conference*, Eindhoven, The Netherlands, 11–14 August 2003, pp123–130. Available from www.ibpsa.org

Capeluto, I. G., Yezioro, A. and Shaviv, E. (2004) 'What are the required conditions for heavy structure buildings to be thermally effective in a hot humid climate?', *Journal of Solar Energy Engineering*, vol 126, pp886–892

Carlson, S. W. (2000) 'GSHP bore field performance comparisons of standard and thermally enhanced grout', *ASHRAE Transactions*, vol 106 (Part 2), pp442–446

Carmody, J., Selkowitz, S., Lee, E. S., Arasteh, D. and Willmert, T. (2004) *Window Systems for High-Performance Buildings*, Norton, New York, 400pp

Carpenter, S. C. (2003a) 'Integrated mechanical systems', in Hestnes, A. G., Hastings, R. and Saxhof, B. (eds) *Solar Energy Houses: Strategies, Technologies, Examples*, 2nd edition, James & James, London, pp52–54

Carpenter, S. C. (2003b) 'The Canadian advanced house in Brampton', in Hestnes, A. G., Hastings, R. and Saxhof, B. (eds) *Solar Energy Houses: Strategies, Technologies, Examples*, 2nd edition, James & James, London, pp83–88

Carpenter, S. C. (2003c) 'The Canadian advanced house in Waterloo', in Hestnes, A. G., Hastings, R. and Saxhof, B. (eds) *Solar Energy Houses: Strategies, Technologies, Examples*, 2nd edition, James & James, London, pp89–94

Carpenter, S. C. and Kokko, J. P. (1998) 'Radiant heating and cooling, displacement ventilation with heat recovery and storm water cooling: an environmentally responsible HVAC system', *ASHRAE Transactions*, vol 104 (Part 2), pp1321–1326

Carpenter, S. C., McGowan, A. G. and Miller, S. R. (1998) 'Window annual energy rating systems: what they tell us about residential window design and selection', *ASHRAE Transactions*, vol 104 (Part 2), pp806–813

Carrié, F. R., Bossaer, A., Andersson, J. V., Wouters, P. and Liddament, M. W. (2000) 'Duct leakage in European buildings: status and perspectives', *Energy and Buildings*, vol 32, pp235–243

Carter, D. J. (2004) 'Developments in tubular daylight guidance systems', *Building Research and Information*, vol 32, no 3, pp220–234

Casas, W. and Schmitz, G. (2005) 'Experiences with a gas driven, desiccant assisted air conditioning system with geothermal energy for an office building', *Energy and Buildings*, vol 37, pp493–501

Caton, J. A. and Turner, W. D. (1997) 'Cogeneration', in Kreith, F. and West, R. E. (eds) *CRC Handbook of Energy Efficiency*, CRC Press, Boca Raton, FL, pp669–683

CEC (California Energy Commission) (1996) *Source Energy and Environmental Impacts of Thermal Energy Storage*. Available from www.energy.ca.gov

CEC (Commission of the European Communities) (1991) *Solar Architecture in Europe: Design, Performance, and Evaluation*, Prism Press, Dorset, UK

CEE (Consortium for Energy Efficiency) (2001) *A Market Assessment for Condensing Boilers in Commercial Heating Applications*. Available from www.cee1.org/gas/gs-blrs/Boiler-assess.pdf.

Cerkvenik, B., Poredoš, A. and Ziegler, F. (2001) 'Influence of adsorption cycle limitations on the system performance', *International Journal of*

Refrigeration, vol 24, pp475–485

Cervero, R. (1998) *The Transit Metropolis: A Global Inquiry*, Island Press, Washington

Cetiner, I. and Özkan, E. (2005) 'An approach for the evaluation of energy and cost efficiency of glass façades', *Energy and Buildings*, vol 37, pp673–684

Chaichana, C., Aye, L. and Charters, W. W. S. (2003) 'Natural working fluids for solar-boosted heat pumps', *International Journal of Refrigeration*, vol 26, pp637–643

Chaichana, C., Charters, W. W. S. and Aye, L. (2001) 'An ice thermal storage computer model', *Applied Thermal Engineering*, vol 21, pp1769–1778

Chalfoun, N. V. (1998) 'Implementation of natural down-draft evaporative cooling devices in commercial buildings: the international experience', in *Proceedings of the 1998 ACEEE Summer Study on Energy Efficiency in Buildings*, vol 3, American Council for an Energy Efficient Economy, Washington, pp63–72

Chameides, W. L., Kasibhatla, P. S., Yienger, J. and Levy II, H. (1994) 'Growth of continental-scale metro-agro-plexes, regional ozone pollution, and world food production', *Science*, vol 264, pp74–77

Chao, C. Y. H. and Hu, J. S. (2004) 'Development of a dual-mode demand control ventilation strategy for indoor air quality and energy saving', *Environment and Building*, vol 39, pp385–397

Charron, R. and Athienitis, A. K. (2003) 'Optimization of the performance of PV-integrated double-façades', in *Proceedings of the International Solar Energy Society World Congress*, Gothenburg, Sweden, June 2003

Chaudhari, M., Frantzis, L. and Hoff, T. E. (2004) *PV Grid Connected Market Potential Under a Cost Breakthrough Scenario*, The Energy Foundation and Navigant Consulting. Available from www.navigantconsulting.com.

Chen, Q. and Glicksman, L. (2003) *System Performance Evaluation and Design Guidelines for Displacement Ventilation*, American Society of Heating, Refrigerating, and Air Conditioning Engineers, Atlanta, Final Reports 949

Chen, T. Y., Burnett, J. and Chau, C. K. (2001) 'Analysis of embodied energy use in the residential building of Hong Kong', *Energy*, vol 26, pp323–340

Chen, Z. D., Bandopadhayay, P., Halldorsson, J., Byrjalsen, C., Heiselberg, P. and Li, Y. (2003) 'An experimental investigation of a solar chimney model with uniform wall heat flux', *Building and Environment*, vol 38, pp893–906

Cheung, C. K., Fuller, R. J. and Luther, M. B. (2005) 'Energy-efficient envelope design for high-rise apartments', *Energy and Buildings*, vol 37, pp37–48

Chiasson, A. D. and Yavuzturk, C. (2003) 'Assessment of the viability of hybrid geothermal heat pump system with solar thermal collectors', *ASHRAE Transactions*, vol 109 (Part 2), pp487–500

Chiasson, A. D., Rees, S. J. and Spitler, J. D. (2000a) 'A preliminary assessment of the effects of groundwater flow on closed-loop ground-source heat pump systems', *ASHRAE Transactions*, vol 106 (Part 1), pp380–393

Chiasson, A. D., Spitler, J. D., Rees, S. J. and Smith, M. D. (2000b) 'A model for simulating the performance of a shallow pond as a supplemental heat rejecter with closed-loop ground-source heat pump systems', *ASHRAE Transactions*, vol 106 (Part 2), pp107–121

Chimack, M. J. and Sellers, D. (2000) 'Using extended surface air filters in heating ventilation and air conditioning systems: reducing utility and maintenance costs while benefiting the environment', in *Proceedings of the 2000 ACEEE Summer Study on Energy Efficiency in Buildings*, vol 3, American Council for an Energy Efficient Economy, Washington, DC, pp77–88

Chirarattananon, S., Chedsiri, S. and Renshen, L. (2000) 'Daylighting through light pipes in the tropics', *Solar Energy*, vol 69, pp331–341

Chow, T. T. (2003) 'Performance analysis of photovoltaic-thermal collector by explicit demand model', *Solar Energy*, vol 75, pp143–152

Chow, T. T., Zhang, G. Q., Lin, Z. and Song, C. L. (2002) 'Global optimization of absorption chiller system by genetic algorithm and neural network', *Energy and Buildings*, vol 34, pp103–109

Chow, T. T., Hand, J. W. and Strachan, P. A. (2003) 'Building-integrated photovoltaic and thermal applications in a subtropical hotel building', *Applied Thermal Engineering*, vol 23, pp2035–2049

Chow, T. T., Lin, Z. and Liu, J. P. (2002) 'Effect of condensing unit layout at building re-entrant on split-type air-conditioner performance', *Energy and Buildings*, vol 34, pp237–244

Chow, W. K. and Chow, C. L. (2005) 'Evacuation with smoke control for atria in green and sustainable buildings', *Building and Environment*, vol 40, pp195–200

Chrisomallidou, N. (2001a) 'Guidelines for integrating energy conservation techniques in urban buildings', in Santamouris, M. (ed) *Energy and Climate in the Urban Built Environment*, James & James, London, pp247–309

Chrisomallidou, N. (2001b) 'Examples of urban buildings', in Santamouris, M. (ed) *Energy and*

Climate in the Urban Built Environment, James & James, London, pp311–380

Chua, H. T., Toh, H. K., Malek, A., Ng, K. C. and Srinivasan, K. (2000) 'A general thermodynamic framework for understanding the behaviour of absorption chillers', *International Journal of Refrigeration*, vol 23, pp491–507

Chua, H. T., Ng, K. C., Malek, A., Kashiwagi, T., Akisawa, A. and Saha, B. B. (2001) 'Multi-bed regenerative adsorption chiller – improving the utilization of waste heat and reducing the chiller water outlet temperature fluctuation', *International Journal of Refrigeration*, vol 24, pp124–136

Chung, M., Park, J. and Yoon, H. (1998) 'Simulation of a central solar heating system with seasonal storage in Korea', *Solar Energy*, vol 64, pp163–178

Chunnanond, K. and Aphornratana, S. (2004) 'Ejectors: applications in refrigeration technology', *Renewable and Sustainable Energy Reviews*, vol 8, pp129–155

Ciampi, M., Leccese, F. and Tuoni, G. (2003) 'Ventilation facades energy performance in summer cooling of buildings', *Solar Energy*, vol 75, pp491–502

Ciampi, M., Leccese, F. and Tuoni, G. (2005) 'Energy analysis of ventilated and microventilated roofs', *Solar Energy*, vol 79, pp183–192

CIBSE (Chartered Institution of Building Services Engineers) (2006) *Guide A: Environmental Design, 7th Edition*, Chartered Institution of Building Services Engineers, London, 336pp

Clarke, J. A. (2001) *Energy Simulation in Building Design*, 2nd edition, Butterworth-Heinemann, Oxford, 362pp

Claridge, D. E., Liu, M., Deng, S., Turner, W. D., Haberl, J. S., Lee, S. U., Abbas, M. and Bruner, H. (2001) 'Cutting heating and cooling use almost in half without capital expenditure in a previously retrofit building', *European Council for an Energy Efficient Economy, 2001 Summer Proceedings*, vol 4, pp74–85

CMHC (Canadian Mortgage and Housing Corporation) (1993) *Commissioning and Monitoring the Building Envelope for Air Leakage*, Report 33127.02, Canadian Mortgage and Housing Corporation, Ottawa, 33pp. Available from www.cmhc-schl.gc.ca

CMHC (Canadian Mortgage and Housing Corporation) (1998) *Tap the Sun: Passive Solar Techniques and Home Designs*, Canadian Mortgage and Housing Corporation, Ottawa. Available from www.cmhc-schl.gc.ca

CMHC (Canadian Mortgage and Housing Corporation) (2000) *Healthy High-Rise: A Guide to Innovation in the Design and Construction of High-Rise Residential Buildings*, Canadian Mortgage and Housing Corporation, Ottawa, 55pp. Available from www.cmhc-schl.gc.ca.

CMHC (Canadian Mortgage and Housing Corporation)

(2004) *Exterior Insulation and Finish Systems (EIFS): Problems, Causes and Solutions*, Canadian Mortgage and Housing Corporation, Ottawa. Available from www.cmhc-schl.gc.ca. 1 vol. (various paginations) + CD-ROM

Cohen, R., Nelson, B. and Wolff, G. (2004) *Energy Down the Drain: The Hidden Costs of California's Water Supply*, Natural Resources Defence Council and Pacific Institute, New York and Oakland, 78pp. Available from www.nrdc.org

Cole, R. J. and Kernon, P. C. (1996) 'Life-cycle energy use in office buildings', *Building and Environment*, vol 31, pp301–317

Collier, R. K. (1997) *Desiccant Dehumidification and Cooling Systems Assessment and Analysis*, Pacific Northwest National Laboratory, Richland, Washington, Report PNNL-11694

Collins, R. E. and Simko, T. M. (1998) 'Current status of the science and technology of vacuum glazing', *Solar Energy*, vol 62, pp189–213

Compagno, A. (1999) *Intelligent Glass Façades*, Birkhäuser, Basel, 182pp.

Compagnon, R. (2004) 'Solar and daylight availability in the urban fabric', *Energy and Buildings*, vol 36, pp321–328

Conde Engineering (2002) *Properties of Ordinary Water Substance at Saturation from the Critical Point Down to −60°C, Equations and Tables*. Available from www.mrc-eng.com

Conde Engineering (2004) *Aqueous Solutions of Lithium and Calcium Chlorides: Property Formulations for Use in Air Conditioning Equipment Design*, 27pp. Available from www.mrc-eng.com

Conroy, C. L. and Mumma, S. A. (2001) 'Ceiling radiant cooling panels as a viable distributed parallel sensible cooling technology integrates with dedicated outdoor air systems', *ASHRAE Transactions*, vol 107 (Part 1), pp578–585

Costelloe, B. and Finn, D. (2003a) 'Indirect evaporative cooling potential in air-water systems in temperate climates', *Energy and Buildings*, vol 35, pp573–591

Costelloe, B. and Finn, D. (2003b) 'Experimental energy performance of open cooling towers used under low and variable approach conditions for indirect evaporative cooling in buildings', *Building Service Engineering Resources Technology*, vol 24, no 3 pp163–177

Courret, G., Scartezzini, J.-L., Francioli, D. and Meyer, J.-J. (1998) 'Design and assessment of an anidolic light-duct', *Energy and Buildings*, vol 28, pp79–99

Coventry, J. S. (2005) 'Performance of a concentrating photovoltaic/thermal solar collector', *Solar Energy*, vol 78, pp211–222

Cox, P. M., Betts, R. A., Jones, C. D., Spall, S. A. and

Totterdell, I. J. (2000) 'Acceleration of global warming due to carbon-cycle feedbacks in a coupled climate model', *Nature*, vol 408, pp184–187

Crawford, R. H., Treloar, G. J., Ilozor, B. D. and Love, P. E. D. (2003) 'Comparative greenhouse emissions analysis of domestic solar hot water systems', *Building Research and Information*, vol 31, pp34–47

Crawley, D. B., Lawrie, L. K., Winkelmann, F. C., Buhl, W. F., Huang, Y. J., Pedersen, C. O., Strand, R. K., Liesen, R. J., Fisher, D. E., Witte, M. J. and Glazer, J. (2001) 'Energy Plus: creating a new-generation building energy simulation program', *Energy and Buildings*, vol 33, pp319–331

Crawley, D. B., Hand, J. W., Kummert, M. and Griffith, B. T. (2005) *Contrasting the Capabilities of Building Energy Performance and Simulation Programs.* US Department of Energy, Washington. Available from www.eere.energy.gov/buildings/tools_directory

Crosbie, M. J. (1998) *The Passive Solar Design and Construction Handbook*, John Wiley and Sons, New York, 291pp

Crowder, H. and Foster, C. R. (1998) 'Building energy codes: new trends', in *Proceedings of the 1998 ACEEE Summer Study on Energy Efficiency in Buildings*, vol 10, American Council for an Energy Efficient Economy, Washington, DC, pp31–40

Curcija, D., Arasteh, D., Huizenga, C., Kohler, C., Mitchell, R. and Bhandari, M. (2001) 'Analyzing thermal performance of building envelope components using 2-D heat transfer tool with detailed radiation modeling', in *Seventh International IBPSA Conference*, Rio de Janeiro, Brazil, 13–15 August 2001, pp219–226. Available from www.ibpsa.org.

Curti, V., von Spakovsky, M. R. and Favrat, D. (2000) 'An environomic approach for the modeling and optimization of a district heating network based on centralized and decentralized heat pumps, cogeneration, and/or gas furnace. Part I: Methodology', *International Journal of Thermal Science*, vol 39, pp721–730

D'Accadia, M. D. (2001) 'Optimal operation of a complex thermal system: a case study', *International Journal of Refrigeration*, vol 24, pp290–301

da Graça, G. C., Chen, Q., Glicksman, L. R. and Norfold, L. K. (2002) 'Simulation of wind-driven ventilative cooling systems for an apartment building in Beijing and Shanghai', *Energy and Buildings*, vol 34, pp1–11

da Graca, G. C., Linden, P. F. and Haves, P. (2004) 'Design and testing of a control strategy for a large, naturally ventilated office building', *Building Service Engineering Resources Technology*, vol 25, no 3,

pp223–239

Dai, Y. J., Sumathy, K., Wang, R. Z. and Li, Y. G. (2003) 'Enhancement of natural ventilation in a solar house with a solar chimney and a solid adsorption cooling cavity', *Solar Energy*, vol 74, pp65–75

Dai, Y. J., Wang, R. Z., Zhang, H. F. and Yu, J. D. (2001) 'Use of liquid desiccant cooling to improve the performance of vapor compression air conditioning', *Applied Thermal Engineering*, vol 21, pp1185–1202

Dalenbäch, J.-O. (1990) *Central Solar Heating Plants with Seasonal Storage – A Status Report*, IEA Solar Heating and Cooling Programme, Task VII. Available from www.iea-shc.org

Daly, A. (2002) 'Underfloor air distribution: lessons learned', *ASHRAE Journal*, vol 44, no 5, pp21–24

Daly, H.E. and Cobb, J.B (1994) *For the Common Good: Redirecting the Economy Toward Community, the Environment, and a Sustainable Future*, Beacon Press, Boston, 534pp

Darley, J. (2004) *High Noon for Natural Gas: The New Energy Crisis*, Chelsea Green, White River (Vermont)

Darwin, R. and Kennedy, D. (2000) 'Economic effects of CO_2 fertilization of crops: transforming changes in yield into changes in supply', *Environmental Modeling and Assessment*, vol 5, pp157–168

Dascalaki, E. and Santamouris, M. (2002) 'On the potential of retrofitting scenarios for offices', *Building and Environment*, vol 37, pp557–567

Davis, B., Baylon, D. and Hart, R. (2002) 'Identifying energy savings potential on rooftop commercial units', in *Proceedings of the 2002 ACEEE Summer Study on Energy Efficiency in Buildings*, vol 3, American Council for an Energy Efficient Economy, Washington, pp79–91

de Carli, M. and Olesen, B. W. (2002) 'Field measurements of operative temperatures in buildings heated or cooled by embedded water-based radiant systems', *ASHRAE Transactions*, vol 108 (Part 2), pp714–725

de Dear, R. J. and Brager, G. S. (1998) 'Developing an adaptive model of thermal comfort and preference', *ASHRAE Transactions*, vol 104 (Part 1A), pp145–167

de Dear, R. J. and Brager, G. S. (2002) 'Thermal comfort in naturally ventilated buildings: revisions to ASHRAE Standard 55', *Energy and Buildings*, vol 34, pp549–561

de Dear, R. J., Leow, K. G. and Ameen, A. (1991a) 'Thermal comfort in the humid tropics – Part 1: climate chamber experiments on temperature preferences in Singapore', *ASHRAE Transactions* 97 (Part 1), pp874–879

de Dear, R. J., Leow, K. G. and Ameen, A. (1991b) 'Thermal comfort in the humid tropics – Part

1: climate chamber experiments on thermal acceptability in Singapore', *ASHRAE Transactions*, vol 97 (Part 1), pp880–886

DEG (Davis Energy Group) (2004a) *Alternatives to Compressor Cooling, Phase V: Integrated Ventilation Cooling*, California Energy Commission Report P500-04-009. Available from www.consumerenergycenter.org/bookstore/index.html

DEG (Davis Energy Group) (2004b) *Development of an Improved Two-Stage Evaporative Cooling System*, California Energy Commission Report P500-04-016. Available from www.consumerenergycenter.org/bookstore/index.html

de Herde, A. and Nihoul, A. (1997) *Improved Solar Renovation Concepts: A Report of Task 20 - Subtask B*, International Energy Agency, Solar Heating and Cooling Programme, Paris

Delp, W. W., Matson, N. E., Tschudy, E., Modera, M. P. and Diamond, R. C. (1998) 'Field investigation of duct system performance in California light commercial buildings', *ASHRAE Transactions*, vol 104 (Part 2), pp722–732

Delsante, A. and Vik, T. A. (eds) (2002) *Hybrid Ventilation: State-of-the-art review*, IEA Annex 35, International Energy Agency, Paris

Demirbilek, F. N., Yalçiner, U. G., Inanici, M. N., Ecevit, A. and Demirbilek, O. S. (2000) 'Energy conscious dwelling design for Ankara', *Building and Environment*, vol 35, pp33–40

Dennis, C. (2002) 'Reef under threat from "bleaching" outbreak', *Nature*, vol 415, p947

de Paepe, M., Theuns, E., Lenaers, S. and van Loon, J. (2003) 'Heat recovery system for dishwashers', *Applied Thermal Engineering*, vol 23, pp743–756

de Salis, M. H. F., Oldham, D. J. and Sharples, S. (2002) 'Noise control strategies for naturally ventilated buildings', *Building and Environment*, vol 37, pp471–484

de Wilde, P. and van der Voorden, M. (2004) 'Providing computational support for the selection of energy saving building components', *Energy and Buildings*, vol 36, pp749–758

de Wilde, P., van der Voorden, M., Brouwer, J., Augenbroe, G. and Kaan, H. (2001) 'The need for computational support in energy-efficiency design projects in the Netherlands', in *Seventh International IBPSA Conference*, Rio de Janeiro, Brazil, 13–15 August 2001, pp513–519. Available from www.ibpsa.org

Dharmadhikari, S. (1997) 'Consider trigeneration techniques for process plants', *Hydrocarbon Processing*, July 1997, pp91–100

Dharmadhikari, S., Pons, D. and Principaud, F. (2000) 'Contribution of stratified thermal storage to cost-effective trigeneration project', *ASHRAE Transactions*, vol 106 (Part 2), pp912–919

Diab, Y. and Achard, G. (1999) 'Energy concepts for utilization of solar energy in small and medium cities: the case of Chambéry', *Energy Conversion and Management*, vol 40, pp1555–1568

Dieckmann, J., Zogg, B., Westphalen, D., Roth, K. and Brodick, J. (2005) 'Heat-only, heat-activated heat pumps', *ASHRAE Journal*, vol 47, no 1 pp40–41

Dieng, A. O. and Wang, R. Z. (2001) 'Literature review on solar adsorption technologies for ice-making and air-conditioning purposes and recent developments in solar technology', *Renewable and Sustainable Energy Reviews*, vol 5, pp313–342

Dinçer, I. and Rosen, M. A. (2002) *Thermal Energy Storage, Systems and Applications*, John Wiley, New York, 579pp

Dixon, A., Butler, D. and Fewkes, A. (1999) 'Water saving potential of domestic water reuse systems using greywater and rainwater in combination', *Water Science Technology*, vol 39, pp25–32

Dobelmann, J. K. (2003) 'Germany's solar success: the 100,000 roofs programme reviewed', *Renewable Energy World*, vol 6, no 6, pp68–79

Dorer, V. and Breer, D. (1998) 'Residential mechanical ventilation systems: performance criteria and evaluations', *Energy and Buildings*, vol 27, pp247–255

Dorer, V., Weber, R. and Weber, A. (2005) 'Performance assessment of fuel cell mirco-cogeneration systems for residential buildings', *Energy and Buildings*, vol 37, pp1132–1146.

Downey, T. and Proctor, J. (2002) 'What can 13,000 air conditioners tell us?', in *Proceedings of the 2002 ACEEE Summer Study on Energy Efficiency in Buildings*, vol 1, American Council for an Energy Efficient Economy, Washington, pp53–67

Dubrous, F. M. and Wilson, A. G. (1992) 'A simple method for computing window energy performance for different locations and orientations', *ASHRAE Transactions*, vol 98 (Part 1), pp912–919

Ducarme, D., Wouters, P., Jardinier, M. and Jardinier, L. (1998) 'Practical experiences with IR-controlled supply terminals in dwellings and offices', *Energy and Buildings*, vol 27, pp275–282

Duer, K. and Svendsen, S. (1998) 'Monolithic silica aerogel in superinsulating glazings', *Solar Energy*, vol 63, pp259–267

Duff, W. S., Winston, R., O'Gallagher, J. J., Bergquam, J. and Henkel, T. (2003) 'Novel ICPC solar collector/

double effect absorption chiller demonstration project', in *2003 International Solar Energy Conference*, Kohala Coast, Hawaii, US, 15–18 March 2003, pp333–338

Duffin, R. J. and Knowles, G. (1981) 'Temperature control of buildings by adobe wall design', *Solar Energy*, vol 27, pp241–249

Duffin, R. J. and Knowles, G. (1984) 'Use of layered wall to reduce building temperature swings', *Solar Energy*, vol 33, pp543–549

Durkin, T. H. and Kinney, L. (2002) 'Two-pipe HVAC makes a comeback: an idea discarded decades ago may be the future of school heating and cooling', in *Proceedings of the 2002 ACEEE Summer Study on Energy Efficiency in Buildings*, vol 3, American Council for an Energy Efficient Economy, Washington, pp93–106

Durkin, T. H. and Rishel, J. B. (2003) 'Dedicated heat recovery', *ASHRAE Journal*, vol 45, no 10, pp18–24

Eames, I. W. and Wu, S. (2000) 'A theoretical study of an innovative ejector powered absorption-recompression cycle refrigerator', *International Journal of Refrigeration*, vol 23, pp475–484

Earp, A. A., Smith, G. B., Swift, P. D. and Franklin, J. (2004a) 'Maximising the light output of a luminescent solar concentrator', *Solar Energy*, vol 76, pp655–667

Earp, A. A., Smith, G. B., Franklin, J. and Swift, P. D. (2004b) 'Optimisation of a three-colour luminescent solar concentrator daylighting system', *Solar Energy Materials and Solar Cells*, vol 84, pp411–426

Edminster, A. V., Pettit, B., Ueno, K., Menegus, S. and Baczek, S. (2000) 'Case studies in resource-efficient residential buildings: the building America program', in *Proceedings of the 2000 ACEEE Summer Study on Energy Efficiency in Buildings*, vol 2, American Council for an Energy Efficient Economy, Washington, pp79–90

Edmonds, I. R. (2005) 'Daylighting high-density residential buildings with light redirecting panels', *Lighting Resources Technology*, vol 37, pp73–87

Edmonds, I. R. and Greenup, P. J. (2002) 'Daylighting in the tropics', *Solar Energy*, vol 73, pp111–121

Edwards, B. (1999) *Sustainable Architecture: European Directives and Building Design*, Architectural Press, Oxford, 277pp

Eftekhari, M. M. and Marjanovic, L. D. (2003) 'Application of fuzzy control in naturally ventilated buildings for summer conditions', *Energy and Buildings*, vol 35, pp645–655

Egan, A. (2001) 'Reasons, results, and remedies for

pump safety factors overuse', *ASHRAE Transactions*, vol 107 (Part 2), pp559–565

Eggen, G. (2000) 'Heat pumps for retrofitting heating systems – an international overview', *IEA Heat Pump Centre Newsletter*, vol 18, no 4, pp10–13

Eicker, U. (2003) *Solar Technologies for Buildings*, John Wiley, Chichester, 323pp

Eiffert, P. and Kiss, G. J. (2000) *Building-Integrated Photovoltaic Designs for Commercial and Institutional Structures: A Sourcebook for Architects*, NREL/BK-520-25272, National Renewable Energy Laboratory, Golden, Colorado, 89pp

Eiffert, P. and Task-7 Members (2002) *Building Integrated Photovoltaic Power Systems: Guidelines for Economic Evaluation*, International Energy Agency, Photovoltaic Power Systems Programme, Task 7, Paris. Summary available from www.iea-pvps.org.

Eijadi, D., Vaidya, P., Reinertsen, J. and Kumar, S. (2002) 'Introducing comparative analysis to the LEED system: a case for rational and regional application', in *Proceedings of the 2002 ACEEE Summer Study on Energy Efficiency in Buildings*, vol 9, American Council for an Energy Efficient Economy, Washington, pp83–98

Eley, C., Syphers, G. and Stein, J. R. (1998) 'Contracting for new building energy efficiency', in *Proceedings of the 1998 ACEEE Summer Study on Energy Efficiency in Buildings*, vol 3, American Council for an Energy Efficient Economy, Washington, pp131–142

Eley Associates (2001) *California 2001 vs. ASHRAE 1999: Comparison of ASHRAE/IESNA Standard 90.1-1999 and the 2001 California Nonresidential Energy Efficiency Standard*, California Energy Commission, 21pp

Elleson, J. and Dettmers, D. (1999) *Arts and Sciences Park Cooling System Evaluation*, HVAC&R Center, University of Wisconsin-Madison, 33pp. Available from www.hvacr.wisc.edu.

Ellis, M. W. and Mathews, E. H. (2002) 'Needs and trends in building and HVAC system design tools', *Building and Environment*, vol 37, pp461–470

Elseragy, A. A. B. and Gadi, M. G. (2003) 'Computer simulation of solar radiation received by curved roof in hot-arid regions', in *Eighth International IBPSA Conference*, Eindhoven, The Netherlands, 11–14 August 2003, pp283–290. Available from www.ibpsa.org

Energy for Sustainable Development (ESD) (2001) *The Future of CHP in the European Market – The European Cogeneration Study, Final Publishable Report*, 88pp. Available from http://tecs.energyprojects.net.

Enermodal Engineering (2003) *Laboratories for the 21st Century: Energy Analysis*. US Department of Energy, Washington, DOE/GO-102003-1694, 80pp. Available from www.labs21century.gov/toolkit/design_guide.htm

Enkemann, T., Kruse, H. and Oostendorp, P. A. (1997) 'CO_2 as a heat pump working fluid for retrofitting hydronic heating systems in western Europe', in *Workshop Proceedings – CO_2 Technology in Refrigeration, Heat Pump, and Air Conditioning Systems*, 13–14 May 1997, Trondheim, IEA Heat Pump Centre, Sittard (The Netherlands), pp79–101

Epstein, G., Abrey, D., Slote, S. and McCowan, B. (2002) 'New energy codes: technical assistance initiatives to aggressively support compliance', in *Proceedings of the 2002 ACEEE Summer Study on Energy Efficiency in Buildings*, vol 4, American Council for an Energy Efficient Economy, Washington, pp121–132

Erb, M. (2005) 'IEA/ECBCS – Annex 39, High performance thermal insulation' in Zimmermann, M. (ed.) *7th International Vacuum Insulation Symposium*, EMPA, Duebendorf, Switzerland, 28–29 September 2005, pp49–56. Available from www.empa.ch/VIP-Symposium.

Erell, E. and Etzion, Y. (1996) 'Heating experiments with a radiative cooling system', *Building and Environment*, vol 31, pp509–517

Erell, E., Etzion, Y., Carlstrom, N., Sandberg, M., Molina, J., Maestre, I., Maldonado, E., Leal, V. and Gutschker, O. (2004) '"SOLVENT": development of a reversible solar-screen glazing system', *Energy and Buildings*, vol 36, pp467–480

Erhorn, H., Beckert, M., Hillman, G., Kluttig, H., Reiss, J., Schmid, H.-M. and Schreck, H. (2003a) 'The German zero heating energy house, Berlin', in Hestnes, A. G., Hastings, R. and Saxhof, B. (eds) *Solar Energy Houses: Strategies, Technologies, Examples*, 2nd edition, James & James, London, pp109–114

Erhorn, H., Beckert, M., Hillman, G., Kluttig, H., Reiss, J., Schmid, H.-M. and Schreck, H. (2003b) 'The German ultra house in Rottweil', in Hestnes, A. G., Hastings, R. and Saxhof, B. (eds) *Solar Energy Houses: Strategies, Technologies, Examples*, 2nd edition, James & James, London, pp115–120

Eriksson, J. and Wahlström, A. (2002) 'Use of multizone air exchange simulation to evaluate a hybrid ventilation system', *ASHRAE Transactions*, vol 108 (Part 2), pp811–817

Erlandsson, M. and Levin, P. (2005) 'Environmental assessment of rebuilding and possible performance improvements effect on a national scale', *Building and Environment*, vol 40, pp1459–1471

Esource (1998a) *Design Brief: Energy Management Systems*, Energy Design Resources, Southern California Edison. Available from www.energydesignresources.com

Esource (1998b) *Design Brief: Integrated Energy Design*, Energy Design Resources, Southern California Edison. Available from www.energydesignresources.com

ESTIF (European Solar Thermal Industry Federation) (2003) *Sun in Action II – A Solar Thermal Strategy for Europe. Volume 1: Market Overview, Perspectives and Strategy for Growth*. Available from www.estif.org/11.0.html

Eumorfopoulou, E. and Aravantinos, D. (1998) 'The contribution of a planted roof to the thermal protection of buildings in Greece', *Energy and Buildings*, vol 27, pp29–36

European Commission (2001) *Green Paper – Towards a European Strategy for the Security of Energy Supply, Technical Document*. Available from ec.europa.eu

European Commission, Directorate General for Energy (1998) *Less is More. Energy Efficient Buildings with Less Installations*, European Communities, Luxembourg, 24pp

Everett, B. (1996) 'Solar thermal energy', in Boyle, G. (ed) *Renewable Energy, Power for a Sustainable Future*, Oxford University Press, Oxford, pp41–88

Faesy, R., Granda, C., Pratt, J. and McNally, M. (2004) 'Screw-based CFLs or pin-based fluorescent fixtures: which path to greatest lighting energy savings in homes?' in *Proceedings of the 2004 ACEEE Summer Study on Energy Efficiency in Buildings*, vol 2, American Council for an Energy Efficient Economy, Washington, pp69–78

Faggembauu, D., Costa, M., Soria, M. and Oliva, A. (2003a) 'Numerical analysis of the thermal behaviour of glazed ventilated facades in Mediterranean climates. Part I: development and validation of a numerical model', *Solar Energy*, vol 75, pp217–228

Faggembauu, D., Costa, M., Soria, M. and Oliva, A. (2003b) 'Numerical analysis of the thermal behaviour of glazed ventilated facades in Mediterranean climates. Part II: Applications and analysis of results', *Solar Energy*, vol 75, pp229–239

Fang, W. and Nutter, D. W. (1999) 'Analysis of system performance losses due to the reversing valve for a heat pump using R-410a', *ASHRAE Transactions*, vol 105 (Part 1), pp131–139

Fanger, P. O. (2001) 'Human requirements in future air-conditioned environments', *International Journal of Refrigeration*, vol 24, pp148–153

Faninger, G. (2000) 'Combined solar biomass district heating in Austria', *Solar Energy*, vol 69, pp425–435

Farrar, S., Hancock, E. and Anderson, R. (1998) 'System interactions and energy savings in a hot dry climate', in *Proceedings of the 1998 ACEEE*

Summer Study on Energy Efficiency in Buildings, vol 1, American Council for an Energy Efficient Economy, Washington, pp79–92

Fechter, J. V. and Porter, L. G. (1979) *Kitchen Range Energy Consumption*, Office of Energy Conservation, US Department of Energy, NBSIR 78-1556

Federal Energy Management Program (FEMP, USA) (2000a) *Technology Profiles: Front Loading Clothes Washers*. Available from www.eere.energy.gov

Federal Energy Management Program (FEMP, USA) (2000b) *Assessment of High-Performance, Family-Sized Commercial Clothes Washers*. Available from www.eere.energy.gov

Fehrm, M., Reiners, W. and Ungemach, M. (2002) 'Exhaust air heat recovery in buildings', *International Journal of Refrigeration*, vol 25, pp439–449

Feist, W., Schnieders, J., Dorer, V. and Haas, A. (2005) 'Re-inventing air heating: convenient and comfortable within the frame of the Passive House concept', *Energy and Buildings*, vol 37, pp1186–1203

Fennell, H. C. and Haehnel, J. (2005) 'Setting airtightness standards', *ASHRAE Journal*, vol 47, no 9, pp26–31

Feuermann, D. and Novoplansky, A. (1998) 'Reversible low solar heat gain windows for energy savings', *Solar Energy*, vol 62, pp169–175

Feustel, H. E. and Stetiu, C. (1995) 'Hydronic cooling – preliminary assessment', *Energy and Buildings*, vol 22, pp193–205

Fiedler, F., Nordlander, S., Persson, T. and Bales, C. (2006) 'Thermal performance of combined solar and pellet heating systems', *Renewable Energy*, vol 31, pp73–88

Financial Times Energy (2001a) *Economizers*, Energy Design Resources, Southern California Edison. Available from www.energydesignresources.com

Financial Times Energy (2001b) *Design Brief: High-Intensity Fluorescents*, Energy Design Resources, Southern California Edison. Available from www.energydesignresources.com.

Fisch, M. N., Guigas, M. and Dalenbäck, J. O. (1998) 'A review of large-scale solar heating systems in Europe', *Solar Energy*, vol 63, pp355–366

Fischedick, M., Nitsch, J., Ramesohl, S. (2005) 'The role of hydrogen for the long term development of sustainable energy systems – a case study for Germany', *Solar Energy*, vol 78, pp678–686

Fischer, S. (2004) 'Assessing value of CHP system', *Journal*, vol 46, no 6, pp12–19

Fischer, J. C., Sand, J. R., Elkin, B. and Mescher, K. (2002) 'Active desiccant, total energy recovery hybrid system optimizes humidity control, IAQ, and energy efficiency in an existing dormitory facility', *ASHRAE Transactions*, vol 108 (Part 2), pp537–545

Fischer, V., Grüneis, H. and Richter, R. (1997) *Sir Norman Foster and Partners Commerzbank, Frankfurt am Main*, Edition Axel Menges, Stuttgart & London, 79pp

Fisher, D., Schmid, F. and Spata, A. J. (1999) 'Estimating the energy-saving benefit of reduced-flow and/or multi-speed commercial kitchen ventilation systems', *ASHRAE Transactions*, vol 105 (Part 1), pp1138–1151

Fisk, W. J., Delp, W., Diamond, R., Dickerhoff, D., Levinson, R., Modera, M., Mematollahi, M. and Wang, D. (2000) 'Duct systems in large commercial buildings: physical characterization, air leakage, and heat conduction gains', *Energy and Buildings*, vol 32, pp109–119

Florides, G. A., Tassou, S. A., Kalogirou, S. A. and Wrobel, L. C. (2002a) 'Measures used to lower building energy consumption and their cost effectiveness', *Applied Energy*, vol 73, pp299–328

Florides, G. A., Tassou, S. A., Kalogirou, S. A. and Wrobel, L. C. (2002b) 'Review of solar and low energy cooling technologies for buildings', *Renewable and Sustainable Energy Reviews*, vol 6, pp557–572

Flourentzos, F., Droutsa, K. and Wittchen, K. B. (2000) 'EPIQR software', *Energy and Buildings*, vol 31, pp129–136

Flourentzos, F., Genre, J. L. and Roulet, C.-A. (2002) 'TOBUS software – an interactive decision aid tool for building retrofit studies', *Energy and Buildings*, vol 34, pp193–202

Flückiger, B., Monn, C., Lüthy, P. and Wanner, H.-U. (1998) 'Hygienic aspects of ground-coupled air systems', *Indoor Air*, vol 8, pp197–202

Ford, B., Patel, N., Zaveri, P. and Hewitt, M. (1998) 'Cooling without air conditioning: the Torrent Research Centre, Ahmedabad, India', *Renewable Energy*, vol 15, pp177–182

Fordham, M. (2000) 'Natural ventilation', *Renewable Energy*, vol 19, pp17–37

Forest, C. E., Stone, P. H., Sokolov, A. P., Allen, M. R. and Webster, M. D. (2002) 'Quantifying uncertainties in climate system properties with the use of recent climate observations', *Science*, vol 295, pp113–117

Foster, S., Calwell, C. and Horowitz, N. (2004) 'If we're only snoozing, we're losing: opportunities to save energy by improving the active mode efficiency of consumer electronics and office equipment', in *Proceedings of the 2004 ACEEE Summer Study on Energy Efficiency in Buildings*, vol 8, American Council for an Energy Efficient Economy, Washington, DC, pp110–123

Fountain, M. E., Arens, E., Xu, T., Bauman, F. S. and Oguru, M. (1999) 'An investigation of thermal comfort at high humidities', *ASHRAE Transactions*, vol 105, pp94–103

Francisco, P. W., Davis, B., Baylon, D. and Palmiter, L. (2004) 'Heat pumps system performance in Northern Climates', *ASHRAE Transactions*, vol 110 (Part 1), pp442–451

Francisco, P. W., Palmiter, L. and Davis, B. (1998) 'Modeling the thermal distribution efficiency of ducts: comparisons to measured results', *Energy and Buildings*, vol 28, pp287–297

Franconi, E. (1998) 'Measuring advances in HVAC distribution system design', in *Proceedings of the 1998 ACEEE Summer Study on Energy Efficiency in Buildings*, vol 3, American Council for an Energy Efficient Economy, Washington, DC, pp153–165

Franklin Associates (2001) *A Life Cycle Inventory of Selected Commercial Roofing Products*, Athena Sustainable Materials Institute, Merrickville, Ontario, Canada, 36pp

Friedrich, H. and Levitus, S. (1972) 'An approximation to the equation of state for sea water, suitable for numerical ocean models', *Journal of Physical Oceanography*, vol 2, pp514–517

Fung, A. S., Aulenback, A., Ferguson, A. and Ugursal, V. I. (2003) 'Standby power requirements of household appliances in Canada', *Energy and Buildings*, vol 35, pp217–228

Furbo, S. and Jivan Shah, L. (2003) 'Thermal advantages for solar heating systems with a glass cover with antireflection surfaces', *Solar Energy*, vol 74, pp513–523

Furbo, S., Andersen, E., Knuden, S., Vejen, N. K. and Shah, L. J. (2005) 'Smart solar tanks for small solar domestic hot water systems', *Solar Energy*, vol 78, pp269–279

Gabrielsson, A., Bergdahl, U. and Moritz, L. (2000) 'Thermal energy storage in soils at temperatures reaching 90°C', *Journal of Solar Energy Engineering*, vol 122, pp3–8

Gage, S. A. and Graham, J. M. R. (2000) 'Static split duct roof ventilators', *Building Research and Information*, vol 28, pp234–244

Gallo, C. (1998) 'Chapter 5: the utilization of microclimate elements', *Renewable and Sustainable Energy Reviews*, vol 2, pp89–114

GAMA (Gas Appliance Manufacturer's Association) (2005) *March 2005 Consumers' Directory of Certified Efficiency Ratings for Heating and Water Heating Equipment*. Available from www.gamanet.org

Gamble, D., Dean, B., Meisegeier, D. and Hall, J. (2004) 'Building a path towards zero energy homes with energy efficient upgrades', in *Proceedings of the 2004 ACEEE Summer Study on Energy Efficiency in Buildings*, vol 1, American Council for an Energy Efficient Economy, Washington, DC, pp95–106

Gan, A. I., Klein, S. A. and Reindl, D. T. (2000) 'Analysis of refrigerator/freezer appliances having dual refrigeration cycles', *ASHRAE Transactions*, vol 106 (Part 2), pp185–191

Gan, G. (1998) 'A parametric study of Trombe walls for passive cooling of buildings', *Energy and Buildings*, vol 27, pp37–43

Gan, G. (2000) 'Numerical evaluation of thermal comfort in rooms with dynamic insulation', *Building and Environment*, vol 35, pp445–453

Gan, G. and Riffat, S. B. (1998) 'A numerical study of solar chimney for natural ventilation of buildings with heat recovery', *Applied Thermal Engineering*, vol 18, pp1171–1187

Gan, G. and Riffat, S. B. (2001) 'Assessing thermal performance of an atrium integrated with photovoltaics', *Building Service Engineering Resources Technology*, vol 22, no 4, pp201–218

Gao, X., McInerny, S. A. and Kavanaugh, S. P. (2001) 'Efficiencies of an 11.2kW variable speed motor and drive', *ASHRAE Transactions*, vol 107 (Part 2), pp259–265

Garg, V. and Bansal, N. K. (2000) 'Smart occupancy sensors to reduce energy consumption', *Energy and Buildings*, vol 32, pp81–87

Garimella, S. (2003) 'Innovations in energy efficient and environmentally friendly space-conditioning systems', *Energy*, vol 28, pp1593–1614

Gasparella, A., Longo, G. A. and Marra, R. (2005) 'Combination of ground source heat pumps with chemical dehumidification of air', *Applied Thermal Engineering*, vol 25, pp295–308

Gaterell, M. R. and McEvoy, M. E. (2005) 'The impact of energy externalities on the cost effectiveness of energy efficiency measures applied to dwellings', *Energy and Buildings*, vol 37, pp1017–1027

Gauzin-Müller, D. (2002) *Sustainable Architecture and Urbanism*, Birkhäuser, Basel, 255pp

Gawlik, K. M. and Kutscher, C. F. (2002) 'A numerical and experimental investigation of low-conductivity unglazed, transpired solar air heaters', in *2002 International Solar Energy Conference*, Reno, Nevada, 15–20 June 2002, pp47–56

Gee, R., Cohen, G. and Greenwood, K. (2003)

'Operation and preliminary performance of the Duke solar power roof: a roof-integrated solar cooling and heating system', in *2003 International Solar Energy Conference*, Kohala Coast, Hawaii, US, 15–18 March 2003, pp295–300

Gee T. A., Cao, J., Mathias, J. A. and Christensen, R. N. (2001) 'Experimental testing and modeling of a dual-fired LiBr-H2O absorption chiller', *ASHRAE Transactions*, vol 107 (Part 1), pp3–11

Genchi, Y., Kikegawa, Y. and Inaba, A. (2002) 'CO_2 payback-time assessment of a regional-scale heating and cooling system using a ground source heat-pump in a high energy-consumption area in Tokyo', *Applied Energy*, vol 71, pp147–160

Genest, F. and Charneux, R. (2005) 'Creating synergies for sustainable design', *ASHRAE Journal*, vol 47, no 3, pp16–21

Georg, A., Graf, W., Schweiger, D., Wittwer, V., Nitz, P. and Wilson, H. R. (1998) 'Switchable glazing with a large dynamic range in total solar energy transmittance (TSET)', *Solar Energy*, vol 62, pp215–228

Gething, W. (2003) 'The environmental building: the Building Research Establishment, Watford', in Edwards, B. (ed) *Green Buildings Pay*, Spon Press, London, pp86–93

Ghiaus, C. (2003) 'Free-running building temperature and HVAC climatic suitability', *Energy and Buildings*, vol 35, pp405–411

Giabaklou, Z. and Ballinger, J. A. (1996) 'A passive evaporative cooling system by natural ventilation', *Building and Environment*, vol 31, pp503–507

Gitay, H. et al (2001) 'Ecosystems and their goods and services', in McCarthy, J. J., Canziani, O. S., Leary, N. A., Dokken, D. J. and White, K. S. (eds) *Climate Change 2001: Impacts, Adaptation, and Vulnerability*, Cambridge University Press, Cambridge, pp235–342

Givoni, B. (1994) *Passive and Low Energy Cooling of Buildings*, Van Nostrand Reinhold, New York, 263pp

Givoni, B. (1998a) *Climate Considerations in Building and Urban Design*, John Wiley, New York, 463pp

Givoni, B. (1998b) 'Effectiveness of mass and night ventilation in lowering the indoor daytime temperatures. Part I: 1993 experimental periods', *Energy and Buildings*, vol 28, pp25–32

Goetzberger, A., Müller, M. and Goller, M. (2000) 'A self-regulating glare protection system using concentrated solar radiation and thermotropic coating', *Solar Energy*, vol 69, suppl, pp45–57

Goldner, F. S. (1999) 'Control strategies for domestic hot water recirculation systems', *ASHRAE Transactions*, vol 105 (Part 1), pp1030–1046

Goldner, F. S. (2000) 'Effects of equipment cycling and sizing on seasonal efficiency', in *Proceedings of the 2000 ACEEE Summer Study on Energy Efficiency in Buildings*, vol 1, American Council for an Energy Efficient Economy, Washington, DC, pp89–100

Goldstein, L., Hedman, B., Knowles, D., Freedman, S. I., Woods, R. and Schweizer, T. (2003) *Gas-Fired Distributed Energy Resource Technology Characterizations*, *NREL/TP-620-34783*, National Renewable Energy Laboratory, Golden, Colorado. Available from www.nrel.gov/analysis/pdfs/2003/2003_gas-fired_der.pdf

Gombert, A., Glaubitt, W., Rose, K., Dreibholz, J., Zanke, C., Blasi, B., Heinzel, A., Horbelt, W., Sporn, D., Doll, W., Wittwer, V. and Luther, J. (1998) 'Glazing with very high solar transmittance', *Solar Energy*, vol 62, no 3, pp177–188

Gómez-Muñoz, V. M., Porta-Gándara, M. A. and Heard, C. (2003) 'Solar performance of hemispherical vault roofs', *Building and Environment*, vol 38, pp1431–1438

Gommed, K. and Grossman, G. (2004) 'A liquid desiccant system for solar cooling and dehumidification', *Journal of Solar Energy Engineering*, vol 126, pp879–885

González, J. E. and Alva Solari, L. H. (2002) 'Solar air conditioning systems with PCM solar collectors', in *2002 International Solar Energy Conference*, Reno, Nevada, 15–20 June 2002, pp97–107

Gordon, J. M. and Ng, K. C. (2000) 'High-efficiency solar cooling', *Solar Energy*, vol 68, pp23–31

Goss, J. O., Hyman, L. B. and Corbett, J. (1998) 'Integrated heating and cooling thermal energy storage with heat pump provides economic and environmental solutions', *ASHRAE Transactions*, vol 104 (Part 1B), pp1598–1606

Granqvist, C. G. (1989) 'Energy-efficient windows: options with present and forthcoming technology', in Johansson, T. B., Bodlund, B. and Williams, R. H. (eds) *Electricity: Efficient End-Use and New Generation Technologies and Their Planning Implications*, Lund University Press, Lund, pp89–123

Granqvist, C. G. (2003) 'Solar energy materials', *Advanced Materials*, vol 15, pp1–15

Granqvist, C. G., Azens, A., Hjelm, A., Kullman, L., Niklasson, G. A., Rönnow, D., Mattsson, M. S., Veszelei, M. and Vaivars, G. (1998) 'Recent advances in electrochromics for smart window applications', *Solar Energy*, vol 63, pp199–216

Gratia, E. and de Herde, A. (1997) 'Technology module 2, Passive solar heating', in *Solar Energy in European Office Buildings*, Altener program: Mid-Career Education, University College Dublin. Available from http://erg.ucd.ie/mid_career/mid_career.html

Gratia, E. and de Herde, A. (2004) 'Optimal operation of a south double-skin facade', *Energy and Buildings*, vol 36, pp41–60

Grätzel, M. (2001) 'Photoelectrochemical cells', *Nature*, vol 414, pp338–344

Green, M. A. (2000) 'Photovoltaics: technology overview', *Energy Policy*, vol 28, pp989–998

Green, M. A. (2003a) 'Crystalline and thin-film silicon solar cells: state of the art and future potential', *Solar Energy*, vol 74, pp181–192

Green, M. A. (2003b) 'General temperature dependence of solar cell performance and implications for device modelling', *Progress in Photovoltaics: Research and Applications*, vol 11, pp333–340

Green, M. A., Emery, K., King, D. L., Igari, S. and Warta, W. (2005) 'Solar cell efficiency tables (version 25)', *Progress in Photovoltaics: Research and Applications*, vol 13, pp49–54

Griffith, B. and Arasteh, D. (1995) 'Advanced insulations for refrigerator/freezers: the potential for new shell designs incorporating polymer barrier construction', *Energy and Buildings*, vol 22, pp219–231

Griffiths, P. W., Di Leo, M., Cartwright, P., Eames, P. C., Yianoulis, P., Leftheriotis, G. and Norton, B. (1998) 'Fabrication of evacuated glazing at low temperature', *Solar Energy*, vol 63, pp243–249

Grorud, C. (2001) 'Energy efficient ventilation filters', *CADDET Energy Efficiency Newsletter*, 13–15 August 2001, International Energy Agency, CADDET Energy Efficiency, Harwell, UK

Grossman, G. (2002) 'Solar-powered systems for cooling, dehumidification and air-conditioning', *Solar Energy*, vol 72, pp53–62

Grut, L. (2003) 'Daimler Chrysler Building, Berlin', in Edwards, B. (ed) *Green Buildings Pay*, Spon Press, London, pp86–93

Gu, L., Swami, M. V. and Fairey, P. W. (2003) 'System interactions in forced-air heating and cooling systems, Part I: equipment efficiency factors', *ASHRAE Transactions*, vol 109 (Part 1), pp475–484

Gugliermetti, F. and Bisegna, F. (2003) 'Visual and energy management of electrochromic windows in Mediterranean climate', *Building and Environment*, vol 38, pp479–492

Guillemin, A. and Molteni, S. (2002) 'An energy-efficient controller for shading devices self-adapting to the user wishes', *Building and Environment*, vol 37, pp1091–1097

Gusdorf, J., Hayden, S., Enchev, E., Swinton, M., Simpson, C. and Castellan, B. (2003) *Final Report on the Effects of ECM Furnace Motors on Electricity and Gas Use: Results from the CCHT Research Facility and Projections*, Canadian Centre for Housing Technology, NRCC-38500

Gutschner, M. and Task-7 Members (2001) *Potential for Building Integrated Photovoltaics*, International Energy Agency, Photovoltaic Power Systems Programme, Task 7, Paris. Summary available from www.iea-pvps.org

Haas, R. (2004) 'Progress in markets for grid-connected PV systems in the built environment', *Progress in Photovoltaics: Research and Applications*, vol 12, pp427–440

Haas, R., Auer, H. and Biermayr, P. (1998) 'The impact of consumer behavior on residential demand for space heating', *Energy and Buildings*, vol 27, pp195–205

Haddad, K. H. and Elmahdy, A. H. (1998) 'Comparison of the monthly thermal performance of a conventional window and a supply-air window', *ASHRAE Transactions*, vol 104 (Part 1B), pp1261–1270

Haddad, K. H. and Elmahdy, A. H. (1999) 'Comparison of the thermal performance of an exhaust-air window and a supply-air window', *ASHRAE Transactions*, vol 105 (Part 2), pp918–926

Hadorn, J. C. (1988) *Guide du Stockage Saisonniere de Chaleur*, SIA/OFEN, Document D 028, Zurich. English Version: 1990. *Guide to Seasonal Heat Storage*. Public Works Canada, Ottawa

Hahne, E. (2000) 'The ITW solar heating system: an oldtimer fully in action', *Solar Energy*, vol 69, pp469–493

Hallé, S. and Bernier, M. A. (1998) 'The combined effect of air leakage and conductive heat transfer in window frames and its impact on the Canadian energy rating procedure', *ASHRAE Transactions*, vol 104 (Part 1A), pp176–184

Haller, A., Schweizer, E., Braun, P. O. and Voss, K. (1997) *Transparent Insulation in Building Renovation*, James & James, London

Halliday, S. P., Beggs, C. B. and Sleigh, P. A. (2002) 'The use of solar desiccant cooling in the UK: a feasibility study', *Applied Thermal Engineering*, vol 22, pp1327–1338

Halozan, H. (1997) 'Residential heat pump systems and controls', *IEA Heat Pump Centre Newsletter*, vol 15, no 2, pp19–21

Halozan, H. and Rieberer, R. (1997) 'Air heating systems for low-energy buildings', *IEA Heat Pump Centre Newsletter*, vol 15(4), pp21–22

Halozan, H. and Rieberer, R. (1999) 'Heat pumps in low-heating-energy buildings', *20th International Congress of Refrigeration, IIR/IIF,* Volume V (paper 499), Sydney

Haltrecht, D., Zmeureanu, R. and Beausoleil-Morrison, I. (1999) 'Defining the methodology for the next-generation HOT2000 simulator', in *Sixth International IBPSA Conference*, Kyoto, Japan, 13–15 September 1999, pp61–68. Available from www.ibpsa.org.

Hamada, Y., Nakamura, M., Ochifuji, K., Yokoyama, S. and Nagano, K. (2003) 'Development of a database of low energy homes around the world and analysis of their trends', *Renewable Energy*, vol 28, pp321–328

Hamilton, S. D., Roth, K. W. and Brodrick, J. (2003) 'Improved duct sealing', *ASHRAE Journal*, vol 45, no 5, pp64–65

Hamlin, T. and Gusdorf, J. (1997) *Airtightness and Energy Efficiency of New Conventional and R-2000 Housing in Canada, 1997*, prepared for CANMET Energy Technology Centre, Ottawa. Available from www.buildingsgroup.nrcan.gc.ca

Hanby, V. I., Loveday, D. L. and Al-Almi, F. (2005) 'The optimal design for a ground cooling tube in a hot, arid climate', *Building Service Engineering Resources Technology*, vol 26, pp1–10

Hansen, J. (2004) 'Defusing the global warming time bomb', *Scientific American*, vol 290, pp68–77

Hansen, J. (2005) 'A slippery slope: how much global warming constitutes "dangerous anthropogenic interference"?', *Climatic Change*, vol 68, pp269–279

Harnisch, J., Höhne, N., Koch, M., Wartmann, S., Schwarz, W., Jenseit, W., Rheinberger, U., Fabian, P. and Jordan, A. (2003) *Risks and Benefits of Fluorinated Greenhouse Gases in Processes and Products under Special Consideration of the Properties Intrinsic to the Substance*, German Federal Environmental Agency (UBA), Berlin, Reference Z 1.6-50422/195

Harriman, L. G. and Judge, J. (2002) 'Dehumidification equipment advances', *ASHRAE Journal*, vol 44, no 8, pp22–29

Harvey, L. D. D. (1993) 'A guide to global warming potentials (GWPs)', *Energy Policy*, vol 21, pp24–34

Harvey, L. D. D. (2000) *Global Warming: The Hard Science*, Prentice Hall, Harlow, UK

Harvey, L. D. D. (2003) 'Impact of deep-ocean carbon sequestration on atmospheric CO_2 and on surface-water chemistry', *Geophysical Research Letters*, vol 30, no 5, p1237 doi:10.1029/2002GL016224

Harvey, L. D. D. (2004) 'Declining temporal effectiveness of carbon sequestration: implications for compliance with the United Nations Framework Convention on Climate Change', *Climatic Change*, vol 63, pp259–290

Harvey, L. D. D. and Kaufmann, R. K. (2002) 'Simultaneously constraining climate sensitivity and aerosol radiative forcing', *Journal of Climate*, vol 15, pp2837–2861

Hasnain, S. M. (1998) 'Review on sustainable thermal energy storage technologies, Part I: heat storage materials and techniques', *Energy Conversion and Management*, vol 39, pp1127–1138

Hassani, V. and Hauser, S. (1997) 'Fundamentals of thermodynamics, heat transfer, and fluid mechanics', in Kreith, F. and West, R. E. (eds) *CRC Handbook of Energy Efficiency*, CRC Press, Boca Raton, FL, pp19–99

Hastings, S. R. (ed) (1994) *Passive Solar Commercial and Institutional Buildings: A Sourcebook of Examples and Design Insights*, John Wiley, Chichester, 454pp

Hastings, S. R. (1995) 'Myths in passive solar design', *Solar Energy*, vol 55, pp445–451

Hastings, S. R. (2004) 'Breaking the "heating barrier" Learning from the first houses without conventional heating', *Energy and Buildings*, vol 36, pp373–380

Hastings, S. R. and Mørck, O. (2000) *Solar Air Systems: A Design Handbook*, James & James, London, 256pp

Haves, P., Linden, P. F. and Carrilho da Graca, G. (2004) 'Use of simulation in the design of a large, naturally ventilated office building', *Building Service Engineering Resources Technology*, vol 25, no 3, pp211–221

Hawkes, D. (1996) *The Environment Tradition: Studies in the Architecture of Environment*, E & FN Spon, London, 212pp

Hawkes, D. and Forster, W. (eds) (2002) *Energy Efficient Buildings: Architecture, Engineering, and Environment*, W. W. Norton, New York, 240pp

Hawlader, M. N. A. and Liu, B. M. (2002) 'Numerical study of the thermal-hydraulic performance of evaporative natural draft cooling towers', *Applied Thermal Engineering*, vol 22, pp41–59

Hawlader, M. N. A., Chou, S. K. and Ullah, M. Z. (2001) 'The performance of a solar-assisted heat pump water heating system', *Applied Thermal Engineering*, vol 21, pp1049–1065

Hay, J. E. (1986) 'Calculation of solar irradiances for inclined surfaces: validation of selected hourly and daily modes', *Atmosphere-Ocean*, vol 24, pp16–41

Hayter, S. J., Torcellini, P. A., Judkoff, R. and Jenior, M. M. (1998) 'Creating low-energy commercial buildings through effective design and evaluation', in *Proceedings of the 1998 ACEEE Summer Study on Energy Efficiency in Buildings*, vol 3, American Council for an Energy Efficient Economy, Washington, pp181–192

Hayter, S. J., Torcellini, P. A., Eastment, M. and Judkoff, R. (2000) 'Using the whole-building design approach to incorporate daylighting into a retail space', in *Proceedings of the 2000 ACEEE Summer Study on Energy Efficiency in Buildings*, vol 3, American Council for an Energy Efficient

Economy, Washington, pp173–184

He, B. and Setterwall, F. (2002) 'Technical grade paraffin waxes as phase change materials for cool thermal storage and cool storage systems capital cost estimation', *Energy Conversion and Management*, vol 43, pp1709–1723

He, J., Okumura, A., Hoyano, A. and Asano, K. (2001) 'A solar cooling project for hot and humid climates', *Solar Energy*, vol 71, pp135–145

Heimonen, I. (2004) 'Outdoor testing and analysis of the thermal performance of supply air window', *Dynamic Analysis Methods Applied to Energy Performance Assessment of Buildings. Warsaw, 13–14 May 2004*, DAME BC / PASLINK EEIG. Available through www.dynastee.info/Warsaw_cdrom.php or www.paslink.org

Heinonen, J. S., Vuolle, M., Heikkinen, J. and Seppänen, O. (2002) 'Performance simulations of hybrid ventilation systems in a five-story office building', *ASHRAE Transactions*, vol 108 (Part 2), pp1241–1250

Heller, A. (2000) '15 years of R&D in central solar heating in Denmark', *Solar Energy*, vol 69, pp437–447

Henderson Jr., H. I. (1999) 'Implications of measured commercial building loads on geothermal system sizing', *ASHRAE Transactions*, vol 105 (Part 2), pp1189–1198

Henderson Jr., H. I., Khattar, M. K., Carlson, S. W. and Walburger, A. C. (2000a) 'The implications of the measured performance of variable flow pumping systems in geothermal and water loop heat pump applications', *ASHRAE Transactions*, vol 106 (Part 2), pp533–542

Henderson Jr., H. I., Parker, D. and Huang, Y. J. (2000b) 'Improving DOE-2's RESYS routine: user defined functions to provide more accurate part load energy use and humidity predictions', in *Proceedings of the 2000 ACEEE Summer Study on Energy Efficiency in Buildings*, vol 1, American Council for an Energy Efficient Economy, Washington, DC, pp113–124

Henderson-Sellers, A. and Robinson, P. J. (1986) *Contemporary Climatology*, Longman, Harlow, UK, 439pp

Hendron, R., Anderson, R., Judkoff, R., Christensen, C., Eastment, M., Norton, P., Reeves, R. and Hancock, E. (2004) *Building America Performance Analysis Procedures Revision 1*, US Deptartment of Energy, Washington, DC

Henning, H.-M. (ed) (2004a) *Solar-assisted Air Conditioning in Buildings – A Handbook for Planners*, Springer-Verlag Wien, Vienna, 150pp

Henning, H.-M. (2004b) 'A breath of fresh air: use of solar-assisted air conditioning in buildings', *Renewable Energy World*, vol 7, no 1, pp94–103

Henning, H.-M., Erpenbeck, T., Hindenburg, C. and Santamaria, I. S. (2001) 'The potential of solar energy use in desiccant cooling cycles', *International Journal of Refrigeration*, vol 24, pp220–229

Henning, H.-M. and Hindenburg, C. (1999) *Economic Study of Solar Assisted Cooling Systems, Bericht TOS1-HMH-9905-E01*, Fraunhofer Institute for Solar Energy Systems, Freiburg, 58pp

Henry, R. and Patenaude, A. (1998) 'Measurements of window air leakage at cold temperatures and impact on annual energy performance of a house', *ASHRAE Transactions*, vol 104 (Part 1B), pp1254–1260

Henry, R. and Dubrous, F. (1998) 'Are window energy performance selection requirements in line with product design in heating-dominated climates?', *ASHRAE Transactions*, vol 104 (Part 2), pp799–805

Hensen, J., Bartak, M. and Drkal, F. (2002) 'Modeling and simulation of a double-skin façade system', *ASHRAE Transactions*, vol 108 (Part 2), pp1251–1259

Henze, G. P., Krarti, M. and Brandemuehl, M. J. (2003) 'Guidelines for improved performance of ice storage systems', *Energy and Buildings*, vol 35, pp111–127

Hepting, C. and Ehret, D. (2005) *Centre for Interactive Research on Sustainability: Energy Performance Analysis Report*. Available from www.sdri.ubc.ca/cirs

Hernández, J. I., Estrada, C. A., Best, R. and Dorantes, R. J. (2003) 'Study of a solar booster assisted ejector refrigeration system with R134a', in *2003 International Solar Energy Conference*, Kohala Coast, Hawaii, USA, 15–18 March 2003, pp343–350

Herold, K. E. (1995) 'Design challenges in absorption chillers', *Mechanical Engineering*, vol 117, pp80–84

Herrera, A., Islas, J. and Arriola, A. (2003) 'Pinch technology application in a hospital', *Applied Thermal Engineering*, vol 23, pp127–139

Herzog, T. (1996) *Solar Energy in Architecture and Urban Planning*, Prestel, Munich, 223pp

Hestnes, A. G. (1999) 'Building integration of solar energy systems', *Solar Energy*, vol 67, pp181–187

Hestnes, A. G. and Kofoed, N. U. (1997) *OFFICE: Passive Retrofitting of Office Buildings to Improve their Energy Performance and Indoor Environment, Final Report of the Design and Evaluation Subgroup*, European Commission Directorate General for Science Research and Development, JOULE Programme: JOR3-CT96-0034

Hestnes, A. G. and Kofoed, N. U. (2002) 'Effective retrofitting scenarios for energy efficiency and comfort: results of the design and evaluation activities within the OFFICE project', *Building and Environment*, vol 37, pp569–574

Hestnes, A. G., Hastings, R. and Saxhof, B. (eds) (2003)

Solar Energy Houses: Strategies, Technologies, Examples, 2nd Edition, James & James, London

Hien, W. N., Poh, L. K. and Feriadi, H. (2000) 'The use of performance-based simulation tools for building design and evaluation – a Singapore perspective', *Building and Environment*, vol 35, pp709–736

Hien, W. N., Liping, W., Chandra, A. D., Pandey, A. R. and Xiaolin, W. (2005) 'Effects of double glazed façade on energy consumption, thermal comfort and condensation for a typical office building in Singapore', *Energy and Buildings*, vol 37, pp563–572

Hildebrandt, E. W., Bos, W. and Moore, R. (1998) 'Assessing the impacts of white roofs on building energy loads', *ASHRAE Transactions*, vol 104 (Part 1B), pp810–818

Hill, D. and Carruth, D. (2000) 'A case study of a successful innovative multi-unit residential building', in *Proceedings of the 2000 ACEEE Summer Study on Energy Efficiency in Buildings*, vol 1, American Council for an Energy Efficient Economy, Washington, DC, pp125–136

Hiller, C. C., Miller, J. and Dinse, D. R. (2002) 'Field test comparison of hot water recirculation loop vs. point-of-use water heaters in a high school', *ASHRAE Transactions*, vol 108 (Part 2), pp771–779

Hiller, M., Holst, S., Knirsch, A. and Schuler, M. (2001) 'TRNSYS 15 – A simulation tool for innovative concepts', in *Seventh International IBPSA Conference*, Rio de Janeiro, Brazil, 13–15 August 2001, pp419–422. Available from www.ibpsa.org

Hinge, A., Bertoldi, P. and Waide, P. (2004) 'Comparing commercial building energy use around the world', in *Proceedings of the 2004 ACEEE Summer Study on Energy Efficiency in Buildings*, vol 4, American Council for an Energy Efficient Economy, Washington, DC, pp136–147

Hirano, T., Kato, S., Murakami, S., Ikaga, T. and Shiraishi, Y. (2006a) 'A study on a porous residential building model in hot and humid regions: Part 1 – the natural ventilation performance and the cooling load reduction effect of the building model', *Building and Environment*, vol 41, pp21–32

Hirano, T., Kato, S., Murakami, S., Ikaga, T., Shiraishi, Y. and Uehara, H. (2006b) 'A study on a porous residential building model in hot and humid regions: Part 2 – reducing the cooling load by component-scale voids and the CO_2 reduction effect of the building model', *Building and Environment*, vol 41, pp33–44

Hirunlabh, J., Wachirapuwadon, S., Pratinthong, N. and Khedari, J. (2001) 'New configurations of a roof solar collector maximizing natural ventilation', *Building and Environment*, vol 36, pp383–391

Hitchcock, R. J. (2003) 'Software interoperability for energy simulation', *ASHRAE Transactions*, vol 109 (Part 1), pp661–664

HMG (Heschong Mahone Group) (2003) *Ceiling Insulation Report: Effectiveness of Lay-In Ceiling Insulation*, California Energy Commission Report 500-03-082-A14. Available from www.consumerenergycenter.org/bookstore/index.html

Hodder, S. G., Loveday, D. L., Parsons, K. C. and Taki, A. H. (1998) 'Thermal comfort in chilled ceiling and displacement ventilation environments: vertical radiant temperature asymmetry effects', *Energy and Buildings*, vol 27, pp167–173

Hoegh-Guldberg, O. (1999) 'Climate change, coral bleaching and the future of the world's coral reefs', *Marine and Freshwater Research*, vol 50, pp839–866

Hoeschele, M., Chitwood, R. and Pennington, B. (2002) 'Diagnostic performance assessment of 30 new California homes', in *Proceedings of the 2002 ACEEE Summer Study on Energy Efficiency in Buildings*, vol 1, American Council for an Energy Efficient Economy, Washington, DC, pp91–102

Hoeschele, M. A., Berman, M. J., Elberling, L. F. and Hunt, M. B. (1998) 'Evaporative condensers: the next generation in residential air conditioning?', in *Proceedings of the 1998 ACEEE Summer Study on Energy Efficiency in Buildings*, vol 1, American Council for an Energy Efficient Economy, Washington, DC, pp147–158

Hogan, J. (2005) '2001 Seattle energy code: striving for 20% total building energy savings compared to Standard 90.1-1999', *ASHRAE Transactions*, vol 111 (Part 1), pp444–456

Hohm, D. P. and M. E. Ropp (2003) 'Comparative study of maximum power point tracking algorithms', *Progress in Photovoltaics: Research and Applications*, vol 11, pp47–62

Holford, J. M. and Hunt, G. R. (2003) 'Fundamental atrium design for natural ventilation', *Building and Environment*, vol 38, pp409–426

Hollands, K. G. T., Wright, J. L. and Granqvist, C. G. (2001) 'Glazings and coatings', in Gordon, J. (ed) *Solar Energy: The State of the Art, ISES Position Papers*, James & James, London, pp29–107

Hollmuller, P. and Lachal, B. (2001) 'Cooling and preheating with buried pipe systems: monitoring, simulation and economic aspects', *Energy and Buildings*, vol 33, pp509–518

Holmes, M. (2001) 'Hybrid ventilation systems: an Arup approach to low energy cooling', in *Cooling Frontier Symposium – The Advanced Edge of Cooling Research and Applications in the Built Environment*, Arizona State University, 4–7 October, 2001, chapter 23

Holst, J. N. (2003) 'Using whole building simulation

models and optimizing procedures to optimize building envelope design with respect to energy consumption and indoor environment', in *Eighth International IBPSA Conference*, Eindhoven, The Netherlands, 11–14 August 2003, pp507–514. Available from www.ibpsa.org

Holton, J. K. and Rittelmann, P. E. (2002) 'Base loads (lighting, appliances, DHW) and the high performance house', *ASHRAE Transactions*, vol 108 (Part 1), pp232–242

Howe, M., Holland, D. and Livchak, A. (2003) 'Displacement ventilation – Smart way to deal with increased heat gains in the telecommunication equipment room', *ASHRAE Transactions*, vol 109 (Part 1), pp323–327

Hu, S., Chen, Q. and Glicksman, L. R. (1999) 'Comparison of energy consumption between displacement and mixing ventilation systems for different U.S. buildings and climates', *ASHRAE Transactions*, vol 105 (Part 2), pp453–464

Huang, B. J., Petrenko, V. A., Chang, J. M., Lin, C. P. and Hu, S. S. (2001) 'A combined-cycle refrigeration system using ejector-cooling cycle as the bottom cycle', *International Journal of Refrigeration*, vol 24, pp391–399

Huang, J., Lang, S., Hogan, J. and Lin, H. (2003) 'An energy standard for residential buildings in South China', in *Proceedings of the 3rd China Urban Housing Conference*, LBNL-53217, 3–5 July, 2003, Hong Kong Special Administrative Region, China

Hughes, H. M. (2001) 'New refrigerants for applied heat pumps', *ASHRAE Transactions*, vol 107 (Part 2), 613–616

Hughes, P. J. and Shonder, J. A. (1998) *The Evaluation of a 4000-Home Geothermal Heat Pump Retrofit at Fort Polk, Louisiana: Final Report*, Oak Ridge National Laboratory, ORNL/CON-460

Humm, O. (1994) 'Dynamic insulation – using transmission losses for building heating', *CADDET Energy Efficiency Newsletter*, 10–12 June, 1994, International Energy Agency, Harwell, UK

Humm, O. (1997) 'Ecology and economy when retrofitting apartment buildings', *IEA Heat Pump Centre Newsletter*, vol 15, no 4, pp17–18

Humphreys, M. A. (1970) 'A simple theoretical derivation of thermal comfort conditions', *Journal of the Institute of Heating and Ventilating Engineers*, vol 33, pp95–98

Hungerford, D. (2004) 'Living without air conditioning in a hot climate: thermal comfort in social context', in *Proceedings of the 2004 ACEEE Summer Study on Energy Efficiency in Buildings*, vol 7, American Council for an Energy Efficient Economy, Washington, DC, pp123–134

Hutson, S. S., Barber, N. L., Kenny, J. F., Linsey, K. S., Lumia, D. S. and Maupin, M. A. (2004) *Estimated Use of Water in the United States in 2000*, USGS Circular 1268. Available from http://water.usgs.gov

Hwang, Y. and Radermacher, R. (1999) 'Experimental investigation of the CO_2 refrigeration cycle', *ASHRAE Transactions*, vol 105 (Part 1), pp1219–1227

IEA (International Energy Agency) (1996) *Solar Collector System for Heating Ventilation air, CADDET Result 228*, OECD, Paris. Available from www.caddet-ee.org

IEA (International Energy Agency) (1999a) *World's Largest Solar Wall at Canadair Facility, CADDET Result 336*, OECD, Paris. Available from www.caddet-ee.org

IEA (International Energy Agency) (1999b) *District Cooling, Balancing the Production and Demand in CHP*, The Netherlands Agency for Energy and Environment, Sittard

IEA (International Energy Agency) (2000a) *Daylighting in Buildings: A Sourcebook on Daylighting Systems and Components*, International Energy Agency, Solar Heating and Cooling Programme, Task 21, Paris. Available from Lawrence Berkeley National Laboratory, Environmental Energy Technologies Division (http://eetd.lbl.gov) as LBNL-47493.

IEA (International Energy Agency) (2000b) *Experience Curves for Energy Technology Policy*, OECD, Paris, 127pp

IEA (International Energy Agency) (2001) *Saving Energy with Daylighting System, CADDET Maxi Brochure 14*, OECD, Paris. Available from www.caddet-ee.org

IEA (International Energy Agency) (2002a) *Ongoing Research Relevant for Solar Assisted Air Conditioning Systems*, International Energy Agency, Solar Heating and Cooling Programme, Task 25, Paris. Available from www.iea-shc.org

IEA (International Energy Agency) (2002b) *Photovoltaics/Thermal Solar Energy Systems: Status of the Technology and Roadmap for Future Development*, International Energy Agency, Photovoltaic Power Systems Programme, Task 7, Paris

IEA (International Energy Agency) (2002c) *The Integrated Design Process in Practice: Demonstration Projects Evaluated*, International Energy Agency, Solar Heating and Cooling Programme, Task 23, Paris. Available from www.iea-shc.org/task23

IEA (International Energy Agency) (2003) *Trends in Photovoltaic Applications: Survey Report of Selected IEA Countries Between 1992 and 2002*, International Energy Agency, Photovoltaic Power Systems Programme, Task 1, Paris. Available from www.iea-pvps.org

IEA (International Energy Agency) (2004a) *Energy Balances of non-OECD Countries 2001–2002*, International Energy Agency, Paris

IEA (International Energy Agency) (2004b) *Energy Statistics of non-OECD Countries 2001–2002*, International Energy Agency, Paris

IEA (International Energy Agency) (2004c) *Electricity Information 2004, with 2003 Data*, International Energy Agency, Paris

IEA (International Energy Agency) (2004d) *Oil Crises & Climate Challenges: 30 Years of Energy Use in IEA Countries*, International Energy Agency, Paris, 211pp

Imbabi, M. S. and Musset, A. (1996) 'Performance evaluation of a new hybrid solar heating and ventilation system optimised for U.K. weather conditions', *Building and Environment*, vol 31, pp145–153

Inoue, T. (2003) 'Solar shading and daylighting by means of autonomous responsive dimming glass: practical application', *Energy and Buildings*, vol 35, pp463–471

Interface Engineering (2005) *Engineering a Sustainable World: Design Process and Engineering Innovations for the Center for Health and Healing at the Oregan Health and Science University, River Campus*. Available through www.interface-engineering.com

IPCC/TEAP (Intergovernmental Panel on Climate Change/Technology and Economic Assessment Panel) (2005) *IPCC/TEAP Special Report on Safeguarding the Ozone Layer and the Global Climate System: Issues related to Hydrofluorocarbons and Perfluorocarbons*, Cambridge University Press, Cambridge

Iribarne, J. V. and Godson, W. L. (1973) *Atmospheric Thermodynamics*, Reidel, Dordrecht

Ismail, K. A. R. and Henríquez, J. R. (2002) 'Parametric study on composite and PCM glass systems', *Energy Conversion and Management*, vol 43, pp973–993

Jaboyedoff, P., Roulet, C.-A., Dorer, V., Weber, A. and Pfeiffer, A. (2004) 'Energy in air-handling units – results of the AIRLESS European project', *Energy and Buildings*, vol 36, pp391–399

Jacobs, P. and Henderson, H. (2002) *State-of-the-Art Review: Whole Building, Building Envelope, and HVAC Component and System Simulation and Design Tools*, prepared for the Air Conditioning and Refrigeration Technology Institute, Arlington, Virginia. Available from www.archenergy.com

Jagemar, L. (1997) *OFFICE: Passive Retrofitting of Office Buildings to Improve their Energy Performance and Indoor Environment, Final Report, Experimental Subtask*, JOULE Programme: JOR3-CT96-0034

Jahn, U. and Nasse, W. (2004) 'Operational performance of grid-connected PV systems on buildings in Germany', *Progress in Photovoltaics: Research and Applications*, vol 12, pp441–448

Jakob, M. (2006) 'Marginal costs and co-benefits of energy efficiency investments. The case of the Swiss residential sector', *Energy Policy*, vol 34, pp172–187

Jakob, M. and Madlener, R. (2003) *Exploring Experience Curves for the Building Envelope: An Investigation for Switzerland for the Period 1970–2020*, CEPE Working Paper No. 22. Available from www.cepe.ch

Jalalzadeh-Azar, A. A., Steele, W. G. and Hodge, B. K. (2000) 'Performance characteristics of a commercially available gas-fired desiccant system', *ASHRAE Transactions*, vol 106 (Part 1), pp95–104

James, P. W., Sonne, J. K., Vieira, R. K., Parker, D. S. and Anello, M. T. (1996) 'Are energy savings due to ceiling fans just hot air?', in *Proceedings of the 1996 ACEEE Summer Study on Energy Efficiency in Buildings*, vol 8, American Council for an Energy Efficient Economy, Washington, DC, pp89–93

Janssen, H., Carmeliet, J. and Hens, H. (2004) 'The influence of soil moisture transfer on building heat loss via the ground', *Building and Environment*, vol 39, pp825–836

Jenkins, D. and Muneer, T. (2004) 'Light-pipe prediction methods', *Applied Energy*, vol 79, pp77–86

Jennings, J. D., Rubinstein, F. M., DiBartolomeo, D. and Blanc, S. L. (2000) *Comparison of Control Options in Private Offices in an Advanced Lighting Controls Testbed*, Lawrence Berkeley National Laboratory, LBNL-43096 REV

Jensen, J. T. (2003) 'The LNG Revolution', *The Energy Journal*, vol 24, no 2, pp1–18

Jensen, K. I., Schultz, J. M. and Kristiansen, F. H. (2004) 'Development of windows based on highly insulating aerogel glazings', *Journal of Non-Crystalline Solids*, vol 350, pp351–357

Jeong, J. W., Mumma, S. A. and Bahnfleth, W. P. (2003) 'Energy conservation benefits of a dedicated outdoor air systems with parallel sensible cooling by ceiling radiant panels', *ASHRAE Transactions*, vol 109 (Part 2), pp627–636

Ji, J., Chow, T., Pei, G., Dong, J. and He, W. (2003) 'Domestic air-conditioner and integrated water heater for subtropical climate', *Applied Thermal Engineering*, vol 23, pp581–592

Johnson, S. (2000) *The Solid State Lighting Initiative: An Industry/DOE Collaborative Effort*, Lawrence Berkeley National Laboratory, LBNL-47589, Berkeley, CA, 5pp

Johnson, S. (2002) *LEDs – An Overview of the State of the Art in Technology and Application*, Lawrence Berkeley National Laboratory, LBNL-49742, Berkeley, CA,

6pp

Johnson, S. and Simmons, J. A. (2002) *Materials for Solid State Lighting*, Lawrence Berkeley National Laboratory, LBNL-49976, Berkeley, CA, 9pp

Johnson, W. S. (2002) 'Field tests of two residential direct exchange geothermal heat pumps', *ASHRAE Transactions*, vol 108 (Part 2), pp99–106

Jones, J. (2005) 'The growth challenge: what will it take for the PV industry to step up to large-scale manufacturing?' *Renewable Energy World*, vol 8, no 4, pp146–157

Jones, P.D. and Mann, M.E. (2004) 'Climate over past millenia', *Reviews of Geophysics*, vol 42, RG2002, doi:10.1029/2003RG000143

Jones, S. H. and Bicker, S. (2004) 'A comparative study of high-efficiency residential natural gas water heating', in *Proceedings of the 2004 ACEEE Summer Study on Energy Efficiency in Buildings*, vol 11, American Council for an Energy Efficient Economy, Washington, DC, pp61–71

Judkoff, R. and Farhar, B. C. (2000) 'Lessons learned: five years of home energy rating systems (HERS) and energy-efficient mortgages (EEMs) in the Pilot States', in *Proceedings of the 2000 ACEEE Summer Study on Energy Efficiency in Buildings*, vol 9, American Council for an Energy Efficient Economy, Washington, DC, pp201–214

Kaarsberg, T., Elliott, R. N. and Spurr, M. (1999) 'An integrated assessment of the energy savings and emissions-reduction potential of combined heat and power', updated and expanded from the version in the proceedings of the ACEEE 1999 Industrial Summer Study, American Council for an Energy-Efficient Economy, Washington, DC, 20036. Available from www.nemw.org

Kalogirou, S.A. (2002) 'Parabolic trough collectors for industrial process heat in Cyprus', *Energy*, vol 27, pp813–830

Kalogirou, S. A. (2004) 'Solar thermal collectors and applications', *Progress in Energy and Combustion Science*, vol 30, pp231–295

Kalogirou, S. A. and Papamarcou, C. (2000) 'Modelling of a thermosyphon solar water heating system and simple model validation', *Renewable Energy*, vol 21, pp471–493

Kammen, D. M. and Pacca, S. (2004) 'Assessing the costs of electricity', *Annual Review of Environment and Resources*, vol 29, pp301–344

Kapur, N. K. (2004) 'A comparative analysis of the radiant effect of external sunshades on glass surface temperatures', *Solar Energy*, vol 77, pp407–419

Karava, P., Stathopoulos, T. and Athienitis, A. K. (2003) 'Investigation of the performance of

trickle ventilators', *Building and Environment*, vol 38, pp981–993

Karlsson, J. and Roos, A. (2000) 'Modelling the angular behaviour of the total solar energy transmittance of windows', *Solar Energy*, vol 69, pp321–329

Karvountzi, G. C., Themelis, N. J. and Modi, V. (2002) 'Maximum distance to which cogenerated heat can be economically distributed in an urban community', *ASHRAE Transactions*, vol 108 (Part 1), pp334–339

Kato, S. and Chikamoto, T. (2002) *Pilot Study Report: Tokyo Gas Earth Port, Tokyo, Japan*, International Energy Agency, Energy Conservation in Buildings and Community Systems, Annex 35. Available from http://hybvent.civil.auc.dk

Katrakis, J. T., Knight, P. A. and Cavallo, J. D. (1994) *Energy-Efficient Rehabilitation of Multifamily Buildings in the Midwest*, Argonne National Laboratory, Argonne (Illinois). Available from www.eere.energy.gov

Kats, G., Alevantis, L. Berman, A., Mills, E. and Perlman, J. (2003) *The Costs and Financial Benefits of Green Buildings: A Report to California's Sustainable Building Task Force*, Sustainable Building Task Force

Kaushika, N. D. and Sumathy, K. (2003) 'Solar transparent insulation materials: a review', *Renewable and Sustainable Energy Reviews*, vol 7, pp317–351

Kavanaugh, S. P. (2001) 'Impact of design simplicity on the economics of geothermal heat pumps', *ASHRAE Transactions*, vol 107 (Part 2), pp481–486

Kavanaugh, S. P. and Lambert, S. E. (2004) 'A bin method energy analysis for ground-coupled heat pumps', *ASHRAE Transactions*, vol 110 (Part 1), pp535–542

Kavanaugh, S. P. and McInerny, S. A. (2001) 'Energy use of pumping options for ground-source heat pumps', *ASHRAE Transactions*, vol 107 (Part 1), pp589–599

Kavanaugh, S. P. and Xie, L. (2000) 'Energy use of ventilations air conditioning options for ground-source heat pump systems', *ASHRAE Transactions*, vol 106 (Part 2), pp543–550

Kaya, Y. (1989) *Impact of Carbon Dioxide Emission Control on GNP Growth: Interpretation of Proposed Scenarios*, IPCC Response Strategies Working Group Memorandum

Kaygusuz, K. (2000) 'Experimental and theoretical investigation of a solar heating system with heat pump', *Renewable Energy*, vol 21, pp79–102

Keoleian, G. A. and Lewis, G. M. (2003) 'Modeling the life cycle energy and environment performance of amorphous silicon BIPV roofing in the US', *Renewable Energy*, vol 28, pp271–293

Keoleian, G. A., Blanchard, S. and Reppe, P. (2001) 'Life-cycle energy, costs, and strategies for improving a single-family house', *Journal of Industrial Ecology*, vol 4, pp135–156

Kerr, R. A. (2005) 'Millennium's hottest decade retains title, for now', *Science*, vol 307, pp828–829

Kessling, W., Laevemann, E. and Kapfhammer, C. (1998) 'Energy storage for desiccant cooling systems component development', *Solar Energy*, vol 64, pp209–221

Khedari, J., Hirunlabh, J. and Bunnag, T. (1997) 'Experimental study of a roof solar collector towards the natural ventilation of new houses', *Energy and Buildings*, vol 26, pp159–164

Khedari, J., Boonsri, B. and Hirunlabh, J. (2000a) 'Ventilation impact of a solar chimney on indoor temperature fluctuation and air change in a school building', *Energy and Buildings*, vol 32, pp89–93

Khedari, J., Waewsak, J., Thepa, S. and Hirunlabh, J. (2000b) 'Field investigation of night radiation cooling under tropical climate', *Renewable Energy*, vol 20, pp183–193

Khedari, J., Ingkawanich, S., Waewsak, J. and Hirunlabh, J. (2002) 'A PV system enhanced the performance of roof solar collector', *Building and Environment*, vol 37, pp1317–1320

Khedari, J., Rachapradit, N. and Hirunlabh, J. (2003) 'Field study of performance of solar chimney with air-conditioned building', *Energy*, vol 28, pp1099–1114

Khudhair, A. M. and Farid, M. M. (2004) 'A review on energy conservation in building applications with thermal storage by latent heat using phase change materials', *Energy Conversion and Management*, vol 45, pp263–275

Kiatsiriroat, T., Tiansuwan, J., Suparos, T. and Na Thalang, K. (2000) 'Performance analysis of a direct-contact thermal energy storage-solidification', *Renewable Energy*, vol 20, pp195–206

Kilkis, I. B. (1998) 'Rationalization of low-temperature to medium-temperature district heating', *ASHRAE Transactions*, vol 104 (Part 2), pp565–576

Kilkis, I. B. (1999) 'Utilization of wind energy in space heating and cooling with hybrid HVAC systems and heat pumps', *Energy and Buildings*, vol 30, pp147–153

Kilkis, I. B. (2002) 'Environmental economy of low-enthalpy energy resources in district energy systems', *ASHRAE Transactions*, vol 108 (Part 2), pp580–588

Kim, S. G., Kim, Y. J., Lee, G. and Kim, M. S. (2005) 'The performance of a transcritical CO_2 cycle with an internal heat exchanger for hot water heating', *International Journal of Refrigeration*, vol 28, pp1064–1072

Kim, T. S. (2004) 'Comparative analysis on the part load performance of combined cycle plants considering design performance and power control strategy', *Energy*, vol 29, pp71–85

Kinney, L. (2004) 'Evaporative cooling for a growing southwest: technology, markets, and economics', in *Proceedings of the 2004 ACEEE Summer Study on Energy Efficiency in Buildings*, vol 11, American Council for an Energy Efficient Economy, Washington, DC, pp72–83

Kiss, G., Kinkead, J. and Raman, M. (1995) *Building-Integrated Photovoltaics: A Case Study*, NREL/TP-472-7574, National Renewable Energy Laboratory, Golden, CO

Kissock, K. (2000) 'Thermal load reduction from phase-change building components in temperature-controlled buildings', presented at *2000 International Solar Energy Conference, A part of SOLAR 2000: Solar Powers Life, Share the Energy*, Madison, Wisconsin, 16–21 June 2000

Klein, G. (2004a) 'Hot-water distribution systems. Part 1', *Plumbing Systems and Design*, March/April 2004, pp36–39

Klein, G. (2004b) 'Hot-water distribution systems. Part 2', *Plumbing Systems and Design*, May/June 2004, pp16–20

Klein, G. (2004c) 'Hot-water distribution systems. Part 3', *Plumbing Systems and Design*, Sept/Oct 2004, pp14–18

Klems, J. H. (2001) 'Net energy performance measurements on electrochromic skylights', *Energy and Buildings*, vol 33, pp93–102

Kleypas, J. (1998) 'Symposium participants assess future of coral reefs', *EOS*, vol 79, no 21, pp249–253

Kleypas, J., Buddemeier, R. W., Archer, D., Gattuso, J.-P., Langdon, C. and Opdyke, B. N. (1999) 'Geochemical consequences of increased atmospheric carbon dioxide on coral reefs', *Science*, vol 284, pp118–120

Knodel, B. D., France, D. M., Choi, U. S. and Wambsganss, M. W. (2000) 'Heat transfer and pressure drop in ice-water slurries', *Applied Thermal Engineering*, vol 20, pp671–685

Knutti, R., Stocker, T. F., Joos, F. and Plattner, G. K. (2002) 'Constraints on radiative forcing and future climate change from observations and climate model ensembles', *Nature*, vol 416, pp719–723

Koch-Nielsen, H. (2002) *Stay Cool: A Design Guide for the Built Environment in Hot Climates*, James & James, London, 159pp

Kolokotroni, M. (2001) 'Night ventilation cooling of office buildings: parametric analyses of conceptual energy impacts', *ASHRAE Transactions*, vol 107 (Part 1), pp479–489

Kolokotroni, M., Robinson-Gayle, S., Tanno, S. and Cripps, A. (2004) 'Environmental impacts analysis for typical office façades', *Building Research and Information*, vol 32, pp2–16

Koomey, J. G., Webber, C. A., Atkinson, C. S., Nicholls, A. and Holloman, B. (2000) 'Buildings sector', in *Scenarios for a Clean Energy Future, ORNL/CON-476*, Oak Ridge National Laboratory, Oak Ridge

Korn, D., Hinge, A., Dagher, F. and Partridge, C. (2000) 'Transformers efficiency: unwinding the technical potential', in *Proceedings of the 2000 ACEEE Summer Study on Energy Efficiency in Buildings*, vol 1, American Council for an Energy Efficient Economy, Washington, DC, pp149–162

Koschenz, M. and Dorer, V. (1999) 'Interaction of an air system with concrete core conditioning', *Energy and Buildings*, vol 30, pp139–145

Koschenz, M. and Lehmann, B. (2004) 'Development of a thermally activated ceiling panel with PCM for application in lightweight and retrofitted buildings', *Energy and Buildings*, vol 36, pp567–578

Kośny, J. (2001) 'Advances in residential wall technologies – simple ways of decreasing the whole building energy consumption', *ASHRAE Transactions*, vol 107 (Part 1), pp421–432

Kośny, J. and Kossecka, E. (2002) 'Multi-dimensional heat transfer through complex building envelope assemblies in hourly energy simulation programs', *Energy and Buildings*, vol 34, pp445–454

Kovács, P., Weiss, W., Bergmann, I., Meir, M. and Rekstad, J. (2003) 'Building-related aspects of solar combisystems', in Weiss, W. (ed) *Solar Heating Systems for Houses: A Design Handbook for Solar Combisystems*, James & James, London, pp93–124

Krampitz, I. (2005) 'A matter of raw material: PV industry expands production', *Renewable Energy World*, vol 8, no 1, pp76–86

Krapmeier, H. and Drössler, E. (eds) (2001) *CEPHEUS: Living Comfort Without Heating*, Springer-Verlag, Vienna

Krarti, M., Erickson, P. M. and Hillman, T. C. (2005) 'A simplified method to estimate energy savings of artificial lighting use from daylighting', *Building and Environment*, vol 40, pp747–754

Krauter, S., Salhi, M. J., Schroer, S. and Hanitsch, R. (2001) 'New façade system consisting of combined photovoltaic and solar thermal generators with building insulation', in *Seventh International IBPSA Conference*, Rio de Janeiro, Brazil, 13–15 August 2001, pp619–626. Available from www.ibpsa.org

Kreider, J. F. and Kreith, F. (1981) *Solar Energy Handbook*, McGraw Hill, New York

Kreider, J. F., Curtiss, P. S. and Rabl, A. (2002) *Heating and Cooling of Buildings: Design for Efficiency*, McGraw Hill, New York, 877pp

Krishan, A. (1996) 'The habitat of two deserts in India: hot-dry desert of Jaisalmer (Rajasthan) and the cold-dry high altitude mountainous desert of Leh (Ladakh)', *Energy and Buildings*, vol 23, pp217–229

Kubo, T., Sachs, H. and Nadel, S. (2001) *Opportunities for New Appliance and Equipment Efficiency Standards: Energy and Economic Savings Beyond Current Standards Programs*, American Council for an Energy-Efficient Economy, Washington, DC

Kuehn, M. and Mattner, D. (2003) 'Solar concepts for building', in Schittich, C. (ed) *Solar Architecture: Strategies, Visions, Concepts*, Birkhäuser, Basel, pp39–55

Kusakari (2005) 'The present status and the future view of the CO_2 refridgerant heat pump water heater for residential use', 8th IEA Heat Pump Conference. Available from www.heatpumpcentre.org

Lam, J. C. (2000) 'Energy analysis of commercial buildings in subtropical climates', *Building and Environment*, vol 35, pp19–26

Lam, J. C. and Li, D. H. W. (1998) 'Daylighting and energy analysis for air-conditioned office buildings', *Energy*, vol 23, pp79–89

Lam, J. C. and Li, D. H. W. (1999) 'An analysis of daylighting and solar heat for cooling-dominated office buildings', *Solar Energy*, vol 65, pp251–262

Lam, K. P., Wong, N. H. and Chandra, S. (2001) 'The use of multiple building performance simulation tools during the design process – a case study in Singapore', in *Seventh International IBPSA Conference*, Rio de Janeiro, Brazil, 13–15 August 2001, pp815–824. Available from www.ibpsa.org

Lam, W. M. C. (1986) *Sunlighting as Formgiver for Architecture*, Van Nostrand Reinhold, New York

Lamp, P., Costa, A., Ziegler, F., Collares Pereira, M., Farinha Mendes, J., Ojer, J. P., Conde, A. G. and Granados, C. (1998) 'Solar assisted absorption cooling with optimized utilization of solar energy', in *Natural Working Fluids '98, IIR – Gustav Lorentzen Conference*, 2–5 June 1998, Oslo, International Institute of Refrigeration, Paris, pp530–538

Lampret, V., Peternelj, J. and Krainer, A. (2002) 'Luminous flux and luminous efficacy of black-body radiation: an analytical approximation', *Solar Energy*, vol 73, pp319–326

Lang, S. (2004) 'Progress in energy-efficiency standards for residential buildings in China', *Energy and Buildings*, vol 36, pp1191–1196

Lang, W. and Herzog, T. (2000) 'Using multiple glass

skins to clad buildings; they're sophisticated, energy-efficient, and often sparklingly beautiful, but widely used only in Europe – at least for now', *Architectural Record*, vol 188, no 7, pp171–176

Laouadi, A., Atif, M. R. and Galasiu, A. (2002) 'Toward developing skylight design tools for thermal and energy performance of atriums in cold climates', *Building and Environment*, vol 37, pp1289–1316

Larsson, N. (2001) 'Canadian green building strategies', in *The 18th International Conference on Passive and Low Energy Architecture*, Brazil, 7–9 November 2001, pp17–25

Law, S. (1998) 'Design tools for low energy cooling technologies', in Tassou, S. (ed) *Low-Energy Cooling Technologies for Buildings: Challenges and Opportunities for the Environmental Control of Buildings*, IMechE Seminar Publication 1998–7, Professional Engineering Publishing, London, pp23–32

Layard, R. (2003) 'Towards a happier society', *New Statesman*, 3 March 2003, pp25–28

Layard, R. (2005) *Happiness: Lessons from a New Science*, Allen Lane, London

Leal, V., Erell, E., Maldonado, E. and Etzion, Y. (2004) 'Modelling the SOLVENT ventilated window for whole building simulation', *Building Service Engineering Resources Technology*, vol 25, no 3, pp183–195

Lee, E. S., DiBartolomeo, D. L. and Selkowitz, S. E. (1998) 'Thermal and daylighting performance of an automated venetian blind and lighting system in a full-scale private office', *Energy and Buildings*, vol 29, pp47–63

Lee, E. S., DiBartolomeo, D. L. and Selkowitz, S. E. (2000) 'Electrochromic windows for commercial buildings: monitored results from a full-scale testbed', in *Proceedings of the 2000 ACEEE Summer Study on Energy Efficiency in Buildings*, vol 3, American Council for an Energy Efficient Economy, Washington, DC, pp241–256

Lee, E. S., Selkowitz, S. E., Levi, M. S., Blanc, S. L., McConahey, E., McClintock, M., Hakkarainen, P., Sbar, N. L. and Myser, M. P. (2002a) 'Active load management with advanced window wall systems: research and industry perspectives', in *Proceedings of the 2002 ACEEE Summer Study on Energy Efficiency in Buildings*, vol 3, American Council for an Energy Efficient Economy, Washington, DC, pp193–210

Lee, E. S., Zhou, L., Yazdanian, M., Inkarojrit, V., Slack, J., Rubin, M. and Selkowitz, S. E. (2002b) *Energy Performance Analysis of Electrochromatic Windows in New York Commercial Office Buildings*, Lawrence Berkeley National Laboratory, LBNL-50096, Berkeley, CA

Lee, E. S., Selkowitz, S. E. and Hughes, G. D. (2004a) 'Market transformation opportunities for emerging dynamic facade and dimmable lighting control systems', in *Proceedings of the 2004 ACEEE Summer Study on Energy Efficiency in Buildings*, vol 3, American Council for an Energy Efficient Economy, Washington, DC, pp177–189

Lee, E. S., DiBartolomeo, D. L., Rubinstein, F. M. and Selkowitz, S. E. (2004b) 'Low-cost networking for dynamic window systems', *Energy and Buildings*, vol 36, pp503–513

Lee, K. H., Han, D. W. and Lim, H. J. (1996) 'Passive design principles and techniques for folk houses in Cheju Island and Ullŭng Island of Korea', *Energy and Buildings*, vol 23, pp207–216

Lee, S. and Sherif, S. A. (2001) 'Thermoeconomic analysis of absorption systems for cooling', *ASHRAE Transactions*, vol 107 (Part 1), pp629–637

Lee, W. L., Yik, F. W. H., Jones, P. and Burnett, J. (2001a) 'Energy saving by realistic design data for commercial buildings in Hong Kong', *Applied Energy*, vol 70, pp59–75

Lee, W. L., Yik, F. W. H. and Burnett, J. (2001b) 'Simplifying energy performance assessment in the Hong Kong Building Environment Assessment Method', *Building Service Engineering Resources Technology*, vol 22, no 2, pp113–132

Lehrer, D. and Bauman, F. (2003) 'Hype vs. reality: new research findings on underfloor air distribution systems', in *Proceedings, Greenbuild 2003*, Pittsburgh PA, November 2003, 12pp. Available from www.cedr.berkeley.edu.

Leigh, S. B., Song, D. S. and Hwang, S. H. (2005) 'A study for evaluating performance of radiant floor cooling integrated with controlled ventilation', *ASHRAE Transactions*, vol 111 (Part 1), pp71–82

Lekov, A., Lutz, J., Whitehead, C. D. and McMahon, J. E. (2000) 'Payback analysis of design options for residential water heaters', in *Proceedings of the 2000 ACEEE Summer Study on Energy Efficiency in Buildings*, vol 1, American Council for an Energy Efficient Economy, Washington, DC, pp163–174

Leemans, R. and Eickhout, B. (2004) 'Another reason for concern: regional and global impacts on ecosystems for different levels of climate change', *Global Environmental Change*, vol 14, pp219–228

Lemar, P. L. (2001) 'The potential impact of policies to promote combined heat and power in the US industry', *Energy Policy*, vol 29, pp1243–1254

Lemire, N. and Charneux, R. (2005) 'Energy-efficiency laboratory design', *ASHRAE Journal*, vol 47, no 5, pp58–64

Lenarduzzi, F. J. and Yap, S. S. (1998) 'Measuring the performance of a variable-speed drive retrofit on a fixed-speed centrifugal chiller', *ASHRAE Transactions*, vol 104 (Part 2), pp658–667

Lenarduzzi, F. J., Cragg, C. B. H. and Radhakrishna, H. S. (2000) 'The importance of grouting to enhance the performance of earth energy systems', *ASHRAE Transactions*, vol 106 (Part 2), pp424–434

Lenzen, M. and Dey, D. (2000) 'Truncation error in embodied energy analysis of basic iron and steel products', *Energy*, vol 25, pp577–585

Lenzen, M. and Treloar, G. (2002) 'Embodied energy in buildings: wood versus concrete – reply to Börjesson and Gustavsson', *Energy Policy*, vol 30, pp249–255

Leslie, R. P. (2003) 'Capturing the daylight dividend in buildings: why and how?', *Building and Environment*, vol 38, pp381–385

Leslie, R. P., Raghavan, R., Howlett, O. and Eaton, C. (2005) 'The potential of simplified concepts for daylight harvesting', *Lighting Resources Technology*, vol 37, pp21–40

Letan, R., Dubovsky, V. and Ziskind, G. (2003) 'Passive ventilation and heating by natural convection in a multi-storey building', *Building and Environment*, vol 38, pp197–208

Levermore, G. J. (2000) *Building Energy Management Systems: Applications to Low-Energy HVAC and Natural Ventilation Control*, E & FN Spon, London, 519pp

Levermore, G. J. (2005) 'Duct loop systems – savings and performance', *ASHRAE Transactions*, vol 111 (Part 1), pp507–514

Levinson, R., Delp, W. W., Dickerhoff, D. and Modera, M. (2000) 'Effects of airflow infiltration on the thermal performance of internally insulated ducts', *Energy and Buildings*, vol 32, pp345–354

Levinson, R., Akbari, H., Konopacki, S. and Bretz, S. (2005) 'Inclusion of cool roofs in nonresidential Title 24 prescriptive requirements', *Energy Policy*, vol 33, pp151–170

Lewis, M. (2004) 'Integrated design for sustainable buildings', *Building for the Future, A Supplement to ASHRAE Journals*, vol 46, no 9, pp2–3

Li, C. H., Wang, R. Z. and Lu, Y. Z. (2002) 'Investigation of a novel combined cycle of solar powered adsorption-ejection refrigeration system', *Renewable Energy*, vol 26, pp611–622

Li, D. H. W. and Lam, J. C. (2000) 'Solar heat gain factors and the implications to building designs in subtropical regions', *Energy and Buildings*, vol 32, pp47–55

Li, D. H. W. and Lam, J. C. (2003) 'An investigation of daylighting performance and energy saving in a daylight corridor', *Energy and Buildings*, vol 35, pp365–373

Li, D. H. W., Lam, J. C. and Wong, S. L. (2005) 'Daylighting and its effects on peak load determination', *Energy*, vol 30, pp1817–1831

Li, Y. and Delsante, A. (2001) 'Natural ventilation induced by combined wind and thermal forces', *Building and Environment*, vol 36, pp59–71

Li, Y., Delsante, A., Chen, Z., Sandberg, M., Andersen, A., Bjerre, M. and Heiselberg, P. (2001) 'Some examples of solution multiplicity in natural ventilation', *Building and Environment*, vol 36, pp851–858

Li, Y., Sumathy, K. and Kaushika, N. (2004) 'Thermal analysis of solar-powered continuous adsorption air-conditioning system', *ASHRAE Transactions*, vol 110 (Part 1), pp33–39

Li, Z., Liu, X., Jiang, Y. and Chen, X. (2005) 'New type of fresh air processor with liquid desiccant total heat recovery', *Energy and Buildings*, vol 37, pp587–593

Li, Z. F. and Sumathy, K. (2000) 'Technology development in the solar absorption air-conditioning systems', *Renewable and Sustainable Energy Reviews*, vol 4, pp267–293

Li, Z. F. and Sumathy, K. (2001a) 'Experimental studies on a solar powered air conditioning system with partitioned hot water storage tank', *Solar Energy*, vol 71, pp285–297

Li, Z. F. and Sumathy, K. (2001b) 'Simulation of a solar absorption air conditioning system', *Energy Conversion and Management*, vol 42, pp313–327

Lian, Z., Park, S. and Qi, H. (2005) 'Analysis on energy consumption of water-loop heat pump system in China', *Applied Thermal Engineering*, vol 25, pp73–85

Liao, S. M., Zhao, T. S. and Jakobsen, A. (2000) 'A correlation of optimal heat rejection pressures in transcritical carbon dioxide cycles', *Applied Thermal Engineering*, vol 20, pp831–841

Lin, F., Yi, J., Weixing, Y. and Xuzhong, Q. (2001) 'Influence of supply and return water temperatures on the energy consumption of a district cooling system', *Applied Thermal Engineering*, vol 21, pp511–521

Lin, Z. P. and Deng, S. M. (2003) 'The outdoor air ventilation rate in high-rise residences employing room air conditioners', *Buildings and Environment*, vol 38, pp1389–1399

Lindenberger, D., Bruckner, T., Groscurth, H. M. and Kümmel, R. (2000) 'Optimization of solar district heating systems: seasonal storage, heat pumps, and cogeneration', *Energy*, vol 25, pp591–608

Littlefair, P. J. and Motin, A. (2001) 'Lighting controls in areas with innovative daylighting systems: a study of sensor type', *Lighting Resources Technology*, vol 33, no 1, pp59–73

Liu, M. and Claridge, D. E. (1999) 'Converting dual-duct constant-volume systems to variable-volume systems without retrofitting the terminal boxes', *ASHRAE Transactions*, vol 105 (Part 1), pp66–70

Liu, M., Claridge, D. E. and Turner, W. D. (2003) 'Continuous commissioning^{SM} of building energy systems', *Journal of Solar Energy Engineering*, vol 125, pp275–281

Lin, Z. and Deng, S. (2004) 'A study on the characteristics of nighttime bedroom cooling load in tropics and subtropics', *Building and Environment*, vol 39, pp1101–1114

Loftness, V., Brahme, R. and Mondazzi, M. (2002) *Energy Savings Potential of Flexible and Adaptive HVAC Distribution Systems for Office Buildings*, Air Conditioning and Refrigeration Technology Institute Report ARTI-CR21/30030-01, 218pp

Löhnert, G., Dalkowski, A. and Sutter, W. (2003) *Integrated Design Process: A Guideline for Sustainable and Solar-Optimised Building Design*, International Energy Agency, Solar Heating and Cooling Program, Task 23 (Optimization of Solar Energy Use in Large Buildings), Paris. Available from www.iea-shc.org/task23

Lomas, K. J., Fiala, D., Cook, M. J. and Cropper, P. C. (2004) 'Building bioclimatic charts for non-domestic buildings and passive downdraught evaporative cooling', *Building and Environment*, vol 39, pp661–676

Lorente, S., Petit, M. and Javelas, R. (1998) 'The effects of temperature conditions on the thermal resistance of walls made with different shapes vertical hollow blocks', *Energy and Buildings*, vol 28, pp237–240

Lorenz, W. (2001) 'A glazing unit for solar control, daylighting and energy conservation', *Solar Energy*, vol 70, pp109–130

Lottner, V., Schulz, M. E. and Hahne, E. (2000) 'Solar-assisted district heating plants: status of the German programme Solarthermie-2000', *Solar Energy*, vol 69, pp449–459

Loudermilk, K. J. (1999) 'Underfloor air distribution solutions for open office applications', *ASHRAE Transactions*, vol 105 (Part 1), pp605–613

Loveday, D. L., Parsons, K. C., Taki, A. H., Hodder, S. G. and Jeal, L. D. (1998) 'Designing for thermal comfort in combined chilled ceiling/displacement ventilation environments', *ASHRAE Transactions*, vol 104 (Part 1B), pp901–911

Loveday, D. L., Parsons, K. C., Taki, A. H. and Hodder, S. G. (2002) 'Displacement ventilation environments with chilled ceilings: thermal comfort design within the context of the BS EN ISO7730 versus adaptive debate', *Energy and Buildings*, vol 34, pp573–579

Lovins, A. (1992) *Energy-Efficient Buildings: Institutional Barriers and Opportunities*, E-Source, Strategic Issues Paper

Lowe, P. R. (1977) 'An approximating polynomial for the computation of saturation vapor pressure', *Journal of Applied Meteorology*, vol 16, pp100–103

Lowenstein, A. and Novosel, D. (1995) 'The seasonal performance of a liquid-desiccant air conditioner', *ASHRAE Transactions*, vol 101 (Part 1), pp679–685

Lowenstein, A., Slayzak, S., Ryan, J. and Pesaran, A. (1998) *Advanced Liquid-Desiccant Technology Development Study*, NREL/TP-550-24688, National Renewable Energy Laboratory, Golden, CO, 43pp

Lowenstein, A., Slayzak, S., Kozubal, E. and Ryan, J. (2005) 'A low-flow, zero carryover liquid desiccant conditioner', *International Sorption Heat Pump Conference*, 22–24 June 2005, Denver

Lstiburek, J. (2002) 'Residential ventilation and latent loads', *ASHRAE Journal*, vol 44, no 4, pp18–22

Lucas, K. (2001) 'Efficient energy systems on the basis of cogeneration and heat pump technology', *International Journal of Thermal Science*, vol 40, pp338–343

Lucas, M., Martínez, P., Sánchez, A., Viedma, A. and Zamora, B. (2003) 'Improved hydrosolar roof for buildings' air conditioning', *Energy and Buildings*, vol 35, pp963–970

Lund, J. W. (2002) 'Direct heat utilization of geothermal resources', in Bundschuh, J. and Chandrasekharam, D. (eds) *Geothermal Energy Resources for Developing Countries*, Balkema, Lisse (The Netherlands), pp129–147

Lund, J. W. (2003) 'The USA geothermal country update', *Geothermics*, vol 32, pp409–418

Lund, J. W., Sanner, B., Rybach, L., Curtis, R. and Hellström, G. (2003) 'Ground-source heat pumps: a world overview', *Renewable Energy World*, vol 6, no 4, pp218–227

Lutz, W., Sanderson, W. and Scherbov, S. (2001) 'The end of world population growth', *Nature*, vol 412, pp543–545

Lutz, J. D., Klein, G., Springer, D. and Howard, B. D. (2002) 'Residential hot water distribution systems: roundtable session', in *Proceedings of the 2002 ACEEE Summer Study on Energy Efficiency in Buildings*, vol 1, American Council for an Energy Efficient

Economy, Washington, DC, pp131–144

Lyons, P. R., Arasteh, D. and Huizenga, C. (2000) 'Window performance for human thermal control', *ASHRAE Transactions*, vol 106 (Part 1), pp594–602

MacRae, M. (1992) *Realizing the Benefits of Community Integrated Energy Systems*, Canadian Energy Research Institute, Calgary, 334pp

Mahdavi, A. and Gurtekin, B. (2001) 'Computational support for the generation and exploration of the design-performance space', in *Seventh International IBPSA Conference*, Rio de Janeiro, Brazil, 13–15 August 2001, pp669–676. Available from www.ibpsa.org

Mahler, B. (2002) 'Sharing the sun – solar district heating in Europe', *Renewable Energy World*, vol 5, no 6, pp91–97

Maker, T. and Penny, J. (1999) *Heating Communities with Renewable Fuels: The Municipal Guide to Biomass District Energy*, Natural Resources Canada and USA Department of Energy. Available from www.nrcan.gc.ca

Malinowski, J. (2004) 'Energy-efficient motors and drives', *ASHRAE Journal*, vol 46, no 1, pp30–32

Maliska, C. R. (2001) 'Issues of the integration of CFD to building simulation tools', in *Seventh International IBPSA Conference*, Rio de Janeiro, Brazil, 13–15 August 2001, pp29–40. Available from www.ibpsa.org

Mangold, D. (2001) 'Active solar heating systems for urban areas', in Santamouris, M. (ed) *Energy and Climate in the Urban Built Environment*, James & James, London, pp199–211

Mann, M. E., Bradley, R. S. and Hughes, M. K. (1999) 'Northern hemisphere temperatures during the past millennium: inferences, uncertainties, and limitations', *Geophysical Research Letters*, vol 26, pp759–762

Mansfield, S. K. (2001) 'Oversized pumps – A look at practical field experiences', *ASHRAE Transactions*, vol 107 (Part 2), pp573–583

Manz, H. (2004) 'Total solar energy transmittance of glass double facades with free convection', *Energy and Buildings*, vol 36, pp127–136

Manz, H., Huber, H. and Helfenfinger, D. (2001) 'Impact of air leakages and short circuits in ventilation units with heat recovery on ventilation efficiency and energy requirements for heating', *Energy and Buildings*, vol 33, pp133–139

Manz, H., Schaelin, A. and Simmler, H. (2004) 'Airflow patterns and thermal behaviour of mechanically ventilated glass double facades', *Building and Environment*, vol 39, pp1023–1033

Marbek Resource Consultants (1992) *City of Toronto Potential for Electricity Conservation, Residential Sector Appendices*, Marbek Resource Consultants, Ottawa

Marshall, J. D. and Toffel, M. W. (2005) 'Framing the elusive concept of sustainability: a sustainability hierarchy', *Environmental Science and Technology*, vol 39, pp673–682

Martin, M. A., Madgett, M. G. and Hughes, P. J. (2000) 'Comparing maintenance costs of geothermal heat pump systems with other HVAC systems: preventative maintenance actions and total maintenance costs', *ASHRAE Transactions*, vol 106 (Part 1), pp408–423

Mast, B., McCormick, J., Vogt, T., Ignelzi, P., Kolderup, E., Berman, M. and Dimit, M. (2000) 'Carrots or sticks? Policy options for building energy standards', in *Proceedings of the 2000 ACEEE Summer Study on Energy Efficiency in Buildings*, vol 9, American Council for an Energy Efficient Economy, Washington, DC, pp261–274

Mastny, L. et al (2005) *Vital Signs 2005: The Trends that are Shaping our Future*, Worldwatch Institute, Washington, 139pp

Mathioulakis, E. and Belessiotis, V. (2002) 'A new heat-pipe type solar domestic hot water system', *Solar Energy*, vol 72, pp13–20

Mavroudaki, P., Beggs, C. B., Sleigh, P. A. and Halliday, S. P. (2002) 'The potential for solar powered single-stage desiccant cooling in southern Europe', *Applied Thermal Engineering*, vol 22, pp1129–1140

Maycock, P. (2005) 'PV market update: global PV production continues to increase', *Renewable Energy World*, vol 8, no 4, pp86–99

Mayo, T. (1993) 'Innovative insulation techniques in Canadian houses', *CADDET Energy Efficiency Newsletter*, December 1993, pp4–7

Mazzei, P., Minichiello, F. and Palma, D. (2002) 'Desiccant HVAC systems for commercial buildings', *Applied Thermal Engineering*, vol 22, pp545–560

McConahey, E., Haves, P. and Christ, T. (2002) 'The integration of engineering and architecture: a perspective on natural ventilation for the new San Francisco Federal Building', in *Proceedings of the 2002 ACEEE Summer Study on Energy Efficiency in Buildings*, vol 3, American Council for an Energy Efficient Economy, Washington, DC, pp239–251

McCowan, B., Coughlin, T., Bergeron, P. and Epstein, G. (2002) 'High performance lighting options for school facilities', in *Proceedings of the 2002 ACEEE Summer Study on Energy Efficiency in Buildings*, vol 3, American Council for an Energy Efficient Economy, Washington, DC, pp253–268

McCullough, J. J. and Gordon, K. L. (2002) 'High hats, Swiss cheese, and fluorescent lighting?' in *Proceedings of the 2002 ACEEE Summer Study on Energy Efficiency in Buildings*, vol 1, American Council for an Energy

Efficient Economy, Washington, DC, pp171–182

McDonell, G. (2003) 'Displacement ventilation', *The Canadian Architect*, vol 48, no 4, pp32–33

McElroy, L. B., Clarke, J. A., Hand, J. W. and Macdonald, I. A. (2001) 'Delivering simulation to the profession: the next stage?', in *Seventh International IBPSA Conference*, Rio de Janeiro, Brazil, 13–15 August 2001, pp831–836. Available from www.ibpsa.org

McEvoy, M. E., Southall, R. G. and Baker, P. H. (2003) 'Test cell evaluation of supply air windows to characterise their optimum performance and its verification by the use of modelling techniques', *Energy and Buildings*, vol 35, pp1009–1020

McGowan, T. (1989) 'Energy-efficiency lighting', in Johansson, T. B., Bodlund, B. and Williams, R. H. (eds) *Electricity: Efficient End-Use and New Generation Technologies and Their Planning Implications*, Lund University Press, Lund, pp59–88

McMahon, J., Toriel, I., Rosenquist, G. J., Lutz, J. D., Boghosian, S. H. and Shown, L. (1997) 'Energy-efficient technologies: appliances, heat pumps, and air conditioning', in Kreith, F. and West, R. E. (eds) *CRC Handbook of Energy Efficiency*, CRC Press, Boca Raton, FL, pp429–443

McMullan, J. T., Williams, B. C. and McCahey, S. (2001) 'Strategic considerations for clean coal R&D', *Energy Policy*, vol 29, pp441–452

McNeil, C. S. L. (1995) 'Applications of Canadian advanced earth energy heat pumps', *CADDET Energy Efficiency Newsletter*, June 1995, pp22–24

Means, R. S. (1999) *Mechanical Cost Data, 22nd Annual Addition*, R. S. Means Company, Inc

Means, R. S. (2005) *Building Construction Cost Data, 63rd Annual Addition*, R. S. Means Company, Inc

Medina, M. A. (2000) 'On the performance of radiant barriers in combination with different attic insulation levels', *Energy and Buildings*, vol 33, pp31–40

Mehling, H., Cabeza, L. F., Hippeli, S. and Hiebler, S. (2003) 'PCM-module to improve hot water heat stores with stratification', *Renewable Energy*, vol 28, pp699–711

Mei, L., Infield, D., Eicker, U. and Fux, V. (2002) 'Parameter estimation for ventilated photovoltaic facades', *Building Service Engineering Resources Technology*, vol 23, no 2, pp81–96

Meier, A., Lin, J., Liu, J. and Li, T. (2004) 'Standby power use in Chinese homes', *Energy and Buildings*, vol 36, pp1211–1216

Meir, M. G., Rekstad, J. B. and Løvvik, O. M. (2002) 'A study of a polymer-based radiative cooling system', *Solar Energy*, vol 73, pp403–417

Meliß, M. and Späte, F. (2000) 'The solar heating system with seasonal storage at the solar-campus Jülich', *Solar Energy*, vol 69, pp525–533

Melet, E. (1999) *Sustainable Architecture: Towards a Diverse Built Environment*, NAI Publishers, Rotterdam

Mendler, S. and Odell, W. (2000) *The HOK Guidebook to Sustainable Design*, John Wiley, New York, 412pp

Menzies, G. F. and Wherrett, J. R. (2005) 'Multiglazed windows: potential for savings in energy, emissions, and cost', *Building Service Engineering Resources Technology*, vol 26, no 3, pp249–258

Mihalakakou, G., Lewis, J. O. and Santamouris, M. (1996) 'On the heating potential of buried pipes techniques-application in Ireland', *Energy and Buildings*, vol 24, pp19–25

Mihalakakou, G., Ferrante, A. and Lewis, J. O. (1998) 'The cooling potential of a metallic nocturnal radiator', *Energy and Buildings*, vol 28, pp251–256

Miller, J. A., Lowenstein, A. and Sand, J. R. (2002b) 'The performance of a desiccant-based air conditioner in a Florida school', *ASHRAE Transactions*, vol 108 (Part 1), pp575–586

Miller, W. A., Loye, K. T., Desjarlais, A. O. and Blonski, R. P. (2002a) 'Cool color roofs with complex inorganic color pigments', in *Proceedings of the 2002 ACEEE Summer Study on Energy Efficiency in Buildings*, vol 1, American Council for an Energy Efficient Economy, Washington, DC, pp195–206

Mills, D. and Morrison, G. L. (2003) 'Optimisation of minimum backup solar water heating system', *Solar Energy*, vol 74, pp505–511

Mills, E. (2003) 'Climate change, insurance and the buildings sector: technological synergisms between adaptation and mitigation', *Building Research and Information*, vol 31, pp257–277

Milne, M. and Givoni, B. (1979) 'Architectural design based on climate' in Watson, D. (ed) *Energy Conservation Through Building Design*, McGraw-Hill, New York

Minca, V. (2003) 'An exhaust-air heat recovery heat pump system optimized for use in cold climates', *IEA Heat Pump Centre Newsletter*, vol 21, no 3, pp14–17

Miyazaki, T., Akisawa, A. and Kashiwagi, T. (2005) 'Energy savings of office buildings by the use of semi-transparent solar cells for windows', *Renewable Energy*, vol 30, pp281–304

Modera, M. P., Brzozowski, O., Carrié, F. R., Dickerhoff, D. J., Delp, W. W., Fisk, W. J., Levinson, R. and Wang, D. (2002) 'Sealing ducts in large commercial buildings with aerosolized sealant particles', *Energy and Buildings*, vol 34, pp705–714

Morbitzer, C., Strachan, P., Webster, J., Spires, B. and Cafferty, D. (2001) 'Integration of building simulation into the design process of an architectural practice', in *Seventh International IBPSA Conference*, Rio de Janeiro, Brazil, 13–15 August 2001, pp697–704. Available from www.ibpsa.org

Morrison, G. L. (2001a) 'Solar collectors', in Gordon, J. (ed) *Solar Energy: The State of the Art, ISES Position Papers*, James & James, London, pp145–221

Morrison, G. L. (2001b) 'Solar water heating', in Gordon, J. (ed) *Solar Energy: The State of the Art, ISES Position Papers*, James & James, London, pp223–289

Morrison, G. L., Budihardjo, I. and Behnia, M. (2004) 'Water-in-glass evacuated tube solar water heaters', *Solar Energy*, vol 76, pp135–140

Motegi, N., Piette, M. A. and Wentworth, S. (2002) 'From design through operations: multi-year results from a new construction performance contract', in *Proceedings of the 2002 ACEEE Summer Study on Energy Efficiency in Buildings*, vol 3, American Council for an Energy Efficient Economy, Washington, DC, pp269–281

Mottillo, M. (2001) 'Sensitivity analysis of energy simulation by building type', *ASHRAE Transactions*, vol 107 (Part 2), pp722–732

Muhs, J. D. (2000) 'Hybrid solar lighting doubles the efficiency and affordability of solar energy in commercial buildings', *CADDET Energy Efficiency Newsletter*, December 2000, pp6–9

Mumma, S. A. (2001) 'Ceiling panel cooling system', *ASHRAE Journal*, vol 43, no 11, pp28–32

Mumma, S. A. and Shank, K. M. (2001) 'Achieving dry outside air in an energy-efficient manner', *ASHRAE Transactions*, vol 107 (Part 1), pp553–561

Muneer, T., Abodahab, N., Weir, G. and Kubie, J. (2000) *Windows in Buildings: Thermal, Acoustic, Visual and Solar Performance*, Architectural Press, Oxford, 258pp

Murakoshi, C. and Nakagami, H. (1999) 'Japanese appliances on the fast track', *Appliance Efficiency*, vol 3, no 3 (Newsletter of IDEA, the International Network for Domestic Energy Efficient Appliances)

Murphy, P. (ed) (2002) *Solar Energy Activities in IEA Countries*, International Energy Agency, Solar Heating and Cooling Programme, Paris. Available from www.iea-shc.org

Nahar, N. M., Sharma, P. and Purohit, M. M. (2003) 'Performance of different passive techniques for cooling of buildings in arid regions', *Building and Environment*, vol 38, pp109–116

Nakagami, H., Ishihara, O., Sakai, K. and Tanaka, A. (2003) 'Performance of residential PV system under actual field conditions in western part of Japan', in *2003 International Solar Energy Conference*,

Kohala Coast, Hawaii, US, 15–18 March 2003, pp491–498

NBI (New Buildings Institute) (2003) *Advanced Lighting Guidelines – 2003 Edition*, 445pp. Available from www.newbuildings.org

Neal, C. L. (1998) 'Field adjusted SEER [SEERFA] residential buildings: technologies, design and performance analysis', in *Proceedings of the 1998 ACEEE Summer Study on Energy Efficiency in Buildings*, vol 1, American Council for an Energy Efficient Economy, Washington, DC, pp197–209

Neathery, J., Gray, D., Challman, D. and Derbyshire, F. (1999) 'The pioneer plant concept: co-production of electricity and added-value products from coal', *Fuel*, vol 78, pp815–823

Nekså, P. (2002) 'CO_2 heat pump systems', *International Journal of Refrigeration*, vol 25, pp421–427

Nekså, P., Girotto, S. and Schiefloe, P. A. (1998) 'Commercial refrigeration using CO_2 as refrigerant – system design and experimental results', in *Natural Working Fluids '98, IIR – Gustav Lorentzen Conference*, 2–5 June 1998, Oslo, International Institute of Refrigeration, Paris, pp270–280

Nekså, P. and Stene, J. (1997) 'Highly efficient CO_2 heat pump water heater', *IEA Heat Pump Centre Newsletter*, vol 15, no 4, pp27–29

New York City Department of Design and Construction (NYC DDC) (2000) *High Performance Building Guidelines*. Available from http://home.nyc.gov/html/ddc/html/ddcgreen/highperf.html

Newman, P. and Kenworthy, J. (1999) *Sustainability and Cities: Overcoming Automobile Dependence*, Island Press, Washington, DC, 442pp

Neymark, J., Judkoff, R., Knabe, G., Le, H. -T., Dürig, M., Glass, A. and Zweifel, G. (2001) 'HVAC BESTEST: a procedure for testing the ability of whole-building energy simulation programs to model space conditioning equipment', in *Seventh International IBPSA Conference*, Rio de Janeiro, Brazil, 13–15 August 2001, pp369–376. Available from www.ibpsa.org

Nguyen, V. M., Riffat, S. B. and Doherty, P. S. (2001) 'Development of a solar-powered passive ejector cooling system', *Applied Thermal Engineering*, vol 21, pp157–168

Nicol, F. (2004) 'Adaptive thermal comfort standards in the hot-humid tropics', *Energy and Buildings*, vol 36, pp628–637

Niles, W. B. (1976) 'Thermal evaluation of a house using a moveable insulation heating and cooling system', *Solar Energy*, vol 18, pp413–419

Niu, J. L., Zuo, H. G. and Burnett, J. (2001) 'Simulation methodology of radiant cooling with elevated air movement', in *Seventh International IBPSA Conference*, Rio de Janeiro, Brazil, 13–15 August 2001, pp265–272. Available from www.ibpsa.org

Niu, J. L., Zhang, L. Z. and Zuo, H. G. (2002) 'Energy savings potential of chilled-ceiling combined with desiccant cooling in hot and humid climates', *Energy and Buildings*, vol 34, pp487–495

NOAA (National Oceanographic and Atmospheric Administration) (2002) *Climatology of the United States No. 81, Supplement 2, Annual Degree Days to Selected Bases 1971-2000*. Available from www5.ncdc.noaa.gov

Nordell, B. and Hellström, G. (2000) 'High temperature solar heated seasonal storage system for low temperature heating of buildings', *Solar Energy*, vol 69, pp511–523

Nordell, B. and Lundin, S.-E. (2001) 'Going underground – solar heat in Sweden', *CADDET Renewable Energy Newsletter*, 4–5 March 2001

Nordmann, T. and Clavadetscher, L. (2003) 'Understanding temperature effects on PV system performance', IEA PVPS Task 2 Activity 2.4, Paper 7P–B3-14, 3rd PV World Conference in Osaka, Japan (WCPEC), May 2003

Norris, G. A. (1998) *An Exploratory Life Cycle Study of Selected Building Envelope Materials*, Athena Sustainable Materials Institute, Merrickville, Ontario, Canada, 75pp

Novoselac, A. and Srebric, J. (2002) 'A critical review on the performance and design of combined cooled ceiling and displacement ventilation systems', *Energy and Buildings*, vol 30, pp497–509

NRCan (Natural Resources Canada) (1997a) *Model National Energy Code of Canada for Buildings 1997*, National Research Council of Canada, Ottawa

NRCan (Natural Resources Canada) (1997b) *Model National Energy Code of Canada for Houses 1997*, National Research Council of Canada, Ottawa.

NRCan (Natural Resources Canada) (2001) *Energy Guide Appliance Directory 2001*, Office of Energy Efficiency, NRCan, Ottawa, 98pp. Available from http://energy-publications.nrcan.gc.ca

NRCan (Natural Resources Canada) (2002) *Operating and Maintaining Your Heat Recovery Ventilator (HRV)*, Office of Energy Efficiency, NRCan, Ottawa. Available from http://energy-publications.nrcan.gc.ca

NRCan (Natural Resources Canada) (2004) *Heating and Cooling with a Heat Pump*, Office of Energy Efficiency, NRCan, Ottawa. Available from http://energy-publications.nrcan.gc.ca

NRCan (Natural Resources Canade) (2005a) *Energy Use Data Handbook, 1990 and 1997–2003*, NRCan, Ottawa. Available from http://oee.brca.gc.ca

NRCan (Natural Resources Canada) (2005b) *EnerGuide Room Air Conditioner Directory 2005*, Office of Energy Efficiency, NRCan, Ottawa. Available from http://energy-publications.nrcan.gc.ca

Nussbaumer, T., Bundi, R., Tanner, C. and Muehlebach, H. (2005) 'Thermal analysis of a wooden door systems with integrated vacuum insulation panels', *Energy and Buildings*, vol 37, pp1107–1113

Oakley, G., Riffat, S. B. and Shao, L. (2000) 'Daylight performance of lightpipes', *Solar Energy*, vol 69, pp89–98

ODPM (Office of the Deputy Prime Minister) (2004) *Proposals for Amending Part L of the Building Regulations and Implementing the Energy Performance of Buildings Directive, A Consultation Document, July 2004*. Available from http://odpm.gov.uk

Oesterle, E., Lieb, R.-D., Lutz and Heusler (2001) Double-skin Facades: Integrated Planning: Building Physics, Construction, Aerophysics, Air-conditioning, Economic Viability, Prestel, Munich

Oke, T. R. (1978) *Boundary Layer Climates*, Methuen, London, 372pp

Oktay, D. (2002) 'Design with the climate in housing environments: an analysis in Northern Cyprus', *Building and Environment*, vol 37, pp1003–1012

Olesen, B. W. (2002) 'Radiant floor heating in theory and practice', *ASHRAE Journal*, vol 44, no 7, pp19–26

Oliveti, G., Arcuri, N. and Ruffolo, S. (1998) 'First experimental results from a prototype plant for the interseasonal storage of solar energy for the winter heating of buildings', *Solar Energy*, vol 62, pp281–290

Olsen, E. L. and Chen, Q. (2003) 'Energy consumption and comfort analysis for different low-energy cooling systems in a mild climate', *Energy and Buildings*, vol 35, pp561–571

Olsen, R., Hart, R., Hatten, M., Ohmart, G. and Brown, G. Z. (2004) 'Ventilative cooling: can business live without mechanical cooling?' in *Proceedings of the 2004 ACEEE Summer Study on Energy Efficiency in Buildings*, vol 7, American Council for an Energy Efficient Economy, Washington, DC, pp260–271

Omer, S. A., Wilson, R. and Riffat, S. B. (2003) 'Monitoring results of two examples of building integrated PV (BiPV) systems in the UK', *Renewable Energy*, vol 28, pp1387–1399

Ong, K. S. and Chow, C. C. (2003) 'Performance of a solar chimney', *Solar Energy*, vol 74, pp1–17

O'Neal, D. L., Rodriguez, A., Davis, M. and Kondepudi, S. (2002) 'Return air leakage impact on air conditioner performance in humid climates', *Journal of Solar Energy Engineering*, vol 124, pp63–69

Oppenheimer, M. and Alley, R. B. (2005) 'Ice sheets, global warming, and Article 2 of the UNFCCC',

Climatic Change, vol 68, pp257–267

O'Regan, B. and Grätzel, M. (1991) 'A low-cost, high-efficiency solar cell based on dye-sensitized colloidal TiO₂ films', *Nature*, vol 353, pp737–740

Ove Arup & Partners Hong Kong (2003) *Implementation Study for a District Cooling Scheme at South East Kowloon Development*, Electrical and Mechanical Services Department, Hong Kong. English summary available from www.emsd.gov.hk/emsd/eng/pee/wacs.shtml

Pahud, D. (2000) 'Central solar heating plants with seasonal duct storage and short-term water storage: design guidelines obtained by dynamic system simulations', *Solar Energy*, vol 69, pp495–509

Papadopoulos, A. M., Oxizidis, S. and Kyriakis, N. (2003) 'Perspectives of solar cooling in view of the developments in the air-conditioning sector', *Renewable and Sustainable Energy Reviews*, vol 7, pp419–438

Parekh, A., Mullally-Pauly, B. and Riley, M. (2000) 'The EnerGuide for Houses Program: a successful Canadian home energy rating system', in *Proceedings of the 2000 ACEEE Summer Study on Energy Efficiency in Buildings*, vol 2, American Council for an Energy Efficient Economy, Washington, DC, pp229–240

Parent, D., Stricker, S. and Fugler, D. (1998) 'Optimum ventilation and air flow control in buildings', *Energy and Buildings*, vol 27, pp239–245

Parent, V., Goldemberg, J. and Zilles, R. (2002) 'Comments on experience curves for PV modules', *Progress in Photovoltaics: Research and Applications*, vol 10, pp571–574

Park, C. S., Augenbroe, G., Sadegh, N., Thitisawat, M. and Messadi, T. (2004) 'Real-time optimization of a double-skin facade based on lumped modelling and occupant preference', *Building and Environment*, vol 39, pp939–948

Park, C. W., Jeong, J. H. and Kang, Y. T. (2004) 'Energy consumption characteristics of an absorption chiller during the partial load operation', *International Journal of Refrigeration*, vol 27, pp948–954

Parker, D. S., Fairey, P. F. and Gu, L. (1993) 'Simulation of the effects of duct leakage and heat transfer on residential space cooling energy use', *Energy and Buildings*, vol 20, pp97–114

Parker, D. S., Sherwin, J. R., Sonne, J. K., Barkaszi, S. F., Floyd, D. B. and Withers, C. R. (1998) 'Measured energy savings of a comprehensive retrofit in an existing Florida residence', in *Proceedings of the 1998 ACEEE Summer Study on Energy Efficiency in Buildings*, vol 1, American Council for an Energy Efficient Economy, Washington, DC, pp235–251

Parker, D. S., Callahan, M. P., Sonne, J. K. and Su, G. H. (1999) 'Development of a high efficiency ceiling fan', FSEC-CR-1059-99. Available from www.fsec.

ucf.edu/bldg/pubs

Parker, D. S., Dunlop, J. P., Barkaszi, S. F., Sherwin, J. R., Anello, M. T. and Sonne, J. K. (2000) 'Towards zero energy demand: evaluation of super efficient building technology with photovoltaic power for residential housing', in *Proceedings of the 2000 ACEEE Summer Study on Energy Efficiency in Buildings*, vol 1, American Council for an Energy Efficient Economy, Washington, DC, pp207–224

Parker, D. S., Sherwin, J. R. and Anello, M. T. (2001) *FPC Residential Monitoring Project: New Technology Development — Radiant Barrier Pilot Project*, Contract Report FSEC-CR-1231-01, Florida Solar Energy Center, Cocoa, Florida. Available from www.fsec.ucf.edu/bldg/pubs

Parker, D. S., Sonne, J. K. and Sherwin, J. R. (2002) 'Comparative evaluation of the impact of roofing systems on residential cooling energy demand in Florida', in *Proceedings of the 2002 ACEEE Summer Study on Energy Efficiency in Buildings*, vol 1, American Council for an Energy Efficient Economy, Washington, DC, pp219–234

Parker, P., Rowlands, I. H. and Scott, D. (2000) 'Assessing the potential to reduce greenhouse gas emissions in Waterloo region houses: is the Kyoto target possible?', *Environments*, vol 28, no 3, pp29–56

Parry, M., Arnell, N., McMichael, T., Nicholls, R., Martens, P., Kovats, S., Livermore, M., Rosenzweig, C., Iglesias, A. and Fischer, G. (2001) 'Millions at risk: defining critical climate change threats and targets', *Global Environmental Change*, vol 11, pp181–183

Parsons Brinkerhoff Asia (2003) *Territory-Wide Implementation Study for Water-cooled Air Conditioning Systems in Hong Kong*, Electrical and Mechanical Services Department, Hong Kong. English summary available from www.emsd.gov.hk/emsd/eng/pee/wacs.shtml

Pasquay, T. (2004) 'Natural ventilation in high-rise buildings with double facades, saving or waste of energy', *Energy and Buildings*, vol 36, pp381–389

Payne, A., Duke, R. and Williams, R. H. (2001) 'Accelerating residential PV expansion: supply analysis for competitive electricity markets', *Energy Policy*, vol 29, pp787–800

Peckler, D. (2003) 'The tankless alternative', *Home Energy*, vol Nov/Dec 2003, p10

Peer, T. and Joyce, W. S. (2002) 'Lake-source cooling', *ASHRAE Journal*, vol 44, no 4, pp37–39

Peixoto, R., Butrymowicz, D., Crawford, J., Godwin, D., Hickman, K., Keller, F. and Onishi, H. (2005) 'Residential and commercial air conditioning and heating', in *IPCC/TEAP Special Report on Safeguarding the Ozone Layer and the Global Climate System: Issues related to Hydrofluorocarbons and Perfluorocarbons*, chapter 5

Pelletret, R. and Keilholz, W. (1999) 'Coupling CAD tools and building simulation evaluators', in *Sixth International IBPSA Conference*, Kyoto, Japan, 13–15 September 1999, pp1197–1202. Available from www.ibpsa.org

Pennsylvania Department of Environmental Protection (PDEP) (1999) *Guidelines for Creating High Performance Green Buildings: A Document for Decision Makers.* Available from www.gggc.state.pa.us/gggc

Perez, R., Hoff, T., Herig, C. and Shah, J. (2003) 'Maximizing PV peak shaving with solar load control: validation of a web-based economic evaluation tool', *Solar Energy*, vol 74, pp409–415

Pessenlehner, W. and Mahdavi, A. (2003) 'Building morphology, transparence, and energy performance', in *Eighth International IBPSA Conference*, Eindhoven, The Netherlands, 11–14 August 2003, pp1025–1032. Available from www.ibpsa.org

Petersdorff, C., Boermans, T., Harnisch, J., Joosen, S. and Wouters, F. (2002) *The Contribution of Mineral Wool and other Thermal Insulation Materials to Energy Saving and Climate Protection in Europe*, ECOFYS, Cologne. Available from www.ecofys.com

Petit, J. R. et al (1999) 'Climate and atmospheric history of the past 420,000 years from Vostok ice core, Antarctica', *Nature*, vol 399, pp429–436

Peuser, F. A., Remmers, K.-H. and Schnauss, M. (2002) *Solar Thermal Systems: Successful Planning and Construction*, James & James, London, 364pp

Pfafferott, J. and Herkel, S. (2003) 'Evaluation of a parametric model and building simulation for design of passive cooling by night ventilation', in *Eighth International IBPSA Conference*, Eindhoven, The Netherlands, 11–14 August 2003, pp1033–1039. Available from www.ibpsa.org

Pfeil, M. and Koch, H. (2000) 'High performance-low cost seasonal gravel/water storage pit', *Solar Energy*, vol 69, pp461–467

Phetteplace, G. and Sullivan, W. (1998) 'Performance of a hybrid ground-coupled heat pump system', *ASHRAE Transactions*, vol 104 (Part 1B), pp763–770

Philip, B. J. H. (2001) 'Three tropical design paradigms', in Tzonis, A., Lefaivre, L. and Stagno, B. (eds) *Tropical Architecture: Regionalism in the Age of Globilization*, Wiley Academy, Chichester, UK, 311pp

Piette, M. A., Kinney, S. K. and Haves, P. (2001) 'Analysis of an information monitoring and diagnostic system to improve building operations', *Energy and Buildings*, vol 33, pp783–791

Piexoto, R., Butrymowicz, D., Crawford, J., Godwin, D., Hickman, K., Keller, F. and Onishi, H. (2005) 'Residential and commercial air conditioning and heating', in IPCC/TEAP Special Report on Safeguarding the Ozone Layer and the Global Climate System: Issues related to Hydrofluorocarbons and Perfluorocarbons, Chapter 5

Pigg, S. (2003) 'The electric side of gas furnaces', *Home Energy*, Nov/Dec 2003, pp24–28

Pigg, S. and Talerico, T. (2004) 'Electicity savings from variable-speed furnaces in cold climates', in *Proceedings of the 2004 ACEEE Summer Study on Energy Efficiency in Buildings*, vol 1, American Council for an Energy Efficient Economy, Washington, DC, pp264–279

Pigg, S., Carroll, E., Nahn, G. and Nagan, J. (2002) 'Ventilation performance in Wisconsin Energy Star homes', in *Proceedings of the 2002 ACEEE Summer Study on Energy Efficiency in Buildings*, vol 1, American Council for an Energy Efficient Economy, Washington, DC, pp247–259

Pilavachi, P. A. (2000) 'Power generation with gas turbine systems and combined heat and power', *Applied Thermal Engineering*, vol 20, pp1421–1429

Plokker, W., Elkuizen, B. and Peitsman, H. (2003) 'Energetic optimal heating and cooling curves (for air supply)', in *Eighth International IBPSA Conference*, Eindhoven, The Netherlands, 11–14 August 2003, pp1047–1052. Available from www.ibpsa.org

Poel, B. (2004) 'Integrated design with a focus on energy aspects', in *Proceedings of the 2004 ACEEE Summer Study on Energy Efficiency in Buildings*, vol 9, American Council for an Energy Efficient Economy, Washington, DC, pp109–120

Poel, B., de Vries, G. and van Cruchten, G. (2002) *The Integrated Design Process in Practice: Demonstration Projects Evaluated*, International Energy Agency, Solar Heating and Cooling Program, Task 23 (Optimization of Solar Energy Use in Large Buildings), Paris. Available from www.iea-shc.org/task23

Poponi, D. (2003) 'Analysis of diffusion paths for photovoltaic technology based on experience curves', *Solar Energy*, vol 74, pp331–340

Popovic, P., Marantan, A., Radermacher, R. and Garland, P. (2002) 'Integration of microturbine with single-effect exhaust-driven absorption chiller and solid wheel desiccant system', *ASHRAE Transactions*, vol 108 (Part 2), pp660–669

Porta-Gándara, M. A., Rubio, E. and Fernández, J. L. (2002) 'Economic feasibility of passive ambient comfort in Baja California dwellings', *Building and Environment*, vol 37, pp993–1001

Prasad, D. and Snow, M. (2005) *Designing with Solar*

Power: A Source Book for Building Integrated Photovoltaics (BiPV), Earthscan, London

Prentice, I. C. et al (2001) 'The carbon cycle and atmospheric carbon dioxide', in Houghton, J. T. et al (eds) *Climate Change 2001: The Scientific Basis*, Houghton, Cambridge University Press, Cambridge, pp183–237

Price, W. and Hart, R. (2002) 'Bulls-eye commissioning: using interval data as a diagnostic tool', in *Proceedings of the 2002 ACEEE Summer Study on Energy Efficiency in Buildings*, vol 3, American Council for an Energy Efficient Economy, Washington, DC, pp295–307

Pridasawas, W. and Lundqvist, P. (2004) 'An exergy analysis of a solar-driven ejector refrigeration system', *Solar Energy*, vol 76, pp369–379

Proctor, J. and Parker, D. (2000) 'Hidden power drains: residential heating and cooling fan power demand', in *Proceedings of the 2000 ACEEE Summer Study on Energy Efficiency in Buildings*, vol 1, American Council for an Energy Efficient Economy, Washington, DC, pp225–234

Proskiw, G. (1998) *Technology Profile: Residential Greywater Heat Recovery Systems*, prepared for CANMET Energy Technology Centre, Ottawa, 67pp. Available from www.buildingsgroup.nrcan.gc.ca

Raeissi, S. and Taheri, M. (1999) 'Energy saving by proper tree plantation', *Building and Environment*, vol 34, pp565–570

Raeissi, S. and Taheri, M. (2000) 'Skytherm: an approach to year-round thermal energy sufficient houses', *Renewable Energy*, vol 19, pp527–543

Raicu, A., Wilson, H. R., Nitz, P., Platzer, W., Wittwer, V. and Jahns, E. (2002) 'Façade systems with variable solar control using thermotropic polymer blends', *Solar Energy*, vol 72, pp31–42

Ramage, J. and Scurlock, J. (1996) 'Biomass', in Boyle, G. (ed) *Renewable Energy, Power for a Sustainable Future*, Oxford University Press, Oxford, pp137–182

Raman, R., Mantell, S., Davidson, J. and Jorgensen, G. (2000) 'A review of polymer materials for solar water heating systems', presented at *2000 International Solar Energy Conference, A part of SOLAR 2000: Solar Powers Life, Share the Energy*, Madison, Wisconsin, 16–21 June 2000

Rashkin, S. (2000) 'If home energy ratings are the solution for energy efficient new homes, what's the problem?' in *Proceedings of the 2000 ACEEE Summer Study on Energy Efficiency in Buildings*, vol 2, American Council for an Energy Efficient Economy, Washington, DC, pp279–286

Ray-Jones, A. (2000) *Sustainable Architecture in Japan: The Green Buildings of Nikken Sekkei*, Wiley Academic, Chichester UK, 190pp

Rea, M. S. (ed) (2000) *Lighting Handbook*, 9th edition, Illuminating Engineering Society of North America, New York

Reay, D. A. (2002) 'Compact heat exchangers, enhancement and heat pumps', *International Journal of Refrigeration*, vol 25, pp460–470

Reay, D. A. and MacMichael, D. B. A. (1988) *Heat Pumps*, 2nd edition, Pergamon Press, Oxford, 337pp

Reddy, A. M. and Narendra, P. (2002) 'India's newest generation – Biomass power in Andhra Pradesh', *Renewable Energy World*, vol 5, no 6, pp77–81

Reddy, B. V. V. and Jagadish, K. S. (2003) 'Embodied energy of common and alternative building materials and technologies', *Energy and Buildings*, vol 35, pp129–137

Reddy, T. A. (1997) 'Active solar heating systems', in Kreith, F. and West, R. E. (eds) *CRC Handbook of Energy Efficiency*, CRC Press, Boca Raton, FL, pp791–847

Redlinger, R. Y., Andersen, P. D. and Morthorst, P. E. (2002) *Wind Energy in the 21st Century: Economics, Policy, Technology and the Changing Electricity Industry*, Palgrave (Basingstoke), 245pp

Reim, M., Beck, A., Körner, W., Petricevic, R., Glora, M., Weth, M., Schliermann, T., Fricke, J., Schmidt, Ch. and Pötter, F. J. (2002) 'Highly insulating aerogel glazing for solar energy usage', *Solar Energy*, vol 72, pp21–29

Reinhart, C. F. (2002) 'Effects of interior design on the daylight availability in open plan offices', in *Proceedings of the 2002 ACEEE Summer Study on Energy Efficiency in Buildings*, vol 3, American Council for an Energy Efficient Economy, Washington, DC, pp309–322

Reinhart, C. F., Voss, K., Wagner, A. and Löhnert, G. (2000) 'Lean buildings: energy efficient commercial buildings in Germany', in *Proceedings of the 2000 ACEEE Summer Study on Energy Efficiency in Buildings*, vol 3, American Council for an Energy Efficient Economy, Washington, DC, pp257–298

Ren, K. B. and Kagi, D. A. (1995) 'Upgrading the durability of mud bricks by impregnation', *Building and Environment*, vol 30, pp433–440

Renzo Piano Building Workshop (1996) 'Debis headquarters, Potsdamer Platz, Berlin', in Fitzgerald, E. and Lewis, J. O. (eds) *European Solar Architecture: Proceedings of a Solar House Contractors' Meeting, Barcelona 1995*, Energy Research Group, University College Dublin, pp190–197

Reppel, J. and Edmonds, I. R. (1998) 'Angle-selective glazing for radiant heat control in buildings: theory', *Solar Energy*, vol 62, pp245–253

Reuss, M. and Mueller, J. (2000) 'Solar district heating with seasonal storage in Attenkirchen', in Benner, M. and Hahne, E. W. P. (eds) *Proceedings Terrastock*

2000, 8th International Conference on Thermal Energy Storage, Stuttgart, 28 August – 1 September 2000, Institut für Thermodynamik und Waermetechnik, Universitaet Stuttgart, Stuttgart, pp221–226

Richard Rogers Architects (1996) 'Three demonstration buildings, Potsdamer Platz, Berlin', in Fitzgerald, E. and Lewis, J. O. (eds) *European Solar Architecture: Proceedings of a Solar House Contractors' Meeting, Barcelona 1995*, Energy Research Group, University College Dublin, pp198–207

Richardson, M. G. (1988) *Carbonation of Reinforced Concrete: Its Causes and Management*, Citis, Dublin

Richter, M. R., Song, S. M., Yin, J. M., Kim, M. H., Bullard, C. W. and Hrnjak, P. S. (2003) 'Experimental results of transcritical CO_2 heat pump for residential application', *Energy*, vol 28, pp1005–1019

Riebesell, U., Zondervan, I., Rost, B., Tortell, P. D., Zeebe, R. and Morel, F. M. M. (2000) 'Reduced calcification of marine plankton in response to increased atmospheric CO_2', *Nature*, vol 407, pp364–367

Riffat, S. B. and Gan, G. (1998) 'Determination of effectiveness of heat-pipe heat recovery for naturally-ventilated buildings', *Applied Thermal Engineering*, vol 18, pp121–130

Riffat, S., Oliveira, A., Facão, J., Gan, G. and Doherty, P. (2000) 'Thermal performance of a closed wet cooling tower for chilled ceilings: measurement and CFD simulation', *International Journal of Energy Research*, vol 24, pp1171–1179

Rigacci, A., Marechal, J. C., Repoux, M., Moreno, M. and Achard, P. (2004) 'Preparation of polyurethane-based aerogels and xerogels for thermal superinsulation', *Journal of Non-Crystalline Solids*, vol 350, pp372–378

Rivard, H., Bédard, C., Ha, K. H. and Fazio, P. (1999) 'Shared conceptual model for the building envelope design process', *Building and Environment*, vol 34, pp175–187

Rivers, N. (2000) 'Unconventional secondary refrigeration in a UK supermarket', *IEA Heat Pump Centre Newsletter*, vol 18, no 1, pp18–19

Robledo, L. and Soler, A. (2000) 'Luminous efficacy of global solar radiation for clear skies', *Energy Conversion and Management*, vol 41, pp1769–1779

Robledo, L. and Soler, A. (2001) 'On the luminous efficacy of diffuse solar radiation', *Energy Conversion and Management*, vol 42, pp1181–1190

Rock, B. A. and Wu, C.-T. (1998) 'Performance of fixed, air-side economizer, and neural network demand-controlled ventilation in CAV systems', *ASHRAE Transactions*, vol 104 (Part 2), pp234–245

Rogdakis, E. D. and Alexis, G. K. (2000) 'Design and parametric investigation of an ejector in an air-conditioning system', *Applied Thermal Engineering*, vol 20, pp213–226

Röing, P. and Avasoo, D. (2001) 'Standby losses caused by low voltage lighting', in *European Council for an Energy Efficient Economy, 2001 Summer Proceedings*, vol 4, pp182–188

Römer, J. C. (2001) 'Simulation of demand controlled ventilation in a low-energy house', in *Seventh International IBPSA Conference*, Rio de Janeiro, Brazil, 13–15 August 2001, pp661–617. Available from www.ibpsa.org

Rosemann, A. and Kaase, H. (2005) 'Lightpipe applications for daylighting applications', *Solar Energy*, vol 78, pp772–780

Rosen, M. A. (1998) 'The use of berms in thermal energy storage systems: energy-economic analysis', *Solar Energy*, vol 63, pp69–78

Rosencrantz, T., Bülow-Hübe, H., Karlsson, B. and Roos, A. (2005) 'Increased solar energy and daylight utilisation using anti-reflective coatings in energy-efficient windows', *Solar Energy Materials and Solar Cells*, vol 89, pp249–260

Rosenfeld, A. H. (1999) 'The art of energy efficiency: protecting the environment with better technology', *Annual Review of Energy and the Environment*, vol 24, pp33–82

Rosenfeld, A. H., Akbari, H., Romm, J. J. and Pomerantz, M. (1998) 'Cool communities: strategies for heat island mitigation and smog reduction', *Energy and Buildings*, vol 28, pp51–62

Ross, J. P. and Meier, A. (2002) 'Measurements of whole-house standby power consumption in California homes', *Energy*, vol 27, pp861–868

Roth, I. F. and Ambs, L. L. (2004) 'Incorporating externalities into a full cost approach to electric power generation life-cycle costing', *Energy*, vol 29, pp2125–2144

Roth, K. W., Westphalen, D., Dieckmann, J., Hamilton, S. D. and Goetzler, W. (2002a) *Energy Consumption Characteristics of Commercial Building HVAC Systems, Volume III: Energy Savings Potential*, Arthur D. Little Inc., Cambridge (MA), 212pp. Available from www.eere.energy.gov

Roth, K. W., Goldstein, F. and Kleinman, J. (2002b) *Energy Consumption by Office and Telecommunications Equipment in Commercial Buildings. Volume 1: Energy Consumption Baseline*, Arthur D. Little Inc., Cambridge, MA. Available from www.eren.doe.gov/buildings/documents/pdfs/office_telecom_vol1_final.pdf

Roth, K. W., Dieckmann, J. and Brodrick, J. (2003a) 'Demand control ventilation', *ASHRAE Journal*, vol

45, no 7, pp91–92

Roth, K. W., Westphalen, D. and Brodrick, J. (2003b) 'Saving energy with building commissioning', *ASHRAE Journal*, vol 45, no 11, pp65–66

Roulet, C. A., Rossy, J. P. and Roulet, Y. (1999) 'Using large radiant panels for indoor climate conditioning', *Energy and Buildings*, vol 30, pp121–126

Roulet, C. A., Heidt, F. D., Foradini, F. and Pibiri, M. C. (2001) 'Real heat recovery with air handling units', *Energy and Buildings*, vol 33, pp495–502

Rousseau, P. G. and Matthews, E. H. (1996) 'A new integrated design tool for naturally ventilated buildings', *Energy and Buildings*, vol 23, pp231–236

Rowlands, I. H. (2005) 'Solar PV electricity and market characteristics: two Canadian case-studies', *Renewable Energy*, vol 30, pp815–834

Rubin, M., Von Rottkay, K. and Powles, R. (1998) 'Window optics', *Solar Energy*, vol 62, pp149–161

Rubinstein, F. and Johnson, S. (1998) *Advanced Lighting Program Development (BG9702800) Final Report*, Lawrence Berkeley National Laboratory, LBNL-41679, Berkeley, CA, 28pp

Rubinstein, F., Jennings, H. J. and Avery, D. (1998) *Preliminary Results from an Advanced Lighting Controls Testbed*, Lawrence Berkeley National Laboratory, LBNL-41633, Berkeley, CA, 19pp

Rudd, A., Kerrigan Jr., P., and Ueno, K. (2004) 'What will it take to reduce total residential source energy use by up to 60%?' in *Proceedings of the 2004 ACEEE Summer Study on Energy Efficiency in Buildings*, vol 1, American Council for an Energy Efficient Economy, Washington, DC, pp293–305

Sabour, S. A. A. (2005) 'Quantifying the external cost of oil consumption within the context of sustainable development', *Energy Policy*, vol 33, pp809–881

Sachs, H. M. and Smith, S. (2004) 'How much energy could residential furnace air handlers save?', *ASHRAE Transactions*, vol 110 (Part 1), pp431–441

Sachs, H. M., Kubo, T., Smith, S. and Scott, K. (2002) 'Residential HVAC fans and motors are bigger than refrigerators', in *Proceedings of the 2002 ACEEE Summer Study on Energy Efficiency in Buildings*, vol 1, American Council for an Energy Efficient Economy, Washington, DC, pp261–272

Sachs, H. M., Nadel, S., Amann, J. T., Tuazon, M., Mendelsohn, E., Rainer, L., Todecso, G., Shipley, D. and Adelaar, M. (2004) *Emerging Energy-Saving Technologies and Practices for the Buildings Sector as of 2004*, American Council for an Energy Efficient Economy, Washington, DC, 247pp

Safarzadeh, H. and Bahadori, M. N. (2005) 'Passive cooling effects of courtyards', *Building and Environment*, vol 40, pp89–104

Saha, B. B., Akisawa, A. and Kashiwagi, T. (2001) 'Solar/waste heat driven two-stage adsorption chiller: the prototype', *Renewable Energy*, vol 23, pp93–101

SAIC (Science Applications International Corporation) (2004) *Solar Seasonal Storage and District Heating, Okotoks Project Phase I Study*, SAIC Canada, Ottawa, 80pp

Saitoh, H., Hamada, Y., Kubota, H., Nakamura, M., Ochifuji, K., Yokoyama, S. and Nagano, K. (2003) 'Field experiments and analyses on a hybrid solar collector', *Applied Thermal Engineering*, vol 23, pp2089–2105

Saitoh, T. S. (1999) 'A highly-advanced solar house with solar thermal and sky radiation cooling', *Applied Energy*, vol 64, pp215–228

Saitoh, T. S. and Fujino, T. (2001) 'Advanced energy-efficient house (Harbeman House) with solar thermal, photovoltaic and sky radiation energies (experimental results)', *Solar Energy*, vol 70, no 1, pp63–77

Saman, W., Bruno, F. and Halawa, E. (2005) 'Thermal performance of PCM thermal storage unit for a roof integrated solar heating system', *Solar Energy*, vol 78, pp341–349

Sánchez, M. M., Lucas, M., Martínez, P., Sánchez, A. and Viedma, A. (2002) 'Climatic solar roof: an ecological alternative to heat dissipation in buildings', *Solar Energy*, vol 73, pp419–432

Sandnes, B. and Rekstad, J. (2002) 'A photovoltaic/thermal (PV/T) collector with a polymer absorber plate. Experimental study and analytical model', *Solar Energy*, vol 72, pp63–73

Sandnes, B., Michaela, M., Rekstad, J. and Mullane, R. (2000) 'Phase change materials for storage of solar heat', in *Proceedings of ISES EuroSun 2000*, 19–22 June, Copenhagen, Denmark

Sanner, B. (1999) *High Temperature Underground Thermal Energy Storage, State-of-the-art and Prospects*, Lenz-Verlag-Giessen, Giessen, Germany

Santamouris, M. and Asimakopoulos, D. (1996) *Passive Cooling of Buildings*, James & James, London, 472pp

Sawhney, R. L., Buddhi, D. and Thanu, N. M. (1999) 'An experimental study of summer performance of a recirculation type underground airpipe air conditioning system', *Building and Environment*, vol 34, pp189–196

Saxhof, B. (2003) 'Transparent insulation', in Hestnes, A. G., Hastings, R. and Saxhof, B. (eds) *Solar Energy Houses: Strategies, Technologies, Examples*, 2nd edition, James & James, London, pp27–30

Saxhof, B. and Carpenter, S. C. (2003) 'High-performance windows', in Hestnes, A. G., Hastings,

R. and Saxhof, B. (eds) *Solar Energy Houses: Strategies, Technologies, Examples*, 2nd edition, James & James, London, pp23–26

Scanada Consultants (1995) *Advanced Houses Technology Assessment Summary Report*, prepared for CANMET Energy Technology Centre, Ottawa, 39pp. Available from www.buildingsgroup.nrcan.gc.ca.

Scartezzini, J. L. and Courret, G. (2002) 'Anidolic daylighting systems', *Solar Energy*, vol 73, pp123–135

Schaetzle, W. J., Brett, C. E., Grubbs, D. M. and Seppanen, M. S. (1980) *Thermal Energy Storage in Aquifers, Design and Applications*, Pergamon Press, New York, 177pp

Schell, M. and Int-Hout, D. (2001) 'Demand control ventilation using CO_2', *ASHRAE Journal*, vol 43, no 2, pp18–29

Schell, M. B., Turner, S. C. and Shim, R. O. (1998) 'Application of CO_2-based demand-controlled ventilation using ASHRAE standard 62: optimizing energy use and ventilation', *ASHRAE Transactions*, vol 104 (Part 2), pp1213–1225

Scheuermann, K., Boleyn, D., Lilly, P. and Miller, S. (2002) 'Measured performance of California buydown program residential PV systems', in *Proceedings of the 2002 ACEEE Summer Study on Energy Efficiency in Buildings*, vol 1, American Council for an Energy Efficient Economy, Washington, DC, pp273–285

Schild, P. and Blom, P. (2002) *Pilot Study Report: Jaer School, Nesodden Municipality, Norway*, International Energy Agency, Energy Conservation in Buildings and Community Systems, Annex 35. Available from http://hybvent.civil.auc.dk

Schittich, C. (ed) (2003) *Solar Architecture: Strategies, Visions, Concepts*, Birkhäuser, Basel

Schlegel, G. O., Burkholer, F. W., Klein, S. A., Beckman, W. A., Wood, B. D. and Muhs, J. D. (2004) 'Analysis of a full spectrum hybrid lighting system', *Solar Energy*, vol 76, pp359–368

Schmidt, D. (2002) 'The Centre for Sustainable Building (ZUB), a case study', presented at *Sustainable Buildings 2002*, Oslo, Norway, International Initiative for a Sustainable Built Environment. Available at www.iisbe.org

Schmidt, K. and Patterson, D. J (2001) 'Performance results for a high efficiency tropical fan and comparisons with conventional fans: demand side management via small appliance efficiency', *Renewable Energy*, vol 22, pp169–176

Schmidt, T., Mangold, D. and Muller-Steinhagen, H. (2004) 'Central solar heating plants with seasonal storage in Germany', *Solar Energy*, vol 76, pp165–174

Schmitz, G. and Casas, W. (2001) 'Experiences with a small gas engine driven desiccant HVAC – system', in *Proceedings of the 2001 International Gas Research Conference*, Amsterdam, The Netherlands, 5–8 November 2001

Schnieders, J. and Hermelink, A. (2006) 'CEPHEUS results: measurements and occupants' satisfaction provide evidence for Passive Houses being an option for sustainable building', *Energy Policy*, vol 34, pp151–171

Schoen, T. J. N. (2001) 'Building-integrated PV installations in The Netherlands: examples and operational experiences', *Solar Energy*, vol 70, pp467–477

Schossig, P., Henning, H.-M., Gschwander, S. and Haussmann, T. (2005) 'Micro-encapsulated phase-change materials integrated into construction materials', *Solar Energy Materials and Solar Cells*, vol 89, pp297–306

Schramm, W., Heckmann, M., Lendefeld, T. and Reichenauer, T. (1994) 'Technology assessment of aerosols', in *Proceedings of the 7th International Meeting on Transparent Insulation Technology*, 21–23 September 1994, Delphth, The Netherlands, Franklin Company Consultant LTD, Birmingham (UK), pp139–143

Schultz, J. M., Jensne, K. I. and Kristiansen, F. H. (2005) 'Super insulating aerogel glazing', *Solar Energy Materials and Solar Cells*, vol 89, pp275–285

Schweigler, C. J., Hellmann, H.-M., Preissner, M., Demmel, S. and Ziegler, F. F. Z. (1998) 'Operation and performance of a 350 kW (100 RT) single-effect/double-lift absorption chiller in a district heating network', *ASHRAE Transactions*, vol 104 (Part 1B), pp1420–1426

Scofield, J. H. (2002) 'Early performance of a green academic building', *ASHRAE Transactions*, vol 108 (Part 2), pp1214–1230

Sekhar, S. C. (1995) 'Higher space temperatures and better thermal comfort – a tropical analysis', *Energy and Buildings*, vol 23, pp63–70

Sekhar, S. C. (1997) 'A critical evaluation of variable air volume system in hot and humid climates', *Energy and Buildings*, vol 26, pp223–232

Sekhar, S. C. and Phua, K. J. (2003) 'Integrated retrofitting strategy for enhanced energy efficiency in a tropical building', *ASHRAE Transactions*, vol 109 (Part 1), pp202–214

Sellers, D. A. (2005) 'Rightsizing pumping systems', *HPAC Engineering*, vol 77, no 3, pp26–36

Sellers, D. and Williams, J. (2000) 'A comparison of the ventilation rates established by three common building codes in relationship to actual occupancy levels and the impact of these rates on building energy consumption', in *Proceedings of the 2000*

ACEEE Summer Study on Energy Efficiency in Buildings, vol 3, American Council for an Energy Efficient Economy, Washington, DC, pp299–313

Sellers, W. D. (1965) *Physical Climatology*, University of Chicago Press, Chicago, 272pp

Sfeir, A., Bernier, M. A., Million, T. and Joly, A. (2005) 'A methodology to evaluate pumping energy consumption in GCHP systems', *ASHRAE Transactions*, vol 111 (Part 1), pp714–729

Shaviv, E. and Capeluto, I. G. (1992) 'The relative importance of various geometrical design parameters in a hot, humid climate', *ASHRAE Transactions*, vol 98 (Part 1), pp589–605

Sherman, M. H. (1987) 'Estimation of infiltration from leakage and climate indicators', *Energy and Buildings*, vol 10, pp81–86

Sherman, M. H. and Jump, D. A. (1997) 'Thermal energy conservation in buildings', in Kreith, F. and West, R. E. (eds) *CRC Handbook of Energy Efficiency*, pp269–303, CRC Press, Boca Raton, FL

Sherman, M. H. and Dickerhoff, D. J. (1998) 'Airtightness of U.S. dwellings', *ASHRAE Transactions*, vol 104 (Part 2), pp1359–1367

Sherman, M. H., Walker, I. S. and Dickerhoff, D. J. (2000) 'Stopping duct quacks: longevity of residential duct sealants', in *Proceedings of the 2000 ACEEE Summer Study on Energy Efficiency in Buildings*, vol 1, American Council for an Energy Efficient Economy, Washington, DC, pp273–284

Shiming, D. and Yiqiang, J. (2003) 'Retrofitting hot water supply in Hong Kong luxury hotel', *ASHRAE Journal*, vol 45, no 12, pp52–55

Shimoda, Y., Fujii, T., Morikawa, T. and Mizuno, M. (2003) 'Development of residential energy end-use simulation model at city scale', in *Eighth International IBPSA Conference*, Eindhoven, The Netherlands, 11–14 August 2003, pp1201–1208. Available from www.ibpsa.org

Shirey, D. B. and Henderson, H. I. (2004) 'Dehumidification at part load', *ASHRAE Journal*, vol 46, no 4, pp42–48

Shonder, J. A., Martin, M. A., McLain, H. A. and Hughes, P. J. (2000) 'Comparative analysis of life-cycle costs of geothermal heat pumps and three conventional HVAC systems', *ASHRAE Transactions*, vol 106 (Part 2), pp551–560

Sick, F. and Erge, T. (eds) (1996) *Photovoltaics in Buildings: A Design Handbook for Architects and Engineers*, International Energy Agency, Paris, 287pp

Simmler, H. and Brunner, S. (2005) 'Vacuum insulation panels for building application: basic properties, aging mechanisms and service life', *Energy and Buildings*, vol 37, pp1122–1131

Simmonds, P. (1994) 'A comparison of energy consumption for storage priority and chiller priority for ice-based thermal storage systems', *ASHRAE Transactions*, vol 100 (Part 2), pp1746–1753

Simmonds, P., Hoist, S., Reuss, S. and Gaw, W. (2000) 'Using radiant cooled floors to condition large spaces and maintain comfort conditions', *ASHRAE Transactions*, vol 106 (Part 1), pp695–701

Simonson, C. J., Ciepliski, D. L. and Besant, R. W. (1999) 'Determining the performance of energy wheels: Part II – experimental data and numerical validation', *ASHRAE Transactions*, vol 105 (Part 1), pp188–205

Simonson, C. J., Shang, W. and Besant, R. W. (2000a) 'Part-load performance of energy wheels: Part I – wheel speed control', *ASHRAE Transactions*, vol 106 (Part 1), pp286–300

Simonson, C. J., Shang, W. and Besant, R. W. (2000b) 'Part-load performance of energy wheels: Part II – bypass control and correlations', *ASHRAE Transactions*, vol 106 (Part 1), pp301–310

Singh, J. B., Foster, G. and Hunt, A. W. (2000) 'Representative operating problems of commercial ground-source and groundwater-source heat pumps', *ASHRAE Transactions*, vol 106 (Part 2), pp561–568

Sirén, K., Riffat, S., Alfonso, C., Oliveira, A. and Kofoed, P. (1997) 'Solar-assisted natural ventilation with heat pipe heat recovery', in *Proceedings 18th AVIC Conference: Ventilation and Cooling*, Athens, Greece, 23–26 September 1997, pp312–315

Skistad, H. E., Mundt, E., Nielsen, P. V., Hagstrom, K. and Ralio, J. (2002) *Displacement Ventilation in Non-Industrial Premises*, REHVA, Federation of European Heating and Air Conditioning Associations, Trondheim, Norway

Skopek, J. (1999) 'BREEAM strategy for reducing buildings' environmental impact', *ASHRAE Transactions*, vol 105 (Part 2), pp811–818

Slayzak, S. (2001) 'NREL's advanced HVAC Project: research to reduce energy use and cost', *Energy Solutions*, spring 2001, pp14–15. Available from www.nrel.gov/dtet/

Sminovitch, M. and Page, E. (2003) *Energy Efficient Downlights for California Kitchens*, California Energy Commission Report 500-04-005. Available from www.consumerenergycenter.org/bookstore/index.html

Smith, B. (2002) 'Economic analysis of hybrid chiller plants', *ASHRAE Journal*, vol 44, no 7, pp42–46

Smith, C. B. (1997) 'Electrical energy management in buildings', in Kreith, F. and West, R. E. (eds) *CRC Handbook of Energy Efficiency*, CRC Press, Boca Raton, FL, pp305–336

Smith, P. F. (2001) *Architecture in a Climate of Change: A*

Guide to Sustainable Design, Architectural Press, Oxford, 214pp

Smith, R. R., Hwang, C. C. and Dougall, R. S. (1994) 'Modelling of a solar-assisted desiccant air conditioner for a residential building', *Energy*, vol 19, pp679–691

Sodec, F. (1999) 'Economic viability of cooling ceiling systems', *Energy and Buildings*, vol 30, pp195–201

Solaini, G., Dall'o', G. and Scansani, S. (1998) 'Simultaneous application of different natural cooling technologies to an experimental building', *Renewable Energy*, vol 15, pp277–282

Song, D. and Kato, S. (2004) 'Radiational panel cooling system with continuous natural cross ventilation for hot and humid regions', *Energy and Buildings*, vol 36, pp1273–1280

Song, D., Kato, S., Kim, T. and Murakami, S. (2002) 'Study on cross-ventilation with radiational panel cooling for hot and humid regions', *ASHRAE Transactions*, vol 108 (Part 2), pp1276–1281

Sonne, J. K. and Parker, D. S. (1998) 'Measured ceiling fan performance and usage patterns: implications for efficiency and comfort improvement', in *Proceedings of the 1998 ACEEE Summer Study on Energy Efficiency in Buildings*, vol 1, American Council for an Energy Efficient Economy, Washington, DC, pp335–341

Sorrell, S. (2003) 'Making the link: climate policy and the reform of the UK construction industry', *Energy Policy*, vol 31, pp865–878

Söylemez, M. S. (2001) 'On the optimum channel sizing for HVAC systems', *Energy Conversion and Management*, vol 42, pp791–798

Sözen, A., Özalp, M. and Arcaklioğlu, E. (2004a) 'Prospects for utilisation of solar driven ejector-absorption cooling system in Turkey', *Applied Thermal Engineering*, vol 24, pp1019–1035

Sözen, A., Arcaklioglu, E. and Ozalp, M. (2004b) 'Performance analysis of ejector absorption heat pump using ozone safe fluid couple through artificial neural networks', *Energy Conversion and Management*, vol 45, pp2233–2253

Späth, M., Sommeling, P. M., van Roosmalen, J. A. M., Smit, H. J. P., van der Burg, N. P. G., Mahieu, D. R., Bakker, N. J. and Kroon, J. M. (2003) 'Reproducible manufacturing of dye-sensitized cells on a semi-automated baseline', *Progress in Photovoltaics: Research and Applications*, vol 11, pp207–220

Springer, D., Loisos, G. and Rainer, L. (2000) 'Non-compressor cooling alternatives for reducing residential peak load', in *Proceedings of the 2000 ACEEE Summer Study on Energy Efficiency in Buildings*, vol 1, American Council for an Energy Efficient Economy, Washington, DC, pp319–330

Spurr, M. (1998) *District Energy Systems Integrated with Combined Heat and Power: Analysis of Environmental and Economic Benefits*, report to the US Environmental Protection Agency, International District Energy Association, Minneapolis, MN

Srikhirin, P., Aphornratana, S. and Chungpaibulpatana, S. (2001) 'A review of absorption refrigeration technologies', *Renewable and Sustainable Energy Reviews*, vol 5, pp343–372

Srinivasan, S., Mosdale, R., Stevens, P. and Yang, C. (1999) 'Fuel cells: reaching the era of clean and efficient power generation in the twenty-first century', *Annual Review of Energy and the Environment*, vol 24, pp281–328

Stec, W. J. and van Paassen, A. H. C. (2005) 'Symbiosis of the double skin façade with the HVAC system', *Energy and Buildings*, vol 37, pp461–469

Stec, W. J., van Paassen, A. H. C. and Maziarz, A. (2005) 'Modelling the double skin façade with plants', *Energy and Buildings*, vol 37, pp419–427

Stein, B. and Reynolds, J. S. (2000) *Mechanical and Electrical Equipment for Buildings*, Wiley, New York

Stetiu, C. and Feustel, H. E. (1999) 'Energy and peak power savings potential of radiant cooling systems in US commercial buildings', *Energy and Buildings*, vol 30, pp127–138

Steven Winter Associates (1998) *The Passive Solar Design and Construction Handbook*, John Wiley and Sons, New York, 291pp

Stewart, W. E., Saunders, C. K. and Dona, C. L. G. (1999) 'Evaluation of service hot water distribution system losses in residential and commercial installations: part 2 – simulation and design practices', *ASHRAE Transactions*, vol 105 (Part 1), pp247–261

Stosic, N., Kovacevic, A. and Smith, I. K. (1998) 'Modelling of screw compressor capacity control', in *Design, Selection, and Operation of Refrigeration and Heat Pump Compressors: Achieving Economic Cost and Energy Efficiency*, IMechE Seminar Publication 1998-15, Professional Engineering Publishing, London, pp9–30

Strand, R. K. (2003) 'Investigation of a condenser-linked radiant cooling system using a heat balance based energy simulation model', *ASHRAE Transactions*, vol 109 (Part 2), pp647–655

Strand, R. K. and Pedersen, C. O. (2002) 'Modeling radiant systems in an integrated heat balance based energy simulation program', *ASHRAE Transactions*, vol 108 (Part 2), pp979–987

Strand, R. K. and Baumgartner, K. T. (2005) 'Modeling radiant heating and cooling systems: integration with a whole-building simulation program', *Energy and Buildings*, vol 37, pp389–397

Strand, R. K., Fisher, D. E., Liesen, R. J. and Pedersen, C. O. (2002) 'Modular HVAC simulation and the future integration of alternative cooling systems in a new building energy simulation program', *ASHRAE Transactions*, vol 108 (Part 2), pp1107–1117

Streicher, W., Jordan, U. and Vajen, K. (2003) 'Heat demand of buildings' in Weiss, W. (ed) *Solar Heating Systems for Houses: A Design Handbook for Solar Combisystems*, James & James, London, pp17–37

Stritih, U. (2003) 'Heat transfer enhancement in latent heat thermal storage system for buildings', *Energy and Buildings*, vol 35, pp1097–1104

Sumathy, K., Huang, Z. C. and Li, Z. F. (2002) 'Solar absorption cooling with low grade heat source – a strategy of development in south China', *Solar Energy*, vol 72, pp155–165

Sun, Z. G., Wang, R. Z. and Sun, W. Z. (2004) 'Energetic efficiency of a gas-engine-driven cooling and heating system', *Applied Thermal Engineering*, vol 24, pp941–947

Sundqvist, T. (2004) 'What causes the disparity of electricity externality estimates?', *Energy Policy*, vol 32, pp1753–1766

Suter, J. M. and Letz, T. (2003) 'Generic solar combisystems', in Weiss, W. (ed) *Solar Heating Systems for Houses: A Design Handbook for Solar Combisystems*, James & James, London, pp38–92

Suzuki, M. and Oka, T. (1998) 'Estimation of life cycle consumption and CO_2 emission of office buildings in Japan', *Energy and Buildings*, vol 28, pp33–41

Sveine, T., Grandum, S. and Baksaas, H. S. (1998) 'Design of high temperature absorption/compression heat pump', in *Natural Working Fluids '98, IIR – Gustav Lorentzen Conference*, 2–5 June 1998, Oslo, International Institute of Refrigeration, Paris, pp539–549

Swedish Energy Agency (2005) *Energy in Sweden: Facts and Figures 2005*, Swedish Energy Agency, Eskilstuna, 40pp. Available from www.stem.se

Swinney, J., Jones, W. E. and Wilson, J. A. (2001) 'A novel hybrid absorption-compression refrigeration cycle', *International Journal of Refrigeration*, vol 24, pp208–219

Syed, A., Maidment, G. G., Missenden, J. F. and Tozer, R. M. (2002) 'An efficiency comparison of solar cooling schemes', *ASHRAE Transactions*, vol 108 (Part 1), pp877–886

Takakura, T., Kitade, S. and Goto, E. (2000) 'Cooling effect of greenery cover over a building', *Energy and Buildings*, vol 31, pp1–6

Tan, K. X. and Deng, S. M. (2001) 'Desuperheater heat recovery hot water heating systems in subtropics: Using water cooling towers to extract heat from ambient air as heat source', *ASHRAE Transactions*, vol 107 (Part 2), pp608–612

Tantasavasdi, C., Srebric, J. and Chan, Q. (2001) 'Natural ventilation for houses in Thailand', *Energy and Buildings*, vol 33, pp815–824

Tao, W. K. Y. and Janis, R. R (1997) *Mechanical and Electrical Systems in Buildings*, Prentice Hall, Upper Saddle River, NJ, 538pp

Tassou, S. A. and Lau, M. M. (1998) 'Low energy cooling technologies – a review', in Tassou, S. (ed) *Low-Energy Cooling Technologies for Buildings: Challenges and Opportunities for the Environmental Control of Buildings*, IMechE Seminar Publication 1998–7, Professional Engineering Publishing, London, pp1–22

Taylor, B. J. and Imbabi, M. S. (1997) 'The effect of air film thermal resistance on the behaviour of dynamic insulation', *Building and Environment*, vol 32, pp397–404

Taylor, B. J. and Imbabi, M. S. (2000) 'Environmental design using dynamic insulation', *ASHRAE Transactions*, vol 106 (Part 1), pp15–28

Taylor, P. B., Mathews, E. H., Kleingeld, M. and Taljaard, G. W. (2000) 'The effect of ceiling insulation on indoor comfort', *Building and Environment*, vol 35, pp339–346

Taylor, S. T. (2002) 'Degrading chilled water plant Delta-T: Causes and mitigation', *ASHRAE Transactions*, vol 108 (Part 1), pp641–653

Tenorio, R. (2002) 'Dual mode cooling house in the warm humid tropics', *Solar Energy*, vol 73, pp43–57

Ternoey, S. E. (2000a) *The Cool Daylighting™ Design Approach Workbook: Volume 1 – Office Buildings*, 34pp. Available from www.ecw.org

Ternoey, S. E. (2000b) *The Cool Daylighting™ Design Approach Workbook: Volume21 – Schools*, 28pp. Available from www.ecw.org

Thanu, N. M., Sawhney, R. L., Khare, R. N. and Buddhi, D. (2001) 'An experimental study of the thermal performance of an earth-air-pipe system in single pass mode', *Solar Energy*, vol 71, pp353–364

Theodosiou, T. G. (2003) 'Summer period analysis of the performance of a planted roof as a passive cooling technique', *Energy and Buildings*, vol 35, pp909–917

Thierfelder, A. (ed) (2003) *Transsolar Climate Engineering*, Birkhäuser, Basel, 229pp

Thomas, A., Shincovich, M., Ryan, S., Korn, D. and Shugars, J. (2002) 'Replacing distribution transformers: a hidden opportunity for energy savings', in *Proceedings of the 2002 ACEEE Summer Study on Energy Efficiency in Buildings*, vol 3, American Council for an Energy Efficient Economy, Washington, DC, pp351–359

Thomas, C. D. et al (2004) 'Extinction risk from climate

change', *Nature*, vol 427, pp133–148

Thomas, R. (1999) *Environmental Design: An Introduction for Architects and Engineers*, E & FN Spon, London, 259pp

Thomas, R. and Fordham, M. (eds) (2001) *Photovoltaics and Architecture*, Spon Press, London, 155pp

Thompson, S. L. and Barron, E. J. (1981) 'Comparison of Cretaceous and present Earth albedos: implications for the cause of paleoclimates', *Journal of Geology*, vol 89, pp143–167

Thormark, C. (2002) 'A low energy building in a life cycle – its embodied energy, energy need for operation and recycling potential', *Building and Environment*, vol 37, pp429–435

Tiwari, P. (2001) 'Energy efficiency and building construction in India', *Building and Environment*, vol 36, pp1127–1135

Tiwari, P., Parikh, J. and Sharma, V. (1996) 'Performance evaluation of cost effective buildings – a cost, emissions and employment point of view', *Building and Environment*, vol 31, pp75–90

Todesco, G. (2004) 'Integrated design and HVAC equipment sizing', *ASHRAE Journal*, vol 46, no 9, ppS42–S47

Tol, R. S. J. (1995) 'The damage cost of climate change toward more comprehensive calculations', *Environmental and Resources Economics*, vol 5, pp353–374

Tombazis, A. N. and Preuss, S. A. (2001) 'Design of passive solar buildings in urban areas', *Solar Energy*, vol 70, pp311–318

Tomlinson, J. J. and Kannberg, L. D. (1990) 'Thermal energy storage', *Mechanical Engineering*, vol 112, pp68–72

Torcellini, P. A., Deru, M., Griffith, B., Long, N., Pless, S., Judkoff, R. and Crawley, D. B. (2004a) 'Lessons learned from field evaluation of six high-performance buildings', in *Proceedings of the 2004 ACEEE Summer Study on Energy Efficiency in Buildings*, vol 3, American Council for an Energy Efficient Economy, Washington, DC, pp325–337

Torcellini, P. A., Hayter, S. J. and Judkoff, R. (1999) 'Low-energy building design – the process and a case study', *ASHRAE Transactions*, vol 105 (Part 2), pp802–810

Torcellini, P. A., Judkoff, R. and Hayter, S. J. (2002) 'Zion National Park Visitor Center: significant energy savings achieved through a whole-building design process', in *Proceedings of the 2002 ACEEE Summer Study on Energy Efficiency in Buildings*, vol 3, American Council for an Energy Efficient Economy, Washington, DC, pp361–372

Torcellini, P. A., Long, N. and Judkoff, R. D. (2004b)

'Consumptive water use for U.S. power production', *ASHRAE Transactions*, vol 110 (Part 1), pp96–100

Tozer, R. (2002) 'Sorption thermodynamics', *ASHRAE Transactions*, vol 108 (Part 1), pp781–791

Treloar, G. J., Love, P. E. D. and Holt, G. D. (2001) 'Using national input–output data for embodied energy analysis of individual residential buildings', *Construction Management and Economics*, vol 19, pp49–61

Tripanagnostopoulos, Y., Nousia, T., Souliotis, M. and Yianoulis, P. (2002) 'Hybrid photovoltaic/thermal solar systems', *Solar Energy*, vol 72, pp217–234

Tripanagnostopoulos, Y., Souliotis, M., Battisti, R. and Corrado, A. (2003) 'Application aspects of hybrid PV/T solar systems', in *Proceedings of ISES Solar World Congress 2003 – Solar Energy for a Sustainable Future*, 14–19 June 2003, Göteborg, Sweden

Tripanagnostopoulos, Y., Souliotis, M., Battisti, R. and Corrado, A. (2004) 'Application aspects of hybrid PV/T air solar systems', in *Proceedings of EuroSun 2004*, 20–23 June 2004, Freiburg, Germany

Trusty, W. and Meil, J. (2000) 'The environmental implications of building new *versus* renovating an existing structure', in *Sustainable Buildings 2000 Conference Proceedings*, Maastricht, The Netherlands, October 2000. Available from www.athenasmi.ca

Tsurusaki, T., Tanaka, A., Nakagami, H. and Ohno, K. (2004) 'Continuous observation and evaluation of photovoltaic power generation systems for residential buildings in Japan', in *Proceedings of the 2004 ACEEE Summer Study on Energy Efficiency in Buildings*, vol 1, American Council for an Energy Efficient Economy, Washington, DC, pp330–340

Tuluca, A. (1997) *Energy-Efficient Design and Construction for Commercial Buildings*, McGraw-Hill, New York, 256pp

Turiel, I., Atkinson, B., Boghosian, S., Chan, P., Jennings, J., Lutz, J., McMahon, J., Pickle, S. and Rosenquist, G. (1997) 'Advanced technologies for residential appliance and lighting market transformation', *Energy and Buildings*, vol 26, pp241–252

Turnbull, P. W. and Loisos, G. A. (2000) 'Baselines and barriers: current design practices in daylighting', in *Proceedings of the 2000 ACEEE Summer Study on Energy Efficiency in Buildings*, vol 3, American Council for an Energy Efficient Economy, Washington, DC, pp329–336

Turnpenny, J. R., Etheridge, D. W. and Reay, D. A. (2000) 'Novel ventilation system for reducing air conditioning in buildings. Part I: testing and theoretical modelling', *Applied Thermal Engineering*, vol 20, pp1019–1037

Turnpenny, J. R., Etheridge, D. W. and Reay, D. A. (2001) 'Novel ventilation system for reducing air conditioning in buildings. Part II: testing of prototype', *Applied Thermal Engineering*, vol 21, pp1203–1217

Tzempelikos, A. and Athienitis, A. K. (2003) 'Simulation for façade options and impact on HVAC system design', in *Eighth International IBPSA Conference*, Eindhoven, The Netherlands, 11–14 August 2003, pp1301–1308. Available from www.ibpsa.org

Ubbelohde, M. S. and Loisos, G. A. (2002) 'Daylight design for multistory offices: advanced window wall design in practice', in *Proceedings of the 2002 ACEEE Summer Study on Energy Efficiency in Buildings*, vol 3, American Council for an Energy Efficient Economy, Washington, DC, pp385–400

Ullah, M. B. and Lefebvre, G. (2000) 'Estimation of annual energy-saving contribution of an automated blind system', *ASHRAE Transactions*, vol 106 (Part 2), pp408–418

UN (United Nations) (1992) *United Nations Framework Convention on Climate Change*, U.N. Doc. A/AC.237/18(Part II)/Add.1:15 May 1992

UN (United Nations (2001) *1998 Energy Statistics Yearbook*. United Nations, Department of Economic and Social Affairs, New York

Ürge-Vorsatz, D., Sroukanska, K. and Asztalos, S. (2002) 'Standing by in central Europe: a survey of Hungarian, Romanian and Bulgarian residences', in *Proceedings of the 2002 ACEEE Summer Study on Energy Efficiency in Buildings*, vol 7, American Council for an Energy Efficient Economy, Washington, DC, pp259–271

US DOE (United States Department of Energy) (2000) *Thermal Energy Storage for Space Cooling*, Federal Technology Alerts. Available from www.pnl.gov/

US DOE (United States Department of Energy) (2001) *Ground-Source Heat Pumps Applied to Federal Facilities – 2nd Edition*, Federal Technology Alerts. Available from www.pnl.gov/

US DOE (United States Department of Energy) (2002a) *National Best Practices Manual for Building High Performance Schools*. Available from www.rebuild.gov/Lawson/attachments/ESSBestPracticesHighPerfSchools.pdf

US DOE (United States Department of Energy) (2002b) *Efficient Lighting Strategies, Technology Fact Sheet*. Available from www.toolbase.org

US DOE (United States Department of Energy) (2003) *Annual Steam-Electric Plant Operation and Design Data*. Available from www.eia.doe.gov/cneaf/electricity/page/eia767.html

US DOE (United States Department of Energy) (2004)

2004 Buildings Energy Databook. Available from http://buildingsdatabook.eere.energy.gov

US EPA (United States Environmental Protection Agency) (1995) *Compilation of Air Pollutant Emission Factors AP-42, Fifth Edition, Volume 1: Stationary Point and Area Sources*. Available from www.epa.gov/ttn/

USGS (United States Geological Survey) (2000) *U.S. Geological Survey World Petroleum Assessment 2000 – Description and Results*. Available from http://pubs.usgs.gov/dds/dds-060

Vaidya, P., McDougall, T., Steinbock, J., Douglas, J. and Eijadi, D. (2004) 'What's wrong with daylight? Where it goes wrong and how users respond to failure', in *Proceedings of the 2004 ACEEE Summer Study on Energy Efficiency in Buildings*, vol 7, American Council for an Energy Efficient Economy, Washington, DC, pp342–357

van Mierlo, B. and Oudshoff, B. (1999) *Literature Survey and Analysis of Non-Technical Problems for the Introduction of Building Integrated Photovoltaic Systems*, International Energy Agency, Photovoltaic Power Systems Programme, Task 7, Paris. Summary available from www.iea-pvps.org

van Zolingen, R. J. (2004) 'Electrotechnical requirements for PV on buildings', *Progress in Photovoltaics: Research and Applications*, vol 12, pp409–414

Vartiainen, E. (2000) 'A comparison of luminous efficacy models with illuminance and irradiance measurements', *Renewable Energy*, vol 20, pp265–277

Vasile, C. (1997) 'Residential waste water heat-recovery system: GFX', *CADDET Energy Efficiency Newsletter*, December 1997, pp15–17

Velraj, R., Seeniraj, R. V., Hafner, B., Faber, C. and Schwarzer, K. (1999) 'Heat transfer enhancement in a latent heat storage system', *Solar Energy*, vol 65, pp171–180

Venta, Glaser, and Associates (1999) *Cement and Structural Concrete Products: Life Cycle Inventory Update*, Athena Sustainable Materials Institute, Merrickville, Ontario, Canada, 90pp

Venta, G. J. and Nisbet, M. (2000) *Life Cycle Analysis of Residential Roofing Products*, Athena Sustainable Materials Institute, Merrickville, Ontario, Canada, 36pp

Vine, E., Lee, E., Clear, R., DiBartolomeo, D. and Selkowitz, S. (1998) 'Office worker response to an automated venetian blind and electric lighting system: a pilot study', *Energy and Buildings*, vol 29, pp205–218

Vineyard, E. A., Sand, J. R. and Durfee, D. J. (2002) 'Performance of characteristics for a desiccant system at two extreme ambient conditions', *ASHRAE Transactions: Symposia*, vol 108, pp587–596

Viridén, K., Ammann, T., Hartmann, P. and Huber, H. (2003) *P+D – Projekt Passivhaus im Umbau* (in German). Available from www.viriden-partner.ch

von Roedern, B. (2003) *Status of Amorphous and Crystalline Thin Film Silicon Solar Cell Activities*, NREL/CP-520-33568, National Renewable Energy Laboratory, Golden, CO, 4pp

Voss, K. (2000a) *Solar Renovation Demonstration Projects, Results and Experience*, James & James, London, 24pp

Voss, K. (2000b) 'Solar energy in building renovation – results and experience of international demonstration buildings', *Energy and Buildings*, vol 32, pp291–302

Waewsak, J., Hirunlabh, J., Khedari, J. and Shin, U. C. (2003) 'Performance evaluation of the BSRC multi-purpose bio-climatic roof', *Building and Environment*, vol 38, pp1297–1302

Wagner, A., Herkel, S., Löhnert, G. and Voss, K. (2004) 'Energy efficiency in commercial buildings: experiences and results from the German funding program SolarBau', presented at EuroSolar 2004, Freiburg. Available from www.solarbau.de

Wakili, K. G., Bundi, R. and Binder, B. (2004) 'Effective thermal conductivity of vacuum insulation panels', *Building Research and Information*, vol 32, no 4, pp293–299

Walker, A., Balcomb, D., Weaver, N., Kiss, G. and Becker-Humphry, M. (2003) 'Analyzing two federal building-integrated photovoltaic projects using Energy-10 simulations', *Journal of Solar Energy Engineering*, vol 125, pp28–33

Walker, D. H. (2000) 'Low-charge refrigeration for supermarkets', *IEA Heat Pump Newsletter*, vol 18, no 1, pp13–16

Walker, I. S. (1999) 'Distribution system leakage impacts on apartment building ventilation rates', *ASHRAE Transactions*, vol 105 (Part 1), pp943–950

Walker, I. S. and Mingee, D. (2003) 'Reducing air handler electricity use: more than just a better motor', *Home Energy*, vol Nov/Dec 2003, pp8–9

Walker, I. S. and Sherman, M. (2005) 'Duct tape and sealant performance', *ASHRAE Journal*, vol 47, no 2, pp34–41

Walker, I. S., Wilson, D. J. and Sherman, M. H. J. (1998) 'A comparison of the power law to quadratic formulations for air infiltration calculations', *Energy and Buildings*, vol 27, pp293–298

Wang, D. C., Xia, Z. Z., Wu, J. Y., Wang, R. Z., Zhai, H. and Dou, W. D. (2005) 'Study of a novel silica gel-water adsorption chiller. Part I. Design and performance prediction', *International Journal of Refrigeration*, vol 28, pp1073–1083

Wang, R. Z. (2001) 'Performance improvement of adsorption cooling by heat and mass recovery operations', *International Journal of Refrigeration*, vol 24, pp602–611

Weiss, I. and Sprau, P. (2002) '100,000 roofs and 99 Pfenning', *Renewable Energy World*, vol 5, no 1

Weiss, W. (ed) (2003) *Solar Heating Systems for Houses: A Design Handbook for Solar Combisystems*, James & James, London, 313pp

Weitzmann, P., Kragh, J., Roots, P. and Svendsen, S. (2005) 'Modelling floor heating systems using a validated two-dimensional ground-coupled numerical model', *Building and Environment*, vol 40, pp153–163

Wellington, G. M., Glynn, P. W., Strong, A. E., Navarrete, A., Wieters, E. and Hubbard, D. (2001) 'Crisis on coral reefs linked to climate change', *EOS*, vol 82, no 1, pp1, 5

Werling, E., Collison, B. and Hall, J. (2000) 'More lessons learned in the Energy Star® Homes Program', in *Proceedings of the 2000 ACEEE Summer Study on Energy Efficiency in Buildings*, vol 2, American Council for an Energy Efficient Economy, Washington, DC, pp335–346

Westphalen, D. and Koszalinski, S. (2001) *Energy Consumption Characteristics of Commercial Building HVAC Systems, Volume 1: Chillers, Refrigerant Compressors, and Heating Systems*, Arthur D. Little Inc., Cambridge (MA), 71pp. Available from www.eere.energy.gov/buildings/info/documents

Whitaker, T. (2003) 'Backlights, airports and vehicles boost LED market', *Compoundsemiconductor.net*. Available from http://compoundsemiconductor.net/articles/magazine/9/12/1/1

White, A., Melvin, G. R. C. and Friend, A. D. (1999) 'Climate change impacts on ecosystem and the terrestrial carbon sink: a new assessment', *Global Environmental Change*, vol 9, ppS21–S30

White, S. D., Yarrall, M. G., Cleland, D. J. and Hedley, R. A. (2002) 'Modelling the performance of a transcritical CO_2 heat pump for high temperature heating', *International Journal of Refrigeration*, vol 25, pp479–486

Wigginton, M. and Harris, J. (2002) *Intelligent Skins*, Butterworth-Heinemann, Oxford, 176pp

Wilkinson, C. R. (1999) 'Global and local threats to coral reef functioning and existence: review and predictions', *Marine and Freshwater Research*, vol 50, pp867–878

Williams, R. H., Bunn, M., Consonni, S., Gunter, W., Holloway, S., Moore, R. and Simbeck, D. (2000) 'Advanced energy supply technologies', in *World Energy Assessment: Energy and the Challenge of Sustainability*, United Nations Development Programme, New York, pp273–329

Winkelmann, F. C. (2001) 'Modeling windows in EnergyPlus', in *Seventh International IBPSA Conference*, Rio de Janeiro, Brazil, 13–15 August 2001, pp457–464. Available from www.ibpsa.org

Withers, C. R. and Cummings, J. B. (1998) 'Ventilation, humidity, and energy impacts of uncontrolled airflow in a light commercial building', *ASHRAE Transactions*, vol 104 (Part 2), pp733–742

Wolfe, D. W. and Erickson, J. D. (1993) 'Carbon dioxide effects on plants: uncertainties and implications for modeling crop response to climate change', in Kaiser, R. U. and Drennen, T. E. (eds) *Agricultural Dimensions of Global Climate Change*, St. Lucie Press, Delray Beach FL, pp153–178

Wong, C. K., Worek, W. M. and Brillhart, P. L. (2002a) 'Use of joint-frequency weather data to determine primary energy consumption of desiccant systems', *ASHRAE Transactions*, vol 108 (Part 1), pp608–616

Wong, C. K., Worek, W. M. and Brillhart, P. L. (2002b) 'Primary energy consumption of ventilation systems – a comparison and rating using joint-frequency weather data', *ASHRAE Transactions*, vol 108 (Part 2), pp563–571

Wong, N. H., Cheong, D. K. W., Yan, H., Soh, J., Ong, C. L. and Sia, A. (2003a) 'The effects of rooftop garden on energy consumption of a commercial building in Singapore', *Energy and Buildings*, vol 35, pp353–364

Wong, N. H., Chen, Y., Ong, C. L. and Sia, A. (2003b) 'Investigation of thermal benefits of rooftop garden in the tropical environment', *Building and Environment*, vol 38, pp261–270

World Alliance for Decentralized Energy (WADE) (2004) *World Survey of Decentralized Energy – 2002/2003*. Available from www.localpower.org

Wright, A. (1999) 'Natural ventilation or mixed mode? An investigation using simulation', in *Sixth International IBPSA Conference*, Kyoto, Japan, 13–15 September 1999, pp449–455. Available from www.ibpsa.org

Wright, J. L. (2001) 'A simplified analysis of radiant heat loss through projecting fenestration products', *ASHRAE Transactions*, vol 107 (Part 1), pp700–708

Wulfinghoff, D. R. (1999) *Energy Efficiency Manual*, Energy Institute Press, Wheaton, MD, 1531pp

Xu, T. T., Carrié, F. R., Dickerhoff, D. J., Fisk, W. J., McWilliams, J., Wang, D. and Modera, M. P. (2002) 'Performance of thermal distribution systems in large commercial buildings', *Energy and Buildings*, vol 34, pp215–226

Xue, H., Chou, S. K. and Zhong, X. Q. (2004) 'Thermal environment in a confined space of high-rise building with split air conditioning system', *Building and Environment*, vol 39, pp817–823

Yamaguchi, Y., Shimoda, Y. and Mizuno, M. (2004)

'Development of district energy supply system for CHP implementation', in *Proceedings of the 2004 ACEEE Summer Study on Energy Efficiency in Buildings*, vol 9, American Council for an Energy Efficient Economy, Washington, DC, pp121–133

Yanbing, K., Yi, J. and Yinping, Z. (2003) 'Modeling and experimental study of an innovative passive cooling system – NVP system', *Energy and Buildings*, vol 35, pp417–425

Yang, H., Burnett, J. and Zhu, Z. (2001a) 'Building-integrated photovoltacis: effect on the cooling load component of building facades', *Building Service Engineering Resources Technology*, vol 22, no 3, pp157–165

Yang, H., Zhu, Z., Burnett, J. and Lu, L. (2001b) 'A simulation study on the energy performance of photovoltaic roofs', *ASHRAE Transactions*, vol 107 (Part 2), pp129–135

Yapici, R. and Ersoy, H. K. (2005) 'Performance characteristics of the ejector refrigeration system based on the constant area ejector flow model', *Energy Conversion and Management*, vol 46, pp3117–3135

Yattara, A., Zhu, Y. and Ali, M. M. (2003) 'Comparison between solar single-effect and single-effect double-lift absorption machines (Part I)', *Applied Thermal Engineering*, vol 3, pp1981–1992

Yavuzturk, C. and Spitler, J. D. (2000) 'Comparative study of operating and control strategies for hybrid ground-source heat pump systems using a short time step simulation model', *ASHRAE Transactions*, vol 106 (Part 2), pp192–209

Yik, F. W. H., Burnett, J. and Prescott, I. (2001) 'A study on the energy performance of three schemes for widening application of water-cooled air-conditioning systems in Hong Kong', *Energy and Buildings*, vol 33, pp167–182

Yonehara, T. (1998) 'Ventilation windows and automatic blinds help to control heat and lighting', *CADDET Energy Efficiency Newsletter*, December 1998, pp9–11

Yorke, P. and Brearley, R. (2003) 'Elizabeth Fry Building, University of East Anglia, Norwich', in Edwards, B. (ed) *Green Buildings Pay*, Spon Press, London, pp180–187

Yoshikawa, S. (1997) 'Japanese DHC system uses untreated sewage as a heat source', *CADDET Energy Efficiency Newsletter*, June 1997, pp8–10

Young, R., Spata, A. J., Turnbull, P. and Allen, T. E. (1999) 'Designing an energy-efficient quick service restaurant', *ASHRAE Transactions*, vol 105 (Part 1), pp1111–1121

Yu, F. W. and Chan, K. T. (2005) 'Experimental

determination of the energy efficiency of an air-cooled chiller under part-load conditions', *Energy*, vol 30, pp1747–1758

Yuan, X., Chen, Q. and Glicksman, L. R. (1999) 'Performance evaluation and design guidelines for displacement ventilation', *ASHRAE Transactions*, vol 105 (Part 1), pp298–309

Yumratuş, R. and Ünsal, M. (2000) 'Analysis of solar aided heat pump systems with seasonal thermal energy storage in surface tanks', *Energy*, vol 25, pp1231–1243

Zalba, B., Marín, J. M., Cabeza, L. F. and Mehling, H. (2003) 'Review on thermal energy storage with phase change materials, heat transfer analysis, and applications', *Applied Thermal Engineering*, vol 23, pp251–283

Zalewski, L., Lassue, S., Duthoit, B. and Butez, M. (2002) 'Study of solar walls – validating a simulation model', *Building and Environment*, vol 37, pp109–121

Zhai, Z. and Chen, Q. Y. (2004) 'Numerical determination and treatment of convective heat transfer coefficient in the coupled building energy and CFD simulation', *Building and Environment*, vol 39, pp1001–1009

Zhai, Z., Chen, Q., Klems, J. H. and Haves, P. (2001) 'Strategies for coupling energy simulation and computational fluid dynamics programs', in *Seventh International IBPSA Conference*, Rio de Janeiro, Brazil, 13–15 August 2001, pp59–66. Available from www.ibpsa.org

Zhai, X. Q., Dai, Y. J. and Wang, R. Z. (2005) 'Comparison of heating and natural ventilation in a solar house induced by two roof solar collectors', *Applied Thermal Engineering*, vol 25, pp741–757

Zhang, L. Z. and Niu, J. L. (2001) 'Energy requirements for conditioning fresh air and the long-term savings with a membrane-based energy recovery ventilator in Hong Kong', *Energy*, vol 26, pp119–135

Zhang, L. Z. and Niu, J. L. (2002) 'Performance comparison of desiccant wheels for air dehumidification and enthalpy recovery', *Applied Thermal Engineering*, vol 22, pp1347–1367

Zhang, L. Z. and Niu, J. L. (2003) 'A pre-cooling Munters environmental control desiccant cooling cycle in combination with chilled-ceiling panels', *Energy*, vol 28, pp275–292

Zhang, Q. and Murphy, W. E. (2000) 'Measurement of thermal conductivity for three borehole fill materials used for GSHP', *ASHRAE Transactions*, vol 106 (Part 2), pp434–441

Zhang, X. and Muneer, T. (2002) 'A design guide for performance assessment of solar light-pipes', *Lighting Research and Technology*, vol 34, pp149–169

Zhang, X. J. and Wang, R. Z. (2002) 'A new combined adsorption-ejector refrigeration and heating hybrid system powered by solar energy', *Applied Thermal Engineering*, vol 22, pp1245–1258

Zhang, Y., Jiang, Y., Zhang, L. Z., Deng, Y. and Jin, Z. (2000) 'Analysis of thermal performance and energy savings of membrane based heat recovery ventilator', *Energy*, vol 25, pp515–527

Zhao, P. C., Zhao, L., Ding, G. L. and Zhang, C. L. (2003a) 'Temperature matching method of selecting working fluids for geothermal heat pumps', *Applied Thermal Engineering*, vol 23, pp179–195

Zhao, Y., Curcija, D. and Gross, W. P. (1999) 'Convective heat transfer correlations for fenestration glazing cavities: a review', *ASHRAE Transactions*, vol 105 (Part 2), pp900–908

Zhao, Y., Ohadi, M. M. and França, F. H. R. (2003b) 'Experimental heat transfer coefficients of CO_2 in a microchannel evaporator', *ASHRAE Transactions*, vol 109 (Part 1), pp533–541

Zhao, Y. and Ohadi, M. M. (2004) 'Experimental study of supercritical CO_2 gas cooling in a microhannel gas cooler', *ASHRAE Transactions*, vol 110 (Part 1), pp291–300

Zhivov, A. M. and Rymkevich, A. A. (1998) 'Comparison of heating and cooling energy consumption by HVAC system with mixing and displacement air distribution for a restaurant dining area in different climates', *ASHRAE Transactions*, vol 104 (Part 2), pp473–484

Zhou, G., Ihm, P., Krarti, M., Liu, S. and Henze, G. P. (2003) 'Integration of an internal optimization module within EnergyPlus', in *Eighth International IBPSA Conference*, Eindhoven, The Netherlands, 11–14 August 2003, pp1475–1482c. Available from www.ibpsa.org

Zimmermann, M. (2004) 'ECBCS building retrofit initiative', *ECBCS News*, October 2004, pp11–14. Available from www.ecbcs.org

Zmeureanu, R. and Peragine, C. (1999) 'Evaluation of interactions between lighting and HVAC systems in a large commercial building', *Energy Conversion and Management*, vol 40, pp1229–1236

Zogg, R., Dieckmann, J., Roth, K. and Brodrick, J (2005) 'Heat pump water heaters', *ASHRAE Journal*, vol 47, no 3, pp98–99

Zondag, H. A., de Vries, D. W., van Helden, W. G. J., van Zolingen, R. J. C. and van Steenhoven, A. A. (2003) 'The yield of different combined PV-thermal collector designs', *Solar Energy*, vol 74, pp253–269

Zuluaga, M. and Griffiths, D. (2004) 'A dynamic ceiling for improved comfort with evaporative cooling', in *Proceedings of the 2004 ACEEE Summer Study on Energy*

Efficiency in Buildings, vol 1, American Council for an Energy Efficient Economy, Washington, DC, pp368–379

Index

Index 697

J

Japan 12, 27–29, 82, 121–122, 135, 148, 169, 172, 208–210, 220–221, 223, 227, 238, 242, 262, 423, 439, 441–442, 445–446, 449–451, 463–464, 473, 481, 494, 511–512, 527, 546–547, 580, 582, 584, 620–621

K

Kenya 218
kerosene heater 135
Korea 12, 198, 318, 347, 442, 449, 509, 527, 562, 588, 620–621
Kuwait 198, 239
krypton 63–64, 69, 78, 82, 87–90, 341

L

laser-cut panel 191, 396–397, 402
latent heat 124, 133, 138, 169, 177–178, 243, 246, 265–267, 282, 290, 457
Leadership in Energy and Environmental Design (LEED™) 538–540
leakage, effective leakage area 102–106, 194, 197, 359
legionnaires disease 289, 576
lifecycle cost 107–109
light pipe 393, 395, 404–405, 413
light shelf 393
light-emitting diode (LED) 383, 387
lighting-guiding shade 398, 402–403
Lithuania 584
liquefied natural gas (LNG) 12
low-e coating, see emissivity
lumen 385, 618
depreciation 410,413
luminescent solar concentrator 401
lux 385, 401–402, 406, 410–411, 415, 618
Luxembourg 584

M

Malaysia 279, 509
mercury 8, 382, 384, 386, 388, 410
meteorological data source 33–34
Mexico 171, 192, 442, 464, 509, 620–621

microchannel heat exchanger, see heat exchanger
Minergie Standard 513
mixing ventilation (MV), see ventilation
model energy code (MEC) 504–506, 508, 516
model national energy code (MNEC) 504–507
Montreal Protocol 499
motor 163, 240, 304–307, 320–323, 450
ECPM, 243, 321–323, 324, 326, 328–329
PSC 320–323

N

nanocrystalline dye cell, see dye-sensitized solar cell
natural gas
supply 12–14
cost 29
externality 8–11, 14
natural gas-fired power plant 564
natural ventilation, see ventilation
Netherlands 29, 97–98, 126, 201, 213, 216, 223, 327, 363, 423, 442, 445, 447, 451, 459, 464, 524, 532, 534, 559, 562, 584, 589, 593, 620–621
New Zealand 29, 423, 464, 492, 509, 620–621
night ventilation, see ventilation
Norway 27–29, 220–221, 238–239, 305, 442, 464, 554, 562, 620–621

O

occupancy sensor 328, 384, 389, 404–405, 410, 526
off-peak cooling 278–285
oven 417, 420
microwave 418, 420, 422
oversizing,
cooling equipment 106, 277–278
cooling tower 286
ducts, pipes 304
fans, pumps, motors 307, 310, 320
heating equipment 106, 140
lighting 410, 413
radiators 578